CONVENTIONAL UNIT CELLS OF THE 14 BRAVAIS LATTICES

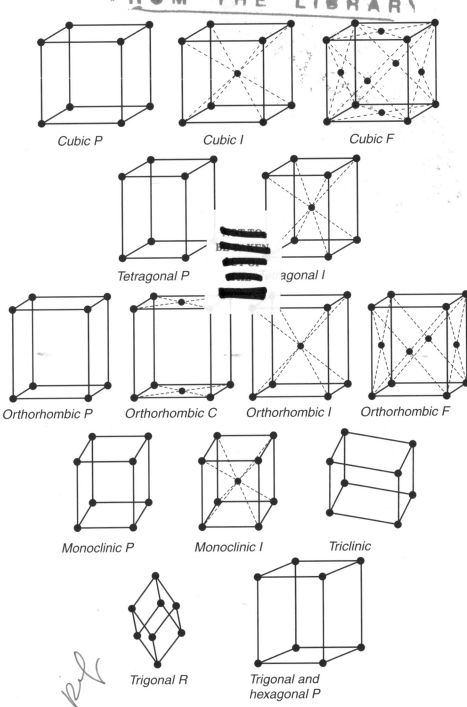

Cubic P Cubic I Cubic F

Tetragonal P Tetragonal I

Orthorhombic P Orthorhombic C Orthorhombic I Orthorhombic F

Monoclinic P Monoclinic I Triclinic

Trigonal R Trigonal and hexagonal P

ENCYCLOPEDIC DICTIONARY OF CONDENSED MATTER PHYSICS

ФИЗИКА ТВЕРДОГО ТЕЛА

Энциклопедический словарь

Главный редактор

В. Г. БАРЬЯХТАР

Киев Наукова думка

FIZIKA TVERDOGO TELA
ENCIKLOPEDICHESKII SLOVAR

Original Edition (in Russian):

Naukova Dumka, Kiev, Ukraine
Volume 1, 1996
Volume 2, 1998
Institute of Physics, Institute of Magnetism

Editor-in-Chief of the Original Edition:

V. G. BAR'YAKHTAR

Department of Physics and Astronomy
National Academy of Sciences of Ukraine

ENCYCLOPEDIC DICTIONARY OF CONDENSED MATTER PHYSICS

VOLUME 2
N – Z

EDITED BY

CHARLES P. POOLE JR.

Department of Physics and Astronomy
University of South Carolina
Columbia, SC, USA

ELSEVIER
ACADEMIC
PRESS

2004

Amsterdam – Boston – Heidelberg – London – New York – Oxford
Paris – San Diego – San Francisco – Singapore – Sydney – Tokyo

ELSEVIER B.V.	**ELSEVIER Inc.**	ELSEVIER Ltd	ELSEVIER Ltd
Sara Burgerhartstraat 25	**525 B Street, Suite 1900**	The Boulevard, Langford Lane	84 Theobalds Road
P.O. Box 211,	**San Diego, CA 92101-4495**	Kidlington, Oxford OX5	1GB London WC1X 8RR
1000 AE Amsterdam	**USA**	UK	UK
The Netherlands			

This is a translation from Russian.

First edition 2004

Library of Congress Cataloging in Publication Data: A catalog record is available from the Library of Congress.

British Library Cataloguing in Publication Data: A catalogue record is available from the British Library.

ISBN 0-12-088398-8 volume 1

ISBN 0-12-088399-6 volume 2

ISBN 0-12-561465-9 set

⊚ The paper used in this publication meets the requirements of ANSI/NISO Z39.48-1992 (Permanence of Paper).

Printed in The United Kingdom.

EDITOR'S PREFACE
TO THE ENGLISH EDITION

At the present time Condensed Matter Physics is the largest of the main subdivisions of physics. For several years the Encyclopedic Dictionary of Solid State Physics has been one of the principal reference works in this area, but unfortunately until now it was only available in the Russian language. Its appearance in English is indeed long overdue. It has been a great pleasure for me to undertake the task of editing the English language edition. This two-volume work contains a great deal of material ordinarily not found in the solid state texts that are most often consulted for reference purposes, such as accounts of acoustical holography, bistability, Kosterlitz–Thouless transition, and self-similarity. It provides information on materials rarely discussed in these standard texts, such as austenite, ferrofluids, intercalated compounds, laser materials, Laves phases, nematic liquid crystals, etc. Properties of many quasi-particles are presented; for example, those of crowdions, dopplerons, magnons, solitons, vacancions, and weavons. The volumes are fully cross-referenced for tracking down all aspects of a topic, and in this respect we have emulated the excellent cross-referencing systematics of the Russian original. Long lists of alternate titles for looking up subjects can considerably shorten the time needed to find specialized information.

I wish to thank the members of the Review Panel of the English Language Edition who read over the initial draft of the translation, offered many suggestions and comments for improving the translation, and also provided valuable advice for refining and updating the contents of the articles: Victor Bar'yakhtar, Alexei A. Maradudin, Michael McHenry, Laszlo Mihaly, Sergey K. Tolpygo and Sean Washburn. Thanks are also due to Horacio A. Farach, Vladimir Gudkov, Paul G. Huray, Edwin R. Jones, Grigory Simin, Tangali S. Sudarshin, and especially to Ruslan Prozorov, for their assistance with the editing. The translators did an excellent job of rendering the original Russian text into English, and the desk editor Rimantas Šadžius did an outstanding job of checking all the many cross references and the consistency of the notation. The Academic Press Dictionary of Science and Technology has been an invaluable source of information for clarifying various issues during the editing.

I wish to thank my wife for her patience with me during the long hours expended in the editing process. The page proofs of this work arrived at my home on the octave day of our 50th wedding anniversary, which we celebrated on October 17, 2003, with our five children and fifteen grandchildren.

Charles P. Poole Jr.

FOREWORD

An Encyclopedic Dictionary of Condensed Matter Physics can serve a diverse set of audiences. Professional researchers and university and college teachers interested in learning quickly about a topic in condensed matter physics outside their immediate expertise can obtain the information they seek from such a source. Doctoral students embarking on research in condensed matter physics or materials science can find it helpful in providing explanations of concepts and descriptions of methods encountered in reading the literature dealing with their research subject and with related subjects. Undergraduate students with no previous background in condensed matter physics will find it a particularly valuable complement to their textbooks as they take their first courses in the field. Educated laypersons can turn to such an encyclopedia for help in understanding articles published in popular scientific journals.

Until the publication of the Russian edition of the Encyclopedic Dictionary of Solid State Physics, with Professor Viktor G. Bar'yakhtar as its General Editor, no such resource was available to the scientific community. Its more than 4300 entries were written by more than 700 contributors, specialists in their branches of condensed matter physics, many of whom were well known to an international audience.

The task of making this encyclopedia accessible to non-Russian readers was undertaken by Academic Press, with Professor Charles P. Poole Jr. in the role of the General Editor for the English translation. Both the publisher and editor deserve the gratitude of the condensed matter community for bringing this project to fruition. All involved in its production hope that that community will find it as useful as its creators intended it to be.

Irvine, California
October 29, 2003

Alexei A. Maradudin

FOREWORD OF THE EDITORIAL BOARD
OF THE ORIGINAL RUSSIAN EDITION

This publication is an attempt at a systematized presentation of modern knowledge within all areas of solid state physics from the most fundamental ideas to applied questions. There has been no similar encyclopedic compilation of solid state physics available in the world until now. The Dictionary contains more than 4000 articles. Among them there are relatively long articles devoted to the basic concepts and phenomena of solid state physics, moderate-length articles on particular problems and physical concepts, and, finally, brief definitions of terms used in the literature. Considerable attention is paid to technical applications of solid state physics. Brief historical information is given in many articles such as the names of authors and the dates of discoveries. In a number of cases the history of the question is covered in the article. Together with the description of classical topics, special attention is paid to those new ideas that arose only recently in solid state physics and, therefore, often still do not have commonly accepted definitions. Many articles are provided with illustrations which supplement the text material to help the reader to acquire more complete and more visual information.

More than 500 scientists participated in the preparation and the writing of the articles of the Dictionary; these articles were written by specialists working in the particular fields that they cover.

The goal of the publication is to provide needed reference material to a wide audience of readers, which will help them to understand the basic principles of solid state physics. Therefore, the authors endeavored to probe deeply into some salient details of the subject, while preserving a broad perspective throughout the text of the article. The content of articles is interconnected by a system of cross-references. The appended Subject Index will help the reader to orient himself easily in the scope of the various articles.

To our deep sorrow, during the period of preparation of the manuscript of this Dictionary, several outstanding scientific members of the Editorial Board Valentin L'vovich Vinetskii, Yakov Evseevich Geguzin, Emanuil Aizikovich Kaner, Boris Ivanovich Nikolin, Kirill Borisovich Tolpygo, Leonid Nikandrovich Larikov and Alexander Bronislavovich Roitsyn have passed from among us. The recently expired well-known scientist and prominent publishing specialist Aleksandr Aleksandrovich Gusev, who shared with us his vast knowledge and encyclopedic expertise, provided invaluable aid in the publication of the dictionary.

Victor G. Bar'yakhtar, Editor-in-Chief
Ernest A. Pashitskii, Vice-Editor-in-Chief
Elena G. Galkina, Secretary

LIST OF CONTRIBUTORS

F.Kh. Abdullaev
A.A. Adamovskii
S.S. Afonskii
I.A. Akhiezer
A.F. Akkerman
A.B. Aleinikov
V.P. Alekhin
A.S. Aleksandrov
L.N. Aleksandrov
B.L. Al'tshuler
E.D. Aluker
S.P. Anokhov
L.I. Antonov
K.P. Aref'ev
I.E. Aronov
M. Ashe
Yu.E. Avotin'sh
V.G. Babaev
Yu.Z. Babaskin
S.G. Babich
Yu.I. Babii
A.S. Bakai
Yu.V. Baldokhin
O.M. Barabash
S.D. Baranovskii
Z.V. Bartoshinskii
V.G. Bar'yakhtar
V.G. Baryshevskii
E.M. Baskin
F.G. Bass
A.N. Bekrenev
I.P. Beletskii
Ya.A. Beletskii
S.I. Beloborod'ko
E.D. Belokolos
V.F. Belostotskii
A.V. Belotskii
M.V. Belous

N.P. Belousov
K.P. Belov
D.P. Belozerov
A.I. Belyaeva
S.S. Berdonosov
B.I. Beresnev
L.I. Berezhinskii
V.I. Bernadskii
I.B. Bersuker
V.N. Berzhanskii
P.A. Bezirganyan
V.F. Bibik
I.V. Blonskii
E.N. Bogachek
A.N. Bogdanov
Yu.A. Bogod
S.A. Boiko
V.S. Boiko
S.I. Bondarenko
V.N. Bondarev
A.S. Borovik-Romanov
A.A. Borshch
Yu.S. Boyarskaya
A.M. Bratkovskii
O.M. Braun
S.L. Bravina
M.S. Bresler
A.B. Brik
M.S. Brodin
V.A. Brodovoi
L.T. Bugaenko
V.N. Bugaev
M. Buikov
L.L. Buishvili
E.I. Bukhshtab
L.N. Bulaevskii
B.M. Bulakh
A.A. Bulgakov

S.I. Bulychev
A.I. Buzdin
A.I. Bykhovskii
A.M. Bykov
Yu.A. Bykovskii
G.E. Chaika
A.A. Chel'nyi
S.P. Chenakin
V.T. Cherepin
N.F. Chernenko
Yu.P. Chernenkov
A.A. Chernov
S.I. Chugunova
B.A. Chuikov
K.V. Chuistov
A.A. Chumak
V.E. Danil'chenko
V.D. Danilov
L.I. Datsenko
A.B. Davydov
A.S. Davydov
S.A. Dembovskii
V.L. Demikhovskii
S.A. Demin
E.M. Dianov
V.V. Didyk
I.M. Dmitrenko
N.L. Dmitruk
V.N. Dneprenko
V.S. Dneprovskii
V.N. Dobrovol'skii
R.D. Dokhner
I.F. Dolmanova
F.E. Dolzhenkov
V.F. Dorfman
S.N. Dorogovtsev
N.V. Dubovitskaya
E.F. Dudnik

V.A. Durov
V.V. Dyakin
M.I. Dykman
A.I. Efimov
A.L. Efros
Yu.A. Ekmanis
M.A. Elango
V.M. Elinson
P.G. Eliseev
V.A. Elyukhin
Yu.P. Emets
V.V. Emtsev
V.Z. Enol'skii
I.R. Entin
I.R. Entinzon
E.M. Epshtein
A.I. Erenburg
A.S. Ermolenko
E.D. Ershov
A.M. Evstigneev
M.I. Faingol'd
V.M. Fal'chenko
V.L. Fal'ko
B.Ya. Farber
M.P. Fateev
I.M. Fedorchenko
O.P. Fedorov
A.G. Fedorus
Ya.A. Fedotov
I.V. Fekeshgazi
A.V. Filatov
B.N. Filippov
V.A. Finkel'
V.M. Finkel'
A.Ya. Fishman
N.Ya. Fogel'
E.L. Frankevich
L.I. Freiman

V.M. Fridkin
I.Ya. Fugol'
B.I. Fuks
A.M. Gabovich
S.P. Gabuda
A.Yu. Gaevskii
Yu.A. Gaidukov
V.P. Galaiko
R.V. Galiulin
Yu.M. Gal'perin
M.Ya. Gamarnik
G.D. Gamulya
E.M. Ganapol'skii
V.G. Gavrilyuk
Ya.E. Geguzin
B.L. Gel'mont
D.S. Gertsriken
I.A. Gilinskii
I.A. Gindin
S.L. Ginzburg
E.I. Givargizov
E.I. Gladyshevskii
A.A. Glazer
L.I. Glazman
A.I. Glazov
M.D. Glinchuk
E.Ya. Glushko
V.M. Gokhfel'd
V.L. Gokhman
V.A. Golenishchev-
Kutuzov
A.V. Golik
M.F. Golovko
T.V. Golub
D.A. Gorbunov
L.Yu. Gorelik
Yu.I. Gorobets
V.I. Gorshkov
V.V. Gorskii
V.G. Grachev
S.A. Gredeskul
V.P. Gribkovskii
P.P. Grigaitis
V.N. Grigor'ev
O.N. Grigor'ev
Yu.M. Grin'
V.V. Gromov
V.V. Gudkov
A.V. Gulbis

A.P. Gulyaev
Yu.V. Gulyaev
M.E. Gurevich
Yu.G. Gurevich
K.P. Gurov
R.N. Gurzhi
M.B. Guseva
A.N. Guz'
V.M. Gvozdikov
A.I. Ignatenko
V.V. Il'chenko
E.M. Iolin
I.P. Ipatova
S.S. Ishchenko
Z.A. Iskanderova
V. Iskra
V.G. Ivanchenko
P.G. Ivanitskii
A.L. Ivanov
B.A. Ivanov
M.A. Ivanov
N.R. Ivanov
G.F. Ivanovskii
P.F. Ivanovskii
V.G. Ivantsov
R.K. Ivashchenko
O.M. Ivasishin
Yu.A. Izyumov
G.A. Kachurin
A.N. Kadashchuk
A.M. Kadigrobov
M.I. Kaganov
E.B. Kaganovich
Yu.M. Kaganovskii
N.P. Kalashnikov
P.A. Kalugin
G.G. Kaminskii
E.A. Kaner
L.N. Kantorovich
A.S. Kapcherin
A.A. Kaplyanskii
A.L. Kartuzhanskii
A.L. Kasatkin
V.N. Kashcheev
N.I. Kashirina
G.A. Katrich
B.V. Khaenko
I.B. Khaibullin
V.A. Kharchenko

E.I. Khar'kov
S.S. Khil'kevich
A.I. Khizhnyak
G.E. Khodenkov
A.R. Khokhlov
G.A. Kholodar'
V.A. Khvostov
K.I. Kikoin
S.S. Kil'chitskaya
O.V. Kirichenko
A.P. Kirilyuk
P.S. Kislyi
A.E. Kiv
Ya.G. Klyava
V.M. Knyazheva
E.S. Koba
L.S. Kogan
V.V. Kokorin
Yu.A. Kolbanovskii
G.Ya. Kolbasov
A.G. Kol'chinskii
V.L. Kolesnichenko
V.V. Kolomiets
V.I. Kolomytsev
E.V. Kolontsova
V.G. Kon
A.A. Konchits
S.N. Kondrat'ev
M.Ya. Kondrat'ko
V.A. Kononenko
V.I. Konovalov
A.I. Kopeliovich
V.A. Koptsyk
T.N. Kornilova
Yu.V. Kornyushin
E.N. Korol'
N.S. Kosenko
A.M. Kosevich
O.G. Koshelev
A.S. Kosmodamianskii
E.A. Kotomin
N.Ya. Kotsarenko
Yu.N. Koval'
V.V. Koval'chuk
A.S. Kovalev
O.V. Kovalev
M.Yu. Kovalevskii
Yu.Z. Kovdrya
I.I. Kovenskii

V.I. Kovpak
V.I. Kovtun
A.V. Kozlov
N.N. Krainik
V.V. Krasil'nikov
A.Ya. Krasovskii
L.S. Kremenchugskii
V.P. Krivitskii
M.A. Krivoglaz
V.N. Krivoruchko
A.A. Krokhin
S.P. Kruglov
I.V. Krylova
A.Yu. Kudzin
L.S. Kukushkin
I.O. Kulik
V.A. Kuprievich
A.I. Kurbakov
A.V. Kurdyumov
V.G. Kurdyumov
G.G. Kurdyumova
M.V. Kurik
M.I. Kurkin
N.P. Kushnareva
V.A. Kuz'menko
R.N. Kuz'min
V.N. Kuzovkov
A.I. Landau
V.F. Lapchinskii
L.N. Larikov
F.F. Lavrent'ev
O.D. Lavrentovich
B.G. Lazarev
L.S. Lazareva
Ya.S. Lebedev
B.I. Lev
A.P. Levanyuk
A.A. Levin
I.B. Levinson
V.A. Likhachev
M.P. Lisitsa
V.M. Lisitsyn
V.A. Lisovenko
F.V. Lisovskii
P.G. Litovchenko
V.G. Litovchenko
V.S. Litvinov
V.A. Lobodyuk
V.M. Loktev
V.F. Los'

D.V. Lotsko	E.V. Minenko	S.V. Panyukov	B.S. Razbirin
V.P. Lukomskii	D.N. Mirlin	I.I. Papirov	A.R. Regel'
G.A. Luk'yanov	I.A. Misurkin	E.S. Parilis	V.R. Regel'
V.A. L'vov	V.Ya. Mitrofanov	A.M. Parshin	V.Yu. Reshetnyak
S.N. Lyakhimets	N.V. Morozovskii	E.A. Pashitskii	Yu.Ya. Reutov
V.G. Lyapin	V.V. Moshchalkov	L.A. Pastur	I.S. Rez
B.Ya. Lyubov	N.P. Moskalenko	R.E. Pasynkov	I.M. Reznik
I.F. Lyuksyutov	S.A. Moskalenko	E.P. Pechkovskii	Yu.A. Reznikov
R.G. Maev	V.N. Murzin	G.P. Peka	D.R. Rizdvyanetskii
V.M. Maevskii	V.P. Naberezhnykh	V.I. Peresada	A.B. Roitsyn
L.L. Makarov	T.A. Nachal'naya	S.P. Permogorov	O.V. Romankevich
N.M. Makarov	V.M. Nadutov	V.G. Peschanskii	A.E. Romanov
V.I. Makarov	E.L. Nagaev	I.S. Petrenko	A.S. Rozhavskii
D.N. Makovetskii	V.E. Naish	E.G. Petrov	S.S. Rozhkov
L.A. Maksimov	N.G. Nakhodkin	V.A. Petrov	V.V. Rozhkov
V.V. Malashenko	V.D. Natsyk	Yu.N. Petrov	E.Ya. Rudavskii
B.Z. Malkin	A.G. Naumovets	G.A. Petrunin	A.V. Rudnev
G.D. Mansfel'd	O.K. Nazarenko	V.Yu. Petukhov	I.N. Rundkvist
V.G. Manzhelii	E.I. Neimark	A.S. Pikovskii	S.M. Ryabchenko
V.G. Marinin	I.M. Neklyudov	G.E. Pikus	P.V. Ryabkov
S.V. Marisova	I.Yu. Nemish	A.N. Pilyankevich	K.P. Ryaboshapka
V.Ya. Markiv	V.V. Nemoshkalenko	V.I. Pipa	V. Rybalka
Yu.V. Martynenko	S.A. Nepiiko	F.V. Pirogov	A.I. Ryskin
T.Ya. Marusii	V.A. Nevostruev	G.S. Pisarenko	V.I. Ryzhkov
S.V. Mashkevich	S.A. Nikitov	E.A. Pisarev	A.M. Sabadash
V.S. Mashkevich	M.Yu. Nikolaev	L.P. Pitaevskii	V.A. Sablikov
T.V. Mashovets	B.I. Nikolin	A.B. Plachenov	A.V. Sachenko
V.V. Maslov	A.B. Nikol'skii	B.T. Plachenov	M.V. Sadovskii
O.I. Matkovskii	G.K. Ninidze	Yu.N. Podrezov	O.G. Sarbei
Z.A. Matysina	A.I. Nosar'	A.E. Pogorelov	V.M. Schastlivtsev
V.F. Mazanko	N.N. Novikov	V.Z. Polinger	Yu.V. Sedletskii
I.I. Mazin	N.V. Novikov	Yu.N. Polivanov	S.I. Selitser
E.A. Mazur	V. Novozhilov	V.V. Polotnyuk	A.V. Semenov
B.V. Mchedlishvili	S.G. Odulov	S.S. Pop	Yu.G. Semenov
M.V. Medvedev	V.I. Okulov	Yu.M. Poplavko	A.I. Senkevich
V.K. Medvedev	E.F. Oleinik	L.E. Popov	T.N. Serditova
A.I. Mel'ker	N.V. Olechnovich	V.E. Pozhar	T.N. Sergeeva
G.A. Melkov	S.I. Olikhovskii	L.I. Pranyavichus	S.E. Shafranyuk
V.S. Mel'nikov	B.Z. Ol'shanetskii	A.I. Prikhna	S.I. Shakhovtsova
F.F. Mende	A.N. Omel'yanchuk	V.E. Primachenko	A.M. Shalaev
Yu.Ya. Meshkov	A.N. Orlov	V.G. Prokhorov	B.N. Shalaev
G.A. Mesyats	S.P. Oshkaderov	G.I. Prokopenko	B.D. Shanina
V.P. Mikhal'chenko	M.E. Osinovskii	P.P. Pugachevich	B.Ya. Shapiro
V.S. Mikhalenkov	S.S. Ostapenko	V.V. Pustovalov	I.G. Shaposhnikov
V.V. Milenin	E.L. Ostrovskaya	V.I. Pustovoit	O.D. Shashkov
V.K. Milinchuk	D.E. Ovsienko	A.P. Rachek	V.G. Shavrov
Yu.V. Mil'man	R.P. Ozerov	V.A. Rafalovskii	D.I. Sheka
M.G. Mil'vidskii	G.A. Pacharenko	V.V. Rakitin	V.I. Sheka
V.N. Minakov	P.P. Pal'-Val'	E.I. Rau	R.I. Shekhter
			V.A. Shenderovskii

S.I. Shevchenko
A.Ya. Shik
V.B. Shikin
Yu.M. Shirshov
S.V. Shiyanovskii
G.N. Shkerdin
O.A. Shmatko
A.S. Shpigel'
T.D. Shtepa
E.I. Shtyrkov
Yu.N. Shunin
A.K. Shurin
K.K. Shvarts
E.A. Silin'sh
A.F. Sirenko
A.I. Sirko
A.A. Sitnikova
F.F. Sizov
E.V. Skokan
V.V. Skorokhod
V.V. Sleptsov
A.A. Slutskin
A.A. Smirnov
L.S. Smirnov
G.A. Smolenskii
O.V. Snitko
V.I. Sobolev
A.I. Sokolov
I.M. Sokolov
A.S. Sonin
E.B. Sonin
A. Sorokin
M.S. Soskin
A.L. Sozinov
V.B. Spivakovskii
I.E. Startseva
A.N. Starukhin
E.P. Stefanovskii
M.F. Stel'makh
I.A. Stepanov
I.A. Stoyanov

M.I. Strashnikova
V.I. Strikha
V.I. Sugakov
A.P. Sukhorukov
S.V. Svechnikov
V.M. Svistunov
I.I. Sych
A.K. Tagantsev
G.A. Takzei
A.P. Tankeev
V.V. Tarakanov
V.V. Tarasenko
G.G. Tarasov
Yu.V. Tarasov
I.A. Tarkovskaya
V.A. Tatarchenko
V.A. Tatarenko
I.I. Taubkin
V.A. Telezhkin
Yu.S. Tikhodeev
L.V. Tikhonov
E.A. Tikhonova
B.L. Timan
S.F. Timashev
B.B. Timofeev
V.B. Timofeev
Yu.A. Tkhorik
K.B. Tolpygo
S.K. Tolpygo
P.M. Tomchuk
V.I. Tovstenko
V.I. Trefilov
O.A. Troitskii
V.A. Trunov
V.A. Tsekhomskii
B.E. Tsekvava
R. Tsenker
B.S. Tsukerblat
V.M. Tsukernik
L.T. Tsymbal
N.V. Tsypin

L.I. Tuchinskii
A. Tukharinov
E.A. Turov
Yu.D. Tyapkin
V.M. Tyshkevich
V.A. Tyul'nin
K.M. Tyutin
A.A. Urusovskaya
A.I. Ustinov
B.K. Vainshtein
S.B. Vakhrushev
M.Ya. Valakh
V.I. Val'd-Perlov
A.A. Varlamov
I.A. Vasil'ev
L.I. Vasil'ev
M.A. Vasil'ev
F.T. Vas'ko
A.V. Vedyaev
B.G. Vekhter
O.G. Vikhlii
V.L. Vinetskii
E.A. Vinogradov
B.B. Vinokur
V.M. Vinokurov
Yu.P. Virchenko
A.S. Vishnevskii
I.M. Vitebskii
N.A. Vitovskii
V.I. Vladimirov
V.V. Vladimirov
N.A. Vlasenko
A.D. Vlasov
K.B. Vlasov
O.G. Vlokh
E.D. Vol
A.F. Volkov
P.Yu. Volosevich
V.A. Voloshin
V.D. Volosov
S.V. Vonsovskii

D.L. Vorob'ev
S.A. Vorob'ev
Yu.V. Vorob'ev
A.M. Voskoboinikov
B.E. Vugmeister
D.A. Yablonskii
Yu.N. Yagodzinskii
L.N. Yagupol'skaya
Yu.I. Yakimenko
B.I. Yakobson
M.V. Yakunin
V.A. Yampol'skii
I.K. Yanson
A.M. Yaremko
A.A. Yatsenko
A.F. Yatsenko
I.A. Yurchenko
Yu.F. Yurchenko
P.A. Zabolotnyi
V.A. Zagrebnov
Yu.R. Zakis
E.Ya. Zandberg
I.M. Zaritskii
E.V. Zarochentsev
I.K. Zasimchuk
E.E. Zasimchuk
S.S. Zatulovskii
E.A. Zavadskii
P.Yu. Zavalii
Yu.S. Zharkikh
I.S. Zheludev
G.N. Zhizhin
A.F. Zhuravlev
S.N. Zhurkov
O.S. Zinets
D.N. Zubarev
M.D. Zviadadze
L.O. Zvorykin
B.B. Zvyagin
I.P. Zvyagin

INTRODUCTION

We introduce to the reader the Encyclopedic Dictionary of Solid State Physics[1]. This Dictionary was planned several years ago, and the aim of its compilers was not only to provide a wide audience of readers with comprehensive and detailed information about different processes, phenomena, and effects in the solid state, but also to enlarge our understanding of the modern terminology in which this information is presented. The latter was very important for the compilers, since they were united by the idea that the achievements of solid state physics during recent decades were due not only to the fantastic development of experimental techniques and the level of experimental skill, but were also due to the creation of a unified system of concepts and terms adequately reflecting the observed regularities and the depth of understanding of the physical processes taking place in solids. The balanced rhythm of the modern language of solid state physics is determined by the theory of symmetry. It governs all theoretical constructions and the coefficients of the equations describing equilibrium and near-equilibrium states of a solid. The dynamic theory of a crystal lattice in the harmonic approximation, which is based on the discrete symmetry of ordered atomic arrangement in a crystal, is a classical example how symmetry considerations are utilized. Another "hidden" symmetry is used, e.g., to describe a number of properties of magnetic, ferroelectric, and superconducting materials. A particular kind of symmetry based on the regularities in diverse phase transitions is apparent in modern scaling theory. The reader will find applications of the theory of symmetry in many of the articles.

The majority of the phenomena of recent interest in solids were associated with low-level excited states of a substance. It appeared that while atoms or molecules are structural elements of a crystal lattice (and, generally, of any condensed state of a substance), certain collective excitations are the elements of every type of motion (elements of dynamics) in a solid. In view of the macroscopic homogeneity of the unbounded model of a solid, or the periodicity of an unbounded crystal lattice, such collective motions must take the form of plane waves, with a dispersion law that satisfies strict symmetry requirements. The quantization of elementary excitations leads to the image of a "corpuscle", or a quasi-particle. The concept of a quasi-particle and the idea of almost ideal quantum gases composed of various quasi-particles, enrich the language of solid state physics and allow one, within a unified system of terms and concepts, to treat physical phenomena that are quite diverse and have, at first glance, little in common. Therefore, it is not surprising that in the articles devoted to phonons, magnons, excitons, and to electrons or polarons as well, we find very similar equations and conclusions. The universality of the physical language reflects the unity of Nature itself.

The passage to the study of highly excited states of a material led to the creation of the nonlinear dynamics of solids. New contributions to the lexical reserve of the language of solid state physics originated from nonlinear dynamics. For example, the concepts of topological and dynamic solitons were put forward, dynamic solitons appearing in solid state physics as a new type of collective excitation of a condensed medium.

Not all parts of solid state physics are presented here with the same completeness and the same amount of detail. The branches reflecting the scientific interests of solid state physicists from the Ukraine and other former Soviet Union republics are the most extensively described. Although general problems of solid state

---

[1] This is the title of the book in Russian (Editor's note).

physics are less systematically presented, the following areas are covered rather deeply and broadly: physical properties of metals, including superconductivity, electrical and optical properties of semiconductors and insulators, magnetism and magnetic materials, mechanical properties of crystals, and physics of disordered systems. One may also notice an emphasis on the applied physics of plasticity and strength. This is because of the fact that, although physics is an objective international science, some of its presentations or overviews found in textbooks, monographs, or encyclopedic publications unintentionally introduce subjective ideas and personal preferences of authors.

The Dictionary includes articles reflecting the latest achievements in solid state physics, sometimes with results that have not yet appeared in monographs. This relates both to experimental discoveries and theoretical ideas. There are articles on high-temperature superconductivity, the quantum Hall effect, Kosterlitz–Thouless phase transitions in two-dimensional systems, autowave processes, fractals, etc. Novel nontraditional materials and techniques for investigation are described: amorphous substances, liquid crystals, quasi-crystals, polymers, low-dimensional systems, etc.

In addition let us point out to the reader the "quantum-mechanical emphasis", which may be noted in many articles devoted to diverse macroscopic phenomena and properties of solid bodies. Traditionally, use is made of the point of view that the quantum statistics of quasi-particles undoubtedly defines such macroscopic low-temperature properties of crystals as specific heat and thermal conductivity, electrical conductivity and magnetism. However, some authors only assign to quantum mechanics the role of describing microscopic laws explaining the dynamics of atomic particles. The discovery of high-temperature superconductivity transferred the quantum manifestations of particle dynamics into the region of moderately high temperatures. State-of-the-art experimental techniques make it possible to observe a number of phenomena involving macroscopic quantum tunneling. Thus, macroscopic solids fairly often exhibit some particular quantum features.

Finally let us stress the role of models in solid state physics. As is well known, a model is a much simplified speculative scheme which allows a researcher, on the one hand, to propose a simple qualitative explanation of a property or phenomenon under study, and, on the other hand, to give a quantitative description of properties and to perform a complete theoretical calculation of characteristics of the phenomenon being studied. Models are utilized very often in solid state physics. Apparently, the most ancient model is that of an "absolutely solid body". The model of perfect gas of quasi-particles is frequently employed; Ising's model is well known in the study of magnetism. The model of a two-dimensional lattice of classical spins in the theory of phase transitions proposed by Onsager allowed him to give the first statistical description of second-order phase transitions. The Frenkel–Kontorova and Peierls models are widely used in dislocation theory. References to them can be found in many of the articles included in this work.

The need for this publication arises from the recent tremendous progress in this field of physics, especially by the extension of its areas of practical applications. There have appeared new fields of electronics (optical electronics, acoustoelectronics); semiconductor electronics has changed and now employs an increasingly wider range of complex structures such as metal–insulator–semiconductor devices and computers with superconducting elements (cryotrons); and memory elements based on new physical principles have been developed.

There are many promising new solid state materials based on garnets, rare-earth metals, magnetic semiconductors, amorphous metals, semiconductors, and, finally, the discovery of high-temperature superconductors in 1986.

The compilers of this Dictionary hope that this publication will be useful for researchers working in solid state and other fields of physics (nuclear physics, plasma physics, etc.) and also for a wide audience of other readers, including, but not limited to, engineers investigating and utilizing solids, as well as doctoral and advanced students at universities.

V.G. Bar'yakhtar

A.M. Kosevich

HOW TO USE
THE ENCYCLOPEDIC DICTIONARY

In this Dictionary, we follow standard fundamental rules generally adhered to in encyclopedic publications.

1. Articles are arranged alphabetically; if a term (typeset in boldface) has a synonym, then the latter is given in the entry after the main meaning of the term (after comma, in usual print).
2. Some terms in the titles of articles adopted from other languages are followed by a brief etymological note (with clear abbreviations: abbr., abbreviated, fr., from; Gr., Greek; Lat., Latin; Fr., French; Germ., German, etc.).
3. The names of scientists appearing in the title or text of an article are commonly followed by the date of obtaining the result (effect, phenomenon) under consideration.
4. We use a system of cross-references (in *Helvetica italics*) to other articles. Additional terms explained within the same article are typeset in *Times italic* and, together with the main entries, are listed in the Subject Index at the end of Volume 2. Many terms synonymous to those in the article titles may be also found in this Subject Index.
5. Units of physical quantities and their symbols are given in the systems of units most commonly used in solid state physics. For the relations which convert these units into SI, see the tables included in the article *Units of physical quantities*.
6. The figures are explained either in their captions or in the text of the article.
7. Usually, the notation used in formulae is explained in the text of that article. However, some symbols have the same meaning throughout the Dictionary (unless otherwise specified): c, speed of light; h and $\hbar = h/2\pi$, Planck constant; T, absolute temperature; k_{B}, Boltzmann constant; λ, wavelength; ν and $\omega = 2\pi\nu$, frequency. The notation of some elementary particles is also standardized: ν, photon and gamma quantum; e, e^-, electron; p, proton; n, neutron; d, deuteron; π and π°, pions; $\mu^\pm$, muons; $K^\pm$ and K°, K-mesons.

Nn

NANO... (fr. Gr. ναυος, dwarf)

Prefix for a physical unit name, which designates a multiple unit equal to 10^{-9} initial units. Symbol: n. Example: 1 nm (nanometer) = 10^{-9} m.

NANOCOMPOSITES

Combined or *composite materials* with sizes of particles of an imbedded phase much smaller than the wavelength of visible light. Nanocomposites consisting of an optically isotropic matrix containing particles of high quadratic or cubic nonlinear electric susceptibility (dielectric constant) exhibit nonlinear optical properties. This allows their use in devices of *integrated optics* for the control of the passage of light beams, including modulation, deflection and optical radiation frequency changes.

NANOCRACK

See *Submicrocracks*.

NANOCRYSTALS

Polycrystalline materials with *crystallite* sizes of 5–10 nm (some authors say 2–50 nm). Nanocrystals are obtained through pressure compacting of ultradisperse powders, which form during vapor condensation in an inert gas atmosphere, and by chemical methods. The density of nanocrystals is 10–15% lower than that of crystalline samples; the density of boundaries is about 600 m^2/cm^3, while about 20% of the atoms belong to boundary or surface layers. *Structure studies* show that boundary layers (about two atomic diameters thick) are significantly disordered and contain many cavities comparable with atomic sizes, as well as larger cavities (10–100 atomic volumes in size). A nanocrystal is a convenient object for the study of solids with inclusions of locally disordered areas. The *specific heat* of nanocrystals is approximately 1.5 times higher than that of *crystals*, and

the *shear modulus* is 25–30% lower. *Internal friction* in nanocrystals is noticeably higher than in crystals. *Diffusion* of atoms takes place mostly at boundaries, with an *activation energy* of self-diffusion 3–4 times smaller than that in crystals. The relatively high *plasticity* is due to diffusive transport of atoms and slippage along the boundaries. High-strength ceramic nanocrystals are obtained by oxidation of metallic samples. Properties of nanocrystals are close to those of amorphous materials, so a possible method for their production is a rapid *crystallization* of amorphous bodies. They can be synthesized with sizes down to 2 nm through methods of colloidal chemistry. Nanocrystal materials may be used as catalysts, high-strength *construction materials*, etc.

NANOELECTRONICS

Subfield of *semiconductor* physics that deals with artificial devices having dimensions or superstructure components (see *Semiconductor superstructure*) equal to or smaller than 10^2 nm in order of magnitude. Such devices, also known as *nanostructures*, can be prepared by high resolution lithography, *molecular beam epitaxy*, and self assembly.

Some variants to the general name "nanostructures" are: *quantum dots, single-electron transistors, Coulomb islets*, and finally *artificial atoms*. Most artificial atoms are created on the basis of *heterojunctions* of the AlGaAs–GaAs type. As the relative dielectric constant of *gallium arsenide* is $\varepsilon = 13$, and the value of the electron effective mass m^* comes to only 7% of the free electron mass m, the effective Bohr radius a^* is $\varepsilon m/m^*$ times greater than a hydrogen atom radius, and equals ≈ 10 nm, whereas the effective Rydberg constant Ry* (*effective Rydberg*) is $m^*/m\varepsilon^2$ times smaller than its standard value, and equals ≈ 6 meV. As a result artificial atoms, like ordinary ones, possess quantum properties at liquid helium temperatures, since their sizes are close to a^* in order of magnitude, and thermal diffusion is small by virtue of $Ry^* \gg k_B T$. The latter statement is confirmed by numerous optical investigations in the IR region of the spectrum, as well as by methods of *tunneling spectroscopy*. In order to enhance the effect, one often works with a large number (up to 10^8) of artificial atoms, embedding them at regular sites of a two-dimensional "crystal lattice", which makes systems under consideration akin to artificial *superlattices*. The positions of spectral levels and, accordingly, of IR absorption lines, may be changed by continuously varying the voltage at the artificial atom gate, as well as by placing nanostructures into external electric and magnetic fields. In particular, when periodic superlattices were placed into a uniform electric field E directed transversely to the layers it was discovered that an electron band split into a periodic set of quantum levels, a phenomenon known as the *Wannier–Stark ladder* (see *Quantum mechanics* in solid state physics); an energetic step of this ladder is equal to $|e|EL$, where e is the electron charge, and L the superlattice constant. Such a spectrum results from the quantization of Bloch oscillations, which were predicted by F. Bloch at the dawn of the development of solid state theory.

Besides a discrete energy spectrum, another quantum property of matter manifests itself in artificial atoms: the discrete nature of charge. This property better matches the alternative name of an artificial atom: *single-electron transistor*. The point is that the addition of charge to an artificial atom increases its energy by $\sim e^2/C$, which, due to the low value of the capacitance (C) of the device, coincides in order of magnitude with the distance between energy levels of an artificial atom. Thus, if there is no voltage at the gate, the transfer of an electron from emitter to collector at $k_B T \ll e^2/C$ is forbidden, since this would entail the climbing of the electron over the barrier, which is $\sim e^2/C$ in height; thus the device carries zero current (the so-called *Coulomb blockade*). The blockade is pierced only at particular gate voltages, which is accompanied by a tremendous increase of conductivity (by 10^3 to 10^4 times). See also *Solid-state quantum electronics*.

NANOTECHNOLOGY, submicrometer technology

A complex of technological processes involved in the development and manufacture of semiconductor *integrated circuits* or other structures with dimensions much smaller than a micrometer. Important processes are X-ray or electron *lithography*, *ion implantation doping*, molecular *epitaxy*, *ion etching*, and plasma *etching*. A pattern on a semiconductor substrate is obtained by exposing a mask, i.e. by irradiating the film made of a special *resist* by either a parallel beam of soft X-ray radiation (*synchrotron radiation*), or an electron beam that is scanned over the resist surface in accordance with a given program. The polymerization of the photoresist film is used for its further removal, and for forming a mask needed either for adding an *insulator* or *metal* layer, or for ion doping. The formation of conductive, semiconducting, or insulating layers is carried out with the aid of particle beams (molecules or ions) in a high vacuum. Related to nanotechnology is *molecular technology* that uses molecular-ordered structures of organic molecules, e.g., *Langmuir–Blodgett films*.

NARROW-GAP SEMICONDUCTORS

Semiconductors which have a narrow *band gap* width. The lower limit to the size of the band gap is represented by semiconductors with a vanishing gap (*gapless semiconductors*), and the nominal upper limit is conventionally taken to be ≈ 0.3 eV. The best known narrow-gap semiconductors are the binary compounds InSb, PbTe, PbSe, *binary alloys* of the general formula $Bi_{1-x}Sb_x$, as well as two classes of isovalent *solid solutions* based on chalcogenides of Groups II and IV: $Cd_x Hg_{1-x}Te$, and also $Pb_{1-x}Sn_xTe$ and $Pb_{1-x}Sn_xSe$. The alloy $Cd_x Hg_{1-x}Te$ behaves as a zero-gap semiconductor for $x < 0.16$; the width of the forbidden zone increasing with increasing x for $x \geqslant 0$. Narrow-band gap semiconductors may be classified as an individual group of *semiconductor materials* owing to their unique physical properties, and their areas of practical application. As a consequence of the narrowness of the band gap, the electron *effective mass* (holes in case of PbTe, PbSe, both pure and alloyed with Sn) is abnormally low, whereas the mobility is extremely high.

Unlike semiconductors with wide gaps, the spectrum of free *current carriers* in narrow-gap semiconductors is highly nonparabolic, with the degree of nonparabolicity increasing with the decrease of the width of the band gap. The electronic properties of $A^{II}B^{VI}$ and $A^{IV}B^{VI}$ compound types are highly sensitive to crystal lattice *defects* involving deviations from *stoichiometric composition*, i.e. to excess quantities of metal or chalcogenide atoms. The predominance of metal *vacancies* results in p-type *electrical conductivity*, whereas dominating chalcogen vacancies give rise to n-type conductivity.

The widespread interest in studies of narrow-gap semiconductors is explained by the fact that these materials are used as infrared (IR) devices such as *photodetectors*, laser *diodes*, and semiconductor radiators operating over the wave range 4–30 μm. Narrow-band semiconductors with a predetermined band gap may be adjusted by varying the ratio of the alloy components. This allows the construction a device for a particular wave band. Binary compounds of elements of Groups III–V (InSb, InAs) are used for producing liquid-nitrogen cooled photodetectors that exhibit high performance over the range of 3–5 μm, which corresponds to the first atmospheric spectral window. Photodetectors based on polycrystalline layers of lead selenide may also be used over this operating range. The latter photodetectors, which either do not require cooling at all or need slight thermoelectric cooling for their operation, are the simplest in design, and readily available for the IR spectral region. The early investigations of mixed crystals HgCdTe and PbSnTe were aimed at creating refrigerated IR detectors for the wave band 8–14 μm, which corresponds to the second atmospheric spectral window, and this is of considerable interest for communication. All the above-listed photodetectors may be used for measuring temperature gradients in the environment, or for medical purposes, for their spectral range corresponds to the *thermal radiation* maximum in the neighborhood of room temperature. *Tunable diode lasers* based on lead chalcogenides are widely used for producing spectral devices that feature high resolving power, and are used in scientific investigations, as well as for analyzing gas mixtures in industry and ecology.

NARROWING OF MAGNETIC RESONANCE LINE

Line width decrease due to modulation of the *paramagnetic center* resonance frequency by dynamic processes within the spin system. A classical example of this line narrowing is the strong decrease of the magnetic dipole line width in liquid and gaseous samples compared to their values in solids at the same concentration of *magnetic moments*. This kind of narrowing is called *motional narrowing* since it is due to the rapid motion of atoms in the environment. A rapid random motion of particles produces fluctuations of the internal *local electric fields* that act on every *spin*. As a result, within a time greater than the fluctuation duration, only the mean value of the local field is effective, a value much smaller than the instantaneous value, so that line narrowing occurs. To produce motional narrowing of the line, the field fluctuations must be rapid compared to the *Larmor precession* in the local field. Since the precessional frequency in the local field is of the order of magnitude of the resonance line width $\delta\omega$ (with static surrounding spins), the condition for line narrowing is $\delta\omega \cdot \tau_c \ll 1$ (here τ_c is the local field correlation time). The width $\Delta\omega$ of the line narrowing can be qualitatively estimated as the inverse of the time interval Δt during which two spins, precessing in corresponding local fields which are initially in phase, become out of phase by the order of $\pi/2$, and the width of the narrowed line becomes $\Delta\omega \sim (\delta\omega)^2 \cdot \tau_c$.

The modulation of local fields which leads to narrowing of the line can also account for the rapid reorientation of neighboring spins. Such a type of narrowing appears, e.g., in NMR pulse experiments, and this is called *pulse line narrowing*. The essence of pulse narrowing consists of exciting the sample by a repeating sequence of radiofrequency pulses which turn the spins in various directions through large angles. This process, like the narrowing of a line due to molecular motion, averages out the *dipole–dipole interaction* and reduces the dipolar width.

Rapid reorientations of neighboring spins which induce line narrowing can also take place as a result of internal interactions in the spin system. Such a phenomenon occurs, in particular, in the presence of a strong *exchange interaction* between spins which induce *exchange narrowing*.

The exchange interaction operator is expressed in the form of a sum of scalar products of *spin operators*; therefore, in the case of a system of identical spins this does not contribute to the second moment of the line M_2 (see *Moments of spectral lines*), the usual indicator of line broadening. However, owing to the presence of spin operators of the type $S_{i+}S_{j-}$ (*i* and *j* denote different spins) in the exchange interaction which induces flip-flop transitions (see *Antiferromagnetism*), a random modulation of the dipole and hyperfine local fields takes place that results in the narrowing of the line. Quantitatively, this exchange interaction narrowing brings about an increase of the fourth moment, so the observed magnitude of the width appears considerably smaller than $(M_2)^{1/2}$, in contrast to the case of the pure dipole–dipole interaction.

NEARLY-FREE ELECTRON APPROXIMATION

Method of *band theory of solids* based on the use of perturbation theory with the crystalline potential (weakly-bound electron approximation) considered to be a small perturbation compared to the free particle kinetic energy. This method finds its explanation in the *pseudopotential method*. It is used for determining the *Fermi surfaces* of metals.

NECK

Local narrowing in a sample appearing under the effect of tensile *strain* arising from an acting stress increase which is not balanced by an increase of material hardening, $d\sigma/de$, where σ is the stress and *e* is the so-called *true strain* (see *Strain hardening*). The sample loses its mechanical stability (first stage) at a maximum stretching effort P_{max}, fixed in a load diagram expressed in the coordinates load versus elongation. By tradition, a neck features relative transverse narrowing $\psi = 1 - (d_2/d_1)^2$, where d_1 and d_2 are the initial and final diameters of the sample, or a *conditional strain* ε (%). In this the true strain is given by $e = \ln[1/(1 - \psi)]$. The magnitude of the strain corresponding to the beginning of neck formation depends on the material structure and chemical composition, the temperature, the rate parameters of the load process, and the geometry of the sample surface. For the case of parabolic hardening satisfying the dependence $\sigma = e_n$, the conditional strain of the neck formation is given by $\varepsilon_{max} \approx n(\sigma_B - \sigma_S)/\sigma_B$, where σ_B is the *ultimate strength*, and σ_S is the *yield limit* of the material. The appearance at the neck of the uniaxial tension leads to the change of a uniform *state of stress* to a triaxial one. At $\varepsilon > \varepsilon_{max}$ (second stage) the *plastic deformation* becomes rather heterogeneous. For conditions of increasing nonuniformity of stress distribution, it is accompanied by the hardening of the neck material, and by the dense local occurrence of pores. At the third stage, in the central part of neck at the point of maximum triaxial tension resulting from the merging of the pores, a microcrack is formed (see *Crack*) which subsequently increases and brings the deformation process to an end by *failure* (*tough failure*, *quasi-brittle failure*). In the neck a texture of the axial type arises.

The word neck is also used to designate narrow connections in adjacent *Fermi surfaces*.

NEEDLE-SHAPED DEFECT

A linear *defect in crystals* corresponding to a closed *Burgers contour*. If $u(x_1, x_2, x_3)$ (where x_1, x_2, x_3 are Cartesian coordinates of all crystal points except those on the defect line) is the *displacement vector* due to a linear defect, and d*l* is the element of a sufficiently large closed contour drawn around the linear defect, then the integral

$$I = \oint \frac{\partial u}{\partial l}\, dl$$

vanishes for a needle-shaped defect. Thus, at the appearance of a needle-shaped defect in an *ideal crystal* with a closed Burgers contour, the Burgers contour remains closed (although its enclosed area may change). In an elastically-isotropic crystal (see *Isotropy of elasticity*), the lattice atom displacements caused by a rectilinear needle-shaped defect are described by the vector

$$u = Cr + \frac{Dr}{r^2},$$

where r is the vector directed from the defect axis to a given lattice point in a plane normal to the defect axis, and C and D are constants dependent on the elastic properties and dimensions of the crystal, properties of the defect, boundary conditions, and so on. A needle-shaped defect can be exemplified by chains composed of *vacancies* or *impurity atoms*. The stability of such chains is low as they easily break up into *point defects* under atomic *diffusion*. Also classified as needle-shaped defects are thin needle-like precipitates of phase components in *alloys*.

NÉEL EQUATION (L. Néel, 1952)

A phenomenological equation describing the displacements of a magnetic *domain wall* (its quasi-stationary dynamics) in a *magnetic substance* during its *magnetic reversal* in the presence of *magnetic viscosity* effects:

$$P(u,t) + R(u) + (M_1 - M_2)B(t) = 0, \quad (1)$$

where $B(t)$ is the external magnetic field; M_1 and M_2 are *magnetization* vectors in adjacent domains separated by the domain wall between them; u is the displacement of the wall; $R(u)$ is the pressure on the wall produced by defects and imperfections in the material plus that from *demagnetization fields*; and $P(u,t)$ is the pressure on the wall due to induced *magnetic anisotropy* and magnetic viscosity effects. In contrast to $R(u)$, the pressure $P(u,t)$ is explicitly time dependent. Moreover, beside its dependence on the displacement u at the instant t, $P(u,t)$ also depends on the earlier displacements at $t' < t$. To account for this, $P(u,t)$ is expressed as

$$P(u,t) = \int_{-\infty}^{t} F\{u(t), u(t'), t, t'\} dt',$$

where $F\{\ldots\}$ is a certain functional depending on the type of domain wall, the nature of the induced anisotropy, etc. The third term in the Néel equation defines the magnetic field pressure on the wall. Thus, the Néel equation describes the domain wall equilibrium condition, which is the condition for zero total pressure affecting the wall. The Néel equation applies when the intrinsic *relaxation* of the magnetic vector to equilibrium proceeds much faster than the relaxation of the induced anisotropy, and when the inertia (mass) of the wall can be neglected (the latter is permissible at low wall speeds). The Néel equation is used to describe the dynamics of *domain walls* in *ferroelectrics*, which involves making the replacements $M \to P$, $B \to E$ in Eq. (1), where P is the polarization vector and E is the electric field.

NÉEL POINT (L. Landau, 1933)

Temperature at which magnetic ordering appears in *antiferromagnets* and *ferrimagnets*. They lose their specific magnetic properties above that point and behave as *paramagnets* whose susceptibilities follow the *Curie–Weiss law*. A maximum of the *magnetic susceptibility*, as well as anomalies in certain other quantities (such as conductivity, sound absorption factor, etc.), are observed in the vicinity of the Néel point. See also *Curie point*, *Magnetic phase transitions*.

NÉEL WALL

A type of magnetic *domain wall* in a thin *magnetic film*, in which its *magnetization* rotates in a

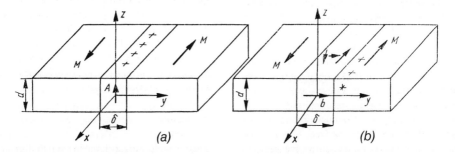

Schematic diagram of Bloch wall (a) and Néel wall (b), where δ is the domain wall width. Arrows A and B show the magnetization M directions in central planes parallel to xz in the Bloch wall and parallel to xy in the Néel wall, respectively; crosses indicate the magnetostatic poles related to the exit of M to the film surface and domain wall surface, respectively.

plane parallel to the film surface. Physically, the reason for the formation of a Néel wall, instead of a *Bloch domain wall*, is a decrease of magnetostatic energy. Indeed, consider a domain wall separating two adjacent *domains* and let the magnetization in it change in a plane parallel to the surface of the wall itself (xz plane in Fig. (a)). *Magnetic charges* will then appear on opposite surfaces of the film in bands of width δ. An additional magnetostatic energy of the crystal is associated with these charges, and it grows while the film thickness, d, decreases. For $d < \delta$ that energy may be lowered if the magnetization vector M rotates in a plane parallel to the film surface (xy plane in Fig. (b)). Thus, there is a critical film thickness d_0 below which the Néel wall is energetically more favorable (see Fig. (b)); a Bloch wall should exist above d_0. The value of d_0 depends on the nature of the film, and may be as large as hundreds of angstroms.

NEGATIVE CRYSTALS

A type of three-dimensional *defect in crystals*, which represents polyhedral cavities with a symmetry corresponding to the microscopic *crystal symmetry*. Such cavities nucleate during crystal growth (in particular, in nature) or at the dissolution of crystals. During the course of growth negative crystals can capture ambient material, which results in the formation of solid, liquid, or gaseous inclusions.

NEGATIVE LUMINESCENCE

The decrease of the radiation by a system compared to its equilibrium *thermal radiation* in the same spectral region. Negative *luminescence* results from the decrease in population of a series of excited energy levels of the system, and the corresponding decrease of the number of radiative transitions at particular frequencies. In this situation the absorption of background thermal radiation by the system predominates over its *spontaneous radiation*, and the resultant radiation energy flux in the indicated frequency range comes from outside to the system surface. The *luminescence power*, defined as the difference between the powers of spontaneous radiation and of absorption, proves to be negative; which is the reason for the term "negative luminescence". Under total suppression of spontaneous radiation,

the absolute value of the luminescence radiation power attains its ultimate magnitude which equals the power of the background radiation absorbed by the system. Negative luminescence in systems with a discrete energy spectrum is usually excited by pumping. In the fundamental absorption band of *semiconductors*, negative luminescence arises through the *depletion* of electrons and holes. The *magnetic concentration effect*, and other methods are used to determine this. Negative luminescence in semiconductors serves as the foundation for a new class of optoelectronic devices, such as high-temperature sources of IR radiation, optical temperature-sensitive elements, etc. (see *Optoelectronics*).

NEGATIVE MAGNETORESISTANCE

A decrease of conductor resistance in a *magnetic field*. The effect is observable in the low-temperature range for a wide class of *disordered solids*, and manifests itself most clearly in systems with reduced dimensionality (in metal *films* and wires, two-dimensional electrons of *heterostructures*, etc.). The explication of this phenomenon involves some difficulties, since both the classical theory of magnetoresistance (which allows for the electron trajectory bending in a magnetic field) and its treatment under *hopping conductivity* (due to the magnetic field contracting the wave functions of localized states) predict an increase of resistance in the magnetic field. The mechanism suggested in the 1960s is related to scattering by magnetic impurities (when the magnetic field suppresses the process of *electron spin-flip scattering*, see *Kondo effect*) and proves to be inapplicable to most of the systems under consideration, where the concentration of such impurities is small. In the early 1980s a theory was put forward to relate the negative magnetoresistance to suppression of the interference quantum correction to the conductivity of a disordered system (see *Weak localization*). This approach explains all specific qualitative features of negative magnetoresistance: (i) the appearance of the effect in weak fields and its square-root (or logarithmic in the two-dimensional case) dependence on the magnetic field strength; (ii) its independence of the angle of the current to the field in the three-dimensional case (while in the two-dimensional case the effect is determined only by the magnetic field component normal to the layer); (iii) its strong temperature dependence caused by suppression of interference during elastic scattering (on account of the electron–phonon and electron–electron processes). A quantitative description of the experiments calls for, as a rule, taking into account the factors that produce a positive contribution to magnetoresistance (spin-flip scattering, quantum corrections due to electron interaction, and so on). See also *Galvanomagnetic effects*.

Materials consisting of alternating nanometer-thick layers of a ferromagnetic material and a nonmagnetic metal experience a very large decrease in resistance in a magnetic field called *giant magnetoresistance*. Examples of such materials are alternating layers of iron and chromium, and multilayers of cobalt and copper. Even larger changes in magnetoresistance called *colossal magnetoresistance* have been found in the perovskite-like material $LaMnO_3$ which has manganese in the Mn^{3+} valence state. Replacing some of the La^{3+} ions by divalent Ca, Ba, Sr, Pb or Cd ions produces the mixed-valence Mn^{3+}/Mn^{4+} system which has many mobile charge carriers. The particular combination $La_{0.67}Ca_{0.33}MnO_x$ displays more than a thousand-fold change in resistance in a 6 T magnetic field.

NEGATIVE PHOTOCONDUCTIVITY in a magnetic field

Weakening of *photoconductivity*, which takes place under *injection* of charge carriers into the central part of a long semiconductor sample placed in a strong *magnetic field B*. For instance, it may be an *n*-type semiconductor where the Hall field strength E_y (see *Galvanomagnetic effects*) is greater than the strength E_x of the electric field producing the current flowing through the sample along the positive *x*-axis (see Fig.). The injected holes enter the sample and drift in the field E_y; the *Lorentz force* shifts them along the OX axis in the direction opposite to that of the force eE_x acting on them, and this produces the negative photoconductivity.

NEGATIVE RESISTANCE

Electric resistance with a negative sign.

(a) For alternating current, it means a negative magnitude of the real part of the complex *impedance* $Z(\omega) = R(\omega) + iX(\omega)$. Under harmonic action with the frequency ω across a subcircuit containing both resistive and reactive impedance, a phase shift δ appears between the current and voltage values, which depends on the circuit parameters and the frequency. In this case, $R(\omega)$ (involving the current amplitude I and voltage V in the relation $IR(\omega) = V\cos\delta$) can be negative over a certain frequency range. Then, *dynamic negative resistance* is spoken about. Across subcircuits that have dynamic negative resistance, the release rather than the absorption of energy takes place.

(b) In the static regime, there is a *negative differential resistance* $R = dV/dI$ that can be negative over a limited part of the *current–voltage characteristic*. The appearance of the negative value can be due to the dependence of the *mobility of current carriers* or their concentration on the electric field. The negative differential resistance leads to instability of the electric circuit, and its spontaneous transition to another state, or to a regime of oscillations.

The negative resistance characteristic is utilized for the operation of various semiconductor devices; see *Gunn diode, Generation type semiconductor diodes, Barrier injection transit time diode, Microwave semiconductor devices*.

NEMATIC LIQUID CRYSTAL, nematic (fr. Gr. νημα, a thread)

A phase of a *liquid crystal* that features long range orientational molecular ordering (see *Orientational order*), but has no translation ordering. Nematic liquid crystals form either during melting of a *molecular crystal* (*thermotropic nematic liquid crystal*), or as a result of the dissolving of

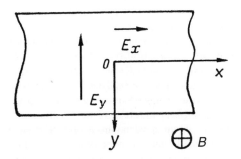

Negative photoconductivity.

molecular compounds in water or other solvents (*lyotropic nematic liquid crystal*). The structural units of the latter are micelles comprising hundreds of molecules, instead of individual molecules (see *Lyotropic liquid crystals*). Nematic liquid crystals are also formed by liquid-crystal *polymers*, in which the preferred direction is specified by ordering either the links of the basic chain or its side branches. Orientational ordering may pertain to either all three principal molecular axes, to two of them (biaxial nematic liquid crystal), or to only one of them (uniaxial nematic liquid crystal). The biaxial type is rarely found. Among the uniaxial types are *calamitics* and *discotics*, with structural elements having the shape of rods and discs, respectively. Orientational ordering of molecules results in a strong anisotropy of the physical properties of nematics. They are birefringent, usually uniaxial media (see *Order parameters in liquid crystals*). Elastic properties of nematic liquid crystals are defined by their orientational strains, which occur when a *director* (prescribing the predominant orientation direction of the molecules) is locally forced out of its equilibrium position (see *Elasticity of liquid crystals*).

Combining a strong anisotropy of their optical and electromagnetic properties with a capability to easily reorient their director when affected by weak external fields, nematic liquid crystals enjoy various uses in systems of recording, processing, and presentation of optical data (see *Frederiks transition*, *Twist effect*, *Guest–host effect*, *Electrooptical effects in liquid crystals*). Nematics owe their name to *disclinations* – linear defects in the director field. Microscopic observations show these to be flexible threads found both within the bulk and at the surface.

NEODYMIUM, Nd

A chemical element of Group III of the periodic system with atomic number 60 and atomic mass 144.24; it belongs to *lanthanides*. Natural neodymium consists of isotopes 142 (27.11%), 143 (12.17%), 144 (23.85%), 145 (8.30%), 146 (17.22%), 148 (5.73%), and 150 (5.62%), the isotope ^{144}Nd is weakly radioactive with half-life of about $5 \cdot 10^{15}$ years, and 17 additional radioactive isotopes are known with mass numbers from 129 to 154. Outer electronic shell configuration is

$4f^4 5d^0 6s^2$. Ionization energy is 5.51 eV. Atomic radius is 0.1816 nm, Nd^{3+} ion radius is 0.102 nm. Oxidation state is $+3$. Electronegativity is ≈ 1.12.

In free form, neodymium is a silvery-gray *metal*. It exists in two modifications (α-Nd and β-Nd). α-Nd is stable below ≈ 1133 K with a double close-packed lattice (alternation of hexagonal atomic layers to ABAC...), space group is $P6_3/mmc$ (D_{6h}^4); $a = 0.36582$ nm, $c = 1.17966$ nm under normal conditions. Stable above 1128 K and up to $T_{melting} = 1291$ K is β-Nd with body-centered cubic lattice, space group is $Im\bar{3}m$ (O_h^9); $a = 0.413$ nm (extrapolated value) after hardening. There are contradictory data that the transition α-Nd $\to$ β-Nd is preceded by its transformation into a (possibly, metastable) allotropic modification with face-centered cubic lattice having conditional unit cell constant $a = 0.480$ nm after hardening, space group $Fd\bar{3}m$ (O_h^5). The density is 7.010 g/cm^3 at 293 K. $T_{boiling} \approx 3340$ K. Binding energy of neodymium is -3.35 eV/atom at 0 K. Heat of melting heat is 7.16–10.9 kJ/mole; heat of sublimation is 288.9 kJ/mole, heat of evaporation is 296 kJ/mole; Debye temperature is ≈ 152 K. The linear heat expansion coefficient of neodymium polycrystal is $6.7 \cdot 10^{-6}$ K^{-1} (at 298 K), coefficient of thermal conductivity is 12.98 W·m^{-1}·K^{-1} (at 299–303 K). Isothermal bulk modulus is 32.6 GPa (at room temperature and zero external pressure); Young's modulus is 37.9 GPa (at room temperature); ultimate tensile resistance is 0.171 GPa (in a cast sample at 293 K); relative elongation is 11%. Vickers hardness (at 293 K) is 35 for cast neodymium and 76 HV for forged Nd. High purity neodymium has high cold deformability (up to 70% reduction) without failure. Application of intermediate annealing allows obtaining thin strips (0.5 mm thick and less). Effective thermal neutron trapping cross-section is 44.6 barn. Electric resistivity of α-Nd polycrystal is 643 nΩ·m (at 298 K), temperature coefficient of electric resistance is 0.00164 K^{-1}. Work function of neodymium polycrystal is 3.3 eV. Magnetic first-order phase transition of the order–disorder type occurs in neodymium at the Néel point $T_N \approx 19.5$ K: the magnetic moments of atoms in neighboring hexagonal layers become antiparallel. At the temperature 7.5 K, the ordering of the magnetic moments of atoms with cubic nearest neighbor symmetry takes place (with a change in the effective number of atoms with ordered magnetic moments). The antiferromagnetic structure of hexagonal layers with cubic environmental symmetry in α-Nd is similar to the structure of layers with hexagonal nearest neighbor symmetry, but with magnetic moment directions in the former turned by $30°$ relative to those of the latter. The magnitudes of magnetic moments of atoms in the layers of both kinds vary as the sine. Magnetic susceptibility of neodymium is $5600 \cdot 10^{-6}$ CGS units; nuclear magnetic moment of isotope ^{143}Nd is 1.25 nuclear magnetons. There are suggestions that the nonequilibrium face-centered cubic modification transforms into the ferromagnetic state at the Curie point $T_C = 29$ K. Neodymium is used as an alloying addition in light alloys (e.g., *magnesium alloys*), as a component for mischmetal, in laser materials, etc.

NEPTUNIUM, Np

An *actinide* chemical element with atomic number 93 and atomic mass 237.048. There are no stable isotopes, and 14 known radioactive isotopes with mass numbers from 228 to 241.

Outer shell electron configuration is $5f^4 6d^1 7s^2$. Atomic radius is ≈ 0.1525 nm; radius of Np^{3+} ion is ≈ 0.0998 nm, radius of Np^{4+} is ≈ 0.093 nm, radius of Np^{7+} is 0.071 nm. Oxidation state is $+5$, more rarely $+2$, $+3$, $+4$, $+6$, $+7$. Electronegativity is ≈ 1.25.

In the free form neptunium is a heavy silvery *metal*. It exists in α-, β-, γ-modifications. Stable at temperatures below ≈ 550 K is α-Np with an orthorhombic crystal lattice (8 atoms in unit cell; space group $Pnma$, D_{2h}^{16}). At 293 K, $a = 0.4722$ nm, $b = 0.4887$ nm, $c = 0.6662$ nm. Stable above 542 K is β-Np with tetragonal crystal lattice (4 atoms in unit cell, space group $P42_1 2$, D_4^2). At 586 K, $a = 0.4896$ nm, $c = 0.3387$ nm. At temperature ≈ 830 K there occurs a polymorphic transformation of β-Np to γ-Np, stable up to $T_{melting} = 910$ K. γ-Np has a body-centered cubic lattice (space group $Im\bar{3}m$, O_h^9); at 873 K, $a = 0.3525$ nm. The density is 20.45 g/cm^3 for α-Np, 19.36 g/cm^3 for β-Np, and 18.0 g/cm^3 for γ-Np; $T_{boiling} = 4174$ K. Bonding energy of neptunium is -4.55 eV/atom at 0 K. Heat of melting is

9.6 kJ/mole, heat of evaporation is 394.8 kJ/mole. Debye temperature of α-Np is 188 K. For neptunium of 99.97% purity the temperatures of polymorphic phase transitions: $\alpha \rightarrow \beta$ at 555 K, $\beta \rightarrow \alpha$ at 542 K, $\beta \rightarrow \gamma$ at 832 K, $T_{melting} = 870$ K (start); the linear thermal expansion coefficient grows from $11.1 \cdot 10^{-6}$ K^{-1} (at 100 K) up to $32.7 \cdot 10^{-6}$ K^{-1} (at 525 K) for α-Np, and from $42.8 \cdot 10^{-6}$ K^{-1} (at 550 K) up to $78.1 \cdot 10^{-6}$ K^{-1} (at 850 K) for β-Np. Calculated value of isothermal bulk modulus is 68 GPa (at room temperature). Electric resistivity of α-Np polycrystal is 35.5 n$\Omega \cdot$m at 4.2 K and 1176.5 n$\Omega \cdot$m at 295 K. The $5f$-electrons in neptunium are delocalized (partially full $5f$-shell is somewhat expanded). Accordingly, neptunium may have a tendency to superconduct for $T_c < 0.4$ K, and Pauli paramagnetism is present in it. On the other hand, as the density of states of $5f$-electrons at the Fermi level is not high, and the criterion of band ferromagnetism is not met, neptunium is an exchange-coupled band paramagnet with magnetic susceptibility about $5 \cdot 10^4$ cm^3/mole. The longest-living isotope ^{237}Np (α-radiator) with lifetime $T_{1/2}$ about $2.14 \cdot 10^6$ years is obtained in nuclear reactors. It finds application in the production of the nuclide ^{238}Pu, which is used in space exploration, nuclear micropower engineering, medicine etc.

NERNST–ETTINGSHAUSEN EFFECT

See *Thermomagnetic phenomena*.

NERNST GLOWER (W. Nernst, 1900)

A rod made from a mixture of refractory *metal* oxides (mainly *zirconium* with *magnesium*) heated by an electric current ($T \sim 1500°$C) that can be used as an IR range radiation source. However it is rarely used for this purpose, since it is only efficient in the short-wave band ($\lambda < 10$ μm). Moreover, because of its poor *electrical conductivity* at room temperature it needs preliminary heating to $\sim 800°$C. The latter circumstance is related to the ionic nature of the electrical transport in ZrO_2, which is a *solid electrolyte*.

NETWORK, PLANATOMIC

See *Planatomic networks*.

NETWORK, RANDOM

See *Random continuous network*.

NEUTRALIZATION, AUGER

See *Auger neutralization*.

NEUTRAL PARTICLE CHANNELING

The motion of uncharged particles along the channels formed by layers or fibers of an inhomogeneous medium. Neutral particle channeling is similar to the *channeling* of charged particles in crystals. In both cases, the transverse motion of a particle is restricted by potential barriers (channel walls) and the particle oscillates between them at grazing angles $\theta < \theta_c$, where $\theta_c = \arcsin(U/E)^{1/2}$ is the *channeling angle*, U is the average barrier height, and E is the particle energy. However, in the case of neutral particles, the barrier is formed by non-Coulombic forces (nuclear, magnetic, etc.). The corresponding potential averaged over the volume $V \sim a^3$, where a is the interatomic distance, is smaller than the Coulomb potential by 5 to 6 orders of magnitude. Therefore, the field of an individual atomic plane or chain is typically too weak to hold a neutral particle in the channel; interfacial tunneling occurs, and neutral particle channeling in the crystal proves to be a rare event. Quite a different situation arises in a *superlattice*, e.g., in the structure of pairwise alternating layers of a material with different physical and chemical properties. With a sufficient thickness of the layers d_1 and d_2, the transverse tunnel drift disappears, and the neutral particles channel between layers with greater U, if $\theta < \theta_c$. The period of a corresponding superlattice is $d = d_1 + d_2 \gg \pi/(k \sin \theta_c)$, where k is the value of the particle wave vector. Since $\pi/(kd)$ is the Bragg angle $\theta_B^{(d)}$ for the period d, this yields the criterion for the possibility of channeling

$$\zeta \equiv \frac{\sin \theta_c}{\sin \theta_B^{(d)}} \gg 1.$$

The dimensionless parameter ζ determines the "channeling" properties of a medium for given particles. If the particle is charged, the criterion can be met even at $d \cong a$, i.e. for an ordinary crystal lattice; in the case of neutral particles, it is met for superlattices with $d = (10\text{–}10^3)a$.

Neutral particle channeling is possible also in a superstructure with amorphous layers. By means of natural or artificially created superstructures, the neutral particle channeling effect can be used, e.g., for effective control of *neutron* or high-energy *photon* fluxes.

NEUTRON DIFFRACTOMETRY

Techniques to study the *crystal structure* and *atomic magnetic structure* of crystalline condensed media, based on the phenomenon of the *Bragg diffraction* of neutrons. These are distinguished as *structural neutron diffractometry* and *magnetic neutron diffractometry*, respectively. Similar to *X-ray structure analysis*, neutron diffractometry is based on analyzing the angular intensity distribution of neutrons that emerge from a crystal after undergoing scattering processes that deflect them from the incoming beam direction. However, in certain aspects, neutron diffractometry differs noticeably from X-ray techniques of studying crystal structure. The cross-sections of neutron scattering by atomic nuclei in a crystal structure depend irregularly on the atomic number of the scatterer. This allows studying structures that include both light and heavy atoms. Since neutrons feature a non-zero *spin* and a *magnetic moment* resulting from it, they may be used to study magnetic ordering in condensed media.

Neutrons have a remarkably high penetration depth. Thermal neutrons carry kinetic energy close to that of thermal excitations, and this property is used to study the dynamics of condensed media. A drawback of neutron diffractometry is the low intensity of available neutron beams, orders of magnitude lower than that of X-ray beams.

NEUTRON DIFFRACTOMETRY, STRUCTURAL

See *Structural neutron diffractometry*.

NEUTRONS in solid state physics

A neutron is a neutral particle with a long lifetime (free state half-life is $6 \cdot 10^2$ s). It has spin $1/2$, rest mass $m_n = 1.6748 \cdot 10^{-27}$ kg, and the *magnetic dipole* moment $-1.91321 \mu_{nuc}$, where μ_{nuc} is the nuclear *magneton*. On entering a solid, a neutrons interacts with the atoms, the predominant interaction depending on the neutron energy. High-energy neutrons may dislodge atoms from their equilibrium positions at *crystal lattice* sites, a process which produces various radiation defects. The capture of neutrons by atoms of a solid results in *nuclear doping* of the solid. Such nuclear reactions are also used in neutron *activation analysis*. Thermal neutrons may be scattered by the atoms of solids, a phenomenon which underlies the *neutron diffractometry* techniques.

In contrast to X-rays and electrons, neutrons are scattered by the atomic nuclei. The basic characteristics of interactions of neutrons with atomic nuclei are the effective *scattering amplitude* and the *cross-section*. Both parameters have been thoroughly studied and tabulated. Neutron scattering by nuclei may be of either a potential or a resonance nature. These two types of scattering govern the amplitude, f, of the scattering of slow (thermal) neutrons which is given by the *Breit–Wigner formula* (G. Breit, E. Wigner, 1936):

$$f = f_n(\mathbf{k}) - \frac{\Gamma_n(2k)}{(E - E_0) + i\Gamma/2}.$$

The first term characterizes potential scattering due to the direct interaction of neutrons with the force field of the nucleus. The second term describes resonance scattering that involves an intermediate excited nucleus formed during the course of the nucleus–neutron interaction. Here f_n is the amplitude of the potential scattering, $\mathbf{k}$ is the wave vector of the incident neutron, E is its energy, E_0 is the resonance energy, Γ is the width of the excited nuclear energy level (proportional to the reciprocal of the excited state lifetime); Γ_n is the width of the resonance energy level (proportional to the probability of composite nucleus decay during elastic neutron scattering). The amplitude of potential scattering is always positive, and proportional to $A^{1/3}$ (A is the mass number of the scattering nucleus). For resonance scattering, the value of f depends on the incident neutron energy, and on the position and half-width of the resonance level of the nucleus. These two inputs combine to produce different amplitudes for the scattering of thermal neutrons by different isotopes of the same element, so that it may be either positive or negative, real or imaginary.

A neutron has a magnetic moment and interacts with the electronic magnetic moment of the atom. The atom can also have a nuclear magnetic moment, but its much lower value makes it difficult to observe by neutron scattering. The scattering of neutrons by atomic electron shells depends on both the spatial and energy distributions of the electrons that generate the magnetic moment (mainly d and f unpaired electrons). The magnetic scattering *form factor*, which depends on the scattering vector (angle), describes such scattering.

Neutron scattering may be classified according to several criteria: (a) the change in the energy of scattered neutrons (elastic versus inelastic); (b) the degree of coherence of scattered neutron waves (coherent versus incoherent). Coherent elastic scattering is structure-sensitive, and takes place when the momentum of the neutron is transferred to the crystal solid as a whole (scattering vector equals site vector of *reciprocal lattice*), resulting in the *thermal neutron diffraction*. This phenomenon underlies the techniques of *structural neutron diffractometry*, and *magnetic neutron diffractometry*. If, during the process of scattering, the neutron exchanges energy with one of the atoms, generating or absorbing either a *phonon* or a *magnon* in the process, there takes place coherent *inelastic neutron scattering*. In that case the scattering vector differs from that of the site of the reciprocal lattice by the wave vector of the phonon or the magnon. *Neutron spectroscopy* techniques are based on this effect. Techniques for studying the dynamics of atom groupings in solids and fluids, which all use inelastic incoherent scattering of neutrons, also belong to the same category. Incoherent elastic neutron scattering takes place in atomically and magnetically ordered systems, and is used correspondingly rarely. Beside the nuclear and the magnetic interactions, there are other weaker (by 4 orders of magnitude) interactions between neutrons and atomic charges (J. Schwinger, L.L. Foldy).

To conduct experiments on the interactions of neutrons with solids, atomic reactors and pulsed sources are used. To slow down fast neutrons produced during fission reactions, slowing materials (moderators) are used. Upon exiting the reactor, the white (broad wavelength range) neutron radiation is monochromated by reflection from a *monocrystal* (*constant wavelength technique*). Modern sources produce monochromatic neutron beams with intensities up to 10^6–10^7 N·cm^{-2}s^{-1}. Fast neutron pulse reactors generate bursts of white neutron radiation (approximately 100 μs long). Electron and proton accelerators are also used to produce neutron fluxes by generating braking γ-quanta (Bremsstrahlung) that strike a target containing heavy nuclei, and release neutrons through photonuclear transformations. The resulting neutron bursts vary within 1–5 μs in duration. When operating pulsed neutron sources, the *time-of-flight method* is employed, where the spectral expansion along the flight path is carried out using a time analyzer and fixing the angle of scattering (or prescribing a set of discrete angles).

NEUTRON SCATTERING, INELASTIC

See *Inelastic neutron scattering*.

NEUTRON SPECTROSCOPY

A set of techniques for studying atomic and molecular vibrations, as well as magnetic excitations (*magnons*) in condensed matter, which are based on inelastic scattering of neutrons (see *Neutrons in solid state physics*). An interaction of a neutron beam with an atom vibrating at a *crystal* lattice site may be treated formally as a collision of a neutron with a *phonon*, so the *conservation laws* of energy and momentum may be applied to that process. Thus one may describe the change in energy and momentum of neutrons during the collision and plot a system of dispersion curves for monocrystal samples, or retrieve the *density of states* of the phonons for polycrystals.

Experimental studies of the spectrum of *inelastic neutron scattering* involve measuring the change in neutron momentum (wavelength) as a function of sample orientation with respect to the neutron beam, and the angle of scattering. There are many schemes available for such measurements involving both stationary and pulsed neutron sources. The most popular among them employs a three-crystal spectrometer, a monochromatic incident beam, and a crystal analyzer positioned behind the sample. By tuning the analyzer to different wavelengths in the scattered emission one measures the so-called *phonon reflections*, their maxima relating to points on the dispersion curve, while their half-widths relate to lifetimes of the phonons. Magnon spectra may be similarly studied. Unlike *optical spectroscopy*, which only yields information on the long-wave part of the spectra of optical phonons in solids, neutron spectroscopy makes it possible to study both optical

and acoustic phonon spectra over the entire range of momenta (wavelengths).

NICKEL ALLOYS

Alloys of *nickel*, alloyed by one or several elements. Nickel forms substitutional solid solutions with Fe, Cr, Cu, Mn, Co, Mo, W and other metallic elements over a wide temperature range. The solutions have a face-centered cubic lattice imparting them with some *plasticity*. The alloys *harden* because of a decrease of the energy of *stacking faults*, reduction of the diffusion mobility of atoms, and changes of electronic structure in the course of *alloying*. Structural nickel alloys (except for *Monel metal*) are low concentration alloys known for their high emission properties, enhanced mechanical *strength* and *corrosion resistance*. They are favorable for all types of machining, even in the cold state. Monel metal (Ni with 1.5% Mn, 2.5% Fe, 28% Cu) is characterized by high grade mechanical properties and corrosion strength in water, acids, strong alkali. In air it retains its strength and plasticity at very low temperatures and upon heating up to 400 °C. *Thermal-electrode and (ohmic) resistance alloys* which are classified as electrical engineering nickel alloys possess high *electromotive force*, electrical resistance and *thermal-environment resistance*. Chromium–nickel alloy (Ni with 9.5% Cr, called *chromel*) and *alumel* (Ni with 1.2% Si, 2.3% Mn, 1.8% Al) are thermal-electrode alloys, used for manufacturing *thermocouples* for measuring temperatures up to 1000 °C (in long term) and up to 1300 °C (in short term). *Nichrome*, *ferronichrome* and also alloys, additionally alloyed with Al, Cu, Mn, Mo, etc., have high ohmic resistance; and are used for manufacturing *resistors*, thermoresistors, tensoresistors and electrical heaters. Nickel alloys with specific physical properties include *soft magnetic materials* (*permalloy*) and corrosion-resistant alloys, e.g., nickel–molybdenum alloys (27% Mo, 1.5% V), and nickel–chromium–molybdenum alloys (15% Cr, 16% Mo, 4% W). They also include alloys with a preset temperature coefficient of linear expansion (*invar*), alloys with high *elasticity* properties over a wide temperature range (e.g., 10% Cr, 10% Mo, 5% W, 4.5% Al), alloys for thermal-bimetallic parts. Some elements, such as Al, Ti, Nb, Ta, Be, Hf, etc. have limited solubility in Ni.

Complex nickel alloys containing these elements are hardened during the decay of oversaturated solid solutions due to *phase* precipitation and to the associated coherent stresses that are caused by the mismatch of the matrix and hardening phase lattices. Besides, the antiphase boundaries and *stacking faults* of the hardening phase somewhat assist hardening. Hardened-matrix nickel alloys containing precipitations of intermetallic (see *Intermetallic compounds*), carbide, boride, and other phases, are useful as *high-temperature materials*. Nickel alloys find their application in different engineering areas such as structural and electrical engineering alloys with specific physical and chemical properties, and as *thermal-environment resistant materials*.

NICKEL, Ni

Chemical element of Group VIII of the periodic table with atomic number 28 and atomic mass 58.69. Natural nickel contains stable isotopes ^{58}Ni (67.89%), ^{60}Ni (26.16%), ^{61}Ni (1.13%), ^{62}Ni (3.66%), ^{64}Ni (1.16%). Outer shell electron configuration is $3d^8 4s^2$. Successive ionization energies are 7.6, 18.15, 35.16 eV. Atomic radius is 0.125 nm; Ni^{2+} ion radius is 0.072 nm. Oxidation state is $+2$, more rarely $+1$, $+3$, $+4$. Electronegativity is 1.8.

In free form nickel is a silvery-white *metal*. It has face-centered cubic lattice, space group $Fm\bar{3}m$ (O_h^5), $a = 0.35243$ nm (at 298 K). Density is 8.897 g/cm^3; $T_{melting} = 1728$ K; $T_{boiling} \sim$ 3100 K. Heat of melting is 16.9 kJ/mole, heat of evaporation is 370 kJ/mole. Specific heat is 445 kJ·kg^{-1}·K^{-1} (at 298 K). Linear thermal expansion coefficient is $13.1 \cdot 10^{-6}$ K^{-1} (at 293 to 373 K). Thermal conductivity is 92.0 W·m^{-1}·K^{-1}. Debye temperature is 465 K. Adiabatic elastic moduli of crystal: $c_{11} \approx 248$, $c_{12} \approx$ 163, $c_{44} \approx 113$ GPa at 298 K. Young's modulus is ≈ 224 GPa (at 298 K), shear modulus is ≈ 86 GPa (at 298 K). Poisson ratio is ≈ 0.30. Tensile strength is ≈ 450 MPa, elastic limit is 80 MPa, yield point is 120 MPa, relative elongation is 40%. Brinell hardness of cast nickel is ≈ 700 MPa (at 293 K). Self-diffusion coefficient is $1.749 \cdot 10^{-21}$ m^2/s (at 700 K). Coefficient of linear term in low-temperature molecular specific heat is 7.02 mJ·mole^{-1}·K^{-2}. Electric resistivity $7 \cdot 10^{-8}$ μΩ·m, temperature coefficient of electrical resistance $292 \cdot 10^{-5}$ K^{-1}. Hall constant is $36 \cdot 10^{-11}$ m^3/C. Polycrystal work function is 4.61 eV.

Nickel is ferromagnetic with the Curie temperature 631 K. Atomic magnetic moment is 0.604 Bohr magnetons, nuclear magnetic moment of ^{61}Ni is 0.746 nuclear magnetons. Total effective high-energy neutron adsorption and scattering cross-section is 3.2 barn (3 to 10 meV), for low-energy neutrons it is 22.1 barn (0.25 eV).

Nickel is mainly used for the production of alloys (with iron, chromium, copper and other metals), distinguished by valuable mechanical, anticorrosion, magnetic and electrical properties, by high temperature strength and heat resistance (see *Nickel alloys*).

NIOBATES

Metallic oxide compounds of *niobium*, including the NbO_3 group with *metals*, e.g., $LiNbO_3$, $KNbO_3$, $Ba_2Na(NbO_3)_5$ ("banana"), $Sr_x Ba_{1-x}$ $(NbO_3)_2$ ("bastron"), etc. They possess nonlinear optical and acoustical properties. They are obtained by the synthetic Czochralski method (see *Synthetic monocrystals*). Uniaxial trigonal monocrystals of *lithium niobate* are *ferroelectrics* which serve as a *piezoelectric material* in *acoustoelectronics* devices for *surface acoustic waves*, and as an electrooptic and optically nonlinear material used in quantum electronics in *electrooptical modulators*, light gates, second harmonic generators of neodymium *laser* radiation (see *Nonlinearoptical crystals*), and in near infrared *optical parametric oscillators*. They possess photorefractive properties (see *Photorefraction*); they exhibit the *photochromic effect* and the *bulk photovoltaic effect*; they are used in optical holographic storage devices with high capacity and short access time (see *Optical techniques of information recording*). Tetragonal *potassium niobate* with the *perovskite* structure is an analog of *barium titanate*, but with $T_c = 708$ K instead of 403 K. Monocrystals with some application in *nonlinear optics* and acousto-electronics are rhombic *barium–sodium niobate* which requires complex mechanical untwinning, as well as *lead–potassium niobate* and some of their analogs, crystallizing in the potassium–tungsten bronze structural type.

NIOBIUM, Nb

A chemical element of Group V of the periodic system with atomic number 41, atomic mass 92.9064, and one natural isotope, ^{93}Nb. The most important artificial radioactive isotope is ^{95}Nb with lifetime of 95 days. Outer shell electronic configuration is $4d^4 5s^1$. Successive ionization energies are 6.88, 13.90, 28.1 eV. Atomic radius is 0.143 nm, radius of Nb^{5+} ion is 0.066 nm. Oxidation state is from +1 to +5. Electronegativity is 1.6.

In a free form, niobium is a refractive light-gray *metal*. It is body-centered cubic, space group $Im\bar{3}m$ (O_h^9); $a = 0.32941$ nm at 298 K. Density is 8.578 g/cm^3, $T_{melting} = 2742$ K, $T_{boiling} \approx 5050$ K. Heat of melting is 26.8 kJ/mole, heat of evaporation is 698 kJ/mole. Specific heat is 273 J·kg^{-1}·K^{-1} (at 373 K), 33.5 J·kg^{-1}·K^{-1} (at 1673 K). Linear thermal expansion coefficient is $7.2 \cdot 10^{-6}$ K^{-1} (at 273 to 373 K). Coefficient of heat conductivity is 13.0 W·m^{-1}·K^{-1} (at 293 K). Debye temperature is 250 K. Adiabatic elastic moduli of monocrystal: $c_{11} = 246.5$, $c_{12} = 134.5$, $c_{44} = 28.7$ GPa. Young's modulus is ≈ 130 GPa (at 298 K), shear modulus is ≈ 50 GPa, Poisson ratio is ≈ 0.365. Tensile strength is 350 MPa (niobium of technical purity). Yield point is 273 MPa. Relative elongation is 50%. Hardness of niobium of technical purity is 1000 to 1600 MPa. Linear low-temperature molecular heat capacity coeficient is 7.79 mJ·mole^{-1}·K^{-2}. Electrical resistivity is $15.24 \cdot 10^{-8}$ Ω·m. Polycrystal work function is 4.01 eV. Ionization potential is 6.77 eV. Superconducting transition temperature $T_c = 9.22$ K. Thermal neutron trapping cross-section is 1.15 barn. Nuclear magnetic moment of ^{93}Nb is 6.144 nuclear magnetons. Niobium is a component of alloys which are heat-resistant, and of structural materials for reactor building, chemical industry, and other areas.

NITRIDES (fr. Gr. $\nu\iota\tau\rho o\nu$, soda, saltpeter, and $\varepsilon\iota\delta o\varsigma$, kind)

Chemical compounds of nitrogen with *metals* and nonmetals. Nitrides of nonmetals are ordinarily covalent compounds; nitrides of the *transition metals* with a deficit of nitrogen are *semiconductors*, and those with a saturated composition possess metallic bonds; nitrides of the alkaline-earth and *alkali metals* are ionic-covalent compounds. Nitrides of transition metals are interstitial phases having their structure based on the lattice of the metal atoms, with nitrogen atoms located at octahedral or tetrahedral sites. Phases of Me_4N (Me is alkali metal) are usually face-centered cubic, Me_2N is generally hexagonal, and MeX can be either face-centered cubic, body-centered cubic, or hexagonal.

All nitrides are brittle materials, they are not as hard as *carbides*, although cubic *boron nitride* is inferior only to *diamond* in its hardness (60 GPa). Nitrides are produced by synthesis from the elements, by interacting metals with ammonia, by reduction of oxides in the presence of nitrogen, by thermal dissociation of compounds containing a metal atom and nitrogen (TiCl$_4$·4NH$_3$), and by deposition from the gas phase. Nitrides are used in various fields of engineering as *refractory materials* (BN, AlN, Si$_3$N$_4$, TiN), as electrical insulators (BN, Si$_3$N$_4$), as radio-frequency transparent materials (Si$_3$N$_4$), and as abrasive and cutting materials (BN$_{cub}$, Si$_3$N$_4$). See also *Aluminum nitride, Silicon nitride, Boron nitride.*

NMR

See *Nuclear magnetic resonance.*

NOBELIUM, No

See *Transuranium elements.*

NOBLE METAL ALLOYS

Alloys obtained by fusion of *silver* (Ag), *gold* (Au), *ruthenium* (Ru), *rhenium* (Rh), *palladium* (Pd), *osmium* (Os), *iridium* (Ir), and *platinum* (Pt), both with each other and with other *metals* (less often with nonmetals). Known are single-phase noble metal alloys (*solid solutions*, chemical compounds) and multiphase ones. The metals form *substitutional alloys* with close-packed crystalline structures (hexagonal close-packed structure for solid solutions based on Ru and Os, and face centered cubic for the remaining ones). *Phase transitions* are possible in noble metal alloys: atomic *alloy ordering*, decomposition of oversaturated solid solutions and *structural phase transitions*. The changes of crystalline and electronic structure occur at the formation of the solid solutions and chemical compounds, and at phase transitions. As a result, some properties of component noble metals may improve and new properties may appear. For instance, the alloys of Pd and Ir in the middle part of their *phase diagram* have the coloration from light yellow to golden and then to pink-violet. Some alloys of *paramagnets* Pd and Rh are ferromagnets (see *Ferromagnetism*). The alloys Au–Cu–Ag are stronger than each individual component (*solid-solution hardening*), while the decomposition of the solid solutions may result in a severalfold increase of their *yield limit*. The use of noble metal alloys is conditioned by their unique physical and chemical properties: high stability to *oxidation, corrosion resistance, high-temperature strength* and *thermal-environment resistance*, beautiful color and others. Noble metal alloys are used in engineering (for electrical contacts, critical-duty *resistors* and potentiometers, for *thermocouples* operating in an oxidizing medium, for *permanent magnets*, for the supports of axes of measuring devices, for their elastic elements, etc.). These alloys are often more efficient than pure *noble metals* as catalysts and *construction materials*.

NOBLE METALS

A group of *metals* with especially low chemical activity.

These are Ag and Au, and metals of the platinum group Ru, Rh, Pd, Os Ir and Pt that belong to Group VIII of the periodic table. Ag and Au are highly plastic; and the other noble metals are notable by their refractory properties ($T_{\text{melting}} \approx$ 1800 °C or higher). Most of the noble metals form *solid solutions* at mutual melting, e.g., Au and Ag (face-centered cubic lattice), Os and Ru (hexagonal lattice). Good *electrical conductivity, corrosion* resistance, high *melting temperature* and high reflectivity of noble metals and their *alloys* determine their wide applications including various contacts, resistors with low thermal coefficient and thermoelectromotive force (paired with copper). *Coatings* of Au 0.01–0.02 mm thick are applied to the outside surface of spaceships and satellites to improve the reflection of the solar electromagnetic radiation. High quality mirrors are made of Ag. Pure platinum and its alloys are used in *thermometry* (resistance thermometers, *thermocouples*). The

wear resistant parts of instruments (e.g., compass needles) are made of Os and Ir alloys. The alloy of Pt (90%) with Ir was used to manufacture the standards of the meter and kilogram.

NOISE in semiconductors

Fluctuations of currents I (or voltages) in semiconductors which are due to the random nature of *current carrier* motion, fluctuations of carrier number, and also the discrete character of the *electrical conductivity* process. Because of their frequent collisions with *phonons* and lattice *defects*, electrons and *holes* move like Brownian particles. Their velocity fluctuations result in inducing electromotive forces associated with *thermal noise*. Often, in addition, *shot noise* occurs in semiconductors. For example, it is observed when current carriers surmount the barriers in *semiconductor junctions*. The individual barrier crossings occur discretely, independently of each other, and lead to uncorrelated current impulses if an external voltage is applied to the barrier.

Generation–recombination noise has a similar character when the number of current carriers, and therefore, the crystal resistance δR fluctuates. In this case there appears *electromotive force noise* $\delta E = I \delta R$. In *many-valley semiconductors, intervalley noise* is observed. It results from the occasional electron transitions from one valley to another. The total number of current carriers is invariant, however, their effective mobility fluctuates, because the mobility of carriers differs in inequivalent valleys.

The resistance of semiconductors depends also on a number of additional physical quantities, the factors which induce the fluctuation of δR, and the excitation of electromotive force noise. Often these are referred to as *current noise*, since in contrast to thermal noise, they exist only for $I \neq 0$. *Flicker noise, telegraph noise* (*popcorn noise*), and also noise due to the fluctuations of current carriers or of the crystal lattice temperature, involve this type of noise. These varieties of noise are spatially nonuniform. Their evolution is described by a diffusion-type equation. Its solution and the spectral density of noise depend on the boundary conditions, and the system dimensions. Also there are local carrier concentration fluctuations in the bulk semiconductor, as well as in the density of the two-dimensional gas adsorbed by its surface, and these have a diffusive character. The investigation of noise is useful for semiconductor diagnostics, for determining their parmeters, for designing low-noise devices, and for selecting optimum operating conditions.

For different varieties of noise, see also *Barkhausen noise, Generation–recombination noise, Telegraph noise, Flicker noise, Intervalley noise, Shot noise.*

NOISE IN ELECTRONIC DEVICES

Voltage (or current) *fluctuations* in electronic devices due to their internal structure and properties. There are also occasional disturbances from external sources, referred to as *interference*. Noise arising in sensitive amplifiers or converters restricts the attainment of maximum sensitivity. Noise may also appear in devices with a high signal level, such as amplifiers or generators. Here the noise may show up as amplitude and phase fluctuations of the main signal. The influence of interferences, and also noise due to the instability of device parameters, may be decreased by selective improvements. At the same time noise such as the thermal type, shot noise (see *Noise*), *generation–recombination noise,* which are due to fundamental random-type processes, affect device operation. As a rule, it is impossible to diminish the amplitude of these noises without changing the device working characteristics. The search for low-noise devices led to the choice of those based on the *Josephson effect,* on principles of quantum electronics (masers), on parametric processes, etc. Such devices have a high sensitivity for measurements of magnetic fields, voltages, currents. Also they can amplify extremely weak signals. See also *Superconductor electronics.*

NOISE TEMPERATURE

Physical quantity T_n, which is a measure of the *noise* power in conductors (resistances). It is defined by

$$S_\omega = 2k_B T_n R(\omega), \qquad (2)$$

where S_ω is the spectral density of the electromotive force *fluctuation* in an electrical conductor, k_B is the *Boltzmann constant,* $R(\omega)$ is the resistive component of the differential *impedance* of the conductor. When a conductor is in *thermodynamic equilibrium* with a thermostat at temperature T, then $T = T_n$. Eq. (1) for S_ω represents the *Nyquist formula* (H. Nyquist, 1928). Consider an ideal resistor with a resistance equal in value to that of the matched load of a system. The noise temperature T_n is the temperature to which the ideal resistor must be raised to generate the same noise as the load. At this temperature the spectral density of the ideal resistor noise equals the spectral density of the load resistor noise under study. This fact is often used for the measurement of noise temperature. A departure of noise temperature from the ambient temperature T demonstrates the nonequilibrium character of the fluctuations. The value of T_n may be used when estimating the extent of the warm-up of current carriers (see *Electron heating*).

NONCENTRAL IONS, off-center ions

Substitutional impurity ions with their equilibrium positions displaced off lattice sites. Such a displacement usually occurs along a *crystal symmetry* direction, and is often accompanied by the appearance of a *dipole moment* oriented along the impurity atom displacement. Thermal or tunnel jumps between different equivalent positions are possible, and there appears an *inversion splitting* between the vibrational levels of the impurity. The simplest possible example is the inversion splitting in a double well potential (see Fig.). The value of such a splitting, Δ, usually lies within the energy range corresponding to $\Delta = 20$ K. Similar to dipolar molecules such as OH^-, noncentral ions may orient themselves in both external and internal electric fields. The presence of noncentral ions significantly alters the properties of crystals (anomalies arise in *specific heat, thermal conductivity,*

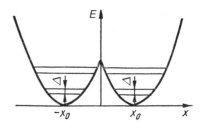

Inversion splitting, Δ, of the vibration levels in a double-well potential ($\pm x_0$ are the positions of the minima).

sound absorption; special features appear in Raman and IR spectra; dielectric, magnetic, and other properties change). These characteristics of a crystal with impurities demonstrate experimentally the noncentral nature of the impurities. The following noncentral ions are known: Li^+ in KCl, RbCl, KBr; Cu^+ in KCl, RbCl, RbBr, KBr, RbI, KI, NaI; Ag^+ in RbCl, RbBr; F^- in NaBr. Because of the presence of certain features in their electron paramagnetic resonance spectra, some other impurities are thought to be noncentral: Mn^+ in KCl; Mn^0 in NaCl; Cu^0 in KCl; Mn^{2+} in BaO; Eu^{2+} in RbCl; Ti^{3+} in LiF; Ag^0 in KCl; Cu^{2+} in SrO, etc. Noncentral ions are actively studied in more complex compounds that undergo *structural phase transitions* or hover near their stability thresholds. For example, some impurity ions in *virtual ferroelectrics* are noncentral (e.g., Li^+ and Na^+ in $KTaO_3$; Ge^{2+} in PbTe). The influence of noncentral ions on physical properties is particularly noticeable in these strongly polarizable lattices. In particular, noncentral ions determine the width of the *central peak* in the *order parameter* fluctuation spectrum, and may induce a *ferroelectric phase transition*. Parameters characterizing the direction, n and magnitude r_0 of the displacement of the best studied noncentral ions are for Li^+ in KCl: $n||[111]$, $r_0 \approx 0.1$ nm; for Li^+ in $KTaO_3$: $n||[100]$, $r_0 \approx 0.1$ nm. An impurity ion can become noncentral when it experiences weaker repulsion and stronger polarization forces compared to those affecting the substituted lattice ion. That is why impurities with similar *ionic radii* and higher polarizability than the substituted ions usually become noncentral. From the point of view of electronic structure, noncentral formations should be expected when one finds energy levels of opposing parities in close proximity to each other. In that case, the displacement of the impurity off the site can be energetically favorable due to the *Jahn–Teller effect*. In highly polarizable *insulators*, such a displacement may take place below a certain temperature T_{loc}. Sometimes, such a situation is conditionally called a *local phase transition* or *local freezing*. Actually no freezing occurs; instead, the ion slowly reorients between the different minima below T_{loc}. Typically the height of the activation barrier is temperature-dependent. This situation may also be encountered in common weakly polarizable lattices.

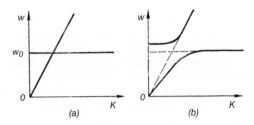

Curve crossing possibilities: (a) intersecting dispersion curves, and (b) degeneracy removed.

NONCROSSING RULE

Crossover (see Fig. (a)) between two dispersion curves (see *Dispersion law*) corresponding to two different types of waves or elementary excitations. The crossing only occurs if the interaction between these waves is neglected, so that a formal degeneracy results, not involving symmetry considerations. Taking into account such an interaction usually lifts this *degeneracy*, and the excitation spectrum rearranges so that the dispersion curves move apart from each other. The new (renormalized) dispersion dependences have no crossings, as shown in Fig. (b). A physical example of a noncrossing rule is given by *Magnetoacoustic resonance*.

NONCRYSTALLINE CLUSTERS, anticrystalline clusters

Connected sets or groupings of atoms from which a crystal structure cannot form because its formation would violate translation symmetry. Polyatomic noncrystalline clusters featuring maximum average atomic binding energies are probable structural elements of amorphous bodies.

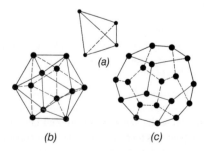

Noncrystalline clusters.

Among the noncrystalline clusters are, e.g., a regular tetrahedron consisting of four atoms (Fig. (a)), an icosahedron containing 12 atoms (Fig. (b)), a dodecahedron containing 20 atoms (Fig. (c)), a Mackay icosahedron (A.L. Mackay, 1952), containing 55 atoms, a Buckyball or fullerene C_{60} containing 60 carbon atoms, and many others.

NONDESTRUCTIVE TESTING TECHNIQUES

Techniques for testing metals, as well as semi-finished and finished items, for their compliance with established specifications to be sure they are suitable to be used for their designated purposes, without harming them. The nondestructive testing methods that are often used are as follows:

(1) a magnetic technique based on analyzing a magnetic field interaction with the sample,

(2) an electric technique based on recording parameters of an electric field interacting with the sample, or such a field generated in it under the effect of some external factor,

(3) an eddy current technique based on analyzing the interaction of the electromagnetic field produced by an eddy current transformer with that of eddy currents induced in the sample,

(4) a radio wave technique based on recording the changes in the parameters of radio band electromagnetic waves interacting with the sample,

(5) a thermal technique based on recording changes in temperature fields of the sample,

(6) an optical technique based on recording the parameters of optical radiation interacting with the sample,

(7) a radiation technique based on recording and analyzing penetrating ionizing radiation after it interacts with the sample,

(8) an acoustic technique based on recording the parameters of elastic waves generated by or originating in the sample, and

(9) a penetration technique based on the penetration of particular substances into the defect cavities of the sample.

Both active and passive nondestructive testing techniques are used. In the former case the sample is exposed to the action of an external (auxiliary) energy source which produces the data reporting signal, and in the latter case the origin of this signal is the sample itself. These techniques involve

recording the measured parameter data either during direct contact of the sensor element with the sample (contact nondestructive testing technique), or while there is no such contact (noncontact nondestructive testing technique). The capabilities of these techniques depend on their sensitivity, resolution, adequacy of data, and reliability of instrumentation and procedures.

Nondestructive testing types can be classed as to the methods (algorithms) used to calibrate the monitored parameter against a prescribed reference; physical phenomena or laws underlying these techniques; the kind of sensors and transducers used; the type of interaction between the monitoring device and the sample; the methods used to recover the values monitored; the kind of instrumentation used; the information parameter of the measured data signal; the properties of the sample used for the purpose. Nondestructive testing also includes *defectoscopy*, structuroscopy and structurometry, thickness gauging, *endoscopy*, etc. Nondestructive testing is used in various industrial technological processes and research practice.

NONEQUILIBRIUM SUPERCONDUCTIVITY

A special state of a superconductor developed under the action of external fields on one of its subsystems – either the electron, the phonon, or the *Cooper pair* condensate. The structure and properties of nonequilibrium superconductivity are determined by both the type of external action that is applied, and the physical parameters and geometry of the superconductor. To induce this phenomenon the following types of external action are commonly used: ultrasonic or electromagnetic fields, tunnel *injection* or *extraction* of monoenergetic quasi-particles, electric current running through either the normal metal–superconductor interface, or through thin superconducting threads. When the fields are weak enough one may apply the following *kinetic equations* for the nonequilibrium *distribution functions* of the electrons f_ε and phonons N_ω:

$$\frac{\partial f_\varepsilon}{\partial t} = I_{ep}\{\varepsilon, f_\varepsilon, N_\omega\} + I_{ee}\{f_\varepsilon\} + I_{ei}\{\varepsilon, f_\varepsilon\}, \quad (1)$$

$$\frac{\partial N_\omega}{\partial t} = L_{pe}\{\omega, N_\omega, f_\varepsilon\} + L_{pi}\{\omega, N(\omega)\}$$

$$+ L_{pp}\{\omega, N_\omega\}, \quad (2)$$

where I_{ep}, I_{ee}, I_{ei} are electron–phonon, electron–electron, and electron–impurity collision integrals; and L_{pe}, L_{pp}, L_{pi} are phonon–electron, phonon–phonon, and phonon–impurity collision integrals, respectively. This system of equations is subject to the self-consistency condition for the nonequilibrium *energy gap*, Δ,

$$I = V_C \sum_k \frac{1 - 2f(E_k)}{2E_k}, \qquad (3)$$

where $E_k = (\xi_k^2 + \Delta^2)^{1/2}$; V_C is the standard electron–electron attraction potential. In a layer of width $\hbar\omega_D$ (ω_D is the *Debye frequency*), the latter has a constant positive value in the vicinity of the *Fermi surface*; ξ_k is the kinetic energy of an electron with momentum P referred to the *Fermi level*. The parameter of the energy gap, which, according to Eq. (3), depends on the function $f(E_k)$, may increase as a result of the external action. The effect of broadening the energy gap (as well as increasing other superconductor critical parameters) experienced under the action of either photons or phonons has been experimentally observed, and was named the *Eliashberg effect*. When the external field is switched off, the energy gap relaxes back to its equilibrium value within the time $\tau_\Delta = 1.2\tau_E/[I - (T/T_c)]^{1/2}$ (here τ_E is the time of the electron energy relaxation, and T_c is the *critical temperature of superconductors*). For $T \to T_c$, the quantity τ_Δ diverges as $(1 - T/T_c)^{-1/2}$. When charged particles are injected into the superconductor (e.g., through the tunnel barrier I of a superconductor–insulator–superconductor *tunnel junction*), the energy gap width does not change, only the phase of the *order parameter* changes. The total charge of the superconductor remains unchanged since the change in the charge of quasi-particles, Q^* is compensated by a change in the charge of the condensate. Essentially, nonequilibrium superconductivity results from an imbalance between the populations of the electron and the hole branches of the spectrum, when the pairs of electron excitations recombine into a superfluid *Bose condensate*, forming Cooper pairs and hole excitation pairs. The quantitative characteristic of such a charge imbalance is given by the difference between the chemical potentials of quasi-particles μ_n and the condensate, μ_S, respectively. The expression for

the *relaxation time* of the charge imbalance has the form $\tau_Q = 4k_B T_c \tau_E/(\pi\Delta)$. For $T \to T_c$, the quantity τ_Q diverges as $(1 - T/T_c)^{-1/2}$, thus impeding the transfer of the charge from quasi-particles to the condensate.

The nonequilibrium distribution function for electrons $f(E_k)$ may be evaluated experimentally using the S–I–S *tunnel junctions* and measuring the current–voltage characteristic $I(V)$:

$$I(V) = \frac{2}{eR} \int_\Delta^\infty \rho(E)\rho(E + eV)$$

$$\times \big[f_1(E) - f_2(E + eV)\big]\,dE, \quad (4)$$

where R is the junction resistance, $\rho(E)$ is the *density of states* in the superconducting electrodes S of the junction. We have for an isotropic superconductor $\rho(E) = |E|/(E^2 - \Delta^2)^{1/2}$. Usually a double tunnel junction of the type S_1–I–S_2–I–S_3 is used, where one of them (e.g., S_1–I–S_2) is low-ohmic and is called the injector (or generator), while the other one is a high-ohmic detector. When a constant bias voltage V is applied to the S_1–I–S_2 junction, this results in an intense injection of monoenergetic electrons (or holes) into the superconducting electrode S_2 resulting in a nonequilibrium change in their distribution function $f_2(E)$. The latter is then recorded at the S_2–I–S_3 junction by measuring its current–voltage characteristic. The explicit form of $\rho(E)$ is determined using Eq. (4). In a similar way the electron distribution function is determined when the superconductor film is bombarded by photons. Additional information about nonequilibrium superconductivity is obtained by studying the effects that result from the nonequilibrium properties of the superconducting order parameter, including its relaxation behaviour. Techniques used to evaluate order parameter relaxation times include subjecting the superconductor to pulses of current, to injection of quasi-particles, and also to laser and thermal pulses. For example, one may exploit the fact that the maximum value of the current I_{dc} that may run through a superconducting sample without bringing it back to its normal state depends on the angular frequency ω and the amplitude of the applied alternating current, I_{ac}. Measurements of the temperature dependence of I_{dc} made it possible to evaluate the relaxation time of the order parameter. The value of τ_Δ thus found for aluminum samples at temperatures close to T_c appeared to be proportional to $(I - T/T_c)^{-1/2}$, which agrees well with theoretical models.

NONERGODICITY

Discrepancy between the results of averaging various physical characteristics of a system over time and over the *Gibbs distribution*. Nonergodicity is typical of *spin glasses* and other *amorphous magnetic materials*; it is found in systems featuring continuous spectra of their *relaxation times* when the duration of an experiment is shorter than the longest relaxation time. A continuous spectrum of relaxation times is due to the infinite degeneracy of the ground state of a spin glass, when averages must be calculated in two stages. First, averaging is done over each of the degenerate states, and then over their ensemble, taking account of the probability of each. Many physical variables (e.g., *magnetic susceptibility, specific heat*) become non-self-averaging with their fluctuations characterizing the degree of the system nonergodicity. That is why, when allowing for the nonergodicity of spin glasses, they are to be described by an infinite number of the *order parameters* that define the overlap of local frozen magnetic polarizations in various states. Nonergodicity results in a large number of irreversible phenomena in spin glasses, in some other amorphous magnetic materials with their lack of equilibrium, and also in memory and aging phenomena, etc., all of them experimentally observed.

NONLINEAR ACOUSTIC EFFECTS

Phenomena originating in an acoustic field with high sonic amplitudes. Among these phenomena are, for example, distortions of the sine profile of longitudinal elastic waves, L, near their source in solids, and the generation of harmonics. This phenomenon (*self-action of acoustic waves*) takes place due to the *anharmonic vibrations* of a crystal lattice, and resultant deviations from *Hooke's law*. If there is no *sound velocity dispersion* and the *sound absorption* remains weak then the nonlinearity, although small in itself, will result in the accumulation of wave shape distor-

tions along its propagation path. The absorption of such a wave is much stronger than that of a linear wave (*nonlinear sound absoption*). Interactions between several elastic plane waves (*phonons*) result, apart from the self-action effect of each individual wave, in the appearance of waves at combination frequencies. If *synchronism conditions* are satisfied in a solid (energy and *quasi-momentum* of phonons conserved), both longitudinal L and transverse T waves propagating at different velocities interact with each other. Examples of nonlinear acoustic effects in solids are *Raman scattering of sound by sound*; generation of harmonics of T waves ("forbidden" by *nonlinear elasticity theory* but appearing because of *dislocations* and internal *mechanical stresses*); nonlinear polarization effects; generation of higher harmonics of *surface acoustic waves*, etc. As far back as in 1937, Landau and Rumer constructed a theory of high-frequency sound absorption ($\sim 10^9$ Hz) at low temperatures (~ 20–70 K) in *insulators* satisfying the condition $kl \gg 1$ (k is the sound wave vector, l is the mean free path of thermal phonons), based on nonlinear interactions of coherent phonons (*sound*), and thermal phonons subject to synchronism conditions. Akhiezer did the same in 1938 for ultrasonic frequencies at room temperature when the condition $kl \ll 1$ is satisfied. The first experiments on coherent phonons to prove the existence of a *three-phonon interaction*, and in particular the *generation of harmonics* in solids, were reported in the early 1960s. In *piezosemiconductors* (e.g., CdS), a strong electroelastic nonlinearity may develop under certain conditions due to the piezoelectric effect (see *Piezoelectricity*), while magnetically ordered crystals (e.g., *antiferromagnets, haematite*) may display a strong *magnetoelastic nonlinearity* due to *magnetostriction*. These "mixed" nonlinearities exceed the lattice (elastic) nonlinearity by 3–4 orders of magnitude, and considerably improve the performance of various nonlinear devices that process radiosignals in *acousto-electronics* and magnetoacoustics (correlators, convolution devices, filters, solid-state *ultrasonic delay lines*). Such devices play a significant role in developing new measurement techniques in solid state physics.

NONLINEAR ELASTICITY THEORY

Part of *elasticity theory* treating large elastic *strains*. Within nonlinear elasticity theory, the strain of a medium is characterized by the *displacement vector* of its elements, u_i, and by the *strain tensor*,

$$u_{ik} = \frac{1}{2}\left(\frac{\partial u_i}{\partial x_k} + \frac{\partial u_k}{\partial x_i} + \frac{\partial u_l}{\partial x_i}\frac{\partial u_l}{\partial x_k}\right),$$

while its *state of stress* is described by the *stress tensor*, σ_{ik}. The energy density of the elastic strain W is considered to be a function of the components of the tensor u_{ik} (the free energy density or some other *thermodynamic potential* is used for W at finite temperatures). By virtue of the requirement of invariance, W depends on algebraic invariants composed from u_{ik} (see *Elasticity theory invariants*). In isotropic media (see *Isotropy*), such invariants are the coefficients I_1, I_2, I_3 of the polynomial $P(x) = |x\delta_{ik} - u_{ik}|$ $= x^3 - I_1 x^2 + I_2 x - I_3$. The defining expressions that relate strains to deformations have the form $\sigma_{ik} = I^{-1/2}(\partial W/\partial u_{ik})$, where $I = |\delta_{ik} + 2u_{ik}|$. These expressions may be written explicitly if one chooses some form for the function of invariants W. The following simplified defining equations are sometimes used in mechanics for nonlinear elastically isotropic media: $\sigma = f(u)$, $\sigma'_{ik} = g(u')u'_{ik}$. Here $\sigma = (1/3)\sigma_{ll}$, $u = (1/3)u_{ll}$, σ'_{ik} and u'_{ik} are the *stress deviator* and the *strain deviator*, respectively. Further, $u' = (2/3)(I_1^2 - 2I_2)^{1/2}$ is the intensity of *shear strain*, while the functions $f(u)$ and $g(u')$ depend on the nature of the medium and are evaluated empirically (in a linear case, we have $f = 3K$, $g = 2\mu$, where K and μ are the *bulk modulus* and *shear modulus*). In the physics of solids, W is usually constructed by the successive approximation method: $W = W_{\text{lin}} + W' + W'' + \cdots$, where W_{lin} corresponds to the linear elastic case, and W', W'', ... are the corrections to W_{lin} of increasing order in u_{ik}. In that case, the relation between σ_{ik} and u_{ik} may also be written down directly from symmetry considerations. Nonlinear defining equations for *crystals* are often used in the form $W = (1/2)c_{iklm}u_{ik}u_{lm} + (1/6)c_{iklmnp}u_{ik}u_{lm}u_{np}$, where c_{iklm} and c_{iklmnp}

are *elastic moduli* of order 2 and 3. The defining equations for isotropic materials in the second order of elasticity theory have the form:

$$\sigma_{ik} = \lambda u_{ll}\delta_{ik} + 2\mu u_{ik}$$
$$+ \left[\frac{\lambda}{2}h_{lm}^2 + \left(\frac{\nu_1}{2} - \lambda\right)u_{ll}^2 + \nu_2 u_{lm}^2\right]\delta_{ik}$$
$$+ 2(\lambda - \mu + \nu_2)u_{ll}u_{ik} + \mu h_{il}h_{kl}$$
$$+ 4(\mu + \nu_3)u_{il}u_{kl},$$

where $h_{ik} = \partial u_i/\partial x_k$; $\lambda, \mu, \nu_1, \nu_2, \nu_3$ are the elastic moduli. Nonlinear elasticity theory is used to calculate the state of stress of highly deformed bodies (e.g., *shells*). It is also used in the physics of solids to analyze the propagation of *shock waves* and strong ultrasonic waves, and to calculate nonlinear effects in crystals containing dislocations (e.g., *dilatations*), etc.

NONLINEARITY, MAGNETODYNAMIC

See *Magnetodynamic nonlinearity*.

NONLINEAR MAGNETOACOUSTIC RESONANCE

A first-order parametric phenomenon, in which uniform oscillations of the *magnetization* occurring in a magnetically ordered medium under the action of external pumping, as a result of a nonlinear *magnetoelastic interaction*, excite a *spin wave* with the frequency ω_k, as well as an elastic vibration of the sample with the frequency ω_a. The pumping frequency, ω, is $\omega = \omega_k + \omega_a$, with $\omega_k \sim \omega \gg \omega_a$. The effect occurs in any clean system, such as a two-dimensional electron gas in the quantum Hall effect regime. It is observed in high-quality *ferrite* monocrystals at a pumping frequency near ≈ 10 GHz. The elastic frequencies of a body depend on its dimensions, and usually are in the range of MHz, with threshold pumping power as high as few mW for perfect monocrystals of yttrium–iron garnet. Other magnetic oscillations may serve for pumping, such as parametrically excited spin waves (*nonlinear secondary magnetoacoustic resonance*, see also *Parametric excitation of spin waves*).

Table 1. Crystal system (syngony) and applications of some common nonlinear-optical crystals

Substance	Chemical formula	Syngony	Applications
Ammonium dihydrogenphosphate	$NH_4H_2PO_4$	Tetragonal	SHG, PG, M
Potassium dihydrogenphosphate	KH_2PO_4	Tetragonal	SHG, PG, M
Barium titanate	$BaTiO_3$	Tetragonal	SHG, M
Iodic acid	HIO_3	Rhombic	SHG
Lithium niobate	$LiNbO_3$	Trigonal	SHG, IO, M, PG
Barium–sodium niobate ("banana")	$Ba_2NaNb_5O_{15}$	Rhombic	SHG
Barium–strontium niobate	$BaSrNb_4O_{12}$	Tetragonal	SHG
Indium antimonide	InSb	Cubic	OBD, SMRS
Gallium arsenide	GaAs	Cubic	IO, M, OBD

Note: SHG, second harmonic generator; PG, parametric generator; M, electro-acoustic modulators and deflectors; IO, integrated optics; OBD, optical bistable devices; SMRS, stimulated magnetic Raman ("spin-flip") scattering.

NONLINEAR MEDIUM

A medium in which wave self-action and wave interaction take place. The principle of superposition no longer applies in a nonlinear medium. Waves do not propagate independently through such a medium; rather their interaction produces changes in the wave field which depend on the parameters of the initial waves. This interaction results from the effects waves themselves have upon the characteristics of the medium. Nonlinear properties of a medium become important when the medium is subjected to highly intense excitation, so that the approximation of *linear response* is no longer applicable, and the reaction of a medium to external factors is determined by its nonlinear susceptibilities. In addition, such nonlinearity may result from the change in the parameters of the medium affected by the wave fields, e.g., when electrons redistribute themselves between their allowed energy levels during excitation, when the medium experiences heating or compression under the action of an external factor, etc. Various nonlinear media may display optical, acoustic, magnetic, and other effects. Waves of a different nature and origin, e.g., acoustic and light waves, etc. may also interact with each other in nonlinear media.

Many *semiconductors*, *ferroelectrics*, and *piezoelectric materials* are essentially nonlinear solids. Nonlinear materials are used in *lasers*, *real-time holography*, and *semiconductor devices*. See also *Nonlinear optics*, *Acousto-optics*, *Acousto-electronics*, *Parametric excitation of spin waves*.

NONLINEAR-OPTICAL CRYSTALS

Crystals used in *nonlinear optics* for both discrete and continuous (parametric) transformation of the frequency of *laser* radiation. Table 1 lists the characteristics of some common nonlinear-optical crystals. These provide transformation efficiencies of tens of percent (even up to 90–95%) for pulse output powers up to hundreds of gigawatts. Nonlinear-optical crystals are most often used for the *second harmonic generation* in neodymium laser radiation (see *Solid-state lasers*).

Concerning their properties, nonlinear-optical crystals are expected to absorb light in their operational optical range, and at the same time yield the highest attainable optical uniformity and durability. See also *Potassium titanylphosphate*.

NONLINEAR OPTICS of solids

Branch of nonlinear optics studying the interaction of intense light with *crystals*. This involves such processes as *second harmonic generation*, parametric generation (see *Optical parametric oscillator*), multiphoton absorption (see *Multiphoton processes*), stimulated Raman scattering (see *Induced light scattering*), *self-action of light waves*, etc. Nonlinear optics became important after the invention of *lasers*. Ordinary optical processes in crystals involve a linear relationship between the polarization, P_a and the field strength E of the electromagnetic wave: $P_a = \chi_{ab} E_b$, where χ_{ab} is the linear *susceptibility*. This follows from the harmonic oscillator approximation that is valid for weak fields, $E \ll E_i$, where E_i is the internal atomic electric field associated with the chemical bonds, $\sim 10^7$–10^8 V/cm. Laser fields reach the levels of $E \sim 10^6$–10^7 V/cm and higher, coming close to E_i. Stimulated oscillations of a bound electron affected by such a field become noticeably anharmonic, which results in a nonlinear dependence of P on E. For $E/E_i < 1$, one may represent this dependence as a power series:

$$P_a = \chi_{ab}^{(1)} E_b + \chi_{abc}^{(2)} E_b E_c + \chi_{abcd}^{(3)} E_b E_c E_d + \cdots$$
$$= P_a^l + P_a^{nl}, \tag{1}$$

where $P_a^l = \chi_{ab}^{(1)} E_b$; P_a^{nl} are higher-order terms: $\chi_{abc}^{(2)}$, $\chi_{abcd}^{(3)}$ are quadratic, cubic etc. nonlinear susceptibilities. As for their order of magnitude, $\chi_{abc} \propto 1/E_i$, $\chi_{abcd} \propto 1/E_i^2$. Typical numerical values for *insulators* in the transparency region are: $\chi^{(1)} \sim 1$, $\chi^{(2)} \sim 10^{-11}$–10^{-12} m/V; $\chi^{(3)} \sim 10^{-21}$–10^{-22} m^2/V^2. These are tensor quantities characterizing the properties of a crystal medium. By symmetry 3rd, 5th and higher odd-rank tensors vanish for crystals with a *center of symmetry*, and χ is nonzero for every rank when there is no center of symmetry.

A nonlinear dependence of the polarization of the medium on the field strength results, accordingly, in a nonlinear dependence of the *electric flux density*, D on the field E (in CGS units):

$$D_a = E_a + 4\pi P_a$$
$$= E_a + 4\pi(\chi_{ab} E_b + \chi_{abcd} E_b E_c E_d + \cdots). \tag{2}$$

This, in its turn, brings about dependences of the *dielectric constant* ε and *refractive index* n on E:

$$n(E) = n_0 + n_2 E^2 + \cdots. \tag{3}$$

The dependence $n(E)$ results in phenomena such as *self-focusing* and *self-defocusing* of a light beam in a nonlinear medium. Terms in Eq. (1), nonlinear in E, describe basic nonlinear optical processes. Thus, the term $\chi_{abc} E_b E_c$ is responsible for generation of the second harmonic (frequency doubling); $\chi_{abcd} E_b E_c E_d$ for the third harmonic, i.e. *parametric oscillation*, and these provide the dependence $n(E)$. In fact the term $\chi_{abcd} E_b E_c E_d$ also contains an intensity-dependent absorption correction accounting for *two-photon light absorption*. Nonlinear susceptibilities

and related optical nonlinearities in solids are pronounced due to the large concentration of particles in a unit volume. They sharply increase as the frequency of the electromagnetic wave approaches either an exciton, or a band-to-band resonance frequency. In addition, there are other mechanisms responsible for the nonlinearity of *refraction of light* in crystals, not related directly to P^{nl}, such as *electrostriction*, laser heating of the medium, input from free electrons, etc. Free carriers excited during either single-photon two-photon absorption, produce a strong nonlinearity of $n(E)$ in *semiconductors*. Diffusive redistribution of charges in photorefractive crystals, such as $LiNbO_3$, $Bi_{12}SiO_{20}$, $BaTiO_3$, etc. results in the generation of an electrostatic field that produces a change in the refractive index via the linear *Pockels effect*.

Applications of nonlinear phenomena in solids include light frequency multiplication, tunable optical parametric oscillators, control of spatial and temporal characteristics of laser beams, etc. Birefringent nonlinear crystals can efficiently match the *phase velocities* of both fundamental and frequency-doubled waves, and several tens of percent efficiency can be achieved in frequency doubling using the crystals potassium- and ammonium-dihydrogen phosphate, as well as $LiNbO_3$, $LiIO_3$, $BaNaNb_5O_{15}$, $ZnGeAs_2$, $CdGeAs_2$, ArGaSe, and others.

The enhanced nonlinearity of refraction in semiconductors and photorefractive crystals (see *Photorefraction*) makes possible highly effective recording of dynamic phase holograms (see *Real-time holography*), for *wave front reversal*, and for producing bistable optical elements (see *Optical bistability*). Other applications include devices to control the parameters of laser radiation, and to develop purely optical communication systems and computers.

NONLINEAR PYROELECTRIC

A pyroelectric substance (see *Pyroelectric materials*) featuring a spontaneous reversible polarization by an electric field (see *Ferroelectricity*). The temperature dependence of this polarization is nonlinear at the *phase transition* point. *Ferroelectrics* which display a strong spontaneous polarization temperature dependence are called nonlinear pyroelectrics. Above the phase transition point the pyroactivity drops sharply or vanishes altogether, and below this point ferroelectrics display their maximum pyroactivity, and are mainly used as sensitive elements in *pyroelectric radiation detectors*. When detecting and measuring weak radiation fluxes in narrow temperature ranges, the pyroresponse is a quasi-linear function of the incident radiation flux. Nonlinear pyroelectric materials are used for high sensitivity receivers.

NONLINEAR SCHRÖDINGER EQUATION
See *Soliton*.

NONLOCAL DENSITY FUNCTIONALS
Quantities used to calculate the exchange-correlation energy of a nonuniform electron gas (see *Correlation energy*). Nonlocal density functionals are characterized by the dependence of the energy density at a given point on the electron density distribution at other points. These functionals are classified into gradient functionals and appreciably nonlocal functionals. Gradient nonlocal density functionals have the density of the exchange correlation energy expressed via the electron density itself and its gradient at the same point. Two types of nonlocal density functionals are used: the *Langreth–Mehl nonlocal density functional* and the *generalized density expansion* (both nonlinear in the gradient squared). Among the appreciably nonlocal functionals the most prevalent involves the so-called *weighted density approximation* which uses the exchange correlation energy expressed via its binary *correlation function*. The weighted density approximation provides high accuracy and yields correct limits for both single-electron systems and a free electron gas, yet lags far behind the local and the gradient functionals in their simplicity of implementation. See also *Density functional theory*.

NONMAGNETIC MATERIALS
Materials with a relative *magnetic permeability* below $\mu = 1.5$, and in most cases below $\mu = 1.1$. Among the most prevalent nonmagnetic materials are nonferrous metals and alloys (*copper, aluminum*, bronzes, *brasses, aluminum alloys*), nonmagnetic *steel* and *cast iron*, as well as *semiconductors*, excluding *ferromagnets* and *ferrimagnets*, insulators like ionic crystals, and most organic compounds.

NONOHMIC CONTACT
A *metal–semiconductor junction* with a nonlinear curent–voltage characteristic: its resistance depends on the value and sign of the current running through the junction. Properties of the nonohmic contact stem from the *Schottky barrier* that forms at the metal–semiconductor interface; they are fully defined by the characteristics of that barrier. Nonohmic contacts may be used to rectify electric current, to transform signals in microwave devices, as a gate in *field-effect transistors*, as a photodetector, or as a *solar cell*.

NONRADIATIVE QUANTUM TRANSITION
Transition of a system between a discrete level to a lower energy state without any *photon* emission. This is in contrast to free atoms where emission is often the only channel for the transition energy release. The transition energy in solids can be transferred to atomic vibrations (*phonons*), to *conduction electrons*, to the nucleation or rearrangement of defects, etc. See *Auger effect, Radiative quantum transition*.

NONRESONANCE ABSORPTION in magnetic materials
Absorption of energy from a low-frequency magnetic field B_1 parallel to a constant magnetic field B. It is used to measure the frequency dependence of the longitudinal *magnetic susceptibility*, $\chi(\omega)$. Nonresonance absorption had been used to study *paramagnetic relaxation* long before the technique of *electron paramagnetic resonance* was discovered. Nonresonance techniques are much less sensitive than resonance ones. Nonresonance absorption is used in combined techniques of *low-frequency enhanced magnetic resonance absorption*. Sometimes it is referred to as the *technique of parallel fields*.

NONSILVER PHOTOGRAPHY
Photographic processes for obtaining an image or recording optical signals on *photographic materials* containing no silver halides (as distinct from classical photography).

Both chronologically and in accordance with the volume of applications, the main nonsilver photographic procedure is *electrophotography* that uses the following substances as the light-sensitive layer: amorphous selenium, chalcogenide vitreous semiconductors, crystalline photoconduction materials in a polymeric host, and *organic semiconductors*. Progress has been achieved in the development of nonsilver photography on the basis of hydrogenated *amorphous silicon*. Also used in nonsilver photography are *photothermoplastic materials* where a development of the latent electrostatic relief is achieved either due to a strain in the polymeric layer that is also a photoconductor, or due to extra layers of inorganic semiconductors. The sensitivity of these materials is high and comparable to that of ordinary photomaterials.

In microphotography, as well as when copying and duplicating technical and text documents, *diazotype materials* and *vesicular materials* are used. The image is produced by light-scattering bubbles of the gas released in a polymer layer under photochemical decomposition of a light-sensitive diazocompound introduced into the layer. There is a possibility of recording images via *photochemical reactions* in the layers of a semiconductor (of PbI_2, BiI_3 type) as well as via stimulated metal *diffusion* in structures containing a metallic layer (Ag, Cu) and a semiconductor (e.g., PbI_2). Their sensitivity is not high but the process can be used to record optical elements and *optical storage disks*. Widely used in nonsilver photography are also *photopigments* where the image recording is based on the phenomenon of *photochromatism* of organic and inorganic substances (see *Photochromic effect*). In view of the irreversibility of light-stimulated changes in coloration, these materials are suitable for repeated image recording with high resolution $v > 10^3$ mm^{-1}, e.g., in *holography*. New materials with similar properties are bacteriorhodopsin layers in a polymeric host. The processes of photopolymerization (*photoresists*), photopretanning (gelatin), and isomerization (stilbene) are used in nonsilver photography in the fields of printing and microelectronics. To record the phase relief of holograms, *reoxan* is used in which the oxidation of the *anthracene* in a polymeric host (polymethylmethacrylate) takes place under the action of light. The sensitivity of this method is low; and fixation of the image is needed to produce a stable phase relief.

Reversible and irreversible photostructural modifications of the chalcogenide vitreous semiconductors, which are accompanied by significant changes of the *refractive index*, are the basis of amplitude–phase recording of images, holograms, and optical data in a bit-by-bit mode in thin layers and bulk plates. It is possible to perform recording with a resolution $v > 10^3$ mm^{-1} in real time. The sensitivity is low, but the materials are applicable in laser systems for the manufacture of memory elements and *integrated optics* elements. *Selective etching* of an exposed layer determines the applicability of the layers of chalcogenide vitreous semiconductors as high-resolution inorganic photoresists.

The phenomenon of *photorefraction*, or photoinduced alteration of the refractive index in electrooptical crystal *ferroelectrics* underlies the application of these crystals to phase recording of holograms.

The *photoinduced phase transformations* take place in crystallizing layers of chalcogenide vitreous semiconductors, in TeO_x and magnetooptical materials MnBi, and offer prospects of producing bit-by-bit recording carriers. Nonsilver photography also uses thermophotographic materials and systems that are sensitive to a temperature change on account of the energy absorbed during the exposure. These are *liquid crystals*, *luminophors* with thermal extinction, and evaporating metallic or organic layers. An extension to the IR region is possible by using the semiconductor photographic processes based on the intrinsic *photoeffect*.

NORMAL DISTRIBUTION
The same as *Gaussian distribution*.

NORMAL VIBRATIONS (normal modes)
Small natural vibrations of a system of several or many particles (including those in a *crystal lattice*) described by such single-frequency time-dependent harmonic functions (*normal coordinates*) that their number coincides with the number of the vibrational degrees of freedom of the system. An arbitrary vibration may be represented in the form of a superposition of normal vibrations. Normal vibrations in a crystal lattice are specified by their *quasi-wave vector* k, and the number of their vibrational branch, α. Together they uniquely define the frequency of such normal vibrations, ω, and the polarization vector, e. The displacement of

the sth atom in the nth unit cell positioned at r_n is written as

$$u_{sn} = \sum_{k\alpha} Q_\alpha(k)e_s(k, \alpha)\exp(\mathrm{i}kr_n),$$

where the normal coordinates $Q_\alpha(k)$ satisfy the harmonic equation

$$\frac{\mathrm{d}^2 Q_\alpha(k)}{\mathrm{d}t^2} + \omega_\alpha^2(k)Q_\alpha(k) = 0.$$

When the equations of motion are linear (*harmonic approximation*), normal vibrations are mutually independent, and the system energy may be represented as a sum of the energies of individual harmonic oscillators corresponding to separate normal modes. When the *anharmonic vibrations* are taken into account the normal mode vibrations interact with each other, so the above expressions become only approximate. In the quantum theory of crystal lattice vibrations the normal vibrations correspond to *phonons*, the elementary excitations of the crystal.

NOTCH

Angular cut or indentation serving as a geometric *stress concentrator* (see *Stress concentration*), used to decrease the coefficient of softness of the *mechanical testing of materials*. Most widely used tests, which involve application of stress concentrators, are tests of samples with a notch for stretching and *flexure*. Samples of circular cross-section are often notched circularly; the preparation of rectangular samples for stretching requires making a symmetrical notch along two sides. Samples for tests of tension are notched along one of the sides of a section, perpendicular to the longitudinal axis. The *stress concentration factor*, the ratio of maximum stress to nominal stress, is used for a quantitative assessment of the notch rigidity. The factor is determined by the geometry of the notch: the deeper and sharper the notch, the higher the factor, and the easier it is to achieve *brittle failure*. For practically every metal, even very plastic ones (see *Plasticity*), one can find an appropriate shape and dimensions of the sample and the notch to produce brittle failure under given test conditions. The criterion of notch sensitivity of a material is the ratio of *ultimate strength* values for intact and notched samples (σ_B/σ_B^n): the

higher this ratio, the more sensitive the material is to the notch. To arrive at comparable results, sample shapes and testing techniques have been standardized.

NOTTINGHAM EFFECT (W.B. Nottingham, 1941)

Heating of the *field-emission cathode* and cathode cooling during *thermal electron field emission*. The energy distribution of electrons at low temperatures is practically identical to the Fermi–Dirac distribution at 0 K (see *Fermi–Dirac statistics*). Therefore, the average energy of autoelectronic emission electrons tunneling through the potential barrier in a strong electric field applied to the surface of the cathode remains lower than the *Fermi energy* in the cathode, and lower than the energy of electrons reaching the cathode from the external electric circuit. This energy difference results in the heating of the cathode. In the case of thermal electron field emission, the electrons escaping from the cathode have energies exceeding the *Fermi level* of the cathode. Their replacement by electrons from the external circuit having average energies lower than that of the escaping electrons results in the cooling of the cathode.

NUCLEAR DOPING, irradiation doping

Method of introducing doping additives into material which is based on nuclear transformations of host nuclei to daughter *impurity atoms* under the action of charged particles (protons p, α-particles, deuterons d, and so on), *gamma rays* or *neutrons* n (in the latter case one speaks of *neutron doping*).

A principal difference between nuclear doping and routine methods of embedding impurities in crystals consists in the direct formation of the impurity in the bulk sample from the atoms of the dopant, instead of introducing the impurity in the *semiconductor* from outside, i.e. from the melt or the gaseous phase, or by *diffusion*. The production of impurity atoms at any point of a bulk sample at a distance smaller than the mean free path distance is equally probable, hence the resulting impurity distribution is essentially isotropic.

The reactions that take place from incoming d, α, and p particles of the types (α, p), (α, n), (p, α), (p, d), etc. are induced by the very high energy of the incident particles which is required to overcome the Coulomb barrier when approaching a nucleus, transfer to it the energy needed to compensate for the breaking of chemical bonds, and again overcome the Coulomb barrier by the charged particle that flies out of the nucleus. This barrier reaches approximately 10, 20, and 30 MeV for light, intermediate, and heavy nuclei, respectively. This process was implemented by doping silicon with phosphorus, e.g., through the reactions ^{28}Si (α, p) ^{31}P; ^{29}Si $(\alpha, 2p)$ ^{31}P; as well as by doping germanium with the impurities Ga, As, and Se and utilizing the actions of deuterons and α-particles. The nuclear reaction ^{28}Si (α, p) ^{31}P corresponds to the silicon isotope ^{28}Si absorbing an α particle to form the excited state sulfur nucleus ^{32}S, which then transforms to the stable phosphorus isotope ^{31}P by emitting a proton p. Due to the high rate of braking (deceleration) of charged particles in matter, the thickness of the resulting doped layer is small, and the distribution of the introduced impurity is nonuniform.

Nuclear reactions which take place under the action of γ-quanta are called *photonuclear reactions*, and reactions of the types (γ, n), (γ, p), $(\gamma, 2n)$, and (γ, pn) are referred to as threshold reactions. A particular feature of these reactions is the dependence of the reaction cross-section σ on the γ-quantum energy E in the range $E \approx 10$ to 25 MeV, where $\sigma(E)$ has a pronounced maximum, the so-called *giant resonance*. The main impurity that is formed in silicon as a result of photonuclear reactions is aluminum. In germanium the gallium atoms are predominantly formed, and the impurities of Se, Ge, and Zn are formed in GaSe. Gamma quanta can penetrate deeply into a semiconductor, so photonuclear reactions are capable of doping large volume regions of the material. Also of interest is nuclear doping through nuclear reactions with thermal neutrons (n, γ). The energy of thermal neutrons $E \sim 0.5$–$5 \cdot 10^{-3}$ eV is comparable with the energy of the atomic thermal vibrations of a moderator, and their velocity is $v \approx 2 \cdot 10^5$ cm/s. The nuclear reaction cross-section is inversely proportion to v ("$1/v$ law"). The main sources of thermal neutrons are nuclear reactors and nuclear fuel, that is ^{235}U. A final product of the (n, γ)-reactions in silicon is phosphorus. In germanium an acceptor impurity and donor impurity, respectively As and Se, are formed. Three acceptor atoms are formed per donor atom. In GaAs

the donor impurities Ge and Se are formed. In semiconductors of the type of InSb, the donor impurity Sn prevails.

The mean free paths of thermal neutrons in various semiconductors are the following (in cm):

Si	Ge	GaAs	InSb
125	4.0	2.8	0.14

The mean free path determines the volume of the doped material within which the impurity is spread uniformly. The indirect negative phenomena that accompany the intentional formation of necessary impurities are induced (bulk and surface) *radioactivity* and radiation defects. The *bulk radioactivity* is due to the radiative decay of an unstable isotope to an atom of the doping impurity. To compensate for this the exposure of the doping material during 10 half-life periods is needed. The source of *surface radioactivity* is the *adsorption* of radioactive elements from the medium of the nuclear reactor by the surface of irradiated ingots. This is eliminated by washing the ingots in special solutions, or by etching the surface layer. Radiation defects are annealed during the *heat treatment*.

The technology of the nuclear doping of semiconductors consists of the following stages: obtaining the initial *monocrystals*, irradiation, decontamination, heat treatment, control of electrical physical parameters, and output radiological monitoring. The main advantages of the nuclear doping of semiconductors include the high degree of uniformity of the introduced impurity distribution (see Fig. 1), the accuracy of the regulation

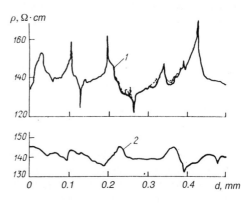

Fig. 1. Distribution of specific resistance in initially undoped (1) and nuclear-doped silicon (2).

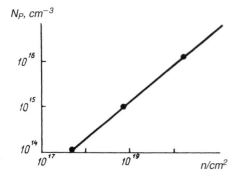

Fig. 2. Phosphorus concentration N_P in nuclear-doped silicon versus neutron dose n.

(see Fig. 2), and the reproducibility of a given concentration. The overall yield of nuclear doping reaches 90–97%.

NUCLEAR ELECTRIC QUADRUPOLE MOMENT

Second-rank tensor q with components

$$q_{ik} = \sum_{\alpha} \left[3x_j^\alpha x_k^\alpha - (r^\alpha)^2 \delta_{jk} \right],$$

arising from the second term of an even-power multipole expansion of the nuclear electric field scalar potential (see *Multipole*), specifying the deviation of the nuclear electric charge distribution from spherical symmetry (the first term is the monopole or total charge). This tensor is symmetric ($q_{ij} = q_{ji}$) and traceless ($q_{xx} + q_{yy} + q_{zz} = 0$). Here α sums over the protons in the nucleus; x_j^α and r^α are, respectively, the coordinates and the length of the αth proton radius-vector in the Cartesian system with its origin at the nucleus. If the nuclear charge distribution has *axial symmetry* then the symmetry axis z is one of the principal axes of tensor q; the component $q_{zz} \equiv Q$ corresponding to this axis is called the *nuclear quadrupole moment*. Deviations from axial symmetry are expressed in terms of the dimensionless asymmetry parameter $\eta = -(q_{yy} - q_{xx})/q_{zz}$, where by convention $|q_{xx}| \leqslant |q_{yy}| \leqslant |q_{zz}|$, and $0 \leqslant \eta \leqslant 1$; for axial symmetry $\eta = 0$. With the help of the *Wigner–Eckart theorem*, the components of tensor q can be expressed using Q and components of the nuclear *spin* vector I. The third term of of

the above-mentioned even-order multipole expansion corresponds to a tensor of the fourth rank, the so-called *electric hexadecapole nuclear moment*, whose components are constructed from the coordinates of the nuclear protons to the fourth power. These moments, together with higher order ones, specify the distortion of a nuclear electric charge distribution from a spherical shape. The nuclear hexadecapole moment causes the nuclear *hexadecapole interaction* which can be reflected in nuclear resonance spectra (although this effect is generally too weak to measure): *nuclear magnetic resonance*, the *nuclear quadrupole resonance*, and *acoustic nuclear magnetic resonance*. This yields opportunities for experimental estimates of hexadecapole moments of interest for nuclear physics.

NUCLEAR FILTERS, Flyorov filters (G.N. Flyorov et al., early 1970s)

Porous walls, *membranes* obtained with the help of methods of nuclear physics; thin (1 to 50 μm) layers or *films* whose system of *pores* is formed by irradiation with accelerated homogeneous multicharged heavy *ions* with subsequent *etching* of the tracks produced through holes (pores). The technology of nuclear filters is based on the *selective etching* of the heavy particle tracks; in this case the etching rate of the material of the nuclear filter is appreciably ($10–10^4$ times) higher along the track than in the untouched material. The following stages of treatment are applied when manufacturing nuclear filters from polymeric films:

(1) obtaining the tracks by irradiating the film with a beam of the cyclotron-accelerated ions with the energy 1–5 MeV, with atomic number 8 to 54 and fluence $10^5–10^{11}$ ions per cm^{-2};

(2) sensitizing of the ion-irradiated film by UV light;

(3) chemical treatment of the irradiated and sensitized film with etchant solutions, e.g., with alkali when using a film made of Lavsan polyester fiber (analogs Therylene, Daeron, Tergal).

Pore sizes in nuclear filters are from 0.01 to tens of μm. Nuclear filters can also be made from thin layers of glass or *monocrystals*.

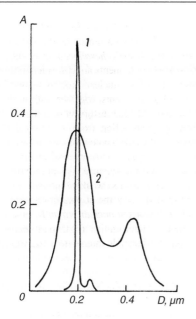

Fig. 1. Pore area distribution A versus pore diameter D in a nuclear filter membrane (curve 1), and in an ordinary random network membrane (curve 2), both with a dominant pore diameter of 0.2 μm.

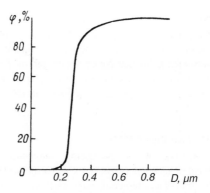

Fig. 2. Percentage φ of detained (not passed) rigid colloidal particles versus the pore diameter D for nuclear filter 1 of Fig. 1.

The main feature which distinguishes nuclear filters from ordinary membranes with a random network structure obtained by colloidal chemistry methods is the regular geometry of the pores, and the excellent uniformity of their diameters. Deviations of the pore sizes from their mean diameter in nuclear filters is only 2–5%, whereas in netlike membranes the deviations are 50% or more (see Fig. 1).

The nuclear filter is a kind of original "membrane knife" that produces a precise purification of liquid and gaseous media. To judge the efficiency of separating colloidal solutions with the help of nuclear filters, a so-called calibrating dependence plot (see Fig. 2) is used. This is a plot of the percentage of detained (held back) particles φ versus the pore diameter D. We see from Fig. 2 that for the nuclear filter 1 of Fig. 1 those particles with diameters d less than 0.2 μm are passed ($\varphi = 0$), while those with $d > 0.2$ μm are blocked from passing ($\varphi \to 100\%$ for $d \gg 0.2$).

Nuclear filters can serve as the basis for obtaining various kinds of secondary structures, e.g., to make points for cold cathodes; to form microstructures of integrated *semiconductor devices* with details smaller than 0.1 μm; to obtain calibrated leaks for vacuum engineering, etc.

NUCLEAR GAMMA RAY RESONANCE
The same as *Mössbauer effect*.

NUCLEAR g-FACTOR SHIFT
A change in the value of the dimensionless g-factor (see *Landé g-factor*) in a metal, g_m, compared to its value in an *insulator*, g_i. The value $\Delta g = g_m - g_i$ is related to the *Knight shift*, $\Delta\omega = \omega_m - \omega_i$, via $\Delta g/g_i = \Delta\omega/\omega_i$, where ω_m, ω_i are the resonance frequencies of the nuclei in *metals* and *insulators*, respectively, measured in the same applied magnetic field, B_0. The g-factor shift Δg, found in every metal, has the following properties: almost always $\Delta g > 0$; the relative shift $\Delta g/g_i$ does not depend on B_0 (for weak fields); the shift is almost independent of temperature, and it increases for larger nuclear charges eZ. These features of Δg are due to the *hyperfine interaction* of the nuclei with the *conduction electrons* that are in an s-state with respect to a particular nucleus. When applying B_0, the spins of the conduction electrons polarize and produce a hyperfine interaction field B' at the nuclei, in the same direction as B_0. Therefore, the field acting upon the nucleus is $B = B_0 + B' > B_0$, so that $\Delta\omega > 0$ and $\Delta g > 0$. Since the frequency shift $\Delta\omega$ is proportional to the electron polarization (magnetization) M_s, it is also

proportional to B_0 or ω_i. Moreover, since M_s does not depend on the temperature (the magnetic moment of a strongly degenerate gas is temperature independent), $\Delta\omega$ and Δg do not depend on T either. Finally, the wave function density at the nucleus grows for larger Z, so it results in a larger Δg shift as well.

NUCLEAR INDUCTION

Phenomenon of precessing *magnetization* (*dynamic nuclear polarization*) which is induced by the action of an alternating magnetic field on a system of nuclear *magnetic moments* in an applied constant magnetic field; also the appearance at the ends of a coil of a voltage induced by the precessing magnetic polarization. The term nuclear induction is taken to denote not only a phenomenon but also a method of detection of *nuclear magnetic resonance* which is based on the measurement of the voltage induced in the coil. Inasmuch as a spin is associated with a magnetic moment, one can employ a more general term such as *spin induction* or *magnetic moment induction* which generalizes the concept of nuclear induction for an arbitrary type of *magnetic resonance*. When a crystal contains particles with non-zero electric *dipole moments*, their precession can be excited by an applied alternating electric field. Their precession is the source of an electric field that can be detected. By analogy with nuclear induction, this phenomenon might be called *electric dipole induction*.

NUCLEAR MAGNETIC RESONANCE (NMR)

Resonance absorption (emission) of electromagnetic waves in the radio-frequency range (10^6–10^9 Hz) related to the precession of nuclear *magnetic moments* in a strong applied magnetic field, a particular case of *magnetic resonance*. It is also a branch *magnetic resonance spectroscopy*. The resonance of nuclei and molecular beams was discovered by I.I. Rabi (1937). NMR in condensed matter was first and independently observed by E. Purcell and F. Bloch et al. (1946).

Overview. In NMR a strong permanent magnetic field B_0 lifts the energy degeneracy with respect to magnetic (and related to it mechanical) nuclear moments, producing a system of energy levels ε_i (*Zeeman splitting*). The energy levels are characterized by their quantum numbers m designating the projection of the nuclear *angular momentum* (spin I) on the magnetic field direction $z \parallel B_0$, where $-I \leqslant m \leqslant I$. The radio-frequency magnetic field B_1 ($B_1 \ll B_0$) induces transitions between these spin sublevels, resulting in the absorption (emission) of a quantum of energy with the frequency $\nu_{ij} = (\varepsilon_i - \varepsilon_j)/h$. In the particular but commonly encountered case of a nucleus with spin $I = 1/2$, there are two levels $\varepsilon_{1,2} = \pm \gamma B_0 \hbar/2$, and the frequency of the transition is $\nu_0 = \gamma B_0$, where γ is the *gyromagnetic ratio*. The applied field B_0 has typical magnitudes from 1 to 10 T. For protons $\gamma = 42.6$ MHz/T, so that for $B_0 = 10$ T we have $\nu_0 = 426$ MHz. At thermodynamic equilibrium the spin system has more nuclei in the lower energy levels that at higher ones. Hence the application of the radio-frequency field initially induces transitions from lower to upper levels, resulting in the absorption of electromagnetic energy. At the same time, there exist situations (see below) when the spin population at a higher level is greater than at a lower one, and in this case, the spin system will emit energy.

Instrumentation. To observe NMR one makes use of an *NMR-radiospectrometer*. Its principal elements are the following: the source of the magnetic field B_0 (permanent magnet, electro magnet, solenoid, etc.); the source of the radio-frequency field B_1 (coil of oscillator); a scanning unit to vary the frequency ν or the applied field B_0 slowly through the resonance condition; a signal detector which includes an oscillatory circuit, a radio-frequency bridge, a receiver and a signal recording device. The radiospectrometer works in the following manner. A sample of the material under investigation is placed in the coil of the signal generation circuit that contains the alternating field B_1 of frequency ν. The coil containing the sample is located in the gap of the magnet that provides the field B_0, oriented so that $B_1 \perp B_0$ to satisfy the selection rules. To observe the NMR, the spin detector transforms the absorbed energy of quantum spin transitions to detectable radio-frequency signals. At the point of resonance ($\nu = \gamma B_0$) a voltage drop takes place in the circuit containing the coil with the sample. This voltage change is amplified and detected in a routine fashion; then it is transformed to a form convenient for display using an oscilloscope or a recorder, or for transmission to a data processor, or a computer, etc.

A particular method for observing NMR is its detection under the conditions of the *electron–nuclear double resonance*, where the detection of the NMR signal is performed indirectly, i.e. through the observation of an *electron paramagnetic resonance* (EPR) signal. In the vicinity of electron paramagnetic centers, the *hyperfine interaction* of electrons and nuclei results in a common system of spin levels. Thus, the nuclear-spin transitions can affect the difference in those *level populations* between which an electron-spin transition takes place. This leads to a change in the EPR signal that serves for detection of the NMR signal.

Hamiltonian. Being a purely quantum effect, NMR is completely described with the help of a Hamiltonian that is called a *spin Hamiltonian* because the Hamiltonian operators all depend on spin variables. This Hamiltonian W has a particular form in every individual case, but in general it can be represented in the manner that takes into account all possible NMR-involved interactions of the nucleus with the environment:

$$W = W_Z + W_Q + W_{ss} + W_P + W_E, \quad (1)$$

where W_Z is the interaction with the applied DC magnetic fields $\boldsymbol{B}$ (*Zeeman interaction*) W_Q is the interaction with the *crystal field gradient* (quadrupole interaction), W_{ss} is the interaction with magnetic moments of surrounding nuclei and electrons (*spin–spin interaction*), W_P is the effective interaction with applied external pressure P (strains), W_E is the effective interaction with external electric fields $\boldsymbol{E}$. Each interaction in Eq. (1), in its turn, consists of one or more terms whose form depends on the nature of the interaction, the number and magnitudes of the nuclear spins, and the *local symmetry* of their sites. The most commonly encountered expressions can be written in the following way:

$$W_Z = -\hbar\gamma(\boldsymbol{I}\,\boldsymbol{B}_0),$$

$$W_Q = \frac{QV_{zz}}{4I(2I-1)}\left[3\hat{I}_z^2 - I(I+1)\hat{e}\right],$$

where Q is the electric quadrupole moment, V_{zz} is the z component of the *electric field gradient* tensor, $\hat{e}$ is the unit matrix; $W_Q = 0$ unless $I > 1/2$,

and the local symmetry at the position of the nucleus is lower than cubic;

$$W_{ss}^{(k)} = \frac{\gamma^2 h^2}{2}\sum_{i>k} r_{ik}^{-3}(3\cos^2\theta_{ik} - 1)$$

$$\times \left(\hat{I}_i\hat{I}_k - 3\hat{I}_{zi}\hat{I}_{zk}\right),$$

where r_{ik} is the distance between the kth nucleus under consideration and the ith nucleus, θ_{ik} is the angle between vectors $\boldsymbol{r}_{ik}$ and $\boldsymbol{B}_0$; W_P and W_E have forms similar to that of W_Q with a single difference in the sense of *spin operators*: in the case of W_P these operators relate to P, and in the case of W_E they relate to the electric field $\boldsymbol{E}$. The term W_E which is linear in respect to $\boldsymbol{E}$, is nonzero only for nuclei in sites without a *center of symmetry*. The terms W_P and W_E can be looked upon as terms of extra-induced quadrupole interactions. When considering the whole system of nuclei, the expression (1) should be supplemented by a summation which takes into account all the nuclei; this is of particular importance in the case of solids. Each term in Eq. (1) can be additively divided into static and dynamic parts. The static part determines the line position in the spectrum, while the dynamic part determines the probabilities of quantum transitions (*line intensities*) and provides for relaxation processes. In this way, noting that $\boldsymbol{B} = \boldsymbol{B}_0 + \boldsymbol{B}_1$, the Zeeman term $W_Z = W_{Z0} + W_{Z1}$ involves two main interactions, a static one W_{Z0} with the large applied field $\boldsymbol{B}_0$ which provides the line positions, and a dynamic one W_{Z1} with the alternating radio-frequency field $\boldsymbol{B}_1$ which induces the transitions. The Zeeman term also includes interactions with local magnetic field inhomogeneities (static) and fluctuations (dynamic) that account for the line width and saturation mechanisms.

Spectrum. The NMR spectrum (i.e. the number of lines, their intensities and angular dependence) is completely determined by the spin Hamiltonian (1). In the case of a nucleus with spin I in a permanent magnetic field $\boldsymbol{B}_0$ (term W_{Z0}), $2I + 1$ equidistant energy levels $\varepsilon_m = -\gamma I m B_0$ appear (here m is the quantum number of operator $\hat{I}_z$ which takes on values $-I, -I+1, \ldots, I-1, I$). The radio-frequency field $\boldsymbol{B}_1$ (time-dependent term W_{Z1}) induces the magnetic dipole transitions between adjacent levels ($\Delta m = \pm 1$) under the condition $\boldsymbol{B}_1 \perp \boldsymbol{B}_0$ (*selection rule*), so that there will appear only a single isotropic line in the NMR spectrum since the transition frequency $v = (\varepsilon_m - \varepsilon_{m-1})/h$ does not depend on m. In fact, owing to presence near the nucleus under consideration of electrons and other nuclei, the magnetic field in atoms, molecules, and condensed matter will differ from the external field $\boldsymbol{B}_0$ due to the polarization of electron shells and the direct nuclear magnetic *dipole–dipole interaction*. As a result, there appear many incremental additions $\Delta \boldsymbol{B}_0$ to $\boldsymbol{B}_0$ which, in the general case, are not parallel to $\boldsymbol{B}_0$. These lead to line broadening and also anisotropic shifts of the NMR lines: *chemical shift*, Knight shift (see *Hyperfine interaction*), and also shifts due to an indirect nuclear spin–spin interaction. In many cases the nuclei in a solid are situated at two or more chemically inequivalent sites, each of which is characterized by its own specific chemical shift. As a result, there appear particular lines in the spectrum which arise from different nuclear groups. The level splittings from equivalent sites are still equally spaced, but broadened.

Other terms of the spin Hamiltonian (1) break the rule of an equidistant arrangement of energy levels. Thus, the operator W_Q adds to a Zeeman level ε_m an extra term which is proportional to $m^2 - I(I+1)/3$. As a result, the lines in the NMR spectra become split into $2I$ lines (*fine structure*). In the case when $B_0 = 0$, the system of levels is determined by operator W_Q (there appear two levels if $I = 3/2$), and transitions between them induced by the radio-frequency field $\boldsymbol{B}_1$ lead to a zero field NMR that is often called *nuclear quadrupole resonance*. A specific line splitting arises under action of operator W_E. There sometimes exist *inversion-inequivalent positions* in crystals which the nuclei of the same type can occupy with equal probability. Since the magnetic field is insensitive to the inversion operation, the field does not distinguish these positions, and the related lines in the NMR spectrum are superimposed. An electric field $\boldsymbol{E}_0$ applied to a crystal, due to the mutual inversion of these inequivalent positions, will be oppositely directed for each of them. As a result, the frequency corrections linear in $\boldsymbol{E}_0$ corresponding to different positions will have different signs, and the shift of these two groups of lines in opposite directions will produce an additional splitting in the NMR spectrum. The term W_E can also

affect the quantum transitions. Hence the electric component of the radio-frequency field $\boldsymbol{E}_1$ can (as well as $\boldsymbol{B}_1$) induce transitions between magnetic sublevels (for the sample positioned between capacitor plates), but with special selection rules (e.g., $\Delta m = \pm 2$ at $\boldsymbol{E}_1 \parallel \boldsymbol{B}_1$).

Line broadening. As a rule, crystals are not perfect. Therefore, NMR is often used for investigating disordered systems (powders, glasses, solutions, crystals with diverse defects). A particular feature of such systems is the inequivalence (nonuniformity) of the conditions at the positions of nuclei, due to: (i) differences of magnetic and electric fields, and deformations induced by structural distortions of the crystal and defects of diverse kinds; (ii) the orientational inequivalence in regard to applied external fields; (iii) the nonuniformity of the external fields, etc. All these lead to dispersion of the parameters of the spin Hamiltonian (1), and as a consequence to *inhomogeneous broadening* of the NMR lines. Despite the existence of a number of methods for describing such broadening (*statistical theory of line shape*, moment method (see *Moments of spectral lines*), theory of randomly or partially disoriented centers, etc.) innomogeneous broadening of all types can be treated from a common viewpoint. In this case, the general expression for the *line shape* takes the form:

$$I(\omega) = \int \varphi\big(\omega - \omega_P(F)\big) V(F)\rho(F)\,\mathrm{d}F, \quad (2)$$

where φ is the initial line shape without taking perturbing factors into account, $V(F)$ is the transition probability per unit time, $\rho(F)$ is the distribution function of the parameters $F(F_1, \ldots, F_k)$ that characterize the broadening mechanisms (components of fields, deformations, angles). So in the case of randomly oriented centers (powders, glasses) one would interpret F as the *Euler angles* that characterize the orientation of a powder particle with respect to the coordinate system related to the external fields.

Relaxation. NMR is accompanied by relaxation processes which act to restore the initial *Boltzmann distribution* equilibrium state which had been disturbed by the transitions induced by the applied electromagnetic radiation. Such relaxation processes are due to the interactions between the nucleus and the lattice, as well as between the nuclei themselves. Therefore, a distinction is made between *spin–lattice relaxation* and *spin–spin relaxation*. If the transitions stimulated by the radio-frequency field are dominant, a *saturation effect* (equalization of level populations) takes place, and manifests itself in the decreased intensity and broadening of the NMR signal. The *relaxation* processes are characterized by their *relaxation times*, and are described by kinetic equations for level populations. In the simplest case of two levels i and j, the equations for populations n_i and n_j take the form:

$$\dot{n}_i = -n_i a + n_j b, \qquad \dot{n}_j = -n_i a - n_j b,$$

where $a = V_{ij}^0 + V_{ij}$; $b = V_{ji}^0 + V_{ji}$; the terms $V_{ij}^0 = V_{ji}^0$ and V_{ij} are the probabilities of transition from level i to level j per unit time under the action of radio-frequency field and relaxation mechanisms, respectively. The relaxation time is determined by the expression $T_r = (V_{ij} + V_{ji})^{-1}$ and characterizes the rate of establishing equilibrium. These relaxation processes shorten the *lifetimes* of particles at spin levels, which leads to their broadening by affecting the width and shape of the NMR lines. This broadening, which is the same for all nuclei, is referred to as *homogeneous broadening*. It can, in particular, determine the shape of φ in Eq. (2).

Spin temperature. For a description of spin systems the concept of "spin temperature" T_S can be introduced. To do this, the relation between the level population and the temperature which determines the Boltzmann distribution is generalized for the case of nonequilibrium populations. The assumption being made is that the spins are in equilibrium with each other, but not with their surroundings (lattice vibrations for a solid and Brownian motion for a liquid). Hence, for arbitrary ratios between the populations of upper (n_u) and lower (n_l) levels, it follows that

$$T_S = -\frac{\varepsilon_u - \varepsilon_l}{k_B \ln(n_u/n_l)}.$$

For $n_u = n_l$ (saturation), $T_S = \infty$; for $n_u > n_l$, $T_S < 0$. The opportunity to produce nonequilibrium populations, including the particular case with $T_S = \infty$ as well as negative spin temperatures

$T_S < 0$, stimulated development of *double resonances* based on NMR. These are characterized by resonance transitions at two frequencies, which in the presence of a multilevel system occur simultaneously or in a certain sequence. The purpose of this is to increase the intensity of the absorption by increasing the difference in population (this principle is used in *electron–nuclear double resonance*), and to obtain a source of electromagnetic radiation through the establishment of a larger population at the upper level than at the lower one (a *maser* uses this principle).

Approximation approaches. The complexity of an analytical description of NMR is due to the enormous number of interacting lattice nuclei. This circumstance complicates diagonalizing the spin Hamiltonian (1), calculating and simulating NMR spectra, and studying the kinetics of spin systems (relaxation, saturation, etc.). In this connection, approximate approaches and methods were developed in order to describe the diverse details of the behavior of the spin system in NMR. First of all, it is worthy to mention the *Bloch equations* (F. Bloch, 1946) which describe the dynamics of nuclear *magnetization* that were introduced at the very beginning of the NMR era. Although these equations are based on a classical description of magnetic moment precession, and hence are not adequate for explaining all experimental data, they played a major role in establishing magnetic resonance as a main branch of science. Subsequently more refined theories were developed which are grounded on the *quantum kinetic equations* for a *density matrix* (*Bloch–Redfield theory*, *Provotorov theory*, and so on). An enormous amount of experimental data have been successfully explained within the frameworks of these theories.

Rotating coordinate system. A widely used approach for analyzing NMR spectra is to transform the motion to a rotating coordinate system, i.e. to examine the NMR resonance in an terms of an effective magnetic field $\boldsymbol{B}_{eff} = \boldsymbol{i}B_1 + \boldsymbol{k}(B_0 - v/\gamma)$ acting upon the nuclear spins, in a coordinate system which rotates with the frequency v about the permanent magnetic field $\boldsymbol{B}_0 \parallel z$, i.e. in the direction of the nuclear precession, where the radio-frequency field $\boldsymbol{B}_1$ is applied normally to $\boldsymbol{B}_0$. The frequency v_{eff} corresponding to $\boldsymbol{B}_{eff}$ is given by $v_{eff} = \gamma B_{eff} = [(v_0 - v)^2 + \gamma^2 B_1^2]^{1/2}$, where

$\nu_0 = \gamma B_0$. This method not only clarified the explanation of many phenomena in the NMR, but it allowed Redfield to predict, and then to observe, a so-called *NMR in a rotating coordinate system*. The essence of this phenomenon involves the following. If an additional alternating field $B_2 \parallel B_0$ with the frequency ν_{eff} is applied to the spin system, then its perpendicular component B_{eff} will induce a new resonance, i.e. NMR in a rotating coordinate system.

High-resolution NMR. Due to the large number of nuclei which influence each other via the dipole–dipole interaction, the width of NMR lines in solids is abnormally large (about 10^{-4} T or tens of kHz), which hinders the observation and investigation of such fine structure effects as chemical and Knight shifts, indirect internuclear interactions, etc. In gases and low-viscosity liquids the thermal motion of atoms and molecules causes the "natural" averaging of the dipole–dipole interaction, diminishing it close to zero. This circumstance makes possible the high resolution of the NMR method which is employed in the majority of the physicochemical and analytical applications of NMR. This branch of NMR designed for detecting and investigating very narrow lines is referred to as high-resolution NMR. For solids, it was necessary to develop special methods for artificially narrowing NMR lines by suppressing the dipole–dipole interaction. Included among them are multipulse methods of various types, and magic angle spinning (MAS) of the sample at the magic angle ($\theta = \cos^{-1}(1/\sqrt{3}) = 54°45'$). These methods have expanded considerably the utility of NMR, particularly in its applications for solids.

Applications. The NMR method is widely employed in various branches of physics, chemistry, biology, medicine, and other areas of science. It has achieved particular success in studies of condensed matter where the need for information is quite extensive, and it involves a variety of diverse interactions of nuclear systems with the environment and other factors. Several examples of areas of application are: establishing the atomic structure of crystals (determination of atomic coordinates); determination of the electronic structure of crystals and the nature of *chemical bonds*; establishing the nature of chemical and physical *adsorption* of atoms and molecules on *solid surfaces*; studies of the attributes of *phase transitions*; determination of spatial ordering in *liquid crystals*, investigating weakly coupled molecular groups in *polymers* and organic crystals; establishing the character of collective motions of atoms (molecules) of the lattice; revealing mechanisms of chemical reactions; *magnetic resonance imaging*. Finally, worthy of mention is the past and present role of NMR in studies of the structure and properties of such recently discovered materials as *fullerenes* and fullerites.

In magnetically ordered materials where NMR is observable in the effective field of a *hyperfine interaction* between the nuclear and electronic spins of a magnet, NMR studies yield data on the *atomic magnetic structure*, its dynamics, the interaction of nuclear and electron *spin waves* (*Suhl–Nakamura interaction*), etc. (see *Indirect nuclear spin–spin coupling*).

NUCLEAR MAGNETIC RESONANCE, ACOUSTIC

See *Acoustic nuclear magnetic resonance*.

NUCLEAR OPTICS

Branch of the theory involving the interaction of *nuclear radiations* with matter where the radiation propagation processes are interpreted using the formalism of optics. In the case of neutron waves, the term *neutron optics* is accepted. In nuclear optics an index of refraction n for a "hard" electromagnetic or electron wave (i.e. high-energy wave) is introduced by the relation

$$n = 1 + \frac{2\pi \rho f(0)}{k^2}, \qquad (1)$$

where $f(0)$ is the forward *scattering amplitude* (elastic coherent scattering amplitude at zero angle), ρ is the density of scattering particles in matter, and k is the wave number of the incident wave. For the high energy (small wavelength) condition $|n - 1| \ll 1$ the geometrical optics laws of refraction (*Snell's law*) and *total internal reflection* are valid. For the case when $f(0)$ depends on the polarization state of the neutron or *photon*, there appears a difference in *refractive index* for right-hand and left-hand circular polarization which yields the *Faraday effect*. Eq. (1) is applicable for the description of matter consisting of randomly distributed scatterers, and for radiation with a wavelength much less than the distance between scatterers. When absorption is taken into account the refractive index n has a complex form.

NUCLEAR ORIENTATION

A collection of methods for producing nonrandomly oriented nuclear *spins*. The extent of their orientation is characterized by two quantities,

$$P_1 = \frac{\langle I_z \rangle}{I}$$

and

$$P_2 = \frac{\langle 3I_z^2 - I(I+1) \rangle}{I(2I+1)},$$

called the *nuclear polarization* and the *nuclear alignment*, respectively. Here I is the spin of the nucleus, and I_z is its projection along the z axis; $\langle \ldots \rangle$ means averaging over the ensemble. P_1 is proportional to the nuclear magnetic polarization (i.e. *magnetization*), it determines the value of the *nuclear magnetic resonance* absorption signal, and in the case of $I = 1/2$ it is the difference in the number of spins directed along and against (antiparallel to) the *magnetic field* direction. Obviously, $-1 \leqslant P_1 \leqslant 1$. For randomly oriented nuclear spins, $P_1 = 0$. The quantity P_2 is a measure of the excess of the number of spins aligned along the z axis (parallel or antiparallel) over those aligned transverse to it. One can see that polarization is accompanied by alignment (when $P_1 \neq 0$, then $P_2 \neq 0$) but not vice versa (when $P_2 \neq 0$ and $P_1 = 0$). The angular distribution of γ- and α-radiation of radioactive nuclei with parity conservation depends on P_2 and perhaps even on higher-order terms. In the case of β-decay parity is not conserved, and the distribution of radiation depends on P_1, P_2 and related higher-order terms, both even and odd.

Methods of nuclear orientation can be subdivided into static and dynamic types. The static method is a "brute force" one since, to attain the maximum values of nuclear orientation, extreme conditions are used such as very high external fields and very low temperatures. Another example of a static method is the *Gorter–Rose method* (C.I. Gorter, M.E. Rose, 1948) which is based on

the action of high internal local fields $\sim$10–100 T relative to the *hyperfine interaction* on the nuclei of paramagnetic ions. Dynamical methods apply *magnetic resonance* in a straightforward manner; high electron spin polarization is "pumped over" to the nuclear spin polarization with the help of the microwave saturation of "allowed" or "forbidden" electron spin transitions. These methods include the *Overhauser effect, solid effect*, electron–nuclear *cross-relaxation*, and dynamic cooling (see *Dynamic nuclear polarization*) of the nuclear spin subsystem which is in (direct or microwave field-induced) thermal contact with the electron spin–spin reservoir that is disturbed from thermal equilibrium by the incomplete saturation of the *electron paramagnetic resonance* line. Examples of dynamic methods are those of chemical and *optical nuclear polarization*. Nuclear orientation methods are employed mainly for obtaining the *polarized nuclear targets* needed for fundamental studies in elementary-particle physics. Another area of application of nuclear orientation is the study of nuclear magnetic ordering at the temperature of the reservoir of nuclear *spin–spin interactions* $T_{II} \leqslant 10^{-6}$ K.

NUCLEAR POLARIZATION

See *Nuclear orientation*.

NUCLEAR QUADRUPOLE INTERACTION

Interaction of nuclei of a crystal with the electric field gradients of their environment caused by the presence of nuclear electric quadrupole moments and a spatial nonuniformity of the fields. The part of the total crystal Hamiltonian which characterizes this interaction is called the *quadrupole Hamiltonian*. Without taking into account the internuclear interactions in the crystal (as a rule, they are very small), the overall quadrupole Hamiltonian is the sum of the quadrupole Hamiltonians of the individual nuclei, each of which is a convolution of the *nuclear electric quadrupole moment* tensor Q_T and the tensor q_T of the *electric field gradient* at the site of the nucleus. By expressing the components of tensors Q_T and q_T in the principal axis system and defining the scalar quadrupole moment Q and the electric field gradient factor q as the components along the main z axis (Q_{zz} and $q_{zz} = -e\partial E_z/\partial z = \partial^2 V/\partial z^2$), the quadrupole nuclear Hamiltonian H_Q is reduced to the form:

$$H_Q = \frac{eqQ}{4I(2I-1)}\left[3I_z^2 - I^2 + \eta(I_x^2 - I_y^2)\right],$$

where e is the electron charge, I is the nuclear spin, $I_{x,y,z}$ are the components of the nuclear spin vector in the principal axes system of tensor q_T, the quantity eqQ is called the *quadrupole interaction constant* (sometimes q_{zz} is defined as $-\partial E_z/\partial z$ and the quantity $e^2 qQ$ is called the quadrupole interaction constant). The asymmetry parameter η in this equation indicates the deviation of the quadrupole moment from axial symmetry ($0 \leqslant \eta \leqslant 1$, $\eta = 0$ for the axial case). The nuclear quadrupole interaction which determines a "pure" nuclear quadrupole resonance (that is without any applied magnetic field) also influences *nuclear magnetic resonance* (NMR): it produces shifts, splittings, and broadening of lines in NMR spectra.

NUCLEAR QUADRUPOLE RELAXATION

Establishment of equilibrium between a *nuclear quadrupole spin system* and a crystal lattice. As a rule, the related processes are studied with the help of the Pauli equations for the *level populations* associated with the nuclear quadrupole Hamiltonian (see *Nuclear quadrupole interaction*) which characterize the interaction of nuclei with the electric field of their electronic environment without taking into account the internal internuclear interactions of nucleons with each other. The roots of the characteristic equation which enter a set of Pauli equations are the *relaxation times* of energy level populations (commonly called *nuclear quadrupole spin–lattice relaxation times*). The interaction of the nuclear quadrupole spin system with the lattice which brings about the nuclear quadrupole relaxation is due to the thermal motion of molecular groups and particles which, in turn, is the cause of the random time dependence of the nuclear quadrupole Hamiltonian terms. The most frequently considered relaxation mechanisms involving thermal motion are the following: the libration mechanism (rotational vibrations of molecular groups containing the nuclei under consideration); the reorientational mechanism (angular jumps of such groups or particles between metastable positions); the modulation mechanism (various movements of particles neighboring those containing the nuclei under consideration). Comparison of results of experimental and analytical investigations of the temperature dependence of the relaxation times is employed in the physics and chemistry of solids for studying the internal dynamics and the *phase transitions* in crystals. A more recent approach to the computation of relaxation times based on the use of the Einstein–Fokker equations is much more general than the earlier conventional approach in the sense that it does not assume that the thermal motions are small, so perturbation theory methods do not apply. The term nuclear quadrupole relaxation is also employed to denote quadrupole interaction mechanisms for the relaxation of nuclear magnetic polarization in an applied external magnetic field.

NUCLEAR QUADRUPOLE RESONANCE (NQR)

Resonance absorption of the energy of a radio-frequency electromagnetic field by a nucleus with a nuclear spin $I > 1/2$ and an electric *quadrupole moment* Q situated in a microscopically inhomogeneous crystalline electric field. This crystal field produces splittings of the nuclear magnetic sublevels which are proportional to the value of the quadrupole moment Q, and to the gradient of the electric field at the nuclear site in the crystal (see *Nuclear quadrupole interaction*). From another viewpoint, nuclear quadrupole resonance (NQR) involves observing magnetic dipole transitions between nuclear spin sublevels of the type studied in *nuclear magnetic resonance* for nuclear spins $I > 1/2$, but NQR involves the case when the quadrupole splitting of the spin sublevels is larger than the Zeeman splitting of these sublevels (see *Zeeman effect*). In this case a nuclear quadrupole resonance can be observed in the absence of an applied magnetic field B. When the magnetic field $B = 0$ and the crystalline electric field E has axial symmetry the quadrupole energy levels E_m are given by

$$E_m = \frac{e^2 qQ}{2I(2I-1)}\left[3m^2 - I(I+1)\right],$$

where $q = \partial E/\partial z$ is the field gradient, and the magnetic quantum number m has the range of values $-I \leqslant m \leqslant I$. Another seldom-used name for this zero field experiment is called "*pure-quadrupole resonance*".

The main parameters which characterize nuclear quadrupole resonance are the following: the quadrupole interaction constant which determines the spectral frequencies; the asymmetry parameter η of the *electric field gradient* which measures deviations from axial symmetry; the spectral *line width*; the spin–spin and *spin–lattice relaxation* times (see *Nuclear quadrupole relaxation*). These parameters are highly sensitive to the chemical and physical structure of solids, the *crystal lattice dynamics*, and external actions such as mechanical, thermal, magnetic, electric, and radiation actions, as well as to the presence of diverse impurities and lattice defects. Owing to this, *nuclear quadrupole radiospectroscopy* is an effective technique for investigating the chemical and physical state of a solid. This technique is employed in nuclear physics to determine quadrupole and hexadecapole nuclear moments, and it is applied in the chemistry and physics of solids for studying intramolecular and intermolecular interactions, molecular dynamics and phase transitions, molecular structure, and the nature of *chemical bonds*. There are also various special applications of nuclear quadrupole resonance (among them a *thermometry* based on nuclear quadrupole resonance).

Tables of nuclear quadrupole resonance frequencies of a variety of chemical compounds are given in monographs devoted to this method, and in handbooks.

The sensitivity of the nuclear quadrupole resonance method is less than that of NMR. As a rule, to observe the signal the sample must be of a few units or even tens of cm^3 in volume.

A weak magnetic field induces an extra Zeeman splitting (in addition to the quadrupole splitting), with the appearance of additional spectral lines. By varying the orientation of this field with respect to the orientation of a monocrystal one can establish experimentally the directions of the principal axes of the electric field gradient tensor with respect to the crystallographic axes. In a strong magnetic field an ordinary NMR transition takes place with a superposed quadrupole splitting (Zeeman splitting > quadrupole splitting). An external electric field causes a shift of quadrupole levels,

and in the case of monocrystals this leads to either a Stark effect shift or a line splitting due to the nuclear quadrupole resonance dependence on the crystal field symmetry, and the localization of the nuclei involved in the resonance (see *Stark effect*). This type of information is useful in studies of the nature of chemical bonds.

NUCLEAR QUADRUPOLE SPIN SYSTEM

A subsystem of a magnetic resonance spin system in a *crystal* whose properties are determined by the isolation of the quadrupole part of the overall Hamiltonian. The word "spin" is used in this term since the nuclear electric quadrupole Hamiltonian can be expressed in terms of the nuclear quadrupole moment, the nuclear spin operator $\boldsymbol{I}$, and the characteristics of the *electric field gradient* $V_{ij} = -\partial E_i/\partial x_j$ of the electronic environment of a nucleus. For example, the quadrupole Hamiltonian can be written in tensor notation as $H = \boldsymbol{I}\boldsymbol{Q}\boldsymbol{I}$, where $\boldsymbol{I}$ is the nuclear spin vector operator, and the quadrupole tensor has the components $Q_{ij} = eQV_{ij}/[2I(2I+1)]$, e is the electronic charge, $Q = \int \rho(r)(3z^2 - r^2)\,dV$ is the scalar quadrupole moment, and ρ is the charge density. Nuclear spins $I > 1/2$ have quadrupole moments (see *Nuclear quadrupole interaction*).

NUCLEAR RADIATION DETECTORS

Devices for detecting charged nuclear particles and *gamma rays*. In a solid-state nuclear radiation detector the *space charge region* of a *p–n* or *p–i–n* junction (*i* is a region of intrinsic conductivity, see *Semiconductor junction*) is the sensitive part. The ionizing particle forms *electron–hole pairs* within the space charge region. An energy ε equal to several widths of the *band gap* (e.g., in *silicon* $\varepsilon \approx 3.6$ eV, in *germanium* $\varepsilon \approx 2.9$ eV) is spent in the formation of each pair, independently of the type of radiation. Electrons and holes are separated by the electrical field, and at the capacitance of the *n–p* junction the charge $\propto E/\varepsilon$ (*E* is the particle energy) accumulates. Thus, the voltage pulse, recorded by the electronic circuit, will be proportional to the particle energy if its path is within the space charge region. This is the operating principle of the so-called *E-detector*. If the path of the particle is longer than the space charge region width then the signal of the detector is proportional to MZ^2/E, where M and Z are the mass and charge of the detected particle (*channeling detector, detector of specific energy losses* or ΔE-detector). Multiplication of the signals from E- and ΔE-detectors provide information about the mass of the particle. In order to determine the location of the particle passage, position-sensitive detectors are used, which simultaneously with the energy measurement provide information about the position of the particle entry into the detector. The charge created by the particle within the detector is distributed between two contacts located at opposite ends of the rear side, and it depends on the resistances R_x from the zero contact to the point of the particle's penetration, and R_y from the penetration point to the second contact. The signal, taken from the front end of the detector, determines the total value of the energy E, and the ratio of these signals determines the location of the particle penetration.

The surface-barrier E-detectors have a fast response ($\sim 10^{-9}$ s) and they are used for the determination of the mass of particles by their time of flight from the target to the detector. For the spectrometry of nuclear particles Si-detectors are used, and for the spectrometry of γ- and X-radiation Ge and Si *diffusion-drift detectors* are used operating at the temperature of liquid nitrogen. The energy resolution of these detectors is some fractions of one percent, which is much better than that for other types of detectors such as ionization chambers, scintillators, etc.

It is possible to detect high-energy (100 to 1000 GeV) nuclear particles by recording their *transition radiation* which appears at the passage of the particle through the interface of two media with different *refractive indices* (e.g., gas–*solid*). In order to increase the sensitivity nuclear radiation detectors are used which contain 10^3 gas–solid layers. The intensity of the forward transitional radiation is proportional to the incident particle's energy, and this permits its determination.

Surface-barrier silicon detectors are also used for detecting *neutrons*. For recording slow neutrons the so-called *converter* (e.g., the layer of ^{6}Li, ^{10}B), within which the (n, α) reaction takes place, is located at the surface-barrier nuclear radiation detector which records the exiting α-particles. For detecting fast neutrons a hydrogen-containing target (e.g., polyethylene film), which is the source of the proton yield upon irradiation by fast neutrons, serves as the detector. Another type of detector of fast neutrons is based on the radiative damage inflicted on a silicon p–n junction. With increasing radiation dose the *silicon* resistance grows and the measured resistance of the p–n junction provides the density of fast neutron flow, and the adsorbed dose within the range 5 to 5000 rad is measured.

NUCLEAR RADIATIONS

Fluxes of nuclear particles and *gamma rays*. Nuclear radiations result from *nuclear reactions* and the decay of radioactive isotopes. The products of nuclear reactions are neutrons, protons, deuterons, α-particles, nuclear fission fragments, high-energy charged atomic nuclei, β-particles including negatively charged β^- (electrons) and positively charged β^+ (positrons), as well as gamma quanta. Gamma rays appear as a result of nuclear capture reactions when a nucleus transforms from an excited state to its ground state, and when high-energy electrons brake (decellerate) in the fields of heavy nuclei.

At their interaction with matter, nuclear radiations bring about ionization and thereby produce positively charged ions and negatively charged electrons. As a result, nuclear radiations are often called *ionizing radiations*. These are divided into the ionizing radiation itself which consists of charged particles with sufficient kinetic energy to ionize atoms at collisions (electrons, positrons, protons, α-particles, etc.), and indirectly ionizing radiation which consists of uncharged particles (neutrons) or gamma rays and X-rays.

At an elastic collision of a nuclear particle with an atom, when the kinetic energy exceeds a certain value atoms can be displaced from lattice sites, with the resultant production of other types of radiation defects.

At an inelastic collision when the energy of the nuclear particle is greater than the nucleon binding energy of the nucleus, then nuclear reactions can accompany the nuclear radiation output.

NUCLEAR SPIN DIFFUSION

Process of transport of nuclear *magnetization* M_z (nuclear spin polarization P_z) caused by dipole–dipole (d–d) *flip-flop transitions* of *spins* of neighboring nuclei. As long as the mutual spin flips do not change the overall Zeeman energy (see *Zeeman effect*) the process of nuclear spin diffusion takes place without involving the lattice. One should stress that nuclear *spin diffusion* does not involve any spatial movement of the nuclei; it only concerns a deviation of the local nuclear magnetization from its value averaged over the sample. In many cases (and always for spin $I = 1/2$) one can introduce the concept of a spatial-coordinate-dependent nuclear spin temperature T_s (see *Level population*), and speak about the equalization of T_s over a sample through the process of spin diffusion. The concept of nuclear spin diffusion was introduced by N. Blombergen (1949) when studying nuclear magnetic relaxation in nonmetallic diamagnetic crystals with small concentrations of paramagnetic impurity ions (see *Magnetic ion*). In such samples the nuclear *spin–lattice relaxation* at low temperatures is mainly due to the magnetic dipole–dipole interaction of the nuclear magnetic moments of the nuclei with the much stronger (by a factor of 10^3) electronic magnetic moments of the impurity ions. Since the nuclei of the host are at various distances from the impurity ions, the nuclei closest to paramagnetic impurities relax at the highest rate. As a result, during the relaxation process there arises a transient variation of the degree of magnetization (e.g., by saturation of *nuclear magnetic resonance*, NMR) throughout the sample, and the gradient of nuclear polarization (spin temperature) M_z becomes position dependent. Thus one might expect a distribution of nuclear *relaxation times*, whereas experimentally a single nuclear relaxation time T_{1n} is observed, that is the nuclei behave as a coherently interconnected system of spins. This process is explained by the efficiency of the process of nuclear spin diffusion in nonmetallic diamagnetic crystals containing paramagnetic impurities, and it is quantitatively described by the *Blombergen equation*

$$\frac{\partial M_z}{\partial t} = D\nabla^2 M_z - c \sum_m |r - r_m|^{-6}(M_z - M_0),$$
(1)

where the first term describes the nuclear spin diffusion directed toward establishing a uniform magnetization M_z, and the second term describes the relaxation itself which brings about transient local nonuniformities of the distribution; r_m is the radius-vector of mth impurity paramagnetic ion,

∇^2 is the Laplacian operator, and c is a known constant. The parameter D is called the *nuclear spin diffusion coefficient*. In the general case, D is a symmetric second-rank tensor whose components depend on the orientation of the external magnetic field B_0 relative to the crystallographic axes. Only for the cases of cubic monocrystals, polycrystalline samples, and a powder belonging to an arbitrary crystal system, is D reduced to a scalar. An exact derivation of Eq. (1) was given by L.L. Buishvili and D.N. Zubarev (1965). At sufficiently low temperatures, the nuclei near paramagnetic impurities have strongly shifted resonant frequencies, and this hinders the flip-flop processes between them. Therefore, there is a diffusion barrier of radius δ near every impurity paramagnetic ion, inside the barrier $D \approx 0$, and outside the barrier $D \approx$ const. It has been proven analytically and experimentally that there is a possibility of breaking the diffusion barrier under the influence of flip-flop transitions of neighboring nuclei which are induced by the spin lattice relaxation or the *spin–spin relaxation* of the paramagnetic ions.

NUCLEAR SPIN RELAXATION IN SUPERCONDUCTORS

A characteristic time is associated with the change of non-equilibrium nuclear spin *magnetization* in a superconductor. The so-called spin–lattice and spin–spin relaxation times, T_1 and T_2, respectively, are distinguished (see *Nuclear spin relaxation time*). Nuclear spin relaxation in metals is caused mainly by the coupling between the nuclear and electronic subsystems via the *hyperfine interaction*. The dependence of T_1 on the temperature calculated in the *Bardeen–Cooper–Schrieffer theory* has a non-monotonic character. As the temperature is lowered below the superconducting transition temperature T_c, the relaxation time T_1 at first decreases, reaches a minimum, and then monotonically increases, due to the exponentially decreasing concentration of normal excitations in the superconductor. The initial decrease of T_1 is caused by the increase (compared with a normal metal) of the *density of electron states* at the edge of the superconductor gap. A high concentration of impurities and a strong magnetic field smear out the singularity in the superconductor density of states, and lead to the smoothing out or disappearance of the minimum in T_1. *Type II superconductors* in the *mixed state* exhibit non-exponential nuclear spin relaxation caused by different values of T_1 within and outside the vortex core.

NUCLEAR SPIN RELAXATION TIME

A characteristic time for the return of a non-equilibrium nuclear spin *magnetization* back to its equilibrium configuration. When a system of nuclear spins in a magnetic field absorbs a pulse of radio-frequency energy at the Larmor frequency a non-equilibrium distribution of nuclear magnetization is established in the system. The spins return to thermal equilibrium by passing this excess energy to the random Brownian motion in the case of liquids, to the lattice vibrations in the case of insulators, and to the conduction electrons in metals (*Korringa relaxation*). The time constant for this transfer of excess energy is called the *spin–lattice* (or *longitudinal*) *relaxation time* T_1. The nuclear spins reestablish thermal equilibrium among themselves in the characteristic time T_2 called the *spin–spin* (or *transverse*) *relaxation time*. For low viscosity liquids $T_1 \sim T_2$, and in typical solids $T_1 \gg T_2$.

NUCLEAR SPIN WAVES

Variety of *spin waves* due to oscillations of nuclear spin directions. The most convenient objects for observation of nuclear spin waves are the easy plane and cubic Mn-based *antiferromagnets* ($MnCO_3$, $CsMnF_3$, $RbMnF_3$, MnO, etc.). The indirect coupling of nuclear spins, which gives rise to a very strong spin–spin interaction, takes place in these materials through the intermediary of *magnons*, which have a very high effective range r_0 of coupling that reaches $10^4 a$ (a is the average distance between spins). Due to this high value of r_0, the nuclear spin waves may exist at helium temperature ($T \leqslant 4K$), which far exceeds the temperature of nuclear spin ordering (about 10^{-2} K for these materials). Experimental investigations of nuclear spin waves involve studying nonlinear effects caused by these waves using the technique of *nuclear magnetic resonance*.

NUCLEAR TARGET, POLARIZED

See *Polarized nuclear targets*.

NUCLEAR ZEEMAN EFFECT

Splitting of energy levels and spectral lines of emission or absorption by atoms, molecules, or crystals under the effect of an external magnetic field B_0, arising from the interaction of a nuclear magnetic moment with B_0. The complete theory of the nuclear Zeeman effect is provided by quantum mechanics. The nucleus with *spin I* has the related dipole *magnetic moment* $\mu_I = \hbar \gamma_I I$ (γ_I is the nuclear *gyromagnetic ratio*). In the field B_0, the nucleus acquires the Zeeman energy $-\mu_I B_0 = \hbar \gamma_I B_0 m$, where the projection m of I along the direction B_0 can assume the $2I + 1$ values $m = -I, -I + 1, \ldots, I$. As a result, the nuclear spin energy level is split into $2I + 1$ equidistant Zeeman sublevels with the spacing $\hbar \omega_I$, where $\omega_I = \gamma_I B_0$ is the nuclear *Zeeman Larmor frequency*. *Nuclear magnetic resonance* arises from quantum transitions between nuclear Zeeman sublevels which are induced by a radiofrequency field perpendicular to B_0, with the frequency $\omega \approx \omega_I$.

NUCLEAR ZEEMAN PSEUDOEFFECT

Phenomenon of splitting of the neutron (n) *spin* $S = 1/2$ energy level into two sublevels upon passage of a flux of slow neutrons through a *polarized nuclear target*. In addition to a weak *dipole–dipole interaction*, there is a stronger magnetic interaction between the neutron *magnetic moment* and the target nuclei, arising from the nuclear forces. The interaction with the target nuclei can be considered as arising from an effective pseudomagnetic field $B^* = \alpha P$ acting upon n from the polarized target, where P is the polarization of the nuclei (see *Nuclear orientation*), and α is a function of the scattering cross-section of the neutron by the nucleus. As a result, the neutron in a constant magnetic field B_0 experiences the effect of the total field $B_0 + B^*$, and its *Zeeman frequency* $\omega_I^* = \gamma_I (B_0 + B^*)$ (assuming $B^* \parallel B_0$) lies in the UHF range, where $B^* \gg B_0$, corresponding to an appreciably large polarization P. Quantum transitions between Zeeman pseudo-levels (so-called *pseudomagnetic resonance*) can be induced by an alternating magnetic field. In this case, the effective alternating field is due mainly to the pseudomagnetic contribution.

NUCLEATING CRACK

An ideally sharp, elastic-equilibrium submicrofracture that forms in a *slip band* of a crystal under the effect of *shear* stress as a result of the merging of n dislocations of a cluster held by a *grain boundary*. According to the *Stroh theory*, the length of an equilibrium nucleating crack is $c \sim n^2 b$ (b is the value of the *Burgers vector*), and a typical dimension at the *yield limit* of polycrystalline *iron* is $(1/70)d$ (d is the grain size). Nucleating cracks are constantly forming during treatment processes accompanied by *plastic deformation* of the material, and they cause the specific volume to increase up to 1%. Nucleating cracks formed under critical normal stress lose their stability, and cause an ideally brittle spalling (*microspalling*), and perhaps total *failure* of a solid at the initial stage of avalanche propagation.

NUCLEATION at phase transitions

A process involving the appearance of small regions of a new *phase* (nucleating center) in a *phase transition* in the near vicinity of the phase transition temperature T_c. Nucleation in solids causes local disturbances because the crystal lattice parameters of the two phases always differ from each other. Therefore, the appearance of a nucleating center is accompanied by the emergence of a field of nonuniform mechanical stresses in its vicinity. If these stresses are sufficiently high, the irreversible *strain* and even the complete *failure* of the sample may occur. Otherwise, additional elastic energy appears that can retard a phase transition. The formation of nucleating centers and the subsequent decomposition of the system can start only in overheated (overcooled) samples, i.e. when a certain degree of metastability (see *Metastable state*) has been attained. This causes a striction contribution to *hysteresis in phase transitions*, which is specific for solids. The sufficiently strong striction effects in some cases bring about complete *striction blocking of nucleating centers*, i.e. they cancel the possibility of nucleation in a phase transition. The most favorable for the formation of nucleating centers are the portions of a sample containing various *defects* of structure and composition, as well as the surface areas. In highly perfect crystals the fluctuation mechanism of nucleation may prevail. At the first stage of nucleating center development (*critical embryos*), their shape is typically close to spherical, which is related to the influence of *surface tension* at the interface of the old and new phases. During the process of further growth (*supercritical nucleating centers*), the *surface energy* becomes insignificant, so the optimum orientation and shape of nucleating centers is completely determined by the requirement of developing toward an elastic energy minimum. Typically, a plate-like shape of nucleating centers proves to be the most favored. In ferroelectric (ferroelastic) phase transitions, the supercritical nucleating centers tend to have a complicated coherent *domain* structure. The rate of nucleating center growth fluctuates over a wide range; it can be limited by such slow processes as *diffusion* and *thermal conductivity*, as well as by rapid processes such as the speed of elastic stress propagation (*sound velocity*).

Oo

OCCLUSION

Gas absorption by solid *metals* or melts, with the adsorbate trapped on the surface or within the bulk, often with the formation of solid or liquid solutions, or chemical compounds (*nitrides, hydrides*, etc.).

OCCUPATION NUMBERS

Numbers n_k associated with various quantum states indicating the number of identical particles in each state k. For a system of fermions (particles with half-integer spin), according to the Pauli exclusion principle the occupation numbers assume values of either 0 or 1. In a system of bosons (particles with integer spin) $n_k = 0, 1, 2, 3, \ldots$. Occupation numbers are normalized by the condition $\sum n_k = N$, where N is the total number of particles in the system. In quantum statistics the averages n_k play an important role. For ideal quantum gases the statistics are determined by the distribution function: Fermi–Dirac for fermions and and Bose–Einstein for bosons (see *Fermi–Dirac statistics, Bose–Einstein statistics*). At $T = 0$ the lowest energy states are occupied. This means that at absolute zero all states are occupied below the Fermi level and all are empty above it for a system of fermions (totally degenerate Fermi gas), and it means that all particles are in the ground state for bosons (*Bose–Einstein condensation*).

The concept of occupation number appears in the formalism of *second quantization* where the number operator $n_\mu(k)$ is the product of a creation operator and an annihilation operator: $n_\mu(k) = a_\mu^+(k)a_\mu(k)$.

OCTAHEDRAL COMPLEX

One of the most widespread coordination polyhedra found in *crystal lattices* containing ions of transition groups with coordination number 6 (see *Coordination sphere*). In a regular octahedral complex, the central metal ion has six identical nearest neighbors (*ligands*; such as ions O^{2-}, F^-, Cl^- etc., molecular ions, and water molecules) equidistant from the center along three orthogonal axes. The octahedron has eight faces in the form of equilateral triangles. The *point group* of symmetry of the central ion is $m\overline{3}m$ (O_h); three fourth-order *symmetry axes* pass through the vertices of the octahedron, four C_3 axes are normal to the faces, and six C_2 axes pass through the centers of edges.

The distortion of an octahedral complex (static or dynamic) caused by lattice vibrations is described by 15 symmetric coordinates that comprise the bases of: one-dimensional Γ_{1g} (A_{1g}), two-dimensional Γ_{3g} (E_g), and three-dimensional Γ_{5g} (T_{2g}), $2\Gamma_{4u}$ ($2T_{1u}$), Γ_{5u} (T_{2u}) irreducible representations of the cubic symmetry group. The normal coordinates are linear combinations of displacements of the central ion and ligands from the equilibrium positions. Under the even-parity (*gerade*, *g*) strain, only the ligands undergo displacements. The central ion can move during odd-parity (*ungerade*, *u*) type vibrations. The totally-symmetric deformation Γ_{1g} corresponds to the expansion or contraction of the octahedron volume while retaining its shape; under the tetragonal (Γ_{3g}) and trigonal (Γ_{5g}) deformations, the octahedron contracts or stretches along one of the axes of fourth or third order, respectively. The electron density distribution in an octahedral complex can be represented by *molecular orbitals* obtained as a result of d^2sp^3-hybridization of six electron states of the central (transition) ion mixed with linear combinations of wave functions (see *Symmetrized functions*) of the outer filled (valence) electron shells of the ligands.

OFF-CENTER IONS

See *Noncentral ions*.

OHMIC CONTACT

Metal–semiconductor junction, or semiconductor–semiconductor contact, with resistance independent of the magnitude of the current running through the contact, and small compared to the remainder of the sample. Good ohmic contacts have resistivity less than 10^{-7} $\Omega\cdot m^2$. In particular, the external terminals of *semiconductor devices* and *integrated circuits* are ohmic contacts. In this case they must satisfy additional requirements concerning both electrical and mechanical stability. In addition, ohmic contacts should not inject minority carriers into a semiconductor. As a rule, ohmic contacts are produced through the formation of a strongly doped layer (with conduction of the same kind) on the semiconductor surface. These function via *diffusion* or *ion implantation*, or else the deposition on the surface and subsequent fusion of an *alloy* with a necessary doping impurity. The best known alloys for ohmic contacts involving Si and Ge are Au–Sn for n-type semiconductors, and Au–In for p-type semiconductors.

OKMAL (a Russian term)

Polycrystalline dense material produced by sintering diamond powders at *high pressures* and temperatures without using binders. By varying the process parameters (pressure, temperature, sintering time), it is possible to obtain okmal with a porosity from 2 to 13% and with different physical properties: *hole conductivity* or *electronic conductivity*, electric resistivity 10^2 to 10^{14} $\Omega\cdot cm$, thermal expansion $(2–4)\cdot 10^{-6}$ K^{-1}, and thermal conductivity up to 260 $W\cdot m^{-1}\cdot K^{-1}$ and above. The temperature dependences of thermo-physical properties of okmal are similar to those of *diamond* monocrystals. Okmal is oxidized on heating in air in the range 200 to 500 °C, which causes reduction of *strength* of the samples and irreversible changes of some physical properties (e.g., increase of electrical resistance by several orders of magnitude). It is used in electrical engineering to manufacture heat sinks.

OKOROKOV EFFECT (V.V. Okorokov, 1965)

Coherent excitation of channeled ions (see *Channeling*). When moving with the velocity v along an ordered atomic chain in a *crystal lattice*, a channeled ion experiences the coherent periodic interaction of the lattice atoms with frequencies

$$\omega_n = \frac{2\pi}{a} \frac{vn}{(1 - v^2/c^2)^{1/2}},$$

where a is the lattice constant; and $n = 1, 2, \ldots$ is the harmonic number. If the frequency ω_n of the external force acting upon a particle emerging from the crystal lattice coincides with the frequency of a transition of the channeled ion from one state to another, $\Delta E = E_e - E_g$, i.e. $\omega_n = \Delta E/\hbar$, a resonance excitation of the particle (similar to the excitation of an atom by the periodic field of a monochromatic electromagnetic wave) may occur. The presence of the *Lorentz factor* $(1 - v^2/c^2)^{-1/2}$ (connected with the relativistic contraction of the interatomic lattice distance in the coordinate system of the moving ion) in the expression for the frequency ω_n allows one to study the coherent excitations of nuclear levels that are characterized by a large value of the transition energy (more than a hundred keV).

The best conditions for the experimental observation of the coherent excitation of a singly charged anion are achieved, if the ion does not change its charge state in passing through a *monocrystal*. If the *resonance* condition $\omega_n = \Delta E/\hbar$ is met, a coherent transition to the excited state takes place.

ONE-DIMENSIONAL METALS

A large class of organic *high molecular weight compounds* and inorganic chain compounds that exhibit a relatively high ("metallic") *electrical conductivity* in one direction, and behave as *insulators* in the transverse plane. The conductivity anisotropy in known one-dimensional disordered metals reaches 10^5. One-dimensional conductors are unstable with respect to the generation of *charge density waves* or *spin density waves*. The emergence of the latter leads to a restructuring of the uniform ground state of the electron system, and to the development of a superstructure with wave number $q = 2p_0$ in the system, where p_0 is the Fermi momentum. This is the so-called *Peierls–Fröhlich instability*. It is suppressed even by a relatively small degree of disorder in the structure of one-dimensional conducting chains. The nature of the disordering of the one-dimensional disordered metals may either be related to randomly located *defects* of the crystal lattice, or pertain to a particular compound. For example, in a unit cell of crystals of the KPC type (abbreviated term for the salt $K_2Pt(CN)_4Br_{0.3}\cdot3H_2O$), the number of equivalent positions for halogen atoms exceeds the number of these atoms, and the positions are randomly occupied. The elastic scattering of one-dimensional electrons by random lattice inhomogeneities leads to the phenomenon of localization of all electron states in a one-dimensional disordered metal, and its transformation into a zero-gap dielectric with a high degree of polarizability. The charge transport in such systems is effected by the hopping of electrons between localized states with energies differing by the magnitude of a quantum of the external field of frequency ω. The low-frequency conductivity of localized electrons at zero temperature is proportional to $\omega^2 \ln^2(\omega\tau)$, where τ is the mean free path time. At a finite temperature, by virtue of inelastic scattering by *phonons*, the electronic states become delocalized, and nonzero static conduction appears. Examples of organic molecular compounds with quasi-one-dimensional metallic type electrical conductivity are salts of 7, 7, 7, 8-tetracyano-p-quinodimethane, called TCNQ for short. Salts of bis(ethylenedithia)tetrathiafulvalene, called BEDT-TTF for short, exhibit low-dimensional electrical conductivity.

ONE-DIMENSIONAL MODELS OF DISORDERED STRUCTURES

Theoretical models that allow one to write closed form equations for one or another dynamic quantity, and thus obtain equations (*Smoluchowski equation* or *Fokker–Planck equation*) for probability densities of these quantities. Such equations can be solved or investigated in detail for many important cases.

The richness in the physical content of one-dimensional models of disordered structures is due to the fact that these models express so clearly the principal difference between *disordered systems* and ordered ones, namely, the presence of a macroscopic number of localized states of an essentially non-Bloch type (see *Bloch theorem*): all states in *one-dimensional disordered structures*

are localized. The simplest quantity to characterize the localization is the *localization length l* that equals the inverse increment (Lyapunov index) of the envelope of the squared magnitude of the wave function. This localization of states drastically changes the kinetic properties of the system. At zero temperature, the low-frequency conductivity tends to zero along with the frequency. The most typical form of such a dependence (*Mott law*) is $\sigma(\omega) \sim (\omega\tau)^2 \ln^2(\omega\tau)$ (with $\tau \sim l/v_F$, where v_F is the *Fermi velocity*). In one-dimensional disordered systems of finite length L, the static electrical conductivity is exponentially small in the parameter L/l.

ONE-DIMENSIONAL SUPERCONDUCTIVITY

See *Superconductivity in quasi-one-dimensional systems.*

ONSAGER METHOD (L. Onsager)

Method for calculating the statistical sum of the *Ising model* for a planar $N \times M$ lattice in the absence of an external field. The Onsager method is based on the representation of the *statistical sum* as $Z = \sum_\sigma (V^N)_{\sigma\sigma}$, where the transfer matrix $V = V_1V_2$. The matrix V_1 describes the spin interactions of a lattice row with a neighboring one:

$$V_1 = \exp\left[K \sum_{m=1,...,M} \sigma_{n,m}\sigma_{n+1,m}\right],$$

and the matrix V_2 describes the interaction of adjacent spins in one and the same row:

$$V_2 = \exp\left[K \sum_{m=1,...,M} \sigma_{n,m}\sigma_{n,m+1}\right].$$

The quantity Z is expressed through the eigenvalues λ_i of the matrix V, i.e. $Z = \sum \lambda_i^N \approx \lambda_{max}^N$ where λ_{max} is the largest eigenvalue. To calculate λ_{max}, the commutation relations of the matrices V_1 and V_2 are used. Due to the presence of translational symmetry along the rows, the problem simplifies through the application of a Fourier transform, and the following relation is found for λ_{max}:

$$\lambda_{max} = (2\sinh 2K)^{M/2} \exp\left(-\int_{-\pi}^{\pi} \frac{dq}{4\pi}\varepsilon_q\right),$$

$$\cosh\varepsilon_q = \cosh 2K \coth 2K - \cos q.$$

The Onsager method is not applicable for calculating the statistical sum of an Ising model in an external field, $B \neq 0$, or this sum for a three-dimensional Ising model, since in these cases the commutation relations of the matrices V_1 and V_2 become too complicated.

ONSAGER THEORY

A theory in the physics of *insulators*, which establishes relations between the static ε_s and high-frequency ε_∞ *dielectric constant* of a medium containing polar molecules (with an intrinsic dipole moment). Account is taken of the difference in the dipole moment of a molecule in a medium and in vacuo, resulting from interactions with neighboring dipoles, and the difference between the electric field in the medium before and after applying an external electric field $\boldsymbol{E}$. A molecule is treated as a sphere with an isotropic polarizability. The Onsager equation is as follows:

$$\varepsilon_s - \varepsilon_\infty = \frac{4\pi \mu_v^2 N_0}{3k_B T} \frac{3\varepsilon_s}{2\varepsilon_s + \varepsilon_\infty} \left(\frac{\varepsilon_\infty + 2}{3} \right)^2, \tag{1}$$

where N_0 is the concentration of polar molecules. The Onsager theory disregards deviations from a macroscopic description of the interaction between nearest neighbor molecules. Accordingly, the correction reduces to adding the factor $(1 + z\langle\cos\gamma\rangle)$ into the right-hand side of Eq. (1). Here, z is the average number of nearest neighbors of the molecule, $\langle\cos\gamma\rangle$ is the mean value of the cosine of the angle between neighboring dipoles. Eq. (1) with the added factor $(1 + z\langle\cos\gamma\rangle)$ is called the *Kirkwood formula* (see *Kirkwood approximation*).

OPPOSED DOMAINS

Domains (magnetic, ferroelectric, etc.), characterized by a jump of the polarization (magnetic, electrical, etc.) component, normal to the *domain wall*, e.g., polarization can reverse direction in opposed domains (see Fig.). Usually such domains are energetically unfavorable because of the appearance of demagnetizing (depolarizing) fields. However in some cases they can form, e.g., in *ferroelectrics* with high conductivity, so that the the bound charges at the walls, creating the high fields, can be compensated by incoming free charges; and

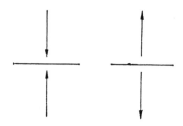

Diagram of the polarization directions of two cases of opposed domains.

in *magnetic substances* they can be compensated by strong nonuniform external magnetic fields.

OPTICAL ABSORPTION SPECTROSCOPY

Measurement of the absorption or reflection of electromagnetic radiation in the infrared (IR), visible, and ultraviolet (UV), frequency ranges to establish the energy level structure of materials.

The absorption spectra of solid *insulators* and *semiconductors* in the UV, visible, and near IR correspond to interband and impurity electron transitions, while in the middle and far IR they arise from *crystal lattice vibrations*, and possibly from the presence of free carriers and impurities. *Metals* are opaque in the visible and transparent in the UV with the cross-over between these two regions given by the plasma frequency $\omega_p = [ne^2/(\varepsilon_0 m)]^{1/2}$ which is typically $\sim 10^{15}$ Hz. Absorption spectra are investigated experimentally with the aid of spectrometers, spectrophotometers, spectrographs, monochromators, etc. Fourier-transform spectrometers (see *Fourier spectroscopy*) which employ a Michelson *interferometer* are widely used in the infrared range. Examples of detectors are photoplates, photocells, photomultipliers, *photoresistors*, thermoelements, *bolometers*, and acoustic-optical receivers. Examples of radiation sources are lasers, heated filaments, gas-discharge and arc-discharge lamps, and laser sparks arising under the action of powerful laser pulses on a material. *Cherenkov radiation* and *synchrotron radiation* are also available. The transmission spectrum of the sample under study is often compared to that of a reference spectrum. The absorption generally follows the *Beer–Lambert law* of exponential decay with material thickness. Typical spectra plot absorption versus radiation frequency (or wavelength). If the absorption spectrum is measured over a wide enough frequency range then the optical conductivity can be determined by the *Kramers–Kronig relations*.

OPTICAL ACTIVITY
See *Rotation of light polarization plane*.

OPTICAL ANALOG OF MÖSSBAUER EFFECT
See *Mössbauer effect, optical analog*.

OPTICAL ANISOTROPY
See *Induced optical anisotropy*.

OPTICAL ANISOTROPY
Dependence of the optical properties on the direction of propagation and the polarization of an electromagnetic wave. Optical anisotropy in *crystals* leads to *birefringence, dichroism of crystals*, changing of ellipticity of light, etc. Optical anisotropy can be natural or induced by an external factor. Noncubic crystals possess natural optical anisotropy.

OPTICAL ANTIRESONANCE
A dip in the *density of states* of excitations in a solid, caused by the interaction of two states coincident or close in energy, one of which is discrete, and the other characterized by a continuous or quite wide-band spectrum. The antiresonance manifests itself in absorption, reflection, *Raman scattering of light*, and *luminescence* spectra (see *Fano resonance*).

OPTICAL AXIS OF A CRYSTAL
Preferential direction in an anisotropic *crystal* where a beam of light propagates without *birefringence*. A crystal can have one or two optical axes, depending on the *crystal lattice* symmetry (see *Crystal optics*).

OPTICAL BISTABILITY
The existence of two (in a multistable system, more than two) stable states for the output optical signal of a system in the presence of a single input signal. This is determined by the totality of static and dynamical processes involved in nonlinear systems with feedback.

Generally the bistable optical element is a *Fabry–Perot interferometer* filled with a *nonlinear medium*. The feedback arises during the reflection of light from the interferometer mirrors. The interferometer transmittance in the absence of absorption is determined by the Airy function:

$$T = \frac{I_T}{I_0} = \frac{1}{I + F \sin^2 \Phi/2}, \tag{1}$$

where I_T and I_0 are the intensities of the transmitted and incident light; $F = 4R/(I - R)^2$ is the interferometer reflectivity coefficient; R is the mirror *reflectance*; $\Phi = 2\pi n L/(\lambda/2)$ is the phase increment; L is the length of the interferometer; and n is the *refractive index* of the material. In the simplest case (for "dispersion" optical bistability), $n(I_c) = n_0 + n_2 I_c$, where I_c is the light intensity inside the interferometer, and n_2 is a constant. Irradiating the interferometer with sufficiently intense light produces a change in the refractive index n, phase Φ and, accordingly, its transmittance which can be expressed via I_c. Disregarding any spatial nonuniformity of the field in the interferometer, we have

$$T = \frac{I_c}{I_0} \frac{(1 - R)}{(1 + R)}. \tag{2}$$

The behavior of an optically bistable system is easy to analyze graphically (see Fig. (a)) by solving Eqs. (1) and (2) simultaneously. The dependences $T(I_0)$ and, accordingly, $I_T(I_0)$ display hysteresis (Fig. (b)) which is in the optical memory regime. By selecting the system parameters one can obtain a differential gain, corresponding to the optical transistor regime (Fig. (c)). Optical bistability in systems with fixed polarization of radiation is subdivided into "dispersion" and "absorption" bistability, depending on whether the dominant process is change in the refractive index or the absorption coefficient of the nonlinear medium. The optical output can involve "intrinsic" optical bistability (signal produced by resonator mirrors, or by the nonlinear process itself) or "hybrid" optical bistability (signal supplied by external electronic device).

Semiconductors are the most promising nonlinear medium for tiny (several micrometers long) devices based on the optical bistability effect; their salient feature being a low energy requirement for the switching mode. Characteristics used are

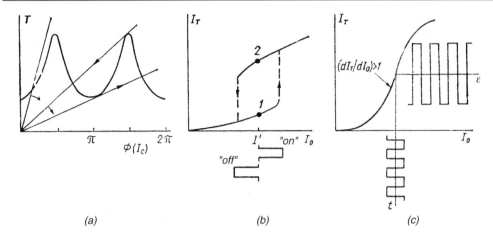

(a)   (b)   (c)

Optical bistability. (a) Graphical solution of Eqs. (1) and (2); (b) hysteresis of optical memory mode: the "on" pulse switches the optically bistable element from state 1 to state 2, whereas the "off" pulse switches from 2 to 1 (I' is the intensity of memory storage); (c) optical transistor mode showing signal amplification.

resonance dynamical nonlinearities in semiconductors, which are determined by specific features of absorption and refraction of laser radiation at resonance transitions, at resonance excitation of *excitons*, *biexcitons*, bound excitons, impurity states, and so on. These nonlinearities (called giant ones) are several orders of magnitude larger than the corresponding classical nonlinearities in non-absorbtive media. The latter nonlinearities determine such processes as the *second harmonic generation*, parametric generation (see *Optical parametric oscillator*), and so on, and manifest themselves in intense light fields comparable to intratomic ones. When using giant resonance non-linearities, one should take into account the dynamical properties and the *relaxation* (lifetime) of excited states, which control the speed of response of devices based on the optical bistability effect. The switching-on time of a device is related to the time τ_c of the field relaxation in the interferometer (if $L = 10$ μm, $\tau_c < 10^{-12}$ s), and to the intensity of the exciting light I_0 (i.e. to the value of the induced nonlinearity), while the switching-off time is determined by the lifetime of excited states.

Optical bistability is of interest both from the practical viewpoint: for the production of optical transistors, ultraspeed selectors, storage elements, and other optoelectronic devices (see *Optoelectronics*) where the light beam replaces an electric current; and for investigating nonlinear opti-

cal processes that determine the bistable behavior of systems: the feedback provides the possibility to detect a nonlinear process under relatively low excitation levels.

OPTICAL BREAKDOWN, light breakdown, optical discharge, laser spark

Transformation of matter to a plasma state as a result of intense ionization under the action of electromagnetic fields at optical frequencies. In optical breakdown a spark seen as a bright flash emerges in the focal point of a lens, and is accompanied by a strong pulse of sound. In irradiated optically transparent *solids* the optical breakdown can occur at the entrance or exit surface, or in the sample bulk, depending on the lens focal point. The latter case requires the highest intensity of focused radiation. The plasma of a spark in the bulk expands in space in such way that the glow front drifts towards the path along the normal to the surface at a speed of 10^7 cm/s. The glow spectrum of the plasma formed in the bulk of the condensed matter is continuous, and that near the surfaces is continuous with a superimposed line spectrum emitted from regions of lower plasma density.

Optical breakdown in condensed matter where power laser radiation propagates can be the cause of a material failure. The evaporation of the solid-state target surface for the case of symmetrical irradiation stimulates uniform compression and heating of the material to high temperatures, with the formation of a high-temperature ($\approx 10^7$ K) dense plasma where thermonuclear reactions might be possible.

OPTICAL COMPENSATOR

A device for introducing a certain optical path-length difference for two coherent light beams, or for reducing an existing path-length difference to zero, or to some constant value. As a rule, an optical compensator is equipped with a reading device to facilitate its use as a path-length difference meter. There are two main types of optical compensators. *Interference optical compensators* are used in double-beam *interferometers* for equalizing the optical path-length difference of interfering beams, e.g., by adjusting the inclination of parallel plates in the beam path to a certain angle. *Polarizing optical compensators* are used for the analysis of elliptically polarized light. The path-length difference in this case is adjusted by varying the thickness of plates cut from a birefringent crystal with differing velocities of the *ordinary ray* and the *extraordinary ray*. Optical compensators are used to study stresses in transparent materials with the help of polarized light; in *crystal optics* in combination with the polarizing microscope, and also to change and analyze light beam polarization.

OPTICAL DEFECTOSCOPY

Application of optical methods to determine *crystal structure* perfection. As distinct from *X-ray defectoscopy* (see *Defectoscopy*) that establishes the nature and concentration of *defects* at the atomic level, optical defectoscopy yields data on the quality of the *crystal lattice* as a whole. The investigation methods used for this purpose are diverse: *luminescence*, *Raman scattering of light*, *light absorption*, as well as shadow, polarization, and interference methods. By analyzing the intensity, line width, and frequency of the lines observed in spectra, and by comparing the crystals under study with standard samples, one can judge the perfection and block (granular) structure of a crystal lattice. The crystal blocks (grains), the traces of slip and other weak local inhomogeneities with clearly pronounced boundaries are easily detected using the *shadow method*, whereby a crystal is inserted in a divergent beam from a point source, and its shadow image is examined on a remote screen. The sensitivity is very high, and allows detecting a heterogeneity that produces a path difference of 0.01λ at the crystal boundaries.

Optical inhomogeneities are easily detectable by interference methods in Michelson, Rayleigh, and Jamin *interferometers*. The essence of interference methods is the fact that the *refractive index* of any inhomogeneity differs from that of the host crystal. The refractive index difference introduces an optical path difference between the interfering beams, causing the interference lines to bend. The magnitude of the bending allows one to judge the geometrical dimensions of inhomogeneities, and to determine the refractive index of impurities. In crystals with microscopic refraction index inhomogeneities, the latter are observed using microinterferometers.

Polarization methods are especially sensitive to stress of any kind in a crystal lattice. To observe stress-induced birefringence, a crystal is positioned between crossed polarizers.

OPTICAL DETECTION

The same as *Pockels effect*.

OPTICAL DETECTION OF MAGNETIC RESONANCE

Methods for detecting *magnetic resonance* optically during microwave saturation of spin sublevels. Applied to solids, this technique is similar to the *optical double magnetic resonance method*. There are two variations of optical detection techniques. One of them uses incident circularly polarized pumping light, and detects effects that are in some way related to the circular polarization of the emergent radiation (*optical pumping method*). Another variety does not necessarily involve the light polarization: these techniques are based on the spin-dependent nature (see *Spin-dependent effects*) of the optical properties of solids (e.g., recombination *luminescence*). When applying the process of optical detection, a sample under study is placed in a resonator of an *electron paramagnetic resonance* spectrometer, and is subjected to the simultaneous action of optical pumping and microwave saturation. As a rule, the microwave power is amplitude-modulated, and the optical detection is performed at this modulation frequency.

An advantage of optical detection relative to electron paramagnetic resonance is the significantly higher sensitivity (approximately by the ratio between the light and microwave frequencies, i.e. $\sim 10^4$). This provides the possibility of studying the radio spectroscopic characteristics of excited states of *paramagnetic centers*, and excited triplet states involving exceedingly small quantities of short-lived intermediate radicals, i.e. *photochemical reaction* products. In the context of semiconductors, the optical detection method allows one to identify unambiguously the pairs of particles (e.g., *donors* and *acceptors*) that are responsible for various bands of recombination luminescence.

The addition of an extra radio frequency source transforms the optical detection technique into that of *electron–nuclear double resonance*. This makes it possible to detect transitions between nuclear spin sublevels in an optical channel with a sensitivity characteristic of the optical detection technique, and a resolution characteristic of the *nuclear magnetic resonance* technique.

OPTICAL EXTINCTION

See *Luminescence quenching*.

OPTICAL FILTERS

Devices that change the intensity of radiation passing through them, and the distribution of intensity over the spectrum. There are absorption, interference, and interference-polarization optical filters, depending on how the light interacts with the working elements of the solid-state filters. The two last types are multilayer systems where the interference of natural or polarized radiation takes place and, as result, a narrow spectral band is isolated.

Solid-state absorptive optical filters can be neutral-gray (reduce light beam uniformly over a broad spectral region) and selective (cut-off or band type). The cut-off optical filters absorb a wide short-wave or long-wave part of the spectrum and pass adjacent regions, while selective filters isolate a certain narrow spectral band. High selectivity is a salient feature of interference-polarization optical filters whose bandwidth can be 10^{-2} nm. Such optical filters are labor-consuming to manufacture, and find use, mainly, in astrophysics for isolating narrow atomic radiation lines arriving from outer space. Among the most widely used optical filters the best parameters are possessed by interference filters with multilayer dielectric mirrors that constitute a half-wavelength layer, in which the transmission band is formed as a result of the interference of the light.

OPTICAL FREQUENCY CONVERTER

Device based on *nonlinear-optical crystals* where laser radiation is converted into light of another frequency. Crystals without a center of symmetry exhibit a nonlinear polarization component proportional to the square of the light wave electric field amplitude. The three-photon interaction occurring in such crystals ($\omega_3 = \omega_1 + \omega_2$) allows one to perform light frequency doubling, and the generation of sum and difference waves. The most efficient transfer of laser radiation energy takes place if the condition of phase synchronism $k_3 = k_1 + k_2$ is met (see *Second harmonic generation*). More than 90% of the laser energy is converted in frequency doublers (*optical harmonic generators*). Nonlinear-optical generators of sum and difference waves find use as coherent radiation sources in the IR and UV regions of the optical spectrum. A pride of place among them belongs to *optical parametric oscillators* that provide a smooth frequency conversion.

When a nonlinear crystal is used to mix laser radiation with IR radiation that carries information on a subject or a process, there results an upward frequency conversion. This can be used to transform the infrared image to a visible spectrum so it is recorded with the help of ordinary methods. Using several successive nonlinear crystals, one can repeat the operations of frequency doubling, addition, and subtraction, and thus obtain sources of coherent radiation at combination frequencies, including high harmonics.

In centrally symmetric crystals with a cubic nonlinearity, the frequency conversion is achieved via four-photon interactions. Such crystals are useful in the direct generation of third-harmonic light waves, and the two-photon pumping of parametric amplifiers and generators. To convert the frequency of intense light waves, use can be made also of the process of *induced light scattering*.

The spectral range of a nonlinear crystal in optical frequency converters is limited by the transparency region of the crystal, and a set of several crystals can be used to cover the overall spectral region from the far-IR to near-UV frequencies.

OPTICAL GATE

A device for one-time or periodic passing or stopping a light beam for a certain time interval Δt_c. In photographic cameras, high-speed cine-cameras, and some optical devices, an optical gate is used to control the exposure time. Optical gates for periodic action provide light modulation. There are mechanical optical gates ($\Delta t_c \geqslant 10^{-4}$ s) based on attenuated total reflection, magneto-optical gates based on the *Faraday effect*, electrooptical gates based on either the *Kerr effect* or the *Pockels effect* ($\Delta t_c = 10^{-13}$–10^{-9} s), and some others. Minimal $\Delta t_c = 10^{-12}$ s is attained by the action of ultrashort light pulses. Optical gates are used in *lasers* for Q-modulation of optical *resonators*. There are *active optical gates* driven by an external signal, e.g., electrooptical gates, and *passive optical gates* driven by the radiation of the laser itself, e.g., bleaching solutions of organic *dyes*, semiconductor crystals, and glasses. Optical gates are used for the generation of nano- and picosecond light pulses, including so-called giant light pulses.

OPTICAL GLASS

An optically homogeneous solid material from a solidified melt of amorphous structure designed for manufacturing transparent elements of optical systems, such as lenses, prisms, plates, and so on. The main source material used for producing optical glass is *silica* (SiO_2). The spectroscopic properties of optical glass (*refractive index, dispersion,* and transparency range) are governed by impurities incorporated in the glass composition. Typically, components made of optical glass are employed in the range 0.4 to 2.5 μm. Impurity-free pure optical glass made of silicon dioxide (fused quartz) is used in the range 0.2 to 4 μm, as well as in the far-infrared ($\lambda > 100$ μm). High-quality optical glass for the infrared range is made of chalcogenides of germanium and arsenic (see *Chalcogenide materials*). Optical glass is the main material for the manufacture of optical fibers and components of *fiber optics*.

OPTICALLY ACTIVE MEDIUM

See *Gyrotropic medium*.

OPTICAL–MAGNETIC RESONANCE

A kind of a *double resonance* involving two frequencies, one in the optical region and the other in the microwave or radio-frequency region. Optical–magnetic resonance serves to enhance the sensitivity and improve the resolution of resonance methods. See also *Optical detection of magnetic resonance*.

OPTICAL NUCLEAR POLARIZATION

A dynamical method for polarizing magnetic nuclei (see *Nuclear orientation*) with the help of optical pumping. The methods of optical nuclear polarization originate from experiments on optical pumping (see *Optical detection of magnetic resonance*) in gases where the polarization mechanism is related to the nuclear-*spin*-dependent *selection rules* for optical transitions. In solid *insulators* and *semiconductors*, the optical nuclear polarization in the external magnetic field arises in the presence of paramagnetic centers whose spin polarization P_e under illumination differs from the equilibrium Boltzmann value P_{e0}. If the paramagnetic centers bring about *spin–lattice relaxation*, i.e. if the time-modulated *hyperfine interaction* of electron and nuclear *magnetic moments* is present, the nuclear polarization $P_n = P_{n0} - \xi f (P_e - P_{e0})$ appears, where $\xi = -1$ for the contact interaction and $0 < \xi < 0.5$ for the dipole–dipole hyperfine one; $f < 1$ is the *leakage factor* that allows for a decrease of P_n due to the presence of extrinsic paramagnetic centers. The difference between P_e and P_{e0} needed to achieve optical nuclear polarization appears in semiconductors (GaAs, Si), e.g., when paramagnetic centers trap conduction electrons that are optically oriented by the circularly polarized light. The excitation of a semiconductor by unpolarized band-to-band light produces equal quantities of electrons with spin projections $+1/2$ and $-1/2$ in the *conduction band*, so $P_e = 0$. If, however, the sample is in a magnetic field, then $P_{e0} \neq 0$, and hence nuclear polarization is possible. In *silicon*, the maximum optical nuclear polarization $P_n / P_{n0} \approx 5 \cdot 10^4$ was observed in samples subjected to irradiation. Under illumination the strongly oriented triplet state paramagnetic centers are excited in these samples, the mechanism of orientation of the centers being related to the selective filling and depletion of triplet state magnetic sublevels. A similar mechanism of polarization takes place in some insulators. Optical nuclear polarization is experimentally observable using the methods of *nuclear magnetic resonance* and recombination *luminescence.*

OPTICAL PARAMETRIC OSCILLATOR

A source of coherent optical radiation, which depends for its operation on the interaction of light waves of various frequencies in a light-transmitting *nonlinear medium.* The optical parametric oscillator performs the transformation of the energy of laser radiation, which is incident upon a *nonlinear-optical crystal*, to light waves of lower frequency. This process involves a special type of three-photon interaction $\omega_p = \omega_1 + \omega_2$, whereby the high-frequency pumping wave ω_p exhibits absolute instability. A crystal without a center of symmetry, which has a quadratic nonlinearity of the polarizability in an electric field of the light wave, serves as the nonlinear medium. Only those waves are excited in the parametric light oscillator, for which the *condition of phase synchronism* $k_3 = k_1 + k_2$ (see *Second harmonic generation*) in the bulk of the nonlinear crystal is fulfilled. Wave *dispersion* obstructs the fulfillment of this condition. The synchronous interaction is realized in anisotropic crystals between an *ordinary ray* and an *extraordinary ray*, which exhibit different frequency dependences of the *refractive index*, and different directions of propagation. The change of dispersion causes a change in the frequencies of excited and amplified modes. This principle provides the basis for the operation of the frequency-modulated optical parametric oscillator. The oscillator is tuned to a particular frequency by changing the angle of synchronism (θ_s). The angle θ_s is changed by the rotation of the crystal, change of its temperature, or application of an external electric field. Positive feedback, which provides the parametric generation, is established by placing the nonlinear crystal between the dielectric mirror plates (M_1, M_2) of a reentrant *resonator* (see the figure). In some instances, the injection of a weak coherent signal (e.g., of a *semiconductor laser*) is used in order to increase the stability of operation of an optical parametric oscillator.

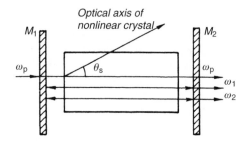

Optical parametric oscillator.

The wavelengths, which are tunable continuously by the optical parametric oscillator, range from 0.4 to 16 μm; this wave band is spanned through the use of a set of optical crystals, which exhibit different transparent regions and different values of nonlinear distortion factors and failure thresholds. The sources of *laser* pulses, as well as the sources which provide continuous or periodic laser radiation and the harmonics of laser radiation, serve as pumps. In certain cases, optical parametric oscillators perform the frequency tuning within 10% of the light frequency. These oscillators are of particular value in the infrared region. Optical parametric oscillators, which are created for the needs of the *optical spectroscopy* of ultrahigh-rate processes, generate laser pulses of pico- and femtosecond duration. Their unique characteristics are coherence, narrowness of generated lines, high power density, continuous frequency tuning. It is due to these characteristics that the optical parametric oscillator is the most widely used (and in certain cases the only one usable) device for performing spectroscopic investigations in the IR region, for the execution of selective actions on certain materials, for control over environments, and for other purposes.

OPTICAL PHONONS
See *Phonons.*

OPTICAL POLARIZATION METHOD
Investigation of the *birefringence* of crystals and of amorphous media under the influence of external (electric) fields. It is based on the dependence of transmission of the interference-polarization signal on the value of birefringence, on the light wave length, on the thickness of the

sample, and on its orientation. This system consists of a crossed polarizer and analyzer with the sample under investigation placed between them. Usually it is not the absolute value of the intensity of the light transmitted by the system that is determined, but rather its variations relative to the center of the interference pattern (see *Conoscopy*), or a study is made of the interference picture which displays the distribution of internal stresses of the object under study. With the aid of the compensator of birefringence, the rotation angle of the analyzer up to the transmission minimum is measured, and using this angle the birefringence is calculated. The optical polarization method is utilized in electro-, piezo-, magneto-, acousto-optical, measurements for investigating stresses in transparent materials.

OPTICAL PROPERTIES OF LIQUID CRYSTALS

Great variety of optical effects related to the *anisotropy of liquid crystals* (LC), which are caused by the orientational ordering (see *Orientational order*) of antisymmetric molecules. The optical axis of a LC (e.g., of a nematic crystal, see *Nematic liquid crystal*) coincides with the direction of a *director*. The orientation of the latter can change easily due to various factors, such as external fields, boundary conditions, and fluctuations. The effects most typical of liquid crystals are as follows:

- Change in optical properties due to rotation of the director under an external action, resulting in birefringence and absorption dichroism (see *Dichroism of crystals*). Dichroism in the optical range is ensured by introducing anisotropic impurity molecules: the *guest–host effect*.
- Visualization of inhomogeneities: *textures, disclinations*, and *point defects*, modulated static and dynamic structures, *wave-like modulation* in liquid crystals (*Kapustin–Williams domains*, dynamic light scattering).
- Scattering of the director by thermal fluctuations in the nematic phase (also in smectic phase C, see *Smectic liquid crystal*) is about 10^6 times that in the isotropic phase. This is related to the fact that the long-wave fluctuations involve negligible energy outlay: the free energy does not change during uniform rotations of the director. Such modes are similar to Goldstone modes (see *Goldstone theorem*), but

prove to be diffusive because of viscosity. The scattering becomes still greater by a few orders of magnitude during the emergence of turbulent flows in a regime of dynamic light scattering.
- Giant optical nonlinearity (roughly 9 orders of magnitude higher than that in carbon bisulfide) is due to director rotation under illumination.

Unique optical properties arise in LC with a helical structure (cholesterics, see *Cholesteric liquid crystal*, and chiral smectics C): (i) enormous optical activity (about 10^4 deg/mm), and (ii) selective light reflection (also seen in *blue phases*) due to *Bragg diffraction* from the helical structure. Since the helix pitch is temperature sensitive, this phenomenon finds wide use in *thermography* in engineering and medicine.

OPTICAL QUENCHING

See *Luminescence quenching*.

OPTICAL RADIATION DETECTORS

Devices for the transformation of optical radiation energy to other types of energy convenient for measurement. *Solids* may be divided into three classes according to the character of the interaction of optical radiation with the material: photonic, thermal, photochemical. These provide the main elements of optical radiation detectors. In the photonic variety individual *photons* are counted, in the thermal types the total energy over the time of observation is integrated, in the photochemical types the optical density of the photolayer, which varies due to the photochemical reaction and to the effect of photographic development, is measured.

Each optical radiation detector is characterized by the following six main parameters:

(1) Threshold sensitivity – minimal optical radiation intensity that gives an output signal at the detector that is equal to the root mean square noise (fluctuation) of the receiving element. At the detector all types of *fluctuations* are reduced to a minimum, and fluctuations of the incoming radiation flux (quantum noise of radiation source and of background) plus the thermal conductivity noise of the sensing element determine the threshold sensitivity (detection ability).

(2) Time constant – time of signal growth up to a specified value, characterizes the rapidity of the response of the optical radiation detector.

(3) The transformation ratio determines the value of the output signal, normalized to the value of detected input radiation flux; it is measured in volts per watt or in amperes per watt.

(4) The spectral sensitivity is the dependence of the output signal of the detector on the wavelength of incoming radiation. Thermal optical radiation detectors usually have a broad spectral sensitivity range, and they are called *nonselective receivers of optical radiation*. Photonic optical radiation detectors operate, as a rule, in narrow ranges of the spectrum, and are very selective.

(5) Dynamic range is the region of values of recorded input optical radiation flux, where the output signal is proportional to the input. Thermal optical radiation detectors and *photodiodes* in the current generator mode possess a wide dynamic range.

(6) Operating temperature of the sensing element.

The main thermal optical radiation detectors are *bolometers, thermocouples* (thermoelements and thermostacks), *pyroelectric radiation detectors* and optico-acoustic (pneumatic) receivers (see *Acousto-optic devices*); they have the threshold sensitivity 10^{-8} to 10^{-12} W/Hz$^{1/2}$ and a time constant 1 to 10^{-3} s. *Superconducting bolometers* are 100 to 1000 times superior according to their sensitivity and fast response. For measurements of very high fluxes of optical radiation calorimeters are used (see *Calorimetry*). The following devices are photonic optical radiation detectors: photocells, photodiodes, photomultiplier tubes, photoresistors with threshold sensitivity up to 10^{-12} W/Hz$^{1/2}$ and time constant up to 10^{-10} s. *Tunnel photodiodes* with microantennae may operate with a time constant 10^{-14} s. *Photoelectronic multipliers* and *avalanche photodiodes* have their threshold sensitivity up to 10^{-17} W/Hz$^{1/2}$.

There are multichannel optical radiation detectors based on arrays of photodiodes combined into shift registers with charge coupling (*charge-coupled devices*), and also receivers of the television tubes type, such as super silicon diode array cameras. Multichannel optical radiation detectors form a new class of multichannel analyzers with improved sensitivity, dynamic range, and fast response in real time compared to the photochemical multichannel optical radiation detectors in the form of different photosensitive layers, as used in photography.

OPTICAL RADIATION, SCANNING

See *Scanning of optical radiation*.

OPTICAL RADIATION SOURCES

Devices for conversion of various kinds of energy into electromagnetic irradiation in the optical range within the conventional wavelength limits (in vacuo) $\lambda = 100$ nm to 1 mm ($3 \cdot 10^{15}$–$3 \cdot 10^{11}$ Hz). Solid state optical radiation sources can be classified into three groups according to their mode of generating radiation:

(1) The *temperature (thermal) mode* converts the heat motion of particles of a solid directly into electromagnetic radiation energy. The commonest are heat sources of optical radiation, viz., incandescent lamp, *globar, Nernst glower*, platinum-ceramic sources heated by an electric current, and electrodes raised to a white glow in a gas discharge (such as a carbon arc). As a rule, thermal optical radiation sources can generate both a discrete and a continuous spectrum with the equivalent brightness temperature of a black body $T_b \leqslant 4000$ K.

(2) The *luminescence mode* uses the recombination radiation of electronic and intramolecular vibrational states of crystals and glasses (*luminophors*). These states can be excited either with electric current (see, e.g., *light emitting diodes*) or a flux of various particles: γ-quanta, photons (*photoluminescence*), electrons (*cathodoluminescence*), etc. Luminescent optical radiation sources can generate both discrete and continuous spectra, and operate in both continuous and pulse modes using single or periodic pulses of flashes with durations longer than several nanoseconds. The equivalent brightness temperature of a black body for continuous luminescent optical radiation sources can reach 10,000 K.

(3) The *laser mode* (see *Laser*) of an optical radiation source possesses temporal and spatial coherence. Laser sources can generate optical radiation at fixed or adjustable wavelengths, and continuous or discrete radiation with a pulse duration less than 10^{-14} s (tens of femtoseconds).

Other optical radiation sources are various *plasma* types (gas discharge and spark) with $T_b \leqslant 10^4$ K, as well as those based on relativistic particle fluxes (e.g., *synchrotron radiation* and undulator radiation with $T_b \sim 10^7$ K for the vacuum ultraviolet range).

OPTICAL RECORDING MEDIA

Light-sensitive materials for *information recording* (including digital data and images). Of widest use are ordinary *photographic materials* with subsequent chemical development. To process and record information in real time, optical recording media with a *photochromic effect* and *photorefraction* are used. Under the action of light, the complex refractive index $\tilde{n} = n - i\kappa$ (here n and κ are the *refractive index* and *absorption index*, respectively) changes in these materials. Depending on the relation between Δn and $\Delta \kappa$, one can differentiate amplitude ($\Delta \kappa \gg \Delta n$) and phase ($\Delta n \gg \Delta \kappa$) optical recording media. In some media, e.g., *photothermoplastic materials* (see *Nonsilver photography*), the medium thickness (profile) changes under the action of light.

Information can be recorded on an optical medium either in a discrete (digital) fashion (point recording), or in the form of images, or two-dimensional or three-dimensional holograms. The recording density in an optical recording medium is limited by *diffraction of light* phenomena, and is as high as λ^{-2} in two-dimensional media and as λ^{-3} in three-dimensional media, which gives for the visible range (wavelength $\lambda = 0.6$ µm) the values 10^8 bit/cm^2 and 10^{12} bit/cm^3, respectively. Usually two-dimensional optical recording media are the main ones used.

Fundamental optical recording processes can be reduced to three kinds of reactions: phototransfer of charge (electrons; protons in some organic compounds); photoinduced *phase transitions*; recordings on *phononless lines*. The phototransfer of charge is carried out in *electroop-

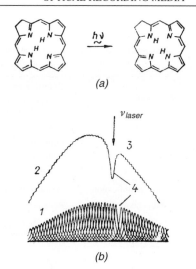

(a)

(b)

Laser hole burning in absorption spectrum of dihydroporphyrin: (a) photoinduced tautomeric reaction in porphyrin molecule; (b) 1, spectral absorption bands for different tautomers; 2, profile of laser radiation absorption band; 3, selective action of laser radiation; and 4, spectral hole in an individual phononless line.

tic materials (LiNbO$_3$, LiTaO$_3$, etc.). The action of light upon these materials is accompanied by an *anomalous photovoltaic effect*, and the establishment of an internal electric field that changes the refractive index (internal *electrooptical effect*). Photoinduced phase transitions take place in amorphous *chalcogenide materials*, e.g., in tellurium and its compounds. Light-induced phase transitions can provide either read–write (*reversible recording*), or read-only (*irreversible*, *one-time recording*) storage. Solid-to-vapor phase transitions (one-time recording) are used also for digital recording, and light-induced phase transitions in *amorphous semiconductors* are used in *optical storage disks*. Recording on phononless lines is performed at molecular centers in organic compounds or *ionic crystals*. The half-width of a phononless line ($\delta_0 \approx 10^{-3}$ cm^{-1}) is smaller by 2 to 3 orders of magnitude than the total half-width of the band (see Fig.). Therefore, this method provides an optical record density significantly beyond the diffraction threshold (due to narrow absorption bands), and achieves in two-dimensional media a 10^{10}–10^{11} bit/cm^2 density. The application of this method is hindered by the necessity to operate at liquid helium temperatures and below.

Table 1. Sensitivity of materials for optical recording

Recording type	Recording material	Sensitivity, mJ/cm^2	Recorded energy, nJ/bit
Irreversible	Tellurium or its compounds	130	1.3
(read only)	Films based on dyes	100	1.0
	Particles of silver in gelatin (Drexon)	60	0.6
Reversible	Films of TeO_x	160	1.6
(read-write)	Films of $Se_{20}In_{35}Sb_{45}$	20–2000	0.2–20
	Chalcogenide semiconductors	10–1000	0.1–10
	Crystals of $KTa_{0.65}Nb_{0.35}O_3$	0.2	$2 \cdot 10^{-5}$
With amplification	Photothermoplastics with CdSe	10^{-5}	10^{-3}
With chemical treatment	Reoxan	2000	–
Silver photomaterials	Silver halides	10^{-2}	10^{-3}

The light sensitivity for direct recording on optical media is lower by several orders of magnitude than that of silver-halide materials with chemical development. Only in photothermoplastic materials (see *Thermoplastic materials*) and photorefractive materials (see *Holography*) recorded in an electric field (enhancement of the primary reaction), does the sensitivity approach that of ordinary photographic media (see Table 1).

Optical recording media find use in recording holograms and discrete information in real time. Controlled spatial light modulators have been developed for optical filtration (see *Optical filters*). Optical recording media are also applied to the production of holographic optical elements such as interference filters, holographic lenses, diffraction gratings, etc. Amorphous chalcogenides are applied as master optical storage disks when manufacturing secondary disks (so-called *compact disks*) for reproduction of audio- and video-recordings. Also, reversible optical disks have been developed for computers (to replace magnetic media in ROM systems), and for playback devices for audio and video recordings.

OPTICAL SOLITON

Self-localized nonlinear excitation (see *Soliton*) of an electromagnetic field. An optical soliton occurs in media that exhibit a linear relationship between the *refractive index* and the intensity of the traveling wave (in fiber-optics tubes or light pipes, *films*, etc.), and in resonant media. The existence of optical solitons in optical single-mode light pipes was predicted by F. Tappert and A. Hasegawa (1973); and was observed by L. Mollenauer et al. (1980). The optical soliton in a light pipe corresponds to the so-called *envelope soliton* of a light wave. Soliton modes may appear in a light pipe when there is mutual compensation for the effect of an increase of slope of the wave front because of nonlinearity, and there is wave packet spreading related to dispersion (see *Wave packet representations of wave functions*). The nonlinear properties of light pipes are related to the deformation of the electronic shells of glass molecules (see *Vitreous state of matter*) under the action of the electric field of the wave. The dynamics of propagation of the envelope soliton is described by the nonlinear Schrödinger equation, in which the role of time is played by the coordinate along the fiber direction. *Dark solitons* arise, if the light pulse propagates over a region of normal *dispersion of light*; *bright solitons* occur if the propagation is over a region of *anomalous dispersion of light*. Dark solitons are dips in intensity of quasi-continuous pumping, or a long pulse; bright solitons are individual localized wave packets. Typical parameter values for the generation of a single bright optical soliton in a light pipe are: wavelength $\lambda = 1.55\ \mu m$, input pulse duration 70 ps, light guide cross-section $S = 100\ \mu m^2$, peak power 1 W. Optical solitons show promise for applications to fiber-optic information cables with extra-high transmission rates, logic gates, soliton *lasers.*

OPTICAL SPECTROSCOPY OF SOLIDS

The branch of spectroscopy concerned with the study of *light absorption, reflection of light, luminescence, Raman scattering of light,* and *photoconductivity* optical spectra of *solids,* the determination of their energy level configurations, and their suitability for various applications. The theory underlying the analysis of the above spectra is the quantum theory of solids and *group theory.* The latter takes into account the symmetry properties of a *crystal lattice* (and, consequently, of wave functions) and establishes *selection rules* for quantum transitions. Spectroscopic data provide an opportunity to judge the nature of the interaction between a light wave and a *crystal,* the energy migration in a crystal, the nature of *defects,* the centers of radiative *recombination* of electrons and holes, the presence of photochemical transformations, the influence of external factors (temperature, pressure, electric and magnetic fields) on the energy bands and discrete levels, and the role of *surface states.* Under high (laser) levels of excitation, *multiphoton processes* and nonlinear optical effects, including stimulated Raman scattering and stimulated radiation emission, become feasible. Optical spectroscopy methods are quite informative for the spectral analysis of solids. Any spectrum depends on the chemical composition, the *crystal structure,* the presence of defects in a crystal lattice, their nature, etc.

In accordance with the *band theory of solids,* quantum transitions between the allowed bands, intraband transitions, those with participation of defects of various kinds, and *quasi-particles,* are possible. Band-to-band transitions produce broad absorption spectra with a sharp cutoff (*intrinsic light absorption edge*) on the long-wavelength side. Adjacent to these spectra are impurity-band, exciton absorption, and luminescence spectra. The latter result from the electron–hole Coulomb interaction. Typical of excitons with a large radius (*Wannier–Mott excitons*) is a set of narrow bands forming a hydrogen-like (Rydberg) series. Its edge coincides with that of intrinsic absorption, while the most outlying band is at a distance correlated with the binding energy of an *exciton,* and determines its ground state.

There are both *direct band-to-band transitions* and *indirect band-to-band transitions.* The former arise without the participation of *phonons,* and are characterized by high values of the *absorption coefficient* (10^4–10^5 cm^{-1}). The latter appear with the participation of either phonons or defects, and have absorption coefficients lower by 2 to 3 orders of magnitude. The interaction of excitons with photons results in the formation of *photoexcitons* (*polaritons*). The edge and exciton absorption, luminescence, reflection, and photoconductivity spectra of *semiconductors* and *insulators* occupy the regions of near-IR, visible, and UV radiation. Electronic transitions in *molecular crystals,* and in particular *dyes,* are found in the same spectral region.

For electronic transitions with the participation of impurity states formed by atoms of rare-earth, transuranium and transition elements, very narrow absorption and luminescence bands can appear even at high temperatures. They have a great number of components due to the splitting of degenerate states in the *crystal field,* and are related with the inner (e.g., $4f^n$ for rare-earth elements) electronic shells of *impurity atoms.*

In addition to quantum transitions between the electronic states that result from vibrations of atoms, ions, or molecules that compose a crystal, there can be a vibration-induced change of the *dipole moment* and the polarizability of the *unit cell.* Also, IR absorption and Raman scattering may appear. For crystals with a center of symmetry in the unit cell some of the lattice vibrations appear in the IR spectrum, and others are found in the Raman spectrum (see *Mutual exclusion*), as permitted by selection rules. The lattice spectra of molecular crystals reveal the rich structure of *intramolecular vibrations.* In the case of anisotropic lattices, all spectra are polarized.

OPTICAL SPECTROSCOPY OF SURFACES

Studies of *solid surfaces* and phase boundaries by optical methods. The optical spectroscopy of surfaces is used mainly to investigate the molecular structure of the near-surface layer, its changes during the course of chemical reactions, and the *adsorption* process at the surface (IR absorption, IR reflection, multiple *total internal reflection*); specific features of *surface quasi-particles* such as *surface plasmons, surface polaritons,* and *excitons* (*attenuated total internal reflection, Raman scattering of light,* surface laser *photoluminescence,* reflection spectroscopy); the rate of *surface recombination, surface state* parameters, and the planar band potential (IR absorption, *electroreflection,* photoluminescence); the structure of energy levels in *heterojunctions, superlattices,* and *quantum wells* (IR absorption, Raman scattering, photoluminescence).

As a general rule, optical spectroscopy of a surface is not associated with a strong interaction there, and it often yields the most direct data on surface parameters (e.g., energy levels in quantum wells and superlattices). This presages the wide use of optical spectroscopy in investigations of physical and physicochemical processes near the surface.

OPTICAL SPIN ORIENTATION

Appearance of anisotropy of the spatial distribution of *spins* of charge carriers excited by circularly polarized electromagnetic radiation in *solids.* The incident circularly polarized radiation possesses a finite momentum, and by virtue of its conservation the charge carriers that emerge after its absorption in a semiconductor should have the same momentum. Since the angular momentum of free electrons in a semiconductor is related to the presence of spin, the spins of the light-generated electrons will have a prevailing orientation along the radiation propagation direction. The *spin–orbit interaction* can also result in the orientation of holes. A similar effect arises when polarized light is absorbed by a system of atoms, and the resultant angular momenta of the atoms are oriented. Experiments reveal an optical spin orientation owing to the presence of circular polarization of the recombination radiation of spin-oriented electrons and holes. In GaAs, the maximum polarization achieved is 90%. This value decreases as a result of relaxation of the oriented electrons and holes during the time delay between excitation and emission of the radiation (see *Paramagnetic relaxation*).

OPTICAL STORAGE DISKS

Information carriers for recording discrete (digital) or holographic data. Storage disks can be read-only or reversible (read–write). The former are used in commercial equipment for the repro-duction of audio and video data or recordings (Fig. 1(a)). The digital master record is usually made with a *semiconductor laser.* The diameter of a single bit (a dot) of a record on an optical storage disk is about 1 μm (Fig. 1(b)), which allows the recording of about one billion data bits on a disk 12 cm in diameter (equivalent to 50,000 pages of text, or a 20 min color video recording with sound). Master recording on the disk usually makes use of *optical recording media* designed for real-time operation. The master disk is copied on to secondary ones that are plastic plates with a system of small holes (about 1 μm in diameter). Substrates for optical storage disks are mostly plastic or glass. Each disk has a required protective *coating.* Those used in computers are 36 cm in diameter. The development of read–write optical storage disks began in 1983–1985 (see *Information recording* in solids).

OPTICAL TECHNIQUES OF INFORMATION RECORDING

Light-induced residual changes of physicochemical properties of materials (see *Optical recording media*). Optical techniques of information recording are applied to the recording of images, discrete (digital) information, holograms and so on, and can be implemented in two-dimensional and three-dimensional media. The limit of the recording density is set by the conditions of diffraction and is as high as λ^{-2} bit/cm^2 in two-dimensional media and λ^{-3} bit/cm^2 in three-dimensional ones (here λ is the wavelength). The theoretical limit of recording density has been attained in two-dimensional systems (at $\lambda = 632.8$ nm it is $2.5 \cdot 10^8$ bit/cm^2).

Optical techniques of information recording use changes in optical constants (*absorption index* κ, *refractive index* n) or profile $\Delta(dn)$ (d is the thickness) of the medium (see p. 928, Fig. 2). Depending on the relation between changes in optical constants, one can distinguish amplitude $|\kappa_1 - \kappa| \gg |n_1 - n|$ and phase $|n_1 - n| \gg |\kappa_1 - \kappa|$ optical recording. The latter provides a high diffraction efficiency (up to 100% in three-dimensional and up to 34% in two-dimensional media) in holographic recording. This high diffraction efficiency is of great practical importance in the manufacture of optical holographic elements (lenses and diffraction gratings).

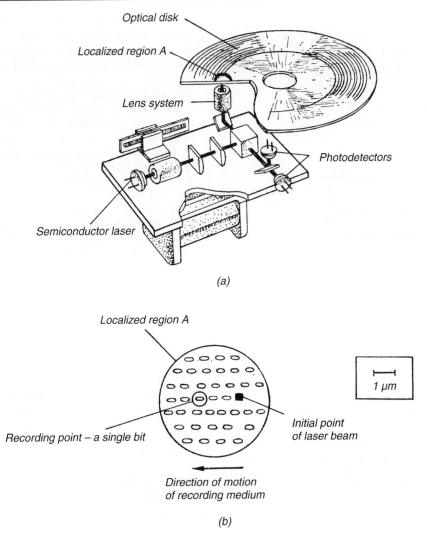

Fig. 1. (a) Optical storage disk, and (b) details of localized region A.

The design of optical information recording techniques is linked to the production of *lasers* and *optical storage disks*. Here, the recording process itself runs in real time (1 to 100 ns for RAM), whereas the information recorded in ROM must be stored for years. These optical techniques are used for both one-time and repeated (reversible) recording. In the latter case, optical erasure is used for a new recording. As a rule, recording media are constructed as complicated multilayer systems containing a light-sensitive layer, various sublayer coatings, and a substrate. Of interest is the so-called *bubble recording* when the light absorbed by the medium also causes heating of the sublayer with subsequent gas liberation (formation of a bubble). This makes it possible to enhance the photosensitivity and improve the detection parameters.

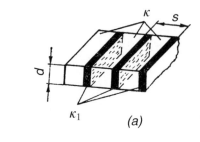

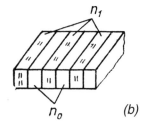

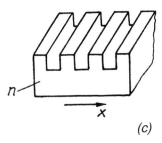

Fig. 2. Three types of optical information recording gratings: (a) amplitude type with thickness d, separation s, and absorption indices κ_1 and κ, (b) phase type with indices of refraction n_1 and n_0, and (c) profile-phase type.

OPTICAL THEOREM

A relation connecting the imaginary part of the *scattering amplitude* $f(k, \Theta)$ for a particle or *quasi-particle* of wave vector k at a zero angle ($\Theta = 0$), with the total (elastic or inelastic) scattering cross-section σ_{tot} of the particle, $\sigma_{\text{tot}} = (4\pi/k)\text{Im}f(k, 0)$.

OPTICAL VIBRATIONS of crystal lattice

Branches of *normal vibrations* that exist in crystals containing more than one atom per *unit cell*. Long-wave displacements of vibrating atoms (or molecules, or ions) occur with no shift in the center of gravity of individual unit cells (with wave

vector $k \to 0$, the atoms move in the "head-on" direction). They are called optical because they are vibrations in *ionic crystals* in the IR part of the spectrum near the visible region. These optical vibrations involve an antiphase motion of ions of the unit cell which produce an alternating crystal dipole moment that interacts with the external electromagnetic field. In contrast to *acoustic vibrations* that always exist in a three-dimensional crystal, the frequencies of optical vibrations do not vanish at $k = 0$. The number of optical vibration branches is $3q - 3$, where q is the number of atoms in a unit cell.

OPTICAL WAVEGUIDE

A dielectric structure along which electromagnetic radiation of an optical spectrum can propagate with low loss. The commonest types involve *fiber optics*, such as fiber light guides, and various kinds of optical waveguides for *integrated optics* (planar, channel, etc.).

OPTICO-ACOUSTICS

See *Acousto-optics*.

OPTICS, NONLINEAR

See *Nonlinear optics* and *Nonlinear-optical crystals*.

OPTICS, NUCLEAR

See *Nuclear optics*.

OPTICS OF METALS

See *Metal optics*.

OPTICS OF SOLIDS

Part of physical optics studying processes of light propagation and light interaction with solids. This includes *polarization of light*, *birefringence*, interference, *diffraction of light*, optical activity (see *Rotation of light polarization plane*), *dispersion of light*, light absorption, *magnetooptics*, electrooptics, etc. The basis of the phenomenological theory of the optics of solids, which reflects the wave aspect of radiation, is Maxwell's equations. The coefficients of these equations are the material constants of matter: the relative *dielectric constant* ε and *magnetic permeability* μ which are related to the *crystal lattice* structure, and determine its *refractive index* $n = (\varepsilon\mu/\varepsilon_0\mu_0)^{1/2}$, where ε_0 and μ_0 are the free space (vacuum) values. All three parameters, n, ε, and μ, are functions of frequency and wave vector. Most crystals are anisotropic, so these three parameters depend on the light propagation direction (see *Anisotropy of crystals*). In nonmagnetic solids $\mu \approx \mu_0$.

All *crystals* are characterized by their intrinsic absorption spectra which are determined by the structure of the *energy bands* of electrons and *phonons*. The electronic absorption spectra are related to the resonance band-to-band transitions of electrons as well as to intraband transitions and, as a rule, are situated in the UV, visible and near-IR regions. The absorption of phonon radiation takes place in the middle and far-IR regions. The strongest absorption over broad optical spectral regions pertains to *metals* owing to their high free-carrier concentration. The band-to-band transitions of metals are, in most cases, in the UV and perhaps visible regions, so many good metals are opaque in the visible. A high transparency of metals is found at frequencies above that of

plasma oscillations, starting at frequencies much higher than those of *highly-doped semiconductors*. The exciton spectra that are usually adjacent to the *intrinsic light absorption edge* are in many cases superimposed on the electron band-to-band spectra. The electronic spectra of molecular crystals reproduce, to a great extent, the spectra of the individual molecules that they contain, and are situated predominantly in the UV and visible regions. Very rich spectra of *intramolecular vibrations* are found in the IR.

OPTICS, VIBRONIC EFFECTS
See *Vibronic effects in optics*.

OPTOELECTRONICS
A technological scientific discipline where a combination of electronic and optical methods and means is used for processing, transmission, recording, storage, and presentation of information. The main feature of optoelectronics is the presence of an optical photon connection to bring

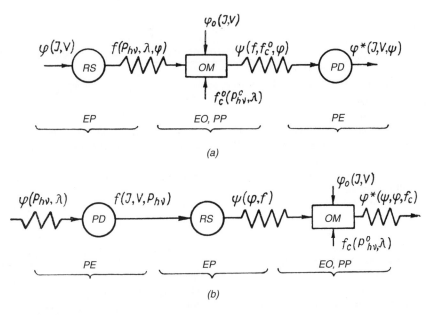

Fig. 1. Diagrams of optoelectronic transformations of: (a) an electrical signal $\varphi(J, V)$ which is a function of current J and voltage V, and (b) an optical signal $\varphi(P_{h\nu}, \lambda)$ which is a function radiation density $P_{h\nu}$ and wavelength λ. The systems include a radiation source (RS), a photon detector (PD) and an optical medium (OM) with electrical $\varphi_0(J, V)$ and optical $f_c^0(P_{h\nu}^0, \lambda)$ input reference signals. The indicated transformations are electron-to-photon (EP), photon-to-electron (PE), electron-to-optical (EO) and photon-to-photon (PP). The intermediate transformed signals are denoted by f and ψ.

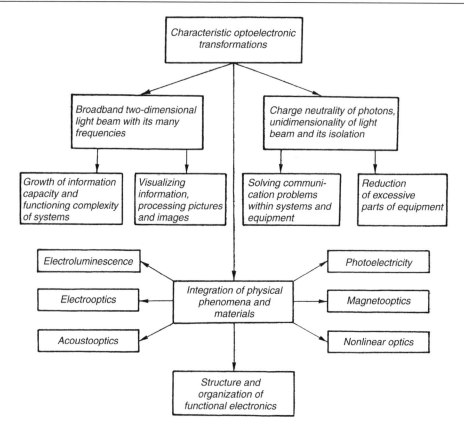

Fig. 2. Qualitative characteristics of optoelectronic transformations and their applications.

about an optoelectronic (either electron–photon or photon–electron) transformation. A characteristic of optoelectronics is charge neutrality, low noise feature, and the accessibility of the information carrier that is a *photon* for visual perception. Besides, optoelectronics provides the possibility of excluding the effect of a frequency-dependent wave impedance from the circuit matching, of suppressing reactances from communication channels. It can also ensure almost complete isolation and autonomy of the channels by combining quantum mechanisms for generating radiation and *current carriers* in the material with electronic methods for controlling these mechanisms. The application of the optoelectronic transformation to electrical (a) and optical (b) signals is illustrated in Fig. 1. The diagrams reflect the duality of optoelectronic circuits, which determined, to a great

extent, the term "optoelectronics". The connections shown can be broken in one or several places. Thus there are several independent branches in optoelectronics: devices for the display and presentation of information, photoelectronics and photoelectronic automatic equipment, *fiber optics, integrated optics, acousto-optics, magnetooptics,* and so on. The block diagram of an optoelectronic transducer can contain any number of electronic elements (e.g., amplifiers, generators, logical isolators, etc.) in its electrical part. This introduces no changes into the algorithm of the optoelectronic transformation, its essence being unambiguously determined by the basic elements of optoelectronics: radiation sources, photodetectors, and optical media. Simple optoelectronic transducers are exemplified by *optoisolators (optrons)* based on an optically connected radiation source and photodetector located in the same enclosure (like integrated circuits); *fiber-optical communication lines* where the optical medium has the form of a long glass-fiber cable; *solid-state amplifiers*; and radiation image converters where the photodetector and radiation source are distributed structures, integrated technologically in a single device. The functional possibilities of an optoelectronic transducer are determined by the nature of the radiation used. Coherent radiation provides more information content, and has a higher Q. The sources of coherent radiation in optoelectronic transducers are *lasers* and *light emitting diodes*. The functionality of incoherent optoelectronics stems, mainly, from the photoelectric transformation. Used to this end in optoelectronics are *photoresistors*, *photodiodes*, phototransistors, and other photoelectric devices and their arrays. The generation, transformation, and detection of the radiation are in most cases carried out by semiconductor elements and structures that lend themselves to microminiaturization and integration. The advantages of optoelectronics that determine its role in solving some technical and scientific tasks, and its relation with various fields of physics, are clearly seen on the diagram of Fig. 2.

OPTOISOLATOR, optron

A *semiconductor device* containing a light source (*light emitting diode*) and a detector (*photodiode*, phototransistor) without direct electrical contacts between input (control) and output signals (see Fig. 1). The signal input current I_1 triggers the generation of radiation by the light-emitting diode. This radiation is detected by the photodiode and then transformed into an electric current I_2 to be applied to the load resistor R_L. Fig. 1 shows a photon $h\nu$ during its transit from the light-emitting diode to the photodetector. Optoisolators or optrons are used for amplifying and converting electrical and optical signals, for switching, for modulation, etc. To increase the current transformation ratio I_2/I_1, which is typically about 10^{-3}, by tens of times, a phototransistor is used in the optoisolator instead of a photodiode. See also *Optoelectronics*.

OPTRON

See *Optoisolator*.

OPW METHOD

See *Orthogonalized plane waves*.

ORBACH–AMINOV–BLUME PROCESS
(R. Orbach, M.K. Aminov, M. Blume)

Establishment of an equilibrium distribution of populations (see *Level population*) E_m and $E_{m'}$ of spin levels of a *magnetic ion* with the nearest excited state energy E_l lower than the maximum *phonon* energy (see Fig. 2). The process results from absorption of a phonon of frequency $\omega_{ph} = \Delta/\hbar = (E_l - E_m)/\hbar$ and the spontaneous or *induced radiation* of a phonon with energy $E_l - E_{m'}$. This process, called *resonant phonon fluorescence*, as a mechanism of *spin–lattice relaxation* involving levels m and m' was considered by Orbach (1961) and investigated independently by Aminov (1962). The rate τ^{-1} of this relaxation process is characterized by its exponential dependence on the reciprocal temperature, $\tau^{-1} \propto \exp[-\Delta/(k_B T)]$, and it is inversely proportional to the level width of E_l determined by the spontaneous phonon emission. The process was observed experimentally with Ce^{3+} ions in cerium–magnesium nitrate, as well as with some ions of the iron group. Blume and Orbach (1962) studied

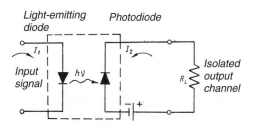

Fig. 1. Sketch of an optoisolator.

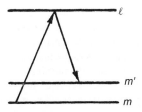

Fig. 2. Energy level diagram of the resonant phonon fluorescence process.

the spin-relaxation process (see *Paramagnetic relaxation*) for the ground state multiplet of a magnetic ion, i.e. when the value of Δ is very small. Such a process results in a temperature-dependent relaxation rate $\tau^{-1} \propto T^5$.

ORBITAL ANGULAR MOMENTUM, orbital moment

The angular momentum of a microparticle arising from its motion in a spherically symmetric force field. According to quantum mechanics, the orbital angular momentum L is quantized; namely, its magnitude as well as its projection upon an axis, arbitrarily chosen in space (z axis), can take on only certain discrete values: $L^2 = \hbar^2 l(l+1)$, $L_z = m\hbar$. Here, $l = 0, 1, 2, \ldots$ is the orbital (azimuthal) quantum number, $m = l, l - 1, \ldots, -l$ is the *magnetic quantum number*. The classification of the states of microparticles in accordance with their values of l is used in the theory of atoms and atomic nuclei, and also in the theory of scattering and collisions.

ORBITAL, HYBRID

See *Hybridization of orbitals*.

ORBITALS

Single-electron wave functions for constructing the wave function of a many-electron system. In the *single-electron approximation*, the state of every electron of the system is described by its own wave function, an orbital, reflecting the existence of an individual state of an electron in a certain effective field of the other electrons. In atoms, the individual electron states are described by *atomic orbitals*, and in molecules they are described by *molecular orbitals*.

A single-electron wave function dependent solely on spatial coordinates of a given electron is called an orbital; or, taking into account spin variables, a spin orbital. The sequence of single-electron orbitals ψ_m corresponds to the spectrum of single-electron energy levels ε_m ($m = 1, 2, 3, \ldots$). If each ith electron of the system is described by its individual spin-orbitals (i.e. $\psi_m(i)\alpha(i)$ and $\psi_m(i)\beta(i)$, allowing for two spin directions: up (α) and down (β)), and its principal quantum number m, then the single-electron approximation lets the overall system wave function

be written in the form of a *Slater determinant*:

$$\psi = \frac{1}{\sqrt{N!}} \times$$

$$\det \begin{bmatrix} \psi_1(1)\alpha(1) & \psi_1(1)\beta(1) & \ldots & \psi_n(1)\alpha(1) & \psi_n(1)\beta(1) \\ \psi_1(2)\alpha(2) & \psi_1(2)\beta(2) & \ldots & \psi_n(2)\alpha(2) & \psi_n(2)\beta(2) \\ \vdots & \vdots & \vdots & \vdots & \vdots \\ \psi_1(N)\alpha(N) & \psi_1(N)\beta(N) & \ldots & \psi_n(N)\alpha(N) & \psi_n(N)\beta(N) \end{bmatrix}.$$

The above representation of the wave function for N electrons ($N = 2n$) ensures its antisymmetric properties, with the Pauli principle automatically satisfied.

ORDER

See *Long-range and short-range order*.

ORDER, ATOMIC

See *Atomic order*.

ORDER DEGREE

See *Degree of order*.

ORDER–DISORDER LAYERED STRUCTURE
(K. Dornberger-Schiff, 1964)

Structures of so-called *order–disorder (OD) layers*, their alternation exhibiting some characteristic features of order–disorder. The OD layers can be adjacent to each other in several equivalent ways, depending on the symmetry of individual layers and neighboring layer pairs. Layers in a pair are related by operations of partial symmetry, which are valid for a given pair, but not necessarily for the system as a whole. Owing to the variation of partial symmetry for different pairs, there could be, in principle, a diversity of OD structures formed from the same layers (of one or several kinds), and symmetrically equivalent pairs of layers (so-called *adjacency condition*), both involving disordered and various strictly periodic alternations of layers.

ORDER–DISORDER PHASE TRANSITION

Phase transition involving an ion (or atom) situated in a potential well with two or more minima, which determines by its position the value of the *order parameter*. These minima are separated by maxima with energy ΔU, considerably higher than the interaction energy J between the different ions of this type (see Fig.).

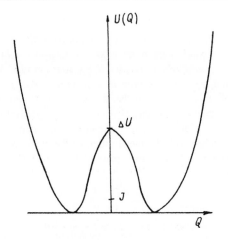

Single-particle potential showing the two minima where the ion is located at the order–disorder transition (Q is a generalized coordinate, and $\Delta U \gg J$).

Above the order–disorder phase transition temperature (T_c) both minima of the single-particle potential are populated with equal probability. In some systems above the transition point the double minimum potential reduces to a simple parabolic form characteristic of only one minimum. Below T_c the interaction J causes an asymmetry of the distributions in the potential wells, and an order parameter different from zero appears. At $T = 0$ all ions occupy equivalent minima. Sometimes, upon the variation of the external conditions, the correlation between ΔU and J becomes reversed, and a *displacive type phase transition* takes place ($\Delta U < J$). Transitions of *alloy ordering*, many *ferroelectric phase transitions* (e.g., KH$_2$PO$_4$ and NaNO$_2$), *structural phase transitions*, etc., are related to the order–disorder type.

ORDERED PHASE

See *Ordered system*.

ORDERED SYSTEM, ordered phase

From the viewpoint of solid state physics, an ordered system is one exhibiting a certain type of long-range order (see *Long-range and short-range order*), such as a *crystal*. In terms of the physics of *order–disorder phase transitions*, the term "ordered system" is used in reference to that of two thermodynamic *phases*, one of which (usually the

low-temperature one) exhibits lower symmetry. In the latter case, the high-temperature phase is called the *disordered phase* or *symmetric phase*. Examples of ordered systems of this type are: *ferromagnets, antiferromagnets, ferroelectrics, superconductors, spin glasses* and *dipole glasses, liquid crystals*, ordered alloys (see *Alloy ordering*). The distinctive feature of an ordered system is that its *order parameter* differs from zero; the generation of an ordered phase is a consequence of *spontaneous symmetry breaking* (in the absence of an external field).

ORDER IN ALLOYS

See *Alloy ordering*.

ORDERING, ISOTOPIC

See *Isotopic ordering*.

ORDERING TEMPERATURE

The temperature of a *first-order phase transition* or of a *second-order phase transition* to a phase of lower symmetry (ordered phase, see *Ordered system*).

ORDER PARAMETER (introduced by L.D. Landau, 1937)

A generalized quantity, which is a measure of the change of structure (change of internal symmetry) of a condensed many-particle system (a *solid*, in particular) in passing through a *phase transition* point. The order parameter differs from zero only in the ordered *phase* (i.e. in the phase of lowered symmetry). It undergoes an abrupt change at a *first-order phase transition* and is continuous at a *second-order phase transition*, where its first derivative with respect to a certain thermodynamic variable (temperature, pressure, density, etc.) has a discontinuity. The nature of the order parameter, its tensor dimension, and the number of its components, are determined by the physical characteristics of the phase transition. In the case of a transition to the ferro- or antiferromagnetic state, the role of the order parameter is played by the vector of spontaneous *magnetization* (ferromagnets) or the *antiferromagnetism vector* (antiferromagnets). In the case of a *ferroelectric phase transition* and ferroelastic transitions, the *polarization vector* and *strain tensor* act as order parameters, whereas in the case of a *superconducting phase transition* the role of the order parameter is played by the complex coherent wave function of the condensate of *Cooper pairs*. The order parameter is a thermodynamic variable and is the statistical average of a certain physical quantity over the state of incomplete thermodynamic equilibrium (see *Anomalous means*). The order parameter is subject to thermodynamic fluctuations, which become abnormally strong in the neighborhood of second-order phase transitions. The *order parameter space* is formed by those values of the order parameter, which correspond to the same (minimal) value of a thermodynamic potential. The geometrical properties of this space are of importance in the analysis of *topological inhomogeneities*. The concept of the order parameter is of crucial importance in the *Landau theory of second-order phase transitions*. See also *Order parameters in liquid crystals, Order parameters in alloys*.

ORDER PARAMETER PHASE

of superconducting and superfluid states

The *order parameter* is a key characteristic of the macroscopic coherence properties of the condensate of *Cooper pairs* in *superconductors* (see *Superconductivity*), or of the superfluid component in quantum Bose and Fermi liquids (see *Superfluidity*). The phase of the order parameter is independent of time and space in the absence of electromagnetic fields and currents in the superconductor, or fluxes and vortex-type motion in superfluid *helium*. This is indicative of the establishment of long-range order (see *Long-range and short-range order*) in the superconducting or superfluid state. This long-range order exhibits overall coherence, and is described by a complex order parameter or wave function $\psi = |\psi| e^{i\varphi}$ (of phase φ and absolute value $|\psi|$). *Macroscopic quantum coherence* is responsible for the presence of persistent (undamped) currents in superconductors, and superfluid flows in liquid helium. The persistent current j is proportional to the space derivative of the phase, that is $j \sim \nabla\varphi$. It also brings about *quantum vortices* in superfluid helium, *Abrikosov vortices* in type II superconductors placed in a magnetic field, the *quantization of flux, Josephson effects*, etc.

ORDER PARAMETERS IN ALLOYS

Parameters, which are a measure of *long-range and short-range order* in alloys. Consider a *binary alloy*, which consists of N_A atoms of type A and N_B atoms of type B. If there are two types of sites 1 and 2 for the atoms A and B, which are taken in amounts $N^{(1)}$ and $N^{(2)}$, respectively, then the quantity that is taken to be the parameter or degree of long-range order, or the *order parameter*, is given by the equation

$$\eta = \frac{p_A^{(1)} - c_A}{1 - \nu}, \tag{1}$$

where $p_A^{(1)}$ is the *a priori* probability of substitution of the site 1 by an atom, $c_A = N_A/N$ (where $N = N_A + N_B = N^{(1)} + N^{(2)}$) is the relative concentration of the atoms of type A, and $\nu = N^{(1)}/N$. The order parameter η ranges from zero (for the case of a *disordered alloy*, i.e. an alloy, which does not exhibit long-range order, when $p_A^{(1)} = c_A$) to the maximum value $\eta_{max} = c_A/\nu$ for $c_A \leqslant \nu$, or $\eta_{max} = (1 - c_A)/(1 - \nu)$ at $c_A \geqslant \nu$. In a fully ordered alloy (see *Alloy ordering*) of stoichiometric composition ($c_A = \nu$), $\eta = \eta_{max} = 1$. The long-range order in an alloy of n components with Q types of sites is defined by independent long-range order parameters, the number of which is equal to $S = (n - 1)(Q - 1)$. The selection of a set of long-range order parameters is determined by various factors, and parameters different from those given by Eq. (1) are used in certain cases (the long-range order parameters are usually proportional to $p_A^{(1)} - c_A$).

The short-range order of an A–B alloy with two types of atomic sites (each site of type 1 having as nearest neighbors only sites of type 2, and vice versa), and $c_A = \nu = 1/2$ may be defined by a short-range parameter (degree) of the form

$$\sigma = \frac{2N_{AB} - N^*}{N^*}, \tag{2}$$

where N_{AB} is the number of pairs of neighboring A–B atoms, and N^* is the total number neighboring atom pairs. In the general case of an n-component alloy with Q types of sites, the short-range order is defined by the *correlation* *parameters* with respect to different *coordination spheres*:

$$\varepsilon_{\alpha\alpha'}^{LL'}(\rho_l) = P_{\alpha\alpha'}^{LL'}(\rho_l) - P_\alpha^L P_{\alpha'}^{L'},$$

$$\alpha, \alpha' = A, B, C, \ldots, \tag{3}$$

$$L, L' = 1, 2, \ldots, Q,$$

where $P_{\alpha\alpha'}^{LL'}(\rho_l)$ is the probability of occupation of a site of type L by an atom of type α and simultaneous occupation of a site of type L' (at the distance ρ_l in lth coordination sphere) by an atom of type α'; P_α^L, $P_{\alpha'}^{L'}$ are *a priori* probabilities of occupation of sites L and L' by atoms α and α' respectively. In the case of alloys, which are described by Eq. (2),

$$\sigma = \eta^2 + 4\varepsilon_{AB}^{12}(\rho_1), \tag{4}$$

i.e. the short-range order parameter σ is determined in the ordered state not only by the correlation parameter, but also by the degree of long-range order η. In the case of disordered $A-B$ alloys ($\eta = 0$), the following short-range order parameters:

$$\sigma_l = 1 - \frac{P_{AB}(\rho_l)}{c_A c_B}, \tag{5}$$

are often used. These short-range order parameters are related to the correlation parameters $\varepsilon_{AB}(\rho_l) = P_{AB}(\rho_l) - c_A c_B$ by the equation $\alpha_l = \varepsilon_{AB}(\rho_l)/(c_A c_B)$ (the indices L and L' are omitted in these equations, because all sites of a disordered alloy are of the same type).

ORDER PARAMETERS IN LIQUID CRYSTALS

Parameters which are representative of *orientational order* and of partial *translational order*. Orientational order is exhibited by the long axes of rod-shaped molecules, and by the normals to the planes of disk-shaped molecules. Partial translational order occurs in *smectic liquid crystals* and columnar phases of *discotics*. In the general case, orientational order is described by the symmetrical traceless tensor $Q_{\alpha\beta}$, which may be used as a measure of a biaxial medium. In the case of, e.g., a plank-shaped molecule, the ordering of both major and minor axes can readily be imagined. This type of ordering may occur in nematics and in smectics C (biaxial lyotropic nematics, see *Nematic liquid crystal*). Furthermore, the fluctuations

of biaxiality may be of importance also in uniaxial *liquid crystals*. In the case of an uniaxial crystal, the tensor *order parameter* is of the form

$$Q_{\alpha\beta} = S\left(n_\alpha n_\beta - \frac{1}{3}\delta_{\alpha\beta}\right), \tag{1}$$

where S is the *degree of order* or *order parameter modulus*, n is a unit vector. The physical meaning of uniaxiality is as follows: there exists an axis n about which the tensor $Q_{\alpha\beta}$ remains invariant under rotation. For instance, in the case of $n = \{0, 0, 1\}$, $Q_{\alpha\beta}$ is diagonal and invariant under rotation about the z-axis. It can be seen from Eq. (1) that replacement of n with $-n$ leaves $Q_{\alpha\beta}$ unchanged, i.e. the states that are described by $Q_{\alpha\beta}$ involving n or $-n$, are equivalent, from which it follows that the medium under consideration is nonpolar (if molecules of the medium exhibit permanent dipole moments, then equal quantities of the dipoles are directed upwards and downwards). The name "order parameter" is often given to the quantity that characterizes the degeneracy of the equilibrium state of a system. In this case, such a quantity is the unit vector n, two directions of which (n and $-n$) are considered to be equivalent (such a vector is called the *director*). The order parameter $Q_{\alpha\beta}$ is the quantity, in terms of which the thermodynamic potentials are expanded in the theory of *phase transitions*. For example, the nematic-to-isotropic liquid transition is described using $Q_{\alpha\beta}$. In the case of *cholesteric liquid crystals*, the order parameter (as given by Eq. (1)) is meaningful only locally, and in the general case consideration must be given to biaxiality, which is related to the helicity of the director (the auxiliary optical axis). The *order parameter space* of a nematic (the domain of values of n) is a sphere with equivalent diametrically opposite points. The order parameter space of a cholesteric is found to possess a more intricate shape, which is why the cholesteric exhibits, e.g., a wider diversity of *defects* and *textures*. A periodic modulation of the density ρ takes place in smectics (as a consequence of one-dimensional translational order). For a smectic A the density modulation is given by the equation

$$\rho(r) = \rho_0\left\{1 + \frac{1}{\sqrt{2}}[\psi(r)\exp(iq_0 n r) + \text{c.c.}]\right\}, \tag{2}$$

where $2\pi/q_0$ is the distance between smectic layers, and the complex modulation amplitude $\psi(r)$ is the order parameter for describing the nematic–smectic A phase transition. In this case, the free energy is set up by invariants, which are derived from $Q_{\alpha\beta}$ and ψ. In the case of a smectic A–smectic C transition, the role of the order parameter is played (at given laminated structure) by the complex function $\psi = \omega(r)\exp\{i\varphi(r)\}$, where ω defines the inclination of the director with respect to the normal to smectic layers in the C-phase, and the phase φ defines the direction of the inclination (C-director). Hence, this phase transition is similar to the transition of ^{4}He to the superfluid state ("*helium analogy*"). Other mesophases and the phase transitions between them are considered in a similar manner. For instance, the density is a two-dimensional periodic function of coordinates for the columnar phases of discotics. As in the case of smectics, the role of the order parameters is now played by the corresponding amplitudes of the given modulation.

ORDER, RADIATION-INDUCED

See *Radiation-induced order*.

ORDINARY RAY, ordinary wave

One of the two light waves arising as a result of the *refraction of light* at the boundary of a *uniaxial crystal*, that is normal to the principal section of the crystal (see *Crystal optics*). The ordinary ray obeys ordinary refraction laws: it lies in the plane of incidence, and its speed of propagation and, consequently, *refractive index n_0*, are independent of the direction of propagation.

ORGANIC CONDUCTORS AND SUPERCONDUCTORS

A series of organic materials has been synthesized that exhibits at high temperatures metallic type *electrical conductivity* proceeding mainly along molecular chains. The *crystals* of such organic conductors, as a rule, consist of planar molecules packed in zigzag stacks that form chains. The extensive overlap of the electron wave functions of neighboring molecules in a stack provides for metallic conduction along the chain. The overlapping of wave functions of neighboring chains is small enough to result in a quasi-one-dimensional

nature of the electron spectrum, as well as a strong anisotropy of crystal electronic properties in directions along and transverse to the conducting chains (see *Quasi-one-dimensional crystals*). Due to the quasi-one-dimensional electron spectrum, at low temperatures these materials are usually *insulators*. In the organic compound TTF-TCNQ (tetrathiofulvalene-tetracyano-quinodimethane) as well as in a number of other organic conductors, the insulating state arises by virtue of a metal–insulator *Peierls transition*. In another class of organic conducting compounds $(TMTSF)_2X$ (TMTSF is tetramethyl-tetraselenium-fulvalene, and X are anions of various types), *metal–insulator phase transitions* are also observable. In compounds $(TMTSF)_2X$ with low-symmetry anions $X = ReO_4$, BF_4, NO_3, this transition is related to the anion ordering that leads to a superstructure with a longitudinal wave vector component $Q_\parallel = 2k_F$ (here k_F is the Fermi momentum). In crystals with symmetric anions AsF_6, PF_6, SbF_6 and TaF_6, the insulating state is related to *spin density waves*. In both cases, the metal–insulator phase transition is caused by the quasi-one-dimensional character of the electron spectrum. As in the case of a Peierls transition, the insulating state in $(TMTSF)_2X$ corresponds to the appearance of a gap in the electron spectrum at the *Fermi surface* (or some part of it). The insulating character of the electron spectrum hinders the superconductivity. The insulating state can be suppressed by reducing the distance between the chains (e.g., using external pressure), since the probability of electron transitions between neighboring chains increases in this case, and the electron spectrum becomes three-dimensional. Studies of compounds $(TMTSF)_2X$ under pressure showed that pressure can be instrumental in suppressing the transition to the insulating state, which is related to the generation of spin density waves or anion ordering in these compounds. Here, the superconducting state appears under pressure $P \approx 1.1$ GPa at $T \leqslant T_c \approx 1$ K. In the compound $(TMTSF)_2ClO_4$ superconductivity appears at $T_c \approx 1.3$ K under pressures as low as one atmosphere. This is attributed to the belief that the so-called *chemical contraction* has taken place in the given compound. Essentially the idea involves the substitution of octahedral anions PF_6, AsF_6 for tetrahedral anions ClO_4

of smaller size. This is equivalent to the action of pressure, and results in enhancing the degree of three-dimensionality of the electronic spectrum. $(TMTSF)_2X$ compounds exhibit superconductivity and spin density waves coexisting in a narrow range of pressures and temperatures.

In addition to $(TMTSF)_2X$, superconductivity has been observed in another class of organic conductors, (BEDT-TTF)X (here BEDT-TTF is bis(ethylenedithiol)tetrafulvalene, X is any anion). The structure of compounds of this class is close to that of $(TMTSF)_2X$ compounds, but the electronic spectrum in the former is more two-dimensional. The compound $(BEDT-TTF)_2I_3$ superconducts with $T_c \approx 6$–8 K.

Apparently the *phonon mechanism of superconductivity* is operative in organic superconductors. The possible realization of the *exciton mechanism of superconductivity*, suggested by W. Little (1964) for obtaining high-temperature superconductivity in organics, is problematic. Still, the great potentialities of organic synthesis sustain the hope of significantly raising their critical temperatures.

ORGANIC POLYMERS

High molecular weight organic compounds whose macromolecules consist of a great number of repeating units called *monomers*. For example, the *polymer* polyvinylchloride $(-CH_2-CHCl-)_n$ is a derivative of the monomer vinylchloride $CH_2{=}CHCl$ which has a double bond that becomes a single bond at polymerization. The degree of polymerization n, called the multiplicity, can be as high as several hundred thousand. If polymer chains contain several kinds of repeat units, the compound is called a *copolymer* (e.g., butadiene–styrene rubber). According to the type of backbone chain, organic polymers are classified into homochain and heterochain types. The backbone of *carbochain polymers* consists solely of carbon atoms. Most carbochain polymers contain saturated (single bonded) polymeric chains, but there are also polymers with unsaturated (multiple) bonds in the chains, such as *raw rubbers* and *polyacetylene*. In *heterochain polymers* the backbone consists of carbon-containing segments joined by heteroatoms. In most cases, oxygen or nitrogen serve as heteroatoms (e.g., in polyesters and polyamides). The backbone of an organic polymer can have a linear or a branching network structure. Linear polymers exhibit some specific properties, including their ability to form anisotropic highly oriented fibers and films, and to undergo extensive reversible deformations, so-called *highly-elastic deformations*. In the progression from linear polymers to branched and cross-linked types, their specific properties becomes less pronounced. Cross-linked polymers (see *Polymer cross-linking*) are insoluble, infusible, exhibit no plasticity, and are often brittle (e.g., phenolic resins). Some polymers are amorphous and some are crystalline. In addition to the highly elastic state, amorphous polymers exist in vitreous and visco-elastic states. Polymers that transform from highly elastic to vitreous below room temperature are classed as *elastomers*; and those with higher transition temperatures are plastics (see *Polymeric materials*). The ability to crystallize is characteristic of polymers with regular sequences of monomers, and a minimal amount of branching. Thus, "low-density" polyethylene with 20–50 branches per 1000 atoms has a 55–70% degree of crystallinity, and "high-density" polyethylene with 5–15 branches per 1000 atoms has 80–90% crystallinity. The density and strength characteristics are higher for the latter. Crystalline polymers are similar to plastics in their behavior. The properties of organic polymers are also determined by intermolecular interactions due to the presence of polar groups. For example, polymethylmethacrylate has a linear structure, and its transition to the highly-elastic state occurs at only 100 °C; cellulose is formed from rigid linear chains, and decomposes before reaching the highly elastic state transformation temperature. The substitution of fluorine atoms for hydrogen produces polymers (e.g., polytetrafluoroethylene, brand name Teflon) with excellent mechanical and electrical insulation properties, as well as high thermal and chemical resistance. Organic polymers can be produced by polymerization reactions (mostly, carbochain polymers) and polycondensation (heterochain polymers).

The mechanical *strength*, elasticity, and electrical insulation properties of organic polymers in combination with a low molecular weight and low production cost are responsible for their wide application. Polymers are the basis for manufacturing plastics, chemical fibers, rubbers, paintwork materials, glues, sealing compounds, ion-exchange resins. *Biopolymers* are the foundation for all living organisms; they participate in all vital function processes, and examples of them are enzymes, nucleic acids, polysaccharides, etc.

Nowadays, the so-called *conducting polymers* are intensively investigated. These are carbochain organic polymers with a backbone of unsaturated polymeric chains containing alternating single and double bonds (polyvinylenes). Owing to their backbone-bond conjugation, these polymers exhibit unique electrical properties. The simplest of them is polyacetylene $(-CH=CH-)_n$ which is a *semiconductor* with conductivity 10^{-3} $(\Omega\cdot m)^{-1}$. Introducing impurities in small quantities, either donors or acceptors (e.g., Li and I_2, respectively), results in an increase in *electrical conductivity* that can reach metallic values. For example, the maximum conductivity of iodine-doped polyacetylene is $1.5\cdot10^7$ $(\Omega\cdot m)^{-1}$. Conducting polymers are promising materials for the creation of current sources, gas-filled capacitors operating on the principle of electrochemical doping, phototransformers, solar cells, etc.

A copolymer is composed of polymer chains containing two or more types of repeat units $(A)_m$ and $(B)_n$. In many cases one polymer component $(A)_m$ is water soluble, and the other $(B)_n$ is not. A polymer nanosphere is formed when one long polymer $(A)_m$ coils up in a spherical shape and develops cross links between adjacent lengths of strands to give rigidity to the sphere. Strands of the other component $(B)_n$ attached to the surface form what is called a "hairy" nanosphere copolymer if they are short compared to the diameter of the sphere, and they form a so-called "star" polymer if they are long compared to the nanosphere diameter. A "brush" copolymer consists of a substrate with long strands of soluble $(A)_m$ projecting on one side, and long strands of insoluble $(B)_n$ on the other side.

ORGANIC SEMICONDUCTORS

A class of materials that belongs, because of their type of bonding, to molecular compounds that exhibit significant *hole conductivity* or *electronic conductivity*, and positive temperature coefficients of *electrical conductivity*. They include *molecular crystals* of organic compounds, organic *dyes*, *charge-transfer complexes*, ion-radical salts, and *polymers* with conjugated bonds. Electrical and photoelectric properties of organic semiconductors are due to their system of conjugated bonds; their *current carriers* are excited π-electrons associated with these conjugated systems. Organic semiconductors exist as *monocrystals*, *polycrystals*, *amorphous films* and powders. The resistivity at room temperature varies from 10^{18} (naphthalene, *anthracene*) to 10^{22} $\Omega\cdot cm$ (ion-radical salts).

Charge carriers in organic semiconductors with weak interactions of the van der Waals type (see *Van der Waals forces*) are able, owing to the pronounced localization of individual molecules, to interact with electron and nuclear subsystems of their environment, and form *polarons*. The *conduction bands* in organic semiconductors are narrow ($\sim$0.1 eV), and the *mobility of current carriers* is typically as low as $\sim$1 cm$^2\cdot$V$^{-1}\cdot$s^{-1}. In addition to the band mechanism of conduction, *hopping conductivity* takes place. The temperature dependence of the mobility of current carriers in model systems of organic semiconductors exhibits a low-temperature behavior similar to that in inorganic *semiconductors*, namely, T^{-n} with $n > 1$.

The trapping centers (*traps*) of charge carriers in organic semiconductors can be of either structural or impurity origin. The structural *defects* that produce a local clustering of molecules in a *crystal lattice* become traps due to the accretion of electron polarization energy of the crystal by a carrier in the defect region. Due to the strong localization of charge carriers at individual molecules in organic semiconductors, structural defects such as *dislocations*, *stacking faults*, orientation and conformational defects, etc., play a more significant role in processes of charge carrier transfer in comparison to inorganic semiconductors. The impurity traps of charge carriers in organic semiconductors are produced by impurity molecules with a greater *electron affinity* or lower ionization potential than those of host molecules. There are charge carrier traps localized in the vicinity of an impurity molecule even when the impurity molecules themselves produce no traps. This kind of trap results from the perturbation of corresponding energy levels of host molecules by a neighboring impurity molecule.

Organic semiconductors find use in *microelectronics* as light-sensitive materials for *non-silver photography* and in other applications. The study of organic semiconductors is important to gain insight into the processes of transfer and transformation of energy in complicated biological systems.

ORGANOPLASTICS

Plastics (see *Polymeric materials*) reinforced with synthetic fibers. Organoplastics possess relatively low *density* and low *thermal conductivity*, high *strength*, good electrical insulation properties, as well as high atmospheric, water, and chemical resistance; they are readily machinable. Organoplastics are used in radio-engineering and for *construction materials* in the aircraft, automobile, and other industries.

ORIENTATIONAL LINE BROADENING

Inhomogeneous broadening that arises due to different orientations of *paramagnetic centers* relative to an external magnetic field B. Randomly oriented paramagnetic centers appear in powders, glasses, frozen solutions, and crystals with a mosaic structure (see *Mosaic crystals*). In many cases the population density of these centers (glasses, powders) directed along any solid angle is the same. There is an additional condition for orientational broadening to appear. The resonance value of the magnetic field B_r should depend on the angles specifying the direction of the field B relative to the principal axes of the paramagnetic centers, i.e. the spectral lines of *monocrystals* should exhibit an *angular dependence of spectra*. In this case the variously oriented paramagnetic centers will have different values of B_r, resulting in orientational broadening with an asymmetrical *line shape* that depends on the type of *spin Hamiltonian* (see the Figure). In some cases of a nonmonotonic angular dependence of B_r, extra absorption peaks appear in electron paramagnetic resonance spectra, which facilitate their analysis. On the whole, the interpretation of spectra with orientational broadening is a much more difficult task than the analysis of the angular dependence

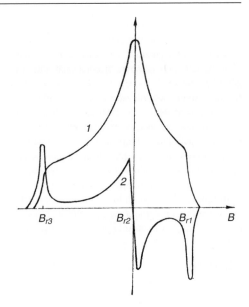

Absorption line shape (1) and its derivative (2) in the case of a spin Hamiltonian H with orthorhombic symmetry of the form $H = \mu_B \sum_{i=x,y,z} g_{ii} S_i B_i$, where $B_{ri} = h\nu/(\mu_B S g_{ii})$ (μ_B is the Bohr magneton, g_{ii} are g-tensor principal value components, S_i are spin operators, and ν is the microwave frequency).

of each individual line in a spectrum from a paramagnetic center in a monocrystal.

ORIENTATIONAL ORDER

A characteristic of a system where, in principle, the spontaneous breaking of rotational symmetry at *phase transitions* is possible (see also *Spontaneous symmetry breaking*). To describe orientational order, the concept of *order parameter* is introduced. As a rule, *short-range orientational order* appears at temperatures above the phase transition point, and *long-range orientational order* appears below this point (see *Long-range and short-range order*). For example, *ferromagnets*, *ferroelectrics*, and nematics (see *Nematic liquid crystal*) display long-range orientational order with the corresponding order parameters *magnetization*, *polarization*, and *director*. In terms of a qualitative description, the orientational order in a ferromagnet and a ferroelectric can be thought of as the aligning of magnetic and electric dipoles along some (particular) axes over the whole (infinite) space. In nematics, the elements ordered along the director are the long axes of rod-shaped molecules or the normals to planes of disk-shaped molecules. True long-range orientational order exists when the average over all orientations of, e.g., magnetic moments or long axes of molecules, differs from zero. However, this can not always be achieved. For instance, in two-dimensional magnets or nematics, true long-range orientational order is impossible, but a quasi-long-range one can exist. The latter is understood as the falling-off of the corresponding correlation functions as a power of distance in contrast to the short-range orientation order (e.g., *paramagnets* or isotropic phase of nematics) when the correlators fall off exponentially.

ORIENTATIONAL PHASE TRANSITION

A *phase transition* where the orientation of the *order parameter* vector η (magnetization, polarization, and so on) changes as the temperature, pressure, or external field changes.

ORIENTATION EFFECTS in irradiated monocrystals

A sharp dependence of the efficiency of various processes during the passage of charged particles or high-energy *photons* through *monocrystals* on the angle ψ of the particle (photon) beam direction with respect to the directions of atomic chains or planes. The following orientation effects are observable in the *channeling* of positive *ions*: abnormally deep penetration of ions into monocrystals; reduction of the nuclear reaction efficiency (e.g., of protons or alpha particles with atoms of the *crystal*); decrease of characteristic X-ray radiation intensity; and decrease of *backscattering* of ions when the the angle of incidence ψ is limited to values below the critical angle ψ_k. Other orientational effects include the *channeling of ion beams* and the related *shadow effect*, the anisotropy of ion–ion and *ion–electron emission* and cathode *sputtering*, and the appearance of *Vener spots* during sputtering (see *Focuson*). During the channelling of electrons and positrons intense hard radiation can appear. An orientational effect at the incidence of a beam of electrons or X-rays at the Bragg angle is their anomalous passage through monocrystals due to interference between transmitted and scattered beams, when the maxima of the resultant beam density are concentrated in interfacial regions of the crystal (see *Anomalous passage of X-rays*).

ORIENTATION RELATIONS

A way to express the geometrical correspondence of two conjugate crystal bodies or *phases*. The orientation relations were established by Heidinger in 1827 for intergrown minerals of hematite and *magnetite*; the basal plane (001) in hematite is parallel to the (111) plane in magnetite. The orientation relations (OR) are expressed in terms of parallel or near-parallel close-packed planes and directions of the *crystal lattices* of these bodies or phases. These relations are found with the help of X-ray, electron diffraction, electron-microscopic, and optical methods. A single relation corresponds to a few crystallographically equivalent orientations of *crystals* of some phase with respect to a crystal of another phase. The *Nishiyama orientation relation* (Z. Nishiyama, 1934–1935) corresponds to 12 equivalent orientationally variant martensitic crystals. For example, a variant of this relation is $(001)_\alpha \parallel (111)_\gamma$, $[0\bar{1}1]_\alpha \parallel [\bar{2}11]_\gamma$; and the remaining 11 variants involve permutation of Miller indices (see *Crystallographic indices*). The *Kurdyumov–Sachs orientation relation* (G.V. Kurdyumov, G. Sachs, 1930) gives the directional interrelation of crystal lattices of *austenite* (A) and *martensite* (M) in *steel* and *iron alloys*. In an austenite monocrystal, the martensite lattice can be oriented in 24 ways, e.g., $(111)_A \parallel (10\bar{1})_M$, $[101]_A \parallel [111]_M$, where the close-packed planes and directions of austenite and martensite are parallel; in this case *twinning structures* may form. An intermediate OR between those of Nishiyama and Kurdyumov–Sachs is the *Greninger–Troiano orientation relation* (A.B. Greninger, A.R. Troiano, 1949), according to which 24 variants of mutual orientations of austenite and martensite exist, e.g., $(111)_A \parallel (011)_M$, $[5, 12, 17]_A \parallel [7, \bar{17}, 17]_M$. Orientation relations are found in practically every diffusion and *diffusion-free transformation* in solids, oxidizing reactions, directional and eutectic *crystallizations*. Orientations often tend to spread (deviations from planar and direction parallelism) due to crystal structure defects and external and internal stresses. The strictest adherence to orientations (within an error of $\sim 1°$) is found in *martensitic transformations*. For the transition from a face-centered cubic (FCC) to a body-centered cubic (BCC) (or body-centered tetragonal) lattice, depending on the composition of the *alloys*, the Kurdyumov–Sachs, Greninger–Troiano or Nishiyama OR is valid. (The latter appears in iron–nickel alloys and differs from the Kurdyumov–Sachs OR by a turn of the martensite lattice through $5°16'$ about the direction $[001]_\alpha$.) A similar transition in *thin films* leads to the *Bain orientation relation* $(001)_\alpha \parallel (001)_\gamma$, $[100]_\alpha \parallel [110]_\gamma$ (E.C. Bain, 1924). At the transition of a BCC phase to a hexagonal close-packed (HCP) phase, *Burgers orientation relation* (W.G. Burgers, 1936) applies most often: $(110)_{BCC} \parallel (0001)_{HCP}$, $[\bar{1}10]_{BCC} \parallel [11\bar{2}0]_{HCP}$.

Orientation relations are used for establishing the crystalline geometrical mechanism of the rearrangement of one crystal lattice into another during a solid state transformation. For example, the rearrangement of a FCC phase into a BCC (or BC-tetragonal) phase with an invariant plane involves the intrinsic (Bain) deformation to rearrange the initial lattice to the final one; a deformation with an invariant lattice by means of *shear* and *twinning of the crystal*; and, finally, the rotation of the rearranged region through an angle that ensures the adaptation of the rearrangement mechanism to the external conditions, in particular, with the elastic action of the surrounding initial phase. It is this rotation that determines the orientation relation. No rotation of the rearranged region is present at the transition of the FCC phase to close-packed martensite phases (in particular, those with the HCP lattice) that are formed by means of the shear of close-packed planes (111). In this case, the ORs are determined by the plane and the shear direction (e.g., $(111)_{FCC} \parallel (0001)_{HCP}$, $[10\bar{1}]_{FCC} \parallel [2\bar{1}\bar{1}0]_{HCP}$). For transformations with a predominance of *surface energy*, those orientation relations are present that correspond to the minimum at the cross-section of a polar diagram of the surface energy, and provide for the best interlinking of the lattices of two crystal phases over the plane of their contact.

ORTHOFERRITE (fr. Gr. $o\rho\theta o\varsigma$, straight, and ferrite)

Ferrites with rhombically distorted variants of the *perovskite* ($CaTiO_3$) structure. The general chemical formula is $RFeO_3$, where R is *yttrium* or a *rare-earth element*. The chemical unit cell coincides with the magnetic one, and contains four formula units. The *space group* is $Pbnm$ (D_{2h}^{16}). The R^{3+} and Fe^{3+} ions are in crystallographic site positions $4c$ and $4b$, respectively. The system of Fe ions experiences magnetic ordering at $T \sim 600$ K. The type of ordering is *antiferromagnetic*, usually with weak *ferromagnetism*. The system of rare-earth element ions is ordered at liquid helium temperatures. With the temperature lowered, spontaneous spin-reorienting *magnetic phase transitions* occur in many orthoferrites.

ORTHOGONALITY, STRONG

See *Strong orthogonality*.

ORTHOGONALIZED PLANE WAVES, OPW method

A *band structure computation method* that uses plane waves, orthogonalized to wave functions of the atomic core, as the basis functions. When the basis of the OPW method is over-determined, use is often made of its modification, the *completely orthogonalized plane wave method*. In the latter case, as many orthogonalized plane waves are removed from the basis set as is necessary to eliminate over-determination. The OPW method was historically the foundation for the *pseudopotential method*.

ORTHORHOMBIC SYSTEM (fr. Gr. $o\rho\theta o\varsigma$, right, and $\rho o\mu\beta o\varsigma$, rhombus)

Crystallographic system determined by the presence of three second-order symmetry axes in crystals; it belongs to the orthorhombic *crystal class*. The rectangular basis coordinate system axes of this system 2 (or $\bar{2}$) are directed along the edges of the *unit cell* with parameters $a \neq b \neq c$, $\alpha = \beta = \gamma = 90°$. The orthorhombic system includes 4 *Bravais lattices*, 3 *point groups*, and 59 *space groups*.

OSCILLATIONS, FRIEDEL

See *Friedel oscillations*.

OSCILLATIONS, KHAIKIN

See *Khaikin oscillations*.

OSCILLATIONS, SELF-

See *Self-oscillation.*

OSCILLATIONS, ZERO SPIN

See *Zero spin oscillations.*

OSCILLATOR STRENGTH

Dimensionless quantity describing the probability of a quantum transition during a radiation process (see *Radiative quantum transition*) and *light absorption*. For electronic transitions between levels of energies E_i and E_j arising from a change of the system *dipole moment* the oscillator strength f_{ij} is given by

$$f_{ij} = -\frac{8\pi^2 m}{3he^2}\frac{1}{g_i}v_{ij}\sum_{\alpha\beta}|\bar{p}_{i\alpha,j\beta}|^2,$$

where m, e are the electron mass and charge, respectively, g_i is the statistical weight of the initial state, $v_{ij} = (E_i - E_j)/h$, and $p_{i\alpha,j\beta}$ is the matrix element of the electric *dipole moment* operator between states $(i\alpha)$ and $(j\beta)$. For $E_i > E_j$ (radiation), the oscillator strength is negative, and for $E_i < E_j$ (absorption) it is positive

$$f_{ij} = -\frac{g_j}{g_i}f_{ji}.$$

The oscillator strength obeys the sum rule $\sum_j f_{ij} = N$, where N is the number of electrons in the quantum system. The term oscillator strength arises from the fact that the contribution of transition $i \rightarrow j$ to the polarizability can be considered as the polarizability of an oscillator multiplied by f_{ij}. Therefore, this term not only applies to electronic but also to other types (in particular, oscillatory) of transitions. The oscillator strength is determined by measuring the integrated absorption, by curves of *anomalous dispersion of light*, and by other methods.

OSCILLISTOR

Oscillations of current in the electron–hole plasma of a semiconductor (see *Solid-state plasma*) placed in parallel electric and magnetic fields. This effect was first detected in *germanium*. The phenomenon of current instability results from the *Kadomtsev–Nedospasov screw instability* (B.B. Kadomtsev and A.V. Nedospasov,

1960). The latter is caused by the Hall drift of the plasma in mutually perpendicular magnetic and electric fields, which arises as a result of the shift of electron and hole screw perturbations of density in a constant electric field. The equality of the Hall and diffusion flows corresponds to the *threshold of excitation of the instability*. Therefore, the instability is excited in sufficiently strong electric and magnetic fields: $\sim$30 V/cm and $\sim$1 T. A related phenomenon is the oscillation of a bar of semiconducting material such as germanium when subjected to a magnetic field and a direct current.

OSMIUM, Os

A chemical element of Group VIII of the periodic system with atomic number 76 and atomic mass 190.2. Natural osmium consists of stable isotopes ^{184}Os (0.018%), ^{186}Os (1.582%), ^{187}Os (1.64%), ^{188}Os (13.27%), ^{189}Os (16.14%), ^{190}Os (26.38%) and ^{192}Os (40.97%). Outer electronic shell configuration is $4f^{14}5d^66s^2$. Successive ionization energies are 8.7, 17.0 eV. Atomic radius is 0.133 nm; radius of Os^{4+} ion is 0.065 nm, of Os^{6+} is 0.069 nm. Electronegativity is 1.52. Oxidation state is +8, +6, +4; less often, +3, +2.

In free form, osmium is a bluish-gray *metal* (heaviest known metal), very hard and brittle. It has a hexagonal close packed lattice with parameters $a = 0.273445$ nm and $c = 0.431745$ nm at 300 K, space group $P6_3/mmc$ (D_{6h}^4). The density is 22.6 g/cm^3, $T_{\text{melting}} = 3318$ K, $T_{\text{boiling}} \sim 5300$ K. Binding energy is -8.11 eV/atom at 298.15 K. Heat of melting is 29.3 kJ/mole; heat of evaporation is 678 kJ/mole, and heat of sublimation is 728.5 kJ/mole. Specific heat is 0.129 kJ·kg^{-1}·K^{-1} at 298 K; Debye temperature is 278.5 K; the linear thermal expansion coefficient is $5.6 \cdot 10^{-6}$ K^{-1} along the main axis $\underline{6}$ and $3.9 \cdot 10^{-6}$ K^{-1} transverse to it (at room temperature), the coefficient of thermal conductivity is 87.1 W·m^{-1}·K^{-1} at 273 to 373 K. Adiabatic bulk modulus is 375 GPa, Young's modulus is 559 GPa, rigidity is 223 GPa, Poisson ratio is 0.25 at 298 K; Vickers hardness is 2.9 to 6.6 GPa for annealed osmium at 298 K and 1.4 to 4.0 GPa at 1473 K (range due to anisotropy of elasticity). Osmium metal is practically impossible to deform and difficult to volatilize in high vacuum (saturated vapor pressure at 2773 K is 0.0133 Pa). Ion-plasma frequency of osmium is 16.127 THz; coefficient of the linear term in low-temperature electronic specific heat is 2.26 mJ·mole^{-1}·K^{-2}; electric resistivity is 96.6 nΩ·m at 298 K; temperature coefficient of electrical resistance 0.0042 K^{-1} from 273 to 373 K; superconducting transition temperature is 0.655 K, critical magnetic field is 6.5 mT (at 0 K). Osmium is paramagnetic with magnetic susceptibility +13.1·10^{-6} cm^3/mole; nuclear magnetic moment of isotope ^{189}Os is 0.651 nuclear magnetons.

Owing to the high hardness, corrosion and wear resistance of osmium, as well as its natural and artificial alloys with other *platinum group metals*, it is employed in various instruments. Osmium (and its compounds) is a good catalyst for many kinds of synthesis.

OVERHAUSER EFFECT (A.W. Overhauser, 1953)

Dynamic nuclear polarization caused by spin relaxation (see *Paramagnetic relaxation*) of an unpaired electron center coupled to nuclear spins by the *hyperfine interaction*. In the case of the Overhauser effect, dynamic nuclear polarization is achieved owing to the electron spin subsystem that was disturbed from thermodynamic equilibrium by external alternating fields. The electron spin *relaxation* involves nuclear spin forbidden relaxation transitions with a certain probability. Since under nonequilibrium conditions the balance of relaxation paths is disturbed, the system is dominated by relaxation processes characterized by a certain change ΔM of the projection M of the electron spin S. The *selection rules* relate the change Δm of the projection m of nuclear spin I with ΔM; this leads to a predominantly nonequilibrium population of particular nuclear spin sublevels. In the case of the Overhauser effect, the subsystem is saturated by the resonance oscillating field, while the electron nuclear relaxation is caused by the contact interaction of the $S \cdot I$ type that does not change the resultant projection of spins: $\Delta M + \Delta m = 0$. The maximum possible dynamic nuclear polarization in the case $S = I = 1/2$ is $p_m = -P_0 = \tanh[\hbar\omega_s/(2k_B T)]$, which is practically coincident in value with the equilibrium polarization P_0 of electron spins in a magnetic field with Zeeman frequency $\omega_S \gg \omega_I$ (ω_I is the splitting of nuclear spin levels). If the electron–nuclear relaxation is controlled by the anisotropic part of the hyperfine interaction with selection rules $\Delta M + \Delta m = \pm 2$, then p_m equals $P_0/2$.

OVERHEATING, METAL
See *Metal overheating*.

OXIDATION, oxide coating

Formation of a solid oxide phase on a metal or semiconductor surface. It proceeds via the electrochemical interaction of ambient oxygen molecules with atoms of the substrate. The rate of formation of oxide decreases as the thickness of the oxide film grows; therefore, the steady-state thickness of the oxide is determined by its transparency to penetration by oxygen molecules. The formation of a dense film (Al_2O_3) promotes *surface passivation*. Loose oxide films (Fe_2O_3, CuO etc.) hinder further oxidation, thereby facilitating surface *corrosion*. The controlled *coating* of a semiconductor surface with oxides is widely used in semiconductor engineering to produce insulating and *passivating coatings*. To this end, thermal and anode oxidation is applied.

OXIDATION, THERMAL
See *Thermal oxidation*.

OXIDE MAGNET

A magnetically hard *ferrite*; a *permanent magnet* made of these materials. Oxide magnets are manufactured from Fe, Ba, Co, and Sr oxides, and most often from barium ferrite $BaO·6Fe_2O_3$ in two modifications, isotropic and anisotropic (the latter obtained by pressing powder in a magnetic field). Some properties of oxide magnets are given in Table 1 of *Hard magnetic materials*.

OXYGEN, O

Chemical element of Group VI of the periodic system with atomic number 8 and atomic mass 15.9994. Naturally occurring oxygen is composed of three stable isotopes: ^{16}O (99.762%), ^{17}O (0.038%), and ^{18}O (0.200%). The tetrahedral atomic radius is 0.066 nm (coordination number 4), the octahedral radius is 0.074 nm (coordination number 6), and the ionic radius of O^{2-}

is 0.136 nm. The electronic configuration of the outer shell is $2s^2 2p^4$. Successive ionization energies are: 13.618, 35.117, and 54.90 eV. The electron affinity is 1.467 eV, and the oxidation state is -2, more rarely, -1.

Oxygen in the free state usually forms the molecule, O_2, more rarely, ozone, O_3 (10^{-4} %). O_2 is one of the few triplet ground state molecules (spin $S = 1$); the dissociation energy is 493.6 kJ/mol (at 0 K), and the distance between the nuclei is 0.120735 nm. The specific heats of O_2 are: $c_p = 29.27$ J·mol^{-1}·K^{-1}, $c_v = 20.5$ J·mol^{-1}·K^{-1} (at 273 K). The density of gaseous oxygen is 1.42897 g/dm^3 (at 273 K and standard pressure), that of liquid oxygen is 1.1321 g/cm^3 (at $T_{boiling}$), and that of solid oxygen is 1.46 g/cm^3 (at 20.3 K); $T_{melting} = 54.43$ K, and $T_{boiling} = 80.15$ K. The triple point parameters are: $T_{triple} = 54.36$ K and $p_{triple} = 146$ Pa; the heat of vaporization is 6.82 kJ/mol, and the heat of fusion is 0.443 kJ/mol. At 273 K, the thermal conductivity is 24 mW·m^{-1}·K^{-1}, the dielectric constant is 1.000547, the viscosity is 18.9 μPa·s, and the susceptibility χ (at 293 K) is $107.8 \cdot 10^{-9}$.

Solid oxygen belongs to the group of *molecular crystals* or cryocrystals. It combines, in a unique fashion, the properties of a molecular crystal and of a *magnetic substance*. Solid oxygen is the only naturally occurring homomolecular magnetic compound with a direct *exchange interaction*, this interaction comprising a substantial portion of the total lattice energy. This fact is responsible for a very close coordination of the magnetic and lattice properties of solid oxygen, as well as for a very rich spectrum of collective excitations: acoustic *phonons*, librons, *magnons*, vibrons, *excitons*, *biexcitons*, and various combinations thereof (multi-particle excitations). Due to the unique character of its properties solid oxygen has been a popular material for study in *low-temperature physics*. At the saturation vapor pressure solid oxygen exists in three crystallographic modifications.

The high-temperature cubic γ-phase (space group $Pm\overline{3}n$, O_h^3) has a high degree of disorder of its molecular orientations. The orientationally ordered rhombohedral (trigonal, $R\overline{3}m$, D_{3d}^5) β-phase ($T_{\gamma\beta} = 43.80$ K) and monoclinic ($C2/m$, C_{2h}^3) α-phase ($T_{\beta\alpha} = 23.88$ K) have a similar pattern of molecular packing since both phases have layered structures. The O_2 molecules in each layer form densely packed planes with the molecular axes perpendicular to these planes. The monoclinic lattice distortion at the $\beta \rightarrow \alpha$ transition arises from the magnetic ordering. The α-O_2 modification is a quasi-two-dimensional *antiferromagnet*. The interlayer exchange in β-O_2 results in a strong antiferromagnetic short-range order, even though there is no long-range magnetic order in this phase. The γ-phase and liquid O_2 are *paramagnets*. Studies of solid oxygen under pressure have revealed several new phases with an orientational structure similar to those of α- and β-O_2.

P p

PAIRING, TRIPLET

See *Triplet pairing*.

PAIRWISE ADDITIVE APPROXIMATION

One of the methods of approximately calculating the total energy of molecular systems. The pairwise additive approximation is used in situations where it is possible to construct the model of the *pairwise interaction of atoms*, in which the many-atomic interactions are effectively taken into account. In its general form the total potential energy of system of N atoms is presented by the sum of double, triple, etc., individual potential energy terms

$$U = \frac{1}{2!} \sum_{i \neq j}^{N} \varphi_{ij} + \frac{1}{3!} \sum_{i \neq j \neq k}^{N} \varphi_{ijk} + \cdots$$

$$+ \frac{1}{m!} \sum_{i \neq j \neq \cdots \neq m}^{N} \varphi_{ij\ldots m},$$

which in some cases can be written in the form

$$U = \frac{1}{2} \sum_{i \neq j}^{N} E_{ij}^{*},$$

where

$$E_{ij}^{*} = \varphi_{ij} \left\{ 1 + \frac{1}{3!} \sum_{k \neq i, j} \frac{\varphi_{ijk}}{\varphi_{ij}} \right\};$$

this includes averaged interactions of higher orders. Introduction of the pairwise additive approximation is especially useful in cases where notions of isolated *chemical bonds* are valid, as, e.g., in covalent structures. The picture of two-centered bonds corresponds to real singularities of electron behavior in saturated systems, where the basic wave functions ψ_i of the ground state can be expressed in terms of one-electron functions χ_i localized at two-center bonds. There is a unitary transformation to express the functions ψ_i as a linear combination of basis functions χ_i localized at the bonds, so the total system energy and charge distribution remain invariant under the transition to the new basis.

PAIRWISE INTERACTION OF ATOMS

An approximation sometimes used to describe the interaction between *atoms* (*ions*) in solids, as well as in liquids and gases. It assumes that the potential energy of a solid (*adiabatic potential*) U within the *adiabatic approximation* may be represented in the form of the sum of potential energies (potentials) of pairs of atoms (ions), the coordinates of which are $\boldsymbol{R}_l$ and $\boldsymbol{R}_{l'}$:

$$U = \frac{1}{2} \sum_{ll'} \varphi(\boldsymbol{R}_l, \boldsymbol{R}_{l'}).$$

This ignores interactions of three or more atoms (ions), a neglect which can not always be justified. Besides that, the approximation of pairwise interaction of atoms fails to take into account energy terms which are not dependent on the solid structure, but which are functions only of its volume. Such a neglect is not warranted in the case of a *metal*, although it does provide good results for crystals with *ionic bonds*, with *covalent bonds* and with van der Waals *chemical bonds* (see *Van der Waals forces*). In the simplest form of the pairwise interaction approximation, the potential φ is assumed to be a function of the interatomic distance $|\boldsymbol{R}_l - \boldsymbol{R}_{l'}|$ only. This approximation restricts the *interatomic interaction potentials* to *central forces* only.

The pairwise interaction approximation is valid when analyzing the properties of solids in the

945

context of the *harmonic approximation*. The inter-atomic interaction potential in this case is inconsistent with a simple two-particle potential, but takes into account the contributions of many-ion potentials. In this connection, the concept of *effective pairwise potential* is introduced, the parameters of which are determined experimentally. This interatomic potential can account for the prominent features of an interatomic interaction in general: short-range repulsion and comparatively long-range attraction.

PALLADIUM, Pd

An element of Group VIII of the periodic table with atomic number 46, atomic weight 106.42. Natural palladium consists of isotopes with following mass numbers: 102 (1.02%), 104 (11.14%), 105 (22.33%), 106 (27.33%), 108 (26.46%), 110 (11.72%). Electronic configuration of the outer shell is $4d^{10}$. Sequential ionization energies (eV): 8.336, 19.428 and 32.92. Atomic radius 0.137 nm. The radii of Pd^{2+} and Pd^{4+} ions are respectively 0.080 nm and 0.065 nm. Oxidation states: +2, +4, less often +3. Electronegativity value ≈ 2.2.

Free palladium is a silvery-white soft *metal*. It exhibits a face-centered cubic structure, $a = 0.38902$ nm (at 296 K). Density 12.02 g/cm³ at 293 K, $T_{melting} = 1827$ K, $T_{boiling} \approx 3210$ K. Heat of melting 16.71 kJ/mol, heat of evaporation 353 kJ/mol; specific heat 0.2445 kJ·kg⁻¹·K⁻¹ (at 298 K); the coefficient of thermal conductivity 71.2 W·m⁻¹·K⁻¹. Coefficient of linear thermal expansion $11.1 \cdot 10^{-6}$ K⁻¹ (at 273–373 K). Resistivity 99.6 nΩ·m (at 298 K). The temperature resistance coefficient 0.00379 K⁻¹ (at 273–373 K). Electron work function 4.99 eV. Palladium is paramagnetic with the specific magnetic susceptibility $5.231 \cdot 10^{-6}$ at room temperature. Saturation vapor pressure is 13.33 MPa (at 1427 K) and 1333 MPa (at 1773 K). Annealed palladium exhibits the elastic modulus of 121.128 GPa, the shear modulus of 49.98 GPa, ultimate tensile strength ≈ 189 MPa, relative elongation from 25 to 40%. The value of Brinell hardness 480.2 MPa. Palladium is amenable to forging, punching, rolling to thinnest sheets, stretching to thin wires, it lends itself readily to polishing and welding. Cold strain hardens palla-

dium noticeably. Thus, 50% reduction increases the ultimate tensile strength to 325.5 MPa; the value of hardness increases by a factor of 2–2.5. The annealing, which follows the reduction, softens the metal again. Alloying elements, particularly *ruthenium* and *nickel*, also increase the value of palladium harness. Palladium exhibits the following values of adiabatic elastic moduli (GPa): $c_{11} = 227.10$, $c_{12} = 176.04$, $c_{44} = 71.73$ at 300 K. The characteristic feature of palladium is its ability to reversibly absorb hydrogen, whereby up to 900 volumes of the gas are taken by a single volume of metal; absorption of hydrogen causes considerable increase of volume and embrittlement of palladium. On moderate heating in vacuum the metal releases the absorbed hydrogen.

Pure palladium is used for the manufacture of chemical equipment; gold-, platinum- and radium-bearing alloys are used in temperature controllers, thermocouples, etc. The chief application of palladium is in medicine, particularly in dental work.

PAPKOVICH–NEUBER SOLUTION
(P.F. Papkovich, 1932; H.Z. Neuber, 1934)

A general solution of the equations of elastic medium equilibrium, which is expressed in terms of the scalar function φ and vector function $\boldsymbol{\psi}$. If the equilibrium equation is written as $\mu \nabla^2 \boldsymbol{u} + (\lambda + \mu) \operatorname{grad} \operatorname{div} \boldsymbol{u} + \rho \boldsymbol{F} = 0$, where $\boldsymbol{u}$ is the *displacement vector*, λ, μ are the elastic *Lamé coefficients*, ρ is the *density*, $\boldsymbol{F}$ is the mass density of the *bulk force*, then the Papkovich–Neuber solution is of the form $\boldsymbol{u} = \operatorname{grad}(\varphi + \boldsymbol{r} \cdot \boldsymbol{\psi}) - 4(1 - \nu)\boldsymbol{\psi}$, where ν is the *Poisson ratio*, and the functions φ and $\boldsymbol{\psi}$ satisfy the equations

$$4(1 - \nu)\nabla^2 \varphi + \frac{\rho}{\mu} \boldsymbol{r} \cdot \boldsymbol{F} = 0,$$

$$4(1 - \nu)\nabla^2 \boldsymbol{\psi} = \frac{\rho}{\mu} \boldsymbol{F}.$$

It follows from the latter equations that the scalar function φ and the components of the vector function $\boldsymbol{\psi}$ are harmonic functions in the absence of bulk forces ($\boldsymbol{F} = 0$).

PARACONDUCTIVITY in superconductors

Increase of conductivity of a normal *metal*, which takes place as the *superconducting phase transition* point $T = T_c$ is approached from above. This increase generally exists, but is quite small. It is due to superconducting fluctuations (see *Fluctuations in superconductors*), and the excess current depends on the magnitude and lifetime of the fluctuations (*Aslamazov–Larkin contribution*, 1968). The excess conductivity depends on the dimensionality of the superconductor, and in the case of *films* is determined by a universal relation $\sigma' = [e^2/(16\hbar d)][T/(T - T_c)]$, where d is the film thickness. This result holds true for temperatures $(T - T_c)/T_c \gg \varepsilon_2$, where ε_2 is the width of the critical region, and is in good agreement with a large number of experimental data. In some cases, however, there is no accord between theory and observation, and it is therefore necessary to consider another mechanism of paraconductivity, which is related to the interaction between normal excitations and superconducting fluctuations: so-called *Maki–Thompson contribution* (K. Maki, 1968; R.S. Thompson, 1970).

PARACRYSTAL

A crystal with a deformed or distorted *crystal lattice* structure that broadens X-ray diffraction lines. A model representation of such an imperfect crystal is formulated in terms of three parameters (basis vectors) a, b and c, which are given independent, perhaps random, values. Statistical relations of the radius vector $R_{m_a m_b m_c}$ may be obtained by adding together and averaging m_a vectors a, m_b vectors b and m_c vectors c, as independent random quantities. The diffraction pattern of a paracrystal is qualitatively analogous with that of a corresponding crystal containing defects, but with broadened diffraction lines (see *Defects in crystals*). The above paracrystal model is used in the analysis of the structure of fibrous crystals, mosaic type crystals, crystals with dispersed *coherent precipitates* of a second phase, those which are correlated to a *macrolattice*, etc.

PARAELECTRIC PHASE

A crystallographic modification of a *ferroelectric* that does exhibit spontaneous polarization, and goes over into the *ferroelectric phase* at a temperature that is called the *Curie point* in analogy with *ferromagnets*. If the Curie point is approached from the paraelectric phase, then the temperature dependence of the dielectric *susceptibility* is described by the *Curie–Weiss law*. If the symmetry of the paraelectric phase is known, then crystallographic considerations allows one to predict the symmetry of plausible ferroelectric phases. The symmetry of the paraelectric phase is a factor in determining the macroscopic properties in the ferroelectric phase. For example, the spontaneous birefringence of ferroelectrics in the ferroelectric phase may vary either quadratically or linearly with the spontaneous polarization depending on whether the ferroelectric crystal is centrally symmetric (e.g., $BaTiO_3$) or noncentrally symmetric (e.g., KH_2PO_4) (see *Ferroelectricity*).

PARAELECTRIC RESONANCE

Resonance absorption of the power of an alternating electric field by a system of tunneling levels of electric dipoles; a method of investigating systems with impurities, which exhibit electric *dipole moments*. Paraelectric resonance was first observed in dipole molecules, and then in a number of crystals with *noncentral ions*. The distance between tunneling levels is ordinarily varied by the application of an external constant electric field, which allows one to "scan" across the *resonance* region by varying the DC field at the constant frequency of the AC field (the latter frequency is usually within the ultrahigh-frequency range). Paraelectric resonance is observed in crystals of $KCl:Li^+$, $RbCl:Ag^+$, $KBr:Li^+$, etc.

The spectrometers used to observe paraelectric resonance are, in many aspects, similar to electron paramagnetic resonance (EPR) spectrometers. The main distinction is in the design of the *resonator*, which accommodates subjecting the sample to uniform alternating and constant electric fields. The intensity of the constant (applied) electric field ranges from zero to the breakdown value. The measurements are usually carried out within the temperature range 1.4–4.2 K, since higher temperatures cause strong broadening of paraelectric resonance lines. *Narrowing of magnetic resonance line* may be achieved by thoroughly cleaning the samples. The concentration of noncentral ions does not usually exceed 10^{17} cm^{-3}. The paraelectric resonance spectrum of Li^+ in KCl is

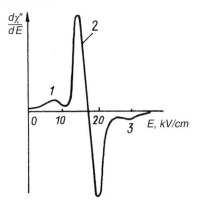

Dependence of the derivative of the ultrahigh-frequency power absorption on the applied electric field E at the noncentral ion ^{7}Li in KCl ($\nu_{klystron} = 63$ GHz, $T = 4.2$ K, $N_{Li} = 7 \cdot 10^{16}$ cm^{-3}): (1, 3) forbidden transitions, (2) allowed transition, χ'' is the imaginary part of the electric susceptibility.

given in Fig. as an example. The locations of the resonance lines provide the magnitudes of the tunnel splitting Δ, and the electric dipole moment d of the impurity. For ^{7}Li$^+$ in KCl $\Delta \approx 32$ GHz, $d \parallel [111]$, $|d| \approx 18.5 \cdot 10^{-30}$ C·m (without taking into account the correction for the *Lorentz field*).

Paraelectric resonance may be also observed in the absence of the applied electric field, a situation called *zero-field paraelectric resonance*. In this case the resonance frequencies, which correspond to the peaks of observed bands, coincide with the tunnel splitting frequencies. For Li$^+$ in KBr they are 62, 65, 141 GHz (for ^{7}Li) and 88, 88, 118 GHz (for ^{6}Li). The 40% difference between magnitudes of the tunnel splitting of the isotopes ^{6}Li and ^{7}Li is consistent with the results of measurements of *thermal conductivity* and *specific heat*, and confirms the tunneling nature of the noncentral ion transport. The study of the location, shape and width of the resonance lines, and of the environmental influences on these factors, provides information on the noncentral ions, their interaction with one another and with the lattice, as well as on the properties of the *solid* under consideration.

PARAELECTRICS

Insulators which exhibit a linear (and hysteresis-less) dependence between the polarization P

and E for weak applied electric fields E. Among paraelectrics are, in particular, *ferroelectrics* in a nonpolar phase in the neighborhood of the *Curie point*.

PARALLELEPIPED, PRIMITIVE
See *Primitive parallelepiped*.

PARALLEL PUMPING OF MAGNONS
See *Spin waves, Parametric excitation of spin waves*.

PARAMAGNET

A material which becomes magnetized proportional to and in the direction of an applied external *magnetic field* B, but does not exhibit magnetic ordering. *Diamagnetism* that may be present is dominated by *paramagnetism* in paramagnets. Classical paramagnets (*Langevin paramagnets*), which have a temperature dependence of the *susceptibility* which satisfies the *Curie law*, include a wide group of materials with intrinsic *magnetic moments*: gases (O_2, NO, metallic vapors), liquids and solids (rare-earth salts, salts of transition metals, solutions of these salts). Many paramagnets are Langevin types, and yet exhibit stronger interactions of their magnetic moments with the matrix and with one another (salts and oxides of transition ions, certain rare-earth ions, including those magnetically ordered above the *Curie point* or *Néel point*, even gases O_2 and NO under elevated *pressure*), do not obey the Curie law, but instead follow the *Curie–Weiss law*. The majority of nontransition metals (all alkali metals; alkaline-earth metals, except for Be and Al, etc.) are paramagnets of the Pauli type with a temperature independent susceptibility $\chi = \mu_B^2 g(E_F)$ arising from the conduction electrons, where $g(E_F)$ is the density of states at the Fermi level. Transition metals, which do not exhibit magnetic order, are clearly defined paramagnets, exhibiting high values of *magnetic susceptibility* χ and rather complicated temperature dependences. This specific group of paramagnets consists of *strong band paramagnets* (see *Strong band paramagnetism*). There exist also *nuclear paramagnets*, nuclear spin systems with spin $I > 0$, such as ^{3}He at low temperatures ($T < 0.1$ K).

PARAMAGNETIC CENTER

A localized state in a solid, which exhibits a nonzero magnetic *dipole moment*; it is the origin of the *paramagnetism* and the source of the *electron paramagnetic resonance* (EPR) signal. Some examples of paramagnetic centers are: *ions* of transition elements with incomplete *d*- or *f*-shells; *impurity atoms*, molecular ions or their associations in natural or synthetic *crystals* (fixed nitrogen in diamond, aluminum in quartz); localized *donor* states and *acceptor* states in doped *semiconductors* (e.g., elements of Groups III and V in *silicon* and *germanium*); *point defects* produced by irradiation, heat or chemical treatment (introduction of impurities), e.g., an electron occupying the *vacancy* of a negative ion (F-center), self-trapped hole (V_k-center) and similar defects of a more complicated nature; intrinsic defects like *dangling bonds* in *amorphous semiconductors*, stable *free radicals*; photoexcited states (e.g., triplet ones) of impurities and defects, which exhibit zero *magnetic moment* in their ground state. The term "paramagnetic center" refers not only to the impurity atom, ion or defect, but also to its close environment, which exerts a strong influence upon the nature of the wave function of the paramagnetic center, and upon its EPR spectrum. One of the challenges of *magnetic resonance spectroscopy* and microwave spectroscopy is the identification of paramagnetic centers, and the determination of their spatial configuration and energy band structure.

PARAMAGNETIC ION

See *Magnetic ion.*

PARAMAGNETIC LIMIT in superconductors

The peak value of a magnetic field B_p in a superconductor with *Cooper pairs* in the singlet state; when the magnetic field reaches the paramagnetic limit, the free energies of the superconducting phase and normal phase become equal in magnitude, taking into account the polarization of electronic spins by the magnetic field, but without regard for diamagnetic undamped currents (*Meissner effect*). The phenomenon of the paramagnetic limit was examined theoretically by A.M. Clogston and B.S. Chandrasekhar (1962). At absolute zero ($T = 0$), the paramagnetic limit

of a weak *electron–phonon interaction* is given by the equation $B_p(0) = \Delta(0)/(\mu_B\sqrt{2})$, where $\Delta(0)$ is the *energy gap* in the spectrum of *quasiparticles* at $T = 0$, and μ_B is the Bohr magneton. For superconductors which exhibit a *critical temperature* $T_c = 10$ K, the paramagnetic limit $B_p(0) \approx 18.5$ T. There are some known materials for which an *upper critical field* $B_{c2}(0)$ exceeds $B_p(0)$. Some examples are laminated dichalcogenides of transition metals, *Laves phases*, etc. The cases of overriding the paramagnetic limit were attributed to the coexistence of superconductivity and electron localization, to the compensation of the external magnetic field by an internal exchange field (*Jaccarino–Peter effect*), to the generation of the nonuniform *Fulde–Ferrell–Larkin–Ovchinnikov state*, etc.

PARAMAGNETIC PROBE METHOD

Investigation of crystals by the *paramagnetic resonance* method, which involves implantation of paramagnetic ions (probes) into the crystalline sample. The method can provide, among other things, the *local symmetry* of the lattice sites, and information on the spatial microscopic *crystal symmetry*. Paramagnetic probes called spin labels (often nitroxide radical ions) are widely used for biological samples.

PARAMAGNETIC RELAXATION

The process of establishing thermal equilibrium in a *paramagnet*; one of the types of *magnetic relaxation*. Also used is the term "*spin relaxation*" when it involves the energy levels of a paramagnetic ion present at ordinary temperatures in paramagnetic crystals (in which intracrystalline interaction and *spin–orbit interaction* are substantial factors) which may be described in terms of the *resultant spin* and *spin Hamiltonian*. Paramagnetic relaxation is of major importance for *electron paramagnetic resonance* and *nuclear magnetic resonance*, for it determines the value of the absorbed power in relation to the radio-frequency field amplitude.

If a paramagnet is placed in a constant magnetic field B_0, then *relaxation* of the longitudinal component (with respect to the field direction), and the transverse components of *magnetization* are distinguished. Accordingly, the terms "*longitudinal relaxation time*" and "*transverse relaxation time*" were introduced by F. Bloch (1946). Longitudinal relaxation is related to the change of Zeeman energy of a paramagnet in an external field B_0, and therefore, may take place only in the presence of a thermal reservoir (e.g., vibrating lattice, Brownian motion); longitudinal relaxation is *spin–lattice relaxation*, and in solids it arises from lattice vibrations. These vibrations are responsible for the change of the intracrystalline electric field, which acts upon the *paramagnetic center*, and influences its *magnetic moment* through the spin–orbit interaction. Transverse relaxation is not related to the change of energy of the spin system, and may take place both through the *spin–lattice interaction*, and through the *spin–spin interaction*. The principal mechanism of *spin–spin relaxation* in dilute paramagnets is the *magnetic dipole interaction*. Spin–spin relaxation generally takes place in a shorter time interval than spin–lattice relaxation, and the spin system reaches a quasi-equilibrium state during the spin–spin relaxation; a state which may be described in terms of a spin temperature, which is distinct from the lattice temperature. In this case, the process of spin–lattice relaxation brings about the equalization of the temperatures of the spin system and the lattice.

One of the techniques for determining the longitudinal relaxation time, which was proposed by C. Gorter (1936) prior to the discovery of the *paramagnetic resonance* phenomenon, consists in measuring the nonresonance absorption of the variable magnetic field energy for a parallel mutual alignment of the constant field and the variable field. Nonresonance absorption is described by an equation, which is similar to the Debye formula for *dielectric loss* in dielectrics, and is characterized by the time parameter τ, which is the relaxation time of the resultant magnetic moment of the system. The parameter τ may differ from the values of the longitudinal relaxation time, which are measured by paramagnetic resonance experiments (pulsed or continuous saturation), when nonequilibrium magnetization is induced by changing the populations of a particular pair of spin levels. All methods of determining the longitudinal relaxation time provide equal values only in the case of systems with an equidistant Zeeman energy spectrum, or when the spin $S = 1/2$. The transverse relaxation time may

also be determined from the experiments on pulse or continuous saturation of electron paramagnetic resonance lines. In systems which exhibit slow *spin diffusion*, information on the spin relaxation times may be gained through the method of electron *spin echo*.

PARAMAGNETIC RESONANCE, also called electron spin resonance

Resonance absorption of microwaves by a *paramagnet*; a particular case of *magnetic resonance*. From the standpoint of physics, the reason for the occurrence of paramagnetic resonance is that the application of a constant *magnetic field* to materials which contain paramagnetic particles gives rise to an energy level system E_l, with each level characterized by a certain orientation of the *magnetic moment* of a particle. The transition from the jth level to the ith level, which involves reorientation of the magnetic moment and absorption of an energy quantum $h\nu_{ij}$, takes place under action of an electromagnetic wave of frequency $\nu_{ij} = (E_i - E_j)/h$. *Zero-field paramagnetic resonance*, i.e. paramagnetic resonance in the absence of external magnetic field, is much less common. In this case, the energy level scheme arises from the interaction between the magnetic moments of the particles and the intracrystalline electric field. Depending on the nature of the paramagnetic particles, *electron paramagnetic resonance* and *nuclear paramagnetic resonance* are distinguished (the latter is ordinarily referred to as *nuclear magnetic resonance*). Frequencies and intensities of paramagnetic resonance lines are sensitive to various fields, which surround the magnetic moments of the particles, and hence, paramagnetic resonance is applied in studies of local properties of solids.

PARAMAGNETIC RESONANCE, ACOUSTIC

See *Acoustic paramagnetic resonance*.

PARAMAGNETISM (fr. Gr. $\pi\alpha\rho\alpha$, near, by, and magnetism)

The property of materials that possess no *atomic magnetic structure* (*paramagnets*) to acquire *magnetization*, which partially align their magnetic moments when placed in an external *magnetic field*. Paramagnetism is the opposite of *diamagnetism.* Paramagnetism was first described by M. Faraday (1847). From the standpoint of microphysics, paramagnetism is characterized by certain particles of paramagnetic substances which possess uncompensated *magnetic moments m* (an exception is provided by *Van Vleck paramagnetism*). The distinction between paramagnetism and other types of magnetism involving uncompensated magnetic moments is that in the case of paramagnetism in the absence of an applied magnetic field these moments are oriented randomly and are independent of one another, i.e. the mean value of the *m* components of each particle is equal to zero (in contrast to *spin glasses*); in addition the magnetization of the body (in contrast to *ferromagnets*) and the mean magnetic moment of any sublattice of the crystal (in contrast to *antiferromagnets*) are also equal to zero. Typical of paramagnetism is the linear relation between the magnetic polarization *M* and the magnetic intensity $H = B/\mu_0$: at low values of the field intensity $M = \chi H$, where χ is the *magnetic susceptibility* (χ may be a tensor quantity in crystals). Another typical feature of paramagnetism is the probability of *magnetic saturation* $M \rightarrow M_s$ when increasing *B* to values which exceed the so-called *saturation field* B_s. The value and the temperature dependence of χ, as well as the value of B_s, are determined by the properties of the particles of nonzero *m* (of electronic or nuclear nature). An electronic magnetic moment is produced by electrons of unfilled shells of atoms and molecules, by collective electrons of conductors, etc.:

$$m = g\mu_B J \qquad (1)$$

where *J* is the angular momentum (of a conduction electron, electronic shell, etc.) in units of *Planck constant* $\hbar$; *g* is the *Landé g-factor*; μ_B is the Bohr *magneton*. As a rule, *m* is different from zero in a condensed medium in the case of inner (*d* and *f*) unfilled shells. Since the orbital moment of the shells in crystals is usually equal to zero ("frozen" or "quenched" because of the *spin–orbit interaction*), the electronic magnetic moment of an atom in a crystal is determined for the most part by the spin. Magnetic moments of nuclei are close in value to the nuclear magneton, and considerably less than electronic magnetic moments. Several types of paramagnetism are distinguished,

and in describing them the interaction of magnetic moments with one another is usually ignored.

If the orientation of the magnetic moments of the particles in the magnetic field is unrelated to their motion, then the phenomenon is referred to as *orientation paramagnetism* or *Langevin paramagnetism* (see *Langevin function*). In this case, paramagnetism is determined by independent orientation of magnetic moments in the external field *B*, and is described on the basis of the *Gibbs distribution* for the projections of the magnetic moments m_z on the direction of the field *B* (on the *z* direction): $m_z = g\mu j_z$. With allowance made for quantization, the j_z projection assumes $2j + 1$ values, and the magnetic moment of *N* atoms is determined by the *Brillouin function*:

$$M = Nm_{\max} B_j \left(\frac{m_{\max} B}{k_B T} \right), \qquad (2)$$

where $m_{\max}$ is the maximum value of the projection of magnetic moment of the particle (for electronic magnetic moment $m_{\max} = g\mu_B J$). Hence it follows that at $m_{\max} B \ll k_B T$ there exists a linear relation between *M* and *B*, and the magnetic susceptibility is governed by the *Curie law*:

$$\chi = \frac{c}{T}, \qquad c = \frac{N}{3} \frac{(j+1)m_{\max}^2}{jk_B}. \qquad (3)$$

It is a common practice to deal with the susceptibility of a unit volume, or the *molar susceptibility*; in these cases *N* is the number of magnetic moments per unit volume or per mole of a material, respectively.

The magnetic saturation ($M \rightarrow M_s = Nm_{\max}$) is observed at $B > B_s$, where $B_s = k_B T/m_{\max}$. Eqs. (2) and (3) adequately describe the paramagnetism of classical paramagnets, namely, liquids and gases, the molecules of which possess magnetic moments, as well as the paramagnetism of localized magnetic moments in solids on the condition that the interactions between these localized moments may be neglected. Interactions between the magnetic moments is responsible for deviations from the Curie law (see *Curie–Weiss law*), and may lead to magnetic ordering at low temperatures.

Pauli paramagnetism is common for a degenerate Fermi gas or *Fermi liquid*, e.g., electrons in metals, liquid ^{3}He, etc. The onset of magnetic polarization (i.e. generation of unequal quantities of fermions with different values of spin projection) in such systems results in an increase of the mean kinetic energy of the gas by virtue of the Pauli exclusion principle. In the case of Pauli paramagnetism the magnetic susceptibility χ_p does not depend on the temperature, and for the electron gas is given by an equation

$$\chi_p = \mu_B^2 D(E_F),$$

where $D(E_F)$ is the *density of states* at the *Fermi surface*. In the case of a spherical Fermi surface, $\chi_p = 3n\mu_B^2/(2E_F)$, where E_F is the *Fermi energy*, n is the electron density. When these equations are compared with Eq. (3), it is apparent that gas degeneracy results in a decrease of susceptibility (by the factor $k_B T/E_F$). Magnetic saturation may arise only in very strong fields ($B_s \propto E_F/\mu_B$). Paramagnetism of the current carriers in nondegenerate semiconductors is characterized by an exponential temperature dependence, which is due to the temperature dependence of the density of carriers. *Van Vleck paramagnetism* or *polarization paramagnetism* is related to quantum-mechanical corrections, which are due to the admixture of excited states with nonzero magnetic moments into the ground (nonmagnetic) state of the atom. This kind of paramagnetism has only a weak dependence on temperature. The study of the paramagnetism of solids and its high-frequency manifestations (*electron paramagnetic resonance* or *nuclear magnetic resonance*) is an important method of investigation of the properties of solids. The phenomenon of paramagnetism is also used for reaching extremely low temperatures (see *Adiabatic demagnetization cooling*).

PARAMAGNETISM, STRONG BAND

See *Strong band paramagnetism*.

PARAMAGNETISM, VAN VLECK

See *Van Vleck paramagnetism*.

PARAMETRIC EXCITATION OF SPIN WAVES

Excitation of *spin waves* in *magnetic materials* (ferro-, ferri- and antiferromagnets), which are exposed to the action of an electromagnetic pumping wave; parametric excitation is observed only when the amplitude h of the magnetic field in the wave exceeds a certain threshold value h_{th}. *Spin waves* arise in the sample at $h > h_{th}$. The frequencies ω_1 and ω_2 and wave vectors k_1 and k_2 of these spin waves are related through *parametric resonance* to the frequency ω and wave vector k of the pumping signal as follows:

$$\omega_1 + \omega_2 = n\omega, \qquad k_1 + k_2 = nk, \qquad (1)$$

where n is the *order of the parametric process*. The amplitude of the oscillating magnetization of the initial spin wave exceeds the thermal level by six or more orders of magnitude, reaching several tenths of a percent of the saturation magnetization M_0 of the material under consideration. Individual phases of the excited spin waves are random variables, and it is only the sum of phases that is single-valued with respect to the pumping wave phase. The experimentally obtained values are $\omega_{1,2} \sim 10^{10}$–$10^{11}$ s^{-1}, $n = 1$ or 2; the length of the pumping wave often noticeably exceeds the dimensions of the sample, i.e. the pumping is uniform, and hence $k = 0$, $k_1 = -k_2$; the value of the wave vector of the excited waves $k_{1,2}$ depends on the pumping frequency and on the intensity of the magnetic biasing field H in which the sample is located, and may range within 10^3–10^6 cm^{-1}.

The first-order parametric excitation of spin waves ($n = 1$) results in the excitation of waves of a frequency that is equal to half the pumping frequency: $\omega_1 = \omega_2 = \omega/2$. Depending on the orientation of h with respect to H, *transverse pumping* ($h \perp H$) (H. Suhl, 1957) and *longitudinal pumping* ($h \parallel H$) (E. Schlömann, 1960) are distinguished. In the former case the electromagnetic field first excites *uniform oscillations of magnetization*, which induce the paramagnetic excitation of spin waves; in the latter case, the immediate cause of the parametric relation between the waves is the longitudinal pumping magnetic field. The threshold value of longitudinal pumping in an isotropic *ferromagnet* is given by the expression

$$h_{th} = \frac{\Delta H_k \omega}{\omega_M \sin^2 \theta_k}, \qquad (2)$$

where $\Delta H_k = \gamma_k/\gamma$, γ_k is the *spin wave damping decrement*; $\omega_M = 4\pi\gamma M_0$, γ is the *gyromagnetic ratio*; θ_k is the angle between the wave vector of the spin waves and H_0. The threshold value of the field in perfect monocrystals of yttrium iron garnet (see *Iron garnets*) amounts to several microtesla at room temperature. Since the uniform precession of magnetic polarization contributes to the process of excitation of spin waves, the threshold of the first-order transverse pumping depends heavily on the degree of closeness of the pumping frequency ω to the *ferromagnetic resonance* frequency ω_0. If $\omega = \omega_0$, then the threshold is particularly low and amounts to $\sim 10^{-8}$ T in yttrium–iron garnet monocrystals. Away from the ferromagnetic resonance frequency, its value is close to the longitudinal pumping threshold value. Second-order parametric excitation of spin waves ($\omega_1 = \omega_2 = \omega$) is observed only in the proximity of the ferromagnetic resonance frequency $\omega = \omega_0$ provided there are no spin waves of frequency $\omega_k = \omega/2$ in the spectrum, which excludes the possibility of first-order parametric spin-wave processes. Its threshold amounts to several nanotesla in these garnet monocrystals. Experimental measurement of the longitudinal pumping threshold h_{th} is the most convenient way of determining the damping decrement of spin waves of wave vectors of the order of 10^4–10^6 cm^{-1} by the application of Eq. (2). Parametric excitation of spin waves has found practical application in the production of *ferrite power limiters*: the damping provided by these devices increases abruptly at $h > h_{th}$ because of the onset of extra losses of signal energy due to the parametric excitation of waves. The maximum operating powers of devices like filters, phase shifters, phase-differential circulators, etc., are determined by the occurrence of the parametric excitation of spin waves: the damping provided by these devices increases for $h > h_{th}$ as in the case of the limiters.

PARAMETRIC LIGHT GENERATION, parametric luminescence

A phenomenon that involves the decay of incident *photons* (of frequency ω_3) in a nonlinear crystal into pairs of photons of lower frequencies ω_1 and ω_2, which are reemitted by the crystal. The production of a pair of photons in a single scattering event is related to generation of parametric *quasi-particles*, which are known as *biphotons*. The sum of the frequencies of the scattered waves is equal to the incident light frequency: $\omega_1 + \omega_2 = \omega_3$. Conservation of momentum of three photons $k_1 + k_2 = k_3$ allows for the effective accumulation (*parametric gain*) of biphotons along coherent directions which are determined by the dispersion and anisotropic properties of the nonlinear crystal. As a result, the scattered waves assume a specific frequency–angular scattering pattern. This circumstance sets the parametric light scattering apart from ordinary *luminescence*. Another distinction between the two phenomena is that the process of parametric light scattering ceases within several periods of light oscillations after the passage of the principal wave ω_3. The correlation between photons, which constitute a single biphoton, allows one to obtain information on the photon ω_1 by detecting photon ω_2. Parametric light scattering is used in the nonlinear spectroscopy of crystals, and may provide a basis for absolute measurements of light intensity. In the case of a high intensity of primary radiation ω_3, parametric light scattering goes over into *parametric superluminescence*. The latter is characterized by an exponential increase in the intensity of waves ω_1 and ω_2 with an increase in the nonlinear crystal axial length, and the power of the wave ω_3.

PARAMETRIC OSCILLATOR

See *Optical parametric oscillator*.

PARAPROCESS, true magnetization

Increase of the absolute value of the *magnetization M_s* of *ferromagnets* and *ferrimagnets* under the action of an external magnetic field B, which takes place on completion of the engineering magnetization processes (see *Saturation magnetization curve*), i.e. when the *magnetic substance* has passed into the single-domain state. The paraprocess is related to the orientation of elementary carriers of magnetism (spin and orbital magnetic moments of atoms or ions) in the field B. It is the final stage of magnetization, during the course of which M_s tends toward its absolute saturation value M_0 (i.e. the magnetization of a ferromagnet at absolute zero, and in the absence of spin *zero-point vibrations*) with increased

B (if $B > B_S$). In the case of magnets with a single *magnetic sublattice* (ferromagnets), the peak of paramagnetism is in the neighborhood of the *Curie point*. The changes of physical properties, which take place in ferro- and ferrimagnets during the course of the paraprocess, exhibit a number of special features (see *Magnetostriction, Magnetothermal phenomena*).

PARENT PHASE

A hypothetical small-volume (microcrystalline) *phase* of a symmetrical structure from which, with the help of minimal (considerably less than interatomic distances) displacements of atoms (ions), a *crystal* of a particular structure may be generated. Macroscopic monocrystals are often grown from seed crystals of the parent phase. A parent phase is convenient for describing the *twinning structure* of some crystals, for describing a *domain structure* which forms from a parent phase, for describing a granular structure, for explaining the specific smallness of some constants of the crystal (e.g., those which are zero in the parent phase due to symmetry requirements), for constructing the theory of some first-order transitions between phases involving symmetry groups which lack a group–subgroup relationship (if the structures of both phases may be derived from the parent phase structure using an *order parameter* transformed according to an irreducible representation of the parent phase symmetry group), etc. Sometimes the term "parent phase" is used with respect to the most symmetrical phase, realized under the experimental conditions, if it is associated with a *phase transition* involving a reduction of the phase symmetry.

PARISER–PARR–POPLE METHOD, zero

differential overlap method (R. Pariser, R.G. Parr, J.A. Pople, 1953)

A popular semiempirical procedure of *quantum chemistry*, which is based on ignoring those matrix elements of the two-particle interaction through localized wave functions, which involve the two-center distributions of the exchange electron density.

PARTIAL DISLOCATION

A *dislocation* with a fractional (non-coincident with a lattice parameter) value of *Burgers vector*. The line of this dislocation delimits a *stacking fault*. As a partial dislocation moves across a *crystal lattice* the lattice does not transform into itself as occurs at a moving *perfect dislocation*. As a result of this partial dislocation motion, a stacking fault appears which is able to break on the other partial dislocation. In so doing, the Burgers vectors b_1 and b_2 of these two partial dislocations sum to the vector b of a perfect dislocation (see Fig.). A system of partial dislocations joined by the area of a stacking fault is a called a *stretched dislocation* (*extended dislocation*).

For example, in face-centered cubic (FCC) crystals a perfect dislocation with Burgers vector $b = a/2\ [\bar{1}01]$ may split up into two partial dislocations with Burgers vectors $b_1 = a/6\ [\bar{2}11]$, $b_2 = a/6\ [\bar{1}\bar{1}2]$. Between these partial dislocations there is a subtraction-type stacking fault, and the equilibrium distance r between the partial dislocation rectilinear areas (see Fig.) amounts to

$$r = \frac{\mu b^2 (2 - \nu)}{8\pi \gamma (1 - \nu)} \left[1 - \frac{2\nu \cos 2\beta}{2 - \nu} \right],$$

where μ is the *shear modulus*, ν is the *Poisson ratio*, β is the angle of the mixed perfect dislocation, γ is the energy of the stacking fault. For an *edge dislocation* r is greater than for a *screw dislocation*.

Top partial dislocations appear at the dislocation splitting accompanied by the transition of the partial dislocation from one slip plane to another, or its interaction with the dislocation located in the other slip plane. The configurations and properties of extended dislocations in other crystalline structures coincide in many respects with that of the FCC lattice type. In hexagonal close-packed crystals, however, unusual varieties of dislocations are observed. In bulk-centered crystals several types

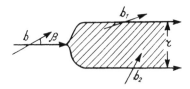

Partial dislocation.

of partial dislocations with high-energy stacking faults are possible, while in *diamond*-type crystals nonequivalent sets of partial dislocations exist, which may be described as having a structure of the face-centered crystal type with a basis composed of two atoms at each site of the lattice. Specific partial dislocations of a superdislocation type are observed in ordered crystals.

Partial dislocations are essential for the explanation of the *twinning of crystals*, phase transformations, *hardening* due to the dislocation barriers that occur at intersecting dislocations, creeping, and *cross slip* processes.

PARTICLE BENDING in a curved crystal

Planar *channeling* involves a forward moving positively charged particle executing stable oscillations between two crystallographic planes, experiencing alternating "total reflections" from each of them. It is possible to turn the direction of a charged particle beam by flexing the crystal planes. The radius of curvature R of a curved crystal may be evaluated from the classical equation of motion:

$$\frac{mv^2}{R} = -\frac{\partial U}{\partial r},$$

where m and v are, respectively, the mass and velocity of the channeled particle, and U is the averaged potential of the plane. In deciding on a particular radius of curvature, it is necessary to take into account that the length of the curved channel $\Delta l = R\Delta\theta$ (where θ is angle of beam bending) must be much less than the *dechanneling* distance l_{dec} (for relativistic particles the dechannelization distance increases proportional to the energy). However, particle beam bending for axial channeling offers several advantages over beam bending with planar channeling: (1) the acceptance angle of axial channeling exceeds that of planar channeling; (2) the number of particles involved in the axial channeling mode is greater than the number accepted by the planar channeling mode; (3) the dechanneling distance for axial channeling is greater by approximately an order of magnitude than that for planar dechanneling. It is difficult to bend low and average-energy particle beams because of the strong particle dechanneling propensity. *Surface channeling* can also be used for the purpose of particle turning. Unlike ordinary channeling, surface-channeled ions of low and average energy lose little energy by reflection from the surface producing the beam bending. *Computer simulation* confirmed the high efficiency of particle beam bending via surface channeling.

PARTICLE EMISSION

Emission of electrons, ions, or neutral particles into a vacuum or gaseous medium by solids or liquids. In most cases the term particle emission is used in names of various types of *electron emission*. As a rule, the kind of action is also added: *field electron emission* (strong electric field), *explosive electron emission* (rapid, very strong electric field), *secondary electron emission* (electron bombardment), *ion–electron emission* (ion bombardment), *thermionic emission* (heating), *photoelectron emission* (illumination), *X-ray photoelectron emission* (X-ray irradiation), *exoelectronic emission* (mechanical action), *autophotoelectronic emission* (strong electric field and illumination), and so on. The emission of molecules from heated solids is called *sublimation*, and from liquids it is called evaporation. In *adsorption* a term "evaporation" is used when describing the particulate emission of adsorbent material; the emission of adsorbed particles is called desorption (*thermal desorption, electron-stimulated desorption, photodesorption, field desorption,* and so on). The emission of ions by incandescent solids has an additional name *surface ionization*. The emission of ions and neutral particles under bombardment of the surface by fast ions and atoms accompanied by surface erosion is called *cathode sputtering* (see *Sputtering*); its particular case which is used in the diagnostics of the surface chemical composition is called *ion-ion emission* (*Secondary ion emission*).

PARTICLE, SMALL

See *Small particles*.

PARTITION FUNCTION, statistical sum

Quantity, which constitutes the reciprocal of the normalization factor of the Gibbs canonical distribution (see *Gibbs distribution*) in quantum statistical physics, and is expressed as the sum of occupation possibilities over quantum states (Zus-tandsumme) Z:

$$Z = \sum_n \exp\left(-\frac{E_n}{k_B T}\right),$$

where E_n is the energy of system in the nth quantum state, and T is the absolute *temperature*. The summation is taken over all allowed n (including states of equal energy). The partition function allows the calculation of all *thermodynamic potentials*. In particular, the Helmholtz free energy given by $F = -k_B T \ln Z$ can be calculated as a function of temperature, volume, and number of particles depending on the particle interaction potential. Given the system Hamiltonian H, $Z = \text{Tr} \exp[-H/(k_B T)]$. In the case of an ideal gas

$$Z = \sum_{\{n_i\}} \exp\left[-\frac{\sum_i \varepsilon_i n_i}{k_B T}\right].$$

The summation in this equation is taken subject to the additional condition that $\sum_i n_i = N$, where N is the total number of particles. The summation in the argument of the exponential function is made over all single-particle quantum states i of energy ε_i; $\{n_i\}$ is the ensemble of possible values of n_i, over which the summation is taken. In the case of *Bose–Einstein statistics*, $n_i = 0, 1, 2, \ldots$, and in the case of *Fermi–Dirac statistics*, n_i is either 0 or 1. In the formalism of classical statistical physics the partition function corresponds to the statistical integral, the integral analogue of the above discrete summations.

PASSIVATION

See *Surface passivation*.

PATENTING

The method of *heat treatment* of medium-carbon low-alloy steel (carbon content 0.45–0.85% C), which involves heating above the transformation temperature, with subsequent controlled cooling. It is used in order to obtain work pieces of high *plasticity*. This circumstance allows the use of patenting prior to the production of cold-formed wires, strips and other articles, which require deep drawing at room temperature. The process of patenting involves heating to 870–920 °C. The

resulting *austenite* is rapidly overcooled to 450–550 °C; at this temperature the austenite undergoes decomposition, yielding thin-laminated *sorbite*. The *steel* with such a structure is substantially hardened under *strain*, reaching the ultimate compression strength of 5000 MPa at total pressing of 99%.

PATH INTEGRAL, functional integral, Feynman path integral (R. Feynman, 1948)

Presentation of the quantum-mechanical time evolution operator and its matrix elements in the form of an integral along a number of classical trajectories.

Suppose that $U_t = e^{it H/\hbar}$ is the unitary operator for the time evolution of a quantum system with a finite number of degrees of freedom (H is the Hamiltonian), q is a point in configuration space, i.e. a set of coordinates, $U_t(q_1, q_2)$ is a matrix element of U_t in the coordinate representation. The integral over the paths has the form

$$U_t(q_1, q_2)$$

$$= \int_{\{q(\tau): q(0)=q_1, q(t)=q_2\}} e^{iS} \prod_{0 \leqslant \tau \leqslant t} dq(\tau),$$

where $q(\tau)$ is the classical trajectory, $0 \leqslant \tau \leqslant t$, the quantity S

$$S = S[q(\tau)] = \int_0^t \left\{\dot{q}^2(\tau) - V[q(\tau)]\right\} d\tau$$

is the *classical action* for the fixed path $q(\tau)$, V is the potential energy; integration is carried out along all paths with a fixed beginning and end. To evaluate the integral the interval $[0, t]$ is divided into sections: $0 < t_1 < t_2 < \cdots < t_n < t$, the continuous path $q(\tau)$ is replaced by a discrete one with vertices at the points of subdivision, the integration is performed sequentially for the variables $q(t_1)$, $i = 1, 2, \ldots, n$, so the result is finite, then the limit of $n \to \infty$ is taken, corresponding to a continuous succession of τ values. The integration is performed with respect to the functions $q = q(\tau)$, whence the origin of the term "functional integral".

In the middle 1950s functional integrals were introduced into the quantum theory of systems with an infinite number of degrees of freedom, i.e. into quantum field theory,

$$U_t = \int e^{iS} \prod d\varphi(x),$$

where x is a point of four-dimensional space–time, $\varphi(x)$ is a classical field, $S = S[\varphi(x)]$ is the action. The path integral method is used in the theory of solids for the description of *phase transitions, polarons, solitons,* and *superconductivity.*

PAULI MATRICES (W. Pauli)

Two-by-two square matrices $\sigma_x, \sigma_y, \sigma_z$,

$$\sigma_x = \begin{pmatrix} 0 & 1 \\ 1 & 0 \end{pmatrix}, \qquad \sigma_y = \begin{pmatrix} 0 & -i \\ i & 0 \end{pmatrix},$$

$$\sigma_z = \begin{pmatrix} 1 & 0 \\ 0 & -1 \end{pmatrix},$$

which are operators for the spin *magnetic moment* of particles of *spin* $1/2$; the magnetic moment operator is $S = (\mu_B/2)\sigma$, where μ_B is the Bohr magneton. The Pauli matrices satisfy the commutation relations

$$[\sigma_j, \sigma_k] \equiv \sigma_j \sigma_k - \sigma_k \sigma_j$$

$$= 2i\varepsilon_{jkl}\sigma_l, \quad j, k, l = x, y, z,$$

where the Levi–Civita symbol $\varepsilon_{jkl} = +1$ for a cyclic permutation of the indices (xyz, yzx, zxy), $\varepsilon_{jkl} = -1$ for an anticyclic permutation of the indices (xzy, yxz, zyx), and $\varepsilon_{jkl} = 0$ if any two indices are the same (e.g., xxy). In addition the Pauli matrices have the cyclic permutation property $\sigma_j \sigma_k = i\sigma_l$, and they anticommute:

$$\sigma_j \sigma_k + \sigma_k \sigma_j = 2\delta_{jk}I, \quad j, k = x, y, z,$$

where δ_{jk} is the Kronecker symbol, and I is the 2×2 unit matrix. Each Pauli matrix has a determinant of -1, and the eigenvalues of the Pauli matrices are equal to ± 1. The Pauli matrices have the unusual property of being both unitary and Hermitian. These matrices are generators of the double-valued representations of the simple unitary group in two dimensions $SU(2)$.

PAULING'S MAXIMUM OVERLAP PRINCIPLE (L. Pauling)

A rule which allows one to predict energetically the most favorable configurations of atoms in molecules or complexes, and to select the atomic *orbitals* which make the maximum contribution to the bonding orbitals of the molecule. Maximum orbital overlaps of two neighboring atoms provide lowering of the energy of the electron of one atom in the field of the other atom; in the case of parallel spins of two atoms the maximum orbital overlap provides the highest value of the exchange integral (involved in the equation for the negative energy of the molecule).

PAULI PARAMAGNETISM

See *Paramagnetism.*

PEACH–KÖHLER EQUATION (M.O. Peach, J.S. Köhler, 1950)

The equation defining the force f that is applied to a *dislocation* of unit length by the field of *internal stresses*: $f_i = e_{ilm}\tau_l \sigma'_{mk} b_k$, where $i = 1,$ 2, 3. Here e_{ikl} is the unit antisymmetric tensor of rank three; τ is the unit vector of the tangent to the dislocation line, which is aligned with the sense of rotation of the dislocation line, b is its *Burgers vector,* and the tensor quantity σ'_{mk} is the *strain deviator.* The Peach–Köhler equation defines a force that is always directed perpendicular to the dislocation line, and is independent of the hydrostatic pressure in the crystal. The Peach–Köhler equation was refined by J. Weertman (1965). The equation is most commonly used for determining the value of the force that acts in the *slip plane* of the dislocation, and is defined as $f = n_i \sigma_{ik} b_k$, where n is the unit vector normal to the slip plane. In the case of a planar *dislocation loop,* this force is determined by a single component of the σ_{ik} tensor (provided two coordinate axes are aligned with the vectors n and b). For the case of an *edge dislocation* which is parallel to the z-axis, and has the Burgers vector aligned with the x-axis, the equation for the force f assumes the simpler form $f = b\sigma_{xy}$.

PEARLITE

A laminar micro-constituent of *steels* and *cast irons*; mechanical mixture of *ferrite* and *cementite.* Under close to equilibrium conditions, pearlite is formed by the eutectoid decomposition (see *Eutectoid alloys, Alloy decomposition*) of *austenite* on slow cooling according to the reaction: austenite $\rightarrow$ ferrite $+ Fe_3C$, where Fe_3C is cementite. The decomposition proceeds at constant temperature $T = 996$ K. The leading *phase* at this pearlite generation is cementite, the nucleation centers of which appear at the boundaries of austenite grains. The austenite region in the neighborhood of the first cementite plates becomes depleted of *carbon*, and as a result, the plate-like ferrite layer develops. As this takes place, the carbon content of the neighboring austenite regions increases and a new nucleation center of cementite appears. All the carbon is concentrated in cementite, for ferrite contains practically no carbon at room temperature. Granular pearlite and lamellar pearlite are distinguished. The ferrite and cementite of *lamellar pearlite* are plate-shaped, the spacing being 0.5–1 μm. Pearlite consists of cementite up to 12%, the balance being ferrite. The ratio of thickness of ferrite laminas to that of cementite ones is approximately 40/3. The structure of pearlite depends to a certain extent on the cooling rate (the higher the cooling rate, the higher the dispersity of structure). Fine-grained modifications of pearlite are *sorbite* and *troostite*. The hardness of pearlite, which is obtained under slow cooling, is $3 \cdot 10^5$ Pa. Construction steels of pearlite structure exhibit adequately high values of *strength* and *plasticity*, and readily lend themselves to machining.

PEARLSTONE

Eruptive volcanic rock of an acidic nature, variation of volcanic glasses; exhibits concentrically conchoidal separation and spheroidal (pearlitic) structure, which is formed due to compression under cooling: the rock is as a whole glassy, and consists of pearl-like balls 1–15 mm in diameter, which are either embedded in the glass individually, or constitute the whole rock. The rock is light gray in appearance, occasionally tinged with blue or yellow, the luster resembles that of wax, enamel, or silk. The rock is brittle. The density 1.3–1.6 g/cm^3 varies with the porosity. It contains up to 3–5% water of hydration. Crushed pearlstone rock swells on rapid heating to 1000–1200 °C, increasing up to 10–20 times in volume. This modification of pearlstone is a sound-absorbing, heat-insulating, and refractory material which is used as filler for concrete, plaster, and paints.

PEIERLS INSTABILITY (after R.E. Peierls)

The phenomenon of the instability of the lattice of a one-dimensional metal with respect to the generation of a lattice deformation (*charge density wave*) with the wave vector $Q = 2p_F/\hbar$ (p_F is the Fermi momentum), which is caused by the interaction between the electrons and *crystal lattice vibrations*. Peierls instability is accompanied by the creation of an energy gap of width Δ at the Fermi surface, and the transformation of the system to the insulating state (see *Peierls transition*).

PEIERLS–NABARRO MODEL (R.E. Peierls, 1940; F.R.N. Nabarro, 1947)

Approximate model which describes the action exerted upon a *dislocation* by forces arising from the discreteness of the *crystal lattice* structure, and from the atomic nature of the dislocation core structure (Peierls–Nabarro forces). The tangential stress τ, which describes the interaction between atomic layers, lying on different sides of the dislocation *slip plane*, is considered to be a periodic function of the local displacement δ of atoms from their positions in an *ideal crystal* lattice; Peierls drew on the simplest periodic relation $\tau = G \sin[2\pi(\delta/b)]$, where b is the magnitude of the *Burgers vector*, the coefficient G is of the order of magnitude of the crystal *shear modulus*. Nabarro applied this model to the evaluation of the *Peierls stress* σ_P, which is required for the motion in the slip plane, and derived the following equation: $\sigma_P = G \exp(-2\pi\omega/b)$, where ω is the dislocation width, whose magnitude is of the order of b. According to this model, the value of σ_P lies within the range from 10^{-5} to 10^{-2} G. The Peierls–Nabarro force, which acts upon the dislocation in its slip plane, is a periodic function of the position of the dislocation center. In the context of a continuum description of dislocations (see *Continuum theory of defects*), the Peierls–Nabarro force, which acts upon every element of the dislocation line in the slip plane, is introduced phenomenologically as a periodic function of the displacement of the dislocation line η in the direction of its motion. This force may be compared to the periodic shift-dependence of the line density of dislocation potential energy (*Peierls relief*)

$$U(\eta) = U_0 + \frac{1}{2}U_P\left[1 - \cos\left(\frac{2\pi}{b\eta}\right)\right],$$

where U_0 is the line density of the intrinsic energy of the dislocation without regard for the discreteness of the structure (the energy of *linear disloca-tion tension*) and $U_P = b^2 \sigma_P/\pi$ is *Peierls energy*.

The chief drawback of the Peierls–Nabarro model is that it assumes a sinusoidal displacement dependence, which is a crude approximation in the case of real crystals.

PEIERLS RELIEF (R.E. Peierls, 1940)

The energy of a *dislocation*, which varies periodically during the course of its movement in the *slip plane*. The minimum value of this energy determines the equilibrium position of the dislocation line when no external and internal forces are applied to the dislocation. The dependence of the dislocation energy on the displacement of the dislocation by an amount $\xi = \alpha b$ along a direction, which is perpendicular to its line, is given by a phenomenological equation $W(\alpha) = W_P \sin^2(\pi\alpha b/a)$, where the quantity W_P, which determines the amplitude of $W(\alpha)$, is commonly called the *Peierls energy*; b is the interatomic distance in this direction; $a = b$ or $a = b/2$, depending on the configuration in the region of the dislocation core. The Peierls relief consists of valleys (grooves), which are separated by energy humps. The maximum value of the stress $\sigma_P(\alpha) = (1/b)(\partial W(\alpha)/\partial\alpha)$ is called the *Peierls stress* ($\sigma_P = \pi W_P/(ab)$). The value of σ_P is considered as a phenomenological parameter, which falls in the range between 10^{-5} and 10^{-2} G (where G is the *shear modulus*); low values of σ_P are typical for close packed *metals*, whereas higher values are characteristic of *covalent crystals* and of, e.g., Si and Ge. In the case of metals with a body-centered cubic structure, σ_P assumes values from the middle of the above interval; for *screw dislocations* σ_P is 6–20 times greater than for *edge dislocations*. Reasonably accurate values of σ_P and W_P are obtained through *computer simulations* of the process of the motion of dislocations in a crystal. Phenomenological curves may exhibit a more complicated nature, having two maxima within the periodicity length. If the external stress σ is smaller than σ_P, then the dislocation lines are located in the valleys of the Peierls relief, and their transport into the neighboring valley may take place either at the expense of *internal stresses*, or under the action of the thermally activated generation and scatter of paired *kinks on dislocations*.

PEIERLS TRANSITION

Metal–insulator phase transition, which takes place in quasi-one-dimensional conductors with congruent regions at the *Fermi surface*. Along with the generation of an energy gap, the Peierls transition brings about a *charge density wave* in the electronic subsystem, and superlattice deformations of the crystal structure with the wave vector $2p_F/\hbar$ (p_F is the Fermi momentum). The Peierls transition takes place in a number of *quasi-one-dimensional crystals*: NbSe$_3$, TaS$_3$, K$_{0.3}$MoO$_3$, in organic conductors TTF-TCNQ (see *Organic conductors and superconductors*), etc. Theoretical and experimental investigations of quasi-one-dimensional metallic systems have made it possible to determine the principal characteristics of the Peierls transition:

- three-dimensional long-range order at the Peierls transition is established in quasi-one-dimensional systems under the condition of a strong enough interaction between one-dimensional conducting chains, which constitute the crystal. The critical region in this case may be rather narrow (*Ginzburg number* $Gi \ll 1$);

- at temperatures above T_P (the Peierls transition temperature), the *phonon spectrum* softens at wave vectors $q \approx 2k_F$, so that $\omega(2k_F) \to 0$ at $T \to T_P$ (the giant *Kohn anomaly*). Below T_P, the charge density wave and superlattice deformations arise (determined by diffraction analysis methods), the constants of which in the general case are incommensurate with the original lattice constant;

- a gap $\Delta(T)$ arises in the electronic energy spectrum below the transition point T_P; the width of this gap increases with decreasing temperature.

The process of "freezing-out" the carriers at the Peierls transition is clearly seen from the temperature dependences of the electric resistance and *magnetic susceptibility*. In the case of a considerably high probability of electron jumps between neighboring chains, the electron spectrum may assume an insulating character, not over the entire Fermi surface, but only in the proximity of the nested (congruent) regions of the Fermi surface. In this case, the Peierls transition gives rise to a metallic phase with a decreased *density of electron states* at the *Fermi level* (a similar transition, which involves a partially insulating behavior of the electron spectrum, is observed in NbSe$_3$).

The Peierls transition involves, along with re-arrangement of the electron system, substantial changes in the phonon spectrum. The collective modes $\omega_{\pm}(q)$ which arise below T_P are related to the amplitude and phase oscillations of the *superlattice*. The low-frequency *Fröhlich collective mode* $\omega_{-}(q)$, which is related to phase oscilla-tions, is optically active and contributes apprecia-bly to the system *polarizability* at low frequencies. At $q \rightarrow 0$, this mode corresponds to the motion of the charge density wave and superlattice along the chains. In a constant current, the contribution of the Fröhlich mode may be rather substantial in electric fields that exceed a certain threshold value E_T, which is determined by the pinning of charge density waves due to structural *defects*, or effects of commensurability with the original lattice (see *Fröhlich conductivity*).

In a quasi-one-dimensional Peierls insulator with a commensurate charge density wave, there may exist excitations, along with ordinary elec-tronic excitations over the Peierls gap, which are related to *solitons* of the charge density wave am-plitude and phase. An example of a system, in which these solitonic excitations play an impor-tant role, is the polymeric material (see *Polymers*) trans-*polyacetylene* (trans-$(CH)_x$).

PEKAR WAVES

The same as *Additional light waves*.

PELTIER EFFECT

See *Thermoelectric phenomena*.

PENDULAR OSCILLATIONS OF INTENSITY

This effect of *dynamic radiation scattering* of X-rays, neutrons, or electrons in *monocrystals* consists of periodical variation of intensities of di-rect and diffracted beams either when the crys-tal thickness increases or when the angle of in-cidence of the beam onto the crystal varies close to the Bragg angle (see *Bragg law*). The reason is interference of two waves with slightly different phase velocities. In a weakly absorbing crystal, the total intensity of the radiation is periodically pumped from the direct beam to the diffracted one and back, similarly to oscillations of two coupled

pendula. The effect of pendular oscillations of in-tensity is used for exact determination of structure factors (see *Structure amplitude*) of crystals and of the *Debye–Waller factor* for real monocrystals con-taining defects.

PENDULUM

An *ideal (mathematical) pendulum* is a mass point connected by a weightless rod (filament) to a fixed point of suspension O, and is in motion due to the force of gravity. A *physical pendulum* is a solid body suspended by a rod at a certain point. The vertical plane motion of an ideal pendulum is described by the equation $\ddot{\varphi} + (g/l) \sin \varphi = 0$, where φ is the angle of deviation of the rod from the vertical, l is the length of the rod, and g is the gravitational acceleration. For a sufficiently small φ, when $\sin \varphi \approx \varphi$, the material point per-forms *simple harmonic oscillations* with period $T = 2\pi (l/g)^{1/2}$. A similar motion is exhibited by a physical pendulum according to the equation $I\ddot{\varphi} + Mgl \sin \varphi = 0$ (I is *moment of inertia* with respect to the suspension point, and l is the dis-tance between the suspension point and the *cen-ter of mass*). The constancy of the oscillation pe-riod of a pendulum makes it convenient to use it as a regulator in timepiece mechanisms. A *torsional pendulum* is a wheel, whose center point is con-nected to a spiral spring that twists and untwists as the wheel rotates back and forth. If angles of twist are not too large, the moment of elastic forces may be considered proportional to the angle of ro-tation. Thus, the motion of the torsion pendulum is also of a simple harmonic nature with period $T = 2\pi (k/I)^{1/2}$, where I is the moment of inertia of the wheel, and k is the coefficient of elasticity of the spring.

PENETRATION DEFORMATION, bursting

A *plastic deformation* effected by a concen-trated load that penetrates through the depth of a crystal within its limited area. A *strain* of this kind produces protuberances at the opposing sur-face, normal to, or oblique with respect to the crys-tal surface undergoing deformation. Such protru-sions are called *bursting figures*. They feature reg-ular geometric shapes that are defined by the el-ements of *slip* in the deformed crystal, and by the orientation of its faces. Bursting was first observed by M. Bauer (1882) in galena (PbS) crystals. The faster the crystal is loaded, and the lower its temperature, the clearer is the shape of the bursting figures. Penetration deformation has been studied most completely in CsCl type crystals. Studies combining a set of comprehensive techniques, such as *optical polarization method*, *X-radiography*, *surface decoration*, and selective *etching* demonstrated that bursting involves such mechanisms as common translation slip, prism slip, *faulting*, and displacement of *point defects*. The actual mechanism depends on the crystal itself, and the conditions of deformation.

PENETRATION DEPTH OF ELECTRIC FIELD

into superconductor

The distance λ_E associated with branch imbalance relaxation, which defines the space scale of nonequilibrium processes in superconductors (see *Nonequilibrium superconductivity*), is called the electric field penetration depth. The simplest case is the passage of current through a galvanic contact between a normal *metal* and a superconductor. In the normal metal the current is carried by electrons, and in the superconductor it is carried by *Cooper pairs*. The transformation of normal electrons to Cooper pairs takes place inside the superconductor within the distance λ_E from the interface with the normal metal. In this region the electric field E is non-zero, so λ_E can be regarded as the penetration depth of the longitudinal "dissipative" electric field into the superconductor, in contrast with its counterpart the usual nondissipative penetration depth λ of the transverse magnetic field (see *Meissner effect*). At a distance close to λ_E, in order of magnitude, a nonequilibrium superconducting state appears which is characterized by an imbalance of the populations of the electron-like ($P > P_F$) and hole-like ($P < P_F$) branches of the spectrum, where P_F is the Fermi momentum, and P is the excitation momentum. The imbalance relaxation is associated with inelastic electron–phonon collisions characterized by the *mean free path* $l_\varepsilon = v_F \tau_\varepsilon$. These collisions have low effectiveness at low temperatures, leading to quite large values of λ_E, up to 10 μm. The theory provides the value $\lambda_E = (D\tau_E)^{1/2}$, where $\tau_E = (4T/(\pi \Delta))^{1/2}\tau_\varepsilon$ is the time constant of the imbalance relaxation, and

$D = (1/3)v_F l_i$ is the electron *diffusion coefficient*. In the vicinity of T_c, the value of λ_E varies as $\lambda_E(T) = \lambda_E(0)/(1 - T/T_c)^{1/4}$, and it determines the specific dimensions of the region of inhomogeneity in the *resistive state* of films.

PENETRATION DEPTH OF MAGNETIC FIELD

into superconductor

A specific temperature-dependent length $\lambda(T)$ that determines the spatial scale of the penetration of an external magnetic field $B(r)$ into a superconductor. Under the assumption $\lambda(T) \gg \xi(T)$ (here $\xi(T)$ is the *coherence length*), the density of the supercurrent (see *Ginzburg–Landau theory of superconductivity*) in the bulk, with the exception of a thin surface layer of thickness ξ, is given by

$$J_s = -\frac{4e^2}{mc}|\Psi_0|^2 A, \qquad B = \nabla \times A, \quad (1)$$

where $2e$ is the charge of a *Cooper pair*. According to the *Ginzburg–Landau–Abrikosov–Gorkov (GLAG) theory* the modulus $|\Psi_0|$ of the spatially uniform, coherent superelectron wave function of the superconduction condensate Ψ_0 (i.e. order parameter) is determined at $T < T_c$ by the equation:

$$|\Psi_0| \cong 3.06\frac{k_B \xi_0}{\hbar}\sqrt{mN(0)T_c(T_c - T)}, \quad (2)$$

where $N(0)$ is the electron *density of states* at the *Fermi surface*, and ξ_0 is the coherence length at $T = 0$. Substitution of Eq. (1) into the Maxwell equation and taking into account Eq. (2) leads to the following expression for the penetration depth:

$$\lambda(T) = \sqrt{\frac{mc^2}{16\pi e^2 |\Psi_0(T)|^2}}$$
$$\cong \frac{\lambda_L(0)}{\sqrt{2(1 - T/T_c)}}, \quad (3)$$

where $\lambda_L(0) = c/\omega_p$ is the *London penetration depth*, and ω_p is the *plasma frequency*. The expression (3) is valid in pure *type II superconductors* when the *Ginzburg–Landau parameter* $\kappa = \lambda_L/\xi_0 > 1/\sqrt{2}$. In *type I superconductors* ($\kappa < 1/\sqrt{2}$) the relationship between J_s and A is nonlocal (A. Pippard, 1953), and in the extreme limit $\lambda_L \ll \xi_0$ it takes the form

$$J_s \approx -\frac{e^2 n_s}{mc}\frac{\lambda_p}{\xi_0}A, \quad (4)$$

where n_s is the superconductor electron concentration, with the *Pippard penetration depth* given by

$$\lambda_P \approx \left(\lambda_L^2 \xi_0\right)^{1/3} = \xi_0 \left(\frac{\lambda_L}{\xi_0}\right)^{2/3} \ll \xi_0. \quad (5)$$

In "dirty" superconductors with a short electron *mean free path* $l < \xi_0$, the coherence and the local correspondence between $\boldsymbol{J}_s$ and $\boldsymbol{A}$ are violated at the distance l, so that

$$\boldsymbol{J}_s \approx -\frac{e^2 n_s}{mc} \frac{l}{\xi_0} \boldsymbol{A}. \quad (6)$$

In this case the effective penetration depth becomes $\lambda_{\mathrm{eff}} \approx \lambda_L (\xi_0/l)^{1/2} > \lambda_L$, and for the condition $l \ll \lambda_L^2/\xi_0$ it significantly exceeds ξ_0, so that in disordered alloys and *amorphous superconductors* $\kappa = \lambda_{\mathrm{eff}}/\xi_0 \gg 1$.

PERCOLATION EFFECTS in superconductors

Particular properties of macroscopically uniform superconducting *composite materials* which depend on the volume fraction p of the superconducting phase. Percolation effects manifest themselves in mixtures of superconducting materials with nonsuperconducting ones, or in the case of the existence of superconducting grids above the percolation threshold (see *Percolation theory*), or in the case of superconducting regions linked by weak Josephson junctions (see *Josephson junction*), as can occur in *granular superconductors*. In the latter case, percolation effects are related to the random scatter of system parameter quantities (in particular, of energies $J_{ij}(T)$ of weak coupling between granules i, j). Taking into account percolation effects, electromagnetic properties are described on the basis of solutions of the Ginzburg–Landau equations (see *Ginzburg–Landau theory of superconductivity*) for grids of thin wires. The selection of grid structures is based on the infinite cluster model of perturbation theory. The fractal dimensionality (see *Fractal*) of the infinite cluster results in the fact that the properties of the grid of an impure superconductor are defined by the quantity $\overline{\xi}_s(T) = [\xi_s(T)]^{2/(2+\theta)}$ (where $\theta \approx 0.8$ for the two-dimensional grid and $\theta \approx 1.7$ for the three-dimensional grid), instead of the ordinary *coherence length* $\xi_s(T)$. Percolation effects are important in the irregular mode, when $\overline{\xi}_s < \xi_p$,

where $\xi_p \propto |p - p_c|^{-\nu_p}$ is the *correlation length* of percolation theory, ν_p is the *critical index*, and p_c is the critical fraction of the superconducting phase. The nature of the temperature dependence of the critical superconducting current I_c of the whole composite varies in accordance with the relationship between $\overline{\xi}_s(T)$ and ξ_p. More particularly, for $\overline{\xi}_s < \xi_p$, $I_c \propto (1 - T/T_c)^{4/3}$ and I_c is independent of $(p - p_c)$, whereas for $\overline{\xi}_s > \xi_p$, I_c is proportional to $(1 - T/T_c)^{3/2}$ and depends on $(p - p_c)$ (T_c is the *critical temperature* of the superconducting transition). The decrease of the size of the granules of a granular superconductor causes its transformation into an *amorphous superconductor*. In the latter case, percolation effects give way to effects related to the *weak localization* of current carriers in superconductors.

PERCOLATION THEORY, seepage theory (fr. Lat. *percolatio*, filtration)

There are two main ways to define percolation theory: (1) it is part of probability theory dedicated to studying problems of the formation and statistical properties of infinite bonded *clusters* on lattices; (2) it is transport theory (of mass, charge, energy) over a random set (cluster), either discrete or continuous. In both cases percolation theory considers bonding of a very large (macroscopic) number of elements on the condition that the bond of each element with a neighbor is of a random nature, but involves a randomness prescribed in quite a specific manner (e.g., by a random number generator with specific properties).

Percolation theory describes well many systems in which a *geometric phase transition* takes place, provided a certain parameter reaches its critical value in such a system. An example is the *metal–insulator transition* in mixtures of conducting and insulating regions, where the concentration of metal *phase* is lower than a certain critical concentration, p_c. Of a similar nature is the development of metallic conductivity in a *semiconductor* when its *doping* exceeds a critical level. Another example is given by rock fracturing when the forces applied to a rock sample exceed a critical value, and microcracks merge into a single major *crack*. Percolation theory is used to describe filtration in porous media, *hopping conductivity* in doped semiconductors, *Anderson localization* in disordered systems, wave propagation through media with a spatially inhomogeneous *refractive index*, elasticity of polymer *gels*, etc.

Many phenomena described by perturbation theory fall into the class of *critical phenomena*. They are characterized by a *critical point* at which the properties of a system change sharply. In percolation theory a change of system properties takes place when an infinite cluster forms from occupied sites bonded together.

The basic statements of percolation theory are usually formulated for lattice problems. Consider a space lattice of nodes or sites, with bonds available for linking together all nearest neighbor sites. In site-percolation theory each site is occupied with a probability p, or empty with a probability $(1 - p)$, and all nearest neighbor occupied sites are connected together by bonds. Bonds only exist between nearest neighbor occupied sites. A cluster is a group of occupied sites such that each such site is bonded to at least one other occupied site in the cluster. For small values of p all clusters are small, each containing only a few occupied sites. An unlimited cluster of occupied sites is called an *infinite cluster*. The *percolation threshold*, p_c, is a particular value of p such that the probability for the existence of an infinite cluster P_∞ (satisfies the conditions $P_\infty(p < p_c) = 0$, $P_\infty(p > p_c) \neq 0$. A geometric *phase transition* takes place when p increases to the point p_c, and isolated clusters merge to form an infinite cluster. For $p > p_c$ the infinite cluster grows in area (or volume) until it fills the entire lattice at $p = 1$. Around p_c many physical properties of the system exhibit critical behavior. Examples are the infinite cluster density, $\rho_\infty(p) \propto (p - p_c)^\beta$, its specific electric conductivity, $\sigma_\infty(p) \propto (p - p_c)^t$, its *diffusion coefficient* for a particle randomly walking through the infinite cluster, $D_\infty(p) \propto (p - p_c)^{\theta v}$ (see *Random walk*). The infinite cluster density $\rho_\infty(p)$ is an analogue of the *order parameter* in *second-order phase transitions*. The analogue of the susceptibility is the average number of points in a finite cluster, proportional to $|p - p_c|^{-\gamma}$. The dynamic characteristics of an infinite cluster (σ_∞, D_∞) have no analogues in the thermodynamic theory of phase transitions.

The dimensionless *critical indices* $\beta, \gamma, t, v, \ldots$ are universal, i.e. they do not depend on the choice of model, and are only defined by the space dimensionality. That statement, tested by a very large number of numerical experiments, serves as the basis of the theory. The system behavior is controlled by the ratio of two spatial scales, the minimum length a (the lattice constant, the average jump distance, the characteristic size, $U(r)$ in continuum problems, see below), and the characteristic spatial dimension of the system – its *correlation length* $L_c \propto a(p - p_c)^{-v}$. System properties fluctuate quite strongly over distances shorter than L_c. At scales far exceeding L_c the system appears homogeneous. It may then be viewed as consisting of blocks of size L_c^d (d is the dimensionality of space into which the percolation system is "immersed"). One may speak about the volume L_c^d as the *representative volume*. This means that practically all the possible combinations of empty and occupied sites are realized within the scale L_c, and self-averaging of the specific characteristics of the system, such as the infinite cluster density ρ_∞, *electrical conductivity, density of states*, etc., takes place there.

When the system approaches the percolation threshold, the size of the region in which self-averaging takes place increases. However, as functions of L_c the properties of the system do not change. For $|p - p_c| \ll 1$ we have $L_c \gg a$, and there exists a range of intermediate asymptotic behaviour $a \ll L \ll L_c$. In the L range all the characteristics of those *percolation clusters* with sizes larger than L are similar to the characteristics of the infinite cluster as a whole. This means that when averaged over many blocks of the size L^d their properties depend on L_c, and moreover, follow the same laws that $\rho_\infty, \sigma_\infty, \ldots$ follow on L_c. In that sense, no self-averaging takes place within the L_c^d volume and the specific characteristics display a universal dependence on the size of the averaging range (the power law). The power indices (β, t, v, θ) are universal characteristics of the physical properties of the system for $|p - p_c| \ll 1$ for a space of a given dimensionality. The reason for such a universality lies with the *statistical self-similarity* (*fractality*) of geometric objects at L scales. In that sense, percolation theory is a particular case of the theory of physical properties of *fractals*.

The universality principle may be violated in those systems whose physical properties are determined by the local microgeometry at scales smaller than a. Among such systems are those described by continuum percolation theory. Continuity problems arise when studying the motion of a particle in the field of a *random potential* $U(r)$. The transport is only possible within the spatial range in which the particle energy E exceeds the potential $U(r)$. The lower limit at which transport through the whole system is possible is the *percolation level* U_c in a continuous problem. The specific electric conductivity in such systems is usually defined by the range around the saddle points of the potential relief, $U(r)$, so that $U_{saddle} \approx U_c$. The characteristic size of such ranges is $b \ll a$. A similar situation is encountered in different classes of continuous problems of percolation theory (so-called Swiss-Cheese, Blue-Cheese), where the resistance of the whole system is defined by the size of the constriction which appears in the conducting matrix when it is saturated with non-conducting inclusions of size a. The characteristic transverse constriction size depends on $(p - p_c)$ and on the condition $b \ll a$.

PERCUSSION CENTER

See *Center of percussion*.

PERCUSSION WELDING in vacuo

Technique of pressure *welding*, which consists in locally heating a thin near-surface layer of the materials to be joined, and applying single- or multi-impulse compression stresses. Percussion welding has its origins in processes of accelerated migration of atoms that are responsible for *interdiffusion* through the joined surface, and in processes of *recrystallization* at the cooling of the weld, and favor the *healing of defects* in the junction area. The technique is used for welding dissimilar *metals*, which form *intermetallic compounds* on being welded together using other methods, as well as for homogeneous materials featuring high thermodynamic stability of corresponding oxides, and high values of vapor pressure. When welding two metals with a tendency toward *chemical bond* formation, e.g., iron and molybdenum, the heating of the area of contact is performed using two ring electron-beam heaters, with the metals to be

welded set apart. The major variables of percussion welding are the temperature and energy of the percussion, which provide the required high-rate *strain* of the near-contact regions of the metals being welded. The duration of the percussion welding process is of the order of a fraction of a second. The total welding cycle time involves a period of preheating to welding temperature, a period of annealing at the welding temperature to ensure uniform heating of components in the welding area, and a time of deformation, which includes welding and cooling time.

PERCUS–YEVICK APPROXIMATION
(J.K. Percus, G.S. Yevick, 1958)

One of the most accurate approximations for the determination of the radial *distribution function* and structure factor (see *Structure amplitude*) of simple Newtonian liquids, which exhibit central *pairwise interactions* between atoms $V(r)$. The *radial distribution function* $g(r)$ is defined from the following relations:

$$n^{(2)}(r_{12}) = \rho^2 g(r_{12}),$$

$$n^{(2)}(r_{12}) = \frac{N(N-1)}{Z_N} \int d^3 r_3 \ldots d^3 r_N$$

$$\times \exp\left[-\frac{1}{k_B T} \sum_{i \neq j} V(r_{ij})\right],$$

where Z_N is the statistical sum, $r_{ij} = |\mathbf{r}_i - \mathbf{r}_j|$, and ρ is the density. The radial distribution function defines the number of particles $dN(r)$ which are to be found within the distance r from a given particle: $dN(r) = \rho g(r) 4\pi r^2 dr$. The energy and *equation of state* of simple liquids can be expressed directly in terms of this distribution function as

$$\frac{U}{N k_B T} = \frac{3}{2} + 2\pi\rho \int_0^\infty dr \, r^2 g(r) \beta V(r),$$

$$\frac{pv}{N k_B T} = 1 - \frac{2\pi\rho}{3} \int_0^\infty dr \, r^2 g(r) \beta \frac{dV}{dr},$$

where U is the energy of a system of N particles, p is the pressure, v is the volume, and $\beta = 1/k_B T$. The structure factor is closely related to the radial distribution function $S_k = 1 + \rho \int a^\beta r e^{-ikr}[g(r) - 1]$. The structure factor

S_k appears in the expression for the cross-section for *diffuse scattering of X-rays* (neutrons) by liquids, which makes it possible to determine this factor directly from experiment.

The direct *correlation function* is introduced through the agency of the Ornstein–Zernike law (L.S. Ornstein, F. Zernike, 1914) for the purpose of analyzing the radial distribution function:

$$h(r_{12}) = C(r_{12}) + \rho \int d^3 r_3 \, C(r_{13}) h(r_{32}),$$

where $h(r) = g(r) - 1$. The characteristic radius of $C(r)$ is smaller than that of $h(r)$, and the Fourier-transformed image of $C(r)$ is related to the structure factor by a simple equation $S_k = 1/(1 - \rho C_k)$. In the limit of low densities, the radial distribution function and its relation with $C(r)$ are determined by application of the *Mayer diagram technique* which leads to a cumulant expansion in powers of the *Mayer function* $f(r) = \exp[-\beta V(r)] - 1$. The simplest diagrams obtained through this technique correspond to the Percus–Yevick approximation $C(r) = f(r) y(r)$, where $y(r) = g(r) \exp[\beta V(r)]$; the *Percus–Yevick equation* is of the form:

$$y(r_{12}) = 1 + \rho \int d^3 r_3 \, f(r_{13}) y(r_{13})$$

$$\times \left[y(r_{23}) \exp\left[-\rho V(r_{23})\right] - 1 \right].$$

A large body of calculations shows that the Percus–Yevick approximation describes the radial distribution function very accurately over broad ranges of parameters. This approximation is widely accepted, since there exists an analytical solution of the Percus–Yevick equation for the physically important case of a hard-sphere liquid (see *Hard-sphere model*), namely:

$$C(r) = 0, \quad r > d,$$

$$C(r) = -a^2 + 6\eta(a+b)^2 \frac{r}{d} - \frac{\eta a^2}{2} \left(\frac{r}{d}\right)^3,$$

$$a = \frac{1 + 2\eta}{(1-\eta)^2},$$

$$b = -\frac{3\eta}{2(1-\eta)^2},$$

where d is the hard sphere diameter, and $\eta = \pi \rho d^3 / 6$ is the degree of packing. In this case, the thermodynamic quantities are functions of the parameter η only. Thus, the equation of state takes the form

$$\frac{p v}{N k_B T} = 1 + 4\eta y(d) = \frac{1 + 2\eta + 3\eta^2}{(1-\eta)^2}.$$

The Percus–Yevick approximation in the case of hard spheres is employed to advantage as the basis system in the thermodynamics of simple liquids in the context of Gibbs–Bogolyubov thermodynamic perturbation theory. This approximation is closely related to the so-called *mean sphere approximation*, which deals with systems with a potential called the "hard spheres plus an attractive tail": $g(r) = 0$ $(r < d)$, $C(r) = -\beta V(r)$ $(r > d)$. The mean sphere approximation is identical to the Percus–Yevick approximation for $r \gg d$, and permits the derivation of analytical solutions for a wide class of potentials, including the Coulomb type.

PERFECT CRYSTAL

A crystal of a perfect shape with crystallographically equivalent faces developed to an equal degree (see *Crystal faceting*). Crystals close to perfect can grow suspended in a gently stirred solution. See also *Ideal crystal*.

PERFECT DIAMAGNETISM

See *Absolute diamagnetism*.

PERFECT DISLOCATION

A dislocation in which the *Burgers vector* is one of the vectors of a crystal *translation*. According to the energetic *Frank criterion* (F.C. Frank) a perfect dislocation with Burgers vector b_1 splits into perfect dislocations b_2 and b_3, if $b_1^2 > b_2^2 + b_3^2$; therefore, energetic stability is exhibited only by *dislocations* with one (or, sometimes, two) possible shortest Burgers vectors.

PERFECTLY RIGID BODY

An idealized concept used in physics and engineering when the *strain* of a body can be neglected, e.g., in the assumed absence of the bending of a rod subjected to torques. Formally, the perfectly rigid body model corresponds to the real body limit with the values of its *elastic moduli* tending to infinity. The model of a perfectly rigid body is in conflict with the special theory of relativity because it allows signal transmission with an infinite speed.

PERIODIC POTENTIAL

The potential of a probe particle, which is a periodic function of its coordinate r. The potential of an *ideal crystal* is periodic as seen by an "excess" electron (or *hole*) moving throughout the crystal. Its period is determined by the translation vector of the crystal lattice. Within the *single-electron approximation*, the description of the motion of an itinerent electron of a crystal in the *self-consistent field* of other electrons also leads to the concept of a periodic potential. The motion of one of the itinerent electrons (the states of other electrons being unaltered) proceeds in the total periodic potential, and the potential of the hole (either free or bound by a *defect*), which is left by this electron. At long distances from the hole, the potential may be treated as approximately periodic (interaction with the hole is a perturbation, which results in scattering). The ideal crystal potential is fully periodic with respect to the motion of the center of gravity of the electron–hole pair (i.e. the *exciton*). The fundamental solutions of the Schrödinger equation in the periodic potential (called the *Hill equation* in mathematics) are *Bloch functions* (see *Bloch theorem*):

$$\psi_{k\lambda} = \exp(ikr)u_{k\lambda}(r),$$

where $u_{k\lambda}(r)$ exhibits translational spatial symmetry, and $\hbar k$ is the *quasi-momentum* of the probe particle. The energy $\varepsilon_\lambda(p)$ of an electron corresponds to this eigenfunction, where λ is the index of the energy band. The energy is a periodic function of the quasi-momentum with the period b of the reciprocal lattice $\varepsilon_\lambda(p) = \varepsilon_\lambda(p + \hbar b)$. This property (periodicity) is a general property of any quasi-particle in the crystal. The particular form of $u_{k\lambda}(r)$ and the energy spectrum of the probe particle are determined by the explicit form of the periodic potential.

A simple model periodic potential is the one-dimensional *Kronig–Penney potential* (R. Kronig, W.G. Penney, 1931) (see Fig.), which admits an exact solution. The form of this solution is relatively simple, if the barrier width $b \to 0$ and

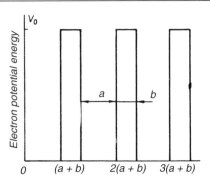

One-dimensional periodic potential energy of an electron in the Kronig–Penney model; the ordinate V_0 is the electron potential energy.

the parameter $\gamma = 2m(V - E)^{1/2}/\hbar^2$ (m is electron mass, E is energy) tends to infinity in such a way that the quantity $p = \gamma^2 ab/2$ remains finite, $(p \sin \beta a)/\beta a + \cos \beta a = \cos ka$, where $\beta = (2mE)^{1/2}/\hbar$, a is the width of the valley between the barriers, V_0 is the barrier height. This relation provides the k-dependence of the energy E. Its graphical solution defines the system of allowed and forbidden *energy bands*. Another model periodic potential is the *sine potential*. The Hill equation with a sine potential reduces to the *Mathieu equation* (E.L. Mathieu, 1868) which has the standard form:

$$\frac{d^2\psi}{dx^2} + \left(b - g^2 \cos^2 x\right)\psi = 0.$$

Its solutions have been studied in detail. A periodic potential for a period far exceeding the *crystal lattice* constant may be produced by artificial means (see *Superlattices*).

PERITECTIC ALLOYS (fr. Gr. περιτηκω, melt around)

Alloys in which crystallization from a liquid (the melt) proceeds through precipitation of the solid *phase* α, followed by formation of the second solid phase β from the α-phase and the remaining liquid. In the case of three-component peritectic alloys, one liquid phase and one solid phase produce two solid phases, or a single solid phase is produced from one liquid phase and two solid phases. The alloys are referred to as *peritectoid alloys*, if the starting phase is not a liquid, but rather a solid phase, which passes into another solid phase on cooling (see also *Phase diagram*).

PERITECTIC REACTION

Phase transformation in which a liquid phase, upon cooling, combines with a solid phase to form a new solid phase. It occurs in compounds which change their composition when they melt.

PERMALLOY (abbr. of permeable alloy)

Iron–nickel based alloy. See *High-permeability magnetic materials*.

PERMANENT MAGNET

A *hard magnetic material* which serves as a source of a constant *magnetic field*. In the simplest cases, a permanent magnet is a material (shaped as a horseshoe, a bar, a washer, etc.) that has undergone an appropriate *heat treatment* and been magnetized up to saturation. In more complex cases, a permanent magnet is a component of a system designed to provide a constant or variable magnetic field. Two parameters associated with the material's magnetization curve or hysteresis loop are: its remanence B_r or the magnetic flux density when the applied field H vanishes, and the coercive field H_c, which reduces B to zero (see Fig.). As a source of magnetic field, a permanent magnet operates under the impact of its own *demagnetization field*, H_d, and its residual magnetic flux density, B_d, is always smaller than the material remanence, B_r. The quantity B_d determines the field in the air gap. It depends upon the permanent magnet shape, on the configuration of the overall magnet system, on the values of H_c and B_r characteristic of the magnetic material, the shape of the

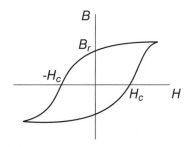

Hysteresis loop showing the remanence B_r and coercive field H_c.

demagnetization curve, viz., the portion of the *hysteresis* loop delimited by the B_r and H_c points. An important characteristic of permanent magnet materials is the maximum value of the product of the coordinates on a demagnetization curve, $(BH)_{max}$, known as the *maximum energy product*. The external magnetic field established by a permanent magnet acquires its maximum energy when the coordinates of its operating point (B_r, H_c) correspond to $(BH)_{max}$. Permanent magnets have $(BH)_{max}$ values amounting to 320 T·kA/m.

Permanent magnets find their application in a diversity of fields. The earliest device utilizing a permanent magnet was a compass. Permanent magnets are employed in electric motors, in electronics, automatics, and robotics, for magnetic couplings, clutches and bearings, in clock-making, in household equipment (e.g., in loudspeakers, for refrigerator door sealing, etc.) and in many other cases. The total industrial output of permanent magnets runs into thousands of tons per year.

PERMEABILITY

See *Magnetic permeability*.

PERMENDUR (abbr. of permeable and durable)

Fe–Co alloy. See *Soft magnetic materials*.

PERMINVAR (abbr. of permeable and invariable)

A *soft magnetic material* based on Fe–Ni–Co alloys (with Cr and Mo additions), which is distinguished by its weak field dependence of the *magnetic permeability* μ over a wide range of magnetic intensities, and low *magnetic losses* due to hysteresis. Cooling perminvar in a magnetic field results in induced *magnetic anisotropy* due to directional ordering of atoms. This results in a right-angle *hysteresis* loop along the field, and a line hysteresis loop in the transverse direction. If perminvar is cooled in the absence of an external field, then the hysteresis loop assumes a specific shape, which is characterized by twisting near the origin of the coordinates, and by a low value of *remanence* ("*perminvar loop*").

PERMITTIVITY

See *Dielectric constant*.

PEROVSKITES (after the mineral perovskite with the formula $CaTiO_3$)

Natural and artificial chemical compounds, which belong to the class of mixed metal oxides, halides and sulfides. The general formula of perovskites is $A^m B^n X^p$ ($m + n = -3p$), where A and B are atoms of metals and X^p is F^-, Cl^-, Br^-, O^{2-}, S^{2-}. Mixed oxides, with valences which satisfy the equality $m + n = 6$, may exist in the form of perovskites of composition $A^+ B^{5+} O_3$, $A^{2+} B^{4+} O_3$ and $A^{3+} B^{3+} O_3$. Mixed halides of the perovskite structure are represented by the formulae $A^+ B^{2+} X_3$ (X is F, Cl, Br) (see Table 1).

The *perovskite structure* is arranged as follows: atoms A (of larger radius) and X form extremely close packing by assuming positions of coordination number 12 (face-centered cubic). Atoms B (of smaller size) occupy the octahedral positions. Octahedra BX_6 are united by the sharing of vertices, thereby forming a three-dimensional octahedral frame (see Fig.), the large cavities of which contain atoms A. The stability of the perovskite structure is determined (in the case of oxides) by the tolerance factor t,

$$t = \frac{d(A-O)}{\sqrt{2} d(B-O)},$$

where d are interatomic metal–oxygen distances. The perovskite structure is realized, if $0.8 \leqslant t \leqslant 1.0$. The perfect arrangement of A and B atoms with an undistorted "oxygen environment" exhibits cubic symmetry ($Pm3m$). The distortion from cubic symmetry is determined by the ratio of the sizes of cations A and B, by charges

carried by these cations, and by the temperature. The decrease of symmetry with decreasing temperature at *phase transitions* occurs in following stages: cubic ($m3m$) → tetragonal ($4mm$) → orthorhombic (mmm) → trigonal ($3m$). This sequence is realized completely only in $BaTiO_3$ and $KNbO_3$. Only a part of the transformation sequence is realized in the case of other compounds ($PbTiO_3$, $PbZrO_3$, $NaNbO_3$). The change in symmetry of the perovskite structure is accompanied by a size change; the volume of the *unit cell* of the newly-generated phase is either slightly different from the volume of initial phase unit cell, or is a multiple of it. The perovskites of formulae $LnBO_3$ (Ln = *lanthanide* or Y; B = Al, Cr, V, Fe, Sc, Ga), $CaRuO_3$ and $SrRuO_3$ are characterized by an orthorhombic distortion due to adaptation of the octahedral frame to cubic octahedra with small-sized Ln^{3+}, Y^{3+}, etc. The degree of orthorhombicity increases with increasing lanthanide atomic number. The high-temperature $BaTiO_3$ may exhibit an hexagonal structure which is close to the perovskite structure. Such a structure is observed in $BaMnO_3$, $BaNiO_3$, $LiSbO_3$. The undistorted octahedral perovskite frame is observed in ReO_3 and WO_3. Filling the cavities of the frame of Ti-, Nb-, W-octahedra with atoms of Li, Na, K, Ca, Sr, Ln results in the structure of cubic bronzes, which may be considered as cation-deficient perovskites of general formula $A_x BO_3$ (A = atom of alkali, alkaline-earth metal, lanthanide, Cu, Al; B = Ti, Nb, Ta, W). There exist oxygen-deficient perovskite phases ABO_{3-x} ($SrTiO_{2.5}$, $SrVO_{2.5}$, $BaFeO_{2.5}$), with

Table 1. Perovskites

Atom	Oxides			Halides	Sulfides
	$A^+ B^{5+} O_3$	$A^{2+} B^{4+} O_3$	$A^{3+} B^{3+} O_3$	$A^+ B^{2+} X_3$	$A^{2+} B^{4+} X_3$
A	Na, K, Rb, Cs, Tl	Ca, Sr, Ba, Cd, Pb, Eu	Sc, Y, Ln*, Pu, Bi	Li, Na, K, Rb, Cs, NH$_4$, Tl, Ag, Hg	Ca, Sr, Ba, Eu
B	Nb, Ta, W, Bi	Ti, V, Mn, Fe, Co, Zr, Mo, Ru, Sn, Ce, Pr, Hf, Nb, Pb, Th, U, Pu	Al, Sc, Ti, V, Cr, Mn, Fe, Co, Ni, Ga, Y, Rh, Bi	Mg, Ca, Sr, Ba, Zn, Cd, Hg, Cu, Cr, Mn, Fe, Co, Ni	Ti, V, Sn, Zr, Hf, Ta

*Ln = La, Ce, Pr, Nd, Sm, Eu, Gd, Dy, Ho, Er, Yb, Lu.

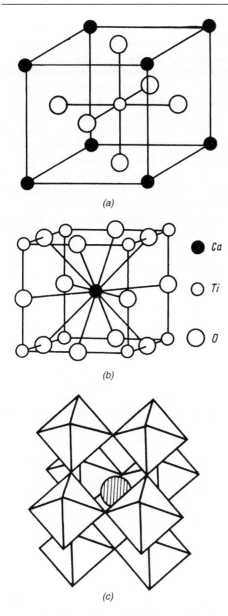

(a)

(b)

● Ca

○ Ti

◯ O

(c)

Crystal structure of perovskites which have the formula $CaTiO_3$: (a) titanium atoms (B-atoms) in the octahedral coordination of oxygen atoms; (b) calcium atoms (A-atoms) in the 12-fold cubic-octahedral coordination of oxygen atoms; (c) polyhedral representation of perovskite structure. Titanium octahedra are located at the vertexes of the primitive cubic cell, whereas the large calcium atoms (shown dashed in (c)) are situated within the cubic-octahedra of the octahedral skeleton.

a homogeneity region which depends on x. Filling the B-positions with atoms of formal valence below $+3$, e.g., Cu^{2+}, results in perovskite-like structures, which do not contain oxygen atoms in the layer of copper atoms (e.g., $YBa_2Cu_3O_6$). On addition of oxygen atoms to this layer, the *high-temperature superconductor* $YBa_2Cu_3O_7$ is formed. Partial or complete separation of octahedral layers of the frame leads to related structures of the type K_2NiF_4, which include, among other compounds, high-temperature superconductors of the general formula $La_{2-x}Sr_xCuO_4$. Another type of closely related structures is presented by *Aurivillius phases*. The perovskite blocks in these phases are separated by double layers of Bi_2O_2. The block height may range from 1 to 4 octahedra. The copper-containing perovskites of similar structure are high-temperature superconductors ($TlBa_2Ca_{n-1}Cu_nO_{2n+3}$, $Bi_2Sr_2Ca_{n-1}Cu_nO_{2n+4}$). The perovskites are characterized by broad ranges of isomorphous substitutions in A and B positions. The position B may be shared by cations of different valence in a stoichiometric ratio (e.g., $SrB_{0.5}^{2+}B_{0.5}^{6+}O_3$).

Perovskites belong to the class of *ferroelectrics* ($BaTiO_3$, $KNbO_3$, $PbTiO_3$, $NaTaO_3$, $KTaO_3$), *antiferroelectrics* ($PbZrO_3$, $PbHfO_3$, $NaNbO_3$), strong ($SrRuO_3$) and weak ($LaFeO_3$) ferromagnets, antiferromagnets ($CaCrO_3$, $LaFeO_3$, $YFeO_3$). The conductivity ranges from the metallic ($LaTiO_3$, $SrVO_3$, $LaNiO_3$, bronzes) to the semiconductor type ($GdTiO_3$, $CaCrO_3$). The conductivity depends on the oxygen stoichiometry, the composition of the *solid solution*, and the presence of impurity elements.

PERSISTENT CURRENT in superconductors

Current density J induced in a superconducting ring (e.g., by a magnetic field) that does not decay with time, provided $J < J_c$, where J_c is the *critical current* density for the material. H. Kamerlingh Onnes (1914) was the first to observe the phenomenon. A persistent current is a metastable state, with its existence explained by the *quantization of flux* (see *Macroscopic quantum phenomena*). The energy barrier preventing the persistent current from decaying with time results from the necessity of destroying *Cooper pairs* in order to suppress the current, and for $J < J_c$ this is disadvantageous in terms of energy.

PERTURBATION MATRIX METHOD,
Koster–Statz method

Group-theoretical method for obtaining matrices that describe changes in the arrangement of paramagnetic ion energy levels under the action of an external disturbance V (constant electric or magnetic field). If the wave functions of an ion in a *crystal* are chosen so that they form basis sets of irreducible representations Γ_α and Γ_β of a *local symmetry group*, and V is resolved into parts that transform as components of an irreducible representations Γ_γ, then the number N of independent matrix elements of V is determined by the multiplicity of irreducible representations Γ_γ^* involved in the direct product $(\Gamma_\alpha^* \times \Gamma_\beta)$ of Γ_α^* and Γ_β. The requirements of being Hermitian, and of invariance under *time inversion* and under the choice of wave function phases reduce N further. The parameters that remain undefined are obtained from comparison of positions of calculated *electron paramagnetic resonance* lines with observed ones. Since the perturbation matrix method is cumbersome, the alternative *method of a generalized spin Hamiltonian* has found much wider use (see *Spin Hamiltonian*).

PERTURBED ANGULAR CORRELATION METHOD

One of the nuclear-spectroscopic methods of investigating *hyperfine interactions* of atomic nuclei with their electronic surroundings in atoms, ions, molecules and condensed phases.

Due to the *conservation law* of momentum the particles radiated by the nucleus during the process of radioactive decay (α-particles, β-electrons, γ-quanta, etc.) are spatially correlated in a specific way. Their angular correlation can be measured by the methods of nuclear spectroscopy. Due to the hyperfine interaction the expected angular correlation of the nuclear emissions can be violated for the condition $\omega_{HF}\tau \geqslant 1$, where τ is the lifetime of the nucleus and ω_{HF} is the characteristic frequency of the hyperfine interaction. Perturbations arise from the precession of the nuclear spin I and its associated magnetic *dipole moment* μ around the direction of the magnetic field B at the nucleus, and from the *nuclear electrical quadrupole moment* Q interacting with the electric field gradient there. The *electric field*

gradient and B are determined by the distribution of the electron density and by its polarization in the closest vicinity of the nucleus, and these are important parameters of the electronic structure of solids. An advantage of the perturbed angular correlation method is the fact that it can be applied at extremely low *concentrations* (10^{-10} and less) of the radioactive impurity atoms in the lattice under investigation, which precludes their interaction with each other. This property has led to, in particular, the successful application of this method for studying the structure and dynamics of *point defects* (including those from radiation damage) in crystals at the microscopic level.

PETA...

Prefix for a physical unit name, which allows obtaining a multiple unit, equal to 10^{15} initial units. Symbol: P. Example: 1 PHz (petahertz) = 10^{15} Hz.

PETCH RELATION, Hall–Petch relation
(E.D. Hall, N.I. Petch, 1953)

The equation, which describes the influence of the crystalline material grain size d on the *yield limit* as $\sigma_3 = \sigma_0 + K_y d^{-1/2}$, where σ_0 and K_y are constants. With allowance made for the orientation factor, the value of σ_0 coincides with the yield point of *monocrystals*, and K_y is characteristic of the force to be applied in order to transfer the sliding across the *grain boundary*. There exist a number of approaches for obtaining the above relation theoretically. The approach that enjoys the widest application consists in the following: the grain size limits the length of the planar agglomeration of dislocations, the *stress concentration* at the tip of which is proportional to $d^{1/2}$. In the event that the metal under consideration exhibits a dislocation substructure, the value for the size of the fragment or the disoriented dislocation cell is substituted for the grain size value d in the above equation (see *Fragmentation*).

PHASE, thermodynamic phase

The state of a substance, which is allowed by general physical and physicochemical laws, and has a particular composition and physical properties which differ from the properties of other possible states of the same substance. Phases realizable in practice are, as a rule, thermodynamically stable (i.e. are at thermodynamic equilibrium), yet non-equilibrium metastable phases are also observed in a number of cases. Examples of thermodynamic phases are: solid, liquid, and gaseous *states of matter*, or paramagnetic (see *Paramagnet*) and magnetically ordered states of a *magnetic material*, or states of the same substance characterized by different *crystal structures* or different electrical properties (metal, insulator, superconductor). In the majority of cases a thermodynamic phase is spatially homogeneous, yet there exist phases in which the order of homogeneity is modulated by a wave, the wavelength of which is in general case not a multiple of the principal structural constant of the substance. The latter phases are referred to as *incommensurate phases*, or incommensurable ones (see *Incommensurate structure*). Examples of incommensurate phases are those involving *spin density waves*, *charge density waves*, waves modulating atomic displacements from initial periodic positions, a number of *modulated magnetic structures*, etc.

Each of the possible phase states of a substance is an equilibrium and thermodynamically stable state over a certain range of values of various quantities (*temperature*, *pressure*, composition, electric and magnetic field intensities, etc.). When these state variables take on values from allowed intervals, the *thermodynamic potential* of the phase becomes minimal relative to other phases. Multidimensional (in accordance with the number of descriptive state variables) geometrical representations of the ranges of values that correspond to equilibrium states of possible phases of a substance form compositional diagrams, or *phase diagrams*. In more common use are cross-sections (plane sections) of these diagrams, which relate a certain pair of state variables: e.g., two-dimensional compositional diagrams relating temperature and pressure, or temperature and magnetic field intensity. The regions of thermodynamic equilibrium of a phase on such a diagram may be closed, i.e. enclosed by a surface (line), the intersection of which (by varying any of the state parameters) causes a *phase transition*, or it may be open or unbounded. In the latter case, the line of phase transition on the diagram ends with a *critical point*, at which the differences between phases disappear. Depending on the kind of phase transition occurring in the system, the coexistence of several phases at the curve (or at the point) of phase transition may be allowed (e.g., *first-order phase transition*) or forbidden (*second-order phase transition*). The number of phases which may coexist in equilibrium is given by *Gibbs' phase rule*.

Usually, it is possible to cross the curve of a first-order transition is such a manner that the phase transition does not have enough time to occur, and the phase remains in a metastable state (*metastable phase*, see *Metastable state*) until its state variables assume those corresponding to a complete lack of stability (e.g., supercooled liquid). In the case of first-order transitions, the formation of a new phase involves the appearance of its nuclei (see *Nucleation at phase transitions*). The presence of factors which permit maintaining the metastable state results in *hysteresis* (e.g., during magnetization reversal). Under the conditions of phase equilibrium, first-order phase transitions are accompanied by heterophase *fluctuations*. The characteristics of spatial phase boundaries in multiphase systems are determined for specified values of state variables, subject to the condition of minimal thermodynamic potential with regard to *surface energy* and the energy of interfaces.

In a number of cases (e.g., a *ferromagnet*, or a *ferroelectric* with uncompensated surface-bound charges), the character of the *order parameter* is such that long-range fields arise (from surface magnetic moments, or surface charges), which increase the energy of the homogeneous phase. In these cases, the requirement that the total energy of the sample and its fields be minimal may result in the generation of special ranges of values of extensive state variables (like, e.g., magnetic field strength or its direction relative to crystal axes), within which the generation of inhomogeneous multiphase states is favored. The latter states are bounded by equilibrium periodic interfaces near first-order transitions across the external field (magnetic or electric). The regions of a homogeneous phase are referred to as *domains*. When a phase transition involves a "change in direction of the vector order parameter" in the neighborhood of zero external field, then phases of different domains exhibit one and the same type of ordering, but the directions of *magnetization* and spontaneous *polarization* are different (most commonly, opposite to each another) in adjacent domains. In the case of an order–order transition in a finite external field, the domains belong to phases of different types of ordering. The fields created by different domains neutralize one another with distance from the sample, and hence contribute less to its energy. When *superconductors* are placed into an external magnetic field that exceeds the critical value of a corresponding state variable (e.g., magnetic field), they show equilibrium multiphase states, the role of phases in the latter being played by the regions that are ordered (superconducting) and disordered (normal). This phenomenon is due to the requirement stating that the total energy of the sample and of external magnetic field, which is distorted because of forcing the magnetic flux to leave the superconducting phase, is to be minimized. In case of *type I superconductors*, the boundary between these phases has a positive energy, and an *intermediate state* of periodic domain structure is realized. In the case of *type II superconductors*, the interface energy becomes negative, and so a *mixed state* is realized, in which the normal phase is represented by cores of vortices of penetrating magnetic flux, which form a periodic vortex lattice structure. (The term "intermediate state" is commonly used in reference to the above-mentioned multi-domain states near a first-order transition in the presence of an external field in magnetics, ferroelectrics.)

The above-mentioned multiphase states are thermodynamically stable and in equilibrium over a certain range of values of extensive state variables. Depending on these variables (and, sometimes, on the sample shape), the number and structure of phase boundaries may change, as well as the amounts of substance related to each of the coexisting phases (e.g., to domains having different directions of magnetization relative to the external field in the case of a ferromagnet in a field too weak to bring about magnetic saturation). On this basis, multiphase states of these types may be treated as spatially homogeneous phases; these states may be studied by constructing phase diagrams, and it is physically correct to discuss phase transitions between multiphase states and other possible phases (both spatially uniform and nonuniform) of a given substance.

Characteristic of a second-order transition is the growth of *critical fluctuations* as the transition region is approached. The phase transition is thereby accompanied by critical deceleration of the dynamics, and an infinite increase of the *correlation length* of order parameter fluctuations. As the system reaches the transition point, the differences between phases vanish (order parameters of different phases undergo continuous interconversions, the symmetry group of one phase being a symmetry subgroup of the other), and the transition takes place coherently over the entire bulk.

When introducing the concept of a thermodynamic phase, it is implied that the properties of a substance in each phase are defined by a complete set of thermodynamic state variables, and are independent of the amount of material. This restricts the concept of thermodynamic phase from being applied to aggregations of atoms (molecules) containing sufficiently small numbers of particles (estimated as 10^2–10^5) that the surface energy becomes comparable to bulk energy.

PHASE ANALYSIS

Analysis of the chemical nature, composition, structure, dispersity, and number of thermodynamic *phases* that comprise a multiphase material under study. A phase analysis is made either at a preliminary separation of phases, without phase separation in the case of equilibrium (or nonequilibrium) systems, or at the onset of a change of state. The phase analysis is carried out using physical and *chemical methods of analysis.*

PHASE DIAGRAM, equilibrium diagram, compositional phase diagram, phase-rule diagram

A diagram describing the equilibrium state of a physico-chemical system as a function of intrinsic and extrinsic parameters. The notion of phase diagram is based on the concepts of *phases*, components and phase equilibrium; it was originally formulated by J. Gibbs. Homogeneous portions of a *heterogeneous system*, or the totality of several such portions separated from each other by interfaces, each according to its own chemical composition, structure and properties, are called phases. The chemical composition of every phase is expressed through the concentration of its *components*, which are individual materials capable of existing in pure form. The most

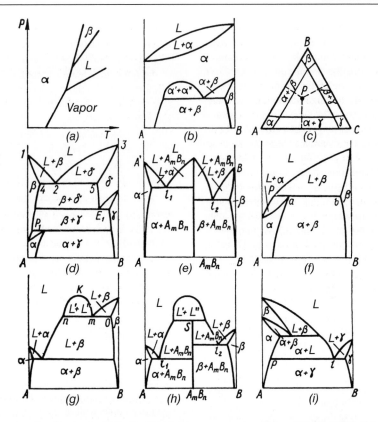

Multiphase diagrams of various types: (a) single-component pressure versus volume plot, (b) and (d)–(i) two-component (binary) temperature versus composition plots, and (c) three-component (ternary) compositional phase diagram. In these plots L denotes liquid, α denotes solid of component A, β is solid of component B, $\alpha + \beta$ is a solid solution, $A_m B_n$ is a particular compound of components A and B.

important parameters are the *temperature T* and *pressure P*. The number of the internal parameters is determined by the number of components. The relation between the number of components c, phases φ and degrees of freedom or *variances f* (the number of independent variables or coordinates which determine the geometry of the phase diagram) is given by *Gibbs' phase rule*: $f = c + 2 - \varphi$. The phase diagram of a condensed system is generally formed from isobaric cross-sections of temperature–composition–pressure diagrams at the standard pressure $P = 101.3$ kPa $= 1$ atm. A small deviation of P from 101.3 kPa does not essentially influence the phase composition of condensed systems, since no gas phase is present. If the pressure is fixed, the phase rule

assumes the simplified form $f = c + 1 - \varphi$. According to the number of components c it is simple (unitary), binary, ternary, etc. Depending on the degrees of freedom f there are the following types of equilibrium: invariant points at $f = 0$, monovariant lines at $f = 1$, divariant surfaces at $f = 2$, etc. The phase diagram of a simple system is plotted in standard coordinates P versus T on Fig. (a). The lines of this diagram show the influence of external factors on the *state of matter*, and also the structure of the solid state if the variation of T or P leads to polymorphous transformations between solids α and β (see *Structural phase transition*). Lines of a *magnetic phase transition* may be shown on a phase diagram.

In order to show the composition of poly-component systems ($c \geqslant 2$) uniform coordinates are used, where the sum of the concentrations of components is equal to a constant number. Thus, e.g., the composition of the alloy of the binary system formed by the two components α and β (Fig. (b)) is determined by the position of a point p along the horizontal length AB, aligned with the abscissa axis of the diagram, whose length is selected equal to 1 or 100. The β-component concentration of the alloy is proportional to the value of the horizontal length Ap, with the percent composition $X = (Ap/AB) \cdot 100\%$. The alloy composition in the ternary system consisting of the components α, β and γ is given by a point within the equilateral triangle ABC (Fig. (c)). There the property is used, that the sum of distances from any point p (composition point) within the triangle to its three sides is a constant value, equal to the height of the triangle. The concentration of the α-component in the alloy at point p varies inversely with the distance from the point p, to the side BC, opposite to the vertex A. The composition of a four-component system is designated by a point within a regular tetrahedron, the vertices of which are labeled by the components. The content of a component is proportional to the distance from the composition point to the face opposite to this vertex. In order to represent equilibrium in the system of five components it is necessary to use a pentatope, or a regular figure with five vertices equally separated from each other in four dimensional space. A system of $(n + 1)$ components is represented by a polytope formed by $(n + 1)$ equally separated vertices in an n-dimensional space.

The basis for a phase diagram classification lies either in (1) the character of phase formation in the solid state, i.e. diagrams with unlimited solubility (Fig. (b)), with boundary solid solutions (Fig. (d)), with intermediate compounds (Fig. (e)), or (2) the type with invariant equilibrium: *eutectic transformation* or the formation of two phases $L \leftrightarrow \alpha + \beta$ from a liquid; *syntectic transformation* or the formation of a solid phase from two liquids $L_1 + L_2 = A_m B_n$ (Fig. (h)); *peritectic transformation*, or the formation of a solid phase from a liquid and another solid phase $L + \beta \leftrightarrow \alpha$ (Fig. (f)); *monotectic transformation* or the formation of a solid phase plus a liquid from a liquid of another composition (Fig. (g)); *metatectic*

transformation or the formation of a liquid and another solid phase from a solid phase $\beta \leftrightarrow L + \alpha$ (Fig. (i)). In the solid state we have the *eutectoid* or the low-temperature decomposition of a solid solution into two solid solution phases, $\delta \leftrightarrow \beta + \gamma$ (Fig. (d)); a *peritectoid* whereby upon cooling two solid phases form a single solid phase, $\beta + \gamma \leftrightarrow \alpha$ (Fig. (d)); and a *monotectoid* transformation $\alpha'' \leftrightarrow \beta + \alpha$ may occur. These differ from eutectic, peritectic and monotectic types since there is no liquid *crystallization* but rather solid state disassociation.

The curve 1–2–3 where the crystallization starts (Fig. (d)) is called the *liquidus curve*, and the curve 1–4–2–5–3 where it ends (beginning of *melting*) (Fig. (d)) is called the *solidus curve*. In three-component systems we have, respectively, the *liquidus surface* and the *solidus surface*. The temperature gap between these curves (surfaces) of liquidus and solidus is called the *crystallization gap*.

The phase diagram of a binary system with liquidus and solidus curves only is called a *melting diagram*. In ternary systems the projection of the solidus surface at the concentration triangle is also called a melting diagram. In the case of systems with one component at room temperature and atmospheric pressure in the liquid state, the term "*solubility diagram*" is used, interchangeably with the term "phase diagram". The phase diagram of a system with $c \geqslant 3$ can be displayed in the form of isothermal cross-sections, or as polythermal cut-away views at fixed concentrations of several components.

Most phase diagrams are plotted experimentally using data from thermal, X-ray structural, resistometric, local X-ray spectral, dilatometric, magnetic, optical microscopy, electron microscopy and other methods of analysis. The complex method of investigating phase balances by combining convential experimental studies with thermodynamic calculations is widely used. It is especially helpful for the study of polycomponent systems. A phase diagram can be the basis for the development of new materials prepared with predetermined properties since it identifies equilibrium states of physico-chemical systems as a function of internal and external parameters.

PHASE DIAGRAM, QUASI-

See *Quasi-phase diagram*.

PHASE EQUILIBRIUM

A *thermodynamic equilibrium* of a system consisting of two or more phases. A closed system in thermodynamic equilibrium has all of its coexisting phases φ at the same *temperature, pressure*, and *chemical potential* in all c of the system components. In addition, the size of each separate inclusion of each phase should be large enough to treat its temperature, pressure and chemical potential as meaningful statistical variables. The number of degrees of freedom f and the number of phases φ in equilibrium with each other in a multicomponent system are related by *Gibbs' phase rule* $f = c - \varphi + 2$. In actual solids exhibiting a defect structure there can appear stresses in the vicinity of appreciable disturbances of the orderly structure of the crystal lattice, and the conditions of phase equilibrium change in these regions. The latter circumstance affects the kinetics of phase transitions, and the structure of a newly formed phase, thus making possible the *thermomechanical treatment* of alloys. A metastable phase equilibrium often develops in solids when a phase persists for a rather long time under conditions in which the other phase is actually more stable (e.g., supercooling). *Metastable phases* form the basis of many metallic materials widely used in technology (e.g., *martensite* in steel). The rate of temperature change alters the conditions of phase equilibrium, which makes plots of the kinetic *equations of state* or *quasi-phase diagrams* necessary for many practical applications.

PHASE EQUILIBRIUM, THERMOELASTIC

See *Thermoelastic phase equilibrium*.

PHASE, HUME-ROTHERY

See *Hume-Rothery phases*.

PHASE INTERFACE

A transition layer which separates two *phases* of the same or different chemical composition which are in contact. It can be a boundary between either *solid* and liquid, solid and gas, or different solids, and so on. As a general rule, a phase interface often has quite a complicated structure. It is heterogeneous both in its structural composition and in its electrophysical and optical properties. Due to the *interdiffusion* of atoms and molecules in the transition layer it is sharply heterogeneous in its specific density, especially in the presence of a gaseous atmosphere. Interface layers in solids and liquids are usually electrically charged (see *Space charge region*), and this charge determines the electrical current that passes through the interface. The *electrical conductivity* of a solid contact can be changed by applying a potential difference, and this property is widely used in solid-state electronic devices.

PHASE MATCHING

See *Second harmonic generation*.

PHASE-MATCHING ANGLE

See *Second harmonic generation*.

PHASER (acronym for PHonon Amplification by Stimulated Emission of Radiation)

Paramagnetic *quantum amplifier* of sound (oscillator). The chief component of a phaser is an active crystal with paramagnetic centers. The structure of the electronic energy spectrum of these centers is required to satisfy the following conditions: on the one hand, the nature of relaxation processes must permit the attainment of an inverted energy *level population*; and on the other hand, the *electron–phonon interaction* between the levels must be strong enough to provide the needed amplification of hypersound (see *Sound*). The active crystal is made in the shape of a sound-conducting rod. Electromagnetic pumping of this rod concentrates in it an electromagnetic field of frequency equal to that of a particular transition, and hypersonic pumping is carried out with an intense hypersonic field of the same frequency. When a phaser is used as a hypersound amplifier an *acoustic wave transducer* is located on one end of the rod, which initiates a hypersonic wave in the crystal. While it propagates the wave builds up, gaining energy from the pumping source. The frequency band of the hypersound amplification signal is determined by the width of the *acoustic paramagnetic resonance* line at the frequency of the transition between levels with the inverted populations. A phaser may serve as the source of coherent hypersound, constituting a *coherent phonon generator* (see *Phonon generation*). This is achieved by filling a high quality factor Q acoustic resonator (rod-shaped crystal with parallel optically flat faces and its axis aligned along a high symmetry crystallographic direction) with active paramagnetic centers. The amplification increases with higher power pumping. Upon reaching the threshold level, when losses are compensated by amplification, the phaser generates coherent phonons at the resonant frequencies of the acoustic resonator. Single-mode phonon generation occurs only if the threshold is slightly exceeded. For pumping far beyond the threshold level, multimode generation of phonons takes place. Phaser amplification and generation are realized by application of paramagnetic centers Cr^{3+} and Ni^{2+} in a crystal of Al_2O_3, as well as Ni^{2+} and Fe^{2+} in MgO. The phenomena of phaser generation and amplification of coherent phonons are applied in hypersonic acoustic paramagnetic resonance spectroscopy to study the dynamics of electron–nuclear spin systems in strong hypersonic fields, and to study nonlinear processes related to ultralong wave propagation of sound in crystals containing active paramagnetic centers.

PHASE SLIP in superconductors

A jump (or change) in the time rate of growth of the phase χ of the complex *order parameter* in superconductors. It is one of the manifestations of *macroscopic quantum coherence* of superconductors in their nonequilibrium state. The spatial gradient $\nabla \chi$ of that phase determines the *superfluid velocity* and the *Meissner effect*, while its time derivative $\partial \chi / \partial t$ prescribes the chemical potential of the *Cooper pairs* in the condensate. This chemical potential $(1/2)\hbar \partial \chi / \partial t$ adds to the electric potential φ to form the electrochemical potential Φ of the superconductor: $\Phi \approx (\hbar/2)\partial \chi / \partial t + \varphi$. The relationship between the phase difference $\Delta \chi$ and the difference between the electric potentials, $\hbar \partial(\chi_1 - \chi_2)/\partial t \approx 2e(\varphi_2 - \varphi_1)$, follows from the equilibrium condition $\nabla^2 \Phi = 0$ outside the region occupied by the field. Due to the *proximity effect* the phase change in a junction between two superconductors separated by a thin layer of dielectric generates a nonstationary *Josephson effect*. The phase difference in the *resistive state* of superconducting films and whiskers occurs at the centers and along the lines of such a slip.

Phase difference centers (PDCs) are microscopic quasi-normal regions in a superconducting channel (a film or a whisker), at which the phase difference of the order parameter occurs in the resistive state. PDCs spontaneously develop as the current grows and provide the electric resistance of the channel. The order parameter in a PDC is close to zero, and the electric field is at its maximum, decreasing in the bulk superconducting regions over large lengths, l_E. Such a decay is due to relaxation of the electric imbalance of the electron and the hole branches of the spectrum (that is, due to mixing). The above length prescribes the size $2l_E$ and the resistance $2l_E\rho_n$ (ρ_n is the normal resistivity) of the discrete *resistive domain*, associated with the PDC. A nonstationary Josephson effect is observed in the vicinity of the PDC.

Phase difference lines (PDLs) constitute a generalization of the concept of PDC to resistive states in wide superconducting films. Such a generalization is based on the experimental observation of specific features of current–voltage characteristics of such films that are similar to features found in the narrow channels. For slightly supercritical currents the resistance of wide films results from the motion of *Abrikosov vortices* across the film. At a certain critical vortex velocity PDLs form along the vortices, and a resistivity mechanism similar to that of channel resistance switches on.

PHASE TRANSFORMATION
See *Phase transitions*.

PHASE TRANSITION HEAT
See *Heat of phase transition*.

PHASE TRANSITION HYSTERESIS
See *Hysteresis in phase transitions*.

PHASE TRANSITION INDUCED PLASTICITY
See *Plasticity induced by phase transitions*.

PHASE TRANSITION, PHENOMENOLOGICAL THEORY
See *Phenomenological theory of phase transitions*.

PHASE TRANSITIONS, phase transformations

The transitions of a substance from one thermodynamic *phase* to another, which are accompanied by an abrupt change of certain physical characteristics of this substance, and a continuous change of external parameters. Not all changes of the phase state of substances involve such phase transitions, e.g., the transition between gaseous and plasma states proceeds in a gradual manner.

The value of temperature, pressure, or some other physical quantity, at which the phase transition occurs, is referred to as the *transition point*. Two kinds of phase transitions are distinguished. A *first-order phase transition* involves an abrupt change of some thermodynamic characteristics of the material, such as the density, or the concentration of components. The free energy remains constant as the temperature is changed through the transition, but its derivative with respect to temperature is discontinuous. The quantity of heat absorbed or evolved by a unit mass during the course of the transition is well-defined, and referred to as latent *heat of phase transition*. For a *second-order phase transition* both the free energy and the temperature derivative of the free energy are continuous as the temperature passes through the transition point. There is no latent heat since heat is neither absorbed nor evolved, although there can be a discontinuity in the specific heat. Other thermodynamic quantities such as the density, which are discontinuous for the first order case, now change continuously across the transition.

The above classification of phase transitions into two groups is not totally precise since there exist first-order phase transitions that feature only slight discontinuities in density and other physical quantities, and low values of the heat of transition, involving pronounced *fluctuations*. A phase transition is a collective phenomenon that occurs at rigidly defined values of temperature and other quantities in a system which, in the limiting case, contains an arbitrarily large number of particles.

See also *Concentration-dependent phase transition, Diffuse phase transition, Ferroelectric phase transition, First-order phase transition, Magnetic phase transitions, Order–disorder phase transition, Orientational phase transition, Second-order phase transition, Structural phase transition, Superconducting phase transition*.

PHASE TRANSITIONS AND DEFECTS

See *Defects in phase transitions*.

PHASE TRANSITIONS AT A SURFACE

Phase transitions that take place either in a *film* adsorbed on a surface, or at a bare surface. During the course of a *surface reconstruction* phase transition, the structure of an unoccupied surface is altered by a change in temperature, under the action of adsorption, etc. The length constants of the resulting *superlattice* structures may be either commensurate or incommensurate with the lattice constants of the free surface (see *Surface lattices*). Adsorbed films exhibit phase transitions between a two-dimensional liquid, gas and crystal, *structural phase transitions*, etc. These transitions take place under changes of temperature, degree of surface covering with adsorbate, chemical potential of the gas in equilibrium with the adsorbate, etc. There exist phase transitions involving a change of orientation of the lattice of adsorbed atoms with respect to the surface: the so-called *oriented epitaxy*. It has been found that during the course of many phase transitions the lattice constants of the adsorbed atoms become incommensurate with those of the substrate. The transformation of an atomically smooth free surface to an atomically rough one occurs during the course of a *loosening-type phase transition*. This transition manifests itself in the disappearance of planar faces from the crystal surface, i.e. by a change of the *crystal* equilibrium shape.

At *adsorption* there can occur a *wetting-type phase transition*, during the course of which there is a change in the thickness an adsorbed film. This thickness may be changed by a single monolayer via a *layering-type phase transition*, or from a finite amount to an infinite one at a *phase transition involving total and critical wetting*. As a rule, phase transitions bring about changes in diffusion, catalytic, electronic, and other surface properties. The types of critical behavior at phase transitions are classified according to the universality class of two-dimensional phase transitions (see *Two-dimensional lattice models*).

PHASE TRANSITIONS IN AN EXTERNAL FIELD

Changes of the characteristics of materials (*magnetization, polarization*, volume, *compressibility*, symmetry, etc.) brought about by an action of a field (electric or magnetic field, elastic stress, gravitation, etc.) When compared with the zero field case, phase transitions in an external field exhibit a number of specific features. An external field may cause a phase transition to die out, may shift the transition temperature T_c, or bring about the generation of new states. Magnetic ordering in a *ferromagnet* can take place as a *second-order phase transition*. At the application of an external *magnetic field B* this phase transition can die out, abrupt changes of characteristic thermodynamic quantities can become smeared since at $B \neq 0$ the magnetization is nonzero, and the symmetry of the magnetic subsystem is the same at various temperatures. However, in ferromagnets exhibiting strong magnetostriction (see *Magnetostriction*), magnetic ordering often occurs as a *first-order phase transition*. This phase transition does not die out at $B \neq 0$, but rather it is characterized by a number of specific features: *hysteresis* with respect to temperature and pressure is unusually high (up to 200 K and 700 MPa respectively; the transition temperature depends strongly on the applied field: ~ 10 K/T); the stability region of the ferromagnetic phase (see *Phase*) may change considerably in strong fields; the action of a magnetic field within the limits of the intermediate region separating the initial and final states allows the onset of an irreversible ferromagnetic phase; the inducing fields are temperature- and pressure-dependent, and may be controlled by changing the sample composition. The phenomenon of *irreversibly inducing metastable phases* is related to the effect of *striction blocking of nuclei* of the nonuniform state (see *Nucleation at phase transitions*). There is a possibility of complete blocking of nuclei, when the generation of these nuclei becomes thermodynamically forbidden over the whole stability region of the initial phase.

Magnetic ordering in *antiferromagnets* is, as a rule, a second-order phase transition, which may occur also at $B \neq 0$. The *Néel point* turns out to be a function of the magnetic field intensity. Various *magnetic phase transitions* related to a change of orientation of the magnetization of sublattices

in an external magnetic field are observed in magnetic materials that have several sublattices (antiferromagnets, ferrites). The strong fields give rise to phase transitions of the type of a *spin-flip transition* (involving the collapse of sublattices, see *Antiferromagnetism*) and magnetic ordering breakdown.

In the case of *ferroelectrics*, an applied electric field brings about the collapse of a first-order transition and a shift of the boundaries of a second-order transition. Since the transitions between ferroelectric and antiferroelectric phases are first-order, one may notice certain parallels between the processes that induce a ferromagnetic phase by a magnetic field, and those that induce a latent ferroelectric phase by an electric field.

PHASE TRANSITIONS IN A RADIATION FIELD

The change of phase composition of a substance under irradiation by charged or neutral particles, or high-energy photons. Irradiation of *metals* and *alloys* with neutrons, electrons, ions, or gamma rays causes the internal energy of system to increase by an amount of the order of 10^8 kJ/mol due to the generation of radiation defects and electronic excitations. This extra energy may bring about *phase transitions* at temperatures lower than those specified by an equilibrium *phase diagram*. Besides that, the radiation field may give rise to a phase transformation that does not proceed under equilibrium conditions. The radiation-induced *segregation* of impurity atoms produces a local increase of their concentration at the sites of *point defects* to the extent that the generation of a new thermodynamic *phase* becomes possible. At the same time, segregation stimulates additional dissolution of the impurity atoms by decreasing their content in the bulk of the crystal. Ion bombardment of a crystal (implantation) brings about the generation of *solid solutions* over a range of concentrations that far exceeds the limits allowed by the compositional diagram for the temperatures at which the irradiation takes place. Irradiation of crystals can also cause ordering or disordering of the alloy crystal lattice (depending on experimental conditions), decomposition of oversaturated solid solutions, *amorphization* of crystal structures, change of stability of the amorphous state of completely disordered structures.

PHASE TRANSITIONS UNDER PRESSURE

Phase transitions under pressure may be either first order or second order. The application of *pressure* shifts the *phase transition* temperature. In the case of a *first-order phase transition*, the change of T_c under the action of pressure is related to the latent *heat of phase transition* L by a simple equation: $L = T_c (dT_c/dP)^{-1} (V_2 - V_1)$, where dT_c/dP is the value of the derivative of temperature with respect to pressure at the phase equilibrium curve, and the difference $V_2 - V_1$ is the change of volume of the material at the transition.

The influence of pressure on the phase transition of *ferromagnets* and *antiferromagnets* can be appreciable, and in the case of *ferroelectrics*, the values of dT_c/dP are of the same order of magnitude. The type of phase transition may also be changed under the action of pressure. Thus, the process of magnetic ordering in the alloy $Mn_{0.91}Ti_{0.09}As$ is second order at pressures below 600 MPa, and first order at higher pressures. The critical pressure, at which the type of phase transition is changed, is 150 MPa for the alloy $MnSb_{0.12}As_{0.88}$. It follows from thermodynamic considerations that the action of a magnetic field B is equivalent in the case of the first-order transition to the shift of the boundaries at the phase diagram relating temperature and pressure (see *Phase diagram*), the isotherms being given by the expression $P \sim B(M_2 - M_1)/(V_2 - V_1)$. Depending on the sign of the volume change at the transformation, the boundary is shifted into the region of either positive or negative pressures. Boundary shifts of more than 1 GPa have been observed experimentally. Within the limits of the intermediate region of the pressure–temperature diagram, the field causes the irreversible onset of the ferromagnetic phase. Materials which exhibit a spontaneous ferromagnetic phase generation boundary in a region of negative pressures, and ferromagnetic phase decomposition boundary in a region of positive pressures, should be assigned to a separate group. In this case, even a short-time application of a magnetic field brings about the irreversible generation of a magnetic phase, which cannot be induced in any other manner. For many *antiferroelectrics*, the ferroelectric phase turns out to be concealed within the region of negative pressures, yet it may be generated under atmospheric pressure by the action of electric field pulses.

PHASE VELOCITY

The velocity of the motion of the phase of a harmonic wave that propagates without changing its shape. The value of the phase velocity is equal to the ratio of the angular frequency ω to the magnitude of the wave vector $k = 2\pi/\lambda$, i.e. $v_p = \omega/k$. For *crystals*, in the general case, the phase velocity is an anisotropic quantity for *quasi-particles*.

PHASON

A *quasi-particle* corresponding to the wave of vibrations of the *order parameter phase* in an *incommensurate structure*. The notion was introduced by A.W. Overhauser (1971). *Fluctuations* cause the amplitude and phase of the stationary periodic distribution of the order parameter corresponding to the incommensurate structure to vary, and these variations may be distributed in a wavelike manner within the crystal. The density of the *thermodynamic potential* of the system depends quadratically on the phase derivatives, but not on the phase itself. Therefore, in an *ideal crystal* the phason spectrum does not have a gap, and its energy depends linearly on the variation of the wave vector with respect to the stationary incommensurate structure.

A particular case of a phason is a *fluctuon* which is formed upon *self-localization* of an electron near the fluctuation region of the second phase. Reduction of the electron energy upon localization makes a phason stable under conditions when it would be unstable in the absence of the electron. Phasons may appear in some temperature ranges near *first-order phase transition* points in nonmetallic systems, and may be 1 nm in size.

PHENOMENOLOGICAL THEORY OF PHASE TRANSITIONS, thermodynamic theory of phase transitions

Theory of *second-order phase transitions*, and of phase transitions close to first order, based on examining the symmetry of the initial and final phases, and on the assumption that the free energy and other *thermodynamic potentials* of the system in the vicinity of the transition point may be expressed as analytical functions of the *order* parameter and of other thermodynamic variables. The phenomenological theory of phase transitions describes well many features of *magnetic phase transitions, ferroelectric phase transitions, structural phase transitions* and *superconducting phase transitions*, but it is not adequate for describing the *critical region* near the transition point. The *Landau theory of second-order phase transitions* is an example of a phenomenological theory.

PHONON AVALANCHE

Phenomenon of a sharp (avalanche) increase of the number of *phonons* upon their interaction with a strongly nonequilibrium (population inverted) physical system. The avalanche process takes place when the probability of phonon generation is proportional to the number of phonons, and radiation exceeds absorption. Such a proportionality is a common feature of *induced radiation* (in our case – of phonons). In fact a phonon avalanche is a realization of the idea of a laser based on phonons (*phonon maser*).

The conditions for phonon avalanche development may be created when a physical system with a finite number of energy levels (a state easily altered by external effects) is brought into contact with a system of phonons. To achieve effective contact it is desirable to have the energy gaps of the system coincide with the phonon frequencies. The higher the extent of the population inversion of the energy levels and the lower the number of phonons, the more favorable the conditions are for the development of a phonon avalanche. For example, a crystal with paramagnetic impurities located in an external magnetic field has an equilibrium distribution of Zeeman energy levels (see *Zeeman effect*). A rapid reversal of the field direction creates an inverted *level population* (opposite to the equilibrium one). Contact with phonons exists due to the spin–phonon coupling. Low temperature, high magnetic atom concentration and fast switching of the field are the necessary conditions for the appearance of the avalanche process in the phonon system.

PHONON BOTTLENECK

Situation which takes place during processes of *relaxation* of an excitation in a solid via phonons, and involving a small group of *phonons* with a narrow spread in frequency. The rate of the relaxation is limited by the fact that these so-called hot phonons are unable to readily transfer the excessive energy, obtained from the excitations, to the remaining vibrations of the *phonon spectrum*. As a result there appears an excessive number of hot phonons compared to the thermal phonons in the normal distribution of the phonon density of states at the temperature of the experiment (there is local overheating of the hot phonons, hence their name), and the process of relaxation is retarded. The spin system and the hot phonons are thus raised to temperatures much higher than that of the thermal phonons (lattice vibrations) which are in direct contact with the thermal reservoir or heat bath (e.g., liquid helium).

An example of phonon relaxation occurs when a high-power pulse of radio or microwave frequency ω raises nuclear or electron spins to an excited energy state in a magnetic resonance experiment, and the spins then pass their excess energy on to the lattice vibrations via the intermediary of phonons. At *cryogenic temperatures* (e.g., 4 K) the direct process of *spin–lattice relaxation* involves the excited spins transferring their energy to the hot phonons which are at resonance (that is at the same frequency ω) with the spins, and then this energy is transferred it to other (nonresonant) phonons which eventually pass it on to the surrounding *heat sink* which has a high *specific heat* (e.g., a liquid helium bath). If the phonons "on speaking terms with" or at resonance with the spins cannot disperse their energy to other phonons rapidly enough then the relaxation path becomes blocked, in other words there is a phonon bottleneck. This can occur because the time constant $\tau_{s\text{-}hp}$ for transferring the spin energy to the hot phonons is much shorter than the time constants $\tau_{hp\text{-}tp}$ for the hot phonons to transfer their energy to the other nonresonant thermal phonons, and is also much shorter than the time constant $\tau_{hp\text{-}hr}$ for the hot phonons to transfer their energy directly to the heat reservoir. In other words: $\tau_{s\text{-}hp} < \tau_{hp\text{-}tp}$ and $\tau_{s\text{-}hp} < \tau_{hp\text{-}hr}$. There is a severe bottleneck when $\tau_{s\text{-}hp} \ll \tau_{hp\text{-}tp}$ and $\tau_{s\text{-}hp} \ll \tau_{hp\text{-}hr}$. Eventually the combined system (excited spins + resonance phonons) relaxes to the reservoir temperature with a longer time constant of the order of $\tau_{hp\text{-}tp}$ or $\tau_{hp\text{-}hr}$, whichever is shorter. For a mild bottleneck the overall spin–lattice relaxation time T_1 is a combination of the three times $\tau_{\text{s-hp}}$, $\tau_{\text{hp-tp}}$ and $\tau_{\text{hp-hr}}$, and for a severe bottleneck it depends mainly on $\tau_{\text{hp-tp}}$ and $\tau_{\text{hp-hr}}$. If one of these two time constants is much shorter than the other then T_1 will be approximately equal to the shorter of the two.

PHONON ECHO

See *Electroacoustic echo.*

PHONON GENERATION

An excitation in a medium of nonequilibrium *phonons* acted upon by an external influence. During the generation process both *coherent phonons* (acoustic waves) and *incoherent phonons* (noise) appear. The main mode of phonon generation is their excitation with the help of either a constant, a radio-frequency, or an optical electric field. High power monochromatic sound sources are called *acoustical generators*. In addition to phonons that propagate in the bulk of the material, surface acoustic waves can also be generated.

For the generation and the amplification of broad-band acoustic noise, an *acousto-electronic effect* is used. This is the excitation of sound waves in *piezosemiconductors* by electrons whose drift velocity under an applied electric field exceeds the sound *phase velocity.*

Many methods of generation are based on the light- (laser-)induced excitation of sound. In thermal methods, the *sound* is emitted by portions of the medium subject to an intensity-modulated or surface scanning laser beam. High power coherent hypersound appears during *Brillouin scattering* induced by the action of a high power laser pulse, with a single spectral component of thermal acoustic noise arising simultaneously.

At low (liquid helium) temperatures, a *phaser* (phonon analogue of a *laser*) can be used for coherent phonon generation. Its working principle is the stimulated radiation of phonons by a *two-level system* of spins with an inverse population established under the conditions of *acoustic paramagnetic resonance.*

The recording of phonons is based on the same principles as their generation. Hence, at low temperatures, *bolometers*, avalanche *diodes*, and detectors using *acoustic paramagnetic resonance* are employed. Optical methods are represented by the diffraction of light on sound (see *Acousto-optic diffraction*). For reception of the acoustic waves *acoustic wave transducers* are used.

PHONON–HELICON RESONANCE

See *Helicon–phonon resonance.*

PHONONLESS SPECTRAL LINE

A narrow line in optical spectra of imperfect crystals due to electron transitions between *defect* levels, which are not accompanied by the creation or destruction of quanta of crystal lattice vibrations (i.e. *phonons*).

The existence of a phononless spectral line was predicted by M.A. Krivoglaz and S.I. Pekar (1953). Depending on the strength of the interaction between the electronic transition and the phonons, the phononless line appears in the spectrum as either a rather isolated peak with weak *phonon wings* on its sides (weak interaction) or as a relatively low-intensity peak on the background of a wide electron-vibrational band (appreciable interaction). The width of the phononless line is considerably less than a characteristic phonon frequency. At very low temperatures for RF transitions, the width approaches the *natural line width* determined by the probability of radiative decay of an excited state per unit time. With increasing temperature, the phononless line intensity decreases, its width increases, and the position of its peak shifts. The nature of these changes is determined by the mechanism of the interaction of the electron transition with phonons, and by the type of phonon spectrum. Therefore, studies of the phononless line provide important information not only on the electron structure of defects (energy spectrum, symmetry, and so on) but also on the *electron–phonon interaction*. For doped disordered systems, phononless lines are investigated with the help of selective laser spectroscopy methods to eliminate the *inhomogeneous broadening*. The phenomenon of phononless lines has wide applications, in particular it underlies the operation of *solid-state lasers* based on transition and rare-earth element impurities, as well as the *Mössbauer effect.*

PHONON MAGNETIC RESONANCE

Change of the kinetic parameters of a *semiconductor* in a *magnetic field* due to resonance inelastic scattering of *current carriers* by phonons. In a resonance magnetic field, the energy difference between adjacent peaks of the *density of states* due to Landau quantization (see *Quantizing magnetic field*) become equal to the a limiting energy, ω_0, of one or several *phonons*. If current carrier transitions occur between states with different spin alignment the phenomenon is termed *spin phonon magnetic resonance*. As distinct from the case of *magnetooptical effects*, contributers to phonon magnetic resonance are current carrier transitions with both small and large changes in the Landau level number, and in the wave vector projection on the direction of the magnetic field. The condition of magnetic phonon resonance for a nondegenerate parabolic band is $N\omega_c = \omega_0$, where the cyclotron resonance frequency $\omega_c = eB/m^*$, $N = 1, 2, 3, \ldots$, and e and m^* are the current carrier charge and effective mass, respectively. Given ω_0, one can determine m^* from the period of the resonance peaks plotted versus $1/B$. Due to the polaron effect, the mass m^* is overestimated in magnetic phonon resonance by a value, according to different data, between $\alpha/2$ and $\alpha/6$, where α is the current carrier–phonon coupling parameter.

Corresponding to phonon magnetic resonance in the ohmic regime are maxima of transverse magnetoresistance (see *Galvanomagnetic effects*). The type of longitudinal magnetoresistance extremum depends upon the relative contributions of elastic and inelastic processes to the scattering. Weak singularities known as *pseudoresonances* can also occur here for the condition $N\omega_c/2 = \omega_0$. Under the conditions of current carrier heating several factors can contribute to the type of the extremum.

PHONON MECHANISM OF SUPERCONDUCTIVITY

A mechanism for the formation of *Cooper pairs* of electrons in metals due to an electron–electron attraction which occurs near the *Fermi surface* in the range of energies of approximate width $\hbar\omega_D$ (ω_D is the Debye frequency) as a result of exchange with *virtual phonons*. This phonon mechanism of superconductivity was initially suggested by H. Fröhlich and J. Bardeen (1950), and

it is characterized by the dimensionless interaction constant:

$$\lambda = 2 \int\limits_0^{\overset{*}{\infty}} \frac{d\omega}{\omega} \alpha^2(\omega) F(\omega), \tag{1}$$

where $\alpha^2(\omega)$ is the square of the *electron–phonon interaction* matrix element averaged over the Fermi surface, and $F(\omega)$ is the phonon *density of states*. The value of λ together with the Coulomb repulsion of electrons determines the *critical temperature of superconductors*. The *isotope effect* is a characteristic feature of the phonon mechanism of superconductivity.

This phonon mechanism in polar (ionic) degenerate semiconductors was taken into account by V.L. Gurevich, A.I. Larkin, and Yu.A. Firsov (1962) on the basis of a retarded interelectronic interaction

$$W(q, \omega) = \frac{4\pi e^2}{q^2 \varepsilon_\infty} \frac{\omega_{TO}^2 - \omega^2}{\omega_{LO}^2 - \omega^2}, \tag{2}$$

where ω_{LO} and ω_{TO} are the frequencies of transverse and longitudinal optical *phonons*, linked by the *Lyddane–Sachs–Teller relation* $\omega_{LO}/\omega_{TO} = (\varepsilon_0/\varepsilon_\infty)^{1/2}$, where ε_0 and ε_∞ are the static and high-frequency *dielectric constants* of the crystal. Eq. (2) follows from the *Fröhlich Hamiltonian* with an electron–phonon interaction constant of the form

$$g_F(q) = \frac{2e}{q} \sqrt{\pi \frac{\hbar\omega_{LO}}{V} \left(\frac{1}{\varepsilon_\infty} - \frac{1}{\varepsilon_0} \right)}$$

obtained from second order perturbation theory based on the Coulomb repulsion term $4\pi e^2/(q^2\varepsilon_\infty)$. In spite of the prevailing repulsion at the *Fermi surface*, $W(q, 0) > 0$ under specific conditions; in particular for $\varepsilon_0 \gg \varepsilon_\infty$ (see *Ferroelectrics*) and a sufficiently high concentration of free carriers the transition is possible to the superconducting state at low temperatures ($T \leqslant 1$ K). M.L. Cohen (1964) considered the phonon mechanism in degenerate *many-valley semiconductors*, taking into account intervalley transitions. Before the discovery of *high-temperature superconductivity* the phonon mechanism was widely considered to be the only acceptable mechanism for explaining the superconductivity of metals, their alloys and compounds, degenerate semiconductors and *semimetals*, and low-dimensional *organic conductors and superconductors*. The role of the phonon mechanism in the superconducting metal-oxide cuprate compounds (see *High-temperature superconductors*) has been discussed in the light of their unusually weak isotope effect.

PHONON OPTICAL SPECTRA

Optical spectra of solids arising from the interaction of electromagnetic radiation with phonon excitations. Such an interaction reveals itself in the spectral dependence of the *light absorption, reflection of light*, light radiation and *light scattering* processes in the solid. Features of IR spectra of *ionic crystals* in the range of frequencies corresponding to the intrinsic transverse ν_{TO} and longitudinal ν_{LO} optical *phonons* are often referred to as phonon optical spectra. The variation of the dipole moment associated with transverse vibrations determines the possibility of a resonance interaction with electromagnetic radiation. According to the quantum theory this interaction leads to the disappearance of the *photon* which entered the crystal, and the generation of a transverse optical phonon of frequency ν_{TO} with a very small momentum (like the photon) $k_{phon} = k_{phot} \approx 0$. This interaction produces an intense absorption band at the frequency ν_{TO}, and in reflection there appears a band at frequencies in the range $\nu_{TO} < \nu < \nu_{LO}$ (*residual ray band*). If the unit cell of the crystal contains more than two ions there may appear several resonances. In crystals with any type of *chemical bonding* two-phonon interaction processes with light can occur, whereby one photon disappears and two phonons are simultaneously created, or one phonon is generated and the other one is annihilated. The intensity of such processes is several orders of magnitude below that of single-phonon resonances, and three-phonon processes of absorption are even less probable. For two-phonon processes the conservation of momentum law allows the participation of phonons with any value of momentum, therefore, the adsorption spectrum extends from $\nu = 0$ to double the maximum frequency of the crystal lattice phonons. Maxima in the spectrum are associated with the participation of phonons corresponding to high-symmetry points of the *Brillouin*

zone (see *Van Hove singularities*). Due to the interrelationship between the absorption and radiation probabilities of materials, processes of photon–phonon interactions contribute to the spectral dependence of crystal radiation in the IR region.

In the phonon optical spectra arising from *Raman scattering of light* there are also resonances associated with the creation and annihilation of long-wavelength ($k \approx 0$) optical phonons, and the structure arising from many-phonon processes. Due to the difference between the *selection rules* of IR absorption and Raman scattering transitions, they generally complement each other. Phonons may also be involved as the third participant in the interaction of photons with an electron system in *intrinsic light absorption edge* spectra, and in *luminescence* spectra. See also *Polariton*.

PHONON–PHONON INTERACTION

Interaction of vibrational excitations (*phonons*) in condensed matter, caused by *anharmonic vibrations*. The phonon–phonon interaction in crystals, as a rule, is weak, and the frequencies obtained in the *harmonic approximation* are weakly renormalized by such an interaction. But due to this interaction there appears a dependence of the phonon frequencies, *sound velocity*, and *elastic moduli* on the temperature, together with a series of singularities in the behavior of the *Debye–Waller factor*, etc. The phonon–phonon interaction plays a fundamentally important role in the *relaxation* processes which lead to the spatial and temporal attenuation of phonons excited by an external action, thus in many cases influencing the value of their lifetimes. As a result, this interaction determines: attenuation of elastic waves (sound), lattice thermal conductivity, and the broadening of spectral lines associated with the scattering and adsorption of various types of radiation when these processes are accompanied by the generation or adsorption of phonons. By taking account of the phonon–phonon interaction it is usually possible to restrict consideration to several of the lowest order terms in the anharmonic potential (the first of such terms determines processes with the participation of three phonons, etc.). A phonon–phonon interaction in an *ideal crystal* conserves the phonon wave vectors within an arbitrary *reciprocal lattice* vector b. For $b = 0$ the processes are called *normal processes*, and for $b \neq 0$ they are referred to as *Umklapp processes*. The latter can provide the main limiting factor of the *thermal conductivity coefficient* of an ideal lattice.

In relaxation phenomena accompanied by scattering processes, or by processes of generation and annihilation of actual phonons, the energy conservation law is always satisfied. The laws of energy and momentum conservation lead to specific restrictions on allowed processes. Three-phonon processes (which determine the attenuation of acoustic branches) in which all three modes of vibration belong to the same branch of the spectrum, appear to be forbidden to a first approximation, but they become possible if the broadening of the highest frequency phonons exceeds the sublinear deviation of the *dispersion law* (in this case they are sometimes called *Simons processes*). Another restriction lies in the fact that at a breakup one phonon should belong to a lower energy branch than the source one, and at a scattering event one phonon is generated in a higher energy branch than that of the disappearing ones.

The probability of relaxation W of phonons caused by the phonon–phonon interaction depends on the phonon frequency ω, and on the temperature T. For long-wave *acoustic vibrations* $W \sim \omega T$ at high temperatures and $W \sim \omega^n T^{5-n}$ at low temperatures (below the *Debye temperature*), where the coefficient n ($1 \leqslant n \leqslant 5$) depends on various factors. For the lowest-frequency vibrations (ω less than the inverse lifetime of the phonons taking part in scattering) $W \sim \omega^2$, and it decreases when the temperature is lowered.

For long-wavelength *optical vibrations* the scattering processes of a single acoustic phonon are forbidden; the decomposition into two (if allowed by energy conservation) or several acoustic phonons is possible, as is also a four-phonon scattering process with the participation of two acoustic modes. Similar processes also determine the broadening of the lines of *local vibrations* in a crystal.

In crystals with impurities or other defects the *quasi-momentum* of the phonon–phonon interaction may not be conserved, which leads to other dependences of the relaxation process probabilities.

PHONONS (fr. Gr. $\varphi\omega\nu\eta$, sound)

The quanta of *crystal lattice vibrations* which appear as a result of quantization of the *normal vibrations*. Phonons are *quasi-particles* which play the role of elementary excitations of a vibrating crystal. As a quasi-particle each phonon possesses the energy $\varepsilon = \hbar\omega$, the momentum $\boldsymbol{p} = \hbar\boldsymbol{k}$, and the velocity $\boldsymbol{v} = \partial\varepsilon/\partial\boldsymbol{p} = \partial\omega/\partial\boldsymbol{k}$, where $\boldsymbol{k}$ is the *quasi-wave vector*, ω is the frequency of the corresponding normal vibration, and the frequency $\omega(\boldsymbol{k})$ is a periodic function of $\boldsymbol{k}$ with the *reciprocal lattice* period. Phonons are Bose particles (bosons) (see *Bose–Einstein statistics*).

In the *harmonic approximation* the energy of the crystal excited state is represented as a sum of independent oscillator energies

$$E = E_0 + \sum_{k\alpha} \hbar\omega_\alpha(\boldsymbol{k})N_\alpha(\boldsymbol{k}), \qquad (1)$$

where E_0 is the energy of *zero-point vibrations*, and the integer $N_\alpha(k) = 0, 1, 2, \ldots$ denotes the number of phonons. The index α differentiates the branches of *acoustic phonons* (see *Acoustic vibrations*) and of *optical phonons* (see *Optical vibrations*). Thus the weakly excited state of a crystal is equivalent to an ideal gas of phonons, and the ground state of the crystal is a *phonon vacuum* ($N_\alpha(k) = 0$). Upon thermodynamic averaging the energy of Eq. (1) provides the lattice part of the internal energy of the crystal, namely $\widetilde{E} = \langle E \rangle$, whereas in the state of thermal equilibrium the average number of phonons is determined by the Bose–Einstein distribution

$$\langle N_\alpha(\boldsymbol{k})\rangle = \left[\exp\left(\frac{\hbar\omega_\alpha(\boldsymbol{k})}{k_B T}\right) - 1\right]^{-1}. \qquad (2)$$

Taking into account the *anharmonic vibrations* of the lattice leads to the *phonon–phonon interaction*.

See also *Ballistic phonons, Biphonon, Intervalley phonons, Magnetic phonons, Virtual phonons*.

PHONON SCATTERING

Processes during which quantized lattice vibration particles of a condensed medium (*phonons*) undergo a change in the direction and perhaps also in the frequency of the wave vector due to their interaction with each other (see *Phonon–phonon interaction*), with lattice *defects*, with electrons, with spin excitations (*magnons*) in magnetic materials, etc. Phonon scattering is the determining mechanism of phonon *relaxation* resulting in their spatial and temporal damping, in broadening of the corresponding spectral lines, etc.

During phonon scattering from defects, only the direction of the phonon wave vector changes, while its frequency, ω remains unchanged (elastic scattering). Moreover, *mode conversion* is possible in which the incoming and outgoing phonons belong to different branches. The probability of scattering W of acoustic phonons (see *Acoustic vibrations*) from *point defects*, such as interstitial or substitutional atoms, is proportional to ω^4 and is temperature-independent. When the phonon frequency approaches the characteristic frequency of the defect itself (in particular, when the wavelength is close to the size of the defect), there takes place a resonance enhancement of W. A noticeable role in scattering from defects is played by *dislocations, grain boundaries*, the crystal surface, etc. The probability of scattering from dislocations is proportional to ω in the long-wave limit. More complex dependences are obtained for phonon scattering from dislocation clusters. Collisions with grain boundaries and the crystal surface may be significant in pure enough crystals at low temperatures. The corresponding probability is $W \propto L^{-1}$ where L is the characteristic size of the grain, or of the crystal as a whole. Phonon scattering from electrons via the *electron–phonon interaction* is accompanied by the absorption of the phonon. In *ideal crystals*, when the *conservation law* of quasi-momentum holds, such a process is observed only if the phonon wave vector does not exceed twice the wave vector of the electron (the corresponding wave vector of an electron is on the *Fermi surface* in a *metal*). For long-wavelength acoustic phonons we have $W \propto \omega$, provided the phonon wavelength is shorter than the electron *mean free path*, and $W \propto \omega^2$ otherwise. As for imperfect crystals, the wave vector is not necessarily conserved during the scattering of phonons off electrons, which explains certain features of the probability of this type of scattering.

PHONON SPECTRUM

The set of natural vibrational frequencies of a crystal. In ideal crystals phonon spectra are described by the dependence of *phonon* frequencies ω_k on the wave vector k for different acoustic and optical branches arising from the vibrations of atoms or molecules interacting with one another in a *crystal lattice*. The electrons and nuclei which form the atoms and molecules in the *crystal* are two strongly interacting subsystems, whose motions are coupled with each other. However, due to the large difference between their masses, their motions may to some extent be treated separately on the basis of the *adiabatic approximation*. For each random configuration of nuclei (with coordinates symbolically designated by the multidimensional variable R) it is possible to introduce the electronic wave function $\varphi(r, R)$, which parametrically depends on the variable R (coordinates of electrons are designated by r). Therefore, the energy of the electrons will depend on the nuclear coordinates R, as if they were constant, i.e. the electron states adiabatically follow the displacements (motions) of the nuclei. Further, it is also possible to write down the classical or quantum-mechanical equations, which describe the motions of the nuclei. In these equations the electron energy $E(R)$ plays the role of the potential energy of the nuclear subsystem. It is called an *adiabatic potential*, and it differs for different states of the electronic subsystem. The expansion of $E(R)$ in terms of the extent of the displacement of the nuclear coordinates from their equilibrium positions $u = R - R_0$ is taken into account up to fourth order. In the most important case, the so-called *harmonic approximation* which is quadratic in u, the quantum vibrational equations lead to the expression for the crystal energy

$$E = \sum_{\mu,k} \hbar\omega_\mu(k) N_\mu(k) + E_0,$$

where k is the wave vector of the vibrations, and μ designates the branches of the frequency dependence on the wave vector k. The individual terms of this sum are the energies a system of non-interacting linear oscillators. In addition, this expression for E may be interpreted as the sum of energies of separate *quasi-particles*, that is *phonons* ($\hbar\omega_{\text{phon}}$) with momenta $\hbar k$. The dependence of ω on k is the *dispersion law* of these quasi-particles, and the factor $N_\mu(k)$ is the number of phonons of the given type. Phonons obey *Bose–Einstein statistics*, so their numbers $N_\mu(k)$

are not restricted, and number of branches μ equals the number of degrees of freedom of the nuclear subsystem per unit cell, i.e. $3p$, where p is the number of atoms within the cell. The totality of the $3p$ functions $\omega_\mu(k)$ encompasses all of the phonon spectra of the atomic system. The inclusion of terms with third and fourth powers of u in the expansion of the adiabatic potential should be interpreted as taking into account interactions between different phonons. These terms contribute much less to the energy than the harmonic ones. From the perspective of perturbation theory the *phonon–phonon interaction* describes processes of interdiffusion of phonons, the decomposition of a phonon into two or three others, and the opposite process of merging two or three phonons into a single one while conserving both the total momentum and the energy.

Besides using perturbation theory, which starts to become inadequate with an appreciable growth of the number of phonons, the average values $N_\mu(k)$ can be determined from the crystal temperature T with the aid of the *Planck formula*. The theory of phonon spectra can be developed via the *self-consistent field* method, whereby the interaction between phonons is taken into account in a self-consistent manner.

A theoretical description of phonon spectra starts with the establishment of the form of the adiabatic potential $E(\boldsymbol{R}) = U(\ldots u \ldots)$, where u embraces the totality of nuclear coordinates. In the harmonic approximation the quadratic expansion has the form

$$U(\ldots u \ldots) = \frac{1}{2} \sum_{\alpha s \beta s'} v^{ll'}_{s\alpha s'\beta} u^l_{s\alpha} u^{l'}_{s'\beta},$$

where l, l' are the numbers of cells, s, s' are the numbers of atoms within the cells, and α, β are the coordinate indices of x, y, z. The constant ratios $v^{ll'}_{\alpha s \beta s}$ have some symmetry properties. *Translational symmetry* is the most important of them: the dependence only on the difference of indices $l - l'$ (i.e. of the coordinates of atoms s, l and s', l'). This permits one to decompose the vibration equations of infinite order into a system of order $3p$ with the aid of the substitution $u^l_{s,\alpha} = u^k_{s,\alpha} \exp(-i\omega t + ik r^l_s)$. Each of these expressions

has the form

$$m_s u_{s,\alpha} \omega^2 = \sum_{s'\beta} A_{s\alpha s'\beta} u_{s',\beta},$$

where

$$A_{s\alpha s'\beta}(k) = \sum_{l'} v^{ll'}_{s\alpha s'\beta} \exp[-ik(r^l_s - r^{l'}_{s'})].$$

The matrix $A_{s\alpha s'\beta}(k)$ of dimensionality $3p \times 3p$ is called the *dynamic matrix*. Solving the problem for the eigenvalues of the determinant $\det(A_{s\alpha s'\beta}(k) - m_s \delta_{\alpha\beta}\delta_{ss'}\omega^2) = 0$ gives the frequencies of the phonon spectrum $\omega = \omega_\mu(k)$ which has $3p$ branches designated by the index μ. Among these branches there are three acoustic ones which start from zero frequency at $k \to 0$. For those acoustic branches in zero order with respect to k, all of the amplitudes $u_{s\mu}(k=0)$ are the same for all the atoms, and the expansion of u^2 according to k starts from the k^2 term. The remaining $3p - 3$ branches are called optical branches (see *Phonon optical spectra*); they have finite (nonzero) frequencies at $k \to 0$ which are called *limit frequencies*. For them $\sum_s m_s u_{\mu s} = 0$ at $k \to 0$, i.e. the center of mass of each cell is not displaced. Equations for the frequencies ω and the amplitudes $u_{s\alpha}$ are usually solved numerically. For positions of the wave vector $\boldsymbol{k}$ that are symmetrical with respect to the crystal axes, finding a solution may be considerably simplified with the aid of *group theory*.

The starting coefficients $v^{ll'}_{s\alpha s'\beta}$ are ordinarily selected on the basis of specific models, e.g.,

(1) the model of *pairwise interaction of atoms*, when the sum of the energies of all the atomic pairs $u = \sum_{ss'll'} \varphi^{ll'}_{ss'}(r^l_s - r^{l'}_{s'})$ is decomposed according to the extent of their displacement,

(2) a *shell model*, whereby the centers of gravity of (spherical) atomic shells are displaced relative to their corresponding atomic nuclei,

(3) *polarized atom models*, wherein besides the displacements u^l_s, the dipole and quadruple moments of shells p^l_s, Q^l_s are introduced, and

(4) *models of rigid atoms (ions)* (e.g., see *Kellerman model*).

See also *Shell model*.

PHONON WIND

Macroscopically ordered motion of nonthermal *phonons* which occurs under the action of an external electric field (or of crossed electric and magnetic fields) and *electron–phonon interactions*. The essence of this phenomenon lies in the fact that the external electric field provides an ordered forward velocity of macroscopic motion to the electrons, and the phonons acquire the same velocity due to electron–phonon collisions. The term is used in the theory of *transport phenomena* in metals and semiconductors.

PHONON WINGS

Bands in the optical spectra of non-ideal crystals, adjacent to the *phononless spectral line* (ideal crystal line), caused by transitions between the energy levels of a *defect* with the simultaneous creation or annihilation of a crystal quantum vibration (*phonon*). At temperatures considerably above the *Debye temperature* the phonon wings are symmetrical, at low temperatures the intensity of the high-frequency (anti-Stokes) wing is much less than that of the low-frequency (Stokes) wing. The shape of the phonon wings depends significantly on the strength of the coupling of the defect level transition with the phonons. For weak coupling the phonon wings result mainly from processes whereby only one phonon is generated or annihilated. In this case the phonon wing width is of the order of the Debye frequency, and the phonon wings reflect the singularities of the *density of states* of the *ideal crystal* phonon spectrum.

PHONORITON

Composite *quasi-particle* which appears at the anti-Stokes scattering of an intense coherent polariton (quantum of coupled phonon–photon transverse wave field) pumping wave, with its frequency in the *transparency* region near an exciton resonance. The dispersion law of a phonoriton reflects the restructuring of a polariton (see *Polariton*), and of *phonon spectra* which include the combining and splitting of levels near the anti-Stokes resonance, in a manner analogous to the formation of polariton dispersion curves. The extent of the phonoriton splitting is determined by the intensity of the pumping wave. Depending

on which *phonons* (optical or acoustic) are involved in the anti-Stokes scattering of the pumping wave, optical and acoustic phonoritons are distinguished. The restructuring of spectra via optical phonoritons is likely a result of the Fröhlich *exciton–phonon interaction*. The observation threshold of optical phonoriton excitations (e.g., in CdS) for an anti-Stokes resonance near the exciton level $n = 2$ corresponds to the pumping intensity $I \approx 10^6$ W/cm^2. Phonoriton restructuring of spectra may be useful in *optoelectronics* as it induces a strong variation of the semiconductor *dielectric constant* near anti-Stokes resonance and exciton transition frequencies.

PHOSPHORESCENCE

Luminescence with long-term afterglow (unlike *fluorescence*). It is characteristic of the majority of crystallophors (see *Luminophors*). The duration of the afterglow of phosphorescence is determined by the lifetime of the intermediate processes, which occur between the act of energy adsorption by the system, and the moment when it reaches the final state from which the *radiative quantum transition* takes place (see *Thermally stimulated luminescence*).

PHOTOCAPACITANCE, SURFACE

See *Surface photocapacitance*.

PHOTOCAPILLARY EFFECT

Acceleration of the *spreading* of a liquid over a *solid surface*, caused by exposure to light. The photocapillary effect has been found to take place in *germanium* of both n- and p-types, *silicon*, antimony and titanium. This effect was observed at the surfaces of these materials for various low-volatile liquids, which do not chemically interact with the *substrate* (high molecular weight alcohols, ethers, etc.). The photocapillary effect is more pronounced for the spreading different liquids over the same substrate, than it is for the case of using different substrates for the same liquid. The heating effect of the light exposure is insignificant compared with the photocapillary effect. The rate of spreading at exposure to light monotonically increases with increased illumination until saturation is reached (e.g., at $E \approx 2.5 \cdot 10^4$ lux for Ge). The photocapillary effect is caused by a decrease of the surface *band bending* at illumination (see *Surface phenomena in semiconductors*), by an increase of the *surface energy* of the substrate, and of the spreading coefficient. The magnitude of this effect is influenced by processes related to adsorption and diffusion, which take place at the surface during the course of spreading. The photocapillary effect is not observed at the surfaces of copper, lead, or tin.

PHOTOCATHODE

Cathode of a vacuum photoelectron device (vacuum tube), at which a flux of *photons* is converted into a flux of electrons (see *Photoelectron spectroscopy*). The main characteristics of a photocathode are its *quantum yield* (ratio of emitted electrons to incident photons), and the relationship between quantum yield and photon energy. Of practical importance are solid effective photocathodes, which provide quantum yields above 10^{-3}. Such photocathodes are usually thin-film p-type semiconductors of low *electron affinity*, which are produced in vacuo on a transparent *substrate*, e.g., Ag–O–Cs, intermetallides Cs_3Sb, dialkaline compounds K_2CsSb, and polyalkaline compounds (Na_2KSb–Cs), which are sensitive to visible, short wavelength infrared, and ultraviolet radiation. The quantum yield of these compounds reaches 0.2–0.3 at λ_{max}. Photocathodes play leading roles in devices such as photoelectric cells, photoelectron multipliers, electron tubes, etc.

Advances in semiconductor physics permit the creation of field-emission photocathodes (see *autophotoelectronic emission*), and photocathodes based on monocrystals of the $A^{III}B^V$ type with negative electron affinities and high quantum yields up to $\lambda_0 \approx 1.2$ μm. Photocathodes with "intervalley transfer" have been developed on the basis of multicomponent *variband semiconductors* of the $A^{III}B^V$ type. Light absorption in these photocathodes takes place within the narrow-band region of the cathode, then the thermalized electrons are transported to the broad-band region of the cathode. In this region they are heated by the internal electric field and move from the Γ-valley to X- and L-valleys of higher-lying energy bands; then these electrons pass into the vacuum through the thin metallic electrode which has a reduced *work function*. The application of rather complicated techniques provides photocathodes with photosensitivity at wavelengths extending to 2.1 μm.

PHOTOCHEMICAL REACTIONS in solids

Chemical and *quasi-chemical reactions*, brought about by the absorption of light. Photochemical reactions take place in various media: in organic materials (*photosynthesis* is a unique type of photochemical reaction); in ionic *insulators* (e.g., in alkali-halide and silver halide crystals). The reactions, which involve transformations of molecules under exposure to light, are often called *photolysis* (e.g., photoionization, photodissociation, photooxidation, etc.). Photochemical reactions are caused by excitation of the electronic subsystem of the solid due to the absorption of *photons*. There is a certain probability that this excitation may lead to a rearrangement of interatomic bonds and the generation of metastable configurations. This process may be local, i.e. it takes place in the vicinity of the atom which has absorbed the photon. In this case there is a high probability of a reverse reaction because the products produced during the course of the primary photochemical event (so-called *genetic pairs*, or *geminal pairs*) are only slightly separated in space. The effect of increased probability of recombination of geminate pairs is called the *cage effect* in photochemical and radiochemical practice. The processes of photoexcitation-induced or photoionization-induced *diffusion* of atoms into solids are also known. Diffusion leads to the capture of one mobile atom by another atom, or by a defect of the solid: in this case the products of the photochemical reactions may not be localized in space. If the energy of the absorbed photon is not enough to overcome the potential barrier of the reaction, then the process may be induced by the absorption of additional photons during the transition to the excited state. In this case the probability of a photochemical reaction increases exponentially with an increase of the temperature of the material exposed to light. The reactions, which are brought about by elastic collisions of high-energy particles with crystal atoms (these displacement collisions can remove atoms from their *crystal lattice* sites) are sometimes overshadowed by photochemical reactions, which result from *nuclear radiations*. Such a situation is observed in the case of *alkali-halide crystals*. Photons (of UV and soft X-ray radiation) cause the generation of excitons in these materials. Extinction of these excitations leads to the generation of *Frenkel defects*. In organic materials, and in certain semiconductors, photochemical reactions proceed under exposure to visible light.

Photochemical reactions are used in *photographic processes, holography*, etc., to record and store information. They may cause aging and degradation of photoelectronic devices.

PHOTOCHROMIC EFFECT

Change of the spectral dependence of the light transmittance of a material by exposure to light, which is perceived by the eye as a change of color. The photochromic effect may be related to the photodissociation or photoinduced polymerization of molecules, the charge exchange of ions (e.g., *transition metal* ions), or of structural *defects* (generation of *color centers*), photoinduced action of individual particles (colloids), appearance of a new phase, etc. The photochromic effect may be either reversible or irreversible. Reversibility (erasure) is often achieved through thermal treatment (*thermal decoloration*), or through irradiation (*photoinduced decoloration*). Most extensively employed are *photochromic glasses*, which owe their color to the presence of metallic colloidal particles, e.g., silver or copper. See also *Photopigments*.

PHOTOCHROMIC GLASSES

Vitreous *photopigments*. Inorganic photochromic glasses are divided into oxide glasses ($SiO_2 \cdot Na_2O \cdot SiO_2$, $SrO \cdot V_2O_5$, etc.), chalcogen glasses (As–S, As–Se, Sb–Se, etc.) and heterophase ones, which consist of a glass matrix with finely dispersed light-sensitive impurities (e.g., oxide glass with 1 μm microparticles of silver halide). There is a tremendous number of organic photochromic glasses among *organic polymers*. Most extensively developed systems of photochromic organic films are based on solid methyl methacrylate. Chalcogenides and organic photochromic glasses are used in *optical storage disks*. See also *Optical techniques of information recording, Optical recording media*.

PHOTOCONDUCTIVITY, photoresistive effect

Increase of the *electical conductivity* of a material, when it is exposed to electromagnetic radiation in the visible, IR or UV region of the spectrum. The phenomenon of photoconductivity is observed in various *semiconductors* and *insulators*. The *photocurrent* is the increment of current which arises in a closed circuit with a voltage source during irradiation of the photoconductor. The value of the photocurrent may exceed the dark current value by many orders of magnitude. The nature of photoconduction is related to the transfer of energy of the absorbed *photons* to the electrons which are located in a completely full *valence band*, or at bound energy states of the *solid*. Thus the electrons become able to move, and more specifically to drift (see *Current carrier drift*) under the action of the applied voltage. Depending on the sign of the *current carriers*, which are produced by the action of light (so-called *photocarriers*), two types of photoconductivity are distinguished: the unipolar (electron or hole) variety, and bipolar (electrons plus holes) photoconductivity. The photoconductivity is called *intrinsic photoconductivity* if the photoexcitation promotes electrons from the valence band to the conduction band. In the case of *impurity photoconductivity*, electrons from a *local electronic level* are excited.

The magnitude and time dependence of the photoconductivity are determined by the intensity I and the spectral composition of the exciting radiation, and also by processes of photocarrier *relaxation*. The energy of the exciting photons must be high enough to promote electrons from the valence band or various bound electron states to the conduction band (or to the state of free *polarons*); each of these transitions is associated with a line or band of wavelengths, which stimulate the photoconduction. The steady-state value of the concentration of photocarriers, which is established under constant illumination, is $n = \eta \alpha I \tau$, where α is the absorption factor (the condition for uniform absorption is $\alpha d \ll 1$, where d is the sample thickness); η is the *quantum yield* of the photoconductivity (portion of absorbed photons responsible for photoexcitation of mobile charge carriers), τ is the effective lifetime of the carriers in the free state, which is determined by the relaxation mechanisms. The carriers of current, which are produced by light (electrons, holes, polarons) possess finite lifetimes, and are either captured by various defects (*traps* of mobile particles), or undergo recombination (see *Recombination*). The influence of defects on the characteristics of the photoconduction depends on the presence of recombination levels, and trapping by them. In particular, the existence of the trapping levels may lead to electric or *thermoelectric instabilities of photocurrent*. There exists *hopping photoconductivity*, which results from the photoinduced redistribution of electrons throughout the local electronic levels by *hopping conductivity*. Along with photoconductivity, which is related to the change of concentration of current carriers at exposure to light (so-called *concentration photoconductivity*), there exists a photoinduced change of current carrier mobility, which is caused by warming-up. This phenomenon is responsible for *mobility photoconductivity*, which is less effective than the concentration type. See also *Anomalous photoconductivity, Negative photoconductivity*.

The phenomenon of photoconductivity provides the basis for the operation of various photoelectric devices (*photoresistors, optical radiation detectors, photoelectron spectroscopy* components, etc.).

PHOTOCONDUCTIVITY, ANOMALOUS

See *Anomalous photoconductivity*.

PHOTOCONDUCTIVITY OF AMORPHOUS MATERIALS

Increase of *electrical conductivity* of an amorphous sample (see *Amorphous state*) at exposure to light. The phenomenon is associated with the process of *photogeneration* of electrons and holes (in concentrations $\Delta n_\pm = G\tau_\pm$), which add to the already present majority equilibrium carriers (present in concentration n_0) responsible for the *dark current* i_d (current in absence of irradiation). The factor G in the above equation is the number of carriers produced in a unit volume per unit time, $\tau_\pm$ is the lifetime of the carriers, which is determined by the processes of *recombination* with free carriers of the opposite sign, and at recombination centers. The photocurrent $i_p = e(\mu_+\Delta n_+ + \mu_-\Delta n_-)E$ (E is electric field, $\mu_\pm$ is mobility) depends on the kinetics of the

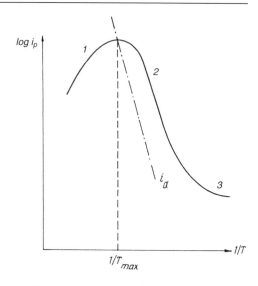

Temperature dependence of the photocurrent i_p.

processes of photogeneration, recombination and thermal activation of carriers from the discrete levels to the bands of continuous states. The temperature dependence of the photocurrent consists of three regions, as shown in the figure:

(1) at high temperatures, $T > T_{max}$, the generated carriers react with thermally produced carriers of the opposite sign ($n_0 \gg \Delta n_\pm$) and $i_d > i_p \propto G$;

(2) at intermediate temperatures with $T < T_{max}$, the recombination of light-produced carriers mainly involves their interaction with one another ($\Delta n_\pm > n_0$) and $i_d < i_p \propto \sqrt{G}$; and

(3) at low temperatures, $T \ll T_{max}$, the photocurrent is proportional to G ($i_p \propto G$) and exhibits a weaker temperature dependence.

The spectral dependence between i_p and the energy of the exciting radiation E correlates with the energy dependence of the optical absorption of amorphous materials $\alpha(E)$: the current begins to rise at the *intrinsic light absorption edge* and then remains approximately constant for higher values of energy. Photoconductive amorphous materials can be used to make inexpensive solar cells.

PHOTOCONDUCTIVITY SPECTROSCOPY

Various methods for determining the physical nature and location of the energy levels of intrinsic and foreign defects in semiconductors, through investigation of the photoconductivity spectra of the sample. The frequency dependence of the photocurrent $I_P(\varphi)$ is obtained by varying the frequency over a narrow spectral excitation band. Each type of photoactive defect has associated with it excited energy levels (or a band) involved in the transfer of an electron, initially bound by the defect, to the conduction band (or the transfer of a hole to the valence band). This process may occur in two stages: first the current carrier of the defect (electron or hole) is transferred to the excitation level located near the band edge, and then it is brought to the band by additional excitation from thermal fluctuations (so-called photothermal ionization). For the purpose of identification of the spectrum the sample is doped with large amounts of impurity atoms of a definite chemical composition, or with large numbers of intrinsic defects, with the subsequent recording of the corresponding bands. The data on structural characteristics of the sample may be obtained through studies of the correlation between photoconductivity spectra and electron paramagnetic resonance spectra, absorption and luminescence spectra, and by investigation of the effect exerted on photoconduction spectra by various perturbations (electric and magnetic fields, elastic stresses, etc.). Not all sites in a crystal lattice provide photoccurrents from impurity atoms situated there, or involve the transformation of an impurity atom to a photoconductively active state (e.g., silicon crystals irradiated with high-speed electrons contain atoms of copper and gold in inactive states). The methods of photoconductivity spectroscopy provide high sensitivity, which permits detection of photocurrents from active centers down to concentrations of 10^{10} cm^{-3}.

PHOTOCONTROLLED MICROWAVE SEMICONDUCTOR DEVICES

Specialized devices incorporated into a microwave transmission-line system which contain a photosensitive semiconductor, and provide light-beam control over electrical signal parameters. These devices include photocontrolled switches, photocontrolled modulators, photocontrolled transparency devices. The control is often exercized through photogeneration of nonequilibrium current carriers in semiconductors such as silicon, gallium arsenide, germanium, cadmium sulfide, etc. Photocontrolled switches are designed for the generation of short pulses. There exist photocontrolled switches in the form of a strip line, in which the role of an insulator is played by a semiconducting material, and one of the metallic strips has a discontinuity in the shape of a narrow slot. Closure of the photocontrolled switch is performed by a laser pulse, whereby the semiconductor surface becomes metallically conducting at the discontinuity region. The minimum duration of the electric pulse (up to 10^{-12} s) is determined by the lifetime of nonequilibrium carriers. Photocontrolled modulators are used for modulation of one or several parameters of the microwaves (amplitude, phase, etc.). There exist photocontrolled modulators in the form of insulating or metallic waveguide, strip line, resonator. Amplitude modulation is often related to the absorption of microwave energy by nonequilibrium carriers in the bulk of the semiconductor. Phase modulation is performed at a considerably higher level of light intensity, which leads to metallic conduction of the semiconductor surface.

Photocontrolled transparency devices are intended for the visualization and conversion of extra-high frequency fields, in particular for exercizing control over the directivity settings of the microwave beam, and for the production of two- and three-dimensional images by means of microwave beams. Photocontrolled transparency devices usually consist of a screen coated with a photosensitive layer; with the dimensions of this screen far exceeding the wavelength. In fact, a photocontrolled transparency device is a matrix of photocontrolled modulators. In order to bring about the conversion of the extra-high frequency field, a certain illumination pattern is created at the surface of the device. This pattern is deposited on the surface with the help of a projector, or by scanning with an optical beam, which is intensity modulated. If the illumination is sufficient for metallization of the semiconductor surface, then this photocontrolled transparency device becomes a phased antenna array. Photocontrolled switches, photocontrolled modulators, and photocontrolled transparency devices are used for exercizing control over electromagnetic waves in the super- and extra-high frequency region.

PHOTOCURRENT INSTABILITY

See *Thermoelectric instability of photocurrent.*

PHOTODESORPTION

Removal of an adsorbate (in form of atoms or ions) from an adsorbent surface (see *Desorption*), which occurs at exposure to light. It is caused by the transfer of the photon energy to the excitation of the desorbing particle. The spectral dispersion and spatial distribution of those particles, which become desorbed through the processes of photodesorption and of *electron-induced desorption*, are identical.

PHOTODETECTOR, SCANNING

See *Sprite.*

PHOTODETECTOR, SEMICONDUCTOR

See *Semiconductor photodetector, Sprite.*

PHOTODIELECTRIC EFFECT

Change of *dielectric constant* ε of an insulator exposed to light. It is caused by phototransitions of electrons of the insulator to higher energy levels, with an increase of the electronic polarizability at these levels. The related *photocapacitance effect* is generally observed, which consists in a radiation-induced change of capacitance C of a capacitor which has insulating material between its plates (the value of C can increase by several orders of magnitude). Since C depends on ε and the geometric dimensions of the condenser (in a plane capacitor, e.g., $C = \varepsilon A / d$, where A is the area of plates, d is the distance between them), both the photodielectric effect (change of ε) and a change of the effective values of A and d (due to spatial redistribution of charges and resistance, which takes place in a dielectric at exposure to light) may cause the photocapacitance effect. Semiconducting light-sensitive capacitors (*photovariable capacitors*, *photovaricaps*) are used in various electronic circuits, e.g., as dynamic capacitors.

PHOTODIODE

Semiconductor diode which serves to convert light energy into an electrical signal. The operation of a photodiode is based on the generation of a reverse current through the barrier layer of the diode (see *Semiconductor junction*) upon exposure to light. The photodiode is designed so that the region of the barrier layer can be illuminated by light. The incorporation in the photodiode of a transistor structure with an illuminated emitter junction provides additional "internal" amplification of the signal (*phototransistor*). This is widely used as a signal receiver in data processing systems, and for measuring light fluxes.

PHOTODOMAIN EFFECT

Effect of exposure to light and concomitant *photoconductivity* on the equilibrium of a *ferroelectric domain structure*, on the kinetics of its generation, and on those properties which are closely related to the domain structure (e.g., electromechanical *hysteresis*). When the *photoferroelectric* has a low-resistivity and the electric field generated by the photovoltaic current in the open-circuit mode is much less than the coercive field, the contribution of the *bulk photovoltaic effect* to the photodomain effect may be neglected. Therefore the photodomain effect is sometimes referred to as a *spontaneous polarization shielding* phenomenon, and is explained in terms of the redistribution of the internal field in the photoferroelectric, which takes place due to surface layers, space screening, and recharging of *surface levels* during exposure to light.

The single-domain state is energetically favorable in the case of ideal short circuiting in the crystal. However, the existence of surface layers in ferroelectrics (equivalent to a dielectric gap) leads to the generation of a certain *depolarization field* in a short-circuited single-domain crystal; this field is caused by surface charges and opposes the external field. Near-surface maxima of the depolarization field are responsible for processes of switching and domain break-up (*polydomainization*) in a short-circuited single-domain crystal. Dark processes of domain breakup are rather slow, because the magnitude of the surface field is insignificant. The screening length decreases at exposure to light, which causes redistribution of the depolarization field: increase of the field at the surface and its decrease in the bulk of the crystal. The increase of the near-surface field accelerates the domain breakup processes. The effect exerted by the nonequilibrium carriers on the domain structure is determined by the ratio between the thickness of the surface layer and the screening length. The photodomain effect was experimentally studied in $BaTiO_3$, SbSI, and other *ferroelectrics*.

PHOTOEFFECT

Redistribution of electrons among their energy levels, which takes place at the absorption of electromagnetic radiation by a solid. A *photoemissive effect* and a *photoconductive effect* are distinguished. During the course of photoemission the electrons, which have absorbed a *photon*, emerge through the *solid surface* into the surrounding medium (see *Photoelectron spectroscopy*). The photoconductive effect is responsible for *photoconductivity* through a photoelectromotive force (see *Photovoltaic effect*), and a *photodielectric effect*. The photoconductive effect may play an important role in nonmetallic crystals. The photoelectric effect may involve electrons, which are located at various quantum levels of the solid, ranging from loosely bound *conduction electrons* to electrons bound at inner atomic shells. Hence the necessary condition for the photoelectric effect is that the photon energy exceeds the binding energy of the electron at the ground level (see *Einstein relation*). The photoemissive effect requires additional energy equal to the *work function* of the solid. This condition determines upper limits to the wavelengths, which result in photoemission from various electron energy levels in a solid.

PHOTOELASTIC TENSOR, elastooptic tensor

The tensor, which establishes the linear relationship between the change of *dielectric constant* ε, which takes place in a body under slight *strain*, and the magnitude of the deformation:

$$\varepsilon_{ik} = \varepsilon_{ik}^0 - \sum_{l,m,n,j} \varepsilon_{il}^0 \varepsilon_{km}^0 P_{lmnj} U_{nj},$$

where ε_{ik} is the dielectric constant tensor of the body under strain, ε_{ik}^0 is the same tensor in the absence of a strain, P_{imnj} is the photoelastic tensor, U_{nj} is the *strain tensor*. The components of the photoelastic tensor are independent of the amplitude for a slight deformation of the object, and as a rule are determined exclusively by the internal structure of the body under consideration, and by the electromagnetic wave frequency. The dependence of the photoelastic tensor components on the deformation amplitude appears under severe strain (*nonlinear elasticity effect*). Typical values of photoelastic tensor components of solids fall in the range from 0.05 to 0.2. The effect of a change of dielectric constant of materials during strain results in the generation of *acousto-optic diffraction*.

PHOTOELECTRIC PHENOMENA

Changes of the electronic properties of materials which take place during exposure to light. Photoelectric phenomena are brought about by the transfer of energy from absorbed *photons* to the electrons of a material. The absorption of a photon raises an electron to a higher energy state (*photoeffect*). If the excitation results in the emission of an electron from the material, then the process is referred to as *photoemission* (or *photoelectron emission*). The energy spectrum of the electrons emitted under irradiation provides information on the energy band structure of the sample (see *Photoelectron spectroscopy*). If the electrons excited by the light remain inside the sample, then the process is referred to as the *photoconductive effect*. This effect is responsible for the phenomenon of *photoconductivity*. If the photon energy is insufficient for the production of free electrons, and electron transitions to bound excited states take place, then a photoinduced change of the *dielectric constant* called the *photodielectric effect*) is observed. Under certain conditions (if the semiconductor contains inhomogeneities, or if inhomogeneities are generated by the exposure to light) a *photoelectromotive force* arises, and a potential difference is produced (*photovoltaic effect*). If the circuit is closed then electric current can flow in the absence of a voltage source. These effects are also observed for uniform illumination of a regular but unipolar crystal; and abnormally high photoelectromotive force values can be reached in this case (see *Anomalous photovoltaic effect*).

PHOTOELECTRIC THRESHOLD

The maximum wavelength for which radiation can still induce a *photoeffect*. The existence of a red limit follows from the quantum nature of the phenomenon: the energy required for the *photon* absorption. This absorption results in a transition between two electronic energy levels, so the photon energy cannot be less than the separation between the two states (at zero temperature). Precise experimental measurements of this red limit remain difficult to make since other factors which contribute to photoexcitation, such as thermal mobility and electric fields, serve to blur the red limit. Therefore its exact position is often extrapolated from the dependence of photoeffect output on the wavelength in the neighborhood of this limit.

PHOTOELECTROMOTIVE FORCE, SURFACE

See *Surface photoelectromotive force.*

PHOTOELECTRON EMISSION

See *Photoelectron spectroscopy.*

PHOTOELECTRON EMISSION, INVERSE

See *Inverse photoelectron emission.*

PHOTOELECTRON SPECTROSCOPY,

photoemission spectroscopy

The branch of spectroscopy which deals with the spectrum of kinetic energies of electrons which are generated as a result of the interaction of electromagnetic radiation with the material, and are emitted from the solid into a vacuum or another medium. Ultraviolet, and synchrotron or X-ray varieties of radiation are often used as the sources of quanta. Under exposure to electromagnetic radiation, the energy absorbed by the material can be transferred to electrons of both the *valence band* and the inner core levels (Fig. (a)), and if the resulting energy of the electrons approaching the surface is sufficient to overcome the potential barrier, then they may escape from the material and generate a flow of free charges in the vacuum.

The general laws of *photoelectron emission* are:

(1) Each material exhibits a certain threshold value of energy: the electron binding energy E_b or minimum *photon* energy above which photoelectron emission takes place (*threshold energy*).

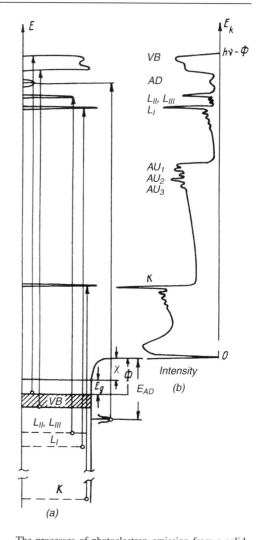

The processes of photoelectron emission from a solid. (a) Sketch representing initial and final (excited) energy states of electrons in the solid: valence band *VB*; inner core levels $L_{I,II,III}$ and K; adsorbate energy E_{AD}, $\Phi = h\nu_0$ is the work function or ionization potential of the highest energy electron of the valence band; χ is the electron affinity; E_g is the width of the band gap. For metals *VB* is the conduction band and $E_g = 0$. (b) Spectrum of photoelectrons in vacuo; *VB* is signal of valence band electrons; *AD* are electrons at the adsorbate level; $L_{I,II,III}$ and K are inner core levels; AU_1, AU_2, AU_3 are lines and bands of Auger electrons, which appear as a result of relaxation of inner core excited states (the background signals belong to scattered primary and secondary electrons; E_k is photoelectron kinetic energy).

(2) The main equation, which describes the process, is *Einstein's equation* $\hbar\omega = E_b + KE$ which connects the energy of incident quantum $\hbar\omega$ with E_b and the electron kinetic energy KE. If the object under investigation is a solid, then it is necessary to include a correction for the *work function*, and for charging the sample (the latter is of highest importance for non-conducting samples).

(3) The number of ejected photoelectrons is directly proportional to the irradiation intensity (*Stoletov's law*).

The absorption of luminous energy results in the emission of electrons which are excited in the bulk of the sample (*bulk photoelectric effect*), the excitation and emission of electrons from the near-surface layer (*surface photoelectric effect*), and the ejection of excited electrons from the surface of the sample (*emission from surface states*). The number of emitted photoelectrons depends on the density of occupied and empty electron states of the solid and its surface, on the probability of optical transitions, on the probability of electron scattering in the course of motion towards the surface, and passing through the potential barrier. The *mean free path* of photoelectrons depends on their energy, and determines the depth of emission. One may control the contributions of the bulk photoelectric effect and surface photoelectric effect to the process of photoelectric emission by varying the excitation energy $h\nu$. Measurable characteristics of photoelectron emission are (1) *quantum yield*, i.e. the number of emitted photoelectrons for a single photon incident upon the surface or absorbed by it; (2) the dependence of the quantum yield on the photon energy (spectral characteristic); (3) the energy and angular distribution of electrons, and (4) the extent of their spin polarization. Photoelectrons exhibit a continuous spectrum from zero energy to $\hbar\omega - \hbar\omega_0$ (Fig. (b)). The fine structure of the spectrum is determined by specific features of the *density of electron states* and the nature of the bulk and surface scattering of electrons. In this connection, photoelectron emission plays a crucial role in the spectroscopy of electronic energy levels of solids, and in the investigation of electron scattering mechanisms.

The method of photoelectron spectroscopy allows the determination of the binding energy of inner core electrons with a high degree of accuracy (10^{-2} eV for UV and 10^{-1} eV for X-ray radiation), and to identify the presence and concentrations of particular atoms. The use of monochromators makes it possible to obtain a resolution of 0.5–0.2 eV, and to investigate various specific features of the structure of a solid. The method permits the study of thin surface layers 2–4 nm thick, and using *synchrotron radiation* it is possible to investigate layers 0.8–1 nm in thickness, and to pass continuously from the surface into the bulk. Photoelectron spectroscopy is used as a method of analysis of the surface and the bulk of a sample. The distinctive feature of photoelectron spectroscopy is that this method provides not only the quantities of the atoms of a given type, but also the *valences* of these atoms. By taking into account the sharp dependence of the photoionization cross-sections on the energy of the incident quanta, which differs in the UV and synchrotron spectral regions, one may determine not only the overall *density of states* in the *valence band*, but also the partial densities of states. Exploration into the angular dependences of the photoelectron spectra is a principal method for investigating of the *dispersion law* of solids. The analytical application of photoelectron spectroscopy is often called *electron spectroscopy for chemical analysis* (*ESCA*) by chemists.

PHOTOELECTRON SPECTROSCOPY EXPERIMENTAL METHODS

See *Experimental methods of photoelectron spectroscopy*.

PHOTOEMISSION SPECTROSCOPY, ULTRAVIOLET

See *Ultraviolet photoemission spectroscopy*.

PHOTOEXCITON

A *quasi-particle* resulting from the admixture of states of a photon and an exciton when they have a strong resonance interaction with each other (see *Polariton*). The term "photoexciton" was introduced to describe the actual elementary excitations in a crystal. Meanwhile the exciton is understood as an idealized excited state of a crystal calculated without taking into account the retardation in the Coulomb interaction between charges, or the curl part of the field that accompanies electromagnetic waves.

PHOTOFERROELECTRICS

Ferroelectrics which exhibit a change of their physical properties (properties related to *ferroelectricity*) when undergoing optical excitation in intrinsic or extrinsic spectral regions. Such crystals include, first of all, *ferroelectrics-semiconductors*, i.e. ferroelectrics, which exhibit semiconducting properties. This class of crystals includes a large number of compounds, among which are broad-band compounds like lithium, barium and strontium *niobates*, compounds of the formula $A_2^V B_3^{VI}$, and narrow-band semiconductors (e.g., compounds of the formula $A^{IV} B^{VI}$). Specific features of semiconducting ferroelectrics consist in the effect exerted by electrons on the thermodynamic functions and their behavior in the neighborhood of the *Curie point*, and in the influence exerted by nonequilibrium electrons produced by the optical illumination on the ferroelectric properties, i.e. *photoferroelectric phenomena*. These phenomena include: (1) shift of the Curie point (often a decrease of T_C during illumination); (2) change of spontaneous *polarization P*; (3) *photorefraction*; (4) change (often decrease) of thermal *hysteresis*; (5) change of spontaneous deformation (*photodeformation effect*), *photodielectric effect*, and other effects.

On the phenomenological level, the description of photoferroelectrics is based on the relationship between the coefficients of expansion of the free energy in powers of the polarization P, and the concentration of photoexcited carriers of current, which are captured by *traps for mobile particles*. The microscopic mechanisms influencing the properties of photoferroelectrics during illumination may result from the *Jahn–Teller effect*, *spontaneous polarization shielding*, recharging of ferroelectrically active impurity centers, etc.

PHOTOGRAPHIC IMAGE

See *Latent photographic image*.

PHOTOGRAPHIC MATERIALS

Light-sensitive layers for image recording or for storage of light signals and other optical information, including binary coded information, applied to a mechanically strong substrate (glass, polymer film, paper, etc.). Photographic materials with a light-sensitive layer of the gelatinous suspension of silver halide (AgHal) microcrystals in gelatin, called *photoemulsions*, have traditionally been the most widespread types in use. In recent years photographic materials with other types of light-sensitive layers having no silver have appeared such as: diazotype (diazo print), photochromic (see *Photopigments*), thermoplastic, electrophotographic, photoresistor, etc. Photographic materials differ according to the type of change that is caused by the light (photochemical, photophysical, photobiochemical), according to their application, or to the presence or absence of amplification of primary photoinduced changes, according to the method of amplification, according to the possibility of reuse, to the type of modulation of reading the light beam (amplitude, phase, mixed), to the ratio of brightnesses in the object and its image (negative, positive), to the possibility of transmission of gradations of the gray tone scale, to the transmission of color (black-and-white, monochrome, color, etc.). For example, the common AgHal's are photochemical photographic materials requiring chemical development, are nonreversible, and are amplitude-modulating. Photographic materials may be negative, positive and reversible; black-and-white or color types, etc.

PHOTOGRAPHIC PROCESSES

Sequence of stages needed for recording optical images and other optical information on *photographic materials*. This sequence includes, along with additional and optional processes, the following main stages: formation of a light sensitive layer, recording the optical information in the form of a *latent photographic image* which is invisible to the eye; visualization of the latent image through amplification of changes induced by light in the recording layer; data reading (automatic or visual). If the recording is performed in a real time mode, then the third stage is either combined with the second or omitted, but the light sensitivity of the process becomes in this case extremely low. At the first stage of this classical photoprocess (in which an emulsion containing silver halide is used) the liquid emulsion is subjected to *chemical nucleation* (warming-up in the presence of chemically active impurities), which leads to the generation of impurity centers (deep electron traps) at the surface of microcrystals of AgHal. At the second stage, the exposure of the sample to light produces small clusters of Ag (latent photographic image) generated at the first stage impurity centers. The third stage (development) involves specific reduction of only those microcrystals which contain the clusters, and which form the latent image; thus the process of reduction is selectively catalyzed. For several atoms of Ag, which are produced under the influence of light, there are up to 10^{10} chemically generated Ag atoms, so the gain factor of the primary photochemical product is very large, $\geqslant 10^9$. It is a common practice to put most of the silver halide layers through spectral sensitization at the first stage of the photoprocess. Spectral sensitization is the process whereby the sensitivity (naturally limited by the edge of the AgHal fundamental absorption band in the blue–violet region of the visible spectrum) of a photographic layer is extended to longer wavelengths (short-wavelength infrared) by the *adsorption* of sensitizing dyes (spectral sensitizers) on the AgHal surface. Besides this development, the third stage (visualization) also includes the removal of unexposed microcrystals from the emulsion layer (photographic fixing).

PHOTOGRAPHY, NONSILVER

See *Nonsilver photography*.

PHOTOINDUCED LIGHT SCATTERING

Quasi-elastic nonlinear *light scattering* in photorefractive crystals (crystals in which the index of refraction depends on the light, see *Holography*). The monochromatic light wave incident upon the *crystal* induces in it a multitude of noise waves, which are scattered by static optical (bulk and surface) inhomogeneities. Each elementary component of the scattered light is coherent with respect to the corresponding elementary component of the incident light; elementary components of scattered and incident light form an interference pattern, through the agency of which a noise holographic grating is formed in the crystal (see *Diffraction grating*). The weak scattered wave may be amplified by virtue of the diffraction of the incident wave by this lattice (*holographic light amplification*). An increase of intensity of the scattered wave brings about an increase of contrast of the interference pattern, and raises the diffraction efficiency of the lattice, which leads to further amplification of the scattering, i.e. the process is nonlinear. Depending on the kind of holographic amplification, which causes photoinduced light scattering, the scattering process may be either stationary or nonstationary. The properties of *photorefractive crystals* manifest themselves in the anisotropy of processes for recording and reading the data of noise holographic gratings; the efficiency of diffraction by these lattices depends on their wave vector, and the polarization of the incident wave. Such anisotropy results in specific angular dependences of the scattered light. The far zone region may exhibit either comparatively large-angle scattering (characteristic of photoinduced light scattering, related to nonlocal mechanisms of photorefractive response and nonstationary types of such light scattering) or cross- and ring-shaped scattering patterns (characteristic of parametric amplification processes at arbitrary local or nonlocal response conditions). Isotropic and anisotropic photoinduced light scattering are distinguished. The scattered light in the isotropic process retains the state of polarization of the incident wave; while in the case of the anisotropic process the scattered light contains a component with orthogonal polarization.

In certain photorefractive crystals ($BaTiO_3$, $LiNbO_3$, Ba–Sr niobate, Te, etc.) photoinduced light scattering takes up most of the energy of the incident beam, for a typical sample thickness of several millimeters; the increments in this process may exceed 10, which can involve the amplification of a primary weak noise wave by a factor of 10^5 or more.

PHOTOLUMINESCENCE

Luminescence stimulated by optical excitation. By virtue of the diversity of processes of energy transfer and conversion, mechanisms of excitation of photoluminescence in solids may differ, depending on the wave length of the exciting radiation. The spectrum of photoluminescence excitation (dependence of *luminescence yield* on wavelength of exciting radiation) exhibits a red-shifted edge located, in accordance with the *Stokes' rule*, in the spectral region before the longest wavelengths of the photoluminescence. *Vavilov laws* also hold for photoluminescence.

PHOTOMAGNETOELECTRIC EFFECT

The same as *Kikoin–Noskov effect.*

PHOTOMECHANICAL EFFECT

The appreciable decrease (by 30–70%) of *microhardness* of a material, which takes place under exposure to light (discovered by L.C. Kuczynski, R.H. Hochman, 1957). The photomechanical effect is observed in monocrystals of *germanium*, *silicon*, various binary and more complex semiconducting materials, in certain metals (*antimony, tellurium*), *alkali-halide crystals*, crystals of yttrium *iron garnets*. Studies have been made to determine the influence of electrically active and gaseous impurities in crystals which exhibit the photomechanical effect. Electrically active impurities reduce the magnitude of this effect in germanium and silicon. The photomechanical effect can even vanish for concentrations of electrically active impurities $\sim 10^{25}$ m^{-3}. The same effect arises from the incorporation of various impurities into lattices of alkali-halide crystals (with the valence of the impurity differing from that of the host lattice ions). Very pure *monocrystals* also fail to exhibit a photomechanical effect. Neutral gaseous impurities (nitrogen, helium) also have no influence on the magnitude of the effect. The spectral range of the photomechanical effect is in the IR, and its magnitude increases with increasing IR light intensity; until at an illumination ~ 1 W/m^2 the microhardness versus illumination curves reach saturation. The exposure time needed to produce the photomechanical effect is between 0.02 and 0.5 s, although it is possible that this time interval does not actually characterize the induction time for generating the effect, but rather the ability to measure the changes of *hardness*. A particular property of this effect is its pronounced decrease in magnitude with an increase in temperature. As a

rule, the effect is not observed even at 100–200 °C. It also does not take place at high microhardness indentations (high loads on *indenter*); this circumstance indicates that the photomechanical effect may occur only in the surface layers of the material. The depth of the layer, which is plasticized by the incident light, is 1–3 μm. Therefore, the photomechanical effect is sometimes called the *surface photoplasticization effect*.

The nature of the photomechanical effect seems to be related to photochemical transformations of hydrogen- and oxygen-containing impurity groups (perhaps OH$^-$) in the surface layers of crystals, although various speculations have been made on this subject at various times. All types of treatments which remove these groups from the surface layers (e.g., vacuum annealing) result in the disappearance of the photomechanical effect. Hydrogen annealing, especially with slight oxidation of the samples, as well as prolonged standing in a damp environment, restores the effect. On the basis of these results the photomechanical effect is classified among the groups of phenomena which are known as the *Rebinder effect*, or the *strength reduction through adsorption* during exposure to a gaseous or liquid medium.

The probability of finding practical applications for the photomechanical effect is related to the deformation of articles of small sizes, produced in brittle materials, and to speeding up the polishing of semiconducting samples. The photomechanical effect plays a significant role in processes involving *friction* and *wear*.

PHOTON

Neutral particle of unit spin, which is the quantum of the electromagnetic field. The photon energy is given by $E = \hbar\omega$, and its momentum is $\boldsymbol{p} = \hbar\boldsymbol{k} = (\hbar\omega/c)\boldsymbol{n}$, where $\hbar = h/(2\pi)$ is the reduced *Planck constant*, c is the *speed of light* in vacuo, $\omega = 2\pi f$ is the frequency of the electromagnetic radiation, and $\boldsymbol{n}$ is unit vector directed along the wave vector $\boldsymbol{k}$. If the number of photons is sufficiently high, then the ensemble of them forms a transverse classic electromagnetic wave $\boldsymbol{E} = \boldsymbol{e}A\exp[-\mathrm{i}(\omega t - \boldsymbol{k}\boldsymbol{r})]$, where $\boldsymbol{e}$ is the *polarization vector* of the wave, and A is its amplitude, $|\boldsymbol{k}| = 2\pi/\lambda = \omega/c$. The following types of electromagnetic waves are distinguished: radio waves,

$f = 10^2$–10^{11} Hz; *infrared radiation*, $f = 10^{11}$–10^{14} Hz; optical (visible) *radiation*, $f = 10^{14}$–10^{15} Hz, ultraviolet radiation, $f = 10^{15}$–10^{17} Hz; X-rays, $f = 10^{17}$–10^{19} Hz; and gamma rays, $f = 10^{19}$–10^{22} Hz.

A *photon in a solid* is a *quasi-particle*, which differs from the photon of free space because of its interaction with excitations of the solid. This interaction, which takes place away from absorption bands, manifests itself in the capture and re-emission of a photon without loss of energy, which results in the retardation of the photon, i.e. the decrease of its velocity $v = c/n$, where $n = \sqrt{\varepsilon}$ is the refractive index of the solid under consideration. Those photons, which exhibit frequencies from the absorption band, are captured by the solid and their energy is transmitted to quasi-particles produced in the solid: electrons, *excitons, phonons*, etc.

A large number of photons of a certain frequency form a classical electromagnetic wave, whose electric field vector brings about a separation of charged particles which are present in the solid. In other words, a polarization wave $P = \chi E = \chi e A \exp[-\mathrm{i}(\omega t - k_0 \cdot r)]$ is induced, where the susceptibility χ is a complex tensor of the system *polarizability*. The electric field of the wave inside the solid has two components: namely, the free electromagnetic field E, and the field of the polarization wave P; taken together, these two components form the electric displacement (*electric flux density*): $D = \varepsilon_0 E + P = \varepsilon_0 (1 + \hat{\chi}) E = \hat{\varepsilon} E$; where $\hat{\varepsilon}$ is the *dielectric constant* (or permittivity) tensor (SI units are being used). In the neighborhood of absorption bands, which involve the production of localized excitations, the $\hat{\varepsilon}$ tensor becomes a complex quantity and the electromagnetic wave decays via $e^{-\alpha z}$ due to the the *absorption coefficient* α, which comes from the imaginary part of $\hat{\varepsilon}$. In the neighborhood of an absorption band, which is associated with the production of nonlocalized excitations of the solid, photons exist in the form of quasi-particles called *polaritons*, which propagate at speeds less than c, and decay as a result of the breakup of excitations of the solid. A polariton is a quantum of a coupled photon and phonon.

PHOTON DRIFT

Directional propagation of *photons* and related drift of nonequilibrium charge carriers (see *Current carrier drift*), which occurs in *variband semiconductors*. Photon drift takes place at high values of internal luminescence *quantum yield* η, and low gradients of the width of the *band gap* E_g as a result of the effects of reemission and anisotropy of the *absorption coefficient* α of the recombination radiation. One may sketch the qualitative pattern of photon drift by assuming (for ease of presentation) $\eta = 1$, and the drift mobility of current carriers $\mu = 0$. In the context of these assumptions, the pairs of carriers, which are generated at the point z_i by external excitation, undergo radiative recombination at the same point. The photons, which move in the direction of decreasing E_g, are absorbed at the distance α^{-1} from the point z_i, thus producing new pairs of carriers. Thus the directional absorption of the recombination emission results in the directional transfer of current carriers from the point z_i to the point $z_i + \alpha^{-1}$. In actual variband crystals this mechanism of nonequilibrium carrier transfer is accompanied by their drift in the applied quasi-electric field (see *Variband semiconductors*); the transfer mechanism of photon drift may play the leading role when the following criteria are satisfied:

$$L^2 \ll \begin{cases} \eta\big[3\bar{\alpha}^2(1-\eta)\big]^{-1} & \text{for } \dfrac{\nabla E_g}{k_B T \bar{\alpha}} \ll 1, \\[2ex] k_B T \big[2\bar{\alpha}^2(1-\eta)|\nabla E_g|\big]^{-1} \\[1ex] & \text{for } \dfrac{\nabla E_g}{k_B T \bar{\alpha}} \gg 1, \end{cases}$$

where L is *diffusion length*, and $\bar{\alpha}$ is the averaged absorption factor for the recombination radiation.

PHOTON ECHO

Coherent radiation of light, which results from the delayed response of a material to irradiation by two (or more) sequential laser pulses; optical analog of an electron *spin echo* or a nuclear (i.e. NMR) spin echo. In an electron spin echo experiment (see Fig.) a high-power pulse I of microwave energy aligns electron spins along the applied magnetic field direction, and then for a time interval τ they precess around the field at slightly different frequencies, gradually getting out of phase

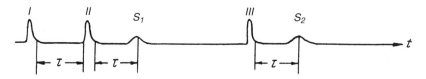

Pulse–echo sequence, where I, II, and III are optical pumping pulses, S_1 is the primary photon echo, and S_2 is the stimulated photon echo.

with each other. After the time interval τ a second microwave pulse II turns around the direction of the precessions, so the spins reverse their motion and gradually get closer together in phase. After another τ interval of time they are all again in phase with each other, and so they radiate a signal S_1 called an echo. This sequence of events is shown on the left side of the figure.

A material which exhibits photon echos must possess *phase memory*, i.e. retain information of optical parameters of the pumping fields over a period of time which exceeds the duration of the pumping pulses, and the time intervals between these pulses.

Photon echo has its origins in matching the phases of oscillations of dipoles with slightly different natural frequencies. The moment of phase matching determines the moment of initiating the photon echo signal, which depends on the pumping conditions (widths of absorption and pumping spectral lines, pulse intensity, etc.). This dependence takes its simplest form in the case of very short and intense pulses, when the delay time τ of the photon echo is equal to the interval between two initial impulses (see Fig., left). Unlike spin echo, the photon echo exhibits sharply defined directionality, which results from the propagation and interference of light waves. The direction of the photon echo wave is determined by the wave vector k_e, and is subject to the conditions of vector alignment (synchronism). In the case of a *primary photon echo* S_1, this condition (given by the equation $k_e = 2k_2 - k_1$) is satisfied only in the case of low angles between k_1 and k_2 (pumping vectors I and II shown in Fig.), i.e. at $k_2 = k_1$. In the case of a *stimulated photon echo* S_2, brought about by a third light pulse III, the alignment $k_e = k_3 + k_2 - k_1$ and phase matching may be observed in several directions (in particular, in Bragg directions). For example, at $k_3 = k_1$ the

wave of the photon echo propagates in the direction of the second pumping beam ($k_e = k_2$), and at $k_3 = -k_2$ the case of a *reversed photon echo* is realized ($k_e = -k_1$).

Applications of photon echo are: spectroscopy of extra-high resolution (determination of dipole moments of transitions, longitudinal and transverse relaxation times, etc.), space–time transformations of wave fields in *real-time holography* (multiplication of phase disturbances, reversal of light wave fronts, optical processing of information).

PHOTOPIGMENTS, photochromic materials

Class of light-sensitive materials which experience a reversible change of color or of optical density under exposure to activating radiation of suitable wavelengths (see *Photochromic effect*). Photopigments may be gaseous (aerosols), liquid, vitreous, or crystalline. Depending on their function, photopigments are produced in various states: in the form of liquid solutions, films, coatings, fibers; their sizes are not limited. Solid-state photopigments are *silicate glasses*, which contain crystals of silver halides, activated *alkali-halide crystals*, oxides, and salts of alkaline-earth metals (CaF_2:La, Ce), *polymers*, etc. The principal characteristics of photopigments include the absorption spectra of initial and photoinduced states, the spectral region and the value of the light sensitivity, rates of darkening processes, reversibility of the photochromic cycle.

Most photopigments undergo photochromic transformations under exposure to UV and short-wavelength visible radiation. Application of various sensitizers or preliminary excitation with UV irradiation (optical sensitization) facilitates the extension of the light sensitivity to longer wavelengths (spiropyranes, *photochromic glasses*).

In theory the limiting light sensitivity of photopigments is 0.25 m^2/J, but actual values of their light sensitivity are smaller by 2 to 3 orders of magnitude. The time constant for the transition of a photopigment from its initial state to the photoinduced state is determined for the most part by the intensity of the exciting radiation, and by the duration of the photochemical processes, the latter being 10^{-13}–10^{-7} s for most photopigments. The lifetime of the photoinduced state of a photopigment varies over a wide range depending on the type of material; and at room temperature it may range from several microseconds to many months and even years. The limits of variations of the lifetime of a given photopigment are temperature dependent.

Certain photopigments exhibit fatigue, i.e. a degredation or loss of photochromic properties during the process of storage and operation, yet most inorganic photochromic crystals survive more than 10^4 darkening–decoloration cycles with no change of properties.

PHOTOREFRACTION, photorefractive effect

Photoinduced change Δn of the *refractive index* n in ferroelectric and piezoelectric materials (i.e. photorefractive materials); it is sometimes called *"optical distortion"*. From the practical point of view, the phenomenon of photorefraction consists in the reversible change of *birefringence* during local illumination of a piezoelectric or ferroelectric crystal with intense transmitted light (focused laser beam). The magnitude of this change reaches 10^{-4}–10^{-3} for certain crystals (LiNbO$_3$, LiTaO$_3$); the memory time of the effect ranges over fairly broad limits (from milliseconds for BaTiO$_3$ to months in the case of LiNbO$_3$). Photorefraction has found an application in the recording of three-dimensional holograms. The recording of a hologram is carried out through volume modulation of Δn, which corresponds to modulation of the recording beam, and the resolution is quite high (see *Holography*).

Photorefraction may be produced by a number of mechanisms:

(1) A *bulk photovoltaic effect* leads to the generation of a strong electric field in an isolated *ferroelectric*, and this field induces the change of Δn by virtue of the linear *electrooptical effect*. This mechanism of photorefraction is the prevailing one in the case of ferroelectrics like LiNbO$_3$:Fe.

(2) Optical recharging of impurity centers, which results in a photoinduced change of the spontaneous polarization (see *Ferroelectricity*).

(3) Generation of local electric fields via nonuniform illumination of the crystal (diffusion mechanism); this mechanism is also observed in nonferroelectric electrooptical photoconductors.

(4) Pyroelectric mechanism, according to which the field arises through the agency of heating via the *pyroelectric effect*.

The change of birefringence Δn, which takes place at illumination of an electrooptical crystal in the external field, is also attributed to photorefractive phenomena. In this case the photorefraction is caused by the screening of the external field applied to the crystal by nonequilibrium charge carriers. This kind of photorefraction has found an application in the production of electrooptical recording of information through the processing of crystals of the sillenite group (Bi$_{12}$GeO$_{20}$, Bi$_{12}$SiO$_{20}$).

PHOTOREFRACTIVE CRYSTALS

See *Holography*.

PHOTOREFRACTIVE EFFECT

See *Photorefraction*.

PHOTORESISTIVE EFFECT

See *Photoconductivity*.

PHOTORESISTORS

Optoelectronic devices (see *Optoelectronics*), whose operation is based on the change of resistance of the active element at exposure to light (see *Photoconductivity*). Unlike electromechanical potentiometers, photoresistors do not have moving mechanical contacts, so there is a sharp increase of the speed of response, a raising of the threshold of sensitivity, and an improvement of the reliability. Along with linear photoresistors, which have a uniform distribution of resistance, there exist widely used functional photoresistors; and to assure that these latter devices have a favorable response in their output characteristics, a nonuniform distribution of resistance is created in them during the course of their manufacture. For this purpose a thin film technique is employed which combines metallic films of good conductivity with photosensitive semiconducting layers. Photoresistors can have a high sensitivity at various regions of the optical spectrum and beyond it, including X-rays and gamma radiations. They are used in automatic devices, computer engineering, systems for information processing, etc.

PHOTOTHERMOPLASTIC MATERIALS

See *Nonsilver photography.*

PHOTOTRONIC EFFECT

See *Photovoltaic effect.*

PHOTOVOLTAIC CELL

A device based on an *semiconductor junction* or a metal–semiconductor structure. The illumination of the rectifying contact by quanta of energy exceeding the *band gap* generates electron–hole pairs in the *semiconductor* which are separated by an electric field inside the contact. This causes electric current to flow in the external circuit of the photovoltaic cell. Such photocells are widely used in electronic devices and in industrial automation systems (see also *Photodiode*).

PHOTOVOLTAIC EFFECT, phototronic effect

Appearance of an electromotive force (emf) upon the absorption of light by an inhomogeneous semiconductor which generates minority carriers called photocarriers. (If the circuit is closed during the light absorption then a current flows.) The reason for the appearance of the *photoelectromotive force* is the separation of the photocarriers by the built-in (internal) electric field, associated with the inhomogeneity. If the boundary of a homogeneous sample plays the role of the inhomogeneity, then the photoelectromotive force exists for differing mobilities of electrons and holes, and it is referred to as a *diffusion photo-emf* (*Dember emf*) (H. Dember, 1931). It is associated with the different velocities of the *diffusion* motions of the nonequilibrium electrons and holes from the illuminated face, where their concentration is high, and also it is associated with the appearance of an *ambipolar diffusion* field.

Upon illumination of an inhomogeneous semiconductor there appears a *bulk photo-emf*, proportional to the dark (before illumination) resistance. The Dember and bulk electromotive forces in semiconductors are low ($V \sim 1$ mV). When light generates charge carriers in the vicinity of a potential barrier, i.e. upon illumination of a *metal–semiconductor junction* or a semiconductor–semiconductor junction, there appears a *gate emf*. The built-in electric field present at the barrier separates the photoelectrons and photoholes, and as a result the height of the potential barrier is reduced, and electric current starts to flow within the circuit. A maximum gate emf appears at *p–n* homo- and *heterojunctions* that may reach several volts. If the *semiconductor junction* is closed by the external load then it plays the role of a transducer of light energy into electrical energy. *Photovoltaic cells* and *solar cells* operate using this principle.

If the illuminated sample is placed in a magnetic field directed perpendicular to the light flux onto one of the sample faces, then even when the electron and hole mobilities are equal there appears a transverse emf perpendicular to both the magnetic field direction and the light flux direction (*Kikoin–Noskov effect*). This electromotive force is associated with the deflection of the diffusing photoelectrons and photoholes to opposite side faces of the sample by the *Lorentz force*. See also *Anomalous photovoltaic effect*, *Bulk photovoltaic effect*, *Circular photovoltaic effect*.

PHOTOVOLTAIC EFFECTS

For different varieties, see also *Anomalous photovoltaic effect*, *Bulk photovoltaic effect*, *Circular photovoltaic effect*.

PHYSICAL ADSORPTION

Type of *adsorption* process, in which the chemical bond between molecules (atoms) that form the *solid surface* (*adsorbent*) and the particles being adsorbed, arises from *van der Waals forces*. Physically the adsorbed molecules preserve their chemical individuality, and the adsorption is reversible so the molecules which become physically adsorbed may be desorbed after some "lifetime" τ at the surface (see *Desorption*). The value of τ depends on the bond energy of the adsorbed particle with the surface, on the type of adsorbant, and on the surface temperature T_s. At room temperature $\tau \sim 10^{-9}$ s, and with the reduction of T_s to a value of the order ~ 100 K, the lifetime τ increases up to times of the order of ~ 1 s. At equilibrium the rates of adsorption and desorption are equal. An equilibrium coating, prepared under ordinary circumstances, amounts to fractions of a monolayer. For an adsorbent surface temperature T_s close to the adsorbate *condensation* temperature, and also under *high pressure*, the equilibrium coating may increase up to a monolayer, or even to a multilayer. The dependence of the monolayer coating on the pressure under constant temperature conditions is described by the *Langmuir adsorption isotherm*.

PHYSICOCHEMICAL MECHANICS

A branch of physical chemistry which studies the relationships between the properties of multiphase (dispersed) systems, and the physicochemical properties of the interface surfaces between the phases. Methods of physicochemical mechanics are used to prepare dispersed systems with pregiven properties (*strength, durability, porosity*, etc.).

PHYSICOCHEMICAL METHODS OF ANALYSIS

Methods used for the qualitative and quantitative determination of the composition of a substance (see *Qualitative analysis, Quantitative analysis*). They are based on the specifics of the individual physical properties of the separate components or phases of the material. The goal of physicochemical methods of analysis is to completely ascertain material compositions over a broad range of *concentration*. In complicated systems this goal is seldom reached by any one method of analysis, but the combined application of several methods allows one to approach it. It is customary to select some specific property of the component under investigation, and to construct a process for bringing about its quantitative determination, or even, in some cases, to separate it from the other components. Typical properties on which physicochemical methods of analysis are based, are: *thermal conductivity, specific heat,* heat of combustion, latent *heat of phase transition, diffusion coefficient* of particles, mass of molecule or ion, size of molecule, *adsorption* energy, ionization potential, *electron affinity,* electrochemical potential for separation at an electrode, *polarizability* and *diamagnetism, light absorption* spectrum, *fluorescence* spectrum, magnetic moment of nucleus and the electron shells, etc. All the methods based on these properties, are, in essence, physical ones. They may be called physicochemical when a chemical reaction is included in the preparation or transformation of the material. For example, the transformation of the material to the gaseous state for mass-spectrometric analysis (see *Mass spectrometry*); the attachment of fluorescent markers to specific molecules for fluorescent analysis.

PHYSICS OF METALS

Division of physics studying the nature, structure, composition and physical properties of *metals* and *alloys*, as well as the processes that take place when metals and alloys are exposed to treatments and other influences. Since physical concepts and physical methods of investigation are widely employed in *metal science* and *metallurgy*, the physics of metals plays an important role in preparing alloys with preset properties.

A major focus of present day research is the study of the electronic structure and characteristics (electronic, magnetic, spectral, etc.) of metals and alloys, their properties associated with the arrangement and motion of atoms (*elasticity, plasticity*, thermal, diffusion properties, etc.), as well as special features and kinetics of *phase transitions*. Of particular importance is investigating the disruption of crystalline regularity that results from defects (see *Defects in crystals*) and imperfectly ordered arrangements of atoms at alloy *crystal lattice* sites. The theory of metals is based on applying the quantum theory of solids, statistical physics, and physical kinetics. Of wide use in the experimental studies of metals are *electron diffraction analysis, neutron diffractometry, X-ray structure analysis*, as well as *electron microscopy*, electron resonance, electrical, magnetic, and other investigation methods.

PICO...

Prefix for a physical unit name, which allows obtaining a multiple unit, equal to 10^{-12} initial units. Symbol: p. Example: 1 pF (picofarad) = 10^{-12} F.

Table 1. Piezoelectric ceramics of functional groups 1 and 2

Functional group	Material type	$\varepsilon_{33}/\varepsilon_0$	k_p	$d_{ij}\cdot10^{12}$, C/N		$g_{ij}\cdot10^3$, V·m/N		Loss tangent		Q_M	T_c, °C
				d_{31}	d_{33}	g_{31}	g_{33}	$E < 25$ kV/m	$E \geqslant 30$ kV/m		
1	PZT-19	1725 ± 325	0.4	100.0	200.0	10.6	24.0	0.035	–	50	290
	PZTNV-1	2250 ± 560	0.45	165.0	400.0	–	–	0.015	–	70	240
	RPC-1	700	0.62	–	–	16.5	38	0.02	–	90	355
2	PZT-23	1075 ± 225	0.43	100.0	200.0	–	–	0.0075	0.04	200	275
	PZTB-3	2300 ± 500	0.45	125.0	320.0	–	15	0.012	0.035	200	180
	RPC-6	2300 ± 500	0.64	195.0	440.0	–	22	0.004	0.01	1100	230

PIEZOELECTRIC CERAMICS

A specific type of ferroelectric ceramic material (see *Ferroelectric ceramics*) with high values of piezoelectric parameters. These ceramics consist of many randomly arranged crystal grains. An individual crystal grain is a *monocrystal* with its volume divided into *domains*. A material with domains arbitrarily oriented in space forms an isotropic *texture* with no piezoelectric properties. Tempering (heat treating) ceramics in a strong electric field ($\sim$30–40 kV/cm) at a temperature around 100 °C results in the appearance of piezoelectric properties, a process called *polarization of piezoelectric ceramics*. During this process the vectors of spontaneous polarization (see *Ferroelectricity*) in separate crystal grains and domains orient themselves along the field. The complex processes of interactions of domains, alligning of crystal grains, ordering of impurities, etc. together with a number of other factors all result in freezing the direction of spontaneous polarization, and the retention of piezoelectric properties in the ceramics after the field is removed. In other words, we obtain a ceramic piezotexture. Not all the polar axes of all the crystal grains and domains align along the direction of the polarizing field, rather they are found within a solid angle that usually does not exceed $\theta \sim$ 50–60°. Therefore, the average values of piezomoduli are about half their respective values for monocrystals. Nevertheless, piezoelectric ceramics are widely used to produce electroacoustic and electromechanical devices. The primary reason of such a popularity is the possibility of shaping piezoelectric elements into various sizes and configurations, the relative simplicity of manufacturing, and the high degree of reproducibility of piezoelectric parameters. Among the most important of these parameters are the *dielectric constant* ε_{ij}, the electromechanical bonding factor k_{ij}, the piezomodulus d_{ij}, the mechanical quality Q_M, the piezoelectric sensitivity g_{ij}, and the relative change of resonance frequency, $\delta f_\theta/f_r$.

A number of piezoelectric ceramic materials have been developed and many are on the market. These are *solid solutions* based on $BaTiO_3$, $PbTiO_3$ and $NaNbO_3$, solid solutions of the $PbTiO_3$–$PbZrO_3$ composition (PZT), and also materials that are called Donetsk piezoelectric ceramics (DPC) and Rostov piezoelectric ceramics (RPC). Piezoelectric parameters of such ceramics depend on their composition, the presence of morphotropic phase transitions, and the specific state and composition of material in the range of that transition. In terms of the possible combinations of piezoelectric parameters the ceramic materials may be classified into the following functional groups:

1. Materials used to produce highly sensitive piezoelectric ceramic elements which function as receptors or emittors. High sensitivity of such materials to mechanical stress is provided by high values of the g_{ij} factor.
2. Materials stable with respect to electric and mechanical actions, used to produce devices that function at high power. The parameters of some of the materials from groups 1 and 2 are presented in Table 1.

Table 2. Piezoelectric ceramics in functional groups 3 and 4

Functional group	Material type	$\varepsilon_{33}/\varepsilon_0$	k_p	$d_{31}\cdot10^{12}$, C/N	$g_{31}\cdot10^3$, V·m/N	Loss tangent, $E = 225$ kV/m	Q_M	$\delta f_\theta/f_r$		T_c, °C
								−60 to +85°C	−10 to +50°C	
3	PZT-22	800 ± 200	0.2	27.0	−	0.025	400	0.435	0.18	320
	NBC-3	1800 ± 400	0.2	45.0	−	0.020	300	1.45	0.60	250
	PZT-35	1000 ± 200	0.38	70.0	−	0.025	550	0.50	0.20	300
	DPC-329	320 ± 50	0.43	40.0	−	−	1000	50 ± 10	−	330
	RPC-15	950	0.45	75.0	−	0.006	2500	0.30	0.10	325
4	PZT-21	550 ± 150	0.2	26.6	5.5	0.025	100	2.90	1.20	400
	RPC-24	480	0.53	64.0	15.1	−	200	3.80	−	−

3. Materials featuring high durability and temperature stability of their resonance frequency, used to build filtering and other selective devices operating with volume waves. Materials with high electromechanical bonding factors are also preferable for this application; the width of their transition band and their mechanical qualities are responsible for the sharp peak of the amplitude-frequency characteristic within the transition band.

4. High temperature materials to produce elements that remain operable at temperatures in excess of 300 °C and feature higher stability of their piezoelectric parameters in the prescribed temperature and mechanical stress ranges. The parameters of some of materials from groups 3 and 4 are presented in Table 2.

PIEZOELECTRICITY (discovered in 1880 by P. and J. Curie in quartz and other crystals)

A phenomenon consisting in the appearance or change of electric polarization as a result of mechanical stress (direct piezoelectric effect), or of mechanical strain under the effect of an electric field (inverse piezoelectric effect) in certain anisotropic insulators and semiconductors (piezoelectrics). The inverse piezoelectric effect should be distinguished from the electrostriction that is observed in all materials. In the piezoelectric effect the strain depends linearly on the electric field, while it is quadratic in that field for electrostriction. The necessary condition for the existence of a piezoelectric effect is the absence of a center of symmetry in the structural elements

of the system. Hence, the piezoelectric effect is observed in crystalline materials belonging to 20 of the 32 symmetry point groups. Among them are nonpolar piezoelectrics (e.g., quartz, potassium chlorate and bromate, nickel sulfate), and polar piezoelectrics or pyroelectrics (see Pyroelectricity), e.g., tourmaline, lithium sulfate, potassium tartrate. A subgroup among the pyroelectrics is comprised of ferroelectrics: monocrystals (e.g., Rochelle salt, barium titanate, ammonium phosphate) and polarized polycrystals – the piezoelectric ceramics; piezosemiconductors (e.g., cadmium sulfide, zinc oxide, antimony sulfoiodite); organic solid piezoelectrics (e.g., wood) and piezoelectric polymers (such as polyvinylidenefluoride), and electrets featuring a preferential direction in their piezoelectric effect, i.e. the so-called piezoelectric texture found in them.

The mechanism of direct piezoelectricity is related to the appearance or change of an electric moment, i.e. to the spatial redistribution of electric charge in the structural elements of solids (e.g., crystal lattice unit cells) that develops under the effect of mechanical stresses produced by compression, extension, shear. When the electric field affects elementary charges in a structure they suffer displacements, the average distances between them change, and a deformation takes place (inverse piezoelectric effect).

Piezoelectricity in various materials is characterized by the values of their piezoelectric constants which are components of third-rank tensors, $d_{i,kl}$, $g_{i,kl}$, $e_{i,kl}$, $h_{i,kl}$ ($i = 1, 2, 3$ and $k, l = 1, 2, 3$) that determine the mutual relation between the components of electric polarization P_i (*electric flux density* D_i) and those of mechanical stress σ_{kl} (e.g., $P_i = \sum_{kl} d_{i,kl}\sigma_{kl}$), of *strain* u_{kl} and the electric field E_i (e.g., $u_{kl} = \sum_i d_{i,kl}E_i$) in the corresponding equations of state. The tensor components $d_{i,kl}$ are symmetric in the last two indices k and l. These components, which are most often used in physics and technology, are called *piezomoduli* (e.g., quartz has $d_{1,11} = 2.31 \cdot 10^{-12}$ C/N, the BaTiO$_3$ piezoelectric ceramics have $d_{3,33} = 10^{-10}$ C/N). Important characteristics of piezoelectricity are the *electromechanical bonding factors*, $k_{i,kl} = d_{i,kl}/(\varepsilon_{ii}S_{nn})^{1/2}$, where $\varepsilon_{ii'}$ and $S_{kk'}$ are, respectively, components of the tensors of the *dielectric constant* and elastic compliance. They define the ratio of the energy of strain due to the piezoelectric effect, to the total energy of the electric field acting on the piezoelectric (e.g., $k_{1,11} = 0.095$ for quartz, and $k_{3,33} = 0.5$ for BaTiO$_3$).

Currently more than a thousand piezoelectrics are known. Among them are natural materials (i.e. found in the natural environment, such as quartz, tourmaline) and artificial piezoelectric materials which are synthesized in the laboratory or processed electrically or mechanically to produce a piezoelectric texture. Such are, e.g., piezoelectric ceramics, piezoelectric polymers, and *piezoelectric composite structures* (the latter combine crystalline piezoelectrics with polymer materials), and also electrets.

Among materials belonging to the 20 groups without centers of inversion the most powerful piezoelectric properties are displayed by ferroelectrics: single crystals and polarized polycrystalline compounds (piezoelectric ceramics) belonging to the group of pyroelectrics (10 such groups are known). Such crystal compounds display abnormally strong piezoelectricity because of a ferroelectric phase transition which differentiates them from the more common, so-called *linear pyroelectrics* (see *Ferroelectricity, Electromechanical effects*). Some examples of these are Rochelle salt, antimony sulfoiodide, piezoelectric ceramics based on barium titanate or lead titanate – zirconate. *Piezoelectric materials* find many technological applications, in particular in electro-acoustics as active elements in piezoelectric transducers (see *Acoustic wave transducer*). Of

significant interest are also piezopolar films and piezoelectric composite structures offering a number of construction advantages, as well as *piezosemiconductors* characterized by a strong *electron–phonon interaction*.

PIEZOELECTRIC MATERIALS

Materials used to manufacture piezoelectric elements for various functional devices in radioelectronic technology: *ceramics, composite materials* and *polymeric materials*. The material may be textured, polarized, made single-domain by superposing a constant electric field or by way of a corona discharge or mechanical deformation. The operations of sound and ultrasound recorders, adapters, microphones, deformation sensors, mechanical-electric generators of electric energy are based on the direct piezoelectric effect, while the operations of sound and ultrasound emitters, piezoelectric motors, atomic force microscope, scanning tunneling microscope, piezoelectric transformers of voltage and current, are all based on the inverse piezoelectric effect (see *Piezoelectricity*).

The operation of resonance frequency stabilizers, sensors of pressure, humidity, chemical composition, etc., and piezoelectric filters is based on the piezoelectric resonance effect. Piezoelectric elements are manufactured preoriented in their crystallophysical directions in such a way as to provide a *quality factor* (QF), optimal for this or that particular application. The value of the QF is given in relative units with respect to piezoquartz (QF = 1). In the reception mode the QF for *piezoelectric ceramics* reaches 2–4; its value is 8 for lithium *niobate*; 10 for lithium tetraborate; 10–12 for polarized polyvinylidenefluoride; 20 for Rochelle salt; 3–30 for piezoelectric composites. In the emission mode, the QF for quartz is 1; it is 13 for Rochelle salt; 15 for lithium niobate; 50–150 for piezoelectric ceramics. The best materials for piezoelectric resonators and other devices using them are piezoquartz, lithium tantalate, selenites, zinc oxide.

PIEZOELECTRIC MODULI

See *Piezoelectricity*.

PIEZOELECTRIC TENSOR

A third-rank tensor defining the interdependence of electric and elastic variables in *piezoelectric materials*. Two piezoelectric tensors are in use: $d_{i,kr}$ and $e_{i,kr}$. The $d_{i,kr}$ piezoelectric tensor relates the components of the *electric flux density* vector D_i to those of the elastic stress tensor, σ_{kr} (*direct piezoelectric effect*), and also the components of the elastic strain tensor u_{kr} to those of the electric field vector, E_i (*inverse piezoelectric effect*): $D_i = \varepsilon_{ik} E_k + 4\pi d_{i,kr}\sigma_{kr}$; $u_{kr} = S_{krmn}\sigma_{mn} + d_{i,kr} E_i$. Here ε_{ik} is the *dielectric constant* tensor, S_{krmn} is the elastic compliance tensor. Summation is done over the repeated indices. When the independent elastic variables are the components u_{kr} then the piezoelectric tensor e_{ikr} is used: $D_i = \varepsilon_{ik} E_k + 4\pi e_{i,kr} u_{kr}$; $\sigma_{kr} = c_{krmn} u_{mn} + e_{i,kr} E_i$. Here c_{krmn} is the *elastic modulus* tensor. Both piezoelectric tensors are symmetric in the last two indices. The maximum number of independent components of the piezoelectric tensor is 18, and the minimum number is one (in the 23 (T) and $\overline{4}3m$ (T_d) cubic classes). The components of the piezoelectric tensor are identically equal to zero in crystals having an inversion center. The following relations link together the $d_{i,kr}$ and the $e_{i,kr}$ piezoelectric tensors: $e_{i,kr} = c_{mnkr} d_{i,mn}$ and $d_{i,kr} = S_{mnkr} e_{i,mn}$. The $e_{i,kr}$ piezoelectric tensor is commonly used in the acoustics of piezocrystals.

PIEZOMAGNETISM (fr. Gr. $\pi\iota\varepsilon\zeta\omega$, to press, and *magnetism*)

A piezomagnetic effect consisting in a change in *magnetization* when elastic stresses are applied to a magnetically ordered crystal. The components of the *piezomagnetic moment* M_i are linearly related to the elastic stress tensor, $\sigma_{\alpha\beta}$, so that $M_i = \sum_{\alpha\beta} \lambda_{i,\alpha\beta}\sigma_{\alpha\beta}$. Piezomagnetism is only possible in those crystals where the magnetic symmetry permits the existence of piezomagnetic tensor components ($\lambda_{i,\alpha\beta}$) which differ from zero. Piezomagnetism in collinear *antiferromagnets* and weak *ferromagnets* results from *spin–orbit interaction* and the *magnetic dipole interaction*, and it is experimentally observed in crystals in which these interactions are strong enough. For example, the experimental value of λ for CoF$_2$ is $\lambda \sim 10^{-9}$ G·cm^2/dyn, while it is $\lambda \sim 10^{-11}$

G·cm^2/dyn for MnF$_2$. In exchange noncollinear antiferromagnets of the type of UO$_2$, Mn$_3$NiN, etc. the piezomagnetism may be of a purely exchange nature.

PIEZORESISTIVE EFFECT, tensoresistive effect

A *strain*-induced change of the resistivity (see *Electrical conductivity*) in a solid. The piezoresistive effect is particularly noticeable in *semiconductors* where it arises from an alteration of the energy spectrum of *current carriers*. Strain produces changes in either the ionization energy of *local electronic levels* or the *activation energy* (for conductivity involving an *impurity band*); there appears a splitting of bands that are degenerate in the absence of directed strain or equivalent energy valleys in *many-valley semiconductors*; the energy gaps between nonequivalent valleys and the widths of *band gaps* can change. The strain-induced changes of band structure parameters lead to changes of the current carrier concentration, the magnitude, and the anisotropy of electron (*hole*) *mobility*. This provides an appreciable piezoresistive effect which makes possible high sensitivity semiconductor *strain gauges* whose resistance can vary greatly for comparatively low strains. The piezoresistive effect in semiconductors is very nonlinear, with the exception of the so-called *range of weak strains* when $\Delta\varepsilon \ll k_B T$ (here $\Delta\varepsilon$ is the strain-induced variation of the band structure energy parameter). In *metals*, owing to some particular features of their *band structure* and a strong electron degeneracy, strain-induced changes are insignificant, and are determined mainly by geometric factors. In wide-gap *insulators* ($\varepsilon_g > 3$ eV, here ε_g is the width of the band gap) the carrier concentration is quite small (with the exception of the breakdown range of electric fields, see *Breakdown of solids*) and varies weakly with the strain because even under high pressure the variation of ε_g is insignificant. The ionic component of conductivity (see *Ionic conductivity*) can vary with strain due to the variation of the ion mobility.

PIEZOSEMICONDUCTORS

Semiconductors exhibiting piezoelectric properties. In the presence of the piezoelectric effect (see *Piezoelectricity, Piezoelectric materials*), acoustic waves in crystals excite an electric field at a frequency coinciding with that of these waves. Such a field affects free current carriers, electrons and holes. Piezosemiconductors play the role of active elements in devices used in electroacoustics and *acousto-electronics*, e.g., to amplify, generate and modify the amplitude and phase of the acoustic waves (in particular, the surface acoustic waves, SAW). They are also used to build electromechanical filters and nonlinear elements for radio-electronic circuitry. The presence of the *inner photoeffect* makes it possible to apply piezosemiconductors to tasks of acousto-optics. Among various semiconductors this piezoelectric effect is more strongly expressed in CdS, CdSe, ZnO, ZnS (hexagonal symmetry crystals), and also in ferroelectric semiconductors and photoferroelectrics such as SbSI.

PIEZOSPECTROSCOPY

Branch of the *optical spectroscopy* of solids involving the study of optical spectra during the directional (anisotropic) elastic *strain* of crystals. The usual objects of such studies are deformations of electron transitions (as manifested in absorption, reflection and excitation spectra, *photoconductivity* or *luminescence*). In a general case the directional elastic deformation lowers the *crystal lattice* symmetry, splits degenerate electron energy levels in crystals (both intrinsic and impurity types), and shifts nondegenerate levels. This entails the reversible splitting (displacement) of the bands of the corresponding spectral transitions, this being the principal spectroscopic effect of a directional deformation. Another piezospectroscopic effect consists in a reversible change of band intensity and polarization (including "excitation" of forbidden transitions) that results from the mixing of electronic states during an anisotropic deformation of the lattice. Piezospectroscopy studies are most informative in the case of cubic crystals, since due to their high symmetry such crystals feature the greatest amount of degeneracy in their energy states.

A static uniform anisotropic elastic deformation is usually produced by the uniaxial compression of a single crystal specimen along its [100], [111], [110] axes, or by the planar extension (or compression) of oriented thin *plates*. The maximum permissible stresses are determined by the mechanical properties of the materials. Specimens are cooled to helium or nitrogen temperatures to narrow their spectral bands, and thereby simplify the detection of their deformational splitting. Such a detection is conducted by direct recording of the spectra in polarized light, or by way of various sensitive techniques from the arsenal of *modulation spectroscopy*. The basic characteristics of the piezospectroscopic effect that are measured are the number of components in the split band, their intensity, their states of polarization, and the frequency shifts of their components.

By comparing experimental characteristics of piezospectroscopic splittings of electronic transitions with the phenomenological theory based on group theoretical calculations (see *Group theory in solid state physics*) and with perturbation theory, one may obtain definitive data on the fundamental symmetry properties of both the impurity and the intrinsic crystal electronic states in the crystals under study. Among them are the point *local symmetry* of impurity centers and defects, the symmetry of their electronic states, their position in *k*-space, and the symmetry of the extrema of exciton and electron bands. Also one may obtain data on the multipole character of optical transitions, and on the phenomenological numerical parameters that characterize the effect of electric deformations on electron states (in band structure calculations *strain potentials* play the role of such parameters). Such determinations were performed for a large number of impurity centers and defects (*color centers*) in *ionic crystals*, of *deep levels* and *shallow levels* in *semiconductors*, *excitons* and band extrema in semiconductors belonging to groups IV, III–V, II–VI, etc.

During piezospectroscopic studies of electron–phonon band transitions (*indirect excitations*, vibronic structure of *impurity atoms*) one may also determine the symmetry of the vibrations taking part in those transitions. Separate observations are also available of direct deformation splittings of bands in the vibrational spectra of crystals (such as *local vibrations* in IR spectra; vibrations active in *Raman scattering of light*), and the resulting data may be utilized to evaluate anharmonic interaction constants.

Studies of the spectra of crystals undergoing anisotropic deformation make it possible to numerically predict the shift of bands during isotropic deformation, i.e. those observed during experiments with bulk (hydrostatic) compression. The changes in the absorption spectra that result from anisotropic electric deformation yield changes in the values of the crystal *refractive index*, that is, changes in the photoelastic constants of the crystal.

PILE-UP OF DISLOCATIONS

A group of *dislocations* which accumulate near an obstacle when exposed to an applied stress. The obstacle for the piling-up of dislocations may be a *grain boundary*, coherent or incoherent inclusions, or barriers formed by extended dislocations. In this case the main dislocation pile-up is fixed because of the interaction with the controlling field of the obstacle whose gradient σ of *internal stresses* is usually rather large, so that at an increase of σ the dislocation displacement is infinitesimal, i.e. it may be considered fixed. At every piled-up dislocation there are present not only external *stresses*, but also interaction forces with other dislocations of the pile-up. As a result, there is a *stresse concentration*, and the main dislocation is acted upon by a force proportional to $N\sigma$, where N is the number of dislocations in the pile-up. If N is large enough then at the pile-up dislocation head there may appear stresses equal to the *theoretical strength* that can initiate plastic flow (see *Plastic deformations*) or induce *cracks* near the obstacle. The process of dislocation pile-up is well approximated by the continuous dislocation distribution approximation. The *dislocation density* $n(x)$ is determined by the condition of force equilibrium resulting from external stresses and interaction forces in the dislocations of the pile-up. In the particular case of a two-sided pile-up, when both sides are symmetrically aligned, there is a pile-up of dislocations with opposite signs, which leads to

$$n(x) = \frac{2(1-\nu)\sigma x}{\mu b[(L/2)^2 - x^2]^{1/2}},$$

where b is the magnitude of *Burgers vector*, μ is the *shear modulus*, and ν is the *Poisson ratio*. The

total number N of dislocations of either sign in a dislocation pile-up is

$$N = \int\limits_0^{L/2} n(x)\,dx = \frac{(1-\nu)L\sigma}{\mu b}.$$

The force acting upon a unit length of the head dislocation is $F = \pi N b\sigma$. For a single-sided pile-up of dislocations clamped to the obstacle by the *shear* stress σ_{xy}^0 the dislocation density $n(x)$ is

$$n(x) = 2(1-\nu)\sigma_{xy}^0 \left[\frac{L_1/2 + x}{L_1/2 - x}\right]^{1/2} (\mu b)^{-1}$$

and

$$N_1 = \pi N.$$

The length of piled-of dislocations is

$$L_1 = \frac{\mu b N_1}{\pi \sigma_{xy}^0 (1-\nu)}.$$

The shear stress σ_{xy} near the piling-up head is

$$\sigma_{xy} \approx \sigma_{xy}^0 \left[\frac{L_1}{x - L_1/2}\right]^{1/2}.$$

The results of the *model of continuous distribution of dislocations* in pile-up are not valid for distances smaller than those separating main dislocations. Pile-up analysis is important for examining the origin and spreading process of *plastic deformations*, *brittle cracks* and *tough cracks*, and also for *strain hardening* theory.

PINCH-EFFECT

See *Magnetic pressure*.

p–i–n DIODE

A *semiconductor diode* that consists of two low-resistivity regions having *n*- and *p*-types of conductivity, and a base region (intrinsic, or *i*-region) of higher resistivity, which exhibits intrinsic conductivity. The base region is so lightly doped that on application of even a small reverse bias it becomes overlapped by the *space charge region* of the *semiconductor junction*. The devices referred to as *p–i–n*-diodes are commonly *p–ν–n*- or *p–π–n*-diodes, where the letters *ν* and *π* designate a lightly doped material having, respectively, *n*- or *p*-type conductivity.

A material with intrinsic conductivity may be obtained by precisely compensating with oppositely charged dopants, but this complicated technological process is rarely used.

Applications of p–i–n-diodes are based on employing the following properties of the non-equilibrium electron–hole plasma in the p–i–n structures: abrupt change of resistance under change of sign of the applied voltage (power rectifiers, detectors, charge-storage diodes, UHF power switches); absorption of electromagnetic waves by electron–hole plasma (absorptive attenuators for controlling the UHF power, optical modulators of IR radiation); injection luminescence of electron–hole plasma (light emitting diodes); change of the lifetime of charge carriers under action of temperature, pressure, irradiation (temperature and pressure transducers, nuclear and X-ray radiation detectors); change of mobility of carriers under the action of a magnetic field (magnetic diodes, magnetic field transducers).

PINNING

See Vortex pinning.

PINNING OF FERMI LEVEL

See Fermi level pinning.

PIPE DIFFUSION

See Tubular diffusion.

PITTING

The destruction of a metal under attack by severe point corrosion involving the formation of small, but sharp, localized cavities (see Corrosion of metals). Materials subject to pitting failure are metals and alloys like iron, carbon and stainless steels, aluminum alloys, nickel alloys, titanium alloys, alloys of zirconium and other alloys in corrosive media containing oxidizing agents (oxygen, nitrates, nitrites, chromates) and activators (chlorine, bromine, and iodine anions). Pitting corrosion occurs most commonly in sea water, in HCl-containing water solutions of alcohols, in cooling brines, in circulating water supply systems of the chemical industry, and other enterprises. Pitting arises at grain boundaries, nonmetallic inclusions (especially sulfide ones), in regions characterized by a lowered chromium concentration. It

also occurs in solid solutions, which have passivating films that exhibit lowered protective ability, and are prone to adsorption of corrosion-activating anions. Corrosion protection involves electrochemical methods (cathodic, anodic, and galvanic protection), introduction of corrosion inhibitors, introduction of corrosion-resisting additives to alloys, e.g., alloying of stainless steels with chromium, silicon, molybdenum. The highest resistance to corrosion in chlorine-containing media is shown by titanium.

PLANAR DEFECT

An inner surface of a crystal lattice which separates crystal regions of different orientation or different phase state. Examples of planar defects are grain boundaries in a polycrystal, in blocks, cells and fragments, dislocation walls, twin boundaries (see Twinning of crystals) and phase interfaces. The separation of planar defects into an independent class of defects is due to the fact that many properties of planar defects are independent of the boundary structure, and are determined solely by the uniform distortion, which transforms the material on one side of the boundary into the state of the material on the another side. The role of such a distortion may be played by rotation, twinning, etc. A theory of planar defects as independent imperfections of a crystal lattice has been developed. In specific cases, the equations of the theory go over into the relations of the theory of dislocations, disclinations and dispyrations. The continuum theory of planar defects (see Continuum theory of defects) allows one to effectively solve various problems of solid state physics: calculating the plasto-elastic fields, analyzing the processes of twinning and of phase transitions, determining the values of stresses in the neighborhood of precipitates, etc.

PLANAR STATE OF STRESS

The state of an elastic body that does not experience the action of normal and shear stresses (see Shear) at elements of area perpendicular to a certain direction (z-axis): $\sigma_{xz} = \sigma_{yz} = \sigma_{zz} = 0$ (σ_{ik} is the stress tensor), so stresses act only in the x, y plane. The state of planar stress is realized, e.g., in the case of a thin plate, which is subjected to surface forces and bulk forces that are perpendicular to the z-axis and uniformly distributed along it. In the general case, the z-dependence of the tensor $\sigma_{\alpha\beta}$ (α, $\beta = 1$, 2) is quadratic. If σ_1, σ_2 are the principal values of $\sigma_{\alpha\beta}$, the peak value of the tangential stress equals $|\sigma_1 - \sigma_2|/2$. The important specific cases of the uniform state of plane stress are: simple shear ($\sigma_1 = -\sigma_2$), *uniform compression* in the x, y plane ($\sigma_1 = \sigma_2 \neq 0$), uniaxial *state of stress* ($\sigma_1 \neq 0$, $\sigma_2 = 0$).

PLANATOMIC NETWORKS (A.V. Shubnikov, 1916)

Planar atomic networks in which every internal atom (node) has an equal number k of linkages or bonds with other neighboring atoms. The network is said to be a *combinatory regular network*, if for each pair of its nodal points there exists a combinatory topologic transformation, which converts the nodes into one another and the network into itself. All possible combinatory differing networks were found by A.V. Shubnikov from consideration of the *Euler equation* for a finite simply connected complex $F - E + N = 1$ where F, E, and N are respectively the numbers of faces, edges (bonds), and nodes (vertices). This relation may be

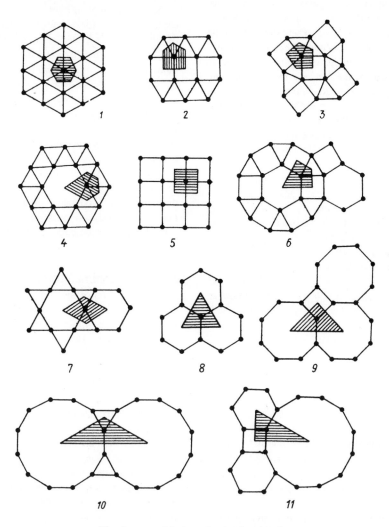

The eleven combinatory planatomic networks.

presented in the form of the following equation:

$$\frac{1}{\alpha_1} + \frac{1}{\alpha_2} + \cdots + \frac{1}{\alpha_k} = \frac{k}{2} - 1,$$

which relates the number of bonds k, which meet at each internal node ($k \geqslant 3$) to the number of vertices α_i of the polygons which meet at that node. The solution of this equation for $k = 3, 4, 5, 6$ (k cannot exceed 6) leads to 11 different combinatory planatomic networks (see Fig.). For example, for network No. 1 we have $k = 6$, $F = 24$, $E = 42$, $N = 19$, and all six α_i values equal 3.

Any combinatory regular network may be transformed without disruption of linkages into a *metrically regular network*. Therefore, any group associated with combinatory topologic transformations of any combinatory regular network is isomorphic with the related two-dimensional *space group* (called Fedorov symmetry group in the Russian literature); in other words, the basis of *two-dimensional crystallography* is purely topological. No analogous statement has yet been obtained for the case of three-dimensional space.

PLANCK CONSTANT, quantum of action, h
(M. Planck, 1900)

One of the fundamental physical constants, which is characteristic of the discrete (quantum) nature of objects of matter. The energy of a quantum particle (*quasi-particle*) ε is related to its frequency v by the equation $\varepsilon = hv$, with the corresponding de Broglie wavelength λ given by the equation $\lambda = hc/\varepsilon$, $p = h/\lambda$ (c is the *speed of light*). The term "reduced Planck constant" is often used in reference to the quantity $\hbar = h/(2\pi)$, $\varepsilon = \hbar\omega$ (ω is the angular frequency, $\omega = 2\pi v$). The accepted value for h is $6.626076 \cdot 10^{-34}$ J·s; hence, the value for h is $1.0545727 \cdot 10^{-34}$ J·s.

PLANCK FORMULA, Planck radiation formula, Planck distribution (M. Planck, 1900)

The spectral distribution of the density of electromagnetic radiation energy in thermal equilibrium with a material is given by the expression:

$$dE_v = \frac{8\pi h v^3}{c^3}\left[\exp\left(\frac{hv}{k_B T}\right) - 1\right]^{-1} dv,$$

where v is the radiation frequency, and T is the temperature.

This Planck formula reflects the nature of the electromagnetic radiation and, in particular, the fact that the *thermal radiation* may be treated as an ideal gas of light quanta (*photons*). It was in this Planck formula that the *Planck constant*, which relates the photon energy ε to the light frequency v ($\varepsilon = hv$), was first introduced. The Planck formula is sometimes written in terms of the density of heat radiation energy per unit wavelength interval.

PLANE PROBLEM OF ELASTICITY THEORY

The plane problem of elasticity theory is concerned with the stress-deformed state of a slab or *plate*, which is loaded symmetrically about its mid-plane (see *Shells*), or of a long (in theory, infinite) cylinder (prism), which is loaded by forces normal to its lateral surface and invariant along its length. Mathematically, the plane problem of elasticity theory is reduced to a boundary value problem for the biharmonic equation (anisotropy adds complexity to this equation) under corresponding boundary conditions. The general solution of these equations is expressed in terms of two analytical functions of complex variables. In the case of dynamic loads (steady-state vibrations, abrupt application of load), the plane problem reduces to solving what can be called metaharmonic equations.

PLASMA-CHEMICAL TECHNOLOGY

The collection of methods for forming materials by the use of chemical reactions, in which the processes occurring in a low-temperature plasma are employed. The specific features of these processes are: high reaction rates, which favor use of miniature equipment; the possibility of employing easily available starting materials of little value (e.g., air); the possibility of obtaining materials, which do not exist at ordinary temperatures (e.g., Al_2O, AlO, *clusters* consisting of 2–9 carbon atoms, etc.). Plasma-chemical technology involves the following stages: production of plasma with plasma generators (electric arc, high frequency and extra-high frequency (microwave) ones), introduction of reagents into the plasma, halting the reaction at the required stage by abruptly lowering the temperature (quenching). The reaction time is controlled with a precision of 10^{-3} s. Plasma-chemical technology is used for obtaining *refractory materials*, fine powders, and *thin films*, for performing directional or "dry" surface *etching* in microelectronics, for the modification of surfaces of various materials, etc.

PLASMA-ELECTRIC EFFECT

Generation of direct current in a *solid-state plasma* as a result of the entrainment of *current carriers* by a longitudinal plasma wave. In semiconductors conforming to a nonparabolic electron *dispersion law*, the external electric field changes the frequency and damping factor of the wave, and may also be responsible for a substantial contribution to the current characterizing the plasma-electric effect.

PLASMA, ELECTRON–HOLE

See *Electron–hole plasma*.

PLASMA OSCILLATIONS, charge density plasma oscillations

Collective charge density oscillations which propagate in a *solid-state plasma* (or in a gaseous electron–ion plasma). The characteristic frequencies of these oscillations are close to the *plasma frequency* $\omega_p = [e^2 n/(m^*\varepsilon)]^{1/2}$, where n is the concentration of free carriers (electrons and *holes*) of *effective mass* m^*, ε is the *dielectric constant* of the crystal (medium). In the long-wave limit, the plasma oscillations exhibit the quadratic *spatial dielectric dispersion*: $\omega_k = \omega_p + k^2 \bar{v}_l^2/\omega_p$, where $\bar{v}_l$ is the characteristic electron velocity, the value of which is close either to the mean thermal velocity (in the case of nondegenerate carriers of current) or to the *Fermi velocity* (in the case of degenerate carriers). For high wave numbers k, the plasma oscillations are attenuated by interaction with resonance plasma particles of velocity $v = \omega_k/k$, or because of the decay into *electron–hole pairs* (classic or *Landau damping*). The spectrum of plasma oscillations exhibits a quasi-classical nature ($\omega_k \to 0$ at $k \to 0$) in the case of a multicomponent electron–hole plasma of many-band (many-valley) *degenerate semiconductors, semimetals*, and *transition metals*, as well as in the case of crystals of low dimensionality (*quasi-two-dimensional crystals*), or chain-type crystals (*quasi-one-dimensional crystals*) exhibiting metallic conductivity of their layers or chains (see *Acoustic plasmon*).

PLASMA POLYCONDENSATION

Synthesis of *solids* (usually in the form of *thin films*), based on the associated processes of *condensation* and polymerization of carbonic radicals in a gas-discharge plasma. The discharge is set up in the static or high frequency mode, the voltage range being typically from 0.1 to 5.0 kV. In contrast to ordinary chemical synthesis, the energy required for the formation of a new structure in the process of plasma polycondensation is brought from the external field directly to the radicals (atoms, ions) to be condensed, whereas the temperature of the substrate may remain rather low. The films are generated as a result of nearly instantaneous (formation times about 10^{-11}–10^{-12} s) *relaxation* of the energy of excited molecular fragments under interaction with a *solid surface*, and the bonding of these fragments to the substrate (thus providing *adhesion*) and to one another (polymerization). The resulting condensate usually exhibits sp (linear chains), sp^2 (graphitic "phase"), and sp^3 (diamond-type "phase") types of chemical bond hybridization. Regulating the plasma parameters exerts control over the proportion of these phases in the film to be synthesized, and hence provides a film with predetermined electrical, optical, mechanical, and physicochemical properties. Using a hetero-organic compound as the parent material, or sputtering a solid-state target (metal, graphite, etc.) synchronously with the process, provides a polycondensate of required composition and properties on nearly any surface, corresponding to radically new fabrication methods (see *Ion-plasma treatment, Plasma treatment*).

PLASMA REFLECTION

The intense reflection of electromagnetic radiation from a certain interface (e.g., between a *metal* and a *semiconductor*), which occurs at frequencies below the frequency ω_p of *plasma oscillations* of free charge carriers,

$$\omega_p = \left(\frac{Ne^2}{\varepsilon_0 m^*}\right)^{1/2}, \qquad (1)$$

where N and m^* are respectively the *concentration* and *effective mass* of free charge carriers; ε_∞ is the *dielectric constant* due to the contribution of band-to-band transitions. In metals, $N \sim 10^{22}$ cm^{-3}, and in most metals the plasma frequency lies far in the UV region, which is responsible for the high reflectivity of metals for visible and *infrared radiation*. The concentration of free charge carriers in semiconductors is lower by at least several orders of magnitude, even under the conditions of heavy *doping* or intense optical excitation. Therefore, the plasma frequency and corresponding increase of the *reflectance* are observed within the IR spectral region. For n-Ge featuring $N \approx 10^{19}$ cm^{-3}, e.g., the region of intense plasma reflection begins at $\lambda \approx 10$ μm, and extends to longer wavelengths.

The abrupt change of the reflectance in the frequency region $\omega \approx \omega_p$ is related to the contribution made by free electrons (or *holes*) to the crystal dielectric constant ε, which determines the *refractive index* n and the reflectance R. Neglecting the attenuation of plasma oscillations and the absorption of radiation by the crystal, we obtain at normal incidence

$$\varepsilon(\omega) = n^2(\omega) = \varepsilon_\infty \left(1 - \frac{\omega_p^2}{\omega^2} \right), \qquad (2)$$

$$R(\omega) = \left(\frac{n-1}{n+1} \right)^2. \qquad (3)$$

Hence it follows that as ω leaves the high-frequency region and approaches ω_p, the value of R decreases and reaches zero as $n(\omega)$ becomes 1, and then abruptly increases again and becomes equal to unity at $\omega = \omega_p$. The study of the plasma reflection spectrum is used for determining ω_p and, hence, N or m^*.

PLASMA, SOLID-STATE

See *Solid-state plasma*.

PLASMA TREATMENT

The physicochemical treatment of a *solid surface* (see *Ion-plasma treatment*). Of primary importance for solid state physics are two types of processes: *etching* and deposition of *thin films*. The isotropic etching in inert or chemically active media (e.g., in freons) is an effective method for surface purification. The etching according to a predetermined pattern (so-called *dry plasma etching*) is one of the tools used in techniques for producing *integrated circuits* with a high degree of integration. The plasma precipitation is used for forming thin dielectric layers for integrated circuits and film *coatings*, which offer a broad spectrum of useful features and applications. The latter trend, which gained wide acceptance in the 1980s, permits multiply increasing the surface hardness, wear resistance, resistance to chemicals, etc., by application of thin films consisting of refractory intermetallics, and particularly of "diamond-like" carbon (see *Plasma polycondensation, Diamond-like carbon films*). These films are applied onto the semiconductor surface at a moderate temperature (down to room temperature). The plasma treatment is carried out in vacuum and continuous plasma reactors, as well as by application of "*plasma torches*", which inject the plasma discharge jet into the atmosphere. Besides etching and deposition of coatings, it is possible in principle to perform the diffusion (implantation) treatment of a solid surface in a plasma. The plasma discharge is also used for processing raw materials and recycling wastes. In this case, the principal role is played by the high-energy reactions, which take place in the plasma.

PLASMA WAVES, MAGNETIC

See *Magnetic plasma waves*.

PLASMON

A *quasi-particle*, the quantum of *plasma oscillations* of characteristic energy $\varepsilon_k = \hbar\omega_p$ (at small values of the *quasi-momentum* k), where ω_p is the plasma frequency. In multicomponent (many-band) systems, or systems of low dimensionality (laminated, chain-type ones), the *dispersion law* for plasmons may assume the quasi-acoustic form, the plasmon energy ε_k tending to zero as $k \to 0$ (see *Acoustic plasmon*). See also *Plasmon in superconductors*.

PLASMON, ACOUSTIC

See *Acoustic plasmon*.

PLASMON MECHANISM OF SUPERCONDUCTIVITY

A hypothetical phononless mechanism of *superconductivity*, which is based on the assumption of Cooper pairing (see *Cooper pairs*) of "light" degenerate *current carriers* (*conduction electrons, holes*) due to the exchange of virtual quanta of *collective excitations* (charge density oscillations) in the system of "heavier" carriers, e.g., *d*- or *f*-shell electrons of transition (rare-earth) *metals*, electrons or holes of higher effective mass in many-valley (many-band) degenerate *semiconductors* (*semimetals*), or in layered (heterogeneous) semiconducting structures. H. Fröhlich and E.A. Pashitsky proposed this mechanism (1968) independently of one another. The spectrum of collective *plasma oscillations* of relatively low frequency is an acoustic one at $v_{Fl} \gg v_{Fh}$ and $\omega_{pl} \gg \omega_{ph}$ (where v_{Fl} and v_{Fh} are *Fermi velocities*, ω_{pl} and ω_{ph} are plasma frequencies of respectively light and heavy fermions), $\omega_q = q(v_{Fl}v_{Fh}/3)^{1/2}$ at $q \to 0$ (see *Acoustic plasmon*), and slightly decays within the frequency range $qv_{Fh} \lesssim \omega \lesssim \omega_{ph}$ (q is the wave number). With regard for the *effects of retardation* (dynamic over-screening) due to the exchange of virtual plasmons, the matrix element associated with the screened Coulomb interaction between electrons may be represented as

$$V_C(q, \omega) = \frac{4\pi e^2}{q^2 + \kappa_l^2}\left(1 - \frac{(\omega_q + i\gamma_q)^2}{(\omega_q + i\gamma_q)^2 - \omega^2}\right),$$

where κ_l^{-1} is the radius of scening by "light" carriers, and γ_q is the plasmon damping factor due to the decay into *electron–hole pairs* (see *Landau damping*) and *current carrier scattering* by the *crystal lattice* defects. It follows from this equation that the retarded interaction is attractive by nature in the region $qv_{Fh} \lesssim \omega \lesssim \omega_q = q\omega_{ph}(q^2 + \kappa_l^2)^{-1/2}$, Re $V_C(q, \omega) < 0$. The efficiency of the *electron–plasmon interaction* increases in polar (ionic) semiconducting crystals due to the hybridization of acoustic plasmons with optical *phonons*, which results in a broadening of the region of attraction up to the energy given by $\hbar(\omega_{ph}^2 + \omega_{LO}^2)^{1/2}$, where ω_{LO} is the frequency of the longitudinal optical phonon. The interest in the plasmon mechanism of superconductivity has been stimulated by the discovery of the phenomenon of *high-temperature superconductivity* in metal-oxide compounds like $La_{2-x}(Ba, Sr)_xCuO_4$; $YBa_2Cu_3O_{7-\delta}$; $Bi_2Sr_2CaCu_2O_{8+\delta}$,

etc. It has been suggested that a plasmon mechanism might play a role in explaining the superconductivity in cuprate compounds with a perovskite-type structure. This hypothesis is supported by the fact that the *isotope effect* in superconducting metal-oxide compounds is generally unusually weak.

PLASMON–PHONON RESONANCE

The closeness of the natural frequency of the *plasma oscillations* of long-wavelength free charge carriers $\omega_p = [e^2 N/(\varepsilon_0 m^*)]^{1/2}$ (ε_∞ is the high-frequency *dielectric constant*; m^* is the *effective mass* of carriers; N is concentration of carriers) to the frequency of longitudinal optical phonons ω_{LO}. In the region of plasmon–phonon resonance, an important role is played by the interaction of corresponding oscillations; this interaction occurs through longitudinal oscillating electric fields, which are associated with both excitations. The contributions of *plasmons* and *phonons* to the crystal dielectric constant are additive:

$$\varepsilon(\omega) = \varepsilon_\infty\left(1 - \frac{\omega_p^2}{\omega^2} + \frac{\omega_{LO}^2 - \omega_{TO}^2}{\omega_{TO}^2 - \omega^2}\right),$$

where ω_{TO} is the frequency of transverse optical phonons.

The frequencies of the coupled oscillations, which result from the resonance, are determined from the condition $\varepsilon(\omega) = 0$, and have the form

$$\omega_\pm^2 = \frac{1}{2}\{\omega_p^2 + \omega_{LO}^2 \pm [(\omega_p^2 + \omega_{LO}^2)^2 - 4\omega_p^2\omega_{TO}^2]^{1/2}\}.$$

The plasmon–phonon resonance modifies the IR spectra of solids, changing the reflection in the frequency region of the *residual ray band* (Reststrahlen). The frequencies of the plasmon–phonon modes and their dependence on the concentration of free charge carriers clearly manifest themselves in the spectra of *Raman scattering of light*.

PLASMONS IN SUPERCONDUCTORS

Quasi-particles which are associated with the *collective excitations* of a system of *conduction electrons*, and which arise from electron–electron Coulomb repulsion. The energy of a plasmon is given by $\varepsilon = \hbar\omega_p$, where ω_p is the *plasma oscillation* frequency. The value of the *plasma frequency* ω_p in homogeneous superconductors coincides with its value in a normal metal, $\omega_p = (ne^2/\varepsilon_0 m)^{1/2}$, where n is the concentration of electrons, and m is the electron mass. Exceptions are provided by superconductors with heavy electrons (*heavy fermions*), anisotropic *superconductivity in quasi-one-dimensional systems*, and *quasi-one-dimensional crystals* featuring a quasi-acoustic *dispersion law* (see *Acoustic plasmon*), as well as by *thin films*, in which the plasma oscillations ("mini-plasmons") are of low frequency. In Josephson tunnel junctions (see *Josephson junctions*), there exists the possibility of the propagation of low-frequency plasma oscillations, the so-called *Josephson plasmons*, the excitation energy of which is $\omega_J = (2eI_c/\hbar C)^{1/2}$, where I_c is the critical Josephson current, and C is the junction capacitance. Typical values of ω_J range from 10^9 to 10^{11} s^{-1}, whereas the plasma frequency of typical metals is $\omega_p \sim 10^{16}$ s^{-1}. The low value of the frequency of Josephson plasma oscillations is due to the low density of charge carriers in the barrier (see *Josephson effect*).

PLASMON, SURFACE

See *Surface plasmon*.

PLASTIC DEFORMATION

The irreversible change of the dimensions and shape of a solid caused by exposure to external actions. From the standpoint of the geometry of deformation, the following types of plastic deformation should be distinguished: *slip*, twinning (see *Twinning of crystals*), *rotation* and *viscous flow*. Twinning takes place in crystals; sliding, rotation, and viscous flow or diffusion *creep* are observed in all solids. Each type of plastic deformation manifests itself in different forms depending on the temperature, the severity and rate of deformation, and each type has a different influence on the resulting structure. The difference of paths of plastic deformation in *monocrystals* and in *polycrystals* are sometimes more important than the change of the deformation conditions. The plastic deformation of real materials proceeds by several mechanisms simultaneously, not by a single one. It proceeds concurrently at several structural

levels: *translation* at one level is accompanied by rotation at a higher level, and vice versa. Depending on the type of elementary deformation carriers, the *dislocation model of plastic deformation* and the *disclination models of plastic deformation* are distinguished, as well as plastic deformation due to the transport of *point defects* (*vacancies, interstitial atoms, crowdions*).

PLASTIC DEFORMATION, DISCLINATION MODELS

See *Disclination models of plastic deformation*.

PLASTIC DEFORMATION LOCALIZATION

See *Localization of plastic deformation*.

PLASTIC DEFORMATION OF AMORPHOUS METALLIC ALLOYS

The plastic deformation of amorphous metal alloys is appreciably different in nature in the temperature intervals above and below the temperature equal to $0.7T_v$, where T_v is the *vitrification* temperature. At low temperatures *plastic deformation* is very localized (*inhomogeneous deformation*), and is concentrated in the shear bands, the width of which is 10–40 nm, according to data provided by *electron microscopy*, the height of the steps at the surface being 100–200 nm. Hence it follows that the severity of the deformation within such a band is very high. On the macroscopic scale, at the same time, the degree of *strain* prior to *failure* is very small. Another specific feature of plastic deformation is either the complete absence of, or the rather small magnitude of, *strain hardening*. One of the mechanisms of strain hardening is the *segregation* of the intrusion elements in regions of increased free volume concentration (see *Free volume model*). At high enough temperatures (above $0.7T_v$), the nature of the deformation of amorphous metallic alloys (see *Amorphous metals and metallic alloys*) changes; the inhomogeneous deformation goes over into a *homogeneous deformation*, no shear bands are generated, and the whole bulk of the material becomes viscously deformed. This takes place with the onset of a pronounced degree of plastic flow. As the temperature further increases ($T_v < T < T_{cr}$, where T_{cr} is the *crystallization* temperature), *superplasticity* arises in amorphous metallic alloys.

By analogy with the mechanisms of deformation of crystalline alloys, several *dislocation models of plastic deformation* have been proposed, among them being the *Gilman model*, according to which sliding is carried out by dislocations with *Burgers vectors* of alternating magnitude and direction. In subsequent studies, the *dislocation lattice model* and the *disclination–dislocation model* were advanced, according to which the structure of amorphous metallic alloys may be represented as a three-dimensional *dislocation grid*. Although the mechanism of plastic deformation still remains to be fully explained, the importance of the role of the free volume is well established. The failure of amorphous metallic alloys under homogeneous deformation proceeds by the opening of a crack along the localized *slip band*.

PLASTIC DEFORMATION OF COVALENT CRYSTALS

The specific kind of *plastic deformation*, which occurs in crystals featuring a very high value of Peierls–Nabarro stress, where the dislocations are practically frozen at low temperatures (see *Peierls–Nabarro model*). As the temperature is increased, the thermally activated motion of dislocations is observed. This motion consists in the generation of a double inflection at the dislocation line (see *Kinks on dislocations*), due to which the dislocation becomes drawn into a neighboring equilibrium position. The low mobility of dislocations under the conditions of the above *strain* mechanism is responsible for high values of the *yield limit* and ultimate *strength*, which decrease with increasing temperature, as well as for the occurrence of a wide temperature interval near 0 K where the failure is brittle by nature (see *Brittle failure*). The starting point of the abrupt increase of the yield point (with decreasing temperature) is determined by the characteristic deformation temperature on the *homologous temperature scale* $t^* = T^*/T_m \approx 0.22[U/(k_B T_m)]^{1/2}$, where U is the *activation energy* of dislocation motion, and T_m is the *melting temperature*. Above T^*, the mechanism of plastic deformation of *covalent crystals* is similar to the mechanism of deformation of plastic metals. In this case the deformation involves the formation of a cellular dislocation structure, and the occurrence of *recovery* and *recrystallization* is possible. The *strain hardening* and the effect of the

structure on mechanical properties above T^* are determined by the same general rules as in metals. Below T^*, no *cellular structure* is formed under deformation, and the failure is brittle by nature. The highest value of T^* ($\sim 0.85 T_m$) is exhibited by purely covalent crystals (Si, Ge, diamond). In partially covalent crystals (III–V and II–VI compound semiconductors, carbides, borides, nitrides, oxides, etc.), the value of T^* decreases with the decrease of directionality of the interatomic bond. Among the group of crystals under consideration, the lowest value of $T^* \approx 0.2 T_m$ is observed in *transition metals* with a body-centered cubic structure.

PLASTIC DEFORMATION RATE

Rate of deformation accompanied by irreversible changes of a sample shape. In terms of the dislocation mechanism the rate of *plastic deformation* of crystalline materials $d\varepsilon/dt = \dot{\varepsilon}$ is defined as the density of mobile *dislocations* ρ times their speed v, that is $\dot{\varepsilon} = b\rho v$, where b is the dislocation *Burgers vector*. For the *diffusion mechanism of crystal creep* the rate of plastic deformation in the elementary case is defined by the gradient of the *vacancy* concentration c times their *diffusion coefficient* D, that is $\dot{\varepsilon} = Aa^3 D\nabla c$, where $A = \text{const}$, and a^3 is the atomic volume.

With the increase of $\dot{\varepsilon}$ the resistance to *plastic deformation* σ, as a rule, also increases. A decrease of the rate of plastic deformation can accompany the development of *strain aging* processes, as observed, e.g., in metallic crystals supersaturated by vacancies. The quantity σ is not a single-valued function of $\dot{\varepsilon}$ at a particular moment of deformation, but rather depends on earlier rates of deformation due to the influence of $\dot{\varepsilon}$ on the dislocation structure. The increase of $\dot{\varepsilon}$ promotes an increase of the homogeneity of plastic deformation (see *Uniform strain*), and the formation of a more fine grained but less flexible structure. There is an increase in the number of active *slip planes* and *slip systems*, fine *shear lines*, and deformation *slip bands*, and a decrease in the cell sizes of the dislocation structure. At very high rates of deformation inertial effects play a pronounced role.

PLASTIC DEFORMATION TEMPERATURE CONDITIONS

See *Temperature conditions of plastic deformation*.

PLASTICITY

Property whereby materials conserve part of the *strain* upon being relieved from the stresses which caused it. When the applied stress exceeds the yield value, plastic (irreversible) deformation occurs. Plasticity is most commonly characterized by the relative elongation prior to failure $\delta = (\Delta l / l_0) \cdot 100\%$ (Δl is elongation, l_0 is the initial sample length). One can distinguish *uniform elongation* δ_u and *localized plastic deformation* (see *Localization of plastic deformation*) in the neck of the sample δ_l, the total value of the relative elongation being the sum of these components: $\delta = \delta_u + \delta_l$. Plasticity may also be characterized by the value of transverse contraction prior to failure. The plasticity of polycrystals is determined mainly by the deformation within grains, which is adequately described in the context of the dislocation mechanism of deformation (see *Dislocation model of plastic deformation*). The *disclination models of plastic deformation* are used for describing the developed *plastic deformation*. The plasticity is largely determined by the difference between the *breaking stress* and the *yield limit*, and hence depends heavily on the mechanism of failure. As the temperature increases, there exists a pronounced correlation between the relative elongation and the mechanism of failure. In the case of covalent and partially covalent crystals, $\delta = 0$ over a broad temperature region near 0 K (see *Brittle failure*). The quantity δ becomes nonzero at the *cold brittleness* threshold temperature, and δ increases in passing from the *quasi-brittle failure* to the tough one. In all cases, the transition from *transcrystalline failure* to the *intercrystalline failure* state results in the decrease of δ. The metals of close-packed structure (Al, Cu, etc.) retain considerable plasticity down to 0 K. The highest plasticity (up to several thousands per cent) is exhibited by metals in the *superplastic* state.

PLASTICITY, ATHERMIC

See *Athermic plasticity*.

PLASTICITY INCREASE

See *Rhenium effect*.

PLASTICITY INCREASE AT SUPERCONDUCTING TRANSITION

A phenomenon that consists in changes in the macroscopic characteristics of plastic deformability of metals and *alloys* during the course of normal-to-superconducting and superconducting-to-normal transitions (see *Superconducting phase transition*). The phenomenon was discovered in 1968 during the course of experiments that consisted in deforming a sample (see *Strain*) while holding the temperature and deformation rates constant, and switching the sample between different magnetic states through variations in the intensity of an external magnetic field. The application of small strains allows comparing the values of *yield limits* of various states, whereas the severe deformation of a sample permits the evaluation of the flow-stress jump at a change of state. The normal-to-superconducting transition brings about a decrease in the yield point (creep limit) (the effect may reach 50%) and the flow-stress; the latter decrease amounts to ~1%, and continues to decrease with the advance of the deformation. These effects exceed the changes in *elastic moduli* at the transition by several orders of magnitude.

More recently it was found that the characteristics related to *creep* and *stress relaxation* are affected during the course of the superconducting transition. The magnitude of creep strain in the superconducting state exceeds that of the normal state by 20%, and the magnitude of stress relaxation in the superconducting state is more than twice that in the normal state. The transition brings about a steep rise of the rate of creep and stress relaxation, which has been observed in all investigated metallic *superconductors*, namely: Pb, Sn, In, Tl, Al, Cd, Nb, Ta, Zn, Mo, Pb–In, Pb–Sn, Nb–Mo, Al–Mg, Pb–Cd, Pb–Tl, Pb–Bi, and has been extensively studied both experimentally and theoretically. The effect under consideration signifies the presence of a factor due to the yield point in the normal state; the presence of this factor arising from the retardation of *dislocations* by electrons (electron drag). Theoretical evaluations indicate that the force due to electron drag may turn out to be a factor in the balance of forces acting on dislocations; the theoretical assessments also predict that this force is reduced by several times at the normal-to-superconducting transition of a metal. The marked role of *conduction electrons* in the macroscopic *plasticity*, which is determined by the conditions of the dislocation motion in the *crystal lattice*, was first shown using the effect under consideration as an example. The presence of this increase of material plasticity is of great importance in various applications associated with the phenomena of *fatigue*, *friction* and *wear*, as well as in the active application of superconducting materials in cryogenics.

PLASTICITY INDUCED BY PHASE TRANSITIONS, transformation plasticity

The occurrence of inelastic (irreversible) mechanical deformation induced by *phase transitions*. Transformation plasticity manifests itself under various conditions: in the first place, it occurs during the course of plastic flow of *metals*, the phase transformation of which may be caused by *mechanical stresses*. In this case, the reaction of the transformation produces additional *strain* in the direction of the applied force. This kind of transformation plasticity is commonly referred to as *transformation-induced plasticity*. In the second place, transformation plasticity occurs during the heating or cooling of stressed crystals, the temperature being changed over the temperature range of the phase transition. In this case, the transformation plasticity is detected by an abrupt increase of inelastic deformation within the temperature range of the phase transition, and sometimes also beyond this range, but it is always caused by the phase transition. The resulting deformation is usually proportional to the stresses applied, which is ruled out in the physics of *plastic deformations*. This is due to the fact that transformation plasticity is determined by forces of a chemical nature, the role of stress reduces to that of a minor disturbing parameter, and the expansion in a power series only involves terms of odd power. Thirdly, the transformation plasticity may be detected under conditions of deformation of a material, when a change of a certain factor, e.g., an abrupt change of temperature, causes the immediate completion of the transition from one phase into another. The typical illustration of this type of transformation plasticity is the immediate decomposition

of the *martensite* of quenched *steel* upon vigorous heating to the *tempering* temperature. During the process of steel tempering, the *yield limit* turns out to be strongly decreased, whereas the value of *plasticity* is considerably increased. The transformation plasticity is characteristic of all kinds of phase transformations: both martensitic and diffusion-types, forward and reverse ones. In the case of *martensitic transformations*, however, the transformation plasticity most often occurs only at the cooling stage, e.g., at the stage of forward transition, whereas only the *shape memory effect* occurs during the heating half-cycle, i.e. during the course of a reverse transformation (TiNi, CuMn, CuAlNi alloys, etc.). The latter effect prevails even under conditions of extremely strong counteraction. In the FeMn and DyCo alloys the transformation $\gamma \leftrightarrow \varepsilon$ features plasticity during both heating and cooling. The phenomenon of the transformation plasticity is attributed to the action of several factors. The main such factor is the *distortion* of the transformation: in accordance with the Clausius–Clapeyron principle, it brings about a greater deformation along the force and a lesser one in the opposite direction. In other cases, transformation plasticity is initiated by the oriented *diffusion* or drift of other *defects* in the direction which favors the propagation of the deformation along the applied force direction, by *microstresses* which arise due to incomplete matching of crystal habits to each other, or by the difference between the specific volumes of the parent phase and the reaction product, etc. This plasticity phenomenon is widely used in physical experiments on creating special types of *steels*, and in metalworking technology.

PLASTICITY MARGIN

See *Margin of plasticity*.

PLASTIC LIMIT

The highest value of *plastic deformation* preceding *failure*. The plastic limit is small for materials with *covalent bonds* (see *Plastic deformation of covalent crystals*). The plastic limit is higher for materials with a face-centered cubic lattice than it is for those with a body-centered cubic lattice. Metals with a hexagonal close-packed lattice have intermediate values. For a given material plasticity the limit depends on the temperature and on the *strain* conditions. Thus, e.g., it grows with a transition from tensile deformation to compression, and especially to *hydroextrusion*. See also *Plasticity*.

PLASTICS
See *Polymeric materials*.

PLASTICS, CONSTRUCTION
See *Construction plastics*.

PLASTICS, GLASS-REINFORCED
See *Glass-reinforced plastics*.

PLASTIC TWISTING DEFORMATION
Process of *plastic deformation* associated with irreversible change of sample shape or of volumes present in it due to the twisting or turning of particular regions of crystallographic directions. The experimental effect of the plastic twisting or bending distortion is detected by an electron-microscopic investigation (see *Electron microscopy*) of the defect structure of deformed materials. Within the defect (dislocation) structure it is possible to identify those formations which cause disorientations in adjacent regions of the material. The disorientation angle may reach tens of degrees. The so-called *cellular structures* are observed, with disorientations of adjacent cells, fragmented regions (see *Fragmentation*), ribbon-like structures, and their various modifications. Information on rotational modes of plasticity may be also obtained by X-ray methods, and with the help of optical microscopy (through analysis of the surface relief).

Plasticity distortions are characteristic of large *strains* or the presence of a nonuniform *state of stress* within the sample. Typical manifestations of *twisting plasticity* include *faulting* or the appearance and development of interlayers with a reoriented *crystal lattice*. The thicknesses of interlayers vary over a wide range: from fractions of micrometers for mesofaults in metals to meters for *faults* in rocks. A typical size of fragments in a fragmented structure is 0.1 to 0.5 μm. These plasticity deformations are present in practically all plastic crystalline materials, and also in *polymers* and *composite materials*.

Microscopic mechanisms of *twisting strain* are associated with the appearance and development of specific collective forms of motion in the dislocation ensemble. These collective effects involve the generation and movement within the material of partial *disclinations*. Thus, twisting modes of plasticity are often called disclination modes. A *partial disclination* moving in an originally monocrystalline region leaves behind boundaries, or points where several boundaries come together (*joint disclinations*). Defects of the disclination type are concentrated in the regions of breaks, or according to their nature, are sources of strong long-range *internal stresses* (internal elastic fields). Thus to bring about the reduction of latent (stored) energy, disclinations join together into groups of defects of opposite sign, e.g., they adapt dipole and perhaps quadruple configurations. The motion within the dipole material of partial wedge disclinations leads to a variation of the shape of the sample, and to the formation of ribbon-like structures and fault bands. Disclination internal stresses play an important role in *strain hardening* at later stages of deformation, and in the *tough failure* of materials.

PLATES, laminae
Solid elastically deformed bodies having the shape of a right-angle prism or a right-angle cylinder, with a thickness h which is much smaller then the transverse dimensions. A plate is said to be thin, if h does not exceed one fifth of the least dimension of the base. Plates are divided into classes according to their *state of stress*. A plate is called rigid if the stresses prevailing in it under transverse loading are flexural stresses, and the tension–compression stresses (see *Flexure*) in the median surface (membrane stresses, see *Shells*) may be neglected. By convention, the deflection of plates classified as rigid does not exceed $h/5$. If membrane stresses are comparable in magnitude to tension stresses then the plate is said to be elastic. An absolutely elastic plate (a *membrane*) is one for which the tensile stresses may be neglected in comparison with the membrane ones; plates featuring a deflection $\geqslant 5h$ are conventionally classed as membranes. The contour of the deflection depends on the distribution of load, and on the boundary conditions (free, built-in, simply supported, or elastically supported edges). The value of the maximal deflection of a stiff plate

under action of normal loads F satisfies (in order of magnitude) the law: $f \propto Fl^2/D$, where l is the characteristic transverse dimension of the plate, $D = Eh^3/12(1 - \nu^2)$ is the so-called *cylindrical rigidity of the plate*, E is *Young's modulus*, and ν is the *Poisson ratio*. In the case of elastic plates and membranes, the dimensionless deflection $\zeta = f/h$ is described by an equation of the form $c_1\zeta^3 + c_2\zeta = \sigma^*$, where c_1 and c_2 are constants, and $\sigma^* = (\sigma/E)(l/h)^4$, $\sigma \propto F/l^2$ is the normal pressure; hence in the case of a strong deflection ($f \gg h$) we have $f \propto F^{1/3}l^{2/3}$.

If the plate is loaded with compressive forces lying in its plane, then at a certain critical value of loading the plate loses stability and buckles (*buckling*). In order of magnitude, $\sigma_k \propto D/hl^2$. See also *Vibrations of rods and plates*, *Plane problem of elasticity theory*.

PLATE VIBRATIONS

See *Vibrations of rods and plates*.

PLATING

See *Chrome plating*, *Electroplating*, *Lead plating*, *Zinc plating*.

PLATINUM, Pt

Chemical element of Group VIII of the periodic table with atomic number 78 and atomic weight 195.08. Natural platinum consists of isotopes ^{190}Pt (0.0127%), ^{192}Pt (0.78%), ^{194}Pt (32.9%), ^{195}Pt (33.8%), ^{196}Pt (25.2%) and ^{198}Pt (7.23%), of which ^{190}Pt and ^{192}Pt are radioactive: half-lives $6.9 \cdot 10^{11}$ and 10^{15} years, respectively. Electronic configuration is $4f^{14}5d^{10}6s^0$. Successive ionization energies are (eV): 8.96, 18.54, 28.5. Atomic radius 0.138 nm, ionic radii: Pt^{2+} 0.080, Pt^{4+} 0.065 nm. Oxidation states $+2$, $+4$, more rarely $+1$, $+3$, $+6$. Electronegativity ≈ 2.2.

Free platinum is a silvery-white *metal*. Exhibits face-centered cubic lattice; lattice constant $a = 0.39233$ nm (at 297 K). Density 21.45 g/cm^3 (at 293 K); $T_{\text{melting}} = 2042$ K, $T_{\text{boiling}} = 4073$ K. Melting latent heat is 21.84 kJ/mole; heat of evaporation 470.4 kJ/mole; specific heat 0.1327 kJ·kg^{-1}·K^{-1} (at 298 K); coefficient of thermal conductivity is 73.047 W·m^{-1}·K^{-1} in temperature range 273–373 K; coefficient of linear thermal expansion is $9.1 \cdot 10^{-6}$ K^{-1} (at 276–373 K). Electric resistivity is 103 nΩ·m, electric resistance temperature coefficient 0.003927 K^{-1} (at 273–373 K). The electronic work function is 5.23 eV.

Platinum is paramagnetic; magnetic susceptibility is $0.971 \cdot 10^{-6}$ CGS units at room temperature. Saturation vapor pressure is 133.3 µPa (at 1773 K) and 13,333 µPa (at 2023 K). The elastic modulus of annealed platinum is 147 GPa, rigidity is 72.4 GPa (at 273 K), ultimate tension 137 MPa. Mohs hardness 4.3; Vickers hardness 37–48 HV (according to different authors). Pure platinum is one of the most malleable and ductile metals; readily lends itself to forging, may be rolled into foil (to 0.0025 mm in thickness) or drawn into wire (0.001 mm in diameter); is amenable to polishing and welding. Incorporation of impurities, even in small amounts, causes a decrease of plasticity and an increase of hardness. Application of cold strain hardens platinum noticeably: relative elongation of cold hardened platinum decreases to 1–2%, whereas the value of HV increases to 90–95. Subsequent annealing restores the softness and plasticity of the metal. All alloy additives, when added within the *solid solution* region, harden platinum: the hardness increases by the factor of 2–2.5. Platinum is one of the most corrosion-resistant materials (see *Corrosion resistance*). Platinum is widely used in various fields of engineering. The metal (and its alloys with rhodium and iridium) are used for production of chemical industry ware. Platinum is one of the most widely used catalysts. Platinum is used for the production of *thermocouples*, resistance thermometers, electric contacts, wires for winding high-temperature furnaces, medical instruments.

PLEOCHROISM (fr. Gr. $\pi\lambda\varepsilon o\nu$, full and $\chi\rho\omega\mu\alpha$, color)

Change of color of a material exposed to transmitted light, which is related to the direction of propagation and polarization of this light. The phenomenon of pleochroism is one of the manifestations of *optical anisotropy*. Pleochroism is brought about by the anisotropy of absorption, which in its turn depends on the polarization of the radiation (see *Polarization of light*), the wavelength, and the direction of wave propagation. The phenomenon of pleochroism is most often observed in crystals. In the case of *uniaxial crystals*, two "principal" (main) color variations are distinguished, which appear when the crystal is observed along its *optical axis*, as well as transverse to it. *Biaxial crystals* feature two principal color schemes to be observed along three directions, which commonly coincide with the principal directions of the crystal (see *Crystal optics*). When viewed from other angles, the crystal exhibits intermediate colorations. Strong pleochroism is found in tourmaline (uniaxial crystal) and copper acetate (biaxial crystal). Variants of pleochroism are the circular and linear *dichroism of crystals*.

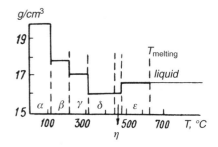

Dependence of the density of Pu on the temperature.

PLUTONIUM, Pu

An artificial radioactive element of Group III of the periodic table with atomic number 94. Plutonium is a *transuranium element*, belongs to *actinides*. Many isotopes ^{232}Pu–^{246}Pu are known: ^{247}Pu and ^{255}Pu were discovered among fission products from thermonuclear explosions. Not easily accessible ^{244}Pu is the most stable plutonium isotope (alpha decay and spontaneous fission, half-life $8.2 \cdot 10^7$ years, atomic weight 244.0642). Of widest application is ^{239}Pu (alpha decay and spontaneous fission, half-life $2.41 \cdot 10^4$ years, atomic weight 239.0522); of practical importance are also ^{238}Pu (half-life 87.74 years), ^{240}Pu (half-life $6.56 \cdot 10^3$ years), ^{241}Pu (half-life 14.34 years) and ^{242}Pu (half-life $3.76 \cdot 10^5$ years). Trace amounts of plutonium, which are generated in uranium ores by various nuclear reactions, are found in nature (0.4–15 parts of Pu per 10^{12} parts of U).

The first plutonium isotope (^{238}Pu) was discovered by G.T. Seaborg, E.M. McMillan, J.E. Kennedy and A.C. Wahl (1940). Outer shell electronic configuration is $5f^6 6d^0 7s^2$. Ionization energy 5.71 eV. Metallic radius 0.16 nm, ionic radii of Pu^{3+}, Pu^{4+}, Pu^{5+} and Pu^{6+} are respectively 0.100, 0.086, 0.074 and 0.071 nm. Electronegativity is ≈ 1.15.

Free plutonium is a brittle silvery *metal*. The specific feature of plutonium is that it exists in the form of six modifications α, β, γ, δ, η (the latter modification is also labeled δ') and ε in the temperature range from room temperature to $T_{\text{melting}} = 640\,^\circ$C. The approximate temperatures of transitions between these modifications are, respectively, $117\,^\circ$C, $195\,^\circ$C, $315\,^\circ$C, $455\,^\circ$C, $478\,^\circ$C. At high temperatures and pressures, the existence of other modifications of metallic plutonium is possible. α-Plutonium exhibits a monoclinic structure with lattice constants (nm): $a = 0.6183$, $b = 0.4822$, $c = 1.0963$ nm and angle $\beta = 101.79^\circ$. The density of α-plutonium is 19.86 g/cm^3 with the temperature dependence shown in Fig., $T_{\text{boiling}} = 3235$–$3350\,^\circ$C; specific heat of α-plutonium is 35.5 J·mole^{-1}·K^{-1}. Melting heat 2.834 kJ/mole. The α-plutonium resistivity is 1.4645 $\mu\Omega$·m (at 273 K). Mean values of thermal expansion coefficients for α-, β-, γ-, δ-, η- and ε-modifications are respectively $42.4 \cdot 10^{-6}$ K^{-1}, $20.9 \cdot 10^{-6}$ K^{-1}, $60.0 \cdot 10^{-6}$ K^{-1}, $-15.4 \cdot 10^{-6}$ K^{-1}, $-27.5 \cdot 10^{-6}$ K^{-1} and $36.4 \cdot 10^{-6}$ K^{-1}. The Vickers hardness of α-plutonium is ≈ 2.55 GPa (at room temperature), elastic modulus of α-Pu is 98.4 GPa (at $20\,^\circ$C), shear modulus 42.7 GPa, bulk modulus 48.4 GPa.

Plutonium is the most chemically active element of all the actinides. The oxidation states, exhibited by plutonium in its compounds, range from $+3$ to $+7$ (most stable is $+4$). Plutonium in compact form is slowly oxidized in air; it self-ignites in air when crushed to a powder.

The most important compounds of plutonium are: PuF$_6$ (low-boiling liquid; thermodynamic stability much less than that of UF$_6$), solid oxide PuO$_2$, carbide PuC and nitride PuN, which may be used as nuclear fuel when mixed with corresponding uranium compounds.

The chief application is the use of ^{239}Pu as a fissionable material in nuclear reactors and nuclear weapons. The critical mass of its α-modification is 5.6 kg (sphere 4.1 cm in diameter). The isotope ^{238}Pu is used in nuclear electrical batteries, which possess long service life. Plutonium isotopes are source materials for synthesis of transuranium elements (Am, etc.). Pu and its compounds are highly toxic.

p–n JUNCTION

See *Semiconductor junction.*

POCKELS EFFECT (F. Pockels, 1894)

A linear *electrooptical effect*, which consists in a change of *refractive indices* under the action of an electric field; a change proportional to the field intensity. The effect was first observed in 1883 in quartz crystals. The effect is possible only in materials without a *center of symmetry*, and occurs in *piezoelectric materials*. A piezocrystal, which is initially isotropic, becomes a *uniaxial crystal* under the action of the electric field; *birefringence*, if it is present in the material, becomes changed. The nature of the Pockels effect is related to the effect of an electric field on the structure of a *crystal lattice* and the electric polarization.

Extensive studies of the Pockels effect began when this phenomenon was observed in crystals of KH_2PO_4, and it was found that it may be used for virtually infinitely fast control over laser radiation. It is used for high-frequency amplitude and phase modulation, deflection and scanning of light, optical systems for transmission and recording of information, and light locators in controllable narrow cut filters. The reverse Pockels effect consists in *optical detection*, i.e. in generation of static polarization of noncentrosymmetric crystals under the action of laser radiation.

POINT CONTACTS

Small-area mechanical, electrical, and heat contacts between bulk conductors which are formed in the vicinity of where they touch. Point contacts are produced by pressing together electrodes of a special kind: needles and planes, sharp edges, etc. As a rule, the area of a mechanical contact is much greater than the spot through which the electric current passes, due to an oxide layer with insulator properties that covers the electrode surface. The electric current passes through the contact either due to electron tunneling (see *Tunnel effect*) in those places where the insulator layer thickness does not exceed a few nanometers, or directly through regions with a direct contact interface between the two *metals*. In experiments

a welding point contact is formed using a short-time current pulse which leads to the formation of a microscopic bridge that shunts the tunnel connection. The point contact resistance with a direct connection is determined by the Sharvin formula (see *Sharvin resistance*) if the contact size d is much smaller than the electron *mean free path l*. In the opposite extreme ($d \gg l$), the contact resistance is of the order of magnitude of the ratio ρ/d (ρ is the metal resistivity). For typical metals such as Na, Cu, Al, etc., a contact spot diameter of 30–3 nm corresponds to a resistance of 1–100 Ω. The point contact current density can reach 10^9–10^{10} A/cm^2, which involves a noticeable deviation of the *current–voltage characteristic* from Ohm's law. Point contacts are employed in microcontact spectroscopy (see *Tunnel spectroscopy*) as local injectors and collectors of nonequilibrium electrons in semiconductors at low temperatures in experiments on the electron focusing in metals by a magnetic field.

Superconducting point contacts are widely used, i.e. the electric contacts formed by touching of a pair of bulk *superconductors* at a small area of interface. The properties of these contacts depend on the ratio between the transverse dimension d of the contact, and microscopic length parameters of the materials such as the coherence length ξ and the electron mean free path l. In the case $d < \xi$, the superconducting point contacts exhibit the *Josephson effect* (see also *Josephson junction*). Owing to the Josephson effect and the *energy gap* Δ in the spectrum of *quasi-particles*, the current–voltage characteristic of these point contacts is essentially nonlinear. For superconducting point contacts with direct conduction there is a substantially larger current at $V \gg \Delta/e$ whose magnitude is of the order of $\Delta/(eR_0)$ (here R_0 is the point contact resistance in the normal state), and it depends weakly on the bias voltage. These superconducting contacts are incorporated into *superconducting quantum interference devices* (SQUIDs) which are widely used for highly precise, supersensitive measurements of magnetic fields. Owing to their small capacity, these superconducting point contacts are used for the detection and transformation of high-frequency electromagnetic radiation.

POINT-CONTACT SPECTROSCOPY

Spectroscopy of elementary excitations in *metals* and *semiconductors* with the help of *point contacts* with dimension d smaller than the length associated with electron energy relaxation. It was applied for the first time to determine the *current–voltage characteristics* of metal–insulator–metal (microbridge across barrier layer) tunnel junctions. Microcontact spectroscopy is specified by the energy duplication of nonequilibrium charge carriers in microcontacts at low temperatures ($k_B T \ll eV$). This phenomenon occurs under the action of an electric bias eV, and consists in the generation of two groups of nonequilibrium carriers with maximum energies differing by the value of the bias. *Relaxation* of this distribution results in a nonlinear current–voltage characteristic; its first derivative is proportional to the frequency of inelastic electron scattering, and the second derivative to the microcontact function of interaction of electrons with other quasi-particles of energy $\hbar\omega = eV$. In the case of the *electron–phonon interaction* at $T \to 0$ and $d \ll l$, we have

$$\frac{1}{R}\frac{\partial R}{\partial V}eV = \frac{8}{3}\frac{ed}{\hbar v_F}g_{\mathrm{pc}}(\omega)\big|_{\hbar\omega=eV}, \qquad (4)$$

where $R = dV/dI(V)$, v_F is the *Fermi velocity*, and $g_{\mathrm{pc}}(\omega)$ is the *point-contact function of the electron–phonon interaction*. The latter differs from the *tunnel function of the electron–phonon interaction* by the presence of a weighting factor to take into account the kinematics of electron scattering processes in a point contact of a certain shape. The point-contact function of the electron–phonon interaction is of the form

$$g_{\mathrm{pc}}(\omega)$$

$$= \frac{(2\pi\hbar)^{-3}}{\oint\limits_{\mathrm{FS}} dS_{\boldsymbol{p}}/v} \oint\limits_{\mathrm{FS}} \frac{dS_{\boldsymbol{p}}\, dS_{\boldsymbol{p}'}}{vv'}$$

$$- \sum_s |M_{\boldsymbol{p}-\boldsymbol{p}',s}|^2 K(\boldsymbol{p},\boldsymbol{p}')\delta(\omega - \omega_{\boldsymbol{p}-\boldsymbol{p}',s}),$$

where $M_{\boldsymbol{p}-\boldsymbol{p}',s}$ is the matrix element of electron transitions from the state with momentum $\boldsymbol{p}$ to that with momentum $\boldsymbol{p}'$ through scattering by a phonon with energy $\hbar\omega$ that corresponds to the sth branch of the phonon spectrum; $K(\boldsymbol{p},\boldsymbol{p}')$ is the *Kulik factor* normalized to its angle-averaged

value; the integration is performed over the states on the *Fermi surface* (FS), and v and v' are the electron velocity components, normal to the Fermi surface, before and after the scattering, respectively. The point-contact function of electron–phonon scattering allows for the kinematics of scattering processes in contacts with a particular configuration, as well as for the electron scattering by static defects in the near-contact region. By analogy to other electron–phonon interaction functions, the point-contact function of electron–phonon scattering is defined by an integral parameter of electron–phonon scattering in a point contact $\lambda_{\mathrm{pc}} = \int_0^\infty g_{\mathrm{pc}}(\omega)\, d\omega/\omega$ that is of the same order of magnitude as other parameters of the electron–phonon interaction in the metal at hand. A relationship similar to Eq. (1) is valid for the interaction of electrons with *magnons*, *excitons*, and other *quasi-particles*. The number of cases investigated by point-contact spectroscopy is constantly increasing, and includes *alloys* and compounds of variable valence, systems with *heavy fermions* (heavy electrons), *Kondo lattices* and Kondo impurities, small-dimension conductors, *superconductors* and other materials.

POINT DEFECT CLUSTERING

A group of *point defects* in some region where their concentration is much greater than the average value throughout the crystal volume. Hence, the point defects in a cluster preserve their individuality by not forming a new *phase*. Clusters of point defects can be formed during *thermal treatment*, *plastic deformation*, and irradiation, as a result of *diffusion* and their interaction with each other and with other large-sized defects, or they can be formed in clusters by, for example, radiation which induces cascades of atom-atom collisions (see *Radiation physics*). The pile-up of electrically active defects strongly influences electrical, optical, and other properties of *semiconductors*. The strongest effect on semiconductor electrical properties arises from the *piling-up of compensated centers*. Such clusters under equilibrium conditions are charged and surrounded by a *space charge region* which is characterized by an effective recombination screening radius ρ, and a potential barrier height φ. The volume in which these point defects are situated (not including the

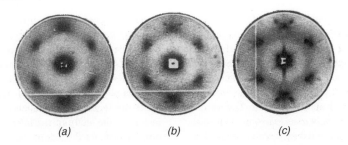

(a)   (b)   (c)

Radiograms (Mo K_α) illustrating the formation of anisotropic defect clusters in monocrystals of LiF under neutron irradiation; (a) non-radiated, (b) irradiation dose $4 \cdot 10^{17}$ neutrons/cm^2, (c) irradiation dose $6 \cdot 10^{18}$ neutrons/cm^2.

space charge region) is called a *nuclear cluster*. Aggregates of point defects not only influence the properties of materials containing them through their net charge Ze, but also by the space charge distribution inside the aggregate, and by the matrix parameters and temperature T. Clusters of point defects with a maximum linear dimension smaller than the de Broglie wavelength of the main charge carriers of the matrix are called quasi-point types, because they are regarded by the charge carriers as *multicharged point centers*. The effective recombination radius (see *Recombination*) of such a cluster is $\rho = [\hbar^2 e^2 \eta /(8m^* \varepsilon k_B T^2)]^{1/3}$, where k_B is the *Boltzmann constant*, η is the effective charge of the cluster at a particular moment (it can vary from 0 up to the equilibrium value $\eta = Z$), m^* is the effective mass of charge carriers in the matrix, ε is the *dielectric constant*, and $\varphi = \eta e/(\varepsilon \rho)$ is the effective height of the recombination barrier. A sensitive method for characterizing point defect clusters in semiconductors is the determination of the kinetics of their nuclear magnetization distribution for a definite nuclear polarization, which detects the presence of small clusters. A method based on long-time *relaxation* and residual *photoconductivity* phenomena is also used. For the investigation of clusters of point defects in different crystals *X-ray diffractometry*, and especially *diffuse scattering of X-rays*, is effective. These methods made it possible to observe changes of size, shape, and orientation of anisotropic clusters of radiation induced point defects as functions of increasing radiation dose (see Fig.)

POINT DEFECT–DISLOCATION INTERACTION

Depending on the *crystal* type and the nature of its defects, there are size, modulus, electrical and chemical interactions of *dislocations* of different types with various *point defects* (*vacancies, impurity atoms*, etc.). In the case of point defects in *metals*, the *size effect* usually dominates. The energy of the elastic interaction of a point defect with a system of *internal stresses* is $U = \Delta V P_0$, where P_0 is the average hydrostatic pressure created by a dislocation which has an edge component; and ΔV is the difference between the effective volumes of the point defect and an atom of the matrix (see *Vacancy–dislocation interaction*). In this case the point defect is considered as a *dilatation center*, and for anisotropic distortions of the lattice the expression for U becomes more complex. A modulus point defect–dislocation interaction occurs when the elastic inclusion, which simulates the point defect, is characterized by elastic moduli that differ from those of the matrix.

The electric point defect–dislocation interaction is stronger in *ionic crystals* than it is in metals, and may become dominant. In this case the effect involves the ability of dislocations to carry or to capture an electric charge. The interaction of an *edge dislocation* with steps (see *Dislocation step*) in an NaCl crystal with positive and negative ion vacancies serves as an example of this interaction. Charged dislocations also interact with *conduction electrons*. The chemical point defect–dislocation interaction occurs between impurity atoms and *stacking faults* in split dislocations.

The presence of one or more of these point defect–dislocation interactions can lead to the formation of *clusters of point defects* of dislocations associated with the nonuniformity of the distribution of the concentration of point defects around a dislocation (see *Cottrell atmosphere*). The formation of atmospheres of other types near the dislocations is possible (see *Snoek atmospheres*). These atmospheres pin the immovable dislocations and make their motion difficult. The formation of an *impurity-atom atmosphere* near an edge dislocation is accompanied by thermal effects. For slowly moving dislocations the atmospheric perturbation increases in proportion to the increase in velocity, but at the same time the interaction of the dislocation with the lagging part of the atmosphere decreases, which can provide a retarding drag force of the atmosphere on the dislocation. The interaction under discussion induces a change in the magnitude and direction of the velocity of point defect *diffusion* (in the presence of the point defect atmosphere there is *ascending diffusion* and *mass transport* under the influence of the gradient of U). There is a probability of generating new phase centers at a *phase transition* in a solid, and of the subsequent dispersity and the manner of separation of the other *phase* in the final structure of the material (see *Disperse structure*). Certain configurations of the *segregation* surface and its location with respect to the dislocation provide a high probability for the appearance of segregation. A cylinder tangent to the dislocation line along its generatrix is the most probable form of the segregation region. The radius of this cylindrical cross-section is not determined by the absolute value of the *surface tension* at the *phase interface* after the appearance of the new phase in the segregation area σ, but rather by its effective value, which is smaller than the value determined from the variation of the total energy of the dislocation interaction with the atoms of the dissolved substance due to their redistribution. In addition, the interaction of the dislocation with impurity atoms influences the rate of growth and the shape of the center of formation of the new phase. These effects provide the physical basis for the so-called *thermomechanical treatment* of alloys.

The point defect–dislocation interaction influences many properties of solids. For example, if a periodically changing external force causes a dislocation to oscillate near its equilibrium position within the point defect atmosphere then sup-

plementary energy dissipation occurs, and therefore, the *internal friction* changes. During *plastic deformation* a special role is played by the interaction of the dislocation with vacancies and *interstitial atoms*. For developing a complete theory of how the physical properties of real solids change under the simultaneous influence of several types of forces, the study of the interaction of dislocations with point defects in the presence of other point defects of more than one type bound to each other is necessary. Thus, in interstitial *solid solutions* vacancies can form non-mobile complexes with impurity atoms, which changes the kinetics of impurity atmospheres involved with dislocation formation.

POINT DEFECTS

Disturbance of an ideal periodic *crystal structure* which is localized in a volume close to atomic size in order of magnitude (a point in the case of a macroscopic description). Fundamental point defects are a *vacancy*, an *interstitial atom*, and an *impurity atom*. For a macroscopic description of point defects, the displacements of atoms, surrounding the defects from their former positions in the *ideal crystal* lattice, would also be also taken into account. A noticeable contribution to the energy of formation of a point defect can be given by including the first, second and further away nearest neighbor atomic spheres (depending on the nature of the bonding forces) from the point defect environment. The relaxation of the atomic environment around the defect can reduce the local crystal symmetry (see *Defect symmetry group*, *Distortion*, *Jahn–Teller effect*). All *real crystals* possess significant concentrations of point defects. The concentration of intrinsic point defects (see *Intrinsic defects*) at sufficiently high temperatures is determined by the conditions of *thermodynamic equilibrium* between a point defect and other defects in the crystal. When cooling the crystal, there remain nonequilibrium intrinsic defects in it, whose concentration depends upon the rate of cooling. In addition, point defects can be generated under the action of *nuclear radiations* (*radiation defects*), or an external static pressure, a *shock wave*, and so on. Impurity atoms generally exist in natural crystals; in artificially grown crystals they are introduced either from the raw material, from the crystal container, or intentionally by special *doping*. The motion of point defects in crystals is determined by thermally-activated diffusion; the *diffusion coefficient* being $D = D_0 \exp[-U/(k_B T)]$, where U is the diffusion *activation energy*, and the coefficient D_0 depends weakly on the temperature. Values of U vary in different crystals, and for different point defects from decimal parts to units of electron-volts (in nonmetallic crystals one must take into account the dependence of parameters D_0 and U on the point defect charge state, and these parameter values can vary by two or more orders of magnitude (see *Donor, Acceptor, Amphoteric center*). Correspondingly, there is a lower temperature limit at which an efficient thermal *diffusion* of given point defects takes place, which varies for different kinds of point defects from far above to far below (less often) room temperature. As a result of this diffusion, the point defects can form quasi-molecules, clusters, and impurity atmospheres around *dislocations* (see *Cottrell atmosphere, Snoek atmosphere, Suzuki atmosphere*). The presence of point defects is reflected in the physical properties of crystals. Their main influence is on electric, photoelectric, and optical properties of semiconductors and insulators, the mechanical and strength properties of metals and alloys based on them, and so on. Diverse types of crystal treatments (*heat treatment*, radiation and mechanical treatment of materials, etc.) lead to various reactions between point defects, and to their redistribution over the possible states in the *crystal lattice*, and over the bulk crystal. This allows one to vary the properties of treated crystals of a given chemical composition over a wide range.

POINT GROUPS of symmetry

Groups of *crystal symmetry* in which the coordinate transformation operators leave at least a single point of the space fixed (rotations about *symmetry axes*, reflections in symmetry planes, and combinations of these elements: rotoflection axes, *center of symmetry*). In crystals rotations are only allowed for particular finite angles $2\pi/n$ ($n = 2, 3, 4, 6$) which are designated as C_n; in the case of a subsequent reflection in the plane normal to the rotation axis $\boldsymbol{n}$ the designation of the axis is S_n (these symbols are usually included in the

group designation symbol). If there is an axis C_2 normal to $\boldsymbol{n}$ then the group is a dihedral type designated as D_n. Auxiliary indices n and v indicate the presence of planes including axis $\boldsymbol{n}$ or normal to it. Particular names are given to the symmetry groups of a cube (or tetrahedron): e.g., the full symmetry group O_h (T_d), and the group composed only of rotations O (T). There are a total 32 point symmetry groups. They are used when describing the spectra of molecules, the dynamics of current carriers in the vicinity of a local center in crystals; they determine the classes of *space groups*. Observable physical quantities are transformed in terms of point group representations (see *Group theory* in solid state physics).

"Rare" point groups of symmetry. Point groups whose particular feature is the presence among their symmetry elements of rotations through the angle 2π divided by 5: $2\pi/5 = 72°$. Such symmetries as C_5, C_{5v}, D_5, D_{5h}, Y, and Y_h did not arouse any interest for a long time because it was commonly believed that they were unable to occur in crystalline materials at the atomic level. However, in recent years there appeared numerous proofs of that fact that such symmetries are not only of mere academic interest, but they reflect real structures which are, in some cases, prospective for applications.

It became clear that small ($\leqslant 10^2$ nm) metal particles called nanoparticles (see *Small particles*) can either take the form of regular pentagons, or possess more complicated structures (icosahedron and pentagonal pyramid). The formation of *clusters* involving several atoms also occurs with other chemical elements, e.g., noble gases; so that in the case of xenon and argon the icosahedral structure is possible.

Quasi-crystals are also notable for this peculiar structure: pentagonal dodecahedrons are clearly seen in their grains. The presence of small regions with icosahedral atomic configurations is inherent in dense supercooled liquids. Five-sided rings are found in the molecule ferrocene $(C_5H_5)_2Fe$ which is a pentagonal antiprism of point group D_{5d}. Icosahedral symmetry has been found in a series of molecules ($B_{12}H_{12}^{2-}$, $C_{20}H_{20}$), as well as in more complicated structures (like viruses). During almost 40 years the model of particular *paramagnetic centers*, so-called *dangling bonds* in diamond-like crystals, has been discussed. There is a model which presupposes the existence of quintuple-coordinated *silicon*; the corresponded state of the uncoupled electron being called a *floating bond*. Similar to this model is a model of another defect which appears at silicon *dislocations*: a quintuple coordination of silicon in addition to a more remote sextuple coordination is clearly seen. Recently a new class of carbon clusters has been found with icosahedra which contain rotations for $72°$ as symmetry elements.

The greatest interest has been aroused by the discovery of *fullerenes*, i.e. molecules C_{60} with icosahedral symmetry, as well as crystals (*fullerites*) built on their basis. The excitement is not only about the particular structure and symmetry of these new molecules, but also their extraordinary properties. For example, it has been shown that when atoms of other elements are embedded in their crystals, such as alkali atoms, then semiconducting, metallic and superconducting properties can emerge. This has been successfully carried out with elements from diverse groups of the periodic system (up to *lanthanides* and even *uranium*).

Thus, various molecular, cluster, and crystalline structures exist whose particular feature is the presence of fivefold *axes of symmetry*. The description of the point groups C_5, C_{5v}, D_5, and D_{5h} is similar to their analogues among the 32 groups. Icosahedral symmetry is exhibited, e.g., by the icosahedron (1), pentagonal dodecahedron (2), and fullerene (3) shown in Fig.

There are four groups of icosahedral symmetry: simple (Y), full (Y_h), ordinary double (Y'), and full double (Y'_h). Group Y consists only of rotations about the axes of symmetry. There are 6 fivefold axes and, hence, 24 rotations about these axes, 10 threefold axes with 20 corresponding rotations; 15 twofold axes associated with 15 rotations. Adding the identity gives the total number of elements in the group, namely 60. Taking into account inversion leads to group $Y_h = Y \times C_i$ (here C_i is the inversion group) which contains 120 elements. Group Y' results from group Y by adding element Q which is a 2π rotation, so that $Y' = Y \times Q$. By analogy with Y_h, $Y'_h = Y' \times C_i$. Groups Y' and Y'_h contain 120 and 240 elements, respectively.

Group Y contains 5 classes and, correspondingly, 5 irreducible representations: $A_g(1)$, $F_1(3)$,

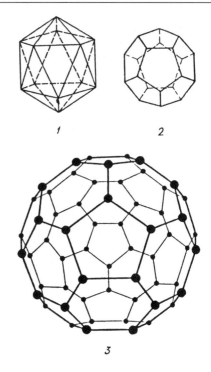

1   *2*

3

Sketches of (1) an icosahedron (20 faces), (2) a pentagonal dodecahedron (12 faces) and (3) a fullerene molecule C_{60} (32 faces), all having icosahedral point symmetry.

$F_2(3)$, $G(4)$, $H(5)$, with their dimensionalities indicated in parentheses. In group Y_h the number of classes and irreducible representations is doubled: to the designations of irreducible representations of group Y the subscripts "g" and "u" are added to indicate the parity or the eveness (g) and oddness (u) of states with respect to inversion. There are 9 classes and, correspondingly, 9 irreducible representations: to the five representations of group Y, four more representations are added: $E'_1(2)$, $E'_2(2)$, $G'(4)$, and $I'(6)$. In group Y'_h the number of classes and irreducible representations is doubled compared to group Y' on account of the added indices "g" and "u".

A character table lists irreducible representations (e.g., matrix representations) of elements of groups of the icosahedron which are used for finding the energy level structures, the selection rules, the description of diverse resonance phenomena, and solving other problems of molecular physics and solid state physics.

POISSON EQUATION

A differential equation of the form $\nabla^2 \psi(r) = -q(r)$, where $\nabla^2 = \partial^2/\partial x^2 + \partial^2/\partial y^2 + \partial^2/\partial z^2$ is the Laplacian operator, and $r(x, y, z)$ is the coordinate vector. When $q = 0$ the Poisson equation becomes the *Laplace equation* $\nabla^2 \psi = 0$. The Poisson and Laplace equations are often encountered in problems of electrostatics, *electrical conductivity, heat transfer, diffusion, elasticity theory*, etc. In electrodynamics the Poisson equation is usually written in the form $\nabla^2 \psi(r) = -4\pi\rho(r)$, where ρ is the distribution of electric charge density, and $\psi(r)$ is the electrostatic potential. In a medium featuring the *dielectric constant* ε, the Poisson equation has the form $\nabla^2 \psi = -4\pi\rho/\varepsilon$. The two-dimensional and three-dimensional versions of the Poisson and Laplace equations can be written in various coordinate systems, and the solutions depend on the boundary conditions of the problem. Often the conditions are such that both $\psi(r)$ and $\rho(r)$ tend to zero for $r \to \infty$, and the solution can be expressed in the form of an integral

$$\psi(r) = \int\limits_{-\infty}^{\infty} dr' \, \frac{\rho(r')}{|r - r'|}.$$

According to the Poisson equation, the potential of a single point charge placed at point r_0, such that $\rho(r) = e\delta(r - r_0)$ (δ is the Dirac δ-function) is described by the *Green's function* of the Poisson equation, which has the form:

$$\psi_0(r - r_0) = \frac{1}{|r - r_0|}.$$

The potential of a single dipole aligned along the unit vector u at the point r_0 is

$$\psi_1(r - r_0) = -(u \cdot \nabla)\psi_0(r - r_0), \quad r \neq r_0.$$

The potential of a multipole of order j located at the point r_0 expressed in polar coordinates is given by

$$\psi_j(r) = \frac{Y_j(\theta, \varphi)}{r^{j+1}}, \quad r \neq 0,$$

where $Y_j(\theta, \varphi)$ is a spherical harmonic function. In many cases a charge density ρ gives rise to a potential function which can be more complex, and often involves nonlinear equations or potentials.

POISSON RATIO (S.D. Poisson)

The ratio v of the relative transverse narrowing (broadening) to the relative longitudinal elongation (compression) during elastic extension (compression) of solids. For example, during the elastic extension of an elastic *rod* of initial length l and diameter d its length reaches l' ($l' > l$) and its diameter shrinks to d' ($d' < d$), and the Poisson ratio is calculated from the expression $v = (-\Delta d/d)/(\Delta l/l)$, where $\Delta l = l' - l$ and $\Delta d = d' - d$. For all material bodies the Poisson ratio lies within the interval 0 to 0.5, and for many of them v has a value close to $1/3$.

POLAR FACETS

Limiting atomic planes of a crystal, at which the *stoichiometric composition* differs from that of the bulk. The facets (0001) of compounds of the wurtzite structure, (100) and (111) of the sphalerite structure, (111) of rock salt, serve as examples. One can designate the polar facets of a binary compound by *crystallographic indices* and the type of atoms which form it (cations A or anions B), e.g., polar facets (111)Ga and ($\bar{1}\bar{1}\bar{1}$)As of gallium arsenide. Unlike polar facets, there are *nonpolar facets*, e.g., (10$\bar{1}$0), (11$\bar{2}$0) of compounds of the wurtzite structure, (110) of the sphalerite structure, (100), (110), (210), (211) of the rock salt structure, at which the bulk stoichiometric composition is preserved.

POLARIMETRY

Methods of investigation based on measuring the degree of radiation polarization (rotation of plane of polarization) and the *polarization of light* in anisotropic and optically active substances. It is used for investigating the optical homogeneity of solids, anisotropic crystal orientations, *rotation of light polarization plane*, the *dispersion* of light, etc.

POLARITON

Quantum of interference between photon and phonon. A quantum of the excited state of an *ionic crystal* involving a combination of electromagnetic wave (*photon*) with optical vibrations (optical *phonon*) or an exciton (see *Photoexciton*). The spectrum of a polariton in the infrared range can be explained within the framework of classical theory. The term polariton became established after

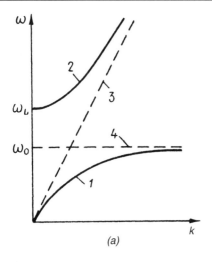

(a)

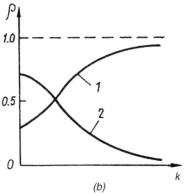

(b)

Polariton.

the first observations by C.H. Henry and J.J. Hop-field (1955) with the help of *Raman scattering of light* at small angles. Fig. (a) shows the dependence of the frequency ω on the wave vector k (*dispersion law*) for a polariton in its low-frequency (1) and high-frequency (2) modes, where the dashed lines characterize noninteracting light (3) and optical lattice vibrations (4). The fraction $\rho(k)$ of mechanical energy (1) and of electromagnetic energy (2) at different wave vectors k is shown in Fig. (b), according to the data of K. Huang.

POLARIZABILITY of atoms, ions, molecules

Coefficient characterizing the ability of particles to acquire an electric *dipole moment* p in an external electric field. Since the local field E_{loc}

effects the particle, its polarizability is given by $\alpha = p/E_{loc}$. The dipole moment per unit volume or *polarization of insulator* P for N_i particles of type i is given by

$$P = \sum_{i=1}^{N} N_i \alpha_i E_{loc}(i). \qquad (1)$$

Specifying the relation between E_{loc} and the mean macroscopic field E, and solving Eq. (1) with respect to P, we obtain the relation of the dielectric susceptibility $\chi = P/\varepsilon_0 E$ with the polarizability in the form of the *Clausius–Mossotti equation* (*Lorentz–Lorenz relation* or *Onsager equation*)

$$\frac{\varepsilon - \varepsilon_0}{\varepsilon + 2\varepsilon_0} = \frac{1}{3} \sum_i N_i \alpha_i, \qquad (2)$$

where SI units are used. (The local field is related to the mean macroscopic field in the Lorentz and Onsager models.)

Several types of polarizability can be distinguished. *Electronic polarizability* is caused by the displacement of atomic electron shells relative to their nucleus; *ionic polarizability* results from the displacement of oppositely charged ions from their equilibrium positions. Molecules or ions with permanent electric dipole moments that can reorient in an electric field possess *orientational polarizability* which is described by the *Langevin–Debye equation*. One of the ways to separate contributions from the different types of polarizability experimentally is to measure the frequency dependence of α: in the optical range the polarizability is almost completely electronic, in the IR range it is ionic, in UHF range it is orientational.

POLARIZABILITY OF DEFECTS

See *Elastic polarizability of defects*.

POLARIZATION CATASTROPHE

Unlimited growth under some critical conditions of the electric polarization (see *Polarization of insulator*). The polarization catastrophe explains the origin of *ferroelectricity* in a number of solids.

The crystal polarization may be approximately written in the form of the sum of products of the *polarizabilities* of the atoms α_i and the corresponding local electrical fields E_{loc}^i (SI units are used):

$$P = \varepsilon_0 \sum_i N_i \alpha_i E_{loc}^i, \qquad (1)$$

where N_i is the number of atoms of ith type in a unit volume. Let us assume for the purposes of simplicity that the *local electric fields* affecting all the atoms are the same, and are given by the *Lorentz formula* $E_{\mathrm{loc}} = E + P/(3\varepsilon_0)$, where E is the macroscopic electrical field within the sample. By substituting E_{loc} into Eq. (1), solving the resultant expression with respect to P, and taking into account that $P = (\varepsilon - \varepsilon_0)E$, we find

$$\frac{\varepsilon}{\varepsilon_0} = \frac{1 + (2/3)\sum_i N_i \alpha_i}{1 - (1/3)\sum_i N_i \alpha_i}. \qquad (2)$$

Thus, the dielectric constant ε becomes infinite, corresponding to a finite polarization for $E = 0$ under the condition

$$\sum_i N_i \alpha_i = 3, \qquad (3)$$

which is the criterion for the polarization catastrophe. From Eq. (2) it is possible to obtain the *Curie–Weiss law*, taking into account that the value of ε is very sensitive to the small deviations of $\sum_i N_i \alpha_i$ from the critical value 3. Actually, if we write $(1/3)\sum_i N_i \alpha_i = 1 - 3s$, where $s \ll 1$, we obtain $\varepsilon \approx 1/s$. If we suppose that near the critical temperature T_c, $s \approx (T - T_c)/C$, we find $\varepsilon = C/(T - T_c)$, which corresponds to the observed dependences of ε on T in the *paraelectric phase* of ferroelectrics.

POLARIZATION FILTER
See *Polaroid*.

POLARIZATION HOLOGRAPHY
A method of recording wave field polarization involving the effect of light-induced anisotropy in recording media: as in the *Weigert effect* in photoemulsions, anisotropic discoloration in *solid solutions* of dyes, reorientation of anisotropic *color centers* in the crystals, etc. Due to the effect of *light-induced anisotropy* the interference of signal and reference beams leads to the spatial modulation of the degree of anisotropy of the medium, i.e. to the storage of a *polarization hologram*. For some recording media the exact restoration of the signal wave polarization is possible through the corresponding selection of polarization of the reference and reading waves. The *dynamic polarization holography* realized in the registering media,

for which the processes of recording and erasing of polarization holograms coincide in time, presents especially interesting possibilities. For example, the possibility to compensate the variation in time of anisotropic inhomogeneities of media due to polarization *wave front reversal*, to control the state of polarization of interacting waves in real time, to investigate the nature of the photon response to the recording media, etc.

POLARIZATION OF INSULATOR
Macroscopically averaged value of the electric *dipole moment* per unit volume P of an *insulator* (sometimes P is called the *polarization vector*). In a *crystal* without a center of inversion (see *Center of symmetry*) the existence of *spontaneous polarization*, caused by the ordered arangement of oppositely charged ions at crystal lattice sites, is possible. In *ferroelectrics* the spontaneous polarization may change substantially under the effect of an electric field E. In the absence of spontaneous dielectric polarization the polarization arises due to an applied electric field. If the inherent atoms (ions, molecules) of an insulator have their dipole moments randomly distributed in the absence of an applied field E, then a macroscopic dipole moment appears at the switching-on of the field due to the partial orientation of microscopic dipoles (so-called *orientational or dipolar relaxation polarization*). In some crystals, there are several different equilibrium positions of ions in the *unit cell*. The mechanism of polarization in insulators, caused by the redistribution of ions under the effect of an external field, is also related to dipolar type relaxation. *Ion polarization* appears in the field E due to the oppositely directed small displacements of ions of opposite sign from their equilibrium positions (at $E = 0$). The field can also induce polarization in the electron shells of the regular atoms or ions of the insulator, which induces the presence of a macroscopic polarization. Other mechanisms of polarization are associated with the presence (perhaps due to external effects) of mobile charges in the insulator, or of impurity ions, vacancies, electrons and holes redistributing within the sample under the effect of a field E (see *Electrets*). In crystals with spontaneous polarization, the presence of mobile charges leads to its screening. Also the process of forming a macroscopic dipole moment in the sample

upon application of an electric field, mechanical stresses (see *Piezoelectricity*), injection of charge carriers, variation of temperature (see *Pyroelectric effect*), and other effects are also called polarization of insulators (the inverse process at the switching-off of the field is called *depolarization*). The characteristic polarization and depolarization times τ vary over very wide limits depending on the mechanism of polarization. For polarization of electron shells $\tau \sim 10^{-15}$ s, for atomic vibrations $\tau \sim 10^{-12}$ to 10^{-13} s, for orientational polarization $\tau \sim 10^{-4}$ to 10^{-6} s at room temperature, whereas processes associated with the transfer of charged particles may develop much more slowly. Therefore, the processes of polarization and depolarization of insulators strongly depend on the frequency of the polarizing field: the mechanisms which are effective are those that are able to follow the field variations (see *Dielectric loss*).

POLARIZATION OF LIGHT

Anisotropy of the optical properties of radiation (see *Optical anisotropy*) in a plane perpendicular to the direction of propagation expressed in terms of definite phase and amplitude correlations between mutually perpendicular components of the electric (magnetic) field intensity. The property of light polarization follows from the transverse nature of an electromagnetic wave. For the complete polarization of light the phase difference and correlation of amplitudes of mutually perpendicular components of the vector E (or H) remain constant for some time period (coherence time). In the general case of an elliptically polarized monochromatic wave the vectors E and H trace out in time and space elliptical cylindrical spirals around an axis along the light beam direction of propagation. *Linear polarization* and *circular polarization* are particular cases of the general (elliptical) type. Linear polarization of light is characterized by the constancy of the planes of vibration of the vectors E and H, and for circular polarization the ends of the same vectors trace out circular spiral paths. For unpolarized radiation there is no definite phase correlation between mutually perpendicular components of the vector E (or H). Sometimes the oscillations of the vector E (or H) will involve partially polarized light.

POLARIZATION OPERATOR

The linear polarizability of a system which depends on the transmitted energy (frequency ν), and momentum (wave vector k); in the general case of an anisotropic medium is designated by $\Pi_{\alpha\beta}(q, \omega)$ and associated with the *dielectric constant* tensor $\varepsilon_{\alpha\beta}(q, \omega) = \delta_{\alpha\beta} - 4\pi(e/q)^2 \times \Pi_{\alpha\beta}(q, \omega)$ (where $\alpha, \beta = x, y, z$; $\delta_{\alpha\beta}$ is the Kronecker delta symbol: $\delta_{\alpha\beta} = 1$ for $\alpha = \beta$, and $\delta_{\alpha\beta} = 0$ for $\alpha \neq \beta$). In particular, for a degenerate electron Fermi gas (see *Fermi–Dirac statistics*) in the high-density random phase approximation without taking account of the spin orientation (see *Spin*), the polarization operator $\Pi(p)$ is displayed graphically with the help of *Feynman diagrams* in Fig. 1(a) (which shows why this operator is often called a *loop*), and it is determined by the expression

$$\Pi(p) = -2\mathrm{i} \int \frac{\mathrm{d}^4 p'}{(2\pi)^4} G(p' - p)G(p'), \quad (1)$$

where $G(p)$ is the *Green's function* of the electrons, and $p = \{q, \omega\}$. In the superconducting state (see *Superconductivity*) (Fig. 1(b)) the operator $\Pi(p)$ has the form

$$\Pi(p) = -2\mathrm{i} \int \frac{\mathrm{d}^4 p'}{(2\pi)^2} \big[G(p' - p)G(p')$$
$$+ F(p' - p)F^+(p') \big], \quad (2)$$

where $F(p)$ is the *Gorkov anomalous function*. The polarization operator describes the dispersion of *plasma oscillations*, and the effects of *electric charge screening*. The spin-polarized polarization operator determines the paramagnetic *susceptibility*, and the exchange effects of magnetic biasing (see *Band magnetism*, *Spin density wave*). In a *Fermi liquid* with a strong Coulomb interaction between particles (electrons) the simple loop is replaced by a polarization operator containing the

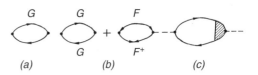

(a)   *(b)*   *(c)*

Fig. 1.

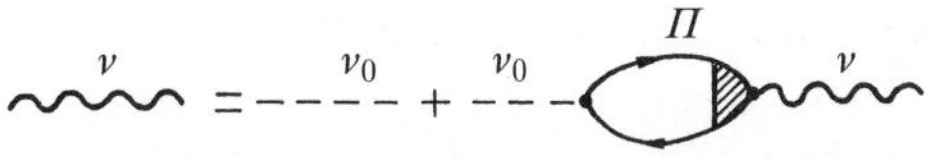

$$\nu \quad = - - - + \quad \nu_0 \quad \nu_0 \quad \Pi \quad \nu$$

Fig. 2.

vertex interaction part (triple-pole) (Fig.1(c)). In the description with the help of Green's functions of Bose type excitations in solids (*phonons, plasmons, excitons, magnons*, etc.), the polarization operator enters the *Dyson equation* as the proper energy part (*mass operator of quantum mechanics*) of the boson field (see Fig. 2).

POLARIZATION, OPTICAL

See *Optical polarization method.*

POLARIZATION PARAMAGNETISM

The same as *Van Vleck paramagnetism.*

POLARIZATION REORIENTABILITY

Capacity of spontaneous polarization P (see *Ferroelectricity*) to change its direction under the effect of an external electric field or mechanical stress by an angle differing from $180°$. *Ferroelectrics* with a vector P, which can reorient to an angle different from $180°$, are called *reorientable ferroelectrics*, unlike *reversible ferroelectrics* for which the vector P can change its direction by only $180°$. This case of reorientability of a ferroelectric is realized when the spontaneous polarization appears along a nonsingular (not unique) crystallographic axis (polar or nonpolar) of the high-symmetry initial phase with n such equivalent crystallographic axes. In this case in the ferroelectric phase there will exist n or $2n$ possible orientations of P, respectively, and the direction of the vector P may change by $180°$, or by an angle that differs from $180°$. Reversibility and reorientability significantly influence the properties of a ferroelectric.

POLARIZATION SWITCHING

The change of direction of spontaneous polarization vector of a *ferroelectric*. All ferroelectrics switch in an oppositely applied electric field. *Ferroelastics* may be partially or completely switched through application of an external mechanical stress. The electric field intensity dependence of the polarization at cyclic switching in an alternating field takes the characteristic form of a *hysteresis* loop. Switching in a pulsed electric field is characterized by a *repolarization current* and a *switching period*. The switching period decays exponentially with the field in a weak applied field and as a power function in a strong field, reaching values of the order of fractions of a microsecond. The displacement current, which arises during polarization reversal of a ferroelectric (see *Ferroelectricity*) is often accompanied by alternating current pulses, which are called *Barkhausen pulses* (by analogy with the *Barkhausen effect* in ferromagnets); the study of these pulses provides information on the *repolarization* mechanism.

Typical switching mechanisms are: generation of wedge-shaped antiparallel domains (reversal centers) in the neighborhood of the sample surface, direct intergrowth of these domains through the sample, and lateral motion of *domain walls*. The characteristics of the repolarization processes are strongly influenced by the actual crystal structure (*internal stresses*, impurities, radiation-induced defects) and the electrodes. Introduction of special dopants (e.g., alanine in *triglycinesulfate* crystals) allows one to achieve polarization, which is resistant to the influence of the applied electric field. The switching process permits the use these ferroelectrics as a *solid-state memory*. It should be noted, however, that repolarization, when repeated many times, causes severe changes of the switching parameter, which is a *fatigue* phenomenon.

POLARIZATION VECTOR

Electric dipole moment per unit volume of a medium. It can change (appear) under the effect of external factors. In a linear *insulator*, in the absence of spontaneous polarization, the polarization vector components P_i and the electric field components E_j are connected by the dielectric susceptibility tensor $P_i = \chi_{ij}\varepsilon_0 E_j$ (SI units). In the principal axis coordinate system the susceptibility tensor χ_{ij} is digonal, with nonzero components only along the principal directions.

POLARIZED NUCLEAR TARGETS

Materials which are highly polarized by the spins of their nuclei (usually protons of deuterons) used for investigations of the spin dependence of

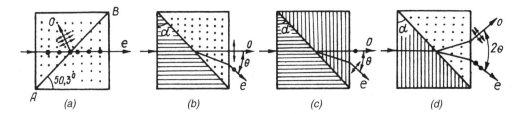

Polarizing prisms: (a) Glan, (b) Rochon, (c) Sénarmont, and (d) Wollaston types.

nuclear forces in elementary particle scattering experiments. Polarized nuclei (see *Nuclear orientation*) may be produced by one of the methods of dynamic orientation using some paramagnetic component. See also *Dynamic nuclear polarization*.

POLARIZING PRISMS

Optical elements for producing linearly-polarized radiation. Polarizing prisms are fabricated from *uniaxial crystals* with *birefringence*. Usually these prisms consist of two (or more) parts with different orientations of the *optical axes* such that the transmitted light beam experiences different conditions of refraction for the *ordinary ray* (o) and the *extraordinary ray* (e). This brings about their separation in space.

The figure shows the following types of prisms: (a) *Glan prism*, where AB is an air interface (ordinary ray is deflected to the side and adsorbed by a special coating applied on the side face); (b) *Rochon prism* (1783), where the angle θ depends on the prism angle α; (c) *Sénarmont prism* (economical to manufacture); (d) *Wollaston prism* provides a symmetrical separation of the rays to an angle approximately twice as large as that of the Rochon prism. In this figure the parallel rulings and dots indicate the directions of the optical axes.

POLAROID, polarization filter

An optical device for obtaining plane-polarized light; one of the main types of optical linear polarizers. The action of polaroid is based on the phenomenon of *linear dichroism*, i.e. of the inequality of adsorption of two linearly polarized mutually perpendicular components of the light falling at the polaroid. Polaroids are often manufactured by pressing a large high number of small similarly oriented crystals, e.g., of iodoquinine sulfate, into the matrix film. Also polaroids are fabricated by stretching polymer films (*dichroism of crystals* appears due to the similar orientation of *polymer* molecules stretched in the same direction). The advantages of polaroids compared to *polarizing prisms* are the ability to obtain wide aperture beams of polarized light, with their compactness, and with the technological effectiveness of their manufacturing process. Drawbacks of polaroids involve the considerable variation of the spectral composition of light passing through them. They are used in the near UV, visible, and near-IR radiation bands.

POLARON

A *quasi-particle* describing the states of a *conduction electron* (*electron polaron*) or hole (*hole polaron*) interacting with the polarization vibrations of a crystal: an electron surrounded by a phonon. In *ionic crystals* the polarization of the ionic framework by the electrical field of the "free" electron may be so pronounced that *self-localization* of the electron takes place: it is localized in the polarization potential well created by the displacements of surrounding ions into new equilibrium positions, i.e. the electron holds the well by its field. Such a self-coordinated state of the current carrier is called a *tightly bound polaron*. The motion of the current carrier takes place together with the polaron cloud of the displacements of its ionic surroundings, hence its effective mass exceeds the *effective mass* of its energy band state. If the potential well is created by small displacements of many ions (its radius significantly exceeds the interatomic distance) it is possible to speak about a strongly bound macroscopic polaron. The term "polaron" was introduced for this model by S.I. Pekar (1946). In strongly bound polarons optical transitions of electrons in the potential well are possible. Quantitatively the bonding force of the electron with the crystal lattice is determined by the parameter $\alpha = R^*/\hbar\omega$, where R^* is the *effective Rydberg*, and $\hbar\omega$ is the characteristic energy of a quantum of the longitudinal optical crystal lattice vibration. The criteria are: strong bond $\alpha > 10$, intermediate bond $10 > \alpha > 1$, weak bond $\alpha < 1$. A *weakly bound polaron* is a state of a band current carrier perturbed by its interaction with the crystal lattice, whose effective mass is increased $1/(1 - \alpha/6)$ times, and whose energy is reduced to the value of $\alpha\hbar\omega$. A *large-radius polaron* for any α is described within the framework of a single Hamiltonian (*Fröhlich Hamiltonian*). A *small-radius polaron* differs from a strongly bound macroscopic polaron by the fact that the *polaron* potential well is formed by the displacements of the equilibrium positions of the small number (e.g., two) of nearest ions. The presence of the polaron well sharply decreases the possibility of electron tunneling (see *Tunneling phenomena in solids*) to adjacent sites, i.e. such tunneling would be accompanied by a simultaneous reconstruction of the polaron well. This implies a large carrier effective mass and a small polaron band width, i.e. a low velocity of spreading of the localized polaron around the crystal. In this case, the charge carrier velocity may be caused by thermally activated electron jumps among adjacent sites. The criterion for the existence of a small radius polaron reduces to $E_p > 2\Delta$, where E_p is the lowering of the electron energy due to the polaron effect, 2Δ is the band gap width.

Together with free polarons, one can also consider bound polarons (e.g., at an anion vacancy: *F-center*), polaron molecules (*bipolaron*, polaron *exciton*, bound electron/hole polarons, with radii smaller than that of an exciton).

Experimental evidence is available for the existence of polarons in various crystals with noticeable *ionic bonding* (e.g., alkali-halide crystals, oxides of transition metals, *perovskites*). Even for the weakly bound case polaron effects show themselves in the width of adsorption lines, in electron-phonon recurrances of resonance transitions, in the nontrivial dependence of the energy on the quasi-momentum, e.g., at the crossing of Landau levels with a phonon energy: *Johnson–Larsen effect*

(D.M. Larsen, E.J. Johnson, 1966). The small radius hole polaron in alkali-halide crystals is a hole localized at two adjacent anion sites. Thus, the hole polaron is the molecular ion X_2^- (X is halogen) with X atoms strongly displaced in the direction toward one another from their positions at adjacent sites. The notion of polarons led to the discovery of other quasi-particles of the polaron type, where the polarization interaction is determined by other types of collective excitations – by the *fluctuons*, piezopolarons, *condensons*, *magnetic polarons*, etc.

POLARON, MAGNETIC

See *Magnetic polaron*.

POLAR PHASE

Term designating the *ferroelectric phase* of a solid. In like manner the *paraelectric phase* or other phase without spontaneous polarization (see *Ferroelectricity*) is called the *nonpolar phase*.

POLE FIGURE

A descriptive image or representation of the main features of a crystallographic *texture*. Direct and inverse pole figures are distinguished. A *direct pole figure* is the stereographic projection of the poles of specific atomic planes $\{hkl\}$ of all the crystallites. An *inverse pole figure* is the standard stereographic projections of poles of the main (with low *crystallographic indices*) planes of the crystal, where an experimentally determined statistical weight is assigned to each pole. Points of equal polar density in the pole figure are shown by isolines. A pole figure is usually normalized with respect to the polar density of a material with an isotropic distribution of crystallite orientations. The method of polar plane plotting is most widely used to display experimental *X-ray diffractometry* results. Any pole figure may be analytically calculated for a known *distribution function* of crystallite orientations, and conversely this function may be reconstructed by a calculation based on one or several measured pole figures. Knowing the pole figure, it is possible under certain conditions to calculate the orientation-dependent (textural) component of the tensor properties of a textured material.

POLISHING

Finishing treatment of articles for purposes of increasing the degree of surface purity, fine tuning articles to required sizes, obtaining certain surface layer properties. Polishing is a set of processes of plastic microdeformation and fine *dispersion* of a surface layer when the latter is acted upon by polishing and finishing materials. Several kinds of polishing are possible: polishing as the limiting or extreme case of *grinding*, in proceeding to a more fine, surface dispersion (fine cutting); utilizing additive adsorbents, which create a soft film of a new substance through chemical reactions of polar groups in their molecules; as this film is formed, it is removed by a mild abrasive, which does not scratch the surface; real polishing involving plastic flow of the layer under treatment while subject to the action of considerable tangential stresses; *electrolytic polishing*.

POLISHING, ELECTROLYTIC

See *Electrolytic polishing*.

POLK MODEL (D.E. Polk, 1971)

A structural model of an amorphous material with a tetrahedral arrangement of atoms in a *random continuous network*. The Polk model allows for all values of dihedral angles. The starting frame, which consists of pentatomic and hexatomic rings, is modified by the addition of atoms which must fulfill the following conditions: (1) there are no free valences inside the model; (2) bond lengths vary through a range of 1%; (3) the deviation of valence angles from the tetrahedral angle ($\approx 109°$) does not exceed $\pm 20\%$; (4) the model is subject to minimum mechanical stresses. The radial distribution function of the Polk model correlates well with experimental data. In principle, it is possible to magnify the sizes of the model indefinitely. The Polk model has been used for the simulation of the structures of *amorphous silicon*, amorphous germanium, etc.

POLONIUM, Po

A radioactive chemical element of Group VI of the periodic table, atomic number 84, most stable isotope is ^{209}Po (half-life 102 years). It was discovered in pitchblende because of its radioactive properties (P. et M. Curie, 1898). Isotopes with mass numbers 210–218 belong to natural radioactive series. Polonium comprises about $2 \cdot 10^{-14}\%$ by weight of the Earth's crust. It resembles *tellurium* in its properties. Electronic configuration of outer shell is $6s^2 6p^4$. Successive ionization energies (eV): 8.43, 19.4, 27.3, 38, 57.1, 73. Most typical oxidation state is +4, less common oxidation states are -2, $+2$, $+6$. The electronegativity is 2.0.

Polonium is a silvery *metal* existing in two modifications (α-Po, β-Po). Low-temperature α-Po forms a cubic crystal lattice, high-temperature β-Po adapts a rhombohedral crystal lattice (phase transition point is 327 K). Density 9.4 g/cm^3; $T_{\text{melting}} = 527$ K, $T_{\text{boiling}} = 1235$ K, heat of melting 12.5 kJ/mole, heat of evaporation 72.8 kJ/mole; specific heat 26.4 J·mole^{-1}·K^{-1}. Coefficient of linear thermal expansion is $2.35 \cdot 10^{-5}$ (at 77 K). Resistivity ≈ 0.42 μΩ·m (at 273 K).

The main application of polonium is the use of the ^{210}Po isotope (half-life 138.38 days). It is a pure alpha particle emitter, employed as an energy source in atomic batteries of satellites and mobile devices. When mixed with *beryllium*, it is used for the preparation of neutron sources. It is produced through irradiation of metallic bismuth with neutrons, and found in the wastes from the processing of uranium ores.

POLYACETYLENE

Organic polymer (CH)$_x$, simplest representative of a large class of organic materials called *conjugated polymers*, that is polymers with alternating single and double bonds (–CH=CH–CH=CH– $\cdots$). Molecular weights range from 400 to several million daltons (atomic mass units). Most extensively studied is polyacetylene of molecular weight 10,000–11,000, which is produced from acetylene (HC≡CH). It has a fibrous structure with a fiber diameter 10–100 nm; and a much greater length. Highly crystalline fibers consist of hexagonally packed chains of monomers

(–CH=CH–)$_n$, oriented along the fiber axis (see Fig.). The degree of crystallinity (up to 90%) of this system is the highest among *polymers*. The overall density is lower than the fiber density (1.16 g/cm^3) and varies over a wide range. A large specific surface (about 80 m^2/g) facilitates incorporation of impurities in polyacetylene, which can cause changes in its electrical properties. After chemical transformations, dopants (I$_2$, AsF$_5$, K, Na, etc.) generate charge carriers (electrons, *holes*) in the polyacetylene. Incorporation of small amounts of charged dopants $D = I_3^-$, AsF$_6^-$, K$^+$, Na$^+$, etc. results in an abrupt increase of specific electrical conductivity of the doped polyacetylene for small concentrations y of dopants ($y_c < y < 10^{-2}$, $y_c \cong 10^{-4}$): electrical conductivity σ increases from 10^{-10}–10^{-6} to 10^2–10^3 ($\Omega \cdot$cm)$^{-1}$, whereby polyacetylene transforms from an *insulator* state to that of an organic *metal*. This property is one of the most interesting and characteristic features of polyacetylene and other conjugated polymers. This behavior and other unusual

properties are due to the fragmented structure of the polymer chains, which are cut by defects into conjugated fragments of finite average length: 30 to 50 *carbon* atoms. The specificity of polyacetylene and other conjugated polymers means that a charge carrier q_1 in the field of a charged dopant D_1 is localized at the strongly anisotropic fragment of continuous conjugation, and may jump from one such fragment to another in response to the temperature, and to electric fields in the direction of a nearby dopant D_2. Like impurity *semiconductors*, at a critical concentration of dopants y_c a *Mott metal–insulator transition* from an insulating to a metallic state can take place. It occurs because of overlap of regions (which contain N_c conjugated fragments in the neighborhood of dopants, where a charge carrier may be situated. In polyacetylene N_c is equal to 15 or 50 (first *coordination sphere*, or first plus second, in the neighborhood of a charged fragment of continuous conjugation). Therefore, according to the assessment given by equation $y_c = B_c/(8N_c l)$, the value of y_c is very small: $y_c \cong y_c$(experimental) $\cong 10^{-4}$, where B_c in the above equation is a universal constant from *percolation theory*. Application of this theory to anisotropic objects leads to an expression which explains the experimentally obtained dependences:

$$\sigma = \sigma_0 + \sigma_3 y^\mu,$$

$$\mu = \frac{\gamma}{l} + 1 - \beta_3 + t_3,$$

where σ_0, σ_3, $\gamma \cong 50$ are characteristic of polyacetylene; $\beta_3 = 0.40$, $t_3 = 2$, $v_3 = 1.7$ are critical indices of percolation theory, $y_c \ll y \ll 10^{-2}$. Optimal packing of chains (see Fig.), which persists at small values of y, exhibits the highest value of N_c, and the highest value of the electron hopping parameter between fragments ($\sim$0.05 eV). This packing provides the highest values of y_c and the lowest values of σ_3 relative to other conjugated polymers. Polyacetylene is widely studied as the simplest quasi-one-dimensional system which is an organic metal and exhibits unique physicochemical properties (see *Fermions in a soliton field*).

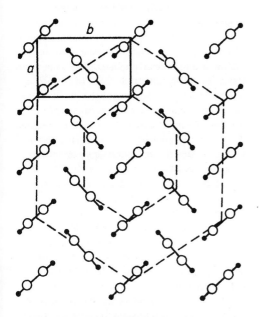

Projection of part of crystalline polyacetylene on a plane perpendicular to the direction of the polymer chain (CH)$_x$. Shown are elements of the honeycomb mesh (solid lines), and chains that are adjacent, nearby and further removed (dashed lines) for an arbitrarily chosen short chain (in center): $a = 0.424$ nm, $b = 0.732$ nm.

POLYALKALINE EFFECT, two-alkali effect

Sublinear relationship between the ionic *electrical conductivity* and the composition of oxide glasses at the equimolar mutual replacement of alkaline components in them, e.g., by varying x from 0 to 1 in the system $(K_2O)_x(Na_2O)_{1-x} \cdot 2SiO_2$. The polyalkaline effect is a special case of *ionic conductivity compensation*.

POLYCLUSTER AMORPHOUS SOLIDS, polyclusters

Condensed amorphous solids, which consist of locally regular *clusters* in contact (i.e. having common boundaries). A *locally regular cluster* is a connected aggregation of regular (locally ordered) atoms. An atom is a *regular atom* if its *coordination polyhedron* belongs to a certain collection of polyhedra of the same type. The number of types and the selection of types are characteristics of the structure. *Polyhedra of the same type* are coordination polyhedra; each having the same number of vertices, edges and faces, with deviations from perfection of edge lengths and angles between edges not exceeding certain critical values. In essence, a locally ordered cluster is a *random continuous network* of atoms with randomly alternating local ordering. Boundaries of locally ordered clusters consist of atoms, which belong to first *coordination spheres* of regular atoms but are not regular themselves. A fragment of a two-dimensional *polycluster* is illustrated in the figure, with an intercluster boundary and an internal cluster boundary shown. Boundaries contain coinciding and noncoinciding sites of regular extensions of locally regular cluster nets (see Fig.). The minimum of the boundary free energy is reached when all coinciding sites and about half of the noncoinciding sites are occupied with atoms. Point defects as well as one- and two-dimensional defects are inherent in polycluster amorphous solids. *Point defects* include vacant sites and atoms in interstices of locally regular clusters, vacant coinciding sites, *impurity atoms*. Vacant noncoinciding sites are partial vacancies (cavity volume approximately twice that of a vacancy in a locally regular cluster). An atom that occupies a noncoinciding site, is a partial interstitial. Two-dimensional defects are boundaries, and one-dimensional defects include edges of internal boundaries, triple junctions

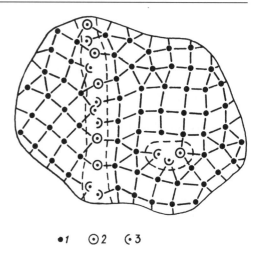

•1   ⊙2   ⊂•3

A fragment of a twofold polycluster with fourfold and fivefold coordinated atoms. Internal (right) and between-cluster (left) boundaries are shown: 1, regular sites; 2, coinciding sites; and 3, noncoinciding sites.

of boundaries, and degenerate one-dimensional boundaries. Generation of a *Volterra dislocation* (see *Somigliana–Volterra dislocations*) in the bulk of a locally regular cluster is related to the formation of a disordered (boundary) layer in a *slip plane*. Prismatic *dislocation loops* of the vacancy or interstitial type also bound the locally disordered nets of vacancies and interstices. Polyclusters possess tunneling states, which are related to atoms in noncoinciding sites. Tunneling states are responsible for *low-temperature anomalies* in amorphous bodies.

The glass–liquid transition (see *Vitrification*) in polyclusters is associated with the "melting" of surfaces, which is caused by the formation of co-operative structural fluctuations at boundaries. *Diffusion* in polyclusters is mostly *boundary diffusion*. Like in crystals, in the bulk of a locally regular cluster the vacancy mechanism of diffusion prevails. Under the action of applied forces, polyclusters may experience *plastic deformation* arising from *slippage along the boundaries*, formation of new slip layers (which possess the structure of boundaries); boundary diffusion. At low temperatures, when diffusion of atoms is quenched, the first and second diffusion mechanisms prevail, which leads to the formation of *slip bands*. At

POLYMER AGING

See *Aging of polymers.*

POLYMER CROSS-LINKING

Formation of transverse *chemical bonds* between macromolecules leading to the development of an interlinked spatial network in the *polymer.* The best known treatment involving the linkage of polymers is the *vulcanization of raw rubber,* the technological process which results in the transformation of *raw rubber* into the *rubber* of commerce. Polymer linkage can be carried out either by a chemical method (e.g., vulcanization of raw rubbers by sulfur with the formation of polysulfide bonds between macromolecules), or under the action of ionizing radiation (*radiation-induced polymer cross-linking*). The trends of the process of polymer linkage are described by the statistical theory of gel formation. A three-dimensional space network is formed when the number of transverse bonds per macromolecule exceeds a certain critical value of the order of unity; simultaneously the polymer loses its *yield* and, as a rule, becomes highly elastic. The polymer linkages can form as a result of *aging of polymers,* e.g., in cases when the degree of *polymer degradation* is less than the extent of radiation cross-linking.

POLYMER CRYSTALLIZATION

See *Crystallization of polymers.*

POLYMER DEGRADATION

Destruction of polymer chains by physical, chemical or biological factors. As a result of polymer degradation the molecular mass of the *polymer* decreases, and its physical and chemical properties change. The following main types of polymer degradation are distinguished: thermal (via increased temperature), thermooxidizing (heat and oxygen jointly), *photodestruction* (light); radiative (γ-radiation, α- and β-particles, neutrons); mechanical (mechanical stresses), biological factors (e.g., fermentation). The majority of polymers break down at 200 to 300°C, but there are some which require much higher temperatures (polytetrafluoroethylene, polyamides, etc.), and these are called thermostable. Oxidation by molecular oxygen considerably accelerates the destruction processes, because under the effect of O_2

free radicals are generated within the polymer, and a chain oxidation process develops. Thermooxidative polymer degradation may be accelerated by autocatalysis through the action of breakdown products. A way to protect from this is to introduce some additives (stabilizers), so-called *antioxidants,* into the polymer. Photodestruction is especially significant for short-wavelength light, and protection from it is provided by *light stabilizers.*

POLYMERIC MATERIALS

Materials containing molecules of *polymers* as main components, which determine the physical and mechanical properties. The polymeric materials of most practical importance are plastics, elastomers (*rubbers*) and fibers. *Plastics* are materials based on polymers. During their period of formation the polymer is in a viscous fluid state or in a highly elastic state, whereas under operating conditions the polymeric material of the plastic is in a vitreous or partially crystalline state. Plastics are classified into *thermosetting plastics* (if processing into articles is accompanied by chemical reactions with the formation of *covalent bonds* between molecules) and *thermoplastic materials* (if no new covalent bonds are formed during processing). Unlike thermosetting plastics, thermoplastics retain the capability of softening and turning into the viscous fluid state under secondary heat treatment. In practice plastics are often loaded with various fillers (organic or inorganic substances in the form of powder, fibers, grains, liquid or gaseous inclusions), which improve and diversify physical and mechanical characteristics. Most frequently used thermoplastic materials contain molecules of polyethylene, polyvinylchloride and polystyrene; the most widely used thermosetting plastics are phenolformaldehyde and epoxy resins.

One of the methods for producing *elastomers* consists in cross-linking (vulcanization) of polymeric melts in the plastic state (see *Polymer cross-linking*). Elastomers may also be obtained directly during the generation of a polymeric substance, if multifunctional monomers are used in the process of synthesis. Formation of links between chains of a polymer, which is responsible for the high elasticity of polymeric materials, may also be caused by hydrogen bonds, specific chain-to-chain interactions, crystalline phase nuclei, etc. The most important property of elastomers, which determines most areas of their practical application, is the property of being able to undergo large elastic strains under moderate stress. *Young's modulus* of typical elastomers is by a factor of 10^4–10^5 greater than that of steel (0.5–8.0 MN/m^2). Under tension, elastomers exhibit substantial deviations from a linear relationship between stress and strain for typical relative strains ($\sim$10%), whereas many rubbers lose elasticity (deformation reversibility) under considerably greater strains ($\sim$100% and higher). Exceeding the *elastic limit* results in rapid rupture of an elastomer.

Polymeric fibers are obtained by extrusion of a strand polymeric solution or melt through spinnerets into a medium which causes polymer solidification. Fibers may be also produced under natural conditions. The most important natural polymeric fibers are: cotton (containing cellulose as chief constituent), wool (chief constituent is keratin), silk (fibroin). Fibers based on chemically modified natural polymers are called artificial fibers (acetate fibers, viscose fibers). Fibers obtained from completely synthetic polymeric derivatives are called synthetic fibers (polyamides (nylons), polyesters, polyacrilonitrile fibers, etc.). Fibers derived from solutions of polymers with stiff chains (e.g., aromatic polyamides) are of a particular *strength*: in such fibers most chains are oriented along the fiber axis. This effect is most pronounced at the formation of a fiber from a nematic liquid crystal polymeric solution.

POLYMERIZATION, SOLID-PHASE

See *Solid-phase polymerization.*

POLYMER MASS DISTRIBUTION

See *Molecular mass distribution.*

POLYMER MODIFICATION

See *Modification of polymers.*

POLYMERS

High molecular weight compounds (macromolecules) which are chain molecules, i.e. consist of a large number of atomic groups united into a chain by *chemical bonds*. Examples of repeated atomic groups (*monomer units*) in polymer chains: [–CH$_2$–]$_n$ (polyethylene), [–CH$_2$–CH$_2$–O–]$_n$ (polyoxyethylene), [–S–]$_n$ (polymeric sulfur). Monomeric units may be arranged in linear chains (*linear polymers*), branched chains (*branched polymers*), or a three-dimensional network constructed from short segments of polymer chains (*polymeric nets*). The number of monomer units in typical linear polymers is 10^2–10^5, and in a number of cases this number reaches 10^8.

In most cases a long linear polymer chain does not assume a linear *conformation*, but exhibits a complicated three-dimensional spatial arrangement of structural units. This is caused by *polymeric chain flexibility*. Polymers containing only identical monomeric units are called *homopolymers*. Chains that contain two or more types of monomeric units are called *copolymers* or *heteropolymers*. According to the nature of the distribution of units in a chain, regular and irregular copolymers are distinguished; notable examples of the latter type are *biopolymers*: macromolecules of DNA and proteins. Copolymers, in which two or more types of structural units form reasonably long continuous sequences, are called *block copolymers*.

The basic (principal) chain of a homogeneous-chain polymer is composed of identical atoms. The most abundant *homogeneous-chain polymers* are *carbochain polymers* with main chains consisting of only *carbon* atoms (with attached hydrogens). Polymers which contain atoms of various elements (e.g., C and S) in their fundamental chains are called *heterogeneous-chain polymers*.

Polymers, which contain only carbon, hydrogen, oxygen and nitrogen, are called *organic polymers*. Polymers, which contain atoms of inorganic elements (most often Si, Al, B, P, Cl) along with organic groups, are called *organoelement polymers*. Polymer macromolecules, which have inorganic main chains and do not contain organic side-chain radicals, belong to *inorganic polymers*.

Polymeric substances may exist in a crystalline or an *amorphous state*. Depending on the strength and the nature of the interactions between their structural units, amorphous polymers may be classified as *polymeric glasses*, high-elasticity polymers, or polymers in the plastic state. *Liquid-crystal polymers*, which are intermediate between crystalline and amorphous polymeric substances, are also studied.

Polymers possess specific physicochemical properties resulting from the chain structure of their molecules, and the long length and flexibility of the polymeric chain. The most important properties are: ability to form very strong anisotropic fibers and films, ability to undergo severe reversible elastic deformations, and relatively simple processing (see *Polymeric materials, Aging of polymers, Polymer degradation*).

POLYMERS AND RADIATION

See *Radiation physics of polymers*.

POLYMERS, INORGANIC

See *Inorganic polymers*.

POLYMER SWELLING

Increase of volume of a polymeric sample caused by absorption of liquid (or vapor) of low molecular weight. Polymer swelling may take place when an amorphous *polymer*, a polymer network, or a partially crystalline polymer, is brought in contact with low molecular weight liquid below the critical temperature of the mixture. Considerable stresses and thermal effects may arise in the sample in the initial stage of polymer swelling.

The kinetics of polymer swelling is, as a rule, determined by cooperative diffusive motion of chains, which make up the polymer body. The average time to attain equilibrium in polymer swelling increases considerably near polymer *phase transition* critical points. In certain cases there is an abrupt decrease of equilibrium swelling for at minor change of the external conditions (e.g., conditions for so called *collapse of polymer networks*). Isotropic *polymeric materials* retain their initial shape on swelling, whereas polymer swelling of anisotropic polymer materials (*films* and, in particular, fibers) is accompanied by a nonuniform change of dimensions of the sample, more change occurring perpendicular to the orientation axis direction than along this direction. In certain cases, even shrinkage of the polymeric material along the anisotropy axis was observed.

POLYMORPHIC TRANSFORMATION

See *Structural phase transition*.

POLYMORPHIC TRANSFORMATION HEAT

See *Heat of polymorphic transformation*.

POLYMORPHISM

Disposition of *crystals* to exist in more than one structural modification of an unchanged chemical composition. Polymorphism is characteristic of many elements (see *Allotropy*), compounds and *alloys*. Polymorphic modifications may differ in the type of interatomic bond. For instance, α-Sn (grey tin) is a *semiconductor* with a cubic diamond-type *crystal lattice* and β-Sn (white tin) is a *metal* with a tetragonal *body-centered lattice*. The transformation α-Sn $\rightarrow$ β-Sn brings about a large change of volume (24%), and leads to corrusion of tin materials (*tin pest*). Stability regions of polymorphic modifications are given by *phase diagrams*. Transformation of one polymorphic modification into another is a *first-order phase transition* which occurs through generation and growth of new phase nucleation centers. If the probability of nucleation of an equilibrium modification is small, then a metastable modification may exist for a long time under normal conditions (e.g., *diamond*). On heating, a metastable transformation is irreversibly converted to a stable one (e.g., *graphite*). Irreversible polymorphic transformations to a final stable form are called *monotropic transformations*, whereas reversible ones between one form which is stable above a certain temperature, and the other is stable below it, are classified as *enantiotropic transformations*. See also *Polytypism*.

Concentration polymorphism is defined as the generation of phases, which are intermediate in chemical composition, in multicomponent systems. Unlike proper polymorphism, concentration polymorphism takes place due to a change of composition of the material under fixed environmental conditions (temperature, pressure). The term "concentration polymorphism" is often used to emphasize structural features of phase diagrams, such as the existence of several different phases in a narrow interval of concentrations; it is also used for the description of an aggregation of phases, which noticeably differ in composition but structurally are all traced back to one initial phase.

POLYNOMIAL BASIS, IRREDUCIBLE

See *Irreducible polynomial basis.*

POLYTYPE, MARTENSITIC

See *Martensitic polytypes.*

POLYTYPISM, (fr. Gr., $\pi o \lambda v \varsigma$, many $\tau v \pi o \varsigma$, pattern).

Ability of a substance of given composition to form a variety of different crystal modifications (*polytypes*); these modifications possess equal *crystal lattice* parameters in two directions and different parameters in the third direction. For example, identical layers (close-packed planes) may be stacked in different sequences. Similar to polymorphic modifications, polytypes have different crystal lattices, therefore, polytypism may be considered as one-dimensional *polymorphism.* Unlike polymorphic modifications, different polytypes are formed at the same temperature and pressure and do not have stability regions; they exhibit nearly equal values of potential energy and possess very similar physical properties. Polytypes are divided into *stable (equilibrium) polytypes* and *metastable (nonequilibrium) polytypes.* Stable polytypes include those of semiconductors, minerals and other substances, which are formed by diffusion during *crystallization,* diffusion annealing (see *Homogenization*). *Martensitic polytypes,* which are generated without diffusion at *martensitic transformations* in metal alloys with a low energy of *stacking faults,* belong to metastable polytypes. Among equilibrium polytypes (e.g., in *silicon carbide*) occur modifications, which contain 4200 close-packed basal planes with lattice spacing $\approx$120 nm in a *unit cell.* For the description of polytypes the symbols A, B, C are most widely used (each of the letters stands for a close-packed plane). Thus, a six-layered polytype SiC-II is written as follows: $-ABC\,ACB-$, and a 15-layered polytype SiC-I is written as $ABC\,BAC\,ABA\ CBC\,ACB$. One of the main challenges of the polytypism theory is the explanation of the generation of ordered (perfect) polytypes of high periodicity. Consideration of such polytypes requires the assumption that there is an interaction (correlation) between atomic layers, which acts through tens and even hundreds of interatomic distances; the range of such an interaction far exceeds the range of known interatomic forces. Some authors do not restrict polytypism to the one-dimensional variety, but consider it the ability of a crystal to occur naturally in more than one form.

POMERANCHUK EFFECT (I.Ya. Pomeranchuk, 1950)

The presence of a minimum on the *melting curve* (the dependence of pressure P on temperature T) for ^{3}He. In liquid ^{3}He the *exchange interaction* leads to a preferable antiparallel orientation of adjacent nuclear spins which reduces the part of the *entropy* associated with them. Due to the smallness of this interaction in solid ^{3}He at the temperatures above $T_N = 1.1$ mK the distribution of spins will be random until the magnetic interaction of the nuclear spins at $T \leqslant T_N$ becomes dominant (see *Helium*). As the entropy of liquid ^{3}He decreases with the reduction in temperature, and at low temperatures the "non-spin" part of entropy is small, then at some temperature T_m ($T_N < T_m < 1$ K) the entropy of the liquid S_L becomes smaller than the entropy of solid ^{3}He, where $S_S = R \ln 2$ per mole in this temperature range. According to the Clausius–Clapeyron equation the slope of the melting curve $dP/dT = (S_L - S_S)/(V_L - V_S)$ becomes negative at $T < T_m$ ($T_m = 0.319$ K, $P_m = 2.9315$ MPa) and it will be less than zero until $T \leqslant T_N$, where the orientation of solid helium ^{3}He nuclei begins, $S_S < S_L$. The molar volume of liquid $V_L > V_S$ of the solid helium in the whole range of temperatures and pressures, so that the sign of dP/dT coincides with the sign of $S_L - S_S$. At the point (T_m, P_m) of the P–T diagram of ^{3}He, where $S_L = S_S$, we find $dP/dT = 0$, which corresponds to the minimum of the melting curve.

POPULATION INVERSION

See *Level population.*

POPULATION OF LEVELS

See *Level population.*

PORES

Gaps or cavities between structural elements of a *solid.* One may distinguish *linked pores* or channels which penetrate deeply into the bulk of the solid, as well as closed or *isolated pores*. The pore volume per unit volume of a solid, or the percentage of its volume that constitutes pores, is called *porosity*. Porosity is also characterized by the *specific surface*, or the ratio of the total surface of a porous body to its volume or mass, often expressed in the units m^2/g.

In *crystals* the effective formation of pores takes place under the effect of nuclear radiation, at high doses of defect-forming particles. Thus, in pure *nickel* irradiated by neutrons with energy $E > 1$ MeV at $T = 400\,^\circ$C, pores are observed at the dose $4 \cdot 10^{21}$ neutrons/m^2, and in reactor steels at 450 to 500 $^\circ$C pores appear at the dose of 10^{26} neutrons/m^2.

Vacancies and *interstitial atoms* created upon irradiation partially recombine with one another, and are partially trapped by different sites. *Dislocations* more readily capture interstitial atoms than vacancies. As a result of this, an excess concentration accumulates, which leads to the generation of pores, and then to their growth (see *Critical pore size*). At low temperatures $T < 0.3T_{\text{melting}}$ (T_{melting} is the melting temperature) the *diffusion* rate of *point defects* is low and there are no processes of pore formation. At the higher temperatures $T > 0.6T_{\text{melting}}$, the diffusion rate is high, defects are effectively annealed, their concentration is small, and there is no pore formation. Pore formation is most likely within the range of temperatures $0.3T_{\text{melting}}$ to $0.6T_{\text{melting}}$. As a result of some nuclear reactions, gases are generated (helium, argon, xenon, etc.), which are practically insoluble in *metals*, so *gas-filled pores* form. The gas pressure in the blisters is so high that atoms of inert gases crystallize in blisters at temperatures above room temperature, i.e. much higher than the *crystallization* temperature at normal pressure. Pore formation leads to the radiation *swelling* of materials.

POROSITY

The presence within the volume of a *solid* of cavities with dimensions significantly exceeding atomic ones. The totality of *pores* forms a system connected to the external surface, which affords access to the bulk of the pores by outside gas molecules. The so-called *sorbents*, used for the uptake of a considerable number of gas molecules, possess the highest porosity: activated *carbon, zeolites*, and silica gels, in which pores are structurally formed elements which are comparable in their dimensions (or larger) to the sizes of gas molecules.

Another type of porosity is determined by the presence of voids in sputtered metallic *films*, in pressed tablets, where the pore size far exceeds an atomic one by orders of magnitude, and may be measured by their fraction of the overall volume (from units to tens of percents).

POROSITY DIFFUSION
See *Diffusion porosity*.

PORPHYRINS, METALLO-
See *Metalloporphyrins*.

POSISTOR (abbr. of Positive Temperature Coefficient Thermosensitive Resistor)

A thermistor or polycrystalline ceramic *ferroelectric–semiconductor* which has a region of abnormally high positive temperature coefficient of resistance $\alpha = R^{-1}(dR/dT)$ (R is resistance, T is temperature). This coefficient has a typical value $\alpha > 10\%$/K, and may reach 30–50%/K. The total resistance jump in the region of positive values (width $\sim 10^2$ K) may reach 5–7 orders of magnitude. On cooling the *posistor effect* becomes reversible. Outside the region of positive values, the temperature resistance coefficient of the posistor is negative (as it is for ordinary semiconductors). The posistor effect includes the following factors:

(1) the existence of internal inhomogeneity: *grain boundaries*; it is observed in *ceramics* and is not observed in *monocrystals*;
(2) ferroelectric properties of the material; observed in the neighborhood of phase transitions of various type in ferroelectrics (ferroelectric–paraelectric transition; ferroelectric–ferroelectric transition involving a change of lattice symmetry; ferroelectric–antiferroelectric transition).

The first patent on an anomalous positive temperature resistance coefficient ($\sim 20\%$/K, resistance increase of three orders of magnitude) in *barium titanate*-based ceramics was issued in 1954.

Of greatest abundance are posistors based on barium titanate (see *Ferroelectric ceramics*) doped with elements of valence 3 or 5. Ceramics based on undoped barium titanate exhibit properties of a good *insulator* ($\rho \propto 10^{10}$ $\Omega \cdot$m). This compound becomes a semiconductor on being doped (see *Doping*) with microamounts (concentration <0.5 at.%) of elements which either possess an ionic radius close to that of Ba^{2+} and a valence +3 (one greater than Ba^{2+}), or possess an ionic radius close to that of Ti^{4+} and valence of +5 (greater by one than Ti^{4+} in barium titanate). Dopants of the first type are usually *rare-earth elements*: La^{3+}, Sm^{3+}, Ce^{3+}, Gd^{3+}. Dopants of second type are Nb^{5+}, Ta^{5+}, Sb^{5+}, Bi^{5+}. Doping results in a considerable decrease of resistivity of the ceramics (up to 0.1–100 $\Omega \cdot$m) and a change of color of the ceramic material, which becomes green. On further increase of concentration of doping impurity (above 0.5 at.%), the resistivity of barium titanate sharply increases again and it becomes an insulator. The posistor effect in barium titanate takes place in the region of a ferroelectric–paraelectric phase transition, with a sharp increase of resistance above the phase transition point. The Curie temperature for pure barium titanate is $T_C = 393$ K. At doping with elements of the same valence as Ba and Ti the Curie temperature may vary over a broad interval: 203 K $\leqslant T_C \leqslant$ 673 K (T_C increases when Ba is replaced with Pb, and it decreases at replacement of Ba with Sr; or Ti with Zr or Sn). The shift of T_C depends on the amount of added dopant and is 2.5 K/at.% for Sr; 4 K/at.% for Zr and Pb; 7.5 K/at.% for Sn. The limiting amounts of *substitutional atoms* at which the doped material still retains semiconducting properties ($\rho < 10^4$ $\Omega \cdot$m) are: 40 at.% for Sr; 25 at.% for Sn and Zr; 70 at.% for Pb.

Along with barium titanate-based semiconductors with a posistor effect, in the neighborhood of the *Curie point* there is a number of ferroelectric semiconductors which exhibit the posistor effect in the region of phase transitions of other kinds: ferroelectric–paraelectric transition and ferroelectric–ferroelectric transition in $KNbO_3$, $KBiTi_2O_6$, Pb_2FeNbO_6; it is in the region of ferroelectric–ferroelectric transitions between rhombic and tetragonal phases, where the posistor effect of $KNbO_3$ and $KBiTi_2O_6$ is most pronounced. This effect at ferroelectric–antiferroelectric phase transition is observed in $NbBiTi_2O_6$.

Posistors find wide application in engineering. The main operating characteristics of posistors are: position of positive temperature resistance coefficient on the temperature scale; value of maximum temperature resistance coefficient in units of %/K; ratio of change of resistance in the region of positive temperature resistance coefficient; switch-over temperature, at which the posistor resistance increases by a predetermined factor; thermal response time which characterizes its transittime effects. Numerous technological applications of posistors include: unbalanced current relays; devices for temperature compensation of resistance, for protection of electric devices from overheating; thermostats; heaters; current regulators, current and voltage stabilizers; device to control the level of a liquid; device for start-up of a single-phase electric motor with starting condenser; automatic demagnetization of color TV sets; time relays and delay elements.

POSITRON

Fundamental particle e^+ which has the same mass and spin as an electron e^-, but differs from it in the sign of the charge and magnetic moment. On penetrating into a solid (usually emitted from a radioactive source) high-speed positrons become thermalized in the short time of about 1 ps through loss of energy by ionization and excitation of atoms. The lifetime of a positron in a material is limited because of positron–electron annihilation. This annihilation gives rise to *gamma rays*; *positron spectroscopy* of solids is based on analysis of the parameters of the gamma quanta. The average lifetime in various solids varies over an interval of 100–1000 ps. The main state of a positron in a solid is the quasi-free state. By virtue of the Coulomb repulsion of positrons by atomic nuclei not all electrons take part in the annihilation of positrons. Therefore, the lifetime is inversely proportional to the "effective" density of electrons, and the angular distribution of annihilation *photons* (e.g., in *metals*) permits one to reproduce the structure and anisotropy of the *Fermi surface*. If there are *defects* in the structure of the solid, then some quasi-free positrons, which are diffused over the material, become trapped by those defects and form with them bound energy states, the so-called *positron centers*.

Certain solids, e.g., MgO, W, possess a negative positron *work function*. As a result of this, after moderation in the bulk of such solid a small percentage of the positrons is emitted from its surface with a kinetic energy of about 1 eV. This phenomenon is used for generating beams of slow positrons (with variable spectrum) and for investigating solid surfaces. Besides quasi-free states, positrons and electrons form hydrogen atom-like bound states in the material, which are called *positronium* (Ps).

POSITRON CENTER

Defect in a solid, at which localization and annihilation of positrons takes place without generation of the positronium state, e.g., *vacancies*, vacancy *clusters*, *dislocations*, complexes of vacancies with impurities. The reason why these defects capture positrons has to do with their excessive negative charge, which is due to the reduced density of positively charged ions. Positronium states are mostly localized at defects in *insulators* (see *Positronium center*). Parameters of electron–positron annihilation (see *Positron spectroscopy*) of a positron, which is localized at a positron center, are different from those of a quasi-free positron; this is due to the difference in the density of valence and inner-shell electrons in the defect relative to those in a crystal lattice. The most useful characteristic of a positron center is the time distribution of positron annihilation: from this distribution one may extract components of lifetimes of positrons at different types of centers, which permits the determination of rates of capture of positrons by positron centers and rates of positron annihilation in the centers. This allows one to determine the concentration of positron centers and the electron density in them.

POSITRONIUM, Ps

Bound state of an electron e^- and *positron* e^+. Free positronium is analogous to a hydrogen atom, in which the proton mass is replaced by the mass of a positron. In the ground state (*parapositronium*) the spins of e^- and e^+ are antiparallel. The level of *orthopositronium* (spins of e^- and e^+ parallel) begins at $8.41 \cdot 10^{-4}$ eV. Parapositronium lifetime with respect to annihilation is 125 ps (two gamma quanta are produced, see *Gamma ray*). Orthopositronium lifetime is 140 ns (three gamma quanta are produced at annihilation). Characteristics of the positronium in a material and in the free state are essentially different. The mechanisms for binding together a positron and electron in a medium at the deceleration of a high-speed positron are specific. According to the mechanism proposed by A. Ore (1949), the formation of Ps takes place at an energy of e^+ in the region of the so-called *Ore gap*, close to the ionization energy of the atoms of the material. According to the alternate track mechanism, the formation of Ps takes place when a positron captures an electron from the region of its track (Spur). The specifics of these mechanisms and the role played by each of them in the generation of positronium are as yet unknown. The so-called *positronium centers* arise, if there are *defects* in the structure of the material. The positronium lifetime in a solid is limited mainly by the so-called pick-off annihilation: annihilation of the positron with other electrons of the substance. This process shortens the positronium lifetime to $\leqslant 1$ ns (for triplet positronium). The most evident manifestation of the formation of Ps in solids is the appearance of narrow peaks in the angular distribution of annihilation phonons. These peaks appear in the region of small angles, and angles which correspond to vectors of the *reciprocal lattice* of the crystal. This is due to the fact that the value of the center of momentum of thermalized (delocalized) Ps is small. Annihilation characteristics of positronium in the material depend to a large measure on the environment; this dependence manifests itself, e.g., in ortho–para conversion of Ps involving conduction electrons or paramagnetic atoms, in the formation of chemical compounds by Ps and atoms of the medium, etc. Delocalized positronium cannot exist in bulk metals. In *insulators* with low concentrations of intrinsic and impurity defects, such as quartz, CaF_2, MgF_2, ice, KCl, NaCl, etc., delocalized positronium has been observed. Positronium localized at the surface is observed in practically all solids.

POSITRONIUM CENTER

Defect, at which localization and annihilation of *positrons*, takes place, giving rise to the *positronium state* (bound electron–positron pair). Positronium centers are observed mostly in ionic *insulators* and molecular *solids*. States of positrons in positronium centers differ from other states in solids by the sensitivity of their annihilation characteristics to the application of a static magnetic field. Positronium centers are often characterized by extensive regions of elevated electron density: *pores*, vacancy *clusters*. The role of positronium centers in molecular solids is also played by interstitials. In ionic insulators, such as *alkali-halide crystals*, the role of positronium centers is played by electronic *color centers*, *vacancies*, and their complexes in the cation sublattice. Localized positronium states may also exist in the neighborhood of vacancy clusters of semiconductors.

POSITRON SPECTROSCOPY

Investigations of characteristics of *solids*, which are based on parameters of *gamma ray* (photons) which are released at the annihilation of *positrons*. Annihilation usually gives rise to two *photons*, which possess the total energy $2mc^2$. The probability of the production of additional photons is small, except when an electron–positron pair forms a bound state (*positronium*, Ps) before annihilation. This state is realized in ionic and molecular *crystals* of *insulators*. The sources are man-made radioactive isotopes ^{22}Na, ^{58}Co, ^{64}Cu, ^{68}Ge. Positron spectroscopy of solids is based on the fact that, in a time of the order of 10^{-12} s, positrons in the material lose energy by scattering, and come into thermal equilibrium with the medium. Therefore, the energy and momentum of the positron at the moment of annihilation are negligibly small compared with corresponding characteristics of an electron, and the emitted photons provide information on the annihilating electrons. This information is obtained through three types of measurements (see Fig.).

1. The positron lifetime τ is determined through measurement of the delay time, i.e. the time interval between the recording of a high-energy gamma quantum, which is produced simultaneously with the positron, and the emission of the annihilation photon. Measurements of τ, which is of the order of 10^{-10} s, are performed with the help of high-speed systems of detecting and recording (Fig. (a)). Since $\tau^{-1} = \lambda = \pi r_0^2 cn$, where λ is the rate of annihilation, r_0 is the classical electron radius, n is electron density in the area of annihilation, the measurement of τ allows one to determine n.

2. The angle of deviation of annihilation photons from collinearity characterizes the component of electron momentum, which is perpendicular to the direction of the scattering of these photons. A measurement of the angular distribution of annihilation photons can provide the density of electron momenta $\rho(p)$. These measurements are performed with the help of coincidence detectors which are situated on each side of the sample; one of the detectors may move over the given interval of angles (Fig. (b)).

3. Doppler broadening of annihilation lines, which is caused by the component of electron velocity that is parallel to the direction of photon scattering, also permits one to determine $\rho(p)$. High-resolution semiconductor detectors are used for the measurements (Fig. (c)).

The *method of electron–positron annihilation* allows one to find $\rho(p)$ and the contour of the *Fermi surface*, to assess structural defects, and to study the near-surface layers. In *metals* there are no filled electron states above the *Fermi level*, therefore, $\rho(p)$ is truncated at this level. The determination of its location for various crystallographic orientations makes it possible to reconstruct the Fermi surface. In modern devices the Fermi surface is drawn with the help of two-dimensional detecting systems, which allow one to sketch outline charts of $\rho(p)$ and find projections of the Fermi surface. The distinctive feature of this method is that the measurements are not restricted to the case of extra-low temperatures and metals of high purity, rather they permit finding Fermi surfaces of high-temperature phases and random alloys.

Diagnostics of defects is based on their ability to capture a *positron*. Centers of localization of positrons are *vacancies*, aggregations of vacancies (vacancy clusters), *dislocations* and other defects, which arise at the generation of localized positrons. Annihilation photons provide information on the type and concentration of defects, as well as on characteristics of the electronic subsystem in the neighborhood of the defect. The interaction of defects, which leads to generation of aggregations of vacancies, complexes of vacancies with impurity atoms and decorated dislocations, manifests itself by a change

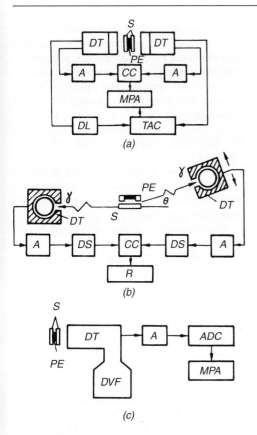

(a)

(b)

(c)

Block diagrams showing the principle of devices for measurements of (a) positron lifetimes; (b) angular distribution of annihilation phonons; (c) determination of the Doppler broadening of an annihilation line. A, amplifiers; ADC, analog-to-digital converter; CC, coincidence circuits; DL, delay; DS, discriminators; DT, detectors; DVF, Dewar vacuum flask; MPA, multi-pulse channel analyzer; PE, positron emitter; R, recording device; S, sample; TAC, time-to-amplitude converter.

of observed annihilation parameters. In this connection positron spectroscopy is used to investigate the evolution of the defect structure of solids, which takes place under exposure to deformation, irradiation, temperature and other factors. It is also used to investigate changes in the defect structure at *phase transitions*, and for *nondestructive testing techniques* of various materials and articles. High-energy positrons from the emitter penetrate into solids to a depth, ranging from tens to thousands of micrometers. Creation of methods of positron moderation and the generation of positron

beams of variable energy permits the study of the electronic structure of the surface and distribution of defects over the depth of the near-surface layer. Generation of positronium is possible at the surface of metals and semiconductors, and in the bulk of insulators, in particular, insulating molecular crystals. The *method of positron annihilation*, applicable to molecular crystals, allows studying both crystal structure and chemical reactions with the participation of Ps.

POTASSIUM, K (Lat. Kalium)

Chemical element of Group I of the periodic system with atomic number 19 and atomic mass 39.0983; it is an *alkali metal*. Natural potassium consists of stable isotopes ^{39}K (93.22%) and ^{41}K (6.77%) and traces of the weakly radioactive isotope ^{40}K (0.00118%). Outer shell electronic configuration is $4s^1$. Successive ionization energies 4.341, 31.820, 46 eV. Atomic radius is 0.236 nm; radius of K^+ ion is 0.133 nm. Work function is 2.22 eV. Potassium is very active chemically. Oxidation state is +1. Electronegativity is 0.78.

In free form, potassium is a soft silvery-white *metal*, quickly dimming in the air, reacting with oxygen, water vapor, and other air components. It has a body-centered cubic lattice, $a = 0.5247$ nm. Density is 0.862 g/cm^3 (293 K); $T_{melting} = 336.7$ K, $T_{boiling} = 1033$ K. Heat of melting is 2.3 kJ/mole; heat of sublimation is 89.4 kJ/mole; specific heat $c_p = 29.6$ J·mole^{-1}·K^{-1}. Debye temperature is 90 K. Electrical resistivity is 0.0623 μΩ·m, temperature coefficient of electrical resistivity is $5.8 \cdot 10^{-3}$ K^{-1} (at 293 K). Hall constant is $-4.2 \cdot 10^{-10}$ m^3/C. Magnetic susceptibility is $\chi = +0.53 \cdot 10^{-9}$. Linear thermal expansion coefficient is $8.4 \cdot 10^{-5}$ K^{-1} (at 273 to 323 K). Brinell hardness is 0.4 MPa. Modulus of normal elasticity upon stretching is 3.5 GPa (at 83 K).

Due to its high chemical activity, applications of metallic potassium are very limited. Liquid alloys (at room temperature) of potassium and sodium (containing 40 to 90% of K) can be used as heat conductors. The mixture of peroxide K_2O_2 and superoxide KO_2, used for air regeneration, is obtained from metallic potassium.

POTASSIUM TITANYLPHOSPHATE

This *nonlinear-optical crystal* KTiOPO$_4$ crystallizes in the orthorhombic space group Pna_1 (C_{2v}^9). Mohs hardness is 5, it is insoluble in water and stable with respect to acids and alkalis, melts with decomposition at temperatures above 1300 K, is transparent from 0.35 to 4.5 μm. It is obtained by hydrothermal synthesis at $T_{max} = 1000$ K and $p_{max} = 100$ MPa, or from the solutions in melts at $T_{max} = 1150$ K. Maximum crystal dimensions are $30 \times 40 \times 20$ mm^3. It is easily machined and polished. It is a *ferroelectric* with $T_C = 1205$ K and $P_C \approx 100$ μC/cm^2; and above 280 K superionic conduction is observed (see *Solid electrolytes*) along the c axis. Nonlinear optical efficiency is 1.5 times higher than that of lithium niobate (see *Niobates*) and 15 times higher than that of potassium dihydrogenphosphate. Dispersion parameters allow *second harmonic generation* from radiation with $\lambda = 1.06$ and 1.32 μm, as well as generation of the sum and difference frequencies within the *transparency* window. It is characterized by low-temperature drift of the synchronism directions and high radiation transparency (up to 1 GW/cm at $\lambda = 1.06$ μm and more) without any of the optical damage that restricts applications of lithium niobate. Conversion efficiency for the second harmonic exceeds 80% at the power of continuous and quasi-continuous radiation up to 20 W at $\lambda = 0.53$ μm. Pumping of the second harmonic from KTiOPO$_4$ of dye *lasers* provides stable tunable generation of picosecond pulses. It is a good electrooptic crystal, three times more efficient than potassium dihydrogenphosphate. It is applicable for *tunable lasers* via *Raman scattering of light*.

POTENTIAL BARRIER

The region of space (medium) where the positive potential energy U of particles (*quasi-particles*) exceeds their kinetic energy E. For classical particles the region $U > E$ is inaccessible, and the potential barrier is impermeable. But due to the quantum (wave) properties of particles their penetration (tunneling) through the potential barrier is possible. The probability of particles passing through the barrier (its transparency), according to the laws of quantum mechanics, is reduced exponentially with the increase in the height and width of the potential barrier. In solids, there may exist both natural (e.g., between electron states of *crystal lattice* atoms in *insulators* and *semiconductors*) and artificially created potential barriers (e.g., in metal–oxide–semiconductor (MOS) and *metal–insulator–semiconductor structures* (MIS), in *tunnel junctions*, etc.).

POTTS MODEL of phase transitions (R.B. Potts, 1952)

Generalization of the *Ising model* of an ideal lattice to an arbitrary number of distinct states. The variable σ_i is specified at each site, assuming the values $\sigma_i = 1, \ldots, q$.

The Hamiltonian of the Potts model has the form $H = -J \sum_{ij} \delta(\sigma_i, \sigma_j) - h \sum_i \delta(\sigma_i, 1)$, where J is the interaction energy; $\delta(\sigma_i, \sigma_i) = 1$, $\delta(\sigma_i, \sigma_j) = 0$ if $\sigma_i \neq \sigma_j$; and h is the "external field". The summation is performed over all pairs of nearest neighbors. For $h = 0$ the Potts model displays a *second-order phase transition* if $q \leqslant q_c$ and a *first-order phase transition* if $q_c < q$. It is known that $q_c(2) = 4$, $q_c(4) = 2$, $q_c(6) = 1$, and $2 < q_c(3) < 3$ (the dimensionality of the space is shown in parentheses). The *order parameter* in the Potts model is the diagonal traceless $q \times q$ tensor $Q_{\alpha\beta}$. The thermodynamic potential of the *Landau theory* of phase transitions of the Potts model as a function of $Q_{\alpha\beta}$ is invariant with respect to the symmetry group of the hypertetrahedron with q vertices in $(q - 1)$-dimensional space. For $q = 2$ the Potts model coincides with the Ising model, in the limit $q \to 1$ it describes the bond problem of *percolation theory*, for $q = 1/2$ in the *critical region* it is equivalent to a dilute Ising *spin glass*, and in case $q = 0$ it models the Kirchhoff lattice of resistances.

The *two-dimensional Potts model* $q = 2.3$ is exactly soluble at the *Curie point* T_C. For the square lattice we have

$$T_C = \frac{J}{k_B \ln(1 + \sqrt{q})}, \quad q = 1, 2, 3, 4.$$

All the *correlation functions* of the two-dimensional Potts model $q = 2.3$ at the phase transition point can be calculated exactly. There exists the hypothesis that the *critical indices* of the two-dimensional Potts model for $q \leqslant 4$ are:

$$\alpha = \frac{2(1 - 2U)}{3(1 - U)},$$

$$\beta = \frac{1+U}{12},$$

$$\gamma = \frac{7-4U+U^2}{6(1-U)},$$

$$\delta = \frac{(3-U)(5-U)}{1-U^2},$$

$$\nu = \frac{2-U}{3(1-U)},$$

$$\eta = \frac{1-U^2}{2(2-U)},$$

$$U = \frac{2}{\pi}\cos^{-1}\frac{\sqrt{q}}{2},$$

where α, β, γ, δ, ν, η are the critical indices of the specific heat, "magnetization", susceptibility, magnetic field, correlation radius, and anomalous dimensionality of the field of fluctuations, respectively. See also *Two-dimensional lattice models*.

POWDER MAGNET

A *permanent magnet* obtained by the compression of ferromagnetic micropowders with a high *coercive force*. Powders of iron, of an iron-cobalt alloy, and also of such highly anisotropic materials as Mn–Bi, SmCo$_5$, etc., are used for this purpose. Sometimes the *pressing* is carried out in the presence of a *magnetic field*. Various bonding agents are used for fixing the particles. See also *Hard magnetic materials*.

POWDER METALLURGY

All of the methods and technological processes involved in the manufacture of polycrystalline single or multiphased materials, and of articles made from them with the help of the *dispersion* of metals, of *pressure* and *heat treatment*. Three groups of technological processes form the basis of *powder metallurgy*: (a) dispersion of metals in solid or liquid states, or direct formation of the dispersed metals with the aid of chemical processes (reduction, decomposition, electrolysis); (b) formation of powders into blanks which approach the shape of final products, by application of pressure or use of fluid suspensions; (c) *sintering* the blanks at high temperatures and transforming the powder conglomerate to a bound polycrystalline body possessing this or that residual *porosity*.

From the physical point of view, the technological operation of caking is the most complex and important. Caking is the process of fusing a powder into a solid mass by pressure, heat, or water. Caking is performed at temperatures from 0.65 to 0.85 of the absolute *melting temperature* of the main *phase*. For a single-phase, and also for many-component systems with closely spaced component melting temperatures, caking takes place in the solid phase and the *mass transport* processes have a diffusion character. If there are low-melt components in the powder mixture, then caking may occur in the presence of a liquid phase and the body preserves its solid framework. The excessive *surface energy* of the dispersed system, which causes the presence of capillary pressures, is the motive force of caking, so upon caking the porous body reduces its volume and its average density increases. The physical theory of this phenomenon is not yet fully worked out.

Powder metallurgy is the progressive highly-organized technology which makes it possible to control the structure and properties of the powder (caked) materials obtained with the help of it. *Metals, alloys, intermetallic compounds*, high-melt compounds of *transition metals* with carbon, nitrogen, boron, silicon, nonmetallic oxygen-free high-melt compounds, and also their combinations with oxides, glass, natural *minerals, polymers* are the objects of powder metallurgy. Powder metallurgy provides the possibility of manufacturing *composite materials* combining substances of different chemical natures and possessing new properties not deducible from the sum of properties of the constituent phases (cermets, metal-glass, metal-polymer materials, *pseudoalloys* etc.). Powder metallurgy plays a special role in producing polycrystalline materials based on nonmelting (sublimating, dissociating) substances (*nitrides* of silicon, aluminum, boron) or metastable *high pressure* phases (diamond-like phases of *carbon* and of *boron nitride*). In addition, the powder metallurgy of amorphous alloys and materials possessing superelasticity and shape memory (see *Shape memory effect*) has been developed.

POWDER METALLURGY MATERIALS

See *Strength and plasticity of powder metallurgy materials*.

PRASEODYMIUM, Pr

A chemical element of Group III of the periodic system with atomic number 59 and atomic mass 140.9077, it is a *lanthanide* (rare earth). 32 isotopes are known, including mass numbers 121, 129, 130, 133 to 151. Natural praseodymium has only one stable isotope ^{141}Pr. Electronic configuration $4f^3 5d^0 6s^2$. Ionization energy 0.04 eV. Atomic radius is 0.182 nm; radius of Pr^{3+} ion is ≈ 0.104 nm, of Pr^{4+} 0.092 nm. Oxidation state +3, less often +4. Electronegativity ≈ 1.14. In a free form praseodymium is silvery-white *metal*, easily oxidized in air and becomes yellow. It has two modifications: α-Pr, β-Pr. Below ≈ 1080 K α-Pr with double close-packed hexagonal lattice is stable; $a = 0.36723$ nm, $c = 1.1834$ nm under normal conditions, interlacing of hexagonal atomic layers according to the type ABAC...; space group $P6_3/mmc$ (D_{6h}^4). Above ≈ 1080 K up to $T_{\text{melting}} = 1206$ K β-Pr is stable with the body-centered cubic lattice with $a = 0.413$ nm (extrapolated value) after hardening, space group $Im\bar{3}m$ (O_h^9). There are contradictory reports that its polymorphous transformation to a (possibly metastable) allotropic modification with face-centered cubic lattice, space group $Fm\bar{3}m$ (O_h^5), precedes (at ~ 860 K) the transition α-Pr $\rightarrow$ β-Pr; $a = 0.488$ (extrapolated value) after hardening. Density is 6.475 to 6.770 g/cm^3 (at room temperature), $T_{\text{boiling}} \approx 3460$ K. Bonding energy of praseodymium at 0 K is 3.9 eV/atom. Heat of melting ≈ 8.5 kJ/mole; heat of sublimation 330.8 kJ/mole; specific heat is 0.192 kJ·kg^{-1}·K^{-1} (at 298 K), Debye temperature is ≈ 110 K; coefficient of the linear thermal expansion of praseodymium polycrystal $6.5 \cdot 10^{-6}$ K^{-1} (in the range of temperatures 100 to 1073 K); coefficient of heat conductivity of α-Pr polycrystal is ≈ 30.4 GPa at zero external pressure and room temperature, Young's modulus is ≈ 34 GPa at room temperature, shear modulus is 13.53 GPa, Poisson ratio is 0.305; tensile strength is 0.110 GPa, relative elongation is 12% (for cast sample at 293 K). Vickers hardness: for cast praseodymium from ≈ 40 HV, for forged praseodymium 76 HV. Activation energy of self-diffusion of atoms in β-Pr at melting temperature is 124 kJ/mole (i.e. 1.29 eV). Effective cross-section of thermal neutron trapping is 11.2 barn. Electrical resistivity of α-Pr polycrystal is ≈ 670 nΩ·m at 296 K, temperature coefficient of electrical resistance is 0.00171 K^{-1} (at 273 K). Electron work function of polycrystal is 2.7 eV. In α-Pr at the Néel temperature $T_N = 24.5$ K the magnetic order–disorder phase transition takes place from paramagnetic state to the sine-wave modulated antiferromagnetic state with collinear position of magnetic moments perpendicular to the principal axis $\underline{6}$ in only half of the hexagonal atomic planes. There are some indications of a transition of nonequilibrium close-packed cubic modification of praseodymium to the ferromagnetic state at the Curie temperature $T_C = 8.7$ K. The magnetic susceptibility of polycrystalline praseodymium $\sim 4000 \cdot 10^{-6}$ CGS units; nuclear magnetic moment of ^{141}Pr isotope is 3.92 nuclear magnetons. Praseodymium is used as an alloying component of steels, of nonferrous metals, e.g., of magnetic alloys with nickel, cobalt, and others.

PRE-BREAKDOWN LUMINESCENCE

See *Electroluminescence.*

PRECIPITATE, COHERENT

See *Coherent precipitate.*

PRECIPITATES in solids (fr. Lat. *praecipitatus*, thrown down)

Extraneous particles within a solid, which consist of large number of molecules. Their appearance is caused by lack of phase equilibrium, e.g., upon decomposition of a solid solution (see *Alloy decomposition*). The *precipitation process* involves the successive appearance and growth of particles of a second phase. Structural defects, and in particular *dislocations*, are loci for precipitation. Precipitates may consist of the matrix material, but possess another structure. Also inclusions of other substances, or compounds of matrix elements with impurities, can also become precipitates. Precipitates may be present in any material – in *metals, semiconductors, ionic crystals, molecular crystals* – independently of their structural state (mono- or polycrystalline, amorphous state). The initial form of precipitates in metals is called a *Guinier–Preston zone*. The shape of precipitates may be of various types: spherical, plate-like, needle-like,

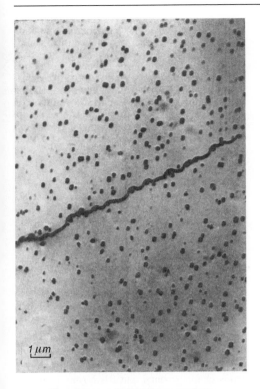

Phosphorus precipitates in germanium.

etc., depending on the system composition, and on the conditions of precipitate formation. Precipitates are sources of *internal stresses* which are minimal for spherically symmetrical shapes, and are maximal for the most anisotropic precipitates. Due to these stresses, precipitates inhibit the motion of dislocations thus causing so-called *precipitation hardening* (see *Precipitation-hardened materials*). The figure presents an electron microscope picture of phosphorus precipitates in germanium. Precipitates distributed within a volume are visible due to diffraction contrast in the distortion field of the associated lattice. The shape of a dislocation is influenced by its retardation at a precipitate.

PRECIPITATION-HARDENED MATERIALS

Materials related to *disperse structures* (composites), for which *hardening* is caused by the presence of particles of *refractory materials* (oxides, nitrides, borides) or particles of high-melt metals (*molybdenum, tungsten*). Unlike aging alloys (see *Alloy aging*), in precipitation-hardened materials the particles of the second phase are introduced artificially. Usually they have a non-coherent *phase interface* with the surrounding medium, and do not interact with it chemically, so the speed of *coalescence* of such particles is low. These materials reach maximum hardening at a particle size of about 10 to 30 nm, with a narrow distribution according size, an optimum distance between them ($\sim$50 to 100 nm), and at bulk contents of about 10%. Nowadays precipitation-hardened materials are obtained by the method of *powder metallurgy*. The material based on *aluminum* containing about 7% of Al_2O_3 and having an *ultimate strength* about 300 GPa at 20°C and a *yield limit* of 70 GPa at 350°C, may be cited as an example. This material is characterized by high *corrosion resistance* and *radiation resistance*.

PRECURSOR, forerunner

A burst of a field that has separated from the main part of a pulse and moves faster than it. It arises from frequencies in the signal spectrum for which the *group velocity* has the maximum value. On the *dispersion law* curve these frequencies correspond to: (1) sections adjacent to the asymptotes $\omega = ck$, where c is the *sound velocity* or *speed of light* in the given medium; (2) the points of inflection ω^* of the dispersion law curve $\omega = \omega(k)$, corresponding to the tiny, high-frequency *Sommerfeld precursor* (A. Sommerfeld), and the larger and lower-frequency *Brillouin precursor* (L. Brillouin). The amplitude of the Sommerfeld precursor, described in the first approximation by a first-order Bessel function, is negligibly small due to the smallness of the specific weight of high frequencies in the pulse spectral function, because the latter has a peak at the carrier frequency ω_0 and drops sharply at its periphery. The velocity of the Sommerfeld precursor is equal to c. After the Sommerfeld precursor the Brillouin precursor arrives with the velocity $d\omega(\omega^*)/dk$. Its envelope is described by the Airy function with a tail on the side of the fundamental signal, with its value reduced in proportion to the spacing of ω_0 from ω^*. In the process of traversing the distance x the Brillouin precursor diffuses slower than the remaining signal ($x^{-1/3}$ instead of $x^{-1/2}$), so at high x and low $\omega_0 - \omega^*$ the Brillouin precursor is the highest spike of the diffused pulse.

Although precursors were predicted in 1914, they were not observed until the sixties with the spreading of ultrasonic waves within a *crystal*, and in coaxial lines filled by a ferromagnetic material. In 1982 the *magnetostatic precursor* was discovered. An *exciton precursor* has been described theoretically. In nonlinear *shock waves* the *elastic precursor* of the main front of a *plastic deformation* wave may be observed.

PREDISSOCIATION OF CHEMICAL BOND

See *Chemical bond predissociation*.

PRESSING, compaction, molding

Process of material treatment by *pressure* in order to bring about its thickening (powder materials, *ceramics*); changing of its initial shape (cast materials); thickening with the introduction of polymerization processes in plastics (see *Polymeric materials*), *rubbers*. The operation of pressing is used in some technological processes: formation of products from powders in press molds in *powder metallurgy*; formation of the materials in plastics and rubber manufacturing in special heated press molds, where cooling of the material under pressure fixes the necessary shape of the products, *extrusion* (squeezing out) of materials from a closed cavity through a die with the shape of a curved hole to provide rods or tubes of a required shape.

The widest areas of pressing applications are the production of parts in machine fabrication, formation in powder metallurgy, and the production of plastics and rubbers. The following processes of pressing are used: isostatic (hydrostatic) (powder metallurgy); dynamic hot (powder metallurgy), explosion (*composite materials*; inducing *phase transitions* in some compounds and alloys); *hydroextrusion* (squeezing out of tools of high-speed steel), *casting* under pressure (obtaining cast parts of accurate size); pressing at room temperature and at higher temperatures (machine fabrication), etc.

In addition, pulsed and high-energy methods of pressing have been developed that help to solve a number of specific problems of metal science, e.g. hot dynamic pressing provides pore-free materials from powders with properties superior to those of cast products: hydroextrusion provides wear resistant cutting material of high speed steel; *explosion pressing* converts the wurtzite modification of *boron nitride* into the densely packed hexagonal one with properties close to those of *diamonds*, and it furnishes reinforced materials based on aluminum with improved mechanical properties; extrusion at elevated temperatures not only forms the necessary profile of the product, but it also improves its properties.

PRESSURE

A scalar (P) equal to the normal component of a force F_n per unit area S

$$P = \lim_{\Delta S \to 0} \frac{\Delta F_n}{\Delta S} = \frac{\mathrm{d} F_n}{\mathrm{d} S}.$$

The pressure characterizes the *state of stress* of a continuous medium. For a medium in motion possessing internal friction the pressure is minus one third of the sum of diagonal components of the *stress tensor*. Pressure, density and temperature are the three basic macroscopic *state variables*. The basic unit is Pascal (Pa) which, in terms of fundamental units, is $\mathrm{N/m^2}$. In meteorology 10^5 Pa $= 0.987$ atm, and for a mercury column 1 mm $= 133.3$ Pa. In modern engineering pressures in use have the ranges: ultrahigh vacuum (10^{-11} to 10^{-7} Pa), high vacuum (10^{-4} to 0.1 Pa), atmospheric ($\sim 10^5$ Pa), high pressure (10^9 to 10^{10} Pa). The latter range is used for the *high pressure* treatment of materials. The pressure at the center of the earth is about 3.4×10^{11} Pa. See also *High pressure cell*.

PRESSURE, HIGH

See *High pressures*.

PRESSURE, MAGNETIC

See *Magnetic pressure*.

PRIMITIVE PARALLELEPIPED, elementary parallelepiped

A parallelepiped of minimal volume, its vertices being *lattice* points. For a given lattice, a primitive parallelepiped may be selected in many ways. To describe lattices unambiguously, one may introduce a *reduced lattice* selected by symmetry considerations. For all crystal systems (with the exception of monoclinic and triclinic ones), the reduced lattice is the *Bravais lattice*. In other cases the reduced lattice is usually built on the three shortest noncoplanar lattice vectors.

PRINCIPAL AXES OF A TENSOR

Axes of a coordinate system in which a tensor is written in its simplest form. Examples are a symmetric tensor A_s of rank two in diagonalized form or an antisymmetric tensor A_a of rank two in the form of a single-nonzero pair of components. An arbitrary second rank tensor $A = A_s + A_a$ in any three-dimensional (e.g., Cartesian) coordinate system can be characterized by either nine components or $3 + 1$ principal values of A_s, A_a, and $3 + 2$ angles which determine the orientation of the principal coordinate system of A_s and of the axial vector A_a, respectively. The symmetric tensor A_s can also be decomposed into an invariant scalar part (its trace) and a two-component zero trace tensor in its principal axis system.

PRINCIPAL MOMENTS OF INERTIA

Moments of inertia relative to principal axes of inertia. In a general case, the *angular momentum* L of a solid relative to a fixed point is a linear function of the components of the angular velocity ω of solid: $L_\alpha = \sum_\beta I_{\alpha\beta}\omega_\beta$, where the summation is over $\beta = 1, 2, 3$. The quantities $I_{\alpha\beta}$ form the *moment of inertia tensor*. In a coordinate system fixed in the body, $I_{\alpha\beta}$ are constants which depend on the body mass distribution and coordinate axes selection. With an appropriate choice of coordinate axes one can rotate the moment of inertia tensor to make the off-diagonal elements $I_{\alpha\beta}$ with $\alpha \neq \beta$ vanish. Then the relationship between L and ω simplifies: $L_\alpha = I_{\alpha\alpha}\omega_\alpha$, $\alpha = 1, 2, 3$. The quantities $I_{\alpha\alpha}$ are called the principal moments of inertia of the body; and the corresponding coordinate axes are called *principal axes of inertia*.

PRINCIPAL SECTION OF A CRYSTAL

See *Crystal optics*.

PRISMATIC DISLOCATIONS

Planar dislocation loops with *Burgers vectors* directed along the normal to the plane of the loop position. Prismatic dislocations may appear upon the collapse of a vacancy disk of single-atom thickness, located along a close-packed plane (*vacancy prismatic dislocations*), or upon the formation of a disk of interstitial atoms of single-atom thickness (*interstitial prismatic dislocations*). Such dislocations are energetically more favorable than corresponding *pores* or inclusions of interstitial atoms when the following condition is satisfied:

$$\left(\frac{R}{b}\right)^{1/3} > \frac{GB}{12\gamma}\pi^{-2/3}(1-\nu)^{-1}\ln\left(\frac{2R}{b}\right),$$

where R is the radius of the prismatic dislocation, b is the modulus of the Burgers vector, ν is the *Poisson ratio*, G is the *shear modulus*, and γ is the pore *surface energy*. For $\gamma \approx Gb/10$, $\nu \approx 1/3$. A prismatic dislocation is stable for $R > 1.3b$. Prismatic dislocations are observed at oversaturation of *point defects*, mainly *vacancies*, which appears during *quenching*, irradiation, *plastic deformation*, etc. The interaction of the rectilinear segment of a *screw dislocation* with a prismatic one during a *slip* process can lead to the formation of a *helical dislocation*.

PRISM, POLARIZING

See *Polarizing prisms*.

PROBE

See *Electrical probes, Electron probe, Microprobe*.

PROGRAMMED HARDENING

One of the procedures used to *harden* crystalline bodies. It is based on the processes of diffusion and microshear *stress relaxation*, which are responsible for improving the degree of structural perfection of the lattice. The distinct features of programmed hardening as compared to deformation hardening are that it achieves a noticeable hardening by virtue of a low degree of residual *strain* (up to 0.2%), while lowering the critical internal energy and increasing the specimen thermodynamic and mechanical stability as a result. All these effects are accompanied by simultaneous growth of the specimen *strength* and *plasticity*. The basic prerequisite for implementing programmed hardening is to load the specimen at a low rate, following a certain temperature and load routine, while maintaining an equality between the growing external forcing and the internal resistance of the material to deformation. In that case the spectrum of weak lattice sites opens up step by step, and they harden due to diffusion processes that develop in them (see *Diffusion*). Also local microshears occur around stress concentrators (see *Stress concentration*) without violating the specimen continuity. Among the principal mechanisms of programmed hardening are changing the individual properties of *dislocations* (their segment length is shortened), and the degree of their splitting (energy of stacking faults is lowered by aggregation of impurity atoms at dislocations). Also both the configuration and the density of the attachments of dislocation centers change. Dislocations redistribute at their accumulation areas near obstacles, and their overall density decreases. Pores and submicrocracks heal and dissolve. The principal parameters of loading in the course of programmed hardening are the temperature, rate of loading and the end value of the stress. The achieved programmed hardening then persists during various subsequent tests within a wide temperature range. The hardening may also result from thermal, radiative, electromagnetic actions that follow certain programs. Steel may also be tempered, that is toughen hardened by softening it and then reheating it to a temperature below the eutechtoid temperature.

PROJECTION OPERATOR

An operator P defined on a linear space, that possesses the property $P \cdot P = P$. The inverse operator P^{-1} does not exist. An expansion of either a vector or a tensor field into invariant components carried out with the aid of projection operators plays an important role in the physics of solids. The simplest projection operator is the x-axis projector: $Pr = ix$. These operators are employed to compute the resonance absorption *line shape* for a paramagnetic system with a multicomponent *magnetic resonance* spectrum. The projection operators P_α, P_β isolate that part of the spin tensor Q which has its matrix elements only between the functions $|\alpha\rangle$, $|\beta\rangle$, so that $\widehat{Q} = \sum_\alpha \sum_\beta P_\alpha \widehat{Q} P_\beta$. Thus a specific resonance line is selected from the total spectrum, and its shape is calculated.

To compute the *band structure* of a solid by the *pseudopotential method* one may use projection operators to select the nonlocal part of the potential acting on band electrons. Expanding the potential into components invariant with respect to the eigenvalues of the *orbital angular momentum l* ($\widehat{V} = \sum_l P_l \widehat{V} P_l$) makes it possible to separate the so-called local part (independent of l) from the nonlocal part (dependent on l) of the potential.

Within the theory of irreversible processes the method of projection operators is used to develop the *kinetic equation* for the *density matrix* of a dynamic subsystem, σ, from the *Liouville–von Neumann equation* for the density matrix of the whole system, ρ. To arrive at the description of the behavior of a dynamic subsystem the projection operator is used to separate either the "significant" part from the tensor ρ, which is diagonal in the coordinates of the dissipative subsystem (*R. Zwanzig technique*, 1960), or the quasi-equilibrium part of the statistical operator. All these operators have been shown to be equivalent to each other.

PROMETHIUM, Pm

A radioactive chemical element of Group III of the periodic system with atomic number 61. It belongs to *lanthanides*. There are 23 isotopes known with mass numbers 132 to 154. [145]Pm is the most stable (half-life $T_{1/2} = 17.7$ years), but it is quite rare, and [147]Pm ($T_{1/2} = 2.6234$ years) is of the greatest practical significance since it forms in considerable quantities in nuclear reactors. Electronic configuration is $4f^5 6s^2$. Atomic radius of promethium is 0.182 nm; radius of Pm^{3+} ion is 0.106 nm; oxidation state is +3. Electronegativity is ≈ 1.17.

Promethium is a light-yellow *metal*. Metallic promethium possesses a double hexagonal close-packed crystal lattice; $a = 0.365$ nm, $c = 1.165$ nm. Density is 7.26 g/cm^3, $T_{\text{melting}} \approx 1400$ K, $T_{\text{boiling}} \approx 3470$ K. Heat of melting is 8.8 kJ/mole; specific heat 27.592 J·mole^{-1}·K^{-1}. Linear thermal expansion coefficient is $9 \cdot 10^{-6}$ K^{-1}.

[147]Pm is introduced into luminescent materials of long-term (several years) persistence, and it is used as a source of radiation in atomic batteries. A promethium battery with the dimensions of a button provides the nominal output power 20 mW; and is not sensitive to variations of temperature, pressure and other external effects.

PROPORTIONALITY LIMIT of stress

Limiting stress producing elastic *strain* of a solid below which the applied stress and the relative strain caused by it are still proportional to each other, i.e. *Hooke's law* is still valid. The limit of proportionality may coincide with the *elastic limit* or it can be below it, when for stresses close to the elastic limit, the deviation from the linear dependence between the applied stress and the relative elastic strain is observed. The basic unit of measurement is the pascal (Pa), which equals $1\,N/m^2$.

PROTACTINIUM, Pa

A radioactive chemical element of Group III of the periodic system with atomic number 91; it belongs to the *actinides*. All the protactinium isotopes (19 are known) are radioactive. ^{231}Pa (half-life $T_{1/2} = 3.276 \cdot 10^4$ years) is the most widely spread in nature, its atomic mass is 231.0359; laboratories of the world have about 150 g of ^{231}Pa. In addition ^{234}Pa ($T_{1/2} = 1.17$ min) and its nuclear isomer ^{234}Pa ($T_{1/2} = 6.70$ h) are contained in negligible quantities in uranium ores. Among artificial isotopes β-radioactive ^{233}Pa ($T_{1/2} = 27.0$ days) formed by the transformation ^{232}Th $\rightarrow$ (n, γ) ^{233}Th $\xrightarrow{\beta^-}$ ^{233}Pa has the widest application. Electron configuration is $5f^2 6d^1 7s^2$. Atomic radius is 0.163 nm. Radius of Pa^{3+} ion is 0.113 nm, of Pa^{4+} is 0.098 nm, and of Pa^{5+} is 0.089 nm. Oxidation states are $+3, +4, +5$. Electronegativity is 1.30.

Protactinium is a silvery-gray *metal* existing in two modifications (α-Pa, β-Pa). Below 1443 K α-Pa is stable with tetragonal crystal lattice; $a = 0.3931$ nm, $c = 0.3236$ nm. Above 1443 K it transforms to β-Pa with body-centered cubic lattice; $a = 0.5019$ nm. Density of α-Pa is 15.32 g/cm^3 (at room temperature), of β-Pa 12.13 g/cm^3; $T_{melting} \approx 1848$ K, $T_{boiling} \approx 4640$ K. Heat of melting is 12 kJ/mole, heat of evaporation is 552 kJ/mole, specific heat is 27.6 $J \cdot mole^{-1} \cdot K^{-1}$, resistivity is 0.12 $\mu\Omega \cdot m$, below 2 K it is superconducting; temperature coefficient of linear expansion is $11.2 \cdot 10^{-6}\,K^{-1}$. Hardness of metallic protactinium is close to that of *uranium*. ^{231}Pa is very toxic. Of potential practical significance is the formation of fissile ^{232}U, a nuclear fuel, by β^--decay of ^{233}Pa which has ^{232}Th as a source material.

PROTON MICROSCOPY, proton diffractometry

Technique for studying atomic structure with the use of accelerated proton beams. It is based on the existence of two types of *orientation effects* observed during the interaction of a beam of charged particles with a crystal: channeling of the proton beam in directions of low Miller indices (see *Crystallographic indices*), that is in the directions of dense packing, and the blocking (see *Channeling of ion beams*) or *shadow effect* in the same directions.

The scattering of a channeled particle in the direction of an atomic row results in a blocking effect that simulates the appearance of characteristic minima of intensity (shadows) in the angular distribution of the proton beam exiting the crystal. Particle outputs in directions with low Miller indices are always lower than in the directions with lower packing density. The appearance of a shadow in the direction of a crystallographic axis (axis shadow) is explained by deviation of the protons that had initially travelled in that axis direction due to the internal atomic electric field of the particles. Flat shadows characterize ranges of lowered scattering intensity for protons propagating in crystallographic plane directions and have the shape of straight lines.

Channeling effects and shadows are observed with the help of a *proton microscope* consisting of a proton accelerator (from 200 keV to 5–10 MeV), a goniometer crystal holder, and a detector of scattered protons. If a photographic plate replaces a detector in the beam path, the so-called *proton image* of a crystal may be obtained by exposing and then processing that plate. It yields a shadow image over a wide solid angle. Dark direct lines in it are the lines along which crystallographic planes of the irradiated *monocrystal* cross the plane of the photographic plate, while spots (points where dark lines cross) are points where the axes with different $\{hkl\}$ indices cross that plane. The distribution of lines and spots in the proton image depends on the crystal structure and the geometric layout of the experiment. The intensity distribution within a single shadow (either an axial or a plane) is determined by many factors, for example the composition and structure of the crystal, the type and energy of the particles, the type and number of *defects* in a crystal. Hence the proton image may yield information on the type of *crystal lattice*, its spatial orientation (to an accuracy of 0.1°–0.5°), on the degree of perfection of both the bulk and surface layers of the crystal. When studying *polycrystals* or *textures* with the help of that technique one may simultaneously obtain proton images from two or three adjacent monocrystals, and deduce their mutual orientation.

PROUSTITE (after J. Proust, 1754–1829)

Ferroelectric-semiconductor, the main representative of the chalcogenide family of the type $A_3^I B^V C_3^{VI}$, which combines sulfur salts of silver – proustite Ag_3AsS_3 and *pyrargyrite* Ag_3SbS_3 known as light- and dark-red silver ores, and also Ag_3AsSe_3 and $AgSbSe_3$, together with sulfur salts of copper – tennantite $Cu_{12}As_4S_{13}$, tetrahedrite $Cu_{12}Sb_4S_{13}$, and also $Cu_{12}As_4Se_{13}$ and $Cu_{12}Sb_4Se_{13}$. *Synthetic monocrystals* of proustite and of its closest isostructural analog pyrargyrite were obtained in the first half of the sixties. At room temperature, the crystallographic space symmetry group of proustite is $R3c$ (C_{3v}^6); the unit cell contains two formula units. The weak bond of Ag^+ ions with groups $(AsS_3)^{3-}$ forming the crystalline frame determines the presence of mixed ion–electron electrical conductivity of proustite and of the members of its family. Pliability and polarization due to the dimensional and electronic properties of Ag and S atoms predetermines the presence of high values of nonlinear optical, acousto-optic and electrooptic coefficients of proustite and pyrargyrite. With the reduction of temperature in Ag_3AsS_3 *phase transitions* are observed, the sharpest being the *first-order phase transition* near 24 to 28 K associated with the symmetry reduction to C_1, accompanied by the stepwise variation of structural and phase parameters. Low-temperature phase transitions are also observed in Ag_3SbS_3 and in Ag_3AsSe_3. *Monocrystals* of proustite and of pyrargyrite have a broad window of *transparency*, from −0.6 to 13 μm at the band gap optical width ∼2 eV. The following applications of synthetic proustite monocrystals are known: optical frequency mixing, transformation of IR radiation into visible, and detection of picosecond radiation pulses of a CO_2 laser. At low temperatures, due to the high value of the pyroelectric coefficient (see *Pyroelectric effect*) to

specific heat ratio, the crystals of Ag_3AsS_3 and Ag_3SbS_3 may be used as materials for highly-sensitive *pyroelectric radiation detectors* in the IR band with operating temperatures of 6 and 11 K, respectively.

PROVOTOROV THEORY (B.N. Provotorov, 1961)

A theory of *nuclear magnetic resonance* that, in contrast to earlier theories (see *Kubo–Tomita method*, *Bloch–Redfield theory*), does not assume weak saturation a of spin system by a radio-frequency magnetic field, B_1, and generalizes the concept of spin temperature (see *Level population*) to the case of several *heat sinks*: the Zeeman Hamiltonian $H_Z = g\mu_B \boldsymbol{B}_0 \cdot \boldsymbol{I}$ and the reservoir Hamiltonian of *spin–spin interactions*, H_d. It is also assumed within this theory that the quasi-equilibrium temperatures of these reservoirs may not necessarily coincide with each other, while the quasi-equilibrium *density matrix* has the form $\sigma = Z^{-1} \exp[-\beta_Z H_Z - \beta_d H_d]$. Saturation by the radio-frequency magnetic field $B_1(t)$ results in a coupled evolution of the two inverse temperatures, $\beta_Z = 1/(k_B T_Z)$ and $\beta_d = 1/(k_B T_d)$. Equations for $\dot{\beta}_Z$ and $\dot{\beta}_d$ are called the *Provotorov equations*. Solving these equations yields a satisfactory description of the change in *magnetization* and *line shape* for the magnetic resonance in a specimen saturated by the high amplitude field $B_1 < (g\mu_B)^{-1} H_d$.

PROXIMITY EFFECT in superconductors

Weakening (suppression) of the superconducting properties of a *superconductor* in the vicinity of its contact with a normal metal, or the appearance of *superconductivity* in a normal metal situated near a superconductor. The proximity effect also refers to superconducting fluctuations induced in a normal metal in contact with a superconductor. This appears in a thin layer with a thickness of the order of a *coherence length* ξ near the interface of a superconductor with a normal metal, or in *thin films* with a thickness $d \sim \xi$. Owing to this effect, in the superconductor the *order parameter* and the bandwidth of the spectrum of *quasi-particles* decrease, while in the metal a gap of finite width is induced. In the vicinity of the *critical temperature* T_c, the proximity effect can be described with the help of the Ginzburg–Landau equations (see *Ginzburg–Landau theory of superconductivity*) by applying the appropriate boundary conditions.

PSEUDOALLOYS

Alloys which cannot be prepared by melting together their components; e.g., *silver* (or *copper*) and *tungsten* (or *molybdenum*) are almost mutually immiscible in both the liquid and solid states. The methods of *powder metallurgy* make it possible to produce pseudoalloys that combine the *hardness* and *strength* of W and Mo with the *electrical conductivity* and *thermal conductivity* of Ag and Cu. Some pseudoalloys acquire a *texture* after *plastic deformation*, and a corresponding anisotropy of properties. For example, the electrical conductivity of pseudoalloys in a direction parallel to the *strain* direction is larger than the corresponding transverse conductivity. This feature is used to produce contacts in relay switches and other instruments. Another group of pseudoalloys combines metals with noble gases. The latter are practically insoluble, however, *ion implantation* of metals makes it possible to obtain metastable substitutional *solid solutions* of He, Ne, Ar, Kr and Xe in various metals. The heating of pseudoalloys results in the decomposition of highly oversaturated solid solutions (see *Alloy aging*), accompanied by the formation of microcrystals of the inert gas. While free Kr has a *melting temperature* of 116 K, its microcrystals in nickel only melt within the interval 825–875 K because of the high *pressure* they produce in the matrix. Heating pseudoalloys beyond that temperature results in the generation of the gas phase. Pseudoalloys containing micropores (see *Pores*) with inert gas in them display anomalous thermal conductivity during the transition of that gas from its liquid to its gas phase. Further heating results in the swelling of pseudoalloys. Triple and quadruple pseudoalloys of such systems as Al–Ar–O and Al–Mn–Ar–O have found their practical application as materials with controlled thermal properties, and also as materials for producing gaskets that feature higher plasticity than sintered *aluminum alloys* due to greater dispersion of their oxide phase.

PSEUDODEGENERACY

The presence of two or several electron states of different energy that are appreciably mixed by the *vibronic interaction* (see *Jahn–Teller effect*).

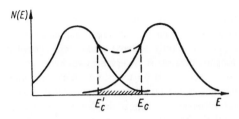

Overlap of electron states in adjacent energy bands producing a pseudogap; truly localized states are shown hatched.

PSEUDOELASTICITY

See *Superelasticity*.

PSEUDOGAP of density of electron states, Mott pseudogap (N.F. Mott, 1966).

A minimum in the *density of electron states* found in the region of the electronic spectrum of a disordered condensed medium, which corresponds to overlaps between the *valence band* and the *conduction band*. Structural perturbations produce *tails in the density of states*, and a combination of levels in the *band gap* that correspond to localized states of electrons, all this resulting in the appearance of a pseudogap, i.e. yielding a nonzero density of states. The position of the pseudogap is fixed by the mobility edges, E_c' and E_c, that separate the localized and the delocalized states of the carriers from each other (see Fig.). See also *Elementary excitation spectra of disordered solids*, *Amorphous semiconductors*.

PSEUDOMONOCRYSTAL (fr. Gr. $\psi\varepsilon\upsilon\delta o\varsigma$, deception, mistake, and monocrystal)

A crystalline sample presenting a combination of monocrystals of a single phase, such that, to the accuracy of a displacement, their mutual orientation corresponds to transformations of a *point group* of symmetry of another phase (matrix). Pseudomonocrystals form in the process of *structural phase transitions* (atomic ordering, peritectoidal and eutectoidal reactions, decay of supersaturated solid solutions, *martensitic transformations*, etc.). A precondition for the formation of a new phase consists in having a relatively defect-free *crystallite* (monocrystal) of the matrix phase (see *Monocrystal growth*). Single phase, two-phase, etc. pseudomonocrystals may be identified. An example of a single-phase pseudomonocrystal is a fully ordered phase with oriented domains present. A two-phase pseudomonocrystal is exemplified by tempered *steel* containing *martensite* and the untransformed matrix (residual *austenite*). In terms of the degree of structural similarity between the matrix and the new phase, and also in terms of the character of the mutual orientation of their crystal lattices, the *reciprocal lattice* of a pseudomonocrystal may be described by the vectors commensurable to the axial vectors of the reciprocal lattice of the matrix that formally corresponds to a monocrystal, or by incommensurable vectors approaching the reciprocal lattice of a textured polycrystal (see *Texture*). A random or an orderly domination of the relative share of crystals of the new phase is possible in several specific orientations, which may result in conversion of the pseudomonocrystal into a twin or into a monocrystal itself (in the limiting case).

PSEUDOPOTENTIAL METHOD

One of the methods used to solve the problem of determining the spectrum of electronic states in a potential produced by a nucleus (nuclei) and screened by electrons. Formally, the pseudopotential method is based on the fact that the spectrum of intrinsic states does not specify a well-defined potential, so that one may find a (weak) potential (the so-called *pseudopotential*) such that its spectrum coincides with the spectrum of the actual potential (which is quite strong close to the nuclei). The weakness of the pseudopotential permits one to apply to this case the powerful apparatus of perturbation theory. The pseudopotential method finds its widest use in the *band theory of solids* (J.C. Phillips, L. Kleinman, 1959). The most comprehensive such method involves orthogonalization of the wave functions of valence electrons ψ_v with respect to wave functions of the electrons of the ion core ψ_c. Specifically, the pseudopotential of orthogonalized plane waves, called the *Phillips–Kleinman pseudopotential*, is defined as the sum of the true potential and the operator W that acts upon ψ_v according to the equation $W\psi_v(r) = \sum_c (E_v - E_c)\psi_c(r)\langle\psi_c|\psi_v\rangle$, where the index c enumerates the electron states of the core. If one treats W as a nonlocal energy-dependent potential then it is always repulsive, i.e.

it compensates the attractive potential of the ion. The spectrum of the Hamiltonian with an added potential W coincides with the actual spectrum of valence electrons. The eigenfunctions of such a Hamiltonian are called pseudowave functions. If φ is a *pseudowave function of the Hamiltonian* of orthogonal plane waves, then the wave function $\psi = \varphi(r) - \sum_c \psi(r)\langle\psi_c|\varphi\rangle$ is an eigenfunction of the true Hamiltonian. The pseudopotential of the orthogonalized plane waves is rarely used in practice. More often empirical model pseudopotentials are used with the operator $W(r)$ of a simple (often local) form. Moreover, it depends on several parameters that are selected to obtain a correspondence of certain calculated values with experiment, e.g., cross-sections of the *Fermi surface*, *elastic moduli*, etc. An important characteristic of pseudopotentials is their transferability. Transferability is considered high when the pseudopotential, adjusted for a given element in a given compound, and for certain properties of that element, describes well the same element in other systems with other physical properties. To improve the transferability one often utilizes an "incomplete" pseudopotential, a "bare" one, i.e. an unscreened pseudopotential. When calculating the electronic structure of a given system one can adapt a model pseudopotential screened in the spirit of *linear response* theory (which is permissible because of the weakness of the pseudopotential), usually with the aid of the *dielectric constant* of a homogeneous electron gas.

The best known model pseudopotentials are:

1. The *Ashcroft pseudopotential* (N.W. Ashcroft), or *empty core pseudopotential*:

$$W_{ps}(r) = \begin{cases} 0, & r < R_c, \\ -\dfrac{z}{r}, & r > R_c, \end{cases}$$

where R_c is a model parameter.

2. The *Heine–Abarenkov pseudopotential* (V. Heine, I.V. Abarenkov, 1964):

$$W_{ps}(r, l) = \begin{cases} -A_l, & r < R_c, \\ -\dfrac{z}{r}, & r > R_c, \end{cases}$$

where R_c and A_l are model parameters. This pseudopotential variously affects electrons with differing angular momenta.

3. The *Shaw pseudopotential* (R.W. Shaw, 1968):

$$W_{ps}(r, l) = \begin{cases} -\dfrac{z}{R_l}, & r < R_l, \\ -\dfrac{z}{r}, & r > R_l. \end{cases}$$

A separate class of pseudopotentials comprise *resonance pseudopotentials*. They are used for transition metals in which the scattering of d-electrons by the ionic potential has a resonance character. In contrast to common pseudopotentials, these are not weak. The corresponding pseudopotential wavefunction is defined by the transformation:

$$\psi = \varphi - \sum_c \langle \psi_c | \varphi \rangle \psi_c - \sum_d \langle \psi_d | \varphi \rangle \psi_d$$
$$+ \sum_d \frac{\langle \psi_d | \Delta | \varphi \rangle}{(E_d - E)} \psi_d,$$

where d denotes the d-electron states and Δ is the so-called *hybridization potential*. The most popular are the so-called *first-principle pseudopotentials* (also called *nonempirical potentials*) that preserve their norm. They are based on *density functional theory*, and are constructed according to the following scheme: a full calculation is made for some system, usually an isolated atom; then an inverse problem is solved, i.e. a pseudopotential is constructed that yields the energy spectrum for the valence electrons under an additional condition that the wave functions and the pseudowave functions coincide for all radii $r > R_c$ (where R_c, so-called core radius, is a free parameter). It is then assumed that the newly found screened potential is a sum of the bare pseudopotential, a Coulomb (Hartree) potential (pseudopotential) of the valence electrons, and their exchange-correlation potential (Hartee–Fock). Starting from this basis the bare pseudopotential is determined self-consistently. The resulting pseudopotential is then used to calculate the electronic structure and total energy of a given element in ions, molecules and solids. A pseudopotential may be utilized with exchange-correlation functionals different from those used to initially construct the pseudopotential. First-principle pseudopotentials feature high transferability and high accuracy, particularly when calculating energy characteristics. Since the pseudopotential method naturally takes into account the nonsphericity of a crystal potential, the first-principle pseudopotential has been applied within the framework of density functional theory to calculate the stability of crystal structures, their electrical properties, and their phonon spectra.

Pseudopotentials may be constructed in a similar manner within the *Hartree–Fock method*, and generally within any method used to describe multielectron systems.

PSEUDO-STARK SPLITTING

See *Stark effect at impurity centers*.

PULSED ANNEALING OF SEMICONDUCTORS

A high temperature treatment of semiconductor specimens and instrumental structures which is characterized by sharp fronts of heating τ_h and cooling τ_c, in practice without holding the temperature at the maximum value. Values τ_h and τ_c vary over a wide range. As a rule, they are set by the conditions of conserving the bulk properties, without any noticeable *diffusion* of impurities, and this provides the possibility of fusing surface layers while conserving the crystal structure of the bulk. In practice, the term "pulsed annealing of semiconductors" covers process durations from 10^{-12} s to several seconds. For pulses shorter than a few microseconds the most efficient results are achieved with exciting pulse energies (light, electrons, ions and SHF) sufficient for melting the surface layer. For longer pulses the target heating is almost uniform due to *thermal conductivity*. Particularly important results were obtained by combining pulsed annealing with *ion implantation* of impurities. The nonequilibrium introduction of impurities together with nonequilibrium *annealing* is the most efficient technique for obtaining highly doped, structurally perfect semiconducting layers, including those saturated by impurities above the ultimate *solubility* limit. The pulse annealing method is widely used for the restoration of a crystal structure disturbed (down to amorphization) during *ion bombardment*. The pulse annealing of semiconductors stimulated the development of prospective technological methods such as *pulse crystallization* and *pulse synthesis*. An advantage of the former is the possibility of growing monocrystalline layers on nonoriented *substrates*;

the latter method provides the opportunity to synthesize compounds from components which differ greatly in vapor elasticity.

PULSE EXPOSURE

A short-time external action upon a material, usually with a duration less than seconds. There are ultrasonic, impact (mechanical), electronic, magnetic pulse, electric spark and other kinds of pulsed exposure. It can cause acoustic, thermal, elastoplastic, plasma and *shock wave* effects. Various kinds of structural and phase modifications occur in materials subjected to pulse exposure; their nature being determined by the type, strength and duration of the exposure. Pulse exposure is applied as a processing technique (*pulse treatment* of metals, *pulse heating* of semiconductors, *pulsed annealing*). *Phase transitions* under pulse exposure exhibit features involving the kinetics of transformations under conditions of short-time heating. For example, pulse heating in semiconductors is used for restoring the surface structure of monocrystals after the amorphization induced by *ion implantation*. The amorphous layer in Si is as thick as 0.5 to 1 μm. Pulse annealing in the range 10^{-2}–10^{-8} s at a pulse energy density 100–1 J/cm^2 causes solid-phase *epitaxy* or *liquid-phase epitaxy* where the *impurity atoms* occupy positions at lattice sites and become electrically active. Comparison of *relaxation times* of all processes running under the interaction of radiation pulses with a semiconductor (electronic subsystem excitation, *electron–phonon interaction*, electron–hole plasma recombination) shows that for pulses of duration longer than 10 ns in monocrystalline Si (or exceeding 0.1 ns in amorphous Si), the phase or structural transformations can be described using a thermal model in accordance with the phase diagram, taking into account the transformation rate.

The figure shows the free energy change of amorphous (A), crystalline (C) and liquid (L) bodies. For $T < T_A$ (where T_A is the amorphous phase melting temperature), transition $A \to C$ (*crystallization*) is possible. At the same time, since the time to initiate crystallization is $t_0^c > 10^{-5}$ s, the amorphous phase heating for pulses $t_p < 10^{-5}$ s transfers it to the temperature range $T >$

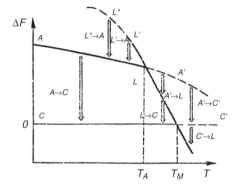

Pulse exposure.

T_A. To start melting, the time $t_0^m < 10^{-8}$ s is sufficient (transition $A' \to L$, i.e. *melting*). The forming melt is supercooled at $T < T_M$ (where T_M is the crystal melting temperature) and then crystallizes (transition $L \to C$). The released latent heat of crystallization can induce self-maintained crystallization in a residual amorphous phase. If the time needed for nucleation of a critical liquid phase is less than t_0, one can obtain an overheated amorphous phase (A') or an overheated crystal (C'). At $t_p \sim 10^{-8}$ s, solidification occurs at the rate 10 to 20 m/s, and either an $L \to C$ or $L' \to A$ transition occurs, unless the atomic structure ordering has time to coalesce and establish long-range order (see *Long-range and short-range order*). At shorter pulses, *vitrification* in the melt with the short-range liquid order retained (L'') may occur. At $T > T_M$, transition $A' \to L$ takes place much faster than transition $A' \to C' \to L$, with times in silicon ~1 fs and 1 ns, respectively. The temperature distribution in a layer and its variation during the pulse heating are described by the *thermal conductivity* equation with sources and sinks that are radiators, including latent heat, heat removal, and intrinsic radiation. At $t_p < 10^{-6}$ s the surface heating is adiabatic so melting the layer of 1 μm thick does not cause changes in lower lying layers. The phase diagrams like that in the figure change under the action of external forces and internal stress, and depend on the rate of heating that stimulates a partial relaxation of strain energy and defects. Shifts of curves A and C affect the values T_A and T_M.

For amorphous Si of various origins, various experiments yield $T_A = (0.7–0.9)T_M$.

PULSE HEATING

A rapid (compared to characteristic times of internal processes) heating of a specimen, or a part or component of it. The concept of pulse heating is used when considering diffusion or *recrystallization* processes in solids. It is here that pulse heating proves to be an efficient instrument to control the processes. Pulse heating is performed using various energy sources: light and electron beams, low-inertia ovens, UHF sources, passage of electric current through conducting solids, etc. All these methods provide large temperature gradients down from the solid surface (up to 10^9 deg/cm using laser pulses). Pulse heating is particularly significant in semiconductor electronics technology due to the high sensitivity of the structural, electrical and physical properties of *semiconductors* to the character of the external action. Particular features of pulse heating under extra short pulses are described by the plasma model of pulse heating.

PULSE SPECTROMETRY

An optical spectrometry of *defects* induced by short pulses of radiation. High-current pulsed electron accelerators with pulse duration τ to 1 ns, or *lasers* with τ to 10 ps, serve as the radiation sources.

There exist pulse spectrometric methods which utilize simultaneous or sequential excitation of crystals by electron or laser pulse fluxes. This allows one to produce, e.g., controlled pre-decay states of *excitons* necessary for gaining insight into the mechanism of the exciton decay into pairs of *point defects*.

The application of pulse spectrometry to *radiation physics of solids* can provide direct information on the processes of radiation-induced defect formation, in contrast to the post-effect studies typical of CW (non-pulsed) research. As a rule (in particular, at high enough temperatures to delocalize primary defect pairs), the structure of spectra induced by the radiation pulse is simpler than that in case of CW irradiation. This is due to the low probability of complex center creation with the participation of diffusion processes. However, metastable electronic excitation, such as *self-localized excitons* in alkali halides, may manifest themselves in short-time absorption spectra. The appearance of some new bands absent in CW spectra characterizes short-time stages of the defect nucleation processes. Pulse spectrometric methods helped to establish that the decay of electronic excitation into structural defect pairs in *alkali-halide crystals* takes $10^{-11}–10^{-10}$ s. Primary excitation decay products could be both neutral and charged defect pairs. The efficiency of defect formation is high in many materials (in particular, in quartz). However, the stability of radiation-induced defects with respect to annihilation is low, which explains the high *radiation resistance* of quartz.

PULSE TREATMENT of materials

A treatment of materials by short-time action of various heat, force and energy sources. Particular features of a pulse treatment are high rates of elastic *strain* and *plastic deformation* (to 10^5 s^{-1}), significant temperature gradients (10^8 deg/cm), and high rates of heating and cooling (to 10^{10} deg/s). Impact (mechanical), ultrasonic, electron beam, laser, magnetic pulse, electrical hydro-pulse, electric spark, explosion, and other kinds of pulse treatment are used. The treatment could be accompanied by *phase transitions* (including fusion and evaporation), and it is characterized by modification of the physicochemical properties of materials and their structure. To produce a pulse regime when processing material with the help of an electron beam, one interrupts either the heater current or supply voltage to the emitting electrode. The pulse duration varies from 10^{-6} to 10^{-1} s and its power density reaches 10^8 W/cm^2. Pulse treatment is applied for *welding*, milling and engraving a surface, making small-diameter holes in plates made of hard materials. For a high-voltage pulse discharge in a liquid the pulse treatment is effected by rapid expansion of the channel under the action of *high pressure* and temperature, accompanied by emission of a pressure pulse (i.e. the compression wave). The discharge channel plasma is characterized by a pressure 10^3 to 10^4 atm, and its life time is 10^{-5} to 10^{-4} s. Electrical hydro-pulse treatment is used for forging, punching, stretching, refining casts of pivotal and molding materials, rock crushing, cleaning wires of scale, etc. Laser treatment pulse duration is 10^{-3} to 10^{-8} s at a power density of 10^5–10^6 W/cm^2. Laser treatment is used for strengthening *metals*, for welding, milling, and punching holes in hard materials. In the case of electric spark pulse treatment, the spark gap is 10–200 μm, the pulse duration is 10^{-6} s. This kind of pulse treatment is used for electric spark *alloying*, surface strengthening, making the holes etc. The pressure at an explosion treatment is 10–10^4 MPa, and the process duration is 10^{-6} to 10^{-3} s; it is used for punching, welding, co-pressing, calibrating, engraving, strengthening of materials, etc. Under conditions of pulse treatment, e.g., vacuum impact welding with a deformation rate above 1 s^{-1}, the rate of *mass transport* in solid metals and *alloys* increases significantly.

Depending on the kind of migrating atoms, either substitutional or interstitial *solid solutions* appear under pulse treatment. In substitutional solid solutions the ultimate solubility shifts toward greater concentrations, and there is an absence of *intermetallic compounds*. In interstitial solid solutions chemical compounds like carbides form in addition to the solid solution; and for the case of insoluble impurities (e.g., krypton in aluminum) the main metal dissolves in the matrix. The greatest mobility of atoms at mass transfer is observable in substitutional solid solutions. The mass transfer process is of a bulk type, and depends weakly on the temperature.

PUMPING, LOW-TEMPERATURE

See *Low-temperature pumping methods*.

PURIFICATION

A process of preparing a material (host) with a minimal impurity content (see *Impurity atoms*). Methods of purification include physicochemical and chemical ones. In physicochemical methods the separation is due to the difference in physicochemical properties between the host and the impurity. Classed among these methods are crystallization from melts (see *Zone refinement*), distillation, redistillation, *crystallization* from a gas (vapor) phase, *thermodiffusion*, and *annealing* in a high vacuum. In chemical methods, the separation is based on the difference in chemical interaction of the host and the impurity with any additional component of the system. These methods include crystallization from solutions with the involvement of coprecipitants or complex-formers,

sorption, ion exchange (see *Ionites*), extraction, synthesis of metal-organic compounds, selective *oxidation* or reduction of impurities, electrolytic refining, and chemical transport reactions.

In order to prepare materials of extra high purity, integrated methods of purification are used. Thus, *indium* of extra high purity was successfully obtained with the help of the synthesis of indium monochloride, fine purification of indium monochloride by rectification and zone *recrystallization*, the isolation of indium from its monochloride in a disproportionation reaction, and then zone recrystallization.

PYROCERAMICS

See *Glass ceramics*.

PYROELECTRIC CURRENT

Electric current, which arises in a circuit involving a pyroelectric crystal in the direction of the polar axis, as the temperature of the crystal is changed. The magnitude of pyroelectric current is related to the spontaneous polarization of the pyroelectric material (see *Pyroelectricity*) by the equation $J = A(dP_s/dt)$, where A is the area of electrodes at opposite faces of the pyroelectric crystal, which is shaped as a parallel-sided plate, with the spontaneous polarization vector P_S directed between the electrodes. Current arises when electrodes absorb periodically modulated or pulsed radiation, as well as when the pyroelectric comes into thermal contact with an alternating or pulsed source of heat. The magnitude of the pyroelectric current is determined by the value of the pyroelectric coefficient (see *Pyroelectric effect*), the area of electrodes, the rate of temperature change in the pyroelectric, or by the density of the absorbed radiation flux. When related to the incident radiation power, the pyroelectric current is expressed as the *ampere–watt sensitivity* of the *pyroelectric radiation detector*.

PYROELECTRIC DETECTION

Detection of electromagnetic radiation by application of the *pyroelectric effect*. The method of pyroelectric detection consists in determining the power or the energy of the radiation, which is incident upon the pyroactive sensing element of the detector, by measuring the output electric signal.

Three types of pyroelectric detection are distinguished: pyroelectric detection of continuous radiation, detection of narrow pulses of radiation, and heterodyne detection. Recording continuous radiation flux calls for modulation of the incident flux, and the best capability for detecting the radiation is achieved at low modulation frequencies. Pulse-type detection can take place when the time constant of the pyroelectric receiver and the input circuit is less than the duration of the pulse of the radiation. In the process of heterodyne detection, the mixing of two optical signals occurs at the surface of the pyroelectric sensitive element: the weak input signal and the strong heterodyne signal. The beating at these optical frequencies results in the heterodyne-amplified difference frequency, which provides information on the signal under investigation. Since the signal to be studied is coherent with the heterodyne input, it is amplified to a considerably greater extent than the incoherent Johnson and thermal noise that accompanies it. The upper limit of the detecting ability of the pyroelectric receiver is set by thermal noise, which is related to the static nature of the heat exchange between the sensitive element of the detector and the environment. For the receiving area of 1 mm^2 at room temperature, the threshold of sensitivity is $5.55 \cdot 10^{-12}$ W·Hz$^{-0.5}$. Sensitivity thresholds attainable under real conditions are an order of magnitude lower. Theoretically, the efficiency of the heterodyne technique is 10^6–10^7 times higher than that of direct detection. See also *Pyroelectric radiation detectors*.

PYROELECTRIC EFFECT

Generation of a charge on the faces of a crystal, cut perpendicular to the polar axis, when it is heated or cooled. The total pyroelectric effect occurring in a uniformly heated or cooled crystal is a complex function of its interrelated electrical, thermal, and mechanical properties. When measured under constant stress, so the crystal may change its shape freely, the pyroelectric effect consists of the *primary ("true") pyroelectric effect* and the *secondary ("pseudo") pyroelectric effect*. The primary pyroelectric effect is due to the direct exchange of thermal energy of the crystal lattice subject to conservation of its shape and volume (ideally mechanically clamped crystal). The

secondary pyroelectric effect is related to the free *strain* of the crystal. The pyroelectric effect may arise in crystals which feature no spontaneous polarization (see *Ferroelectricity*), in the case of nonuniform heating. The generation of a temperature gradient may affect the symmetry (see *Crystal symmetry*) of crystals, which have not shown pyroelectric properties before heating, in such a way that they become pyroactive. The pyroelectric effect arising in this case is referred to as the *tertiary pyroelectric effect ("pseudo" pyroelectric effect of the second kind)*. The *pyroelectric coefficient* is an index of the magnitude of the pyroelectric effect exhibited by a given material at a given mechanical pressure, electric field strength and temperature. The pyroelectric coefficient is determined by the density of charge for a change of temperature by one degree. The following ferroelectrics exhibit the highest pyroelectric coefficients: strontium–barium niobate, *triglycinesulfate*, lithium tantalate, and ferroelectric ceramics (see *Ceramics*) like lead titanate–zirconate within the region of a *phase transition*. For triglycinesulphate at room temperature, the pyroelectric coefficient is $3.5 \cdot 10^{-4}$ C·m^{-2}·K^{-1}. In the case of a bulk sample, the temperature at which the pyroelectric coefficient and dielectric constant have maxima, coincides with the phase transition point. The effect observed in thin-layer ferroelectrics consists in a shift of the temperature of the maximum pyroelectric coefficient relative to the temperature of the dielectric constant maximum.

The pyroelectric coefficient may be determined using dynamic or static methods. Its value may vary over a wide range, particularly for ferroelectric ceramics, depending on the method of measurement. The pyroelectric coefficient measured by application of a static method depends upon both the "fast" and "slow" components of the total polarization. When the *dynamic method of pyroelectric coefficient determination* is used, the major contribution to the coefficient comes from fast polarization processes.

PYROELECTRICITY

Generation of electric charges at the faces of a number of *pyroelectric crystals*, which lack a center of symmetry (see *Pyroelectric effect*) on heating or cooling. The term "pyroelectricity" was in-

Table 1. Comparative characteristics of pyroelectric materials

Material name	Detecting capacity, arb. units	Noise equivalent power, arb. units	Coefficient of performance of pyrovidicon, arb. units	Coefficient of performance of point source receiver
Triglycinesulphate	1.00	1.00	1.00	1.00
Triglycinesulphate with alanine	1.18	1.04	1.60	1.68
Lithium tantalate	0.91	2.20	0.03	0.16
Polyvinylfluoride	0.18	4.15	0.01	–
Lithium sulphate monohydrate	0.64	3.15	0.01	1.00
Barium strontium niobate	0.28	8.3	0.005	0.5
Lead/lanthanum zirconate/titanate ceramic	0.14	14.3	0.015	0.33

troduced by D. Brewster (1824). Of 21 noncentrally symmetric *crystal classes*, 10 have a principal polar axis and exhibit pyroelectric properties. Pyroelectrics are spontaneously polarized in the absence of external actions such as an applied electric field. However, the spontaneous polarization under fixed environmental conditions cannot be detected by the presence the charges on the crystal faces, for these charges are compensated by the free carriers due to bulk and surface conductivity, and from the environment. Nevertheless, since the spontaneous polarization is temperature dependent, its change may be detected during the process of heating or cooling the crystal.

The highest pyroelectric activity is exhibited by ferroelectric (see *Ferroelectrics*) monocrystals and *ceramics* in the regions of *phase transitions*. The temperature dependence of the pyroelectric activity of these materials is nonlinear. The thin-layer ferroelectrics undergo an increase of coercive fields and become harder. The anomalous behavior of the pyroelectric coefficient dies out in the phase transition region, and the ferroelectric degenerates into a *linear pyroelectric* as the layer thickness decreases.

PYROELECTRIC, LINEAR AND NONLINEAR

See *Linear pyroelectric*, *Nonlinear pyroelectric*.

PYROELECTRIC MATERIALS

Pyroelectric materials, such as ferroelectric barium titanate, tourmaline and cane sugar, which develop a small potential difference across their crystals when they undergo a change in temperature. Pure *triglycinesulfate* is taken as a primary standard of pyroelectric materials. The detecting capacity (D) and noise equivalent power (NEP) of this substance are respectively $D \approx 1.1 \cdot 10^9$ cm·Hz$^{1/2}$·W^{-1} and NEP = $4.8 \cdot 10^{-11}$ W·Hz$^{1/2}$. The comparative characteristics of some basic pyroelectric materials are given in Table 1.

Pyroelectric materials are capable of providing a temperature resolution of up to 0.1 °C under the conditions of television resolution, the data integration time per frame being less than 200 μs. The time resolution obtained when recording high-rate processes is 1 ps, within the range of measurable power density from μJ/cm^2 to kJ/cm^2.

Pyroelectric materials are widely used in equipment for thermal physics measurements, to provide panoramic thermal imaging for various scientific and engineering purposes (including manufacturing), and for medical diagnostics. These materials have been employed as uncooled targets for television camera tubes which convert IR radiation into visible radiation (pyrovidicons), and as nonselective receiving elements of point and linear detectors (see *Pyroelectric target*).

PYROELECTRIC PYROMETERS

Devices for measuring temperature using a sensitive element made of a pyroelectric material. The operation of pyroelectric pyrometers is based on the *pyroelectric effect*, which consists in the change of spontaneous polarization of pyroelectric crystals with a change of temperature. The magnitude of the polarization charge is proportional to the volume-averaged temperature of the pyroelectric material. In the typical pyroelectric crystal *tourmaline* the field of 400 V/cm arises from a temperature change of 1 K. The pyrometer rate of response depends on the density and *specific heat* of the material, and attains 10^{-7} s. A pyroelectric material in an electric circuit plays the role of a variable-capacitance transducer, so the determination of the temperature involves measuring its capacitance (ordinarily using the compensation method).

Pyroelectric pyrometers are used for a non-contact measurement of temperature (see *Pyrometry*), for controlling thermal processes of short duration, and for recording *thermal radiation*.

PYROELECTRIC RADIATION DETECTORS

Receivers that depend for their operation on the *pyroelectric effect*. Pyroelectric radiation detectors belong to the class of rapid-response thermal detectors of radiation. Their sensitive elements are capacitors, which may be given complex shapes in order to attain high values of receiving area. In distinction to *bolometers*, pyroelectric detectors require no additional power sources for their operation. These radiation detectors have a sensitivity and time constant that may be changed by many orders by varying the load resistance. Their output signal is proportional to the derivative of the temperature with respect to time, so these devices exhibit a high rate of response which is comparable to that of photodetectors. Featuring high detection capability, pyroelectric radiation detectors are the only high-speed thermal radiation detectors capable of working at room temperature.

When operating by self-absorption in the IR region, the limiting time of pyroelectric detectors is determined by the phonon relaxation time, and lies within the range of 10^{-11}–10^{-13} s. In the event that radiation is absorbed by an electrode, the pyroelectric response time is determined by the thermal diffusion time, and amounts to 10^{-6}–10^{-8} s. The "detecting ability" of pyroelectric radiation detectors is 10^9 cm·Hz$^{0.5}$·W^{-1}.

Pyroelectric radiation detectors are used within the spectral region ranging from ionizing radiation wavelengths to millimeter waves (10^{-3}–10^4 μm).

There are many types of pyroelectric radiation detectors, with the planar single-element detectors the most common type. The area of a sensitive detector element may range from 10^{-3} to 100 mm^2, with a high degree of uniformity and sensitivity maintained over this range. See also *Pyroelectric detection*.

PYROELECTRIC TARGET

The sensitive element of a *pyrovidicon*, i.e. a transmitting cathode-ray tube which converts IR radiation into visible radiation. The pyroelectric target is a pyroactive crystal, which is cut perpendicular to the polar axis. The target is given a layer of conducting coating on the side to be irradiated, a layer of electron beam resistant dielectric *coating* on the opposite side, and is mounted on metallic rings. The IR image under study is projected by a lens on the pyroactive target, thereby creating a thermal pattern which is transformed via the *pyroelectric effect* into a charge pattern, which in turn is read by the scanning electron beam. The signal generated in this process modulates the brightness of the developer of the television-type image. The pyroelectric target exhibits low values of transverse *thermal conductivity*, *specific heat*, *dielectric constant*, and has a high pyroelectric coefficient. To increase their sensitivity pyroelectric targets are constructed in the form of 15–20 μm thick plates; the targets must be stable, retaining the single-domain state responsible for their uniform band sensitivity. Of the materials used as working media in pyroelectric targets the highest sensitivity is exhibited by *triglycinesulfate* and substances isomorphic to it, in particular by deuterated triglycinesulfates and triglycinefluoroberyllate. Lead germanate and titanate-zirconate, as well as polymeric films based on polyvinylfluoride and polyvinylidenfluoride are less commonly used. Pyroelectric targets are given a cellular structure (*mosaic crystals*) to decrease the *diffusivity*. Intercellular gaps of structured pyroelectric targets are filled with a material that has a low temperature conductivity factor, e.g. with araldite, and then the target is ground to a 15 μm thickness. Methods of laser and photolithographic generation of a mosaic structure have been developed.

The material used as dielectric coatings to be applied on the backside of a pyroelectric target is a mixture of arsenic selenide and tellurium selenide. In pyrovidicons, which convert the IR radiation without scanning, the pyroelectric target is the principal part, determining the major parameters. Pyrovidicons are very promising IR–visible radiation converters since they are faster, and have a higher sensitivity than evaporographs, edgeographs, and liquid crystal converters.

PYROMAGNETIC EFFECT

An analog of the *pyroelectric effect*, based on the temperature dependence of the spontaneous *magnetization* of certain materials. A pulse of radiation incident on a pyromagnetic material changes the magnetization of the sample because of the heating. The electric signal that arises in a detecting coil placed in the magnetic field of the sample is proportional to the pyromagnetic coefficient, and the rate of the change of the sample temperature. The pyromagnetic effect may be used for detecting radiation over a broad spectral range. *Pyromagnetic radiation receivers*, developed on the basis of the pyromagnetic effect, are high-speed temperature detectors, with a time constant ranging from 10^{-6} to 10^{-9} s. These radiation receivers use ferromagnetic materials (see *Magnetic materials*) as working media. The method of pyromagnetic detection has not gained wide acceptance in IR and laser engineering.

PYROMETRY (fr. Gr. $\pi \upsilon \rho$, fire and $\mu \varepsilon \tau \rho \omega$, am measuring)

Method of determining the *temperature* of bodies by measuring their *thermal radiation*. The corresponding devices are referred to as *pyrometers*. Pyrometry has its origins in the relationship between the intensity of heat radiated by bodies and their temperature. The quantities measured in pyrometry are the intensity of monochromatic radiation from the object (*brightness pyrometry*), the ratio of the radiation intensities of a body at two different wavelengths (*color pyrometry*), or the total radiant intensity (*radiation pyrometry*). At high temperatures (10^3–10^4 K) brightness pyrometers are effective, and at low temperatures (~ 100 K) radiation ones are preferable (also called *total radiation pyrometers*). Sensitive receiving devices in pyrometry are *thermocouples*, bolometers, photoelectric and *pyroelectric radiation detectors*.

QUADRUPLE CRITICAL POINT

See *Tetracritical point*.

QUADRUPOLE

A system of charged particles that has zero net electric charge and no electric *dipole moment*; it is a *multipole* of the second order. The main characteristic of a quadrupole is its *quadrupole moment* Q. For the quadrupoles shown in the figure, $Q = 2ela$, where e is the absolute value of electron charge, and l is the size of the component dipoles. At large distances R from the quadrupole the strength E of its electric field decays in inverse proportion to R^4. The traceless quadrupole tensor has three principal components $Q_{ZZ} = -(Q_{XX} + Q_{YY})$, with principal axes chosen so Q_{ZZ} has the largest magnitude. For lower than axial symmetry ($|Q_{XX}| < |Q_{YY}|$) we define the dimensionless asymmetry parameter $\eta = (Q_{XX} - Q_{YY})/Q_{ZZ}$, where $0 \leqslant \eta \leqslant 1$, and $\eta = 0$ for the common case of axial symmetry. The quadrupole moment determines the energy of the quadrupole in an external smoothly nonuniform electric field, such as an electric field with a uniform gradient $\partial E/\partial x$.

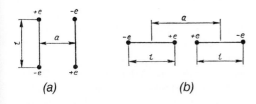

(a) Transverse quadrupole, (b) axial quadrupole.

QUADRUPOLE HAMILTONIAN

See *Nuclear quadrupole interaction*.

QUADRUPOLE INTERACTION

See *Nuclear quadrupole interaction*.

QUADRUPOLE MOMENT

See *Nuclear electric quadrupole moment*.

QUADRUPOLE RELAXATION

See *Nuclear quadrupole relaxation*.

QUADRUPOLE RESONANCE

See *Nuclear quadrupole resonance*.

QUADRUPOLE SPIN SYSTEM

See *Nuclear quadrupole spin system*.

QUADRUPOLE SPLITTING

The splitting of energy levels caused by the interaction of the electric quadrupole moment (see *Quadrupole*) of a *paramagnetic center* or of a nucleus with a *crystal field gradient*. Sometimes this is referred to as an *initial splitting* or a *zero-field splitting* (i.e. with no external fields present). The resonant absorption (or radiation) of electromagnetic energy for transitions between the resulting levels is referred to as *quadrupole resonance*, and sometimes it is called *zero-field resonance* (see *Nuclear quadrupole resonance, Zero-field magnetic resonance*).

QUALITATIVE ANALYSIS

Identification (detection) of the components of a substance under analysis. Qualitative analysis may be an independent aim of a study, or may merely precede a *quantitative analysis*. The qualitative analysis can be elementary, molecular, functional, isotopic, *phase analysis*. To detect the components, one uses chemical, physicochemical, and physical methods. Depending on the quantity of the substance being analyzed, we can distinguish macro (>0.1 g), semi-micro (0.01–0.1 g), and micro (0.001–0.01 g) methods of qualitative analysis. The detectable quantity of a component is characterized by its detection limit (c_{min}), i.e. the least quantity (or concentration) which can be measured in a sample.

QUANTITATIVE ANALYSIS

Determination of the amount of individual components (or their ratios) in a specimen. Several types of analysis are: elemental, molecular, functional, isotopic, and *phase analysis*. Quantitative analysis is based on the dependency of an analytical signal (mass, volume, electrode potential, etc.) on the content or concentration of a component. One possible classification of analytical methods is based on the type of analytical signal: gravimetry (see *Gravimetric analysis*), titrimetry, potentiometry, etc. The trend in quantitative analysis is the development of automated instrumental methods. Quantitative analysis can be conventionally divided into determining a main component (1–100% of the sample mass), a micro-component (0.01–1%), and a trace amount ($<0.01\%$). It is characterized by the following parameters: accuracy, reproducibility, range of component distribution, and errors at separate stages of analysis.

QUANTIZATION OF ELECTRON ENERGY in crystals

Formation of a system of alternating allowed and forbidden bands of electron energies characterized by discrete quantum numbers (band number) and continuous quantum numbers (*quasi-momentum* of electron). The physical foundation for establishing the *energy bands* is the motion of electrons in a *periodic potential* of ions forming the *crystal lattice*; the allowed bands arise as a result of the splitting of the electron energy levels corresponding to individual atoms. In a narrower (but more accepted) sense, the term "quantization of electron energy" in crystals is applied to the motion of free *current carriers* in *metals* and *semiconductors* in external fields (electric or magnetic fields) or potentials created by impurities, defects, or boundaries of *crystals*.

The most important case of such quantization is *Landau quantization* in a strong magnetic field causing a great variety of phenomena: diamagnetism of *conduction electrons*, quantum oscillations of thermodynamic and kinetic coefficients of metals and degenerate semiconductors (*de Haas–van Alphen effect, Shubnikov–de Haas effect*), Azbel–Kaner *cyclotron resonance* in metals, *cyclotron (diamagnetic) resonance* in semiconductors, etc. In *thin films* of metals and semiconductors the *size quantization* of the energy of the free current carriers occurs. Size quantization in semiconductors determines the features of electron spectra in *superlattices* and *quantum wells*, and in combination with Landau quantization in a magnetic field perpendicular to the surface of a semiconductor, it provides the conditions for the observation of the *quantum Hall effect*.

A rather peculiar manifestation of electron state quantization in solids is the interference of closed electron orbits formed during consecutive elastic scattering of electrons from impurities in "dirty" metals or alloyed semiconductors (phenomenon of *weak localization*). See also *Quantum interference phenomena* in metals.

QUANTIZATION OF FLUX in superconductors

A phenomenon of superconductivity occurring in multiply connected samples (e.g., rings, hollow cylinders) whereby the amount of magnetic flux Φ in a region can take on only discrete values

$$\Phi = \oint A\,\mathrm{d}l + \frac{mc}{n_s e^2} \oint j_s\,\mathrm{d}l = n\Phi_0,$$

where n is an integer. Here A is the vector potential of the external source field, m the is electron mass, e is its charge, j_s is the superconducting current density, n_s is the concentration of super-electrons (Cooper pairs), and Φ_0 is the *flux quantum*. The integration is carried out over a closed contour. In bulk superconductors the current density is equal to zero, and the flux $\Phi = \oint A\,\mathrm{d}l$ enclosed within the contour is equal to an integer number of flux quanta $\Phi = n\Phi_0$. In thin-walled cylinders and rings ($d < \lambda$, d is the thickness of the walls and λ is the *penetration depth of magnetic field*) the quantization of flux displays itself in the appearance of a circulating super-current flowing on the sample surface, with its magnitude and direction changing as a function of Φ with a period of Φ_0. The direct consequence of the flux quantization in such systems is a periodic changing of the superconducting transition temperature with an applied magnetic field (*Little–Parks effect*). The quantization of flux is a macroscopic quantum phenomenon exhibited by the *Aharonov–Bohm effect*, and by the existence of long-range magnetic order (see *Long-range and short-range order*). The quantization of flux was predicted by F. London (1950) and discovered by R. Doll and M. Näbauer, as well as by B.S. Deaver Jr. and W.M. Fairbank (1961). The value $\Phi_0 = h/2e$ found in experiments confirmed the main assumption of the theory of superconductivity, namely that the supercurrent is carried by *Cooper pairs*.

QUANTIZING ELECTRIC FIELD

An electric field that leads to quantizing the motion of a band *electron in a crystal*. Under the effect of the electric field E, in the general case, the electron traverses a complex trajectory in k-space. If *interband tunneling* is neglected, and the electron motion within the crystal is considered along some constant direction with the lattice constant a, the electron performs periodic motion in k-space following Newton's equation. During this motion, e.g., along the x coordinate axis, in the reduced band scheme the component k_x of the wave vector varies from some value k_{0x} to π/a, where the electron experiences reflection with a change of k_x to $-\pi/a$, whereupon it again begins to move from $-\pi/a$ to π/a, and so on. As a result, its spectrum takes on the form

$$\varepsilon_n(\boldsymbol{k}_\perp) = eEan + \frac{a}{2\pi} \int\limits_{-\pi/a}^{\pi/a} \widetilde{\varepsilon}(\boldsymbol{k})\, \mathrm{d}k_x,$$

where, for a crystal with a *center of symmetry*, the quantity $\widetilde{\varepsilon}(\boldsymbol{k})$ is the energy of an electron in this band, and $n = 0, 1, 2, \ldots$. The *dispersion law*, $\varepsilon_n(\boldsymbol{k}_\perp)$, and the wave functions

$$\varphi_{n,\boldsymbol{k}_\perp}(\boldsymbol{r}) = \left(\frac{a}{2\pi}\right)^{1/2} \int\limits_{-\pi/a}^{\pi/a} \mathrm{d}k_x\, \psi_{\boldsymbol{k}}(\boldsymbol{r})$$

$$\times \exp\left\{\frac{\mathrm{i}}{eE} \int^{k_x} \left[\varepsilon_n(\boldsymbol{k}_\perp) - \widetilde{\varepsilon}(\boldsymbol{k})\right] \mathrm{d}k_x\right\},$$

where $\psi_{\boldsymbol{k}}(\boldsymbol{r})$ is a Bloch function (see *Bloch theorem*), were first found by E.O. Kane (1959). The discrete energy levels are often referred to as the *Wannier–Stark ladder*. The latter name becomes clearly understandable, if one schematically depicts the energy band in an electric field as a function of the x coordinate in the shape of an infinite tilted band: corresponding to each lattice site will be an energy level, these levels making up steps of height eEa. The electric field in a *semiconductor* should be sufficiently strong to be a quantizing one, since an electron must reach the top of an allowed band before being scattered, i.e. the criterion $eEa/\hbar > \nu$ (ν is the characteristic frequency of collisions) should be satisfied. There are several experimental situations under which the quantized nature of electron motion makes its appearance. For instance, the phenomenon of *electric phonon resonance* exists, similar to the *phonon magnetic resonance* under Landau quantizing conditions (see *Quantizing magnetic field*).

QUANTIZING MAGNETIC FIELD

A sufficiently strong magnetic field capable of changing the electron energy spectrum of a *semiconductor* or a *metal*, and thereby causing various oscillation effects, such as the *de Haas–van Alphen effect*, the *Shubnikov–de Haas effect*, and others. The quantum mechanical problem of electron motion in a uniform magnetic field was solved by L.D. Landau (1930). In the case of a quadratic *dispersion law* of electrons in a crystal $\varepsilon = p^2/2m^*$ (where $\boldsymbol{p}$ is the *quasi-momentum*, m^* is the *effective mass* of an electron) in the presence of a uniform magnetic field $\boldsymbol{B} = \{0, 0, B\}$, the electron energy spectrum assumes the form $\varepsilon_\nu = (n + 1/2)\hbar\omega_c + p_z^2/(2m^*)$ (disregarding the electron spin), where $\omega_c = |e|B/m^*$ is the *cyclotron frequency*, $\nu = \{n, p_z\}$, $n = 0, 1, \ldots$. This spectrum describes the quantum nature of the transverse (relative to $\boldsymbol{B}$) orbital motion of an electron (harmonic oscillator) and its free longitudinal motion along $\boldsymbol{B}$. The energy ε_ν represents a set of sublevels $p_z^2/(2m^*)$ spaced by the value $\hbar\omega_c$, referred to as *Landau levels* or sublevels. The electrons in a crystal collide with *phonons*, impurities, and one another, which causes the broadening of the Landau levels. For the quantization to actually take place, the electron has to make several (or many) revolutions in a cyclotron orbit between two consecutive collisions; therefore, the cyclotron frequency must be much greater than the average electron collision rate with scatterers. In addition, the strong quantization condition requires that a greater number of electrons must be at the lower Landau levels (for degenerate electrons, counted from the *Fermi level*), i.e. $k_B T < \hbar\omega_c$. The case $k_B T \ll \hbar\omega_c$ is referred to as the *quantum limit*. Hence, there is a limit on the value of a magnetic field for it to be a quantizing one. The most favorable conditions for this are attained in pure crystals with a small electron effective mass at low temperatures when collisions are sufficiently rare and $k_B T < \hbar\omega_c$.

QUANTUM ACOUSTICS

Acoustical investigations of various phenomena in solids which are theoretically interpreted using concepts and principles of quantum mechanics. Among them there are investigations of hyper-sound waves in *insulators* at low temperatures when the mechanism of *sound absorption* in crystals involves the three-phonon process of scattering of hyper-sound quanta off long-lived *phonons*; as well as hyper-sound investigations of the electron energy spectrum structure and the *electron–phonon interaction* on the basis of *acoustic paramagnetic resonance*. This field includes *acoustic nuclear magnetic resonance*, investigations of magnetically-ordered crystals by

acoustic methods based on hyper-sound, investigations of *magnetoacoustic resonance*, and the softening of the phonon modes at *orientational phase transitions*.

QUANTUM AMPLIFIER, maser

A device with extremely low level inherent noise which amplifies electromagnetic waves based on the effect of *induced radiation*. The term "quantum amplifier" or *maser* (acronym for Microwave Amplification by Stimulated Emission of Radiation) refers to devices operating in decimeter, centimeter and millimeter wavelength bands. The amplification of the input signal in a quantum amplifier proceeds via the interaction of an electromagnetic wave with the active substance where an energy level population inversion takes place (see *Level population*). As active substances one uses insulator crystals with paramagnetic impurities, among which the most popular are corundum Al_2O_3 and rutile TiO_2 with impurities of Cr^{3+} or Fe^{3+}, as well as beryl $Al_2Be_3Si_6O_{18}$ and andalusite Al_2SiO_5 with impurities of Cr^{3+} or Fe^{3+}, respectively. In thermal equilibrium $n_1^0 < n_2^0 < n_3^0$. The population inversion for a pair of levels 1 and 2 (see Fig.) is produced by a pumping field which equalizes the populations of levels 1 and 3 due to *saturation effects*. The essential characteristic of the active substance is its *inversion coefficient* $J = -(n_2 - n_3)/(n_2^0 - n_3^0)$, where n_i and n_i^0 are the populations of the levels i un-

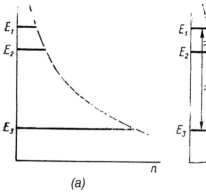
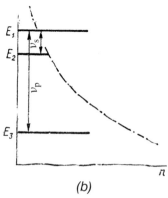

(a)   (b)

System of three energy levels (a) in an equilibrium state, and (b) in the presence of a pumping field, where E is the energy, n is the level population, ν_p and ν_s are the pumping and signal (observed) frequencies; and the dashed lines indicate equilibrium populations.

der the pumping field action, and in the thermodynamically equilibrium state, respectively. The coefficient J depends on the energy level splitting
(arising from intracrystalline electric and external magnetic fields acting on the impurity center),
on the relationship between probabilities of relaxation transitions, processes of *cross-relaxation* in
the spin system, and other factors. The possibility of J being large ($J > 10$ was achieved) is determined by checking crystals using the *electron
paramagnetic resonance* method. Since the intrinsic probabilities of transitions being initiated by
the signal input field from level 1 to 2 and from
level 2 to 1 are equal ($W_{12} = W_{21}$), and since
$n_1 > n_2$ under pumping conditions, it follows that
the number of downward transitions $1 \rightarrow 2$ per
unit time $W_{12}n_1$ (emission) exceeds the number
of upward transitions $2 \rightarrow 1$ per unit time $W_{21}n_2$
(absorption). As a result, an input signal at the frequency ν_s is amplified. To improve the amplification one can increase the time of interaction of
the input wave with the active substance. To do
this the active substance is placed in a system of
the linked *resonators* or in a particular type of
waveguide (slow wave system).

The successful operation of a quantum amplifier (in particular, ensuring a low inherent noise
level) calls for cooling the active substance to liquid helium temperatures, which severely restricts
the applicability of these amplifiers. At present
microwave quantum amplifiers (masers) are used
in systems of long-range space communications
and in radio astronomy where the extremely high
effectiveness of ground-based radio-receiving devices is required. The development of the principles of quantum amplifier operation had preceded
and inspired the creation of *lasers* which are analogous light amplifiers.

QUANTUM-CHEMICAL SIMULATION

Computer simulation of the dynamics of atomic processes in non-metallic materials, used pri-
marily for *solids* with covalent *chemical bonds*.
The quantum-chemical simulation in its original
version was based on using effective *interatomic
interaction potentials* constructed by extrapolation of a number of phenomenological parameters to the domain of large displacements of atoms
from their equilibrium positions. The equations

of atomic motion were written and solved using
interatomic interaction potentials. The quantum-
chemical simulation is typically performed in the
cluster approximation using the *pairwise additive
approximation.*

In distinction from the classical version of
quantum simulation, in the case of quantum-
chemical simulation the refinement of the angular
and radial wave functions of the *cluster* atoms, and
the calculation of corrections to the interatomic interaction potential, are carried out at every step of
the computer simulation. The quantum-chemical
simulation is used to describe dynamic and static configurations of *defects* in covalent structures
arising during the damaging of materials by radiation and other influences.

QUANTUM CHEMISTRY of solids

Applications of computational methods developed in the quantum mechanics of atoms and
molecules to the theory of *solids*. These numerical computations are sometimes carried out from
first principles (e.g., by solving the Schrödinger
equation for systems of electrons and nuclei forming the solid), and sometimes they employ semiempirical methods such as the Hückel method (see
Hoffmann method), the Roothan et al. method, and
others. Typically the scope of the calculations does
not involve the entire *crystal*, but rather only part
of it (a cluster), often including some *defects* and
interstitial and substitutional impurities.

While in classical solid state theory one often
considers only one or two electrons of a defect,
and the overall crystal is treated as a "medium"
including a *periodic potential* (frequently excluded
via the *effective mass* method) with screening of
the defect potential (typically by dividing the potential by the dielectric constant), in the quantum chemistry approach the problem of finding the wave function and energy levels of the
multielectron system of a cluster is stated and
solved. Typically the system of Hartree–Fock
equations (see *Hartree–Fock method*) is solved,
and the individual wave functions of electrons
$\psi_i(r)$ plus the corresponding energy eigenvalues are found. These equations follow from the
variational principle with the multielectron wave
function $\psi(r_1, \ldots, r_n)$ found as a determinant
of individual spin-orbitals $\psi(r)$. On the basis of

Koopmans theorem the change of the whole system energy resulting from the replacement in the determinant of one of the functions $\psi_i(r)$ by an excited one $\widetilde{\psi}_i(r)$ is equal to the difference of the corresponding energies: $\Delta E = \widetilde{\varepsilon}_i - \varepsilon_i$. To simplify the problem the exchange operator in the Fock equation is replaced by an exchange potential which has a common form for all electrons. One commonly employs the Slater exchange potential (see also *Exchange interaction*) which is proportional to $\rho^{1/3}$, where $\rho(r)$ is the electron density given by $\sum_{i=1}^{n} |\psi_i(r)|^2$.

The methods of *Green's functions* and *density functional theory* have been used with increasing frequency in quantum chemistry.

An advancement over the *self-consistent field* method is the *many-configuration approximation*, whereby the function $\psi(r_1, \ldots, r_n)$ is found in the form of a linear combination of some determinants, i.e. a superposition of different configurations of electrons. If these configurations are equal or very close in energy, a so-called *quantum resonance* appears. The lowering of the total energy in comparison to the energy of a *one-configuration approximation* is brought about by the off-diagonal matrix elements $M_{ij} = \int \psi_i^* H \psi_j \, d\tau$, where ψ_i and ψ_j are many-electron wave functions corresponding to different configurations. In semiempirical theories the resonance energy plays the role of a parameter of the theory, and it can be estimated from experimental results.

QUANTUM CHEMISTRY INTEGRALS, molecular integrals

Specific matrix elements of operators of the energy of electrons interacting with each other and with atomic nuclei. Quantum chemistry integrals typically emerge during computations of the energy states of many-atom (many-electron) systems as components of more complicated matrix elements of the form $M_{ij} = \int \psi_i^* \widehat{H} \psi_j \, dV$, where ψ_k and $\widehat{H}$ are many-electron wave functions and the Hamiltonian, respectively, and V is the domain of integration. If ψ_k is represented by a linear combination of products of one-particle (atomic) functions, φ, and $\widehat{H}$ is represented by a sum of one- and two-particle interactions $\widehat{h}$, then one can represent M_{ij} as a linear combination of elementary one- and two-electron integrals, the latter being called

quantum chemistry integrals. A general expression for the *one-electron integral* has the form

$$[ab|c] = \int \varphi_a^{(1)}(\nu) \widehat{h}_c(\nu) \varphi_b^{(2)}(\nu) \, dV_\nu,$$

where ν is the coordinate of the νth electron; a, b and c are the coordinates of nuclei on which the functions φ and the operator $\widehat{h}$ (superscript indicates the possibility of different forms of the φ_a functions) are centered. The Coulomb operator of the interaction between electron ν and nucleus c is designated by $\widehat{h}_c(\nu)$. In this case the integral is usually referred to as a *direct integral*. If the operator $\widehat{h}$ is independent of ν, the integral $S = \int \varphi_a^{(1)}(\nu) \varphi_a^{(2)}(\nu) \, dV_\nu$ is called an *overlap integral*. One-electron integrals are triple integrals. A general expression for a two-electron integral has the form

$$[ab|cd] = \int \varphi_a^{(1)}(\nu) \varphi_b^{(2)}(\nu) \widehat{h}(\nu, \mu) \varphi_c^{(3)}(\mu)$$
$$\times \varphi_d^{(4)}(\mu) \, dV_\nu \, dV_\mu.$$

The Coulomb interaction between electrons ν and μ is an example of $\widehat{h}(\nu, \mu)$. In this case the integral is called a *repulsion integral*. Two-electron integrals are sextuple integrals.

Integrals are also classified as to their centering. If all positions a, b, c, and d are different, we have a *four-center integral*, if there are three positions, the integral is a *three-center integral*. In a similar way, there may be *two-center integrals* and *one-center integrals*. The three and four-center ones are known as *multicenter integrals*. Particular two-center integrals are the Coulomb $[aa|bb]$, the hybrid $[aa|ab]$, and the exchange $[ab|ab]$ integrals. Accordingly, the three-center integrals include Coulomb-hybrid $[aa|cd]$ and hybrid-exchange $[ab|ad]$ integrals.

As distinct from the elementary integrals, there are quantum chemistry integrals with φ and $\widehat{h}$ represented by a linear combination of functions or operators. Thus, the matrix elements of the effective one-electron complete Hamiltonian $\widehat{h}_{\text{eff}}$ introduced in the LCAO method (*Linear Combination of Atomic Orbitals*) are as follows:

$$m_{ij} = \int \varphi_i(\mu) \widehat{h}_{\text{eff}}(\mu) \varphi_j(\mu) \, dV_\mu.$$

Here the so-called group integral m_{ij} is called a *Coulomb integral* for $i = j$, and a *resonance* integral for $i \neq j$. When using *group theory* to deal with symmetric systems, it is convenient to utilize *symmetrized functions* that transform according to a particular irreducible representation of the symmetry group, and are linear combinations of atomic functions of *ligands*. In this case the matrix element m_{ij} is called a *group integral*. A special case is the *overlap group integral* $S_{\text{group}} = \int \varphi_i{}^{\text{sym}}(\mu)\varphi_j{}^{\text{sym}}(\mu)\,dV_\mu$. One can express the Coulomb, resonance, and group integrals through elementary integrals.

The evaluation of one- and two-center integrals involves no special difficulties. They can generally be expressed in a so-called closed form, i.e. as a finite linear combination of elementary and special functions. For a long time no one has succeeded in expressing multicenter integrals in a similar way. Therefore, these integrals are evaluated with the help of approximation formulae, representing expansions in infinite series; on the basis of semiempirical expressions; by the method of integral transformations; by the use of Gaussian functions; and by direct numerical integration of the initial expression. The most natural, exact analytical calculations lead to closed-form expressions that are difficult to evaluate. This is related both to the absence of a general systematic approach for the evaluation of multicenter integrals, and to the cumbersome nature of the analytic procedures. Only in recent years, owing to the development of methods for the calculation of multicenter integrals, and the use *computer algebra*, have solutions for some of these integrals been attained, including the most complicated four-center quantum chemistry types. These quantum chemistry integrals underlie the microscopic theory of atoms, molecules, and condensed media.

QUANTUM CRYSTALS

Crystals in which the translational and rotational motion of the constituent atoms and molecules cannot be properly described in the classical approximation, but is determined by quantum-mechanical laws. The existence of quantum crystals is a consequence of *Heisenberg's uncertainty principle*, according to which the indeterminacies of positions of particles lead to the presence of finite kinetic energies in the ground state. The ground state translational and rotational motion energies are comparable with the potential energies of their central ε and non-central ε_φ interactions. Quantitative features of a quantum crystal are the dimensionless quantum parameters of translational motion $\lambda_t = \hbar/[\sigma(m\varepsilon)^{1/2}]$ (*de Boer parameter*) and rotational motion $\lambda_r = \hbar^2/(I\varepsilon_\varphi)$ (where m, I, and σ are the mass, moment of inertia and diameter of the particles), which to a first approximation are proportional to the potential energy. Some known values of these translational parameters are: ^{3}He ($\lambda_t \approx 0.5$), ^{4}He ($\lambda_t \approx 0.4$), H_2 ($\lambda_t \approx 0.3$), HD ($\lambda_t \approx 0.25$), D_2 ($\lambda_t \approx 0.2$). Classical calculations give imaginary frequencies for at least one and in some cases all of the acoustic branches of the energy spectrum of these crystals. Examples of rotational motion parameters are: H_2 ($\lambda_r \approx 5$), HD ($\lambda_r \approx 4.5$), D_2 ($\lambda_r \approx 4$), CH_4 ($\lambda_r \approx 1$).

Dynamic crystal lattice features of quantum crystals are anharmonicity (see *Anharmonic vibrations*) of translational and orientational vibrations even at $T = 0$ K, and the coordinated motion of neighboring particles in the lattice (short-range correlations). These features change the relations between particle parameters and macroscopic equilibrium properties, but the properties themselves only undergo quantitative changes.

The large amplitude of *zero-point vibrations* and low potential barriers result in a comparatively high tunneling probability (see *Tunneling phenomena in solids*) of particles to nearby sites in the crystal, or to equivalent orientational positions at the same site. This leads to the translational and orientational delocalization of particles, and the former causes *defects* of the crystal (impurities, vacancies) to transform into *quasi-particles* (impuritons, vacancions). Some important effects in quantum crystals (*quantum diffusion*, localization in solutions of ^{3}He in ^{4}He, *crystallization waves*, nuclear, magnetic ordering in ^{3}He) are associated with tunneling processes.

The rotation of the molecules in the crystals (or in sublattices, as with CH_4) is almost unrestricted. A consequence of this is the significant influence of the total nuclear spin of molecules on the spectrum of their rotational motion. As a result, the different nuclear-spin modifications of quantum *molecular crystals* differ dramatically in their properties. This circumstance, together with the tunneling processes, leads to the strongest manifestation of the quantum properties of these crystals, quantization relative to the rotational motion of molecules (quantum diffusion in hydrogen, negative thermal expansion in methane, etc.). The quantum nature decreases as the pressure rises.

QUANTUM CYCLOTRON RESONANCE

Resonant absorption of an electromagnetic wave by a *metal* in a magnetic field B parallel to the surface of a sample. It occurs because the quantization of electron energy levels by the magnetic field produces a well-resolved discrete spectrum of cyclotron frequencies Ω_n ($n = 1, 2, \ldots$), spaced at a value $\geqslant \tau^{-1}$, on the *Fermi surface* (τ is the characteristic electron lifetime). The values $\Omega_n = \Omega(\varepsilon_{\mathrm{F}}, p_{zn})$, where $\Omega(\varepsilon_{\mathrm{F}}, p_{zn})$ is the *cyclotron frequency* corresponding to the given value of the longitudinal momentum p_z of the electron, ε_{F} is the *Fermi energy*, and p_{zn} are the values of p_z on the Fermi surface, are determined by the *Lifshits–Onsager quantization rule* $S(\varepsilon_{\mathrm{F}}, p_z)/(e\hbar B) = 2\pi(n + 1/2)$, where S is the area of the Fermi surface section. This means that the separation between neighboring Ω_n is of the order of $\delta\Omega = (e\hbar B/S_0)\Omega_0$, where Ω_0 and S_0 are the characteristic values of Ω and S. Typically, $\delta\Omega \leqslant (10^{-3}-10^{-4})\Omega$. If $\delta\Omega \gg \tau^{-1}$, then narrow windows of resonant absorption appear $|\Omega - l\Omega_n| \leqslant \tau^{-1}$ (Ω is the frequency of the electromagnetic wave, l is the multiplicity of the resonance). Pertinent to quantum cyclotron resonance are the particular *giant quantum oscillations* of impedance in B^{-1} with period slowly changing with B, the characteristic value of the latter being $\sim e\hbar/S_0$. At $\delta\Omega \ll \tau^{-1}$, the amplitude of quantum cyclotron resonance oscillations is exponentially small, and the ordinary case of classical *cyclotron resonance* takes place. The smearing of the Fermi distribution edge also causes damping of the oscillations according to the exponential law $\exp[-2\pi^2 k_{\mathrm{B}} T/(\hbar\overline{\Omega})]$ (the parameter $\overline{\Omega} \sim \Omega_0$). The observation of quantum cyclotron resonance in fields $B \sim 10$ T on the main electron groups of a typical metal requires values $\tau \geqslant 10^8$ s. Quantum cyclotron resonance was predicted by I.M. Lifshitz in 1961.

QUANTUM DIFFUSION

Low-temperature *diffusion* occurring under conditions for which quantum tunneling is significant (see *Tunneling phenomena in solids*). It can occur when *defects* interact with each other, with *phonons*, and with other defects. It was predicted by A.F. Andreyev and I.M. Lifshitz (1968) and observed experimentally by V.N. Grigoryev, B.N. Yeselson, V.P. Mikheyev and Yu.E. Shulman (1972).

The coefficient of quantum diffusion has an anomalous temperature dependence, and is very sensitive to the concentration of impurities. For a small defect concentration the delocalization of the diffusing particle occurs as the temperature decreases; and it transforms to the state of band motion when its *mean free path* is determined by the interaction with phonons, and the coefficient of quantum diffusion $D \sim T^{-9}$, where T is the absolute temperature of the crystal. If the concentration of defects is considerable then at $T = 0$ K the localization of the diffusing particle occurs in the inhomogeneous elastic fields of defects, and its motion is possible only due to the interactions with phonons, and hence $D \sim T^9$. A similar temperature and concentration dependence of quantum diffusion is observed in *helium* for ^{3}He atom transport in a solid ^{4}He matrix.

Quantum diffusion sometimes occurs along particular lines in a crystal containing *dislocations*, acquiring thereby a one-dimensional character. It is observable in deformed solid hydrogen.

QUANTUM ELECTRONICS

See *Solid-state quantum electronics*.

QUANTUM FERROELECTRICS

Ferroelectrics with their *Curie points* at low-temperatures where *zero-point vibrations* dominate. The term has been introduced to stress that the temperature corresponding to the stability limit of the *ferroelectric phase* is determined by the balance between long-range dipole–dipole forces (see *Dipole–dipole interaction*) and short-range repulsive forces associated with zero-point vibrations. Quantum ferroelectricity may be induced by the experimental introduction of a small amount of impurities or by applying pressure. Low-temperature *phase transitions* were induced in a range of *virtual ferroelectrics* by impurities

($K_{1-x}Li_xTaO_3$, $KTa_{1-x}Nb_xO_3$, $Pb_{1-x}Ge_xTe$, $x > x_c$, $x_c \geqslant 0.01$ is the critical concentration), or by axial pressure P ($KTaO_3$, $P > P_c$, $P_c \approx 0.56$ GPa is the critical pressure). The shift of T_c to the quantum range was obtained in normal ferroelectric materials in some range of hydrostatic pressure $P > P_c$ where $\partial T_c / \partial P < 0$ ($P_c \cong 1.3$ GPa for KH_2PO_4). Dielectric, hypersound and pyroelectric measurements show that a characteristic feature of quantum ferroelectrics is their dependence on T and P: $T_c \sim (y - y_c)^{1/2}$; $P'(T = 0 \text{ K}) \sim (y - y_c)^{1/2}$; $\varepsilon^{-1}(T = 0 \text{ K}) \sim y - y_c$, where y is the impurity concentration or pressure, y_c is its critical value, P' is the polarization, and ε is the dielectric constant. The properties of quantum and normal type ferroelectrics coincide outside the temperature range of the quantum domain.

QUANTUM GENERATOR

See *Solid-state quantum electronics*, *Laser*, *Solid-state laser*, and *Optical parametric oscillator*.

QUANTUM GYROSCOPE

A device for determining the angular speed of rotation relative to a specified axis. Its operation is based on the quantum-mechanical properties of particles forming the working material of a sensor. One distinguishes two groups of quantum gyroscopes: devices oriented by the field lines of external magnetic fields (interplanetary and Earth's), and devices oriented relative to an inertial coordinate system. To the first group belong *nuclear and electronic quantum gyroscopes* whose sensitive element is the precession frequency of intrinsic magnetic moments (nuclear or electronic spin) depending on the angle of the device axis relative to the external magnetic field strength vector. The second group comprises *optical* and *superconducting quantum gyroscopes* with their working principle based on the difference of the transit time of electromagnetic waves and quasi-particle excitations (see *Quasi-particle*) moving along a closed circuit (ring resonator) to meet each other in the rotating coordinate system (analogue of Michelson's experiment). At this there is a mixing of the oscillations arriving at the oscillator from opposite directions, and their difference frequency is proportional to the rotation speed of the ring resonator. In optical quantum gyroscopes one uses *lasers* as oscillators, and in superconducting gyroscopes one uses *Josephson junctions*.

QUANTUM HALL EFFECT

A quantum phenomenon of the transverse (Hall) conductivity $\sigma_{xy} \equiv \sigma_H$ and its inverse the transverse (Hall) resistance $\rho_{xy} \equiv R_H$ in a *two-dimensional electron gas*, or 2D-electron gas (e.g., in the *inversion layer* of semiconductor *field-effect transistors* (M–O–S structures like metal–SiO_2–Si) and in *heterostructures* (as GaAs–$Ga_{1-x}Al_xAs$)), present in a strong transverse magnetic field $\boldsymbol{B} \parallel z$ with the intensity $B \geqslant 10$ T (100 kG) at low temperatures $T \leqslant 1$ K. The effect was discovered by K. von Klitzing et al. (1980).

The *integer (normal) quantum Hall effect* (QHE) is the series of plateaus observed on the dependence R_H or σ_H on B, or the surface concentration of 2D-electrons N_s (see Fig.) when values of R_H or σ_H correspond to integer values of the filling factor $v = N_s/N_0$, where $N_0 = eB/h$ is the number of states on Landau levels (see *Quantizing magnetic field*) per unit area and plateau, $\sigma_H = (e^2/h)n$ ($n = 1, 2, 3, \ldots$), e is the electron charge, h is Planck's constant. The deep minima (practically to zero) of diagonal components

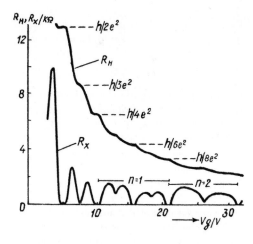

Quantum Hall effect in a silicon metal–oxide–semiconductor field-effect transistor (MOSFET) showing the dependence of the transverse Hall resistance R_H (upper curve) and the longitudinal resistance R_x (lower curve) on the voltage V_g applied to the gate. This voltage V_g is proportional to N_s. The magnetic induction $B = 18.9$ T, and $T = 1$ K.

of the conductivity tensor σ_{xx} and the resistance tensor $\rho_{xx} = \sigma_{xx}/(\sigma_{xx}^2 + \sigma_{xy}^2) = R_x$ correspond to all the plateaus, so $R_H = 1/\sigma_H = h/e^2 n = R_0/n$. The quantity $R_0 = h/e^2 = 25{,}812.8056\,\Omega$, called the *quantum of resistance*, or the *Hall resistance*, is expressed through the fundamental physical constants h and e. The qualitative interpretation of the integral quantum Hall effect is based on concepts concerning the structure of the single-particle electron spectrum of a degenerate 2D-system in a transverse magnetic field when a random potential from *defects* (impurities) is present. Due to this potential all the states in the gaps of the discrete spectrum $E_n = \hbar\omega_H(n + 1/2)$ are localized, where $\omega_H = eB/m^*$ is the *cyclotron frequency*, m^* is the *effective mass*, $\hbar = h/2\pi$, and $n = 0, 1, 2, 3, \ldots$ (see *Anderson localization*). Delocalized (extended) states remain only in the vicinity of Landau levels. Under the influence of the electric field $E \parallel x$ these states carry the Hall current $j_y = \sigma_H E$ (in the y direction) and the dissipative current $j_x = \sigma_{xx} E$ (in the x direction), if the *Fermi level* is near one of the Landau levels. When N_s or B changes, the Fermi level shifts into the region of localized states, and the dissipative conductivity disappears ($\sigma_{xx} = 0$, $\rho_{xx} = 0$ at $T = 0$). These facts explain the deep dip in the dependence of $R_x = \rho_{xx}$ on N_s (see Fig.). In this case $\sigma_H \neq 0$, since due to the drift of carriers in the crossed fields E and B the delocalized states on the totally full Landau levels below the Fermi level contribute to $\sigma_H = 1/R_H$. The quantum drift arises from the mixing (hybridization) of wave functions of neighboring Landau levels under the influence of the electric field E. The filling of localized states does not change when the Fermi level is shifted into the localized states (in gaps between Landau levels), so the value σ_H remains constant in the corresponding regions where B or N_s change. This explains the presence of plateaus on the dependences of R_H on B and on N_s.

The *fractional quantum Hall effect* was was discovered by D.C. Tsui, H.L. Störmer and A.G. Gossard (1982). This effect is characterized by the presence of anomalies (breaks) in the dependences $R_H(B)$ and $R_H(N_s)$ at fractional values of the filling factor $\nu = p/q$, where p and q are integers, and q is an odd number ($q = 3, 5, 7, \ldots$). According to modern conceptions, the fractional quantum Hall effect results from electron–electron interactions, forming new phases in the 2D electon system, the so-called *incompressible Fermi liquid* (see *Fermi liquid*), having finite *energy gaps* in the spectrum of elementary excitations. These excitations involve so-called intermediate statistics, i.e. they are *anyons*. The fact that q is an odd number is associated with the asymmetry of the wave functions of the 2D-system degenerate state in a magnetic field. The requirement of translation and gauge symmetry (*invariance*) restricts the possible occupation numbers of Landau levels to the discrete multiplicity of the rational numbers. The fractional Hall quantum effect is observed in perfect 2D-structures, where the effects of interelectron correlation involve the scale of energies $e^2/\varepsilon\lambda$ (ε is the *dielectric constant*, $\lambda = (\hbar/eB)^{1/2}$ is the *magnetic length*, i.e. the orbit radius of the lowest Landau level) which exceeds the quantum uncertainty of energy $\hbar/\tau$ (τ is the relaxation time), related to electron scattering by defects, and causing the appearance of localized states.

The quantum Hall effect is used in metrology as the source of an *absolute standard of resistance*, and also for a more accurate determination of the constants h and e (if an independent accurate resistance standard is available).

QUANTUM INTERFERENCE PHENOMENA
in metals

The aggregate of macroscopic effects determined by the interference of quasi-classical electron waves scattered at static centers, or on the surface of a sample. One should distinguish two main types of quantum interference phenomena in normal *metals*, those in disordered metals and those in pure metals.

The first type involves quantum interference phenomena in disordered metallic structures, which arise from the interference of quasi-classical electron waves scattered off randomly dispersed impurities. Two main spatial scale sizes exist: the size of the sample L and the *phase coherence length* L_φ – the distance over which the electron diffuses during the time of quasi-classical electron wave phase loss due to inelastic scattering, or *electron spin-flip scattering*, i.e. L_φ is the characteristic size of the domain in which all the impurities take part in forming a composite interference pattern. In the case $L_\varphi \ll l$ (l is the mean free path) the *phase coherence* is totally destroyed, and the kinematics of the metal are described on the basis of a classic *kinetic equation*. For $L \gg L_\varphi \gg l$ small quantum-interference additions to classical *kinetic coefficients* arise. These additions depend strongly on the frequency, external field, temperature, and weak magnetic field, and therefore, they can be detected on a background of the main effect. The quantum interference addition to the static conductivity $\Delta\sigma$ is always negative, i.e. the interference leads to the enhancement of a metal's resistance. The theory describing the properties of $\Delta\sigma$ is referred to as the theory of *weak localization*. In low-dimensional samples the value of $\Delta\sigma$ can become of the order of the conductivity itself. In this case one talks about *strong localization* (*Mott–Anderson localization*). At $L_\varphi \gg L$ the interference pattern is formed by all impurities and essentially depends on the configuration of their distribution. This leads to strong dispersion of physical properties of samples prepared in a similar manner, i.e. kinetic characteristics become *self-averaging quantities*. The last statement relates not to microscopic objects, but to systems consisting of many atoms. This branch of the physics of metals is referred to as *mesoscopics*.

The second type of quantum interference in pure metals arises, first, in the presence of a *magnetic breakdown* as a result of interference of quasi-classical electron waves scattered off the centers of the magnetic breakdown (called *coherent magnetic breakdown*); and second, at the *multichannel specular reflection* from the surface of a sample, where they are determined by the interference of quasi-classical waves of different channels of the multichannel specular reflection. The anomalous sensitivity to weak external influences which do not change the classical dynamics of electrons, but rebuild the interference pattern, is characteristic of quantum interference phenomena in pure metals. It involves a broad range of non-linear effects (for more details see *Magnetic breakdown*, *Quantum size effects*).

QUANTUM KINETIC EQUATION

An equation of quantum statistical physics describing the time evolution of the *distribution function* $\rho_\alpha(t)$ of the probabilities of occupancy by the particles of the full set of states with quantum numbers α. This equation for the diagonal elements of the statistical operator (see *Density matrix*) of the system, called the *master equation*, has the form

$$\frac{d}{dt}\left[\rho_\alpha(t)\right] = \sum_\beta (\rho_\beta W_{\beta\alpha} - W_{\alpha\beta}\rho_\alpha),$$

where $W_{\alpha\beta}$ is the intensity of transitions of particles per unit time from state α to state β. W. Pauli initially derived this equation by perturbation theory for the interaction of particles on the assumption that the phases of the unperturbed system during their evolution in time are incoherent. This assumption is analogous to the molecular chaos hypothesis used by L. Boltzmann for deriving the classical kinetic equation (see *Boltzmann equation*). In this sense the quantum kinetic equation has the same physical meaning as the Boltzmann equation; i.e. it connects changes of probabilities $\rho_\alpha(t)$ in a small time interval with the balance of the probability flux entering the quantum state α and exiting from it. Another approach for describing the evolution of nonequilibrium quantum systems of many bodies has been developed on the basis of the idea of a reduced description introduced by N.N. Bogolyubov. In the framework of this approach a method has been created that allows the construction of quantum kinetic equations describing the time evolution of a single-particle distribution function in the form of asymptotic series in powers of a small parameter. In particular, if the small parameter is the ratio of the interaction radius to the mean de Broglie wavelength, then in the first non-vanishing approximation the kinetic equation for a gas of *quasiparticles* appears, which plays an important role in solid state physics, and which, in addition to the Boltzmann equation, accounts for pair-only collisions:

$$\frac{df_p(t)}{dt}$$

$$= \pi \sum_{p',p_1 p_1'} \left| T^+_{pp',p_1 p_1'}(\varepsilon_p + \varepsilon'_p) \right|^2$$

$$\times \delta(\varepsilon_p + \varepsilon_{p'} - \varepsilon_{p_1} - \varepsilon_{p_1'})$$

$$\times \left\{ f_{p_1} f_{p_1'}(1 \pm f_{p'})(1 \pm f_p) \right.$$

$$\left. - f_p f_{p'}(1 \pm f_{p_1})(1 \pm f_{p_1'}) \right\}.$$

Here ε_p is the energy spectrum of quasi-particles, $T^{+}_{pp',p_1p_1'}$ is calculated on the mass surface scattering matrix depending on the momenta of collided and departing quasi-particles, and the signs $+$ and $-$ correspond to *Bose–Einstein statistics* and *Fermi–Dirac statistics*, respectively. This equation agrees with the *Pauli equation* if the intensities $W_{\alpha\beta}$ are calculated in representations of occupation numbers, and the transitions associated with pair collisions are taken into account.

QUANTUM LIQUID

The special state of a liquid at which the *de Broglie wave* length for the thermal motion of its constitutent particles becomes comparable with the distance between them. This is possible only at very low temperatures (see *Low-temperature physics*) when the existence of the liquid state is associated with quantum effects (*zero-point vibrations*). If the zero-point vibrational energy is comparable with the energy of particle interactions then the body will remain liquid until an infinitely low temperature. This condition (due to the weakness of the atomic interaction and the small atomic mass) is satisfied for the stable isotopes ^{3}He and ^{4}He of *helium*. According to the statistics of its particles, a quantum liquid is a *Bose fluid* (^{4}He), a *Fermi liquid* (^{3}He, *conduction electrons* in normal metals, *neutron liquids*), or a mixed type liquid (^{3}He–^{4}He solutions).

A quantum liquid possesses a number of specific properties, the most remarkable of which is *superfluidity*. This is typical of a Bose-liquid, and is associated with the phenomenon of Bose-condensation (see *Bose–Einstein condensation*) resulting in the appearance of a superfluid condensate in a quantum liquid. These results were explained by the phenomenological theory of quantum liquids (in particular, the two-fluid hydrodynamics of a quantum liquid) constructed by L.D. Landau, and by the microscopic theory of a Bose quantum liquid developed by N.N. Bogolyubov.

A Fermi quantum liquid can be either superfluid or normal (non-superfluid). The distinctive feature of a normal type is the possibility of propagating undamped *zero sound* (oscillations of the *Fermi surface* relative to its equilibrium value).

In a Fermi-liquid with attractions between particles, coupled states of two particles are created (*Cooper pairs*) below some temperature. These pairs are described by Bose statistics, and after their appearance the quantum liquid transforms to a superconducting state. The coupling may occur with zero-valued (see *Superconductivity*) and non-zero *orbital angular momenta* (*superfluid phases of* ^{3}He).

QUANTUM MAGNETOMETER

See *Magnetometry*.

QUANTUM MECHANICS in solid state physics

Basic theory introduced in the early 20th century which explains the physics of condensed systems consisting of a huge number of particles ($\sim 10^{22}$ per cm^{3}). The quantum theory of *crystals* has been developed in detail, while that of amorphous bodies is more qualitative (see *Amorphous state*).

One of the main results of this quantum approach was the concept of *quasi-particles*. The energy of an excited state of a crystal which is close to its ground state can be represented as a sum of the energies of individual quasi-particles. This allows one to introduce the concept of a "gas" of quasi-particles, and to use the methods of the kinetic theory of gases, for investigating thermal, magnetic and other properties of solids. The macroscopic characteristics of solids are expressed through those of quasi-particles (*mean free path*, speed, etc.). Quasi-particles do not exist in free space, but they are present in a crystal lattice whose structure is reflected in the properties of quasi-particles. Applications of the quasi-particle approach for the description of excited states of amorphous materials has not been very successful.

There are several characteristic features of solids considered as physical objects containing many particles.

1. Atoms, molecules, and ions are the structural units of a solid, and their interaction energy is small compared to the energy needed to destroy the structural unit itself. At the same time, this interaction energy is not small compared to their energy of thermal motion, hence a solid body is a system of strongly interacting particles.

2. Classically the mean energy of thermal motion of particles is $\sim k_B T$. At high temperatures the thermal energy of a solid is $\varepsilon \sim 3N k_B T$, where N is the number of particles. When lowering the temperature the energy of a solid decreases faster than expected from classical physics since the discrete (quantum) character of the energy spectrum leads to "freezing out" motion at $T \to 0$ K. The greater the energy separations between the levels, the higher the temperature of the "freezing-out" of the corresponding motion, so different motions in solids freeze out at different temperatures.

3. Because of the variety of forces acting between the particles, properties associated with gases, liquids and plasma are all displayed in crystals under particular conditions. For example, a *metal* can be considered as an ionic lattice immersed in an electron liquid; a *ferromagnet* at $T \gg T_c$ behaves as a gas of magnetic arrows (the *magnetic susceptibility* of a solid *paramagnet* has same temperature dependence as a gaseous paramagnet); under the action of a high-frequency electromagnetic field an electron gas of a metal or *semiconductor* behaves like a plasma (see *Solid-state plasma*).

4. The motions of atomic particles are diverse, especially in view of their diversity of masses. Since ions are a thousand times heavier than electrons, the speeds of ions are small compared those of electrons. In some approximations (e.g., *adiabatic approximation*) when considering the electron motion the ions may be assumed to be stationary at their equilibrium positions.

5. All motions of atomic particles in solids may be divided into four types.

A. *Diffusion* of intrinsic and foreign atoms. During the course of *vibrations, fluctuations* can cause the kinetic energy of a particle to move it beyond the range of the potential well which confines it, and the particle is able to "break away" from its equilibrium position. The probability W of such a process at room temperature is extremely small, and rises with the temperature: $W = \nu_0 \exp[-U/(k_B T)]$ where $\nu_0 \sim 10^{12}$–10^{13} s^{-1}, and U is the order of the binding energy per particle. The life time of an atom at its site is much longer than the time of its transit, so it makes rare and random jumps. The *diffusion coefficient* is proportional to W. It rises near the *melting temperature* and depends on the condition of the crystal surface. *Plastic deformation* "loosens" a crystal, reduces potential barriers separating equilibrium positions of atoms, and enhances the probability of their "jumps". Diffusion is a rare example of classical motion of atoms in solids.

B. In particular cases, in e.g., solid He under pressure, the tunneling "percolation" of atoms from one equilibrium position to another is possible (see *Tunnel effect*). This process, which is referred to as quantum diffusion, involves a diffusion coefficient that is not equal to zero at $T = 0$ K. The possibility of tunneling converts *impurity atoms* and *vacancies* into unusual quasi-particles (vacancions, impuritons) which determine the properties of so-called *quantum crystals*.

C. There exist collective motions of particles on an atomic scale such as *crystal lattice vibrations*. The simplest such motion is a wave with a particular wave vector and corresponding frequency. Another example of collective motion on an atomic scale is the electronic excitation of an atom (e.g., by absorption of a photon or at an increase in temperature) which is not localized at a particular site of the crystal lattice, but rather jumps from site to site (*Frenkel exciton*). The energy of this process is of the order of the excitation energy of an individual atom. Collective motions on an atomic scale have a discrete structure; e.g., the energy of atomic vibrations associated with the frequency ω can be equal to $\hbar\omega$, $2\hbar\omega$, etc. The quasi-particles that can be used to describe atomic vibrations are called *phonons*. In *ferromagnets* and *antiferromagnets* close to $T = 0$ K deviations from magnetic order can propagate in the form of a wave (*spin wave*), and the resulting quasi-particle is a *magnon*. Different types of particle motion in solids are typically uncoupled, but sometimes there is a resonant interaction between unrelated wave processes, and their frequencies coincide. This leads to "mixing" of the motions; e.g., atomic vibrations (sound) can be excited by precessing atomic magnetic moments in an alternating magnetic field, and a sound wave can spontaneously transform to a spin type (see *Magnetoacoustic resonance*).

D. At low temperatures (e.g., close to $T = 0$ K) many *metals* become superconducting (see *Superconductivity*). The conduction electrons in a superconductor attract each other in pairs through the intermediary of an electron–phonon interaction, and couple together as so-called Cooper pairs which form a *Bose condensate*. The escape from the condensate requires some energy to overcome an *energy gap* which makes the superconducting motion stable. The superconducting state is characterized by the total absence of electrical resistance, and by anomalous magnetic properties (*absolute diamagnetism*).

6. To describe various phenomena and properties of solids one uses the concept of quantum gases of quasi-particles. For example, the thermal motion of atoms of a crystal lattice is described by phonons, and the *electrical conductivity* by a gas of *conduction electrons* and *holes*. The electrical resistance of metals and semiconductors is determined by the scattering of conduction electrons and holes at phonons and lattice defects (see *Current carrier scattering*). All quasi-particles (especially phonons) transfer heat, and according to the kinetic theory of gases the contribution of each gas to the *thermal conductivity* can be written as $\kappa = \beta c l \langle v \rangle$, where β is a numerical coefficient; c, $\langle v \rangle$, and l are the *specific heat*, mean thermal speed, and mean free path of the quasi-particles, respectively. Magnons contribute to magnetic and thermal properties of magnetic materials. The temperature dependence of the magnetization of ferromagnets and of the magnetic susceptibility of *antiferromagnets* at $T < T_C$ results from the "freezing out" of spin waves as the temperature decreases. Light absorption in semiconductors and *insulators* results from the presence of *Wannier–Mott excitons*.

7. At a particular temperature all degrees of freedom of atomic particles in a solid can generally be subdivided to two groups. For one group the interaction energy U_{in} is small compared to $k_B T$, while for another group it is large. If $U_{in} \ll k_B T$, then the corresponding degrees of freedom behave as an aggregate of gas particles, and if $U_{in} \gg k_B T$, then the corresponding degrees of freedom are ordered, and the motion may be described by a system of quasi-particles weakly interacting with each other. Therefore, in both limiting cases the "gas approximation" is valid. A good example is the magnetic moments of atoms: at $T > T_C$ they are a gas of magnetic arrows localized at crystal lattice sites, at $T < T_C$ they form a gas of magnons. Close to a *second-order phase transition* the "gas approximation" is inapplicable. The solid body behaves as a system of strongly interacting particles or quasi-particles, with the motions of atomic particles (atoms) correlated. The correlation has a special (not power law) nature: the probability of collective motion is as large as the probability of individual motions. It is displayed by the growth of fluctuations in anomalies of the *specific heat*, magnetic susceptibility, and others. Due to the variety of motions inherent to the particles of a solid, the temperature dependence of most characteristics is very complicated, and becomes further complicated at *phase transitions*, which are followed by abrupt changes of many quantities (specific heat, for example).

QUANTUM OF ACTION
The same as *Planck constant*.

QUANTUM RADIO-FREQUENCY GENERATOR
A device producing electromagnetic waves on the basis of the *induced radiation* effect. The physical processes taking place in quantum-mechanical generators are similar to those in *quantum amplifiers*, and the principal distinctive feature of these devices is the presence of positive feedback. Quantum generators are characterized by high frequency stability of the generated signal, which justifies their application as *quantum frequency standards*, devices to obtain electromagnetic oscillations with a frequency that is stable in time.

QUANTUM SIZE EFFECTS

Change of thermodynamic and kinetic properties of a *crystal* when at least one of its geometrical sizes is comparable with the de Broglie wavelength of electrons. In *metals* one finds macroscopic effects in *films* determined by the quantization of the finite motion of *conduction electrons* across the film (*size quantization*). I.M. Lifshitz and A.M. Kosevich predicted quantum size effects (1955), although the influence of size quantization on the electronic *specific heat* of metals had been noted earlier by Fröhlich (1937). The first experimental report was in 1966. Size quantization leads to the splitting of an energy band of conduction electrons into two-dimensional subbands $\varepsilon_n(P_\parallel)$ (n is the band number) depending on the longitudinal component $P_\parallel$ of the electron *quasi-momentum* relative to the surface of the film, and conserved in reflection. The characteristic distance between subbands is $\delta\varepsilon \sim \hbar v_F / L$ (L is the film thickness, v_F is the electron velocity at the *Fermi surface*). In the simplest case of *single-channel specular reflection* of electrons from the metal surface, when on the *isoenergetic surface* $\varepsilon(P) = E$ (where $\varepsilon(P)$ is the *dispersion law* of electrons) there are only two quasi-momenta $P_{1,2}$ with the given projection $P_\parallel$, the length $\Delta P(E, P_\parallel)$ of the segment connecting P_1 and P_2 is quantized: $\Delta P = 2\pi\hbar(n + \gamma)/L$ (the dimensionless constant $\gamma \sim 1$). Due to the anisotropy and periodicity of the dispersion law, $2m$ ($m \geqslant 2$) quasi-momenta on the Fermi surface correspond to the given $P_\parallel$. In such a case multichannel specular reflection arises: an electron incident on the boundary of a metal transforms to a quantum superposition of $m \geqslant 2$ states with a given energy and $P_\parallel$ value.

Most quantum size effects are based upon oscillations (on L and other external parameters) of the electron density of states $g(\varepsilon_F)$ on the edge of the Fermi distribution at ε_F. These oscillations have the same nature as in the *de Haas–van Alphen effect*. For a single-channel specular reflection the oscillation mechanism involves the fact that the size quantization terms $E_n(P_\parallel)$ cross the *Fermi level* ε_F sequentially as L (or another parameter) changes. When this takes place $g(\varepsilon_F)$ passes through a sharp maximum every time ε_F coincides with an extremal value of one of the

functions $E_n(P_\parallel)$. As a result all thermodynamic and kinetic features of a metal film oscillate together with $g(\varepsilon_F)$. Their periods along L are equal to $\Delta L = 2\pi\hbar/\Delta_{\text{ext}}$, where Δ_{ext} are the extremal values of $\Delta P(\varepsilon_F, P_\parallel)$. The amplitude of the *size oscillation* is small in accordance with the quasi-classical parameter $\hbar/L P_F \ll 1$ (P_F is the Fermi momentum). Thermal broadening at the edge of the Fermi distribution decays according to the exponential law $\exp(-k_B T/\Delta)$, where the parameter $\Delta \sim \delta\varepsilon$. Size oscillations also exist for the case of multichannel specular reflection (in spite of the disorder in the size quantization spectrum).

The imperfections of the sample surface which are responsible for the diffuseness of electron scattering also diffuse the levels of a size quantized spectrum, and hence a weakening of the quantum size effects takes place. The volume *electron scattering* on impurities, dislocations, and phonons causes the same effect. These quantum size effects are most pronounced in *semimetals* where the electron wavelength is greater than interatomic distances. To observe these effects in metals either a high perfection of the sample surface or a small thickness of the film is required ($l \approx 10$ nm). Quantum size effects have also been observed in films of semiconductor alloys. In a good metal (*tin*) *quantum size oscillations* of the superconducting transition temperature were detected.

Quantum size effects give important information regarding features of the electron energy spectrum in films, and concerning the nature of electron scattering from the boundary of a sample (see also *Size effects*).

QUANTUM STATES OF CHANNELED PARTICLES

The main result of the quantum theory of *channeling* which characterizes the motion of channeled particles in a periodic one-dimensional (*plane channeling*) or two-dimensional (*axial channeling*) continuous potential. By analogy with the *band theory* of crystals the particle motion is described by Bloch wave functions associated with discrete one-dimensional or two-dimensional energy bands. The observed values depend essentially on the populations of the quantum states of the channeled particles, which are determined by the angle of incidence of the particle on the crystal, and the transitions between different states which depend on the extent to which the actual interaction potential deviates from being continuous. The specific *channeling emission* is a result of radiative transitions of channeled particles.

QUANTUM VORTICES

Topologically stable extensive (linear) *defects* in macroscopic quantum systems (including *solids*) with *spontaneous symmetry breaking* and a complex (sometimes multi-component and anisotropic) *order parameter* with a coherent phase (see *Macroscopic quantum coherence*) in an axial vector gauge-invariant field slowly changing in space and time. The following systems are relevant: superfluid *helium* (^{4}He and ^{3}He) in a rotating container (see *Superfluidity*, *Superfluid phases of ^{3}He*), a *type II superconductor* in an external magnetic field, superfluid nuclear matter in rapidly rotating neutron stars (pulsars) possessing superstrong magnetic fields, the ground state (vacuum) of nonlinear (non-Abelian) gauge fields, and others.

Quantum vortices are represented by stretched or curved filaments with a normal state core of radius $r \sim \xi$ (*coherence length*), inside which the order parameter is suppressed, and outside it the undamped encircling super current or superfluid flux circulates, localized in a domain with radius $r \sim \lambda > \xi$. For instance, in type II superconductors λ corresponds to the *penetration depth of magnetic field*. The quantum vortices (see *Abrikosov vortices*) due to the repulsion of the encircling currents form a hexagonal *vortex lattice*.

A single quantum vortex in a superconductor carries one *flux quantum* with the value $\Phi_0 = h/2e = 2.0678 \cdot 10^{-15}$ T·m^2, and in superfluid helium – one *quantum of velocity circulation*. In non-homogeneous (heterogeneous) superconducting systems with weak Josephson bonds (see *Josephson effects*), in particular in layered crystals and ceramics (referred to as *Josephson media*), in weak magnetic fields *Josephson vortices* can exist which are quantum vortices of a macroscopically large radius. In systems with an anisotropic and multicomponent order parameter (^{3}He, nuclear matter) the structure and configuration of quantum vortices can be quite complex.

QUANTUM WAVES

Electromagnetic waves propagating in a *metal* in the presence of a strong magnetic field B. Quantum waves can appear in pure metals ($\nu \ll \Omega$) in the ultra-quantum limiting case $k_B T \ll \hbar\Omega$ (ν is the relaxation frequency of the electrons, $\Omega = eB/m$ is the cyclotron frequency), when the quantization of electron levels in a magnetic field (*Landau levels*) becomes manifest. The onset of quantum waves is closely connected with the phenomenon of *giant quantum oscillations* of a collision-free absorption. The spectrum of the quantum waves is acoustic, and they can be treated as "electron sound" in a degenerate electron gas.

L. McWhorter and M.G. May first pointed out the possibility of quantum waves (1964). Transverse quantum waves have a small *phase velocity*, while longitudinal ones have a speed of the order of the *Fermi velocity* (v_F). The spectrum of longitudinal quantum waves, in contrast to the transverse one, does not depend on the electron concentration.

QUANTUM WELL of a semiconductor structure

A semiconductor *heterostructure* including a thin (comparable to the effective *quasi-particle* de Broglie wavelength) layer of semiconductor with a small *band gap* sandwiched between wider-gap materials. In this case a potential well is created in the *conduction band* (as well as in the *valence band*, or in only one of these bands (see Fig.)) in which the transverse motion of the charge carriers is quantized, while their motion along the layer remains free. In addition to such cases, a more complicated energy spectrum can be realized. For example, if the conduction band bottom of the quantum well appears lower than valence band extrema of the neighboring layers (a situation realized in GaSb–InAs–GaSb), then a two-dimensional *semimetal* could be present, and by decreasing the quantum well thickness a semimetal-to-semiconductor transition can take place. In the case of a quantum well with nonsymmetric barriers the spin degeneracy of the energy spectrum is removed, and along with the kinetic energy $p^2/2m$ a contribution linear in the two-dimensional *quasi-momentum* p appears. One often experimentally investigates the structure of quantum wells. If the distance between the

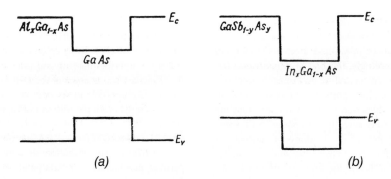

Energy diagrams for two types of heterostructures (a, b), where E_C and E_V are the edges of the conduction and valence bands, respectively.

atoms is sufficiently small so that the tunneling between size-quantized levels begins, then another type of heterostructure – a *superlattice* – is formed. The formation of a quantum well requires the production of *herojunctions* which lack any appreciable surface charge so the *band bendings* caused by the electric potential are small. This is possible for materials with a small disparity of the lattice constant d (for the structure GaAs–Al$_x$Ga$_{1-x}$As the lattice constants differ by only 0.3%) using modern methods of *heteroepitaxy*. If the difference between values of d is larger (up to 5%, making many more materials available), then one can realize only thin quantum wells (or superlattices) in which the disparity of lattices is compensated by the deformation of layers without the generation of *dislocations* at phase boundaries.

Modern technology not only provides sharp heterojunctions, but also can change the profile of the content of a superconductor alloy, e.g., to produce a parabolic or triangular quantum well. These possibilities have prompted the onset of a new semiconductor *materials science – band engineering* – ways to intentionally alter the content and geometry of a material to obtain a needed energy spectrum. The simplest feature of a quantum well energy spectrum based on GaAs consists in enhancing the energy gap by size quantization. In this case the edge of the region of optical transitions shifts to red or near-infrared frequencies. Based on such a structure, a visible band *injection laser* is produced, matched with fiber light guides (see *Fiber optics*), and used in modern systems

of information processing and optical communications (on the basis of In$_x$Ga$_{1-x}$As). The successes of quantum wells suggests creating other *semiconductor devices*, in particular integrated electron and optoelectronic circuits.

The *selective alloying* of a quantum well (when donor impurities are in the wide-band material remote from the heterojunction so "two-dimensional" electrons are ineffectively scattered) enables considerably enhanced low-temperature mobility of electrons in such structures. The enhancement of mobility up to $(2–8) \cdot 10^6$ cm$^2 \cdot$V$^{-1} \cdot$s^{-1} permits the use of such structures, as well as those of isolated heterojunctions with selective alloying, for new physical investigations of two-dimensional electron systems and their applications. The fractional *quantum Hall effect* was discovered on this system. *Field-effect transistors* based on heterostructures with selective alloying possess high switching speed, as well as enhanced high frequency and noise parameters.

QUANTUM YIELD

Ratio of the number of products of photoreactions (e.g., electrons of a *photoeffect*, quanta of *photoluminescence*) to the number of *photons* that enter the irradiated medium.

QUASI-AMORPHOUS STATE

The latent crystalline state of a solid where the long-range order (see *Long-range and short-range order*) cannot be resolved by accepted methods of *structure studies*. Depending on the physical principle and resolving apparatus (optical or *electron microscopy, electron diffraction* or *X-ray diffraction*, neutron diffraction, etc.), one can detect internally ordered grains of solid phase from 10^3 to 1.0 nm in size. Accordingly, the line of demarcation between evidently crystalline and quasi-amorphous materials will shift. Only the highest-resolution apparatus enables one to assert reliably that the body under study is really amorphous. Still, some macroscopic properties (isotropy of mechanical properties, character of fracture) and microrelief sometimes can reveal the fine structure of a substance. The quasi-amorphous state is especially typical of *thin films* obtained by *condensation* from molecular and ionic-molecular beams, as well as from plasma on cold substrates. With fast particles (energy 10–100 eV and higher) present in the condensing flux, the implantation of the high-energy component into forming grains occurs alongside the surface-diffusion mechanism of structure formation, which is typical of growth from uncondensed phases. As a result, secondary structure with block sizes of the order of 1–10 nm forms inside the grains with diameters of the order of 10^2–10^3 nm, which complicates the structural diagnostics of the films. In contrast to the truly *amorphous state*, the quasi-amorphous one is rather stable even in the case of metallic films, and finds its use in *microelectronics*.

QUASI-BRITTLE FAILURE

Type of *failure* of a solid, intermediate between ideally *brittle failure* and *tough failure*. Quasi-brittle failure is most common for *metals* with a body-centered cubic lattice at low temperatures. An ideally brittle (with no traces of *plastic deformation*) failure of metals never takes place, since the emergence of the failure source itself, a *nucleating crack*, the creation of a group of lagging *dislocations* in the active *slip band*, is required. In the framework of the *microspalling* model, the quasi-brittle failure of solids is describable with a quantitative criterion $j = K_{Ve}$. Here, $j = \sigma I / \sigma_i$ is the *strained state rigidity*, $K_{Ve} = R_{MSe} / \sigma_e$ is the *viscosity* coefficient for *iron* or *steel*, plastically deformed to the deformation extent e (σI is the major principal stretching stress, σ_i is the intensity of the stress, R_{MSe} is the *microspalling resistance*, and σ_e is the flow *strain* of a material deformed with the extent e). This criterion is met in iron

and steel with a body-centered cubic lattice with a moderate amount of deformation $e \approx 0.5$–0.8. At a higher level of deformation the plastic failure arises through the merging and growing of *pores* that form a *viscous crack* at the failure site. Such a crack is observable in the *neck* of plastic metals immediately before the moment of rupture.

QUASI-CHARACTERISTIC RADIATION

Electromagnetic radiation of a fast charged particle moving in a *monocrystal*, which arises from radiation transitions between different energy levels (zones) of lateral or transverse motion. Charged particles moving in a monocrystal under the conditions of planar and axial *channeling* are characterized by energy levels (bands) of bound transverse motion. Such quasi-bound motion in the regime of planar or axial channeling can be considered as a one- or two-dimensional *atom*. In the coordinate system moving relative to the laboratory one at the longitudinal component of the particle speed $\beta = p_{\parallel} c / E$, radiation transitions between levels (bands) of different parity are accompanied by electromagnetic emission at a frequency equal to the difference between two energies of the transverse motion. In the laboratory coordinate system the frequency of quasi-characteristic radiation is determined in accordance with the *Doppler effect* formula

$$\omega_{nm}(\theta_\gamma) = \frac{1}{\hbar} \frac{(E_n - E_m)\sqrt{1 - \beta^2}}{(1 - \beta \cos \theta_\gamma)},$$

where E_n and E_m are the energies of the initial and final states of the transverse motion of the particle, and θ_γ is the emission angle. The mechanism associated with these radiation emission transitions has a number of special features. The maximum of the spectral distribution depends markedly on the energy of the incident particles. If the particles move parallel to crystallographic planes the radiation is linearly polarized, and the emission takes place within a narrow cone close to the direction of the incident particle momentum. The intensity of the radiation into a unit increment of solid angle $\Delta \Omega$ and frequency $\Delta \omega$ is much higher than that of *Bremsstrahlung*. It will be noted that the radiation characteristics (intensity, angular and frequency distribution, polarization) are functions of the crystal structure, and the parameters of the charged particle beam. This quasi-characteristic radiation has several important properties from the viewpoint of a source of radiation: the distribution of *photon* energy, its monochromatic nature, linear polarization, directivity within a narrow cone, and high intensity compared to other radiation sources for energies above 10 keV.

QUASI-CHEMICAL REACTIONS in solids

Reactions with the participation of *point defects*: *impurity atoms*, *vacancies*, and *interstitial atoms* of the host material. They comprise the formation of point defects, their rearrangement, including *segregation* of *point defects*, creation of *clusters*, *pores*; *sticking* (trapping with subsequent thermal release), trapping at extended defects; and *recombination* reactions (annihilation of complementary defects). Quasi-chemical reactions in solids have a number of typical features. For one, components of quasi-chemical reactions can be vacancies as well as atoms. Then the reaction of creating a *Frenkel defect* in a crystal is an analogue of creating an atom–antiatom pair in a vacuum, but differs from it by the smallness of the amount of energy involved (one to several eV), and the high probability of a reaction occurring through thermal mechanisms or excitation by high-energy particles. Therefore, quasi-chemical reactions in solids often occur with a variable total number of particles or reaction components.

Quasi-chemical reactions include not only those that change the chemical composition of a defect, but also those that change the positions of an atom in a crystal lattice. For example, transitions of a particular atom between substitutional sites in different sublattices of binary or multicomponent crystals, between positions at lattice sites and interstitial sites of different types (tetrahedral, octahedral, hexagonal), etc. Such transitions often cause changes in the atom mobility, and the efficiency of different channels of quasi-chemical reactions.

A potential barrier exists between components of many quasi-chemical reactions in solids due to the deformation of the closest atomic surroundings of a defect. This leads to a considerable reduction of the effective *cross-section* σ of the interaction of nearby defects, compared to the geometrical cross-section and its additional (caused by diffusion of a defect) strong dependence on the temperature: $\sigma \propto \exp[-\varepsilon/(k_B T)]$, where ε is the energy barrier height. In nonmetallic crystals, the barrier height can vary considerably due to a changing charge state of a defect. Due to small values of σ determined by the presence of the barrier many cases feature, along with reactions limited by the *diffusion* of reagents, those limited by an elementary act of trapping. A frequent distinction of the kinetics of quasi-chemical reactions in solids is the essential role of *unstable Frenkel pairs*. Due to their relatively small spatial separation, the components of such a pair have an enhanced probability of recombination by first-order kinetics compared to the probability of quadratic recombination (with the involvement of "another" pair of defects). In some nonmetallic crystals, the quasi-chemical reactions take place under the action of low-energy excitations of the electron subsystem.

Due to the wide variety of impurity atoms and intrinsic defects in *real crystals*, there is quite a large number of quasi-chemical reaction channels. Their relative effectiveness depends on the concentration of point and extended defects in the particular sample, on the temperature and external factors, and it varies with time as the defects are created (see sample aging or degradation under irradiation).

QUASI-CLASSICAL APPROXIMATION in
quantum mechanics, Wentzel–Kramers–Brillouin (WKB) approximation

A method of approximate solution of the Schrödinger equation in the case when the *de Broglie wave* length of a particle λ is much smaller than the characteristic length for the variation of the potential $U(r)$. In the one-dimensional case the stationary state Schrödinger equation has the form

$$\frac{d^2\psi}{dx^2} + k^2(x)\psi = 0, \tag{1}$$

where ψ is the wave function of a particle, $k^2 = p^2/\hbar^2$, and $p(x) = \{2m[E - U(x)]\}^{1/2}$ is the classical momentum of a particle with mass m and energy E. The solutions of Eq. (1) for $k^2 = $ const prompt the substitution of the form $\psi(x) = \exp[i\varphi(x)]$. Then the *quasi-classical condition* is the slowness of changes in the phase φ in accordance with the inequalities

$$\hbar \left| \frac{dp}{dx} \right| \ll p^2 \quad \text{and} \quad \frac{1}{2\pi} \left| \frac{d\lambda}{dx} \right| \ll 1. \quad (2)$$

To first order in the small parameter determined by inequality (2), the solution of Eq. (1) is given by

$$\psi(x) = \frac{C_{\pm}}{\sqrt{p(x)}} \exp \left\{ \pm \frac{i}{\hbar} \int p(x)\, dx \right\}, \quad (3)$$

where $C_{\pm}$ are constants.

The motion of a particle in a potential well is described as follows. The wave functions are calculated by Eq. (3) in the regions where $U(x) < E$ and $U(x) > E$, and are joined at the turning points determined from the equation $U(x) = E$. Outside the region of classically admissible motion (i.e. for $U(x) > E$), the wave functions must decay. At the joining, one should take into account that condition (2) fails to hold near the turning points; therefore, one has to track the variation of the wave function phase during passage through these points. To do this, the approximate solutions to the left and right sides of the turning points can be compared to the exact solutions of Schrödinger equation in the neighborhood of these points. Thus, two expressions can be obtained for the wave function in the region $U(x) < E$ (from joining at the turning points), which should coincide with each other. The last leads to the following condition:

$$\int_a^b p(x)\, dx = \left(n + \frac{1}{2} \right) \pi \hbar. \quad (4)$$

Here a and b are the turning points (where the potential $U(x)$ is regular), and $n = 0, 1, 2, \dots$ is the quantum number identifying the energy levels E_n found from Eq. (4).

In some cases (*harmonic oscillator*, hydrogen atom, and others) the expressions for E_n obtained from Eq. (4) coincide with the exact values derived from the Schrödinger equation. The integer n is equal to the number of zeros passed through by the wave function, and the distance between zeros is of the order of λ; so the quasi-classical criterion is satisfied for values of n that are not too small. The condition (4) is referred to as the *Bohr–Sommerfeld quantization rule*. In

1913, N. Bohr postulated this rule to interpret the absorption spectra of hydrogen atoms.

The quasi-classical approach also provides a satisfactory explanation for the tunneling of a particle through a potential barrier (see *Tunnel effect, Interband tunneling*), which is forbidden for a classical particle. In 1928, G. Gamow used this approach to explain α-particle decay. This quasi-classical method is easily generalized to cases with many degrees of freedom.

QUASI-CRYSTALS

Solids which, like crystals, possess long-range order (see *Long-range and short-range order*), but have a *point group of symmetry* that is not allowed for a periodic lattice. They were discovered by D. Shechtman et al. (1984) in the cooled melt of composition $Al_{0.86}Mn_{0.14}$. At present, scores of quasi-crystals are known, they are mostly intermetallides of aluminum (see *Intermetallic compounds*).

Quasi-crystals exhibit unusual properties in X-ray or electron diffraction patterns (see *X-ray structure analysis, Electron diffraction analysis*). The diffraction pattern, similar to that of a crystal, consists of distinct spots positioned with a symmetry forbidden for crystals, e.g., that of a regular decagon (ten-sided polygon; see *Crystal symmetry*). Being spatially uniform phases, quasi-crystals differ from symmetrical twin crystals (see *Twinning of crystals*) that may yield the same types of diffraction patterns.

The majority of known quasi-crystals are metastable, and may be obtained only under highly nonequilibrium conditions (e.g., during extra fast cooling from the melt, laser melting, incomplete annealing of *metallic glasses*, or alloy decomposition). The size of quasi-crystal particles formed under these conditions is of the order of 0.1 to 1 μm. However, there are stable quasi-crystals, e.g., substances of the compositions: $Al_{65}Fe_{15}Cu_{20}$, Al_6CuLi_3, Al_6CuMg_4. The grains of these quasi-crystals, grown under conditions close to equilibrium, have well-defined facets, a size of the order of 1 mm, and icosahedral symmetry which is forbidden for crystals.

The theoretical concept of the structure of quasi-crystals dates back to the so-called *Penrose lattices* (see Fig.). Whereas this lattice is not a periodic one, the Fourier image of a function defined

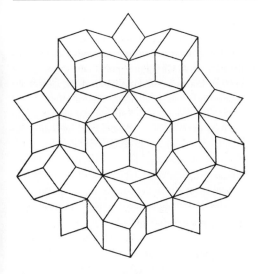

Sketch of a Penrose lattice.

at its sites consists, like that of a periodic function, of sharp δ-function peaks. The Penrose lattice (R. Penrose) is a particular case of patterns resulting from the limitation of periodic structures within a space of a larger dimensionality to a two- or three-dimensional cross-section of the latter. If in this case the intersecting hyperplane transforms into itself for some transformations from the point group of the symmetry of the multidimensional periodic crystal lattice, the corresponding symmetry persists in the cross-section as well. For instance, the symmetry of the cyclic permutation of the basis vectors of a five-dimensional cubic lattice transforms (with a rotation by $72°$) some two-dimensional planes into themselves. Accordingly, the section of the periodic structures by this plane produces quasi-crystalline patterns with rotation symmetry of the fifth order, in particular, the Penrose lattice. The majority of known quasi-crystals have the icosahedral symmetry that appears in the three-dimensional cross-section of a six-dimensional crystal lattice, although quasi-crystals are known with symmetry axes of orders 8, 10 and 12. In their physical properties, quasi-crystals are closest to *metallic glasses*. *Dislocations* that exist in them apparently cannot move under the effect of elastic stresses, so that quasi-crystals are incapable of *plastic deformations* and fail with a conchoidal fracture, typical

of brittle materials (see *Brittle fracture*). The *electrical conductivity* of quasi-crystals is low and increases weakly with increasing temperature, as is the case for *amorphous metals*. The similarity to glasses may be explainable by the high density of defects in real quasi-crystals; and because of the latter, the long-range order persists over distances, not exceeding 20 to 60 nm.

QUASI-EXPANDED UNIT CELL

A segment of a *crystal* which is selected for ease of computing the electronic structure of the crystal in the *cluster approximation*. A particular case of a quasi-expanded unit cell is the *periodic cluster model* which uses cyclic boundary conditions. It ensures the equality of equivalent atoms inside the volume of the molecular fragment, and on its boundary. The quasi-expanded unit cell is used for computations of a system corresponding to the region of a *Bravais lattice* with a volume which is an integer times that of a primitive *unit cell*.

QUASI-GAP in vibrational spectrum

A feature in the frequency spectrum of *crystal lattice vibrations* in the presence of a considerable concentration of *point defects*, which arises when an isolated defect creates *quasi-local vibrations*. It manifests itself by the appearance near the frequency of a quasi-local vibration of a frequency range in which the vibrations of a plane wave type having particular values of the wave vector cannot propagate. Then the *dispersion law* of those long-wavelength crystal vibrations that have the form of slightly damped plane waves possesses a forbidden frequency gap that is a quasi-gap. The frequencies that fall in the quasi-gap may correspond to *local vibrations* fortuitously arising from *clusters* of point defects. The quasi-gap appears also in spectra of other types of elementary excitations (magnons, electrons) when the energy of an isolated defect falls inside the continuous spectrum of elementary excitations, and the concentration of impurities is sufficiently large.

QUASI-HARMONIC APPROXIMATION

An approximation involving the *crystal lattice dynamics* of slightly anharmonic crystals, when the anharmonic terms in the Hamiltonian are replaced by effective harmonic terms, and the coefficients of the latter are found in a self-consistent way. In particular, the quasi-harmonic approximation is used to describe displacive *ferroelectric phase transitions*. An example is a Hamiltonian of the type

$$H = \sum_l \left(\frac{1}{2} a \xi_l^2 + \frac{1}{4} b \xi_l^4 \right) - \frac{1}{2} \sum_{ll'} V_{ll'} \xi_l \xi_{l'}. \quad (1)$$

The variable ξ_l describes the displacement of an active atom in cell l, $V_{ll'}$ is the interaction potential of atoms located in different cells (a and b are constants). The quasi-harmonic approximation corresponds to the substitution $\xi_l^3 \to 6\xi_l \langle \xi_l^2 \rangle$, where $\langle \xi^2 \rangle$ is the statistical average. Such an approximation corresponds to disregarding the interaction of the Fourier components of fluctuations with different wave vectors. The equation of motion for the Fourier components $\xi(q)$ and a particle of unit mass ($m = 1$) assumes the simple harmonic form (see *Harmonic approximation*):

$$\ddot{\xi}(q) = -\Omega^2(q)\xi(q) \quad (2)$$

with the quasi-harmonic frequency Ω

$$\Omega^2(q) = a + 3b \langle \xi_l^2 \rangle - V(q). \quad (3)$$

The self-consistent formalism is closed with the equation

$$\langle \xi_l^2 \rangle = \frac{1}{N} \sum_q \langle \xi(q)\xi(-q) \rangle$$

$$= \sum_q \frac{\hbar}{2\Omega(q)} \cot \left[\frac{\hbar \Omega(q)}{2k_B T} \right], \quad (4)$$

where q is the wave vector of a vibration. Here the statistical average is taken with the harmonic Hamiltonian $H = \sum \hbar \Omega(q) C_q^+ C_q$, and $C_q^+ C_q$ are the operators for the creation and annihilation of phonons. Eqs. (2)–(4) in the quasi-harmonic approximation determine the frequency of the *soft mode* as a function of temperature.

QUASI-LOCAL VIBRATIONS

The vibrations of a *real crystal*, which are related to a point defect or an extended *defect*. Ordinarily such vibrations spread across the entire crystal, but here the amplitude of the defect vibration is much greater than that of atoms in the bulk. The frequencies of quasi-local vibrations, in contrast to those of *local vibrations*, lie within the frequency bands of an *ideal crystal*. Their positions are usually close to the edges of these bands, where the density of vibrations of a defect-free crystal is small. In the case of a *point defect*, the density of vibrations has a narrow resonant peak at the *quasi-local frequency*.

QUASI-MOMENTUM

A characteristic of a particle (or a *quasi-particle*), moving in the periodic field of a *crystal*, which corresponds to the momentum of a free particle $p = \hbar k$, where $k = 2\pi/\lambda$ (λ is the de Broglie wavelength); and k also defines the wave function of a free particle, $\psi_k = \exp(ik \cdot r)$, where r is the coordinate. For a particle in a periodic field, k is a quasi-wave vector; it defines the wave function of a particle $\psi_k = u_k(r) \exp(ik \cdot r)$, where $u_k(r)$ is a periodic function with a period equal to the crystal lattice constant (see *Bloch theorem*). In both cases, the force F acting upon an electron is found via the derivative of k (or p) with respect to time t:

$$F = \hbar \frac{dk}{dt} = \frac{dp}{dt},$$

while the velocity of the particle equals $v = \hbar^{-1} \nabla_k E(k) = \nabla_p E(p)$, where E is the particle energy. In view of the periodic dependence of $E(k)$, it suffices to consider the energy spectrum of the particle for k values within the first *Brillouin zone*. However, the *conservation law* of momentum, which expresses the uniformity of space, is replaced in the periodic field with that of quasi-momentum: changes of $\hbar k$ by an amount $\hbar G$ are allowed, where G is a *reciprocal lattice* vector. The concept of quasi-momentum is applicable not only to electrons in the framework of *band theory*, but also to any elementary excitations in the crystal, including those interacting with each other.

QUASI-ONE-DIMENSIONAL CRYSTALS

Compounds with a chain or filament structure (see *Thread-like crystals*) with weak overlapping of the electron wave function of atoms (molecules) in adjacent chains. The electronic spectrum of quasi-one-dimensional crystals is highly anisotropic: the *conduction band* width for the motion of an electron along the chains is significantly greater than in the case of motion perpendicular to them. Quasi-one-dimensional crystals include several classes of compounds:

(1) Flat-square complexes with variable *valence* of the type of $K_2Pt(CN)_4Br_{0.3}\cdot 3H_2O$, where the motion of electrons across the band formed by atomic wave functions of Pt (extended along the chain) proves to be almost free, but electron jumps between the chains of Pt atoms are hindered due to the large inter-chain separation;

(2) *crystals* of *polymers*, e.g., of polyacetylene $(-CH{=}CH{-})_x$ and of polysulphurnitride $(SN)_x$ with conjugated bonds;

(3) ion-radical salts with charge transfer, which consist of stacked flat organic molecules of the type of tetracyanquinodimethane (TCNQ), tetrathiofulvalene (TTF) or tetramethyltetraselenofulvalene (TMTSF); stacks of charged molecule-anions $TCNQ^-$ alternating with those of molecule-cations TTF^+ (see Fig. 1); the conjugated chains within the molecules, and the overlapping of the π-electron wave functions of neighboring molecules, allow electrons to move freely along the stack, but the electron jumps between the stacks are hindered because of the large distance between the electrons;

(4) trichalcogenides of *transition metals* (TaS_3, $NbSe_3$) with the cotton wool type filament (fibrous) structure and with a quasi-one-dimensional large anisotropy of electronic properties.

Many quasi-one-dimensional crystals are *metals* at room temperature, but they transform to the insulating state with decreasing temperature. This may result from a *Peierls transition* with the emergence of a *charge density wave* (CDW) or a *spin density wave* (SDW), and with the appearance of an energy gap in the *band gap* at the *Fermi surface* (see Fig. 2). Another reason can be the *Anderson*

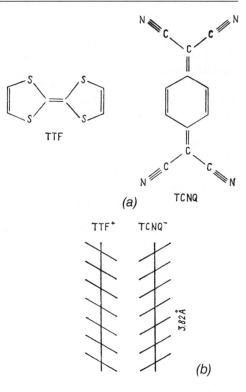

Fig. 1. (a) Structural formulae of the molecules tetrathiofulvalene (TTF) and tetracyano-p-quinomethane (TCNQ); (b) side view of the stacks of the molecules.

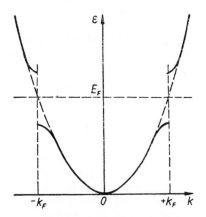

Fig. 2. The one-dimensional band of electrons in the presence of Peierls displacements with period π/k_F. The energy gap is shown at the Fermi surface, i.e. at the points with wave numbers $k = \pm k_F$.

localization of electrons due to the disordering of the structure, or due to the strong Coulomb electron repulsion (*Hubbard transition*, see *Hubbard model*).

A Peierls transition with the emergence of a charge density wave has been detected in many organic crystals (e.g., TTF-TCNQ) or in trichalcogenides (TaS_3). There are quasi-one-dimensional compounds that are *Peierls insulators* already at room temperature, e.g., platinum complexes and polyacetylene. The transition with the formation of a SDW was detected, e.g., in $(TMTSF)_2PF_6$. Still, some quasi-one-dimensional crystals with weak anisotropy remain metals at any temperature, and may transform to the superconducting state upon cooling (see *Superconductivity*). Classed among such systems are *organic conductors and superconductors*, e.g., $(TMTSF)_2CeO_4$, $(SN)_x$ compounds, and $TaSe_3$. Quasi-one-dimensional crystals which are Peierls insulators exhibit specific electron excitations, or *solitons*. They may carry a charge without a spin or a spin without a charge. It is these excitations that determine, e.g., electronic kinetic and magnetic properties of polyacetylene. Peierls insulators exhibit electrical conductivity related to the motion of a CDW with respect to the host *crystal lattice* in strong electric fields (*Fröhlich conductivity*). Conduction of such types is accompanied by generation of low-frequency noise, which was also observed in trichalcogenides. In an alternating high-frequency electric field the dipole active mode of the CDW is observed (so-called *Fröhlich collective mode*), its existence leading to the abnormally high *dielectric constant* of a Peierls insulator, to strong nonlinear effects in high-frequency conductivity, etc.

QUASI-PARTICLE

A concept in the quantum physics of *condensed state of matter* which includes the quanta of *elementary excitations* of macroscopic systems of particles with strong interactions: *crystals* and *quantum liquids*. In a condensed medium the atoms, which are the structural units of a body, cannot serve as structural units of its energy. Indeed, in solids and liquids the total energy does not even approximate the sum of the kinetic energies of particles that make up the body. A condensed body (especially a solid) in a slightly excited state can be treated as a system of elementary excitations (quasi-particles) above a background of the ground (unexcited) state. The excitation energy approximates the sum of the quasi-particle energies, and a system of strongly interacting particles is reduced to one of weakly interacting quasi-particles, with the macroscopic properties of bodies described in terms of a gas of quasi-particles. The similarity of a gas of quasi-particles to one of particles permits the use of concepts and terms of statistical thermodynamics and kinetics of gases, considering the *cross-sections* of various scattering processes, *mean free path*, etc., for the calculation of thermodynamic (see *Thermodynamic potentials*) and kinetic properties (see *Kinetic coefficients*) of solids.

Dynamical properties of quasi-particles. The motion described by an elementary excitation involves all the particles of a body. Typical elementary excitations (quasi-particles) are a wave of displacements of atoms from their equilibrium positions (*phonon*); a wave of moving optical excitation of a molecule through the crystal (*Frenkel exciton*); *electrons* and *holes* in a *semiconductor* or *metal*, and others. The periodicity of the ground state of crystals (and the homogeneity of the ground state of quantum liquids) allows characterizing the elementary excitations by *quasi-momentum* (in quantum liquids – by momentum). This makes elementary excitations similar to quantum particles in a vacuum, their state being specified by the momentum, so elementary excitations are often referred to as *single-particle excitations*. The mathematical formalism for expressing the transition from elementary excitations to quasi-particles is *second quantization* which introduces the occupation numbers n_λ of elementary excitations in state λ to describe the body state. The index $\lambda \equiv (s, \boldsymbol{p})$ includes, in addition to the quasi-momentum $\boldsymbol{p}$, the designation s of the type of elementary excitation.

Like quantum particles, the elementary quasi-particle excitations are of two types: *boson* ($n_\lambda = 0, 1, \ldots$ – any integer) and *fermion* ($n_\lambda = 0$ or 1). Every fermion has an antiparticle (the term antiquasi-particle is not used) with which it can annihilate. Boson quasi-particles appear (are created) and disappear individually, while fermion quasi-particles do so in pairs (particle plus antiparticle).

An example of a fermion antiparticle is a hole – the antiparticle for an electron in a metal or semi-conductor. The total energy near the ground state (with energy E_0) is approximately equal to

$$E = E_0 + \sum_{s,p} \varepsilon_{s\,p} n_{s\,p}, \qquad (1)$$

where $\varepsilon_{s\,p}$ is the energy of the sth quasi-particle in a state with quasi-momentum (or momentum) p.

The similarity between a quasi-particle and a particle is underscored by the fact that the velocity of a quasi-particle can be written $v_{s\,p} = \partial \varepsilon_{s\,p} / \partial p$. Hence, the energy $\varepsilon_{s\,p}$ has a meaning similar to the kinetic energy of a particle. The energy of an anti-quasi-particle with a certain value of p does not equal that of a quasi-particle with the same p. This is one of the main distinctions between "actual" particles and quasi-particles.

According to Eq. (1), to describe the energy spectrum of a body, one needs to know which quasi-particles can exist, whether they are fermions or bosons, as well as the dependences of the energies of quasi-particles on the quasi-momentum (*dispersion laws*). The values of the energy of quasi-particles in crystals are confined to finite intervals (*energy bands*). Therefore, the theory of energy spectra of solids is referred to as *band theory*. The energy of atomic particle motion in condensed bodies is only approximately reducible to the sum of elementary excitation energies. The expression (1) fails to account for the interactions of elementary excitations (collisions of quasi-particles). The decays and transmutations of quasi-particles are possible through their collisions. The quasi-particles under nonrelativistic conditions resemble high-energy elementary particles. The energy and quasi-momentum *conservation laws* hold in the collisions of quasi-particles. In collisions with the *defects* of a crystal (or with *impurities* in a quantum liquid), quasi-particles typically experience elastic collisions. However, transmutations are possible as this takes place.

The statistics of quasi-particles. The mean number of quasi-particles in a state λ is described by *Bose–Einstein statistics* for bosons, and by *Fermi–Dirac statistics* for fermions. The chemical potential of the gas for the same type of quasi-particles is zero. This is a corollary of the absence of a conservation law for the number of particles:

there are no quasi-particles in the ground state (at $T = 0$), and their number grows as the temperature rises. If the energy of a quasi-particle $\varepsilon_\lambda \gg k_B T$, then quasi-particles are described with the formulas of classical statistics taking into account, however, the fact that their number is not conserved. At low temperatures, the main role in the statistical thermodynamic properties of condensed bodies is played by those quasi-particles that have zero mean minimal energy (*Goldstone excitations*).

The nature of quasi-particles. The ground state of a condensed system ($T = 0$) is the vacuum for quasi-particles. This representation makes quasi-particles closer to ordinary (elementary) particles, if one treats them as the quanta of excitations of a physical vacuum. From this point of view, the principal difference between a quasi-particle and a particle lies in the properties of the space where they reside: quasi-particles – in macroscopic bodies (crystals, quantum liquids), particles – in a physical vacuum. If macroscopic vibrations (waves) with wave vector k and frequency $\omega(k)$ can be excited in a condensed system, then there should exist boson quasi-particles with quasi-momentum $\hbar k$ and energy $\hbar \omega$. Their aggregation in the state with a particular k provides for the classical nature of the wave motion.

Electrons moving in condensed media are fermion quasi-particles. An elementary excitation of the electron subsystem is affected by the transition of an electron from an occupied state to a free one. Then, two quasi-particles appear, a *conduction electron* and a hole.

The introduction of quasi-particles is based on singling out the types of motion or degrees of freedom that are approximately independent of one another, e.g., lattice motion, i.e. motion of atoms at lattice sites, and that of electrons (orbital and spin separately), and this is an approximate procedure. Resonant mixing of heterogeneous elementary excitations often permits introduction of new quasi-particles (*polaritons*, e.g., the result of mixing bosons – an optical phonon with a *photon*). The interaction of fermion quasi-particles can cause new types of quasi-particles to appear. An example of the interaction of a fermion with bosons is an electron plus induced lattice polarization forming a *polaron*. A particularly striking illustration of a quasi-particle interaction is the rearrangement of the energy spectrum at the transition of metal to the superconducting state (see *Superconducting phase transition*). The conduction electrons are attracted to one another by phonon exchange and form *Cooper pairs*. The existence of a *Bose condensate* of Cooper pairs and a gap in the electron energy spectrum of the metal (a gap due to the binding energy of a Cooper pair, see *Energy gap in superconductors*) underlies the transition of a metal to its superconducting state. Indeed, the coherent motion of the electrons cannot, due to this gap, dissipate by passage to the thermal motion of the quasi-particles. Similarly, the photon–roton spectrum in He II provides a possibility of superfluid motion of He atoms without viscosity (see *Superfluidity*). Although some properties are common, various quasi-particles differ significantly from each other in their dispersion laws $\varepsilon = \varepsilon_s(\boldsymbol{p})$, lifetimes $\tau_{s\,p}$ (see below), and their role in the behavior of condensed bodies.

Quasi-particles and defects. The role of defects in the dynamics of quasi-particles is not limited to the scattering of the latter by them. Tunneling of defects (*vacancies*, in particular) in *quantum crystals* transforms them to quasi-particles such as *defectons, vacancions, impuritons*. The ever present defect of any solid – its boundary – gives rise to two-dimensional surface quasi-particles: Rayleigh waves and *Tamm levels*. Along linear defects such as *dislocations* one-dimensional sonic waves propagate (one-dimensional phonons), and in *magnetic substances* – one-dimensional *magnons*.

Phase transitions and quasi-particles. At phase transitions the energy spectrum of a condensed system changes; hence, its constituent quasi-particles change also. Sometimes the phase transition proceeds as if prepared by quasi-particles: the *soft mode* arises in the phonon spectrum of the crystal on approaching a second-order *structural phase transition*, the energy of one of the phonons changes sign at the transition point, and the *phase* loses its stability.

Limits of quasi-particle concept. Explanations involving the dynamical properties of gases of quasi-particles can be informative, although their interactions with each other and with defects are relatively weak, and the lifetime τ_λ of an individual quasi-particle lasts long enough so that

$$\varepsilon_\lambda \tau_\lambda \gg \hbar. \qquad (2)$$

In the neighborhood of the ground state (e.g., at a low temperature), the number of quasi-particles in a body is small, the mean distance between them is large, and they rarely collide. Thus condition (2) holds and Eq. (1) is valid to a high accuracy, so quasi-particles can be treated as close to an *ideal gas*. As the temperature rises, their number and their interaction becomes stronger. The highest temperatures where the concepts of quasi-particles are still valid differ for different condensed media, and also for different quasi-particles types of the same material. A gas of magnons can be treated as almost ideal only at temperatures far below the *Curie point* or *Néel point*, while phonons interact with one another slightly under almost all conditions in a solid body. There are situations when a transition from strongly interacting particles to quasi-particles fails to provide any simplification, such as close to a second-order phase transition (in the *critical region*), or in the case of *disordered solids* for short-wavelength elementary excitations (see *Amorphous state, Vitreous state of matter*). Nevertheless, some aspects of condensed matter do not lend themselves to a quasi-particle descriptions, such as the diffusive motion of an atomic particle at high temperatures which is better treated classically without resorting to the concept of quasi-particles.

The quasi-particle concept has also had applications in nuclear physics, astrophysics, plasma physics and turbulence of liquids. If one proceeds from the idea that particles are the quanta of excitation of a physical vacuum, then it becomes clear that quasi-particle is a more general concept than particle.

QUASI-PHASE DIAGRAM

A diagram that characterizes the state of a solid as a function of (changing) external parameters, and is drawn in the coordinates of a state variable versus a kinetic parameter (as distinct from a compositional *phase diagram* drawn against a thermodynamic parameter such as, for example, a pressure versus temperature diagram).

As *quasi-phases* in contrast to ordinary phase states, one can consider nonequilibrium (including disordered) structure forms, different types of *crystal faceting* and their morphological features, nonuniform distributions of impurities, nonequilibrium defects, etc. Quasi-phase diagrams are of an approximate nature since kinetic ranges of quasi-phase formation are usually diffuse, and the stronger they are the farther the process stays from equilibrium. Examples are the conditions of epitaxial growth, as well as polycrystalline and amorphous *condensation* from molecular or ion-molecular beams. Sometimes the collective mechanism of ordering only provides the conditions of local equilibrium, even if the overall process is a highly nonequilibrium one. Thus, under the said condensation from beams with the relative over-saturation as high as 10^{10} and more, the role of a collective ordering factor is played by a "*two-dimensional gas*", namely the ensemble of mobile particles on the surface of a growing crystal or a film. The two-dimensional gas density is characterized by a number of critical values separating nonequilibrium structural-morphological states of a film, which correspond to values associated with kinetic *phase transitions*. The latter may be similar to an ordinary *first-order phase transition* and *second-order phase transition*. First-order kinetic phase transitions can be exemplified by the development of grains of the "second phase" in the formation of a crystal from the gaseous phase or from beams; those of second order – by the changes in the structure of the *crystallization* front during the growth from the melt. In their turn, changes of the front can affect all parameters of the growing crystal. Since quasi-phase diagrams are plotted with allowance for the nonequilibrium nature of real processes, they are important indicators for selecting conditions for the synthesis of solids (see *Crystallization, Crystal morphology, Epitaxy*).

QUASI-TWO-DIMENSIONAL CRYSTALS

Layered compounds with very strong anisotropy of the electronic spectrum, when the movement of electrons along two directions (in the plane of the layers) is close to free motion, and is strongly hindered along the third direction (perpendicular to layers) due to the relatively large distance between the layers, with the energy of the electron weakly dependent on the projection of the *quasi-momentum* along this direction. Quasi-two-dimensional crystals include, e.g., *layered crystals* of dichalcogenides of *transition metals* (of

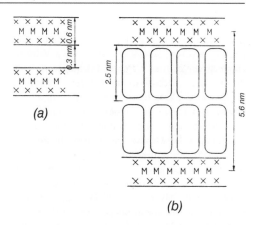

(a)

(b)

Sketch of the structure of layered crystals MX_2 in the direction along the layers: (a) without intercalation, and (b) after intercalation with molecules of octadecylamine.

the type $NbSe_2$, TaS_2), intercalated compounds of these crystals of the type TaS_2-(pyridine)$_{1/2}$, organic conductors of the type $(BEDT-TTF)_2I_3$ (BEDT-TTF is abbreviated notation for the molecule bis-diethylenedithio-tetrathiofulvalene), cuprate metal-oxide compounds of the type of La_2CuO_4, and others. Crystals of dichalcogenides of transition metals are built of close-packed layers, each of them being a sandwich of two layers of halogen atoms X with a monatomic layer of a *metal* M between them. There is a strong *covalent bond* between atoms M and X in the sandwich, whereas MX_2 layers are bound within the crystal by weak *van der Waals forces* (see the figure). Electrons move freely along MX_2 layers, but the overlap of electron wave functions between adjacent layers is small. Anisotropy of the electrical conductivity (ratio $\sigma_\parallel/\sigma_\perp$ of conductivity along, $\sigma_\parallel$, and across, $\sigma_\perp$, the layers) in dichalcogenides of transition metals is as high as several dozen. It is possible to introduce large organic molecules between the layers of dichalcogenides, e.g., the cyclic molecules pyridine or octadecylamine, which draw apart the MX_2 layers from 0.3 nm to 1.2 nm or 5.6 nm, respectively. The compounds obtained in such a way are called intercalated (see *Intercalated structures*). The inserted molecules hinder the jumping (tunneling) of electrons between the layers still more, and the anisotropy of the conductivity grows up to hundreds of thousands ($\sigma_\parallel/\sigma_\perp \geqslant 10^5$). The conducting layers in organic crystals $(BEDT-TTF)_2I_3$ are composed of flat molecule-cations $BEDT-TTF^+$ that are separated by layers of anions I_3^-, where the density of conduction electrons is low, and the anisotropy of the electrical conductivity there $\sigma_\parallel/\sigma_\perp \sim 10^2$.

Structural phase transitions of the *charge density wave* (CDW) type are observed in many layered metals upon their cooling. The ions of the lattice under these transitions are periodically displaced off their old equilibrium positions. This produces an energy gap at the *Fermi surface* (or at part of it), which lower the electronic energy of the system. The quasi-two-dimensional nature of the electronic spectrum promotes the formation of the CDW state. The origin of the CDW transitions is believed to involve the matching of sections of the Fermi surface; it is at these sections that the energy gap appears. Many layered metals are *superconductors*, and the quasi-two-dimensional nature of their electronic spectrum leads to a strong anisotropy of their superconducting properties (e.g., London *penetration depth of magnetic field*, upper and lower critical magnetic fields). In the limit of very strong anisotropy, the theory predicts the transition to a qualitatively new behavior of layered superconductors, to a Josephson interaction of the layers (see *Josephson effects*). The CDW state and superconductivity may coexist within the same crystal, with their interaction being antagonistic. Application of pressure at the coexistence stage reduces the temperature of the CDW transition (due to the convergence of layers) and the value of the energy gap, and this raises the *superconducting phase transition* temperature. In layered cuprate metal-oxide compounds *high-temperature superconductivity* is observed with critical temperatures $T_c \approx 90$ to 130 K.

QUASI-TWO-DIMENSIONAL SUPERCONDUCTORS

Superconducting compounds with layered structures (see *Quasi-two-dimensional crystals*) and strongly anisotropic electronic properties. The electronic anisotropy in them is associated with weakly overlapping electronic wave functions in alternating layers. If the corresponding width of the band for the electron motion transverse to the layers significantly exceeds the superconducting transition temperature energy $k_B T_c$ (see *Critical temperature of superconductors*) then the superconducting

properties become strongly anisotropic, but there are no qualitatively new effects. In the other limiting case one can speak about the Josephson interaction between the layers (see *Josephson effects*) with such new effects as the appearance of Josephson vortices (lacking a normal state core) and Josephson oscillations in an electric field. Strong anisotropy of electronic properties is observed in layered compounds of dichalcogenides of *transition metals* of the type of $NbSe_2$ and TaS_2 and of their intercalated compounds; in *graphite* intercalated with metallic atoms; in *organic conductors and superconductors* based on the molecules BEDT-TTF, and in high-temperature superconductors such as $YBa_2Cu_3O_7$ (see *High-temperature superconductivity*). The closest approach to Josephson junctions between the layers are dichalcogenides of transition metals, intercalated by large organic molecules of the pyridine type.

QUASI-WAVE VECTOR

One of two quantum numbers which characterize, according to the *Bloch theorem*, the single particle states in a *periodic potential*. The quasi-wave vector equals the *quasi-momentum* of a particle up to a factor (*Planck constant*). A corollary of the Bloch theorem is the fact that a quasi-wave vector is defined to an accuracy of a *reciprocal lattice* vector. The term "quasi-wave vector" is sometimes taken to mean, in a narrower sense, the quasi-wave vector within the first *Brillouin zone*. It should be noted that it is not accepted terminology in *band theory* to denote as degenerate those electron states with different quasi-wave vectors and the same energy. Here the term degenerate refers only to states with the same quasi-wave vectors, but with different band numbers.

QUENCHED (FROZEN) ORBITAL ANGULAR MOMENTUM

Orbital angular momentum L with an average quantum-mechanical value close to or equal to zero. The concept of quenched orbital angular momentum is used when considering an atom or ion (usually of the iron group) introduced into the crystal. If the ground state energy term is an orbital *singlet* arising from raising the orbital degeneracy by the *crystal field*, then the orbital angular momentum is frozen, and the magnetic properties become determined by the *spin degeneracy*. In other cases, one can speak about partial quenching of the orbital angular momentum, depending on its contribution to different properties.

QUENCHING, also called hardening

Heat treatment of materials (*metals*, *alloys*, glasses, etc.) that consists of heating them to temperatures above critical points and holding (if necessary) at these temperatures, with a subsequent fast cooling (quenching). The purpose is to obtain a high-temperature state that is metastable at low temperatures (quenching without polymorphic conversion) or an intermediate transitional state differing from both the low temperature and the high-temperature one (quenching with polymorphic conversion). The quenching without polymorphic conversion is exemplified by quenching of duralumin or some brands of *steel* with the aim of obtaining a supersaturated *solid solution*. The quenching of carbon steel to a new transitional state is an example of quenching with polymorphic conversion. The metastable structures obtained by quenching tend to decompose upon subsequent heating.

QUENCHING OF LUMINESCENCE

See *Luminescence quenching*.

Rr

RADIATION-ABSORBING MATERIALS

Materials featuring high *absorption coefficients* (either linear or mass *absorption* types). Absorption indices of materials depend on their characteristics and the type of incident radiation. That dependence prescribes the choice of materials for radiation protection screens.

RADIATION BY CHARGED PARTICLES

Photon emission during the motion of charged particles. Every time a charged particle undergoes an acceleration, that is changes either the magnitude or direction of its velocity, the so-called electromagnetic *Bremsstrahlung* (braking radiation) arises. A charge undergoing uniform rectilinear motion in vacuo does not radiate. There also exists a class of phenomena where the radiation is caused by an alteration of the properties of the medium through which the charged particle moves. Here, the particle velocity can remain constant both in magnitude and direction. If a charge (electric monopole or *magnetic dipole*) moves with a constant velocity v exceeding the phase velocity of light $c/\sqrt{\varepsilon}$ in the medium with relative *dielectric constant* ε, this gives rise to *Cherenkov radiation* departing at the angle $\cos \theta = c/(v\sqrt{\varepsilon})$. In addition, the change of the dielectric constant, resulting from the particle crossing an interface between two media, leads to emission of *transition radiation*. In general, the parameter that determines the radiation is the ratio of charged particle velocity, v, to the speed of light in the refracting medium. Cherenkov and transition radiation can, generally, coexist and interfere with the radiation resulting from the charge acceleration. The electromagnetic radiation of charged particles at high energies in *monocrystals* differs essentially from that of an isolated atom. If a fast charged particle enters a single crystal at a small angle, θ, with respect to some crystallographic plane (or axis), it interacts coherently with all atoms found in a certain *zone of radiation formation*. In the case of relativistic particles, the formation zone is large relative to the interatomic distance. Therefore, the details of the arrangement of atoms along this length are immaterial, and one can replace the actual atomic arrangement by an averaged one, i.e. introduce the averaged (plane or string) potential. The energy of transverse motion of a particle in the averaged potential is proportional to the square of the angle between the incident particle momentum and the crystal plane (or axis). Before the entry into the crystal, this energy is positive. Zero energy of transverse motion corresponds to a charged particle travelling precisely parallel to a crystallographic plane or axis. This particle begins populating states belonging to levels (bands) of quasi-bound motion (*subbarrier occupation*) and to continuum bands (*superbarrier occupation*).

Radiative transitions between different energy states of a charged particle determine the electromagnetic radiation intensity

$$dI = dI_{cc} + dI_{cb} + dI_{bb},$$

where dI_{cc} is the radiation intensity between different states of the continuous spectrum; dI_{cb} is the radiation intensity from the continuous spectrum states to the levels of the quasi-bound transverse motion (*channeling* regime); and dI_{bb} is the radiation intensity between different levels of the quasibound transverse motion. The first term, dI_{cc}, is related to the Bremsstrahlung mechanism of electromagnetic radiation in a single crystal, and represents *coherent braking radiation*. The second term, dI_{cb}, describes the radiation accompanying the process of charged particle trapping into the channeling regime, involving the rearrangement of the incident wave function of the particle into a superposition of fundamental (Bloch) functions (see *Bloch theorem*) in the averaged plane (string) potential. Thus, the rearrangement of the charged particle's wave function at the crossing of a single crystal boundary is accompanied by radiation independent of the crystal thickness. The intensity of such transition radiation can be expected to increase in proportion to the number of single crystals in a layered radiator. The third term, dI_{bb}, describes the radiation of channeled particles and represents *quasi-characteristic radiation* under radiative transitions between different levels of the quasi-bound transverse motion. The angle at which the maximum of coherent braking radiation is observed at high frequencies (the high frequency radiation corresponds to the transitions between high superbarrier states, i.e. is determined by the coherent braking radiation mechanism) exceeds the critical channeling angle. Hence, for relativistic particles coherent braking and quasi-characteristic radiation in the high frequency range are "set apart" in angles with respect to the direction of the particles' entry trajectory.

Other types of radiation of charged particles undergoing acceleration are not specific for solids. For example, when a relativistic charged particle in a magnetic field is moving along a circumference it emits pulses, on a nanosecond time scale, of electromagnetic radiation (so-called *synchrotron radiation*) that are highly polarized, and an undulator imparts an *undulating motion* to relativistic electrons in a linear accelerator that causes them to emit microwave radiation.

RADIATION CATALYSIS

Phenomenon of acceleration of chemical reactions at *solid surfaces* in the presence of ionizing radiation. In a *solid*–gas (liquid) system in which radiation-induced catalysis takes place, ordinary catalytic processes are either not observed at all or need considerably higher temperatures to take place. The characteristic temperatures of radiation catalysis do not exceed $100\,°C$. The solids most often used for radiation catalysis are either *insulators* or "poor" *semiconductors* (SiO_2, TiO_2, Al_2O_3, ZrO_2, ZnO, CdO, etc.). The active catalytic centers formed in the solid under the effect of radiation are often paramagnetic as reflected in their paramagnetic resonance spectra.

The termination of the irradiation results in a lowering of the rate of catalysis. However, it never stops abruptly (*post-irradiation effect*). The slowing down of this rate during the post-irradiation period is explained by the thermal *annealing* of the active catalytic centers. Many reactions available and known from ordinary catalysis may also be triggered by radiation catalysis techniques: e.g., hydrogen–deuterium exchange, oxidation of hydrogen, etc. But parallel to such reactions it becomes possible to carry out certain specific processes, e.g., the radiation catalysis for synthesis of higher monocarboxylic acids of normal structure with lengths of C_{200}, from ethylene and carbon dioxide at the surface of γ-Al_2O_3 in a gamma ray flux.

RADIATION CHEMISTRY

Branch of chemistry that studies chemical and physico-chemical transformations that result from the action of *nuclear radiation* on materials (*radiolysis*). When radiation affects intrinsic or impurity atoms (ions) in a solid matrix they are raised to excited states; processes of relaxation and interaction ensue, new chemical states of atoms (ions) develop, the character of *chemical bonds* changes, and the formation of new *phases* becomes possible. Radiation chemistry of *solids* specifies those features of radiation-chemical reactions both within the bulk and at the surface of solids that affect the *radiation resistance* and radiation sensitivity of materials. Important directions of radiation chemistry research include studies of short-lived excited states in solids (an efficient method of pulse radiolysis); of stable and metastable *point defects*; heterogeneous oxidation/reduction processes at *solid surfaces*, that develop under the effect of radiation (*radiation-induced adsorption, desorption* and catalysis) and the possibilities of applying those to chemical technologies. Studies have been made of the mechanisms of radiolysis that result in changing the properties of solids since these are needed to produce radiation-stable *construction materials* for instrument building, as well as radiation-sensitive memory elements for *information recording*. A separate area of research is studying the effects of radiation on polymers (see *Radiation physics of polymers*).

Targets and study techniques for radiation chemistry of solids are practically the same as those encountered in the *radiation physics of solids*.

RADIATION DETECTOR, PYROELECTRIC
See *Pyroelectric radiation detectors*.

RADIATION, INDUCED
See *Induced radiation*.

RADIATION-INDUCED ADSORPTION
Binding of particles from either a gas or a liquid phase by a *solid surface* in the presence of ionizing radiation. The level of radiation-induced adsorption is defined by the number of particles additionally taken up per unit surface area of a solid after the dark adsorption–desorption equilibrium has been established.

Radiation-induced adsorption is stimulated by the absorption of radiation energy by a solid during which new adsorption-active defects (often of a paramagnetic nature) form. Substances most susceptible to radiation-induced adsorption are highly porous materials such as γ-Al_2O_3 and η-Al_2O_3, silica gel, *zeolites*, etc. Radiation-induced adsorption of H_2O_2 and other gases is known to take place at surfaces.

Radiation-induced adsorption has been used to improve the vacuum in enclosed volumes. To accomplish this, e.g., silica gel is placed into a sealed vessel which is routinely evacuated. Then the vessel is subjected to γ-irradiation, which results in a significant improvement of the vacuum due to the radiation-induced adsorption.

RADIATION-INDUCED CREEP of metals
A radiation-stimulated acceleration of *creep* and its effects at lower temperatures, when regular thermal creep is almost absent. The appearance of a radiation-stimulated component of creep results from damages to the crystal produced during its bombardment by high-energy particles (*neutrons*, electrons, ions, etc.). *Clusters*, micropores, and *dislocation loops* formed in the process of irradiation turn into barriers in the pathway of *dislocations*. However, simultaneously radiational *point defects* promote the creep of edge dislocations, and hence accelerate the process of creep under irradiation. Theoretical concepts of mechanisms for radiation-induced creep are based on accounting for the anisotropic character of the interaction between the radiation-induced point defects and dislocations in the field of an applied stress. In addition, nonuniformly distributed dislocation loops appear during the process of irradiation. Another factor to be taken into account is the acceleration of the diffusive surmounting of obstacles by dislocations. All of these effects serve to control the rate of creep.

RADIATION-INDUCED DEFECTS
See *Radiation physics of solids*.

RADIATION-INDUCED DIFFUSION
Observed acceleration of the rate of *diffusion* of impurity atoms or intrinsic defects under the influence of *nuclear radiations* in various materials. One of the most efficient mechanisms of radiation-induced diffusion is the interaction of *impurity atoms* with radiation *defects* (see *Radiation physics of solids*). The appearance of mobile *vacancies V* speeds up the diffusion of *substitutional atoms* A_s, both during the one-time jump of an atom to a neighboring vacant site with the ensuing spatial separation between V and A_s, and the formation of quasi-molecules ($A_s V$) from A_s and V located at adjacent sites. When *substitutional atoms* shift into interstitial sites under the effect of irradiation (e.g., due to expulsion of A_s by an intrinsic *interstitial atom* forced into that site by radiation, as occurs with Group III elements in Si and Ge) then diffusion speeds up as well. Meanwhile the formation of vacancies may slow that process for interstitial impurity atoms, A_i, since the A_i atoms scavenged by V need additional activation energy for their further thermal release and ensuing diffusion. Diffusion may also go faster due to radiation stimulation when external high-energy particles elastically interact with the impurity atoms. Diffusion in *semiconductors* may be effectively accelerated by the excitation of their electron subsystem. One of the possible mechanisms of such acceleration consists in *recombination-induced diffusion*. During this process a free electron or a hole is captured by a *point defect*, accompanied by the jump of an atom, its *activation energy* having been lowered by the energy (or by part of it) released during such a capture. Repeated successive captures

of electrons and holes open a possibility of the long-range spreading of radiation-induced diffusion. Other possible mechanisms involving diffusion are the following: a change in the charge state of the "defect" atom occurring when it captures either an electron or hole, or the possibility of a transition of an electron located at an atom to an excited state, which changes the potential energy of the system. As a result, the barrier height for a thermally activated atom is lowered, thus increasing the probability of an electron jump. When there are two types of interstitial sites in a crystal (P_1 and P_2), with P_1 corresponding to a potential energy minimum and P_2 to a potential energy maximum, an *inversion* becomes possible when atoms at P_1 and P_2 interchange positions.

The effects of the acceleration and the slowing down of diffusion of impurity and intrinsic atoms are observed during irradiation of various materials with various nuclear radiations. The effective *diffusion coefficient D^** corresponding to radiation-induced diffusion may exceed by several orders of magnitude the value of D that exists during to room temperature equilibrium in the absence of irradiation. The effects of radiation-induced diffusion play an important role in radiation technologies (see *Ion implantation, Nuclear doping*) as well as in processes controlling the *radiation resistance* of materials.

RADIATION-INDUCED FILM GROWTH

A process taking place under the effect of irradiation by high-energy electrons, *ions*, and hard *photons*. Irradiation accelerates *film* growth, increases specific yield from the initial material, makes it possible to localize the condensed material within the irradiated field, to control the structure and properties of generated films. Irradiation of the surface produces *point defects* both at the surface itself, and in the subsurface layer, which serve as condensation centers. By varying their concentration and spatial distribution one may significantly modulate the rate of embryo formation that controls the kinetics of film growth and the features of its structure, and hence the overall properties of the resulting films. *Point defects* at the surface of *ionic crystals* may act as orientation centers, i.e. centers of epitaxial growth. Surface aggregations of identical point defects can exhibit

such properties. In particular, they form during the process of *annealing* point defects when these defects reach the surface from the bulk. Besides, the capture of point defects by phase interfaces between the growing islands of the newly formed film serves to aid the *epitaxy* processes during the growth. Radiation-induced excitation of adsorbed atoms stimulates their mobility and chemical activity, thus affecting the kinetics of embryo generation and growth. The latter effect may alter the mechanism of *condensation* and stimulate *crystallization* under conditions otherwise favorable for forming an amorphous condensate. The interaction of radiation with the condensed phase results in a significant change of the film morphology at its early stage of growth: film islands change their shape, merge (*coalescence*), further break up the grid structure to form new islands (provided coverage coefficients are low), etc. In addition, metastable phases form in films, and polymorphic transformations take place (see *Structural phase transition*).

RADIATION-INDUCED JOLT (term introduced by V.L. Indenbom)

Large amplitude vibrations and possible displacements of atoms of *solids* around their equilibrium positions, produced by the effect of *nuclear radiation*. Atoms acquire significant energy during collisions with external particles flying by, and can be jolted far off their lattice sites. This energy transfer is sufficient to cause a *vacancy V*, and an *interstitial atom I* associated with the same lattice site, to become spatially separated from each other. The separated V and I centers can be detected by various experimental techniques. However, most of the collisions between the atoms of an irradiated target and the external particles striking it, as well as their collisions with each other, involve the transfer of very small amounts of energy, insufficient for forming stable Frenkel pairs (see *Unstable Frenkel pair*). After the initial event some of the atoms surrounding this displaced atom undergo subsequent collisions, and induce the ensuing relaxing vibrations of an excited microvolume of the crystal. Their amplitude may, at the initial stages, significantly exceed that of thermal vibrations, so that radiation-induced jolt takes place. During that process certain atoms may temporarily reach outside the limits of their *unit cell*, but eventually the excited part of the crystal returns to its state of minimum of potential energy corresponding to the initial state of the crystal as a whole. When the irradiated crystal is monatomic, the absorbed radiation energy goes exclusively to the generation of *phonons*. As a result of the radiation-induced jolt, certain atoms may exchange their positions in the lattice. The crystal region around the vacancy such that a jolted-out atom entering it cannot produce a stable perturbation of the initial crystal structure is called the *instability zone*. Provided the initial pair V and I are separated beyond this distance, a bound Frenkel pair is formed. Even further separation results in the appearance of isolated (i.e. noninteracting) V and I. When irradiated crystals feature structure defects or electric excitations, the energy of radiation-induced jolt (which is practically undetectable in a *perfect crystal*) becomes detectable through the changes in the system of defects or of electric excitations. This is explained by the fact that these changes occur after the transfer of increments of energy that are too small to jolt an atom out of its lattice site. To offer an example, computer simulations make it possible to model the displacement of a dumb-bell configured interstitial atom during the jolt. *Alkali-halide crystals* serve as a basis for experimentally studying the process of extinguishing *exciton luminescence* during radiation-induced jolt.

RADIATION-INDUCED ORDER

An increase in the degree of order of atoms at sites in the *crystal lattice* of alloys that sometimes results from the interaction of high-energy charged and neutral particles or *photons* with *atoms* and electrons of the crystal. Depending on the irradiation conditions (temperature, parameters of radiation field) and the type of crystal, such a radiation flux works to produce new ordered or disordered local regions that appear in the *alloys*. The mechanisms of ordering–disordering at work in this case are *radiation-induced diffusion* and the generation of *thermal spikes*. Direct experiments show that disordering is largely localized in the vicinity of thermal spikes. As a rule (see *Long-range and short-range order*), long-range order is established via mechanisms of radiation-induced diffusion. In disordered systems, irradiation often increases the

degree of ordering, while in fully ordered systems it induces disordering.

RADIATION-INDUCED REARRANGEMENT OF STRUCTURE

Change of *crystal structure* under the effect of irradiation by *neutrons*, *ions*, electrons, X-rays, observed for high radiation doses. Radiation-induced rearrangement of structure manifests itself in *amorphization*, decomposition of initial compounds into their constituent elements, formation of a different modification, and change of phase composition of *solid solutions* and *alloys*. As a rule, such processes are mutually competitive. The appearance and dominance of one particular type of structural change is prescribed by the irradiated material and conditions of irradiation: the type of radiation, its energy, dose, flux magnitude and temperature. Radiation-induced heterogeneous amorphization is possible locally, within the size ranges of a few interatomic distances. Under the effect of higher radiation doses it develops into homogenous amorphization which may spread throughout the entire irradiated crystal. Amorphization may be further radiation-stimulated into a *first-order phase transition*, and the reverse phase transition from an amorphous to a crystalline state has also been observed. During radiation-induced decomposition of a compound, its products are not necessarily limited to its constituent elements. Certain compounds may form from these and other elements available from the crystal environment. Changes in phase composition may also involve changes in the *phase diagram* of the system. Rearrangement of the crystal structure is also observed, ending with the formation of a new stable modification. The symmetry of such a structural state may correspond to a high-temperature modification of a non-irradiated crystal (see figure on next page).

Such structural changes are found in crystals with *ionic-covalent bonds*, in intermetalloids (see *Intermetallic compounds*), *polymers* and complex compounds of *rare-earth elements*. Restructuring goes either via the formation of embryos of high temperature modifications at *displacement spikes*, followed by their growth in the process of irradiation, or via correlated displacements of atoms produced by radiation point defects. By varying the

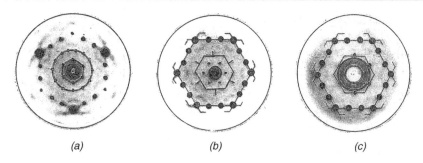

(a)   (b)   (c)

Radiograms (K_α Mo) illustrating changes in the crystal structure of quartz during irradiation by neutrons and heating: (a) α-phase ($T = 20\,^\circ$C); (b) β-phase ($T = 600\,^\circ$C); (c) phase after irradiation at $\sim 6 \cdot 10^{19}$ n/cm^2.

dose and temperature of irradiation, the concentration and type of impurities, and also the degree of the initial crystal perfection, it is possible to control the crystal *radiation resistance*.

RADIATION-INDUCED SHAPE CHANGE,
radiation-induced growth

Change of shape of crystalline bodies during exposure to radiation, in the absence of any externally applied load or forcing, and unaccompanied by any significant change of volume. The effect of radiation-induced growth is particularly evident in materials with anisotropic structures (*uranium, zirconium, graphite*, etc.). Irradiated *monocrystals* of uranium elongate along the [010] crystallographic direction, and shrink along the [100] direction at the same rate, while undergoing no changes along the [001] direction. Rods of polycrystalline uranium with [010] type texture elongate by a factor of 2 to 3 when irradiated by *neutrons*, all due to radiation-induced growth.

RADIATION, NUCLEAR
See *Nuclear radiations*.

RADIATION PHYSICS OF METALS AND ALLOYS

Subfield of *radiation physics of solids*. What is specific for *metals* (compared to *semiconductors* and ionic *insulators*) is their high *radiation resistance*, as determined from changes in their mechanical strength and electric properties under the effect of irradiation. Therefore, radiation physics of metals and alloys pays particular attention to studying defects that are formed by large

radiation doses, as well as the high concentrations of *point defects* that those actions introduce. Merging of such defects into *clusters* also plays a significant role. As the number of defects increases, clusters of *vacancies* and of intrinsic *interstitial atoms* form. These have the arrangement of vacancy and interstitial *dislocation loops*, and planar accumulations of vacancies and interstitial atoms. Further growth of vacancy clusters results in the formation of three-dimensional radiation defects that are called *vacancy pores*. The impurity atoms of gases that either form in the specimen or are implanted during irradiation, diffuse subsequently into these vacancy pores to form *gas-filled pores*. Gas pores significantly differ from vacancy pores in their physical properties (mobility, growth capacity, *coalescence*), and are treated as a separate type of radiation defect. Gas pores that form during the course of bombardment of metals by α-particles and protons at small depths below the metal surface (down to 1 µm) produce *blisters* (see *Blistering*) at the metal surface. Eventually these blisters break to produce *scabbing* (see *Ion-stimulated surface modification*).

Clusters of impurity atoms may, during their growth, induce new *phase formations*. The same type of radiation defects may develop in *alloys* of the *solid solution* type. These result in a redistribution of alloy components in the region of the initial perturbation, whereby one of the components of that alloy breaks out of its solubility limit. Another possible result is the onset of *radiation-induced phase transitions* associated with a change of lattice structure. Intense irradiation may also result in their *amorphization*, i.e. a full or a partial destruction of the crystal structure and the disruption of long-range order. Direct observations of point radiation defects are possible with the help of the techniques of field ionization microscopy (see *Field ion microscopy*). For that, one needs a microscope power up to 10^7 and a resolution at least 0.1 nm. *Electron microscopes* permit one to observe dislocation loops, pores, and new phase formations at powers of 10^5–10^6 and resolutions at least 1 nm. To study energy characteristics of radiation defects (their energies of formation, migration, and mutual interaction) indirect methods are in use that are based on relationships between macroscopic characteristics of irradiated materials known from direct observation, and such factors as the concentrations and the well-known temperature dependences of the rates of *annealing* of their defects. Indeed, when heating the specimen such defects often annihilate, coalesce, or migrate to traps at the *solid surface, grain boundaries, pores*, or *dislocations*.

Radiation defects that form as a result of irradiation can bring about serious adverse changes in the properties of metals and alloys, particularly those used as *construction materials* for the needs of the atomic power industry. For example, the development of a dislocation structure and changes in the phase composition result in a deterioration of such characteristics as *plasticity* and *strength*, while the developing porosity expands the overall construction geometrically. Meanwhile blisters erode construction surfaces. Therefore, a necessary condition for the successful development of atomic and thermonuclear power installations consists in producing irradiation stable construction materials, i.e. materials which can hamper deterioration and withstand the effects of radiation defects.

RADIATION PHYSICS OF POLYMERS

A branch of physics studying changes in the electrical, dielectric, mechanical, optical, and other properties of *polymers* under the effect of *nuclear radiations*. During the course of absorbing a dose D of nuclear radiation energy there takes place both reversible and irreversible increases in the *electrical conductivity* of polymers. The (reversible) irradiation-induced electrical conductivity is produced when extra *current carriers* (free and bound electrons, holes, ions) are generated in the polymers. One may differentiate between stationary radiation-induced electrical conductivity (during continuous irradiation), and the pulsed type. The dependence of steady state irradiation-induced electrical conductivity σ on the rate of absorbing a dose dD/dt is a power law function: $\sigma = A(dD/dt)^{\Delta}$, where A and Δ are characteristic constants. In particular, we have $0.5 < \Delta \leqslant 1$. The irradiation-induced conductivity does not fall off abruptly after termination of the irradiation, but lingers depending on the temperature of the polymer. This persistence may extend from minutes to tens of hours. Irradiation pulse-induced electrical conductivity is differentiated into immediate pulse conductivity, determined by free electrons that had not suffered scavenging, its characteristic time scale being about 10^{-10} s, and delayed conductivity resulting from electrons thermally released from their *traps*. The level of this induced conductivity is proportional to D. Irreversible changes in the initial electrical conductivity (which may change by a factor of several units) are produced by changes in the physical structure and chemical composition of the polymers. Such changes result from radiation-induced processes of interlacing, *polymer degradation, crystallization, amorphization*, disorientation, plasticization, etc.

Radiation-induced changes in the *dielectric loss* tangent, $\tan \delta$, and the *dielectric constant* ε' may be either reversible or irreversible. Reversible changes depend on the rate of the radiation dose, the ambient temperature, and other parameters which affect the dynamics of generation and decay of the unstable products of *radiolysis* (*free radicals*, electrons, ions, low molecular weight compounds, etc.). Irreversible changes of $\tan \delta$, ε', and the value of electric field strength develop from the processes of *oxidation*, interlacing, destruction, crystallization, and amorphization of polymers. For example, oxidizing nonpolar polymers during their irradiation in air, and amorphization of polar polymers, both result in increasing $\tan \delta$. A reversible increase of the *creep* rate in polymers takes place during irradiation. Their durability under load decreases, hence the *sound velocity* increases. Note that these variables depend on the dose rate, the level of mechanical stress, and the polymer temperature. Irreversible changes in the mechanical properties of polymers (such as their *ultimate strength*, ultimate strain, *elastic modulus*, etc.) are mainly a result of the absorbed dose. The most general dependences are the following: the specimen ultimate strength remains unchanged up to a certain irradiation dose, and then gradually diminishes; the ultimate strain falls off monotonically with the dose; with increasing temperature and radiation power the ultimate strength and strain both decrease. Radiation-induced changes in the thermo-physical properties of polymers (such as their thermal conductivity κ, specific heat C, density ρ, etc.) result from the processes of interlacing, destruction, crystallite disintegration, the formation of gaseous products, etc. Radiation interlacing routinely increases κ and ρ, and lowers C, while destruction results in the opposite changes.

Various active agents, such as excited molecules, charged particles, free radicals, etc., absorbing at longer wavelengths than the initial macromolecules, are responsible for reversible changes of the optical properties of polymers. Their *extinction* coefficients lie within the range 10^2 to 10^7 $l \cdot mole^{-1} \cdot cm^{-1}$. The processes responsible for reversible changes include the formation and destruction of chromophore groups and low molecular weight impurities, as well as crystallization or amorphization of the polymers.

RADIATION PHYSICS OF SEMICONDUCTORS

Branch of physics that studies processes of forming lattice *defects* induced by *nuclear radiation*, and the effects of radiation defects on semiconductor properties. Among various solids, the radiation physics of *silicon* is the most studied. It was found experimentally that the *activation energy* for the diffusion of isolated *vacancies* in crystals of silicon, U_g, reaches $\sim$0.2–0.3 eV (depending on the vacancy charge state). The value of U_g obtained previously from the temperature dependence of self-diffusion is of the order of 2 eV. Such a large discrepancy, referred to as a *vacancy paradox*, might be related to the scavenging of diffusing vacancies by various defects, thus putting in doubt the classical interpretation of data on the high temperature *diffusion* of vacancies in silicon and other semiconductors. It was found experimentally that at low temperatures (including liquid helium temperatures) *interstitial atoms* of silicon diffuse independently of any activation, and may force certain *impurity atoms* (e.g., elements from Group III of the periodic system) from their substitutional positions; the so-called *Watkins effect* (G. Watkins). The existence of potential barriers for reactions involving intrinsic and impurity *point defects* has been established. Accounting for the exponential factor in these barriers, of the form $\exp[-E/(k_B T)]$ where E is the barrier height, explains the significant difference found between the cross-sections of such reactions for atoms of various chemical types. Studies were conducted on the behavior of bound Frenkel pairs. The energy spectrum of secondary radiation defects in silicon strongly depends on the initial impurity content (not always under control) in the irradiated specimen. Scores of radiation defects, differing in their energies, have been experimentally identified. The most common ones are the so-called A-centers (interstitial oxygen atom, plus vacancy at a nearby site); E-centers (atom of Group V plus vacancy at a neighboring site), and bivacancies. The possible number of defects of various types is noticeably larger in semiconductors compared to monatomic crystals.

The appearance of radiation defects often results in changes in the observed properties of the sample. Semiconductors are specific in their high sensitivity to radiation effects, compared to metals, particularly with respect to changes in the lifetimes of their minority *current carriers*. As for pure substances, the same is true with respect to their dark conductivity. A specific case of this high sensitivity to radiation which takes place in appreciably imperfect crystals is designated as the *small-dose effect*. Changes in the mobility of electrons and holes are noticeable, as well as changes in the spectra of *electron paramagnetic resonance*, in optical and thermal properties, in *internal friction*, etc.

The effects of radiation on *semiconductor materials* find their use in radiation technology, including *ion implantation doping*, *nuclear doping*, and improvement of material parameters with the help of controlled radiation doses.

RADIATION PHYSICS OF SOLIDS

The study of interactions of *nuclear radiations* (*neutrons*, fast charged particles, hard *photons*) with *solids*. One of the most important areas in radiation physics is studying *radiation defects*, which are products of atom displacements in irradiated materials, as well as electron excitations produced by ionization processes in them. The character of radiation-induced processes depends on the properties of the target, in particular on the spectrum of its excited states, which differ for various types of solids. They also depend on the parameters of the radiation, such as the particle mass, charge and energy, the photon energy, the intensity and dose of radiation, as well as the conditions of irradiation, such as temperature, presence and level of elastic, electric and magnetic fields. The principal mechanism for the formation of *defects* in a crystal lattice of an ionic *insulator* is the relaxation (decay) of electronic excitations: *excitons* or electron–hole pairs (see *Semiconductor junction*). In other materials, *metals* and *semiconductors* in particular, such defects mainly form as a result of elastic collisions of external particles with atoms of the solid. The *primary radiation defects* formed, such as *vacancies V*, and *interstitial atoms I*, exhibit significant mobility over a wide temperature range, which results in diverse reactions between the primary defects and the *impurity atoms*. *Secondary radiation defects* that are stable at irradiation temperatures, form during the course of such reactions.

When bombardment energies remain low and radiation doses remain moderate, most primary defects are *point defects*. There exists an *energy threshold* for their formation, E_a. If the energy transferred to an atom is less than E_a, the probability of generating a defect sharply decreases. Provided the initial energies are sufficient, *collision cascades* follow, resulting in the formation of *clusters*, that is regions of higher defect density (see *Seeger zone, Displacement spike, Thermal spike*). For high radiation doses point defect coagulation becomes efficient as these defects diffuse more readily. Radiation-induced *phase transitions* become possible then, such as, e.g., *amorphization* of crystal layers or, for a different irradiation regime, *crystallization of amorphized layers*. Further annealing may bring about even further restructuring of defects, including annihilation of V and I pairs.

Introducing radiation defects changes the basic physical properties of materials; varying the conditions of irradiation and annealing makes it possible to control such changes to a certain extent. In that sense bombarding solids with *ions* is quite specific because chemically foreign atoms are then imploded into the target (see *Ion bombardment, Ion implantation doping, Ionic synthesis* and also *Nuclear doping*).

An important area of radiation physics of solids involves studying the behavior of high-energy particles and photons during their passage through the crystal, in particular their *orientation effects* (see *Channeling, Hyperchanneling, Particle bending*). Methods for studying crystal structure have been developed on that basis. *Radiation by charged particles* during their passage through solid targets, and in particular through crystalline targets, is widely studied (see *Quasi-characteristic radiation, Coherent Bremsstrahlung, Okorokov effect*).

The applied aspects of radiation physics of solids include such applications as increasing the *radiation resistance* of materials, developing radiation technology, improving solid state nuclear radiation detectors (see also *Radiation physics of polymers, Radiation physics of semiconductors, Radiation chemistry, Radiation physics of metals and alloys*).

RADIATION, QUASI-CHARACTERISTIC
See *Quasi-characteristic radiation*.

RADIATION RESISTANCE
A qualitative concept characterizing the degree to which a given parameter of an irradiated target changes under the influence of this or that radiation dose. Sometimes it is identified quantitatively as the radiation dose that results in a significant change in the monitored parameter, but one which still permits an acceptable degree of normal functioning of the target. One may differentiate between radiation resistances to light (its absorption by elements of optical devices), to equilibrium or nonequilibrium conductivity (in elements of electric devices), to strength (in *construction materials*), etc. The level of radiation resistance depends on the type of radiation, the energy of particles or *photons*, and in many cases on the intensity of the radiation, the ambient temperature, the contents

of various *impurity atoms* (often remaining uncontrolled), the intrinsic *defects* in the irradiated material, and other conditions. To compare radiation resistances of a given target to various types of radiation for one of its parameters one may introduce the concept of *"equivalent" doses*. These are doses of different radiation types bringing about identical changes to the selected parameter. Such an "equivalence" may be violated, however, if the conditions of the irradiation change, since in that case the processes of formation and restructuring of radiation effects can develop differently. The often observed lack of additivity between the doses of various radiation types that impact a target simultaneously proves that there is really no universal approach to such "equivalence" assignments of radiation doses.

RADIATION, STREAMER

See *Streamer radiation*.

RADIATION, THERMAL

See *Thermal radiation*.

RADIATION TRANSPARENCY

Capability of a material to transmit the radiation flux incident on it in the microwave or the UHF regions (10^5–10^{12} Hz). A commonly used measure of radiation transparency is the inverse *absorption coefficient* determined by the imaginary parts of the complex *dielectric constant* and *magnetic permeability*. Materials with very low values of the *dielectric loss* tangent and its magnetic counterpart have a high radiation transparency. The effect of radiation transparency is weakened due to reflections of the emitted radiation from medium interfaces. Such reflections may be lessened or eliminated by antireflection or matching *coatings*.

RADIATIVE QUANTUM TRANSITION

A transition of a quantum system to a lower energy state with the simultaneous emission of a *photon*. Radiative quantum transitions can occur spontaneously (*spontaneous emission*) or under the impact of external photons (*stimulated emission*). The formation of corresponding quantum states in solids can involve several interacting particles. This results in a diversity of transition mechanisms and *selection rules*. Under this radiative transition photons differing in energy can be emitted with different probabilities. Here the energy and quasi-momentum of the system are conserved due to changes in the states of neighboring particles (changes in the number of phonons, *Auger effect*, and so on). Experimentally, this is observed as broad (in the spectral sense) bands of radiation, which can change under the influence of external factors (see also *Nonradiative quantum transition*).

RADICAL RECOMBINATION LUMINESCENCE

See *Luminescence during radical recombination*.

RADII OF ATOMS

See *Atomic radii*.

RADIOACTIVITY

A property of nuclei of certain isotopes of chemical elements to transform spontaneously into nuclei of other elements, through the emission of α-particles, electrons, positrons; this transformation is sometimes accompanied by the emission of neutrinos, antineutrinos, γ-rays (photons). The radioactivity of isotopes that is found in nature is called natural. There is a large number of artificial radioactive isotopes including many not found on Earth in their natural form. The process of radioactive decay is characterized by an exponential reduction of the number of original radioisotopes with time. The time span within which the number of original isotopes is reduced to one half of its initial number is called the *half-life* ($T_{1/2}$). Half-lives of nuclei of various elements vary over extremely wide limits, from practically instantaneous decay to many millions of years.

RADIO-FREQUENCY SELECTIVE SATURATION

A technique to measure the *hyperfine interaction* in inhomogeneously broadened *electron paramagnetic resonance* (EPR) lines (see *Inhomogeneous broadening*). Consider an electron spin system with a hyperfine interaction being saturated with high microwave power at an EPR resonance in a magnetic field. Radio-frequency *selective saturation* involves applying to this spin system a high power radio-frequency (NMR) saturating signal that induces resonance transitions in the nuclear subsystem. Depending on the saturation mechanism the holes burned in the spectral line either fill in or become deeper, new holes may appear, etc. The mechanisms involved are similar to those of *electron–nuclear double resonance* (ENDOR). Ordinarily a selective NMR saturation unsaturates the EPR resonance so that the depth of the "burned hole" decreases.

Radio-frequency selective saturation is often explained in terms of changes in the hyperfine *level populations* and the redistribution of saturation throughout the EPR resonance line under the influence of the high power radio-frequency field. Since the value of the hyperfine interaction is determined during the NMR selective saturation by measuring the radio-frequency at resonance, this technique approaches ENDOR in its accuracy for measuring hyperfine coupling constants.

RADIO-FREQUENCY SIZE EFFECTS

Anomalies in the dependence of the *surface impedance* of a metal *plate* on the magnetic field that appear when the plate thickness is a multiple of the extremum length of the electron trajectory. Radio-frequency size effects result from the ballistic features of the *anomalous penetration* of an electromagnetic field into a metal. They essentially involve a finite number of bursts of anomalous penetration that fit into a plate of thickness d. The magnetic field B is varied so that the condition $d = nD$ is satisfied ($n = 1, 2, 3, \ldots$), where D is the characteristic size of the electron trajectory (i.e. the distance between bursts), and the subsequent burst is brought to the opposite face of the plate. The electromagnetic field that appears is emitted into space, and may be measured. When B changes further, the condition $d = nD$ is violated, and the specimen becomes opaque for radio waves. Radio-frequency size effects can also be related to the cut-off of the cyclotron orbits (see *Resonance cutoff*). An isolated singularity then appears in the plot of the impedance as a function of B, which depends on the shape of the extremal electron trajectories. Radio-frequency size effects offer an effective method to measure the characteristics of the *Fermi surface*, the temperature dependence and the anisotropy of the mean free path, etc. See also *Gantmakher–Kaner effect*.

RADIO-FREQUENCY SPECTROSCOPY

See *Magnetic resonance spectroscopy*.

RADIO-FREQUENCY SPECTROSCOPY, VIBRONIC EFFECTS

See *Vibronic effects in radio-frequency spectroscopy*.

RADIOGRAPHY

A method of studying *solids* and biological specimens based on obtaining images produced by emissions from radioactive elements that are detected by sensitive recording materials. Radioactive elements may either be introduced into a specimen artificially, be present naturally, or be produced by irradiation from an external source. The α-, β-, γ-, and neutron types of radiography techniques are available. Radiography makes it possible to investigate the distribution, redistribution and diffusion properties of radioactive elements present in the specimen under study (*autoradiography*, *microautoradiography*), and also the presence of defects, impurities, structural, phase, and chemical inhomogeneities in them. To obtain radiographic images (*radiograms*) various *photographic materials* can be used (special X-ray films, nuclear emulsions, etc.), as well as particle track detectors (for α-particles, fission fragments, etc.). Depending on the exposure procedures one may employ contact or remote radiography. Contact radiography involves a peal-off emulsion. Depending on the analysis techniques, one may differentiate between contrast and track radiography. After exposure, one measures the optical density of the photographic materials.

RADIOGRAPHY, MICRO-

See *Microradiography*.

RADIOLUMINESCENCE

Luminescence excited by various types of *nuclear radiation*: γ- and X-rays (*X-ray luminescence*), electrons (*cathodoluminescence*), protons and ions (*ionoluminescence*), α-particles, neutrons, etc. During the interaction of radiation with materials, their atoms and molecules undergo ionization and excitation, followed by the formation of secondary particles with lower energy. In their turn the latter trigger further ionization and excitation, all eventually resulting in the formation of

energy states which are responsible for the final emission transitions. The energy output from radioluminescence is significantly less (1–30%) than the input because part of that energy is expended forming radiation defects, exciting *phonons* in the material, and inducing radiationless transitions. The principal mechanisms for stimulating radioluminescence include the following. In *ionic crystals* (CsI:Tl, NaI:Tl) *excitons* that form under the effect of the radiation then transfer their energy to emission centers. In semiconductors (ZnS:Ag) electron–hole pairs form (see *Semiconductor junction*), after which one of the carriers is localized at an emission center where it undergoes emission during *recombination* with a carrier of opposite sign. In organic crystals (*anthracene*, etc.) and *solid solutions* of organic materials (such as those used for plastic scintillators), either molecules of the basic material or those from an intentionally introduced luminescent admixture transform to an excited state. Luminescence may be observed as *fluorescence* (immediate emission) and *phosphorescence* (delayed emission). It is used for various detection and measurement tasks, and for spectroscopic studies of ionizing emissions. Some substances are also used in self-sustaining sources of light.

RADIOLYSIS

Chemical transformations of matter brought about by ionizing radiations. During radiolysis, there take place both the destruction of the initial chemical compounds and the formation of new compounds. The change of chemical composition that results from radiolysis can bring about variations in the physico-chemical, mechanical, optical, electrical, magnetic, and other properties of materials. Radiolysis is the process involved in *radiation chemistry*.

RADIOMETRIC ANALYSIS

A combination of physical techniques for measuring the activity of radionuclides. Radiometric analysis is often used to provide a relative measurement of *radioactivity*. Its absolute values may then be determined by comparing the activity of a radioactive specimen against that of a radioactive standard (i.e. standard radioactive source). Absolute measurements of activity involve gas discharges, scintillation counters, ionization chambers, and semiconducting detectors. During radiometric analysis one needs to account for the type of radioactive emission, since the respective instruments may only be able to effectively detect and record a particular type of ionizing radiation. Radiometric analysis is used to solve various problems, from those encountered during studies involving radioactivity monitoring, to those that concern the age of archeological and geological specimens. Layer-to-layer radiometric analysis is applied to determine concentration profiles of radioactive elements through a specimen depth.

RADIOTHERMOLUMINESCENCE

A type of thermal luminescence (see *Luminescence*) which develops when one heats to moderate temperatures substances preexposed to *nuclear radiations* before such heating. As a rule, radiothermoluminescence is observed in materials capable of long-term phosphorescence. The phenomenon is used to study *metastable states* and charge-carrier trapping centers (see *Traps for mobile particles*), both intrinsic and formed under the effect of irradiation.

RADIUM, Ra

A radioactive element of Group II of the periodic system with atomic number 88 and atomic mass 226.0254; it is an alkaline-earth element. All isotopes are radioactive, and four of them, ^{223}Ra, ^{224}Ra, ^{226}Ra and ^{228}Ra, are present in very small quantities in nature, because they are formed by the decay of long-lived ^{238}U, ^{235}U and ^{232}Th. The most stable isotope ^{226}Ra found in uranium ore (≈ 1 g per 3 T of uranium) is the most significant (half-life $T_{1/2} = 1.62 \cdot 10^3$ years). Electronic configuration of outer shell is $7s^2$. Atomic radius is 0.235 nm. Radius of Ra^{2+} ion is 0.143 nm. Oxidation state is $+2$. Electronegativity is 0.93.

In a free form, radium is a silvery-white lustrous *metal*; it quickly becomes dim in the air. It has a body-centered cubic lattice, $a = 0.5148$ nm. Calculated density 5.5 g/cm^3, $T_{melting} \approx 1100$ K, $T_{boiling} \approx 1770$ K. Heat of melting 7.9 kJ/mole, heat of sublimation is 162 kJ/mole, standard entropy is $S^\circ_{298} = 69.0$ kJ·mole^{-1}·K^{-1}; specific heat $c_p = 29$ kJ·mole^{-1}·K^{-1}. It is paramagnetic with magnetic susceptibility $1.05 \cdot 10^{-6}$ CGS units.

All the compounds of radium as well as metallic radium in the air possess a dull-bluish glow. As a result of the radioactive decay of 1 g of ^{226}Ra, about 550 J of heat per hour is released. Radium is used for the preparation of neutron ampoule sources (in mixtures with Be), and for manufacturing permanently luminescent materials (in mixture with ZnS).

RAMAN LINE, ANTI-STOKES

See *Anti-Stokes Raman lines*.

RAMAN RELAXATION PROCESS

A magnetic resonance relaxation transition between spin sublevels with the splitting $\hbar\omega$. Such a transition is accompanied by the virtual absorption of a phonon with energy $\hbar\omega_1$ and the emission of a *phonon* with energy $\hbar\omega_2$, provided $|\omega_2 - \omega_1| = \omega$ (a *two-phonon process*). It is called a *Raman process* in analogy with *Raman scattering of light*. In contrast to the *direct relaxation process*, phonons from every part of their overall frequency spectrum may take part in the Raman process. Thus, the Raman relaxation process starts to dominate direct relaxation at $T \geqslant 10$ K despite the fact that it is essentially a second-order process. The *spin–lattice relaxation* rate for the Raman relaxation process is given by $\tau_1^{-1} \propto T^n I_{n-1}$, where

$$I_{n-1} = \int\limits_0^{\theta/T} \frac{x^{n-1} \exp(x)}{[\exp(x) - 1]^2} \, dx,$$

and θ is a parameter that is generally smaller than the *Debye temperature* θ_D: $\theta \leqslant \theta_D$. We have $I_{n-1} \sim (n-1)!$ for $T \leqslant \theta/3$, so that $\tau_1^{-1} \propto T^n$, where $n = 7$ for ions with an even number of electrons, and $n = 5$ for a basic multiplet state featuring minor splittings. We have $\tau_1^{-1} = bT^9 + b' B^2 T^7$ for ions with an odd number of electrons, and for transitions within *Kramers doublets*. Here b and b' are coefficients that do not depend on T and B. For the limit $T/\theta \geqslant 1$ we have $\tau_1^{-1} \propto T^2$ for all the above cases.

Raman processes involving more than two phonons are possible, e.g., three- and four-phonon processes of spin–lattice relaxation. However, their probability is quite low. Such a probability grows for $T > \theta_D$ when the whole phonon spectrum is excited.

RAMAN SCATTERING

See *Enhanced Raman scattering of light*.

RAMAN SCATTERING OF LIGHT, Raman effect; called combinational scattering in the Russian literature

Scattering of optical radiation by matter accompanied by a change of the frequency of the scattered light. Raman scattering was discovered by G.S. Landsberg and L.I. Mandelshtam (1928) in crystals and, simultaneously, by Ch. Raman and K. Krishnan (1928) in liquids. Upon the irradiation of a material with monochromatic light, additional lines or bands appear in the scattered radiation spectrum alongside the non-shifted line of elastic *Rayleigh scattering of light*. These lines are symmetrically positioned on the low-energy (*Stokes component*) and the high-energy (*anti-Stokes component*) sides. The magnitudes of the frequency shifts characterize the energies of excitations of the medium where the process of inelastic scattering takes place. Unlike *Brillouin scattering* from intrinsic elastic *vibrations* of condensed media, which exhibits only relatively small shifts ($\Delta h\nu \leqslant 10^{-4}$ eV), the type of excitations is more diverse for Raman scattering, and typical shifts lie in the range $10^{-3} \leqslant \Delta h\nu \leqslant 1$ eV.

For many years only the vibrational and rotational excitations of constituent molecules (atoms, ions) were considered as having a bearing on Raman scattering in gaseous, liquid, and solid media. According to quantum theory, Raman scattering involves two inter-connected processes (see Fig.). By interacting with the electron system of a

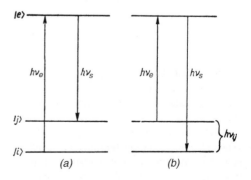

Transitions of Raman.

molecule in the initial vibration-rotation state $|i\rangle$, the primary *photon* of energy $h\nu_e$ transforms it into an intermediate (virtual) excited state $|e\rangle$. From this, the molecule makes a transition to another vibration-rotation state $|j\rangle$ by emitting a secondary light quantum $h\nu_s$. The energy *conservation law* $h\nu_e - h\nu_s = h\nu_{ij}$ defines the value of the vibrational (rotational) quantum $h\nu_{ij}$ generated in the Stokes process. If the initial level corresponds to an excited vibrational state then an anti-Stokes process can take place (see *Anti-Stokes Raman lines*) with the elimination of a vibrational quantum $h\nu_e - h\nu_s = h\nu_{ij}$. For solids under these conditions one ordinarily speaks of Raman scattering from *phonons* $h\nu_{ij}$. Raman scattering can also be brought about by other types of elementary excitation in crystals: *magnons*, surface and volume *polaritons*, *plasmons* and bound plasmon-phonons; single-particle excitations of free *current carriers*, from electrons and *holes* bound at impurity centers; transitions with spin flips, etc. In all these cases the cause of the Raman scattering involves space and time modulation of the electronic contributions of these excitations to the crystal polarizability.

The experimental detection and use of the great majority of Raman scattering processes in the solid state became possible only after the introduction of laser light sources (1962) for the excitation of the Raman scattering. In combination with high-sensitivity methods of photon counting and information gathering, the use of lasers permits the detection of very weak Raman scattering lines, some lower in intensity compared to that of the exciting radiation by 6 to 7 orders of magnitude. There is a pronounced increase in the scattered intensity when the exciting radiation frequency is very close to a dipole allowed electronic transition in the analyzed material. Such *resonant Raman scattering* is utilized for the study of the energy *band structure* of crystals, and the *electron–phonon interaction*. New opportunities for the analysis of very small quantities of material opened up after the discovery of *giant Raman scattering*.

Using the high intensity radiation of pulsed *lasers* enhances the probability of the scattering processes, and *induced light scattering* can occur whose intensity is comparable with that of the exciting light.

RAMAN SCATTERING OF SOUND BY SOUND

Appearance of a scattered acoustic field of combination frequencies beyond the region of intersection or overlap of incoming acoustic beams. A non-linear interaction of the acoustic waves takes place in the region of their intersection, and sources arise which radiate the scattered field. For instance, if two acoustic beams of frequencies ω_1 and ω_2 intersect, then a scattered field of sum and difference frequencies $|\omega_1 \pm \omega_2|$ appears (in the second-order perturbation approximation of the three-wave or three-phonon interaction) outside the interaction region, provided that synchronism or phase coherence requirements are satisfied (see *Nonlinear acoustic effects*), and provided that there is a finite probability for such an interaction to take place in accordance with the selection rules. Since the velocities of elastic waves in solids, e.g., longitudinal, transverse and *surface acoustic wave* types, differ from one another, the Raman scattering of sound by sound occurs at oblique intersections of acoustic beams. The angles of scattering are determined by coherence requirements, and depend upon the type of interacting waves, on their frequencies, and on the elastic parameters of the medium. The displacement amplitude in the scattered waves is proportional to the product of the amplitudes of the interacting waves, and to the cube of the wave number, $k = \omega/c$, where c is the *sound velocity*; viz., the Raman scattering of sound by sound is most pronounced at high-ultrasound and hypersound frequencies. Experimentally, this scattering is observed in the MHz range of ultrasound. The interaction of longitudinal and transverse waves allowed by the selection rules results in scattering when the frequencies of the interacting waves differ by no more than one order of magnitude. The Raman scattering of sound by sound in the case of three-wave (three-phonon) interactions is a basis for the kinetics of phonon systems; in particular, the theories of *sound absorption* in insulating crystals, of *thermal conductivity* and other collective *kinetic phenomena* in solids are based on this acoustic Raman scattering.

RANDOM CLOSE-PACKING MODEL (introduced by J.D. Bernal, 1959)

Model of liquid structure, also used for modeling *amorphous metals* and *metallic alloys*. The

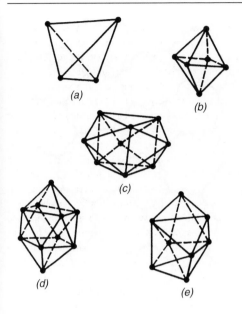

Fig. 1. Types of interatomic cavities: (a) tetrahedron, (b) octahedron, (c) trigonal prism covered by three semioctahedra, (d) Archimedean antiprism covered by two semioctahedra, (e) tetragonal dodecahedron.

random close-packing model has no formal definition, and it involves an algorithm with an arbitrary choice. The general rule for this random model structure building is the incorporation of the most possible choices for the locally close-packed interpositions of spheres. There are mechanical models made by mixing spheres with identical and differing diameters inside elastic covers, in rubber bubbles, and numerical models constructed by computer calculations involving interactions in accordance with some law (e.g., *Lennard-Jones potential*), using atoms with random positions and velocities, followed by "quick hardening" or quenching which means the abrupt removal of atomic kinetic energy from the system. There are also other algorithms for numerical random close-packing model building. Experiments have shown that the structural properties obtained by different approaches to random close-packing modeling are similar, i.e. they are not sensitive to the method of model building.

The main properties of random close-packing model structures are the following. The structure consists of five types of interatomic cavities (see

Fig. 1): tetrahedron, octahedron, trigonal prism covered by three semioctahedra, Archimedean antiprism covered by two semioctahedra, and tetragonal dodecahedron. These cavities are called *Bernal cavities* or *Bernal cells*. In a typical case tetrahedra occupy 48% of the volume, semioctahedra 27%, trigonal prisms 8%, Archimedean antiprisms 2%, and tetragonal dodecahedra occupy 15%. Structural defects of the random close-packing model have been studied by numerical modeling. There are regions of high density with pressure stresses concentrated inside (p-defects), regions of low density with tension stresses (n-defects), and regions of relatively high shear stresses (τ-defects). Each of these defect types includes, as a rule, nearly 10 atoms. *Vacancies* in the model are found to be unstable, so their volumes are redistributed between surrounding atoms. *Edge dislocations* are also found to be unstable, while *screw dislocations* have some stability. Atomic *diffusion* in the model structures proceeds by cooperative mechanisms due to correlated motions of nearest neighbor atomic groups, or through vacant cavities with volumes equal to those which result from fluctuations in the atomic structure.

Random close-packing models are used for studying structural and physical properties of amorphous metals and metal alloys. The *free volume model* is generally applied for investigating kinetic and mechanical random close-packing properties. Numerous experiments have shown that actual amorphous states have more perfect short-range order (see *Long-range and short-range order*) than random close-packing models predict, so the model is inadequate for simulating real amorphous structures.

RANDOM CONTINUOUS NETWORK (introduced by W.H. Zachariasen, 1932)

A model of amorphous solid state structure (see *Amorphous state*) according to which every atom of a given type has a definite local order and coordination, but interatomic bonds admit some distortions (bends, shears, contractions, extensions). The random continuous network model is used to describe nonmetallic materials with *covalent bonds*. Fig. 2 shows (a) a two-dimensional network with one type of atom with coordination 3, and (b) one with two types of atoms with

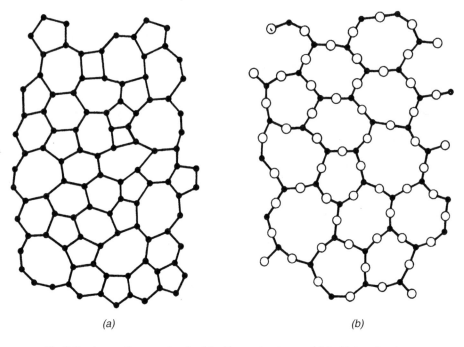

Fig. 2. Random continuous networks: (a) with one atom type, and (b) with two atom types.

coordinations 2 and 3. This network lacks long-range, but not short-range order (see *Long-range and short-range order*). Typical defects of this network are *dangling bonds* whose appearance is associated with local order violation. There may also be present extended defects of the type found in *polycluster amorphous solids*.

RANDOM FIELD CORRELATION FUNCTION

See *Correlation function of random field*.

RANDOM PHASE APPROXIMATION

Approximation used in the quantum theory of a gas of interacting Coulomb particles (*electrons in crystals, solid-state plasma*) according to which each particle is looked upon as free, and electrons interact with a mean overall potential. In this case, for the random phase approximation two-particle bound states (*electron–hole pairs, excitons*) participate in charge density formation only to the extent of their charge and polarizability. This approximation originated from processes of *electric charge screening* and electron liquid collective be-

havior, and is based on the physically reasonable supposition that under certain conditions one may neglect the exponent of the sum over states with randomly changing phases in comparison with the number N of electrons (hence the name of the effect). So, with this neglect, we have for the Fourier component of the charge density:

$$\rho_{k-q} = \sum_i \mathrm{e}^{\mathrm{i}(q-k)x_i} \approx N\delta_{k,q}.$$

The random phase approximation is used for investigations of solid state electronic properties, in particular it permits one to take into account electron interactions when analyzing quasi-particle properties, and it is also used for introducing collective coordinates for the description of plasmons (see also *Plasma oscillations*). Using the framework of this approximation the *dielectric constant* in quantum shielding theory is calculated. The approximation is also used for calculating collision integrals of the *quantum kinetic equation* for an electron gas, and for *Fermi liquid* theory.

RANDOM POTENTIAL

Potential energy $V(r)$ of a *quasi-particle* in a *disordered solid* state environment which is changing in a random manner in space, and is described by a random function of three-dimensional coordinates, i.e. by definition it is a *random field*. The random potential is usually established by parameterizing families of functions $V = V(r, \omega)$, where $\omega = (\omega_1, \omega_2, \ldots, \omega_n, \ldots)$ is the parameter set in which ω_i are the values with given *distribution functions* $p(\omega)d\omega$. The *Gaussian random field* (see *Gaussian distribution*) and the *Poisson random field* are often used in random potential theories. Gaussian field parameters are the coefficients u_k of the Fourier expansion $V(r) = \sum_k u_k \exp(ikr)$ which are associated with the distribution $p(u_k) = N \exp[-|u_k|^2/(2a_k)]$, where N is a normalization factor, a_k are material positive quantities corresponding to the mean values of the u_k coefficients squared: $\langle|u_k|^2\rangle = a_k$. A Poisson field is a superposition of potentials $v_i = v(r - R_i)$ with centers situated at points designated by coordinates $(R_1, R_2, \ldots, R_N, \ldots)$ that parameterize the field: $V(r) = \sum_{i=1}^{N} v(r - R_i)$. Usually coordinates R_i are considered homogeneously distributed in the system, so $p(R_i)dR_i = dR/\Omega$, where Ω is the system volume. Many physical values in disordered systems are expressed with the help of pair correlation functions $\psi(r_1, r_2) = \langle V(r_1)V(r_2)\rangle$. For a homogeneous isotropic random potential this function depends only on $r = |r_1 - r_2|$, i.e. $\psi(r_1, r_2) = \psi(r)$. Random potentials are often classified in terms of pair correlation functions, and when the correlation function is of the form $\psi(r) = \psi_1 \exp(-r/r_c)$, where ψ_1, r_c are constant values, we have a *Coulomb random potential*. A Poisson field with the potential $v_i(r) = Ze^2\varepsilon^{-1} \times |r - R_i|^{-1} \exp[-|r - R_i|/r_0]$ (where Ze is ion charge, r_0 is *screening radius*, ε is the static *dielectric constant*, $i = 1, \ldots, N$), is a Coulomb type, and in this case $r_c = r_0$, $\psi_1 = 2\pi Z^2 e^4 r_0/\varepsilon^2$, and $n = N/\Omega$ is the density of scattering centers. A delta-function random potential is one in which $\psi(r) = \Phi_0 \delta(r)$, where Φ_0 is a constant. An example of such a random potential is Poisson's field expressed as a superposition of short-range interaction potentials approximately described by the expressions $v_i(r) = v_0\delta(r - R_i)$, $i = 1, \ldots, N$.

For such a field $\Phi_0 = nv_0^2$, $n = N/\Omega$. If the pair correlation function is continuous together with all its derivatives at $r \to 0$, then the random potential is called a smooth one. Using a random potential facilitates tasks of finding spectral *densities of states* of electrons and phonons, and describing *transport phenomena* in disordered systems (see also *Correlation functions of random fields*).

RANDOM WALK

Model of particle transport in ordered and *disordered solids* describing the thermal motion of classical particles (*atoms*, molecules), or incoherent transport of quantum particles (e.g., *triplet excitons*). The simplest such model is a *discrete random walk on a lattice* in which at every time increment the wandering particle jumps from an occupied site to one of the neighboring unoccupied sites with equal probability. Over a long time interval the random walk process resembles a diffusion type behavior: the mean square distance covered in time t is $\langle r^2(t)\rangle = Dt$, where the *diffusion coefficient* is $D = \langle a^2\rangle/\tau$, $\langle a^2\rangle$ is the mean square particle displacement per step, and τ is the time interval between successive steps (see *Diffusion, Migration*). A more commonly used model is a *random walk with continuous time* in which the probability density for the waiting time until the next step is set by some (normalized) function $p(t)$. If there is a mean *waiting time* $\tau = \int_0^{\infty} tp(t)\,dt$, then a random walk with a continuous time evolution will, over a long time span, lead to the same result as a discrete random walk with step time τ. If the first moment $\langle r(t)\rangle$ of the expectation time distribution vanishes, then the nature of the time dependence of the second moments $\langle r^2(t)\rangle$ is determined by the specific form of $p(t)$. In this case the particle behavior involves *anomalous diffusion*. Many other random walk models have been studied in detail: models with log-range jumps over regular lattices, models with correlated sequential steps, random walks in disordered systems with different life time distributions at sites, or different probabilities for jumps between sites.

Random walk mathematical theory permits the calculation of the probability of a given site occupancy, the time distribution of reaching sites at various distances, the distribution moments of the motion, and other values. An important property of a random walk is its possibility of intersecting itself. The wandering is called *unrestricted random walk* if there is a probability of repeatedly visiting the same sites of the lattice. The opposite case in which each site can be visited only once is called a *self-avoiding random walk*, and it involves a wandering path that never crosses itself. A self-avoiding random walk is a useful model for connecting the monomers of a long chain polymer. An unrestricted random walk on a *fractal* system involves assignments of *fracton dimensionality* values.

RAPID ELECTRIC HEAT TREATMENT

One of the most widely applied types of *rapid heat treatment*, which is characterized by the use of high-frequency electric currents or electrical contact heating. Under these conditions the latent heat required for phase and structural transformations is generated directly in the reacting metal volume. With rapid electric heat treatment the kinetics of a transformation does not depend on heat transfer processes (unlike heating by an external source), which essentially influence the diffusion redistribution of elements, the formation of a grain structure, and may lead to an alteration of the *phase transition* mechanism. The original state of a metal acquires a special meaning with rapid electric heat treatment, because the heating of alloys, which are in a *metastable state*, causes relaxation processes to take place, which result in the formation of intermediate structures with a complex of physical properties, that cannot be obtained through conventional *heat treatments*.

During rapid electric heat treatment the temperature ranges of phase transitions, *dehardening* processes, *restoration of crystals*, and *recrystallization* are displaced toward higher temperatures. In deformed steel relative to the annealed state, the lowering of the austenitizing temperature, changes of phase transition kinetics, and volumetric effects of this transition can occur. The latent *work hardening* energy can be partially retained up to the phase *recrystallization* onset temperature. The lowering of the critical point is proportional to the part of the latent cold-working energy relaxation at the phase transition. The austenite formation at heating for hardening depends on the dispersion of the initial structure (see *Disperse*

structure): the rougher it is, the higher the *homogenization* temperature. This property can be applied in the heat treatment concepts based on the limited homogenization effect. When high-carbon steels are heated the limited dissolving of *cementite* in austenite can cause local microfusion in accordance with a eutectic type reaction. The heating of ferrito-perlite structures can induce the alpha to gamma conversion to perlite regions and structurally free ferrite, and this is associated with the inhibition of the diffusion concentration equalization process in the two-phase region of the compositional *phase diagram*. This is widely used for hardening with the formation of so-called *duplex structures*. The hardening of high-temperature phases being formed at rapid heating allows one to change the structure and morphology of hardened products over a broad range. By heating strengthened structures for annealing or aging, the morphology and phase composition of decay products (see *Alloy decomposition*) can also be altered over a broad range: e.g., during the *tempering* of alloyed steels the formation of special carbides can be inhibited. At rapid *electric tempering* the following are possible: changing the decay sequence as a result of different kinetics at all annealing stages, the formation of particular structures proper only for electroannealed steels, with a special favorable complex of physicochemical properties (*strength*, *plasticity*, *hardness* etc.) that is associated with the generated fine structure of a matrix and carbide phase. Electric tempering and heating for aging are used as the terminal operation of thermal processing, and also as a preparation technique for preparing a structure suitable for subsequent mechanical and thermal processing. The ultrarapid heating for annealing inhibits the pre-recrystallization processes that determine the recrystallization kinetics. As a result of ultrarapid heating, the rate of initiating recrystallization centers, and the rate of their growth, vary noticeably, and their concentration increases, thus promoting the formation of a special fine-grained homogeneous structure. Increasing the heating rate shifts the recrystallization toward higher temperatures, and can reach a phase transition point. The influence of the rate is more strongly pronounced for the heating of *heterogeneous structures* than for the heating of homogeneous alloys and fine metals.

A rapid electric heat treatment has particular technological advantages in comparison with burning processing. In most cases the properties of the processed metal are superior, with momentary heating the metal surface is less oxidized, and the overheating structure formation or metal burning is eliminated. The rapid electric heat treatment method offers wide opportunities for automation and robotizing the operation, and solves problems involving environmental concerns.

RAPID HEAT TREATMENT, rapid thermal annealing

Heat treatment at a heating rate of tens and hundreds of degrees per second. *Pulse heating* should be applied in the cases when the cross-sections of the heated metal or the heating method allow one to neglect the *heat transfer* into the depth of the metal, which during conventional heat treatment considerably influences the kinetics of *phase transitions* and structural transformations (see *Structural phase transition*). Heating that involves pulsed thermal treatment using external sources, e.g., due to radiation, can be carried out with the help of resistance furnaces, induction-muffle furnaces, salt tanks and lamps. Heating by electron or laser beams is widely used. These methods are most effective for the surface heat treatment of metals and alloys, and are used for the purpose of *quenching* and *tempering* during a strengthening treatment, and also for an overall or recrystallization *annealing* during a softening treatment of near-surface layers.

RAPID THERMAL ANNEALING

See *Rapid heat treatment.*

RARE-EARTH ELEMENTS, lanthanides

Fifteen chemical elements from ^{57}La to ^{71}Lu, filling the $4f^n$ electronic shell with $n = 0$ to 14. Each element has a Xe rare gas inner core of electrons ($2 + 8 + 18 + 18 + 8 = 54$). The outer valence electrons are $6s^2$ and sometimes $5d^1$. A salient feature of rare-earth elements is their close association in nature (e.g., the monazite *mineral* RPO$_4$ contains R = Y, La, Ce, Nd and sometimes phosphates of other rare-earth elements and Th). In addition, they are very close to each other in their chemical properties. Rare-earth elements form thermally very stable oxides practically insoluble in water. Many rare earths exhibit unusual optical and mechanical properties, which provide wide possibilities for their applications (see *Magnetism of rare-earth metals*). In the older literature sometimes *yttrium* (^{39}Y) and *scandium* (^{21}Sc) are mentioned as rare earths, but the term lanthanides is always restricted to the $4f^n$ series of elements.

RARE EARTH MAGNETISM

See *Magnetism of rare earths metals.*

RASHBA EFFECT (E.I. Rashba, 1957)

An increase in *light absorption* by an impurity center as the impurity level approaches the bottom of the exciton band with an extremum at $K = 0$. The effect has been observed in *molecular crystals* and *semiconductors*. The Rashba effect is explained by the fact that, as the energy of a localized state approaches the band edge, exciton states "mix" with it more and more, its radius correspondingly grows, and the *dipole moment* of the impurity center transition increases.

RASTER (fr. Lat. *rastrum*, rail)

A track traced out on a surface by a scanning electron beam, or the pattern formed by a sequence of such tracks. For example, the raster in a *scanning electron microscope* presents details of lines formed by a focused electron beam scanning along the surface of a target (a specimen or an electron beam tube screen). In that case the raster delineates the analyzed area at the specimen surface, which as a rule is rectangular. In an imaging tube the electron beam scans along a similar but larger scan raster, each point in that scan corresponding to a particular position of the beam at the solid surface of the specimen. In optics a raster is a grid of transparent, opaque or reflecting elements, placed on a substrate, its purpose being to transform or modify the structure of a light beam incident upon the raster. Rasters are used in cathode ray tubes, television, photography, and in printing.

RASTER ELECTRON MICROSCOPE

See *Scanning electron microscope.*

RAW RUBBERS

Polymers consisting of macromolecules that are similar to isoprene ($CH_2=C(CH_3)-CH=CH_2$) and have a chain structure. They form a complicated topologically disordered system. At ordinary temperatures they are highly elastic; they transform to a viscous fluid state as the temperature T increases, and to a *vitreous state of matter* as the temperature decreases below that of *vitrification*. The elasticity of raw rubbers is due to untwisting and aligning (in the direction of stretching) macromolecules accompanied by a change in the valence angles and interatomic distances. To remove existing *plastic deformation*, consisting in a relative displacement of the centers of gravity of individual molecules, the latter are cross-linked together with transverse chemical bonds, i.e. vulcanized (see *Rubbers*). Raw rubbers with molecules exhibiting a regular structure crystallize at certain temperatures, especially during the the the course of stretching. Irregular raw rubbers are more stable and do not crystallize.

RAYLEIGH DIFFRACTION, MÖSSBAUER

See *Mössbauer radiation Rayleigh diffraction*.

RAYLEIGH LAW (for magnetization)

(J.W. Rayleigh, 1887)

Parabolic dependence of the *magnetization M* of *ferromagnets* on the external *magnetic field* strength in weak magnetic fields (below the *coercive force H_c*). The Rayleigh law has the form:

$$M = \chi_a H \pm bH^2.$$

Here $\pm$ relates to the cases $H > 0$ and $H < 0$, respectively, χ_a is the initial *magnetic susceptibility*, and b is the *Rayleigh constant*. According to modern concepts, the nature of χ_a may be associated with reversible *magnetization rotation processes* and *domain wall displacement processes* in the magnetization, whereas b is associated with irreversible processes. The values of χ_a and b can sometimes be determined by the *remanence*. For many materials (Fe, Co, Ni, etc.), the Rayleigh law is valid over a wide range of temperatures up to the *Curie point*.

RAYLEIGH SCATTERING OF LIGHT

(J.W. Rayleigh, 1871)

Coherent scattering (without change of frequency) from optical inhomogeneities, whose dimensions are much smaller than the wavelength of the exciting light. In a solid, Rayleigh scattering occurs from the short-wavelength fluctuations of static defect concentrations, and from microinhomogeneities of various types.

RAYLEIGH WAVES

See *Surface acoustic waves*.

REACTANCE, reactive impedance

The part of the total *impedance* of an electrical (alternating current) circuit that comprises a *capacitance C* and an *inductance L*. According to Ohm's law we have for a series circuit: $V = IZ$, $Z = (R^2 + X^2)^{1/2}$. Here Z is the total impedance of the circuit, V is the voltage amplitude, I is the current amplitude, R is the *resistance* (dissipative impedance), and $X = \omega L - 1/(\omega C)$ is the reactance, where ω is the angular frequency of the harmonic oscillations of the current. The quantity ωL is called the *inductive impedance*, and $1/(\omega C)$ is the *capacitive impedance*. If $R \ll \omega L$, $1/(\omega C)$ then the series circuit exhibits *resonance* properties at the frequency ω of an applied alternating voltage which coincides with the natural resonance frequency $\omega_0 = 2\pi f_0 = 1/(LC)^{1/2}$ of the circuit.

REACTION, CHAIN

See *Chain reactions*.

REACTION, PERITECTIC

See *Peritectic reaction*.

REACTION, PHOTOCHEMICAL

See *Photochemical reactions*.

REACTION, QUASI-CHEMICAL

See *Quasi-chemical reactions*.

REACTION RATE CONSTANT, specific reaction rate

The rate of a chemical reaction with the *concentration* of each reacting substance equal to unity. A reaction rate constant, strictly speaking, is a value for a simple reaction which follows the law of mass action. In solid bodies it is meaningful only for true monomolecular and bimolecular reactions; thus the rate constant is the reciprocal of the mean lifetime of a molecule before entering into the reaction. In other cases the effective reaction rate constant is a combination of the true rate constant with the concentrations of the reagents. The temperature dependence of a reaction rate constant usually follows the *Arrhenius law*: $k = k_0 \exp[-E/(k_B T)]$, where E is the *activation energy*, and k_0 is the preexponential factor. Exceptions are *tunneling reactions* for which the rate constant either has no dependence, or only a weak dependence on the temperature.

REACTIVE DIFFUSION, reaction diffusion, reagent diffusion

Diffusion process accompanied by a reaction between the diffusing components that results in the formation of new *phases* or chemical compounds. For example, reactive diffusion appears during the growth of a metal oxide film at a metal surface that forms in the course of diffusion of metal atoms through that film to the surface enriched with absorbed oxygen. *Intermetallic compounds* may also form and accumulate during the *interdiffusion* of two metals. The kinetics of reactive diffusion may be controlled either by the rate at which the diffusion feeds the reacting components, or by the rate of reaction between them. The kinetic law describing the growth of the new phase depends on the rate-controlling step in the process.

REACTIVE IMPEDANCE

See *Reactance*.

REACTIVE STRESSES

Mechanical stresses that form during the course of implementing the *shape memory effect*, provided the material recovering its shape after heating and deformation is held fixed in some restraining device. Reactive stresses manifest themselves when heating brings the material through the temperature interval of the inverse *martensitic transformation*. The temperature at which such stresses start to form *en masse* coincides with the temperature at which *austenite* begins to form, and the temperature of its completion coincides with the temperature at which the inverse martensite transformation terminates. Note that the temperature of the end of this transformation depends on the level of stresses generated, increasing for higher levels of stress. The respective shift in the temperature scale may reach 100–150 K. For strong materials reactive stresses may reach ~600 MPa, while record cases such as doped titanium nickelide attain ~1200 MPa. The kinetics of generating reactive stresses repeats that of the martensite–austenite transformation in general, but in certain cases reactive stresses with heating may appear far earlier than in the martensite transformation case. Sometimes reactive stresses grow over a very broad temperature range. For example, they cover the range from 77 K to 850 K for ferromagnetic alloys. In other cases, reactive stresses are found both during heating and cooling. A *reverse effect of reactive stresses* is known for titanium nickelide, in which, as heating proceeds, reactive stresses pass through a maximum, and then repeat their kinetics in an inverse order during the ensuing cooling. Moreover, this effect occurs over and over again. As a rule, however, reactive stresses only generate during the heating half-cycle, while during the cooling half-cycle they relax within the temperature range of a direct martensite transformation. The process of generation–relaxation of reactive stresses may be repeated in successive thermal cycles.

The physical nature of reactive stresses is apparent: a redistribution of elastic and inelastic *strains* takes place in the regime of repeated shape memory effect and transformation of metal *plasticity*. The effect of reactive stresses is widely used to produce stressed units, e.g, when joining pipelines by mufflers made of alloys with shape memory.

REAL CRYSTALS

Crystals characterized by deviations of their *crystal lattice* from an *ideal crystal* periodicity because of the presence of *defects* of various types. The latter form during *crystallization* from the melt or the gas phase, during *recrystallization*, or under the effect of various external actions, e.g., *plastic deformation*.

The number and nature of the defects that form during the growth of real crystals are controlled by the conditions in which the crystal grows, and also by the impurities and inclusions that it contains. Faces of the growing real crystals capture impurities under certain conditions. Moreover, this process is not uniform in nature since various crystallographic faces exhibit differing *adsorption* energies. This can result in a segmented crystal structure with a corresponding chemical inhomogeneity. The latter also develops when the temperature changes periodically during *monocrystal growth*, and the crystal may develop a zonal structure. An inhomogeneous distribution of impurities in a crystal results in local changes of the crystal lattice parameters, providing an additional source of *internal stresses* and of generated defects (*dislocations*) that may produce *cracks*. Defects contained in real crystals are always in mutual relation to and interaction with each other. For example, the appearance of *point defects* promotes the formation of dislocations, while the surfacing of *screw dislocations* is accompanied by the formation of a surface defect, the so-called *growth spiral* at a face of the growing crystal, etc.

REAL STRENGTH, actual strength

Real *strength* of materials while in use. The actual strength of polycrystalline materials widely used in practice (see *Polycrystals*) is 2–3 orders of magnitude lower than their *theoretical strength*. Such a discrepancy is explained by the *Griffith theory* as resulting from the development and growth under load of microcracks (see *Cracks*) already present initially in the specimen. For *metals* of low plasticity with body-centered cubic lattices, the actual strength may be determined within the framework of the *Stroh theory*, which takes into account that a microcrack is generated as a result of *plastic deformation* prior to the breakdown itself, while the development of that microcrack is treated within the framework of the Griffith theory. For highly plastic metals in which their breakdown is viscous in nature (see *Tough failure*) the low actual strength is explained by the high mobility of the *dislocations* and a low value of the *yield limit*. Strength in that case is equal to the fluidity limit plus *strain hardening*. Levels of actual strength close to the theoretical strength can be achieved in dislocation-free *thread-like crystals*.

REAL SURFACE, nonideal surface

State of a *solid surface* found under ordinary atmospheric conditions. The real surface forms as a result of the interaction of surface atoms of a *solid* with the atmosphere, first of all with molecules of oxygen and water, that can result in the formation of an oxide (hydroxide) film up to several ångströms thick. The state of the real surface of a *semiconductor* also depends on the chemical processing of the respective specimens after they are cut out of a more massive crystal so as to eliminate structural *defects*. Chemical processing usually consists in *etching* a semiconductor with various acids and bases, followed by washing in distilled (deionized) water. An oxide film forms at the surface of a semiconductor after etching, its composition and structure depending on the conditions of the etching, and on the subsequent interaction of the etched surface with the ambient medium. Since the natural oxide surface film is not very stable, there results variations of the electrophysical properties of the surface with time. In most cases these electrophysical properties are studied in vacuo where the surface is more stable. To improve the surface stability certain techniques for passivation are available (see *Surface passivation*). This is achieved by growing various *coatings* of *films* at the surface or coating it with them.

REAL-TIME HOLOGRAPHY, dynamic holography

A method of recording and reconstructing the wave field of coherent light beams. In the usual static *holography* process a recording is made of the pattern formed from the interference of the object beam with a reference beam. The information is stored within the medium of the *hologram* (system of recorded interference lines) by an optical recording that employs the spatial variation of the *refractive index n* or of the *absorption coefficient κ*. The *phase holograms* associated with variations of *n* are the brightest and most effective. These variations provide the data needed for reading the stored image. In real-time holography the recording occurs through the direct effect of the incident beams without an intermediate stage of development, so the recording waves are to some extent distorted during the course of this process itself. There are different schemes used in information processing systems for redistributing the intensities and phases of the recording beams interacting through the holographic medium, e.g., energy pumping between the beams (attenuation of donor beam and amplification of acceptor beam), *wave front reversal*, etc. The nature of the beam variation also depends on how quickly the changes of n or κ induced within the medium exert their effect, and on the so-called *reversibility of the medium*.

In solids there are different mechanisms bringing about the variation of the index of refraction (and absorption) that affect the light beams. In materials where optical nonlinearity is not associated with real light-induced transitions between the states, the non-inertial recording of the interference field of the light beams is carried out. Particular mechanisms of recording are associated with light absorption, owing to variations of state of the medium, or to the appearance of free *current carriers* or *excitons* in *semiconductors*, to the heating, etc. In each of these cases the photoinduced variation of n and κ occurs due to the variation of the electronic redistribution of charge carriers between *traps*, due to polarizability variations, due to the appearance of an electric field (*electrooptical effect*), due to the dependence of n on the temperature, or on the magnitude of the elastic stresses, etc. During the recording of the interference field these variations of n are spatially-modulated with respect to the distribution of intensity in the interference field.

The shift in the recorded lattice of the refractive index relative to the recorded interference picture, whose value is substantially reflected in the properties of the dynamic transformation of the beams, is the important parameter of *holographic recording*. Most of the techniques for the photoinduced variation of n act on the local recording, upon which the phases of the interference field and of the recorded hologram coincide. For the *electrooptical method of recording* phase mismatching (nonlocal recording) occurs. Such a shift may be artificially obtained through local mechanisms: e.g., in the lattice recording of free electrons with the help of crossed electric and magnetic fields, if the *Lorentz force* is directed across the strokes of the holographic lattice.

In many non-metallic crystals the above listed mechanisms of the photoinduced variation of n

are realized with n variations of the order 10^{-3} to 10^{-4}, which is quite sufficient for the effective dynamic transformation of light beams in optically homogeneous media. The following materials are applicable: ferroelectric and pyroelectric crystals, in particular, lithium and potassium *niobates, barium titanate*; semiconductor materials – silicon, germanium, gallium arsenide, cadmium sulfide. These are good sensitive media for applications using methods of acoustic wave dynamic holography. Methods for real-time holography of long-wavelength electromagnetic radiation and neutron beams have been developed. See also *Optical techniques of information recording*.

REARRANGEMENT OF STRUCTURE, RADIATION-INDUCED

See *Radiation-induced rearrangement of structure*.

RECIPROCAL LATTICE

A lattice in reciprocal or k-space constructed from the basis vectors of a lattice in coordinate space (*Bravais lattice*). The three basis vectors of a reciprocal lattice have the form

$$b_1 = a_2 \times a_3/v,$$

$$b_2 = a_3 \times a_1/v,$$

$$b_3 = a_1 \times a_2/v,$$

where a_1, a_2, a_3 are the basis vectors of the direct lattice, $a_i \times a_j$ is the vector product, and $v = a_1 \cdot (a_2 \times a_3)$ is the volume of a *unit cell* of the crystal direct lattice. An arbitrary vector g of a reciprocal lattice can be written

$$g = 2\pi n_1 b_1 + 2\pi n_2 b_2 + 2\pi n_3 b_3,$$

where n_1, n_2, and n_3 are integers. The volume of a unit cell of the reciprocal lattice is equal to $(2\pi)^3/v$. All physical characteristics of a crystal that depend on the *quasi-wave vectors* k are periodic functions of k with the period of the reciprocal lattice. The convenience of introducing the reciprocal lattice is that any function of coordinates (that is periodic in the direct crystal lattice) can be represented in the form of a Fourier series summed over all the vectors of the reciprocal lattice:

$$f(r) = \sum_g A_g \exp(i\,g \cdot r).$$

For any vector $\boldsymbol{R}$ of the crystal lattice (direct lattice) and $\boldsymbol{g}$ of the reciprocal lattice, the equation $\exp(i\,\boldsymbol{g}\cdot\boldsymbol{R}) = 1$ holds. This means that each vector of the reciprocal lattice is normal to a certain set of planes of the direct lattice, and allows the planes of the direct lattice to be designated in terms of the reciprocal lattice, or by using the *Miller indices* n_1, n_2, and n_3 introduced above (see *Crystallographic indices*).

As a symmetric unit cell for the reciprocal lattice, it is often convenient to select the so-called *Brillouin zone* rather than a parallelepiped (primitive cell) constructed from the vectors $\boldsymbol{b}_i$. The Brillouin zone is useful when presenting the dispersion relations of elementary excitations in a crystal.

RECIPROCAL LATTICE CELL

A parallelepiped constructed of vectors of the *reciprocal lattice* (k-space), which can replicate and fill this space to generate a reciprocal lattice.

RECIPROCAL SPACE

The same as k-*space*.

RECOIL ATOM

An *atom* of a material which has received a considerable momentum due to an impact from an incident particle (e.g., *ion*, *neutron*, high-energy electron, etc.), or as a result of a nuclear reaction (see *Nuclear doping*).

During the bombardment or irradiation of a *solid* with particles, the recoil atoms contribute to its *sputtering* by forming the high-energy part of the spectrum of sputtered atoms. When the recoil atoms that were knocked out from the surface of a solid body or from within a *thin film* are excited or ionized in their inner shells they can emit electromagnetic radiation and Auger electrons (see *Auger electron spectroscopy*) with spectra having Doppler broadening (see *Doppler effect*) and other features that provide information on the crystalline and electronic structure of a solid, and the chemical composition of its surface. Under the ionic bombardment of a solid coated with a thin layer of another substance, the recoil atoms induce the implantation of atoms from the coating into the solid (*implantation of recoil atoms*). In the bulk of a solid, recoil atoms form radiation *defects*. See also *Hot atoms*.

RECOMBINATION of current carriers

The phenomenon of the annihilation of an electron and a *hole*. Recombination occurs in either a semiconductor or an insulator during a direct interaction called *band-to-band recombination*. This involves a transition of an electron from its energy state in the *conduction band* into a vacant state in the *valence band*.

Energy of the order of the width of the *band gap* E_{g} is released during the recombination of *current carriers* in semiconductors or insulators. Recombination is classified according to the mode of the energy release; it is: *radiative (emission) recombination* when a quantum of light is emitted, and *nonradiative recombination* when the energy is transferred to a third carrier or to lattice vibrations (see *Auger effect*). Radiative recombination is effective in direct-gap semiconductors.

Radiative processes can take place both in the bulk of a semiconductor, and at its surface. In many cases (particularly during nonradiative recombination) the process goes through several stages in which *local electronic levels* involving impurity centers or *defects* of crystal structure take part. The *Shockley–Read–Hall theory* specifies the dependence of the recombination rate on the defect concentration, the temperature, and the parameters describing the *trapping* of carriers into the local states for that case.

An important role in recombination processes is played by the binding of electrons and holes into *excitons* and exciton complexes (see *Electron–hole liquid*). The phenomenon of *recombination waves* is closely associated with recombination processes.

RECOMBINATION, SURFACE

See *Surface recombination*.

RECOMBINATION WAVES

Recharging waves in *traps* in *semiconductors*, which develop under certain conditions when a strong electric field is applied to the material. For large enough *hole* trapping cross-sections in n-type semiconductors, the local fluctuations in conduction electron concentration induce changes in hole concentrations at traps. Provided the electron trapping capture cross-sections are low, such a combined fluctuation dissipates slowly, and under the influence of an external electric field, it drifts through the specimen in the direction opposite to that of the field. The condition of current instability with respect to the formation of recombination waves envisages simultaneously satisfying the conditions involving a high enough concentration of secondary current carriers, holes. First, one should have $p_0/n_0 > \tau_p/\tau_n$, where n_0 and p_0 are concentrations, τ_n and τ_p are capture times for electrons and holes ($\tau_n \gg \tau_p$), respectively; second, the drift length for holes in the external field should exceed the diffusion length; $n_0 < N_e$, N_h, where N_e and N_h are the concentrations of centers capturing electrons and holes, respectively. The phenomenon of recombination waves in semiconductors was theoretically predicted and experimentally studied in *germanium* crystals doped with certain elements (e.g., Sb, Mn) that have *deep levels*, in *silicon* (Zn), etc.

RECOVERY

Totality of processes increasing the structural perfection of a distorted (containing many *defects*) crystalline material without change of dimension or shape of grains, limited by large-angle *grain boundaries*.

In the course of recovery of the deformed crystals two stages are distinguished: *restoration of crystals* and *polygonization*. The repose stage involves the redistribution and reduction in concentration of *point defects*, and also the redistribution and partial *annihilation of dislocations* without forming new boundaries. Repose is the main low-temperature stage of increasing the perfection of distorted crystal structures. During the investigation of the recovery process the most useful information is obtained with the help of electrical resistance measurements and calorimetric analysis (see *Calorimetry*), as well as the method of internal friction, and others. The *Van Buren classification* distinguishes five stages of recovery of properties upon increasing the annealing temperature (with different values of activation energies). The first three stages are related to repose: (I) *recombination* of *interstitial atom–vacancy* pairs, (II) *diffusion* of the interstitial atoms and groups of vacancies, (III) diffusion of the single and dual vacancies. Stage IV is associated with the migration of vacancies to *dislocations* which helps the

polygonization process, when metastable dislocations become aligned, and stage V directly involves the thermally activated motion of dislocations in the processes of *recrystallization* and nucleus formation. Mechanical properties of crystals (*yield limit*, *plasticity*) at the stage of repose vary only weakly because the dislocational structure itself also varies weakly. The more pronounced changing of mechanical properties appears at the stage of polygonization.

RECOVERY, ELASTIC
See *Elastic recovery*.

RECOVERY OF AGED ALLOYS

The phenomenon of the regeneration of properties of aged alloys (see *Alloy aging*) due to the passing into solution of the decomposition products of the supersaturated solid solution (see *Alloy decomposition*). Alloys subjected to continuous decomposition of the solid solution can achieve almost total recovery of their structure and properties. In alloys aged according to the cellular (interrupted) mechanism, a *fragmentation* of the grain structure occurs, and the source structure is not regenerated.

RECRYSTALLIZATION

Three meanings of this term are in use.

1. A process which consists in dissolving, melting, or evaporating crystalline material, followed by a repeated *crystallization* from solution (melt, gaseous phase). Recrystallization is used with the object of removing impurities from starting materials.

2. Another use of the term is for the change of the structure of a material in the solid state due to a polymorphic transformation (see *Structural phase transition*). The nuclei of a new *crystal structure* can arise inside the initial phase on heating (or cooling) over a certain temperature range. The growth of the new *phase* consists in the consequent increase of sizes of nuclei consistent with orientational and dimensional relations (see *Orientation relations*). As a result, the crystals of the new phase possess a certain orientation with respect to the grains of the initial phase. Recrystallization in the solid state may either involve or not involve diffusion; the process is accompanied by elastic stresses, which arise because of the difference between the volumes of the initial and final

phases. If the temperature of recrystallization is close to the temperature of thermodynamic equilibrium, then the generation and growth of the nuclei are slow; this results in the generation of a fine-grained structure which has a favorable effect on the mechanical properties of *alloys*; the procedure is used for increasing the strength and other characteristics of steels through *annealing* and *quenching*.

3. Process of replacement of a perturbed structure of crystalline material by new polycrystal grains (see *Polycrystal*) within the same phase, such grains having a more perfect *crystal structure* and lower free energy. Recrystallization may be observed both in deformed crystals and crystals perturbed through other techniques (e.g., during *monocrystal growth* while being bombarded by high-energy particle beams, etc.).

Recrystallization is a stage of structural changes during the process of heating the deformed crystal. It follows the stage of *recovery* of crystalline properties (*restoration of crystals* and *polygonization*) that results in the formation of a polycrystalline structure with a low density of *dislocations* and other structural *defects*. Recrystallization occurs through the formation of embryos, and their subsequent growth involves the migration of large angle boundaries.

Three principal stages of recrystallization may be identified: initial, collective, and secondary. The *initial recrystallization* stage is the process of generating recrystallization centers and their growth up to contact with neighboring new grains in the recrystallized regions. The new grains have a more perfect structure than the initial matrix, and are separated from it by large angle boundaries. The *collective recrystallization* stage consists in the growth of newly crystallized grains at the expense of others, via expansion or migration of their large angle boundaries. The average grain size then increases, but no major scatter in grain size is observed. The driving force of the initial recrystallization consists in the depletion of energy stored during *plastic deformation* related to dislocations and other defects of the crystalline structure. Meanwhile the driving force of collective recrystallization is the lowering of the free energy of *grain boundaries* due to their shrinkage per unit volume. *Secondary recrystallization* differs

from the collective type in that only a few of the recrystallized grains posses the capacity for further growth, and they may reach a very large size relative to the fine grain matrix. Secondary recrystallization occurs in the absence of normal continuous grain growth. It is observed if grain boundaries are fixed by the dispersed particles of a second *phase*. In that case, when heated, the grain boundary may move by way of either "dragging" these particles with it (*slow collective recrystallization*), or by tearing itself from the particles (provided the moving force of recrystallization is strong enough). In the latter case the torn-off boundary begins to quickly migrate while the other boundaries are still held pinned by the particles of the second phase. This is when secondary recrystallization develops, resulting in a widely discordant material granularity. This latter process acquired the designation of *quick or anomalous recrystallization*. Certain *textures* may also stabilize the structure (hamper grain growth), thus causing secondary recrystallization. The separate stages are thermally activated processes, so that their rate grows exponentially at higher temperatures, while their onset is usually preceded by an incubation period. The values important in practice are the initial and the final temperature of primary recrystallization (t_r^i and t_r^f), and the starting temperatures of the collective and the secondary recrystallization. These temperatures depend on the presence of impurities, the initial structure (*strain* ε), and the duration of *annealing* τ. For example, t_r^i and t_r^f decrease when ε and t grow. The presence of dispersed particles of a second phase may sharply slow down recrystallization, since t_r^i then increases due to blocking of the large angle boundaries of dislocation units (see *Fragmentation*). It also serves to increase the temperature of the collective recrystallization. The latter effect is widely used for technical applications. The value of t_r^i significantly depends on the type of interatomic bonding. The condition $t_r^i \geqslant t^*$ (see Table 1), where t^* is the characteristic temperature of crystal strain, holds for covalent and partially covalent crystals, the mobility of dislocations significantly decreasing below that temperature (see *Plastic deformation of covalent crystals*). This effect is caused by the need to have dislocation transformations before a recrystallization nucleus can form. When *cold brittleness* is present,

Table 1. Characteristic temperature t^* and onset temperature t_r for recrystallization of crystallized materials on a homologous scale T/T_{melting}

Material	Ge	Si	TiC	WC	$NbC_{0.76}$	TiB_2	ZrB_2	Cr	Ta	Mo
t^*	0.83	0.82	0.61	0.49	0.48	0.42	0.29	0.20	0.20	0.17
t_r	0.89	0.84	0.51–0.60	0.52–0.55	0.53	0.51–0.52	0.47	0.3–0.4	0.3–0.4	0.3–0.4

recrystallization of a deformed metal with a *cellular structure* results in a higher temperature of brittle failure, while its *plasticity* simultaneously increases above the temperature of brittle failure and the *yield limit* decreases in accordance with the *Petch relation*. The tendencies for *localization of plastic deformation* and for the formation of *necks* both diminish after recrystallization. To select an optimal processing temperature for a metal, the concept of a *homology recrystallization temperature*, $t_r^i = T/T_r^f$, is introduced, where T is the deformation or *annealing* temperature, and T_r^f is the final temperature of the initial recrystallization, both expressed in kelvins. For example, the optimal plasticity of *molybdenum* and *tungsten* alloys is reached at $t_r^i \sim 0.95$. Crystallographic texture significantly affects recrystallization. Various texture components in deformed metals feature various t_r^i, which may result in a partial recrystallization during annealing. Annealing textures formed after recrystallization either correspond to a deformation texture or noticeably differ from it. Textures may also vanish: e.g., a transition to a random distribution of orientations may take place.

Recrystallization may occur directly during the process of hot plastic deformation (see *Temperature conditions of plastic deformation*). This is the so-called *dynamic recrystallization*. In that case, an equilibrium sets in between the processes of *strain hardening* and *dehardening*, and the grain size is controlled by the deformation temperature and rate. That is why dynamic recrystallization may be used for obtaining extremely finely grained metals with low grain variability.

RECRYSTALLIZATION ANNEALING

Heating a solid to a temperature that provides full *recrystallization* within a given time period. Recrystallization annealing is used to lower the *dislocation* density, to change *texture*, to form a polyhedral *polycrystal* structure, and thereby to bring the physical and chemical properties of a solid to a level characteristic of an unhardened annealed state (see *Hardening*). Conditions for this annealing of deformed materials are chosen using recrystallization diagrams which provide the dependences of the temperatures for the beginning and end of the initial recrystallization on the degree of *strain* for a given processing duration.

RECRYSTALLIZATION AT LOW TEMPERATURES

See *Low-temperature recrystallization*.

REENTRANT SUPERCONDUCTORS

Compounds where *Cooper pairs* appear at the critical temperature T_{c1}, then disappear upon further cooling to the point $T = T_{c2} < T_{c1}$ which has been called the *second critical temperature*.

Such a transition from the superconducting state back to the normal state at decreasing temperature has been called a *reentrant transition*. This destruction of superconductivity in reentrant superconductors is associated with the appearance of ferromagnetic ordering in the regular lattice of the magnetic atoms. The basic mechanism of suppression of superconductivity in a *ferromagnet* is the strong exchange field that effects the conduction electrons via their magnetic moments. The spins of the Cooper pair electrons are antiparallel, and the exchange field acts to make them parallel, thereby destroying the singlet superconductivity. Unlike the case of ferromagnetic superconductors (see *Magnetic superconductors*), antiferromagnetic superconductors do not exhibit a reentrant transition because rapid spacial oscillations of the exchange fields (at the atomic scale) average to zero over the distance scale characteristic for the Cooper pair, which is a coherence length ξ_0.

The destruction of superconductivity in reentrant superconductors always takes place via a phase transition which distinctly exhibits hysteresis. Both a direct transition from a paramagnetic superconducting state into a normal ferromagnetic state, and the emergence of a nonuniform magnetic structure in the superconducting state, the so-called *coexistence phase*, are possible. The latter gives way to the normal ferromagnetic state under further lowering of the temperature. The first of these possibilities is realized in reentrant superconductors with closely spaced transition temperatures T_{c1} and T_{c2}. For example, in $Tm_2Fe_3Si_5$ $T_{c1} = 1.3$ K and $T_{c2} = 1.1$ K. Direct transition from the paramagnetic superconducting phase into the ferromagnetic state is also observed in reentrant superconductors with disorder in the magnetic subsystem, i.e. in the pseudo-ternary compounds $(Er_{1-x}Ho_x)Rh_4B_4$ over a wide range of concentrations $0.3 < x < 0.9$, and also in other series of pseudo-ternary compounds. The occurrence of the intermediate phase of coexisting magnetism and superconductivity has been observed in the reentrant superconductors $ErRh_4B_4$ ($T_{c1} = 8.7$ K, $T_{c2} = 0.8$ K) and $HoMo_6S_8$ ($T_{c1} = 1.8$ K, $T_{c2} = 0.65$ K).

In the framework of the model of the *indirect exchange interaction* of moments through conduction electrons (*Ruderman–Kittel–Kasuya–Yosida interaction*, RKKY interaction), the temperature of the appearance of magnetic ordering is determined by the paramagnetic susceptibility of the electrons χ_s, and the wave vector q of the magnetic structure corresponding to the maximum of $\chi_s(q)$. In the superconducting state at $T < T_{c1}$ the value of $\chi_s(q = 0)$ is noticeably reduced due to the appearance at the *Fermi surface* of a gap in the electronic spectrum, whereas for the wave vectors $q \gg \xi_0^{-1}$ the electronic *susceptibility* approaches its value in the normal metal. The value $q_0 \approx (a^2\xi_0)^{-1/3}$, where a is on the order of the interatomic distance, corresponds to the maximum of $\chi_s(q)$, while the temperature of the emergence of a nonuniform magnetic phase with wave vector q_0 practically coincides with the *Curie point* θ which should have been observed in the absence of superconductivity (1 K in $ErRh_4B_4$ and 0.74 K in $HoMo_6S_8$). *Magnetic anisotropy* leads to a magnetic structure with a sinusoidally modulated magnetic moment directed along the *easy magnetization axis*. With the reduction in temperature,

the sinusoidal modulation very quickly transforms into a domain-type structure with magnetization changes localized in the *domain walls*. The electromagnetic interaction appears to cause the domain structure to be transverse with the period $d \sim (a\xi_0)^{1/2}$, but the remaining parameters are determined by the *exchange interaction*. Neutron scattering data indicate $d \sim 10$ nm in $ErRh_4B_4$, and in $HoMo_6S_8$ it is about 20 nm. The *magnetic domain structure* in the superconducting phase can exist down to zero temperature ($HoMo_6S_8$), but only the small gain in energy arising from Cooper pairing, stabilizes it. When this gain becomes less than the excessive magnetic energy caused by nonuniformity, the first-order transition to the normal ferromagnetic phase occurs, and this determines the reentrant transition temperature of $ErRh_4B_4$ and $HoMo_6S_8$.

REFLECTANCE, reflectivity, reflectance factor

The ratio of the flux of energy (particles) reflected by a body to that incident on it. In the case of electromagnetic radiation, the reflectance depends on the angle of incidence, the degree of optical homogeneity, and the state of the surface of the reflecting medium (characteristic size d of the surface roughness compared to the wavelength λ). For $\lambda \gg d$, *specular reflection* takes place, whereas for $\lambda \ll d$, *diffuse reflection* occurs. To calculate the reflectance one can use the *Fresnel equations*. In particular, the specular reflection factor for a homogeneous medium at normal incidence is

$$R = \frac{(n - 1)^2 + \kappa^2}{(n + 1)^2 + \kappa^2},$$

where n is its *refractive index*, and $\kappa = k''\lambda/(4\pi)$ characterizes the absorption (k'' is the *absorption coefficient*). In other words, the intensity falls to $[\exp(4\pi\kappa)]^{-1}$ of its initial value when the light beam travels a distance λ through the medium.

REFLECTION OF ACOUSTIC WAVES

Reradiation of waves by an obstacle back into the medium where the incident wave originated. The term "reflection of acoustic waves" is applicable in those cases when the approximation of *geometrical acoustics* is valid; otherwise, one deals with the diffraction (see *Acoustic diffraction*) or scattering of sound (see *Acoustic wave scattering*). As a rule, at the interface of two media, the *acoustic wave refraction* takes place simultaneously with their reflection. In the general case, an incident monochromatic elastic plane wave excites three reflected (one quasi-longitudinal, and two quasi-transverse) and three refracted waves. For two media attached rigidly across the interface, the boundary conditions reduce to the continuity of displacements and mechanical stresses at every point of the interface. In this case all waves, i.e. incident, reflected, and transmitted, are taken into account. The solution of the problem of reflection and refraction of harmonic plane waves demonstrates that all wave vectors lie in the plane of incidence that is defined by the normal to the interface and the wave vector direction of the incident wave. In reflection and refraction of acoustic waves, the frequency remains unchanged. The wave vectors can be used to determine the directions and the rates of energy transfer for each wave (see *Phase velocity*, *Group velocity*). There can appear, at the interface between two anisotropic media, phenomena that are identical to those arising at the interface between isotropic media at so-called critical and supercritical angles of incidence, the latter in the anisotropic case being defined with respect to the wave vectors. Since the direction of energy transport in *crystals* does not, in the general case, coincide with that of the wave vector, the reflection of energy (sound beam) from the boundary can occur coincident with the refraction of the wave vector. The intensity of reflected and refracted waves is characterized by reflection and refraction factors that depend on the properties of both media, the wave polarization, and the angles of incidence, reflection, and refraction.

REFLECTION OF LIGHT

A phenomenon that occurs during the arrival of light at the interface between two media with different indices of refraction, and involves the formation of a secondary light wave that propagates from the interface back onto the medium of incidence. There are specular, diffuse, and mixed kinds of light reflection. *Specular light reflection* is observable when the sizes of irregularities or inhomogeneities of the reflecting surface are much smaller than the light wavelength λ. The reflected beam lies in the same plane as the incident one, and the normal to the interface (plane of incidence). The angle of reflection ψ (angle of reflected beam to the normal) equals the angle of incidence φ (angle of incident beam to the normal). The *reflectance R* depends on the angle of incidence, light wave polarization, and *refractive indices* (n_1, n_2) of the adjacent media. The incident light beam can be thought of as two beams plane-polarized in two mutually perpendicular directions: one in the plane of incidence (*p*-component) and the other in the plane perpendicular to it (*s*-component). In the case of reflection from an *insulator*, for $n_2 > n_1$, the reflectance R_p for the first component decreases down to the so-called *Brewster angle* φ_B, where $\tan\varphi_B = n_2/n_1$ (D. Brewster, 1815), as the angle of incidence increases; R_p vanishes at the Brewster angle and then sharply increases to 1 as φ grows to $\varphi = \pi/2$. The reflectance R_s for the second component increases gradually, and at $\varphi = \pi/2$ it also attains unity. Thus, at the angle equal to φ_B, the interface totally reflects polarized light. This is used in reflective radiation polarizers. If the beam is incident from an optically denser insulator ($n_1 > n_2$), then, starting with a certain angle φ_{cr} called the critical angle, determined by the relation $\sin\varphi_{cr} = n_2/n_1$, the entire incident beam reflects at the interface for $\varphi > \varphi_{cr}$ (*total internal reflection*).

The reflectance of many highly-absorbing media (in particular, *metals*) is large. Since the refractive index for absorbing media is complex-valued, there is no Brewster angle. The high reflectance of metals is explainable by the large concentration of free carriers. There is a pronounced sensitivity of the state of polarization of the reflected wave to the nature of the metal, and its surface treatment finds wide use in *metal optics* where the technique for studying the optical parameters is the analysis of intensity, phase difference, and polarization of the reflected beam.

Diffuse light reflection involves scattering in all directions. This takes place when the sizes of the interface irregularities exceed λ. In the case of ideal diffuse reflection, the reflected light intensity distribution is described by the *Lambert law* (J. Lambert), which states that the brightness of a diffusely reflecting surface is the same in all directions.

REFLECTION, PLASMA

See *Plasma reflection*.

REFRACTION OF ACOUSTIC WAVES

See *Acoustic wave refraction*.

REFRACTION OF LIGHT

Phenomenon at the interface between optically differing media which consists, in the general case, in a change of direction, amplitude, polarization and velocity of an electromagnetic wave upon its passage from one medium to the another. The refracted wave is related to the waves incident at the interface and reflected from it. The problem of *reflection of light* and refraction of light is solved for particular boundary conditions imposed on the electric E and magnetic H field vectors of the three waves. For transparent isotropic *insulators* the solution is provided by the *Fresnel equations*. For incidence on a *uniaxial crystal*, two refracted waves result, an ordinary ray (o) and an extraordinary ray (e) (see *Birefringence*), which have different polarizations and velocities of propagation. The problems associated with the reflection and refraction of light at boundaries of transparent crystals and of isotropic media are important for the calculation and design of optical systems containing crystalline components such as lenses, prisms and mirrors.

Another use of the term is for bending of a light beam associated with the variation of the *refractive index* of the medium in which it propagates. For a continuous variation of the refractive index a smooth bending of the light beam is observed. A smooth variation of the refractive index across the cross-section from the center to the periphery of a glass optical fiber allows the concentration of the propagating radiation along the fiber, and thus facilitates the optical transmission of information (see *Fiber optics*).

REFRACTIVE INDEX

The ratio between the velocity of electromagnetic waves in vacuo and the *phase velocity* of the radiation in the medium under consideration. Thus defined, the refractive index n is also called the *absolute refractive index*. The value of the refractive index in nonmagnetic media is determined by the *dielectric constant* ε: $n = \sqrt{\varepsilon}$; and depends on the frequency of the radiation. In anisotropic media, in particular in other than cubic crystals, n depends on the direction of propagation and the *polarization* of the light.

REFRACTOMETRY

A branch of technical optics which includes various methods for determining the *refractive index n* of light in a gas or condensed phase. According to the value of n, it is possible to obtain some information about polarizability and the effective charges of atoms of solids, or to identify the properties of the *band structure* of a crystal.

REFRACTORY MATERIALS, high melting temperature materials

Materials characterized by high *melting temperatures* (usually above $1800\,^\circ$C). Refractory materials based on *refractory metals* and their alloys (*chromium, molybdenum, niobium, tantalum, tungsten,* vanadium, etc.) and ones based on refractory compounds (*carbides*, oxides, borides, etc.) are distinguished. *Chromium, niobium, tantalum,* and *vanadium* serve as base metals for producing alloys with an appreciable content of the minor alloy component. The majority of elements are poorly soluble (see *Solubility*) in molybdenum and tungsten, and hence low concentration alloys (see *Alloying*) based on these two elements are for the most part used. High-melting metals exhibit *cold brittleness*, a property more pronounced in alloys based on chromium, molybdenum and tungsten, than in those based on vanadium, niobium and tantalum. The reduction of the cold brittleness temperature is brought about by purifying the alloy from interstitial elements, as well as by the generation of a fine-grained dislocation *cellular structure*. The hardening of alloys is accomplished by *solid-solution hardening* at alloying, as well as through precipitation hardening (see *Precipitation-hardened materials*) with high-melting compounds. In order to achieve effective *hardening* the particle size of the second *phase* must not exceed 0.1 μm. In wide use is alloying molybdenum and tungsten with considerable amounts of *rhenium* (see *Rhenium effect*), which simultaneously increases both the *high-temperature strength* and low-temperature *plasticity*, at the expense of decreasing the energy of *stacking faults*, and reducing the Peierls–Nabarro stress (see *Peierls–Nabarro model*). Metallic refractory alloys are most commonly used as structural *high-temperature materials*. *Tungsten alloys* exhibit the highest heat resistance at high temperatures, and the highest *elastic modulus*. However, in view of the high density and scarcity of these alloys, they are replaced, whenever possible, by *molybdenum alloys*. Niobium alloys and vanadium alloys offer promise as high-temperature heat-resistant materials for nuclear power engineering. *Chromium alloys* exhibit the highest heat resistance and *corrosion resistance*. *Tantalum alloys* are used most extensively in electronic equipment.

There exist both oxygen-containing (metal oxides) and oxygen-free *refractory compounds*. Refractory materials based on oxygen-containing high-melting compounds are distinguished for high refractoriness, and they are used for lining high-temperature furnaces; those based on oxygen-free compounds (borides, carbides, nitrides, etc.) belong to interstitial phases or to compounds of diamond-like structure. A number of binary refractory compounds exhibit a high mutual solubility of their components; e.g., components of the systems TiC–ZrC, VC–TaC, etc. are completely soluble in one another, whereas most other systems exhibit only partial mutual solubility. This provides a variety of compositions and properties of refractory materials based on such binary systems. These above materials are obtained, for the most part, by methods of *powder metallurgy*. The mechanical properties of refractory compounds are conditioned mainly by the covalent component of the interatomic bond (see *Plastic deformation of covalent crystals*). Materials based on refractory compounds find a wide range of applications: cutting tools, *abrasive materials*, superconducting materials, materials for jet-propelled devices and nuclear power engineering, phases for dispersion hardening of steels and high-melting metals, etc.

REGULAR SOLID SOLUTIONS

Solid solutions whose thermodynamic properties may be described within the framework of the *model of regular solutions*. This statistical-mechanical model assumes that the solution consists of almost spherically shaped particles of sim-

ilar size. The potential energy of the solution is assumed to equal the sum of the energies of interaction between immediate neighbors of atoms. The change of enthalpy (see *Thermodynamic potentials*) accompanying the formation of a single mole of such a binary solution is given by $\Delta H_m^M = \Delta U_m^M = N_A Z x (1 - x)\omega_{12}$, where N_A is the *Avogadro number*, Z is the coordination number, x is the mole fraction, $\omega_{12} = U_{12} - (U_{11} + U_{22})/2$ is the parameter called the *mutual exchange energy*; U_{11}, U_{22}, U_{12} are interaction energies for nearest neighbor particles: identical (11, 22) and differing (12) in nature. The change of *entropy* during the formation of regular solid solutions is given by an expression that holds for ideal solutions: $\Delta S_m^M = -R(x_1 \ln x_1 + x_2 \ln x_2)$, where R is the gas constant. The model of regular solid solutions is used for calculating thermodynamic properties of solid and liquid solutions, describing ordering phenomena (see *Alloy ordering*) and stratification in solutions (see *Alloy stratification*).

REGULAR SYSTEM OF POINTS (L. Sopke, 1874)

A set of points that are equivalent with respect to symmetry transformations of a crystallographic *space group*. Each system point may be obtained from a single initial point by repeated symmetry operations of the space group. An atom or a group or atoms associated with each point is called the basis, and when repeated in space it forms a crystal structure. The crystal structure is a replication in space of a finite number of atoms in a unit cell, the replication being carried out by symmetry operations (e.g., translations) of the space group. There are 7 crystal systems and 14 Bravais lattices in three-dimensional space. Bravais lattices can have one (primitive, P), two (body centered, I; base centered, C), or four (face centered, F) sites. The collection of transformations of the regular system of points leaving one of its points fixed in its place constitutes a symmetry *point group*, and the number n of these transformations is called the order of the point group. Each crystal structure conforms to one of the 32 *crystal classes*. If a point in the unit cell generates a new point for each symmetry operation of the point group then the resulting regular system of points is called a *general regular system of points*, otherwise it is a *particular regular system of points*. The former corresponds to a *Wyckoff general position*, and the latter to a *Wyckoff special position*. The number of sites for a Wyckoff general position equals the order n of the point group times the number of sites in the Bravais lattice. These Wyckoff positions are listed for each of 230 space (Fedorov) groups in the *International Tables for Crystallography*.

As an example consider the tetrahedral point group 23 (T) which has 12 symmetry operations, namely an identity, three $180°$ rotations, and eight $\pm120°$ rotations. In the primitive tetrahedral space group $P23$ (T^1) there are 12 atoms at a Wyckoff general position and 1, 3, 4 or 6 at the Wyckoff special positions. In the body-centered tetrahedral space group $I23$ (T^3) there are 24 atoms at a Wyckoff general position, and in the face-centered tetrahedral space group $F23$ (T^2) there are 48 atoms at a Wyckoff general position.

RELATIVISTIC EFFECTS IN THE BAND THEORY OF METALS

These effects play an essential role at high electron densities, in a homogeneous electron gas with a density typical of a *metal* when the relation $v_F/c \approx 0.01$ is satisfied, where v_F is the electron velocity at the *Fermi surface*. Near an ionic nucleus the electron density is much higher and relativistic effects have a significant indirect influence on the the valence electrons through their effect on the wave functions. The electron density functional method (see *Density functional theory*), which allows a many-electron system to be reduced to a single-electron system, fails to be logically consistent because of relativistic effects. In this case the single-particle approach is unjustified due to the effects of vacuum polarization, i.e. of the generation of particles (real or virtual) involving electrons with high energy. Nevertheless, in practice the relativistic generalization of the electron density functional allows one to obtain single-particle *Dirac equations* (P.A.M. Dirac, 1928) with a relativistic exchange-correlation potential including the *effects of retardation* of Coulomb repulsion and magnetic interactions of moving electrons. The relativistically invariant Dirac equation provides four-component (bispinor) single-particle wave functions, and the spectrum of electron energies in the crystalline potential. For an approx-

imate account of relativistic effects in the Dirac equation the wave function is expanded in powers of $1/c$. In the first order of expansion the *Pauli equation* is obtained (W. Pauli, 1927), which differs from the Schrödinger equation by taking into account the spin magnetic moment of the electron. In the second order of expansion there appear three terms describing the following: a dependence of the electron mass on its velocity; a Darwin term, different from zero at those points where charges are located (it influences only electrons of s-symmetry); a *spin–orbit interaction*, i.e. the interaction of the moving *magnetic moment* with the electric field. In practical calculations it is relatively simple to take into account the first two effects without the decomposition in powers of $1/c$ (*semirelativistic approach*), and when this is done the symmetry of the Schrödinger equation is preserved.

Taking into account *spin* and the spin–orbit interaction changes the symmetry of the wave functions, and requires switching to crystallographic double *point groups* which brings about the splitting of the energy bands with different values of total angular momentum, and the lifting of symmetry degeneracies at many points of the *Brillouin zone*. In this case the total energy of the crystal depends on the *magnetization* direction, which introduces the important effect of the *magnetic anisotropy* of metals. The Darwin term and the term involving the dependence of mass on the velocity lead to the lowering of s-bands relative to the *Fermi energy*, and the spin–orbit interaction splits the p-, d- and f-peaks in the *density of states* to two states with different values of the total angular momentum. In some cases relativistic effects must be taken into account to obtain agreement between calculated and experimental data on the topology of the Fermi surface. Relativistic effects are particularly important for *band structure* calculations of crystals containing elements of the fifth and sixth periods of the periodic system.

RELATIVISTIC EFFECTS IN SOLIDS

Changes of *energy bands* and equations of motion that arise from the *spin–orbit interaction*, whose role increases with an increase in the number of electrons (Z, atomic number) forming the atoms of a *crystal*. The characteristic or effective *fine structure* constant in a *solid* $\alpha_{\text{eff}} = Ze^2/(2\varepsilon_0 hc)$ is large, and a large spin–orbit interaction energy, comparable to or exceeding the gap E_g, leads to a qualitative change of the Hamiltonian (e.g., *Luttinger Hamiltonian*, *spin Hamiltonian*); in particular, its dimensionality varies due to the *Kramers theorem*. It brings about a change of the spectrum of *quasi-particles* and local centers (see *Local electronic levels*). *Narrow-gap semiconductors* stand out in this respect, where besides other features, the electronic *g*-factor (see *Landé g-factor*) has a pronounced dependence on the spin–orbit interaction, sometimes reaching magnitudes appreciably above 2. In addition, the spin–orbit interaction is responsible for adding new terms to the Hamiltonian, which leads to the appearance of new optical transition lines (*combined resonance*); similar effects are observed in *electron paramagnetic resonance* and *Raman scattering of light*.

RELAXATION (fr. Lat. *relaxatio*, relaxation, easing)

A process of establishing *thermodynamic equilibrium* in macroscopic physical systems (solids, liquids, gases). The state of a system is controlled by numerous parameters, and the equilibrium state of each may be attained differently. Each rate of approach to equilibrium is associated with a characteristic time called a *relaxation time*. Having attained an equilibrium state for certain parameters a physical system may remain in a nonequilibrium state with respect to other parameters, so the overall state is one of partial equilibrium. Some relaxing systems successively pass through several states of partial equilibrium.

A microscopic theory of the *relaxation of excitation* in solids is based on the *kinetic equations* that prescribe certain relations between the *kinetic coefficients*, and the collision characteristics for particles and quasi-particles (see *Boltzmann equation, Quantum kinetic equation*). Quasi-particles have finite *lifetimes* that may serve to provide estimates of relaxation times in solids.

Relaxation may be accompanied by a *phase transition*. During a *first-order phase transition* from a nonequilibrium to an equilibrium state a system first enters a *metastable state* from which it may exit via growth of *critical embryos* of a stable

phase. Since such nuclei need to achieve macroscopic dimensions to be effective, the process of relaxation from a metastable to a stable phase may become so slow that metastable phases behave as stable ones (see *Amorphous state, Crystallization*).

When approaching a *second-order phase transition* the order parameter that characterizes the difference between the properties of the *phases* tends to zero, and the relaxation times for such a system tend to become long (see *Critical phenomena*). The *spins* of atoms and subatomic particles being quite closely related to each other (via *crystal lattice vibrations*, orbital moments of *conduction electrons*), the spin system becomes a quasi-independent subsystem in the solid, with its own *magnetic relaxation* time.

As a rule, relaxation manifests itself experimentally in various indirect ways: through excitation of macroscopic motion, limitation of particle and heat fluxes that develop in various materials subject to external actions, and also through the dependence of the kinetic coefficients (of *electrical conductivity, internal friction*, etc.) on the frequency, ω, provided the external action undergoes periodic motions itself. This frequency dependence (dispersion) of the kinetic coefficients is one of the immediate consequences of relaxation processes. If a static force f_i brings about a deviation $\Delta x_i = \tau_i f_i$ from an equilibrium position x_i, then an alternating force of the same amplitude $f_i(t) = f_i \cos \omega t$ produces the deviation $\Delta x_i = \tau_i f_i (1 + \omega^2 \tau_i^2)^{-1/2} \cos(\omega t + \kappa)$, where $\tan \kappa = \omega \tau_i$, and τ_i is the relaxation time of the variable x_i. The effective reduction of the effect with increasing frequency ω and the phase shift between f_i and Δx_i leads to a nonmonotonic dependence of the periodic adsorbed energy $Q(\omega) \propto \omega \tau_i / [1 + (\omega \tau_i)^2]$. The presence of a maximum of $Q(\omega)$ at $\omega \tau_i = 1$ is called *kinematic (relaxation) resonance*. Measuring a kinematic resonance provides a convenient way to determine the relaxation time. The presence of several maxima of $Q(\omega)$ is evidence for the presence of several paths or mechanisms of relaxation (see *Resonance*).

See also *Dipole relaxation, Direct relaxation process, Ion–lattice relaxation, Magnetic relaxation, Maxwell relaxation, Paramagnetic relaxation, Nuclear quadrupole relaxation, Raman relaxation process, Snoek–Köster relaxation, Spin–lattice relaxation, Spin–spin relaxation, Surface relaxation, Zener relaxation.*

RELAXATION KINETICS METHODS

Methods of studying temporal changes of the parameters of various systems (electronic, nuclear, spin, molecular, structural) which take place due to the tendency of a system, driven out of its state of thermodynamic equilibrium by a certain forcing action, to spontaneously return to that equilibrium state. Such a forcing may be thermal, electromagnetic, mechanical, chemical, radiative, laser, electric, etc. Relaxation processes develop differently in various solids (*metals, semiconductors, polymers*, etc.), and depend on the external conditions (temperature, pressure, magnetic and electric fields, etc.).

A feature common for all methods of relaxation kinetics is the need to measure the characteristic *relaxation times* which depend on the properties of a given system, on its particular state, and on external conditions. For example, electrons in metals relax to equilibrium within 10^{-13}–10^{-14} s, while the electronic excitations in molecules do so in 10^{-10}–10^{-12} s. Relaxation times for solid polymers occupy a wide range from 10^{-8}–10^6 s and depend on the macromolecule structure, the molecular mass distribution, and the composition of the mixture. Rapid relaxation processes proceed via the motion of separate molecular groups in the segments of the polymer chains, and the slow ones do so via the motion of the larger segments of the chains.

Methods of relaxation kinetics in the physics of solids are used to retrieve data on the electronic and chemical structure of molecules (*nuclear magnetic resonance, electron paramagnetic resonance*, double electron–electron and *electron–nuclear double resonance*, acoustic resonance, *nuclear quadrupole resonance, optical spectroscopy* in the UV, visible and IR ranges, *luminescence, Raman scattering of light*, laser spectroscopy, X-ray photoelectron spectroscopy, etc.). These techniques also yield data on the structure and molecular dynamics of solids (nuclear magnetic resonance, mechanical and *dielectric loss, light scattering*, X-rays, neutrons, polarization luminescence, nuclear γ-resonance, positron annihilation,

etc.). Another application is to acquire data on the kinetics of processes with the participation of various intermediate state active particles, and various defects (using electron paramagnetic resonance, pulsed *radiolysis* and photolysis, low temperature spectroscopy, luminescence, conductometry, electron *spin echo, cyclotron resonance*, etc.).

RELAXATION OF EXCITATION in crystals

Process involving the return of a *crystal* to the state of *thermodynamic equilibrium* after its excitation. Unlike macroscopic nonequilibrium states (of temperature, fields, etc.) this case concerns the elementary excitations of microscopic states. Mechanisms of relaxation are very diversified. In *radiative relaxation* the energy of an elementary excitation is expended in photon release, while in *nonradiative thermal relaxation* the energy dissipates into low-energy thermal excitations (*phonons, magnons*). The time of radiative relaxation for an electric dipole transition $i \rightarrow k$ is $\tau_{ik} = 2.3 \cdot 10^{-8} \, M/(E_{ik}^2 f_{ik})$ (in s), where M is the particle mass in electron mass units, E_{ik} is the transition energy in eV, and f_{ik} is the oscillator strength. With increasing E_{ik} the speed of radiative relaxation increases and the speed of thermal relaxation exponentially decreases in the region of E_{ik} corresponding to the generation of a large number of phonons in one act of thermal relaxation.

Conduction electrons and *excitons* relax via the *electron–phonon interaction* when the thermal relaxation rate is low due to the small size of the parameter m/M (ratio of electron to nucleus mass). If the electron kinetic energy is sufficient to ionize atoms then it is transferred to secondary electrons (δ-electrons), relaxing along the thermal relaxation channel. If the elementary excitation is associated with a spatial separation of the produced particles (of electron and *hole*, of fragments of molecules, etc.), then thermal recombination relaxation takes place, and high τ values are possible. Excitons and local excitations relax along the radiative relaxation channel (for not too small f_{ik}) as well as the thermal channel (especially in the presence of a lower-lying close or intersecting term). Competition between radiative and thermal relaxation determines the population ratios of different excited levels, and the *quantum yield* of luminescence. Raising the temperature accelerates thermal relaxation, but usually does not influence radiative relaxation. In the case of vibrational excitations (in particular molecular *vibronic states*) the radiative relaxation rate, in view of the large M and small E_{ik}, is negligibly small, and thermal relaxation is caused by their anharmonic coupling to the lattice. The weakened bonding of some electron-excited molecules to inert cryomatrices retards the thermal relaxation and facilitates *hot luminescence* from the high vibrational levels. When there are freely rotating molecules in the lattice then the transfer energy to their rotational degrees of freedom precedes its dissipation as vibrational energy (as also occurs in a gas). Thermal relaxation of phonons in *insulators* is caused by their anharmonic interactions, mainly by the decomposition or decay of one phonon into two phonons, and by the inverse process of their coalescence. Phonons with the frequency $\omega \sim \omega_D$ (ω_D is Debye frequency) decompose during several lattice vibration periods. With reduction of the average value of ω thermal relaxation is retarded due to reduction of the spectral *density of states*. With increasing temperature T the influence of the density of states and the role of induced processes become more pronounced. The relaxation of a spin subsystem (magnons) is mainly associated with the scattering of magnons by magnons, caused by their *exchange interaction*. The role of exchange scattering sharply increases with increasing magnon *quasi-momentum*. Magnons with low values of quasi-momentum relax due to the processes of magnon decay into two magnons, as well as via inverse processes, and also through processes involving the participation of phonons (cf. *Magnon–phonon interaction*).

RELAXATION OF NUCLEAR SPINS
See *Nuclear spin relaxation time*.

RELAXATION OF STRESS
See *Stress relaxation*.

RELAXATION TIME
Time interval τ of a non-equilibrium process during which the deviation of some parameter(s) of a system (or the *distribution function* of this parameter) decreases from its initial value by an amount comparable with this initial value, usually by the factor of $e = 2.718$ in the case of single exponential relaxation. The simplest example of a relaxation time is the case of *linear relaxation* when $|X - X_0| \ll X_0$, where X is a system parameter, X_0 is its equilibrium value, and the time dependence has the form $dX/dt = -\tau^{-1}(X - X_0)$. Therefore, $X(t) - X_0 = [X(0) - X_0] \exp(-t/\tau)$. The value τ^{-1} is called the *relaxation frequency* (or rate). A relaxation time is a measure of the efficiency of a *relaxation* mechanism. The values of relaxation times in solids depend on the external and internal conditions, and lie in wide ranges: 10^{-8}–10^{-13} s for *conduction electrons*, and up to many hours for *spin–lattice relaxation* of nuclei in solids.

RELAXATION TIME, ELECTRON
See *Electron relaxation times*.

RELIABILITY OF MATERIALS
The capacity of materials to perform their functions while retaining specified characteristics within the limits required for this performance. The main advantage of high reliability is to preclude premature or unforeseen failure, i.e. loss of properties that are necessary for serviceability. The degree of reliability is determined by indicators associated with failure, which can take place either suddenly or gradually. Sudden failure is characterized by an abrupt change of properties, e.g., by *brittle failure*. Gradual failure is preceded by a progressive and regular change of properties of the material, e.g., *creep*, *fatigue*, *wear*, aging (see *Alloy aging*). The reliability of various types of materials is gauged by different properties: failure-free performance, longevity, retaining quality, maintainability. These properties are of varying significance when applied to specific materials. In this connection, reliability is assessed with the help of indicators chosen to take into account specific features of the material, operating conditions, and consequences of failure. The main ways to evaluate reliability are experimental studies, methods of mathematical statistics, and statistical modeling. A necessary condition for assuring the reliability of materials is through thorough and comprehensive tests of samples under various conditions, including especially unfavorable circumstances that may arise during realistic operation,

with subsequent detailed analysis, and followup of the results.

RELIEF

See *Peierls relief*.

RELUCTANCE

The ability of a material to resist the passage of a *magnetic flux* through a section of a *magnetic circuit*. For a uniform magnetic circuit where no *magnetomotive force* (mmf) is applied, the reluctance R_m is the ratio of the magnetic potential drop ΔV_m across this section to the passing magnetic flux Φ. This relationship, $\Delta V_m = \Phi R_m$, is the magnetic circuit analogue of Ohm's law $\Delta V = I R$ for electric circuits. If the specimen is shaped as a toroid and a magnetic coil is wound uniformly round it, then the magnetic potential drop equals the magnetomotive force, while the reluctance is equal to the ratio of the mmf to the magnetic flux. The reluctance of a uniform magnetic circuit section is be computed, in the SI system, as $R_m = l/(\mu\mu_0 A)$, where l and A are, respectively, the length and cross-sectional area of the magnetic circuit section, μ is the relative *magnetic permeability* of the circuit material, and $\mu_0 = 4\pi \cdot 10^{-7}$ N/A^2 is the *magnetic permeability of vacuum*. In the case of a nonuniform magnetic circuit (comprising uniform sections in series with differing l, A, and μ), the reluctance is the sum of R_m for the sections. Computing reluctance with the above equation is only an approximation, since it takes no account of the *magnetic loss* in the materials, the magnetic field nonuniformities in the circuit, the nonlinear dependence of reluctance on the magnetic field, etc. For an alternating applied magnetic field the reluctance is a complex-valued quantity, as μ depends on the frequency of the magnetic field oscillations.

Ferromagnets (electromagnets, magnetic flux concentrators, etc.) can amplify magnetic flux, and give it a direction. Materials with large reluctance are often used to reduce magnetic flux in a corresponding magnetic circuit section. Some materials, when included in a magnetic circuit magnetized with a direct current, change their reluctance periodically. They can be used to gauge the magnitude of a magnetic flux passing through a circuit section, and make this procedure automated,

for instance, in ferroprobe coercitometers and differential magnetic field measurement devices (see *Magnetometry*) for *magnetic structure analysis* of ferromagnetic materials, etc.

REMAGNETIZATION

See *Magnetic reversal*.

REMANENCE, residual magnetization

Magnetization M_r that is inherent in a premagnetized *ferromagnet* at zero applied *magnetic field* strength. The value of the remanence depends on many factors, including the magnetic properties of the material, its magnetic history, and its temperature. Remanence increases with the increase of the magnetizing field strength, and tends to a limiting value that is taken to be the remanence of a given ferromagnet. The remanence of a material is to be distinguished from the remanence of a body of a particular shape made from that material. Since the body experiences the effect of its own *demagnetization field*, which depends on its shape, its remanence is almost always smaller than that of the material itself: the greater the demagnetizing factor of the body, the smaller its remanence. To determine the material remanence, conditions must be established so that the internal field strength in a sample is zero. It is convenient to compare the relative remanence $m_r = M_r/M_s$ for different materials, where M_s is the magnetization of engineering saturation (see *Saturation magnetization curve*). In some materials, $m_r \approx 1$, which is accomplished by creating a *magnetic texture* in them. The remanence is one of the main parameters to be taken into account when establishing a ferromagnet as a *hard magnetic material* for *permanent magnets*. See also *Magnetic hysteresis*.

REMANENT POLARIZATION

The part of the polarization (see *Polarization of insulator*) of a *ferroelectric* that remains after the applied electric field is removed. It is the electric analogue of remanent magnetization. In a strong electric field a ferroelectric becomes polarized. If the field exceeds the value of the coercive field (see *Coercive force*) and its action is sufficiently prolonged, the ferroelectric crystal can pass into a *single-domain state*. After the field has been turned off, the *depolarization field* remains in the sample (see *Photodomain effect*). If no special steps are taken the single-domain state eventually decays. However, in most ferroelectrics the orientation of a majority of the *domains* is conserved. The value of the polarization P is then determined by the *unipolarity factor* φ; $P = \varphi P_0$, where P_0 is the saturation polarization. The remanent or residual polarization is easily detectable by its *dielectric hysteresis* loop under cyclic repolarization of a ferroelectric in a strong alternating electric field (see *Polarization switching*). The polarization on the loop, which corresponds to zero applied field, is the remanent polarization. As a rule, it is observable in ferroelectrics with a large value of the coercive field, its magnitude depending on the crystal perfection, the presence of internal mechanical stresses, and impurities. It can be increased by carrying out the polarization in a permanent field while decreasing the temperature from the *Curie point* to room temperature. This method is often used for polarizing *piezoelectric ceramics*.

RENNER EFFECT (R. Renner, 1934)

Totality of effects which characterize linear molecules in degenerate electronic states; they arise from quadratic terms of the *vibronic interaction* in the bending vibration displacements of nuclei Q (in the linear approximation there is no effect). The weak Renner effect involves states degenerate at a point Q_0 in which a branch of the *adiabatic potential* splits in such a way that their minima remain at the point Q_0, as shown in Fig. (a), and in the strong Renner effect the ground state at this point becomes unstable, as shown in Fig. (b). In the latter case the Renner effect has characteristics somewhat similar to those of the *Jahn–Teller effect*.

RENORMALIZATION GROUP METHOD

Statistical method developed in quantum field theory and widely used for studying materials near critical points. The basic idea of the method lies in the investigation of transformations, in which the effective Hamiltonian of the system evolves subject to the variation of particular spatial and/or temporal scales, with these transformations exhibiting some group properties, and there is very often a kind of invariance principle at work (see

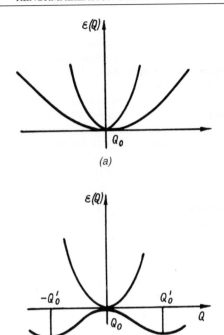

(a)

(b)

Behavior of the adiabatic potentials $\varepsilon(Q)$ in the space of nuclear displacements Q near the point of degeneracy Q_0 of the Renner effect: (a) weak effect with no instability; (b) strong effect with dynamic instability.

Group theory). The renormalization transformations do not, however, actually form a mathematical group because there is no inverse transformation. The evolution of Hamiltonian parameters under transformations involving infinitely small length scales is described by differential equations, the *renormalization group equations*. The singular points (*fixed points*) of these equations determine the parameters of the asymptotic behavior of the system, e.g., the values of the *critical indices* (*critical exponents*) of thermodynamic quantities near points of *second-order phase transitions*, the ratios of critical amplitudes, etc. The variety and range of modifications of the renormalization group method extend from field-theory types to versions that are very exotic with respect to their mathematical form. There are some versions formulated for computer applications, e.g., *Monte-Carlo renormalization group method*. The renormalization group method is a basic working tool in the modern theory of *critical phenomena*, where with its assistance many qualitative and quantitative results in conformity with experiment have been obtained. It has been applied to problems involving random walks, self-similarity, fractal dimensionality, chaos, the Ising model, the *Kosterlitz–Thouless transition*, and the Migdal–Kadanoff transformation. It is also widely used in localization theory, in *percolation theory*, in the theory of phase transitions in disordered systems, in the theory of hydrodynamic instabilities, in the theory of the *quantum Hall effect*, etc.

REOXAN (acronym for REcording OXidizing medium with ANthracene)

A polymeric light-sensitive medium. Reoxan consists of a *dye*-sensitizer and *anthracene* dispersed in the polymer polymethylmethacrylate. Before exposure the layer is saturated with oxygen, which, upon collisions with the dye molecules, converts them to the singlet state and oxidizes anthracene. This causes the adsorption spectrum in the UV region to reflect the variation of the *refractive index* in the visible region ($\Delta n \sim 10^{-3}$). Reoxan is used for the creation of holographic optical elements (gratings, filters, lenses, etc.) of high efficiency ($\sim$90%).

REPLICA

An exact copy of the surface relief of bulk samples or small particles in the form of a *thin film*, intended for investigation of the *surface atomic structure* using an *electron microscope*. Replicas are obtained by the application and subsequent separation of a thin layer of organic mixtures, of carbon, of silicon oxide, of light metals directly on the sample or on initially manufactured prints. To increase the contrast of the image, use is made of the *method of shadowing* the inclined *spraying* of a thin layer of a heavy metal (gold, platinum, tungsten, etc.) or carbon. To achieve high resolution, replicas made from sprayed metals (silver, platinum, aluminum, etc.) are used.

RESIDUAL MAGNETIZATION

See *Remanence*.

RESIDUAL RAY BAND, Restrahlen

A region of the infrared reflection spectrum of an *ionic crystal* situated between the frequencies of the long-wavelength transverse optical *phonons* ω_{TO} and the long-wavelength longitudinal optical phonons ω_{LO} (the *reflectance* in this region is close to 100%). After multiple reflection from the crystal the only remaining radiation corresponds to the residual ray band. See also *Infrared spectroscopy*.

RESIDUAL STRESSES

Internal stresses that remain in a solid after external factors, such as force, heat, etc., cease acting upon it. The residual stresses can be of three kinds, according to the size of the volume where these stresses are balanced (see *Stress scales of defect structures*). Stresses of the first kind (*macrostresses*) are balanced in the macroscopic bulk of the entire sample, or its component parts. Stresses of the second kind (*microstresses*) are balanced in individual grains or parts of grains. Stresses of the third kind (*submicrostresses, nanostresses, static lattice distortions*) are balanced within one or several *unit cells* of the crystal lattice. Residual stresses are due to nonuniform elastic *strain* in solids (i.e. by distortions of the *crystal lattice*). Lattice distortions are caused by crystal defects and the nonuniformity of their distribution. Distortions of the first and second kinds can be related to *dislocations, grain boundaries*, phase boundaries, microcracks (see *Cracks*), and so on. Distortions of the third kind, as a rule, are due to *point defects* (*vacancies*, dissolved atoms); yet, they can be observed in thin near-surface regions, as well as in regions of dislocation cores. In many cases, defects of one kind cause distortions of other kinds. The distortions emerge during *plastic deformation, phase transitions*, cooling of one- and many-phase systems due to a difference in heating temperatures, and different linear expansion coefficients. Residual stresses can produce both positive (*hardening*) and negative (warping, cracking) effects on the properties of products. The most common means for removing residual stresses is *annealing* with subsequent slow cooling. The *stress relaxation* near stress concentrators like microcracks or inclusions can be achieved by means of minor plastic deformations. Residual stresses are calculated from the physical parameters of the material, and measured using mechanical, X-ray, electromagnetic, polarization-optical, and other methods.

RESIST

Material sensitive to light (*photoresists*), X-rays (*X-ray resists*), electron beam (*electron resists*). Resists are used in *integrated circuit* technology as protection mask coatings for processing chip surfaces (e.g., during chemical *etching*, or *ion implantation*). They are usually made of particular *polymeric materials*. Photoresists are applied to the surface, then the latter is illuminated through a *mask*, that is a transparency with a particular pattern drawn on it. Processing the plate then produces a thin protective layer of the required pattern on its surface. Under the effect of light, negative photoresists form insoluble patches of the pattern, and positive photoresists form soluble ones. Either a positive or negative image of the pattern is projected, respectively, upon the semiconductor plate from the mask. An important characteristic of resists is their resolution, the possible number of equally wide lines separable per unit length of the surface, which is achievable by *lithography* techniques. It may reach thousands of lines per millimeter (submicron resolution). *Two-layer resists* and *three-layer resists* are also used.

RESISTANCE of an electric circuit

A quantity R characterizing the rate of energy loss during the passage of an electric current: $R = P/I^2$, where I is the current amplitude, and P is the power dissipated by the circuit. For an alternating current an average over the period of oscillation, provides the transformation of electromagnetic field energy into thermal, light, mechanical, or other energy types. For a sinusoidal current and voltage the power loss is $P = IV \cos\varphi$, where V is the voltage amplitude and φ is the phase shift between the current and voltage. If $\varphi = 0$, then *Ohm's law*, $V = IR$, is valid.

RESISTANCE TO MICROSPALLING

See *Microspalling resistance*.

RESISTANCE, THERMAL

See *Thermal resistance*.

RESISTANCE TO CORROSION

See *Corrosion resistance*.

RESISTIVE DOMAINS in superconductors

Normal regions in a *superconductor*, that is regions in the *resistive state* in which energy dissipation takes place. This state appears for currents exceeding the critical value, and is characterized by an electrical resistance different from zero, while superconducting order is still present. Specific realizations of the resistive state are related to microscopic mechanisms of resistance, and depend on the temperature of the superconductor, and on the cross-section of the specimen. In bulk *type I superconductors* or in bulk *type II superconductors* the role of the resistive state is, respectively, played by the normal layers of the *intermediate state*, or by *Abrikosov vortices* Φ_0 that move under the action of the Lorentz force $\boldsymbol{J} \times \boldsymbol{\Phi}_0$ induced by the transport current density $\boldsymbol{J}$ (see *Dynamic mixed state*). The resultant transport of magnetic flux Φ causes an electric voltage V to be generated in the material. Resistive states in superconducting films and whiskers are associated with centers and lines of *phase slip*. Their appearance in the specimen induced by current growth is experimentally identified as voltage jumps in the *current–voltage characteristic*, and as multiple discrete slopes of differential resistance within the regions between these jumps. Thermal resistive states form during local overheating of the specimen above the superconducting critical temperature.

RESISTIVE STATE of superconductors

A state in which the *order parameter* and the electric field E within the bulk of a *superconductor* both differ from zero. The presence of an E field results in complex spatial–temporal variations of the order parameter. The overall electric current then consists of a superfluid and a normal component. There appear ranges of *nonequilibrium superconductivity* characterized by an imbalance between the charges of electrons and *holes*, by a splitting of chemical excitation potentials and condensed (Cooper) pairs, by a lack of equilibrium in the distribution of excitations that builds up at higher frequencies of the nonstationary processes, etc. In particular, the resistive state features a complex nonlinear response to electromagnetic actions (that serves as the basis for using superconductors in electronics), and the spontaneous generation of HF and UHF (microwave) currents and electromagnetic fields (see *Josephson effect*). The resistive state develops under the action of a transport current (*kinetic uncoupling*, centers and lines of *phase slip*) during the combined effect of the current and a magnetic field (*dynamic mixed state, intermediate state*) within the range of fluctuation superconductivity. The resistive state may exist within a broad range of currents and voltages separating the superconducting state from the normal one.

RESISTIVITY OF ORDERED ALLOYS

The resistivity of ordered alloys ρ depends essentially not only on the *alloy* composition but also on the *long-range and short-range order* parameters. As a rule, the appearance of long-range order results in a decrease of ρ. The figure illustrates typical dependences ρ versus temperature T for alloys in an equilibrium state, for the two cases when an *order–disorder phase transition* is a *second-order phase transition* (curve 1), and when it is *first-order phase transition* (curve 2). The jump in resistivity in the latter case is associated with a latent heat.

At the *ordering temperature* T_0, curves 1 and 2 demonstrate a break and a discontinuous change of ρ (examples are β-brass and the alloy AuCu$_3$, respectively). At *alloy ordering*, a crystal lattice restores its potential periodicity, the scattering of *conduction electrons* decreases, and hence the residual electric resistance ρ_0 is reduced. Quantum perturbation theory takes this into account (without correlation), and leads to the following

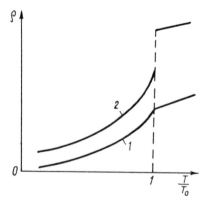

Typical temperature dependence of the resistivity ρ for ordered alloys. Curve 1 is for a second-order phase transition, and curve 2 is for the first-order type.

dependence of ρ_0 in A–B alloys on the concentration c_A, and on the long-range *order parameter* η:

$$\rho_0 = c_r \left[c_A(1 - c_A) - \nu(1 - \nu)\eta^2 \right],$$

where ν is the relative concentration of sites which can be occupied by atoms A, and c_r is a constant in most cases. This formula explains the above-mentioned experimental trends. The value of c_r depends on the nature of the electronic energy spectrum. In some particular cases when the ordering affects this spectrum strongly, one could take into account a dependence of c_r on η, or develop a more complicated theory with more complex expressions for ρ_0; this could result in effects which increase ρ_0 at ordering. This is possible, in principle, even when the appearance of the split in the spectrum leads to a *metal–insulator transition*. The theory of the electric resistance of ordered alloys is developed by taking into account *correlation in alloys*, as well as with the application of a more perfect *coherent potential method*. In some alloys an anomalous increase of ρ at the onset of the ordered state is observed. The explanation of these effects in various alloys has been correlated with the influence of ordering on the electronic spectrum, the existence of short-range order, heterogeneous structures of different kinds, and with other causes.

RESISTOMETRY

Branch of physics and technology, which develops the theory and techniques of measuring electrical resistance (see *Electrical conductivity*). The most widely accepted measurement technique employs an ammeter and voltmeter when the current through the specimen and the potential across it are measured directly. For a more accurate resistance measurement particular techniques are used that are based on comparing a measured *resistor* with a standard (see, e.g., *Quantum Hall effect*) using potentiometer and bridge circuits. To measure electrical resistance of low ohmic specimens, and in bulk materials, electrode-free methods were developed that are based on exciting eddy currents in the material. Commercial digital bridges for measuring the absolute electrical resistance and its deviations from nominal values find wide use. To develop supersensitive measurement techniques one may make use of the *Josephson effect*.

RESISTOR

A component with a constant or variable electrical resistance. The resistance may be sensitive to various actions: temperature (*temperature-sensitive resistor*), electromagnetic emissions (*photoresistor*), electric voltage (*varistor*); mechanical stress (*resistance strain gauge*); magnetic field (*magnetoresistor*). Resistors may be used to stabilize parameters of electric circuits, as a contact-free adjustable resistance, to measure and control temperature, etc. *Functional resistors* transform the output voltage according to a prescribed law, so they are manufactured with a specified distributed resistance. Semiconducting temperature-sensitive resistors offer a high *resistance–temperature factor* that may be achieved through an exponential temperature dependence of the resistivity: $\rho = \rho_0 \exp[U/(k_B T)]$, typical for semiconductors, where U is the *activation energy*. The temperature factor is given by the expression $d\rho/dT = -U\rho/(k_B T^2)$, so that for $U > 0$ it is negative. Semiconducting devices controlled by an electric field (*transistors*) are also used as resistors.

RESONANCE

A sharp selective response of an oscillatory system to a periodic external action when the frequency ω coincides with one of the system's *natural resonance frequencies*, ω_0. The dependence of the magnitude of the response on the value of ω is called the *resonance curve (line)*, its width $\Delta\omega$ denoting the frequency range within which the resonance amplitude is greater than half of its maximum value. Initially the concept of resonance was introduced in 1583 by Galileo Galilei, who applied it to describe mechanical systems (pendulums, weights on springs, etc.). Mechanical resonance develops when the amplitude of forced oscillations grows as the frequency ω of the forcing action approaches the frequency ω_0 prescribed by the mechanical properties of the system itself. Resonance is also found in electrical circuits when the frequency of an applied alternating voltage coincides with the resonance frequency of a circuit with a capacitance and an inductance. When the frequency applied to unpaired spins in a magnetic field is periodically varied then *paramagnetic resonance* absorption can take place. The concept

of resonance (resonance energy absorption, resonance phenomena in general) entered the physics of solids from quantum mechanics in relation to the behavior of microparticle systems (molecules, atoms, electrons, atomic nuclei). According to the laws of quantum mechanics, such a system possesses discrete energy levels, ε, with gaps between them, these gaps prescribing a set of natural frequencies $\nu_{ij} = (\varepsilon_i - \varepsilon_j)/h$, where i and j are the indices of the respective energy levels, and h is Planck's constant. When the external action frequency (usually that of an electromagnetic field) coincides with ν_{ij}, quantum transitions can occur between the ith and the jth levels, which leads to resonance absorption or emission of an electromagnetic field. Since in thermal equilibrium the number of particles in the lower level is larger than that at the upper one, while the probabilities of induced transitions to the upper (and lower) levels are equal to each other, energy absorption usually takes place at such resonances. If, however, the upper level is metastable and acquires an excess population then emission of quantized photons occurs at resonance. Absorption or emission of electromagnetic energy during transitions between the levels produce the resonance lines of the observed spectrum. The amount of energy that may be applied to a system of microparticles during resonance is limited by *relaxation* processes, that is by spontaneous (either radiative or nonradiative) transitions from upper to lower levels, that return the system to its equilibrium state. Together with various processes of interaction between the particles, these transitions determine the *line width*.

Depending on the types of the levels and the nature of the external actions, one may identify optical emission and absorption lines which correspond to transitions between the electronic levels in atoms and molecules. *Electron paramagnetic resonance* and *nuclear magnetic resonance* spectra correspond to transitions between Zeeman sublevels (see *Zeeman effect*) under microwave or radiofrequency pumping. *Nuclear quadrupole resonance* energy states are related to various orientations of the quadrupole moment of the nucleus in an electric field gradient. Acoustic electron and *acoustic nuclear magnetic resonance* involve transitions between levels formed under the effect of elastic waves (phonons). *Ferromagnetic resonance* and *antiferromagnetic resonance* represent *magnetic resonances* in systems with a strong *exchange interaction*. *Spin-wave resonance* describes the selective absorption of energy of an alternating magnetic field in ferromagnetic films (see *Magnetic films*) that develops when *spin waves* of certain frequencies are stimulated in them. *Cyclotron resonance* describes the transitions between diamagnetic levels of free *current carriers* in metals and semiconductors located in magnetic fields. *Paraelectric resonance* takes place during transitions between the tunnelling states of a system. When atoms coalesce into a solid (e.g., a *crystal*) their energy levels form separate bands of allowed energies, with resonance transitions between them also being possible. Many phenomena in solid state physics, such as light reflection and scattering, *photoconductivity*, *electrooptical effects*, etc., are of a resonance nature.

Resonance absorption and emission of energy is used to study the state and composition of materials and their *defects*, *phase transitions* in crystals, to measure *magnetic moments* and quadrupole moments (see *Quadrupole*) of microparticles, and to build *lasers* and *masers*.

RESONANCE CUTOFF (E.A. Kaner, 1958; M.S. Khaikin, 1960)

Disappearance of *cyclotron resonances* in thin parallel-sided metal plates of thickness d in magnetic fields B, smaller in strength than the "cut-off" field $B_c = \Delta p_F/(ed)$, where e is the electron charge, and Δp_F is the diameter of the resonance section of the *Fermi surface* (in momentum units) in a plane normal to B. In accordance with the resonance condition, $\omega = n\Omega$ (here ω is the external wave frequency, $\Omega = eB/m^*$ is the *cyclotron frequency*, m^* is the extremal cyclotron effective electron mass, $n = 1, 2, 3, \ldots$), the nth harmonic of the cyclotron resonance arises in a magnetic field $B_n = \omega m^*/(ne)$. It is sustained by electrons that move in the plane normal to B along a trajectory (cyclotron orbit) with diameter $D_n = \Delta p_F/(eB_n)$. If $D_n \geqslant d$, then the orbit fails to fit in the plate (and is cut off), i.e. cyclotron resonance in fields $B_n \leqslant B_c$ is unobservable. Thus, those harmonics of cyclotron resonance become cut off that have $n \geqslant \omega m^* d/\Delta p_F$. The resonance cutoff results in the appearance of a singularity in

the dependence of the *surface impedance* on the magnetic field at $B = B_c$. Using data from the resonance cutoff, the diameters of sections of the Fermi surface as well as their anisotropy can be found experimentally.

RESONANT CAVITY
See *Resonators*.

RESONANT MAGNETIC FIELD

The strength B_r of the applied constant magnetic field B at which the energy difference between two spin levels and the energy $\hbar\omega$ of the absorbed UHF (or microwave) energy quantum of an applied oscillating (electro-) magnetic field $B_1 \ll B_r$ become equal to each other. It is measured in electron paramagnetic resonance (EPR) and nuclear magnetic resonance (NMR) spectrometers during a scan of the magnetic field B or frequency ω. Ordinarily the field is scanned in EPR and the frequency in NMR. The value of B_r depends on the observation frequency, and in *crystals* it also depends on the orientation of B relative to the crystallographic axes (see *Angular dependence of spectra*). Comparing the observed B_r with those calculated using the *spin Hamiltonian* or a more generalized Hamiltonian makes it possible to determine the value of the *spin*, the *local symmetry group* of its Zeeman and *hyperfine interactions* (see *Zeeman effect*), the *spin–spin interactions*, and interaction with the crystal field, as well as the parameters of those interactions (such as $2D$ and g). These are the characteristics of a particular electron *paramagnetic center* or a nuclear spin in a given crystal. The figure presents the energy levels (E) of a paramagnetic center plotted versus the applied magnetic field B for the EPR case of an electronic spin $S = 3/2$ in an axial crystal field (D term) with the applied magnetic field B parallel to its axis, and no hyperfine interaction present. The triplets of vertical lines between energy levels indicate the values of B_r for two particular frequencies ω: dashes above the abscissa (B) axis correspond to B_r for the three transitions at the higher (microwave) frequency, and dashes below this axis correspond to the three B_r transitions at the lower (UHF) frequency. For spin $S = 1/2$ the distance between the levels with spin projection quantum numbers $M = 1/2$ and $M = -1/2$

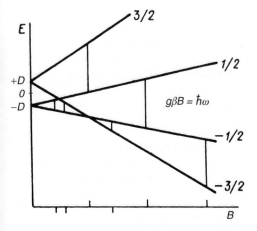

EPR energy levels with resonant magnetic fields indicated.

depends on the g-factor alone, since $D = 0$ for electronic spin $1/2$.

RESONATOR, ACOUSTIC

See *Acoustic solid-state resonator*.

RESONATOR, DISPERSION LASER

See *Dispersion laser resonator*.

RESONATORS, electromagnetic resonators, resonant cavities

Radio, optical, quasi-optical, etc., devices capable of sustaining free or forced (induced) electromagnetic oscillations. The latter feature a certain spectrum of resonant frequencies, and their electromagnetic field has a stationary distribution within the resonator which ideally can be sustained with minimal energy dissipation. Tuned circuits are used as resonators in low-frequency radio technology, resonant cavities play that role in microwave technology, and quasi-optical devices are used at wavelengths $\lambda \approx 1$ mm. A *resonant cavity* is bounded by a conducting surface on every side. The simplest resonant cavity is a section of a waveguide closed at both ends by flat conductor plates normal to its longitudinal axis.

Natural oscillations of such a resonator are classified as waveguide modes ($TE_{l,m,n}$ or $TM_{l,m,n}$ where the l, m, n indices define the number of half-waves fitting along the three dimensions of the cavity). To excite the cavity, various rods, loops, slits, and openings are used. The most widely used resonant cavities are rectangular, cylindrical, or perhaps toroidal in shape. Oscillations with spatially separated electric and magnetic fields may be excited in them, which is convenient for various experiments in *magnetic resonance spectroscopy*. The principal parameter of a resonant cavity is its *quality factor* or *Q-factor*, that is the ratio of energy stored to losses within a single oscillation period. It may reach values of 10^3–10^6. One may differentiate between the intrinsic and the loaded quality factor (the latter depends on coupled external circuits, and objects placed into the cavity). Various *double resonance* cavities have been developed for the radiospectroscopy of solids at both high and low temperatures, and attachments permit optical illumination of samples, operation at *high pressure*, etc. Besides, cavities are used with wave meters, *masers*, as reference elements to stabilize the frequency, etc. In the millimeter, submillimeter, and IR ranges cavities become too small to operate at their dominant mode, so many resonant modes are present simultaneously when their volume is increased. Therefore, conventional cavities are often replaced by open resonators in these ranges, the spectrum of natural frequencies in the latter being considerably reduced due to irradiation of modes with larger l and m indices through slots or openings in their side surfaces.

RESTORATION OF CRYSTALS

A subsequent partial recovery of physical and mechanical properties of *crystals* after their *hardening* by means of external factors (*irradiation hardening*, *plastic deformation*, and *quenching*). The restoration corresponds to the initial stages of the *recovery* of crystal properties. During the course of restoration a partial *healing of defects* takes place, including a redistribution and annihilation of *point defects* (*vacancies* and *interstitial atoms*) as well as *dislocations* of opposite signs. The rate of restoration is highest at the starting moment, after which it decreases more gradually. Formally, this is described by equations of chemical reaction kinetics: $dC/dt = -kC^n$, where C is the defect concentration in the crystal, n is the order of the reaction, and k is the *reaction rate constant*. The dependence on the temperature T follows the *Arrhenius law* $k = k_0 \exp[-E/(RT)]$, where E is the *activation energy* of the process. As a result of restoration, both the *electrical conductivity* and *thermal conductivity* at helium temperatures change. The *mechanical properties of solids* at restoration are recovered to a lesser degree. Minor variations in *density* and enthalpy are also observed.

RESTRAHLEN

See *Residual ray band.*

RESULTANT SPIN

Net sum of spins S_i of individual electrons of an ion, or the total spin of individual ions (nuclei) $S = \sum_i S_i$, $i = 1, 2, \ldots$. For a pair of spins the quantity S has the range of values $|S_1 - S_2| \leqslant S \leqslant S_1 + S_2$. For several spins, $n > 2$, it is necessary when summing to take into account the possible intermediate angular momenta and coupling schemes. A group of levels with the same S can be considered independently from other levels, and can be described by an individual effective *spin Hamiltonian*. The resultant spin of electronic configurations determines the *exchange splitting*, and the resultant spin of the surrounding nuclei determines the hyperfine structure (see *Hyperfine structure*) of the *electron paramagnetic resonance* (EPR) lines, as well as the structure of second-order lines of *electron–nuclear double resonance*.

RETARDATION EFFECTS

See *Effects of retardation.*

REUSS–VOIGT–HILL APPROXIMATION

(A. Reuss, W. Voigt, R. Hill)

An approximate method for calculating *elastic moduli* (*bulk modulus* K, *shear modulus* G) of single-phase isotropic *polycrystals* using the elastic parameters of *monocrystals*. The method consists in calculating moduli in the Reuss approximation (K_R, G_R) and the Voigt approximation (K_V, G_V, see *Voigt averaging*), and subsequently determining the moduli (K_x, G_x) as the arithmetic (or geometric) average from the values K_R, K_V and G_R, G_V, respectively. Since $K_R \leqslant K \leqslant K_V$ and $G_R \leqslant G \leqslant G_V$, the elastic moduli calculated in this approximation (K_x, G_x) are often very close to the values of K and G experimentally measured for polycrystals.

REVERSIBLE PROCESSES

Processes in a *thermodynamic system* involving conservation of *entropy*. In the case of a reversible process, it should be possible to return the system to its initial state by applying the earlier actions in reverse order. Examples of reversible processes are adiabatically slow ones.

RHENIUM, Re

A chemical element of Group VII of the periodic system with atomic number 75 and atomic mass 186.207. Natural rhenium has 2 isotopes: stable ^{185}Re (37.07%) and weakly radioactive (long-living) ^{187}Re (62.93%). Electronic outer shell configuration $4f^{14}5d^56s^2$. Successive ionization energies are 7.87, 16.6 eV. Atomic radius is 0.1373 nm; radius of Re^{4+} ion is 0.072 nm, of Re^{7+} ion 0.056 nm. Oxidation state is from -1 to $+7$. Electronegativity is 1.5.

In a free form rhenium is lustrous light-gray *metal*, very refractory, and at the same time plastic in the cast and recrystallized state. Rhenium is deformable in the cold state, and it may be strongly hardened. Due to its high elastic properties, rhenium is distinguished by high deformation strength and fast wear hardening upon pressure treatment, and also by high long-term strength. Rhenium does not corrode under normal conditions in humid and aggressive media. It has a hexagonal close-packed lattice, space group $P6_3/mmc$ (D_{6h}^4), with $a = 0.27615$ nm, $c = 0.44568$ nm at 300 K. Density is 21.03 g/cm^3. $T_{\text{melting}} = 3453$ K, $T_{\text{boiling}} = 5915$ K. Binding energy is -8.11 eV/atom at 0 K. Heat of melting is 33.1 kJ/mole, heat of sublimation is 712 kJ/mole, heat of evaporation is 783 kJ/mole, specific heat is 134 J·kg^{-1}·K^{-1}; Debye temperature is 416.2 K; linear thermal expansion coefficient is $4.7 \cdot 10^{-6}$ K^{-1} along the principal axis $\underline{6}$ and $6.1 \cdot 10^{-6}$ K^{-1} perpendicular to it (at 300 K); coefficient of thermal conductivity is 71.1 W·m^{-1}·K^{-1} at room temperature; adiabatic elastic moduli of rhenium monocrystal are: $c_{11} = 617.73$, $c_{12} = 274.91$, $c_{13} = 205.57$, $c_{33} = 682.82$, $c_{44} = 160.54$ (GPa) (at 298 K); bulk modulus is 364.6 GPa; Young's modulus is $\approx$461 GPa; shear modulus is 179 GPa, Poisson ratio is $\approx$0.275 (at 298 K); ultimate strength of annealed rhenium is 1.08 GPa, and that of strained rhenium is 2.21 GPa (for high-purity rhenium tensile strength is 490 MPa); relative elongation at the moment of rupture of annealed rhenium is 25%, of strained rhenium 2%. Vickers hardness of annealed samples of rhenium is 1.23, of cast samples 1.42, of strained ones 7.84 GPa; temperature of the onset of recrystallization at the strain of 10% is 1823 K, at 20% 1623 K, at 60% 1473 K. Ion-plasma frequency of rhenium is 27.919 THz; Sommerfeld linear low-temperature molar electronic specific heat coefficient is 2.3 mJ·mole^{-1}·K^{-2}. Electric resistivity is 189.0 nΩ·m (at 295 K); temperature coefficient of electrical resistance 0.00455 K^{-1} at room temperature; work function of polycrystal is $\approx$4.95 eV; critical temperature T_c to superconducting state is 1.698 K, critical magnetic field is $\approx$20 mT (at 0 K); molar magnetic susceptibility of the paramagnetic rhenium is +68.7·10^{-6} cm^3/mole at room temperature; nuclear magnetic moment of ^{187}Re is 3.176 nuclear magnetons.

High-temperature and refractory alloys of rhenium with tungsten, molybdenum, tantalum are used for manufacturing parts of supersonic aircraft and rockets. Rhenium *coatings* (rhenizing) serve for corrosion and wear protection. Rhenium and its compounds are used as catalysts in oil refining.

RHENIUM-BEARING ALLOYS

Alloys in which the element *rhenium*, Re, is a component. The following rhenium-bearing alloys are distinguished: dilute *solid solutions* of rhenium in a *transition metal* X; solid solutions located in the middle part of the compositional *phase diagram* with comparable amounts of Re and X; dilute solutions of X in Re. The first group of rhenium-bearing alloys is of greatest practical interest. Rhenium has a high *solubility* in transition metals with body-centered cubic lattices, which may reach 65%. *Alloying* with Re leads to essential changes in the physical and chemical properties of *molybdenum*, *tungsten*, *chromium*, *nickel*, *tantalum*, *niobium*, *titanium*, *cobalt*, and other *metals*, and provides *construction materials* with favorable technical parameters. By alloying Re with Mo, W and Cr, the anomalous phenomena of the "*rhenium effect*" were observed (alloying not only increases strength, but also improves low-temperature plasticity). The effect is associated with the variation of the electronic structure during the course of the alloying, which leads to a decrease of the Peierls–Nabarro strength (see *Peierls–Nabarro model*), to a decrease of the *stacking fault* energy, to increasing the solubility of *interstitial atoms*, and to *strain* via *twinning of crystals*. The most widespread are *molybdenum alloys* and *tungsten alloys* with Re, which are distinguished by high *strength*, and *high-temperature strength* combined with high plasticity and technological effectiveness, with high emission characteristics, with increased optical resistance, and with a reduced coefficient of electrical resistivity. Rhenium-bearing alloys located in the middle part of the compositional phase diagram are chemical compounds of the *sigma phase, chi phase* or *Laves phase* types, which possess high *hardness* and brittleness (see *Alloy brittleness*). Rhenium-bearing alloys with a dominant Re content are characterized by a sharp variation of their properties upon alloying. Rhenium-bearing alloys of the first group have found their application in electronic engineering (*thermocouples* and heaters in vacuum technology, etc.). Rhenium-bearing alloys of the second and third groups are used as semiconductors, superconductors, and other materials with specific physical properties.

RHENIUM EFFECT of low-temperature plasticity increase

A phenomenon of simultaneously increasing the *strength* and *plasticity* of transition metals of Group VI of the periodic system (Cr, Mo, W) by *alloying* them with Re and other transition elements (Fe, Co, Ru, Os) located to the right of Group VI in the periodic table. This effect is observed in *alloys* with a wide range of *solubility* when the alloying brings about essential changes in the electronic structure, thus disturbing the conditions of resonance *covalent bonding*, and varying the *energy band* population, but it does not change the type of *crystal lattice*. The observed increase of the *density of states* at the *Fermi surface* $N(E_F)$ is accompanied by a reduction of Peierls stress, by reduction of the energy of *stacking faults*, by increasing the *activation volume* and the velocity of *stress relaxation* near the concentrators, by increasing the solubility of additions of *doping atoms*, and also by the involvement of the additional mechanism of *strain* via the *twinning of crystals*. The increase of $N(E_F)$ should be gradual enough to allow the reduction in energy of the stacking faults to accommodate the additional mechanism of strain or twinning, but it should not involve the *cross slip* of helical components of *dislocations* (see *Screw dislocations*). The increasing solubility of the added interstitial elements reduces their *segregation* at the dislocations. During the alloying growth of *elastic moduli*, in particular of the *shear modulus* directly associated with the *breaking stress*, there is observed an increase in the low-temperature plasticity.

RHEOLOGICAL EQUATION (fr. Gr. $\rho\varepsilon\omega$, flow, and $\lambda o\gamma o\varsigma$, knowledge)

Relation between *mechanical stress* and *strain* in a continuous medium where the processes of aftereffect, *relaxation*, ordering, etc. are possible, whereby the mechanical state of the medium is characterized by stress and strain not only at the moment under consideration, but also at previous times. The rheological equation is the basic (determining) equation of *rheology*, the science of deformation and *yield* of materials simultaneously displaying the elastic and viscous properties (see *Viscosity*). The rheological equation is used to describe viscoelastic bodies, *stress relaxation*, *internal friction*, dynamic *hysteresis*, etc.

Different models of rheological media were originally considered by Lord Kelvin (1865), the first rheological equation was suggested by J.C. Maxwell (1868) and W. Voigt (1892). L. Boltzmann (1874) originally formulated the theory of *isotropic viscoelasticity*, and V. Volterra (1909) the theory of *anisotropic viscoelasticity*. The rheological equation can be expressed in either a differential or an integral form. The Maxwell form is an example of a differential rheological equation: $\partial_t \sigma_{ik} + \tau^{-1} \sigma_{ik} = 2\mu \partial_t u_{ik}$ ($i \neq k$), where σ_{ik} and u_{ik} are *stress tensor* and *strain tensor* components, μ is the *shear modulus*, ∂_t is the partial derivative with respect to time, and τ is the *relaxation time* which describes the isotropic body (see *Isotropy*). This time τ characterizes for short time intervals the properties of an elastic medium, and for long time spans the properties of a viscous liquid. In solid state physics, the rheological equation of a standard solid, suggested by C. Zener (1938),

is often used: $\sigma_{ik} + \tau_u \partial_t \sigma_{ik} = M(u_{ik} + \tau_\sigma \partial_t u_{ik})$, where τ_u and τ_σ are the relaxation times at constant strain and constant stress, respectively, and M is the *elastic modulus*. The most general rheological equation of this type may be written in the form

$$\left(\sum_{n=0}^{N} p_n \partial_t^n \right) \sigma_{ij} = \left(\sum_{m=0}^{M} q_m \partial_t^m \right) u_{ij},$$

where p_n and q_m are constants. The following expression in the form of Stieltjes' integral is an example of a rheological equation in integral form:

$$\sigma_{ij}(t) = \int_0^\infty u_{kl}(t - s)\, dG_{ijkl}(s).$$

Here the tensor $G_{ijkl}(t)$ characterizes the mechanical properties of the material, and is called the *relaxation tensor*. The Fourier-transform image of the tensor $G_{ijkl}(t)$ is called the tensor of *complex elastic moduli*. Its components are connected by the dispersion relations. For an isotropic body, the relaxation tensor has the form

$$G_{ijkl}(t) = G_1(t)\delta_{ij}\delta_{kl} + G_2(t)(\delta_{ik}\delta_{jl} + \delta_{il}\delta_{jk}).$$

Here $G_1(t)$ and $G_2(t)$ are independent relaxation functions, δ_{ij} is the unit tensor. In its general representation the linear rheological equation expresses the stress tensor in the form of a linear uniform tensor functional of the stress tensor at the moment under consideration, and at all preceding times. A nonlinear rheological equation expresses the stress tensor in the form of a nonlinear function of the strain tensor, at the particular moment and for all previous times, whereupon this function is often expressed in the form of a polynomial expansion in linear functionals. The rheological equations used in practice are either postulated in a phenomenological manner, or deduced on the basis of kinematic theory and nonequilibrium thermodynamics.

The investigation of the mechanical state of a rheological medium is reduced to the solution of a rheological equation together with the dynamical equation of the medium

$$\frac{\partial}{\partial x_j} \sigma_{ij}(t) + \rho F_i(t) = \rho \partial_t^2 u_i(t),$$

where $F_i(t)$ is the mass density of the *bulk force*, $u_i(t)$ is a *displacement vector*, and ρ is the medium *density*. If a rheological equation is linear, then we can perform a Laplace transformation according to time for both equations to reduce the problem under consideration to a static problem of *elasticity theory* for the Laplace transforms of stresses and strains. This relationship between elasticity theory and rheology is called the *elastic-rheological similarity*.

RHODIUM, Rh

Chemical element of Group VIII of the periodic system with atomic number 45 and atomic mass 102.9055. Natural rhodium consists of one stable isotope ^{103}Rh. Electronic configuration of outer electronic shells is $4d^85s^1$. Ionization energies are 7.46, 18.07, 31.05 eV. Atomic radius is 0.134 nm; radius of Rh^{2+} ion is 0.086 nm, of Rh^{3+} is 0.068 nm. Oxidation state is $+3$, more rarely $+1, +2, +4, +6$. Electronegativity is ≈ 1.9.

Rhodium is a silvery-white *metal*. It has face-centered cubic lattice: $a = 0.3803$ nm (at 299 K). Its density is 12.41 g/cm^3, $T_{melting} = 2233$ K, $T_{boiling} = 3973$ K. Specific heat is 0.2466 kJ·kg^{-1}·K^{-1} (at 298 K); coefficient of thermal conductivity is 150.1 W·m^{-1}·K^{-1} (from 273 to 373 K), linear thermal expansion coefficient is $8.3 \cdot 10^{-6}$ K^{-1}, electrical resistivity is 43.3 nΩ·m (at 273 K); temperature coefficient of electrical resistance 0.00457 K^{-1} (from 273 to 373 K). Specific paramagnetic susceptibility is $0.9903 \cdot 10^{-6}$ CGS units (at 293 K); thermoelectromotive force in thermocouple with platinum is 25.35 mW (at 1723 K). Rhodium is barely volatile in high vacuum (saturated vapor pressure at 1723 K is 133.322 µPa). Mechanical properties of rhodium, especially its plasticity, strongly depend on its degree of purity. Normal elastic modulus of annealed rhodium is $3864 \cdot 10^8$ Pa, shear modulus is $153 \cdot 10^9$ Pa; tensile strength is $42.0 \cdot 10^7$ Pa, relative elongation is 9 to 15%. Brinell hardness is 989.8 MPa. Rhodium is subjected to plastic deformation with difficulty. It may be forged and drawn to wire of diameter 1 mm at temperatures above 1073 K. Sheet rhodium is manufactured by hot rolling up to the thickness of about 0.75 mm, after which metal becomes sufficiently flexible for cold rolling, during which multiple intermediate annealing is used. Zone-refined monocrystalline rhodium may be subjected to cold deformation with reduction of up to 90% without intermediate annealing. It is used for protective coatings of mirrors, electrical contacts, reflectors, manufacture of crucibles for melting of laser materials, etc.

RHOMBIC SYSTEM
See *Orthorhombic system*.

RICHARDSON EQUATION (O. Richardson, 1901), Richardson–Dushman equation (S. Dushman)

Equation which expresses the dependence of the density of the *thermionic emission* saturation current i on the material temperature T

$$i = AT^2 \exp\left(-\frac{\varphi}{k_B T}\right),$$

where A is a constant characteristic for a given substance, called the *Richardson constant* or *emission constant*, φ is the *work function* for removing an electron to the vacuum, and k_B is the *Boltzmann constant*. In the free electron model

$$A = \frac{emk_B^2}{2\pi^2\hbar^3} = 1.2 \cdot 10^6 \text{ A·m}^2 \cdot \text{K}^{-2},$$

where m and e are the mass and charge of the electron. The derivation of the Richardson equation employs the equilibrium *distribution function* of electrons within the crystal (*Fermi–Dirac statistics* or *Boltzmann distribution*). Since the thermal emission process is a nonequilibrium one there is a flux of electrons from the material to the vacuum outside; electrons with energy of the order of φ, significantly exceeding the average energy in the material, contribute to the thermal emission current. Taking this into account leads to the dependence of the Richardson constant on the value of the emitted current, which becomes stronger, the lower the work function. For work functions of about 1 eV, the Richardson constant deviates from its equilibrium value by about 10%. Quantum-mechanical reflection of the electron after passing through the potential barrier also leads to the deviation of the Richardson constant from its equilibrium value by several percent. The Richardson equation is not applicable to semiconductors for thermionic emission in a strong electric field when the electron distribution function deviates from its equilibrium form.

RIEDEL SINGULARITY of tunneling current
(E. Riedel, 1964)

Sharp increase of the amplitude of superconducting Josephson tunneling current at the voltage V equal to $V_c = (\Delta_1 + \Delta_2)/e$, where Δ_1 and Δ_2 are the *energy gaps* of the two *superconductors* that participate in the tunneling (see *Tunnel effects in superconductors*). In the ideal case the Josephson current has a logarithmic singularity

$$I(V) \approx \frac{(\Delta_1 \Delta_2)^{1/2}}{4eR} \ln\left(\frac{V_c}{|V - V_c|}\right),$$

where R is the contact resistance in the normal state. In real samples the logarithmic singularity at $V = V_c$ is smoothed by the anisotropy of the gap, and by the finiteness of the *quasi-particle* lifetime. The Riedel singularity is caused by the singularity in the superconductor *density of states*. For Josephson generation (see *Josephson effect*) the Riedel singularity leads to the appearance of a tunneling current at

$$V = V_n = \frac{\Delta_1 + \Delta_2}{e(2n + 1)},$$

where $n = 1, 2, 3, \ldots$, for the so-called odd series of gap subharmonics. The Riedel singularity has been observed both for the tunneling and for the *point contact* type of junction.

RIEMANN TENSOR

The same as *Curvature tensor*.

RIGHI–LEDUC EFFECT

See *Thermomagnetic phenomena*.

RIGID BODY, PERFECT

See *Perfectly rigid body*.

RIGIDITY

See *Shear modulus*.

RIGID MAGNETIC BUBBLE DOMAIN

A *magnetic bubble domain* containing, within the *domain wall*, vertical magnetic *Bloch lines* of the same topological sign, having a total width comparable to the bubble domain perimeter. It is

characterized by a large value of the *topological index S* (several tens). Magnetic bubble domains that differ in energy and have different S values may coexist if there are no reasons preventing the formation of Bloch lines. Vertical Bloch lines preclude the contraction of rigid magnetic bubble domains, so collapse of the latter occurs in greater displacement fields as against "normal" domains. The collapsing mechanism seems to involve the annihilation of Bloch lines with the assistance of *Bloch points* (sometimes, chemically activated). With the decrease of the displacement field a rigid magnetic bubble domain turns into a *dumb-bell domain* due to the redistribution of the Bloch lines along the bubble domain perimeter. In some cases, a rigid magnetic bubble domain with preset S has two states, stable in diameter, in the same displacement field. The dynamics of rigid magnetic bubble domains exhibits strong gyrotropic deviation from the direction of the moving force. Their mobility is many times smaller than that of magnetic bubble domains with small values of S. For these latter reasons, rigid magnetic bubble domains are unacceptable in conventional devices based on magnetic bubble domains. Usually they are suppressed in *iron garnet* materials by *ion implantation* of the film surface. In view of the above properties, rigid magnetic bubble domains are sometimes referred to as *anomalous magnetic bubble domains*.

RKKY INTERACTION

See *Ruderman–Kittel–Kasuya–Yosida interaction*.

ROCK SALT, common salt, halite, sodium chloride

A *mineral* with chemical composition NaCl. Density is 2.1 to 2.3 g/cm^3; Mohs hardness is 2. Pure crystals are transparent and colorless. *Refractive index* is 1.5446 (for $\lambda = 0.59$ μm at $T = 287$ K) and decreases with increases in temperature. The crystals form a face-centered cubic lattice with $a = 0.5628$ nm. Space group is $Fm\bar{3}m$ (O_h^5) with 4 molecules in a unit cell. Pure crystals of rock salt are used to manufacture prisms for IR spectrographs in the range of 0.950 to 1.550 nm. In nature, rock salt crystallizes from water solutions, in laboratory practice from the melt.

RODS

Elastically-deformable *solids* of uniform cross-section, with lengths that far exceed the cross-section diameter. The following simple types of static and dynamic loading of rods are possible: axial tension, torsion about an axis (see *Rod twisting*), *flexure*. When a linearly elastic isotropic (see *Elasticity theory, Isotropy*) rod is stretched by applying the force F, the elongation per unit length is given by *Hooke's law* $\Delta l / l = F / (EA)$ (l is length, A is cross-sectional area, E is *Young's modulus*); the ratio of the lateral contraction to the longitudinal elongation equals *Poisson ratio* ν. When the rod is twisted by the torque N, directed along the rod axis, the angle of twist per unit length τ is related to N by an equation of the form $N = D\tau$, where D is referred to as the *torsional rigidity* of the rod. In order of magnitude, $D \propto A^2 G$ (G is the rigidity or the *shear modulus*); in the case of a circular rod of radius R, the torsional rigidity is given by $D = (\pi/2)GR^4$. In the case of a rod slightly bent by two parallel and oppositely directed torques N, which are applied at the ends of the rod and oriented at right angles to its axis, the radius of curvature of the axis r is determined by the relation $N = EI/r$ (I is the appropriate *moment of inertia*). The following types of waves may travel in long rods: *longitudinal waves* with velocity $c_k = (E/\rho)^{1/2}$ (ρ is the *density* of the rod material); *torsional waves* with velocity $c_k = (D/\rho I_0)^{1/2}$ (I_0 is the moment of inertia of a cross-section with respect to its center of inertia), for a circular rod, $c_k = (\mu/\rho)^{1/2}$; and *flexural waves*. Rods are widely used, e.g., in *acousto-electronics*, as an important component of delay lines, in high-frequency piezoelectric pressure transducers, in musical instruments, etc.

ROD-SHAPED DEFECTS

A disruption of the regular structure in semiconducting crystals with a diamond-like lattice. This imperfection is characterized by its rod-like shape. Rod-shaped defects arise when, e.g., *silicon* and *germanium* monocrystals are exposed to heat and radiation, when these materials undergo *ion implantation* followed by *annealing*, when crystals are irradiated with high-energy electrons at elevated temperatures, when oxygen-rich silicon samples are subjected to a *heat treatment*.

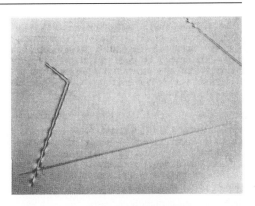

Rod-shaped defects in silicon.

The figure is a photograph, obtained by transmission *electron microscopy*, which presents a rod-shaped defect in silicon. The propagation of the defect took place mainly in the $\langle 110 \rangle$ direction. It follows from an analytical treatment of the principal characteristics of rod-shaped defects that they exhibit some diversity in their propagation. Their generation and growth along the $\langle 110 \rangle$ direction are observed in all cases, but the lateral growth may take place in different planes. This results in the existence of a certain set of *displacement vectors*. It has been found from examining the contrast in electron microscopic images that the following translational vectors are met with more commonly than others: $(a/x)\langle 110 \rangle$, $(a/11)\langle 113 \rangle$, $(a/25)\langle 166 \rangle$, where a is the *crystal lattice constant*, x is an integer. Rod-shaped defects are formed mostly by *interstitial atoms*, but in single crystals some impurities may be incorporated into them.

ROD TWISTING

A deformation state in *rods* brought about by *twisting torques* applied to the ends of the rods. As a rule, the *hypothesis of planar cross-sections* is used to describe the twisting of round rods: it is assumed that the cross-sections of the rods remain planar and their radii are linear. When subjected to twisting moments the cross-sections rotate with respect to each other by a certain angle φ, experiencing a shear $\gamma = \varphi r / l$, where r is the distance to the rod axis and l is its length. Tangential strains τ result when $\tau = \mu\gamma$ lies within the range of elastic *strains*, where μ is the *shear modulus.* The strongest tangential strains develop in the outermost fibers of the cross-sections. Sections tilted relative to the rod axis also undergo normal strains σ that are strongest at the surface in the planes tilted by $45°$ to that axis. During the free twisting of rods of non-circular cross-section the lateral planes bend freely in the same manner relative to each other, and no normal strains develop, while the tangential strains remain identical in every cross-section of the rod. Rod twisting is used in the physics of solids for the *mechanical testing of materials,* and for the determination of their static shear moduli and strength characteristics. *Torsional vibrations* are employed in various physical instruments, torsional pendula for example, to study *internal friction.*

ROD VIBRATIONS

See *Vibrations of rods and plates.*

ROENTGEN (R)

Obsolete unit of exposure dose of X-rays, gamma rays, α- and β-particles, neutrons, which involves total ionization adsorption in 1 cm^3 of dry atmospheric air (1.2393 mg dry air at $0°C$ and 101 kPa) such that ions are formed with a total charge of 1 CGS of either sign. Named after W.K. Roentgen. 1 R $= 2.57976 \cdot 10^{-4}$ C/kg of air. At the dose of 1 R, $2.082 \cdot 10^9$ ion pairs form in 1 cm^3, or $1.61 \cdot 10^{12}$ ion pairs form in 1 g of air. Measuring radiation dose according to its ionizing ability for a particular material allows establishing the physical equivalent of a roentgen (rep). For air, 1 rep $= 0.85 \cdot 10^{-2}$ J/kg. To quantize radiation by its biological effect, the biological equivalent of a roentgen (roentgen equivalent man, rem) is used; the adsorbed radiation energy, biologically equivalent to one roentgen, is 1 rem $= 10^{-2}$ J/kg. What is considered safe for a human being is a dose per unit time that is 250 times the the background of cosmic radiation plus the radioactive radiation from the depths of Earth.

ROENTGEN-EQUIVALENT

A non-system unit measuring in *roentgens* the dose of ionizing radiation of α-, β-particles and neutrons, under which in 1 cm^3 of dry air at the temperature of $0°C$ and atmospheric pressure 101 kPa, there appears $2.082 \cdot 10^9$ pairs of ions of opposite sign (subject to an exposure dose of X-ray or gamma radiation equal to 1 roentgen).

ROSETTON (fr. Fr. *rosette*)

Stable motion of a negatively charged particle in the axial *channeling* mode. For *axial channeling,* motion in the transverse direction (in plane perpendicular to crystal axis) is determined by the effective potential U_{eff}, i.e. by the sum of the accelerated continuous potential of the atomic chain, and by the centrifugal energy of the particle:

$$U_{\text{eff}}(\rho) = U(\rho) + \frac{M_z^2}{2\mu\rho^2},$$

where $M_z = \hbar m$ is the particle angular momentum projection on the crystallographic axis, $m = 0, \pm 1, \pm 2, \ldots$, ρ is the distance between the particle and the atomic chain, μ is the relativistic mass. Thus, the energy of the levels of transverse motion is determined not only by the "principal" quantum number n, as it is in the case of the planar channeling, but by two numbers, including the angular momentum projection m. For the designation of the levels, a notation similar to atomic spectroscopic terms $1s$, $2p$, $3d$, etc., is used. With the growth of the channeled particle energy, the number of discrete levels in the effective axial potential grows in proportion to the *Lorentz factor* (ratio of total energy to rest energy). The motion of a negatively charged particle in the rosetton mode is presented in the figure as a centered helical trajectory labeled (a), on an axis delineated by a chain of atoms, coincident with the corresponding crystallographic axis; (b) is the projection of this trajectory on a plane perpendicular to the crystallographic axis. A rosetton is similar to the state of a two-dimensional atom.

ROTATING COORDINATE SYSTEM

An auxiliary Cartesian coordinate system rotating around the direction of a constant uniform applied magnetic field B_0 with the frequency ω of an additionally superimposed radio frequency magnetic field $B_1(t) = B_1 \cos \omega t$, with $\boldsymbol{B}_1 \perp \boldsymbol{B}_0$. In *magnetic resonance* the introduction of the rotating coordinate system provides a way to describe the evolution of a system of spins interacting with each other, which

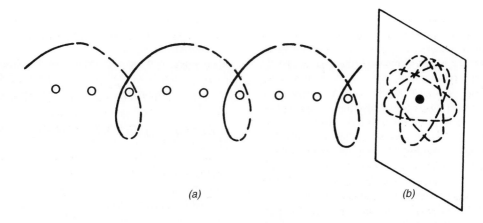

(a)

(b)

Rosetton showing (a) centered helical trajectory, and (b) projection of trajectory on a plane.

have been drawn out of alignment by the alternating magnetic field through an effective truncated Hamiltonian H*. The transformation to the rotating coordinate system is performed by the rotation operator $U = \exp\{i\omega t s_z\}$, where s_z is the operator of the z-component of spin, and the angle ωt is around the axis z directed along $\boldsymbol{B}_0$ so that an arbitrary operator A in the rotating coordinate system is replaced by the operator $\tilde{A} = U A U^+$. As a result, the evolution of the *density matrix* σ in the rotating coordinate system is described, with a high degree of accuracy, by the equation $i\dot{\tilde{\sigma}} = [\mathrm{H}^*, \tilde{\sigma}]$. If ω_0 and ω_1 are the Larmor frequencies (see *Larmor precession*), corresponding to the fields $\boldsymbol{B}_0$ and $\boldsymbol{B}_1$, then the effective Hamiltonian H* corresponds to the Hamiltonian of a system of spins (see *Spin Hamiltonian*) in the constant magnetic field directed at the angle $\theta = \arctan[\omega_1/(\omega_0 - \omega)]$ to $\boldsymbol{B}_0$ and having the Larmor frequency $\omega^* = [(\omega_0 - \omega)^2 + \omega_1^2]^{1/2}$. In the operator for the *spin–spin interaction* H_{ss} the non-secular part of H_{ss} (i.e. the part of H_{ss} that does not commute with s_z), can be neglected for the case $|\omega_0 - \omega| \ll \omega$. The spins in the rotating coordinate system will reach a state of internal equilibrium characterized by a spin temperature (see *Level population*), which differs from the temperature of the crystalline lattice.

ROTATING CRYSTAL METHOD

A basic X-ray diffraction method for investigating atomic *crystal structure*. It is based on the diffraction images of a monochromatic X-ray beam reflected from a sequence of crystal lattice planes of a rotating *monocrystal*. The crystal itself is uniformly rotated about one of its crystallographic axes, and when it does so it continuously exposes a sequence of lattice planes from which the X-rays can be coherently reflected to produce images on a cylindrical photographic film coaxial to the rotation axis, or the detector might perform a rotational movement relative to one of the crystallographic axes that does not coincide with the direction of primary beam (see *X-ray structure analysis*). The detected X-ray diffraction rays correspond to reflections which arise from particular points of the *reciprocal lattice*, and the resulting data can be Fourier analyzed to provide the atomic coordinates of the direct lattice. The diffracted rays give reflections and characteristic extinctions or missing reflections which permit one to determine the *lattice* period along the rotation axis, and accordingly to deduce the *space group* and the presence or absence of a *center of symmetry*. X-ray goniometric methods have been developed for the rotating monocrystal method which permit the synchronized rotation of the monocrystal and the detector, and thus to provide *X-ray diffraction patterns* with distorted (*Weissenberg method*) or non-distorted (*precession method*) reciprocal planes of the monocrystal reciprocal lattice. Automated *X-ray diffractometers* have been developed for monocrystals.

ROTATION MODES

Rotational motion of molecules or atomic groups within *crystals*. As a rule, rotational modes have the nature of retarded rotation, because the energy of a molecule in the *crystal field* depends on its orientation. But for large angular momenta (rapid turning) the rotation differs but slightly from free rotation, with the energy $E = \hbar^2 I(I + 1)/(2M_0)$, where M_0 is the moment of inertia, and the rotational quantum number I assumes the values $I = 1, 2, \ldots$. The crystal field causes the ground state of the above-mentioned groups of atoms to be oriented in a specific way relative to the symmetry directions of the crystal. If the anisotropy of the energy of a molecule in the crystal field is sufficiently high, compared to $\hbar^2/(2M_0)$, then the low-frequency excitations are *librations*.

ROTATION OF ACOUSTIC WAVE POLARIZATION PLANE

Rotation of the acoustic wave polarization plane was originally predicted by Ch. Kittel (1958), and experimentally observed by H. Matthews and R. Le Craw (1962) in yttrium *iron garnet* at the frequency 528 MHz. It appears during the propagation of a transverse acoustic wave in matter. Unlike the analogous phenomenon in light (see *Rotation of light polarization plane*) it occurs only in *solids* in the following manner. The linear-polarized acoustic wave is incident from an elastically anisotropic medium I at the interface with an elastically anisotropic medium II, propagates in the latter, and then is incident on the interface with another elastically isotropic medium III. As a result, in medium III the polarization plane is found to be rotated through the angle φ with respect to the polarization plane of the wave in the medium I. In the general case the differences of absorptions and *phase velocities* of the transverse waves in medium II bring about the rotation of the acoustic wave polarization plane. This rotation is often accompanied by *acoustic wave ellipticity*. In this case the rotation angle of the main axis of the ellipse, characterizing the wave polarization in medium III, is identified as the angle φ. This ellipticity and rotation are observed only for a wave propagating along a low symmetry direction, and can arise from the internal structure of the material, or be caused by the interaction of the material

with an applied magnetic field (see *Acoustic analog of Faraday effect*, *Acoustic analog of Cotton–Mouton effect*). The ellipticity and this acoustic rotation are used for investigations of the physical parameters of solids: of the electronic structure of metals, of magnetoelastic constants of magnetic materials, etc.

ROTATION OF LIGHT POLARIZATION PLANE, optical activity

Ability of some materials to rotate the plane of polarization of linearly polarized light passing through them (see *Polarization of light*). It was discovered by D. Arago (1811) in quartz. Some materials exhibit this rotation only in the solid phase (quartz, CdP_2, ZnP_2, $NaBrO_3$, $NaClO_3$, etc.), and some do so for any *state of matter* (sugar, turpentine, carbon disulfide). In the first case this rotation is a result of the interaction of the *crystal lattice* as a whole with the light (see *Crystal optics*). In the second case the rotation is associated with the absence of an *center of symmetry* in the molecules. L. Pasteur (1948) had shown that the rotation can occur in two directions (right- or *dextro* (d) and left- or *levo* (l) rotating) with the same rotation capability. The corresponding molecules or crystal lattices are mirror images of each other.

For transparent *crystals* the rotation angle φ is determined by the relation $\varphi = \alpha d$, where d is the thickness of the sample, and α is the rotational constant, characterizing the nature of the material. J.B. Biot (1931) established the law $\varphi = [\alpha]cd$ for solutions, where c is concentration of the active material, and $[\alpha]$ is the rotational constant per unit concentration. He also discovered optical rotary *dispersion*, i.e. the dependence of the constants α and $[\alpha]$ on the wavelength λ of the incident light.

The first phenomenological theory of linear polarization rotation was proposed by A.J. Fresnel (1823) when he suggested that the plane-polarized beam entering the crystal is decomposed into two circularly polarized beams with opposite directions of rotation, propagating with different *phase velocities*. When recombined at their exit from the crystal, these waves form a plane-polarized wave with its polarization direction rotated through the angle $\varphi = \pi d(n_+ - n_-)/\lambda_0$, where λ_0 is the wavelength in vacuo, and n_+, n_- are the *refractive indices* of the slower and faster moving waves, respectively.

Polarized light can also be rotated by applying an external magnetic field (*Faraday effect*) or electric field (*electrogyration*) to the crystal. See also *Self-induced light polarization change*.

The phenomenon of polarization plane rotation has played a significant role in physics, chemistry, and biology, especially in the area of stereochemistry. It has led to interesting advances in medicine, such as the discovery that optical isomers of the same substance can differ sharply in some properties. For instance, left-rotating l-mycetin is a good antibiotic, but right-rotating d-mycetin is not.

ROTATION SYMMETRY AXIS

An axis about which a rotation C_n of a body through an angle of $\varphi = 2\pi/n$ (where n is an integer) preserves the appearance of the body. The number n, which is the quotient of 2π and the minimum angle of rotation φ, is called the *order of the axis*. Crystallography permits 1-fold, 2-fold, 3-fold, 4-fold and 6-fold rotation symmetry axes in crystals (see *Crystal symmetry*), denoted by the figures 1, 2, 3, 4, 6, respectively. Some molecules exhibit other orders of rotational axes (e.g., C_5 in ferrocene, $(C_5H_5)_2Fe$). Descriptions of properties of continuous media and physical fields may include rotation symmetry axes C_∞ of infinite order (∞) (see *Limiting symmetry groups*).

ROTATION TENSOR

A tensor of *elasticity theory* characterizing the rotation of a body volume element. Two close points, separated by the vector dx_i without *strain*, have after the deformation a separation by the vector

$$dx_i' = (\delta_{ik} + u_{ik} + \omega_{ik})\,dx_k, \qquad (1)$$

where δ_{ik} is unit tensor, u_{ik} is the *strain tensor*, and ω_{ik} is the *rotation tensor*. The second term u_{ik} on the right-hand side of Eq. (1) describes the deformation, and the third ω_{ik} gives the rotation of the element of volume. In the case of small deformations

$$\omega_{ik} = \frac{1}{2}\left(\frac{\partial u_i}{\partial x_k} - \frac{\partial u_k}{\partial x_i}\right) \qquad (2)$$

is asymmetric, and it is possible to select the vector $\boldsymbol{\omega} = (1/2)\nabla \times \boldsymbol{u}$ (according the cyclic rule $\omega_1 = -\omega_{23}$, etc.), along the local axis of rotation with its modulus equal to the rotation angle.

ROUGHNESS OF SURFACE

See *Surface roughness*.

RUBBERS

Rubber itself is a natural (*cis*-polyisoprene) or synthetic high polymer material characterized by its high elasticity. Rubbers are products of *raw rubber* vulcanization, and they have the capability for large reversible *strain*. They are divided into the following groups: general use rubbers (operational temperatures from -50 to $150\,^\circ$C), heat-resistant (operational temperatures above $150\,^\circ$C), frostproof (operational temperatures below $-50\,^\circ$C); rubbers stable in hostile media (e.g., gasoline- and oil-proof); dielectric, electrically conducting, frictional, etc. rubbers. They are used to produce tires (more than 50% of all the world output of rubber), conveyor belts, driving belts, footwear, etc. Biological protein rubbers include abductin, elastin and resilin.

RUBBERS, RAW

See *Raw rubbers*.

RUBIDIUM, Rb

A chemical element of Group I of the periodic system with atomic number 37 and atomic mass 85.4678; it is an *alkali metal*. Natural rubidium consists of two isotopes: stable ^{85}Rb (72.165%) and radioactive ^{87}Rb (27.835%), which is a β-emitter with half-life $T_{1/2} = 4.88 \cdot 10^{10}$ years. It is not found in nature in its free form. Electronic configuration of outer shell is $5s^1$. Ionization energies are 4.176, 27.501, 39.68 eV. Atomic radius is 0.243 nm; radius of Rb^+ ion is 0.147 nm. Oxidation state is $+1$. Electronegativity is 0.77.

Rubidium is a silvery-white *metal*, it ignites spontaneously in the air, and reacts with water explosively. It has a body-centered cubic lattice; $a = 0.570$ nm. Density is 1.5348 g/cm^3; $T_{melting} = 311.8$ K, $T_{boiling} = 976$ K. Heat of melting is 2.310 kJ/mole; heat of evaporation is 76.02 kJ/mole; specific heat is 29.61 kJ·kg^{-1}·K^{-1} (at 273 K) and 31.206 kJ·kg^{-1}·K^{-1} (at 323 K); coefficient of the thermal expansion is $9.0 \cdot 10^{-5}$ K^{-1} (273 to 311 K). Electrical resistivity is 0.1125 μΩ·m. Mohs hardness is 0.3; Vickers hardness is 0.022 in HV units. Metallic rubidium is a paramagnetic material with specific magnetic susceptibility $0.2 \cdot 10^{-6}$ CGS units (at 290 K). It is used in *lasers*, magnetometers, fuel elements, photomultipliers, and also as a getter in vacuum electronic engineering.

RUDERMAN–KITTEL–KASUYA–YOSIDA INTERACTION, RKKY interaction
(M.A. Ruderman, Ch. Kittel, T. Kasuya, K. Yosida, 1954–1957)

Indirect exchange interaction of spins of paramagnetic impurity atoms in *metals* with each other through the *conduction electrons*. Electron spin-flip scattering by a paramagnetic atom leads to the polarization of their spin near the atom. This disturbance spreads over distances far exceeding the *screening radius* of the impurity potential (*Debye radius*) due to the mechanism of *Friedel oscillations*. As a result, the spins of other paramagnetic atoms located at distances smaller than the electron *mean free path* l are also polarized, which leads to the effective interaction of atoms with each other. This interaction is described by the effective Hamiltonian

$$H_{\text{eff}} = \sum_{ij} J_{\text{eff}}(R_{ij}) S_i S_j,$$

where

$$J_{\text{eff}}(R_{ij})$$

$$= \left(\frac{J}{n}\right)^2 \nu(\varepsilon_F) \frac{p_F^3}{\pi \hbar^3}$$

$$\times \left[\left(\frac{2p_F R_{ij}}{\hbar}\right)^{-3} \cos \frac{2p_F R_{ij}}{\hbar} \right.$$

$$\left. - \left(\frac{2p_F R_{ij}}{\hbar}\right)^{-4} \sin \frac{2p_F R_{ij}}{\hbar} \right]$$

$$\times \exp\left(-\frac{R_{ij}}{l}\right) \exp\left(-\frac{2p_F R_{ij}}{\hbar} \frac{k_B T}{\varepsilon_F}\right),$$

where J is the exchange constant of the electron spin interaction with the paramagnetic impurity, n is the density of atoms, $\nu(\varepsilon_F)$ is the *density of electron states* at the *Fermi surface*, p_F is the Fermi momentum (see *Fermi–Dirac statistics*), S_i and S_j are spin operators of paramagnetic impurities, located at the distance R_{ij} from one another, and ε_F is the Fermi energy (the dependence of J_{eff} on T is appreciable only in *semimetals* and degenerate semiconductors). Many physical phenomena are associated with the RKKY interaction. In particular, at a sufficiently low temperature, it may lead either to the ordering of the paramagnetic atom spins (see *Magnetic phase transitions, Curie point*), or to the transition of the paramagnetic atom system to the *spin glass* state.

RUSSELL–SAUNDERS COUPLING,
LS-coupling (H. Russell, F. Saunders, 1925)

A situation in a many-electron atom where the electrostatic interaction of electrons (responsible for the addition of their orbital angular momenta, $L = \sum_i L_i$) significantly exceeds the corresponding *spin–orbit interaction*, responsible for forming the total angular momentum $J = L + S$ (where $S = \sum_i S_i$ is the total spin angular momentum). In that case the quantities L and S are approximately conserved, each separately from the other, and $2S + 1$ is the multiplicity of a level with a given S, its value depending on L and S. The energies of separate states of the multiplet depend on the value of the total angular momentum J. As a rule, levels lower in energy correspond to larger L and larger S, because of the greater number of negative exchange integrals for electron pairs with parallel spins (see *Hund rules*). For lighter elements, such as those of the first transition series, the ground state has a lower J value when the d-shell is less than half full, and a higher J value when it is more than half full (10 electrons fill the $3d^n$ shell). As an illustration, the ion Cr^{3+} with the $3d^3$ configuration has a ground state with $L = 3$, $S = 3/2$ and $J = 3/2$, whereas Co^{2+} with the $3d^7$ configuration has a ground state with $L = 3$, $S = 3/2$ and $J = 9/2$.

For heavier atoms, as the nuclear charge grows and hence the spin–orbit interaction, which is proportional to Z^4, also increases, there develops a deviation from Russell–Saunders coupling since the role of the electrostatic interaction of electrons in the outer shells correspondingly decreases (particularly in excited atoms). The total angular momenta of individual electrons are approximately conserved: $J_i = L_i + S_i$, and there develops the so-called jj-coupling scheme, whereby the total angular momentum of a configuration is given by the sum over individual total angular momenta: $J = \sum_i J_i$. *Selection rules* for allowed optical transitions are $\Delta L = 0, \pm 1$, $\Delta J = 0, \pm 1$. Transitions between the terms of differing multiplicity, $\Delta S \neq 0$, that have a low probability in Russell–Saunders coupling, become possible in jj-coupling.

RUTHENIUM, Ru

A chemical element of Group VIII of the periodic system with atomic number 44 and atomic mass 101.07. Natural ruthenium has 7 stable isotopes ^{96}Ru (5.57%), ^{98}Ru (1.86%), ^{99}Ru ($\approx$12.7%), ^{100}Ru ($\approx$12.8%), ^{101}Ru (16.98%), ^{102}Ru (31.61%) and ^{104}Ru ($\approx$18.5%). Electronic configuration of outer shells is $4d^7 5s^1$. Ionization energies are 7.364, 16.76, 28.46 eV. Atomic radius is 0.132 nm; radius of Ru^{4+} ion is 0.067 nm. Oxidation state is $+8, +6, +5, +4, +3$ (more rarely $+7, +2, +1$). Electronegativity is 1.45.

In free form, ruthenium is a silvery-white high-melting *metal* (in powder form, it is gray), it is very hard and brittle; has high corrosion resistance. It has a hexagonal close-packed lattice; $a = 0.2706$ nm, $c = 0.42816$ nm at 300 K, space group $P6_3/mmc$ (D_{6h}^4). For ruthenium, polymorphous transformations are possible at 1308, 1463 and 1773 K. Density is 12.45 g/cm^3; $T_{\text{melting}} = 2607$ K (for ruthenium monocrystal with purity of 99.98%), T_{boiling} is around 4370 K. Binding energy is 6.615 eV/atom at 0 K. Heat of melting is 24.3 kJ/mole, heat of evaporation is 620 kJ/mole, heat of sublimation is 670 kJ/mole. Specific heat is 0.238 kJ·kg^{-1}·K^{-1} at 298 K. Debye temperature is 580 K; linear thermal expansion coefficient is $11.1 \cdot 10^{-6}$ K^{-1} along the principal axis $\underline{6}$ and $6.7 \cdot 10^{-6}$ K^{-1} perpendicular to this axis (at 300 K); coefficient of thermal conductivity is 105.1 W·m·K^{-1} (at 293 K); adiabatic elastic moduli of ruthenium monocrystal: $c_{11} = 562.6$, $c_{12} = 187.8$, $c_{13} = 168.2$, $c_{33} = 624.2$, $c_{44} = 180.6$ (in GPa) at 298 K; bulk modulus is 284.49 GPa, Young's modulus is $\approx$454 GPa, shear modulus is $\approx$171 GPa, Poisson ratio is 0.25 (at 298 K); tensile strength of the annealed ruthenium is 0.49 GPa, Vickers hardness is $\approx$230 to $\approx$430 in HV units (range of values is caused by anisotropy). Polycrystalline ruthenium is practically undeformable metal, hardly subjected to plastic deformation at temperatures above 1770 K. Zone-refined monocrystalline ruthenium is plastic at room temperature (it holds flexure of more than 90°). Ruthenium is hardly volatile in high vacuum (pressure of saturated vapor is 0.0133 Pa at 2273 K). Coefficient of linear (in temperature) term in low-temperature molar electronic specific heat is 2.97 to 3.35 mJ·mole^{-1}·K^{-2}. Electrical resistivity is 74.27 nΩ·m (at 298 K); temperature coefficient of electrical resistance 0.00458 K^{-1} at room temperature; Hall constant is $+2.25 \cdot 10^{-10}$ m^3/C; transition temperature to superconducting state is 0.493 K; critical magnetic field is 6.6 mT (at 0 K); molar magnetic susceptibility of paramagnetic ruthenium is $+39.2 \cdot 10^{-6}$ cm^3/mole; nuclear magnetic moment of ^{101}Ru isotope is -0.69 nuclear magnetons.

Ruthenium alloys are noted for their hardness and wear resistance. Ruthenium forms broad ranges of *solid solutions, sigma phases* and *Laves phases* with transition metals.

RYDBERG, EFFECTIVE

See *Effective Rydberg*.

SADOVSKI EFFECT (A.I. Sadovski, 1898)

The appearance of a mechanical moment of rotation in a solid under the effect of elliptically polarized light. The Sadovski effect is caused by the transfer of the *angular momentum* of an elliptically polarized electromagnetic wave to a solid interacting with the wave. The experiment can involve the use of a 1/4-wave plate of an *uniaxial crystal* cut parallel to the crystal *optical axis*. Historically the Sadovski effect was one of the direct proofs of the applicability of the conservation of angular momentum law to processes involving the interaction of electromagnetic radiation with matter.

SAINT-VENANT PRINCIPLE, principle of
statically equivalent loads (A. Saint-Venant, 1855)

A general principle in mechanics whereby a set of forces that add vectorially to zero, acting on a region of a solid, will only be felt near that region, and will decay to become negligible sufficiently far beyond that region. When applied to *elasticity theory* the principle states that the replacement of a system of loads imposed on a small surface site with the characteristic size l by another statically equivalent one (that is having the same principal vector and principal moment of force) results in only a localized change of the deformation mode, and has only a negligible effect at large distances $r \gg l$. A.E.H. Love formulated the Saint-Venant principle as follows: "If the forces impressed on a small surface portion of a body are equivalent to zero force and zero torque, then the stress decreases away from the place of load application, and is negligible at distances large compared to the linear dimension of the loaded part of the body." The Saint-Venant principle allows one to modify the boundary conditions, to simplify them, and to derive approximate analytical solutions to elasticity problems by replacing some configurations of surface loads with others that are more convenient for the analysis. A general proof of the Saint-Venant principle is unavailable; but it has been verified for many special cases. For example, a load system that is statically equivalent to zero, and is distributed on a small area of an elastic half-space, exhibits rapidly decreasing stresses, $\propto r^{-4}$ or $\propto r^{-3}$, depending on the direction of the *surface forces*.

SAMARIUM, Sm

A chemical element of Group III of the periodic system with atomic number 62 and atomic mass is 150.4; it belongs to *lanthanides*. Its content in the Earth's core is $7 \cdot 10^{-4}\%$. In the natural state it has six stable isotopes and one α-radioactive isotope with half life $T_{1/2} = 1.3 \cdot 10^{11}$ years. There are 10 radioactive isotopes obtained artificially by the fission of *uranium* or *plutonium*. Due to the large neutron trapping cross-section, samarium is a "reactor nucleus". Outer shell electronic configuration is $4f^6 6s^2$, successive ionization energies are 5.6 and 11.2 eV. Atomic radius is 0.1804 nm; radius of Sm^{3+} ion is 0.104 nm. Oxidation state is $+3$, more rarely $+2$. Electronegativity is 1.1.

Samarium is silvery-white soft *metal*; in the air it quickly grows dim, being coated by gray oxide film. It has two modifications: α-Sm, β-Sm. At room temperature α-Sm has a rhombohedral crystal lattice; $a = 0.8996$ nm and $\alpha = 23°13'$. At temperatures over 1190 K α-Sm transforms to β-Sm with a body-centered cubic lattice; $a = 0.407$ nm. Density of α-Sm is 7.536 g/cm^3 (at 293 K), of β-Sm is 7.40 g/cm^3 (at 1190 K); $T_{\text{melting}} = 1345$ K, $T_{\text{boiling}} = 2073$ K. Heat of melting is 8.6562 kJ/mole; heat of evaporation is 19.32 kJ/mole; specific heat is 0.1764 kJ·kg^{-1}·K^{-1}; temperature coefficient of linear expansion is $1.04 \cdot 10^{-5}$ K^{-1}. Resistivity is

0.9 μΩ·m; temperature coefficient of resistivity is $1.48 \cdot 10^{-3}$ K^{-1}. Normal elastic modulus is 34.104 GPa, shear modulus is 12.603 GPa, Poisson ratio is 0.352. Vickers hardness is ≈0.427 GPa. Samarium is paramagnetic up to the Néel point $T_N = 14$ K. Upon alloying with metals, it forms compounds. Samarium is obtained by reduction from its oxide by lanthanum. It is used in screens for protection from neutron irradiation, because samarium oxide Sm_2O_3 strongly adsorbs thermal neutrons. An intermetallic compound of samarium with cobalt (see *Samarium–cobalt magnet*) is used for the manufacture of very strong permanent magnets.

SAMARIUM–COBALT MAGNET

A *permanent magnet* made of the *intermetallic compound* $SmCo_5$ having a high *magnetic anisotropy*. Fine powders of this compound have a very high *coercive force* (~800 kA/m). Samarium–cobalt magnets may be obtained from this powder by cold *pressing* in a magnetic field with or without the aid of a binding material (see *Magnetodielectric, Powder magnets*). In technology, the caking of blanks premanufactured by cold pressing is used to obtain *cermet magnets*. There are other modifications of samarium–cobalt magnets, e.g., on the basis of the compound Sm (Co, Fe, Cu, Zr)$_8$. Samarium–cobalt alloys are the most power-consuming and highly-coercive among the hard magnetic materials (for its properties, see Table 1 in *Hard magnetic materials*).

SASAKI EFFECT, Sasaki–Shibuya effect
(W. Sasaki, M. Shibuya, 1956)

Variation of the angle between the directions of the electrical field E and the current density j as a result of the *hole* and *electron heating* in a semiconductor. It is described by the redistribution of *current carriers* according to the state of the energy band that creates or changes the anisotropy of the *electrical conductivity*. In the case of *many-valley semiconductors* of cubic symmetry (*germanium, silicon*) the conductivity caused by each valley is anisotropic, and the electric current j_D in the general case is noncollinear to E. As a result of the symmetry of the location of the valleys

in the space of quasi-wave vectors at low (non-heating) electrical fields, the total electric current in all the valleys $j = \sum j_D$ is parallel to E, and the conductivity is isotropic. If the valleys are oriented inequivalently with respect to E, then due to the differing field direction values there are differing contributions from the *effective masses* of the current carriers in the nonequivalent valleys, the warming-up of electrons is different, and this leads to an *intervalley redistribution* of electrons different from the above values of j_D and their contribution to j. In such a situation an angle different from zero appears between the directions of j and E. In long samples where the direction of j is fixed, the Sasaki effect involves a change in the direction of E, in short samples the direction of E is fixed and the vector j rotates.

At an initially equivalent orientation of valleys with respect to E, due to fluctuations the stable redistribution of electrons between the valleys, which leads to a transverse field, may occur spontaneously. In silicon the direction along $\langle 110 \rangle$ or $\langle 111 \rangle$ corresponds to this phenomenon, in germanium those directions are along $\langle 110 \rangle$, $\langle 111 \rangle$ or $\langle 100 \rangle$. Redistribution probabilities leading to transverse fields with the same absolute magnitudes, but opposite directions, are equal to each other. This is called the *multivalued Sasaki effect*, or the *multivalued distribution of electrons over valleys*.

SATELLITE LINES in spectra

Low-intensity lines accompanying a main spectral line that is usually more intense. They appear due to weaker types of interaction. Thus satellite lines in *magnetic resonance* spectra are low-intensity lines at frequencies $0, 2\nu_0, 3\nu_0, \ldots$ (ν_0 is the frequency of the fundamental line), caused by the mixing of the states of isolated spins by means of the *dipole–dipole interaction*. Taking them into account is important when comparing the moments of experimental spectra with those calculated theoretically. In electron paramagnetic resonance, superhyperfine satellite lines can appear which arise from the interaction of the electron spin with nearby isotopes having nonzero nuclear spins.

SATURATION EFFECTS in magnetic resonance

A phenomenon whereby the power absorbed by a spin system fails to increase further despite increased power pumped into it by the incident radio-frequency (rf) electromagnetic field used to produce the resonance transition. Saturation effects are described by a *saturation factor* $S = (n_1 - n_2)/(n_1^0 - n_2^0)$, where n_i^0 and n_i are *level populations* for energy levels 1 and 2 in thermodynamic equilibrium and during resonance absorption, respectively. For the simplest case (when high-frequency modulation, coherence effects, etc. are absent), we have $S = (1 + 2T_1 W)^{-1}$, where T_1 is the longitudinal or *spin–lattice relaxation* time, and W is the probability of a transition induced by the rf field. Sometimes the value $x = 2T_1 W$ is called the saturation factor (*saturation parameter*). Saturation effects take place when $2T_1 W > 1$, and are absent when $2T_1 W \ll 1$. The power absorbed by a spin system is given by the expression $P = \hbar\omega(n_1 - n_2)W(1 + 2T_1 W)^{-1}$, where $\hbar\omega$ is the energy of a single rf field quantum. When $2T_1 W \gg 1$, the value P gives the maximum power that the spin system can transfer to the lattice (note that here $n_1 = n_2$). For particles with spin $1/2$ that are in exact resonance, one has $x = \gamma^2 B_1^2 T_1 T_2$, where B_1 is the rf magnetic field strength, T_2 is the transverse or *spin–spin relaxation* time, and γ is the *gyromagnetic ratio*. A magnetic resonance line saturates when $\gamma^2 B_1^2 T_1 T_2 > 1$. In such a case the intensity U, the shape, and the width, of the resonance signals all become complicated functions of the conditions under which the spectra are recorded (i.e. rf field strength B_1, amplitude and frequency of modulating fields, etc.). By studying the dependence of U on B_1 (as well as on other conditions of signal recording), one may determine the *relaxation times* for the spin system, and perhaps deduce mechanisms of line broadening. As a rule, $T_2 \ll T_1$ in solids ($T_2 \approx T_1$ in low-viscosity liquids), so the condition $\gamma^2 B_1^2 T_1 T_2 > 1$ can be met in solids for $\gamma B_1 T_2 < 1$. When $\gamma B_1 T_2 > 1$, describing saturation effects becomes much more difficult, since one then needs to take into account the effects of coherence.

SATURATION, MAGNETIC

See *Magnetic saturation*.

SATURATION MAGNETIZATION CURVE

Curve describing the process of *magnetization* of *ferromagnetic* materials under the effect of either an internal, B, or an external, $B_0 = \mu_0 H_0$, *magnetic field*, while being driven to saturation (range 3 in Fig.). *Saturation* means that the *magnetization* M approaches the spontaneous magnetization (*true magnetization*) M_s at a given temperature. Further increases in the applied magnetic field may result in a weak linear growth of M_s (called a *paraprocess*), as shown in the figure. The difference of B from B_0 is explained by the demagnetizing effect of the sample surfaces. The true field B affecting M inside the ferromagnet is a sum of the external and the *demagnetization fields*. Thus, the $M(B)$ versus B curve characterizes the properties of the material itself. In many cases, e.g., while magnetizing a crystal along its *easy magnetization axis*, saturation may be reached in quite weak applied fields B_0, fractions of a millitesla. Generally saturating magnetic fields vary from 10^{-2} to 1 T.

The nature of magnetization may vary from material to material. In multidomain ferromagnets it proceeds via displacement of the domain walls (DW) (see *Domain wall displacement processes*). In many cases the initial stage of magnetization involves reversible DW displacement, corresponding to the beginning of the $M(B)$ versus B curve, range 1 in Fig. As B increases further, irreversible displacements become important (steeper range 2 in $M(B)$ curve). After these displacement processes are completed then *magnetization rotation processes* operate. To rotate M towards B

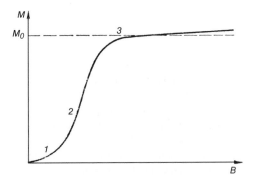

Saturation magnetization curve.

magnetic anisotropy forces must be overcome, and the growth of M at higher B is slower. In practice it is not always possible to separate the processes of rotation and displacement. In a more general case these processes may either develop in parallel, in sequence, or a combination of these.

SATURATION, RF

See *Radio-frequency selective saturation*.

SATURATION, SELECTIVE

See *Selective saturation*.

SCABBING, flaking, chipping-off

A type of *failure*, induced by a short-time intensive action that results in the separation of a part of a sample (*chipped-off plate*). Scabbing or flaking is observable at collisions of *solids*, under action of powerful laser radiation, at polymorphic transformations (see *Structural phase transition*), and so on. The *flaking fracture* was first observed experimentally by B.E. Hopkins (1914) during the collision of metal rods. The mechanism for this, in brief, is as follows. Under the impact-wave conditions of loading, compression waves start to propagate from the contact zone. Upon reaching the free surface the waves reflect back from it. In the region where incident and reflected waves meet, high-intensity tensile stresses appear, which cause thermal-fluctuation-induced nucleation and the development of microscopic damage. For a sufficient intensity and duration of the tensile stress, the growth of microscopic damage leads to the formation of the main *crack* and eventually sample break-up.

To describe the flaking, use is made of the *instant flaking* model whereby the failure is regarded as a loss of continuity, which occurs at reaching a certain magnitude of load (*breaking stress*), and lasts for a time close in order of magnitude to the period of interatomic oscillations (10^{-12} s), i.e. practically instantly. This scheme of instant flaking is applicable in the range of stresses comparable in order of magnitude to those of interatomic bonds (for *metals*, decimal fractions of *Young's modulus*), when the flaking surface is sufficiently smooth (*specular flaking*). At lower tensile stress levels, the regions near the flaking surface are characterized by the presence of a large amount of microscopic damage. Under these conditions, there is no single value of critical stress that can specify a failure of the same sample in a series of experiments involving different loading regimes. Then, when performing calculations, one should use the temporal criteria of failure, which allow for the time-dependent accumulation of damage. The phenomenological criteria of flaking failure are mostly threshold criteria with respect to accumulated damage, and involve some kind of kinetics of damage growth depending on the material rheology. According to the form of microscopic damage, one can distinguish *tough flaking* involving spherical regions of damage, and *brittle flaking* involving damage with the shape of flat disks.

SCABS

Thin slag or oxide *film*, which disturbs the continuity of a metal. Scabs are generated mostly on the surface of the melt, jet, or casting mold. The strength properties of the material, which experiences the attack of scabs, are several times decreased, up to the complete loss of *strength* and impact *viscosity*. Deep-lying scabs are detected by application of methods of magnetic *defectoscopy* and infrared defectoscopy. The generation of internal scabs is prevented by protecting the alloys from aggregations of nonmetallic (slag) inclusions, cavities, and *pores*.

SCALE EFFECT (G. Sharby, 1912)

Deviation from the *law of similarity* in the deformation of an elastic solid body. This law implies that equal relative *strains* are produced by equal stresses. The scale effect was first observed during measurements which showed that, when notched, geometrically similar specimens are subjected to an impact, the *failure* work decreases with an increase of the specimen size. In tension tests, the scaling effect shows up in the decrease of the ultimate characteristics of *plasticity* (relative constriction) and *strength* (true *ultimate strength*) with an increase in size. The physical nature of the scale effect is related to the greater degree of *localization of plastic deformation* in large specimens, and to an increased probability of finding large *defects* and sharp stress concentrators in these specimens (see *Stress concentration*). In cold-strained materials (wire, strip), the scale effect shows itself in a decrease of the plasticity margin (of allowable total reduction) with an increase in diameter, which is due to nonuniform distribution of strain across the section. In measurements of *hardness* and *microhardness*, the scale effect shows itself in their increase as the load on the *indenter* decreases.

SCALING INVARIANCE HYPOTHESIS, similarity hypothesis, scaling

Assumption that certain features, typically the most fundamental ones, of a system's behavior are independent of temporal or spatial scales (static similarity hypothesis). The scaling invariance hypothesis is mathematically formulated in terms of similarity relationships that connect values of physical quantities corresponding to different scales.

Most generally the scaling invariance hypothesis assumes homogeneity of the functions under consideration. In *phase transition* theory, such functions are *thermodynamic potentials* and *correlation functions* in the *critical region*. For instance, considering a function of two arguments $f(x, y)$, this means that for arbitrary x, y and λ we may write $f(\lambda^a(x), \lambda^b(y)) = \lambda^p f(x, y)$, where a, b and p are similarity parameters. It follows from this relationship that $f(x, y) = y^{p/b} F(x/y^{a/b})$, where $F(z) = f(z, 1)$. The latter means that the function $f(x, y)$ has a rather specific form. The scaling invariance hypothesis in phase transition physics evolved on a semiempirical basis, and initially it provided a rationale for summarizing and classifying a large body of experimental data on magnetic, structural, superfluid, and other *second-order phase transitions*. In the late 1960s the scaling invariance hypothesis received microscopic substantiation. Since the early 1970s it has been considered an integral part of *critical phenomena* theory based on the *renormalization group method*. *Scale invariance* becomes manifest for a second-order phase transition at distances much greater than interatomic ones; at $T \neq T_c$ the pattern of critical fluctuations of the order parameter is scale invariant only as long as the distances do not exceed the *correlation radius*. The term *scaling relationships* often refers to formulae that connect critical indices of various thermodynamic and kinetic quantities. They are exemplified by $\alpha + 2\beta + \gamma = 2$, and $\alpha = 2 - d\nu$, where $\alpha, \beta, \gamma, \nu$ are indices of *specific heat, order parameter, susceptibility,* and *correlation length,* respectively, and d is the system dimensionality. The scaling concept is widely applied not only in phase transition physics, but also in disordered system theory (*percolation theory, localization theory*), in *polymer* physics, in hydrodynamic instability theory, etc. (see *Dynamic scaling*). The scaling invariance hypothesis has been verified in works beyond the scope of the *phenomenological theory of phase transitions*, which depend upon such concepts as the renormalization group method, and feature *self-similarity* and the Gibbs distribution at the *critical point*.

SCALING THEORY OF LOCALIZATION

A theory intended to explain the behavior of disordered systems (in the single-particle approximation at zero temperature). The theory proceeds from the concepts of similarity theory (*scaling invariance hypothesis*) and the *renormalization group method*. These have their origin in *phase transition* theory and in quantum field theory. In accordance with them, the scaling theory of localization proceeds from the assumption that the properties of a disordered system are those of an insulator or a conductor, changing from one to the other at the passage of the *Fermi level* through the *mobility edge* (*Anderson localization*), and are describable in a sufficiently large sample L with a unified dimensionless *electrical conductivity* parameter (*electrical conductance*) $g(L)$. It follows then that the change of g under varying dimensions depends only on g itself, i.e. may be described by an equation of the form: $(\mathrm{d}\ln g/\mathrm{d}\ln L) = \beta(g)$. Deep in the fluctuation region of the spectrum, i.e. for $g \ll 1$, the states decay exponentially; therefore, $g \propto \exp(-L/l)$ in this region, where l is the localization length (*localization radius*). On the other hand, when $g \gg 1$, the system should exhibit excellent conducting characteristics, and in such a situation for a d-dimensional system $g \propto \sigma L^{d-2}$, where σ is the microscopic conductance. That is why $\beta(g) \propto \ln g$ when $g \ll 1$, and $\beta(g) \propto d - 2$ when $g \gg 1$. More accurate argumentation based on a modified perturbation theory (see *weak localization*) shows that $\beta(g) = d - 2 - a/g - b/g^2 - \cdots$ when $g \gg 1$. Under the additional assumption that $\beta(g)$ is a smooth monotonic function, the above asymptotic expressions indicate that at $d = 1, 2$ total localization takes place (although at $d = 2$ and higher energies, the states may happen to decay as a power function and not exponentially), while a mobility edge that is defined by the equation $\beta(g_c) = 0$ exists at $d = 3$. In the neighborhood of this boundary, the conductivity goes continuously to zero: $g \propto (E_F - E_c)^\nu$, with the slope of the curve $\beta(g)$ at $g = g_c$ acting as the *critical index*. The same index, as is always the case in one-parameter scaling, determines the divergence of the localization length as the Fermi energy approaches E_c from the insulator side: $l \propto (E_c - E_F)^{-\nu}$.

SCANDIUM, Sc

A chemical element of Group III of the periodic system with atomic number 21 and atomic mass 44.95591. Natural scandium has one stable isotope ^{45}Sc. Outer shell configuration is $3d^1 4s^2$. Successive ionization energies are 6.7, 12.8, 26.19 eV. Atomic radius is 0.162 nm; radius of Sc^{3+} ion is 0.081 nm. Oxidation state is +3, less commonly +2, +1. Electronegativity is ≈ 1.26.

In the free form scandium is a soft *metal* of a silvery color. It exists in two modifications (α-Sc, β-Sc). Below 1607 K α-Sc is stable with hexagonal close-packed crystal lattice, space group $P6_3/mmc$ (D_{6h}^4), with parameters $a = 0.33088$ nm, $c = 0.5268$ nm (at 298 K). Above 1607 K up to $T_{\text{melting}} = 1812$ K β-Sc is stable with body-centered cubic lattice, space group $Im\bar{3}m$ (O_h^9), with parameter $a = 0.4541$ nm. Binding energy is -3.93 eV/atom (at 0 K); above 1870 K it is volatile. Density of α-Sc is 2.99 k/cm^3 at 293 K; $T_{\text{boiling}} = 3123$ K. Heat of melting is ≈ 16.9 kJ/mole, heat of sublimation is 293.1 kJ/mole, heat of evaporation is 329 kJ/mole, specific heat is 0.560 kJ·kg^{-1}·K^{-1}. Debye temperature is 359 K, linear thermal expansion coefficient is $15.1 \cdot 10^{-6}$ K^{-1} along the principal axis $\underline{6}$ and $7.61 \cdot 10^{-6}$ K^{-1} in perpendicular to this axis at 300 K; adiabatic elastic moduli of α-Sc monocrystal: $c_{11} = 99.3$, $c_{12} = 45.7$, $c_{13} = 29.4$, $c_{33} = 109.6$, $c_{44} = 27.7$ (in GPa) at 293 K; isothermal bulk modulus is 43.5 GPa at room temperature; tensile strength is 0.245 to

0.294 GPa, compression strength is 0.981 GPa. Brinell hardness is 75 to 100, cross-section of thermal neutron trapping is 13 barn; Sommerfeld term of linear low-temperature electronic specific heat is ≈ 10.5 mJ·mole^{-1}·K^{-1}. Electrical resistivity of α-Sc is $(47-71) \cdot 10^{-8}\,\Omega$·m, temperature coefficient of the electrical resistivity is 0.000815 K^{-1} at room temperature, work function is 3.23 eV; molar magnetic susceptibility of paramagnetic scandium is $+315 \cdot 10^{-6}$ CGS units at room temperature (value along the principal axis $\underline{6}$ in α-Sc is higher than in perpendicular direction), nuclear magnetic moment of isotope ^{45}Sc is 4.749 nuclear magnetons. Elements of fast-response memory for computers are manufactured from scandium *ferrites*. Other areas of application are *metallurgy*, rocket and aircraft building.

SCANNING ELECTRON MICROSCOPE (SEM), raster electron microscope (introduced by M. Ardenne, 1938)

An electrooptical device designed for observation and analysis of *solid surfaces*. Its spatial resolution, which reaches ~ 1.5 nm, is much closer to that of a transmission *electron microscope* than it is to an optical microscope. In an SEM a narrow *electron probe* beam with diameter 1–5 nm is scanned across a solid surface using a scanning system. The interaction of the electron beam with a sample produces secondary, reflected Auger electrons (see *Auger effect*), X-rays, etc. Depending on the type of the detector, the radiation emitted by the sample is transformed into electrical signals which, after amplification, are used to modulate a cathode-ray tube display where an image of the sample surface is formed (see Fig. 1). During the investigation of a semiconductor surface structure by an SEM there appears a *charge-induced emf* which arises from nonequilibrium charge carriers generated by the scanning electron probe beam in the subsurface layer, and these induce surface *band bending* and a redistribution of *surface states*, leading to variations of the surface charge and potential. The local surface potential gives rise to a charge-induced emf in a metallic ring-shaped detecting electrode located above the sample, which is then amplified and displayed on the microscope screen. The video image carries information about local states on the

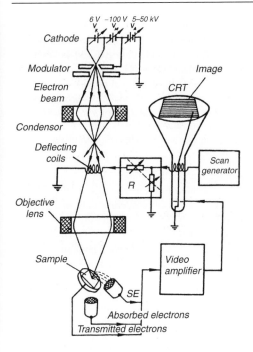

Fig. 1. Block diagram of a scannning electron microscope (SE, secondary electrons; CRT, cathode-ray tube).

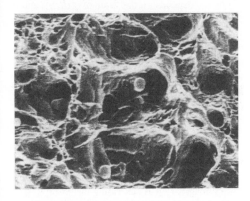

Fig. 2. SEM picture of the structure of a fracture in α-Fe deformed at 293 K (magnification 18,000).

surface, and the distribution of electrophysical parameters in the subsurface semiconductor layer. The charge-induced emf method is effective for the nondestructive, noncontact detection of defects in semiconductor materials and devices (see *Defectoscopy*).

The advantage of SEM is the possibility for examining large samples without prior sample preparation, with sufficiently high resolution and significant depth of focus (see Fig. 2). An apparatus combining SEM and an X-ray microanalyzer offers the best experimental potentiality, and allows not only the examination of sample surface structure, but also can provide the chemical analysis of structural components with the help of a *crystal diffraction spectrometer*, and a spectrometer with energy dispersion and a spatial resolution of ~1 μm. Together with a computer this apparatus allows the quantitative analysis of the experimental data.

SCANNING ELECTRON MICROSCOPY

A microprobe analysis technique for studying the microstructure and the local composition of *solid surfaces*. It is based on scanning a focused electron beam over the target surface (see *Raster*). It provides a significant improvement in the resolution of the image of a solid surface compared to light *luminescence*. The resolution of modern scanning electron microscopes reaches several nanometers. To obtain an image of the surface topography of a bulk solid specimen scanning electron microscopy uses secondary or reflected electrons that are collected together (see *Scanning electron microscope*). The structure of thin targets may also be studied in the transmission mode. Recording absorbed electrons provides one with information on the local subsurface properties of the target. Detecting X-rays and cathode luminescence emissions from the target makes it possible to conduct elemental and chemical analysis of the target composition. Scanning electron microscopy techniques find applications in the study of both electric and magnetic microfields, and the electrophysical properties of semiconducting structures, as well as for fundamental and applied investigations in the physics of solids, physics of metals and material science, mineralogy, biology, and other branches of science and technology.

SCANNING OF OPTICAL RADIATION

Spatial movement of an optical beam on or across a particular path or area on a surface. Scanning is performed using *light deflectors*, devices that deviate the direction of a light beam. The simplest light deflectors are opticomechanical types.

Solid-state light deflectors depend on changes of *refractive index* of a crystal under exposure to electric or magnetic fields, as well as on their interaction with electromagnetic and acoustic waves (see also *Kerr effect*, *Faraday effect*).

Scanning of optical radiation is also used in laser TV systems, aerial surveillance, detection and tracing of targets, image sensing, *optical techniques of information recording* and other technologies.

SCANNING SEMICONDUCTOR PHOTODETECTOR

See *Sprite*.

SCANNING TUNNELING MICROSCOPY

Technique for examining *solid surfaces* based on the measurement of the current I arising from the tunneling of electrons through the potential barrier in a gap of width l (tens of nanometers) between the material surface and a scanning electrode shaped in the form of a sharp tip. If the current I and the bias voltage V are maintained constant, the topographic image of the surface, which represents the surface distribution of electron density, can be recorded, thereby making possible the determination of the surface atomic structure. The very sharp (exponential) dependence of I on l guarantees a high spatial resolution: 0.01 nm normal to the surface and up to 0.1 nm laterally. By modulating l it is possible to determine the distribution of the *work function* across the surface of a sample. Measuring the *current–voltage characteristics* $I(V)$ can provide information on the spectrum of *surface electron states*. The change of l and the scanning across the surface with an electrode are performed using piezoelectric elements. Scanning tunneling microscopy is useful for examination of *atomically clean surfaces*, as well as *real surfaces* of metals and semiconductors, adsorbed layers on these surfaces, and various biological objects. The technique uses low (of the order of several volts) operating voltages, is non-destructive (see *Nondestructive testing techniques*), and does not always require a vacuum. The *Scanning Tunneling Microscope* inventors G. Binnig and H. Rohrer (1982) shared the 1986 Nobel Prize in physics with the electron microscope inventor E. Ruska.

SCATTERING AMPLITUDE

Amplitude $f(k')$ of a spherical wave $e^{ik'r}/r$ scattered by a particle during its excitation by an incoming plane wave e^{ikr}, where $k' = |k'|$. It has the dimension of length and provides the magnitude (or effective differential cross-section) of the scattering with the wave vector k':

$$d\sigma(k') = |f(k', \theta, \varphi)|^2 \sin\theta \, d\theta \, d\varphi,$$

where θ is the angle between k' and k, and φ is the azimuthal angle. Integration over all angles yields the *total scattering cross-section* in the channel $k \rightarrow k'$ (for elastic scattering only one channel $k' = k$ exists). The total scattering cross-section in all channels is expressed via the imaginary part of $f(k, 0)$ (*optical theorem*). The scattering amplitude can provide the angular and energy distribution of the scattered particles, including when multiple scattering is taken into account. The propagation of the beam of particles in a medium is determined by the interference of the incoming and all elastically scattered waves. In an amorphous medium (see *Amorphous state*), the only waves that are in phase are those scattered in the forward direction ($\theta = 0$), so that the result of interference depends on the value of $f(k, 0)$, and it is expressed by the change of the resulting *phase velocity* of the wave, i.e. in terms of an index of refraction $n \neq 1$. Therefore, the value of n is determined by the scattering amplitude in the forward direction. For example, for a homogeneous medium with a weak enough scattering interaction, we have

$$n^2 = 1 + \frac{4\pi}{k^2} N f(k, 0),$$

where N is the number of scatterers per unit volume. Because $f(k, 0)$ is a complex quantity, the n has an imaginary part describing the absorption of the beam. In crystalline solids, due to the spatially periodic arrangement of their atoms, the scattered waves can constructively interfere in a discrete number of directions, which depend on the crystal lattice geometry. Under certain conditions (*Bragg law*), the diffracted beams can be observed in these directions, with an intensity and phase that depends on the individual scattering amplitude $f(k)$. The concept of scattering amplitude can be applied to a scatterer with a complicated spatial structure, e.g., a system of atoms arranged regularly along a line or a plane.

SCATTERING OF CURRENT CARRIERS

See *Current carrier scattering*.

SCATTERING OF ELECTRONS

See *Current carrier scattering*.

SCATTERING OF LIGHT

See *Light scattering, Induced light scattering, Light scattering in liquid crystals, Photoinduced light scattering, Raman scattering of light, Rayleigh scattering of light*.

SCATTERING OF PARTICLES

Change in direction of incoming particles due to their interactions with particles in a target medium. One may identify *elastic scattering* (without change in particle internal state or total kinetic energy) and *inelastic scattering* (e.g., with a change in total kinetic energy). The scattering is characterized by its cross-section and scattering angle. The latter is the angle between the direction along which the particle traveled prior to the interaction with the scattering particle, and the direction of motion after the interaction. A *differential scattering cross-section* $d\sigma/d\Omega$ or $d\sigma/dE$ characterizes the elastic and inelastic scattering processes. These two quantities correspond to the ratio of the number of particles scattered within a unit solid angle Ω, or those which have their energy E changed by a unit energy increment during the act of scattering, relative to the density of the incident particle flux, nv (n is concentration of incident particles, v is their velocity). The *total scattering cross-section* σ is obtained by integrating the differential cross-section over angles or energies, respectively. A quantum description of the scattering of particles yields a distribution of probabilities of scattering into various angles, presented as a *scattering amplitude* $f(\theta)$ that is related to the differential scattering cross-section: $d\sigma/d\Omega = |f(\theta)|^2$. The angle of scattering, θ, is measured in a coordinate system with its origin at the center of mass of the colliding particles. A popular approximation technique for calculating the scattering amplitude is the *Born approximation* that uses first-order perturbation theory,

with the perturbation itself as the particle interaction potential $V(r)$:

$$V(r)f(\theta) = 2\pi \int_0^\infty V(r) \frac{\sin qr}{qr} r^2 \, dr,$$

$$q = 2k \sin\left(\frac{\theta}{2}\right),$$

where k is the wave number in the coordinate system of the particles' center of mass, and the energy of the particle is large enough, $\lambda_{deB} \ll R_0$. At low particle energy (de Broglie wavelength $\lambda_{deB} \gg R_0$, where R_0 is the interaction force range of action) when bound or quasi-bound states of the colliding particles (those having a finite lifetime) exist, a *resonance scattering* can take place. The scattering cross-section is a maximum when the energy of the relative motion of the particles equals the energy of the bound (or quasi-bound) state. The scattering of *quasi-particles* (*electrons, holes, phonons*, etc.) off each other from *defects* of the crystal lattice controls various kinetic processes, such as *electrical conductivity, thermal conductivity, galvanomagnetic effects*, and others. See also *Current carrier scattering*.

SCATTERING OF PHONONS

See *Phonon scattering*.

SCATTERING OF X-RAYS

See *X-ray scattering, Diffuse scattering of X-rays, Small-angle scattering of X-rays*.

SCHER–MONTROLL THEORY (H. Scher, E.W. Montroll, 1975)

A model of *hopping conductivity* in *amorphous semiconductors*. It starts from the assumption that the time distribution of jumps is broadly spread out, and the peak concentration of the *current carriers* injected by current or induced by light tends to remain in the initial region where it is generated. In practice a typical semiconductor film of width 50 μm is commonly enclosed in a structure of a "sandwich" type with two attached electrodes. Carriers may be injected from either electrode by an electrical impulse, or be produced as *electron–hole pairs* by means of an absorbed electron beam or light. The time of flight of carriers to the opposite electrode is determined by observing the arrival a transient current of rectangular shape, or an increase of charge. The transient current in the Scher–Montroll theory arises from the number of carriers reaching the opposite electrode from the forward edge of the hopping time distribution. Physically, it means that the injected carrier moves to the second electrode by a hopping mechanism. The *activation energy* of these hops is taken as constant, but sometimes a charge carrier lands on a site and remains there fairly long before the next jump. This causes a blurring of the forward moving fronts, and a broadening of the current pulse.

SCHLIEREN-METHOD

See *Optical defectoscopy*.

SCHOENBERG EFFECT (D. Schoenberg, 1962)

Instability of the uniform distribution of magnetic induction B in a metallic sample at low temperatures. The Schoenberg effect is due to a nonmonotonic dependence of the sample magnetization M on the magnitude of the magnetic induction B (e.g., *de Haas–van Alphen effect*). See also *Diamagnetic domains*.

SCHOTTKY BARRIER, depletion layer
(W. Schottky, 1939)

Potential barrier in the vicinity of a metal–semiconductor contact arising from differing *work functions* of the materials in contact, and the presence of *surface states* at the boundary. The Schottky barrier involves *band bending*, with the depletion of majority charge carriers near the surface (see *Semiconductor surface*). In this region of width L (*Schottky layer* width) the charge is due only to ionized impurities, so charges of free electrons and holes may be neglected. For the approximation of a uniform impurity distribution in the semiconductor the band bending of the energy barrier varies with the distance x from surface as $\varphi(x) = e^2 n_0 (L - x)^2 / (2\varepsilon_0 \varepsilon)$, and the electric field varies as $E(x) = -e n_0 (L - x)/(\varepsilon_0 \varepsilon)$, where n_0 is the impurity concentration, L is the width of Schottky layer, and ε and ε_0 are the *dielectric constants* of the semiconductor and free space, respectively. In 1939, W. Schottky developed a theory of semiconductor diodes with such a barrier (see *Schottky diode*). See also *Nonohmic contact*.

SCHOTTKY DEFECT, Schottky pairs
(W. Schottky, 1930)

A *defect*, a missing atom or *vacancy* in a crystal lattice, commonly occurring in a monatomic lattice, or in a binary crystal MX. In the latter case two vacancies often occur in different sublattices M and X. If the pair of vacancies are bound by their mutual attraction and are located at adjacent sites, then the defect is called a *double vacancy*. The energy of Schottky defect formation is defined as the energy required for the pair of atoms M and X to depart from the bulk of the crystal and move to its surface. It often amounts to one or more eV in energy. In many crystals (e.g., *alkalihalide crystals*, oxides MgO, SrO, etc.) Schottky defects are the main type of equilibrium defects at high temperatures in the presence of thermal *fluctuations*.

SCHOTTKY DIODE

A *semiconductor diode* containing a *Schottky barrier* at a metallized semiconductor boundary. The voltage dependence of the *barrier capacitance* was put to practical use when Schottky diodes with nonlinear capacitances were developed: *varicaps*, *varactors*, frequency doubling *diodes*. In Schottky rectifiers the sharp dependence of direct electric current through the barrier on the applied voltage is utilized. The same principle is applied in mixer Schottky diodes, where the decrease of super high-frequency radiation is handled by mixing two signals with nearby frequencies to produce their difference frequency.

SCREENING DISTANCE

See *Screening radius*.

SCREENING, ELECTRIC

See *Electric charge screening*.

SCREENING LENGTH

See *Screening radius*.

SCREENING RADIUS, screening distance, screening length

A characteristic length beyond which there is a significant (exponential) decrease of the Coulomb interaction (either attractive or repulsive) between electric charges in various media (e.g., solids, liquids, gases). The phenomenon arises from the presence of free mobile particles, such as *conduction electrons* and *holes* in metals, semimetals, semiconductors, *ions* in electrolytes, electrons and ions in gas discharge plasmas, etc.; see *Electric charge screening*. In a classical (nondegenerate) plasma in a semiconductor (or gas discharge) as well as in a weak electrolyte solution, the screening radius is $r_{\mathrm{DH}} = [k_B T/(8\pi e^2 n_0)]^{1/2}$, where n_0 is the concentration of free charged particles (electrons or ions) with the charge e. This radius is called the *Debye–Hückel screening radius* (see *Debye–Hückel theory*). When computed within the Thomas–Fermi approximation (L. Thomas, E. Fermi, 1928) the effective screening radius of a static point charge in a degenerate electron gas will be $r_{\mathrm{TF}} = [E_F/(6\pi e^2 n_0)]^{1/2}$. Here E_F is the *Fermi energy*, and n_0 is the electron concentration. The value of r_{TF} is called the *Thomas–Fermi screening radius*.

SCREW AXIS, called helical axis in Russian literature

An element of crystallographic space group symmetry denoted by n_p (see *Crystal symmetry*) combining an n-fold *rotational symmetry axis* (rotation by $360/n$ degrees) and a *translation* axis parallel to it. The pitch of the screw motion is p/n and the translation is p/n times the shortest lattice translation vector along the axis. The screw axes of symmetry are: 2_1, 3_1, 3_2, 4_1, 4_2, 4_3, 6_1, 6_2, 6_3, 6_4, 6_5. The pairs 3_1 and 3_2, 4_1 and 4_3, 6_1 and 6_3, 6_2 and 6_4 are called enantiomorphic (see *Enantiomorphism*), because for the same value of translation they are distinguished only by the direction of the rotation. The axes with smaller indices in these pairs are called right-handed, and those with larger indices are left-handed. The screw axes 2_1, 4_2 and 6_3 are called neutral because right- and left-handed rotations are equivalent.

SCREW DISLOCATION, called helical dislocation in the Russian literature

Rectilinear *dislocation* with a *Burgers vector* parallel to the dislocation line. In an infinite elastically isotropic medium (see *Isotropy of elasticity*) a screw dislocation creates anti-plane *strain* inversely proportional to the distance from it. By this the displacements of the elements of the medium

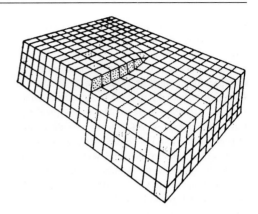

Sketch of a screw dislocation.

are parallel to the dislocation line and depend only on the coordinates of the point in the plane, perpendicular to it. In a *crystal* the Burgers vector of a screw dislocation coincides with one of the lattice vectors. Successive lattice planes passing the dislocation are transformed into the surface of a helix. The figure shows a model of a screw dislocation in a simple cubic lattice.

SCREW ROTATION

See *Crystal symmetry*.

SECONDARY ELECTRON EMISSION

Emission of secondary electrons from *solids* and liquids (emitters) upon their bombardment with primary electrons. Secondary electrons can be emitted from the surface facing the primary beam, and from the back surface if the emitter thickness is less than the penetration range of the primary electrons. Quantavely this secondary electron emission is characterized by the *secondary electron emission coefficient* $\delta = i_2/i_1$, where i_1 and i_2 are the current flows of primary and secondary electrons, respectively. The coefficient δ depends on the primary electron angle of incidence, on the nature and structure of the material, on the surface state, and on the primary electron energy (E_n), reaching its maximum in most materials at $E_n = 100$ to 1000 eV.

Secondary electrons have a continuous energy spectrum from 0 to E_n. They consist of elastically (usually called back-scattered) and inelastically reflected electrons, Auger electrons, and slow, truly secondary electrons with the most probable energy of the order of 1 eV for non-metals, and 2 to 4 eV for *metals*. The fine structure of the electronic spectrum of secondary electrons is determined by the *characteristic energy losses*, by the *Auger effect*, and by the *electron diffraction* on crystalline emitters. It is used for diagnostics, and allows one to make a judgement about the chemical composition and electronic state of the atoms of the surface layer. Most materials have small values of δ ($\delta_{max} \propto 1$ to 2), but in *insulators* with wide gaps and small *electron affinity* χ, $\delta_{max} = 20$ to 40 at E_n of several hundred electron-volts. In semiconductor emitters with a negative electron affinity ($\chi < 0$) δ increases up to 10^2–10^3. Secondary electron emission has found wide practical use in the *secondary electron multiplier*, a device for the measurement of very small electron currents. Secondary electron emission is also the basis for obtaining a high-resolution image of the surface in scanning electron microscopes by utilizing a so-called *secondary electron detector*.

SECONDARY HARDNESS, secondary hardening

The transformation of residual *austenite* to *martensite* (not decomposed at a high *annealing* temperature) during its subsequent cooling. The causes of this phenomenon are the depletion of the residual austenite of *carbon*, they can involve the alloying elements (see *Alloying*) at the annealing, and therefore exert their influence during the temperature increase at the start of the *martensitic transformation*. The effect of relaxation processes is also not excluded. To achieve complete or almost complete transformation of the residual austenite to martensite it is necessary to repeat the high-temperature annealing several times. The secondary hardness displays itself especially prominently in high-speed and high-chromium tool *steels*.

SECONDARY ION EMISSION, ion–ion emission

Emission by a *solid surface* or liquid to the vacuum of positive or negative secondary ions in the form of charged atoms, molecules, atomic complexes (clusters) during bombardment of the surface by a flux of primary ions. Quantitatively secondary ion emission is characterized by the *secondary emission coefficient* (K), which equals the ratio of the secondary to the primary ion flux, and also by the energy distribution of the secondary ions. The K value depends on the chemical composition and temperature of the target, on the physico-chemical condition of the surface under bombardment, and also on the kinetic energy, angle of incidence, and density of the primary ion beam.

In most cases the secondary ions of metals have a single positive charge, and those of non-metals have a negative one, but the emission of multicharged ions is also possible. For metals $K = 10^{-2}$ to 10^{-5} ions per ion. The secondary ion distribution has been characterized by *mass spectrometry*, as containing ions of the basic target material, from impurities, from ions of the primary beam, as well as of atomic complexes formed during physical and chemical processes at the target surface both before the bombardment and after it. This secondary ion emission can be enhanced by bombarding the target with ions of chemically active gases, e.g., of oxygen, or by a surface coating with an adsorbed layer of oxygen (positive secondary ions) or cesium (negative secondary ions). Secondary ion emission is a physical process which forms the basis of the method of analysing the composition of matter via *secondary ion mass spectrometry*.

SECONDARY ION MASS SPECTROMETRY

(SIMS), mass spectrometry via secondary ions

A technique for investigating materials through determination of masses and quantities of *atoms*, molecules and atomic *clusters* formed as a result of *secondary ion emission* by bombarding the sample under study with a flux of primary *ions*. Secondary ion mass spectrometry exhibits a high absolute (up to 10^{-14} g) and relative (up to $10^{-7}\%$) concentration sensitivity, and a broad dynamic range of measurement of concentrations. It is used for the analysis of surface composition of solid and liquid bodies, for *layer-by-layer analysis* of near-surface layers and *thin films*, for *local analyses* of substances of heterogeneous composition. The technique makes it possible to carry out analyses of all elements and their isotopes, as well as of compounds, in particular of large organic and bioorganic molecules. It is carried out using specialized instruments called *secondary ion mass spectrometers.* The composition of a surface and surface physicochemical processes are analyzed by using a primary ionic flux of low density (10^{-7} A/cm^2) (*method of statistical mass spectrometry of secondary ions*). The layer-by-layer analysis is performed by measuring the quantity of impurity ions from deeper layers revealed as a result of sputtering the surface by a primary ion flux. Layer-by-layer resolution reaches several nm. Local analysis is performed either through bombardment of the sample by a focused ion beam (*ionic microprobe*), or through formation of magnified ionic images of the surface, which are separated according to mass (*ionic mass-spectrum microscope*). In the first case, the spatial resolution reaches 30 nm, and in the second case about 1 μm.

SECONDARY LUMINESCENCE

Total light response of a system with a complex spectrum (*crystal* or polyatomic molecule) to optical excitation. Secondary luminescence includes the coherent (transmission and *reflection of light, photon echo*, optical harmonics, etc.) as well as incoherent (*light scattering*, hot and ordinary *luminescence*) components. This luminescence approach is useful for the description of non-coherent components of emission when the lines of luminescence and of light scattering have similar or coinciding spectral and polarization characteristics, and are also of comparable intensity. There are several radiation processes which are differentiated by the time constant involved. Those processes, for which there is no time delay between optical excitations and secondary luminescence, are related to scattering. The radiation, which appears during the course of energy *relaxation* of an excited state, is called *hot luminescence*. The most common type of luminescence occurs after the completion of the relaxation, when the energy distribution between the sublevels of the excited states has attained a quasi-equilibrium character described by an effective temperature.

SECOND HARMONIC GENERATION
in nonlinear optics (P. Franken, 1961)

The doubling of a light frequency by a *nonlinear medium*. The generation of a second harmonic became possible with the appearance of *lasers* which provide high power, monochromatic, phase coherent radiation. In classical electrodynamics this generation is explained on the basis of an *anharmonic oscillator model*, i.e. by a nonlinear dependence of the polarization of atoms and molecules on the electric field strength of the light wave. The principle of superposition which is a basis of linear optics is violated, the waves interact with each other, and this results in distortion and the appearance of harmonics. In quantum mechanics second harmonic generation is looked upon as the coalescence of two phonons with the frequency ω. When the thickness of the *crystal* in which the generation takes place is much greater than the wavelength, the second harmonic intensity depends on the phase relations between the main wave and the second harmonic wave. As a result, in the general case, the intensity of the second harmonic has a pulsating (interference) nature with an efficiency of 10^{-10}–10^{-12}. The maximum transfer of energy from the wave of frequency ω to the that of frequency 2ω is possible when there is *phase matching*, i.e. equal *phase velocities* at frequencies ω and 2ω. In quantum mechanics this condition corresponds to the conservation of momentum in the coalescence of photons. The phase matching takes place in anisotropic crystals as a result of the interaction of differently polarized waves. For example, in KH_2PO_4 crystals, a plane ordinary ray of frequency ω propagating at a certain angle relative to an optical axis (*angle phase-matching*) is phase-matched to an extraordinary ray of frequency 2ω. The phase matching determined by the dispersion and the *anisotropy of crystals* is only a necessary but not a sufficient condition for efficient second harmonic generation in a thick crystal. A second condition is the vanishing of a component of the *nonlinear susceptibility tensor* χ (a polar tensor of rank three) for a given kind of interaction. The second harmonic generation is used for the transformation of long-wavelength laser radiation into the short-wavelength range. The maximum efficiency for the transformation to a second harmonic in the presence of phase matching is approximately 0.8.

SECOND-ORDER PHASE TRANSITION

The *phase transition* of a material from one thermodynamic *phase* to another, during the course of which the state of the material varies continuously. At a second-order phase transition both the free energy and its derivative with respect to the temperature are continuous, and there is no latent heat. It may involve growth from zero of particular characteristics such as: spontaneous electric polarization or magnetization, spontaneous deformation of the crystal lattice, concentration of superconducting electrons, etc. Certain thermodynamic quantities, such as the *dielectric constant* in *ferroelectrics* and the *magnetic susceptibility* in *ferromagnets*, show an anomalous rise when approaching the transition point from the region of a symmetric as well as from an ordered phase (see *Ordered system*). The second-order transition point is a singular point of the *thermodynamic potential* of the system when this potential is expressed as a function of corresponding variables, e.g., of temperature and external field. In the neighborhood of the transition the characteristic quantities are proportional to $|(T - T_c)/T_c|^{-\varepsilon}$, where ε is a dimensionless critical exponent. The critical exponents α for specific heat, β for magnetization, γ for susceptibility, and ν for coherence length obey the scaling law $\alpha + 2\beta + \gamma = 2$. For the two-dimensional Ising model these exponents have the values $\alpha = 0$, $\beta = 1/8$, $\gamma = 7/4$, and $\nu = 1$.

A second-order phase transition involves a change of the system symmetry and the onset of a certain new system property (e.g., magnetic or ferroelectric order, superconductivity). In its most general form, this fact was realized by L.D. Landau (1937), and forms the basis of his *phenomenological theory of phase transitions* (see *Landau theory of second-order phase transitions*). This theory is widely used for interpreting experimental data. Characteristic features of second-order phase transitions are the increase of the *correlation length* and the *relaxation time* of thermodynamic fluctuations of the *order parameter*, as well as of the fluctuation amplitude, as the transition point (in the *critical region*) is approached. These features manifest themselves as anomalies in the scattering of light and neutrons, in EPR and NMR spectra, in Mössbauer spectra, etc.

SECOND QUANTIZATION

A quantum method of treating identical particles using, instead of wave functions, the operators $\widehat{\psi}$, $\widehat{\psi}^+$:

$$\widehat{\psi} = \sum_i a_i \psi_i, \qquad \widehat{\psi}^+ = \sum_i a_i^+ \psi_i^*,$$

where ψ_i with $i = 1, 2, \ldots$ is a complete set of single-particle wave functions with complex conjugates ψ_i^*, and a_i^+, a_i are the creation and annihilation operators of particles in the ith state, satisfying the relations:

$$a_i^+ |N_1, \ldots, N_i, \ldots\rangle$$
$$= \sqrt{N_i + 1} |N_1, \ldots, N_i + 1, \ldots\rangle,$$
$$a_i |N_1, \ldots, N_i, \ldots\rangle$$
$$= \sqrt{N_i} |N_1, \ldots, N_i - 1, \ldots\rangle,$$

where the symbol $|N_1, \ldots, N_i, \ldots\rangle$ describes the state of the system with $N_1, N_2, \ldots$ particles when $i = 1, 2 \ldots$,

$$[a_i^+, a_j^+] = [a_i, a_j] = 0, \qquad [a_i, a_j^+] = \delta_{ij}$$

for *Bose–Einstein statistics*, and

$$\{a_i^+, a_j^+\} = \{a_i, a_j\} = 0, \qquad \{a_i, a_j^+\} = \delta_{ij}$$

for *Fermi–Dirac statistics*, where δ_{ij} is the Kronecker symbol, $[\hat{b}, \hat{c}] = \hat{b}\hat{c} - \hat{c}\hat{b}$ is a commutator, $\{\hat{b}, \hat{c}\} = \hat{b}\hat{c} + \hat{c}\hat{b}$ is an anticommutator. A single-particle Hamiltonian $H^{(1)}$ in this representation has the form $H = \sum_{ik} H_{ik}^{(1)} a_i^+ a_k$, where $H_{ik}^{(1)} = \int \psi_i^*(x) H^{(1)} \psi_k(x) \, dx$. Interactions of particles of a particular type with each other and with other types of particles can be described with the help of second quantization operators. For example, the Hamiltonian for harmonic vibrations (see *Harmonic approximation*) of the crystalline lattice has the form $H_\varphi = \sum_{k,s} \hbar\Omega_s(k)(b_{ks}^+ b_{ks} + 1/2)$, where the indices k, s determine the values of the wave vector $\boldsymbol{k}$ of the sth branch of vibrations, $\Omega_s(k)$ is the frequency of vibration, b_{ks}^+, b_{ks} are the creation and annihilation operators of the corresponding *phonons*. The cubic term of the expansion of the potential energy in powers of atomic displacements provides the anharmonicity operator $W = \sum V_{ss't}(q, k, k')\widehat{\varphi}_{q,t}\widehat{\varphi}_{ks}\widehat{\varphi}_{k's'}$, where the summation is carried out over all repeated indices,

$\widehat{\varphi}_{ks} = b_{ks} + b^+_{-k,s'}$, and the coefficients $V_{ss't}$ are expressed in terms of the third derivative of the potential energy of the atoms with respect to their displacements. The Hamiltonian of a free electron in the crystal has the form $H = \sum_\lambda E_\lambda a^+_\lambda a_\lambda$, where E_λ is the energy of the electron in a *periodic potential* in the state λ (the index λ denotes the *quasi-wave vector* and *spin* of the electron).

The interaction of *conduction electrons* with *crystal lattice vibrations* is described by the *Fröhlich Hamiltonian*

$$H = \frac{1}{\sqrt{N}} \sum_{k,s,q} F(k) a^+_{q+k} a_q \big(b_{ks} - b^+_{-ks} \big),$$

where index q numbers the values of the electron wave vector q, a^+_q and a_q are creation and annihilation operators of the electron in the qth state, and N is the number of unit cells in the crystal. In an *ionic crystal* for long-wavelength optical *phonons* $F(k) = e|k|^{-1}(2\pi\hbar\Omega c/V)^{1/2}$ (here Ω is the optical phonon frequency, V is the crystal volume, $c = \varepsilon_\infty^{-1} - \varepsilon_0^{-1}$, ε_∞ and ε_0 are the high-frequency and static *dielectric constants*).

The interaction with long-wave acoustic vibrations leads to $F(k) = -i\sigma[\hbar|k|/(2Mc_a)]^{1/2}$, where σ is a parameter of the *strain potential*, c_a is the velocity of the longitudinal acoustic vibrations, and M is the total mass of the atoms in one *unit cell*). Interaction of the long-wave optical phonons with photons in an ionic crystal is described by the Hamiltonian

$$H = \sum_{k\alpha} \hbar D_k \big(a_{-k\alpha} - a^+_{k\alpha} \big) \big(b_{k\alpha} + b^+_{-k\alpha} \big),$$

where $a^+_{Q\alpha}$, $a_{Q\alpha}$ are creation and annihilation operators of photons with the wave vector Q and polarization α,

$$D_k = -\frac{i}{2} \sqrt{\frac{\Omega_t \omega_Q (\varepsilon_0 - \varepsilon_\infty)}{(\varepsilon_0 \varepsilon_\infty)^{1/2}}},$$

$k = Q$, Ω_t is the frequency of the transverse *optical vibrations*, ω_Q is the photon frequency, etc.

The second quantization method is convenient for applications to linear transformations involving interacting particles. Such canonical transformations, which satisfy the canonical equations of mechanics, but adopt the Hamiltonian form (see *Canonical transformation method*), have been suggested by N.N. Bogolyubov for use with non-ideal *Bose gases* and *Fermi gases*, and they are widely used in the theory of Bose gas *superfluidity* and of *superconductivity*.

SECOND SOUND

Temperature waves in the gas of elementary excitations in solids and quantum liquids, e.g., *phonons* in crystals or *quasi-particles* in liquid ^{3}He. Second sound can be considered as ordinary sound (*first sound*) in a gas of quasiparticles. In superfluid phases of helium second sound involves weakly damped antiphase oscillations of the density of the normal and superfluid components. Such antiphase oscillations produce negligible changes in the overall density and pressure, and correspond to temperature waves. The speed and attenuation of second sound have been investigated in detail in ^{4}He and in ^{3}He–^{4}He mixtures. Second sound has also been found in ^{3}He (see *Superfluid phases of ^{3}He*), and in *quantum crystals* of ^{4}He and ^{3}He. In crystals second sound can propagate in the so-called hydrodynamic regime, i.e. when its wavelength and period of oscillation are much longer than, respectively, the mean free path and mean free path time of the quasiparticles (phonons, *rotons*). The velocity of second sound in a gas of phonons has the same order of magnitude as that of ordinary sound.

SEEBECK EFFECT
See *Thermoelectric phenomena.*

SEEGER ZONE (A. Seeger)
A region with a high (about 0.2) concentration of *vacancies* that forms as a result of a *collision cascade* of atoms in a crystal irradiated by high energy particles. This zone forms after the *recombination* and escape of atoms displaced from the cascade region to interstitial sites during the course of collisions. The lifetime of a Seeger zone is determined by the rate of the diffusional escape of vacancies. The presence of such zones contributes to the radiation hardening of irradiated metals and alloys (see *Irradiation hardening*).

SEEPAGE THEORY
See *Percolation theory.*

SEGREGATION (fr. Lat. *segregatio*, separation of lamb from the flock)

The formation of regions enriched by individual components (impurities, *defects*) under the influence of *heat treatment* or other factors in compounds with a complex composition. Segregation can involve nonequilibrium composition differences in the material. One basis for segregation is a *phase transition* to a state with a different equilibrium composition that involves a distinct binding energy for separation into the individual phases. Segregation is often observed at the surfaces of multicomponent *solid solutions*. The main reason for surface segregation is the availability of free valence bonds on the surface. Therefore, the *grain boundaries*, *dislocation walls*, and other defects are centers of preferential detached localization. In the case of solidification of a metal alloy, the segregation is sometimes called *liquation*. Segregation is used in *metallurgy*, in the production of high purity compounds (*zone melting*), in crystal growth (see *Crystallization*); it affects the strength of alloys. Porous compounds ordinarily obtained by the *powder metallurgy* approach can contain a considerable amount of impurities, e.g., due to the adsorption of impurities by the compressed particles. *Auger electron spectroscopy* and other methods of *local analysis* find the preferential segregation of impurities close to various crystal structure defects in alloys with a body centered cubic lattice. Segregation is observed on grain boundaries and at interparticle contacts that form during *sintering*, but the most acute segregation is observed close to surface of *pores*, involving both both interparticle and intraparticle types. The segregation zone of impurities close to the pore surface is usually about 10 nm wide; thus, the *competitive segregation* region is found in the segregation zone itself, so the impurities most responsive toward segregation (sulfur and *potassium* in *molybdenum*) are located close to the surface (within several atomic layers), forcing out oxygen, carbon, and other elements. The impurity segregation taking place close to the pore surface reduces the *surface energy*, and this usually results in decreasing the sintering rate, and healing the pores (see *Healing of defects*). The simultaneous lowering of the surface energy on many extended sites (e.g., on a porous grain boundary) results in a decrease of

the *breaking stress*. Impurity segregation on grain boundaries which inhibit the processes of collective *recrystallization* can be considered as a positive phenomenon for *refractory materials*, as the *high-temperature strength* rises and the metal becomes slightly more brittle (see *Alloy brittleness*) as a result of high temperature exposure.

SELECTION RULES

Rules determining the system states (e.g., energy levels) between which transitions are possible under the effect of a particular disturbance (e.g., incoming radiation). The selection rules depend on the symmetry of the states and the perturbation that induces the transition. In solid state physics the selection rules usually assume no displacement of the atoms (molecules) of the *crystal* from their equilibrium positions. Such displacements can cause violations of the selection rules, and lead to the appearance of "forbidden transitions" or weak lines of *light absorption* and *light scattering*. See also *Optical spectroscopy* of solids.

SELECTIVE ETCHING

Special processing of a crystal *solid surface* that results in the formation of a surface contour, or an increase of optical contrast between surface sites with different physicochemical properties. Depending on the surface processing technique the following types of selective etching are distinguished:

(1) *Chemical etching* based on the difference of chemical activity and the dissolving rate of different crystal sites.

(2) *Electrolytic etching* consisting of nonuniform anodic dissolving of separate surface sites; the combination of chemical and electrolytic *etching* methods which can enhance the selective exposure on the surface.

(3) *Color etching* based on differences in phase oxidizability. During the etching oxide layers of different widths are formed for differing phases, which imparts color contrast to these sites.

(4) *Ion etching* which consists in extracting the compound from the crystal surface by ion bombardment. The selective effect of ion etching consists in the fact that the atomization

rate of the compound from the surface under exposure to the ion beam is unequal for crystal sites with different structural inhomogeneities.

Selective etching is used in metallographic *phase analysis* to identify the crystal microstructure boundaries of blocks and grains, and for detection of the direction of the *rotation of light polarization plane* in a crystal. Selective etching is widely used in the study of semiconductor devices to establish the location and depth of *semiconductor junctions* by etching a polished section, and to determine the color contrast of crystal sites of different types of conductivity. Selective etching is one of main research techniques for characterizing *dislocations*.

SELECTIVE SATURATION (T. Sanadze et al., 1966)

The appearance of one or more dips (holes) on an *electron paramagnetic resonance* (EPR) spectral line during fast passage after saturation by a short high power microwave pulse. The passage through resonance is fast compared to the *spin–lattice relaxation* and *cross-relaxation* rates. The phenomenon of selective saturation is observed in diamagnetic non-metallic crystals containing *paramagnetic centers* if the Zeeman energy of the nuclei (see *Nuclear Zeeman effect*) E_0 surrounding the paramagnetic center, and the *hyperfine interaction* energy of these nuclei E_{hf} are of the same order. For such systems the probability of an allowed (only the electron *spin* reorients) and of a forbidden (the nuclear spins also flip) have about the same values.

The appearance of selective saturation can be explained in terms of two electronic spins $S = 1/2$ and two equivalent nuclear spins $I = 1/2$ with the energy level diagram shown in the figure. The systems of nuclear sublevels 1, 2, 3, (with energy spacings ε_+) and sublevels $1'$, $2'$, $3'$, (with spacings ε_-) are labeled with the electron spin projections $+1/2$ and $-1/2$. The values ε_+ and ε_- depend on E_0 and E_{hf}. In the state of thermodynamic equilibrium the *level populations* of levels 1, 2, and 3 are less than those of the levels $1'$, $2'$, and $3'$. If a short high power microwave pulse (solid vertical line in Fig.) saturates and thereby equalizes the populations of levels 2 and

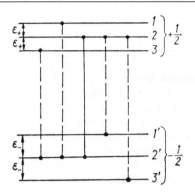

Selective saturation.

$2'$ of the allowed transition, then observation of the EPR spectrum at a low (non-saturating) microwave power level causes a hole to appear at the allowed saturated 2–$2'$ transition and dips to appear on the spectrum at the forbidden transition positions indicated by dashed vertical lines on the figure. This occurs because the corresponding differences of the populations are less than their equilibrium values so the forbidden transitions at these frequencies strongly influence the EPR line shape. The forbidden transition dips are located on both sides of the hole, separated from it by $\pm\varepsilon_+$ and $\pm\varepsilon_-$. Measurement of the distances between the dips provides values of hyperfine interaction constants.

SELECTIVE SPUTTERING

The predominant *sputtering* of one component of a target compound compared to the other components. The selectivity A is the ratio of sputtering rates (or factors) of compound B, sputtered at its maximal rate, to that of the other components D:

$$A = \frac{V_B}{V_D} = \frac{S_B}{S_D},$$

where V_B and V_D are the rates, S_B and S_D are the sputtering factors for compounds B and D, respectively. Selectivity at physical sputtering is insignificant, it does not exceed 2–3; at chemical sputtering, it can be much higher than 20. This difference is determined by the fact that the chemical activity of bombarding particles in relation to the target components, and the volatility of interaction products, can differ by many orders of magnitude. At the same time the difference in the binding energy that determines the selectivity of physical sputtering cannot vary by nearly as much.

SELENIUM, Se

A chemical element of Group VI of the periodic system with atomic number 34 and atomic mass 78.96. It is related chemically to S, Te and Po; in nature it is found associated with sulfur compounds. Outer shell electronic configuration $3d^{10}4s^24p^4$. Successive ionization energies are 9.75, 21.2, 32.0, 42.9, 68.3 eV. Atomic radius is 0.116 nm, radii of Se^{2-}, Se^{4+}, Se^{6+} are 0.2 nm, 0.05 nm, 0.042 nm, respectively; oxidation states are -2, $+4$, $+6$. Electronegativity is 2.48. Electron affinity is 2.02 eV.

Solid selenium has several allotropic modifications, among which the γ-modification is the most stable (so-called *gray selenium*) with a hexagonal crystal lattice and parameters $a = 0.4363$ nm, $c = 0.4959$ nm. The density of gray selenium is 4.807 g/cm^3. Besides gray selenium, also known are the red modification with a monoclinic lattice, which exists in an α-form ($a = 0.9054$ nm, $b = 0.9083$ nm, $c = 1.601$ nm, $\beta = 90°42'$) and in a β-form ($a = 0.931$ nm, $b = 0.807$ nm, $c = 1.285$ nm, $\beta = 93°08'$); there is also vitreous (amorphous) selenium black in color, and amorphous selenium of red color. All these modifications, upon long-term storage and holding at temperatures 373 to 423 K, transform to the hexagonal modification. For the hexagonal modification, $T_{melting} = 494$ K, $T_{boiling} = 958.4$ K. Heat of melting is 6.7 kJ/mole, heat of evaporation is 30 kJ/mole, specific heat $c_p = 25.3$ J·mole^{-1}·K^{-1}, Debye temperature is 89 K.

Selenium is a *semiconductor*. The band gap width for gray selenium is 1.8 eV, for an amorphous selenium film it is 2.25 eV. Electrical resistivity of gray selenium is $8 \cdot 10^{-2}$ Ω·m (at 0 °C), temperature coefficient of electrical resistance is $0.6 \cdot 10^{-3}$ K^{-1}. Solid selenium is a diamagnet. Dielectric constant of amorphous selenium is 6.24 (at 290 K). Brinell hardness of gray crystalline selenium is about 750 MPa; normal elastic stretching modulus is 10.2 GPa, shear modulus is 6.6 GPa; Mohs hardness of selenium is 2.0.

Selenium is brittle, its plasticity grows upon heating above 373 K. Selenium is used in rectifiers and as a semiconductor material. The ability

of selenium to change its electrical resistance upon illumination is used in electrical engineering (selenium *photoresistors*). Introduction of selenium into glass provides a material opaque to IR radiation. Compounds of selenium with the elements of zinc, germanium, and the arsenic subgroup (e.g., the *selenides* ZnSe, GeSe, As_2Se_3, etc.) possess semiconductor properties. In addition selenium is used in *metallurgy*, in chemical synthesis (as a catalyst), and for other purposes.

SELF-ACTION OF LIGHT WAVES

Broad class of *nonlinear optical* phenomena where the influence of intense electromagnetic radiation on the optical parameters of a *nonlinear medium* causes variations in the propagation characteristics of the radiation itself in the medium. There are two main groups of self-action effects of light waves, namely coherent phenomena displayed in the case of short light pulses (compared to characteristic *relaxation times* of nonlinear medium polarizability), and effects which occur under quasi-stationary conditions. The phenomenon of *self-induced transparency* is included in the first group, whereas saturation of absorption, self-focusing, *self-diffraction of light*, formation and propagation of *optical solitons*, self-curving, *self-induced light polarization change*, etc., are associated with the second group.

The phenomenon of *saturation of absorption* involves the equalization of energy level populations through resonant light absorption by the nonlinear medium; the absorption is reduced and the corresponding *transmittance* grows. This was predicted by R. Karplus and J. Schwinger (1948) for the case where resonance absorption is caused by phototransitions in two-level centers. During the growth of intensity of the radiation the *level populations* (ground and excited) are equalized, and the absorption is correspondingly reduced. In crystals, saturation of absorption is observed at the excitation of impurity centers, as well as near the *intrinsic light absorption edge*. The *Burstein–Moss effect* associated with the generation of a large number of *electron–hole pairs* in the intensive resonance field is one of the mechanisms activated in the latter case. The effect of absorption saturation is used in laser engineering for modulation of the quality factor of *resonators*, etc.

The phenomenon of *self-focusing of light* involves the reduction of the divergence of the light beam (compared to the diffraction case) in proportion to its propagation in the nonlinear medium. This was predicted by G.A. Askarian (1962) and observed by N.F. Pilipetski and A.R. Rustamov (1965). As the density of energy flow across the cross-section of the light beam is not uniform, the *refractive index*, depending on the light intensity, will vary across the cross-section of the beam. Depending on the mechanism of nonlinearity, the sign of the nonlinear addition to the refraction index $n_2 E^2$ (E is electric field intensity) may be either positive or negative, If $n_2 > 0$, then for a beam with its cross-section intensity distribution falling toward the periphery, the velocity of propagation in the outer regions is greater than it is toward the center. As a result, the phase front of the beam, which is uniform at the input, transforms during its propagation in the nonlinear medium to a concave phase front, and self-focusing takes place.

The path length traversed in the nonlinear medium by a Gaussian beam with input radius a and field intensity along the axis E_0, during which the beam converges (or "collapses") toward the axis, is called the *effective self-focusing length*:

$$z_{\mathrm{sf}} = \frac{a}{2}\left(\frac{n_0}{n_2 E_0^2}\right)^{1/2}.$$

If the convergence angle of the beam, determined by nonlinear refraction $\theta_{\mathrm{nl}} \approx a/z_{\mathrm{sf}}$, is equal to the angle of the diffraction divergence $\theta_{\mathrm{d}} \approx 0.3\lambda/a$ (λ is wavelength of light in the nonlinear medium), the nonlinear refraction completely compensates for the diffraction spreading of the beam in this medium. Thus the beam creates for itself an effective waveguide along which it propagates without divergence, an effect called *self-channeling*.

When the nonlinear addition to the refraction index is negative, $n_2 < 0$, then *self-defocusing* appears, or an increasing divergence of the light beam during its propagation. *Self-curving of a light beam* is the deviation of its trajectory from a rectilinear one upon its propagation in a nonlinear medium. It was observed by M.S. Brodin and colleagues (1969); it is exhibited by beams with an unsymmetrical intensity profile across their cross-section. Due to the corresponding unsymmetrical variation of the refraction index of the medium trajectory the beam curves toward the side of greater refractive index. The angle of curvature φ is proportional to the intensity gradient ∇E^2 of the beam:

$$\varphi \sim \frac{2\pi}{\lambda} a n_2 \nabla E^2.$$

The self-curving has no threshold, and it appears in the nonlinear medium with any sign of the nonlinear added factor n_2.

SELF-AVERAGING QUANTITY

A characteristic of a disordered system that is valid in the macroscopic limit. The existence of a self-averaging quantity is the result of two fundamental properties of disordered systems: spatial uniformity in the mean and reduction of correlation at large distances. For each self-averaging quantity f it is possible to find some small volume V_{c}, such that for $V_{\mathrm{c}} \ll V$ its distribution becomes Gaussian (see *Gaussian distribution*) with the mean value $\langle f \rangle$, independent of V, with a root mean square deviation $(\langle f^2 \rangle - \langle f \rangle^2)^{1/2}$, proportional to $(V_{\mathrm{c}}/V)^{1/2}$. In systems with a volume smaller than V_{c} fluctuations of the (mesoscopic) values f become pronounced, due to which various mesoscopic effects are displayed. Simple examples of self-averaging quantities are specific extensive (additive) physical values: density of states, specific free energy, electrical conductivity, diagonal elements of single-particle Green's function in k-space, etc. In mesoscopic materials these quantities do not self-average. The notion of self-averaging quantity was introduced by I.M. Lifshits who proved its applicability to the dipole moment of a unit volume of a crystal with isotropic disorder.

SELF-CONSISTENT FIELD

An averaged potential field of many interacting particles which determines the state of each particle. The self-consistent field method permits the reduction a many-particle problem to that of a single particle moving in the self-consistent field. The self-consistency lies in the fact that every particle of the system makes some contribution to the average potential field, and its state is determined by this field. The varieties of self-consistent field approaches depend on the method used for averaging the particle interactions. The dependence of the energy of a particle in the field of all the other particles on the average number of particles having a particular energy is the condition for self-consistency. The self-consistent field approximation is used in the quantum and statistical theories of solids, e.g., to calculate and describe the states of electrons in *metals* and *semiconductors* (*Hartree–Fock method, Hartree method* and *Thomas–Fermi method,* see *Atom*), and spin systems in magnetic materials (see *Molecular field*), to study atomic *alloy ordering,* to deal with various *phase transitions* (see *Landau theory*), including the transition to the superconducting state (in this case the interpretation is far from simple, see *Self-consistent field method in superconductivity theory*), etc. The introduction of a self-consistent field simplifies the calculation of physical properties of solids, but this method does not give qualitatively correct results in situations when long-wave *fluctuations* have a pronounced effect (see *Critical region*).

SELF-CONSISTENT FIELD METHOD
in superconductivity theory

An approximate method for calculating the ground state energy and the spectrum of elementary excitations (*quasi-particles*) in superconducting (superfluid) Fermi systems. It is based on the partial replacement of *Cooper pair* creation (annihilation) operators by their average values, which is an introduction of the so-called *Bogolyubov anomalous mean values*

$$\langle a_{p\sigma}^+ a_{-p,-\sigma}^+ \rangle \neq 0; \qquad \langle a_{-p,-\sigma} a_{p\sigma} \rangle \neq 0. \quad (1)$$

Here $a_{p\sigma}^+$ ($a_{p\sigma}$) is the fermion creation (annihilation) operator with momentum p and spin σ, angular brackets $\langle \dots \rangle$ mean quantum-mechanical averaging in the ground state of the system at $T = 0$, and thermodynamic averaging at a finite temperature $T \neq 0$. This approach simplifies the Hamiltonian of the *Bardeen–Cooper–Schrieffer theory* by converting it to a quadratic form with respect to $a_{p\sigma}^+$ ($a_{p\sigma}$):

$$H = \sum_{p,\sigma} (E_{p\sigma} - \mu) a_{p\sigma}^+ a_{p\sigma}$$

$$+ \frac{1}{2} \sum_{p,\sigma} (\Delta_p a_{p\sigma}^+ a_{-p,-\sigma}^+$$

$$+ \Delta_p^* a_{-p,-\sigma} a_{p\sigma}). \quad (2)$$

Here $E_{p\sigma}$ and μ are the fermion energy and chemical potential, and Δ_p is a complex *order parameter* that describes the superconducting (superfluid) condensate of Cooper pairs, and is determined by the expressions

$$\Delta_p = -\sum_{p',\sigma} V_{pp'} \langle a_{-p',-\sigma} a_{p'\sigma} \rangle,$$

$$\Delta_p^* = -\sum_{p',\sigma} V_{pp'} \langle a_{p'\sigma}^+ a_{-p',-\sigma}^+ \rangle, \quad (3)$$

where $V_{pp'}$ is a Fourier component of the fermion pair interaction. The simplified (reduced) Hamiltonian (2) allows exact diagonalization with the help of the *canonical transformation method* (see also *Bogolyubov method*), and leads to a spectrum of quasi-particles of the form

$$E_p = \sqrt{(E_{p\sigma} - \mu)^2 + |\Delta_p|^2}, \quad (4)$$

which is similar to the spectrum of quasi-particles in the Bardeen–Cooper–Schrieffer theory. The self-consistent field method is asymptotically accurate in the thermodynamic limit when the number of particles $N \to \infty$ and the volume of system $V \to \infty$, with the density N/V remaining finite. In coordinate space this method is used to describe spatially nonuniform states in superconductors (see *Bogolyubov–de Gennes equations*).

SELF-DIFFRACTION OF LIGHT

A phenomenon observed upon the interference in a *nonlinear medium* of at least two mutually coherent beams converging at an angle θ, and resulting in the appearance in transmitted and reflected radiations of additional beams propagating at angles that are multiples of θ with respect to each of the source beams. The periodic nonuniformity of the complex *dielectric constant* of the nonlinear medium created within the interference field of the source beams, or a dynamic hologram (see *Real-time holography*), and the diffraction of the source beams which leads to the appearance of additional beams, is the cause of the self-diffraction of light. The condition for observing the self-diffraction of light is $2\theta \sin(\theta/2) < \lambda/d$, where λ is the wavelength of the source radiation, and d is the thickness of the nonlinear medium.

SELF-DIFFUSION

See *Diffusion*.

SELF-INDENTING

Appearance of a source of *plastic deformation* in the region of contact of two similar bodies limited by sections of convex surfaces. A model situation is the contact of two spheres (with radius R). The *indenting force* is the force F_c of capillary origin, which appears where two free surfaces belonging to the osculating spheres (*surface energy* α) disappear in the zone of contact in some area S of the contact circle with radius x ($S = \pi x^2$), and one boundary surface (surface energy α_b) is formed in its stead. The force F_c may be found also in such phenomena as the merging of two liquid drops that touch each other, or the mutual depression of two contacting soap bubbles. The reduction of energy associated with the surface upon indenting is $\Delta W_s = -S(2\alpha - \alpha_b) = -\pi x^2 \Delta\alpha$. Since for $x \ll R$ the approaching of the centers of spheres $h = x^2/(2R)$, $\Delta W_s = -2\pi R h \Delta\alpha$ and therefore $F_c = d\Delta W_s/dh = 2\pi R \Delta\alpha \sim R$. The indenting stress $\sigma_c = F_c/(\pi x^2) = 2R\Delta\alpha/x^2$ for small x may exceed the *elastic limit* and cause plastic self-indenting. For this, the condition of *multiplication of dislocations* should be met: $\sigma_c > Gb/x$ (G is *shear modulus*, b is *Burgers vector*); this yields an estimate of the maximum value of x/R, at which plasticity is possible in the zone of contact: $(x/R)_{max} = [2\Delta\alpha/(Gb)]\cdot 10^{-1}$. The effect of self-indenting of crystals is observed by formation of dislocation aggregations, the so-called *dislocation rosettes*, in the contact zone. Self-indenting is detected in the well-known process of *caking* (self-pressing) of disperse powders (see *Disperse structure*).

SELF-INDUCED LIGHT POLARIZATION CHANGE

An effect of the *self-action of light waves* which involves the variation of its polarization as it propagates in an optically *nonlinear medium*. The effect is due to the anisotropy of the *refractive index*, *absorption coefficient*, and optical activity (see *Rotation of light polarization plane*) induced by the light. The value and nature of the self-induced change of the light polarization depend on the medium symmetry, the optical nonlinearity, and the light intensity. The most detailed investigations of self-action are performed for systems where the polarization of weak light beams either does not change in the absence of optical activity (isotropic media, cubic crystals, *uniaxial crystals* with light propagation along the *optical axis of a crystal*), or it changes slightly and does not mask the effect.

In isotropic non-optically active media a self-induced light polarization change is possible only for elliptically polarized light which violates "right" and "left" symmetry within the medium, and where circular birefringence or *dichroism of crystals* is present. In transparent media a self-induced light polarization change appears as a rotation of the polarization ellipse, the so-called *Maker–Terhune effect* (P.D. Maker, R.W. Terhune, 1964). With nonlinear absorption present, the shape of the polarization ellipse varies with the change in light propagation.

In non-optically active cubic and uniaxial crystals, due to their lower symmetry (compared to isotropic media), a self-induced light polarization change also appears for linearly polarized incident radiation. However, there are preferential directions for which this polarization change is absent. In the general case they are determined by the symmetry (in particular, planes of symmetry of the medium). As a rule, there are several such directions in commonly used experimental geometries where the radiation propagates along one of the symmetry axes.

Self-induced light polarization oscillations and *self-induced light polarization plane rotations* are characteristic manifestations of the self-action of light waves. The former case involves oscillations of the orientation of the polarization ellipse with respect to one of the selected directions (depending on the incident light polarization), accompanied by oscillations of the degree of ellipticity up to the limit of linear polarization. This effect is displayed in regions of comparatively weak absorption. The latter effect involves the rotation of the polarization plane to one of the selected directions (also depending on the initial radiation polarization). The effect shows itself in the region of nonlinear resonance absorption when the refraction nonlinearity is low.

In optically active *gyrotropic media* the light intensity dependent nonlinear optical activity is superimposed. A self-induced light polarization change in *liquid crystals* has a specific nature since the nonlinear response is essentially nonlocal.

Due to the high sensitivity of polarization measurements, the study of self-induced light polarization changes, its spectral dependence, and its kinetics, allows one to determine the parameters of various physical systems (impurity centers, biexcitons, electron–hole systems, etc.), which induce the appearance of optical nonlinearity.

SELF-INDUCED TRANSPARENCY

A phenomenon of light pulse propagation without attenuation in an absorbing medium (selftransmission of light pulse). It appears in the coherent interaction of radiation with a resonantly absorbing medium when the duration of a strong light pulse is shorter than the phase memory time (dephasing time) of the corresponding transition, and its area $\Theta = (\mu/\hbar) \int_{-\infty}^{\infty} E \, dt > \pi$ (E is the envelope of the electric field pulse at its entry into the material, μ is the dipole moment matrix element of the transition). A light pulse that coherently excites the medium loses energy at its front edge; and this energy returns to it at its rear edge due to the induced re-radiation of the medium. As the duration of the exciting pulse is shorter than the dephasing time, the pumping of the light energy to the medium, and vice versa, occurs more rapidly than the relaxation processes destroy the coherence of the interaction. The processes of energy pumping lead to the decrease of the speed of the pulse motion (as a whole), so the pulse deforms and acquires some fixed shape. Selfinduced transparency differs from the clearing up of the medium as a result of absorption saturation (see *Self-action of light waves*). Self-induced transparency was discovered by S.L. McCall and E.L. Hahn (1967) during the resonance excitation

of ruby (Cr^{3+} doped Al_2O_3) cooled to liquid helium temperature by strong *laser* pulses with a modulated Q-factor.

SELF-LOCALIZATION, also called
autolocalization (fr. Gr. $\alpha \upsilon \tau o \varsigma$, self)

The localization of an electron, *exciton*, or other electronic excitation in some part of an *ideal crystal*.

By virtue of the translation *crystal symmetry*, the electron excitation is described by Bloch wave functions (see *Bloch theorem*), and the corresponding electron density is periodically distributed over the entire *lattice*. However, because of the interaction between electronic reexcitation and the atoms of the *crystal lattice*, the former is capable of causing a local deformation of the regular atomic order (the latter creates the potential for localizing the electronic excitation sustaining this deformation). In a polar crystal, if the electronic excitation involves a *conduction electron (hole)*, such an autolocalized state is called a *polaron*. A self-localized state can move through the crystal as an integral whole (the electronic excitation moves along with the wave of deformation). However, the *effective mass* of such a *quasi-particle* can be much greater than that of the "bare" electronic excitation. Self-localization was predicted by L.D. Landau in 1933. See also *Self-localized exciton, Soliton*.

SELF-LOCALIZED EXCITON, also called
autolocalized exciton

An electronic excitation in a crystal, formed as a result of the *self-localization* of a band *exciton*.

Under self-localization, the exciton makes a transition from a band to a localized state of a potential well arising from the *crystal lattice* deformation brought about by the selflocalizing *quasiparticle* itself. The term *self-localization* (or *self-trapping*) underscores that the phenomenon is due to the interaction of the particle with the ideal lattice, and is not due to its trapping by an impurity or a structural defect (see *Defects in crystals*). The microstructure of a self-localized exciton is mainly determined by the properties of the initial exciton and its ability to deform the lattice. Self-localization is energetically favorable if the increase in the quasi-particle's kinetic energy, and the energy of the lattice *strain* of the self-trapping is compensated by the decrease of the potential energy during the formation of the potential well, as occurs for the case of a sufficiently strong *exciton–phonon interaction*. Self-localized excitons are experimentally observed mainly in *semiconductors*. *Frenkel excitons* (*small-radius excitons*) are self-localized in a single-site state within a single unit cell (*molecular crystals*). The self-localization of *Wannier–Mott excitons* (*large-radius excitons*) depends on the band structure and the type of inter-atomic bond. A typical example of self-localization occurs when the ratio of the electron to hole effective mass is small. In this case, the region of deformation of the crystal is determined mainly by the interaction between the lattice and the heavy hole, and it can be small, while the electron cloud has a large radius and interacts only weakly with the lattice (*alkali-halide crystals*). The self-localization of excitons of intermediate radius is highly sensitive to the symmetry of the exciton–phonon interaction. Such a self-localized exciton can be formed both as a single-site excitation of the quasi-atomic type and as a two center structure of the excimer type of molecule (see *Excimer*), in which two neighboring lattice atoms are coupled by a resonant bond (crystals of inert elements, and alkali-halides). In three-dimensional crystals self-localization requires overcoming an energy barrier which is higher, the lower the effective mass of the exciton. In the case of wide-band *insulators* (crystals of inert gases and alkali-halides), a high barrier height indicates the concurrent existence of free and self-localized excitons. Theoretical and experimental investigations of this coexistence provide information on the energy barrier, the structure of the self-localized exciton, and the magnitude of the exciton–phonon interaction. An *energy band* corresponds to the translation of self-localized excitons through the crystal. However, the band is narrow so the *diffusion coefficient* of the self-localized excitons is determined by a jump mechanism, and is smaller by several orders of magnitude than that for initial free excitons. As a result, a self-localized exciton substantially retards the *energy transfer by excitons* and the energy *relaxation of excitation* in a crystal.

SELF-OSCILLATION, also called auto-oscillation

Undamped nonlinear oscillation in a dissipative system where the *frictional loss* is compensated by an external energy source (term introduced by A.A. Andronov, 1928).

Self-oscillations arise in nonlinear systems over a range of parameter values which is specific for each system. The amplitudes and frequencies of self-oscillations are determined by the values of the system parameters only. The time-varying energy input from the applied source is controlled by the system itself, and is determined by its set of natural frequencies and eigenfunctions. Its nature is complicated, especially in systems with several degrees of freedom. Self-oscillations may be of different forms, including both strictly periodic and stochastic types. Self-oscillating systems are referred to as *systems with hard excitation* when some critical value of the initial amplitude must be exceeded to initiate them, or as *systems with soft excitation* when they can arise from the infinitesimal *fluctuation* of a variable. Self-oscillations are found in mechanical systems (pipes, buildings, or bridge supports swaying in the wind, flutter-type vibrations of airplane wings associated with bending and torsion of the wing; ship yaw, pitch, and roll, etc.); in electric circuits (current and electromagnetic field oscillations); in optical systems (interaction of light beams in *real-time holography*, and light intensity oscillations in *lasers*); and in acoustic, acousto-electric, thermal, and other systems. In many cases, self-oscillations are caused by physical processes in solid-state *active media*. In *semiconductors*, there are different mechanisms of bound auto-oscillations associated with nonequilibrium concentrations of *current carriers* in *conduction bands* and at *local electronic levels*; as well as auto-oscillations of the temperature, the electric current, and the electric field. A more general phenomenon than self-oscillations is *autowaves*.

Self-oscillations find wide use in generators and amplifiers of electromagnetic and acoustic oscillations, lasers, etc.

SELF-SIMILARITY

The similarity of system characteristics at different moments of time (see *Similarity* of phenomena).

System characteristics can be expressed through a set of independent dimensionless parameters Π_i $(i = 1, 2, \ldots)$ (*similarity parameters*), some of which involve the time t. It is the invariability of the similarity parameters with changes in time (a possibility in the presence of time dependence of the dimensional parameters entering Π_i, usually coordinates) which indicates that the system exhibits self-similarity. A transformation to dimensionless variables in the differential equations describing the system in the case of self-similarity should correspond to the initial and boundary conditions for the original dimensional variables. Such a situation, corresponding to a complete self-similarity, very seldom occurs. However, in many cases it proves possible to treat the system as approximately self-similar in certain spatial-temporal domains (so-called *intermediate asymptotic*) which considerably simplifies the description of the system. The property of self-similarity is helpful in solving problems of *diffusion*, *heat transfer*, propagation of elastic and other waves in a medium, etc. For example, the diffusion equation $\partial \rho / \partial t = D \partial^2 \rho / \partial x^2$ (where ρ is the density of diffusing particles, D is the diffusion coefficient, t is time, and x is a coordinate) admits the introduction of a single dimensionless parameter $S = x^2/(Dt)$. Assume the initial condition at $t = 0$ to be $\rho = \rho_0 = \mathrm{const}(x)$, $x \leqslant 0$, and the boundary condition at $x = 0$ to be $\rho = 0$ (such conditions describe, in particular, the *gettering* of impurity atoms through the plane $x = 0$ from the impurity phase located at $x < 0$). The two conditions at $t = 0$ and at $x = 0$ lead to consistent conditions for S both at $S \to \infty$ and at $S = 0$, i.e. the system has the property of self-similarity. The substitution $\rho = \rho_0 f(S)$ provides a differential equation for $f(S)$, namely $4Sf'' + Sf' = 0$, i.e. it involves the transition from a partial differential equation to an ordinary differential equation. The solution of this latter equation is $\rho(x, t) = \rho_0 \Phi([S/2]^{1/2})$, where Φ is the probability integral.

SELF-SUSTAINING CRYSTALLIZATION

Crystallization of an amorphous film which spreads from the initial crystallized area, and is maintained by the release of the latent heat of the *phase transition*. Other names for self-sustaining crystallization are autocatalytic, impact, explosion, avalanche crystallization. There are two types of self-sustaining crystallization: solid-phase (direct transition of amorphous into crystalline phase) and liquid-phase (initial melting of amorphous phase, then overcooled melt crystallizes). The possibility of this type of crystallization in amorphous films of metals, semiconductors and insulators is determined by the relation $W_c \leqslant \beta Q_c$, where W_c is the activation energy, Q_c is the latent heat of crystallization, and β is a coefficient taking into account the heat loss from radiation and heat conduction.

Initially self-sustaining crystallization was observed in amorphous films of Sb, then in films of Ge, Bi, Si, CdTe, and of the insulators SiO_2, Si_3N_4, beginning from a thickness adequate for the accumulation of heat, upon heating and conservation of the internal stresses within the films. The start of self-sustaining crystallization requires the establishment of conditions at the film by shock, puncture, laser or electron-beam pulse, whereby crystal nuclei are formed. Then the growth of *crystallites* continues with the formation of either a radial structure, or of several concentrically located ring-like sections with crystals of various sizes. During the transverse motion of the flat front of self-sustaining crystallization in amorphous films on an unorientable substrate (of the type α-Si at SiO_2) it is possible to obtain, depending on the initial temperature and heating pulse, either coarse-grained (periodic) or fine-grained structures. *Pulse heating* of an amorphous film on an orienting substrate promotes the process of *epitaxy* under the conditions of self-sustaining crystallization. The self-sustaining crystallization velocity of propagation is 1 to 30 m/s, which corresponds to the velocity of heat conduction or of phase transition kinetics at the crystallization temperature.

SELF-WAVES

See *Autowaves*.

SEMICONDUCTING SUPERCONDUCTOR

See *Superconducting semiconductors and semimetals*.

SEMICONDUCTOR DEVICES

Electronic devices, whose action is based on control of electronic processes by *semiconductors*. In electronics semiconductor devices are used for the transformation, processing, and storage of information contained in electrical, acoustic, electromagnetic and other signals; in power engineering they are used for the direct transformation of one form of energy into another.

The effects caused by electronic processes in semiconductors were discovered in the second half of the 19th century, such as the variation of selenium column resistance upon its illumination (1873), and the unilateral conductivity and evident nonlinearity of the current–voltage characteristic of a semiconductor crystal junction (1874). *Transistors* appeared in the late 1940s. In subsequent years different semiconductor devices were developed: *tunnel diodes, Gunn diodes, impact ionization avalanche transit time diodes, barrier injection transit time diodes, varicaps, thyristors*, etc. Semiconductor optoelectronics then provided *photodiodes, light emitting diodes, photothyristors* (switch controlled by radiation), *optoisolators, semiconductor lasers*. In the 1960s came *microelectronics, integrated circuit* engineering, *microwave semiconductor devices*, and *field-effect transistors*. The reduction of the temperature of semiconductor devices leads to reduction of inherent noise, increasing fast response, improving sensitivity, and the possible application of high-temperature superconducting materials in integrated electronics (see *High-temperature superconductors*). Dielectric and magnetoelectric devices using physical effects in *piezoelectrics materials, ferroelectrics, ferromagnets*, find their application in semiconductor devices.

Industry worldwide produces more than 300 billion transistors and approximately the same quantity of integrated microcircuits at the average integration level of more than 10,000 transistors at one crystal. Thus, the general number of transistors produced in the world annually (including also those within integrated microcircuits) is more than 50 thousand times greater than the number of vacuum tubes during the peak years of the late 1980s. The use of semiconductor devices in electronics increased the reliability of devices and systems due to the dramatic reduction (by 4 to 5 orders of magnitude) of the incidence of defects in a single active element (of lamp and transistor, respectively). Similarly, the mass-weight indices and energy consumption have been reduced by

tens or hundreds of times. For manufacturing devices and integrated microcircuits a wide range of semiconductor materials is used: *germanium, silicon*, compounds of elements of Groups III and V, and also II and VI, of the periodic system (*gallium arsenide*, indium antimonide, lead–tin–tellurium, etc.).

The so-called *planar technology* includes the application of *thin films* of dielectric (mainly silicon dioxide) and metallic materials, photolithography (electron lithography, X-ray lithography), different types of *thermal treatments*, and other types of chemical treatments. Integrated microcircuits have achieved extremely high accuracy and resolution by endowing their main and auxiliary materials with maximum achievable degrees of purity from foreign inclusions and impurities.

In the 1980s several tens of thousands of transistors had connections at a crystal with an area of about 50 mm^2, and at present there is the possibility of the placement of up to 10 to 20 million transistors on a single crystal with an area up to 100 mm^2. Manufacturing difficulties and a series of technological and physical restrictions cause the reliability and fast response of integrated microcircuits to be limited, not by the parameters of the active semiconductor devices, but rather by the interconnections between them. The search for an alternative solution led to the appearance of *charge-coupled semiconductor devices* which developed within the framework of the important area of integrated electronics called *functional electronics*. Thus, semiconductor devices which evolved from discrete devices to active elements of integrated microcircuits, have now reached the stage of functional electronics. This is a non-circuit engineering approach in integrated electronics which uses the integration of different physical effects in solids as the basis of action principles of functional electronics devices. In power engineering, semiconductor devices are used for the conversion of alternating current energy to that of direct current, for the transformation of light (mainly solar) energy to electrical energy (solar batteries), of thermal energy into electrical energy (thermogenerators) and for the inverse transformation (refrigerators), etc. See also *Diode, Generation type semiconductor* diodes, *Charge coupled devices, Varactor, p–i–n diode, Semiconductor memory, Semiconductor photodetectors, Schottky diode.*

SEMICONDUCTOR DIODE

A two-electrode device with unipolar conduction, with a nonlinear dependence of the current passing through the *diode*, or of its capacitance, on the voltage across it. These properties of a semiconductor diode are caused by nonlinear properties of the doped semiconductor used to construct it, by the electric properties of its p–n-junction (see *Semiconductor junction*), or by the properties of the Schottky barrier of the diode (see *Schottky diode*). Semiconductor diodes may be used as active elements of generators and amplifiers in the UHF range (see *Generation type semiconductor diodes, Varactor*), for rectifying and transforming electromagnetic signals with frequencies from low (50 Hz) to the optical range (see *Photodiode*), and as controlling diodes (see *p–i–n-diodes*).

SEMICONDUCTOR–ELECTROLYTE BOUNDARY

A boundary between a semiconductor and an electrolyte (solution containing ions that carry electric current) characterized by the presence of an *electric double layer* formed by charges of opposite sign in the two phases in contact (see Fig. (a)). In the semiconductor a charge accumulation in the near-surface region results from the redistribution of electrons and holes; and in the electrolyte it arises from the redistribution of ions which form an ionic coating at the double layer; under equilibrium conditions the charge magnitudes on the facing double layer coatings are equal to each other.

The double layer comprises the three regions sketched in Fig. (a). A dense region called the *Helmholtz* layer lies in the electrolyte immediately adjacent to the semiconductor surface formed by ions attracted to the surface, as well as by solvent molecules; this layer is very thin, approximately one ion radius in thickness. On the electrolyte side of the Helmholtz layer there is a diffuse region with a charge gradient called the *Gouy–Chapman layer* which is clearly evident in dilute electrolytes ($\leqslant 10^{-1}$ mole/l). On the other side of the Helmholtz layer there is a *space charge region*

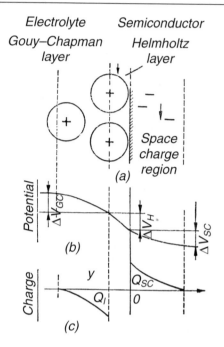

(a) Structure of the electric double layer at a semiconductor–electrolyte boundary, (b) variation in potential, and (c) variation of charge across the boundary, where ΔV_{GC} is the potential drop in the Gouy–Chapman layer, ΔV_H is the potential drop in the Helmholtz layer, ΔV_{SC} is the potential drop in the space charge region of the semiconductor, Q_I is the ionic charge adhering at the interface of the double layer, and Q_{SC} is the charge in the space charge region of the semiconductor.

arising from with the accumulation of charge at the surface inside the semiconductor, and Fig. (c) sketches the charge variation on these two sides of the Helmholtz layer. There is a continuous potential drop across the semiconductor–electrolyte boundary, sketched in Fig. (b), that is associated with the capacitance of the space charge region of the semiconductor C_{SC} being less than the the capacitance C_H of the of the Helmholtz layer ($C_H \approx 6$ to $20\ \mu F/cm^2$). The rate of electrochemical reactions at the semiconductor–electrolyte boundary is determined by the free *current carrier* concentration in the semiconductor, by the ion concentration in the electrolyte, and by the positions of the energy bands and the *Fermi level* in the semiconductor with respect to the energy levels of ions in solution. The stability of the boundary depends on the rate of chemical and electrochemical reactions on the semiconductor surface, which are associated with the irreversible processes of *oxidation* (reduction) of ions and molecules in the electrolyte, the *adsorption* of products of the reactions, and the degradation of the semiconductor.

The properties of the semiconductor–electrolyte boundary are widely used in the chemical *etching* and *electrolytic polishing* of semiconductors, and for obtaining metal contacts in semiconductors in the instrument-making industry. This boundary can be used for the determination of the impurity concentration and the impurity distribution over the depth within semiconductors (method of volt–capacity differential profiling during etching of semiconductors). The high photosensitivity of the semiconductor–electrolyte boundary allows one to employ it for the photoelectrolysis of water.

SEMICONDUCTOR JUNCTION, *p–n* junction

Transition region in a contact between two *semiconductors*, one with electron-type (*n*) and the other with hole-type (*p*) conductivity. The junction can be formed either in a semiconductor having a homogeneous chemical composition by nonuniform *doping* (*p–n homojunction*), or as a contact between two semiconductors with differing chemical compositions (anisotropic *heterojunction* or *p–n heterojunction*). As a rule, the term semiconductor junction ordinarily denotes a *p–n* homojunction.

When forming a contact, the equilibrium between the semiconductors is established by the fluxes of *current carriers* and corresponding differences in *work functions*. The electrons from the semiconductor with a smaller work function pass to the semiconductor with a larger work function, leaving behind a non-compensated positive charge; *holes* pass in the reverse direction and leave behind a negative charge. As a result, a space charge builds up in the transition region; its electric field induces a potential barrier which is so high that the *Fermi levels* in the two semiconductors become equal, and the movement of the carriers comes to a halt. The barrier height is determined by the *contact potential difference* V_k. In the case of a *p–n* homojunction, the *n*-type-semiconductor becomes depleted

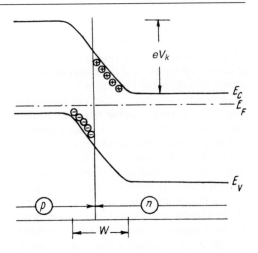

Fig. 1. Energy diagram of semiconductor junction in equilibrium: E_c is the conduction band bottom, E_v is the valence band top, E_F is the Fermi level, $\oplus$ denotes uncompensated charges of donors, and $\ominus$ indicates uncompensated charges of acceptors.

and hence positively charged as electrons diffuse into the *p*-region, while the *p*-type-semiconductor become depleted of holes and charged negatively. A space charge in the transition region is formed by impurity ions which are donors in *n*-type-semiconductors, and acceptors in *p*-type-semiconductors. The contact electric field induces the potential barrier for the majority (most numerous) charge carriers. Its value is equal for electrons in the *conduction band* and for holes in the *valence band* (see Fig. 1). In a *p–n* heterojunction, the depletion of the transition region of majority charge carriers also takes place, but the heights of the potential barriers for electrons and holes are not the same. This is determined not only by the contact field, but also by the misalignment of the edges of the valence band and conduction band.

As long as the *space charge region* is depleted of majority carriers the potential drop significantly influences the passage of electrical current. The external voltage changes the barrier height and disturbs the equilibrium between the charge carrier fluxes through the junction. If the polarity of the voltage is opposite to the contact electric field (forward bias), the potential barrier decreases, and there is an uncompensated flow of majority charge carriers through the barrier to the other semiconductor, i.e. a flux of holes from the p-type to the n-type semiconductor (where these carriers are minority ones). The *injection* of minority carriers takes place, and their concentration increases with the applied electric potential (which drops in the space charge region) in accordance with an exponential law. For example, the hole concentration at the edge of the space charge region in an n-type semiconductor is equal to $p = p_n \exp[eV/(k_B T)]$ where p_n is the equilibrium concentration of holes in the n-type semiconductor. Simultaneously with the injection of minority carriers through the current-conductive contacts, an equal quantity of majority carriers enters the n- and p-regions. Within the *Maxwell relaxation* time the majority carriers compensate the charge of the minority carriers, and an *injection plasma* is produced. The current through the semiconductor junction is determined by the total number of *recombination* events involving electrons and holes over the entire bulk per unit time. The recombinations occur in three space regions: in the regions where there exists a quasi-neutral plasma in n-type and p-type-semiconductors, and in the semiconductor junction itself. Corresponding to them are two components of current density: $J = j_D + j_R$. Here j_D is the diffusion current associated with the diffusion and recombination of charge carriers in quasi-neutral regions

$$j_D = j_S \exp\left(\frac{eV}{k_B T}\right), \qquad (1)$$

where

$$j_S = e\left(\frac{D_p}{L_p} p_n + \frac{D_n}{L_n} n_p\right), \qquad (2)$$

D_n and D_p are the *diffusion coefficients* of electrons and holes, respectively; L_n and L_p are the *diffusion lengths* of carriers in semiconductors of n- and p-type, respectively; n_p is the equilibrium concentration of electrons in p-type semiconductors; p_n is the equilibrium concentration of holes in n-type semiconductors. The recombination current related to the recombination of carriers in the semiconductor junction is

$$j_R \approx \frac{n_i}{2\tau} \frac{k_B T}{V_k V} W \exp\left(\frac{eV}{n k_B T}\right), \qquad (3)$$

where n_i is the intrinsic carrier concentration in the semiconductor, τ is the lifetime of carriers in

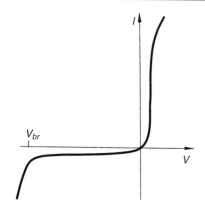

Fig. 2. Current–voltage characteristic of a semiconductor junction.

the semiconductor junction, W is the semiconductor junction width, and n is a dimensionless number close to 2. In the case of a large forward bias when $V \sim V_k$, the current increases so sharply that the potential drop in n-type and p-type regions becomes substantial, and the *current–voltage characteristic* deviates significantly from the exponential law given by Eqs. (1) and (2) (see Fig. 2).

In the case of a reverse bias (negative potential in p-type semiconductor, $V < 0$), the potential barrier increases, the majority carrier current through the semiconductor junction drops, whereas the minority carrier current (which experiences no barrier) remains almost unchanged. As a result, the semiconductors in the vicinity of the junction are depleted of minority charge carriers (*depletion*), and the current through the junction is due to thermally generated *electron–hole pairs*. The pairs beyond the semiconductor junction diffuse to the barrier and are separated by its field, thus inducing the current j_S independent on the electric potential (saturation current). The current carriers generated inside the junction are also separated by the contact field: electrons and holes go to n-type and p-type semiconductors, respectively. The resulting so-called *generation current* is

$$j_G = e\frac{n_i}{\tau_e} W(V), \qquad (4)$$

where n_i is the natural concentration of carriers in the semiconductors, and τ_e is the life time for the generation of carriers in the semiconductor junction. The total backward current is equal to the sum of j_S and j_G. The generation current grows with the increase in potential due to the widening of the semiconductor junction. The generation and recombination of charge carriers in it are usually associated with deep impurity centers (see *Deep levels*).

The width of a semiconductor junction and its dependence on the applied potential are determined by the concentration of doping impurities and their distribution in space. In the case of a stepwise impurity distribution (sharp semiconductor junction),

$$W = \left[\frac{\varepsilon}{2\pi e} \frac{N_A + N_D}{N_A N_D} (V_k - V) \right]^{1/2},$$

where N_A is the acceptor concentration in a p-type semiconductor, N_D is the donor concentration in an n-type semiconductor, and ε is the *dielectric constant*. If the impurity concentration varies linearly with the distance (gradual semiconductor junction), then $W \sim (V_k - V)^{1/3}$. In the case of a reverse bias the semiconductor junction widens, while for a direct bias it narrows.

The current–voltage characteristic of the junction is sharply asymmetric (see Fig. 2). For a direct bias the current is higher than the inverse current by several orders of magnitude. At a sufficiently high inverse voltage $-V_{br}$ a sharp current buildup in magnitude (*breakdown*, negative in direction) takes place. This is related to extra mechanisms of carrier generation whose intensity depends strongly on the electric field in the junction, or on the current. There are three main mechanisms of breakdown: *avalanche breakdown* due to *impact ionization* of the crystal lattice by electrons and holes in the junction; *tunnel (Zener) breakdown* related with carrier tunneling through a barrier (see *Interband tunneling*); and *thermal breakdown* induced by lattice heating with an increase in the rate of thermally generating carriers. A tunnel current is significant only for a sufficiently small barrier layer width, i.e. for strong doping. When a semiconductor is doped up to degeneracy, the tunnel current is also large for a forward bias. As the potential grows, the tunnel current increases, and it drops at reaching its maximum.

This results in N-shaped current–voltage characteristic for a direct bias (see *Tunnel diode*).

In an alternating current circuit, the semiconductor junction possesses a capacitance which in the equivalent circuit is in parallel with a resistance. There are two mechanisms for the formation of the capacitive component current with two corresponding capacitances: a *charging capacitance* due to the charge variation of the barrier layer (regarded as a capacitor with the distance W between plates) and a *diffusion capacitance* associated with the inertia of the carrier diffusion in quasi-linear regions. The diffusion capacitance is proportional to the nonequilibrium carrier concentration, and for this reason it is effective only for a forward bias. The diffusion capacitance depends on the frequency ω, and falls as ω^{-1} when ω exceeds the inverse carrier lifetime. In the case of a reverse bias, the determining factor is the charging capacitance which is weakly dependent on the frequency, and is effectively controlled by the potential (see *Varactor*).

Semiconductor junctions form the basis of many *semiconductor devices*: *diodes* of various types, *transistors*, *thyristors*, *light emitting diodes*, *injection lasers*, *p–i–n-diodes*, *stabilizers*, *varistors*, *varactors*, etc. The high sensitivity of these junctions with respect to external actions (thermal and mechanical actions, action of light and ionizing radiation) is the reason for their application in detectors of various kinds. Semiconductor junctions are also used for light energy transformation in *solar cells*. The junction can be produced by diffusion or implantation of doping impurities (e.g., acceptors in n-type semiconductors), introduction of a metal plus some needed impurities into a semiconductor, epitaxial growth, and monocrystal growth from the melt with a change of the impurity composition (see *Monocrystal growth*).

SEMICONDUCTOR LASERS

Lasers with semiconductors as the *active medium*, using *radiative quantum transitions*, and *conduction electron* and *hole* mobile charge carriers. Induced radiative transitions occur between levels in the bands, and also between the levels of small impurity centers. In order to obtain optical amplification it is necessary to invert the population density of the working levels (see *Semiconductors*),

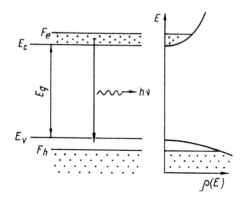

Fig. 1. Energy level diagram (left) showing the band gap E_g and a radiative transition $h\nu$ from the inverted population of the occupied bottom of the conduction band (E_c) and the empty top of the valence band (E_v). The electron and hole Fermi levels F_e and F_h, respectively, are indicated. Right: the distribution of the densities of states $\rho(E)$ (horizontal axis) in the semiconductor bands.

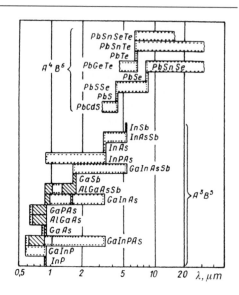

Fig. 2. Spectral overlapping for semiconductor lasers constructed on the basis of binary compounds, and of solid solutions formed from them.

and the condition for this may be expressed in the form

$$F_e - F_h > h\nu,$$

where F_e, F_h are, respectively, the *Fermi quasi-levels* of electrons and holes; $h\nu$ is the photon energy. This condition for intensive pumping is fulfilled at the band edges, therefore $h\nu$ is usually close to the width of the *band gap* E_g (see Fig. 1). In semiconductor lasers the so-called *direct-gap semiconductors* are used, which have the bottom of the *conduction band* and the top of the *valence band* at the same point ($k = 0$) in the phase space. Many binary compounds of the type $A^{III}B^V$, $A^{II}B^{VI}$ and $A^{IV}B^{VI}$ (see *Semiconductor materials*), e.g., GaAs, CdS, PbTe, and also the related series of *solid solutions* of these compounds, form the basis on which these lasers were made. The wavelength ranges spanned by various semiconductors, and their extent of overlap, are shown in Fig. 2.

According to their methods of pumping, semiconductor lasers are subdivided into *injection lasers* (injection pumping via direct current through an electron–hole junction), *electron beam pumped lasers* (pumped by high-energy electrons, e.g., with energy 20 to 50 keV), lasers with optical

pumping, and those with pumping by breakdown in an electrical field, including *streamer lasers*.

Semiconductor lasers are characterized by their small dimensions (especially injection lasers), by their economic efficiency, by the simplicity and rigidity of their design, by the possibility of tuning their frequency, by the availability of active media with broad frequency ranges, by the possibility of direct internal modulation of radiation within the band of 1 to 10 GHz, by the possibility of generating of ultrashort pulses, by their compatibility with semiconductor circuits, and by the possibility of monolithic integration with other *semiconductor devices* and integrated-optical elements. In its simplest version a semiconductor laser is a crystal *diode* with two-sided flat-parallel mirror faces forming the optical *resonator*. The laser radiation generated in the active layer located along the axis of resonator, i.e. perpendicular to the specular faces, exits through a face. In more complex modifications with many-element resonators, structures with Bragg reflectors and coupled resonators are used. Heterolasers, where the active area is a potential well for the carriers and a dielectric waveguide for the generated photons, is an important variety of semiconductor lasers. In the

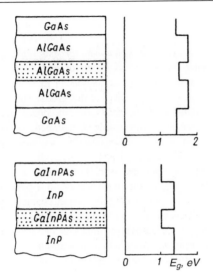

Fig. 3. Diagrams of some laser double heterostructures (left), and their corresponding band gaps (right).

so-called *double heterostructure* the active layer is located among the passive emitter layers of the broader-band semiconductor, and due to the *total internal reflection*, radiation is channeled along it at its boundaries. The circuits of some double heterostructures are shown in Fig. 3. If the active layer thickness is comparable or smaller than the de Broglie electron wave length, then due to the quantization of electronic states in the potential well the radiative properties of the medium are significantly changed (*quantum-size lasers*). Within the limits of one wave-guide layer of the laser heterostructure there may exist several quantum-size layers. If they are strictly periodic, the laser is called a *superlattice laser*. Reduction of the radiative layer thickness and optimization of the waveguide geometry permit the reduction of the threshold current density in quantum-size lasers to the level of 100 to 200 A/cm^2, and lower, at 300 K.

Stripline injection lasers where the active region is restricted to lateral directions within the range of a narrow strip (from 1 to 10–20 μm) are of practical significance. In order to restrict the current outside the strip the planar technology of oxide isolation is used, or a covering of a poorly conducting material, e.g., one containing

"counter" (i.e. locking) *p–n*-junctions. In overgrown stripline lasers the threshold current is reduced to 5 to 50 mA. A typical radiation power level is 10 mW in a stripline semiconductor laser, and up to several watts in monolithic multistrip lattices; in the radiation pulse mode the peak power of an injection laser radiation attains 10 to 50 W, and in lasers with electronic or optical pumping, up to several megawatts (due to a high active volume and a higher laser beam cross-section compared to an injection laser). Injection lasers operating in the wavelength ranges 0.78, 0.85, 1.3, and 1.55 μm have found wide practical applications, in particular in systems for the storage and reading of information, including audio and video disc devices, in fiber-optical communication systems, in automation. Semiconductor lasers are used for optical pumping of other types of *solid-state lasers*. Additional areas of semiconductor laser application are: high-resolution spectroscopy, monitoring of the atmosphere and other media, projection television, night vision, optical detection and ranging, telemetry, and metrology.

SEMICONDUCTOR MATERIALS

Materials characterized by values of *electrical conductivity*, intermediate between those of *metals* and *insulators*, i.e. 10^4 to 10^{10} $(\Omega \cdot cm)^{-1}$ (300 K). Unlike metals, the electrical conductivity of a semiconductor increases with an increase in temperature. The *band gap* of a semiconductor varies within the range 0 to 5.6 eV. For these materials the following is characteristic: the presence of two oppositely charged *current carriers* (electrons and holes), and also a high sensitivity of electrophysical properties to the *impurity atom* content, to structural defects, and to external conditions (heating, radiation, deformation, etc.) (see *Semiconductors*).

Classification. Semiconductor materials are subdivided into several groups, and the principal ones are:

1. *Elemental semiconductor materials*, the most important representatives being *silicon* and *germanium*, the basic materials of modern semiconductor electronics. Having four valence electrons, atoms of Si and Ge form a *diamond* type crystal lattice with *covalent bonds*

between the atoms; they are indirect-gap semiconductor materials; they form a continuous series of *solid solutions* one with another.

2. Binary compounds of the type $A^{III}B^V$, that is compounds formed from elements of Groups III and V of the periodic system. They generally have a sphalerite (zinc-blende) structure, with the bonding of their atoms in the crystalline lattice mainly of a covalent nature (see *Covalent bond*) with some admixture (up to 5 to 15%) of *ionic bonding*. They melt congruently with a narrow region of homogeneity. The most important representatives *gallium arsenide* (GaAs), InP, InAs, InSb are direct-gap semiconductor materials, although GaP has an indirect band gap. Many $A^{III}B^V$ compounds form continuous binary solid solution series with one another, as well as some ternary or more complex series.

3. Compounds of elements of Group VI (O, S, Se, Te) with elements of Groups I to V of the periodic system, and also with *transition metals* and *rare-earth elements*. In this general group of semiconductor materials $A^{II}B^{VI}$-*type compounds* are of greatest interest, i.e. compounds of elements from Groups II and VI of the periodic system. They have a crystal structure of either the sphalerite (zinc-blende) or the wurtzite types, less often of the NaCl type. The bond between the atoms has a mixed (covalent-ionic) character (ionic component reaches 45 to 60%). For the compounds $A^{II}B^{VI}$ the phenomenon of *polymorphism* is common, as well as the presence of polytypes (see *Polytypism*) of cubic and hexagonal modifications. They are mainly direct-gap semiconductor materials. The most important representatives are cadmium sulfide (CdS), CdTe, ZnO, ZnS, ZnSe, ZnTe, HgTe. Many $A^{II}B^{VI}$ compounds form continuous solid solution series with each other.

4. Ternary compounds of the type $A^{II}B^{IV}C_2^V$ formed from elements of the Groups II, IV and V of the periodic system. They generally crystallize in the chalcopyrite lattice. In many aspects they are electronic analogues of the compounds $A^{III}B^V$. Typical representatives are $CdSnAs_2$, $CdGeAs_2$, $ZnSnAs_2$.

5. *Silicon carbide* SiC exists in several structural modifications, the most important being: α-SiC (hexagonal structure), β-SiC (sphalerite structure).

6. *Noncrystalline semiconductor materials.* The vitreous (see *Vitreous state of matter*), chalcogenide and oxide semiconductor materials are typical representatives of this group. The solid solutions of the series of *amorphous semiconductors* with hydrogen, hydrated noncrystalline semiconductor materials, are important: α-Si:H, α-Si$_{1-x}$C$_x$:H, α-Si$_{1-x}$Ge$_x$:H, etc. Atomic hydrogen is highly soluble in these materials, and it attaches itself to a considerable number of *dangling bonds*, characteristic of amorphous semiconductor materials. As a result, the *density of states* in the *band gap* sharply decreases, and there appears the possibility of creating *semiconductor junctions*. Also many *ferrites*, *ferroelectrics*, and *piezoelectrics materials* are semiconducting. Some organic materials also exhibit semiconductor properties, and form a large group of *organic semiconductors*.

Obtaining semiconductor materials. A thorough purification from foreign impurities is a necessary condition for the achievement of highly refined electrophysical characteristics of semiconductor materials. In the cases of Ge and Si this problem is solved by syntheses from their volatile compounds (chlorides, hydrides), after carrying out a thorough purification of these compounds by the methods of fractional distillation, sorption, partial hydrolysis, and special thermal treatments. Chlorides of specific purity are then subjected to high-temperature reduction by hydrogen, after undergoing an intense preliminary purification. Ge and Si are separated from their purified hydrides by thermal decomposition. Semiconductor compounds are obtained using starting components which have passed a thorough purification. The most widespread method for obtaining *monocrystals* of semiconductors is their *crystallization* from the melt using the Czochralski method. Techniques of directed and zone crystallization of the melt in the container, and also floating *zone melting*, are widely used. The growth of monocrystalline epitaxial *films* on different single crystal *substrates* is widespread. For epitaxial growth the crystallization processes from the melt, from the gas phase, and also *molecular beam epitaxy* are employed.

For intentionally varying the electrophysical properties of semiconductor materials they are *doped* by the introduction of specific impurities. These impurities are added either during the course of *monocrystal growth* and epitaxial structures, or in the course of their subsequent *diffusion* or *ion implantation*. Doping may be brought about by the effect of radiation which induces nuclear reactions which form electrically neutral impurities. In order to obtain semiconductor materials with *n*-type electrical conductivity that varies with the concentration of the *current carriers* (electrons) over wide limits one ordinarily uses donor impurities, which form *shallow levels* in the band gap (see *Donor*). For hole or *p*-type semiconductor materials the analogous task is solved by the introduction of acceptor impurities (see *Acceptor*). The atoms of these impurities at room temperature are almost completely ionized so the concentration of the charge carriers coincides almost exactly with the concentration of the introduced impurity.

Applications of semiconductor materials. *Microelectronics* is the fundamental area of application of semiconductor materials. These materials are the basis of modern large-scale and superlarge-scale integrated circuits. At present *integrated circuits* are mainly manufactured on the basis of Si. Further progress in increasing the fast response and reducing the power consumption may result from the development of integrated circuits based on GaAs, InP, and their solid solutions with other $A^{III}B^{V}$ compounds. On a broader scale, semiconductor materials have been prepared for the manufacture of "power" electronic devices (gates, thyristors, powerful transistors). Here Si is also the main material, and further progress in the area of higher operating temperatures should be associated with the use of GaAs, SiC and other broad-band semiconductor materials. The application for solar batteries is promising. Semiconductor materials are widely used for creating sources of radiation (lasers and light emitting diodes), and also photodetecting devices. Semiconductor lasers and photodetectors are especially important components of fiber-optic communication lines. These materials are widely used for the creation of various ultrahigh frequency generators, and of detectors of nuclear radiation. Thermal refrigerators, strain gauges, highly sensitive thermometers, sensors of magnetic fields, modulators and wave guides of infrared radiation, "optical windows", and a series of other devices are fabricated on the basis of these materials.

SEMICONDUCTOR MEMORY

Devices for storage of digital information, fabricated on the basis of semiconductor integrated circuit technology. They are distinguished by their manufacturing technology: bipolar memory, metal–insulator–semiconductor (M–I–S) elements, memory based on *gallium arsenide*; according to their functional designation: random access memory (RAM), read-only memory (ROM); according to the method of information storage: dynamic memory, static memory; according to the method of information sampling: RAM, associative sampling. An integrated circuit contains memory elements and control circuits. A memory storage unit with random sampling consists of a memory element array connected to a system of buses, the control of which facilitates the sampling of the separate elements of the array for the purpose of reading, writing, and storing information. The control circuit transforms the address code into the signal at one of the control buses, and performs the exchange between the memory element and the information buses of the integrated circuit.

Integrated circuits of RAM are the most frequently used, and among these are dynamic integrated circuits of M–I–S *transistors*, which possess high information storage capability. An M–I–S capacitor connected with the array buses by the key M–I–S transistor is their memory element. Information is stored in the form of a capacitor charge, its discharge being refurbished by periodic reading and re-recording, through regeneration. In the integrated static random access memory of M–I–S transistors the symmetric trigger of 4 to 6 transistors serves as the memory element. In terms of information capacity they are four times inferior to dynamic ones, but they have a faster response and low dissipated power (memory with 256 kbyte capacity in the storage mode consumes only a few milliwatts). A bipolar RAM exceeds an M–I–S type RAM in its fast response, but it is inferior in terms of information capacity. Read-only memories are energy-independent, i.e. information is stored with the power supply switched off. In addition, a combined memory may be used, or an energy-independent memory, which involves a complicated memory element combining the functions of random access memory and reprogrammable read-only memory.

In 1997 the leading manufacturers of dynamic RAM devices released memory microcircuits with a capacity of 64 Mbyte; we now have 256 Mbyte capacity using a 0.25 μm technology, and eventually we expect to have integrated circuits of dynamic RAM with 1 Gbyte capacity, with an access time of 30 ns (using a 0.18 μm technology). A change in the industrial standard for the diameter of the silicon plate, on which the crystals of integrated circuits are formed, from 200 to 300 mm, has been proposed, which would reduce the cost of integrated circuits. See also *Information recording in solids*.

SEMICONDUCTOR MICROWAVE DEVICE

See *Microwave semiconductor device*.

SEMICONDUCTOR PHOTODETECTORS

Semiconductor devices for the transformation of information contained in optical radiation to an electrical signal. *Thermal photodetectors* and *photoelectronic semiconductor photodetectors* may be distinguished. In thermal or heat sensitive photodetectors, *bolometers*, the adsorption of electromagnetic waves leads to a rise in temperature which causes an increase of the *current carrier* concentration and of the *electrical conductivity* of the sensitive element. In photoelectronic detectors the adsorption of radiation leads to excitation of the semiconductor electronic subsystem accompanied by an external *photoeffect* (*photoemission photodetectors*), or by an internal photoeffect (*photoelectric semiconductor receivers*). The main types of photoelectric semiconductor receivers are *photoresistors, photodiodes, metal–insulator–semiconductor structures*. The main parameters of semiconductor photodetectors are the spectral range of sensitivity and the threshold sensitivity (minimal recordable optical signal). In order to improve their threshold sensitivity the photodetectors are cooled with the help of different cryogenic devices. The development of an integrated technology leads to the design of integrated photosensitive microcircuits, providing receiving, preamplification, and processing of the signals entering photodetectors. Charge-coupled device (CCD) cameras read the relative brightness of each picture element, while a self-scanning semiconductor imaging device varies the intensities of the incoming light.

SEMICONDUCTOR, PULSED ANNEALING

See *Pulsed annealing of semiconductors*.

SEMICONDUCTORS

Broad class of materials, in which the concentration of mobile *current carriers* is much lower than the concentration of atoms, and may vary over wide limits in response to temperature, exposure to light, and addition of relatively small amounts of impurities. These properties of semiconductors set them apart from *metals*: thus there is a qualitative difference between semiconductors and metals. The difference between semiconductors and *insulators* is much less pronounced and is, in fact, established by convention: it is customary to classify as insulators the materials with resistivity $\geqslant 10^{11}$–10^{12} $\Omega{\cdot}$cm at room temperature (300 K).

According to structure, semiconductors are divided into solid, amorphous, vitreous and liquid types. A special class is represented by *solid solutions* in which atoms of different types are randomly distributed over the crystal lattice sites. Crystalline semiconductors are discussed below.

According to the chemical composition, semiconductors are divided into single-element (elemental) semiconductors (Ge, Si, Se, Te), and two-element, three-element, four-element, etc. compounds. *Organic semiconductors* are also known. It is common practice to classify semiconducting compounds according to the numbers of the groups of the periodic table to which the semiconductor constituent elements belong. For example, $A^{III}B^V$ compounds contain elements of the Groups III and V (GaAs, InSb, etc.). The elements Ge, Si, together with $A^{III}B^V$ compounds and their solid solutions play the most important role in semiconductor electronics. The $A^{II}B^{VI}$ and $A^{II}B^V$ semiconducting compounds are also widely studied (see *Semiconductor materials*).

Semiconductor band structure. Electrical and optical properties of semiconductors are related to the fact that filled electron states are separated

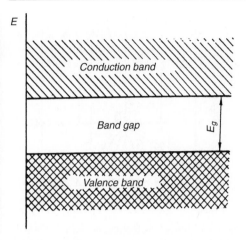

Fig. 1. Energy band diagram of a semiconductor: electron-filled states are designated by shading. The conduction band, band gap, and valence band are indicated.

Table 1. The width of the band gap E_g (eV) for some semiconductors at $T = 300$ K

Material	E_g	Material	E_g
Ge	0.65	InP	1.26
Si	1.10	GaSb	0.67
Se	1.89	GaAs	1.35
InSb	0.17	GaP	2.24
InAs	0.35	AlSb	1.60

from vacant energy states by a forbidden band called a *band gap*, which represents a range of energy that cannot be acquired by electrons (see Fig. 1). Impurity atoms and structural defects can give rise to *local electronic levels* in the band gap, but the number of these local levels is relatively small, and the concept of band gap retains its validity. The lower (almost) completely filled band is called the *valence band*, and the upper (almost) empty band is called the *conduction band* (see *Band theory*). The width of the gap E_g is an important characteristic of a semiconductor, and many of its electronic properties are to a great extent determined by it. This width varies over a wide range, and Table 1 lists the band gap width of several semiconductors in eV at 300 K.

There exist *gapless semiconductors* with $E_g = 0$ (e.g., α-Sn, HgTe, HgSe), and in solid solutions which include these semiconductors the value of E_g may acquire very low values (see *Narrow-gap semiconductors*).

The state of an electron in a semiconductor is characterized by the band number s and the *quasi-momentum* $p = \hbar k$, where k is the wave vector. The band structure is determined by the *dispersion law* $E(p)$. If the valence band is completely full of electrons, then there are no excitations in it. If for some reason one electron is absent in the valence band, then the valence band is said to have an ex-

citation in the form of a positively charged *quasiparticle*, i.e. a *hole*. Current carriers in semiconductors are the electrons in the conduction band (conduction electrons) and the holes in the valence band.

Nondegenerate bands. Conduction bands of typical semiconductors (Ge, Si, $A^{III}B^V$) do not have degenerate states near the minimum of the $E(p)$ function (except for the Kramers degeneracy with respect to spin). Certain semiconductors exhibit a minimum value of $E(p)$ at $p = 0$, i.e. at the center of the *Brillouin zone*. In the local neighborhood of this point $E(p)$ may be expanded in a power series of p. For crystals of cubic symmetry this expansion may be limited to the first two terms, to give

$$E(p) = E_c + \frac{p^2}{2m}, \qquad (1)$$

where E_c is the energy of the bottom of conduction band, and m is a constant with the dimensionality of mass. Eq. (1) is applicable to electrons of moderately high energy; with the quantity m the electron mass or, more generally, the *effective mass*. If, e.g., an electron is placed in a potential field and the characteristic distance over which the field changes is considerable compared with the lattice constant a_0, then the energy levels and wave functions of this electron may be found with the help of the Schrödinger equation. Performing this calculation, one does not have to take any account of the periodic potential generated by the crystal atoms: all one needs to do is to replace the mass of a free electron in vacuo m_0 with an effective mass m (*effective mass method*). Thus, the effective mass $m = \hbar^2/(\partial^2 E/\partial k^2)$, which is a measure of the curvature of $E(k)$, determines the kinetics of electrons at low energies, and several effective mass values are listed in Table 2.

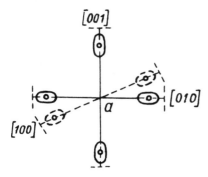

Fig. 2. Arrangement of electron isoenergy surfaces in the Brillouin zone for Si: Brillouin zone edges are indicated by dashed lines. Coordinate axes are p_x along the [100] direction, p_y along [010], and p_z along [001].

Table 2. The effective mass m of conduction electrons for several semiconductors with the minimum of $E(p)$ at the center of the Brillouin zone (at $T = 300$ K)

Material	m/m_0
InSb	0.01
InAs	0.02
InP	0.08
GaSb	0.05
GaAs	0.07

The dispersion law (1) is quadratic and isotropic, and is sometimes called the standard law. *Isoenergetic surfaces* in the *momentum space* $E(p) = $ const in the vicinity of the point $p = 0$ are spheres centered at the point $p = 0$. If the minimum value of $E(p)$ is not located at the center of Brillouin zone but rather at $p \neq 0$, then the constant energy surfaces in its neighborhood are ellipses (see Fig. 2). For this case the effective mass m depends on the direction with respect to crystallographic axes (crystal symmetry axes) even in crystals of cubic symmetry. An energy band in the neighborhood of a minimum is called a *valley*; and a semiconductor with several equivalent minima is called a *many-valley semiconductor*.

Degenerate bands. The valence band of a typical semiconductor (Ge, Si, $A^{III}B^V$) is sixfold de-

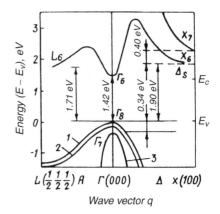

Fig. 3. Energy bands located directly above and directly under the band gap of gallium arsenide ($T \approx 300$ K): 1, heavy hole band; 2, spin orbit split-off band; 3, light hole band.

generate at the point $p = 0$ ignoring the *spin–orbit interaction*. However, the valence band splits into bands, which are twofold and fourfold degenerate at the point $p = 0$ by virtue of the spin–orbit interaction (see Fig. 3). The energy interval Δ between these bands is called the *spin–orbit splitting energy*. For $p \neq 0$ the fourfold degeneracy is removed and two doubly degenerate bands arise: a band of light holes (E^l) and a band of heavy holes (E^h). The dependence of the energies of these bands on the momentum is given by the following expression:

$$E^{l,h}(p)$$

$$= -\frac{1}{2m_0}\left\{ \gamma_1 p^2 \pm \left[4\gamma_2^2 p^4 + 12\left(\gamma_3^2 - \gamma_2^2\right) \right.\right.$$

$$\left.\left. \times \left(p_x^2 p_y^2 + p_y^2 p_z^2 + p_x^2 p_z^2\right)\right]^{1/2}\right\},$$

$$(2)$$

where the plus sign is related to the band of light holes, the minus sign denotes the band of heavy holes; $\gamma_1, \gamma_2, \gamma_3$ are dimensionless parameters (Luttinger parameters, see Table 3) (see *Luttinger Hamiltonian*).

The surfaces $E(p) = $ const, which are described by Eq. (2), do not possess spherical symmetry, they are slightly corrugated. In a number of semiconductors, including Ge, the anisotropy of isoenergetic surfaces is low. Therefore, the bands

Table 3. Luttinger parameters and spin–orbit splitting energy Δ (eV) for Ge and Si

Material	γ_1	γ_2	γ_3	Δ
Si	4.22	0.39	1.44	0.04
Ge	13.35	4.25	5.69	0.29

of light and heavy holes (E^l and E^h) are approximately described by the equations

$$E^l = -\frac{p^2}{2m^l},$$
$$E^h = -\frac{p^2}{2m^h}, \tag{3}$$

where $m^l = m_0(\gamma_1 + 2\gamma)^{-1}$ is the light hole effective mass, $m^h = m_0(\gamma_1 - 2\gamma)^{-1}$ is the heavy hole mass, and $\gamma = (3\gamma_3 - 2\gamma_2)/5$. For Ge we have $m^l = 0.04m_0$, $m^h = 0.3m_0$. In the absence of transitions between bands of heavy and light holes, m^l and m^h describe the dynamics of light and heavy holes, respectively. This pattern of hole bands holds good for Si and Ge crystals, which possess a *center of symmetry*. A more complicated form of the dispersion law is observed for crystals of semiconductors of type $A^{III}B^V$ for small values of p.

Kane model (E.O. Kane, 1956). The kinetic energy $E - E_g$ of an electron or a hole shows a quadratic dependence on the momentum provided that it is small compared with E_g. This condition is violated for narrow-gap semiconductors (small E_g). However, for $E > E_g$ it is also possible to obtain simple equations for the dispersion law, which will hold good when the electron wavelength far exceeds the lattice constant a_0, and the distance to other allowed energy bands exceeds the electron energy. In this case only the overlap of wave functions of valence band electrons and conduction band electrons should be taken into account: the interaction with other bands is insignificant. This approximation is called the Kane model. Except for the quantities E_g and Δ, only one parameter P appears in it, which characterizes the overlap of wave functions. It is expressed in terms of the effective mass of an electron at the bottom of the conduction band E_c. For very small values of

the momentum p, when $E \ll E_g$, the Kane model provides the following parabolic expressions for the energy of electrons $E^e(p)$, the energy of light holes $E^l(p)$, the energy of heavy holes $E^h(p)$, and the energy of the holes of the band that is split off as a result of the spin–orbit interaction $E^{so}(p)$:

$$E^e = E_g + \frac{p^2 P^2}{3\hbar^2}\left(\frac{2}{E_g} + \frac{1}{E_g + \Delta}\right),$$
$$E^l = -\frac{2p^2 P^2}{3\hbar^2 E_g},$$
$$E^h = 0, \tag{4}$$
$$E^{so} = -\Delta - \frac{2p^2 P^2}{3\hbar^2(E_g + \Delta)}.$$

As may be seen from Eq. (4), this approximation does not provide the spectrum of heavy holes. If $E_g \ll \Delta$, then by comparing Eq. (4) with Eqs. (1) and (3) we find that an electron and a light hole have identical masses m

$$m = \frac{3\hbar^2}{4P^2}E_g. \tag{5}$$

If therefore $p \ll (2m\Delta)^{1/2}$, then the energies of both the electrons and the light holes are given by the following equations:

$$E^e = \frac{E_g}{2} + \frac{E_g}{2}\left(1 + 4\frac{p^2}{2mE_g}\right)^{1/2},$$
$$E^l = \frac{E_g}{2} - \frac{E_g}{2}\left(1 + 4\frac{p^2}{2mE_g}\right)^{1/2}. \tag{6}$$

Eqs. (6) show the energies of electrons and of light holes deviating from quadratic behaviour when the kinetic energy of an electron or a hole is of the order of E_g.

Impurities and defects in semiconductors. Electrically active and electrically inactive impurities are recognized. Electrically active ones can acquire a positive or negative charge which is compensated by the creation of an electron in the conduction band or a hole in the valence band, respectively. Electrically inactive impurities remain neutral and have only a slight effect on the electrical properties of a semiconductor. As a rule, electrical activity is related to the fact that the *impurity atom* and the atom that is substituted for in the process of introducing an impurity, often have different valences (i.e. different numbers of valence electrons), and the order of arrangement of nearest neighbors around the foreign atom is determined by the crystal lattice of the doped semiconductor. Thus, e.g., when a Group V element enters the lattice of Si, which exhibits tetrahedral symmetry, it rearranges its valence electrons in such a way, that four of them make up a stable tetrahedral configuration, and the fifth one becomes relatively loosely bound to the impurity atom. As a first approximation, assume that this fifth electron is an "extra" one experiencing only the force of electrostatic attraction to the impurity ion, which is weakened by the dielectric constant of the lattice (see *Electrically active extended defects*). In the simplest case of a nondegenerate standard band the equation of motion of the extra electron appears identical to that of an electron of a hydrogen atom. The binding energy is given by the expression

$$E_0 = \frac{me^4}{2\varepsilon^2\hbar^2} = \frac{m_0 e^4}{2\hbar^2}\left(\frac{m}{m_0}\right)\frac{1}{\varepsilon^2}, \qquad (7)$$

where e is the electron charge, and ε is the lattice dielectric constant. For typical semiconductor values ($m/m_0 = 0.1$, $\varepsilon = 12$) E_0 is smaller than the hydrogen atom binding energy (13.6 eV) by a factor of approximately 10^3. The thermal motion easily strips this electron off the impurity atom, after which it is available to take part in the conduction of electric current. Such impurity atoms are called *donors*.

On being added to a lattice of tetrahedral symmetry, Group III elements capture an electron from the valence band and form a stable tetrahedral configuration with its help. A hole, which is thereby created in the valence band, becomes attracted to the negatively charged impurity atom, and remains in the bound (localized) state at low temperatures. In the case of a standard band the binding energy of a hole is also defined by Eq. (7), where m is the hole effective mass. When this hole breaks away from the impurity atom it can also take part in the conduction of current. Impurity atoms, which produce holes, are called *acceptors* (acceptor impurity). At interatomic distances the potential created by an impurity ion differs substantially from that of a point charge potential, and depends on the chemical nature of the impurity. This short-range component of the impurity

potential is responsible for an additional (with respect to Eq. (7)) shift of the impurity level, which is called the *chemical shift*. By virtue of the chemical shift the levels of various impurities differ from each other. This difference is much sharper for *s*-states than for *p*-states, because the wave function of a *p*-state vanishes at the impurity center. If there are several equivalent extrema in the band (e.g., the band consists of several equivalent ellipsoids) then the impurity levels exhibit an additional degeneracy, the degeneracy multiplicity being equal to the number of equivalent ellipsoids. This degeneracy (e.g., fourfold in Ge, sixfold in Si, see Fig. 2) is partially removed by the short-range component of the impurity potential: the ground impurity level of Ge splits into three levels, and the ground impurity level of Si splits into two levels. In the presence of degeneracy, valence band acceptor states have characteristic features. If the value of the spin–orbit splitting Δ is large compared with the acceptor binding energy E_0, then the split-off degenerate band can be neglected. Ignoring corrugation of isoenergetic surfaces, the acceptor states are rated in order of their values of total angular momentum J and its projection on the quantization axis. A fourfold degenerate state with $J = 3/2$ turns out to be the ground state.

In many semiconductors $m^h \gg m^l$. In this case, the wave function of an impurity electron exhibits two different scales, which are the lengths of the de Broglie waves for particles with equal energy but different effective masses. As the distance to the impurity center increases, the wave function is determined first by the smaller scale, which corresponds to the heavy holes, and then by the larger scale, which corresponds to the light holes. The binding energy, which is determined by the heavy hole mass, may be derived from Eq. (7) by replacing m with m^h and adding the multiplier 4/9.

Impurity states, whose binding energy E_0 is small compared to E_g, are called *shallow levels*. The *deep levels* arise, as a rule, if the main contribution to the binding energy is made not by electrostatic attraction, which is weakened by the dielectric constant ε, but by the short-range potential, which is determined by the chemical nature of the impurity. It may be said that shallow donor states are split off from the valence band, and shallow acceptor states are split off from the conduction band. Deep states belong to a comparable extent to both the valence and the conduction bands, and may partake the nature of both donor and acceptor. The relationship between the concentrations of electrons and holes may vary depending on the quantity and nature of the impurities. Those (*e* or *h*) which are most numerous are called *majority current carriers*, the remaining ones being called *minority current carriers*. The addition of impurities provides semiconductors with specified properties (see *Doping*). If the impurity atom and the substituted atom belong the same group of the periodic system (isovalent substitution), then this impurity atom generally does not form any localized electron state (see *Isovalent impurities*), i.e. it is electrically inactive. These inactive impurities may enter the lattice in large concentrations and form solid solutions. The arrangement of lattice sites in solid solutions exhibits long-range order, but substitutional atoms are randomly distributed over these sites. *Solid solutions* are of vital importance in semiconductor electronics, because it is possible to vary their E_g values by varying their composition. It is therefore feasible to obtain a number of crystals, in which the value of E_g varies continuously, and even crystals in which E_g varies from one point to another. However, solid solutions are *disordered solids*. Their composition is spatially uneven, which results in the smearing of band edges, and the presence of variations in *current carrier scattering* (see also *Heterojunction, Heterostructure*). Lattice defects in semiconductors also may be electrically active or inactive. An important role is played by *vacancies, interstitial atoms, dislocations*.

The behavior exhibited by impurities in noncrystalline and *liquid semiconductors* differs from the behavior of impurities in their crystalline counterparts. The absence of a regular crystal structure results in the fact that an impurity atom, which has a valence different from that of the substituted atom, may saturate all its valence bonds is such a way that it will be disadvantageous for it to accept an extra electron, or to lose one of its own electrons. As a result, the impurity atom turns out to be electrically inactive. This circumstance prevents the use of doping to change the type of conduction, which is necessary, e.g., for the creation of *semiconductor junctions*. Certain *amorphous semiconductors* change their properties under doping, but the effect of doping is considerably weaker than in the case of crystalline semiconductors. The doping sensitivity of the amorphous bodies may be increased by a processing treatment.

Statistics of electrons in semiconductors. In the state of thermodynamic equilibrium the concentrations of electrons and holes are uniquely determined by the temperature, by the concentration of electrically active impurities, and by the band structure parameters. For the calculation of concentrations of electrons and holes an electron may be located in the conduction band, at a donor level, at an acceptor level, or it may leave the valence band (causing the generation of holes in it) because of thermal transfer, or some other factor.

Electrons obey *Fermi–Dirac statistics*, and the distribution of the electrons in energy levels E is described by the Fermi function. The parameters of this function are the temperature T and the chemical potential $\mu = E_F$ (sometimes called the *Fermi level*). The probability of filling a level of energy E is equal to

$$\frac{k_B T}{1 + \exp(E - E_F)}.$$

If the concentration of impurities is not too high, then the Fermi level is in the gap (see Fig. 4). The behavior of mobile electrons and holes obeys the laws of classical statistics (see *Maxwell distribution*). The concentration of electrons in the conduction band n and the concentration of holes in

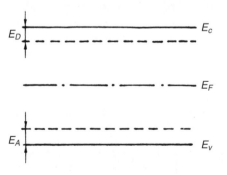

Fig. 4. Impurity levels (dashed lines) in a semiconductor: E_c is the bottom of the conduction band, E_v is the top of the valence band, E_D, E_A are the binding energies for donors and acceptors, respectively, and E_F (dash-dotted line) is the Fermi level.

the valence band p are given by the following relations (E_F is measured from the bottom of the conduction band):

$$n = N_c \exp\left(\frac{E_F}{k_B T}\right), \qquad (8)$$

$$p = N_v \exp\left(-\frac{E_g + E_F}{k_B T}\right), \qquad (9)$$

where N_c and N_v are respectively characteristic concentrations of electrons and holes, which are determined by the spectrum of electrons and holes. For the standard spectrum we have (m^e and m^h are the effective masses of electrons and holes, respectively):

$$N_c = \frac{(2\pi m^e k_B T)^{3/2}}{4\pi^3 \hbar^3},$$

$$N_v = \frac{(2\pi m^h k_B T)^{3/2}}{4\pi^3 \hbar^3}. \qquad (10)$$

In the case of ellipsoidal isoenergetic surfaces m^e should be replaced by $(m_1 m_2 m_3)^{1/3}$, where the effective masses m_1, m_2, m_3 correspond to the principal axes of the ellipsoid. In the case of a degenerate valence band the equations for N_c and N_v acquire a more complicated form; however, if the masses of the heavy holes far exceed the masses of light holes then Eq. (10) may be used (m^h in it must be replaced by the heavy hole mass).

The concentration of electrons, which are located at donor levels, is given by the equations

$$n_D = N_D \left\{1 + g_D^{-1} \exp\left(-\frac{E_D + E_F}{k_B T}\right)\right\}^{-1}, \qquad (11)$$

where g_D is the degeneracy multiplicity of the lowest donor band (taking into account the spin degeneracy); N_D is the concentration of donors; E_D is the donor binding energy ($E_D > 0$). The concentration of holes, which are trapped at the acceptor levels, i.e. the concentration of neutral acceptors, is given by the equation

$$p_A = N_A \left\{1 + g_A^{-1} \exp\left(-\frac{E_A - E_F - E_g}{k_B T}\right)\right\}^{-1}, \qquad (12)$$

where g_A is the degeneracy multiplicity of the acceptor level; N_A is the concentration of acceptors; E_A is the acceptor binding energy ($E_A > 0$). The Fermi level E_F is determined from the *electrical*

neutrality condition, which states that the concentration of negative charges (electrons and charged donors) must be equal to the concentration of positive charges (holes and negative acceptors):

$$n + N_A - p_A = p + N_D - n_D. \qquad (13)$$

In order to determine the concentrations of electrons n and holes p, one should substitute Eqs. (8)–(12) into Eq. (13), solve the resulting expression for E_F, and then insert the value of E_F into Eqs. (8) and (9). It is seen from Eqs. (8) and (9) that the product $n_i^2 = np$ of the concentrations of electrons n and holes p is independent of the concentrations of impurities:

$$np = n_i^2 = N_c N_v \exp\left(-\frac{E_g}{k_B T}\right). \qquad (14)$$

In the standard case

$$n_i = \frac{(2\pi k_B T)^{3/2}}{4\pi^3 \hbar^3} (m^e m^h)^{3/4}$$
$$\times \exp\left(-\frac{E_g}{2k_B T}\right). \qquad (15)$$

Intrinsic and extrinsic semiconductors. A semiconductor with a negligibly small impurity content is called an *intrinsic (pure) semiconductor*. It contains an equal number of electrons and holes $n = p = n_i$. These electrons and holes were generated by thermal electron transfer from the valence band to the conduction band. The Fermi level of the intrinsic semiconductor is located approximately at the center of the band gap and is defined by the following equation:

$$E_F = -\frac{E_g}{2} + \frac{3}{4} k_B T \ln\left(\frac{m^h}{m^e}\right). \qquad (16)$$

At sufficiently high temperatures, a semiconductor may be intrinsic even if it contains large numbers of impurities. The necessary condition is $n_i \gg N_D$ and $n_i \gg N_A$. The temperature region over which a given semiconductor may be considered intrinsic is determined by the width of the gap E_g, the concentration of impurities, and spectrum of electrons and holes. In Ge $n_i = 2 \cdot 10^{13}$ cm^{-3}, in Si $n_i = 1.5 \cdot 10^{10}$ cm^{-3} (at 300 K).

The semiconductor is called an *extrinsic (impurity) semiconductor* if N_D or N_A far exceeds n_i. The main property of an extrinsic semiconductor is the sharp difference between the concentrations of electrons and holes. If the dominant carriers are electrons, then the material is an *n-type semiconductor*. If most of the current is carried by holes, then it is a *p-type semiconductor*. The prevailing impurities are donors in the first case, and acceptors in the second case.

If the semiconductor contains only donor impurities and the temperature is high enough for the ionization of all the impurities, but at the same time low enough to neglect the ejection of electrons from the valence band ($n_i \ll N_D$), then the concentration of electrons $n \approx N_D$, and E_F is defined by the following equation:

$$E_F = k_B T \ln\left(\frac{N_D}{N_c}\right). \qquad (17)$$

For $N_D < N_c$ the Fermi level is located somewhat below the bottom of conduction band E_c. In this case the concentration of holes is negligible compared to concentration of electrons. In the case of acceptor impurities an analogous temperature interval is observed: the concentration of electrons over this interval is negligibly small and E_F is in the neighborhood of E_v:

$$E_F = -E_g - k_B T \ln\left(\frac{N_A}{N_v}\right). \qquad (18)$$

If there are both donors and acceptors and $N_D \gg N_A$, every acceptor traps an electron from a donor. Then for the total ionization of donors the concentration of electrons $n = N_D - N_A$. In the same way, at $N_A \gg N_D$, $p = N_A - N_D$. Thus, the impurities compensate each other. Therefore, semiconductors, which contain both donor and acceptor impurities, are called *compensated semiconductors*; the ratio between the concentration of the minor impurity and the concentration of the major impurity is called the *degree of compensation* (K); $0 \leqslant K \leqslant 1$.

At sufficiently low temperatures there will be only a small density of electrons in the conduction band of an *n*-type semiconductor. In this case, the concentration of electrons in the conduction band depends exponentially on the temperature:

$$n = \frac{1}{\sqrt{2}} (N_D N_c)^{1/2} \exp\left(\frac{E_D}{2k_B T}\right). \qquad (19)$$

Eq. (19) holds only for a weakly compensated semiconductor. E_F is located approximately in the middle of the gap between the donor level and E_c:

$$E_F = -\frac{E_D}{2} + \frac{1}{2} k_B T \ln \frac{N_D}{2N_c}. \qquad (20)$$

There exist analogous equations which are valid for *p*-type semiconductors. In this case the Fermi level E_F is located between the acceptor level and E_v, and the concentration of holes depends exponentially on the temperature T. In compensated semiconductors of *n*-type at low temperatures E_F practically coincides with the donor level, and the dependence $N(T)$ at $n \ll N_D$ has the form

$$n = N_c \frac{N_D - N_A}{2N_A} \exp\left(-\frac{E_D}{k_B T}\right). \qquad (21)$$

Fig. 5 schematically shows the dependence of $\ln(1/n)$ on $1/T$ in *n*-type semiconductors. The steep slope region I corresponds to the semiconductor itself. According to Eq. (15) the activation energy characterizing the slope of the straight line in this region is $E_g/2$. In region II all the donors are ionized and $n = N_D - N_A$. In the lowest-temperature region III almost all the electrons are located at the impurities and, according to Eq. (21), the activation energy is equal to E_D. In the weakly compensated semiconductors where $K \ll 1$, between regions III and II there is a region where, according to Eq. (19), the activation energy equals $E_D/2$.

Thus, the concentration of mobile electrons and holes in semiconductors decreases exponentially with the temperature, vanishing at $T = 0$ (see Fig. 5). This phenomenon is called the *"freezing-out" of carriers*, and can be explained by the localization of carriers at impurities. This phenomenon disappears at a sufficiently high impurity concentration.

Heavily doped semiconductors. In semiconductors with a sufficiently high impurity concentration there is some residual concentration of mobile electrons (or holes) approximately equal to the concentration of the impurities, and weakly depending on T at low temperatures. This leads to the appearance of a residual *electrical conductivity* of a metallic type, i.e. weakly dependent on T. For instance, in *n*-Si with the addition of P the residual

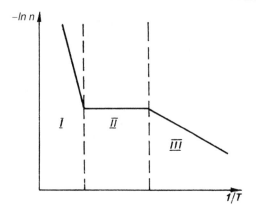

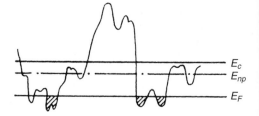

Fig. 6. Energy level diagram of a compensated semiconductor: the irregular curve characterizes the bending of the bottom of the conduction band, the upper solid line labeled E_c is the bottom of the conduction band in the absence of an impurity potential, the lower solid line labeled E_F is the Fermi level, and the dash-dotted line E_{np} is the level of percolation. The shaded areas are occupied by electrons (electron droplets).

Fig. 5. Dependence of the logarithm of the electron concentration, $\ln(1/n)$, on $1/T$ in an n-type semiconductor. In region I the slope of the line corresponds to an activation energy of $E_g/2$. In region II all the donors and acceptors are ionized, so n is constant, i.e. $n = N_D - N_A$. In region III almost all the electrons of a compensated semiconductor are situated at impurities with the activation energy equal to E_D.

conductivity is observed at $N_D > 3.7 \cdot 10^{18}$ cm^{-3}, and in n-Ge with the addition of Sb it is observed at $N_D > 1.5 \cdot 10^{18}$ cm^{-3}.

The transition to metallic electrical conduction is explained by the overlapping of the adjacent impurity levels, which causes an *impurity band* to form and eventually overlap with the conduction band. The critical concentration N_{cr} at which the metallic-type electrical conductivity is displayed can, as a rule, be described by a correlation of the following type:

$$N_{cr}a^3 \approx 0.02, \qquad (22)$$

where a is the radius of the impurity state, corresponding to the particular type of impurity under conditions of weak doping. At the concentrations of donors N_D satisfying the inequality $N_D a^3 > 1$, the electron gas at $T = 0$ is degenerate. The Fermi level is within the conduction band, and for the usual case is expressed by the formula

$$E_F = \frac{(3\pi^2)^{2/3}\hbar n^{2/3}}{2m}, \qquad (23)$$

whereas in the absence of compensation ($N_A = 0$), $n = N_D$. At $N_D a^3 \gg 1$ the Fermi energy E_F is higher than the interaction energy of electrons

with impurities and with each other, and hence one can consider the electron gas to be ideal. Thus, the statistics of electrons in *highly-doped semiconductors* are the same as in metals, although the concentration of carriers is much lower than in the metallic state. At sufficiently high T ($k_B T \gg E_F$) a Fermi degenerate electron gas disappears, it becomes a Maxwell gas, and E_F is given by Eq. (17).

If the n-type semiconductor also has acceptors, then it is necessary to substitute $n = N_D - N_A$ into Eq. (23). At exact compensation, when N_D and N_A are sufficiently close, the electron gas is not ideal. Electrons are located in the field with a random potential, created by donors and acceptors. The random potential may be considered as the bending of the bottom of the conduction band E_c. At a very exact compensation the characteristic amplitude of the random potential becomes higher than E_F determined by Eq. (23). In this case electrons are only in the deepest places of the potential relief, thus forming droplets, isolated from one another (see Fig. 6). At $T = 0$ such a system becomes an insulator. The electron flow results from the thermal excitation of electrons to the so-called *percolation level* (see *Percolation theory*).

Transport processes. Conduction. Besides electrons, ions also may be charge carriers, but ionic *electrical conductivity* in typical semiconductors is low (*ionic superconductors* are an exception). In semiconductors three main mechanisms of electron transfer are fulfilled: band transport (electron motion associated with variation of energy within the limits of a single, allowed, energy band); jump or hopping transport along localized states (see *Hopping conductivity*), and polaron transport (see *Polaron*). The electrical conductivity of semiconductors varies over a very wide range with variations of temperature and impurity concentration. The variation occurs due to changes of the free carrier concentration n, as well as due to the nature of their dissipation. In order to separate these two factors, it is convenient to write the electrical conductivity σ in the form

$$\sigma = en\mu, \qquad (24)$$

where μ is the *mobility of current carriers*, which in a nondegenerate semiconductor does not depend on n (or only weakly depends on it). Mobility is defined by the ratio of the drift velocity v_{drift} of the carriers under the effect of an electric field to the field intensity E:

$$\mu = \frac{v_{\text{drift}}}{E}. \qquad (25)$$

The value of μ and its temperature dependence are determined by the state of the carriers (band, impurity or polaron states), and by the mechanism of their scattering. For band conduction of semiconductors high values of μ are inherent. Thus in the weakly doped n-Ge at $T = 77$ K, $\mu = 10^4$ cm$^2 \cdot$V$^{-1} \cdot$s^{-1}. If $\mu < 1$ cm$^2 \cdot$V$^{-1} \cdot$s^{-1}, it usually means that there is a polaron or hopping mechanism of electric conduction.

An electron whose energy lies in the allowed band of an ideal crystal lattice may move without scattering, thus preserving its quasi-momentum. Scattering is caused by deviations from the ideal periodic structure associated with the thermal vibrations of atoms (scattering by *phonons*), and by impurities and structural defects. Besides this, the carriers may scatter off one another (see *Current carrier scattering*).

The most important mechanisms which determine the mobility of carriers within the range $T \leqslant 300$ K are scattering by acoustic phonons and by charged impurities. In nondegenerate semiconductors: upon scattering by acoustic phonons $\mu \sim T^{-3/2}$, and upon scattering by charged impurities $\mu \sim T^{3/2}$. At the higher temperatures one mechanism prevails, and at lower temperatures the other dominates, so the dependence $\mu(T)$ has a maximum. If the energy of the thermal motion of carriers $(k_B T)$ is comparable to the energy of an optical phonon or exceeds it, then the scattering by optical phonons, especially strong in polar semiconductors, plays an important role. In solid solutions, the scattering from the variations of composition, for which $\mu \sim T^{-1/2}$, is important. In heavily doped semiconductors at low temperatures the scattering by charged impurities, screened by free electrons, plays the main role. In this case also the mobility μ and the electric conductivity σ depend weakly on T, and it is possible to talk about the electrical conductivity $\sigma(0)$, resulting from the extrapolation of the function $\sigma(T)$ to $T = 0$ K. For a concentration of impurities N smaller than N_{cr}, the low-temperature electrical conductivity has an activated character, because the concentration of free carriers decreases exponentially with decreasing temperature. At $N > N_{\text{cr}}$, $\sigma(0) = 0$, which means that electrons are localized at the impurities. At low impurity concentrations the separate impurities are centers of localization, and at concentrations approaching N_{cr} the region of electron localization includes many impurity centers. According to the theory the value of $\sigma(0)$, as a function of N, vanishes at $N \rightarrow N_{\text{cr}}$ via the power law

$$\sigma(0) \sim (N - N_{\text{cr}})^t, \qquad (26)$$

where $t > 0$ is a number called the *critical index*. The transition from electrical conductivity of the metallic type to conductivity of the activation type is called the *Mott metal–insulator transition*.

Conduction in a strong electric field. Deviations from Ohm's law in a strong electric field in semiconductors is mainly associated with heating of the carrier gas. Energy obtained by the carriers from the field is transferred upon collision to the phonons and leads to the release of Joule heat. However, the power obtained from the field may be so high that the carriers do not succeed in transferring it to the phonons, so their temperature appears to be higher than the lattice temperature. In this case it is possible to speak about *hot carriers* (see *Hot electrons*). The heating occurs if the amount of energy obtained by a carrier from the field during the period between collisions exceeds the energy transferred to the phonon during the course of one collision.

If the carrier temperature depends on the electric field then Ohm's law is not obeyed, and the shape of the semiconductor *current–voltage characteristics* is determined by several factors. The heated carriers, e.g., may be located in another region of the energy spectrum, and thereby sharply change their mobility. This may lead to instability, exemplified by the *Gunn effect* (see also *Solid-state plasma*). *Avalanche breakdown* is another type of instability. Electrons in the electric field acquire a kinetic energy comparable to the gap width E_g and thereby knock electrons out of the valence band to the conduction band. In their turn they are accelerated by the field and knock out new electrons, etc. The so-called *impurity breakdown* (see *Frenkel effect*), which appears in a significantly weaker field, is specific for semiconductors. In this case the electrons are not knocked out from the valence band, but from the impurity levels.

Galvanomagnetic phenomena in semiconductors provide ways to experimentally investigate band structure parameters and impurity composition. Measuring the *Hall constant* R_H in a weak magnetic field is the simplest method of obtaining the charge carrier sign and concentration. For a single type of carrier we have

$$R_H = \frac{r}{en}, \tag{27}$$

where r is a coefficient depending on the mechanism of the carrier scattering. If both electrons and holes serve as carriers, and it is possible to neglect their interaction, then the electrical conductivity is given by

$$\sigma = en\mu_e + ep\mu_h, \tag{28}$$

where μ_e, μ_h are the mobilities of electrons and holes, respectively. The Hall constant is associated with μ_e and μ_h by the expression

$$R_H = \frac{e^2(p\mu_h^2 - n\mu_e^2)}{\sigma^2}. \tag{29}$$

The sign of R_H in semiconductors of n- and p-types is different. It is possible to determine the carrier concentration by measuring the Hall effect in a strong magnetic field, when the *cyclotron frequency* of the carriers is high compared to the frequency of collision for electrons as well as for

holes. A useful expression for the Hall constant R_H is

$$R_H = \frac{1}{e(p-n)}. \tag{30}$$

The so-called *quantum Hall effect* plays a specific role. It appears in two dimensional systems, and occurs, e.g., in the inversion layer of an *metal–insulator–semiconductor structure*, or in a *quantum well*. If a strong magnetic field is directed normal to the layer the dependence of the Hall electrical conductivity σ_H on the magnetic field contains "steps" described by the formula

$$\sigma(H) = \frac{\nu e^2}{h}, \tag{31}$$

where ν can have integer or fractional values. The accuracy to which Eq. (31) is satisfied is so high that the quantum Hall effect constitutes a method for measuring and checking fundamental constants. Besides, measurements of *negative magnetoresistance* in a weak magnetic field may play an important role. A magnetic field destroys the quantum interference of electron states, and thus increases the electrical conductivity of the system (see *Galvanomagnetic effects*, *Weak localization*).

Thermoelectric effects in semiconductors are important both as a means of defining semiconductor parameters, and also for practical applications. The *thermal electromotive force* of semiconductors is significantly higher than that of metals. That of a degenerate electron gas is of the order of $(k/e)(k_B T/E_F)$, whereas for metals the factor $k_B T/E_F$ is very small. The thermal electromotive force of a nondegenerate semiconductor does not contain this factor, and therefore, it is considerably larger. In connection with this a semiconductor is used for the creation of thermoelements. The measurement of thermoelectric phenomena in a magnetic field (see *Thermomagnetic phenomena*) plays an important role.

Optical properties of semiconductors. Fundamental or *inherent light absorption* in semiconductors is associated with the transition of electrons from the valence band to any of the upper unfilled bands. These transitions may be direct or indirect. Only an electron and a photon (see Fig. 7) take part in a *direct transition*. If the extrema

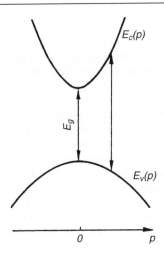

Fig. 7. Two direct transitions for the case when both the bottom of the conduction band and the top of the valence band are located at the point $p = 0$ in the center of the Brillouin zone.

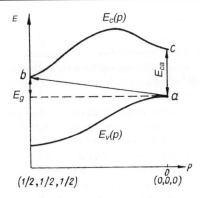

Fig. 8. Direct ($a \rightarrow c$) and indirect ($a \rightarrow b$) transitions involving the "indirect gap" band structure of Ge.

of both bands are at the same point of k-space, then the threshold of direct transitions (*absorption edge*) coincides with E_g. Photons with $\hbar\omega < E_g$ may be adsorbed only through considerably less probable processes, and the transparency is significantly enhanced for $\hbar\omega < E_g$.

Transitions in which, besides an electron and a photon, phonons or impurity centers also take part, are called *indirect transitions*. These indirect transitions are less probable, but they determine the light absorption coefficient when $\hbar\omega > E_g$, when the extrema of the two bands are at different locations in phase space. For instance, for Ge the minimum of the conduction band is at the point b (see Fig. 8) which is the boundary of the Brillouin zone. The maximum of the valence band lies at point a where $p = 0$. The conduction band has a higher minimum at point c at $p = 0$, and the difference of energies between the points c and a is denoted by E_{ca}. Direct transitions are possible only for $\hbar\omega > E_{ca}$. In the region of energies $E_g \leqslant \hbar\omega < E_{ca}$ only indirect transitions can take place (sloping line).

Exciton. The structure of the *intrinsic light absorption edge* band is complicated by the interaction of the electron in the conduction band with the hole in the valence band at the absorption of

a quantum of light. The electron and hole may form a bound state called a *Wannier–Mott exciton*, whereby the energy of the photon is reduced to the value of the binding energy of the exciton. Because the exciton also has excited states, the fundamental absorption edge has a structure reminiscent of the Balmer series of a hydrogen atom, but at much lower energies. For a sufficiently high light intensity an appreciable number of excitons may be formed in the semiconductor. With increasing concentration they may condense and form an *electron–hole liquid*.

Influence of external fields. Size quantization. The structure of the fundamental adsorption edge changes under the influence of applied electric and magnetic fields. An electric field "slopes" the bands and makes possible a tunneling transition for $\hbar\omega < E_g$ (see *Franz–Keldysh effect*). A magnetic field causes the quantization of electron and hole energies, i.e. the appearance of equidistant *Landau levels* (see *Quantizing magnetic field*). The *density of states* near Landau levels grows, and there appear oscillations of the absorption as a function of the light frequency. A maximum of absorption corresponds to transitions between Landau levels. The investigation of these oscillations helps to determine the spectra of electrons and holes.

In addition, the so-called *size quantization*, which appears if the sample is a *thin film* or has small dimensions in all directions, influences the fundamental absorption edge. The corresponding energy levels are also involved in the interband light absorption (see *Quantum size effects*). For $\hbar\omega < E_g$ the *intraband absorption* plays an important role. Quantization in a magnetic field or size quantization may significantly increase the intraband adsorption at frequencies associated with this quantization, which also makes possible the study of the spectrum of current carriers. *Cyclotron resonance* appears to be the most important phenomenon of this type: electrons in a strong constant magnetic field B move along closed trajectories, with their period of revolution depending on the type of energy spectrum of the semiconductor, on the value of magnetic field B, and on its direction relative to crystallographic axes. Experimentally the sample is placed in a high-frequency magnetic field, and a study is made of how the energy absorption depends on the value of B. Resonance appears when the applied magnetic field frequency coincides with the electron cyclotron frequency.

Nonequilibrium charge carriers. Generation of nonequilibrium carriers. The concentration of nonequilibrium electrons and holes is strongly influenced by the sample temperature. Many of the important properties of semiconductors are associated with the nonequilibrium carriers which may be created by various methods, e.g., during excitation by light, and by *injection* through the junctions. Irradiation by light with $\hbar\omega > E_g$ generates electrons and holes which are not in equilibrium. For stationary illumination their concentration is independent of time, and depends on the light intensity and the carrier *lifetime* (in a free state). They are responsible for the phenomenon of *photoconductivity*, or the increase of electrical conductivity brought about by the effect of light. Sometimes the electrical conductivity under illumination is many orders greater than the so-called *dark electrical conductivity*.

When the illumination is turned off, the carrier concentration returns to its equilibrium value during a time of the order of the lifetime of the nonequilibrium carriers. The small inertia of this phenomenon makes possible the design of sensitive devices for recording the light radiation, including that in the IR range.

When current passes through the contact of a semiconductor with a metal or with another semiconductor, the nonequilibrium electrons and holes

fill the region near the contact, so their concentration is determined by the value of the current, and the thickness of the region filled by the nonequilibrium carriers is determined by the distance over which they diffuse during their lifetime (see *Injection*).

Recombination of electrons and holes. The carrier lifetime is determined by recombination processes as a result of which electron–hole pairs disappear, i.e. electrons return from the conduction band to the valence band. *Recombination* of nonequilibrium carriers may be accompanied by the emission of light (*luminescence*). Luminescence may be caused by light (*photoluminescence*) or by an electrical current (*electroluminescence*). The operation of most semiconductor light radiators is based on the phenomenon of electroluminescence. Due to the nonequilibrium carriers, a *population inversion* may appear in semiconductors, when the number of electrons in upper energy levels exceeds that in lower ones. Under such conditions the radiation of light exceeds its absorption, i.e. there occurs light amplification, which is observed only in the so-called active region of a semiconductor. At other places the population inversion is absent, and the absorption of light prevails. If light amplification in the active region is so high that it compensates for the losses in the passive area, as well as the release of light energy to the outside, then light generation takes place. In *semiconductor lasers* the population inversion is usually achieved by injection of nonequilibrium carriers through the junctions.

Due to *nonradiative recombination* the released energy eventually returns to the lattice. There are various mechanisms of nonradiative recombination. At low carrier concentrations recombination through an intermediate state in the gap, formed by an impurity, or by a lattice defect, is the main mechanism. At first the impurity traps a carrier of one sign (e.g., an electron), and then of the other sign (i.e. a hole). As a result of this, the electron and hole disappear, and the impurity or defect returns to its original charge state. *Surface recombination*, which occurs in the presence of *surface states*, has a similar mechanism. At high concentrations the so-called *Auger recombination* (see *Auger effect in semiconductors*) plays an important role, when the energy is transferred to a

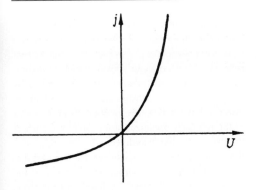

Fig. 9. Current–voltage characteristic of a p–n-junction.

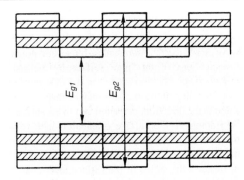

Fig. 10. Energy level diagram associated with a superlattice; minibands are shaded.

third carrier. Auger recombination is caused by the interactions of the electrons. In *light emitting diodes* and *lasers* nonradiative recombination is undesirable.

Semiconductor structures. The simplest semiconductor structure is an electron–hole junction (see *Semiconductor junction*). It is obtained by doping the sample in such a way that donor impurities prevail in one of its regions, and acceptor impurities dominate in the other. The main property of a p–n-junction lies in the fact that the absolute value of the current j which flows through it depends strongly on the polarity of the applied voltage V (see Fig. 9). If the junction is switched to the forward direction, the electrons and holes move toward the boundary of the interface, and recombine in its vicinity. This mechanism provides a relatively high current. If the junction is switched to the reverse direction, then the carriers move away from the interface. In this case the current flows only due to the generation of electron–hole pairs near the boundary, and the current is much lower than that in the forward direction. Thus, a p–n-junction may operate as a rectifier. In addition solar cells, light emitting diodes, lasers, etc., operate on the principle of p–n-junctions (see *Semiconductor devices*). Two p–n-junctions switched towards each other form a *transistor*.

For applications of semiconductor electronics the so-called p–i–n-*structures* are manufactured, where p- and n-regions are separated by an intermediate region with intrinsic conductivity (i), and also there are periodic structures, which consist of a large number of p- and n-regions (p–n–p,

etc.). All the above-listed structures are obtained by doping particular materials with donors and acceptors (see *Doping*). *Heterostructures* and *heterojunctions*, which are contacts formed from different semiconductor materials, are used for the creation of semiconductor lasers and other semiconductor devices.

The methods of molecular beam *epitaxy* and chemical vapor deposition provide different types of heterostructures, e.g., *superlattices* which are formed by the periodic interleaving of semiconductors with different gaps E_g. By this effect there emerge in the conduction and valence bands periodically spaced potential wells and barriers with dimensions which may be of the order of several interatomic distances. As a result, there appear in the conduction band and in the valence band so-called *minibands* separated by forbidden energy gaps (see Fig. 10). As a result of this superlattices possess properties which have important applications in solid-state electronics.

The surface of semiconductors. A surface is understood to be a region several atomic layers thick near the boundary of a semiconductor. A surface possesses some properties that differ from those within the bulk. The presence of the surface violates the translational symmetry of the crystal, and leads to surface states for electrons, and also to special electromagnetic waves (*surface polaritons*), as well as vibrational and *spin waves*. Due to its chemical activity, the surface, as a rule, is covered by a layer of foreign atoms or molecules, adsorbed from the environment. These atoms modify the physical properties of the surface, thus masking the states inherent to a clean surface. The development of superhigh vacuum engineering permits one to obtain and to preserve an *atomically clean surface* for several hours. Investigations of this pure surface by methods of *low-energy electron diffraction* (LEED) showed that crystallographic surfaces may be displaced as a unit in the direction, perpendicular to the surface. Depending on the surface orientation with relation to the crystallographic axes, this displacement may be directed into the semiconductor or out of it. Besides, the atoms of the surface layer change their equilibrium positions in the plane, compared to their positions at the same plane located far below the surface (*surface reconstruction*). On the surface there can appear ordered structures with lower symmetries than in the bulk, or incompletely ordered structures. The former are in thermodynamic equilibrium, with a symmetry that depends on the surface orientation. See also *Semiconductor surface*.

See also *Amorphous semiconductors, Degenerate semiconductor, Doped amorphous semiconductors, Gapless semiconductors, Highly-doped semiconductors, Liquid semiconductors, Magnetic semiconductors, Many-valley semiconductors, Narrow-gap semiconductors, Semimagnetic semiconductor, Variband semiconductors.*

SEMICONDUCTORS AND RADIATION

See *Radiation physics of semiconductors*.

SEMICONDUCTOR SUPERSTRUCTURE

Crystallographic semiconductor structure involving a substitutional solid solution with long-range order (see *Long-range and short-range order*). The superstructure (enlarged unit cell) is formed as a result of a *phase transition* which converts a random substitutional compound into a *crystal structure* of reasonably high symmetry. The resulting superstructures may be of various types and dimensionalities. The type depends on the symmetry and on the composition of the initially disordered substitutional *solid solution*. In triple solid solutions based on the binary structure of ZnS, e.g., $A_x^{III}B_{1-x}^{III}C^V$, semiconductor superstructures are formed in the sublattice which consists of the atoms of two varieties, and they may be of the following types: CuAu I, CuAu II, based upon the ratio of the molecular components 1 : 1,

or of a more complex type such as Cu_3Au or Al_3Ti based upon the ratio 3 : 1. One of the main causes of the *order–disorder phase transition (superstructure)* in a semiconductor substitutional solid solution is the internal (inherent) stresses caused by the presence of the substituted components in the initial *crystal lattice*. The internal strain energies of the regularly organized solid solution are lower than those for the disordered solution, and this is the cause of the phase transition. The greatest difference between the internal strain energy of the disordered and completely ordered states in triple solid solutions with the ZnS structure occurs with $In_{0.75}Ga_{0.25}P$ and $GaSb_{0.75}As_{0.25}$. The presence of external stresses applied to the solid solution may considerably raise the phase transition temperature. In externally unstressed Si_xGe_{1-x} solid solutions with the *diamond* structure the phase transition usually does not occur because of the low rate of self-diffusion of atoms at the transition temperature. However, in externally stressed epitaxial layers of Si_xGe_{1-x} in a *heterostructure* the phase transition temperature exceeds 700 K and a semiconductor superstructure is formed, similar to the double superstructure of the CuPt type, where pairs of (111) planes interlace, with preferential filling by the atoms of Si and Ge (see Fig.). The influence of external stresses on the phase transition temperature is explained by the fact that the lattice

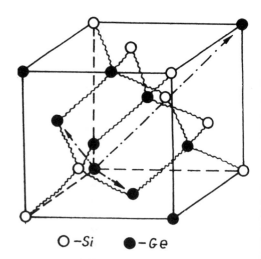

O –Si   ●–Ge

Semiconductor superstructure.

constants for the disordered solution and those of the solution with a superstructure do not coincide.

SEMICONDUCTOR SURFACE

The boundary between a *semiconductor* and other phases (gaseous, liquid, solid). The distinctive property of a semiconductor surface is the existence of *surface electron states* on it. A *physical model of a semiconductor surface* is a model of an electronic spectrum of free and localized states at the surface, providing an explanation for its structural, mechanical, optical, electric, etc., properties, which are observed at the investigation of surface-sensitive phenomena in *semiconductors*. This model may be adapted according to the physicochemical state of the semiconductor surface, and the material in contact with it.

The following types of semiconductor surfaces are recognized: (1) *atomically clean surface*; (2) *real surface*; (3) surface coated with an insulating film; (4) surface of the boundary between a semiconductor and a conducting medium (metal, electrolyte, degenerate semiconductor, etc.); (5) heterojunction; (6) internal semiconductor surfaces: boundaries of bicrystal, of two-dimensional dislocations (*stacking faults*).

An atomically clean surface is achieved through spalling or through bombardment with ions of inert gases in extra-high vacuum. It is characterized by a high concentration of surface electron states (10^{14}–10^{15} cm^{-2}), which cluster into two surface bands. The high concentration of surface electron states results in the stabilization of the Fermi level at the surface near the middle of the forbidden band gap of the semiconductor (see *Fermi level pinning position*). Another characteristic feature of an atomically clean surface is the existence of a *surface superlattice*.

A real surface is often generated by treating with a chemical etching-agent. Such a surface is coated with a thin (2–7 nm thick) amorphous layer of its own oxide.

The donor or acceptor nature of *impurity atoms* at the semiconductor surface is determined by the chemical electron affinity of the particular element with respect to the matrix material. It is therefore possible to make an unambiguous prediction of the type of center for the elements, which exhibit characteristic values of *electron affinity* (e.g., F and Cl

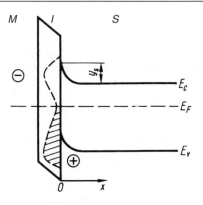

Band bending at a semiconductor surface: M, I and S denote metal, insulator and semiconductor; E_c, E_v and E_F are the energy at the bottom of the conduction band, the energy at the top of the valence band, and the Fermi energy, respectively; y_s is the band bending potential. The dashed curve gives the surface density of states, with occupied states cross hatched.

are always acceptors, whereas hydrogen and alkali elements always act as donors).

Surface levels of *heterostructures* are close to the above-discussed surface structures. They are determined mostly by distortion of *chemical bonds*, complexes of structural defects, and by the presence of foreign atoms at the surface.

Surface electron states are responsible for the properties of so-called *surface phenomena*, such as: *surface electrical conductivity*, *surface capacitance*, *surface recombination*, *work function*, electron emission, adsorption phenomena, etc. The real surface model includes two groups of surface electron states, which differ radically from each other in their times of exchange with bulk *energy bands*: "fast" surface electron states, which are situated at the boundary between the semiconductor and the oxide, and "slow" surface electron states, which are localized at the internal surface of the oxide. A high proportion of long-term conduction relaxation in the *field effect* method may also be related to ionic processes, which take place in the oxide layer. The existence of charged surface electron states is responsible for the generation of a near-surface double layer of space charge (see *Space charge region*), whose thickness is of the order of a Debye *screening radius*, and for the existence of an electric field which is directed normally to the surface. The onset of an electric field in the near-surface region is accompanied by a change in the distribution of an electrostatic potential $\psi(x)$ over the width of the region, and consequently, by a change of the potential energy of electrons at the bands and local levels $\Delta E(x) = e\psi(x)$. In accordance with what has been said above, bands bend near the surface, forming the so-called *band bending* $y(x)$. At the surface $y = y_s = \psi_{x=0} - \psi_0$, where ψ_0 is the potential in the bulk (see Fig.). The potential distribution in the near-surface region, as well as the values of band bending and the surface potential, can be calculated with the aid of the Poisson equation.

A semiconductor surface coated with an insulating film has found wide application in devices, in particular in the production of *integrated circuits*.

There is a thin interfacial layer at the boundary between a semiconductor and a conducting medium: it is called the *Helmholtz layer* if the materials in contact are a semiconductor and an electrolyte; for the case of a contact between a semiconductor and a metal, the interfacial layer consists of chemical substances based on the semiconductor and metal. A *Schottky barrier* often arises at the boundary between a semiconductor and a conducting medium. In a number of cases its existence is combined with the presence of an inversion layer.

SEMIMAGNETIC SEMICONDUCTOR, magnetic mixed semiconductor, dilute magnetic semiconductor

Solid solution of a *magnetic semiconductor* in a nonmagnetic one. Typical semimagnetic semiconductors are compounds of the general formula $A^{II}_{1-x}Mn_x B^{VI}$ or $A^{IV}_{1-x}Mn_x B^{VI}$, which contain manganese ions as magnetic components and exhibit the *crystal lattice* symmetry of $A^{II}B^{VI}$ or $A^{IV}B^{VI}$, where A^{II} is Zn, Co or Hg; A^{IV} is Pb, Sn; B^{VI} is S, Se, Te. These semimagnetic semiconductors have electrical properties close to those of the corresponding *solid solutions* of nonmagnetic *semiconductors*. Semimagnetic semiconductors exhibit an *exchange interaction* of band electrons (holes) with the ensemble of localized electrons at unfilled $3d$-shells of *magnetic ions*. Unlike magnetic semiconductors, the role of the

direct *spin–spin interaction* of Mn ions (over a broad composition range of the magnetic component x) decreases to the point where the *magnetization* of a semimagnetic semiconductor is governed by magnetic fields which can easily be experimentally attained. The difference between semimagnetic semiconductors and semiconductors with low concentrations of magnetic ions is the capability of the former to create a strong internal "molecular" field $B_{MF} = -J \langle S_{loc} \rangle$ (where $\langle S_{loc} \rangle$ is the thermodynamic average of the spin of localized moments of the magnetic component), which acts upon the band electron spin S_{loc} in addition to the external magnetic field B, but considerably exceeds this external field in magnitude (*effect of giant spin splitting*). Since the exchange constant of the *magnetic impurity interaction* for the above-mentioned semimagnetic semiconductors is $J = 0.1$–0.2 eV for the *conduction band* and 0.5–1.5 eV for *valence bands*, at sufficiently high spin polarization of magnetic ions $|\langle S_{loc} \rangle|/S_{loc} \leqslant 1$ the effect of giant spin splitting values is realized beginning with concentrations of Mn, which correspond to $x = 0.001$–0.01. As the field B_{MF} is not associated with the vector potential, the orbital motion of electrons is subject only to the action of the external field B. Hence, the relatively weak field B influences the physical properties of semimagnetic semiconductors more sharply than in the case of ordinary semiconductors.

SEMIMETALS

Materials with an even number of *conduction electrons* per *unit cell*, which are intermediate in their concentration of mobile charge carriers between *metals* and *semiconductors*. Unlike pure (intrinsic) semiconductors, which at zero temperature have a completely filled *valence band* separated from an unoccupied *conduction band* by a band gap, semimetals are characterized by a slight overlap of the energy bands. In the absence of impurities the total number of electrons n_e in a semimetal is equal to the total number of holes n_h. The difference between semimetals and metals is expressed in terms of the conventional concept of "the amount of overlap", so the line of demarcation is not rigidly drawn between these two. Bi, Sb, As (and binary alloys like BiSb) and *carbon* C in its *graphite* modification are classed with semimetals. Ca and Sr are also sometimes classified with semimetals at pressures which give rise to *strain* of several tens of a percent. The crystal structure of Bi, Sb, As is the trigonal (see *Trigonal system*) *space group R3m* (D_{3d}^5) with two atoms per unit cell. The lengths of the edges (nm) are, respectively, 0.474, 0.450, 0.412 nm, while the angles between the edges are $57°14'$, $57°6'$, $54°10'$. Graphite is hexagonal (see *Hexagonal system*), has the space group $P6_3mc$ (C_{6v}^4) with four atoms per unit cell, and the ratio $c/a \approx 2.4$. The mobile charge carrier concentration, overlap energies $\varepsilon_{overlap}$, cyclotron masses m^*, and electronic *specific heat* temperature coefficient γ in semimetals are, in order of magnitude, 10^{17}–10^{20} cm^{-3}, 10^{-2}–10^{-1} eV, 10^{-2}–$10^{-1}m_0$, 10^{-2}–10^{-1} mJ·mole^{-1}·deg^{-2}, respectively (for graphite $\gamma \sim 13$ mJ·mole^{-1}·deg^{-2}, and the *effective mass* of electrons along the c axis $\approx 6.7 m_0$). In combination with the condition $n_e = n_h$ and the presence of several isolated regions of the *Fermi surface*, the above parameter values determine the features of the electronic properties of semimetals: relatively slow growth of electric resistance at low temperatures, strong magnetoresistance, high amplitudes of quantum oscillations of thermodynamic and kinetic characteristics in a magnetic field, realization of diffusive *size effect* and *static skin-effect* (Bi, Sb), transition to ultra-quantum limit in comparatively weak magnetic fields (~ 1 T in Bi), and the opportunity to observe bulk *phonon generation* by the ultrasonic drift of charge carriers (Bi, C).

The best studied representative of the semimetal group is Bi ($n = 3 \cdot 10^{17}$ cm^{-3}, i.e. 10^{-5} electrons per atom; $\varepsilon_{overlap} = 3.7 \cdot 10^{-2}$ eV, $m_{min}^* \approx 10^{-2} m_0$; $\gamma \approx 2.1 \cdot 10^{-2}$ mJ·mole^{-1}·deg^{-2}), which also exhibits a high coefficient of specular *surface scattering* of electrons (~ 1), high values of the g-factor ($\approx 2 \cdot 10^2$), and of the static *dielectric constant* ($\sim 10^2$), and similar values of the characteristic energies in electron, *phonon* and *plasmon* spectra. Because of the unique properties of bismuth, the investigation of this metal has played an important role in the development of solid state physics: bismuth was the first metal, in which strong magnetoresistance, oscillation of physical parameters caused by quantization of the *density of electron states* in a magnetic field, undamped extra-high frequency waves, *cyclotron resonance*, generation of acoustic noise in metals, *quantum size effect*, and lateral electron focusing in a magnetic field were observed. The first thorough investigations of *magnetic surface levels* and geometric oscillations of ultrasound were carried out in bismuth.

SEMIMETAL SUPERCONDUCTOR

See *Superconducting semiconductors and semimetals*.

SHADOW METHOD

See *Optical defectoscopy*.

SHAKE-OFF

A multielectron process of excitation of an electron subsystem of atoms (including those that compose molecules or solids) resulting from a sudden change of the central atom potential with the excitation of an electron to non-bonding states of the continuous energy spectrum. In contrast to a *shake-up*, a shake-off process includes electron transitions to the states of the continuous spectrum. The sudden change of the potential occurs as a result of a change of either the nuclear charge in the case of β-decay, or the effective charge of the external electron shell due to the formation of an electron vacancy at the core or the valence level during *photoelectron emission*, *secondary electron emission*, or *ion–electron emission*. Additional energy is expended for the excitation of these electrons. Therefore, in energy spectra of photo- or secondary electrons, diffuse lines spread from the energy at the ionization limit of a given level to the high energy loss region (lower electron kinetic energy). The probability of shake-off depends on the kind of atom, the structure of its electron shells, and the excitation energy. The shake-off produces a considerable contribution to the excitation process (e.g., its probability is about 20% per absorbed X-ray quantum). The theory of shake-off was developed in 1941 by A.B. Migdal and Ye.L. Fainberg.

SHAKE-UP, monopole excitation

A process of excitation of an electronic subsystem of atoms, which forms part of a molecule or a *solid*, by the sudden variation of the central potential, or upon the transition of an electron into a bound state. Sudden changing of the potential may occur either as a result of a change in the nuclear charge, or of the electronic charge due to the formation of *vacancies* at the valence electron levels under the effect of *photoelectron emission, secondary electron emission,* or *ion–electron emission.* Some additional energy is expended for the shake-up. Therefore, in the energy spectra of photo- and secondary emissions, besides the spectral lines (monochromatic groups of electrons), satellites with less kinetic energies appear. Their position and intensity are characteristics for each set of molecular *orbitals.* The allowed transitions occur between states with the same symmetry. In solids there is high common background noise, and the *characteristic energy losses* of the main spectral line mask the satellites. The satellites manifest themselves most prominently in electronic spectra of *transition metals,* of *rare-earth elements* and their compounds. Deciphering the shake-up spectra can provide a more accurate understanding of the electronic state of the system.

SHALLOW LEVELS

Local electronic levels with binding energy of 15–100 meV formed by substitutional impurities of *valence* different from that of atoms of the crystal lattice. Atoms of lower valence are *acceptors* (Al in Si, Be in GaAs, etc.), while impurities of higher valence are *donors* (P in Si, S in GaAs, etc.). The properties of shallow levels are determined by the long-range nature of the Coulomb potential of an impurity, whose radius of action far exceeds the crystal lattice constant a_0. As a result, details of the potential configuration of the crystal have no noticeable effect on characteristics of shallow levels, and their properties may be described using the approximation of an *effective mass.* In this approximation, the characteristic wave function scale of localized electrons and their binding energy are described by an effective Bohr radius $a_B^* = a_B \varepsilon m / m^*$ and effective Rydberg constant $R^* = R m^* / m \varepsilon^2$ (see *Effective Rydberg*), respectively. Here, R is the Rydberg constant, ε is the relative dielectric constant, m is the free electron mass, and m^* is the effective mass of carriers near the bottom of the band from which the shallow level has split. The shallow level wave

function can be written approximately as

$$\psi_i(r) = F(r) U_{n0}(r),$$

where $U_{n0}(r)$ is the wave function of an electron at the bottom of band n, from which the level is split, $F(r)$ is a "hydrogen-like" envelope described by the Schrödinger equation with a screened Coulomb potential and effective mass m^*. For various bands in crystals Si, Ge and compounds $A^{III}B^V$ and $A^{II}B^{IV}$, the ratio m^*/m varies over a range of 0.01–0.15 eV, and ε is in the range of 10–12. Accordingly, the effective Bohr radius covers up to 10^2 *coordination spheres* around the *impurity atom,* and R^* is $10^{-3} R$. In the approximation of a spherical Coulomb potential of an impurity, the spectrum of shallow levels forms a hydrogen-like series with binding energies $E_i = R^*/n^2$ ($n = 1, 2, \ldots$). In the ground state, the impurity electron exhibits s-symmetry. Deviations from the simple model of hydrogen-like shallow levels are explained by anisotropy of the effective mass, divergences of the impurity potential from the Coulomb potential at short distances from impurity centers, the spatial dependence of the dielectric constant, the presence of several conduction band valleys in Si, Ge, GaP, orbital *degeneracy* of valence bands, and the *spin–orbit interaction* contribution to the spectrum of shallow acceptor levels.

SHAPE MEMORY EFFECT

The tendency of *metals* and *alloys* to regain their original shape on being heated after considerable inelastic deformation (in the martensitic state or in the *martensitic transformation* temperature interval). Depending on the mechanism of the deformation of the *martensite* or of the high-temperature phase, the shape may be recovered either totally or partially. The complete shape restoration is observed when inelastic *strain* takes place through the generation of preferentially oriented martensite crystals, reorientation of already existing martensite crystals, through deformational *twinning of crystals.* When being heated during the process of a reverse martensitic transformation through reversible motion of interphase, intermartensite, and intramartensite boundaries, the high-temperature phase regains its original orientation, which results in the elimination of inelastic deformation. The magnitude of the inelastic deformation reaches saturation at a critical value ε_c. For different alloys, ε_c is determined by the characteristics of crystal geometry and by the ability to change the shape under deformation only by the martensitic mechanism, and not by ordinary *plastic deformation* mechanisms (see *Dislocation model of plastic deformation, Disclination models of plastic deformation*). In the case of alloys, which undergo a thermoelastic martensitic transformation, which is characterized by high mobility of boundaries, the reversible deformation may reach 5–10%. The alloys, which undergo a nonthermoelastic martensitic transformation, exhibit complete shape recovery at $\varepsilon_c \leqslant 1$–3%. A slight shape memory effect may take place in alloys which do not undergo a martensitic transformation. Joint action of the plastic deformation, and the deformation which proceeds through the martensite mechanism, results in a *reversible shape memory effect*: the disposition of metals and alloys to undergo spontaneous changes of shape within the interval of forward and reverse martensitic transformation, as well as in initiating a *reversive shape memory effect*: a twofold spontaneous change of shape. A characteristic feature of the shape memory effect is the essential difference between the force, under the action of which the accumulation of inelastic deformation takes place, and the stresses, which are generated in the material on heating under conditions which counteract the shape recovery (see *Reactive stresses*).

Applications involving alloys allow one to solve various applied problems on a new basis. Alloys are used in engineering and medicine as temperature-sensitive elements, load-bearing members and final-control elements, new kinds of thread-less fasteners, as heat-to-work converters, for suppression of vibration, etc.

SHAPIRO STEPS

Discontinuities on the *current–voltage characteristic* of weakly coupled superconducting junctions which appear as a result of the interaction of electromagnetic radiation (external, or generated within the contact) with the alternating Josephson current. They appear at regions where the current can change within some range (step width) at the following fixed values of voltage (see Fig. 3 in *Josephson effects*):

$$V_n = \frac{n\hbar\omega}{2e},$$

where $n = 1, 2, 3, \ldots$. Processes involving the passage of *Cooper pairs* through a junction with absorption or stimulated emission of quanta of electromagnetic radiation are responsible for the Shapiro steps. One distinguishes steps induced by external electromagnetic radiation, and the inherent Shapiro steps arising from the self-detection of radiation generated within the contact. The positions of the latter are determined by the resonance frequencies of the junction as an electrodynamic system. The height of the steps is determined by the value of the constant magnetic field B applied to the contact, and the controlling *phase velocity* of the traveling density wave of ac Josephson current. The externally induced Shapiro steps are used in metrology for the creation of quantum standards of voltage of high accuracy and stability.

SHARVIN RESISTANCE (after Yu.V. Sharvin)

The resistance of a *point contact* with direct contact for electrical conduction, having a size much less than an average electron *mean free path*. If the contact has a hole shape with an area S in an infinitely thin partition, opaque for electrons and separating two metallic half-spaces, then the Sharvin resistance is as follows:

$$R = \left[\frac{e^2 S S_F}{(2\pi\hbar)^3} \langle \cos\theta \rangle_{v_z > 0} \right]^{-1},$$

where S_F is the projected *Fermi surface* area, $\langle \cos \rangle_{v_z > 0}$ is the cosine of the angle between the contact axis z and the electron velocity $\pm v$, averaged over the Fermi half-surface.

SHEAR

A result of *strain* of a solid produced by opposing tangential stresses. During such a deformation, a parallel displacement of parts of the solid with respect to its other parts takes place; i.e. the surfaces in contact slide against each other. The shear may be either elastic or plastic, depending on the stress applied. See also *Shear modulus, Shear strain*.

SHEAR LINES, slip lines (G. Luders 1854; D.K. Chernov, 1885)

The result of *shear strain*, observed on the surface of solids as a system (or systems) of parallel lines or bands. Carpenter and Elam (1920) were the first to explain them. The lines of strain localization at concentrators (see *Localization of plastic deformation, Stress concentration*) of stress (*Luders–Chernov lines*) involve the development of shear processes. Shear lines are observable with an optical microscope at rather weak magnification. Finer shear lines are identified with the help of transmission *electron microscopy* and the use of *replicas*, as well as by scanning electron microscopy. The presence of shear lines in crystalline bodies is explained by the surfacing of *dislocations* that move in the *slip plane*. Observing shear lines makes it possible to identify *slip systems* in crystals and determine their number, orientation and length, which depend on the crystal lattice type, the temperature and manner of deformation, and also by the stage of deformation that the crystal is under (see *Strain hardening*). At the stage of easy slip (see *Linear hardening of crystals*), a single system of parallel straight shear lines is found on the surface of the crystal. At the later stage of multiple slip the number of systems of shear lines increases, their distribution becomes nonuniform, and the shear lines group into bands (see *Slip bands*). Distortions of shear lines are observed, among which *fault bands* and *bands with secondary shear* are identified. Fault bands are explained as rotation effects (see *Plastic twisting deformation*) within the framework of *disclination models of plastic deformation*. Shear lines in amorphous metal alloys (see *Metallic glasses*) are slightly wavelike, and are located approximately in the plane of action of the maximum tangential stresses. They result from the localization of plastic deformation that develops during the motion, and the growth of the free volume (see *Free volume model*).

SHEAR MODULUS, modulus of rigidity

An *elastic modulus* of isotropic bodies μ, which characterizes the ratio between shear stresses σ_{ik} ($i \neq k$) and the corresponding shear strains u_{ik}: $\sigma_{ik} = 2\mu u_{ik}$, $i \neq k$. In stable systems, $\mu > 0$. The rigidity has the particular values: $\mu = 0.25$ GPa for Al, 1.5 GPa for W, 0.77 GPa for Fe, 0.3 GPa for quartz (see *Shear*).

SHEAR STRAIN

The *strain* of a solid caused by tangential stresses (see *Stress tensor*). The shearing strain γ is determined as $\tan\alpha$, where α is the angle of shear. Through the process of *shearing* an arbitrary rectangle transforms to a (non-rectangular) parallelepiped. In this case α is the rotation angle due to the shearing of the rectangle sides, perpendicular to the shearing stress τ. The elastic shearing strain obeys the *Hooke law* $\tau = \gamma G$, where G is the *shear modulus* (see *Elasticity*). When the stress τ exceeds the *critical shearing stress* then *plastic shearing strain* is observed. The motion of *dislocations* is the elementary act of *plastic deformation* that results from shearing in a crystal.

SHEAR STRAIN INTENSITY

A value equal to twice the square root of the second invariant of the *strain deviator* u_{ik} (see *Elasticity theory invariants*):

$$\Gamma = 2\left[I_2(\text{Dev}\, u)\right]^{1/2}$$
$$= \left(\frac{2}{3}\right)^{1/2}\left[(u_{11} - u_{22})^2 + (u_{11} - u_{33})^2 \right.$$
$$\left. + (u_{22} - u_{33})^2 + 6(u_{12}^2 + u_{23}^2 + u_{13}^2)\right]^{1/2}.$$

It is proportional to the *shear strain* ε in the plane equally inclined to the principal axes of the *strain tensor* (octahedral plane): $\varepsilon = (2/3)^{1/2}\Gamma$.

SHELL

A nonplanar thin layer; that is a body with one dimension (thickness d) small compared to the other two. It can be closed, like the surface of a sphere, closed in one dimension like a cylindrical tube, or open like the surface of a hemisphere. It has two surfaces, an inner and an outer one (or a top and a bottom one). If the shell is not uniform in thickness then its shape or configuration is characterized by a fictitious *middle surface* which is equidistant from the inner and outer ones. The *shell thickness* is the length of a normal to the middle surface stretching between the two outside surfaces. From the geometrical viewpoint, closed shells are defined by the shape of the middle surface, and the law of its variation in thickness. Nonclosed shells are additionally defined by the contour shape (end surface) confining the middle surface at its edges. If the middle surface is a plane the configuration is called a *plate*; plates, like shells, can be of variable thickness.

Shells are basic and indispensable units in various branches of engineering: in rocket, aircraft and ship building, in electrical and chemical and mechanical engineering, in building structures, etc. As structural elements, shells are combined with other structural elements such as *rods*, plates, shells of different shapes, and other three-dimensional bodies. In this connection, there arise problems of studying shells acted upon by other structural elements. Shells can possess a complicated shape of their middle surface and the confining contour, interact with mobile loads and shock waves in liquids and gases, operate at variable temperatures and in electromagnetic fields. Under these actions and situations shells develop static and dynamic stress and *strain* (elastic, plastic, and viscous) (see *Elasticity*, *Plasticity*, *Viscosity*) fields. Shells are made of *metals*, *alloys*, *composite materials*, and others. *Laminated shells* are often used, in certain cases with layers of variable thickness. The internal layers (fillers) are often of a complicated structure and shape. The mechanics of shells should provide the possibility for studying various fields in shells operating under complicated conditions, as indicated above.

The mechanics of shells deals with two essentially different classes of problems. In the first case, the geometric parameters and material properties of shells as well as external loads and fields are given, and it is necessary to determine the distribution of stress, strain, and other fields (including the temperature and electromagnetic fields). In the second case, the shell material properties and external load and fields are given with certain constraints imposed on the geometrical parameters or the internal field magnitudes, and it is necessary to determine the geometrical parameters of the shell (problems of this type are found in design optimization theory). In theoretical research, including the application of numerical methods and computer calculations, the composition of the initial set of equations and corresponding boundary and initial conditions seems to be the most essential feature. The most general factor of this approach is the application of variational principles. Typical for shell theory is the application of two-dimensional equations and corresponding boundary and initial conditions (two-dimensional in the spatial variables correspond-

ing to the middle surface). The transition to two-dimensional equations employs various hypotheses (*Kirchhoff–Love hypothesis*, *Timoshenko hypothesis*, "broken line" hypothesis, etc.). These concern the nature of the distribution of fields over the thickness, algorithms connected with representations of fields over thicknesses as a series in terms of special functions (power series, Legendre polynomials, etc.), or asymptotic methods taking into account the boundary layers. In principle, two-dimensional theories can describe such fields in shells where quantities change noticeably over distances much greater than the shell thickness. This means that two-dimensional theories cannot describe all classes of phenomena occurring near the end surfaces (within the limits of the thickness). Along with the general theory of shells, there are particular or individual theories (the moment-free theory of shells, the theory of shells of revolution, the theory of easy-gradient shells, the theory of shells with a great index variability, and so on).

SHELL MODEL of crystal lattice

A model of atoms in a *crystal* used to describe *crystal lattice vibrations*. In contrast to the model of rigid atoms (ions), the shell model treats their motion as vibrations of a negatively outer charged shell of valence electrons, and a positively charged inner ion core with a quasi-elastic force acting between them. Empirical parameters used in the shell model are determined by comparing the results of calculations of *elastic moduli*, *dielectric constant*, and vibrational frequency spectra with experimental data.

SHIELDING, MAGNETIC

See *Magnetic shielding*.

SHIELDING OF ELECTRIC CHARGE

See *Electric charge screening*, *Screening radius*.

SHIELDING, POLARIZATION

See *Spontaneous polarization shielding*.

SHOCK HARDENING, explosion hardening

Hardening of a material by means a of a *shock wave* initiated by an *explosion*. Explosion hardening depends on the loading scheme, on the previous history of the material, on its *strain*, and especially on the shock wave parameters (pressure, form and duration of a pulse). All the materials at the enhanced pressure region within the shock wave display a particular dependence of explosion hardening on the pressure. There always exists a pressure range of intensive hardening (for metals up to 20–30 GPa), followed by a range with fairly constant hardening. A further increase of pressure (∼50 GPa) induces a sharp reduction of explosion hardening. Single-phase *metals* and *solid solutions* are significantly hardened (to 100–150%) by explosion hardening, whereas polyphased stable *alloys* are hardened less (30–40%). Materials with a developed mosaic structure (see *Mosaic crystals*) are either not hardened or are hardened weakly. Metastable alloys (see *Metastable state*) are hardened and dehardened at the explosion, depending on the character of the *phase transition* caused by the explosion. *Strain hardening* at the explosion depends on the energy of *stacking faults*, and is determined by the type of dislocation structure being formed, as well as by its parameters (see *Dislocations*).

SHOCKLEY LEVELS (W. Shockley, 1939)

Energy levels relating to *surface electron states* in crystals with mainly *covalent bonds*, where an interpretation of the energy spectrum formation in terms of chemical bonds is the most descriptive and fruitful one. Shockley levels are considered to be based on concepts of dangling external hybridized sp^3 atomic orbitals for surface atoms in diamond-type crystals. In this case the energy level of an electron occupying the *dangling bond* proves to be higher than a maximum energy of an electron located in the crystal *valence band*, and falling within the energy gap it is related to the *Tamm levels*. The number of surface atom dangling bonds depends on the crystallographic orientation of the *solid surface*. They influence the *surface level* spectrum, the formation of new bonds, and the reorganization of the surface.

SHOCK, MICRO-

See *Microshock loading*.

SHOCK WAVES in solids

Traveling discontinuities of a particular characteristic of a medium. In the case of an electromagnetic field, the *electromagnetic shock waves* arise due to a discontinuity of field intensity, and in case of discontinuities of density the concept of *density shock wave* is used.

Consider the velocity u of a shock wave that executes uniform rectilinear motion. The total derivative of the velocity with respect to time du/dt in this case equals zero. When stated in terms of Cartesian variables (x is coordinate, t is time, $u = \partial x/\partial t$), this condition assumes the form:

$$\frac{\partial u}{\partial t} + u\frac{\partial u}{\partial x} = 0. \tag{1}$$

The general solution of the Cauchy problem with respect to this equation is given by a relation of the form

$$u = f(x - ut), \tag{2}$$

where $f(x)$ is the velocity profile at the initial instant of time $t = 0$.

The profile contour may be such that the particles in the upper region of the contour start to outrun those in the lower one with increasing values of the argument. As this takes place, the velocity profile becomes steeper, and as the upper particles leave the lower ones behind the so called overlap begins, i.e. the dependence of the velocity (in Cartesian variables) on the coordinate and time is no longer unique. In this case, three different values of velocity correspond to a single value of the coordinate or time, which is physically meaningless. In actuality, a velocity shock wave forms which undergoes uniform rectilinear motion. The coordinate and time of the discontinuity generation are determined from Eq. (2), and the condition that the velocity becomes infinite at the discontinuity point. On evaluating the velocity and time of the discontinuity, the magnitude of the discontinuity may be found from Eq. (2).

It follows from Eq. (1) that the generation of a discontinuity requires nonlinearity of the system and the absence of dissipation, temporal dispersion (see *Temporal dielectric dispersion*), and spatial dispersion (see *Spatial dielectric dispersion*). Due to the smearing caused by dissipation and inversion, the discrete step that corresponds to the original discontinuity becomes flattened into a gently descending slope. This pattern is realized in all media, but the mechanism of initiating the nonlinearity is different in different media. Most typical for solids are electromagnetic and elastic shock waves, as well as hybrid modes.

Electromagnetic shock waves in *insulators* and *semiconductors* are related to the nonlinear dependence of the *dielectric constant* on the electric field intensity, particularly for very high field strengths. The relationship between the dielectric constant and the electric field strength in insulators is often determined by the electrostriction effect (see *Electrostriction*). The nonlinearity of this dependence in semiconductors may arise from the warm-up of current carriers by the electric field, the non-quadratic momentum-dependence of the energy of electrons and holes (see *Electron heating*), and the effect of light (photon) pressure forces on the carrier concentration. Shock waves may also arise in transition layers of semiconductors. Typical of magnetically ordered nonconducting media are shock waves related to the dependence of the *magnetic permeability* on the magnetic field.

Elastic shock waves in semiconductors owe their existence to the nonlinearity of the elasticity equations; these waves are the discontinuities in density and velocity of the material constituting the solid. In order for a shock wave to arise, the time of its formation must be much less than the characteristic times defined by dissipation and dispersion wave processes. Thus, the time of the onset of a shock wave in an insulator should be short in comparison with the time related to dielectric losses, whereas in semiconductors it is to be small compared to the Maxwell relaxation time (see *Maxwell relaxation*) and the current carrier *relaxation time*.

Shock waves have been repeatedly studied experimentally, and are used in engineering. Shock waves in *ferrites* are used for the generation of ultrashort pulses.

SHORT-RANGE ORDER

See *Long-range and short-range order*.

SHOT NOISE

Insignificant random deviations of the anode current of vacuum tube and *semiconductor devices* from its mean value caused by nonuniformity of the electron emission from a cathode and the *injection of current carriers* into a semiconductor. The fluctuation magnitude depends on the working regime of the device. If all the emitted electrons reach the anode, the emission fluctuations are exactly reproduced in the anode current. Otherwise, in the vicinity of the cathode a negative charge cloud can form which possesses the damping capability to smooth the anode current fluctuations.

Shot noise accompanies any processes connected with the formation of charged or neutral particles, such as *photoelectron emission*, *secondary electron emission*, and so on. The shot effect had been predicted analytically by W. Schottky (1918) (see *Noise in semiconductors, Noise in electronic devices*).

SHUBNIKOV–DE HAAS EFFECT
(L.V. Shubnikov, W.J. de Haas, 1930)

An oscillating dependence of the static electrical conductivity of *metals*, *semimetals* and *degenerate semiconductors* on the applied magnetic field B due to its quantizing effect on the electron energy spectrum (see *Quantizing magnetic field*). As is the case with the *de Haas–van Alphen effect* of quantum oscillations, quantum *Shubnikov–de Haas oscillations* are determined by the oscillating dependence on the magnetic field of the density of states $g(\varepsilon_F)$ of electrons with an energy equal to the *Fermi energy* ε_F. Oscillations of electrical conductivity appear due to the fact that along with $g(\varepsilon_F)$, the probability of *electron scattering* per unit time (on impurities, phonons, etc.) oscillates with a dependence on the field B. Periods of Shubnikov–de Haas oscillations with variations in $1/B$ coincide with periods of the de Haas–van Alphen effect. A boundary blurring the Fermi distribution (see *Fermi–Dirac statistics*) and the finiteness of the electron life time τ diminish the magnitude of the Shubnikov–de Haas effect, as well as that of the de Haas–van Alphen effect. At zero temperature in the limit $\omega_c \tau \gg 1$ (ω_c is the characteristic *cyclotron frequency*), the relative amplitude of the Shubnikov–de Haas oscillations is $\propto (e\hbar B/p_F^2)^{1/2}$, where p_F is the characteristic Fermi momemtum. Also other *kinetic coefficients*: thermoelectric power, thermal conductivity, coefficient of sound absorption, impedance, etc., undergo quantum oscillations with magnetic field dependences similar to the Shubnikov–de Haas ones. The Shubnikov–de Haas effect should be distinguished from the oscillations of kinetic coefficients arising at the onset of *magnetic breakdown*. These latter ones are not expressed through $g(\varepsilon_F)$, and have a quantum-interference origin (see *Quantum interference phenomena*).

SHUBNIKOV GROUP OF MAGNETIC SYMMETRY

See *Magnetic symmetry group*.

SHUBNIKOV GROUPS, color groups, black-and-white groups

Point groups and space groups of *antisymmetry* which describe crystals containing atoms with magnetic moments. Making use of the antisymmetry operation called antiidentity increases the number of *Bravais lattices* from 14 to 50, the number of point groups goes from 32 to 122, and then taking into account the additional antisymmetry operation of *antitranslation* expands the number 230 of ordinary *space groups* to 1651 Shubnikov groups. An example of an actual physical situation described by means of Shubnikov groups is the space lattice of spin $S = 1/2$ atoms with *spin* projections that take on two values ("up" and "down") in a magnetically ordered crystal (see *Magnetism*). The antiidentity operation reverses the spin projection directions of the atoms. Crystals with various types of magnetic ordering (ferromagnetism, antiferromagnetism, etc.) belong to different Shubnikov space groups. Shubnikov groups are often called color groups when the antiidentity operation is associated with a reversal of the colors black and white of, e.g., otherwise identical balls of these two colors arranged on a lattice.

SHUBNIKOV PHASE (called mixed state in the Western literature) (L.V. Shubnikov, 1937)

Nonuniform state of a *superconducting alloy* (i.e. *type II superconductor*) existing in the range of fields between the *lower critical field* B_{c1} and the *upper critical field* B_{c2}. The Shubnikov phase represents a *mixed state* with its internal region permeated by quantum vortices of magnetic flux (*Abrikosov vortices*). The vortex density n (vortices per m^2) determines the magnitude of the internal magnetic induction B according to the equation $B = n\Phi_0$, where $\Phi_0 = h/(2e)$ is the *flux quantum*. Theory predicts the dependence of n on the applied field B_{app}, and the shape of the magnetization curve for a homogeneous type II superconductor. In heterogeneous alloys the Abrikosov vortices are pinned at defects (see *Vortex pinning*), giving rise to flux bundles involving a large number ($\sim 10^3$–10^5) of vortices. The properties of such alloys (occasionally called *type III superconductors*) differ from ideal hard superconductors by the great magnitude of their *hysteresis* involving the dependence of B (or the magnetization M) on B_{app}, and by the large values of their *critical currents*, associated with the depinning of the bundles of flux from defects and inhomogeneities.

SIGMA PHASE (σ-phase)

Phase representing an *intermetallic compound* of variable composition with a composite tetragonal structure (see *Tetragonal system*) falling into space group $P4_2/mnm$ (D_{4h}^{14}). This phase usually forms in systems containing atoms of *transition metals*. Its unit cell holds 30 atoms with a stacked arrangement, and the structure is very similar to an hexagonal close-packed lattice. There are five inequivalent positions in the unit cell at which particular atoms are distributed statistically, although some data suggest an ordered arrangement of atoms on several inequivalent positions. An important role in their formation is played by dimensional and electronic factors. A minor difference of atomic size of the components is apparently a necessary, but is not a sufficient requirement for their formation. The temperature and concentration intervals for the appearance of a σ-phase differ for different systems. A σ-phase is characterized by high hardness and brittleness (see *Alloy brittleness*). Their *melting* in technical alloys is undesirable due to the decrease of the *viscosity*. Some σ-phases, e.g., particular concentrations of the alloys MoRe, MoTc, NbRh, TaRh, and WOs, are superconducting.

SILENT LATTICE VIBRATIONS, mute lattice vibrations

Optical vibrations of crystal lattices in several symmetry classes (see *Crystal symmetry*) which, according to *selection rules*, are forbidden as single-phonon processes (see *Direct relaxation process*) when the crystal interacts with incident light. Therefore, silent or mute vibrations do not appear in spectra of IR absorption, or *Raman scattering of light*, i.e. they are neither IR nor Raman active. Silent vibrational frequencies can be determined from *hyper-Raman scattering*, from the spectra of two-phonon absorption, from *light scattering* (see *Phonon optical spectra*), or from experiments of electron, neutron or X-ray scattering by crystals.

SILICA, silicon dioxide, SiO_2

A chemical compound of *silicon* with *oxygen*, present in different crystalline and *amorphous states*. It is the most abundant compound on the Earth. It crystallizes in different modifications: cristobalite, tridymite, *quartz*; artificial crystals include coesite, kytite, stishovite (the latter is the most dense form of crystalline silica with density of 4,350 g/cm^3, Mohs hardness of 9). Most *minerals* are silica crystals containing various impurities providing different coloring, e.g., violet amethyst, green chrysoprase, black morion, yellow citrine, and so on.

Applications make use of heat resistance, thermal stability, mechanical strength (for ceramic protective coatings in space technology); pyroelectric properties (in acoustoelectronics, in radio engineering, as piezoelectric oscillators, including those in quartz clocks). Quartz glass is distinguished by its high thermal and chemical resistance, good transparency in the visible and UV range, so it is applied in optics, including waveguides, also as an *active medium* in lasers with activators introduced in the glass. Technical glass is another application. Good insulating properties (resistivity of quartz is about 10^{18} Ω·cm) promote the use of silica for integrated circuits in microelectronics, where thin (fractions of a micrometer) insulating or protective layers of SiO_2 are prepared by oxidizing pure silicon. Many other applications could be mentioned.

SILICON (Lat. *silicium*), Si

A chemical element of Group IV of the periodic system with atomic number 14 and atomic mass 28.086. The three stable isotopes are ^{28}Si (92.28%), ^{29}Si (4.67%), ^{30}Si (3.05%), other radioactive isotopes are also known. Silicon shows the second largest abundance (after oxygen) in the Earth's crust, constituting 26% of its mass. It is found in nature only in chemical compounds: *silica* or silicates. Crystalline and *amorphous silicon* are prepared synthetically. Outer shell electron configuration is $3s^2 3p^2$. Successive ionization energies are 8.15, 16.34, 33.46, 45.13 eV. *Electron affinity* energy is 1.8 eV, atomic radius 0.134 nm, ionic radii are 0.039 nm for Si^{4+} and 0.198 nm for Si^{4-}. Work function is 4.8 eV. Silicon demonstrates low chemical activity, oxidation state +4 (less common +2 etc.).

Crystalline silicon is a dark-gray brittle substance. It has *diamond*-like face-centered cubic lattice with four nearest neighbors located at equivalent sites at tetrahedron vertices (coordination number 4, distance 0.235 nm), with the total of eight atoms in the unit cell: $a = 0.5431$ nm, space group $Fd3m$ (O_h^5). Density of solid silicon is 2.329 g/cm^3 (at 300 K), that of liquid samples 2.530 g/cm^3 (at melting); $T_{melting} = 1690$ K, $T_{boiling} = 2873$ K. Debye temperature is 628 K. Heat of melting is 46.47 kJ/mole, heat of evaporation is 297.26 kJ/mole, specific heat is 0.8 kJ·kg^{-1}·K^{-1}, thermal conductivity coefficient is 125.6 W·m^{-1}·K^{-1} (at 298 K). The linear thermal expansion coefficient +2.5·10^{-6} K^{-1} (at 120 K) reverses its sign and acquires the value of −0.31·10^{-6} K^{-1} at 100 K. Mohs hardness is 7.0, Brinell hardness is 2.35 GPa/m^2, elastic modulus 106.79 GPa, compressive strength is 92.87 MPa, compressibility factor is 3.314·10^{-12} Pa^{-1}. Silicon becomes notably plastic deformable at 1073 K. It is a diamagnet, with atomic magnetic susceptibility of −0.13·10^{-6}.

Electrical resistivity of high-purity silicon is 2.3·10^3 Ω·m (at 300 K), intrinsic concentration of charge carriers (electrons and holes) is 1.6·10^{10} cm^{-3}. Band gap width is 1.166 eV (at 0 K) to 1.119 eV (at 300 K). It decreases with temperature first as a square law and then linearly (within 250–400 K with the thermal coefficient of −2.4·10^{-4} eV/K). Mobilities of electrons and holes scattered by lattice phonons behave with temperature as $4.0 \cdot 10^9 T^{-2.6}$ and $2.5 \cdot 10^8 T^{-2.3}$, respectively, and at 300 K they equal to 1450 and 500 $m^2 \cdot V^{-1} \cdot s^{-1}$, respectively. In this case diffusion coefficients of electrons and holes are respectively 37.6 cm^2/s and 13 cm^2/s. The lifetime of secondary *current carriers* in high-purity silicon is $2.5 \cdot 10^{-3}$ s, the breakdown electric field strength is $3 \cdot 10^3$ V/cm. Silicon forms a continuous series of solid solutions with *germanium* for any ratio of the components, and the band gap of the solution varies between the values corresponding to pure constituents.

The *conduction band* of silicon involves six equivalent valleys located at points along $\langle 100 \rangle$ directions corresponding to the wave vector $k = 0.8 k_{max}$, where k_{max} corresponds to *Brillouin zone* boundaries in these directions. The constant energy surfaces of the conduction band as a function of the wave vector are ellipsoids of rotation with their symmetry axes directed along $\langle 100 \rangle$. The longitudinal and transverse components of the electron *effective mass* are $0.98m$ and $0.19m$, respectively (m is the free electron mass). The anisotropy coefficient is 5.15, ohmic effective mass is $0.26m$, effective mass of the density of states is $0.33m$. The valence band top is located at the center of the first Brillouin zone; it is doubly degenerate at point $k = 0$. The scalar effective mass of heavy holes is $0.49m$, that of light holes is $0.16m$. Due to *spin–orbit interaction* the energy of the third band is lowered by 0.035 eV, and the effective mass of holes in the split-off band is $0.245m$.

Silicon is optically isotropic, transparent in the IR range with wavelengths λ from 1 to 9 μm. In this range the reflectivity is 0.3, refraction index is 3.42 (for $\lambda = 6$ μm). The dielectric constant is 11.7.

Elements of Groups III and V can be electrically active impurities in silicon, presenting shallow hydrogen-like *acceptors* and *donors*, respectively. *Lithium* may also be a shallow donor. Among deep acceptors, zinc, beryllium, copper, magnesium, thallium, cobalt, nickel, indium are known; deep donors are sulfur, chromium, tellurium, manganese, iron, etc. Amphoteric impurities in silicon are silver, gold, mercury, platinum, tungsten, oxygen. Silicon itself may be present as an amphoteric impurity in a number of $A^{III}B^V$ compounds.

Silicon reacts violently with alkalis. At elevated temperatures in oxygen (above 873 K) it forms the dioxide, which could then be reduced to monoxide. At room temperatures silicon reacts with fluorine (yielding SiF_4), and upon heating with other halogens, yielding $SiCl_4$, $SiBr_4$, etc. At 2273 K it forms silanes (SiH_4, Si_2H_6) by reacting with hydrogen, and at 1373–1573 K it forms nitride Si_3N_4 or carbide SiC as a result of direct reaction with nitrogen or carbon, respectively. When heated with almost all metals, it will form silicides that often violate classical valence rules (MeSi, $MeSi_2$, and so on).

Heat treatment of silicon may induce thermal defects of different kinds, including electrically active ones. Irradiation with γ-quanta, neutrons, or charged particles leads to appearance of different radiation defects, both primary and secondary. A nuclear reaction with slow neutrons

$$Si_{14}^{30}(n, \gamma) Si_{14}^{31} \xrightarrow[2.624]{\beta^-} P_{15}^{31}$$

brings about the formation of phosphorus. This process underlies *nuclear doping* of silicon.

Silicon is a raw material in the manufacture of *microelectronics* devices (e.g., integrated circuits) and individual components (diodes, transistors, etc.).

SILICON, AMORPHOUS

See *Amorphous silicon.*

SILICON CARBIDE

A chemical compound of *silicon* with *carbon*, SiC. There are many polytypes of silicon carbide (see *Polytypism*). All polytypes have the same dimension of a hexagonal *unit cell* $a = b = 0.3078$ nm and a variable parameter $c = 0.2518n$ nm, where n is the number of layers in a period which varies from 2 to an indefinitely large number. The majority of known polytypes of SiC have a hexagonal unit cell (H), and others have rhombohedral (R) cells. One polytype called β-SiC has a diamond-type cubic lattice (C), and the others are called α-SiC. Examples are $6H$-SiC and $3C$-SiC, where the digit stands for the number n of layers within the period. The cubic modification is metastable and transforms to hexagonal above 2100 °C. Under pressure above 3.0 GPa at temperatures above 1200 °C the reverse transformation α-SiC $\rightarrow$ β-SiC is observed. The density is 3.214 g/cm^3 for α-SiC and 3.166 g/cm^3 for β-SiC. *Thermal conductivity* of α-SiC monocrystals is 490 W·m^{-1}·K^{-1}, temperature coefficient of linear expansion is 5.94·10^{-6} K^{-1} (at 250 to 2500 °C); for β-SiC, it is 3.8·10^{-6} K^{-1} (200 °C). *Debye temperature* is 1200 K for α-SiC and 1430 K for β-SiC. Knoop hardness under 100 g load is 29.17 GPa for α-SiC (face 0001); for β-SiC, it is 28.15 GPa (face 111) and 31 to 34 GPa (polycrystal). *Elastic modulus* of α-SiC is 392 GPa, *shear modulus* is 171 GPa, bulk modulus is 98 GPa.

Silicon carbide is a *semiconductor*, usually of *n*-type, *band gap* width of α-SiC is 3.3 eV (2*H*) and 2.86 eV (6*H*, at 300 K); thermoelectromotive force of α-SiC is 70 μV/K (293 K) and 110 μV/K (1273 K). The observed width of the band gap of β-SiC is 2.4 eV (at 1.6 K) and corresponds to *indirect transitions* of electrons from a *valence band* maximum to a *conduction band* minimum, displaced to the point $x(0, 0, \pi/a)$. *Effective masses* of electrons are $m_\perp = 0.24m_e$, $m_\parallel = m_e$, where m_e is the free electron mass.

Silicon carbide does not decompose in mineral acids at temperature above 1000 °C. It can be prepared by the interaction of *silica* with coal, by synthesis from the elements, by *crystallization* from solutions, melts, gaseous compounds. Silicon carbide is used for manufacturing abrasive, refractory, abrasion-resistant, electric engineering materials. As a semiconductor, silicon carbide is used to manufacture electroluminescence devices, photoresistors, detectors of visible and ultraviolet radiation. See also *Semiconductor materials*.

SILICON DIOXIDE

See *Silica*.

SILICONIZING

The thermodiffusion saturation by *silicon* of the surfaces of metallic and nonmetallic products with the purpose of increasing their resistance to corrosive attack (see *Corrosion resistance*), to thermal, wear, and to acid contamination in aggressive media. Siliconizing is realized from solid, fluid, steam, or gas phases using as silicon-containing compounds elemental silicon, ferrosilicon, silicon carbide, hydrolyzed ethylsilicate, halides of silicon, etc. Electrolytic siliconizing from the melt of salts at high temperature is also utilized.

SILICON NITRIDE, Si$_3$N$_4$

Chemical compound of *silicon* with nitrogen. It has two modifications, α-Si$_3$N$_4$ and β-Si$_3$N$_4$. The α-Si$_3$N$_4$ compound is hexagonal with space group $P31c$ (C_{3v}^4), crystal lattice parameters $a = 0.7765$ nm, $c = 0.5622$ nm; the β-Si$_3$N$_4$ is also hexagonal with space group $P6_3/m$ (C_{6h}^2), $a = 0.7606$ nm, $c = 0.2909$ nm. Both structures are formed by SiN$_4$ tetrahedra, but β-Si$_3$N$_4$ is stable, while α-Si$_3$N$_4$ transforms into β-Si$_3$N$_4$ at temperatures above 1400 °C. Silicon nitride does not melt, it vigorously evaporates (sublimes) with disproportionation to nitrogen at temperatures over 1600 °C. Its density is 3.19 g/cm^3, thermal conductivity is 62.8 W·m^{-1}·K^{-1} at 300 °C; *microhardness* is 34.5 GPa; elastic modulus is 298 GPa. It is an insulator with electric resistivity $\rho > 10^{15}$ Ω·m (300 K); *band gap* width is 4.0 eV. Silicon nitride is obtained by the interaction of silicon oxide with *carbon* in the presence of nitrogen, by synthesis from the elements, or by interaction of silane with nitrogen or ammonia. Silicon nitride is the basis of structural *ceramics*; it also a cutting ceramic material, and an *abrasive material*.

SILVER (Gr. $\alpha\rho\gamma\upsilon\rho\sigma\varsigma$, Lat. *argentum*), Ag

A chemical element of Group I of the periodic system with atomic number 47 and atomic mass 107.8682. The natural element silver has two stable isotopes ^{107}Ag (51.35%) and ^{109}Ag (48.65%). Outer shell electronic configuration is $4d^{10}5s^1$. Successive ionization energies are 7.58, 21.50, 35.79 eV. Atomic radius is 0.144 nm; radius of Ag$^+$ ion is 0.089 nm. Oxidation state is +1, rarely +2, +3. Electronegativity is $\approx$1.5.

In a free form, silver is a white *metal*. It has a face-centered cubic lattice with parameter $a = 0.408624$ nm. Density is 10.5 g/cm^3; $T_{melting} = 1233.9$ K, $T_{boiling} = 2457.1$ K. Heat of melting 11.34 kJ/mole, heat of evaporation is 255.024 kJ/mole; thermal conductivity coefficient is 452.87 W·m^{-1}·K^{-1} at 0 °C; specific heat $c_p = 0.2387$ kJ·kg^{-1}·K^{-1}. Temperature coefficient of linear expansion is 19.51·10^{-6} K^{-1} (at

300 K). Resistivity is 14.7 nΩ·m (at 293 K). Compared to other metals, silver possesses the highest electrical conductivity $62.97 \cdot 10^6$ $(\Omega \cdot m)^{-1}$, thermal conductivity, and reflectivity. Depending on the mechanical treatment, the breaking strength of silver is 98 to 147 MPa upon its elongation of 48%. Brinell hardness is 245 MPa, Mohs hardness is 2.5 to 3.0. Adiabatic elastic moduli of silver: $c_{11} = 122.2$, $c_{12} = 90.7$, $c_{44} = 45.4$ (GPa) at room temperature.

It is mainly used in the form of alloys (photography, electrical engineering, electronics, etc.). Silver is a component of several solders; it is used for the low-temperature soldering of steels, copper-based alloys, and other alloys.

SILVER, GERMAN

See *German silver*.

SIMILARITY of phenomena

Similarity is an approach which allows one to obtain characteristics of one phenomenon using known characteristics of another through a change-over to another system of units. The concept of similarity lies at the basis of simulations of various phenomena, which are difficult to investigate experimentally, using more available models. For instance, equilibrium of elastic structures is determined by *Young's modulus E* and the dimensionless *Poisson ratio* σ, by the density of the material ρ, the loading P and the characteristic size d. Three dimensionless parameters may be derived from these five quantities: σ, $E/\rho d$, and P/Ed^2. For corresponding values of these three parameters all *strains* of the natural structure and of the model will be similar, and the amount of displacement of arbitrary sections of the structure may be obtained from the model. The so called Π-*theorem* provides mathematical justification for such a simulation. A certain feature X of the system under study is determined by a set of system parameters x_l (so-called *governing parameters*): $X = X(x_1, \ldots, x_k, x_{k+1}, \ldots, x_N)$. Part of the system characteristics, including X, may be expressed as dimensionless quantities Π with the help of several (from x_1 to x_k) governing parameters (e.g., with dimensionality of length, time, mass): $\Pi = \Pi(x_1, \ldots, x_k, \Pi_{k+1}, \ldots, \Pi_N)$. According to the Π-theorem, this relation may

be represented as $\Pi = \Phi(\Pi_{k+1}, \ldots, \Pi_N)$, where the function Φ does not contain the parameters $x_1, \ldots, x_k$. The equality of the values of all $N - k$ dimensionless parameters $\Pi_{k+1}, \ldots, \Pi_N$ of systems under consideration is a necessary and sufficient condition for the similarity of these systems; the values of the dimensions of governing parameters are different. The quantities $\Pi_{k+1}, \ldots, \Pi_N$ are called *similarity parameters* (*similarity criteria*).

Simulation is also possible if the similarity of a system is incomplete (i.e. it only holds for certain governing parameters). The concept of similarity and simulation is useful for solving various problems of *elasticity theory*, propagation of waves in media, motion of solids in liquids, etc.

SIMPLE CRYSTAL FORMS

Combinations of *crystal* faces that are equivalent to each other (to within a parallel displacement) under the symmetry transformations of the corresponding *crystal classes*. To the accuracy of a similarity transformation (homothety) each simple crystal form is characterized by an *isohedron*, i.e. a polyhedron (either open or closed), with faces equivalent in the respective symmetry class. Two simple crystal forms belong to one and the same type if the isohedrons corresponding to them are combinatorily identical and have identical complete *symmetry groups*. The figure shows 47 types of simple crystal forms.

SIMPLE STATE OF STRESS

A *state of stress* in which the *stress tensor* σ_{ij} is a linear function of the Cartesian coordinates of the body x_k: $\sigma_{ij} = A_{ij} + B_{ijk}x_k$, where A_{ij} and B_{ijk} are constant tensors. The principal property of the simple state of stress is that it arises for one and the same boundary conditions in all homogeneous elastic bodies of identical shape, independent of the elastic properties of the bodies themselves (see *Elasticity*). A simple state of stress during which the stress tensor does not depend on the coordinates is called a *uniform state of stress*. Particular cases of this uniform state are: *uniform compression*, $\sigma_{ij} = -p\delta_{ij}$; *uniaxial strain*, $\sigma_{ij} = -sn_in_j$ (here $\boldsymbol{n}$ is a unit vector); *shear*, $\sigma_{ij} = -\sigma(n_im_j + n_jm_i)$ (here $\boldsymbol{n}$, $\boldsymbol{m}$ are orthonormal vectors). Nonuniform simple states of stress

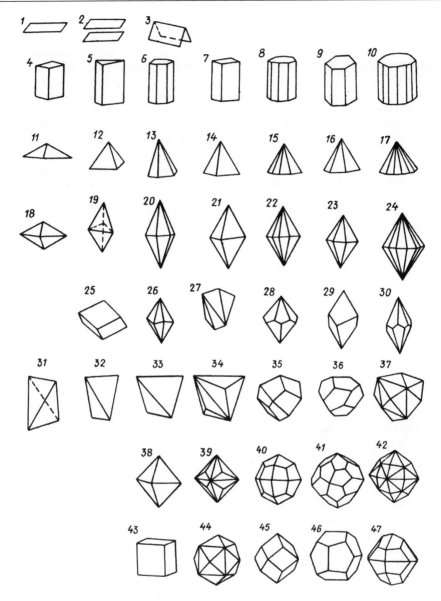

Simple crystal forms: 1, monohedron; 2, pinacoid; 3, dihedron; 4, rhombic prism; 5, trigonal prism; 6, ditrigonal prism; 7, tetragonal prism; 8, ditetragonal prism; 9, hexagonal prism; 10, dihexagonal prism; 11, rhombic pyramid; 12, trigonal pyramid; 13, ditrigonal pyramid; 14, tetragonal pyramid; 15, ditetragonal pyramid; 16, hexagonal pyramid; 17, dihexagonal pyramid; 18, rhombic dipyramid; 19, trigonal dipyramid; 20, ditrigonal dipyramid; 21, tetragonal dipyramid; 22, ditetragonal dipyramid; 23, hexagonal dipyramid; 24, dihexagonal dipyramid; 25, rhombohedron; 26, trigonal scalenohedron; 27, tetragonal scalenohedron; 28, tetragonal trapezohedron; 29, trigonal trapezohedron; 30, hexagonal trapezohedron; 31, rhombic tetrahedron; 32, tetragonal tetrahedron; 33, tetrahedron; 34, trigon-tristetrahedron; 35, tetragon-tristetrahedron; 36, pentagon-tristetrahedron; 37, hexatetrahedron; 38, octahedron; 39, trigon-trioctahedron; 40, tetragon-trioctahedron; 41, pentagon-trioctahedron; 42, hexoctahedron; 43, cube; 44, tetrahexahedron; 45, rhombic dodecahedron; 46, pentagon-dodecahedron; and 47, didodecahedron.

develop during *flexure* of a rectangular prism, during *rod twisting*, and in several other cases.

SINE-GORDON EQUATION

See *Soliton*.

SINGLE CRYSTAL

See *Monocrystal*.

SINGLE-DOMAIN PARTICLES

Small particles of ferro- or ferrimagnetic materials so small that the formation of a *magnetic domain structure* is not energetically favorable. As a result, the particles are uniformly magnetized (each is a single *magnetic domain*). The condition for the existence of single-domain particles (*criterion for single domain formation*) is that their size D is smaller than the critical size D_{cr} for the existence of a single domain. This value D_{cr} depends on the particle shape, as well as on the nature of the *magnetic anisotropy* and its magnitude k, the magnitude of the nonuniform exchange energy A, and the saturation magnetization M_{s}. The inequalities for values of the magnetic field $B > B_k$ and other quantities also depend on the single-domain criterion. In the case of weak anisotropy $k \ll 2\pi M_{\mathrm{s}}^2$, the value of D_{cr} essentially depends on the particle shape and demagnetization factor N (see *Demagnetization fields*) along the *easy magnetization axis*

$$D_{\mathrm{cr}} \approx C \left(\frac{A}{N M_{\mathrm{s}}^2} \right)^{1/2}, \qquad (1)$$

with $C \approx 5$–6. For spherical particles, D_{cr} proves in this case to be smaller than a *magnetic domain wall* thickness Δ; e.g., for Fe or Ni $D_{\mathrm{cr}} = 3.3 \cdot 10^{-6}$ cm or $7.6 \cdot 10^{-6}$ cm, respectively. In the strong anisotropy case, D_{cr} is weakly dependent on the particle shape:

$$D_{\mathrm{cr}} \approx \frac{9(Ak)^{1/2}}{\pi M_{\mathrm{s}}^2}. \qquad (2)$$

In order of magnitude, D_{cr} coincides with a *characteristic length of magnetic material* l, $D_{\mathrm{cr}} > \Delta$, and increases as the anisotropy grows. For example, D_{cr} is about $4 \cdot 10^{-5}$ cm for MnBi and $1.3 \cdot 10^{-4}$ cm for barium hexaferrite.

Single-domain particles and their clusters embedded into weakly-magnetic hosts have some characteristic magnetic properties. Their *remanence* coincides with M_{s}, and *magnetic reversal* occurs solely by irreversible *magnetization rotation processes*. If the single-domain particle anisotropy is high (e.g., for elongated particles, or due to crystalline magnetic anisotropy), the *coercive force* B_{c} for their ordered aggregate is high. Hence such systems are *hard magnetic materials*, and are used to manufacture *permanent magnets*. Liquid emulsions of single-domain particles also have some specialized magnetic properties (see *Ferrofluid*). As the size of single-domain particles decreases, their magnetic reversal due to thermal fluctuations becomes possible, and the system acquires the property of *superparamagnetism*.

SINGLE-ELECTRON APPROXIMATION

Introduction of single-electron wave functions for each electron of a many-electron quantum-mechanical system ($i = 1, 2, \ldots$), which depend on the coordinates r_i of only one electron, $\psi(r_i)$, and the replacement of the many-electron function $\psi(r_1, \ldots, r_i)$ with a product of single-electron functions (Hartree approximation), or a linear combination of such products (see *Hartree–Fock method*). In this case, the energy of interaction of an arbitrary lth electron with all others is expressed by a sum of the form

$$V_l(r_l) = \sum_{i \neq l} \int V_{il}(r_{il}) |\psi_i(r_i)|^2 \, \mathrm{d}r_i,$$

$$V_{il} = \frac{e^2}{\varepsilon r_{il}},$$

where ε is the *dielectric constant*, and $r_{il} = |r_i - r_l|$ is the distance between electrons. The wave functions $\psi(r_i)$ are found from the Schrödinger equation with a Hamiltonian that involves $V_i(r_i)$, i.e. the single-electron approximation leads to a nonlinear Schrödinger equation (see *Soliton*) for single-electron wave functions. The functions obtained from the Schrödinger equation are used to calculate this V_l (Hartree or Hartree–Fock *self-consistent field*). The system energy in this approximation does not equal the total energy of all the electrons.

In solid state physics the one-electron approximation is one of the most widely used methods for investigating the quantum states of a system.

If the potential of a single electron is regarded as periodic with a period equal to the crystal lattice constant, the solution obtained is self-consistent: the Bloch wave functions (see *Bloch theorem*) that are solutions of the equation with a *periodic potential* give, upon substitution into the expression for V_l, a periodic dependence $V_l(r_l)$. As a general rule, the single-electron approximation is believed to provide a high degree of accuracy in the description of the quantum states of a many-electron system. However, in some cases, the single-electron approximation appreciably distorts the true nature of the spectrum of states. For example, the interaction of two electrons in a polarizable medium, in the case of a continuum description in the single-electron approximation $\psi(r_1, r_2) = \psi(r_1)\psi(r_2)$, is unable to provide for the formation of their spherically symmetric bound state, i.e. a *bipolaron*. Through the use of a two-electron function $\psi(r_1, r_2) = \psi(r_1)\psi(r_1)(1 + \beta r_{12})$ (here β is a variational parameter) that allows for the correlation in the motion of electrons, the bipolaron state proves to be explainable.

SINTERING

The process of *heat treatment* of porous structures, which takes place through the *pressing* of powders, resulting in a decrease of the *porosity* of the compressed material. Sintering causes fine particles of a material to bond chemically at a temperature high enough to induce atomic diffusion. Sintering is one of the principal stages of *powder metallurgy* fabrication, and the manufacture of refractory materials. *Mass transport*, an essential process for sintering, may take place spontaneously by various diffusion mechanisms under the influence of gradients of the chemical potential μ, arising from gradients of the local curvatures K of free surfaces, which limit the grains of the powder: $\nabla \mu_k = \alpha \omega \nabla K$, where α is the *surface tension*, ω is the atomic volume. These flows decrease the area of free surfaces, hence decrease the free surface energy, and make spontaneous powder compression thermodynamically advantageous. The compression rate may be considerably accelerated through application of *uniform compression*, which causes and maintains the plastic (dislocation) flow of material into *pores*.

The compression, which takes place in the pressure sintering of powder mixtures, is accompanied by the process of diffusion *homogenization*, which involves the formation of *solid solutions*, intermetallide phases, etc. The processes of diffusion compression (removal of cavities) and diffusion homogenization are interrelated by their common diffusion mechanism, and in many cases may be competitive with each other because of the inequality of opposing diffusion flow rates of components (see *Diffusion porosity*). Thus, the compression process may become decelerated and even reversed in sign in the formation of solid solutions. The process of sintering is accompanied in many mixtures by the formation of some liquid phase. The presence of a liquid phase accelerates all the mass transport processes which take place in the compressed material during sintering.

SIZE EFFECTS

Dependence of physical properties of a material on the specimen size and shape, when its thickness d is comparable to the *de Broglie wave* length, or to the mean free path l for *quasi-particles*, or to the diameter $2r$ of the electron orbit in a magnetic field B, etc. The principal cause of size effects in solids is the additional dissipation of thermal flux and electric current, induced, in turn, by the scattering of *conduction electrons*, *phonons* and *magnons* by the specimen boundary. Size effects depend on the state of the specimen surface. When colliding with an ideally smooth defect-free surface the energy $\varepsilon(p)$ and the tangential component of the quasi-particle momentum are conserved (*specular reflection*). Such collisions are nondissipative so the *thermal conductivity* κ and *electrical conductivity* σ remain proportional to l. The presence of roughness or of impurity atoms at the surface perturbs the correlation between the momentum of the incident, p_-, and the reflected, p_+, quasi-particle. In case of a total loss of such correlation (*diffuse reflection*, when all velocity directions of the reflected quasi-particle are equally probable), the effective transport mean free path l_{eff} in a bulk specimen $(d \gg l)$ is given by the relation $l_{\text{eff}}^{-1} = l^{-1} + d^{-1}$, while in a thin specimen $(d \ll l)$ the principal mechanism of thermal flux and electric current dissipation is the collisions of quasi-particles with the boundary.

Heat transport in *insulators* occurs via the *diffusion* of phonons. When d is much larger than l_N, the bulk mean free path of phonons undergoing normal phonon–phonon scattering (elastic, in which the overall momentum is conserved), but is smaller than $l_R = l$, where l_R is the bulk mean free path for resistive *phonon scattering* by phonons, defects, or impurities in the crystal (inelastic, in which quasi-momentum is not conserved in the process), then for significantly nonspecular reflections from the boundary, their effective transport mean free path is d^2/l_N, provided $d^2 < l_N l$. In other words, it decreases with decreasing temperatures T. The latter effect is explained by the fact that the number of phonons is proportional to T^3, while the Brownian step length, equal to l_N, grows without limit when $T \to 0$. Further lowering of the temperature results in the l_{eff} becoming equal to l_R and then, when $l_N > d$, it reaches a value of $l_{eff} = d$. This combination of effects results in a minimum in the temperature dependence of the thermal resistance.

Thermal conduction of *ferrites* results from the magnon (*spin wave*) drift, so that one expects a specific dependence of κ on the temperature in a pure enough relatively thin specimen (about 10^{-4}–10^{-3} cm thick). That dependence is related to a change of spin wave *relaxation* mechanisms at lower temperatures.

Size effects in the conduction of *metals* were first observed by J.J. Thomson (1900). K. Fuchs (1938) calculated the dependence of the electrical conductivity σ on d by assuming that a fraction q of the electrons undergoes specular reflection, and a fraction $(1-q)$ undergoes diffuse reflection. If $q = 1$, size effects are absent under isotropic conditions, and σ coincides with the bulk conductivity σ_0. These results are significantly changed when one takes into account the fact that the specular reflection parameter, q, depends on the angle of incidence of electrons at the surface (specular reflection is more probable for grazing incidence (electron sliding along specimen surface) than for incidence at a steep angle). A more realistic account of charge carrier interaction at the metal surface provides a logarithmic dependence of σ on l in wires when $d \ll l$, while plates with appreciable surface roughness yield $\sigma = (d/l)^{1/2}$. For anisotropic conductors, in which *multichannel specular reflection* takes place, that is when

the equation $\varepsilon(\boldsymbol{p}_+) = \varepsilon(\boldsymbol{p}_-)$ has several solutions for $\boldsymbol{p}_+$ for a given value of $\boldsymbol{p}_-$, we find that $\sigma(d) < \sigma_0$, even at $q = 1$, although $\sigma(d)$ still remains proportional to l. The behavior of $\kappa(d)$ is similar.

Proceeding to *semimetals* and *semiconductors*, size effects manifest themselves in a sufficiently thick specimen when their thickness is comparable to the diffusion length, L, which is explained by a weak *intervalley scattering* of charge carriers at the surface. The two downward trends in the $\sigma(d)$ dependence for *bismuth* that appear in the plot when the specimen thickness is decreased were explained by E.I. Rashba et al. by assuming the development of an electric field in the specimen which compensates for the cross-currents produced by the curvature of the energy bands close to the surface at distances $L \gg l$. Intervalley scattering and electron–hole *recombination* at the surface in *many-valley semiconductors* result in a spatially inhomogeneous distribution of charge carriers throughout the specimen bulk. An excess of charge carriers belonging to a certain group (valley) builds up at one of the plate faces as a result, while the other face experiences a deficit of such carriers. Under these conditions size effects appear not only in the dependence of σ on d for $d < L$, but also in the development of an electric field tangential to the electric current, even in cubic crystals. Experimental studies of the dependence of σ on d make it possible to determine the state of the conductor surface. However, it is difficult to prepare a set of samples of differing thicknesses with identical surfaces. It seems more reliable to study the nature of the reflection of charge carriers by measuring the dependence of σ on the temperature T, and on the magnetic field strength. The higher the degree of specular reflection off the surface, the less the $\sigma(T)$ dependences differ between bulk and thin film values In pure enough conductors with rough surfaces, when one may neglect electron–impurity scattering and the dissipation of electron fluxes via small angle electron–phonon scattering and by collisions of electrons with the surface of a thin conductor, there appears a specific dependence of σ on d and T at $T < T_D$. Here T_D is the activation temperature of intervalley processes, which coincides with the *Debye temperature* in its order of magnitude.

The resistivity, $\rho = \sigma^{-1}$ of conductors with finite sizes grows with temperature slower than the *Bloch law* prescribes ($\sigma \propto T^5$), and depends significantly on the form of the $q(\theta)$ function (θ is the angle of incidence for electrons striking the surface). For small d, in a model assuming $q = \mu \cos \theta$, where $\mu = $ const, the asymptotic behavior of σ in plates is given by

$$\sigma = c\sigma_0 \left(\frac{d}{\mu l} \right)^{3/4} \sim d^{3/4} T^{-5/4}, \quad c \sim 1, \quad (1)$$

while in wires it is

$$\sigma = \frac{3}{4} \sigma_0 \frac{d}{\mu l} \ln \frac{l}{d} \ln \frac{T_D}{T}. \quad (2)$$

In the *Fuchs model* ($q = $ const) we have $\sigma \propto \ln(T_D/T)$ in plates, and $\Delta \rho(T) = \rho(T) - \rho(0) \propto d^{-2/3} T^{5/3}$ in wires.

Size effects are most pronounced and are more informative in magnetic fields. Among such effects, one should mention the *Sondheimer oscillations* related to the focusing role of the magnetic field, the *static skin-effect*, and a large number of *high-frequency size effects*. Their experimental studies yield detailed information on the spectrum of charge carriers in conductors, and on their relaxation properties.

See also *Quantum size effects, High-frequency size effects, Magnetoacoustic size effects, Radio-frequency size effects*.

SKIN-EFFECT

The attenuation of an alternating electromagnetic field as it penetrates into a bulk conducting medium. A high-frequency field in a bulk conductor concentrates in its boundary *skin-layer*. The greater the conductivity σ and frequency ω the smaller is the skin depth δ (depth of field penetration). If the electric current obeys Ohm's law locally then the value of the field in the skin layer decreases exponentially $e^{-x/\delta}$ (*normal skin-effect*), and in an isotropic conductor it is lowered by the factor $1/e = 0.368$ over the distance $\delta = c/(\omega \sigma \mu/2)^{1/2}$ (μ is the *magnetic permeability*, SI units are used) inward from the surface. The quantity δ at the frequency $\omega/(2\pi) = 1$ MHz and room temperature in a number of *metals* takes on values from 60 up to 230 µm, and in bismuth $\delta = 545$ µm. *Ferromagnets* at low frequencies exhibit rather small δ values because of their

large permeability μ. At frequencies ω exceeding the collision frequency of the current carriers (infrared range) δ is defined by the *high-frequency conductance*, and is equal to c/ω_L where ω_L is the plasma frequency, and c is the speed of light. In pure metals at low temperatures the values of the penetration depth in the radio frequency and microwave ranges can be considerably smaller than the electron *mean free path* so that strong spatial dispersion (conductivity dependence on wave vector) is exhibited. The skin-effect under such conditions is called the *anomalous skin-effect*, and it exhibits a non-exponential field decrease in a skin-layer width defined by the quantity $\delta_a = [c^2 v_F/(\omega \omega_L^2)]^{1/3}$ where v_F is the electron *Fermi velocity*.

In conductors located in a strong constant magnetic field at a low temperature the nature of the skin-effect can change (depth of penetration increases) due to excitation of weakly decaying waves, e.g., *helicons, magnetic plasma waves*, Alfvén waves (see *Solid-state plasma*), *dopplerons* etc.

SKIN-EFFECT, ANOMALOUS

See *Anomalous skin-effect*.

SKIN-EFFECT, STATIC

See *Static skin-effect*.

SKYRMIONS

Fermion (or boson) particles with finite radii, composed from boson fields; named after English scientist T.H.R. Skyrme (1922–1987). In solid state physics they are macroscopic spin excitations (*quasi-particles*) of a vortex type in two-dimensional electronic systems in a *quantizing magnetic field* with odd filling of the Landau level under the conditions for the occurrence of the *quantum Hall effect*. Theoretically skyrmions correspond to nonuniform rotations of *second quantization* Ψ-operator spinors for fermions (electrons, holes) with the help of a nonuniform rotational displacement matrix $U(r)$ that translates initial spinors Ψ into new spinors $X(r)$ in accordance with the relation $\Psi(r) = U(r)X(r)$. The matrices $U(r)$ are parameterized by the three *Euler angles* with the help of the *Pauli matrices*. The size of the skyrmion vortex core (see *Quantum vortices*) is determined by the competition between Coulomb electronic repulsion, which tends to increase the region of charge density perturbation, and the Zeeman magnetic energy which tends to reduce the region with unfavorable spin orientation.

SLATER–CONDON PARAMETERS (J. Slater, E. Condon)

Empirical parameters in atomic spectra. In the $L - S$ coupling case (see *Russel–Saunders coupling*) the electrostatic energy of non-repeating terms (energy term sums of electron configuration) are expressed via linear exchange (J) and coulomb (K) integral combinations (see *Exchange interaction*) which are defined by single-electron atomic states ψ_{nlm}:

$$J(\psi_{nlm}, \psi_{n'l'm'}) = \sum_k a^k(lm, l'm') F^k(nl, n'l'),$$

$$K(\psi_{nlm}, \psi_{n'l'm'}) = \sum_k b^k(lm, l'm') G^k(nl, n'l'),$$

where a^k and b^k are tabulated numerical coefficients, and the Slater–Condon parameters F^k, G^k are radial integrals R^k involving the wave functions ψ_{nlm} and components of the expansion of $1/|r - r'|$ in Legendre polynomials. These parameters have the form:

$$F^k(nl, n'l') = R^k(nl, n'l', nl, n'l'),$$

$$G^k(nl, n'l') = R^k(n'l', nl, nl, n'l').$$

The quantities F^k and G^k are considered as empirical parameters which are estimated from the energy separations between levels corresponding to optical spectral lines. In many cases only a few Slater–Condon parameters are needed to specify the energy of a particular spectral state. The atomic spectra designation in terms of Slater–Condon parameters is not adequate when the *spin–orbit interaction* is strong, or when the *configuration interaction* is appreciable.

SLATER DETERMINANT (J. Slater, 1929)

Antisymmetric wave function of N electrons (or other fermions) constructed from N linearly

independent and orthonormal single-particle functions (*spin-orbitals*) $\psi_k(\xi_l)$ expressed in the form of a determinant:

$$[\psi_1 \psi_2 \ldots \psi_N] = \frac{1}{\sqrt{N!}} \begin{vmatrix} \psi_1(\xi_1) & \cdots & \psi_N(\xi_1) \\ \vdots & \ddots & \vdots \\ \psi_1(\xi_N) & \cdots & \psi_N(\xi_N) \end{vmatrix},$$

where ξ_i is the set of space and spin coordinates of electron i. Any antisymmetric N-electron function for a particular system may be represented by an expansion (linear combination) over all ordered Slater determinants: $[\psi_{k_1} \psi_{k_2} \ldots \psi_{k_N}]$, $k_1 < k_2 < \cdots < k_N$, which are constructed from an arbitrary but complete set of spin-orbitals of the system.

SLATER RULES (J. Slater)

The set of rules making it possible to write wave functions and electronic energies of many-electron atoms in a rough but very simple approximation. According to the Slater rules, the wave functions for many-electron atoms are similar to hydrogen atom *orbitals*. *Slater's functions* have the form:

$$\psi_n = A r^{n^*-1} \exp\left[-\frac{Z^*(n^*)}{n^*}\right] r,$$

where A is a normalization factor, n^* is the effective principal quantum number, $Z^*(n^*)$ is the effective charge acting on an electron in an orbit with definite n^*. According to the Slater rules, the value of n^* is related to the principal quantum number n in the following way:

n	1	2	3	4	5	6
n^*	1	2	3	3.7	4	4.2

where the value of Z^* is determined from the condition $Z^* = Z - \gamma$, where Z is the atomic charge, and γ is the screening constant which has the same particular values for electrons within each of the following groups: $(1s)$, $(2s, 2p)$, $(3s, 3p)$, $(3d)$, $(4s, 4p)$, $(4d)$, $(4f)$, and so on. It is assumed during the determination of γ that all additional electron orbitals beyond these groups do not influence γ; and every electron from a group has a γ value of 0.35 (except in the $(1s)$ group $\gamma = 0.30$); if an electron is in an (s, p) group with the principal quantum number n then electrons in the group with quantum number $n-1$ increase γ by 0.85, and more remote inner groups increase γ by 1.00. For groups (d) and (f) all electrons of an inner group increase γ by 1.00. In accordance with Slater rules the electron energy is $E_{n^*} = E_0(Z^*/n^*)^2/2$. Slater functions with varied parameters are widely used in quantum-chemical calculations.

SLATER–TAKAGI MODEL (J. Slater, 1941; U. Takagi, 1948)

Quantitative description of the *ferroelectric phase transition* in potassium dihydrogen phosphate KH_2PO_4 (KDP). Historically the first model comes from the idea that the KDP dielectric properties are determined mainly by the proton (hydrogen atom) configurations. Each proton is considered as having two equilibrium positions along the bond connecting two PO_4 adjacent tetrahedral groups. If one takes into account only the interaction of the closest protons then one must consider the proton configurations near each PO_4 tetrahedron, as shown schematically in the figure.

Here PO_4 tetrahedra are indicated by squares, and protons are shown by small circles. There are $2^4 = 16$ possible positions for four protons surrounding every PO_4 group, $1+1=2$ for configuration A, $4+4=8$ for B, 4 for C and 2 for configuration D. Thus only six of them corresponding to H_2PO_4 groups with two nearest protons have the lowest energies (neutral C and D configurations in Fig.). The A and B type configurations are charged, and their energies W and W' are large, as indicated. In the Slater–Takagi model such configurations are neglected because the probability of their contribution is small. The phase transition here is a *first-order phase transition* with the transition temperature $T_c \propto E/(k_B \ln 2)$ (E is type C configuration energy); the dielectric constant above T_c satisfies the *Curie–Weiss law* $\varepsilon(T) \propto (T - T_c)^{-1}$. Takagi has shown that a calculation involving the charged configurations A and B produces a *second-order phase transition*. For a quantitative description of the phase transition properties on a modern level, *cluster* theory is used which, unlike the *mean field approximation*, permits one to take into account strong short-range proton correlations.

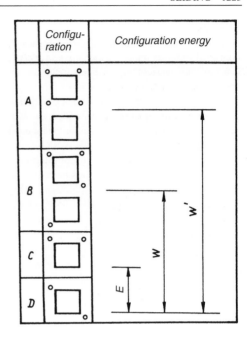

Configuration	Configuration energy

Slater–Takagi model.

SLAVUTICH

A superhard *composite material* composed of a tungsten–cobalt hard *alloy* and natural *diamonds* (0.2 to 0.8 mm). To provide *adhesion* splicing of the components, slavutich is caked by hot *pressing* in a protective medium under conditions which exclude saturation of the interface, with *carbon* up to *stoichiometric composition*. By changing the content of diamonds in slavutich and of *cobalt* in the alloy, the qualities of different types of slavutich are obtained. The most widespread slavutich variety has a wear resistance which exceeds that of hard alloys by 30 to 200 times. Slavutich has a somewhat reduced *strength* ($\sigma_{comp} = 820$–930 MPa) and significantly increased *thermal conductivity* (125 to 130 $W \cdot m^{-1} \cdot K^{-1}$) compared to a hard alloy. Slavutich is used in the form of elements of any pregiven shape and dimensions in boring and dressing tools.

SLIDING

See *Slip*.

SLIP, sliding

Relative displacement of two solid bodies in contact. In the region of contact there is *friction*. The force arising from the roughness in the contact region that opposes the onset of motion is called the *static friction*. After motion begins the friction force decreases to a lower value, called the *sliding or kinetic friction*, which must be maintained by the applied force to sustain the motion. Sliding is accompanied by the response of intermolecular interactions that act at various points of the contact to oppose the motion. The losses arising from these forces are expressed in terms of the dimensionless parameter τ_{mol}/σ_S, where τ_{mol} is the shear resistance of the molecular bonds, and σ_S is the *yield limit*. The molecular shear resistance is given by $\tau_{mol} = \tau_0 + \beta P_{cp}$, where τ_0 is the shear *strength* of a point of contact in the absence of a compressing load, β is the *hardening* ratio, and P_{cp} is the *pressure* at the contact point. The mechanical component of sliding friction depends on the dimensionless parameter h/R, which is the ratio of the depth h of insertion of a unit unevenness to the radius R of the spherical segment inscribed in it. The tangential stress $\tau_{mech} = k\alpha(h/R)^{1/2} P_{cp}$, where k is a coefficient that depends on the distribution of roughness in elevation, and α is the *hysteresis* loss ratio. The overall sliding friction coefficient f_s is the sum of the molecular and mechanical components: $f_s = \tau_0/P_{cp} + \beta + k\alpha(h/R)^{1/2}$. The dependence of f_s on the pressure for constant roughness, and on roughness for constant pressure, causes it to pass through a minimum. After the settling down of the contact surfaces by sliding, f_s attains and sustains its minimum value. A further reduction of friction can be brought about by introducing lubricant between the sliding surfaces.

SLIP BANDS

Narrow bands, visible through an optical microscope, which appear at the surface of single crystal samples as a result of *plastic deformation*. Plastic deformation of *crystals* is nonuniform, and is focused at isolated regions. Slip bands are traces of these regions at the surfaces of deformed samples. In their turn, slip bands are also nonuniform and exhibit fine structure: electron-microscope photographs show that the slip bands

are split into narrower *shear lines* (slip lines). On the atomic level, slip bands and slip lines make up an aggregation of steps at the crystal surface; the heights of these steps being multiples of the *crystal lattice* constant. These steps are generated as a result of numerous shear displacements (see *Shear*) along crystallographic planes of close packed atoms (*slip planes*). As a rule, many of the microscopic shears in the region of *strain* remain unfinished, and give rise to dislocation lines. Application of various methods of observation of *dislocations* reveals a complicated dislocation structure. Depending on the type of dislocations that intersect the surface under consideration, slip bands are divided into *edge slip bands* and *helical slip bands*.

SLIP LINES

The same as *Shear lines*.

SLIP OF DISLOCATIONS

See *Dislocation slip*.

SLIP PLANE in a crystal

It is a plane along which the displacement of crystal layers takes place, as *plastic deformation* through slippage occurs. The *slip plane* and the *slip direction* (*slip vector*) parallel to it (displacements of an *edge dislocation*) are given the name *slip elements*, and constitute the *slip system* which depends on the structure of the crystal lattice, and is characterized by the minimal amount of work done in performing a shift by one *translation* vector. This work is less, the weaker the bonds between the displacing planes, and the smaller the displacement of these planes with respect to one another. A minimal amount of work is required for shifting close-packed planes along the close-packed directions. The systems of planes, which correspond to the minimum work, are referred to as *major slip systems* (or *primary slip systems*). In a crystal of NaCl, e.g., the slip occurs over the planes {110} along the directions $\langle 1\bar{1}0 \rangle$, the total number of planes of this type in the crystal being six.

SLIP SYSTEM

This involves a plane and a direction of *dislocation slip* in *crystals*. Each crystal lattice has its own preferential *slip plane* which is usually associated with closest-packed planes and directions. The basic slip schemes are: in crystals with a face-centered cubic (FCC) lattice $\{111\}\langle 011\rangle$, in crystals with a body-centered cubic (BCC) lattice $\{110\}\langle 111\rangle$, in crystals with a hexagonal close-packed (HCP) lattice $\{0001\}\langle 11\bar{2}0\rangle$ or $\{11\bar{2}0\}\langle 0001\rangle$, in crystals with a *diamond* lattice $\{111\}\langle 110\rangle$, in *ionic crystals* with a NaCl lattice $\{110\}\langle 110\rangle$, in crystals with a CsCl lattice $\{110\}\langle 001\rangle$. Other things being equal, the slip begins in that system in which, for a given orientation of the crystal relative to the applied stress, the critical stress displacement is minimal.

SLONCZEWSKI EQUATIONS (J.C. Slonczewski, 1972)

A system of differential equations involving magnetic *domain wall* coordinates u, and the angle ψ of the exit of the *magnetization* out of the plane of the wall:

$$\frac{\partial \psi}{\partial t} - \frac{\alpha}{\Delta_0}\frac{\partial u}{\partial t} = -\frac{\gamma}{2M_0}\frac{\delta \sigma}{\delta u},$$

$$\frac{1}{\Delta_0}\frac{\partial u}{\partial t} - \alpha\frac{\partial \psi}{\partial t} = \frac{\gamma \Delta_0}{2M_0}\frac{\delta \sigma}{\delta \psi},$$

where Δ_0 is the thickness of a domain wall at rest, α is the dimensionless *Gilbert relaxation constant*, $\sigma = \sigma(u, \psi)$ is the domain wall energy calculated from the magnetization distribution that corresponds to a resting domain wall with a given shape and value of ψ. The Slonczewski equations are adequate for *ferromagnets* with strong uniaxial *magnetic anisotropy*, e.g., materials with *magnetic bubble domains*. For a ferromagnet with a 180° domain wall in an external magnetic field $B = B_z$, parallel to *easy magnetization axis*, and with low anisotropy in base plane, the Slonczewski equations have the form:

$$\frac{\partial \psi}{\partial t} - \frac{\alpha}{\Delta_0}\frac{\partial u}{\partial t} = \gamma B_z + \frac{\gamma \sigma_0}{2M_0}\nabla_\perp^2 u;$$

$$\frac{1}{\Delta_0}\frac{\partial u}{\partial t} - \alpha\frac{\partial \psi}{\partial t} = \frac{\gamma B_a}{2q}\sin 2\psi + \frac{\gamma \sigma_0 \Delta_0}{2M_0}\nabla_\perp^2 \psi,$$

where B_a is the *anisotropy field*, q is the quality factor, σ_0 is the resting domain wall energy, $\nabla_\perp$ is the gradient operator in the wall plane. In this case, when $B_z = $ const, the Slonczewski equation describes the *Walker solution*; when $B_z = B_z(t)$ the

Slonczewski equation treats forced vibrations of a domain wall; and when $B_z = 0$, it describes *flexural waves* of the wall. These equations are used for an approximate (to the extent of $1/q$ smallness) description of certain effects of *magnetic domain wall dynamics*.

SLONCZEWSKI LIMITING VELOCITY

This defines the "plateau" associated with the dependence of the velocity of a twisted domain wall (see *Domain wall twisting*) on an applied external *magnetic field*, exceeding a prior value related to the *domain wall peak velocity*. It has also been referred to as the *saturation velocity*. The Slonczewski limiting velocity v_{SL} is determined by stationary dynamic processes of initiation and elimination of horizontal Bloch lines, as well as by the peak velocity which is inversely proportional to the thickness of the *magnetic film*, v_{SL} being, however, smaller than the peak velocity. Its value is often derived from the experimentally measured saturation velocity, reaching about $1/5$ of the Walker velocity (see *Walker solution*).

Alongside the Slonczewski mechanism, an important role in the dynamic mode can be played by the formation processes of Bloch lines with more complex configurations, and by the participation of *Bloch points*; bends of boundary planes, etc. For the determination of the Slonczewski limiting velocity it may not be adequate to employ the simplified *Slonczewski equations*, but rather preferable to proceed from the more basic *Landau–Lifshits equation*, carefully taking into account *relaxation*, and irregular boundary motion.

Experimental research measurements require the fulfillment of the following conditions: homogeneity of the film, stability of the applied magnetic field, as well as quality control of the surface of the films and *crystals*.

SMALL-ANGLE SCATTERING OF X-RAYS

Diffuse scattering of X-rays localized near the origin of the *reciprocal lattice* and measured in directions that form a small angle with the incident primary beam. Here, the maximal angle subtends no more than half of the angle of observation of the first Bragg maximum. Such diffuse scattering is caused by the presence in the sample of concentrations of inhomogeneities with respect to lattice elements (differing in X-ray scattering *atomic form factors* from average values). They are inclusions (with elemental content differing from average concentrations) or *pores*. In real crystal materials, one can observe scattering at small angles resulting from *Bragg double reflections*. The angular dependence of the diffuse scattering intensity carries information on parameters defining the concentration of inhomogeneities (shape, size, chemical constitution, and spatial location). The angular dependence of Bragg double reflection intensity is controlled by the structural imperfections (angular disorientation of grains, grain fragments, etc.) of the specimen. Small-angle scattering of X-rays is also used to study phase composition fluctuations near *phase transition* points.

SMALL-DOSE EFFECT

An effect in solid state *radiation physics* involving the qualitative difference between radiation-induced changes of various physical properties of a material for rather small doses Φ relative to those for typical larger radiation doses. For low doses, for instance, the mobility of *current carriers* increases, the reverse outflow of channeled particles (see *Channeling*) decreases, the phonon absorption of light decreases, etc. As Φ becomes larger changes opposite to these take place. The small-dose effect is clearly evident in imperfect *crystals*, which is related to the fact that changes in a system of defects are taking place for much lower values of incident energy transfer than those needed to dislodge an atom from a site in a *crystal lattice*. The difference between cross-sections for dislodging an atom out of a lattice site with the creation of a stable defect, and that needed for a jump of an *interstitial atom* to a neighboring site under the action of a highly energetic external particle, may amount to several orders of magnitude. Therefore, when the concentration of atoms in highly imperfect crystals in various *metastable states* is several orders of magnitude smaller than that at lattice sites, the irradiation may result in a more effective elimination of defects (e.g., escape of interstitial atoms to the surface, their entrapment by *dislocations*, etc.), rather than in the generation of Frenkel pairs in regions of an ideal lattice. With an increase in dose, when the concentration of metastable atoms in the crystal decreases considerably, the process of accumulation of *vacancies*, interstitial atoms, and products of reactions of metastable atoms with one another and with *impurity atoms* becomes predominant. Thus, in the above described examples of the small-dose effect, the early stages of the irradiation can bring about radiation-induced improvements of the structure of an initially imperfect crystal, a process which later degrades. At sufficiently large Φ, when defect accumulation prevails, some radiation-induced partial defect elimination still takes place (so-called *irradiation annealing* of defects). The efficiency ratio of defect formation to radiation annealing, along with the dose range where the small-dose effect is observed, depends on the energy of the incident particles.

SMALL PARTICLES, small clusters

Clusters of dimensions smaller than certain characteristic physical lengths (light wavelength, *mean free path*, *de Broglie wave* length, etc.). Features of small clusters differ considerably from those of a bulk sample. Thus, infrared absorption by *small metallic particles* depends strongly on their shape. When the dimensions of metallic clusters are smaller than the electron mean free path there is a significant decrease of their electron-lattice energy exchange. This favors the creation of *hot electrons* by putting power into small metal islands (see *Insular films*). As the cluster size becomes less than the de Broglie wavelength the *quantization of electron energy* occurs. In the *phonon spectrum* for small clusters, the input of Rayleigh surface modes (see *Surface acoustic waves*) grows, while the long-wave spectral region (wavelengths exceeding cluster dimensions) is cut off. The fraction of surface atoms in small clusters may reach 10% which favors the clear expression of all *surface state* features.

Ferromagnetic particles smaller than *magnetic domains* exhibit increased values of the *coercive force* and saturation *magnetization*. Electric and magnetic polarizabilities of small clusters depend strongly on their shape, and on the presence of other nearby clusters (*local field effect*). Ensembles of small clusters exhibit many unique physical properties.

Nanoparticles, with dimensions in the range from 1 to 100 nm, have a large percentage of their atoms at the surface, so surface effects can dominate. Metallic nanoparticles have conduction electrons which cannot effectively delocalize in the conventional sense because of the small size, so they exhibit confinement effects in a quasi-zero-dimensional structure. Their density of states, which in a bulk conductor or semiconductor serves to determine electrical, thermal, and other properties, becomes a series of delta functions, characteristic of a zero-dimensional system. A nanoparticle of this type is often called a *quantum dot*.

SMECTIC LIQUID CRYSTAL, smectic
(fr. Gr. $\sigma\mu\eta\gamma\mu\alpha$, soap)

A *liquid crystal* formed from rod-like molecules, having a one-dimensional layer structure. It is classified by the number of molecules in the layers and other factors, using the notation: A, C, B, D, E, F, G, H, I. Most smectic liquid crystals that are encountered are types A, B, and C. In types A and C there is no translational order of molecules inside the layers, but there is *orientational order*, with the long molecular axes parallel to each other. In smectic A liquid crystals the molecules are perpendicular to the layer planes (see Fig. (a)), and in the smectic C variety they have some slope in the direction described by the *C director* (see Fig. (b)). Both phases are solid crystals in one dimension and liquids in the other two directions (compare with *discotics*). All this can be explained by the specific molecular structure having a rigid main group and two flexible hydrocarbon end groups. Liquid layers may slide relative to each other, and easily adopt curved configurations, but only those which maintain constant interlayer spacings.

Smectic B liquid crystals have a special order inside the layers, and there are two types of them known. The first is characterized by the presence of a two-dimensional crystal lattice in the layers; this type is essentially a solid three-dimensional crystal with large *anisotropy of elasticity*. The second type of smectic B liquid crystals called *hexadic smectics* have no translational order in the layers, but as a rule have hexagonal order in bond orientations among molecules. The less commonly investigated smectic liquid crystals D, E, F, G, H, I resemble by their structure smectics of the first B variety, and are three-dimensional solid crystal types. Smectic liquid crystals are used as

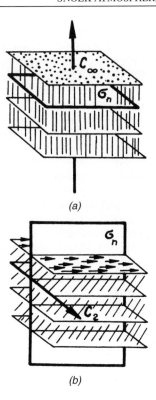

(a)

(b)

Structures and symmetry elements of smectics of types A (a) and C (b).

long-term memory displays for *information recording* (for several months). Laser light warms some smectic liquid crystal regions to the isotropic liquid state, and thereby changes the optical display characteristics. Smectic liquid crystals serve as a model for biological *membranes*.

SMOOTHING
See *Surface smoothing*.

SNELL'S LAW
At an interface of two media with indices of refraction n_1 and n_2 Snell's law for an incident light beam is $n_1 \sin\theta_1 = n_2 \sin\theta_2$.

SNOEK ATMOSPHERE
Region where *doping atoms* are regularly distributed at interstitial octahedral sites in the stress

of a *dislocation* field in body-centered cubic (BCC) metals. The Snoek atmosphere of impurity atoms is brought about by the inequivalence of octahedral interstices arising from the presence of stress, and this lies along three crystallographic directions for the occupancy by *impurity atoms*. This distribution occurs via stress-activated jumps of these atoms to more favorable near-neighbor interstices, during a time several orders of magnitude shorter than that needed for forming a *Cottrell atmosphere*. The jump activation enthalpy for forming a Snoek atmosphere of *carbon* in α-iron is nearly 0.5 eV. The break away dislocation stress on the atmosphere is proportional to the impurity concentration, and independent of temperature. The retardation of moving dislocations by a Snoek atmosphere is a maximum for the condition $v_d = vR$, where v_d is the dislocation speed, v is the jump frequency, and R is the radius of the atmosphere. For α-iron at 293 K we have $v_d = 20$ interatomic distances per second, so the corresponding rate of *strain* is $\approx 10^{-6}/\text{s}$. The blocking of dislocations by the Snoek atmosphere substantially influences the value of the lower *yield limit*, and the phenomenon of *microcreep*.

SNOEK–KÖSTER RELAXATION (I. Snoek, 1941; W. Köster, 1954)

Damping of free *vibrations* in strained *metals* which contain interstitial *impurity atoms* (e.g., in Fe–C, Fe–N *alloys* at 473 K and a frequency 1 Hz). The phenomenon was discovered by Snoek and extensively investigated in *iron* alloys by W. Köster et al. in the 1950s. The phenomenon was also observed in Nb–O, Ta–O, Nb–N, Mo–N alloys, in hydrogenated α-Fe, Mo, and W. The following basic facts were determined experimentally:

(1) the process of Snoek–Köster relaxation requires the presence of *dislocations* and *doping atoms*;

(2) the magnitude of the *relaxation* depends on the lattice type, impurity type, distribution and nature of dislocations;

(3) the enthalpy of activation of Snoek–Köster relaxation far exceeds the activation enthalpy of doping impurity *diffusion*;

(4) the relaxation peak is broad and corresponds to a range of *relaxation times*.

The mechanism of Snoek–Köster relaxation is still a debated topic. One of the popular theories is the *Schoeck model* (G. Schoeck, 1963) which attributes the relaxation to the motion of dislocations together with impurity atom atmospheres. This model fails to explain the occurrence of Snoek–Köster relaxation in hydrogenated metals, and the concentration dependence of relaxation in other alloys. According to the theory proposed by Seeger (A. Seeger, 1979), Snoek–Köster relaxation is caused by the generation of double kinks on *screw dislocations* $(a/2)\langle 111 \rangle$, and by diffusion of impurity atoms in response to the dislocation motion. This model is in qualitative agreement with experimentally observed activation enthalpies, but is inconsistent with data on the concentration dependence of the magnitude and temperature of the relaxation peak, and the influence of the distribution and type of dislocations on Snoek–Köster relaxation. The most adequate interpretation of the available data is provided by the model which suggests that Snoek–Köster relaxation results from the motion of geometric kinks across the secondary *Peierls relief*, which is accompanied by *tubular diffusion* of interstitial impurity atoms. Possible practical applications of Snoek–Köster relaxation are the evaluation of relaxation resistance of *steels*, and the determination of the diffusion coefficient.

SOAKING

A technical operation of filling microcavities and/or *pores* in a solid by melted metals or alloys that are lower melting than the bulk material itself. The aim of soaking consists mainly in obtaining dense, poreless materials, or in consolidating different phases in *composite materials*. Using that operation one may obtain the so-called *soaked materials* for various needs (antifriction, instrumental, construction, electric contact, etc.) with favorable functional characteristics. In certain cases (e.g., during processing of timber, soaking it with a liquid preservative) such soaking is called *impregnation*.

SODIUM, Na

Chemical element of Group I of the periodic system with atomic number 11 and atomic mass 22.9898; it is an *alkali metal*. Natural sodium has only one isotope ^{23}Na. It is widely distributed in the Earth's crust (2.5%), does not occur in free form in nature; but forms compounds (NaCl, Na_2CO_3), as has been well-known from ancient times. Outer shell electron configuration is $3s^1$. Ionization energies are 5.138, 47.29, 71.65 eV. Atomic radius is 0.189 nm; ionic radius of Na^+ is 0.098 nm. Oxidation state is +1, electronegativity is 1.01.

Sodium is a lustrous silvery-white *metal*. It has body-centered cubic lattice; $a = 0.4282$ nm. Density is 0.9727 g/cm^3; $T_{melting} = 371$ K, $T_{boiling} = 1156$ K. Heat of melting 2.605 kJ/mole; heat of evaporation is 106.01 kJ/mole. Linear heat expansion coefficient is $7.21 \cdot 10^{-5}$ K^{-1} (273 to 323 K); specific heat 132.72 $W \cdot m^{-1} \cdot K^{-1}$. Electrical resistivity is 4.288 $n\Omega \cdot m$ (at 273 K). Brinell hardness is 0.6867 MPa; Mohs hardness is 0.4. Metallic sodium is plastic, easily cut by a knife, drawn through dies, but becomes brittle if deeply cooled.

Metallic sodium and its alloys are used as conductors of heat in many technological processes, and also in nuclear reactors. Sodium vapors are used in gas-discharge lamps. *Modification of aluminum alloys* by sodium (up to 0.05%) improves their strength; alloys with lead are used for the manufacture of bearings. Metallic sodium is used as a reducing agent during the course of refinement of niobium, tantalum, titanium, zirconium, and hafnium.

SOFTENING

See *Dehardening*.

SOFT MAGNETIC MATERIALS, rarely called soft magnetics

Magnetic materials that do not retain appreciable magnetization after the removal of an applied magnetic field, and are easily demagnetized. Soft magnetic materials have a narrow *hysteresis* loop with a *coercive force*, B_c, smaller than 1 mT, low *magnetic loss*, and high *magnetic permeability*: both the initial one, μ_a, and the maximum one, μ_{max}. During the remagnetizing of soft magnetic materials with a high overall *magnetic anisotropy*, *domain wall displacement processes* prevail, with *magnetization rotation processes* in the magnetization curve being dominant for low anisotropy. Soft magnetics are subjected to different kinds of *heat treatment* to obtain the best magnetic properties. This provides upgrading of the chemical purity, improvement in the material structure, changes in the *long-range and short-range order* of atoms, the crystal grain moderation, and the formation of a favorable crystalline (see *Texture*) and *magnetic texture*. The material properties may vary over a wide range, depending on treatment procedures.

Soft magnetic materials are classified by their chemical composition and application, and major classes are:

(1) Pure *iron* (*Armco iron*) and low-carbon *steels* have a high saturated magnetic flux density, $B_s = 2.15$ T, and low electric resistivity, $\rho = (10–12) \cdot 10^{-6}$ $\Omega \cdot cm$; they are used to manufacture cores in electrical engineering devices operating under the conditions of quasi-static *magnetic reversal* at low frequencies.

(2) Fe–Si alloys with Si content of up to 6.5% (electrical steels) exhibit their *magnetostriction* constant, λ_s, decreasing down to zero, and ρ increasing up to $80 \cdot 10^{-6}$ $\Omega \cdot cm$ with increasing Si content. Yet B_s falls off to 1.8 T at such an increase, while *hardness* and brittleness become higher. With Si contents below 3.5%, sheets are produced with cold rolling, whereas hot rolling is used at higher Si contents. A method has been developed to manufacture thin sheets of silicon-rich alloys: *quenching* right from the melt yields a fine-grain flexible strip. Good magnetic properties can be obtained by *annealing* at 1200 °C in vacuo, but at the cost of *embrittlement*.

(3) Fe–Al and Fe–Si–Al alloys exhibit high values of $\rho \sim 100 \cdot 10^{-6}$ $\Omega \cdot cm$, μ_a and μ_{max}, with λ_s varying between 0 and $40 \cdot 10^{-6}$. These alloys are very responsive to heat treatment, including a *thermomagnetic treatment* and a *thermomechanical treatment*. Alloys possessing high μ are used for recording heads, and those with high λ_s for magnetostrictive transducers.

(4) Alloys based on Fe–Ni (35 to 90% Ni) have their electric and magnetic properties variable over wide ranges by alloying them with Mo, Cr, Cu and Co and by heat treatment: μ_a and μ_{max} may reach values of up to 10^5 and 10^6 (accordingly, B_c changes between 0.002 and 10 G, or between 0.0002 and 1 mT) (see also

High-permeability magnetic materials). These alloys are very ductile, so one can roll them as thin as few micrometers. Alloys based on Fe–Ni include those with high permeability and those with a rectangular or slanted hysteresis loop. The latter show a weak $\mu(B)$ dependence (see *Perminvar*), and are used in radio engineering, instrument making, automation, etc.

(5) Alloys based on Fe–Co are marked for their enhanced values of $B_s = 2.4$ T and T_C (the *Curie point*) of up to 980 °C; they are used for transformer cores with a high *magnetic flux* density and reduced mass operable at elevated temperatures. The alloy of 49% Co, 2% V, the remainder Fe (*permendur*) possesses $\lambda_s = 70 \cdot 10^{-6}$; it is employed for magnetostrictive transducer cores (see *Magnetostriction*).

Also classified among soft magnetic materials are some *ferrites, thermomagnetic materials, magnetostrictive materials*, alloys with high *corrosion resistance*, and *metallic glasses* (amorphous ferromagnets).

SOFT MODE

Mode of *collective excitations* (quasi-particles) with a frequency that tends to zero at the instability point of a given phase of the material, most often at the point of a *second-order phase transition*. In the latter case, the soft mode describes oscillations of the *order parameter*, and the *phase transition* may be represented as a *soft mode condensation*. In this case the soft mode frequency ω_0 follows the *Cochran relationship* (W. Cochran, 1960) $\omega_0 \propto (T - T_c)^{1/2}$, where T is an external parameter, ordinarily the temperature, and T_c is the value of this parameter at the phase transition point, e.g., at the *Curie point*. One may also observe soft modes in the cases of *first-order phase transitions*, which are close to second-order ones. In this case, the soft mode frequency, while failing to vanish, becomes abnormally small at the transition point. For *structural phase transitions*, the soft mode is one of the optical phonon modes; in particular, for displacive transitions in *ferroelectrics*, the soft mode is a transverse mode of optical *phonons*, and describes oscillations of electric polarization. For *magnetic phase transitions* between commensurate phases, the soft mode frequency becomes zero at the center of the *Brillouin zone*. For transitions involving enlargement of the unit cell (e.g., for certain *antiferroelectric* materials and *antiferromagnets*), one may observe a zero value of the soft mode frequency at the boundary of the Brillouin zone. For transitions to incommensurate phases, e.g., to *modulated magnetic structures*, the soft mode frequency may vanish at arbitrary points of the Brillouin zone.

SOL (fr. Germ. *sol*, colloid solution)

A colloidal system where the solid particles of the dispersed phase (micelles) are small enough to pass through filter membranes, and undergo Brownian motion in the liquid dispersing medium. We can distinguish liquid *lyosols* and gaseous *aerosols*. *Micelles*, which range in size from 10^{-7} to 10^{-9} m, consist of a nucleus and an adsorption-solvate shell which includes a double layer at the surface. Sol stability is determined to a large extent by the structure of the *electric double layer*, especially by its thickness, which in turn depends on the ionic force of the dispersing medium. Sols are divided into lyophilic (stable systems) and lyophobic, with the latter often characterized by thermodynamic instability, aging with time, coagulation, and structure formation processes. One principal feature which distinguishes a sol from a true solution is its capability of opalescence. Sols with a disperse phase (see *Disperse structure*) that consists of conducting particles (*metal sols*) are intensely colored, with the color determined by the particle size. Metal sols are used to manufacture dyed glasses, e.g., ruby glass is a colloidal solution of *gold*.

SOLAR CELL

A semiconductor device which brings about photovoltaic conversion of sunlight. The solar cell depends for its action on the photoconductive effect (see *Photoeffect*), which is responsible for the generation of free *current carriers* in bands under illumination. The separation of electrons and holes in space, which induces an *electromotive force*, is achieved through the action of the electric field, which exists in a homogeneous *semiconductor junction*, in a *heterojunction*, or in surface barrier structures based on a *metal–semiconductor junction* or a *metal–insulator–semiconductor structure*. The losses of solar thermal power conversion are due to the following factors: (1) photons of energy less than the *band gap* pass through the junction; (2) surplus energy due to photon absorption of energy which exceeds the gap width is converted to heat; (3) the reflection of photons from the solar cell surface; (4) *recombination* of charge carriers in the bulk semiconductor, in the *space charge region*, and at interfaces. The maximal efficiency, which is 27.5% for a single-transition solar cell, may be attained using a semiconductor with the band gap width 1.5 eV. Materials with a gap width ranging from 1 to 2 eV, such as Si, α-Si:H, GaAs, InP, CdTe, CuInSe$_2$, are considered promising materials for the construction of solar cells. The material most often used for this purpose is Si; the efficiency of silicon cells reaches 20%. The GaAs solar cell is 21–22% efficient. Of wide application are *thin-film solar cells* in amorphous Si, with an efficiency in excess of 11%.

A further increase in efficiency may be achieved through application of cascade solar cells, which consist of a series of *semiconductor materials* with different band gap widths arranged one above another in such a way that the band gap width decreases away from the illuminated surface. Solar cells are used as independent energy sources both on Earth and in space.

SOLAR CELL, THIN FILM

See *Thin-film solar cells.*

SOLDER

A substance or alloy used to fill the gap between joined surfaces in order to obtain a monolithic soldering seam. Solders feature lower *melting temperatures* than the joined metals. One may distinguish between soft solders with melting temperatures up to 400 °C (such are the Pb-, Sn-, Cd-, Bi-based alloys), and hard solders with melting temperatures above 550 °C. The latter typically display greater *strength.* Hard solders are based on Cu, Ag, Ni, Zn.

SOLID

A *state of matter* characterized by the stability of shape and the nature of the thermal motion of atoms which perform small vibrations about their equilibrium positions (see *Thermal displacements of atoms*). There exist crystalline and amorphous solids. Crystals are characterized by the regularity of the spatial arrangement of the equilibrium positions of atoms at the sites of the *crystal lattice*. In amorphous solids the atoms vibrate about more randomly distributed points. Particular types of solids are *quasi-crystals* and gaseous crystals (see *Gaseous crystalline state*). The stable state of a solid (corresponding to minimum internal energy) is a crystalline state. From the thermodynamic viewpoint, an amorphous solid (see *Amorphous state*) is in a *metastable state*, and must eventually crystallize in the course of time (see *Crystallization*). All substances in nature (except for *helium*) solidify at atmospheric pressure and temperatures $T > 0$ K.

Studies of properties of solids are part of the field of condensed matter physics whose development is stimulated by the needs of technology. Solid state physics has been a source of new materials; and many new physical concepts that first appeared in solid state physics have influenced and contributed to the progress of high-energy physics, nuclear physics, astrophysics, biophysics, and other branches of science.

The properties of solids can be explained in terms of their atomic-molecular structure and the laws of motion of their atoms (*atoms, ions*, molecules) and subatomic particles.

Atoms in a solid. Interatomic bonds. Atoms, molecules and ions serve as structural units of solids. The *crystal structure* of a solid depends on the forces acting between the atoms. The same atoms can form diverse structures, e.g., gray and white *tin, graphite* and *diamond*, and so on (see *Polymorphism*). By changing interatomic distances with the help of external pressure, it is possible to change the crystal structure and properties of solids. A great number of various crystalline modifications were found to form under *high pressures*. Under extreme pressures when the volume per atom becomes smaller than an ordinary atomic size the atoms lose their individuality, and the matter transforms into a compressed electron–nuclear plasma. The study of such a condition of matter is of importance, in particular, in order to gain insight into the structure of stars. Alterations of the structure and properties of solids (*phase transitions*) can also take place at particular temperatures, brought abut by the action of *magnetic fields* and other external actions (see *Phase transitions in an external field*, *Phase transitions under pressure*, *Phase transitions in a radiation field*).

Solids are divided into five types, each characterized by its spatial electron distribution: (1) *ionic crystals* (NaCl, KCl, etc.); (2) *covalent crystals* (diamond, Ge, Si); (3) *metals* (Cu, Ag, Pt); (4) *molecular crystals* (N_2, O_2, Ne, Ar, etc.); (5) *hydrogen bonded* crystals (*ice*). Although the forces acting between atomic particles in solids are quite diverse, their fundamental origin is electrostatic attraction and repulsion. The formation of a stable solid from atoms and molecules demonstrates that at a distance of $\sim 10^{-8}$ cm the attraction forces are counterbalanced by repulsion forces (the latter are quantum-mechanical in nature, and decay rapidly with distance). The *melting temperature* $T_{melting}$ of solids of different types is different: for hydrogen $T_{melting} = -259.1\,°C$; for tungsten $T_{melting} = 3410\,°C$; for graphite $T_{melting} = 4000\,°C$. Helium remains liquid down to absolute zero at atmospheric pressure.

Mechanical properties of solids are determined by the bonding forces acting between its structural elements and particles. The diversity of these forces leads to a variety of mechanical properties: some solids are plastic and others are brittle (see *Alloy brittleness*). As a rule metals are more plastic than *insulators*. *Plasticity* usually grows with an increase in temperature. Under light loading all solids exhibit elastic *strain*. The crystal strength does not reflect the power of the interatomic bonds. The low strength observed in *real crystals* is explained by the presence of macroscopic *defects* and *dislocations*. Under appreciable mechanical load, the crystal reaction depends on the presence or the absence of dislocations. In most cases these determine the plasticity of a solid. The mechanical properties of solids depend on their pretreatment which introduces or removes defects. *Diffusion* in solids depends substantially on the presence of *point defects* (*vacancies* and *interstitial atoms*).

Crystal lattice dynamics. The vibrational motion of atoms and ions in solids persists up to the melting temperature $T_{melting}$. Even at $T = T_{melting}$, the mean amplitude of the atomic vibrations is much smaller than interatomic distances, and the melting is due to that fact that the *thermodynamic potential* of the liquid at $T = T_{melting}$ is less than the thermodynamic potential of the solid. The quantization of atomic vibrational motion involves the concept of a *phonon*, and the description of the thermal properties of solids in terms of a *quasi-particle* (phonon) gas. See also *Crystal lattice dynamics*.

Electrons in solids. The quantum-mechanical description the motion of electrons in a crystal-lattice periodic field provided an explanation of the motion of electrons in a crystal, and stimulated the creation of the *band theory* that is the basis of the modern electronic theory of the solid state. In terms of the *band structure* picture of the filling the electron bands by electrons, solids are naturally divided in *metals* (conductors), *insulators* (dielectrics), *semiconductors* and *semimetals*. The presence of defects and impurities in a crystal results in the formation of additional (impurity) energy levels (see *Donor*, *Acceptor*, *Color centers*, *Local electronic levels*, etc.). It seems that there are no strictly forbidden energy bands (see *Band gap*) in amorphous solids, however, there are quasi-band gaps (see *Pseudogap*) where the density of states is considerably less than in allowed bands. The existence of an analogue of band structure in amorphous solids justifies their division into metals, insulators, and semiconductors. Modern ideas about electron motion, their interaction with phonons, each other, and defects provide an explanation of the thermodynamic, kinetic, and high-frequency properties of solids of various types. In particular, the nature of *superconductivity* has been clarified.

Magnetic properties. At sufficiently high temperatures all solids are diamagnetic or paramagnetic. *Diamagnetism* is a common feature of all atomic systems. The *conduction electrons* (in metals and semiconductors) also contribute to the diamagnetic *susceptibility* (see *De Haas–van Alphen effect*) owing to their quantized motion in the plane normal to a magnetic field. *Paramagnetism* is a consequence of the orientation of *magnetic moments* of atomic and subatomic particles by a magnetic field. The splitting of energy levels in a magnetic field produces *electron paramagnetic resonance* (EPR). The structure of magnetic levels is highly sensitive to the environment of a particle. Therefore, EPR can be a source of information about the arrangement of atoms in a crystalline *unit cell*, *chemical bonds*, defects, and so on. By decreasing the temperature, at T_C (i.e. at the *Curie point*) some *paramagnets* (insulators and *transition metals*) pass to either a ferromagnetic or antiferromagnetic state (see *Ferromagnetism, Antiferromagnetism*) for which an ordered orientation of atomic magnetic moments is specific in the absence of an external magnetic field. Non-transition metals, as a rule, remain paramagnets or diamagnets down to the absolute zero.

The role of atomic nuclei is not restricted to the fact that they are the principal contributors to the mass of a solid. A quantum quenching of most motions in a solid can bring to light the contribution of nuclear magnetism (if nuclei possess magnetic moments). Nuclear magnetic levels manifest themselves in the resonance absorption of electromagnetic energy (see *Nuclear magnetic resonance*). NMR is one of the methods used in studies of solids since the structure of nuclear magnetic levels depends substantially on the local surroundings of the nuclei, and in particular on the properties of their atomic electron shells. Many nuclear processes in solids acquire some specific features that allow one to use them for studying the properties of solids. For example, studies of electron–positron annihilation (*Positron spectroscopy*) provide information on the properties of the electron system in solids; the resonance absorption of γ-quanta (see *Gamma ray*) by nuclei provide information on the local intracrystalline fields (see *Mössbauer effect*), and many other fine details of the solid state.

Interaction of fast charged particles with solids. The regular arrangement of atoms influences the energy transfer from an incoming fast particle to atoms of a solid. On the one hand, there is a strong dependence of the *mean free path* of the fast particle on the direction of its motion with respect to the crystallographic axes (see *Channeling*). On the other hand, the irradiation of a solid by fast particles and *photons* affects the properties of the solid (see *Radiation physics of solids*).

Role of the surface. Each solid possesses a surface by means of which the body comes in contact with the environment. The surface of a solid plays the determinative role in such phenomena as catalysis, *corrosion*, crystal growth (see *Crystallization*), and so on. See *Solid surface*.

SOLID EFFECT

One of the methods effective for obtaining *dynamic nuclear polarization*. If there are *paramagnetic centers* in a *solid* and the nuclei of the matrix possess *magnetic moments*, then the polarization of nuclei, which are close to paramagnetic centers, may be achieved through the intermediary of the electron–nuclear *magnetic dipole interaction* with the microwave saturation of a *flip-flip transition* ($\Delta M = \Delta m = \pm 1$) and a *flip-flop transition* ($\Delta M = \pm 1$, $\Delta m = \mp 1$), which are not completely forbidden in this case. Here M, m are respectively the projections of the electronic ($S = 1/2$) and nuclear ($I = 1/2$) spins. This produces the nuclear polarization $P_1 = \pm g\beta B/(2k_B T)$ (the sign depends on which transition is saturated), which exceeds the thermodynamic equilibrium nuclear polarization $P_1^0 = g_n \beta_n B/(2k_B T)$, where $\beta/\beta_n = 1836$, so the overall factor $g\beta/(g_n\beta_n) \sim 10^3$ (β is the Bohr *magneton*, β_n is the nuclear magneton, g and g_n are, respectively, the electronic and nuclear g-factors). This does not take into account a correction for the "*leakage factor*" of the nuclear polarization. The polarization subsequently propagates into the bulk of the crystal by *nuclear spin diffusion*. Activation of this mechanism of nuclear polarization allows one to create highly polarized nuclear targets (usually proton-containing) for elementary particle scattering experiments; there is no need for the extreme conditions (very low temperatures, strong magnetic fields), necessary for static methods, such as the "brute force" approach (see *Nuclear orientation*).

SOLID ELECTROLYTES, superionic conductors

Crystalline or amorphous materials whose *ionic conductivity* σ is comparable with σ of electrolytic melts, and is characterized by a low *activation energy* (~ 0.1 eV). An example of a classical solid electrolyte is the *ionic crystal* AgI which at 420 K passes from a low-temperature *beta phase* with $\sigma \leqslant 10^{-4}$ (Ω·cm)$^{-1}$ to a high-temperature superionic *alpha phase* with $\sigma \geqslant 1$ (Ω·cm)$^{-1}$. Solid electrolytes are divided into cationic (e.g., Ag^+ ions move in AgI and $RbAg_4I_5$, and protons in $CsHSO_4$) and anionic (e.g., PbF_2 where F^--based high conductivity is reached without a clear transition). In some solid electrolytes σ is highly anisotropic, it is bidimensional in

β-alumina, unidimensional in β-eucryptite, and so on. Many solid electrolytes such as α-AgI are, in fact, purely ionic conductors, although there exist solid electrolytes with a noticeable electron contribution to σ (e.g., α-Ag$_2$S). Superionic glasses (Ag–Ag$_2$O–B$_2$O$_3$, and so on) are also known. The superionic conduction is a consequence of the "melting" of one of the sublattices of the *solid electrolyte*. Thus, in α-AgI, I$^-$ ions are situated in the framework in the form of a body-centered cubic lattice while two ions Ag$^+$ in every unit cell are distributed statistically over 42 crystallographic positions [so-called *Strock model* (L.W. Strock, 1934)] and can easily migrate through channels that pass over saddle points of the crystal potential. The analytical description of solid electrolytes is grounded on the idea of movable *defects* with various ways of specifying their interaction through the following models: the hydrodynamic model, the model of a *lattice gas*, versions of the *Debye–Hückel theory*, etc. Solid electrolytes are used for making electrochemical current sources, capacitors of high capacity, chemical transducers, and so on.

SOLIDIFICATION

Transition of a substance from a liquid to a crystalline solid (or amorphous) state. The transformation of the liquid phase to a crystalline one is accompanied by the release of the latent heat of *crystallization*, and is characterized by two stages, the formation of the crystal nucleating centers followed by their growth. With increasing supercooling ΔT or supersaturation ΔC of the melt (T is temperature, C is solute concentration), which is the motive force of the process, both the rate of nucleating center formation J and their growth rate v increase. The crystal growth rate becomes noticeable at $\Delta T \approx 0.1$ K. Meanwhile, the formation of the crystal nucleating centers requires large (so-called threshold) values of supercooling, ~ 10–10^2 K, depending on the purity and nature of the material, the presence of solid particle impurities, their catalytic activity, etc. Upon deep supercooling of the melt, both parameters J and v become small, which is related to intense reduction of the mobility of molecules and, accordingly, to the increase of the melt *viscosity*. As a result, neither the formation of new crystal nucleating centers, nor their growth, are possible. Under such conditions of suppression of the crystallization processes, the liquid passes to a solid *amorphous state*, which corresponds, in its structure, to a quenched liquid with short-range but not long-range order (see *Long-range and short-range order*). Cooling rates that preclude nucleation processes can bring about transitions from the liquid state to an amorphous one. These cooling rates depend on the degree of purity, and on the nature of the material. Thus, for the materials with long-chain molecules (salol, etc.) having high viscosities, the transition to the amorphous state occurs at slow cooling rates 1–10 K/s; whereas in the case of metals with low viscosity of the melt, much faster cooling rates are needed, 10^7–10^8 K/s. The reverse transition requires the same heating rates.

SOLID-PHASE POLYMERIZATION

Polymerization of monomers in the crystalline (or vitreous) state. The fixed mutual arrangement of monomer molecules in a *crystal* results in certain kinetic and structural characteristics of solid-phase polymerization which distinguish it from other polymerization processes.

As a rule, solid-phase polymerization is initiated by radiation (γ-rays, X-rays, fast electrons). The measure of performance of solid-phase polymerization depends significantly on the ratio between the parameters of the *crystal lattice* of the solid monomers and those in the related polymer chain. If the formation of macromolecules requires substantial displacements of the monomer molecules from their equilibrium positions in lattice sites, then the polymer phase is formed as localized *domains* with the orientation of macromolecules uncorrelated with the orientation of any monomer crystal axis. In some cases this results in the termination of the polymerization long before the completion of the conversion of monomers (e.g., this occurs at low-temperature polymerization of polyacrylonitrile). However, if the formation of polymer phase domains is accompanied with an increase of the number of defects involved in the new onset of nucleation then the process of solid-phase polymerization may exhibit auto-acceleration with an increase of the conversion depth (e.g., solid-phase polymerization of acry-lamide). The rate of solid-phase polymerization of this type grows with the increase in temperature, and rises sharply in the vicinity of *phase transition* points.

Another situation emerges in those cases when only a comparatively small change of their positions in crystal lattice sites is needed for the linkage of monomers by *chemical bonds*. The macromolecules formed as a result of the solid-phase polymerization are oriented along a certain axis of the original crystal along which the mutual arrangement of monomers is optimal for the formation of chemical bonds between them (*topotactic process*). The rate of the of the polymerization process grows when approaching the *melting temperature*, and then drops in the close vicinity of the melting point where the phenomena of monomer crystal disordering becomes substantial.

At sufficiently low temperatures, the electronic mechanism of chain growth of solid-phase polymerization can be stimulated by *tunnel effects* in spite of the thermal activation; and this leads to a diversity of the features of low-temperature polymerization.

SOLID-SOLUTION HARDENING

An increase of the strength of *steels* and *alloys* as a result of *alloying* (without the formation of new *phases*) which arises from an increase of the *internal friction* forces during *dislocation* motion in crystals, and the blocking of dislocations by atmospheres of *impurity atoms*. There exist three kinds of solid-solution hardening induced by the interaction of dislocations with the atoms of dissolved elements: (1) the dislocations are immovable, the atoms of dissolved elements are movable, and this results in friction resistance; (2) the dislocations are movable, the atoms of dissolved elements are immovable, and this results in blocking the dislocations by impurities with the formation of impurity "atmospheres"; (3) the dislocations and the atoms of dissolved elements both move; and this results in the phenomenon of dynamic *strain aging*.

The friction resistance caused by the atoms of dissolved elements (*substitutional alloy*), $\tau_{im} = ZG\varepsilon^{3/2}c^{1/2}$ where $Z = 1/760$, G is the *shear modulus*, ε is the misfit parameter resulting from

both the mismatch of the *crystal lattice* parameters between the dissolved elements and the host, and the mismatch of their *elastic moduli*; c is the concentration of atoms of dissolved elements. The elastic interaction energy of dislocations with *doping atoms* is higher than that with *substitutional atoms*; the atmospheres formed by atoms of carbon and nitrogen can effectively block the dislocations in steels and alloys with a body-centered cubic lattice. The atmospheres due to chemical interactions of impurity atoms with the *stacking faults* of split dislocations (*Suzuki atmospheres*), which are specific for *solid solutions* with a face-centered cubic lattice, are more thermally resistant then the elastic *Cottrell atmospheres*. The formation of a solid solution can result not only in an increase of the strain resistance at low temperatures ($T \leqslant (0.2\text{–}0.25)T_{melting}$), but also in an increase of the *high-temperature strength* of alloys.

SOLID SOLUTION, IDEAL

See *Ideal solid solution*.

SOLID SOLUTIONS

Homogeneous crystalline materials which consist of two (or more) components that preserve their homogeneity under a changing concentration ratio of the components. There are solid solutions of substitution, interstitial intrusion, and subtraction types. In *substitutional solid solutions* the atoms of dissolved components occupy solvent site positions in the *crystal lattice*. The *alloys* of copper with nickel, copper with silver, of tungsten with rhenium, etc., serve as samples of substitutional solid solutions. If the components substitute for each other in all possible proportions, a continuous series of solid solutions results. The solid solutions are called limited if their range of existence in a *phase diagram* is restricted to a certain region adjacent to pure components. In *interstitial solid solutions* the atoms of dissolved components are at interstices in the solvent lattice. Ordinarily in this case the atomic sizes of components differ significantly one from another. As a rule, such solutions are formed by H, B, C, N, and O. *Silicon* (atomic diameter 0.235 nm) forms substitutional solid solutions in *metals* with atomic size ≈ 0.25 nm, while in niobium (atomic diameter 0.285 nm) it forms an interstitial solid solution. Atoms of components of solid solutions can be either randomly distributed over the lattice sites (interstices), or partially or completely ordered with a regular arrangement of atoms of different kinds relative to each other. A fully ordered lattice forms only at certain (stoichiometric) concentration ratios of components. *Subtraction solid solutions* can exist in ordered solutions of non-stoichiometric composition. For example, in the case of an excess of aluminum atoms in solid solution based on the compound CoAl (CsCl-type structure), these atoms occupy sites in the Al sublattice with resulting unoccupied sites in the Co sublattice. Ordered solid solutions can exist if atoms of different components interact more strongly than identical atoms. As the temperature rises, the degree of the long-range order (see *Long-range and short-range order*) decreases, and the *order parameter* goes to zero at a certain critical temperature. Solid solutions in equilibrium contain *point defects* (*vacancies* and *interstitial atoms*) which lower the crystal free energy. There are rules concerning solid state solubility limits (see *Hume-Rothery rules*). One of the rules points out that if the difference in atomic sizes of components exceeds 14–15%, the *solubility* would be limited. If the difference in sizes is less than 15% then the solubility is determined by other factors, such as the alloy electron concentration, and the chemical affinity of components. The homogeneity of solid solutions is violated by concentration nonuniformities (clusters and *segregations* of the same atoms) which form when the interaction between identical atoms is stronger than that between different ones. In this case, the temperature decrease often leads to the decomposition of the solid solution (see *Alloy decomposition*) with the appearance of new *phases*. If one rapidly cools (quenches) a solution from a temperature where it is thermodynamically stable (region of homogeneous solid solutions in the phase diagram of the related alloy) down to a temperature at which the rate of diffusion processes is insignificant, then it is possible to obtain a *supersaturated solid solution*. The heating of a supersaturated solid solution accelerates its decomposition. At early stages of decomposition clusters of impurity atoms are formed (called *Guinier–Preston zones* in solid solutions based on Al), while at final decomposition stages particles of phases in excess are precipitated and isolated. These processes are known as *decomposition of supersaturated solid solutions* (an engineering term is *alloy aging*), and they can be utilized for improving the metal material properties.

SOLID STATE CHEMISTRY

Branch of chemistry concerned with the chemical properties and composition of *solids*, general rules and specific features of chemical and other reactions in solids, as well as methods of synthesis, and practical applications of various solids. This field began with studies of *chemical bonds* and *crystal structure*. After the discovery of *defects* in crystals, the determination of the role played by these defects in chemical reactions, and the effects of *diffusion* and other processes, a new trend in solid state chemistry appeared: the chemistry of imperfect crystals. This discipline studies the structure of defects, interactions of defects with one another and with the *crystal lattice*, participation of defects in chemical and physicochemical transformations. An important part of solid state chemistry is the thermodynamics of the solid state, which includes processes of *phase transitions* and heterogeneous equilibria. Solid state chemistry also investigates chemical reactions in solids, the characteristics of which are strongly influenced by the high degree of order of the lattice structure, the presence of various types of crystal lattice defects (including *solid surface* defects), *phase interfaces* between initial substances and products, and the slow rate of transport processes in solids. A distinguishing feature of solid state reactions is the nonstationary nature of these processes, which is due to the heterogeneity of systems which undergo transformations: disappearance of phases of initial substances, formation of phases of products, generation of a number of intermediate phases. The procedures of *melting* and *crystallization, diffusion, adsorption* and *desorption* have significant effects on the rates of solid state reactions, and determine the dispersity and structure of the reaction products. Reactions in solids begin at particular sites on the surface or in the bulk of the crystal (so-called reaction centers). If only particular centers localized at a surface separating the phases of a solid reagent and a product are capable of initiating a reaction, then the process is called a *topochemical* type. Typical topochemical reactions are: reduction of metals from ores, roasting, and alkali treatments.

SOLID-STATE LASER

Generator of coherent radiation in the optical range using an activated *crystal*, glass, or a semiconductor to serve as the active medium (see *Semiconductor laser*). Among the more widely used solid-state *lasers* are the ruby (Cr doped Al_2O_3), yttrium–aluminum garnet (YAG), and neodymium glass systems. The first one operates as a three-level system, while the latter two employ four energy levels. The pumping systems are optical. Gas discharge sources operating in either the pulse or the continuous wave (CW) mode induce electronic transitions that transform active centers into excited states, and these centers quickly relax to a metastable state. Then, due to feedback, these same centers in the three level system return to the ground state in a forced mode by emitting coherent monochromatic radiation. Feedback is provided by two mirrors, one of them semitransparent, and the other totally reflects the generated radiation. These mirrors form a resonator of the *Fabry–Perot interferometer* type with the active material placed within it. In systems operating in a four-level mode, the quantum radiation transition does not take place to the ground state, but rather to a higher energy level that is depleted of electrons. Therefore, the energy threshold is much lower for producing an inverted population of active centers whose concentration at the upper operating level exceeds that at the lower one (see *Level population*). Such solid-state lasers may operate in a continuous mode even at room temperature, while three-level systems require cooling of their active media.

Besides the active media mentioned above, numerous dielectric crystals are used for this purpose, and among them are oxides of rare and alkaline earth elements (niobates, molybdates, tungstates, etc.), some of which are simple and others quite complex in their chemical composition and *crystal structure*. *Rare-earth elements* are more often used as activators, and all such systems function via electronic transitions. *Color center lasers* have been developed utilizing *alkali-halide*

crystals with *color centers* (F_2^+, F_2^-, F_A, F_B, etc.) and molecular ions. Lasers of the first group also operate using electronic transitions between energy levels of their color centers, while those of the second group employ vibrational transitions of an impurity ion such as CN^-. Depending on the parameters of the active center, solid-state lasers emit coherent beams within the wavelength range 0.88–3 μm, or lower.

In contrast to solid-state lasers that use dielectric crystals and glasses, with active bodies tens of centimeters long, semiconductor lasers are much smaller-scale systems. The utmost achievement in semiconductor laser technology is the generation at superlattices (*quantum wells*) that consist of alternating *thin films* of semiconductor materials (each about tens of nanometers thick) with broader or narrower *band gaps*. Typically they having higher *quantum yields*, consume very little energy, have a lower generation threshold, and are easily controlled.

Emissions produced by modern semiconductor lasers cover ranges across the optical spectrum from the UV to several microns into the IR. The materials most effective for use in semiconductor solid-state lasers are compounds of the compositions A_3B_5, A_2B_6, A_4B_6, and mixed crystals grown on their bases (see *Semiconductor materials*).

SOLID-STATE MEMORY

Crystalline and film devices for data storage in computers. The physical principles of solid-state memory elements can be classified in the following fashion:

(1) rapid reversible electronic processes in semiconductors, cryoelectronic, and perhaps in the future molecular-electronic and molecular-optical apparatus and devices designed for a computer main memory;

(2) electron transitions between *shallow levels* and *deep levels* in *semiconductors*, mainly, in *metal–insulator–semiconductor structures* with a double-layer *insulator* designed for nonvolatile alterable memory units with an electronic *information recording* and an electronic or optical (UV) erasable memory;

(3) controlled transport of *magnetic domains* or *magnetic domain walls* (including *magnetic*

bubble domains) in nonvolatile semipermanent ready-access memory units;

(4) magnetic domains with immovable (with respect to carrier) boundaries at movable carriers such as magnetic disks, tapes, and magnetic drums, i.e. auxiliary computer storage;

(5) magnetooptical systems (see *Magnetooptical effects*) with laser recording and reading of information: *optical storage disks* for auxiliary
high-capacity (to 10^{10} bit/cm^2) storage with
reloading (see *Optical recording media*);

(6) optical single-recording systems using irreversible media;

(7) optical systems without reloading: the physical principle of recording (the same as in
item (6)) is the formation of microrelief or an
irregular pattern at the disk surface.

The above classification is not exhaustive. Any
nonlinear electromagnetic phenomenon in solids
can serve, in principle, as a basis for designing
main-memory devices. Any electromagnetic phenomenon that leads to the formation of long lived
states (10^{-6}–10^9 s) can be a basis for permanent
or semipermanent memory units. Therefore, almost all known phenomena in solids have been
proposed for this purpose. However, very few of
the proposed memory devices satisfy the necessary requirements in the areas of technology, reliability, and energy saving. In the sequence of items
(1)–(7) the memory capacity grows with decreasing operating speed. An absolute limit of a solid
storage unit is about 10^{11} bit in the case of surface
recording, and 10^{16} bit for volume recording. The
limiting speed (switching time) of a *solid memory device* is 10^{-12} s. The storage capacity 10^{11}
bit has now been achieved in optical laser storage units, and 10^{16} bit is available in experimental
tests of cryoelectronic (see *Microelectronics*) and
semiconductor devices. The actual limiting rate of
solid-state memory in computers is equivalent to
$(1–2) \cdot 10^{-11}$ s.

SOLID-STATE PLASMA

A subsystem consisting of mobile *current carriers* of both signs (electrons and holes). A neutral *equilibrium plasma* (in which the concentrations of electrons and holes are equal) occurs
in *semiconductors* of very high purity at a high
enough temperature, in *semimetals*, and in a number of *metals* (e.g., in *tungsten, molybdenum*, etc.).
A neutral *nonequilibrium plasma* may be produced in a solid by application of double *injection*,
impact ionization, exposure to light. Examples of
a charged plasma (i.e. a plasma having unequal
concentrations of electrons and holes) are a *metal
electron plasma*, and an electron or hole plasma
of semiconductors having *donor* or *acceptor* centers. The crystal including the charged plasma subsystem is electrically neutral, for the ion matrix of
the *crystal lattice* (in metals) or charged *defects* (in
semiconductors) compensates the plasma charge.

A *monopolar plasma*, which consists of carriers of like sign, may in its turn include groups of
carriers having different mobilities. This specific
feature of plasma is due to the many-valley structure (see *Many-valley semiconductors*) of the *conduction band* (or the overlapping of hole subbands)
in a number of semiconductors and to the complicated shape of the *Fermi surface* in the case of metals. The concentration of current carriers ranges
from 10^{22}–10^{23} cm^{-3} in metals and down close
to zero in semiconductors. Therefore, the plasma
in metals is always degenerate and is described using *Fermi–Dirac statistics*. In a number of semiconductors, the degeneracy occurs at low enough
temperatures even at a concentration of current
carriers $n \sim 10^{14}$ cm^{-3} because of the small *effective mass* of the current carriers $m^*(m_e^*/m \approx
0.01$ in InSb and alloys of the types of $\mathrm{Bi}_{1-x}\mathrm{Sb}_x$,
where m is the free electron mass).

A characteristic solid-state plasma property
is the occurrence of collective excitations called
plasma oscillations (or *Langmuir oscillations*) with
the frequency given by (using CGS units)

$$\omega_\mathrm{p} = \left(\frac{4\pi n e^2}{\varepsilon_\mathrm{L} m^*} \right)^{1/2},$$

where ε_L is the *dielectric constant* of the lattice.

In metals, the energy of *plasmons* $\hbar\omega_\mathrm{p} \approx$
10 eV ($m^* = m$, $\varepsilon = 1$) and the excitation of
plasma oscillations is observed in experiments on
the passage of electrons through thin foil sheets (in
this process, electrons lose energy in quantized increments $\hbar\omega_\mathrm{p}$). Semiconductors exhibit a smaller
value of n and hence a smaller plasmon energy:
$\hbar\omega_\mathrm{p} \approx 0.01$ eV; quanta of this energy manifest
themselves in the process of *Raman scattering of*

light. At the boundary between the solid and the vacuum, the *surface Langmuir waves* of frequencies $\omega_n = \omega_p/(1 + \varepsilon_L)^{1/2}$ ($\omega_n = \omega_p/\sqrt{2}$ in the case of metals) may travel in the plasma. These oscillations are also observed in the spectrum of electron beam losses. The *plasma frequency* ω_p determines the boundary between the regions of transparency and of specular reflection when light is incident on the metal surface: the metallic film is transparent to light at $\omega > \omega_p$ and reflects the light at $\omega < \omega_p$. It is due to this fact that metals exhibit their characteristic lustrous appearance for frequencies of the visible band that are lower than the metal plasma frequency. The plasmons are clearly defined (i.e. decay only slightly) as elementary excitations for $\omega_p \gg \nu_{coll}$, where ν_{coll} is the frequency of collisions of electrons (holes) with the lattice, and for wave numbers $k \ll \omega_p/v$, where v is either the thermal or the Fermi electron velocity (the latter restriction stems from the fact that strong *Landau damping* occurs at $k \approx \omega_p/v$).

In the presence of a strong magnetic field, *helical waves* can travel in a monopolar plasma. These waves are electromagnetic oscillations referred to as *helicons* (O. Konstantinov, V. Perel, 1960). If the wave propagates along the external magnetic field H_0, then the plasma electrons, which are drifting at the rate $v = -icE/H_0$ under the action of the electric field of the wave $E = -i\omega H/(ck)$, produce the Hall current $j = env$, which sustains the alternating magnetic field H:

$$-kH = \frac{4\pi}{c} j = -\frac{4\pi}{c} \frac{en\omega}{kH_0} H.$$

Hence follows the dispersion relation for helicons: $\omega = [cH_0/(4\pi en)]k^2$, where c is the *speed of light*. The helicon frequency may range from 10 Hz in metals (because of the high values of n) to 10^{11} Hz in semiconductors. The ratio between the damping factor γ and frequency ω for these waves is $\gamma/\omega \approx (\Omega/\nu_{coll})^{-1}$, where $\Omega = eH_0/(m^*c)$ is the *cyclotron frequency*, and they decay slightly at $\Omega \gg \nu_{coll}$ (even if $\omega \ll \nu_{coll}$). Figuratively speaking, the discovery of helicons "removed the skin-depth coating from metals": until helicons were discovered, it was believed that waves are unable to propagate in metals because of the *skin-effect*. Measuring the frequency and attenuation of helicons allows determining the concentration

of plasma particles, the collision frequency, and the effective mass of current carriers. Helicons are used in metals for the reconstruction of the shape of the Fermi surface by application of the *Doppler-shifted cyclotron resonance*. As a result of the *Doppler effect*, the current carriers are affected by the helicon wave at the frequency $\omega + kv < kv_F$, where v_F is the *Fermi velocity*. While decreasing the magnetic field, the onset of the cyclotron absorption of helicons starts as the condition $kv_F = \Omega$ is satisfied. Having determined the critical values of the magnetic field, below which strong absorption starts ($H_{critical} \propto \omega^{1/3} v_F^{2/3}$), one is enabled to evaluate the Fermi velocity at different orientations of the magnetic field with a high degree of accuracy. The conduction becomes substantially nonlocal in the neighborhood of the indicated resonance, which brings about a strong spatial dispersion of the plasma dielectric constant (see *Spatial dielectric dispersion*), and the occurrence of new solutions of the dispersion equation, which are referred to as *dopplerons* (waves characteristic of a highly degenerate metal plasma). Unlike helicons, dopplerons may propagate also in the neutral plasma of metals, for the nonlocal conductivity is different for electrons and holes in the neighborhood of the resonance region (the Hall current differs from zero).

Metals placed in a strong magnetic field also exhibit other types of resonance absorption of waves by electrons: *paramagnetic resonance* (*spin resonance*) and Azbel–Kaner *cyclotron resonance*. *Spin waves* and *cyclotron waves* are initiated in the neighborhood of regions of these resonances. There exist branches of spin (in nonmagnetic metals) and cyclotron waves, which are due to the Fermi liquid type interaction between electrons (discovered by V.P. Silin, 1957).

In the case of a neutral plasma in a strong magnetic field ($\Omega \gg \nu$), propagation of linearly polarized electromagnetic waves is possible. These waves are called *Alfvén waves* (after H. Alfvén, 1942). The frequency of these waves is defined by an equation of the form

$$\omega = kV_a,$$

where

$$V_a = \frac{H_0}{[4\pi n(m_e^* + m_h^*)]^{1/2}}.$$

The condition for the low damping of these waves ($\omega \gg \nu$) is stricter than in the case of helicons, because the Hall currents of the electrons and holes compensate one another. These waves were first observed in *semimetals*. By virtue of the controllability of their *phase velocities*, these waves are used for studying the relations with other wave types (e.g., with sound waves).

Of great importance for applied research concerned with the creation of electromagnetic radiation generators is the study of the *solid-state plasma instability*. The instability of the *electron–hole plasma*, namely *helical instability*, was first observed in semiconductors by J.L. Ivanov and M. Ryvkin (1958). They reported that low-frequency oscillations of current (~ 10–15 kHz) arose in *germanium* test samples which were placed into a strong magnetic field ($H_0 \sim 1$ T) with an electric field parallel to it ($E_0 \sim 30$ V/cm). Later it was shown that the amplitude of the oscillating current component amounts to 70% of the constant quantity, and the oscillations are sinusoidal over many hours. The samples (Ge, InSb, Si, etc.) in which such an instability may be set up were given the name *oscillistors*. The phenomenon of current instability owes its occurrence to the evolution of the *Kadomtsev–Nedospasov helical instability* (B.B. Kadomtsev, A.B. Nedospasov, 1960), the theory of which was first developed to describe a gas-discharge plasma. The instability is caused by the Hall drift of a plasma in a constant magnetic field, and transverse to it electric fields, which arise when the electron and hole excitations of density are shifted in the constant electric field. The threshold of instability initiation corresponds to the equality between the Hall and diffusion currents. Therefore, the instability is set up in strong enough fields E_0 and H_0. The development of an oscillistor is accompanied by nonlinear effects: an anomalous increase of the sample resistance due to the increased frequency of the pulse relaxation of current carriers on scattering by helical fluctuations of field and density; and strong *hysteresis* of the threshold conditions of initiation and quenching of the hysteresis effect. Under the conditions of uniaxial deformation of a semiconductor, which exhibits a many-valley structure of the conduction band, the principal oscillistor characteristics (frequency, excitation threshold) are severely changed even under slight deformations because of the *intervalley redistribution* of electrons. This principle provides the basis of operation of *strain gauges* with a frequency output of the load signal. The phenomenon of the existence of oscillistors is used for modeling the behavior of a gas-discharge plasma.

In a semiconductor plasma, the generation of a static *negative differential conductivity* σ_d (with $\sigma_d < 0$) is possible. The field dependence of the current $j(E_0)$ may be N-shaped in this case (the electric field is not a single-valued function of the current). Under these conditions, the initial current fluctuation in the sample increases, for the run-off of electrons from the fluctuation region decreases with increasing field intensity. Thus a dipole layer ("$-$" at the cathode side and "$+$" at the anode side) of space charge arises, which moves in the anode direction. At fixed voltage, the drift velocity of electrons beyond the *domain* decreases with the increase of field intensity in it. The build-up of the field stops when the domain velocity equals the velocity of electrons outside the domain. Such instability is exemplified by high-frequency current oscillations in gallium arsenide and indium phosphide of n-type in strong electric fields in the absence of an external magnetic field (discovered by Gunn, see *Gunn effect*).

If the dependence $j(E)$ is S-shaped (the current density is a multi-valued function of the field) then the sample may exhibit a stratification of current. A broad class of instabilities in semiconductors is due to the generation of a dynamic negative differential conductivity ($\sigma_{d\infty} < 0$) at certain frequencies. The classical example of an ultrahigh-frequency oscillator operating under the condition $\sigma_{d\infty} < 0$ is the *impact ionization avalanche transit time diode* (IMPATT diode). According to the traditional IMPATT diode circuit design, the attainment of $\sigma_{d\infty} < 0$ takes place in the drift space, where the plasmoid injected by the avalanche band (the strong field band, where the impact ionization occurs) is retarded by the oscillating electric field $\widetilde{E}$, and gives up its energy to this field. The avalanche band provides the phase shift of $\pi/2$ between the plasmoid conduction current $\widetilde{j}$ and $\widetilde{E}$ at the moment of avalanche injection into the drift space. The IMPATT effect occurs at saturation of the drift velocity of current carriers in the electric field both in the avalanche band and in the drift space (*Maxwell relaxation* does not occur). The Gunn effect and IMPATT effect provided the basis for the creation of oscillators operating at frequencies up to 300 GHz.

At high rates of *current carrier drift* in the external electric field, resonance excitation of bulk sound and cyclotron waves may take place in a semiconductor (*Cherenkov excitation mechanism*). Excitation of the surface waves of this type is brought about by bunches of high-speed electrons, which pass parallel to the sample surface.

In a nonequilibrium degenerate semiconductor plasma, a *population inversion* between the *conduction band* and the *valence band* may arise. The quanta of energy $\hbar\omega < \mu_e + \mu_h + E_g$ (μ_e, μ_h are the Fermi levels of electrons and holes, which are measured from the bottoms of their corresponding bands, E_g is the width of the *band gap*) do not get absorbed, but bring about the *radiative quantum transition* of an electron from the conduction band into the valence band. This causes the build-up of a photon avalanche. Population inversion in a monopolar plasma, e.g., between the subbands of light and heavy holes (*p*-Ge), arises in strong crossed electric and magnetic fields even in the absence of pumping into the subband of light holes. If the cyclotron frequency far exceeds the hole collision frequency, then the light and heavy holes drift at the same rate $V_D = cE_0/H_0$. Since the mass of heavy holes m_h^* is appreciably greater than that of the light ones m_L^* ($m_h^*/m_L^* \approx 9$ in *p*-Ge), the heavy hole distribution function in *momentum space* is shifted to regions of higher momenta ($p_h = m_h^* V_D$) relative to the distribution of light holes ($p_L = m_L^* V_D$), and the population inversion occurs in the neighborhood of the value $p = p_L$. The frequency of the direct transition from the maximum of the light hole distribution into the subband of heavy holes is given by a relation of the form

$$\omega = \frac{p_L^2}{2\hbar}\left(\frac{1}{m_L^*} - \frac{1}{m_h^*}\right) \approx \frac{m_L^* V_D^2}{2\hbar},$$

where V_D is the hole drift velocity, $V_D = cE_0/H_0$.

In *p*-Ge ($m_L^* = 0.04m$) at $E_0 = 3$ kV/cm, $H_0 = 1$ T, this frequency corresponds to the sub-millimeter band ($\lambda = 100$ μm). Inversion occurs at sufficiently low temperatures $k_B T < \hbar\omega$ (see also *Semiconductor lasers*).

SOLID-STATE QUANTUM ELECTRONICS

A branch of quantum electronics with deals with the physics and technology for the generation and transformation of coherent radiation by devices based on solids (both *crystals* and amorphous materials such as semiconductors, glasses, and so on). In some cases *semiconductor lasers* and transformers are considered separately, and the term solid-state quantum electronics acquires a narrower meaning. Active media in the devices of solid-state quantum electronics include those with impurity activators, doped glasses, numerous binary (e.g., types $A^{III}B^V$ and $A^{II}B^{VI}$) and ternary compounds of *semiconductor materials*, and *solid solutions* based on them. The devices of solid-state quantum electronics can be divided into two classes. The first class includes oscillators and amplifiers whose operation is based on *induced radiation* (lasers). The main types of *solid-state lasers* are the following: (1) semiconductor lasers; (2) lasers made from doped crystals and glasses, e.g., ruby, garnet, neodymium–glass, etc., lasers; *optical fiber lasers* and *microlasers* (mainly employing mixed salts); (3) solid-phase *dye lasers*, e.g., *film lasers*; (4) lasers based on media with stimulated scattering such as Raman scattering, including magnetic-Raman scattering (*"spin-flip" lasers*), and so on. The second class includes devices for transforming and utilizing coherent radiation on the basis of nonlinear optical phenomena (see *Nonlinear optics*). The main kinds of nonlinear devices are (1) nonlinear resonators which function as bistable optical devices, differential amplifiers ("transphasors"), and optical limiters; (2) harmonic oscillators and *optical parametric oscillators* which transform the incident radiation frequency, e.g., "up-converters" for visualization of IR radiation, and so on; (3) nonlinear mirror-filters designed for wave front reversal; (4) modulators and selectors based on *electrooptical effects* including fiber- and integrated-optical units; (5) *light deflectors* and other scanning electrooptical devices, and so on. Some of these devices are employed in laser and optoelectronic apparatus (see *Optoelectronics*), e.g., high-power light sources at the wavelength 0.63 μm (e.g., based on *yttrium–aluminum garnet* with neodymium impurity), and frequency doublers involving a nonlinear crystal (e.g., on $LiNbO_3$). In many cases the nonlinear elements are introduced inside the laser resonator cavity.

Solid-state quantum electronics also includes techniques for obtaining inverted populations (see *Level population*) in laser systems and solid-state optics under high radiation intensity, including such phenomena as self-focusing (see *Self-action of light waves*), optical damage and *failure* of solids, *ablation* and photochemical processes, laser annealing, multiphoton processes, etc. The technical aspects of solid-state quantum electronics consist in designing new versions of solid-state lasers for various applications, such as *laser technology*, optical coherence, data processing and storage systems (see *Optical techniques of information recording*), atmosphere monitoring, control devices, navigation, medicine, designing integrated optics, optical-fiber technique, designing laser systems for controlled thermonuclear fusion (mainly, on the basis of neodymium–glass lasers), application of lasers in audio and video disk technique, designing various transducers and telemetric devices.

SOLID SURFACE

Type of interface or area of contact which exists between a *solid* and the surrounding gaseous, liquid phase or solid phase of another composition. A solid surface has certain features which differ from the bulk properties of the same material. The atoms of the surface layer are bound to the parent substance by a lesser number of chemical bonds, than atoms in the bulk. Some chemical bonds for surface atoms (which originate from underlying atomic layers) either do not exist (broken, or *dangling bonds*), or are reoriented to neighboring surface atoms (atomic bridges), and thus weakened.

This leads to a change in the mutual arrangement of the surface layer atoms, involving a swelling of the surface layer, and as a result, the generation of a more rarefied structural atomic configuration both along the surface and transverse to it. This swelling is considerably decreased at the second monolayer, and almost vanishes at the depth of 4–5 monolayers. That is how *surface superlattices* arise: periodic structures with lattice

constants which differ appreciably from those in the bulk of the material, e.g., 2×2 (the spacing is double that in the bulk), 2×1; 5×5; 7×7, etc. These figures show how much the spacing of the surface periodic structure (superlattice) exceeds the spacing of the underlying three-dimensional structure. However, there are many examples of the reverse situation, when atoms of the first surface layer are bound more firmly and arranged more tightly, than atoms of underlying layers. Such situations may take place in the case of *atomically clean surfaces* and specifically, in the case of surfaces which border chemically active adhesive substances of the external medium. In the latter case the external substance fastens the solid surface atoms together through binding them by its chemical bonds, which are stronger than the ones intrinsic to the solid under consideration. This "surface reinforcement" is caused by the appearance of an additional constituent of the chemical bond of the surface configuration, or by strengthening of this constituent. For example, covalent (or almost covalent) substances of the $A^{III}B^{V}$ type exhibit an increase of the Coulomb interaction force F in the neighborhood of the surface because of the decrease of the effective dielectric constant ε^* in the region adjacent to the medium, which causes strengthening of the polar chemical bond. Thus, we have for this force

$$F = \frac{q_1 q_2}{\varepsilon^*(z) z^2} \exp\left(-\frac{z}{L_D}\right),$$

where z is distance between the atoms, which carry charges q_1 and q_2, and L_D is the Debye length.

A general thermodynamic property of surfaces is the excess of the *surface energy* E_S with respect to the volume, and the presence of a *surface tension* σ. These features are responsible for the rearrangement of the crystal (molecular) structure of the surface (see *Surface reconstruction*), generation of superlattices, microrelief of the surface layer (its corrugation, etc.), as well as for broadening of the *phonon spectrum*, reduction of the *melting temperature* of the lattice at the surface, and lowering of the crystal strength. A solid surface exhibits elevated chemical activity, the ability to orient molecules, to accumulate (getter) foreign molecules and, as a result, to generate additional electric charge (surface *electric double layer*).

The condition of minimum free surface energy E_S at constant volume determines the parameters of equilibrium solid phases at their generation, and the shape of the *crystal* at *crystallization* (if anisotropic forces of molecular interaction, wetting, gravitational and convection forces are taken into account). The surface energy E_S and surface tension σ of various faces of anisotropic crystals exhibit considerable differences.

The surface is the site of the interaction between the solid under consideration and various other phases (gaseous, liquid, solid), which is responsible for many *surface phenomena*. The most important of these are *cohesion, adhesion, wetting, friction, spalling*, generation, and development of new phase nuclei at the precipitation of the substance on a substrate, *adsorption, chemisorption, gettering*, which includes transport of the substance along the surface (*migration, surface diffusion*, surface electron or gravitational drift) followed by adsorption of foreign atoms and defects by the surface layer, which exhibits either a specific (with respect to the bulk of the substance or other surface regions under cleaning) or a different phase composition.

Features of the atomic and electron structure of the surface cause the generation of characteristic modes of surface excitations (see *Surface quasiparticles*). Propagation of longitudinal (with respect to the surface) atomic vibrations will be, of course, much different from propagation of atomic vibrations, which are transverse to the surface. For instance, a longitudinal wave (Rayleigh wave) may be initiated along the surface (see *Surface acoustic waves*), which acquires a geometrical relief in the form of ripples in a standing wave. Superlong-wavelength flexural waves may also be generated, whereas in the bulk waves of this type either cannot be excited at all, or are heavily damped. Hence the frequency dispersion of surface modes differs considerably from that of bulk modes.

Surface optical phenomena, which are caused by these modes (molecular, plasmon, exciton scattering) provide the physical basis for the technique in optics called *surface polariton spectroscopy*. This approach deals with electromagnetic waves, which are localized to the surface with phonons and propagate along it; the intensity of these electromagnetic waves decays sharply in both directions away from the surface. This circumstance causes the generation of the so-called surface-enhanced effects, which include *enhanced Raman scattering of light* and the enhancement of surface *photochemical reactions*.

There are, as a rule, electric potential jumps and intrinsic and structural defect energy levels, the so-called *surface levels*, at the *phase interface*. The existence of these levels determines a rich variety of emission phenomena: thermionic emission, *field electron emission*, field ion emission. *Characteristic energy loss spectroscopy* and spectroscopy of characteristic surface X-ray radiation, which is stimulated by electronic (ionic) beams, owe their origin to the fact of the excitation of surface phonons and electrons.

There are many methods for investigating physical properties of a solid surface. These methods may be classified according to the type of characteristics to be measured. Optical methods evaluate the state of a solid surface (or of thin near-surface layers) using *optical spectroscopy* techniques. These methods are: *ellipsometry, electroreflection*; spectra of light reflection at excitation of an extremely thin surface layer (by laser radiation or by a grazing, slightly divergent light beam); measurement of surface *photoluminescence* and *electroluminescence*; methods of surface polariton spectroscopy, in which the probing beam grazes the solid surface, providing information on parameters of surface light scattering, *current carriers*, plasma state, etc.; *Raman scattering of light* and other investigation methods.

Emission methods of solid surface investigation deal with surface emission of electrons (see *Auger electron spectroscopy, Field emission microscopy*) and ions (see *Field ion microscopy, Secondary ion mass spectrometry*) or of various kinds of radiation (UV, visible, X-rays); in a number of cases the spectral analysis of a certain characteristic radiation is used: ESCA (Electron Spectroscopy for Chemical Analysis), characteristic electron loss method, *photoelectron spectroscopy*, etc. (see *Emission spectral analysis*).

Three very widely used methods for investigating electrical properties of a solid surface are: (1) the field effect which, in particular, provides the basis for the operation of *field-effect transistors*; (2) the method of *capacitance–voltage char-*

acteristics, which consists in measuring a reactive quantity: transverse capacitance which allows one to calculate capacity parameters of the *space charge region* (*band bending*); (3) methods of *surface recombination* determination, which are based on the measurement of the photoconduction or nonequilibrium conduction of minority carriers, generated by an electric field, and on the measurement of the saturation current of a *semiconductor junction*, which is localized in the neighborhood of the solid surface. These methods are applied mostly to the study of *semiconductor surfaces*.

Investigation of adsorption characteristics of solid surfaces is performed through the measurement of adsorption isotherms (by recording changes of gas pressure in a limited volume, or by determination of the *work function* using methods of thermionic emission, field electron and field ion emission, or contact potential with the help of a microbalance (e.g. a quartz beam balance). Adsorption methods are related to catalytic investigation methods, in which yields of surface reactions are recorded. Catalytic methods include mass microscopy, and mass spectrometry of secondary ions and *clusters*. Of wide application are also the following methods: *low-energy electron diffraction* (LEED), *X-ray diffraction, electron microscopy, scanning tunneling microscopy*, atomic force microscopy, high-sensitivity methods of *electron paramagnetic resonance* and *nuclear magnetic resonance*.

SOLITON (fr. Lat. *solus*, only or alone)

Stable, spatially localized disturbance of a *nonlinear medium*. A soliton exhibits certain corpuscular properties, e.g., an interaction between solitons may be represented as elastic scattering, with retention of internal structure, or as soliton–*antisoliton* annihilation, etc. The concept of a soliton is fundamental for characterizing nonlinear distributed oscillatory systems in terms of mathematics and physics. Soliton-like or *particle-like excitations* contribute to a description of nonlinear ordered media.

From the standpoint of mathematical physics, solitons are solutions of nonlinear differential equations (in particular, wave equations), which describe localized, solitary waves that do not decay in their interaction with each other (*asymptotic superposition principle*). The soliton property of asymptotic stability is related to the so called *exact integrability* of the corresponding equations. The most familiar equations of this kind are: the *Korteweg–de Vries equation* (D.J. Korteweg, G. de Vries, 1895), which describes gravitational waves on the surface of a liquid in a long shallow canal, the *Boussinesq equation*, the *sine-Gordon equation* in the angular variable φ:

$$\frac{\partial^2 \varphi}{\partial t^2} - \frac{\partial^2 \varphi}{\partial x^2} + \sin \varphi = 0 \qquad (1)$$

and the *nonlinear Schrödinger equation* for a complex function ψ:

$$i\frac{\partial \psi}{\partial t} + \frac{\partial^2 \psi}{\partial x^2} + |\psi|^2 \psi = 0. \qquad (2)$$

Eqs. (1) and (2), written in dimensionless variables x and t, are widely used in the physics of condensed media. The *Landau–Lifshits equation* is also exactly integrable under certain conditions. There exist two-dimensional (the independent variables are two coordinates and time) exactly integrable equations, e.g., the *Kadomtsev–Petviashvili equation* (B.B. Kadomtsev, V.I. Petviashvili, 1970) in the real scalar variable u. The latter equation is an extension of the Korteweg–de Vries equation to two dimensions:

$$\frac{\partial}{\partial x}\left(\frac{\partial u}{\partial t} - 6u\frac{\partial u}{\partial x} - \frac{\partial^3 u}{\partial x^3}\right) = 3\frac{\partial^2 u}{\partial y^2}. \qquad (3)$$

However, related nontrivial three-dimensional equations are as yet unknown. The so-called *many-soliton solutions* are obtained for exactly integrable equations; these solutions describe an interaction of an arbitrary number N of solitons. It is feasible to solve the Cauchy problem, and to construct periodic and almost periodic solutions.

In condensed matter physics, the term "soliton" is sometimes used in reference to any stable or sufficiently long-lived excitation of a nonlinear medium; the existence of a soliton is not directly related to exact integrability. In terms of solid state physics examples of solitons are: *dislocations* and *disclinations* in crystals, including *liquid crystals*; dynamic *crowdions, Josephson vortices, magnetic domain walls*, certain solitary *magnetic domains* in magnetically ordered media, etc. Solitons are essential to the thermodynamic description of low-dimensional systems (*quasi-one-dimensional crystals* or *quasi-two-dimensional crystals*), *incommensurate structures* and phases, highly excited states of matter. *Topological solitons* differ from other soliton states because the *order parameter* distribution $\Phi(r)$, which fits this soliton state, is not reducible to a uniform distribution through continuous deformation of the function $\Phi(r)$ (see *Topological inhomogeneity*).

The equations, which describe the dynamics of the order parameter are reducible to exactly integrable ones in the case of certain ordered media. For example, the dynamics of vortices in long *Josephson junctions* and certain models of *elasticity theory* (see *Frenkel–Kontorova model*) and magnetism theory are described by the sine-Gordon equation (1); macroscopic perturbations in a slightly imperfect Bose gas (see *Quasi-particle*), the phenomenon of self-focusing in nonlinear optics (see *Self-action of light waves*), and other phenomena are described by the nonlinear Schrödinger equation (2); a number of models of quasi-one-dimensional *ferromagnets* are expressed in terms of the exactly integrable Landau–Lifshits equation (see *Magnetic soliton*). Although the reduction of real problems to integrable equations always involves approximations, the familiar many-soliton solutions may be used to describe systems with solitons; from these solutions it follows that the interaction between solitons is non-reflexive: solitons regain their initial shapes and velocities after leaving the interaction region. The difference between real systems and idealized exactly integrable ones, which can lead to, e.g., the break-up and relaxation of solitons (see *Domain wall drag*), is described on the basis of soliton perturbation theory. Tidal waves have been treated as solitons.

SOLITON FIELD FERMIONS

See *Fermions in a soliton field*.

SOLITON, MAGNETIC

See *Magnetic soliton*.

SOLITON, OPTICAL

See *Optical soliton*.

SOLITONS IN MOLECULAR (PROTEIN) SYSTEMS

The bound state of intrapeptide excitation of a protein chain, and a local *strain* in it. Such *solitons* are generated by two additional types of interaction, which take place on the excitation of intrapeptide vibrations. The first one is the resonance *dipole–dipole interaction* between neighboring peptide groups which results in a delocalization of the excitation, e.g., a spread of excitation along the length of the peptide molecule. The second type of deformation interaction is responsible for the displacement of the equilibrium positions of peptide groups, which are neighbors of the excited group. *Collective excitations* due to these interactions are described by *nonlinear Schrödinger equations* whose solutions are solitons in molecular systems (*Davydov solitons*).

Linking together the intrapeptide vibration and local deformation results in the liberation of binding energy. This is one of the reasons for the high stability of the soliton. An amount of energy equal to the binding energy must be spent to break up a soliton. Another important reason for the high soliton stability is that the speed of their motion is always less than the *sound velocity* so solitons do not emit *phonons*. In other words, the kinetic energy of solitons is never converted into the energy of thermal vibrations in the formalism of the exactly integrable model. The *effective mass* of a soliton far exceeds that of an *exciton*, in which the transport of internal excitation does not involve local deformation. Only excitons, and not solitons, are excited in a protein molecule on absorption of far-IR radiation, because the duration of the light absorption is not long enough to displace the equilibrium positions of peptide groups. The excitation of a soliton is possible only through exposure to some local action, like a chemical reaction, e.g., in the hydrolysis of the adenosine triphosphate (ATP) molecule, which is attached to the end of a polypeptide α-helix. It is possible that solitons, which move along polypeptide helixes and cross the plasma *membrane* of a cell, are responsible for the transport of information into a cell from the environment.

SOLUBILITY

Capability of materials to form homogeneous mixtures or solutions. For solids this means the capacity of different elements to mix in various proportions, forming a common *crystal lattice* typical for one of the components of the solution (the solvent). Solubility may be either limited or unlimited. *Limited solubility* involves particular concentration ranges of the constituents within which the solution can form, quantitatively characterized by concentrations of the individual substances entering the solution. Exceeding these concentration ranges results in the separation of new phases from the solution. A necessary but not sufficient condition for *unlimited solubility* at a given temperature consists in an identical *state of matter* of all the components of the solution. Such a condition for *solid solutions* implies an identical *crystal structure* of all the components, the latter also being a necessary but not a sufficient condition. Lowering the temperature of continuous solid solutions may result in a transition from unlimited solubility to a limited one, by way of forming intermediate phases (Fe–Cr, Cr–Ti), *superlattices of alloys* (Cu–Au, Mg–Cd), as well as *alloy stratification* (Cr–Mo) or a polymorphic transformation (see *Structural phase transition*) in one of the components (Fe–Ni). Lowering the temperature of a limited solubility system is accompanied, as a rule, by restricting the ranges of allowable concentrations of the constituents in the solvent. This phenomenon is used in technology to control physical and mechanical properties of materials through their *heat treatment*. As long as solid solutions may range widely in both their elemental and compound composition, it is possible to produce materials with a wider spectrum of physical properties.

SOLUTION HEAT

See *Heat of solution*.

SOMIGLIANA–VOLTERRA DISLOCATIONS

(C. Somigliana, 1905, 1909; W. Volterra; 1907)

Specific singular fields of intrinsic *internal stresses* and *strains* of an elastic continuum, with singularities grouped along a certain line L. Somigliana–Volterra dislocations are constructed in accordance with the following algorithm: assume that there is a cut in an arbitrary surface S,

which is stretched over a contour L, the edges of the cut being separated by an arbitrary distance. The resulting opening is filled with a certain material, surplus material is removed from overlapping regions, and then the whole system is bonded together and given an opportunity to relax. The result is a "defect", a source of internal stresses, which is commonly called a *Somigliana dislocation*. The deformation and stress fields of this dislocation may have discontinuities over the surface S. In a specific case, when the opposite edges of the cut remain congruent, the defect is called a *Volterra dislocation*. There are no discontinuities of stress and deformation fields over the surface of a Volterra dislocation, but these fields always exhibit singularities over the line L, which borders the surface S. Because of this, Volterra dislocations may be generated, strictly speaking, only in doubly connected regions. The concept of Somigliana–Volterra dislocations is widely used in the *continuum theory of defects* in crystalline structures. Volterra dislocations are the continuum analog of lattice *dislocations* and *disclinations*. The double connectedness of a crystal is established naturally by the atomic composition of the medium.

SONDHEIMER OSCILLATIONS

(E.H. Sondheimer, 1950)

Periodic variation (oscillation) of kinetic parameters of thin film conductors as a function of the magnetic field. Sondheimer oscillations are related to the focusing role of the magnetic field. A beam of electrons, having the same energy ε and momentum projection $p_B = \boldsymbol{p} \cdot \boldsymbol{B}/B$ along a magnetic field $\boldsymbol{B}$ normal to a plate (film) surface, will, once started from a surface point, be collected at another surface point if it takes an integral number of periods $2\pi/\Omega$ for the electrons to pass through the thickness d of the sample. This condition will be met if the magnetic field changes by the value

$$\Delta B = \frac{1}{ed} \frac{\partial S}{\partial p_B},$$

where S is the area enclosed by the intersection of the *Fermi surface* with a plane $p_B = $ const. The period of oscillation can be used to determine $\partial S/\partial p_B$ for these electrons, and studies of the amplitude (sensitive to the nature of the charge carrier reflection by the sample boundary) for different orientations of **B** clarifies the nature of the *surface scattering* of conduction electrons. In a magnetic field parallel to the plate, the *Sondheimer effect* only occurs for open cross-sections of the Fermi surface. Displacements of electrons downwards into the sample during the period are the same in the entire range of open cross-sections. There are no special cases, all electrons participate in the oscillations, their amplitude being greater the more specular the reflection of charge carriers by the faces of the plate. A necessary condition for Sondheimer oscillations is nonuniformity of an electrical or thermal field. During the propagation of sonic or electromagnetic waves through a thin conductor, the length scale of the electric field nonuniformity is provided by the wavelength λ. Hence, Sondheimer oscillations exist even in those cases when there is no *size effect* in the static *electrical conductivity*. In a high frequency field under conditions of the *anomalous skin-effect*, these oscillations can be amplified owing to the appearance of lightly-damped waves. The Sondheimer oscillations were first observed by J. Babiskin and P. Siebenmann in thin bismuth wires, and further studies involved *monocrystals* (Cd, Ga, Cu, etc.). These oscillations can be used to refine the energy spectrum of conduction electrons. The high resolution of Sondheimer oscillations permits one to differentiate contributions of electrons with very similar oscillation parameters, and this makes it possible to investigate local changes of Fermi surface geometry caused by an external factor such as the *pressure*.

SORBITE (after H.C. Sorby)

A structural constituent of *steels* and *cast irons*; a fine-grained mixture of *ferrite* and *cementite*. Sorbite has a pearlitic-type structure, but its dispersity is higher than that of *pearlite* (a volcanic glass). The thickness of ferrite laminas in sorbite ranges from 0.2 to 0.4 µm (in perlite 0.5–1.0 µm). Sorbite is formed by overcooling *austenite* decomposition over its region of lowest stability 723–823 K, both isothermally and during continuous cooling for several tens of degrees. The main characteristic of sorbite, which is responsible for the *strength* of steel, is the length of free slip Δ_s within the ferrite interval, which is proportional to the thickness of the ferrite lamina Δ_f.

The *ultimate strength* of a steel of the sorbite structure is defined by the equation $\sigma_u = \sigma_0 \Delta_f^{-1/2}$. The value of Brinell *hardness* of sorbite is 230–360, the strength $\sigma_u = (830-1420) \cdot 10^6$ Pa depending on the carbon content of the steel (0.3–0.8%). Hardened sorbite and tempered sorbite are distinguished. *Quenched sorbite* is obtained immediately from overcooled austenite (see *Quenching*), whereas *tempered sobite* is generated at the high-temperature (723–923 K) *tempering* of hardened steel. Tempered sorbite contains cementite (carbide) in the form of short laminae with rounded edges, or it can even have a spheroidal shape. Steels of the sorbite structure are strong, plastic (see *Plasticity*) and wear resistant (see *Wear*); they are used for production of cold-drawn high-strength wires (hardened sorbite), and heavy-duty units (tempered sorbite).

SORET EFFECT

See *Thermodiffusion*.

SORPTION

Uptaking by a *solid* (*sorbent*) from its surrounding gas or liquid (*sorbate*). The term "sorption" is used in reference to a combination of processes that include *adsorption* (adhesion of particles to a solid surface) and *absorption* (solution of particles in the bulk of a solid) phenomena. The adsorption proceeds with a gain in energy so the solid surface is often covered with an adsorbed film at sufficiently low temperatures, with the amount of adsorbed substance proportional to the area of the surface. An adsorbed particle (atom or molecule) has to overcome an activation barrier (separating adsorption and absorption states) to pass into the bulk of the solid. The absorption process is favorable for only a limited number of gas–solid pairs, e.g., H–Pd, H–Ti, O–Nb, O–Ta, O–Ag (see *Occlusion*). The sorption power of certain materials is enhanced by their porous structure; such materials are used in practice for the purification of gases or liquids, and the removal of gases from mixtures.

SOUND in solids

Elastic *strains* of a wave type in solids. Such elastic strains of a medium can be represented in the form of a set of propagating *bulk acoustic waves* that are characterized by their frequency, wave vector, and damping.

Sound waves in solids have some fundamental differences from sound in liquids and gases. The presence of resistance to *shear* of the medium induces the excitation of transverse (shear) modes of *acoustic vibrations*. The ordered internal structure of crystals gives rise to the anisotropy of elastic properties (see *Anisotropy of elasticity*), in particular, anisotropy of the *sound velocity*. The parameters of the sound wave are determined by the structure and state of a solid, and by internal processes. In particular, the damping of sound is due to its scattering by inhomogeneities, and by absorption caused by the interaction with electrons, *phonons*, *magnons* and other types of excitations. The sensitivity of the acoustic wave parameters to the medium underlies the development and the wide applicability of *acoustic methods of studying solids*.

SOUND-ABSORBING MATERIALS

Materials with high damping ability with respect to mechanical oscillations of different frequencies and amplitudes. As a rule, they have a moderate density (up to 0.70 g/cm^3) and are manufactured from fibrous and porous materials. They find wide use in architectural and industrial acoustics to create optimal sonic fields in rooms: cinemas, concert halls, sound-recording studios, etc. Another field of application is the reduction of the noise level at factories, for transportation, and in aviation. In mechanical engineering, in order to suppress harmful vibrations of polymer, rubber and plastic materials, *alloys* with high damping properties are employed. In terms of their basic mechanism of absorbing oscillatory energy and their chemical composition, *highly damping alloys* can be classed into four groups: those with *elastic twinning*, with a reversible *martensitic transformation*, with magneto-mechanical damping (see *Magnetomechanical phenomena*), and with a highly distinct *heterogeneous structure*. Besides these alloys, coatings with a damping effect, as well as multilayer and *composite materials* are also used.

SOUND ABSORPTION in crystals

The decrease of acoustical wave energy with time, caused by dissipation processes.

Sound absorption is for the most part determined by the interaction of sound with elementary crystal excitations, crystal imperfections, and with the *order parameter* at *second-order phase transitions*. If the sound frequency ω is much less than the reciprocal of the characteristic excitation lifetime τ^{-1}, then the sound absorption coefficient $\alpha \propto \omega^2$, where α is expressed in terms of macroscopic characteristics of the crystal: *viscosity*, *thermal conductivity*, and *electrical conductivity*. The physical nature of the absorption of such low-frequency sound is related to relaxation processes of the nonequilibrium part of the *distribution function* of elementary excitations. Absorption of high-frequency sound ($\omega\tau \gg 1$) may be treated as a quantum-mechanical process involving an acoustic quantum of energy $\hbar\omega$ and several elementary excitations. Sound absorption in an ordinary *insulator* is determined by its interaction with *phonons*: at $\omega\tau \ll 1$ the mechanism is related to relaxation of the phonon distribution function (*Akhiezer mechanism*, A.I. Akhiezer, 1937); at $\omega\tau \gg 1$ the absorption mechanism is associated with the interaction between a sound quantum and two phonons: if both phonons belong to the same branch, then the *Landau–Rumer mechanism* (L.D. Landau, J.B. Rumer, 1937) takes place and $\alpha \propto \omega$; if phonons belong to different branches, then the *Herring mechanism* (C. Herring, 1940) operates and $\alpha \propto \omega^2$ or ω^3. The interaction with *magnons* is important in *magnetic substances*, whereas in *semiconductors* and *metals* the main interaction is with *conduction electrons*. If there is a linear relation between *strain* and the order parameter at a second-order phase transition, then sound absorption exhibits a strong anomaly:

$$\alpha \sim \frac{\omega^2\tau}{1+\omega^2\tau^2},$$

where the *relaxation time* τ of the order parameter increases indefinitely as the transition is approached.

Among crystal imperfections, polycrystallinity (see *Polycrystal*) and defects with internal degrees of freedom (e.g., *two-level systems*) are of primary importance from the point of view of sound absorption. The study of sound absorption is one of the most important methods for the investigation of physical processes that take place in crystals.

The *absorption of acoustic waves by amorphous bodies* (glasses) differs markedly from the corresponding process in crystals, both in magnitude, and in the dependence on the temperature T and the acoustic wave frequency ω. The most distinct feature of the sound absorption process in dielectric glasses (fused quartz at $T \approx 50$ K; see *Vitreous state of matter*) is that the temperature dependence of the *absorption coefficient* Γ has a maximum. This maximum value of Γ is greater by approximately two orders of magnitude than the absorption coefficient of a crystal sample, and $\Gamma \propto \omega$. For the high-temperature branch $\Gamma \propto \omega^2$; for the low-temperature branch there is a so-called shoulder at $T \approx 10$ K, and on further decrease of temperature Γ monotonically decreases. This pattern is only slightly affected by the composition of the glass. For $T < 1$ K the factor Γ exhibits a universal behavior in all glasses. The value of Γ is of the same order for various glasses. Γ may be represented as a sum of two contributions: the first component of the sum is proportional to $\omega \tanh[\hbar\omega/(2k_\mathrm{B}T)]$ and sharply decreases with an increase of the intensity I of the acoustic wave; the second contribution exhibits a different behavior in metallic and nonmetallic glasses: in metallic glasses it is of the order of T^3 and depends very little on ω and I; in nonmetallic glasses it is of the order of ω and almost independent of T, whereas its I-dependence appears at considerably higher intensities than for the first contribution. Such a behavior has been explained in terms of a concept called *two-level systems*, which is specific for a vitreous type of excitation.

SOUND BY SOUND SCATTERING

See *Raman scattering of sound by sound*.

SOUND DIFFRACTION

See *Acoustic diffraction*.

SOUND GENERATION

See *Electromagnetic generation of sound*.

SOUND, SECOND

See *Second sound*.

SOUND VELOCITY

The velocity of elastic wave propagation in solids, liquids and gases. Phase and group velocities of sound are distinguished. The *phase velocity* $v_\mathrm{p} = \omega/k$ is the speed of the displacement in the medium of the fixed phase $\varphi = \omega t - kx$ of the elastic wave (ω is the frequency, k the wave number). The *group velocity* $v_\mathrm{g} = \partial\omega/\partial k$ is the speed of energy transport by a wave packet with frequencies close to the mean value ω_i. In liquids and gases only longitudinal waves propagate (amplitude u_0 and wave vector k are parallel), and in solids there are three types of waves with different polarizations. Therefore, in the solid state there are three different group v_{gi} and three phase v_{pi} ($i = 1, 2, 3, \ldots$) velocities. The group and phase velocities for small wave vectors k ($\lambda \gg a$, where $\lambda = 2\pi/k$ is the wavelength, a is the lattice constant), coincide with each other, so in this wave vector range sound propagation in a solid can occur with three possible velocities. An isotropic solid has only two velocities of sound, a longitudinal $v_l(u_0\|k)$ and a transverse $v_t(u_0\perp k)$ one. The velocities of transverse sound waves with different polarizations coincide with each other (are degenerate). For an isotropic solid

$$v_{lp} = v_{lg} = v_l = \left(\frac{E(1-\sigma)}{\rho(1+\sigma)(1-2\sigma)}\right)^{1/2}$$

$$= \left(\frac{K + (4/3)G}{\rho}\right)^{1/2},$$

$$v_{tp} = v_{tg} = v_t = \left(\frac{E}{2\rho(1+\sigma)}\right)^{1/2} = \left(\frac{G}{\rho}\right)^{1/2},$$

for $v_t < v_l$,

where E is *Young's modulus*, σ is the *Poisson ratio*, G, K are the *shear modulus* and *bulk modulus*, respectively, and ρ is the material *density*.

SOUND VELOCITY DISPERSION

The dependence of the *phase velocity* v_ph of harmonic acoustic waves on their frequency ω. In a broad sense this notion is applicable also to other types of elastic waves. There are two main types of sound velocity dispersion:

(1) The relaxation type, caused by the effects of after-action in the medium, where the acoustic wave is propagated. The *relaxation dispersion of sound* is always accompanied by *sound absorption* (see *Kramers–Kronig relations*),

(2) Sound velocity dispersion caused by the waveguide nature of the sound wave propagation. This type of dispersion is not associated with sound absorption.

Dissipation of the elastic wave energy in the *solid* occurs in the case when the stress and *strain* are not connected with one another by an analytic dependence during the period of vibrations and the strain lags behind the stress. Of importance in this case is the ratio of the time τ of *relaxation* of the subsystem of the solid excited by the sound wave, to the wave period $T = 2\pi/\omega$. The greatest sound velocity dispersion is present in the frequency region $\omega\tau \sim 1$. The main types of interactions which produce a strong effect on the sound velocity adsorption and dispersion in solids are the following: thermal or *thermoelastic effects*; dislocation *friction*; *acousto-electronic interaction* in metals; *magnetoelastic interactions* in ferromagnets, responsible for the motion of the *domain walls* and for the *spin–lattice interaction*; *phonon–phonon interactions*; interactions with nuclear spins; interaction with the electronic spins of the *paramagnetic centers*; interaction with the *current carriers* in semiconductors; and *acousto-electric effects* in semiconductors.

Besides the relaxational variety of sound velocity dispersion in microscopically nonuniform media there also exists spatial sound velocity dispersion caused by the v_{ph} dependence on l/λ, where l is the characteristic size of the micro-nonuniformities (defects), and λ is the acoustic wave length. The *spatial dispersion of sound* is observed in crystals at the frequencies of hypersound, when the space periodicity of the *crystal lattice* leads to the dispersion of the elastic properties of the crystal, and conditions the so-called *acoustic activity*, or the ability to rotate the polarization plane of the acoustic wave. Sound velocity dispersion of this type (sometimes called "geometrical" dispersion) is observed also for the propagation of waves within *plates* and *rods*. The parameter which determines the value of the sound velocity dispersion in this case is the relation of

the wavelength to the transverse dimension of the sound guide. The presence of this dispersion in plates and rods results because the *elasticity* of such bodies towards *flexure* is proportional to the size of the bent section. Thus at high frequencies, when the size of the bent section, determined by λ, is reduced, the dynamic elasticity increases and the speed of wave propagation grows.

The dispersion of normal waves during sound propagation along an *acoustic waveguide* is an important type of sound velocity dispersion. For instance, the layer in a lamellar-non-uniform medium, where different layers have different *acoustic impedances*, may play the role of such a waveguide. The sonic field in the waveguide may be presented in the form of a superposition of normal waves of different mode numbers, spreading with different phase speeds. Due to the dependence of the phase speed of each normal wave on its mode number and frequency, there appears a specific type of the sound velocity dispersion.

SPACE CHARGE in an insulator, bulk charge

An electrically charged region in an *insulator*. A bulk charge emerges under the conditions of thermodynamic nonequilibrium of an insulator as a result of an external action: irradiation by charged particles of ionizing radiation, *injection of current carriers* through a contact under the action of an electric field, and so on. The effects also take place under the conditions of *thermodynamic equilibrium*, as a result of the redistribution of a charge carriers near an interface between the dielectric and other media, or between regions of spatially inhomogeneous (e.g., nonuniformly doped) insulator (see *Space charge region*). An isolated insulator may be charged as a whole. This can be accomplished, e.g., by irradiation with fast electrons with a mean free path of a few centimeters or longer. Owing to the low electron mobility in an insulator, the excess electrons accumulate in the bulk of the sample. In this case, the redistribution of the intrinsic electrons of the insulator is possible due to a complicated spatial arrangement of space charge regions with different signs in the irradiated insulator. A natural limit to the accumulation of bulk charge is set by reaching the *breakdown* voltage when an electric discharge appears in the bulk of the material, and the bulk charge falls off to practically zero. As a result of such spontaneous discharges, the insulator can be damaged, or some traces of the breakdown, *Lichtenberg figures* (G. Lichtenberg, 1777), would remain in its interior. After the irradiation ceases, the bulk charge can remain for a long time in insulators with resistivity of at least 10^{10} Ω·m. For example, in a pane of glass irradiated by electrons, the bulk charge remains stable with almost no change for 3 to 4 months. Eventually the charge dissipates, and the insulator discharges. The bulk charge relaxation takes place within the time of the *Maxwell relaxation* $\tau = \varepsilon\varepsilon_0/\sigma$, where ε is the dimensionless *dielectric constant* of the material, ε_0 is the permittivity of free space, and σ is the *electrical conductivity*. Bulk charge stability and relaxation are controlled not only by the electrophysical properties of the material but also, to a great extent, by particularities of the spatial distribution of the charge. The most stable structures are laminated ones with alternating positive and negative space charge regions. A nonuniform spatial distribution of charge is often observable under irradiation of various insulators with charged particles. This was also found in radioactive insulators, e.g., beta-radioactive high-ohmic materials. The bulk charge strongly affects the mechanical, optical and electrophysical properties of insulators, and influences the course of various heterogeneous processes such as mass transfer, dissolution, adsorption, catalysis, etc.

In metal–insulator–semiconductor systems, in *semiconductor junctions*, in *heterojunctions*, and so on, a bulk charge emerges due to the redistribution of charge carriers in near-contact regions, since the values of the *work function* of electrons from the system components in contact or those of the charge carrier concentration differ from each other. Such bulk charge determines the shapes of current–voltage plots, and other characteristics of these systems. See also *Space charge region.*

SPACE CHARGE IN A SEMICONDUCTOR

Aggregation of excessive charge carriers of the same sign, localized in some region of a semiconductor. The space charge occurs at the boundary of two differently doped semiconductors, of different semiconductors, of a semiconductor and a metal (see *Semiconductor junction, Heterojunction,*

Schottky barrier). This charge plays an important role in the operation of practically all semiconductor devices. It can be formed also by an external field (e.g., in the surface area of a semiconductor or at its boundary with a dielectric). By means of an external field control it is possible to induce the directed motion of the charge cluster that is used in *charge coupled devices*. Space charge in a semiconductor sometimes forms by *ionized* impurities, and this generally has a negative effect on the operation of devices and integrated circuits, but it can also be explicitly utilized for the development of electrochemical, biochemical and other sensors and detectors. The case when the excessive carriers are fixed at deep *traps*, is especially important since their life time can reach 10^6 to 10^9 s, and often even longer. In the vicinity of such space charge the mobile *current carriers* also redistribute to form a depleted or enriched region, and this finds an important application in nonvolatile memory elements of computers (see *Solid state memory*). In such elements the space charge is formed by the tunneling of electrons through thin (several nanometers thick) dielectric layers under the effect of a strong field. Then the electrons are held at deep traps at the boundary of thin and thick dielectric layers or within the "floating gate" *thin film* of a *metal* or polycrystalline *silicon* between these layers. The erasing of the space charge in a semiconductor is performed by a field of opposite polarity, or by ultraviolet light.

SPACE CHARGE REGION

Region of space containing a dense electric charge in the vicinity of interfaces between *solids*, or of those between semiconductors and other media, as well as in gas-filled and vacuum tubes, etc. A space charge region may arise from various causes: the presence of a *contact potential difference*, of *surface electron states*, of an applied potential difference, of variations in doping impurity concentration (see also *Space charge in an insulator*). Concerning semiconductors, one may identify space charge layers depleted of majority charge carriers (*Schottky barrier*) and those enriched with them, as well as *inversion layers*, i.e. layers rich in minority charge carriers. Space charge regions play an important role in the functioning of *semiconductor devices*.

To describe the distributions of charge density, electric field, and potential within a space charge region, the concepts of Debye screening length and surface screening length were introduced. The *Debye screening length* L_D is the distance over which a potential, as small as the thermal value $(k_B T/e)$, diminishes by a factor of $1/e$. The *surface screening length* is defined as the distance over which the potential relative to its surface value decreases by the amount of the thermal potential.

SPACE GROUP, space symmetry group, Fedorov symmetry group (name used in the Russian literature)

Crystallographic symmetry groups formed by the totality of transformations (rotations, reflections, translations or their combinations), which make an infinite ideal *crystal lattice* coincident with itself. They were formulated independently by E.S. Fedorov and A. Schoenflies in 1890. The total number of space symmetry groups corresponding to the variety of *crystal structures* that are possible in nature is 230. They are classified according to their point groups into seven *crystal systems* or *syngonies*: triclinic, monoclinic, orthorhombic, tetragonal, trigonal, hexagonal and cubic (isometric). Each of these systems, in its turn, is subdivided into a finite number of classes (2, 3, 3, 7, 5, 7, 5, respectively), so the total number of classes is 32. Each class corresponds to one of 32 *point groups*. Further subclassification of space groups within individual point groups is determined by the allowed operations of translational symmetry (see *Crystal symmetry*). Space groups have extensive applications in various areas of solid state physics (see *Group theory*).

SPALLING

1. Process of development of brittle or quasi-brittle *fracture* in crystalline material. Spalling usually propagates on the surface with the least effective *failure* energy γ_{eff}, accompanied by thin fragments breaking off in layers parallel to the surface. In a *monocrystal* the spalling occurs along those particular crystallographic planes that have a minimal *elastic modulus* perpendicular to the planes, the largest interplanar distance, and the largest atom density in the plane. In most *metals* with a body-centered cubic structure the spalling occurs along the {100} planes which have a lower atom packing density than the {110} planes. This diversion from the indicated regularity is explained for BCC metals by the large contribution of interatomic interaction in the second *coordination sphere*. In mineralogy planes of spalling are called *cleavage planes*. In polycrystals spalling occurs within grains along the indicated crystallographic planes, or along the *grain boundaries* if the *surface energy* γ_0 there is sharply reduced as the result of impurity *segregation*; and in the last case *intercrystalline failure* can occur. At *brittle failure* $\gamma_{\text{eff}} = \gamma_0 + \gamma_s$, where γ_s is the energy of the spalling step formation that occurs at the intersection by the spalling front of *screw dislocations* and twist boundaries. The spalling steps result in the formation of a stream-like surface pattern of the spalling. In *quasi-brittle failure* it is necessary to add to the indicated relation the term γ_p describing the energy consumption on the *plastic deformation* adjacent to the region of spalling. The quasi-brittle failure type spalling is described by the *Stroh theory*.

2. Method of preparing *atomically clean surfaces* of semiconductors. Spalling is carried out in a high vacuum to decrease the rate of surface coating by residual gases. Surfaces with electrophysical parameters close to those being realized on the spalled surface in a high vacuum are prepared by spalling the sample in the presence of chemically inert condensed gases such as nitrogen or helium.

SPALLING, MICRO-

See *Microspalling*.

SPATIAL DIELECTRIC DISPERSION

The dependence of the components $\varepsilon_{\alpha\beta}$ of the *dielectric constant* tensor of a medium on the wave vector $\mathbf{k}$, which stems from the nonlocal relation between the vector of electric induction $\mathbf{D}$ and the electric field strength $\mathbf{E}$. Such a nonlocal relation produces certain physical phenomena that are effects of spatial dispersion, such as the *rotation of light polarization plane*, *optical anisotropy* of cubic crystals, etc., where an explanation requires taking into account the dependence of $\varepsilon_{\alpha\beta}$ on $\mathbf{k}$. Spatial dispersion also manifests itself by the possibility of propagation of *additional light waves* in crystals. The theory of spatial dispersion effects is closely related to that of *excitons*. One should take into account spatial dispersion when studying such problems as the *anomalous skin-effect* in metals, *crystal lattice vibrations*, light scattering, and the behavior of certain optical vibrations of crystals around their *second-order phase transition* points, etc.

SPECIFIC HEAT, heat capacity

A *susceptibility* of a body being heated. It is determined as the amount of heat needed to raise the *temperature* of one kilogram of a body by one degree kelvin (in SI units), or more exactly, it is a ratio of received heat ΔQ to temperature change ΔT (at $\Delta T \to 0$). Thus the specific heat c is given by $c = dQ/dT$. As a quantitative characteristic of matter the terms *specific heat*, *volume heat capacity*, and *molar heat capacity* are used, i.e. the heat capacities of a unit of mass, a unit of volume, and one mole. The corresponding units of measurements are $J \cdot kg^{-1} \cdot K^{-1}$, $J \cdot m^{-3} \cdot K^{-1}$, and $J \cdot mole^{-1} \cdot K^{-1}$. Heat capacity is a more general, nonspecific term. The unit calorie is often substituted for joule in practical applications. The body heat capacity depends on the conditions of measurement. As a rule, measurements are made of specific heat at constant volume c_V and at constant pressure c_P. Values c_V and c_P are expressed through *entropy* S and the *thermodynamic potential* of a body:

$$c_V = T\left(\frac{\partial S}{\partial T}\right)_V = \left(\frac{\partial E}{\partial T}\right)_V = -T\left(\frac{\partial^2 F}{\partial T^2}\right)_V,$$

$$c_P = T\left(\frac{\partial S}{\partial T}\right)_P = \left(\frac{\partial H}{\partial T}\right)_P = -T\left(\frac{\partial^2 G}{\partial T^2}\right)_P,$$

where E is the internal energy, F is the Helmholtz free energy, H is the enthalpy, G is the Gibbs thermodynamic potential. It follows from the second law of thermodynamics that $c_P, c_V > 0$; the third law gives $S \to 0$ at $T \to 0$, which means that specific heat goes to zero at $T = 0$. In isotropic bodies the following relation holds:

$$c_P - c_V = -T\frac{(\partial V/\partial T)_P^2}{(\partial V/\partial P)_T}, \qquad (1)$$

where $(\partial V/\partial T)_P$ is the temperature volume expansion coefficient, $(\partial V/\partial P)_T$ is the *compressibility*. Since for all stable bodies $(\partial V/\partial P)_T < 0$, it follows that $c_P > c_V$. For anisotropic solids (see *Anisotropic medium*)

$$c_P - s_V = T\sum_{ijkl}\alpha_{ij}f_{ijkl}\alpha_{kl}, \quad i,j,k,l = x,y,z,$$
$$(2)$$

where α_{ij} is the second-rank tensor of the thermal expansion coefficient, f_{ijkl} is the fourth-rank *elastic modulus tensor*. For cubic crystals, Eq. (2) reduces to Eq. (1). In crystals far from the region of a *phase transition*, the values c_V and c_P differ very little from one another, and in an ideal gas $c_P - c_V = R$, the gas constant.

In the classical limit, i.e. at sufficiently high temperatures, the law of equipartition of energy is valid. In this case, the specific heat is equal to $k_B/2$ per translational or rotational degree of freedom, and k_B per vibrational degree of freedom, where k_B is the *Boltzmann constant*; $k_B = 1.3807 \cdot 10^{-23}$ J/K. This results in the validity of the *Dulong and Petit law* (P. Dulong, A. Petit, 1819) for simple (non-molecular) crystals at high temperatures: $c_V = 3N\nu k_B$ where N is the number of *unit cells*, ν is the number of atoms in a cell; this means that the specific heat per atom is the universal constant with the value $3k_B$ (each atom has 3 vibrational degrees of freedom). The molar specific heat of a monatomic crystal in this limit is $c_V \approx 3R \approx 25$ J·mole^{-1}·K^{-1}, where $R = N_A k_B$ is the gas constant, and N_A is the *Avogadro number*. In accordance with the third law of thermodynamics, the specific heat decreases at low temperatures. Such a behavior is explained within the framework of the quantum theory: the specific heat of a single vibrational degree of freedom (of an oscillator) has the form:

$$c_V^0(T) = k_B f\left(\frac{\hbar\omega_0}{k_B T}\right),$$

$$f(x) = x^2\frac{e^x}{(e^x - 1)^2},$$

where $\hbar = h/(2\pi)$, and ω is the angular frequency of a given oscillator. It is seen that $c_V^0 = k_B$ at high temperatures ($k_B T \gg \hbar\omega_0$) with an exponential decrease ($f(x) \approx x^2 \exp(-x)$, $x \gg 1$) at low temperatures ($k_B T \ll \hbar\omega_0$). In a *harmonic approximation* the *crystal lattice vibrations* can be represented in the form of $3N_v$ independent oscillators. In this case the lattice specific heat is

$$c_V = \sum_{n=1}^{3N_v} c_{nV}(T), \qquad (3)$$

where n is the number of the harmonic oscillators with frequency ω_n. In the simplest quantum theory of solids put forward by Einstein (*Einstein approximation*) it was assumed that all frequencies ω_n are the same ($\omega_n = \omega$). Therefore, at low temperatures the specific heat would go to zero according to an exponential law. In contrast to this, experimental data follow the law $c_V \sim T^3$ in this temperature range. For describing such behavior it is necessary to take into account the actual vibrational spectrum of the solid (see *Crystal lattice dynamics*). To accomplish this Eq. (3) for crystals may be rewritten in the form:

$$c_V = 3k_B N_v \int g(\omega) f\left(\frac{\hbar\omega}{k_B T}\right) d\omega,$$

where $g(\omega)$ is the density of states of phonons in the crystal normalized to unity. Among the $3v$ branches of the crystal lattice vibrations three of them are *acoustic vibrations* (their frequency $\sim |q|$ at $q \to 0$), and the other ($3v - 3$) branches are called *optical vibrations* (their frequencies remain finite at $q \to 0$). To describe the specific heat in the low-temperature range it is sufficient to consider only the acoustical branches. A good description of the specific heat of simple crystals over a wide temperature range is provided in the framework of the *Debye model of specific heat*. This theory assumes that over the entire range of wave vector values for the acoustical branches, the linear *dispersion law* applies: $\omega_{qj} = s|q|$, where the *sound velocity* s does not depend on either the polarization or the wave vector direction. A *Brillouin zone* is replaced by a sphere of the same volume with radius $q_D = (6\pi^2/v_0)^{1/3}$ called the *Debye radius* where v_0 is the unit cell volume. The highest possible vibrational frequency of the acoustical spectrum $\omega_D = s q_D$ is called the *Debye frequency*. As a result, the specific heat of a monatomic crystal takes the form:

$$c_V = 3k_B N \frac{v_0}{2\pi^2} \int_0^{q_D} f\left(\frac{\hbar s q}{k_B T}\right) q^2 \, dq$$

$$= 3N k_B f_1\left(\frac{T}{\Theta}\right); \qquad (4)$$

$$f_1(x) = 3x^3 \int_0^{1/x} \frac{t^4 e^t}{(e^t - 1)^2} \, dt;$$

$$\Theta = \frac{\hbar\omega_D}{k_B} = \frac{\hbar s q_D}{k_B}.$$

Thus, in the Debye theory the temperature dependence of the crystal specific heat is described by a single parameter, the *Debye temperature* Θ, which is related in a simple manner with the Debye frequency $\omega_D = k_B \Theta / \hbar$. At high temperatures when $T \gg \Theta$ (function $f_1(x) \to 1$ at $x \to \infty$) the Dulong and Petit law ($c_V = 3N k_B$) is valid; at low temperatures ($T \ll \Theta$) $c_V = (12/5)\pi^4 N k_B (T/\Theta)^3$, i.e. the specific heat varies as T^3 in agreement with experiment. At intermediate temperatures, deviations from the behavior predicted by Eq. (4) can occur. Such deviations are sometimes erroneously interpreted as a temperature dependence of Θ. The contribution of the optical branches to the specific heat is described fairly adequately in the framework of the Einstein theory, i.e. disregarding dispersion. In *quasi-two-dimensional crystals* and *quasi-one-dimensional crystals* at not very low temperatures, the specific heat appears to be proportional to smaller power indices of the temperature (smallest power index $T^{1/2}$ is for chain crystals) with the T^3 law valid only for the lowest temperatures.

The *conduction electrons* make an extra contribution to the specific heat of *metals*. The associated electronic specific heat depends linearly on the temperature:

$$c_V^e = \beta T, \qquad \beta = \frac{\pi^2}{3} N(\varepsilon_F),$$

where $N(\varepsilon_F)$ is the electron density on the *Fermi surface*. At high temperatures the electronic specific heat is negligibly small compared to the lattice portion, but at very low temperatures its contribution becomes appreciable. Hence studies of the electronic specific heat provide important information about the properties of conduction electrons. A transition to the superconducting state (see *Superconductivity*) is accompanied by a jumplike rise of the electronic specific heat; and further lowering of the temperature decreases the specific heat according to an exponential law.

In *magnetic materials* the magnetic excitations (*magnons* or *spin waves*) contribute to the specific heat. In *ferromagnets* with a small *magnetic anisotropy*, the related specific heat at temperatures below the *Curie point* varies as $T^{3/2}$ due to a quadratic dependence of the spin wave frequencies on the wave vector. Owing to a linear dispersion law, the extra specific heat contribution in *antiferromagnets* is described by the same law ($c_V \sim T^3$) as that due to acoustic lattice vibrations. At extremely low temperatures, below the minimal magnon energy ε_0, the magnon specific heat decays at $T \to 0$ as the exponential function, $c_V \propto \exp[-\varepsilon_0/(k_B T)]$.

A substantial contribution to the specific heat of crystals can be introduced by a high density of impurity centers (see *Donor, Acceptor*), which is of particular importance if their characteristic excitation frequencies are much less than the Debye frequency. As an example, heavy *impurity atoms* whose classical limit is found at significantly lower temperatures than the overall crystal, and two-level centers whose specific heat (*Schottky specific heat*) is described by a curve with the maximum

$$c_V^{\text{Sh}} = \left(\frac{\Delta E}{2k_B T} \right)^2 \coth^{-2} \left(\frac{\Delta E}{2k_B T} \right),$$

where ΔE is the energy difference between levels. In the most disordered systems such as amorphous materials and glasses (see *Amorphous state, Vitreous state of matter*) the specific heat at the lowest temperatures is a linear function of temperature due to the presence of so-called two-level states (see *Two-level systems*) with randomly distributed transition energies whose *density of states* $g(\Delta E)$ remains finite at $\Delta E \to 0$. The same type of specific heat temperature dependence takes place in *spin glasses*, and also in crystals at the presence of impurities of certain kinds: Jahn–Teller atoms (see *Jahn–Teller effect*), reorienting centers, etc.

Particular features in the temperature dependence are observable in the neighborhood of *second-order phase transitions*. In accordance with the *Landau theory*, a jump of the specific heat would take place at the transition point. Such jumps can readily be isolated if one considers the asymptotic dependencies of the specific heat far from a transition temperature T_c. In the vicinity of the transition point (*critical region* where *order parameter* fluctuations are significant) the specific heat exhibits a divergent character:

$$c_P \sim \left| \frac{T - T_c}{T_c} \right|^{-\alpha},$$

where α is the *critical index* of the specific heat.

In the framework of the now accepted *scaling invariance hypothesis*, critical indices are regarded as universal values dependent on the space dimensionality (d) and the number of components (n) of the order parameter. Experimental values of α are insignificant, close to 0.1. In the framework of the Wilson ε-expansion (see *Epsilon expansion, Renormalization group method*), the first two members of the expansion of α assume the following form:

$$\alpha = -\frac{n-4}{2(n+8)}\varepsilon - \frac{(n+2)^2}{4(n+8)^3}(n+28)\varepsilon^2 + \cdots,$$

where $\varepsilon = 4 - d$.

A comparison of theoretical and experimental values of α for the present case has very low accuracy due to the smallness of α.

Experimental methods for the determination of the specific heat are quite diverse, and depend on the process employed, the sample properties including dimensions, and the temperature range covered in the experiment.

SPECIFIC HEAT JUMP in superconductors

The difference of *specific heat* between superconducting and normal phases at the *superconducting phase transition* temperature $\Delta C = C_S(T_c) - C_N(T_c)$. The specific heat jump is the difference between the heat capacity values on both boundaries of this transition region, and it is associated with an unusually narrow *critical region* in the majority of superconductors. Outside of this region the *Landau theory of second-order phase transitions* is valid. Thus, the specific heat jump is understood as the difference of specific heat values at the two limits of the critical region. The majority of superconductors come close to satisfying the relation $\Delta C / C_N(T_c) = 1.43$ predicted by the *Bardeen–Cooper–Schrieffer theory*. However, in some cases significant deviations from this result are observed, e.g., $\Delta C = 2.65$ for Pb and 2.3 for Hg. Some of these values are explained by the theory of strongly coupled superconductors (see *Eliashberg equations*), whereby the enhancement of the specific heat jump is attributed to an increase of the electron–lattice interaction.

SPECIFIC HEAT, SURFACE

See *Surface specific heat*.

SPECTRAL ANALYSIS, EMISSION

See *Emission spectral analysis*.

SPECTRAL ANALYSIS, MOLECULAR

See *Molecular spectral analysis*.

SPECTRAL DENSITY OF STATES

The same as *Density of states*.

SPECTRAL DIMENSIONALITY

See *Fracton dimensionality*.

SPEED OF LIGHT in a solid

The speed of propagation of electromagnetic waves with frequencies for which the medium is transparent (that is absorption is negligibly small). In crystals the electric polarization P induced by the field E of the light wave, and consequently the *electric flux density* $D = \varepsilon_0 E + P$ (ε_0 is *dielectric constant* of free space) in a general case do not coincide in direction with E at a given point. The simplest relation between D and E along the principal dielectric axes (see *Principal axes of a tensor*) of the crystal x_α ($\alpha = 1, 2, 3$) has the form $D_\alpha = \varepsilon^{(\alpha)} E_\alpha$, that is the coefficients of proportionality between D_α and E_α vary for different principal directions α. This gives rise to two important features of light propagation:

(a) The *group velocity* of light v (speed of electromagnetic energy transport) which is perpendicular to E and H (H is the *magnetic field* intensity) differs from the *phase velocity* U (perpendicular to D, H) not only in magnitude but also in direction.

(b) For each direction $s = U/|U|$ there are generally only two definite (mutually perpendicular) planes of possible linear *polarization of light*, and two different speeds of propagation denoted by $U_{1,2}(s)$.

The set of numerical values $U_{1,2}(s)$ for all s form a double self-intersecting surface described by the *Fresnel equation*

$$
\begin{aligned}
s_1^2 \big(U^2 - \overline{U}_{(2)}^2\big)\big(U^2 - \overline{U}_{(3)}^2\big) \\
+ s_2^2 \big(U^2 - \overline{U}_{(1)}^2\big)\big(U^2 - \overline{U}_{(3)}^2\big) \\
+ s_3^2 \big(U^2 - \overline{U}_{(1)}^2\big)\big(U^2 - \overline{U}_{(2)}^2\big) = 0.
\end{aligned}
\tag{1}
$$

Here $U_\alpha \equiv c/[\varepsilon^{(\alpha)}]^{1/2}$ is the speed of waves polarized along x_α (*principal speeds of light* in the crystal). The group velocity v is characterized by a similar surface.

If all $\varepsilon^{(\alpha)}$ are identical (crystals of *cubic system*) then $U_{(1)} = U_{(2)} = U_{(3)}$, and the surface (1) reduces to a sphere with $U = $ const in all directions. If $\varepsilon^{(1)} = \varepsilon^{(2)} \neq \varepsilon^{(3)}$ (*uniaxial crystals*), then $U_{(1)} = U_{(2)} \neq U_{(3)}$; and the surface (1) looks like a sphere and an ellipsoid of revolution touching each other at two points which define the direction of the *optical axis of a crystal*. The sphere is associated with the light polarized perpendicular to the optic axes (*ordinary ray*, o-ray), and the ellipsoid corresponds to the light with its polarization plane parallel to the optic axis (*extraordinary ray*, e-ray). Finally, if all the $\varepsilon^{(\alpha)}$ differ (*biaxial crystals*) then surface (1) has intersections at four points, defining two different optic axes. The infinite set of ray directions generating the conical surface (*conical refraction of light waves*) corresponds to the speed of light U along such an axis. When a light beam is incident on a crystal with two values $U_{1,2}(s)$, there are two refracted beams, the o-ray and the e-ray, each of which is linearly polarized (*birefringence*). The properties of light propagation in crystals are studied in *crystal optics*, and find various applications, in particular for the correlation of the refraction coefficients of different harmonics (see *Refraction of light*) in second harmonic generation in *nonlinear optics*.

SPERIMAGNETISM

See *Speromagnetism*.

SPEROMAGNETISM (fr. Gr. $\sigma\pi\varepsilon\iota\rho\omega$, spread, scatter)

A magnetic state of an *amorphous magnetic substance* in which the equilibrium and temporally constant orientations of atomic magnetic moments of individual magnetic atoms (ions) are fixed in position and randomly distributed over space (over lattice sites) in such a way, that the overall *magnetization* of the atoms (ions) is zero; in addition the correlations between the directions of magnetic moments of neighboring atoms die out within several interatomic distances.

Based on specific features of their magnetic properties and their type of *atomic magnetic structure*, speromagnets are special cases of *spin glasses* in a "frozen" state. The lack of correlation between the directions of local magnetic moments in traditional spin glasses is most often attributed to random interchanges of the magnitudes and signs of *exchange interaction* constants of nearest neighbor atoms (ions), or to frustration of antiferromagnetic linkages. In some speromagnets there are additional reasons for the lack of correlation: random distribution of local directions of *easy magnetization axes* in an amorphous material, and alignment of magnetic ions along directions where the anisotropy energy dominates over exchange energy. The latter mechanism is characteristic of *metallic glasses* of the type rare-earth metal (of nonzero orbital moment)–noble or transition metal, e.g., Dy–Cu, Tb–Ag, etc.

The distribution of the directions of local moments may be slightly aspherical. In this case, the term "*asymmetric speromagnetism*" is used. *Asperomagnets* are characterized by a certain mean macroscopic magnetization in combination with typical varieties of spin glass states (*magnetic viscosity*, irreversible change of properties upon *magnetization*). The regular component for local directions of the easy magnetization axis may arise in asymmetric speromagnets due to external or spontaneous deformation (related to magnetoelasticity). Antisymmetric speromagnets are sometimes called asperomagnets, although the term *asperomagnetism* more properly refers to an antisymmetric speromagnetic state.

If an amorphous material includes two or more types of magnetically inequivalent ions (ions which differ chemically in valence electron configuration), then several magnetic subsystems may arise in this material. Each of these subsystems can be speromagnetic, and related to a certain distinct kind of magnetic ion. Such a state resembles a randomized *ferromagnet*. If at least one of these subsystems is asymmetric (asperomagnet), then the material is called a *sperimagnet*. This material will also exhibit both spin glass properties and a certain nonzero mean magnetization. The typical representatives of sperimagnets are amorphous *solid solutions* of the type "rare-earth metal (of nonzero atomic orbital moment)–ferromagnetic metals of the ferrous group", e.g., Nd–Co, Nd–Fe, Dy–Co, etc.

SPHERICAL HARMONICS

Eigenfunctions $Y_{lm}(\theta, \varphi)$ of the orbital angular momentum operator (θ and φ are angular coordinates). Spherical harmonics characterize the electron density distribution in a centrally symmetric field for an electron with orbital *angular momentum l*, the square of which is equal to $l(l+1)$, and whose projection on the axis of quantization is m. The functions Y_{lm} are solutions of the Legendre differential equation. The infinite series of spherical functions Y_{lm} with $l = 0, \pm1, \pm2, \ldots, \pm\infty$ and $m = -l, \ldots, l$ is a complete set of orthonormal functions, i.e. functions which satisfy the conditions of orthogonality and normalization:

$$\int Y_{lm}(\theta, \varphi) Y_{l'm'}(\theta, \varphi)\, d\Omega = \delta_{ll'}\delta_{mm'}.$$

This property allows one to expand any function of the difference between radius vectors of two particles in a series of spherical harmonics as

$$f(\boldsymbol{r}_1 - \boldsymbol{r}_2) = F(|\boldsymbol{r}_1|, |\boldsymbol{r}_2|)$$

$$\times \sum_{l=-\infty}^{+\infty} \sum_{m=-l}^{l} a_{lm} Y_{lm}(\theta_1, \varphi_1) Y_{l,-m}(\theta_2, \varphi_2).$$

The expansion of the Coulomb interaction energy of point charge ions in terms of spherical harmonics is, in particular, used in *crystal field theory* for deducing the effective *spin Hamiltonian*.

SPIKE, DISPLACEMENT

See *Displacement spike*.

SPIN

The intrinsic angular momentum of an elementary particle (electron, nucleon, meson, etc.), which is quantum in nature, and is unrelated to the spatial motion of the particle. Numerous phenomena of solid state physics are related to spin (see, e.g., entries beginning with the word "spin").

SPIN COMPLEXES

Many-particle *magnon bound states*. In the case of a one-dimensional system, spin complexes of arbitrary length exist for any nonzero value of the total *quasi-momentum* of the magnons. There exist explicit expressions for the wave function and energy levels of spin complexes of the spin chain of the *xyz-model*. The spin complexes of a chain, which exhibits a "ferromagnetic" interaction for the indicated number of *magnons* and a fixed value of total quasi-momentum, have the lowest energy. There exists a relationship between spin complexes and *magnetic solitons*, which are solutions of the corresponding *Landau–Lifshits equation*.

SPIN DEGENERACY

Degeneracy of energy levels, due to spin states (spin wave functions). Degenerate quantum states are states with the same energy. The ground state of *paramagnetic centers* in crystals is often an orbital singlet due to splitting of atomic terms by intracrystalline electric fields. In this case, the multiplicity or degeneracy of a level with spin S is given by the quantity $2S + 1$.

SPIN DENSITY

The square of the amplitude of the wave function at the point r, i.e. $|\psi_{km}(r)|^2$, where m is the magnetic spin quantum number, and k is the set of other quantum numbers. The term "spin density" is occasionally used in reference to squared coefficients of the linear combination of atomic orbitals in the LCAO method (see *Linear combination of atomic orbitals*). The concept of spin density is of crucial importance in studying the *hyperfine structure* in spectra due to magnetic resonance phenomena, since this hyperfine structure is determined by the total spin density $\sum_{kk'} [|\psi_{k\uparrow}(r)|^2 - |\psi_{k'\downarrow}(r)|^2]$ localized at the nuclear position r. The total spin density of filled electron shells may differ from zero ($k = k'$) due to the fact that electrons in full shells are subjected to exchange polarization by electrons of unfilled shells, which have zero spin density in the proximity of the nucleus. In line with the above relations, the description of the magnetic properties of materials sometimes involves the introduction of the *spin density operator* with respect to the *spin angular momentum* $S(r') = \sum_l S_l \delta(r_l - r')$, where the coordinate r_l locates lth spin (nucleus), and r' is the running coordinate.

SPIN DENSITY WAVE

A stable, static, periodic in space, distribution of spin density (and correspondingly of *magnetic moment*) of charge carriers (electrons and holes) in solids. The appearance of a spin density wave means a *phase transition* to a magnetically ordered state of the antiferromagnetic type. Accordingly, macroscopic and spin-wave properties of such an *antiferromagnet* are indistinguishable from the antiferromagnetism of localized electrons described by a *Heisenberg Hamiltonian*. However, in the spin density wave (SDW) case other microscopic models, based on *band magnetism* (see *Antiferromagnetic metals*), are used.

The first model of *band antiferromagnetism* was developed by Overhauser in the late 1950s and early 1960s. It involves a one electron band, where a SDW appears due to correlations between electrons and holes at opposite regions of the *Fermi surface*. Such correlations resemble those of *Cooper paired* electrons in the *Bardeen–Cooper–Schrieffer theory* of superconductivity, and also the electron–hole correlations responsible for the *Peierls instability* which brings about a *metal–insulator transition* with the formation of a *charge density wave* (CDW). A difference between a SDW and a CDW is that for the former the *electron–hole pair* that appears as a result of correlation is in the spin triplet state, while in the latter case it is in the singlet state. The efficiency of the electron–hole interaction responsible for the formation of a charge density wave, as well as of a spin density wave, is determined by how well the opposite regions of the *Fermi surface* line up (i.e. are nested), separated in k-*space* by the wave vector Q, which determines the period of the CDW or SDW. For good nesting straight sections of a Fermi surface are required, so the best conditions for the formation of these waves in a single band are found in *quasi-one-dimensional crystals*. There are also two-band models of SDW formation where correlations appear between electrons and holes from the different bands. The efficiency of such correlations depends on how well the Fermi surface electron bands nest through the displacement of one of them by the Q vector. The first *two-band model* of a SDW was considered by W.H. Lomer (1962) with application to *chromium* and *chromium alloys*, which are well investigated materials with SDWs. The two-band model is also a particular case of the *exciton dielectric* model, suggested by L.V. Keldysh and Yu.V. Kopayev (1964) for a *semimetal* with overlapping *conduction band* and *valence band*. In this model, at the non-coinciding band extrema in *k*-space, the formation of a singlet electron–hole pair brings about the appearance of a CDW, while the triplet pair produces a spin density wave. The SDW vector Q can be commensurate (or incommensurate) with vectors of the *reciprocal lattice* of a crystal; in the first case it is possible to speak about a *commensurate* SDW, and in the second case about an *incommensurate* one. These two types of waves and the phase transition between them have been experimentally observed, in particular, in antiferromagnetic chromium by varying its degree of *alloying*. Spin density waves are distinguished by the polarization of the magnetic moment in the wave. Circular polarization (*helicoidal waves*) is possible, and also two types of the linear polarization: parallel and perpendicular with respect to the Q vector (see *Modulated magnetic structures*). Since the Q vector is parallel to some selected crystalline direction, the longitudinal SDW is the analog of an *antiferromagnet* with anisotropy of the "easy axis" type, and a transverse SDW is an analog of an antiferromagnet with anisotropy of the "easy plane" type.

SPIN DENSITY WAVE IN SUPERCONDUCTOR

See *Superconductor with charge (spin) density waves*.

SPIN-DEPENDENT EFFECTS

Effects related to the fact that certain mobile and/or localized interacting particles (e.g., electrons and holes) have spin (angular) momentum. A great variety of phenomena involves manifestations of spin-dependent effects, which may be of either a resonance or a nonresonance nature. In the former case the spin-dependent effects show themselves as a resonance change of a certain measured quantity, e.g., the *photoconductivity* of a sample under electron paramagnetic resonance (EPR) conditions (in an external magnetic field, and subjected to microwave saturation at a resonance frequency).

Most spin-dependent effects are described in terms of the concept of pairs of particles, which have *spins*; the interactions of these particles (e.g., electron–hole *recombination*) results in a nonequilibrium distribution of these pairs between states with parallel and antiparallel spins. This deviation from an equilibrium distribution of spin orientations may be due to, e.g., preferential recombination of "singlet pairs" relative to a recombination of triplet states. Application of resonance microwave pumping disturbs the statistical equilibrium, and hence changes the recombination rate, causing, e.g., resonance decrease of photoconductivity or amplification of recombination *luminescence*. Spin-dependent effects are involved in a wide diversity of phenomena, which occur extensively in semiconductors (particularly *amorphous semiconductors*) and organic solids. Phenomena in semiconductors accompanied by spin-dependent effects are: *transport phenomena*, Hall effect, magnetoresistance (see *Galvanomagnetic effects*), photoconductivity, *photoluminescence*, *trapping* and ejection of current carriers by paramagnetic centers (the latter processes appear as distortions in corresponding EPR spectra obtained through optical excitation), photoelectromotive force (photovoltaically detected *magnetic resonance*), and photoelectric absorption (both phenomena observed in *deep level transient spectroscopy*). Organic solids exhibit, along with phenomena of a similar nature (photoconduction, *fluorescence*), the resonance change of the rate of photoprocesses (magnetic resonance, which is detected by the change in the reaction rate); the latter effect permits one to detect 10^5–10^7 particles of lifetimes $\sim 10^{-8}$–10^{-9} s in the sample under study. The techniques, which employ spin-dependent effects, are extraordinarily sensitive compared with the ordinary EPR method. This sensitivity is high because the magnitude of the effect does not depend on the concentration of the particles involved, but is determined by the relative contribution of the spin-dependent energy to the phenomenon under study.

SPIN DIFFUSION

The process of spin excitation transfer across the sample, due to *spin* flip-flop transitions (see *Cross-relaxation*). The spin diffusion in dilute magnetic electron spin systems and nuclear paramagnets is caused by the dipole–dipole *spin–spin interaction*. Based on rather general assumptions, the process of spin excitation transfer is described by the bilinear *kinetic equation*:

$$\frac{\partial P_i}{\partial t} = \sum_j W_{ij}(P_j - P_i), \qquad (1)$$

where W_{ij} is the probability of flip-flop transitions of spins i and j, and P_i and P_j are spin polarizations.

The term "spin diffusion" implies that polarization redistribution is described by the *diffusion* equation. However, the change-over from Eq. (1) to an approximate description in terms of diffusion is possible only when the characteristic scale of the polarization spatial nonuniformity exceeds the mean distance between spins. The latter situation is almost always the case in nuclear paramagnets, which is where the term "*nuclear spin diffusion*" comes from. The characteristic nonuniformity scale is in practice determined by the paramagnetic impurities present in the sample. Since the magnitude of the *dipole–dipole interaction* between nuclear spins and paramagnetic impurities far exceeds the magnitude of the dipole–dipole interaction of nuclear spins, the polarization nonuniformity scale far exceeds the distance between spins. Yet the situation may be different for spin systems in which the interaction constants of spins with one another and with impurities may be of the same order of magnitude. In this case, the process of spin polarization redistribution is described by integral equations (not by differential equations),

and *spin migration* dominates over spin diffusion. In the case of diluted magnetic solids, the migration process is also strongly affected by the disorder in the arrangement of paramagnetic atoms.

The special case of polarization transfer is represented by the so-called *hopping polarization mechanism*, which is the opposite of the diffusion mechanism. This mechanism takes place when the polarization nonuniformity scale is smaller than the distance between spins.

SPIN ECHO

Appearance of *magnetic resonance* emission signals within a certain time t after the application of radio-frequency electromagnetic radiation pulses to a sample. Usually, two pulses are applied sequentially separated by an interval τ. The spin echo signal is observed at the time τ after application of the second pulse (see Fig.). Spin echos appear in both *electron paramagnetic resonance* and in *nuclear magnetic resonance*. From the standpoint of physics, the reasons for the spin echo are as follows. At the initial instant of time ($t = 0$), the spins are preferentially aligned along the constant magnetic field $B_\parallel$ (z-axis). The first ($\pi/2$) pulse of the field $B_\perp$ turns the total magnetization into the plane perpendicular to the z-axis (e.g., along the y-axis). During the time interval between the first and second pulses the magnetization spreads out in the x–y plane into individual spin packets which precess at different rates (due to differing local fields, which cause *inhomogeneous broadening* of the magnetic resonance line). The second (π) pulse specularly reflects the entire spin alignment pattern in the z–x-plane. As a result of this

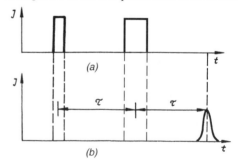

Scheme of formation of a spin echo: J is intensity in arbitrary units, (a) shows a pair of input pulses ($\pi/2$ and π), and (b) is the echo signal.

phase inversion the "phase lagging" spin packets become leading, but their motion is still slow as before, so the spin wave packets from behind will catch up with them. All the spin wave packets will be again in phase at the time τ after the second pulse, and this produces the spin echo signal.

If the crystal under consideration has *defects* with nonzero *dipole moments*, then the *electric dipole echo* effect may take place along with the *electric flux density* effect. The electric dipole echo effect arises upon excitation of the system of electric dipoles by a series of pulses of electric or magnetic components. The electric dipole echo signal may appear at times, which differ from the time intervals, at which a magnetic dipole (spin) echo signal appears.

SPIN ECHO, ACOUSTIC

See *Acoustic spin echo*.

SPIN EFFECTS in solid-phase chemical reactions

Change of rate of a nonequilibrium process in a material associated with interactions involving electronic spins, sometimes occurring in response to the application of an external *magnetic field*. This rate change results in either an increase or a decrease in the product yield of the process. Chemical changes in solids (*molecular crystals, polymers*) often take place with the participation of electronic excitations (*excitons*, ion-radicals, etc.), the nonequilibrium concentrations of which are formed upon exciting the substance with light, ionizing radiation, etc. Spin effects manifest themselves in processes involving the formation of intermediate pairs of paramagnetic particles, with the yield of the end product depending on the spin state of these particles after their recombination. The spin state of the pair, which is initially determined by precursor spins, undergoes a certain evolution, tending to an equilibrium state. Under actual conditions, when the energy of interaction between paramagnetic particles, which have a magnetic moment M, and an external magnetic field B is much smaller than the thermal energy ($E = M \cdot B \ll k_B T$), the spin effects are not affected by thermal motion because changes of spin state of the pair, caused by the external magnetic field, may proceed much quicker (in a time $\approx 10^{-9}$ s) than the process of *relaxation* of pairs

to their equilibrium spin state (characteristic time $\tau_I \sim 10^{-7}$ s).

A typical reaction, which involves a spin effect, is the photochemical excitation of an electronic transition in a molecular crystal of the *anthracene* type (molecule D), which contains impurity molecules A. Exposure of the crystal to exciting radiation brings about the generation of ion-radical pairs ($\dot{D}^+ \cdots \dot{A}^-$) in accordance with the scheme: $^1D + h\nu \to {}^1D^*$, $^1D^* + {}^1A \to {}^1(\dot{D}^+ \cdots \dot{A}^-)$. The initial state of the pair is a singlet. The spin multiplicity of the pair becomes a triplet due to the evolution of its spin state, which may related to the the Zeeman interaction (see *Zeeman effect*) and/or *hyperfine interaction. Recombination* reactions of $^1(\dot{D}^+ \cdots \dot{A}^-)$ and $^3(\dot{D}^+ \cdots \dot{A}^-)$ give different products: singlet excitons $^1D^*$ are formed in the recombination of a singlet pair, whereas a triplet pair recombines giving *triplet excitons* $^3D^*$. The probability of dissociation of pairs into free *current carriers*, which are responsible for *photoconductivity*, is also changed. The magnetic effect manifests itself in a change of *fluorescence* intensity and photoconductivity (change of up to 20%) on the application of a magnetic field $B = 1$–10 mT.

Additional examples of processes involving spin effects (ordinarily not characterized as chemical) are: triplet–triplet annihilation of excitons, quenching of triplet excitons by paramagnetic impurities, recombination of charge carriers in semiconductors, and the jump motion of charge in substances with paramagnetic centers.

In *ionic crystals*, the paramagnetic pairs involve crystal lattice defects, e.g., F-centers (see *Color centers*) of spin $1/2$. The photoexcitation of F-centers brings about paired electron tunneling, and results in the generation of uncharged singlet defects. The change of spin state of a pair due to an external magnetic field results in a change of the ratio between the rate of electron transfer, and the rate of a competitive radiation process.

SPINEL, FERRITE

See *Ferrite spinels*.

SPIN-FLIP TRANSITION

See *Antiferromagnetism, Magnetic phase transitions*.

SPIN-FLOP TRANSITION

See *Antiferromagnetism, Magnetic phase transitions*.

SPIN GLASS

Disordered state of a magnetic system, which involves random, alternating-sign, spin–spin interactions. The *spins* in this state become "frozen" in a certain direction which, however, varies randomly from point to point, so that there is no long-range order in the system (see *Long-range and short-range order*). The term "spin glass" implies lack of long-range order, as in the case of simple glasses (see *Vitreous state of matter*). The freezing of spin moments distinguishes spin glasses from *paramagnets*, in which individual spin moments are subject to rapid fluctuations with the resultant spin moment averaged to zero. The spin glass state appears when the temperature T is lowered below the *glass freezing temperature* T_f. The temperature dependence of the dynamic *magnetic susceptibility* in zero *magnetic field* B shows a pronounced inflection at T_f, which becomes smoothed with increased field strength; certain physical quantities show specific behavior at T_f, e.g., a nonlinear magnetic susceptibility $\chi_3 = \partial^3 M / \partial H^3 \propto (T - T_f)^{-\gamma}$, and a *critical index* value $\gamma = 3.5$ for the CuMn alloy. The resistivity and *specific heat* of spin glasses do not exhibit anomalous behavior at the freezing temperature. On approaching the freezing point, the range of spin relaxation times (see *Paramagnetic relaxation*) broadens abruptly at $T > T_f$. The maximal relaxation time is given by the *Vogel–Fulcher law* (I. Vogel, 1921; G. Fulcher, 1925): $t_{max} \propto t_0 \exp\{E_A / [k_B (T - T_f)]\}$, where E_A, T_f are experimentally determined parameters. The range of *relaxation times* is unlimited at $T < T_f$, which results in a *nonergodicity* in the behavior of spin glasses. The cause of the destruction of ergodicity is the presence of an infinitely large number of nearly degenerate minima of the system energy. Therefore, the description of equilibrium characteristics of the system requires specifying an infinitely large number of *order parameters*, which characterize different spin configurations of the spin glass state. The "classic" spin glasses are *noble metal alloys* with *transition metals* (e.g., AuFe, CuMn), in which transition metal content is several mole percent.

The atoms of these alloys interact with one another via the *Ruderman–Kittel–Kasuya–Yosida interaction* through the intermediary of *conduction electrons*; the distinguishing features of this interaction are its long range and alternating-sign nature, which depends on the relative positions of the interacting atoms. Therefore, there is a competition between ferromagnetic and antiferromagnetic interactions, which are randomly distributed due to the disordered arrangement of magnetic atoms. The spin glass state is also observed in concentrated alloys of transition elements, e.g., in polycrystalline $Fe_x Ni_{80-x} Cr_{20}$ ($0 < x < 70\%$), and amorphous $(Fe_x Ni_{1-x})_{77} \cdot B_{13} Si_{10}$. Another class of materials that exhibits a spin glass state are insulating compounds of *rare-earth elements*, e.g., $Eu_x Sr_{1-x} S$. The *exchange interaction* of Eu ions in this compound is of short range, but extends over the first and second *coordination spheres*; the exchange interaction has different magnitudes and signs in the two coordination spheres.

SPIN GLASS FREEZING

Phase transition in *amorphous magnetic substances*, when a random frozen *magnetization* $m_i = \langle S_i \rangle_T$ appears, different from zero over the entire space (the average spin is obtained from Gibbs averaging). The value of m_i is a random function of the coordinates. Its simplest parameters are first moments $M = \langle m_i \rangle_c$ and second moments $q = \langle m_i^2 \rangle_c$, where the averaging is over configurations, and q is called the *Edwards–Anderson order parameter* (S.G. Edwards, P.W. Anderson, 1975) (see *Spin glass*).

SPIN HAMILTONIAN

Energy operator of spin subsystem of atoms, ions, molecules and solids, which is expressed in terms of *spin operators* of electrons and solids comprising these quantities (see also *Heisenberg Hamiltonian, Ising model, Exchange interaction*).

The *spin Hamiltonian of the magnetic ion in a crystal* is a matrix, which is composed of the components of a transition ion *spin* operator; this matrix being used for calculating the splittings of its dominant energy term ^{2S+1}L (L is the value of the total *orbital angular momentum*) in an external constant magnetic field B. The spin Hamiltonian is obtained from the complete Hamiltonian, which accounts for interaction of electrons with a crystal field $H_C \sim 10^4$ cm^{-1}, *spin–orbit interaction* $H_{LS} = \lambda L \cdot S \sim 10^2$ cm^{-1}, *spin–spin interaction* $H_{SS} \approx \rho(L \cdot S)^2 \sim 1$ cm^{-1}, and Zeeman interaction $H_Z = \beta(L + 2S)B \sim 1$ cm^{-1}, by applying two or three perturbation theory approximations. The matrix elements $H_{LS} + H_{SS} + H_Z$, which are calculated using the coordinate wave functions of the unperturbed Hamiltonian H_C and determine the sought-for splittings, are polynomials of degrees 1 and 2 with respect to S. In this approach, the parameters of spin operators (g- and D-tensors) are expressed in terms of the known constants λ and ρ, matrix elements of the operator L, and energy intervals between the dominant term and excitation terms. In the case of ions in crystals, the lowest energy state of which is an orbital singlet (Cr^{3+}, V^{2+}, Ni^{2+}) or doublet (Cr^{2+}, Mn^{3+}, Cu^{2+}), the spin Hamiltonian of their actual spin S is used as the electron spin Hamiltonian. The spin orbital multiplet of $3(2S + 1)$ orbital triplet sublevels is split by LS-coupling into groups, the *effective spin* $\widetilde{S}$ of which is either $S + 1$, or S, or $S - 1$. Each of these groups can be described with a spin Hamiltonian or generalized spin Hamiltonian, which is constructed of components of $\widetilde{S}$. The spin Hamiltonian is applicable to ions with D- and F-terms and adequately describes the spectra of these ions. The spin Hamil-

tonian fails to explain the splittings of an S-state term, which are observed in spectra of ions in $3d^5$- and $4f^7$-configurations (e.g., Mn^{2+}, Gd^{3+}) with $L = 0$. The spectra of the latter ions are usually described in terms of the phenomenological generalized spin Hamiltonian, obtained by application of methods of *invariants*, the *perturbation matrix method*, and *irreducible tensor operators* of the form

$$H = \sum_{lml_1l_2\ldots} a_{lm}^{l_1l_2} \cdots T_{lm}^{l_1l_2} \cdots (S, H, \ldots),$$

where T_{lm} are irreducible tensor operators, l_1, $l_2, \ldots$ are powers of the SH-components and other vectors and tensors, which appear in T_{lm} (nuclear spin I, electric field E, spin S_2 of second ion of *exchange-coupled pair, strain tensor*, etc.); a_{lm} are arbitrary independent parameters determined by comparing calculated spectra with experiment. The component of H characterized by $l_2 = 0$, $l_1 = 2, 4, 6$ ($l_1 \leqslant 2S$) describes the interaction between electrons of the paramagnetic ion and the crystal field, the components, which are characterized by $l_2 = 1$, $l_1 = 1, 3, 5, 7$, are related to a Zeeman interaction linear in B. The admissible values of lm for all possible spin Hamiltonian symmetry groups are given in Table 1, which gives the most abbreviated form of a spin Hamiltonian.

Table 1. Collection of lm operator combinations for all eleven symmetry groups σ_i of the spin Hamiltonian

lm	σ_1	σ_2	σ_3	σ_4	σ_5	σ_6	σ_7	σ_8	σ_9	σ_{10}	σ_{11}
00	c	c	c	c	c	c	c	c	c	c	c
20	c	c	c	c	c	c	c	c	c		
22	cs	cs	c								
40	c	c	c	c	c	c	c	c	c	c	c
42	cs	cs	c								
43	cs					cs	c				
44	cs	cs	c	cs	c	c	c	c	c	c	c
60	c	c	c	c	c	c	c	c	c		c
62	cs	cs	c								c
63	cs					cs	c				
64	cs	cs	c	cs	c					c	c
66	cs	cs	c			cs	c	cs	c	c	

*Notes: c (or s) indicate an allowed combination of inequivalent tensor operators, and of transformations involving cosines (sines) of the azimuthal angle. In addition to the terms included for σ_1 there are also cs-operators (not shown) with lm, equal to 21, 41, 61, and 65. For the two cubic groups σ_{10} and σ_{11} linear groupings of some lm operators form individual combinations, as indicated by vertical lines.

The *generalized spin Hamiltonian* was introduced for the description of EPR spectra of ions in the *S*-state, for which the ordinary electron spin Hamiltonian is inapplicable (the generalized spin Hamiltonian applies also to spectra of other ions, due to its versatility). The calculation of the energy spectrum is based on the fact that a group of states relatively close together in energy, which are far removed in energy from all other states, have energy eigenvalues which can be calculated with the aid of matrix elements expressed in terms of a small number of parameters through the introduction of an effective spin $\widetilde{S}$, and the application of symmetry considerations and the Wigner–Eckart theorem (E.P. Wigner, 1927; C. Eckart, 1930). The important characteristic of the generalized spin Hamiltonian is its *symmetry group*, i.e. the *point group* of the symmetry of the magnetic and electric interactions of a paramagnetic defect present in a molecule, in a crystal, etc. In the presence of an external electric field E (crystal field) the symmetry group coincides with one of 32 possible *local symmetry groups*. For $E = 0$ there are only 11 fundamentally different generalized spin Hamiltonians because the magnetic field is invariant under the inversion operation. Each of these 11 generalized spin Hamiltonians corresponds to one of the Laue sets σ_i of the point groups (see *Laue classes*):

$$\sigma_1(C_1, C_i),$$
$$\sigma_2(C_2, C_s, C_{2h}),$$
$$\sigma_3(C_{2v}, D_2, D_{2h}),$$
$$\sigma_4(C_4, S_4, C_{4h}),$$
$$\sigma_5(C_{4v}, D_4, D_{2d}, D_{4h}),$$
$$\sigma_6(C_3, S_6),$$
$$\sigma_7(C_{3v}, D_3, D_{3d}),$$
$$\sigma_8(C_6, C_{3h}, C_{6h}),$$
$$\sigma_9(C_{6v}, D_6, D_{3h}, D_{6h}),$$
$$\sigma_{10}(T, T_h),$$
$$\sigma_{11}(O, T_d, O_h).$$

The symmetry group of the generalized spin Hamiltonian should be distinguished from the *symmetry group of the spectrum*, for the latter is one of the symmetry groups of the generalized spin Hamiltonian only for a partial spectrum (of equivalent *paramagnetic centers*), whereas for the case of a complete spectrum it contains additional operators of microscopic point *crystal symmetry*, which convert equivalent centers into each other.

Furthermore, an *effective spin Hamiltonian* can also be designated. It is constructed from the effective spin operators (used in EPR spectroscopy and its modifications), and is intended for calculating the nuclear splitting in a particular electron state M. This Hamiltonian is obtained by averaging the electron–nuclear component of the spin Hamiltonian using electron spin functions. In the averaging process *hyperfine interactions* are transformed into an M-dependent magnetic field, which affects nuclear spins, and into *indirect nuclear spin–spin coupling*. An effective spin Hamiltonian is used for the description of *electron–nuclear double resonance* spectra.

To describe the phenomena of *nuclear magnetic resonance, nuclear quadrupole resonance,* and *Mössbauer effect,* the *nuclear spin Hamiltonian* is used, which includes the: Zeeman interaction between nuclear moments and the applied magnetic field B; the interaction proportional to B of the electron shell of a molecule or crystal with the screening field (*chemical shift* of resonance); the interaction between quadrupole moments of nuclei and gradients of electric fields produced by intrinsic electrons of the incomplete shells of the ion and the charges of other ions; spin–spin interactions between nuclei (direct dipole–dipole and indirect ones). See also *Heisenberg Hamiltonian.*

SPIN HAMILTONIAN PARAMETERS, spin Hamiltonian constants

General name for the parameters of an electron (nuclear, generalized) *spin Hamiltonian*, which are characteristic of the Zeeman, *exchange interactions, hyperfine interactions,* quadrupole interactions, plus interactions with the *crystal field*. The representation of spin Hamiltonian parameters involves Cartesian *g*-, *D*-, *A*-, *Q*-tensors.

The *g-tensor* is the set of numbers that characterize the Zeeman splitting of the energy levels of a paramagnetic defect in an external constant magnetic field. Ordinarily the *g*-tensor appears in the Hamiltonian as a symmetric 3×3 matrix with 9 components, six of which are independent because of the symmetry $g_{ij} = g_{ji}$. In the principal axis system in which the tensor is diagonal it has three principal values $g_{ii} = g_i$, where $i = x, y, z$, and three Euler angles to describe its orientation. In the absence of hyperfine interactions the value of g in an arbitrary direction expressed in polar coordinates is given by $g = [(g_x^2 \cos^2 \varphi + g_y^2 \sin^2 \varphi) \sin^2 \theta + g_z^2 \cos^2 \theta]^{1/2}$. The principal values g_i are only slightly different from 2.0 for ions in an S-state. In the case of ions in other states, g_i are expressed through parameters of the *spin–orbit interaction* and energy splittings, and can exhibit large anisotropies at some crystallographic sites.

The *crystal field splitting tensor* or *D-tensor* is a traceless symmetric second-rank tensor, which describes the indirect interaction of the resultant electronic spin of a paramagnetic ion S in terms of the spin–orbit interaction with the electric field of the crystal (molecule) under consideration. Five parameters of the *D*-tensor may be transformed into two principal values (involving constants D and E) and three Euler angles, which characterize the orientation of the principal axes with respect to the crystal axes.

The *quadrupole tensor* or *Q-tensor* is also a traceless symmetrical second-rank tensor, which describes the interaction of the nuclear quadrupole moment (see *Quadrupole*) with the *electric field gradient*, the latter being produced by electrons of the intrinsic ion, and charges of ions in the nearby environment. The *Q*-tensor is often characterized by Q_{zz} and the asymmetry parameter $\eta = (Q_{xx} - Q_{yy})/Q_{zz}$, where $0 \leqslant \eta \leqslant 1$, in the system of *principal axes of a tensor*.

The *hyperfine tensor* or *A-tensor* is a second-rank tensor of general form, which characterizes the interactions between the resultant electronic spin of a paramagnetic ion with the spin of its own nucleus or of nuclei of nearby atoms. In its principal axis system it is diagonal with three principal components.

In accordance with the above tensor nomenclature, the corresponding interactions are written as products of a certain tensor with two vectors (βBgS, SDS, IQI, SAI), where β is the Bohr magneton, B is the magnetic field, and S and I are the electronic and nuclear spin vector operators, respectively. However, the representation of a spin Hamiltonian in terms of *irreducible tensor operators* $T_{lm}(S, H, I, \ldots)$ is more general, and facilitates theoretical calculations and algorithms for deriving the spin Hamiltonian parameters from experimental data.

Also of importance are *low-symmetry spin Hamiltonian parameters* which are present for transition ions at crystallographic sites of low symmetry. Sometimes these components are small because there is, e.g., a tetragonal, rhombic, trigonal or other small distortion of a dominantly octahedral or tetrahedral site symmetry. If the components of the spin Hamiltonian are represented as $\beta BgS + SDS + IQI + SAI + \cdots$, then another cause for the presence of low-symmetry spin Hamiltonian parameters can be a lack of alignment between the principal axes of the g-, D-, A- and Q-tensors. These parameters can be determined from the position, *line width*, and *line intensity* of the allowed and forbidden transitions in electron paramagnetic resonance, electron–nuclear double resonance, and nuclear magnetic resonance spectra. The said parameters manifest themselves in the features of the *angular dependences of spectra* (*low-symmetry effects*), as well as in *spin–lattice relaxation*, broadening of spectral lines, and the occurrence of *forbidden quantum transitions* with respect to the orientation of B along the *crystal symmetry* axes.

SPIN INDUCTION

See *Nuclear induction*.

SPIN–LATTICE INTERACTION, spin–phonon interaction

Direct (or indirect) interaction of localized (or free) *spins* of atoms of a solid with crystal ion displacements (see also *Magnetoelastic interaction*). Electric and magnetic forces acting on a *paramagnetic center* are changed under the influence of *crystal lattice vibrations* (*phonons*). The spin–lattice interaction is a perturbation, which causes *spin–lattice relaxation*, induces changes of the static *spin Hamiltonian* constants according to temperature and external pressures, brings about resonance absorption of sound waves by paramagnetic centers (*acoustic paramagnetic resonance*), and an indirect *spin–spin interaction* of paramagnetic centers by means of the phonon field. The magnitude of the spin–lattice interaction is characterized by the constants involved in the *spin–phonon Hamiltonian*.

SPIN–LATTICE RELAXATION

The process of establishing thermal equilibrium between the *spin subsystem* (electronic or nuclear) and the phonon or lattice vibration subsystem. The common enough case is the one when the times τ_s and τ_{ph} for the establishment of quasi-equilibrium thermodynamic states within the individual spin and phonon subsystems, respectively, are much shorter than the time $\tau_{s\text{-}ph}$ for the establishment of joint thermodynamic equilibrium between the spin and phonon subsystems which are in loose (weak) contact with each other. It is customary to call $\tau_{s\text{-}ph}$ the spin–lattice (or longitudinal) *relaxation time* and to denote it by the symbol T_1, and to call τ_s the spin–spin (or transverse) relaxation time denoted by the symbol T_2. The time T_1 of spin–lattice relaxation is determined by the mechanisms involved in the *spin–lattice interaction*, and these depend strongly on the external and internal conditions and parameters of the system. The values of T_1 and T_2 for a particular spin system may be determined experimentally by using methods of continuous saturation and pulse saturation (see *Saturation effects*), *spin echo*, *electron paramagnetic resonance*, as well as by studies of resonance broadening of electron paramagnetic resonance lines. For most solids $T_1 \gg T_2$, and for low-viscosity liquids $T_1 \sim T_2$. For relaxation in low-viscosity liquids Brownian motion plays the role that lattice vibrations play in solids.

Ordinarily the phonon subsystem is so large that it serves as a thermal reservoir or heat sink with a stable temperature, and τ_{ph} need not be taken into account. An exception occurs for high-power saturation studies involving the phenomenon of "hole burning" when a narrow frequency range of the phonon subsystem is raised in temperature, and τ_{ph} plays a role in the delaxed reestablishment of phonon thermodynamic equilibrium.

Studies of spin–lattice relaxation provide information on the mechanisms of the corresponding interactions, and this is needed for an understanding of the dynamic properties of the spin and phonon subsystems in a solid. See also *Magnetic relaxation*.

SPINODAL

Curve on the *phase diagram* of a decomposing alloy (see *Alloy decomposition*). This curve consists of points, at which the second derivative of the free energy with respect to the composition assumes negative values. Within the region bounded by the spinodal, alloy decomposition proceeds without generating new thermodynamic *phase* nuclei, and this is referred to as a *spinodal decomposition*. Chemical and coherent spinodals are distinguished to describe, respectively, incoherent and coherent precipitation.

In the field of metallurgy there are structures called spinodal structures which have two phases, each of high homogeneity.

SPINODAL DECOMPOSITION

Decomposition or breakdown of a *solid solution* (see *Alloy decomposition*) within the region bounded by a *spinodal*. The spinodal decomposition proceeds at a fast rate (in a matter of milliseconds) which is determined by the low value of the diffusion *activation energy*.

Spinodal decomposition proceeds by the *ascending diffusion* mechanism. According to the phenomenological theory, this decomposition involves the formation of a spectrum of concentration waves. In the beginning of the process, these waves have low amplitudes and long wavelengths. The spectrum includes a wave that increases at a maximum rate mainly along one of the crystallographic directions ($\langle 100 \rangle$ or $\langle 101 \rangle$), depending on the sign of the *anisotropy of elasticity* of the alloy. The spinodal decomposition results in the formation of a *modulated structure* in a crystal. This breakdown process has been observed in both metallic (Al–Zn, Cu–Ni–Fe, Fe–Ni–Al) and nonmetallic (PbO–Al_2O_3, PbS–PbTe, $CoFe_2O_4$–Co_3O_4) decomposing alloys. As a result of this decomposition, the alloys acquire improved mechanical and physical properties (magnetic, electric).

SPIN OPERATOR REPRESENTATIONS

Notation for *spin* operators with the help of Bose (a^+, a) and Fermi (b^+, b) creation and annihilation operators of *quasi-particles*, used in the quantum physics of magnetism. The *Holstein–Primakoff representation* (T. Holstein, H. Primakoff, 1940) and *Dyson–Maleyev representation*

(F. Dyson, 1956, S.V. Maleyev, 1957) may be written in the unified form:

$$S_+ = a^+ \sqrt{2S}\left(1 - \frac{a^+a}{2S}\right)^\gamma,$$

$$S_- = \sqrt{2S}\left(1 - \frac{a^+a}{2S}\right)^{1-\gamma} a,$$

$$S_z = -S + a^+a,$$

where $\gamma = 1/2$ for the Holstein–Primakoff spin operator representation and $\gamma = 1$ for the Dyson–Maleyev spin operator representation, S is the atomic spin value, $S_\pm = S_x \pm iS_y$, and the operators a, a^+ for these spin operator representations have the sense of operators for the creation and annihilation of *magnons*. These two spin operator representations allow one to take into account systematically the dynamical interaction of magnons, and to develop the theory of high-frequency and thermodynamic properties of *magnetic substances* at temperatures $T \ll T_c$ (T_c is the *Curie point*). The main difficulty of their application is the necessity of introducing the projection operator of the infinitely-dimensional space of the Bose particle state in the $(2S + 1)$-dimensional space of the action of spin operators S_x, S_y, S_z, the subspace of the magnon physical state. According to Dyson the finite dimensionality of the physical subspace may be taken into account by the introduction of the kinematic interaction of magnons, specific for the spin system of magnetically ordered materials. The problem of taking into account the kinematic interaction of magnons systematically was solved by V.G. Baryakhtar, V.N. Krivoruchko, D.A. Yablonski (1982), who introduced a new spin operator representation:

$$S_+ = a^+ \sqrt{2S}\left(1 - \frac{a^+a}{2S}\right)$$

$$- \sqrt{\frac{2}{S}}(2S + 1)a^+b^+b,$$

$$S_- = a\sqrt{2S},$$

$$S_z = -S + a^+a + (2S + 1)b^+b.$$

Here, the kinematic interaction of magnons (a^+, a) is described by their interaction with fictitious Fermi particles, the so-called *spurions*

(b^+, b). In addition to the above spin operator representations in the theory of magnetically ordered materials, there are representations of J. Schwinger (1952); of N.N. Bogolyubov (1949), and an oscillator representation introduced by Baryakhtar and Yablonski (1975).

SPIN OPERATORS

Matrix operators that act on the spin component of an ionic (or nuclear) wave function. The rank of the spin operator matrix is $2S + 1$, where S is the value of the *spin* (or effective spin). Under space–time inversion spin operators act like angular momentum operators. The spin operators for electronic wave functions ($S = 1/2$) were introduced by W. Pauli (1927) and are called *Pauli matrices*. The simplest operators S_x, S_y, S_z (or S_+, S_-, S_0) may be used to build up polynomials of higher degree called *irreducible tensor operators* (in particular, *Stevens operators*). Since the matrix elements of electrostatic interactions (interaction of electrons and nuclei with a crystal field, *exchange interactions, hyperfine interactions* and *spin–spin interactions*) are expressible in terms of these operators, the spin operators serve as a basis for describing energy splittings in terms of *spin Hamiltonians*.

SPIN–ORBIT INTERACTION

The interaction between an electron spin and the orbital angular momentum that is second order in v/c (v is the velocity of the electron, c is speed of light), and is linear in the spin S. When moving in a Coulomb field at the speed v, an electron experiences the effective magnetic field $\boldsymbol{B} = -c^{-2}[\boldsymbol{v} \times \nabla U]$ due to the Coulomb field gradient $\nabla U = (\boldsymbol{r}/r)(\mathrm{d}U/\mathrm{d}r)$. The interaction of the intrinsic magnetic moment (spin) with the field B is the classical spin–orbit interaction. By virtue of the so-called *Thomas relativistic correction* (L.H. Thomas, 1926), the spin–orbit interaction for an electron in an atom is half as strong as the classical spin–orbit interaction. The same results are provided by the systematic quantum relativistic theory developed by P.A.M. Dirac (1928); in particular, the spin–orbit interaction energy is given by $H_{so} = \lambda(\boldsymbol{L{\cdot}S})$, where $L = [\boldsymbol{r} \times m\boldsymbol{v}]/\hbar$ is the *orbital angular momentum*, the constant $\lambda = -[e\hbar^2/(4m^2c^2)](\mathrm{d}U/\mathrm{d}r)r^{-1}$ is referred to as the *spin–orbit coupling constant*. The spin–orbit interaction determines the *fine structure* of atomic levels, leads to the splitting of degenerate energy bands in solids, and is responsible for the *g*-factor shift (see *Landé g-factor*) of the splitting in *electron paramagnetic resonance* spectra relative to the Landé g_e-factor of a free electron ($g_e = 2.0023$).

SPIN-PAIRED COMPLEXES

See *High-spin complexes*.

SPIN–PHONON HAMILTONIAN

Hamiltonian of the *spin–lattice interaction*. The spin–phonon Hamiltonian H_{s-ph} has the form of the sum of products of *spin operators* O_μ with lattice operators U_μ. In the majority of problems it is sufficient to choose for U_μ a linear or bilinear form of the operators of the displacements of *crystal lattice* sites from their equilibrium positions U_σ, and the rates of these displacements $\dot{U}_\sigma$ ($\sigma = x, y, z$). For this case, the spin–phonon Hamiltonian has the form

$$H_{s-ph} = \sum O_\mu \left(a_\mu^\sigma U_\sigma + b_\mu^{\sigma\sigma'} U_{\sigma\sigma'} + c_\mu^\sigma \dot{U}_\sigma \right.$$
$$\left. + d_\mu^{\sigma\sigma'} \dot{U}_\sigma \dot{U}_{\sigma'} + e_\mu^{\sigma\sigma'} U_\sigma \dot{U}_{\sigma'} \right),$$

where the operators $O_\mu = O_\mu(S_1, S_2, \ldots, I_1, I_2, \ldots)$ are functions of one or several spin matrices for *magnetic ions* S_i and nuclei I_j; the form of the operators O_μ is determined by the values of the *spins* of the interacting particles, and by the total symmetry of the system, including the *invariance* of H_{s-ph} under *time inversion*. The spin–phonon coupling constants a and b are descriptive of the influence of the static displacement of crystal lattice sites on the *spin subsystem*, the terms c and d determine the dynamic (non-adiabatic) mechanisms, and the term e represents a mixed mechanism of the spin–phonon interaction. It follows from the invariance of H_{s-ph} under an arbitrary uniform displacement $U = const$, and the Galilean principle, that H_{s-ph} is, in fact, an expansion in terms of the differences of atomic shifts and their rates of change. The spin–phonon Hamiltonian is used in *magnetic resonance* studies for describing *spin–lattice relaxation, acoustic paramagnetic resonance*, and pressure-related effects in *electron paramagnetic resonance*. Another

approach to the spin–phonon interaction is the *dynamic spin Hamiltonian* which is expressed in terms of the crystal *strain tensor* $e_{\alpha\beta}$, the spin matrices S_σ, and a number of phenomenological parameters $J_{\alpha\beta\gamma\delta}$; the number of nonzero parameters is determined from *crystal symmetry* considerations. Restricting consideration to the part of the interaction between the spins and the lattice strains that is linear with respect to the tensor $e_{\alpha\beta}$ and bilinear with respect to S_α ($S > 1/2$), the following expression is obtained for the dynamic spin Hamiltonian: $H = \sum J_{\alpha\beta\gamma\delta} S_\alpha S_\beta e_{\gamma\delta}$ ($\alpha, \beta = x, y, z$). Since in the case of $S = 1/2$ the spin–phonon interaction involves a change of *g*-factor (see *Landé g-factor*), this Hamiltonian is defined within the same approximation by the expression $H = \sum J'_{\alpha\beta\gamma\delta} S_\alpha B_\beta e_{\gamma\delta}$ linear in S_α. A form of dynamic spin Hamiltonian that allows for higher powers of S_α ($S > 1/2$) and the tensor $e_{\alpha\beta}$ can be introduced in a similar way.

SPIN–PHONON INTERACTION

See *Spin–lattice interaction*.

SPIN PHONON MAGNETIC RESONANCE

A variety of *phonon magnetic resonance*, which involves inelastic *electron scattering* between Landau levels (see *Quantizing magnetic field*) of opposite spin orientation. Available theoretical works do not explain the amplitudes of spin phonon magnetic resonance peaks, although these amplitudes are considerably smaller than those of magnetic phonon resonance. Experimental features, which are presumably due to spin phonon magnetic resonance, were detected in the magnetoresistance of *n*-InAs under ohmic conditions in a magnetic field of 48 T, as well as in the *electron heating* or warm-up mode of magnetoresistance (see *Galvanomagnetic effects*) and *photoconductivity* of *n*-InSb.

SPIN RELAXATION

See *Paramagnetic relaxation*.

SPIN, RESULTANT

See *Resultant spin*.

SPIN–SPIN COUPLING, INDIRECT

See *Indirect electron spin–spin coupling*, and also *Indirect nuclear spin–spin coupling*.

SPIN–SPIN INTERACTION

General name for *dipole–dipole interactions* and *exchange interactions* involving spins; sometimes this term is also used in reference to *hyperfine interactions*. In concentrated spin systems, a spin–spin interaction determines the processes of establishing equilibrium, and influences the shape and width of *magnetic resonance* lines. The secular (diagonal) part of the spin–spin interaction is an independent *heat sink* exhibiting its own values of *temperature* (different from the spin temperature) and *specific heat*. The occurrence of a spin–spin interaction between the nuclei of different isotopes of a material allows the study of low abundance isotopes by application of double nuclear–nuclear resonance (see *Double resonances*).

SPIN–SPIN RELAXATION

The process of establishment of equilibrium within a spin system. In magnetic resonance experiments the spin–spin relaxation is conditioned by the secular component of the spin–spin *dipole–dipole interactions*, which brings about transverse *relaxation* (relaxation of those *magnetization* components that are perpendicular to the constant magnetic field) due to both the dephasing of precessions of different *spins*, and the precession phase of every spin caused by modulation of static local fields created by neighboring spins, and spin-flop transitions (see *Cross-relaxation*). The spin–spin relaxation manifests itself also in the redistribution of the Zeeman energy of individual spins within the spin system.

SPIN SUBSYSTEM

The system of interacting spin (magnetic) moments, which are isolated or decoupled from the lattice. The spin subsystem behaves as a unit if the *spin–spin relaxation* time is much shorter than the *spin–lattice relaxation* time. The spin subsystem Hamiltonian used in magnetic resonance experiments involves the Zeeman interaction with a constant applied magnetic field, and the spin–spin dipole–dipole interaction. The alternating microwave or radiofrequency magnetic field is considered as a small perturbation. The Zeeman energy (see *Zeeman effect*) generally far exceeds the *dipole–dipole interaction* energy,

and hence only the secular contribution of the dipole–dipole interaction, which commutes with the Zeeman Hamiltonian, is taken into account for interpreting the energy levels. The nonsecular part of the dipole–dipole interaction is responsible for the spin–spin relaxation. Due to this relaxation, the spin system may be considered as in quasi-equilibrium over sufficiently long time intervals, and is characterized only by conserved macroscopic quantities (integrals of motion): mean values of Zeeman energy H_Z and secular dipole–dipole interaction energy H_{ss}. The quasi-equilibrium state is described by a *density matrix* of the form

$$\rho \sim \exp\left[-\frac{H_Z}{k_B T_Z} - \frac{H_{ss}}{k_B T_{ss}}\right],$$

where the parameters T_Z and T_{ss} represent the temperatures of the Zeeman and the spin–spin energy reservoirs, respectively.

SPIN TEMPERATURE

See *Level population*.

SPIN WAVE EXCITATION

See *Parametric excitation of spin waves*.

SPIN WAVE INSTABILITY

See *Kinetic instability of spin waves*.

SPIN WAVES

The collective motion of *spins* of magnetic atoms (ions) in magnetically ordered media. The existence of spin waves is due to the strong electrostatic exchange interactions (see *Exchange interaction*) and weak *spin–orbit interaction* and *dipole–dipole interaction* between magnetic atoms, which results in the shifting of intrinsic magnetic moments (spins) of atoms from their equilibrium orientations, and their propagation in the form of waves.

A spin wave is an elementary excitation of a subsystem of atomic magnetic moments in a magnetically ordered medium; the *quasi-particles*, which correspond to spin waves, are called *magnons*. The existence of spin waves in ferromagnets was predicted by F. Bloch in 1930. Like any wave in a crystal, spin waves are characterized by a *dispersion law*, which follows from the equations of motion of the magnetization of the magnetic sublattices (*Landau–Lifshits equation*), and from Maxwell's equations, and depends on the structure of the equilibrium state of the *magnetic ions*. The latter is determined by its magnetic structure, which depends on the temperature T, the external magnetic field B, and the shape and relative dimensions of the sample. Spin waves in uniform magnetic materials are characterized by their wave vector k and frequency ω, and the number of spin wave branches is equal to the number of *magnetic sublattices* n. Out of these n branches, several (s) branches have a low activation energy $\hbar\omega_0$ (ω_0 is the frequency for $k \to 0$); the activation energy of these latter branches is low because of the weakness of the spin–orbit and dipole–dipole interactions, and vanishes in the exchange approximation; these s branches are called *acoustic spin waves*. The activation energy of the remaining $n - s$ branches, called *optical spin waves*, is of the order of the exchange integral.

The number of acoustic spin waves is determined by the magnetic structure of the material: $s = 1$ for *ferromagnets* and *ferrimagnets*; $s = 2$ in the case of collinear *antiferromagnets*; $s = 3$ in the case of antiferromagnets, which possess more than two sublattices and a noncollinear magnetic structure. The law of spin wave dispersion in an isotropic ferromagnet is of the form

$$\omega = \left[\omega_B + \omega_E (ak)^2\right]^{1/2}$$
$$\times \left[\omega_B + \omega_E (ak)^2 + \omega_M \sin^2 \theta_k\right]^{1/2},$$

where $\omega_E = \varepsilon/\hbar$ (ε is the exchange interaction energy between neighboring atoms; $\omega_B = \gamma B$; $\omega_M = \mu_0 \gamma M$, γ is the *gyromagnetic ratio*, μ_0 is the *magnetic permeability of vacuum*, M is the *magnetization* of the material at the temperature T), a is the lattice constant, θ is the angle between B and k. In the long-wave limit, when the nonuniform exchange interaction may be neglected ($\omega_E a^2 k^2 \ll \omega_B, \omega_M$), the spin waves are referred to as *magnetostatic spin waves*. In the case of ferromagnets, which exhibit low *magnetic anisotropy* constants (e.g., in yttrium *iron garnet* $Y_3Fe_5O_{12}$), the low-frequency (acoustic) branch of the spin wave spectrum corresponds to a spatially nonuniform precession of the total magnetization, and is also described by the above equation. The remaining (optical) branches belong to

the submillimeter and IR wave bands ($\omega \propto \omega_E$). The spin wave spectrum of ferromagnets depends on the ground state structure. Consider the case of a uniaxial crystal with two magnetic sublattices, the magnetic moments of which are collinear to the crystal principal axes and the external magnetic field direction: such a crystal exhibits two branches of acoustic spin waves, the dispersion law having the form

$$\omega s_{1,2} = \sqrt{\omega_E \omega_A + \omega_E^2 (ak)^2} \pm \omega_B,$$

where $\omega_A = \gamma B_A$, B_A is the anisotropy field, $\omega_B < (\omega_E \omega_A)^{1/2}$. Spin waves may also exist in *nonmagnetic metals*. In this case, they exist in the form of *collective excitations* of spin moments of the system of strongly interacting *conduction electrons* (see *Spin waves in nonmagnetic metals*). There exist *nuclear spin waves*: the collective oscillations of nuclear spins. *Surface spin waves* can also exist near the magnetic–vacuum interface, or in the neighborhood of the boundary between two magnetic materials; the law of dispersion of these waves has a pronounced dependence on the interface properties.

Experimentally, the $k \neq 0$ spin waves are excited either through application of special converters (antennae, strip-line waveguides, etc.) or by the technique of *parametric excitation of spin waves* (*parallel (longitudinal) pumping*), when a single pumping photon breaks up into two spin waves. Standing spin waves of finite wavelength may be excited in thin *magnetic films* ($d \leqslant 10^{-7}$ m), and they appear as peak signals of *spin-wave resonance* on the electromagnetic energy absorption curve.

SPIN WAVES IN NONMAGNETIC METALS
(V.P. Silin, 1960)

Oscillations of the *spin density* of conduction electrons due to the *exchange interaction*. The study of spin waves in nonmagnetic metals has played a special role in preparing the way for treating conduction electrons as a charged *Fermi liquid*. In particular, the observation of spin waves in nonmagnetic metals in an external magnetic field provided, for the first time, experimental evidence that an electron gas exhibits properties of a Fermi liquid.

In the absence of an external magnetic field H the spin waves in a charged Fermi liquid are completely identical with spin waves in a neutral Fermi liquid. This is because such spin density oscillations do not give rise to any noticeable electric field, hence the conditions for their existence are the same as those of a neutral Fermi liquid. Thus, in the simplest case when the spin components of the Landau correlation function are independent of momenta, it is also necessary for the Landau parameter B_0 to be positive (see *Fermi liquid*). Estimates indicate that spin waves of this variety make an insignificant contribution to the *surface impedance*, and they have not been observed experimentally.

The equation of motion of the nonequilibrium *magnetization* $M(r, t)$ of an electron liquid in an external magnetic field involves the Fermi liquid Landau parameters B_0 and B_1. When the nonuniformity can be ignored the differential equation for M is a *Bloch equation*, which describes *electron paramagnetic resonance*. The resonance frequency $\omega_{0,0} = 2\mu_B H/\hbar$ does not depend on Fermi liquid effects; μ_B is the Bohr magneton. With allowance for *spatial dielectric dispersion*, one of the branches $\omega_{0,0}(k)$ of the spin waves that travel in the metal has the activation frequency $\omega_{0,0}$. The *dispersion law* $\omega_{0,0}(k)$ involves the parameters B_0, B_1 as well as the angle α between the wave vector k and the field H. Under the conditions of generating a standing spin wave ($kL = n\pi$) in a metal plate of thickness L, a number of extra transparency peaks arise in the neighborhood of the principal peak of the paramagnetic resonance. The positions of these additional peaks with respect to the principal one depend on the angle α and the parameters B_0 and B_1. The observation of this effect was instrumental in the determination of the constants B_0 and B_1 for the *alkali metals* Na and K.

SPIN WAVES, NUCLEAR

See *Nuclear spin waves*.

SPONTANEOUS POLARIZATION

See *Ferroelectricity*.

SPONTANEOUS POLARIZATION SHIELDING

Reduction of the electrostatic field of a spontaneously polarized pyro- (ferro-) electric sample produced by either free charges, charges adsorbed at the sample surface from the environment, or internal charges. Spontaneous polarization shielding by the internal carriers takes place not only in ferroelectric semiconductors (see *Photoferroelectrics*) but also in wide-gap materials. An essential, or even decisive, contribution to the shielding is brought about by *surface states*.

This shielding reduces the energy of the electrostatic field of a spontaneously polarized sample, and thereby weakens the main cause for the appearance of a *domain structure*. In particular, the existence of an equilibrium single-domain state is possible.

An interesting representation of spontaneous polarization shielding is the appearance of a periodic phase structure in a ferroelectric semiconductor at a *first-order phase transition*. The type of ferroelectric domain structure is, to a great extent, determined by the conditions of shielding. The Debye length (see *Debye–Hückel theory*) characterizes the *screening radius* in real *ferroelectrics*, due to the high value of the *dielectric constant*. During the illumination of ferroelectric semiconductors the conditions of shielding can change, resulting in a rearrangement of the domain structure (see *Photodomain effect*).

SPONTANEOUS SYMMETRY BREAKING

The transformation of a physical system to a symmetry state lower than that of the Hamiltonian and equations of motion which describe of the system. An example of spontaneous symmetry breaking in classical mechanics is the behavior of a particle in a symmetric one-dimensional potential with two minima: the particle, which is originally placed in the center (unstable maximum point) falls into one of the two potential wells, hence the initial symmetry, which exhibits invariance under reflections, is spontaneously broken. A spontaneous breaking of symmetry in quantum and thermodynamic systems may take place when the system under consideration has an infinite number of degrees of freedom, which is the case, e.g., at a *phase transition* (see *Landau theory*). The broken symmetry may be discrete, like in a uniaxial *ferroelectric*, or continuous, like in a *Heisenberg magnet* or a *superconductor* (see *Continuous symmetry transformation groups*). In the latter case, *Goldstone excitations* (see *Goldstone theorem*) (zero-mass, zero-spin boson excitations) may arise in the energy spectrum of the system. In order to study theoretically the spontaneous breaking of symmetry in systems of continuous symmetry, N.N. Bogolyubov derived (1963) the *quasiaverage method*, which consists in analyzing an object in a weak external field, which forces the breaking of symmetry; and then letting the value of the field strength go zero in the resulting equations.

SPONTANEOUS TRANSITION

See *Radiative quantum transition*.

SPRAYING

The process of deposition of a layer of material onto the surface of another material called the *substrate*. It may consist of a sequence of several processes: forming a flux of the material to be sprayed; transferring that material from its source to the substrate; *condensing* the material vapor when it reaches the substrate. Table 1 subdivides spraying methods according to techniques used to generate the flux of material towards the substrate.

Thermal vacuum deposition consists in heating the material in a deep vacuum to such a temperature that the *pressure* of its vapor exceeds that of residual gases by several orders of magnitude. The breaking of bonds between surface atoms of the material to be evaporated results in thermal *evaporation*. Nominally, the evaporation temperature is assumed to be such that the vapor pressure at the surface is $P = 1.33$ Pa.

For the ionic case, when a solid or a liquid target surface is bombarded by individual atoms, molecules, or ions with a kinetic energy exceeding that of atom-to-atom bonds (which determines the heat of *sublimation* for the target material), the atoms of the *lattice* move to new positions. As a result, atoms migrate across the surface and escape off it into the gas phase. Such a process is called physical *ionic sputtering*. Since cathode material is sputtered by a subnormal discharge, the term "*cathode sputtering*" is used (see *Sputtering*).

Explosion techniques of spraying produce effective fluxes of sprayed material that far exceed the background atmospheric level. At the same time, the detrimental roles of parasitic chemical reactions and of *diffusion* which need a longer time to develop can be largely suppressed. The explosion techniques provide particularly short spraying times, within fractions of a second. During *explosion*, the metal undergoes almost no evaporation. Instead, most of it is sputtered into a finely dispersed mist consisting of droplets (not vapor). This results in forming *films* from the flux, rather than large agglomerated high-energy particles: the higher the energy, the finer the particles.

Chemical techniques of spraying presuppose that a specific medium containing one or several components or reactants is formed around the substrate. This medium includes constituents of the material to be sprayed. A chemical reaction mainly occurring on the substrate itself then results in forming a solid condensate of the required composition. The chemical reaction itself may be

Table 1. Spray-coating techniques

Thermal vacuum deposition	Ionic case (cathode sputtering)	Explosion spraying	Chemical type	Plasma-jet spraying
Resistive	Diode	Electric	Deposition	Plasma generator
Electron beam	Triode	Gasoline	Pyrolysis	Arc
Laser	Ionic plasma	Acetylene	Activation techniques	Flame
Burst	Jet	Explosives	Hydrolysis	
Discrete	Subnormal discharge		Electron beam decomposition	
High frequency	High-frequency sputtering		Deposition to electron beam track	
Electric arc	Ionic sputtering in arc discharge		Electron beam polymerization	

activated in various ways: applying either heat or a high-frequency field, irradiating the substrate with visible light or X-rays, by an electric arc discharge, by electron or *ion bombardment*, or by using the catalytic effect of the substrate surface itself. The nature of the chemical reaction and of its activation mechanism may noticeably affect the resulting morphology of the sprayed layer.

One runs into difficulties trying to draw a clearcut distinction between physical and chemical deposition when considering such processes as thermal evaporation or cathode sputtering in a chemically active atmosphere, combined evaporation of interacting material, or vacuum deposition followed by transformation of the deposited material via a chemical reaction initiated through any of the techniques available.

During *plasma–jet spraying*, the material to be deposited is introduced into a high-temperature *high-pressure* zone (in excess of atmospheric values). That zone may be formed by an electric arc or using an arc plasma generator. Next, the material to be deposited is evacuated from the hot zone by a plasma-forming gas and is delivered onto the substrate.

SPREADING

A process of spontaneous dispersal of liquid in the form of a single-phase layer across a *solid surface* or liquid surface. Spreading occurs because it diminishes the system free energy, in particular, its surface *free energy* (see *Surface energy*). The driving force of the spreading of liquid across a smooth surface in a medium, calculated per unit width of the spreading liquid wave front, is equal to $\Delta\sigma = \sigma_{12} - \sigma_{13} - \sigma_{23}\cos\theta_D$, where σ_{ik} is the phase-to-phase tension between the ith and the kth phases, and θ_D is the *dynamic edge angle* (see *Wetting*). The value of $\Delta\sigma$ for $\theta_D = 0$ is called the *spreading factor*. When a liquid spreads across a rough surface, $\Delta\sigma = k(\sigma_{12} - \sigma_{13} - \sigma_{23})$, where k is the *roughness factor*, equal to the ratio of the actual surface roughness to that of an ideally smooth one (see *Surface roughness*). During spreading, resistance forces concentrate either along the wetting line (kinetic resistance), or within the bulk liquid (*viscosity*, inertial forces, etc.). In the first case the liquid position $x = vt$, where v is the velocity of spreading (*kinetic spreading regime*). In

the second case, we have a *hydrodynamic spreading regime*. During the initial stage of spreading, called the *inertial spreading regime* (a fraction of a second long), inertial forces play an important role. The main stage of the process involves viscous spreading. To study the kinetics of spreading one may consider either an infinite or a limited source of liquid. In the case of an infinite source we have $x^2 = At\Delta\sigma$ during the principal stage of the process, where x is the propagation distance and t is time. For a limited source of liquid $x^3 = Bt\Delta\sigma$ for one-dimensional spreading, and $x^4 = Ct\Delta\sigma$ for a two-dimensional case. At higher temperatures the *kinetic coefficients* A, B, C grow. Spreading is affected by the surface profile, it is sensitive to the illumination of semiconducting surfaces (*photocapillary effect*), to the electrochemical polarization of a metal in an electrolyte, and to the temperature gradient at the surface.

SPRITE (acronym for Signal PRocessing In The Element)

Scanning *semiconductor photodetector* (line scanning device), which may be considered as a solid-state analog of a *charge coupled device*. In this device, the light-produced profile of the distribution of minority carriers moves at a constant rate along the photodetector due to the drift of charge in the electric field, which is applied to the photodetector; the light beam, which established this distribution, is set in motion at the same rate, and in the same direction. There is a reading element (usually a *semiconductor junction*) at the output of the photodetector, which produces a current proportional to the concentration of the minority carriers in close proximity to the reader; as a result, the time dependence of the output current reproduces the profile of the charge distribution, and consequently of the irradiance of the input beam of light. A photodetector of the sprite type may be constructed only on the basis of a semiconductor featuring high charge carrier mobility. The device has a number of disadvantages, among which are the necessity of scanning the beam of light, and the narrow range of adjustment of the scan rate: the image can be blurred due to *diffusion* at low rates of scanning, and at high rates there is a drop in sensitivity caused by the effects of the strong electric field. The principal advantages of sprite are its simple construction, high sensitivity (reached by accumulation of the photocurrent signal during the motion of the charge carrier packet in the detector), and the fact that, relative to other scanning photodetectors, the operation of sprite is much less affected by the heterogeneity of the material. A high-resolution small computer animation object that can be shifted across a screen, independently of images and text material, is also called a sprite.

SPUTTERING

Release of atoms from the surface layers of a *solid* during its bombardment by fast particles: ions, atoms, electrons, neutrons. It was first observed as "evaporation" of metal electrodes, mainly cathodes in gas tubes at temperatures below the *melting temperature* of the metal, which is why *ion sputtering* has been called *cathode sputtering*. Sputtering occurs because of energy transfer by ions to the atoms of the target in the process of their collisions in the surface layers of a solid target. The process of sputtering occurs under conditions far removed from thermodynamic equilibrium, and in this respect sputtering differs from *evaporation*. The products of sputtering are atoms, ions, and *clusters* (see *Cluster sputtering*).

Sputtering is characterized by a *sputtering coefficient S* that is equal to the number of atoms knocked out of their positions by a single incoming ion. Values of S may vary over wide limits, and can even far exceed unity, depending on the properties of both the bombarding ions (mass, energy, angle of incidence) and the target (structure, ion mass, bonding energy, temperature). When bombarding the solid with high-energy particles ($E > 100 \, keV$) atomic collisions rarely takes place in the surface layer, and the sputtering coefficient is low. When bombarding a specimen with particles of larger mass and lower energy, sputtering becomes a dominant factor: mean free paths are then short, and the energy is mainly transferred to atoms in the surface layer.

During the initial stages of sputtering of multicomponent systems, one portion of their components may be evacuated faster than another, to produce an enrichment of the surface with the remaining components. Selectivity of sputtering (e.g., depletion of austenite steels in nickel, carbon, manganese) may result in a $\gamma \rightarrow \alpha$ transformation, i.e.

a lowering of thermal stability, change of *magnetic permeability*, etc.

Along with *physical sputtering*, produced by the transfer of energy and momentum from high-energy particles to atoms of the target, there exists *chemical sputtering* as well. This occurs in cases when the bombarding particle forms a highly volatile compound with the target material. For example, when bombarding graphite with accelerated ions of hydrogen, gaseous methane CH_4 can form.

Ionic sputtering is used for cleaning and etching surfaces (*ion etching*), for growing *films* (by precipitation of sputtered particles), for analyzing the surface via the technique of *secondary electron emission*, etc.

The diminishing of the wall depth due to sputtering is one of the problems to be overcome on the way to building a thermonuclear reactor.

SPUTTERING, CLUSTER

See *Cluster sputtering*.

SPUTTERING, SELECTIVE

See *Selective sputtering*.

SQUID

See *Superconducting quantum interference device*.

STABILITY CONDITIONS OF CRYSTALS

See *Crystal stability conditions*.

STACKING FAULTS, stacking defects

Two-dimensional *defects in crystals* of a crystalline structure associated with the disordered interlacing of close packed layers of the *crystal lattice*. Stacking faults have atomic sizes in directions normal to their plane, and macroscopic dimensions in the two other directions. The nature of the distortions of the crystal structure is the same along the entire defect surface. They are subdivided into *defects of interstitial packing* (added atomic plane among normally interrelated lattice layers, e.g., in the face-centered cubic lattice – ABCACBCAB) and *defects of subtraction packing* (missing atomic plane in a closely packed crystal, e.g., in the face-centered cubic lattice – ABCACABC). These structures do not necessary extend throughout the whole crystal, but may stop at a *partial dislocation*. A system of two partial dislocations connected by a stacking fault is called a *split dislocation* (see also *One-dimensional models of disordered structures*).

STAINLESS STEEL

Solid solution of a large group of alloying elements in *iron* with the surface having enhanced resistance against rusting under atmospheric conditions, and against *corrosion* in hostile media. According to chemical composition, stainless steels are subdivided into chromium, chromium–nickel, and chromium–manganese–nickel types, so *chromium* is the main alloying element with its content varying between 12 and 20%.

Together with the elements accompanying Fe (C, Si, S, P), other elements (Ni, Mn, Ti, Nb, Co, Mo, N) are introduced into stainless steel to provide the desired level of mechanical and anti-corrosion properties. *Corrosion resistance* of stainless steel is caused by the formation of a thin protective film of oxides and insoluble compounds at the surface. According to the phase composition, chromium stainless steels are divided into:

(1) martensite, semiferrite and ferrite types with predominantly body-centered cubic lattices;
(2) chromium–nickel and chromium–manganese–nickel stainless steels which comprise austenite and austenite–carbide steels with face-centered cubic lattices, and also
(3) austenite–ferrite and austenite–martensite steels with body- and face-centered cubic lattices.

If the alloys tend to form *epsilon phases*, their phase composition will be represented also by hexagonal close-packed lattices. Depending on the type of *crystal structure*, stainless steels may possess the magnetic properties of either paramagnets or ferromagnets.

STARK EFFECT at impurity centers (after German physicist J. Stark)

Shift and (or) splitting of energy levels of impurity centers in an external electric field E. The dependence of the energy spectrum on E depends on the presence or absence of inversion symmetry at the sites of the local centers. When inversion is included in the *point group* of the site symmetry the electron states of the center have a definite parity, and the square-law Stark effect may occur in the spectra of these centers (shift and splitting of levels proportional to E^2). A weak *quadratic Stark effect* has been observed experimentally in absorption and luminescence spectra of a number of crystals (alkaline-earth fluorides, diamond, etc.) with *color centers* and impurity ions of the "inversion" type.

When the inversion operation is not included in the point symmetry group of the center then the *linear Stark effect* is possible. This linear effect in spectra of centrally symmetric crystals with color centers lacking inversion is of particular interest. Such centers possess spontaneous electric *dipole moments* which are oriented along some equivalent crystallographic directions. In an applied electric field, the linear shift of the bands depends on the projection of E on the dipole moment of the center, which differs for centers having different orientations. It causes the splitting of certain bands in the spectrum into symmetric multiplets, with each component arising from a shifted transition from a center with a particular orientation (so-called *pseudo-Stark splitting*). The latter was observed in corundum Al_2O_3 (D_{3d}) crystals activated by $3d^n$ transition metal ions, in alkali halides (O_h) with dipolar rare-earth centers, in diamond (all these crystals possess inversion symmetry), and other systems. An analysis of the pseudo-Stark band splitting spectrum provides information about the local symmetry of the center, and about the multipole nature of the optical transitions. Experimentally the splittings are studied in a static external field, or by a more sensitive differential procedure using a modulated applied electric field.

STARK EFFECT OF EXCITONS

Shift and (or) splitting of energy levels and spectral lines or bands in *exciton* optical spectra in an external electric field E. For crystals with ordinary isotropic electron energy bands, the electric field affects the hydrogen-like *Wannier–Mott excitons* in a manner similar to the electric field action on a hydrogen atom. The energy states of excitons with principal quantum number $n \geqslant 2$ are split into sublevels, and the sublevel shift with respect to the initial position in weak fields depends linearly on E (*linear Stark effect*). For the state $n = 1$ only a square law (*quadratic Stark effect*) is possible. The ratio between the magnitudes of the Stark effect for an exciton and that for a hydrogen atom is proportional to the ratio of their Bohr radii $(a_{eff}/a_H) \approx 10^2$, where a_{eff} is the effective exciton Bohr radius, and $a_H = 0.052918$ nm is the Bohr radius of the hydrogen atom.

The features of the exciton Stark effect depend upon the properties of the individual crystal. In *uniaxial crystals* (CdS, CdSe, GaSe, etc.), the internal *crystal field* partially removes the degeneracy of the states with a given orbital angular momentum quantum number. In weak fields the dependence of the splitting on E is linear. With a further increase of E the excitation of *forbidden quantum transitions* occurs, which permits the determination of the location of optically inactive exciton states. The splitting in an anisotropic crystal depends on the orientation of the field E relative to the crystallographic axes. In centrally symmetric crystals (e.g., Cu_2O), the effect of the external electric field is like that on alkali metal atoms which exhibit no linear Stark effect, namely the excitation of forbidden transitions and a square-law shift of the levels with increasing field. An electric field $E \sim E_n/(ea_e)$ easily brings about the ionization of an exciton state with a binding energy E_n. That is why the highest terms of the exciton Rydberg series are not observed in fields ~ 100 V/cm.

STATE OF STRESS

Internal *mechanical stresses* produced by external fields (such as electric, magnetic, mechanical, thermal, etc.) and those remaining in the material after removal of the field (see *Internal stresses*). States of stress may be subdivided into: macroscopic (neutralized within a range comparable to the sample size), microscopic (neutralized within the volume of a single *crystallite*) and submicroscopic (neutralized in a volume comparable to a unit cell or single atom). Quantitatively all states of stress may be evaluated using radiographic techniques (see *X-radiography*). X-radiography techniques of evaluating *macrostresses* are all based on the relationship of elastic strain (see *Elasticity*) to stress via *Hooke's law*, and the angle of diffraction of X-rays for a specified reflection from an $\{hkl\}$ reflecting plane.

Using Debye X-ray imaging, one may calculate macrostresses from line shifts in the corresponding radiograms. As for the *microstresses*, these are calculated from line broadening in the *X-ray diffraction patterns* and are characterized by a value $\Delta d/d$, where Δd is the maximum deviation of the plane-to-plane distance for a given interference line from its average value, d. Measuring the line half-width in radians in a sample (b), and a standard (B), one may find the broadening, β, as $\beta = (b^2 - B^2)^{1/2}$, while the averaged value of microstresses is given by

$$\frac{\Delta d}{\Delta d_{ave}} = \pm \frac{(b^2 - B^2)^{1/2}}{2R \tan\theta}.$$

Here R is the goniometer radius in the diffractometer, θ is the Bragg angle for the chosen diffraction line $\{hkl\}$ (see *Bragg law*). *Submicrostresses*, related to *defects* in crystals that cause displacements of atoms from their equilibrium positions, are determined from the decreased intensity of X-ray interference lines against those obtained from an undistorted crystal.

STATES OF MATTER

Gaseous, liquid, and solid states.

Atoms and molecules in the *gaseous state* are not bound to each other; but rather they move about as free particles limited only by collisions with each other, and with the vessel walls. In the *liquid state* they are bound by interatomic forces and vibrate about randomly distributed and continuously changing equilibrium positions. Therefore a liquid has a fixed volume and possesses boundaries like a *solid*, but it acquires the shape of a vessel in which it is placed, like a gas. The atoms in crystalline solids (*crystals*) are bound to equilibrium positions arranged strictly periodically; and they undergo small oscillations about these positions in the *crystal lattice*. The atomic positions in solids exhibit long range order, those in liquids have short range order, and atomic positions in gases lack order. The equilibrium positions in amorphous solids (see *Amorphous state*) are arranged irregularly somewhat as in liquids, but without the continuous rapid motional rearrangements of liquids. Gases and liquids are referred to collectively as *fluids*, and liquids and solids constitute *condensed states of matter*. The state of matter in which a material is found depends on the temperature (T) and pressure (P), and regions of different states are represented by *phase diagrams* which are distinctive for each material. Sometimes a *plasma*, or electrically neutral collection of charged particles, is spoken of as the fourth state of matter. The concept of states of matter is also applied in solid state physics to describe the state of *quasi-particle* quantum systems (e.g., *Bose gas*, *Fermi gas*, *Bose fluid*, *Fermi liquid*, etc.).

STATE VARIABLES, state parameters

Physical quantities, which are characteristic of the state of a *thermodynamic system*. Extensive and intensive state parameters are distinguished. *Extensive state parameters* (internal energy, *entropy*, *enthalpy*, *Helmholtz free energy* or simply *free energy*, *Gibbs free energy*) are proportional to the volume (or mass) of the system, whereas *intensive state parameters* (*pressure*, *temperature*, *concentration*, *magnetic flux density*, etc.) are mass-independent. Since not all state parameters are independent, the equilibrium state of a system may be uniquely determined by specifying only a limited number of them, such as pressure and temperature for an ideal gas (see *Equations of state*, *Gibbs' phase rule*).

STATIC DISPLACEMENTS OF ATOMS

Changes of equilibrium positions (for thermal vibrations) of atoms relative to their corresponding positions in an *ideal crystal* with an ideal infinite lattice. These changes take place on introducing *defects* into the crystal, or on applying external distorting stresses to it. The stochastic field of static atomic displacements arises in a crystal as a result of the superposition (in the general case, nonlinear) of displacement fields produced by various defects. The fields of static displacements of atoms are conveniently calculated using the *lattice statics method*, assuming a specific model of crystal structure. The fields of static displacements of atoms produced by individual defects are in the general case anisotropic, with the exception of a defect with cubic symmetry in an elastically isotropic medium (see *Isotropy of elasticity*). In the latter case, the static displacements of crystal atoms vary as $u = u(r) = Ar/|r|^3$ (A is the *elastic potential of a defect*) at distances $|r|$

from the defect long compared to the period of a crystal *translation*. At short distances from the defect the discreteness of the crystal becomes appreciable, and in all cases this fails to conform to the above relation for $u(r)$, an effect which is most pronounced in crystals with high values of the ion–ion interaction radius.

STATIC SKIN-EFFECT, steady-state skin-effect

A pronounced increase in the constant electric current density near the surface of a metallic sample located in a strong magnetic field, compared to the current density within the bulk. For the Larmor orbit radius condition $r < l$ (l is electron *mean free path*), the bulk conductivity σ_b is extremely low, for it is determined by the *random walk* of the center of closed electron orbits assuming infrequent collisions of this center with lattice *defects*: $\sigma_b = \sigma_0(r/l)^2$, where σ_0 is the conductivity of a bulk conductor in the absence of a magnetic field. Another situation arises within a region of width r in the neighborhood of the surface when the closed orbits can be disrupted by frequent collisions of electrons with the surface.

The contribution of the surface to the total current may be analyzed most easily in the case of a slab or *plate* placed into crossed electric and magnetic fields that are both parallel to its surface. If the reflection of electrons from the surface is specular, then these electrons follow periodic open (non-closed) paths in the magnetic field, with their mean velocity along the electric field direction of the order of the *Fermi velocity* v_F. Therefore, the conductivity in the near-surface layer and σ_0 are of the same order of magnitude, and the contribution of the surface to the thickness-averaged conductivity is given by $\overline{\sigma}_s = \sigma_0 r/d$, where d is the plate thickness. A similar situation is observed in the case of diffuse reflection (all reflection angles equally likely) in the absence of transitions between electron and hole states, although the trajectory is not periodic, the mean velocity being of the order of v_F. If reflection results in transitions between electron- and hole-type charge carriers (*surface recombination*), then a zero value of the mean velocity is possible: the transformation of an electron into a hole changes the direction of the motion of a *quasi-particle* along the electric field. The role of the mean free path in this process is played by the quantity r/w, where w is the probability of recombination, and hence $\overline{\sigma}_s \cong \sigma_0 r^2/(wld)$. The above expressions are valid for *metals* with equal numbers of electrons and holes ($n_e = n_h$). If $n_e \neq n_h$, then the bulk conductivity turns out to be of the order of σ_0 because of the appearance of a strong Hall field, and the static skin effect does not take place. If, however, the applied field is inclined with respect to the surface, then the static skin effect also occurs at $n_e \neq n_h$, and determines the value of the dissipative component of the conductivity. The static skin effect in *semimetals* is of a specific type because the probability of *recombination* as a result of collisions of electrons with impurities in the bulk is small; the skin width in this case is on the order of the *diffusion length*, which far exceeds r.

STATIC STRENGTH

The value of *strength* determined by static *mechanical testing of materials* carried out during a relatively slow change of the test piece loading corresponding to a low rate of its *strain* (strain rate $<10^{-2}$ s^{-1}). Within this range of rates the mechanical properties of materials only change slightly. The static strength is determined by mechanical tests for tension, compression, *flexure*, twisting. When undergoing mechanical tests the material exhibits a lower value of the *yield limit* and higher *plasticity* than under dynamic tests, in which higher loading rates are used (see *Impact strength*). This is because the relaxation processes (see *Stress relaxation*) in static deformation are more complete than those in dynamic deformation.

STATISTICAL OPERATOR

The same as *Density matrix*.

STATISTICAL SUM

See *Partition function*.

STATISTICAL THEORY OF LINE SHAPE

Theory of the spectral distribution of the resonance frequencies of atoms (*paramagnetic centers*) under the conditions of a random scatter of these frequencies. This spread in resonance frequency values may be caused by, e.g., internal inhomogeneities of the crystal due to the presence of

defects: impurities, *dislocations*, etc. This statistical approach to line shape theory is used in studies concerned with the phenomenon of *magnetic resonance* for estimating the broadening due to the *dipole–dipole interaction*.

The principle of the method is as follows. It is assumed that the resonance frequency shift $\Delta\omega$ due to defects may be expressed as a sum of contributions from individual defects $\Delta\omega = \sum \Delta\omega_i(\varepsilon_i, r_i)$, where r_i is the coordinate of a defect, and ε_i is the characteristic of its internal state of (e.g., of its orientation, if the defect has an elastic or electric *dipole moment*). The *line intensity* of absorption $I(\omega)$ at the frequency ω is proportional to the probability of the absorption frequency lying within the interval from ω to $\omega + d\omega$, i.e.

$$I(\omega) = \left\langle \delta\left(\omega - \sum_i \omega_i\right)\right\rangle$$

$$\equiv \frac{1}{2\pi}\int_{-\infty}^{\infty} dt\, e^{i\omega t}\langle e^{-i\omega t}\rangle^N, \qquad (1)$$

where the angle brackets $\langle \ldots \rangle$ designate averaging over the variables r_i and ε_i, and N is the total number of defects. The possibility of factorization in the averaging in Eq. (1) is related to the lack of correlations between the positions and orientations of individual defects, which is the principal reason for the applicability of the statistical method. The product of integrals in Eq. (1) is easily calculable in the thermodynamic limit $N \to \infty$, $V \to \infty$, $n = N/V = $ const (V is crystal volume), which leads to the final equation

$$\langle e^{-i\omega_i t}\rangle$$

$$= \exp\left[-n\int dr \int \varphi(\varepsilon)\,d\varepsilon\left(1 - e^{-i\omega(r,\varepsilon)t}\right)\right], (2)$$

where $\varphi(\varepsilon)$ is the distribution density of different values of ε.

The statistical theory of line shape was elaborated by H. Margenau (1936), and used to explain the broadening of lines (see *Line (level) width*) in very dense gases. P. W. Anderson (1951) employed this method for calculating the electron paramagnetic resonance (EPR) line shape due to dipole–dipole interactions (see *Line shape*) in magnetically dilute solids. This calculation consisted in determining the corrections to the resonance frequency that arise from pairs of interacting spins, and summing over all the spin moments of the system. For this case of dipole–dipole broadening the statistical line shape approach predicts the *Lorentzian shape* of an EPR signal. The application of the statistical theory of line shape broadening to the magnetic resonance lines arising from defects in solids was developed by A.M. Stoneham (1965). In recent years this theory has not found widespread use.

STATISTICS

See *Bose–Einstein statistics*, *Fermi–Dirac statistics*.

STEADY-STATE SKIN-EFFECT

See *Static skin-effect*.

STEEL

An *alloy* of iron with carbon (2.0% C). It often contains technological additives (Mn < 1%, Si < 0.5%) and, in smaller amounts (<0.1%), permanent and foreign impurities; such steel is called *carbon steel*. If particular elements are added to steel (Cr, Ni, W, Mo, Co, etc.) to give it special properties, then it is referred to as *alloy steel*. The properties of steel may be varied over a broad range by *heat treatment*. The wide application of steel is due to its good mechanical properties, and the especially efficient combination of *strength*, *plasticity*, and *viscosity*. All these factors as a whole provide high *construction material strength* to materials and articles made of steel.

Alloy steels are classified according to the presence of various alloying elements (nickel, chromium, nickel–molybdenum, etc.), or by their structure (ferritic, austenitic, etc.; see *Ferrite*, *Austenite*, etc.). Steels are designated according to the content of major elements. The first two numbers of the marking are indicative of carbon content (in hundredths of %), the following letters of the Russian alphabet indicate the presence of one or another alloying element (H indicates Ni, X = Cr, Γ = Mn, M = Mo, C = Si, etc.), and the numbers placed after the letters denote the percentage of that element in steel; e.g., steel marked 30X2H4M1 contains 0.3% C, 2% Cr,

4% Ni, and 1% Mo. In terms of classification according to functions performed, there are four classes of steels, namely: structural, tool, special purpose, and stainless types.

Structural steel is used for making major parts of machines and constructions. The essential feature required of this class of steel is that it must exhibit an appropriate combination of strength, plasticity, and viscosity. The principal structures (for building, etc.) are made of the so-called *low-alloy (construction) steel*. This steel includes enhanced amounts of Mn and Si and small additives of other elements (N, Ti, V, Al, Nb, etc.). The carbon content of this steel is low (less than 0.15%) in order to provide good weldability. The class of construction steels includes spring steel, ball bearing steel, etc.

The *tool steels* are carbon, alloy and high-speed steels (the latter used to make high performance cutting tools). On being subjected to *quenching* by cooling from high temperatures (1200–1250 °C) and triple *tempering* at 560 °C, the tool steel attains the property of *red-hardness*, i.e. it retains high hardness at red heat (600 °C), which makes it capable of cutting metal. There are also other types of tool steels: stamp steels, steels for fabricating measuring tools, etc.

Of widest application among alloy *special purpose steels* are the following two types: heat-resistant steels and high-temperature ones. The alloys of the former type are usually high in chromium content, and exhibit high resistance to oxidation at elevated temperatures (up to 900 °C), whereas alloys of the second type feature satisfactory heat resistance and exhibit enhanced strength at temperatures up to 800 °C.

Acid-resistant (stainless) steels are used for working in aggressive media. Alloying with *chromium*, the content of the latter reaching 12–30%, provides *corrosion resistance*. Purely chromium steels (ferritic ones) are unweldable, whereas chromium–nickel steels are weldable by all *welding* techniques, which defines their field of application. Smelting steels to a very low carbon content (0.03%) eliminates the tendency toward *intercrystalline corrosion*. *Precision (machinery) steels* include those featuring various specific properties: nonmagnetic steels, steels with special magnetic properties, steels exhibiting features of thermal expansion, etc.

STEEL, HADFIELD
See *Hadfield steel.*

STEEL, STAINLESS
See *Stainless steel.*

STEPANOV EFFECT (A.V. Stepanov, 1933)

Generation of an electric charge (potential difference) at the surface of *alkali-halide crystals* under *plastic deformation.* The sign and magnitude of the charge depend on many factors: the type of crystalline material under study, presence of impurities, subjection of the crystal to preliminary *heat treatment* and irradiation, and conditions of the plastic deformation. As the degree of plastic deformation and temperature increase there may be a reversal of the sign of the charge caused by a change in the character of, respectively, the *dislocation* motion and the state of impurities. The Stepanov effect has a complex physical nature, and the following phenomena may contribute to it: the transport of charge to the surface by mobile dislocations, which occurs because the cores of these dislocations trap excessive amounts of *vacancies* of a similar type (sweep-up mechanism); the drag of vacancies by dislocations outside the core, the generation of a bulk charge due to the unbalanced flow of dislocations of unlike sign, the shift of dislocations with respect to the surrounding cloud of impurities at the initial stages of the quasi-static or alternating-sign plastic deformation; the preferential displacement in a certain direction of *interstitial atoms* or vacancies of certain types. The Stepanov effect is observed not only in alkali-halide crystals, but also in $A^{II}B^{VI}$ type semiconductor crystals, which feature charged *edge dislocations* along the edges of excess half-planes {111}, composed of atoms (ions) of types A^{II} and B^{VI} (α- and β-dislocations) which are oppositely charged.

STEVENS OPERATORS
See *Irreducible tensor operators.*

STOICHIOMETRIC COMPOSITION

A chemical compound of the general formula $A_l B_m C_n \dots$ is of stoichiometric composition, if the ratios of *concentrations* $C_A : C_B : C_C : \cdots$ of its components to each other are in exact correspondence with its formula:

$$C_A : C_B : C_C : \cdots = l : m : n : \cdots ,$$

where l, m, n are integers. Deviations from a stoichiometric composition of a crystal arise due to introducing excess amounts of atoms of one of its components, e.g., as *doping atoms* or as *substitutional atoms*, or because of the deficiency of atoms of a certain type compared with stoichiometry, which is detected by the presence of vacant lattice sites. The deviation from stoichiometry Δ_A of the component A of a binary non-stoichiometric crystal $A_{n+\alpha} B_m$ relative to its stoichiometric analogue $A_n B_m$ is given by $\Delta_A = \alpha/n$. In like manner $\Delta_B = \beta/m$ is the deviation from stoichiometry for the non-stoichiometric crystal $A_n B_{m+\beta}$. Variations from stoichiometry α and β can be due to the exchange of atoms between the sample and its environment. Many compounds (oxides, chalcogenides, etc.) exhibit considerable deviations from stoichiometry, e.g., the composition of TiO crystals ranges from $TiO_{0.69}$ to $TiO_{1.35}$. In such cases there is no rigid line of demarcation between the concepts of a non-stoichiometric compound and that of a *solid solution*. These deviations are responsible for various properties of crystals; it should be noted that *stoichiometric defects* and impurities can be present in the crystal concurrently. Sometimes, the stoichiometric composition is defined in a more restricted way; by designating only the concentrations of corresponding atoms in normal positions at their *crystal lattice* sites. In this context, the transitions of atoms between the lattice sites and interstitial positions are treated as variations of stoichiometry, which makes the present interpretation of stoichiometry less appropriate in many cases.

STOICHIOMETRIC DEFECTS

Intrinsic defects in a crystal lattice, which consist in the general case of two spatially separated components: *vacancies* and *interstitial atoms*. The presence of stoichiometric defects does not disturb the regular overall *stoichiometric composition* of the crystal. The designation "stoichiometric defects" is used in reference to *Frenkel defects*, *Schottky defects*, and *anti-Schottky defects*. If the components of stoichiometric defects are electrically inactive (e.g., in *metals*), then their equilibrium concentration is determined by the relation $N_1 N_2 = N^2 \exp[-W/(k_B T)]$, where $N_1 = N_2$ is the concentration of stoichiometric defect components, N is the effective concentration of lattice sites, W is the formation energy of stoichiometric defects. The stoichiometric defects in nonmetals are commonly electrically active (easily ionized), hence their equilibrium concentration depends on the equilibrium concentration of electrons n:

$$N_1 N_2 = N^2 \left(1 + \frac{n}{Q_d}\right)\left(1 + \frac{Q_a}{n}\right)\exp\left(-\frac{W}{k_B T}\right),$$

where

$$Q_d = Q \exp\left(\frac{\varepsilon_d}{k_B T}\right),$$

$$Q_a = Q \exp\left(\frac{\varepsilon_a}{k_B T}\right),$$

Q is the effective *density of states* in the *conduction band*, ε_d and ε_a are the energies of the *local electronic levels* of, respectively, donor and acceptor components of stoichiometric defects, measured from the bottom of the conduction band (positions of levels in the energy gap satisfy the inequalities $\varepsilon_d < 0$, $\varepsilon_a < 0$). The reversible change of the equilibrium concentration of electrically active stoichiometric defects with changes of the crystal temperature is responsible for the *intrinsic conductivity* of semiconductors, and for *conductivity self-compensation*.

STOKES' RULE (G.G. Stokes, 1852), Stokes' shift

An empirical rule which states that the wavelength of emitted *luminescence* light is longer than that of the incident exciting light responsible for the luminescence. Stokes' rule is representative of the quantum nature of the process of photoexcitation and light emission: in the absence of other energy sources, the energy of the emitted quantum does not exceed the energy of the prior absorbed quantum. Interaction with the environment causes, as a rule, loss of a certain part of the excitation energy (*Stokes' losses*), due to which the emission becomes shifted to longer wavelengths. It is also to be noted that Stokes' rule may be slightly violated when some part of the energy of thermal

motion is transferred to the emitted light. Therefore, there exists also another statement of Stokes' rule, according to which both the radiation spectrum as a whole and its peaks are shifted toward longer wavelengths, compared to the absorption spectrum and its peaks (E. Lommel). There are cases in which Stokes' rule violated. For example, the whole broad band of CdS luminescence, which peaks at 541 nm, may be initiated by light of wavelength 633 nm. In *Raman scattering of light* Stokes' rule is obeyed by Stokes lines, and violated by anti-stokes lines. See also *Anti-Stokes luminescence*.

STONER CRITERION (F. Stoner, 1936)

The possibility of delocalized electrons to exhibit the phenomenon of *ferromagnetism*. See *Band magnetism*.

STORED ELASTIC ENERGY

Energy spent for the formation of crystal lattice *defects* which are stable at the defect formation temperature (e.g., during *nuclear radiation*). The subsequent *annealing* at a higher temperature results in the healing or annihilation of defects. Different annealing temperatures correspond to different types of defects, which produce peaks of heat release at different temperatures, typical for particular types of defects. For some types the release of stored energy during annealing is observed also in the form of peaks of activated *electrical conductivity*, *luminescence*, etc. The temperature positions of the corresponding peaks can help to identify the defects.

STRAGGLING

A notion sometimes used for the mean-square spread of the values of a certain randomly changing quantity about its mean value. The term is commonly employed to characterize the passage of charged or neutral particles through a condensed medium, e.g., straggling of *charged particle paths* in solids, *charged particle energy loss*, etc.

STRAIN, deformation

Variation of the volume of a body due to the effect of external fields (thermal, gravitational, mechanical, electrical, magnetic, etc.). Deformation may be elastic (see *Elasticity*), i.e. disappearing after the cessation of the field effect, or *plastic deformation*, i.e. with residual effects. In the elastic case there is a linear relationship between the stress and the deformation (generalized *Hooke law*). For isotropic materials (see *Isotropy*) the connecting coefficients are constants (*elastic moduli*), and when anisotropy is present they depend on the coordinates (see *Anisotropy of elasticity*). Plastic deformation of crystalline solids is usually associated with the motion of *dislocations* or translational linear *defects* which are observed at the early stages of deformation. As the deformation proceeds the free dislocations that are capable of moving begin to disappear, and the crystal structure becomes a system of components (cells, fragments) disoriented with respect to each other (see *Cellular structure, Fragmentation*). In such a system plastic deformation may proceed by other paths, e.g., by the movement of *grain boundaries* as independent defects. Plastic deformation is generally understood as a process which occurs at different structural levels, from the atomic to the macroscopic.

STRAIN AGING

Variation of mechanical and physical properties of strained *metals* and *alloys* as a result of their storage at room temperature (*natural strain aging*) and of their heating (*artificial strain aging*). The physical nature of strain aging lies in the formation of atmospheres of *impurity atoms* around *dislocations*, as a result of which the strength parameters (e.g., *yield limit*) are increased and the *plasticity* is reduced. The most valuable strain aging is in metals with a body-centered cubic lattice characterized by low solubility of the impurity *interstitial atoms*. The combination of heating and *strain* brings about *dynamic strain aging*, conditioned by the interaction of the impurity atoms with moving dislocations. Dynamic strain aging is displayed through the toothed character of "stress–deformation" curves (*Portevin–Le Chatelier effect*). Strain aging is used in technological schemes for hardening metals and alloys, e.g., during the manufacture of springs (see *Strain hardening*).

STRAIN AXES, axes of strain

Principal axes of the *strain tensor*. This tensor U_{ik} is symmetric; therefore, using a linear transformation of coordinates at each point, it can be reduced to the *principal axes of a tensor*, i.e. to a diagonal form. The corresponding diagonal components are called *principal strains*.

STRAIN, BREAKING

See *Breaking strain*.

STRAIN COMPATIBILITY

Restrictions which are imposed on *strains* so the strained elements of the sample will fit tightly with one another, and not interfere with each another after the deformation. From the standpoint of mathematics, the necessity of imposing these restrictions is due to the fact that six components of the *strain tensor* $u_{ik} = (1/2)(\partial_i u_k + \partial_k u_i)$, where $\partial_i = \partial/\partial x_i$, are expressed in terms of three components of the *displacement vector* u_i and therefore cannot be chosen arbitrarily, but must comply with certain conditions called strain compatibility conditions, or *continuity conditions*. The corresponding six conditions are of the following form (no summation with respect to repeated indices):

$$\partial_{ii} u_{kk} + \partial_{kk} u_{ii} = 2\partial_{ik} u_{ik},$$

$$ik = 12, 23, 31;$$

$$\partial_{ik} u_{ll} = \partial_{ii} u_{kl} + \partial_{kl} u_{ii} - \partial_{ll} u_{ik},$$

$$ikl = 123, 231, 312.$$

STRAIN, CRITICAL

See *Critical strain*.

STRAIN DEGREE

The quantity, which characterizes the change of geometric dimensions of a sample under the action of an external stress (force) or mechanical load. The most commonly used measure of strain is the value of elongation or compression per unit length: $\varepsilon = \Delta L/L_0$ (ΔL is the absolute change of length in the direction of the applied load, $\Delta L = L - L_0$, L_0 is the initial effective length, L is final length). As the stress is increased the strain increases, and the material can undergo a sequence of successive structural changes. The nature of the resulting dislocation structure and *strain hardening* are a direct function of the *stacking fault energy*, the rate and temperature of deformation, and the type of applied *stress*.

STRAIN DEVIATOR, strain deformation tensor

A tensor characterizing, in *continuum mechanics*, the variation of the shape of an object as a result of *strain*. The strain deviator $\tilde{u}_{ik}$ is determined through the *strain tensor* according to the formula $\tilde{u}_{ik} = u_{ik} - u_0 \delta_{ik}$, where $u_0 = (1/3)u_{ii}$ is the average strain, and δ_{ik} is the unit tensor (Kronecker delta). From here the decomposition of the strain tensor follows: $u_{ik} = u_0 \delta_{ik} + \tilde{u}_{ik}$, where the first term describes the even all-round strain of an element of volume without any variation of its shape. After reduction to its *principal axes* the tensor $\tilde{u}_{ik}$ takes on a diagonal form with its principal values $\tilde{u}_1, \tilde{u}_2, \tilde{u}_3$, satisfying the zero trace condition $\sum \tilde{u}_{ll} = \tilde{u}_1 + \tilde{u}_2 + \tilde{u}_3 = 0$. Thus the strain deviator describes the combination of local stretches or compressions along three mutually perpendicular axes, without any accompanying change of overall volume.

STRAIN ENERGY

See *Elastic energy*.

STRAIN GAUGE

A sensitive element of a *strain measuring device* intended for measuring stresses and *strains* in solids under static and dynamic loads (see *Strain measurement*). Strain gauges are designed for the transformation of the magnitude of the strain into a signal (usually an electrical signal) that is used for further detection and treatment. Depending on the type of transformation and the field of application, strain gauges are divided into resistance strain gauges, and capacity, induction, electromechanical, and optomechanical varieties.

The most widely used strain gauges are based on the *piezoresistive effect* whereby an electrical conductor changes its resistance R when subjected to a strain ε. The functional dependence has the form $\Delta R(\varepsilon) = k\varepsilon$ where k is the *piezosensitivity coefficient*. The basic requirements for a material to be suitable for use in a resistance strain gauge are the following: as great as possible a magnitude of k (for wire conductors $k \leqslant 2$–2.5, for semiconductors $k \sim 200$), a small temperature coefficient of electric resistance, a large specific resistance (resistivity), and high mechanical *strength*.

The most widely utilized materials are the alloys constantan, nichrome, manganin, chromel, etc.

STRAIN HARDENING

Increasing stress of crystalline material flow as a result of *plastic deformation*. Strain hardening results from increasing the density of crystal structure *defects*, especially of *dislocations*. According to the *dislocation model of plastic deformation* the hardening results from the motion and reproduction of dislocations; it is associated with the effects of interactions of individual dislocations and dislocation ensembles.

It is possible to estimate the value of the strain hardening with the aid of *deformation diagrams* or τ–γ *hardening curves*, where τ is the applied shear stress along the *slip plane*, and γ is the shear deformation along this plane. In crystals oriented for slipping in a particular system usually three stages of *hardening* are observed, characterized by the corresponding *hardening coefficients* Θ_{I}, Θ_{II}, Θ_{III}.

At stage I the hardening process results from interactions of "primary" dislocations initially present within the crystal moving with respect to each other. Toward the end of this stage dislocations of other "secondary" systems begin to appear, thus forming with the primary ones barriers that are called "sedentary" dislocations (*Lomer–Cottrell barriers* in face-centered cubic lattices). At the second stage the density of these barriers increases (according to the *Seeger theory*), thereby increasing the number of flat frozen groups of dislocations within the crystal, and bringing about a decrease of the mean free path the dislocations of the primary system.

At stage III the hardening coefficient decreases, which in the Seeger theory is considered as the result of easing the transverse slipping of *screw dislocations* of the primary system, leading to the relaxation of stresses of the flat aggregations, pressed against the Lomer–Cottrell barriers. Electron microscope studies have shown that this stage of hardening involves the formation of fragmentary structure in the form of interlacing regions, weakly disoriented with respect to each another (see *Fragmentation*). The appearance of such substructure is a result of the collective dislocation interactions. The deformation curve of a crystal oriented for multiple slipping has a parabolic shape.

The coefficient of strain hardening is determined by the slope of the deformation diagram; it depends on the material composition and structure, and on the test conditions. It decreases with increasing degree of strain, of material purity, with the test temperature, and also with reduction of the deformation speed. The following effects are observed in material subjected to strain hardening: decreasing *plasticity* and increasing *hardness*, electrical *resistance* and *magnetic permeability*.

STRAIN INTERACTION OF DEFECTS in a
crystal structure

A form of indirect interaction of *defects in crystals*. It is conditioned by the interference of the fields of static distortions, and of mechanical stresses associated with them and created by the different defects within the crystal. The strain interaction of defects contains a contribution from the *elastic interaction of defects* which decreases rapidly, as a rule, with increasing distance r between them (e.g., for defects restricted in all directions). This strain interaction also contains a component which varies smoothly with distance, and is associated with mirror reflection forces in analogy with electrostatics, specified by the individual defects (see *Image forces*). In many crystals this strain interaction appears to be the only essentially long-range one, and is dominant compared with the other types of interactions between defects. Thus, the energy values characteristic of the strain interaction of defects are 0.1 to 1 eV/atom (less often 0.01 eV/atom) for *impurity atoms* depending on their type and location (*doping atoms*, *substitutional atoms*), on their elastic properties, and on the symmetry of the surrounding lattice (see *Crystal symmetry*). Thus, to a large extent this interaction determines the structural mechanisms and quantitative dependencies of some physico-mechanical processes in interstitial *solid solutions*. The *interstitial atoms* of the impurity cause the *uniform strain* of the crystal lattice, and also induce the static displacements of the surrounding atoms from their lattice sites. The effects specified by this become especially important in the close vicinity of a *second-order phase transition*, or of a critical point associated with the decomposition curve of an elastically-anisotropic solid solution (see *Anisotropy of crystals*). The sequential (noncontinuous) theory of the strain interaction of defects is based on the *lattice statics method* (or *fluctuation waves* of defect concentration or of static displacements see *Fluctuations of atomic positions*). With the help of numerical calculations the following characteristic features of this strain interaction within crystals of different structures have been identified: the nonmonotonic ("oscillating") character of the dependence of the interaction energy on the distance between defects (as a rule adjacent similar *point defects* experience an indirect attraction, while widely separated defects may experience repulsion as well as attraction); the sharp non-central (anisotropic) character resulting from the dependence of the strain interaction energy on the orientation of the radius-vector, connecting the defects, with respect to the crystallographic axes of the matrix. For a fixed distance r in a non-ideal crystal this interaction energy depends on the temperature.

STRAIN MEASUREMENT

Techniques developed for studying the strain state of a solid acted on by an applied force and heat loads. The *strain* value in a solid is determined with the help of *strain measuring devices*.

The operating principles of several strain gauges are as follows. A *mechanical* type is based on measuring the distance between two points of a solid before and after loading. An *electric* device uses the *piezoresistive effect* which is present in some materials. An *optical* one is based on polarized light (see *Polarization of light*) transmission through a transparent sample made of an optically active material in the loaded state. An *X-ray* device is based on measuring the distance between the atoms of the crystal lattice. An *ultrasonic* one involves measuring the *sound velocity* in the solid. A *Moiré band method* uses the appearance of dark and light bands resulting from the superposition of two networks with identical or close parameters; a *method of varnish coatings* is based on the determination of loading-stimulated cracks at a varnish coating applied to the wear surface. Strain measurement is referred to as "tensometria" in Russian (fr. Lat. *tensus*, tension and Gr. $\mu\varepsilon\tau\rho\varepsilon\omega$, I am measuring).

STRAIN, MICRO-

See *Microstrain*.

STRAIN POTENTIAL

An operator characterizing the dependence of energy states of *conduction electrons* on the *strain* $\varepsilon(p, r, t)$ of the crystal lattice. The passage of a sonic wave through a *metal* adds to the Hamiltonian of the electron the additional strain tensor deformation potential term $\delta\varepsilon(p, r, t)$ which is linear in the *strain tensor*

$$u_{ik} = \frac{1}{2}\left(\frac{\partial u_i}{\partial x_k} + \frac{\partial u_k}{\partial x_i}\right),$$

namely $\delta\varepsilon = \sum_{ik}\lambda_{ik}(p)u_{ik}(r, t)$; where $\lambda_{ik}(p) = L_{ik}(p) - m_0 v_i v_k$, p is the *quasi-momentum*, $v = \partial\varepsilon/\partial p$ is the conduction electron speed, m_0 is mass of the free electron, and $u(r, t)$ is the displacement of the lattice ions. The value $L_{ik}(p)u_{ik}$ describes the average incremental energy input to an electron with quasi-momentum p resulting from a variation of the corresponding lattice potential due to the strain. For a free electron $L_{ik} \to 0$. The value $m_0 v_i v_k$ is the transfer of momentum to the free motion of the electron. The tensor $\lambda_{ik}(p)$ satisfies the symmetry condition $\widehat{\lambda}(p) = \widehat{\lambda}(-p)$. The strain potential tensor plays the role of the "phonon charge", because it describes the interaction of electrons with *phonons*.

STRAIN, SHEAR

See *Shear strain*.

STRAIN TENSOR, deformation tensor

A dimensionless, symmetric tensor u_{ik} which connects the distance between two infinitely close points of a solid after a deformation dL with that before the deformation dl: $dL^2 = dl^2 + 2u_{ik}\,dx_i\,dx_k$. The diagonal elements of u_{ik} are the *deformations of stretching (compression)*, and the off-diagonal ones are those of shear.

The deformation tensor may be written as a function of the Cartesian coordinates x_k of the points of the body before deformation:

$$u_{ik} = \frac{1}{2}\left(\frac{\partial u_i}{\partial x_k} + \frac{\partial u_k}{\partial x_i} + \frac{\partial u_l}{\partial x_i}\frac{\partial u_l}{\partial x_k}\right),$$

where u_k are components of the *displacement vector*. In terms of the coordinates X_k after the deformation this tensor, called the *Almansi–Hamel deformation tensor* (E. Almansi, G. Hamel), has the form

$$U_{ik} = \frac{1}{2}\left(\frac{\partial u_i}{\partial X_k} + \frac{\partial u_k}{\partial X_i} - \frac{\partial u_l}{\partial X_i}\frac{\partial u_l}{\partial X_k}\right).$$

It is possible to represent the tensor u_{ik} as a sum of the *strain deviator* and the spherical part of the *strain* $u_{ik} = (u_{ik} - u_{ll}\delta_{ik}/3) + u_{ll}\delta_{ik}/3$. The initial deviator term represents a pure shear and the spherical part ($u_{ll}\delta_{ik}/3$) represents a *uniform compression*. For weak strains of "three-dimensional" bodies the linear approximation of the *theory of elasticity* is used. In this case the derivatives with respect to X_k and x_k differ up to the second order of smallness so all the differentiations may be performed along the coordinates x_k, and we obtain $u_{ik} = ((\partial u_i/\partial x_k) + (\partial u_k/\partial x_i))/2$. The deformation tensor completely determines the strained state of a nonmagnetic body, and this state is called *uniformly strained* if u_{ik} is independent of the coordinates.

STRAIN, UNIFORM

See *Uniform strain*.

STRATIFICATION-TYPE FAILURE

Failure of a sheet, a wire, or any other prefabricate that proceeds via easy development of a *crack* along planes specifically oriented with respect to the direction of *strain*. Stratification develops as an *intercrystalline failure* in the presence of the *structural texture of metals*. The following types of stratification-type failure are known:

(1) Stratification along the boundaries of dislocation cells compressed during the process of deformation is most often observed in strongly deformed metals with a body-centered cubic lattice.

(2) Stratification along *grain boundaries* in deformed metals occurs in the presence of *segregations* or film formations of a second *phase* in the vicinity of grain boundaries.

(3) Stratification along grain boundaries in recrystallized metal is observed when the structural texture is conserved during the process of *recrystallization*.

The first type of stratification, that is most widespread and most dangerous, is triggered by large elastic stresses related to *dislocations* in *subboundaries*. These stresses fully relax (see *Stress relaxation*) when dislocations "fall" into the growing crack, which leads to an easy spread of that crack.

Two other types of stratification are produced by segregation effects when a structural texture is present. Stratification limits the rigid deformation regime for metals with a body centered cubic lattice, and produces *anisotropy of mechanical properties*. *Strength* in the direction normal to the plane of possible stratification may be an order of magnitude smaller than in a "good" direction. The tendency to stratify in a deformed metal augments as the deformation temperature drops (rolling, *drawing*) significantly below the recrystallization temperature.

STREAMER BREAKDOWN

Incomplete *electric breakdown* (see *Breakdown of solids*), when narrow current-carrying channels emanating from a point of high potential intergrow into the crystal, but no shorting of electrodes takes place. The phenomenon of streamer breakdown has been observed since 1930 in *alkali-halide crystals*, glasses, rosin, celluloid, etc.; the streamer breakdown causes irreversible thermal and chemical transformations. This breakdown in high-ohmic ($\rho \geqslant 5 \cdot 10^3$ $\Omega \cdot$cm) direct-band *semiconductors* is initiated by applying high-voltage pulses ($V \geqslant 6$ kV) to a needle electrode and detecting it in the form of brightly glowing tracks, which propagate along 48 (ZnSe, ZnS), 36 (CdS), or 6 (ZnO) definite crystallographic directions, which are temperature-dependent (coefficient of bending being $\sim(4–0.7) \cdot 10^{-2}$ deg/K) at the rate of $1.2 \cdot 10^7$–$4 \cdot 10^9$ cm/s. The direction of streamer breakdown propagation often coincides with the direction of *phonon avalanche*. The free carriers are established in the streamer breakdown channel (the so-called *streamer*) with the concentration of 10^{17}–10^{19} cm^{-3} by application of an electric field pulse to the streamer head. For each semiconductor, there is a certain temperature within the interval 290–470 K, at which the streamer breakdown pattern is most pronounced and developed. At helium temperatures and above 500 K, the streamers can be excited only slightly. In defect-containing crystals, like in amorphous bodies, corona or diffuse breakdown occurs, instead of streamer breakdown. The breakdown occurring at a clean semiconductor surface exhibits a herring-bone or branch-shaped pattern. The streamer breakdown is used in *streamer lasers* for establishing an inverted population (see *Level population*), for initiation of incoherent *streamer radiation*, for obtaining information on the structure of a solid and processes occurring in it, for elucidation of the general features of electric breakdowns. There exist streamer-based techniques for rapid evaluation of the orientation of crystals, crystallographic characteristics of surfaces of *plates*, the level of defects in test samples, etc.

STREAMER LASER, laser of the streamer type

Semiconductor laser with the inverted population state (see *Level population*) established by application of incomplete electric breakdown (see *Breakdown of solids*) of the working material. The generation occurs behind the ionization front in the streamer channels in regions of characteristic size 2–3 μm, which move throughout the crystal at a rate reaching $4 \cdot 10^9$ cm/s. The generation characterized by the peak power $\sim$15 kW and pulse duration of $\sim$2 ns has been established in the direction transverse to the streamer axis at 300 K in thin ($\sim$20 μm) CdS samples, the major faces of which are covered with dielectric *coatings* of 100% and 97% reflectivity. Using an iris for the near field radiation allows one to obtain picosecond radiation pulses. The generated power increases with voltage V as $V^{1.8–2.5}$. In the absence of an optical resonator, the stimulated radiation of a clearly expressed mode structure with principal mode angle 10–15° is observed in the discharge propagation direction, and also in the opposite direction. The streamer laser design is based on semiconducting *monocrystals* in which *streamer radiation* is observed.

STREAMER RADIATION in solids

Radiation, which arises under the conditions of an incomplete electric *streamer breakdown* in a solid. The streamer is an electrical discharge, resembling a bolt of lightning, which arises at a point of high potential in a charged body, and emanates along narrow paths in the material. This type of radiation is observed in the form of thin (2–5 μm) fluorescent streams, which propagate into the depth of the sample, or in its near-surface layer, at the rate of 10^7–$4 \cdot 10^9$ cm/s along certain crystallographic directions. The phenomenon of streamer radiation has been studied since 1930 in *alkali-halide crystals*, and since 1973 in high-ohmic $(5 \cdot 10^3 – 10^{12}$ Ω·cm) highly luminescent *semiconductors* of group II–VI featuring *direct transitions* (CdS, CdSe, CdTe, ZnO, ZnS, ZnSe), of group III–V (GaAs, InP), etc. The streamer radiation spectra of semiconductors practically coincide with the spectra of *photoluminescence* under the conditions of intense two-photon excitation. The bands dominant in these spectra at ~300 K are due to *recombination* in an electron–hole plasma. The shift of streamer radiation bands by 1 nm to longer wavelengths is caused by absorption in the bulk crystal, and the contraction of the *band gap* in the streamer channels, in which the concentration of nonequilibrium carriers can reach $5 \cdot 10^{19}$ cm^{-3}.

STREAMING

Distribution of current carriers, which is markedly extended along an applied electric field. This type of current carrier distribution is realized in *semiconductors* at low temperatures. In intermediate electric fields the carriers acquire the energy of optical phonons $\hbar\omega_0$, and have no time to become scattered in the passive region of *momentum space* $(\varepsilon(p) \leqslant \hbar\omega_0)$; when passing to the active region $(\varepsilon(p) \geqslant \hbar\omega_0)$ the carriers revert back to the low-energy range due to the spontaneous emission of optical phonons. The ultimate needle-shaped distribution is realized when the scattering is insignificant, and the emission of optical phonons is an instantaneous process. The distribution of carriers in the passive region is concentrated in the neighborhood of the main trajectory, which is determined from the equations of motion in the field, and passes through $p = 0$. The motion of carriers involves cycles of free acceleration to the optical phonon energy range. The trajectory becomes curved in crossed electric and magnetic fields, and may close upon itself within the passive region. The conditions of streaming provide an opportunity for creating an electromagnetic wave amplifier, in particular, based on

heavy holes in germanium with negative *effective masses*, or *lasers* employing hot holes.

STRENGTH of a material

Capability of solids to resist *failure* or irreversible change of shape (*plastic deformation*). The material strength is defined by the stress that corresponds to the maximum bearing capacity of the specimen (the maximum load, P). During *brittle failure* below the *cold brittleness* threshold the material strength corresponds to failure stress. Plastic materials usually have a maximum in their P versus ε *deformation diagram* (ε is strain). The stress corresponding to that maximum is the *ultimate strength* or the *temporal resistance*, σ_t of plastic materials. Material strength characteristics are also commonly understood in terms of the conditional *yield limit*, σ_ε corresponding to a certain $\varepsilon = \text{const}$ (e.g., $\sigma_{0.2}$), with upper and lower yield limits when the deformation diagram contains a *yield cusp*, failure stress, and also the coefficient of *strain hardening* defined by the $d\sigma/d\varepsilon$ derivative. The theoretical *Orowan strength* estimate (1949) (see *Theoretical strength*), which does not envisage a possibility of plastic deformation, yields the value $\sigma_T = (E\gamma/a)^{1/2}$, where E is *Young's modulus*, γ is the *surface energy*, and a corresponds approximately to the lattice parameter. Currently the levels of strength achieved in *thread-like crystals* of various materials are close to these theoretical values. Bulk crystalline materials, however, exhibit actual strengths approximately two orders of magnitude smaller than the corresponding theoretical values. On the assumption that microcracks are already present in a crystal (see *Cracks*), one of them opening under stress can fracture the crystal. The *Griffith theory* yields an expression of the Orowan type in which the length of the crack c stands in place of a. Prescribing $c = 1$ μm the Griffith theory yields values of strength for failure of solids that are close to those actually achieveable. During *tough failure* the strength is determined by the sum of the yield limit and the value of strain hardening until the value σ_T is reached. The yield limit for crystalline materials is defined by the resistance of the crystal lattice to dislocation motions at the temperature 0 K (the *Peierls–Nabarro strength*, see *Peierls–Nabarro model*), and also by the capability of dislocations to penetrate in a thermally active manner the potential barriers of the crystal lattice as the temperature grows. The Peierls–Nabarro strength appears to be higher, the larger the covalent component in the interatomic bond (see *Covalent bond*). The concept of dislocations makes it possible to satisfactorily describe the temperature dependence of the yield limit. The dependence of strength on the structure factor (the size of *crystal the grains d* in case of *quasi-brittle failure* of a crystal) is satisfactorily described by the *Mott–Stroh equation* (N.F. Mott, 1956; A.N. Stroh, 1957), whereby $\sigma \propto d^{-1/2}$. During tough failure when the strength is controlled by the value of σ_v, the same type of dependence of strength on the grain size still prevails. During brittle, in contrast to tough or quasi-brittle failure, a significant scatter is found in the values of strength, explained by the difference in *defect* ensembles from specimen to specimen. In order to process experimental results adequately in that case, one may use a statistical theory of strength, the most developed being the *Weibull theory* (W.A. Weibull, 1939), that is based on the concept of the weakest link. The main parameter of the theory, the *Weibull modulus m*, characterizes the homogeneity of the defect distribution in the material (flaw size distribution under uniform stress). The higher the value of *m*, the narrower is the scatter in the values of strength, and the weaker is the dependence of strength on the scale factor. Metal strength may be increased by *alloying*, by dispersion strengthening (see *Precipitation-hardened materials*), and by developing certain special structural and substructural states (such as dislocation *cellular structures*).

See also *Construction material strength*, *Cyclic strength*, *High-temperature strength*, *Impact strength*, *Long-term strength*, *Real strength*, *Static strength*, *Theoretical strength*, *Ultimate strength*.

STRENGTH AND PLASTICITY OF POWDER METALLURGY MATERIALS

Mechanical properties of materials produced by *powder metallurgy* techniques that are defined by their *porosity*, and also by the structure and *strength* of particle-to-particle contacts. Increasing the volume porosity of baked materials usually results in a monotonic decrease of their strength and plasticity characteristics. To describe the *ultimate strength* dependence on porosity a number of empirical relations is used, but unfortunately they do not take into account of the effect of *pore* size and shape, or the particle-to-particle contact strength. Therefore, they may only be applied when a streamlined uniform technology is used to produce specimens of different porosity. Expressions of higher complexity were suggested to account for the parameters of pore structure, and the particle-to-particle contacts. The size of particles in the initial powder obtained by chemical reduction does not usually affect the mechanical properties of baked plastic materials in any significant way, since the effective length of a *slip plane* is defined by finer fragments, limited by interparticle pores and separation interfaces. In the case of monocrystal powder particles, particularly in brittle materials, their strength grows for smaller particles. The dependence of the *yield limit* on porosity may be described by the same expressions, however, the effect of porosity appears to be weaker than in case of the strength limit. With growing porosity the material *plasticity* decreases even faster than the fluidity limit. Baked *iron* features a parabolic dependence of its *strain hardening* on the degree of *strain* similar to high-density iron. The coefficient of deformation strengthening then decreases at a higher porosity, following the same dependence as the fluidity limit.

STRENGTH OF COMPOSITE MATERIALS

This characteristic is defined by the composition, properties of the components, binding strength and structure of *composite materials*. Reinforced composite materials characteristically display a strength anisotropy that is expressed not only as the dependence of strength on the applied stress direction, but also as the difference in strength with respect to distension and compression in each direction, and also as the dependence of *ultimate strength* for *shear* on the sign of the tangential stresses. The strength of composite materials reinforced with fibers oriented in a single direction is defined, to the first approximation, by the mixing rule. Generally speaking, the dependence of the strength of composite materials on the spatial density of fibers is nonmonotonic. This effect is explained by the existence of critical fiber densities at which the mechanism of *failure* changes. The strength of composite materials reinforced with discrete fibers decreases as the ratio of fiber length to fiber diameter decreases.

Precipitation-hardened materials, with their specific strengthening mechanism, present a special case of composite materials. The *Long-term strength* of dispersion strengthened composite materials depends on the thermodynamic stability of the dispersed inclusions of the second phase. The nature of the failure of *pseudoalloys* featuring a two-phase structure depends on the nature and concentration of the corresponding phases. When the content of the harder phase increases then the *yield limit* of these pseudoalloys grows, and their *fracture toughnes* diminishes. A favorable combination of the yield limit with the resistance to *crack* propagation results in a maximum strength of composite materials which is usually found at a certain ratio of phase concentrations.

STRENGTH REDUCTION THROUGH ADSORPTION, Rebinder effect (P.A. Rebinder, 1928)

The development of deformation and failure of *solids* due to adsorption of *surface-active agents* on their outer and inner bounding surfaces that appear during the deformation. The effective strength reduction is brought about through adsorption of organic acids and alcohols, or melts of salts and *metals*. The physical reason for the strength reduction is related to the decrease of the *surface energy*. Adsorption is important in the technology of metal processing by cutting and pressure, drilling of rock soil, and crushing various solids.

STRESS, BREAKING

See *Breaking stress*.

STRESS CONCENTRATION

A local increase in the *stress tensor* components of a loaded body. Stress concentration is caused by *stress concentrators* such as *cracks*, holes, *notches*, scratches, internal *defects*, and other local changes of the condition (form) of the body. The stress distribution in the stress concentration zone is characterized by increases of the magnitude and the gradient of stress with the approach to the concentrator. A convenient quantitative index is the *stress concentration coefficient*

α_σ, which equals the ratio of the maximum local stress σ_{max} to its nominal value σ_n calculated without taking into account the stress concentration: $\alpha_\sigma = \sigma_{max}/\sigma_n$. For example, in the elongation of a *plate* with a circular hole whose diameter is much smaller than the plate dimensions, the maximum normal stresses arise on the ends of the diameter normal to the elongation direction with $\alpha_\sigma = 3$. The following types of stress concentration coefficients are distinguished: theoretical (σ_{max} and σ_n calculated by *elasticity theory* methods), technical (actual σ_{max} and σ_n values determined by calculation or by measurement taking into account inelastic deformations), effective (determined by the relation of tolerance limits of the bodies with and without stress concentration). Zones of stress concentration are the most probable points of origin of *plastic deformation* and *failure* of a sample or machine components. Stress concentration (in solid state physics) refers to the local increase of mechanical stresses caused by *defects* of body structure or their clusters. Especially large stress concentrations are associated with flat *pile-up of dislocations*. In this case α_σ is approximately equal to the number of dislocations in the cluster, and this helps to explain the difference between theoretical and technical *strength* values of crystal phases.

STRESS, CONTACT

See *Contact stresses*.

STRESS CORROSION

See *Corrosion under stress*.

STRESS DEVIATOR, stress deformation tensor

A tensor of *continuum mechanics* characterizing the shear portion of the stresses (see *Shear*). The stress deviator $\widetilde{\sigma}_{ik}$ is determined through the *stress tensor* σ_{ik} according to the expression $\widetilde{\sigma}_{ik} = \sigma_{ik} - \sigma_0 \delta_{ik}$, where $\sigma_0 = (1/3)\sigma_{ll}$ is its average which corresponds to the *hydrostatic stress*, and δ_{ik} is the unit tensor (Kronecker delta). This equality determines the decomposition $\sigma_{ik} = \sigma_0 \delta_{ik} + \widetilde{\sigma}_{ik}$ of the stress tensor into hydrostatic ($\sigma_0 \delta_{ik}$) and shear ($\widetilde{\sigma}_{ik}$) portions. In the *principal axes* system the stress deviator is a diagonal tensor diag $[\widetilde{\sigma}_1, \widetilde{\sigma}_2, \widetilde{\sigma}_3]$ which satisfies the zero trace condition $\sum \widetilde{\sigma}_{ll} = \widetilde{\sigma}_1 + \widetilde{\sigma}_2 + \widetilde{\sigma}_3 = 0$ or

$\tilde{\sigma}_3 = -(\tilde{\sigma}_1 + \tilde{\sigma}_2)$. From this it follows that $\tilde{\sigma}_{ik}$ can be decomposed into two diagonal tensors diag $[\tilde{\sigma}_1, 0, -\tilde{\sigma}_1]$ and diag $[0, \tilde{\sigma}_2, -\tilde{\sigma}_2]$ which describe pure shears along planes at $45°$ with the principal axes. Thus, the general *state of stress* is locally the superposition of a *uniform compression* (stretching) plus two *shears*.

STRESS INTENSITY

See *Tangential stress intensity*.

STRESS INTENSITY FACTOR

A factor characterizing the *mechanical stress* concentration (intensity) in the vicinity of a *crack* tip, which follows from solving the problem of stress distribution near this tip. Depending on the nature of the load distribution and the load application mode, one can single out the stress intensity factors of normal tear K_1, longitudinal shear K_2, and transverse shear K_3. The *stress tensor* components $(\sigma_{r\theta}, \sigma_{\theta\theta}, \sigma_{rr})$ in the vicinity of the tip of a normal tear crack at a point with coordinates r and θ is determined by the formula

$$\sigma_{ij} = \frac{K_1}{2\pi r} f_{ij}(\beta) + \mathrm{O}(r),$$

where $f_{ij}(\theta)$ are known angular functions; $\mathrm{O}(r)$ is a small quantity for $r \to 0$. The current value of this factor depends on the applied load, the crack size, and the body configuration, but it is independent of the coordinates r and θ. For an infinite *plate* of a finite thickness with a central crack of dimension $2a$, $K_1 = \sigma\sqrt{\pi a}$, where σ is the applied stress. The critical value K_{10} corresponding to the start of an unstable crack propagation is a force criterion of *failure*. The critical factor K_{10} is a constant of the material reflecting its ability to resist the crack propagation, and hence it is called the crack resistance (see *Fracture toughness*). To determine these characteristics special procedures have been developed based on *mechanical testing of materials* with cracks.

STRESS, INTERNAL

See *Internal stresses*.

STRESS, MACRO- AND MICRO-

See *Macrostresses* and *Microstresses*.

STRESS, MECHANICAL

See *Mechanical stress* and *Mechanostriction*.

STRESS PROPORTIONALITY LIMIT

See *Proportionality limit of stress*.

STRESS, REACTIVE

See *Reactive stresses*.

STRESS RELAXATION

Phenomenon of reduction, during the course of time, of *stress* in elastically stressed solids for a constant preestablished *strain*. As in the related phenomenon of *creep* which involves an increase of strain at constant load, stress relaxation is a consequence of *plastic deformation* at low loads. The velocity of stress relaxation is generally reduced by a decrease in temperature or effective stress, by increasing the binding forces within the lattice, by creating in the material obstacles for the motion of *dislocations*, e.g., due to solid-state *hardening* and to *dispersion hardening* (see *Precipitation-hardened materials*) with increasing uniformity and stability of the structure. Stress relaxation of fasteners, spring units, stressed fitted joints, and other components of machines is undesirable. The ability of a material to oppose the reduction of stresses is called *relaxation stability*. In engineering the stress σ_τ is determined by the amount that remains within the material after a preassigned time τ, or by the decrease in stress $\Delta\sigma_\tau = \sigma_0 - \sigma_\tau$ during that same time interval (σ_0 is the initial stress at $\tau = 0$).

In some cases it is necessary to measure the relaxation of undesirable stresses, which appear in the material during the course of manufacturing and processing of articles by the method of cold plastic deformation, *welding* etc. For fast stress relaxation the material is *annealed* (see *Residual stresses*).

The phenomenon of stress relaxation is used to study mechanisms of plastic deformation. With the help of this method the parameters of thermally activated dislocation motion (*activation volume*, *activation energy*, etc.) can be determined.

STRESS SCALES OF DEFECT STRUCTURES

(N.N. Davidenkov, the 1930s)

Classification of *internal stresses* σ_{int} by their structure scale. Under *heat treatment* and *plastic deformation*, solids develop internal (residual) stresses. By virtue of *crystal stability conditions*,

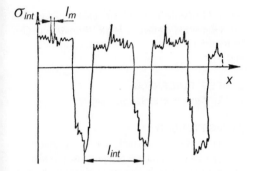

Variation of internal stresses along the x axis in a solid section $(0, L)$, where l_{int} is the characteristic scale of internal stress changes, and l_m is the small-scale relief of internal stresses, for the case of two scale levels.

average values of internal stresses across every section of a solid are equal to zero. The average extent of internal stress variations is called the *characteristic scale of internal stresses* (see Fig.). Both average values of internal stresses

$$\langle \sigma_{int} \rangle = \left(\frac{1}{s} \int_{(s)} \sigma_{int}^2 (r) \, ds \right)^{1/2},$$

and the scale of these stresses have considerable influence on material strength characteristics.

Three types of stresses are distinguished. *Stresses of the first kind* are internal stresses with a scale of the order of the body dimension L, those of the second kind have a scale of the order of crystal grain size l_g, and those of third kind are of the order of an atomic size a. *Strain* of plastic materials usually begins with motion of individual *defects* (usually *dislocations*) and small groups of these defects. As their density increases the intrinsic defect energy becomes smaller than that of their interaction, while external forces become smaller than those of interactions. The collective effects begin, and ordered defect structures, e.g., block boundaries, are generated. The characteristic dimension of *block structures* $l_b \sim 1$ μm is far greater than that of isolated dislocations $l_d \approx 0.01–0.1$ μm. At the same time, defects of ordered structures (second order defects) arise, e.g., partial *disclinations*. The deformation growth causes a transition to *plastic twisting deformation*, to the increase of disclination density, to collective

effects in their ensembles, and to ordered disclination structures. The latter are often called fragmented; a standard terminology has yet to be established (see *Fragmentation*). Their scale for *metals* is $l_f \sim 10$ μm and more; accordingly, such structures develop third order defects. Possibilities in solids with a larger grain size are collective effects of the third-order defects, formation of structures and defects of the fourth order, and so on. In parallel with the multiscale structures, ensembles of defects form at the initial boundaries that interact with them. Their characteristic sizes are on the order of those of phase precipitation l_{ph}, grain l_g, and grain groups l_{gg} (which undergo collective motion such as rotation). Thus, the strain causes the formation of an hierarchy of defect structures with scales $a, l_d, l_b, l_f, l_{ph}, l_g, l_{gg}, L$. In particular cases, some of these scales may be absent.

Point defects, e.g., *impurity atoms*, produce internal stresses of a scale $\sim a$. A dislocation with *Burgers vector* b generates stresses $\sigma_{int}(r) = (Gb/r) f(v)$, where G is the *shear modulus*, r is the distance from the dislocation axis, $f(v) \propto 1$ is a function of the angle v. In a multipole structure with dislocation density ρ, they are effective at distances $l_D \approx \rho^{-1/2}$. At larger values of r the sum of stresses from multipoles with roughly equal numbers of dislocations with different signs, $N_+ \approx N_-$, yields negligible values of internal stresses.

Generation of ordered structures of defects on all scale levels is accompanied by partition of multipoles into dislocation groups of the same sign $N \rightarrow N_+ + N_-$, their spacing being of the order of the structure scale. Therefore, structures produce internal stresses of the same scale. If the structure extends along b (a kind of *pile-up of dislocations*), then the stresses are of the dislocation type, i.e. $\sigma_{int} \approx (Gb/r) f(v) \Delta N$; if the structure is extended perpendicular to b (a kind of *dislocation wall*), then internal stresses of the dislocation type are generated due to structural defects, i.e. by its ends and joints, $\sigma_{int} \propto Gb/d$, where d is the spacing of neighboring dislocations ΔN along the structure. Thus, at plastic deformation there arises a spectrum of stress scales of defect structures $l_d, \ldots, l_{gg}, L$, the most important scale l_{int} increasing with deformation. Attainment of the critical value of l_{int} (which depends on material and structure type) signals "plasticity exhaustion" and *failure*.

STRESS STATE

See *State of stress, Planar state of stress, Simple state of stress*.

STRESS–STRAIN CURVE

The same as *Deformation diagram*.

STRESS SURFACE (introduced by A. Cauchy, 1827)

Characteristic second-order surface given by the equation $\sigma_{ij} x_i x_j = \pm 1$, where σ_{ij} is the *stress tensor*. It provides a geometric representation of the *state of stress*. Depending on the sign of the principal values of the stress tensor, the stress surface may be either a real or an imaginary ellipsoid or hyperboloid. Assume that a vector V of unit length specifies the direction; corresponding to this vector there is a ray which passes from the origin of coordinates along the vector V; there is a plane tangent to the characteristic surface at the point p where the ray intersects the characteristic surface (see Fig.). The normal stress σ_N is directed along the vector V and is proportional in magnitude to r^{-2}, where r is the distance between the origin of coordinates and the point p of intersection of the ray and the characteristic surface. The tangential stress σ_T is perpendicular to the vector V; it lies in the plane generated by V, and is normal to the plane which contains V. In absolute value $\sigma = (r^2 h^{-2} - 1)^{1/2}/r^2$: the meaning of r is given above, and h is the distance between the

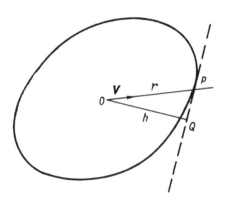

Stress surface.

origin of coordinates and the tangent plane. Besides the stress surface, a *Mohr diagram* may also be used for the determination of the quantities σ_N and σ_T.

STRESS, TANGENTIAL

See *Tangential stress intensity*.

STRESS TENSOR

Symmetric second rank tensor σ_{ik} that completely defines the *state of stress* of a body. An internal force, which acts on an element of an arbitrarily chosen surface inside the body ds (ds_i are Cartesian coordinates of the surface element vector, which is directed along the outward normal to it), is equal to $\sigma_{ik}\, ds_k$. An internal force, which acts on a unit volume of a strained body, is equal to $\partial \sigma_{ik}/\partial X_k$ (X_k are Cartesian coordinates of points of the body after deformation). The component σ_{ik} of the stress tensor is the ith component of the force acting on a unit surface perpendicular to kth axis. The three diagonal components σ_{ik} are *normal stresses*, and the three off-diagonal components are *tangential (shear) stresses*. At every point of the body there is a Cartesian coordinate system (of axes) in which σ_{ik} is diagonal, with the diagonal values called *principal stresses*. The force applied by *internal stresses* to the entire surface S of the deformed body, equals $\oint_S \sigma_{ik}\, dS_k$ (dS is directed along the outward normal to S). A state of stress is said to be uniform if σ_{ik} is independent of coordinates. In the case of *uniform compression*, $\sigma_{ik} = -P\delta_{ik}$ (P is the pressure). The tensor σ_{ik} may be represented as the sum of a *stress deviator* and a spherical part: $\sigma_{ik} = (\sigma_{ik} - \sigma_{ll}\delta_{ik}/3) + \sigma_{ll}\delta_{ik}/3$. The deviator consists of pure *shear* stresses, the spherical part consists of a uniform compression. The average value of the stress tensor of a deformed sample is

$$\sigma_{ik} = \frac{1}{2}V \int_V (X_i F_k + X_k F_i)\, dV$$

$$+ \frac{1}{2V} \oint_S (X_i P_k + X_k P_i)\, dS,$$

where F and P are the forces that act on a unit volume of the deformed body, and on its unit surface area, respectively. In particular, in nonuniform heating in a nonlinear approximation, $F =$

$-K\alpha'\nabla T$ and $P_i = K\alpha(T - \langle T\rangle)n_i$, where K is the *bulk modulus*, α is the volume coefficient of *thermal expansion*, n is the outward normal to the surface, and $\langle T\rangle$ is the average temperature over the volume of the body. See also *Mohr diagram*.

STRIPE DOMAIN STRUCTURE

See *Magnetic films*.

STROH THEORY (A.N. Stroh, 1954)

A physical model representing the generation of *nucleating cracks* in polycrystalline solids, followed by the deleterious propagation of these cracks, leading to the onset of brittle failure of the metallic material; provides the basis for the physical theory of *brittle failure* (*cold brittleness*) of body-centered cubic materials. According to the Stroh theory, the nucleating crack is generated at the head of the conglomeration of n line *dislocations*, with the Burgers vector nb, which are inhibited by the *grain boundary*; and the concentration of stress $n\tau$ established at the point of crack initiation eventually reaches the level of the theoretical breaking stress (see *Theoretical strength*). According to Stroh, the critical stage of the process is the confluence of two initial dislocations at the head of the *pile-up of dislocations*, after which all the remaining dislocations flow spontaneously into the *submicrocrack*, thus forming a wedge-shaped cavity of equilibrium length $c \approx n^2 b$. By assuming that the normal stress at the instant of completion of the microcrack generation equals the Griffith stress, Stroh obtained the following equation for the brittle failure stress σ_p of a polycrystalline metal of grain size d:

$$\sigma_p = \sigma_0 + \left(\frac{6\pi\gamma G}{1-\nu}\right)^{1/2} d^{-1/2},$$

where σ_0 is the resistance to the dislocation motion, γ is the *surface energy*, G is the *shear modulus*, and ν is the *Poisson ratio*.

In the Stroh theory, the first nucleating crack alone causes complete *failure* of a metal, which is inconsistent with the experimental evidence. The refinement of the Stroh theory introduces an additional stage of elastic equilibrium to the mechanism of the generation of nucleating cracks. At this stage, only a moderate fraction of all n dislocations (approximately one fourth) falls into the crack. This refinement provides us with a quantitative description of the conditions of failure of polycrystalline iron and steel in the context of the *microspalling* model (see *Microspalling resistance*). The Stroh model is the basis of the first consistent theory of failure of metals, which relates the earlier energetic *Griffith theory* of cracks (that fracture in brittle materials occurs because of preexisting flaws) to modern dislocation concepts. The first physical model of cold brittleness of metals was developed in the context of the Stroh theory. Analyzing the conditions of the crack opening at the end of the arrested *slip plane*, Stroh assumed that the growth of the crack may be suspended if the sources near the pile-up were set into operation by the *stress concentration* of the pile-up. Given a certain combination of structural and thermal-rate parameters of loading, the *stress relaxation* may be suppressed, and hence the crack opening will cause brittle failure. The Stroh theory has provided the initial basis for the development of the dislocation theory of cold brittleness of metals. This theory is ideally suited for perturbation calculations. At present the Stroh theory is mainly of interest from the standpoint of providing a methodology for studying the physics of failure of metals.

STRONG BAND PARAMAGNETISM

Paramagnetism of delocalized electrons which is characterized by a strong and nonlinear dependence between the *magnetization* $m(B, T)$ and the external field $\boldsymbol{B}$; this dependence is analogous to the Wohlfarth isotherm (see *Ferromagnetism of metals and alloys*) in *band magnetism*,

$$\frac{m^2(B, T)}{m_n^2} = \frac{\chi_n B}{\mu_0 m(B, T)} - \left(1 + \frac{T^2}{T_n^2}\right),$$

where χ_n is the longitudinal *magnetic susceptibility* at $T = 0$, and m_n and T_n are parameters. Strong band paramagnetism is theoretically described by the Stoner relations (see *Band magnetism*, Eqs. (1)), and was first observed experimentally in a Cu–Mn alloy by B.K. Ponomarev et al. (1977). Strong band ferromagnetism is sometimes incorrectly referred to as "non-Stoner ferromagnetism".

STRONG ORTHOGONALITY

In a many-particle quantum system the *orthogonality condition* of different states has the form

$$\int \cdots \int \varphi_i^*(r_1, r_2, \ldots, r_N) \varphi_j(r_1, r_2, \ldots, r_N)$$
$$\times d^3 r_1 \ldots d^3 r_N = \delta_{ij}.$$

In some cases to simplify the calculations a stronger restriction is applied, strong orthogonality, whereby

$$\int \varphi_i^*(r_1, r_2, \ldots, r_k, \ldots, r_n)$$
$$\times \varphi_j(r_1', r_2', \ldots, r_k, \ldots, r_N') d^3 r_k = 0$$

when $i \neq j$ for arbitrary values of the coordinates $r_1, r_1', r_2, r_2', \ldots, r_N, r_N'$ of all the other particles except the kth one. The condition of strong orthogonality is easy to meet if φ_i, φ_j are approximated by multiplicative wave functions of the form $\prod_k \psi_k^{(i)}(r_k)$, $\prod_k \psi_k^{(j)}(r_k)$, or by a determinant of functions, assuming the individual orthogonality of all factors

$$\int \psi_k^{(i)}(r_k) \psi_k^{(j)}(r_k) d^3 r_k = \delta_{ij}.$$

As the system Hamiltonian contains only single-particle and two-particle terms, e.g., Δ_k, e^2/r_{kl}, the strong orthogonality condition reduces all integrals to single-particle and two-particle types. At the same time, strong orthogonality, which does not follow from the postulates of quantum mechanics, does not allow one to obtain a true minimum of the system energy by the variational method, and thus it limits the accuracy of the approximation.

STRONTIUM, Sr

Chemical element of Group II of the periodic table with atomic number 38 and atomic weight 87.62. Natural strontium consists of four stable isotopes, among which the isotope ^{88}Sr is dominant (82.56%). The radioactive isotope ^{90}Sr (half-life 27.7 years) is a product of uranium fission. Electronic configuration of outer shell is $5s^2$. Successive ionization energies 5.692, 11.026, 43.6 eV. Atomic radius 0.214 nm; radius of Sr^{2+} ion is 0.112 nm. Oxidation state +2. Electronegativity 1.0.

Strontium is a silvery-white soft *metal* that occurs in several polymorphic modifications (α-Sr, β-Sr, γ-Sr). The modification stable under standard conditions is α-Sr, with a face-centered cubic structure and lattice constant $a = 0.6085$ Nm. The polymorphic transformation α-Sr $\rightarrow$ β-Sr takes place at 488 K; β-Sr exhibits a hexagonal lattice; at 878 K it transforms into γ-Sr (body-centered cubic lattice). Density 2.63 g/cm^3 (at 293 K), $T_{\text{melting}} = 1043$ K, $T_{\text{boiling}} = 1653$ K, heat of melting 9.24 kJ/mole, heat of evaporation 142.0 kJ/mole; specific heat 0.7392 kJ·kg^{-1}·K^{-1}; coefficient of linear thermal expansion of polycrystal is $23 \cdot 10^{-6}$ K^{-1}. Debye temperature 148 K. Electronic work function is 2.74 eV. The electric resistivity is 0.228 $\mu\Omega$·m. The atomic magnetic susceptibility is $91.2 \cdot 10^{-6}$. Elastic modulus is 15.6 GPa, modulus of compression is 11.956 GPa. The value of Mohs hardness is 1.8. Metallic strontium is used for the deoxidization of copper and bronze, and as a getter in electric vacuum engineering; strontium salts are used for the production of luminous compounds, in the fabrication of glazes and enamels; isotope ^{90}Sr is used as a β-radiation source.

STRUCTURAL COHERENCE

An identity of the atomic arrangement and a closeness of the interatomic distances of two or more phases over all *phase interfaces*, or a part of it which is not a dislocation boundary. A structural coherence can be one-, two-, or three-dimensional. If the interatomic distances in two phases at the phase interphase differ slightly then the structural coherence causes a coherent elastic *strain*. Substantial coherent deformations which exceed a material *yield limit* disrupt the structural coherence and account of the formation of structural *dislocations* at the phase interface. The presence or absence of structural coherence can be established with the help of electron microscopy, by the methods of X-ray diffuse scattering, auto-ion microscopy (see *Field ion microscopy*), and *internal friction*. In *precipitation-hardened materials*, the presence of structural coherence between the disperse inclusions and the host causes an especially sharp increase of hardness and yield point, with a concurrent reduction of *plasticity*.

STRUCTURAL DOMAIN

A *domain* which differs from other domains in the nature of its atomic or structural component arrangement. Structural domains can emerge as a result of *structural phase transitions* accompanied by the reduction of the *crystal symmetry*. Other examples are the *Kapustin–Williams domains* (A.P. Kapustin, R.J. Williams, 1963) in *nematic liquid crystals*, the structural domains in proteins that are regions of tertiary structure with a certain autonomy of structural composition, and the domains of *ferroelastics*. Domains in *ferroelectrics* can also be regarded as structural domains. In recent times, for describing the structure of glassy *amorphous films*, to counterbalance the model of the *vitreous state of matter* as a *random continuous network*, a model based on structural domains has been introduced. The accumulation of strain energy during the growth of islands hinders their *coalescence*, and leads to the formation of a broken film structure. This is broken into structural domains of a certain size separated by narrow grooves that form the domain boundaries. The most pronounced domain morphology has been observed by *electron microscopy* in films of As_2S_3.

STRUCTURAL GLASSES

Solids which contain atomic groups with electric *dipole moments* or *multipole* moments having interactions randomly varying in sign between the dipoles or other multipoles. The structural glasses are divided into: dipolar glasses (of the type $K_xLi_{1-x}TaO_3$) (see *Dipolar glass*), *proton glassses* that constitute a low-temperature state in *mixed crystals* of the kind $Rb_{1-x}(NH_4)_xH_2PO_4$, quadrupolar glasses, and so on. Structural glasses, in regard to their properties, resemble *spin glasses* with the difference that the role of spin is now played by a dipole or higher-order multipole. A *phase transition* in structural glasses is associated with the presence of an infinite number of quenched dipoles, and this is related to the appearance of long-range order (see *Long-range and short-range order*). A structural glass is referred to as a *non-ergodic system* (see *Nonergodicity*) with a continuous spectrum of *relaxation times*. To describe structural glasses, it is necessary to introduce an infinite number of *order parameters* characterizing the overlapping of quenched dipoles.

The experimental consequences of nonergodicity of structural glasses are lack of equilibrium, diverse irreversible processes, *memory phenomena*, aging, etc.

STRUCTURAL INSTABILITY OF SUPERCONDUCTORS

A phenomenon of *lattice destabilization* of superconductors which appears in the form of structural transformations with a lowering of the lattice symmetry, an anomalous decrease of elastic moduli with the decrease in temperature, a softening of the high-frequency branches of the *phonon spectra*, etc. due to a strong *electron–phonon interaction*.

According to modern theories of strong coupling in superconductors, increasing the electron–phonon coupling constant λ could lead to an increase of T_c, and this could affect crystal lattice instability at $T_m > T_c$. The change of the crystal lattice symmetry leads to a perturbation of not only the phonon but also the electron spectrum of a superconductor. The increase of elastic energy resulting from the crystal lattice *strain* is compensated by the reduction in energy of the perturbed electron subsystem.

A structural instability of superconductors was revealed in superconducting compounds with crystal lattices of the A15 type (e.g., Nb_3Sn, V_3Si), the CsCl type (VRu), *Laves phases* (HfV_2, ZrV_2), and so on. As a rule, the structural transformations are *first-order phase transitions* (see *Structural phase transition*).

STRUCTURAL NEUTRON DIFFRACTOMETRY

Collection of methods for investigating the atomic structure of materials (mainly *crystals*) based on measuring the nuclear scattering of neutrons during their diffraction by a sample. The geometric aspects of neutron and X-ray scattering have many common features. When comparatively low intensity neutron sources are used, however, the neutron diffractometry experiment (see *Neutron diffractometry*) is much more complicated, and requires using much larger samples than in the X-ray case (~ 1 mm^3 in studies of *monocrystals* and ~ 1 cm^3 for *polycrystals*). The elementary interaction events of the neutron and X-ray beams with an atom differ from each other.

The specificity of structural neutron diffractometry is related to the absence of any dependence of the nuclear *scattering amplitudes* f on the charge Z and mass number A of nuclei, and the fact that some of the scattering amplitudes are negative or complex. Owing to these circumstances, the methods of structural neutron diffractometry give more reliable and more exact data on the position and the nature of thermal vibrations of light elements in the presence of heavy ones (hydrogen-containing compounds are the extreme case), and this allows one to study the structure of compounds with different atoms close together in their Z values, and to distinguish two isotopes of the same element. It is possible to exclude the coherent elastic neutron scattering in some systems by the creation of so-called zero matrices, to study the absolute configuration of molecules in crystals containing abnormally absorbing nuclei, and so on. In all these cases other types of *structure studies* are either inefficient, or do not provide any result al all. Structural neutron diffractometry has been successfully applied for the study of the crystal chemistry aspects of hydrogen-containing compounds such as hydrates, hydrides, organic and biological molecules (myoglobin, vitamin B_{12}), and so on.

A structural neutron diffractometry experiment is performed with the help of computer-controlled *diffractometers* (both poly- and monocrystalline units) located close to the neutron source. Information on the diffraction reflection intensities (from tens to many thousands of reflections) is collected and processed with a computer. To reconstruct the crystal structure, the known modes and methods of *X-ray structure analysis* are employed. The methods of structural neutron diffractometry continue to be modified by designing special apparatus for high resolution and wide apertures (used for small samples), those for *high-pressure*, high- and low-temperature studies, and so on (see also *Neutrons in solid state physics*).

STRUCTURAL PHASE TRANSITION,
polymorphic transformation

A *phase transition* accompanied by a rearrangement of the crystal structure under variation of an external parameter (temperature, pressure, field,

concentration of components, etc.). The most widespread are *displacive type structural phase transitions* when the atoms shift from their original positions described by a space group G to form a new configuration with the space group G_D. As a rule, the atomic displacements are small, so that the structure of the new phase is a minor distortion of the original phase, and the transformation is called a *distortion transition*. There is a subgroup–group relation $G_D \subset G$ that exists, so one can use the concepts of a higher symmetry (e.g., tetragonal) original phase (G), and a lower symmetry (e.g., orthorhombic) derived phase (G_D). In almost all cases the symmetric phase is the stable higher-temperature one. Transitions in *perovskites* ($BaTiO_3$, $KMnF_3$, etc.) serve as examples of distortion transitions. Distortion transitions are *second-order phase transitions* (no latent heat), or are close to them. There are also structural phase transitions that involve atomic displacements without a subgroup–group relation. The atomic displacements here are quite pronounced, and the crystal structure undergoes a much more substantial reconstruction. This *reconstructive transition* is a *first-order phase transition* with a latent heat. Included among these transitions are specific *metal* and *alloy* transformations from a body-centered cubic phase into either a face-centered cubic or a hexagonal close-packed phase.

Another class of structural phase transitions are *order–disorder phase transitions*, or those involving some kind of ordering. These are observable in ordered alloys (CuZn, CuAu, Fe_3Al, etc., see *Alloy ordering*), in metal hydrides, etc. Ordering takes place when cooling a crystal down to the temperature of the structural phase transition. The transition consists of the appearance of a regular alignment of atoms of various kinds, while remaining at their regular lattice of sites; their settling into a disordered phase is random. The ordering might involve forming a magnetic (e.g., ferromagnetic) phase, or a *superlattice*, that involves an additional order over the already existing structure of sites; a subgroup–group relation $G_D \subset G$ can also be present here. A further type of structural phase transition is an *orientational phase transition*. The transition in $NaNO_2$ can serve as an example; in this case a certain regular orientation of the triangular complexes NO_2 appears below a transition point, with the orientation being random in the higher temperature disordered phase. A similar situation takes place for the dumb-bell-like O_2 molecules in solid *oxygen*.

Numerous *ferroelectric phase transitions* which produce in the crystal a spontaneous electric polarization P ($BaTiO_3$, $KNbO_3$ and so on) are also structural phase transitions. From a macroscopic viewpoint, the appearance of P is caused by a mutual displacement of ions of different sign. In some cases the transition is accompanied by the appearance of a spontaneous crystal strain, and it is called a *ferroelastic transition*.

A thermodynamic description of a structural phase transition employs the concept of an *order parameter* η. In the general case, the order parameter is multicomponent, with the number of components related to the dimensionality of that irreducible representation of the group G according to which the transition occurs (see *Landau theory of second-order phase transitions*). In many cases when describing a structural phase transition, a macroscopic quantity serves as the order parameter: the polarization vector P for ferroelectric transitions, the strain tensor U_{ik} for ferroelastic transitions, and so on. To describe the symmetry of the phases their *point groups* are sufficient. There are so-called *isostructural phase transitions* in which overall G is not changed, but the lattice constants (e.g., in Ce), or the atomic position parameters in a space group (as in La_3S_4) change in a jumpwise fashion. Many phase transitions of the electronic kind (e.g., the metal–insulator transition in VO_2, and the Jahn–Teller transition in $KCuF_3$, see *Jahn–Teller effect*) can also be regarded as structural phase transitions. There are experimental examples of such structural phase transitions ($NaNO_2$, K_2SeO_4, and so on) where the new phase differs from the original phase by a spatial modulation with a period incommensurate with the lattice constants. The principal experimental methods of studying structural phase transitions are X-ray, neutron, and electron diffractometry. Structural phase transitions are sometimes accompanied by diverse anomalies of macroscopic physical properties (e.g., mechanical, electric, magnetic, optical, and others). Their studies yield important information on the nature of the phase transitions. The temperature range of *critical phenomena* that take plane at structural phase transitions is small, as a rule.

STRUCTURAL TEXTURE OF METALS

A nonequiaxial shape of grains and subgrains in a metal which emerges as a result of *plastic deformation*. The structural texture of metals is also determined by the type of arrangement of grains (see *Polycrystal*) after a *strain*. If the arrangement permits a direct (without turns) propagation of the main *crack* along the *grain boundaries* (or subgrain boundaries), the metal strength in some directions is appreciably reduced. The concept of structural texture also includes the condition of grain (subgrain) boundaries of a deformed metal which is determined by such parameters as elastic *internal stresses* related to the presence of *dislocations* in subboundaries, by the presence at the boundary of islands of the second *phase* rolled during the deformation process, and the *segregation* of impurities at the interfaces and in the near-interface region. In porous sintered metals, *pores* under plastic deformation elongate along the stress directions. The nonequiaxiality of pores is considered as their structural texture, and the nonequiaxiality of grains takes place if the plastic deformation occurs below the *recrystallization* temperature. The degree of nonequiaxiality can be calculated on the basis of the *Taylor and Polany principle* (G.I. Taylor, 1938; M.Z. Polany, 1925): during the plastic deformation, each grain changes its shape in the same manner as the sample as a whole. Nonequiaxiality of dislocation cells is smaller than that of grains in connection with processes of *polygonization* and *recovery*. As the temperature decreases (when these processes are impeded), the nonequiaxiality of cells increases and approaches that of grains. The structural texture of metals substantially affects the micromechanism of *failure*, thus stimulating a transition from *transcrystalline failure* to *intercrystalline failure*. The structural texture of metals increases the tendency of the metal toward breakdown of the lamination type, and grain size changes in various directions. Therefore, the structural texture of metals together with the crystallographic texture is responsible for the *anisotropy of mechanical properties* of deformed metals.

STRUCTURE AMPLITUDE

In general, a complex quantity $F(hkl)$ which characterizes the dependence of the scattering amplitude of the X-ray, the electrons, and other radiations incident on a crystalline *solid* on the number and the arrangement of atoms in an *unit cell*. The absolute value $|F(hkl)|^2$ is called the *structure factor*. The structure amplitude of a unit cell can be written in two different forms:

$$F(hkl) = \sum_{j=1}^{N} f_j \exp\left[2\pi i(hx_j + ky_j + lz_j)\right], (1)$$

$$F(hkl) = \int \rho(x, y, z)$$

$$\times \exp\left[2\pi i(hx + ky + lz)\right] dV, \quad (2)$$

where f_j is the atomic scattering factor (*form factor*) for the jth atom; x_j, y_j, z_j are the coordinates of the jth atom in the unit cell measured in fractions of the unit period; h, k, and l are the *crystallographic indices* (Miller indices) of the reflecting plane; $\rho(x, y, z)$ is the distribution of the factor that determines the scattering (such as the electron density, the electron potential, etc.). The summation in Eq. (1) is performed over all N atoms of the unit cell, and the integration in Eq. (2) is over the unit cell volume V_0. Eqs. (1) and (2) refer to a static lattice; to take into account the atomic thermal motion that decreases the intensity of scattered X-rays it is necessary to make the replacement $f_j \rightarrow f_j \exp(-W_j)$, where $\exp(-W_j)$ is the *Debye–Waller factor*.

The values of the structure amplitudes for a given crystal depend on the *space group* to which its *crystal structure* belongs, i.e. on the type of *Bravais lattice* and on the set of symmetry elements inherent in the crystal. For example, if the crystal structure belongs to the space group $P\bar{1}$ (a primitive lattice with a single symmetry element that is a *center of symmetry*) the following expression for the structure amplitude is obtained:

$$F(hkl) = 2 \sum_{j=1}^{N/2} f_j \cos\left[2\pi(hx_j + ky_j + lz_j)\right].$$

Due to phase factors $\exp[2\pi i(hx_j + ky_j + lz_j)]$, in crystals with a complicated unit cell some reflections are weakened or totally suppressed (e.g.,

in body-centered cubic crystals, the reflections hkl with $h + k + l$ equal to an odd number do not appear).

STRUCTURE ANALYSIS

See *Magnetic structure analysis, Microprobe X-ray defect structure analysis, X-ray structure analysis*.

STRUCTURE FACTOR

See *Structure amplitude*.

STRUCTURE MEMORY (structure heredity in Russian)

A phenomenon observable at repeated heating of quenched (see *Quenching*) or quenched and tempered *steels*: at the *structural phase transition*, when grains of *austenite* (*gamma phase*) newly formed from the *alpha phase* (see *Polycrystal*) reflect their orientational and microstructural relations with the prior austenite grains obtained as a result of the preceding heating. The structure memory manifests itself most clearly when the newly-forming grains of gamma phase reproduce in size, shape, and crystallographic orientation the primary austenite grains.

STRUCTURE REARRANGEMENT

See *Radiation-induced rearrangement of structure*.

STRUCTURE STUDIES

Studies of *crystal structure* and *defects* in solids based on using the following phenomena: the diffraction of X-rays (see *X-ray structure analysis*), electrons (*electron diffraction analysis* and *electron microscopy*), neutrons (*neutron diffractometry*), Mössbauer γ-quanta (*Mössbauerography*); the phenomenon of the ionization of atoms in a non-uniform electric field (*field ion microscopy*), and the orientation effects of the scattering of charged particles in crystals (*channeling*) are also used. To study the microstructure of solids (including grain and subgrain structure), an optical or *scanning electron microscope* is utilized. Indirect methods widely used in structure studies are based on the sensitivity of certain physical properties of solids to the presence of crystal structure defects, or to the variation of their composition (*magnetic structure analysis, resistometry, dilatometric analysis, Mössbauer spectroscopy, positron spectroscopy, nuclear magnetic resonance*, relaxation effects, *acoustic emission*, and so on). To determine the crystal structure (*unit cell*) beam scattering methods (X-ray, neutrons, or electrons) are generally employed. In this case, the *crystal lattice* is implied to be perfect. The diffraction of radiation from the lattice produces sharp Bragg maxima whose position and intensity determine the crystal lattice type, the dimensions of its unit cell, the amount, variety and arrangement of atoms in it.

In actual crystals with defects, the positions and the intensities of the Bragg maxima may change, and in addition an extra *diffuse scattering of X-rays* appears. From its quantitative analysis it is possible to determine the kind of defects, their spatial distribution and concentration in the crystal, and the magnitudes of the atomic displacements. Additional information about the actual crystal structure is available from the method of *small-angle scattering of X-rays* or neutrons. This method gives the dimensions, the shape, the size and concentration distributions of defects such as pores and disperse inclusions of a second phase. The morphology of individual defects (dislocations, stacking faults, segregations, and clusters of a second phase) are studied with the help of diffraction electron microscopy, or field ion microscopy. A method of "direct evaluation" of the crystal lattice has been developed which provides the determination of the magnitude of interplane distances in regions ~ 1 nm in size.

Neutron diffractometry allows one to study not only the crystal structure but also the *atomic magnetic structure* of solids, as well as to distinguish different isotopes of the same element. The method is efficient in structural studies of solids containing light elements or their compounds.

In studies of the localized state of an individual atom or atomic group, gamma resonance spectroscopy or *nuclear magnetic resonance* are often used. These methods allow one to determine the internal magnetic field at a nucleus, the changes in the symmetry of the environment around an atom, and the local chemical and phase composition.

A particular place among the methods of structural studies is occupied by *field ion microscopy* which gives the patterns of individual atoms on surfaces, and enables one to determine their type (atomic probe) and the atomic structure of defects, but only at the sharp end point of the object under study with a radius of curvature of only a few nanometers.

SUBBOUNDARIES

These are boundaries of subgrains (see *Block structure*) disoriented by an angle of $\theta \approx 1°$ with respect to each other, consisting of randomly arranged *dislocations* of various kinds with one prevailing *dislocation sign*, which provides the resulting disorientation of the subgrains. Subboundaries arise during *plastic deformation* of monocrystals or polycrystals at moderate temperatures. During *annealing*, the subboundaries rearrange into *polygonal boundaries* due to *annihilation of dislocations* with opposite signs.

SUBCASCADES

Branches of a *collision cascade* in a *solid*. When elastic collisions of atoms incident on the atoms of a solid occur with an energy exceeding a certain threshold energy they split the cascade of successive atom–atom collisions into subcascades ($\geqslant 10$ keV), and these subcascades are divided into two classes. In the majority of collision events, the incident high-energy atom is scattered into small angles; a substantially lesser number of collisions involves scattering at a wide angle. A *graph* serves as a mathematical model of such a process, with arcs that are close to linear trajectories of incident atoms (see Fig. on next page) that scatter into small angles (arcs AB, BC, and BD), and apices (B, C, D) that are the branching points of the cascade in which the high-energy incident atom is scattered at a wide angle. A *vacancy* is formed at every branching point, and to every arc there corresponds a high-energy branch with the following structure: there are vacancies and associated athermic *clusters* produced by the scattering atom along its trajectory; the momenta of atoms knocked out of the lattice sites are normal to the trajectory; the starting points of the ejected atoms serve as origins of high-energy cascade branches. One high-energy branch in combination with adjacent branches forms a subcascade.

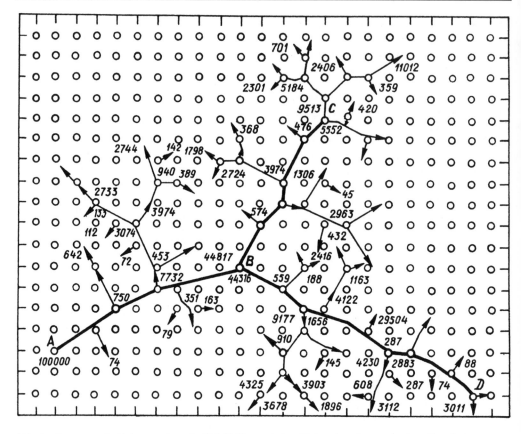

Calculated structure of a high energy branched 0.1 MeV cascade in α-Fe (numbers denote energies of knocked-out atoms in eV, and arrows indicate directions of motion).

SUBLIMATION

A type of *first-order phase transition* whereby a solid transforms directly to a gas, bypassing the liquid phase. Sublimation takes place in a certain range of temperatures and pressures, and is accompanied by a latent *heat of sublimation*. When applied to an atom in a monatomic solid, this quantity of heat energy has been considered as corresponding to the potential well depth in an *interatomic interaction potential*. Owing to this, the heat of sublimation can be expressed through the *compressibility*, the Grüneisen constant (see *Thermal expansion*), the *specific heat*, and so on. A kinetic characteristic of sublimation is its rate (i.e. amount of mass passing from solid to gaseous state per unit time) which depends on the type of

solid (*metal, insulator*), its surface temperature, its saturation vapor pressure, and the rate of vapor removal from the surface. Sublimation is employed for cleaning solids (sublimation with subsequent cooling). It is of importance when intensive vaporization takes place (e.g., thermonuclear reactors).

SUBMICROCRACKS in surface layers, nanocracks

Cracks with dimensions from 40 to 200 nm that nucleate in crystals during *plastic deformation* (both creep and active loading) under conditions of low and moderate temperatures ($T > 0.4 T_{\mathrm{melting}}$). Cracks with dimensions smaller than the Griffith size (see *Griffith theory*) are called *microcracks*; their indedependent growth is energetically unfavorable. Their dimensions vary from

0.1 μm to a few mm depending on the properties of the solid and the loading conditions (in rocks they can be up to tens of cm). Submicrocracks (or *nucleating cracks*) are those that nucleate as a result of a single event within a time of the order of 10 μs at a *pile-up of dislocations*, or at disclination structures (see *Disclinations*). Reconstructions of defect structures provide the energy gain for the nucleation of submicrocracks; and the latter process can be stimulated by thermal *fluctuations*. At moderate temperatures the rate of development of dislocations (see *Dislocations*) and disclination structures is higher in the surface layers of 10–100 μm thickness (depending on crystal size and loading conditions). Therefore, at $T < 0.4T_{melting}$ the submicrocracks nucleate mainly in these layers. Submicrocracks can be detected as early as at the stage of less than 10% *strain*. In quasi-brittle solids (see *Quasi-brittle failure*) the submicrocracks remain open and are able to grow during increasing deformation; submicrocracks transform into microcracks, and then into macrocracks. In plastic solids (see *Plasticity*) the submicrocracks rapidly become blunted and transform into *pores*. Their subsequent growth is due to the confluence of a high concentration of submicrocracks. On reaching the critical submicrocrack concentration, the avalanche stage of their confluence begins; eventually ending with a transition from a *failure* process to the next stage of macroscopic crack nucleation, or to *neck* formation. At the critical concentration of submicrocracks there is a critical *dilatation* from 0.5 to ~2% (depending on the submicrocrack mean size). *Annealing* the submicrocracks at the intermediate stage results in a significant improvement of the plastic properties, and an increase of *durability* and *strength*.

SUBMICRON TECHNOLOGY

See *Nanotechnology*.

SUBMILLIMETER SPECTROSCOPY

A field of spectroscopy which deals with the interaction of electromagnetic radiation with matter in the submillimeter spectral range ($\lambda \approx 50$ to 1000 μm) that occupies an intermediate position between the middle IR range and the high-frequency microwave-range. This is characterized by specialized instrumentation techniques due to the inability to utilize experimental methods developed for neighboring spectral ranges. The submillimeter range covers the energies of *phonons* in crystals; the energies of *Cooper pairs* (energy gap) in *superconductors*; the energies of *phase transitions* in *ferroelectrics, ferroelastics, ferromagnets*, and so on; specific binding energies and the energies of intrinsic excitations of impurity centers, *excitons* and more complicated bound and collective states of current carriers (*impurity atoms* and *biexcitons*, *clusters*, H^--like impurity states, and the *electron–hole liquid*) in semiconductors; the frequencies of *cyclotron resonance*, and plasma and magneto-plasma resonances in the system of free current carriers in *semiconductors*. The methods of submillimeter spectroscopy are useful for studying processes involving the delocalization of electrons, *photoconductivity*, and photothermal impurity ionization, as well as phase transitions in solids.

The main instruments of submillimeter spectroscopy are diffraction spectrometers, interference Fourier-spectrometers (see *Fourier spectroscopy*), gas *lasers*, and spectrometers based on backward-wave tubes. There are semiconductor generators based on *impact ionization avalanche transit time diodes*, as well as band-to-band transition *solid-state lasers*, and semiconductor-hot-plasma *tunable lasers*.

SUBSTITUTIONAL ALLOYS

Alloys based on substitutional *solid solutions*, where the atoms of the dissolved element replace solvent atoms at *crystal lattice* sites. When the differences between the radii of the atoms exceeds 14–15% then only restricted ranges of composition of these solid solutions can form; otherwise unlimited solubility of the components is possible. Many two-component or more complex substitutional alloys are known, which form on the basis of crystal lattices of metals or intermediate compounds.

SUBSTITUTIONAL ATOM

An *impurity atom* replacing a regular atom in a *crystal lattice*, i.e. substituting for an inherent atom that has been removed. The term substitutional atom also refers to an inherent atom of a crystal (chemical compound), occupying the position at a site of another sublattice, e.g., an M atom at a site of sublattice X in the crystal M_pX_q. An inherent substitutional atom is referred to as an *antisite defect*.

SUBSTRATE

A surface of a certain material, on which a layer of the same or a dissimilar substance is built up through a particular technique (see *Thin film growing*). Monocrystalline and amorphous (glass, *ceramics*) substrates are used, depending on the function and the method of layer deposition. For good *adhesion* and a high degree of structural perfection, the surface of the substrate must be thoroughly cleaned prior to the deposition process to avoid mechanical or structural imperfections. The following methods are used for this purpose: chemical and mechanical polishing, chemical *etching*, degreasing with organic solvents, ultrasonic, ionic and plasma cleaning. In *gas phase epitaxy* and *liquid-phase epitaxy* cleaning of the substrate surface is brought about through the removal of the surface layer by certain chemical reactions with substances in the gaseous phase, or by dilution in the melt, which is achieved by exposure at an elevated temperature. Control over the quality of the substrate surface is carried out through *electron diffraction analysis* and Auger spectroscopic analysis (see *Auger electron spectroscopy*); wherever possible, the control is accomplished in a vacuum chamber just before starting to precipitate the *film*. The following additional requirements are imposed upon the substrate for producing epitaxial structures: (1) it must have a low density of *dislocations*, (2) the discrepancy between parameters of the *crystal lattice* of the substrate and the new layer must not be too pronounced. Besides that, the lower side of the substrate may be put through a defect-generating treatment, which forms discharge channels for defects that arise during the process of layer deposition.

SUBSURFACE VOLUME ACOUSTIC WAVES (SSVAW)

Horizontally polarized shear volume waves generated by opposing rod transducers (see *Acoustic wave transducer*). These waves propagate through a *piezoelectric material* near its surface. SSVAWs differ from *surface acoustic waves* (SAW). The latter localize immediately under the surface of a piezoelectric while SSVAWs propagate at a small angle to the surface and follow the $(\lambda/R)^{1/2}$ law in their extinction (λ is the wavelength and R is the distance from the input transducer). The basic material for SSVAW devices is a rotated I-cut quartz. The SSVAW velocity at that cut exceeds the respective SAW velocity by a factor of approximately 1.6. Therefore, employing SSVAWs in *acousto-electronic* signal processing devices makes it possible to expand their frequency range to 2–3 GHz, and also to improve their temperature and overall long-term stability, as compared to the corresponding surface wave devices.

SUM RULE

See *Friedel sum rule*.

SUPERALLOYS

Complex *alloys* based on *nickel*, *cobalt* or *iron*, that comprise numerous components (up to 13 in number and 5–18% by mass), such as Cr, Al, Ti, Mo, W, Ta, etc., and they often include 10–12 impurities and alloying elements specifically introduced (C, B, O, Mn, Hf, Zr, etc.). Superalloys exhibit certain favorable features: high-temperature *long-term strength*, good *plasticity*, high *impact strength*, resistance to high and low temperature *fatigue* and high-temperature *corrosion*, good molding properties (see *Casting*), stable microstructure at high temperatures. All these properties result from the specific selection of basic doping elements that ensure *solid-solution hardening* and the formation of *intermetallic compounds*, *carbides*, etc. These feature respective *crystal structure*, shape, size, distribution and overall properties that ensure their stability during long term operation. Superalloys are capable of withstanding long-term loads, $(50–60) \cdot 10^3$ Pa, for $10^4–10^5$ h at temperatures of 800–1100 °C. Superalloys are used in aviation and space technology, chemical industry, etc.

SUPERANTIFERROMAGNETISM

A particular magnetic state of a tiny well-cut *antiferromagnet* slab with an even number N of atomic planes parallel to the surface, belonging alternatively to one of the two *magnetic sublattices* of the antiferromagnet. The slab retains its basic antiferromagnetic properties, but since the planes at the edges (boundaries) are in weaker effective antiferromagnetic *exchange fields* than those deep inside the plate, the related *magnetic moments* become stronger along the direction of an external magnetic field B. As a result, in an applied field there appear local variations of *antiferromagnetism vector* ($L = M_1 - M_2$) directions which are nonuniform over the plate thickness, but are symmetric in respect to its central plane. The *magnetic susceptibility* of this thin plate in weak fields grows up to twice its value in the corresponding bulk antiferromagnet. Cf. *Superparamagnetism*.

SUPERCONDUCTING ALLOYS

Disordered and ordered *alloys* and *solid solutions* of various transition and simple *metals*, and also of metals and nonmetals that exhibit superconducting properties (e.g., NbTi, PbBi, etc.). Superconducting alloys belong to the so-called "dirty" *type II superconductors*. They find wide practical applications in electric current carrying and magnetic superconducting systems (see *Superconducting magnets*).

According to the *Anderson theorem*, electron scattering by atoms of different elements (or by impurities) in an alloy does not result in changes of thermodynamic properties of a superconductor (its transition temperature, T_c, energy gap Δ, specific heat jump ΔC at the transition point, etc.), provided that this superconductor is invariant with respect to time inversion. However, electromagnetic properties of alloys differ from those of pure superconductors, the values of their critical magnetic fields and currents change, a transition from the situation of an ideal *Meissner effect* to a *mixed state* takes place, and characteristic microscopic superconductor lengths (*coherence length* ξ, penetration depth of magnetic field λ) also change. Violation of the Anderson theorem is explained, if one takes into account *electron spin-flip scattering* in alloys with paramagnetic impurities and anisotropic effects. A strong dependence of T_c on the structural state and stoichiometry is also observed in *intermetallic compounds* with complex lattices, and those containing features of low dimensionality.

In an isotropic alloy, the characteristics of its superconducting state are smooth functions of the ξ_0/l ratio, where $\xi_0 = \hbar v_F/(\pi \Delta_0)$ is the coherence length for a pure superconductor at $T = 0$ K, l is the electron scattering length (mean free path). A pure superconductor corresponds to the criterion $l \gg \xi_0$, and an impure one to $l \ll \xi_0$. The concentration of superconducting electrons, which controls the diamagnetic response of a superconductor, is expressed by the formula $n_s = n\chi(l/\xi_0, T/T_c)$. Here n is the number of electrons per unit volume, $\chi(x, t)$ is the Gor'kov function with its asymptotes $\chi(x, 0) \approx x$ for $x \to 0$ and $\chi(x, 0) \approx 1$ for $x \to \infty$. The concentration of superconducting electrons around the critical temperature point is

$$n_s = \frac{\pi \Delta^2 \tau n}{2T} \chi(2\pi \tau T),$$

where $\chi(x) = 1$ and $\chi(0) = 0.85/x$ for $x \gg 1$ ($\tau = l/v_F$ is mean free path time). The magnetic field penetration into a pure superconductor is given by the London penetration depth $\lambda_0 = [m/(\mu_0 n_s e^2)]^{1/2}$. In a dirty superconductor the penetration depth λ increases relative to the London value λ_0 as $\lambda = \lambda_0(\xi_0/l)^{1/2}$. The characteristic displacement of an electron within its correlation time $t_0 = \hbar/\Delta_0$ in an alloy is given by the relation $\Delta = (2Dt_0)^{1/2}$, where $D = v_F l/3$ is the *diffusion coefficient*. The value $\Delta x = (\xi_0 l)^{1/2}$ is significantly smaller than ξ_0; in other words, the radius of a *Cooper pair* in an alloy decreases. Around the critical temperature, T_c, the value of ξ follows an asymptote $\xi(T) \approx 0.74\xi_0/(1-t)^{1/2}$ in a pure alloy, and $\xi(T) \approx 0.86(\xi_0 l)^{1/2}/(1-t)^{1/2}$ in a "dirty" alloy, where $t = T/T_c$.

According to the *Ginzburg–Landau–Abrikosov–Gorkov (GLAG) theory*, superconducting alloys become type II superconductors at high enough impurity concentrations. The critical concentration value is given by the condition $\kappa = 1/\sqrt{2}$, where κ is the *Ginzburg–Landau parameter*, $\kappa = \lambda/\xi$. In an impure alloy, the value of κ may be expressed as $\kappa = \kappa_0 + 0.23(ec/k_B)\gamma^{1/2}\rho$, where ρ is the residual electric resistivity, and γ is the coefficient of the linear term in the electronic *specific heat*. In alloys containing paramagnetic impurities the transition temperature decreases in proportion to the impurity concentration (see *Gapless superconductors*).

SUPERCONDUCTING CRITICAL TEMPERATURE

See *Critical temperature of superconductors*.

SUPERCONDUCTING GROUP SWITCH

(commutator)

A set of *superconducting switches* such that a prescribed sequence of turning them on results in a prescribed direction and magnitude of current in the switched circuits. Such a group switch can make use of cryotrons in which current (below the critical value) in one or more superconducting input circuits controls the superconducting to normal transition in one or more output circuits. The most important parameter of a superconducting group switch is its switching time. When wire cryotrons are used to build such a group switch, this time may reach 10^{-6}–10^{-5} s. Film cryotrons provide switching times within 10^{-7}–10^{-6} s, and Josephson elements may reach 10^{-9} s. Wide application of the latter using classical superconductors has been limited in the past by the complex technology, and the need to have a cryogenic base for them. The use of *high-temperature superconductors* with liquid nitrogen cooling should facilitate the adoption of *Josephson junction* switching.

SUPERCONDUCTING INTERMETALLIC COMPOUNDS

A wide class of superconducting materials, with their *crystal lattice* formed by atoms of heterogeneous metals. Among the superconducting intermetallic compounds are superconductors with the A15 structure, the *Laves phases*, *delta phases*, and compounds of the α-Mn type.

Superconductors with the A15 structure are compounds with an A_3B chemical formula, their crystal lattice having body-centered cubic symmetry (β-W structure). Their B atoms are positioned at the apices and the center of a cube, and the A atoms pair-off at cube faces, forming families of mutually perpendicular infinite chains along the principal crystallographic directions. The *unit cell* contains 8 atoms and belongs to space group $Pm\bar{3}n$ (O_h^3). The A atom is usually a transition metal atom: V, Nb, Cr, etc. (metals of Groups IV–VI of the periodic system). The B atom is a nontransition (Al, Sn, Si, Ge, Ga, ...) or transition metal (Co, Ni, Ru, ...) element from among

those positioned to the right of Group VI of the periodic system. Prior to the discovery of the cuprate *superconductors*, the A15 type had been the compounds with the highest *critical temperatures*, reaching 23 K and 20 K for such pseudobinary compounds as Nb_3Ge and Nb_3Ga, respectively, and 24 K for Nb_3Ge films. The A15 compounds are *type II superconductors*, featuring high critical magnetic fields B_{c2} ($B_{c2}(0) \approx 47$ T for $Nb_3(Al, Ge)$), and high current densities. The critical current density in the diffusion layers of Nb_3Sn reaches $j_c \approx 3 \cdot 10^{10}$ A/m^2 in a magnetic field $B = 5$ T, while ribbons made of V_3Ge are capable of carrying currents up to 10^7 A/m^2 in a magnetic field $B = 20$ T at 4.2 K.

Because of their chain crystalline structure, the A15 compounds exhibit certain unusual physical properties. For example, V_3Si, Nb_3Sn and Nb_3Ge display anomalous temperature dependences of their electron and phonon characteristics. The spin component of their electronic *magnetic susceptibility* and their Knight shift (see *Hyperfine interaction*) both grow with lowering temperatures, the more rapid the growth the higher the T_c for the respective material, because of the presence of a peak in the *density of electron states* in the vicinity of their *Fermi level*. The temperature dependences of their *specific heat*, Hall constant (see *Galvanomagnetic effects*) and electrical *resistance* are all unusual. The latter tends to saturate at high temperatures. With lowering temperature, V_3Si and Nb_3Sn show a softening of their elastic shear moduli: $C_s = (C_{11} - C_{12})/2$, so at a temperature T_m that somewhat exceeds T_c ($T_m = 21$ K, $T_c = 17$ K for V_3Si; $T_m = 45$ K, $T_c = 18$ K for Nb_3Sn), a *structural phase transition* and a transformation from a body-centered cubic crystal lattice to a tetragonal one becomes possible. Such a phase transition manifests itself by the instability of the acoustic *phonon* spectrum with wave vector $q \parallel \langle 110 \rangle$ and polarization $P \parallel \langle 110 \rangle$. The velocity of these phonons, proportional to $C_s^{1/2}$, tends to zero at $T = T_m$. Experimental data indicate a possibility of low-temperature structural transitions in the $Nb_3Al_xGe_{1-x}$, V_3Ga and Nb_3Al alloys. No structural phase transitions were observed in any A15 material with low T_c (note that the *electron–phonon interaction* is weak in that case). Both types of phase transitions in A15, namely the transition to a superconducting state at high $T_c = T_m$ and the lattice instability result from strong a electron–phonon interaction that, in its turn, results in changes of the lattice symmetry accompanied by transformation of the A15 electron spectrum. The nature of the latter may be related to both the *Peierls instability* and the *Jahn–Teller effect*. In both cases, the loss in the elastic energy accompanying the appearance of a tetragonal deformation of the crystal lattice is compensated by a lowering of the overall energy of the transformed electron subsystem. When T_m decreases (e.g., under external pressure), T_c grows. Introducing defects, impurities, as well as radiation disordering always results in lowering of both T_m and T_c in A15 materials with high T_c (Nb_3Sn, V_3Si, Nb_3Al, Nb_3Ga, Nb_3Ge, V_3Ge), while A15 compounds with low T_c (Mo_3Ge, Mo_3Si) may display an increase of these temperatures.

Superconductors with a B1 structure are chemical compounds with an AB chemical formula, having a crystal lattice structure similar to that of NaCl. Atoms of metal A form a face-centered cube lattice, and atoms of metal B occupy octahedral sites. Atoms of type A usually belong to the transition metals of subgroups III–VI of the periodic system, and atoms of type B are nontransition elements of the same subgroups. This B1 group includes borides, nitrides, oxides of the transition metals. Available vacancies characteristic of these compounds are ordered, forming *superlattices*. The highest transition temperature $T_c = 10$ K is found in nitrides and carbides of Zr, Nb, Mo, Na, W. The value of T_c decreases for larger deviations from stoichiometry and increasing number of vacancies. The highest critical parameters are reached in niobium carbonitrides, NbN_xC_{1-x}, where T_c and the *upper critical field* B_{c2} both peak as functions of x at $x = 0.7$ ($T_c^{max} \approx 18$ K, $B_{c2}^{max}(0) \approx 16$ T). The critical current density then increases to $j_c \approx 8 \cdot 10^{10}$ A/m^2 at $B = 10$ T and $T = 4.2$ K.

Among the B1 substances are palladium compounds with hydrogen, PdH, and deuterium, PdD. No superconductivity is found in either pure Pd or its alloys, but saturating Pd with either H or D results in superconductivity at $T = 2$ to 9 K, while the H/Pd ratio varies from 0.8 to 1. When substituting deuterium for hydrogen the value of T_c increases to 11 K, in other words, an inverse *isotope*

effect develops. The absence of superconductivity in pure Pd is explained by the strong spin *fluctuations* that are suppressed in PdH due to the filling of the narrow palladium d-band by the hydrogen electrons.

Superconductors with a C15 structure are compounds with a face-centered cubic crystal lattice (as in MgCu) that belong to the $Fd3m$ (O_h^7) space group (see *Crystal symmetry*). Most C15 substances have both their superconduction transition temperatures T_c and their upper critical magnetic field $B_{c2}(T)$ relatively low. An exception is the $Hf_{1-x}Zr_xV_2$ alloys, with their maximum $B_{c2}(4.2$ K$) = 23$ to 24 T, $j_c(4.2$ K$) = 10^5$ to 10^6 A/m^2, and $T_c = 10.1$ to 10.2 K attained for $x = 0.5$ to 0.6. At $T_m = 100$ to 150 K these alloys undergo a structural phase transition from cubic to body-centered orthorhombic ($x < 0.40$) or rhombohedral ($x > 0.45$) symmetry. The T_m temperature is sensitive to the conditions of thermal processing of these materials, and decreases for deviations from stoichiometry, and when doping admixtures are introduced. The $T_m(x)$ dependence exhibits a minimum, coinciding with the maximum in $T_c(x)$. Such a correlation is related to the electronic nature of the structural phase transition. It also manifests itself in the peaks found in the vicinity of T_m in the temperature dependences of the electric resistance, magnetic susceptibility and specific heat of the C15 materials. The *band structure* of the electron energy spectrum contains plateaus around its L- and X-points, and in the H–C direction of the *Brillouin zone*. If the *Fermi level* passes in the vicinity of these regions, the strong electron–phonon interaction results in a transformation of the spectrum of elementary excitations of current carriers, and initiates a structural phase transition that follows the Peierls instability mechanism. As a result, an insulating gap opens at flat regions of the *Fermi surface*, with its presence responsible for certain features in the temperature dependences of electron characteristics of the C15 materials. The critical temperature T_c in C15 compounds is only weakly sensitive to radiation damage.

SUPERCONDUCTING MAGNET (solenoid)

Devices used for generating constant or alternating magnetic fields that employ solenoids made of superconducting materials. Superconducting magnets and solenoids with magnetic fields up to 10 T are manufactured from multiple conductor cable of the Nb–Ti alloy. The Nb_3Sn alloy solenoids have produced fields above 20 T, and their operating volume ranges from about 10 cm^3 to several tens of m^3. Such magnets are widely used in laboratory research, in *magnetic resonance imaging* (MRI), in accelerators, in plasma traps built for the control of thermonuclear synthesis installations, as induction energy capacitors, etc. The value of the maximum operating magnetic field is determined, first of all, by the current capacity of the superconductor, i.e. by the *critical current* density for a given magnetic field, and, second, by the structural and strength qualities of the solenoid material and the unit as a whole.

SUPERCONDUCTING MAGNETIC SHIELD

A device used to screen away a magnetic field, or stabilize its value in a prescribed volume. Superconducting magnetic shields are usually fabricated from *lead* or *niobium*, and have the shape of hollow cylindrical shells. They function in liquid helium ($T = 4.2$ K) in external magnetic fields B_e with magnitudes below the critical field for the respective superconductor. Since a superconductor pushes the magnetic field outside its own volume and "freezes" the magnetic flux with a field strength B inside the shell, the magnetic flux may penetrate partially inside the shell only through its openings. At a distance l into the shell from the edge of its opening of diameter d, the longitudinal component of a variable field $B_\parallel$ is screened by a factor of $S_\parallel = B_{e\parallel}/B_{l\parallel} \approx 7.6l/d$, and the transverse component $B_\perp$ is screened by the factor of $S_\perp = B_{e\perp}/B_{l\perp} \approx 3.8l/d$. The longitudinal component of a constant field may either be screened (by the same factor or less) or become "frozen" at a lesser gradient. This is one of the advantages of superconducting magnetic shields over the ferromagnetic and electromagnetic types. Another advantage consists in the independence of the screening that they provide for ($S_\parallel$, $S_\perp$) from the frequency (f) of the external field in the range $f \sim 0$–10^{10} Hz. Superconducting magnetic shields are used to protect magnetometric transformers from magnetic interference, to provide a

high-grade magnetic "vacuum", to build superconducting sources of strong magnetic fields, and to generate highly homogeneous magnetic fields.

SUPERCONDUCTING MICROBRIDGES

Short, narrow superconducting bridges (e.g., weak links) that couple together two superconducting electrodes of macroscopic dimensions. Electric and physical properties of superconducting microbridges depend on the relation between their characteristic dimensions (length L, width w, and thickness t) and the parameters of the superconducting material from which they are fabricated: *penetration depth of magnetic field* λ and *coherence length* ξ. Provided $\max(L, w, t) < \xi, \lambda$, *Josephson effects* develop in the superconducting microbridges (see also *Josephson junction*). When the transverse dimension of a microbridge or its length exceeds ξ, then its *resistive state* involves either the motion of *Abrikosov vortices* through the bridge, or the formation of centers of *phase slip*. The *current–voltage characteristics* of superconducting microbridges exhibit strong nonlinearity, so the bridges are able to detect and transform electromagnetic radiation, and may also serve as switches. Thin film superconducting microbridges link together superconductors made from *thin films*. One should differentiate between constant thickness film superconducting microbridges and those with a variable thickness. In the first case the microbridges are the same thickness as film electrodes, and in the second they are thinner than such electrodes, thus providing for a three-dimensional spread of current through electrodes which results in less Joule heating of the regions where the current concentrates.

SUPERCONDUCTING MICROWAVE ELECTRONIC DEVICES

Devices in which electron beams or plasma interacts with electromagnetic fields generated by microwave instruments that consist of *superconductors* in whole or in part. The basic advantages offered by this application of superconductors, their alloys and compounds in microwave electronic devices stem from the fact that the active component of the *surface impedance of a superconductor* may be several orders of magnitude lower than in normal metals. That feature is very efficiently used to produce high-Q resonator cavities (see *Superconductor electronics*), highly stable generators, band pass and band stop filters, wide-band transmission lines, delay lines, analogue elements for signal processing, various components in accelerator technology, etc. Since the energy loss in a superconductor is many orders of magnitude less than in normal metals in the microwave range, one can produce various units and devices with parameters unattainable by common technologies. For example, the Q of microwave cavities may exceed 10^{11}, the transmission band in coaxial transmission lines may be as wide as 100 GHz at an attenuation of 10^{-2}–10^{-3} dB/m, band filter amplitude-frequency characteristics may have their edges as steep as 30 dB/octave, and the relative instability of the frequency generator may be about 10^{-16}. Researchers are seeking ways and means to use superconductors in elementary particle accelerators. The low losses of superconductors at these operating frequencies make it possible to improve their efficiency by several orders of magnitude, and to proceed from pulse to quasi-continuous and even continuous operation acceleration modes when running accelerators. Small-size superconducting accelerators that operate in a continuous mode already produce gradients in their accelerating fields equal those of the best modern pulse accelerators. However, there still remain some problems in building accelerating units (mostly of a technical nature) that await solutions.

SUPERCONDUCTING PHASE TRANSITION,

phase transition from normal to superconducting state

Transition to a phase with zero electric resistance, which takes place in certain materials at low enough temperatures. It was first observed close to 4 K in mercury by H. Kamerlingh-Onnes (1911). It was later found to occur in a wide class of metallic, semiconducting and polymer materials. Transition temperatures T_c for several classical superconductors are: 3.7 K for *tin*, 7.2 K for *lead*, 9.5 K for *niobium*, 18 K for Nb_3Sn. Prior to the late 1980s T_c could not be raised above 23.2 K (Nb_3Ge). In recent years high-temperature superconductors, which are copper-containing *metal oxides* called *cuprates* with a layered structure,

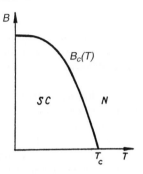

Phase diagram of a type I superconductor in a magnetic field (B) versus temperature (T) coordinate plane. The superconducting phase is denoted by SC and the normal phase by N; $B_c(T)$ indicates the critical magnetic field line where the phase transition occurs.

have been found with their T_c reaching 133 K. The value of T_c for the most popular cuprate superconductor, $YBa_2Cu_3O_{7-\delta}$, is $T_c = 90$–95 K (see *High-temperature superconductivity*).

An applied magnetic field B destroys the superconducting state of a *type I superconductor* if it exceeds a certain critical level, B_c, which is a function of temperature. In other words, there is a *first-order phase transition* that takes place along the $B_c(T)$ line in the *phase diagram* of superconductors (see Fig.). *Type II superconductors* have more complicated phase diagrams with two critical fields $B_{c1}(T)$ and $B_{c2}(T)$. Physically, the superconducting phase transition takes place when a single multiparticle quantum-mechanical state develops for a macroscopically large number of *conduction electrons* when these electrons with antiparallel spins and opposing momenta form *Cooper pairs* in the vicinity of the *Fermi surface*. Long-range ordering then spreads throughout the system, and an *order parameter* which is complex appears, in contrast with, e.g., the case of a *magnetic material*. The characteristic size of Cooper pairs, i.e. their *coherence length* ξ, plays the role of the *correlation length* of fluctuations of the superconducting order parameter, which grows without limit for $T \rightarrow T_c$. The classical low-temperature superconductors have a *critical region* which is extremely narrow, and their superconducting phase transition is very well described by the phenomenological *Ginzburg–Landau theory of*

superconductivity, as well as by the more fundamental microscopic *Bardeen–Cooper–Schrieffer
theory*, both of which employ the rationale of the
self-consistent field approach.

SUPERCONDUCTING QUANTUM
INTERFERENCE DEVICE (SQUID),
superconducting quantum interferometer

A superconducting current loop containing
either a single (one-junction) or a pair (two-
junction) of weak links (*Josephson junctions*).
SQUID is an acronym for Superconducting QUantum Interference Device. It appears more correct
to envisage a SQUID as a superconducting quantum interferometer combined with electronic circuitry used for processing the resulting signal information. Depending on the type of SQUID used,
its *critical current*, *impedance* or actual loss display a periodic dependence on the external magnetic flux Φ. The quantum interference of wave
functions of superconducting electrons, which develops as a result of combining *quantization of
flux* with *Josephson effects*, controls that periodicity and also the discrete quantum states of
the SQUIDs: their total magnetic flux Φ being
equal to an integral number n of quanta (flux-
ons), $\Phi = n\Phi_0$. Modern electronic devices are capable of recording 10^{-4}–10^{-6} $\Phi_0/\mathrm{Hz}^{1/2}$, which
makes it possible to exploit the unique capabili-
ties of SQUIDs as a new class of measuring instruments. Despite the impressive capabilities of two-
junction SQUIDs, the one-junction SQUIDs are
more widely used. A simple scheme of such a one-
junction SQUID circuit is presented in the figure.
The superconducting quantum interferometer circuit (SQUID) is inductively connected on the left
to the input solenoid of a superconducting transformer (ST). The latter transforms the flux from
the magnetic antenna (MA). A resonant circuit
(RC) is also present. The amplitude of oscillations
in the latter is determined by the periodic parameters of the SQUID. Depending on a dimensionless
parameter $l = LI_k/\Phi_0$, where L is the inductance
and I_k is the critical contact current, there may be
two regimes in which the HF SQUIDs operate: the
hysteresis-free ($l = 1$) and the hysteretic ($l > 1$)
regime. When $l < 1$, the circuit reactance is modulated and when $l > 1$ the losses in the RC are modulated. The pumping generator (G) frequency is
close to the RC intrinsic frequency, and is usually
within the range 30–400 MHz. With the pumping frequency growing, the SQUID sensitivity increases (*microwave SQUIDs*). A low-noise amplifier (A) and a power amplifier (PA) provide signal detection and recording, often in the feedback
mode, so the SQUID operates as a null-indicator.
Various types of antennae makes it possible to
measure the magnetic field, the components of the
tensor of field gradients, and its derivatives of second and higher orders.

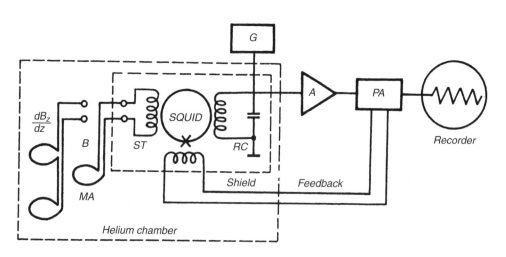

Circuit for measuring a magnetic field B or field gradient $\mathrm{d}B/\mathrm{d}z$ using a single-junction SQUID.

SQUIDs are the most sensitive low-frequency measurement devices available, making it possible to reach the quantum limit at frequencies that are significantly below $k_B T/h$. The sensitivity of SQUIDs already realized is 10^{-16} T/Hz$^{1/2}$ in a magnetic field, 10^{-16} T$\cdot$m$^{-1}\cdot$Hz$^{-1/2}$ in a field gradient, 10^{-13} A/Hz$^{1/2}$ in current, 10^{-15} V/Hz$^{1/2}$ in voltage (such instruments are called femtovoltmeters). SQUIDs are capable of measuring magnetic susceptibility in low fields (10^{-4}–10^{-10} T) at a sensitivity of 10^{-12} SI unit, and at reaction times down to 10^{-5} s (at 30 MHz). When conducting *nuclear magnetic resonance* spectroscopy and *tomography* SQUIDs make it possible to detect a single spin, in principle. The energy sensitivity of SQUIDs reaches 10^{-28}–10^{-29} in commercial and 10^{-31}–10^{-34} J/Hz in laboratory instruments. The best result achieved by means of a two-junction SQUID is $4.8\cdot10^{-34}$ J/Hz $\approx 3.2\hbar$. In other words, the fundamental threshold has been almost reached. The application range for SQUIDs embraces such fields as scientific instrumentation, geophysics, search for magnetic anomalies, extremely low frequency communications, biomagnetism, computer technology, etc. SQUIDs first initiated quantitative studies of the quantum behavior of macroscopic systems with dissipation (*macroscopic quantum tunneling*) at temperatures about 1 K, and of the quantum electrodynamics of superconducting circuits.

SUPERCONDUCTING SEMICONDUCTORS AND SEMIMETALS

Superconducting materials with a degenerate electron spectrum and a density of current carriers n that is low in comparison with that found in the more conventional superconducting metals and alloys. Free electrons or *holes* in superconducting semiconductors and semimetals are generated by way of an admixture *doping* or via the equilibrium excess or deficiency of one of the components. All of these materials are *type II superconductors*. Several of the $A^{IV}B^{VI}$ structure are known (see *Semiconductor materials*) with hole conductivity. Among them are GeTe with its maximum superconducting *critical temperature* $T_c^{max} \approx 0.53$ K. Its concentration range of current carriers in which the superconductivity exists is $8.5\cdot10^{20}$ cm^{-3} $< n < 6\cdot10^{21}$ cm^{-3}.

Next are SnTe, with $T_c^{max} \approx 1.1$ K and the respective concentration range $4\cdot10^{20}$ cm^{-3} $< n <$ $7.5\cdot10^{21}$ cm^{-3}; PbTe doped with Tl ($T_c^{max} \approx$ 1.4 K, $n < 10^{20}$ cm^{-3}); and SnTe alloys with GeTe or PbTe ($T_c^{max} \approx 2.75$ K, 10^{19} cm^{-3} $<$ $n < 1.5\cdot10^{21}$ cm^{-3}). Of particular interest is superconductivity in Tl-doped PbTe that develops due to the formation of quasi-local states in the Tl *impurity band* (up to 1.25 at.% of Tl). Adding 0.5 at.% of Na raises the critical temperature T_c of that system to 1.6 K. There are also materials based on In: InSn$_{1-x}$Te$_{1+y}$ ($y \ll 1$) with $T_c^{max} =$ 3.5 K ($n \geqslant 5\cdot10^{20}$ cm^{-3}, p- and n-type conductivity) and an n-type *inversion layer* in p-type InAs, its $T_c = 2.7$ K. The La–X systems (X = S, Se, Te) display relatively high values of T_c, their T_c^{max} reaching 3.8 K for La$_3$X$_4$, X = Te, 8.25 K for X = S, and 10 K for X = Se at 10^{21} cm^{-3} $<$ $n < 6\cdot10^{21}$ cm^{-3} (n-type conductivity). The *upper critical field* B_{c2} ($T = 0$) for La$_3$S$_4$ is 16 T.

Two additional electron conductivity superconductors in this group have a perturbed *perovskite* structure. These are SrTiO$_3$ with $T_c^{max} =$ 0.65 K, $8.5\cdot10^{18}$ cm^{-3} $< n < 3\cdot10^{20}$ cm^{-3} and a nonmonotonic $T_c(n)$ dependence, and BaPb$_{1-x}$Bi$_x$O$_3$ (BPB) with its $T_c^{max} \approx 13$ K at $x = 0.25$ and $n^{max} \approx (1–5)\cdot10^{21}$ cm^{-3}. BPB features one of the highest T_c values among all the materials that contain no transition metal atoms. Although the $T_c(x)$ dependence is similar to $T_c(n)$ for SrTiO$_3$, in the case of BPB $T_c(n)$ monotonously increases with n, and its $B_{c2}^{max}(0) \approx 4.4$ T. The *Bardeen–Cooper–Schrieffer theory* relation $\Delta(0) = 1.76k_B T_c$ between T_c and the energy gap Δ is approximately satisfied in all the superconducting semiconductors and semimetals for which the *energy gap* $\Delta(T)$ was measured.

Superconductivity in all the above-mentioned types of materials has been variously explained by interaction of their current carriers with either longitudinal or transverse optical phonons, by the multivalley nature of their band spectrum, by the *plasmon mechanism of superconductivity*, by two-phonon exchange between the electrons, by the *Bose–Einstein condensation* of bipolarons, and by various combinations of these and other factors. Most theoretical studies assume that either superconducting semiconductors and semimetals are *ferroelectrics*, or that their *crystal lattice* is on the verge of transforming into this state with spontaneous polarization.

SUPERCONDUCTING SUPERLATTICES

Either natural or synthesized solid state systems, such that their structure is characterized by a periodicity (or a quasi-periodicity) in one or several directions. Moreover, such a period (quasi-period) should significantly exceed unit cell (or interatomic) distances. Modern methods of crystal growth (see *Monocrystal growth*) and nanotechnology (see *Molecular beam epitaxy*) make it possible to produce artificial crystals with interfaces between layers that are smooth on an atomic scale, while the layers themselves may be spatially oriented in some desired sequence. When the superconducting superstructure is periodic in a single direction its is called one-dimensional. Two-dimensional and three-dimensional superconducting superlattices are characterized by superperiods along either two or three axes, respectively, of the coordinate system. The mutual positioning of layers may be either periodic, quasi-periodic (i.e. following a certain algorithm), or disordered, so one may differentiate between these three types of superconducting superlattices. Among the artificial superlattices are the following types: superconductor/insulator (Al/Ge, Nb/Ge, Nb/Si, V/Si, Mo/Si); superconductor/normal metal (Nb/Cu, V/Ag, Nb/Ta); superconductor/magnetic atom (V/Cr, Mo/Ni, V/Ni). Intercalated layered superconductors (see *Quasi-two-dimensional superconductors*) and transition metal dichalcogenides of the type Nb/Se_2, Ta/S_2 also form one-dimensional superconducting superlattices since their transverse periods across the layers may significantly exceed interatomic distances (5–6 nm) as a result of the *intercalation*. In the same sense, one may speak about layered *high-temperature superconductors* formed on the basis of Y, Bi, Tl, etc. in which their cuprate layers $(-CuO_2-)_n$ or double layers are spaced at 1–3 nm, while other elements controlling their chemical composition form regular layers of an "atomic intercalate" free of atomic disordering. One may also produce superconducting superlattices with layers one *unit cell* thick that are composed of high temperature superconducting crystals, e.g., $YBa_2CuO_x/PrBa_2Cu_3O_x$. Replacing Y by Pr does not particularly affect

the size of the corresponding unit cell, but makes the layer of $PrBa_2Cu_3O_x$ nonconducting. Therefore, varying the layer thickness at a discrete step of a single unit cell, one may study *superconductivity* in monolayers of pure $YBa_2Cu_3O_x$. Superconducting superlattices in cuprates based on Bi are also known. In them Bi is partially replaced by Pr, so that by gradually varying the concentration of Pr the superconducting superlattice is gradually brought to the superconductor/insulator type. Finally, one may speak about superconducting superlattices in high temperature superconductor crystals because a quasi-regular mesh of planes of *twinning of crystals* is present in some of them. Such a superlattice is interesting because of the effect of local amplification of superconductivity that may occur in the vicinity of twinning planes, similar to regular *superconductors*; the so-called *Khaikin–Khlyustikov effect* (M.S. Khaikin, I.N. Khlyustikov, 1981). In that sense, the PbTe/PbS superconducting superlattice is also unusual. First, it consists of layers of PbTe and PbS, both non-superconducting when separated from each other. However, the *critical temperature* of such a joint superconducting superlattice featuring the semiconducting concentration of electrons 10^{19} cm^{-3} is $T_c = 5.5$ K. Second, beside the one-dimensional alternating layer period, the PbTe/PbS superlattice also features a two-dimensional rectangular mesh of *epitaxial dislocations* positioned in the vicinity of layer-to-layer interfaces.

Since such quantities as the *coherence lengths* along and across the layers, the *penetration depth of magnetic field*, the thickness of the layers and the intervening interfaces, the period and the quasi-period may all vary quite widely in the above superconducting superlattices, their features and properties are also as variable. Among the critical characteristics of superconducting superlattices, those studied in greater detail are the superlattice critical temperatures, *critical currents* and *critical magnetic fields*. All of them are necessary for the development of an elemental basis for the next generation computer components.

One of the most interesting phenomena in superconducting superlattices is *dimensional crossover*. It appears as a smooth transition from three-dimensional distributions of characteristic variables (typical for the *Bardeen–Cooper–Schrieffer theory*) to the corresponding two-dimensional distributions typical for *thin films* at lower temperatures. Such a crossover must be clarified in finer detail to understand the temperature dependence of the *upper critical field*. Studies of superconducting superlattices are pursued quite actively, although they are still far from their final stage of maturity. See also *Superlattices*.

SUPERCONDUCTING SWITCH

A switch which utilizes the property of superconductors to increase their electrical resistance dramatically during the transition to the normal state. A superconducting switch consists of a bulk superconductor (a piece of wire, a band) or a superconducting film converted to its normal state by heating or applying a magnetic field. An example is a superconducting current carrying wire or film (e.g., *cryotron*) that is flipped back into its normal state by the magnetic field of a control device. These switches are used in superconducting magnetic systems to obtain persistant currents by short circuiting the ends of the current carrying *superconducting magnet* coil. When the switch is in its normal state, in other words it has a finite resistance, a prescribed current in injected into the superconducting solenoid. After that, the switch is flipped into its superconducting state, so that a persistent current flows continuously through the short-circuited solenoid.

SUPERCONDUCTIVITY

The phenomenon of a sharp drop of electric resistance to zero of certain *metals, alloys* and *intermetallic compounds* at some *critical temperature, T_c*. Critical temperatures may vary from well below 1 K in some elemental superconductors to appreciably above 100 K in *high-temperature superconductors*. To give a few examples: Al, $T_c = 1.2$ K; Nb, $T_c = 9.2$ K; Nb_3Ge, $T_c = 23$ K; $La_{2-x}Sr_xCuO_4$, $T_c = 36$ K; $YBa_2Cu_3O_{7-x}$, $T_c = 92$ K; $HgBa_2Ca_2Cu_3O_x$, $T_c = 135$ K. Superconductivity was detected by H. Kamerlingh-Onnes (1911). In 1986–1987 J.G. Bednorz, K.A. Müller and C.W. Chu discovered *high-temperature superconductivity* in metal oxide compounds. Besides their ideal (infinite) conductivity,

superconductors display *absolute diamagnetism* (the *Meissner effect*) whereby they expel magnetic flux from their interior. In other words, we have $B = 0$ in bulk superconductors within a certain range of magnetic field strength from 0 to B_{max}. The value $B_{max} = B_c$ is called the *thermodynamic critical magnetic field* for *type I superconductors* (pure metals). Another value, $B_{max} = B_{c1}$ is the *lower critical field* for the case of *type II superconductors* (alloys, intermetallic compounds). Type II superconductors are in their *mixed state* within the range $B_{c1} < B < B_{c2}$ (B_{c2} is the *upper critical field*). In this mixed state the field penetrates into the bulk superconductor in the form of quantized *Abrikosov vortices*. We have $B = n\Phi_0$ for the magnetic field (n is the density of vortices, and Φ_0 is the *flux quantum*, the amount of flux in one vortex). The fields B_c and B_{c1} are usually of the order of 10^{-2}–10^{-1} T, while B_{c2} can be quite high (10–100 T).

When the applied magnetic field is zero, the *superconducting phase transition* is a *second-order phase transition* accompanied by a jump in the *specific heat* (it increases below the transition point T_c). The nature of superconductivity consists in the formation of macroscopic large-scale ordering of the wave function $\psi(r)$ into a coherent superposition of wave functions of separate particles (electrons or holes coupled into *Cooper pairs*). The phase of that function is a macroscopic variable associated with so-called quantum coherence effects such as *quantization of flux*, and the dc and ac *Josephson effects*.

According to the microscopic *Bardeen–Cooper–Schrieffer theory* of superconductivity (BCS theory) the macroscopic quantum state arises from the *Bose–Einstein condensation* of Cooper pairs. Because of the *electron–phonon interaction* electrons in common superconductors bind into pairs that have a total momentum $P = 0$ and spin $S = 0$ (singlet state), provided an electric current is completely absent. Due to exchange of *virtual phonons*, electrons develop a common attraction in the vicinity of the *Fermi surface*. In the common case when the pairs form via the electron–phonon interaction mechanism the critical temperature depends on the mass M of its atoms (the so-called *isotope effect*). Some compounds including *heavy fermions* (e.g., $CeCu_2Si_2$, UBe_{13}) may form triplet Cooper pairs with a spin of $S = 1$ (see *Triplet pairing*).

As a result of pairing there develops an *energy gap* Δ in the energy spectrum of a superconductor. That gap yields the minimum value of excitation energy, so that $\varepsilon_p = (\Delta^2 + \xi_p^2)^{1/2}$, where ξ_p is the electron energy as a function of the momentum p in a normal metal, measured relative to the Fermi surface. The gap is zero at T_c, and at $T = 0$ it reaches its maximum value $\Delta(0)$, which for most superconductors has the approximate value $\Delta(0) \approx 1.76 k_B T_c$. Since, during the interaction of a superconductor with an electromagnetic field, excited (unbound) electrons are released in pairs, a threshold of energy absorption from that field appears, $\hbar\omega = 2\Delta \approx 3.5 k_B T_c$. A superconducting state with zero gap is also possible (see *Gapless superconductors*). However, the *density of states* still exibits an anomaly within a range of energies of the order of $k_B T_c$ in the vicinity of the *Fermi level* (i.e. it is significantly lower than its normal state value).

Superconductivity finds numerous practical applications such as the generation of superstrong magnetic fields (*superconducting magnets*), devices for measuring superweak magnetic fields (*Superconducting QUantum Interference Devices*, SQUIDS), measuring devices, and computer elements and components.

See also *Dislocation superconductivity, High-temperature superconductivity, Nonequilibrium superconductivity, Surface superconductivity, Thin film superconductivity, Twinning plane superconductivity, Weak superconductivity*.

SUPERCONDUCTIVITY, HIGH-TEMPERATURE

See *High-temperature superconductivity*.

SUPERCONDUCTIVITY IN QUASI-ONE-DIMENSIONAL SYSTEMS

The phenomenon of *superconductivity* in *quasi-one-dimensional crystals* that present a system of weakly interacting metal filaments (chains), such that their electronic energy spectrum is basically one-dimensional:

$$E_\pm(p) = \pm v_F(p_\parallel \pm p_F) + E_\perp(p_\perp),$$

$$E_\perp(p_\perp) \ll v_F p_F,$$

where p_F, v_F are the Fermi momentum and velocity; $p_\parallel$ and $p_\perp$ are the components of *quasi-momentum* of the electron along and transverse to the chain, respectively. In case of a single electron band, the Fermi surface of such a system consists of two weakly corrugated planes that superimpose as a result of transposition by a wave vector $2p_F/\hbar$. If the transverse dispersion of the electron spectrum is low ($E_\perp \to 0$) the *condition of nesting* $E_+(p) = -E_-(p - 2p_{F\parallel})$ is satisfied for the Fermi surface. Other instabilities may develop along with the Cooper instability in a quasi-one-dimensional system superconducting state, all generated by interactions of electrons. Examples are instabilities with respect to the formation of *charge density waves* (CDW) or *spin density waves* (SDW). Their wave vector is equal to $2p_F/\hbar$, and both result in an insulating state of the system (see *Peierls transition*). In a quasi-one-dimensional crystal the above instabilities compete against each other and affect each other. Theoretically, such a competition was studied in detail using a one-dimensional model of a metal in which the electron–electron interaction is described by constants g_1 and g_2. These correspond to electron scattering accompanied by strong ($\sim 2p_F$) or weak (~ 0) transfer of longitudinal momentum, respectively. In a commensurable case ($2p_F = \pi n \hbar/a$) one has to account for a third interaction constant, g_3, which is responsible for *Umklapp processes* (see *Conservation laws*). The form of the ground state in such a metal is prescribed by the relationships between the above constants. In particular, if $g_1 - 2g_2 > 0$, then Cooper instability is the important factor and the ground state is superconducting. When this inequality has a negative sign then instability with respect to a CDW dominates, and the basic state of the system is a *Peierls insulator*. When the transverse dispersion of the electron spectrum is low then *fluctuations* also play a considerable role, preventing the formation of long-range order in the system and lowering the temperature of a three-dimensional phase transition. The role of three-dimensional effects becomes more important when the probability of an electron transition between the chains increases. The level of Fermi surface corrugation then increases, generally resulting in a suppression of instabilities with respect to the formation of CDW (or SDW). Besides, the role of *order parameter* fluctuations diminishes so that the transition of a quasi-one-dimensional metal into its superconducting state becomes possible (see *Organic conductors and superconductors*), provided the values of $E_\perp$ are not too low.

SUPERCONDUCTIVITY MECHANISM

For different varieties, see *Anderson mechanism of high-temperature superconductivity, Bipolaron mechanism of superconductivity, Cooper effect, Cooper pairs, Exciton mechanism of superconductivity, Magnon mechanism of superconductivity, Phonon mechanism of superconductivity, Plasmon mechanism of superconductivity.*

SUPERCONDUCTOR CHARACTERISTIC LENGTH

See *Characteristic lengths in superconductors.*

SUPERCONDUCTOR COLLECTIVE EXCITATIONS

See *Collective excitations in superconductors.*

SUPERCONDUCTOR CURRENT STATES

See *Current states in superconductors.*

SUPERCONDUCTOR ELECTRONICS

A branch of cryogenic *solid-state quantum electronics* which employs nonlinear quantum phenomena of superconductors, their zero resistance and *absolute diamagnetism*, to transform signals and to solve various problems of radio electronics and physics. The most important result achieved by superconductor electronics lies with the greater sensitivity of various instruments. Deep freezing (to 0.1–4.2 K) lowers thermodynamic noise in such instruments. However, various types of noise appear in the *resistive state* of superconductors, and quantum fluctuations set fundamental limits on the possibilities of further sensitivity increases.

Most of the active elements used in superconductor electronics employ the phenomena of quantum coherence: the *Josephson effect*, the *quantization of flux*, and quantum interference. Superconducting magnetic screens make it possible to lower constant magnetic fields (shield the Earth's field) to 10^{-12} T, and variable fields by a factor of 10^{12}. Superconducting *resonators* (cavities) in the VHF and microwave bands achieve

quality factors Q up to 10^{12}–10^{14}. Superconducting *bolometers* can reach threshold sensitivities of 10^{-10}–10^{-16} W/Hz$^{1/2}$ and time constants of 10^{-10}–10^{-1} s at temperatures of 4.2–0.1 K. *Josephson junctions* with direct conductivity are used as detectors, mixers, parametric amplifiers with self-pumping, etc. The sensitivity of Josephson sensors reaches 10^{-15}–10^{-16} W/Hz$^{1/2}$ in various regimes. Due to certain features of their current–voltage characteristics, *tunnel junctions* are promising memory and logic elements for new computer technologies. Single-particle current–voltage characteristics of small area tunnel junctions (S–I–S contacts, weak links) and those of *Schottky superdiodes* (see *Schottky diodes*) make it possible to actually develop new types of video and heterodyne detectors and mixers, with noise levels approaching the quantum threshold at frequencies from tens of GHz to the near IR band. One may expect to see new technologies which employ systems with multiple superconducting junctions that can operate synchronously, perhaps produced by electron or *ion lithography* at submicron resolution (1–100 nm). The widest use in measurement technologies is offered for *superconducting quantum interference devices* (SQUIDs), their energy sensitivity having almost reached the quantum threshold already at the level of 10^{-28} J/Hz in mass produced and 10^{-33} J/Hz in laboratory SQUIDs. Resolution in magnetic flux lies within 10^{-4}–10^{-6} Φ_0, where $\Phi_0 = h/2e$ is the *flux quantum*. Measurement systems based on SQUIDs can record magnetic fields of 10^{-16} T, voltages down to 10^{-16} V, and other variables transformed into variations of magnetic flux. These achievements provide new classes of instruments for scientific research, geophysics, medical science, metrology, and other technical fields. Several countries have established their national standard for the unit volt using the Josephson effect expression $f/V = 1/\Phi_0 = 483.6 \cdot 10^{12}$ Hz/V, where f is the frequency. Certain other standards, comparators, measurement systems to conduct fundamental physical experiments followed suite, all of them employing SQUIDs, superconducting cavities, etc.

Superconducting cavities are electronic resonator systems in which their current conducting elements and other parts operate in a superconducting state. Their principal property is a quality factor which is far higher ($Q \sim 10^{11}$) than that of resonators built using normal metals. Superconducting resonators feature extremely high operational quality factors as standards since at their operative temperatures metals have a very low linear expansion coefficient. The concept of "superconducting resonators" usually includes both cavity microwave resonators and resonating systems that operate in various frequency ranges. The gain in the quality factor may vary from several units to several factors of ten, depending on their design, operating temperature, and the material from which the cavity is manufactured. To produce superconducting cavities such pure metals as *niobium* and *lead* are sometimes used, with critical temperatures of 9.25 K and 7.18 K, respectively. A15 *intermetallic compounds* such as Nb_3Sn, Nb_3Ge and others with somewhat higher transition temperatures (15–20 K) have been under consideration for such applications. Superconducting resonators are effectively applied to solve various problems in electronics, radio physics, measurement techniques, and particle acceleration installations. Their use can improve instrumentation sensitivity by several orders of magnitude, as well as its accuracy and efficiency. Widespread application of superconductor resonators is so far hampered by the difficulties of liquid helium cooling under continuous operation. One may expect a breakthrough in this field when new *high-temperature superconductors* operating at liquid nitrogen temperatures enter routine use.

SUPERCONDUCTOR ENERGY GAP

See *Energy gap*.

SUPERCONDUCTOR INTERMEDIATE STATE

See *Intermediate state of superconductor*.

SUPERCONDUCTOR NUCLEAR SPIN RELAXATION

See *Nuclear spin relaxation in superconductors*.

SUPERCONDUCTOR PARAMAGNETIC LIMIT

See *Paramagnetic limit*.

SUPERCONDUCTOR PROXIMITY EFFECT

See *Proximity effect*.

SUPERCONDUCTOR RESISTIVE DOMAIN

See *Resistive domains in superconductors*.

SUPERCONDUCTOR RESISTIVE STATE

See *Resistive state of superconductors*.

SUPERCONDUCTORS

Solid materials (*metals, alloys, solid solutions,* chemical compounds, *organic polymers,* doped *semiconductors, semimetals*) that exhibit the property of *superconductivity* at low enough temperatures. Among the pure metals (Hg, Pb, Sn, Zn, Al, Tc, Nb, etc.) the highest *critical temperatures* T_c of the *superconducting phase transition* are found in Nb ($T_c \approx 9.46$ K) and Tc ($T_c \approx 7.75$ K), while among the metal alloys and compounds the highest is the *intermetallic compound* Nb_3Ge ($T_c \approx 23.2$ K). Degenerate *semiconductors* doped with atoms of metals (Nb, Te), such as GeTe, PbTe, SnTe, InTe, $SrTiO_3$, exhibit low critical temperatures: $T_c < 5$ K. Layered dichalcogenides, NbX_2, TaX_2 and trichalcogenides NbX_3, TaX_3 of the transition metals (X = S, Se) feature somewhat higher critical temperatures, $T_c \approx 2$–8 K. Values of $T_c \approx 8$–10 K were reached by organic compounds with chain (quasi-one-dimensional) and layered (quasi-two-dimensional) structures and strongly anisotropic conductivity. Among the superconductors available by 1986 that did not contain any transition metals the highest critical temperature was found for the solid solution $BaPb_{1-x}Bi_xO_3$ ($T_c = 13$ K at $x = 0.25$). Its critical temperature T_c sharply dropped to zero at higher contents of Bi ($x \geqslant 0.3$), because of the metal–insulator transition during which alternating valence states of Bi^{3+} and Bi^{5+} ions appear at adjacent sites of the *crystal lattice*. In 1988 the value of $T_c \approx 30$ K was reached for the solid solution $Ba_{1-x}M_xBiO_3$ (M = K, Rb).

In 1986, *high-temperature superconductors* based on layered metal oxide cuprates were discovered, with $T_c > 90$ K by 1987, and $T_c > 130$ by 1993.

See also *Amorphous superconductors, Gapless superconductors, High-temperature superconductors, Magnetic superconductors, Organic conductors and superconductors, Quasi-two-dimensional superconductors, Reentrant superconductors*.

SUPERCONDUCTORS AND PLASMONS

See *Plasmons in superconductors.*

SUPERCONDUCTORS AND TUNNELING

See *Tunnel effects in superconductors.*

SUPERCONDUCTOR SPECIFIC HEAT

See *Specific heat jump in superconductors.*

SUPERCONDUCTORS OF TYPE I AND TYPE II

See *Type I superconductors, Type II superconductors.*

SUPERCONDUCTOR SURFACE ENERGY

See *Surface energy of superconductors.*

SUPERCONDUCTOR SURFACE IMPEDANCE

See *Surface impedance of a superconductor.*

SUPERCONDUCTORS WITH CHARGE (SPIN) DENSITY WAVES

Anisotropic metal systems in which one finds coexisting *superconductivity* and *charge density waves* (CDW) or *spin density waves* (SDW), formed as a result of *phase transitions* at higher temperatures. As a rule, superconductors with these density waves are low-dimensional *metals*, their *Fermi surface* spanning a reciprocal lattice vector Q (Q is the CDW or SDW wave vector).

The transition to a CDW or SDW state results in the appearance of an insulator gap in the electron spectrum in the vicinity of nested regions (regions spanned by reciprocal lattice vectors) of the Fermi surface; that, in its turn, results in a decrease of the *density of electron states* at the *Fermi level*. The critical temperature, T_c of the superconducting phase transition is smaller than the value T_{c0} that would exist in the absence of either a CDW or a SDW. Suppressing a CDW (SDW) by the introduction of impurities, or by applying an external pressure, makes it possible to increase the transition temperature up to the value $T_c \approx T_{c0}$. Such a situation in which the superconducting state coexists with a CDW or SDW is observed in double and triple chalcogenides of the *transition metals*, i.e. the compounds of the form MX_2 and MX_3 (M = Nb, Ta; X = S, Se), and in the case of SDW in *organic conductors and superconductors* of the form $(TMTSF)_2X$ ($X^- = AsF_6^-$, ClO_4^-, FSO_3^-, etc.).

A number of theoretical studies treat, along with the competition between superconductivity and a CDW (SDW), the possibility of increasing the critical temperature of superconducting transition in systems with a CDW or SDW. A theoretical model suggested by Yu.V. Kopayev et al. puts forth such a mechanism for raising T_c. The model assumes that doping may result in shifting the Fermi level in CDW systems from the vicinity of nested areas of the Fermi surface to the edge of the energy gap in the electron spectrum. The density of electron states then develops a feature of the form $g(\varepsilon) \propto 1/(\varepsilon^2 - \Delta^2)^{1/2}$, where Δ is the width of the insulating gap. All this should result, under certain conditions, in increasing the superconducting transition temperature.

Currently known superconductors with CDWs (SDWs) have certain particularities in their physical properties. Among them, there can be a significant restructuring of their *phonon spectrum* during the transition to the superconducting state, that results in a change in the frequency of the amplitude mode of CDW oscillations. Using *Raman scattering of light* such a feature has been observed in transition metal chalcogenides.

SUPERCONDUCTORS WITH LOCAL PAIRS

A theoretical model that explains macroscopic quantum properties of superconductors by the formation, in the actual physical space and due to a strong *electron–phonon interaction*, of electron pairs that belong to one and the same or adjacent sites of a *crystal lattice*. Superconductors with local parameters are often described using the phenomenological *Hubbard Hamiltonian* (see *Hubbard model*) with an effective attractive interaction potential $(-U)$:

$$\widehat{H} = \sum_{i\sigma} \varepsilon_i a_{i\sigma}^+ a_{i\sigma} + t \sum_{\langle ij \rangle \sigma} a_{i\sigma}^+ a_{j\sigma}$$

$$- U \sum_i n_{i\uparrow} n_{i\downarrow} + V \sum_{\langle ij \rangle \sigma \sigma'} n_{i\sigma} n_{j\sigma'},$$

$$n_{i\sigma} = a_{i\sigma}^+ a_{i\sigma},$$

where $a_{i\sigma}^+$ ($a_{i\sigma}$) are the creation (annihilation) operators of an electron at the ith site; V is the

Coulomb energy of interaction of electrons belonging to different sites, t is the site-to-site transition integral, and the summation $\langle ij \rangle$ is limited to adjacent sites. At low temperatures, when $T \ll U/k_B$, the Hamiltonian may be transformed into a different one that operates in the space of electron pairs and is, in its turn, equivalent to the anisotropic spin *Heisenberg Hamiltonian*. The *bipolaron mechanism of superconductivity* results in the same Hamiltonian if the treatment is brought to the limit at which the binding energy of the bipolaron, Δ, significantly exceeds the width of the polaron band, W (bipolarons of small radius). The random phase treatment demonstrates that the spectrum of elementary excitations is acoustic, and below the transition temperature T_c the *specific heat* $C(T)$ follows a power law. The dependence of the *upper critical field* B_{c2} on T, taking into account scattering from impurities, has a positive slope. The *penetration depth of magnetic field* entering superconductors with local parameters, λ_P, significantly exceeds the London value, λ_L, while the specific heat jump at $T = T_c$ may be abnormally small, and remain experimentally unobservable. The effect of disorder in the material significantly lowers T_c. The drawback of the model of a superconductor with local parameters is that it does not account for the large-scale Coulomb repulsion. The *Anderson model* with localized centers of electron attraction against the background conduction electron gas that screens the Coulomb interaction overcomes that deficiency.

SUPERCONDUCTOR THERMAL CONDUCTIVITY

See *Thermal conductivity of superconductors*.

SUPERCONDUCTOR THERMODYNAMICS

See *Thermodynamics of superconducting transition*.

SUPERCONDUCTOR TUNNELING SPECTROSCOPY

See *Tunneling spectroscopy of superconductors*.

SUPERELASTICITY, pseudoelasticity, anomalous elasticity

A property whereby some *alloys*, under the effect of applied stress σ, undergo considerable strain ε (up to 25%), then, without any noticeable

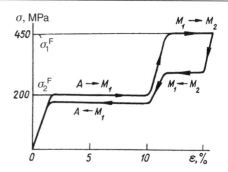

Stress (σ) versus strain (ε) curves showing the plateaux of the regions of superelasticity.

residual effects, they return to their initial state when the stress is lifted. Superelastic deformation was first observed by A. Ölander in Au–Cd alloys (1932), and is known to exist in many metals and alloys (such as Cu–Al–Ni, Cu–Al–Zn, In–Tl, Ni–Ti, etc.). Superelasticity may arise from diverse causes: it may develop when predominantly oriented crystals of *martensite* form under applied stress in the range of temperatures where the martensite phase does not form without stress. These crystals vanish when the stress is lifted, so that an inverse *martensitic transformation* takes place. Intermartensite transformations may also occur in a field of external tensions. They consist in the successive formation of close-packed phases that differ in their *crystal structure*. When the stress is lifted, martensitic phases disappear in the subsequent inverse transformation. Stress may also affect the motions of twinning boundaries in martensite (see *Twinning of martensite*). In other words, certain twins may grow at the expense of others. When the stress is lifted a reverse motion of twinning boundaries occurs (the martensite crystal structure does not change its symmetry).

Depending on the alloy chemical composition, the temperature of the test, the crystallographic orientation of the initial phase, etc., one or several mechanisms of superelasticity may be at work. Superelasticity appears mainly when *monocrystals* are stretched along the [001] axis of the cube. In those monocrystals in which superelasticity is produced by martensite and the intermartensite transitions, the respective stress versus strain (σ–ε) curves typically exhibit several plateaus which correspond to the stress $\sigma_i^F = $ const. Apparently these are transitions during which martensite (M) forms from the high temperature phase (A), and one martensite phase forms from another (see Fig.). The effect of superelasticity is closely related to the *shape memory effect*, since both of them are based on direct and inverse martensitic transformations. In certain cases the residual deformation found during superelasticity tests may be removed by additional heating.

Superelasticity is sometimes observed in alloys that do not undergo martensitic transformations, e.g., in Fe$_3$Al monocrystals. They feature a 1.5% reversible strain under shear loading at room temperature (see *Shear*). In that case, the possible mechanisms include *twinning of crystals* and disordering (see *Alloy ordering*). Alloys with a superelasticity effect are used in various branches of technology to produce sensors and activating elements (such as microcoolers, clamps for solid-state circuitry assembly operations, etc.).

SUPERFLUIDITY

The phenomenon of the vanishing of viscous *friction* in *quantum liquids* at low temperatures. Superfluidity in liquid *helium* (^{4}He) below the *lambda point* ($T_\lambda = 2.17$ K) was discovered by P.L. Kapitza (1938). The transformation from the natural (He I) to superfluid (He II) state is a *second-order phase transition*.

Superfluidity is a universal property of quantum fluids provided the spectrum of their elementary excitations satisfies the *Landau superfluidity criterion* (L.D. Landau, 1946). Superfluidity develops when a macroscopically large number of particles transforms into a single coherent quantum state (a condensate). In liquid ^{4}He which is a quantum *Bose fluid*, superfluidity is often associated with the phenomenon of *Bose–Einstein condensation*, which is typical for an ideal or weakly non-ideal *Bose gas* (N.N. Bogolyubov, 1947; S.A. Belyaev, 1959). However, due to the strong interaction between the atoms in liquid ^{4}He, the single particle *Bose condensate* becomes depleted, in other words particles with zero energy and momentum almost disappear. Thus, at $T < T_\lambda$, the superfluid component in He II forms a correlated multiparticle quantum state (an effective condensate) that is not directly related to Bose condensation.

Superfluidity is also typical of quantum *Fermi liquids*, both neutral (e.g., liquid ^{3}He, see *Superfluid phases of ^{3}He*) and charged (*conduction electrons* in metals). Pairs of linked (correlated) fermions with half-integer spin couple together to form atom-like entities of ^{3}He or electrons under the effect of some attraction (*van der Waals forces, electron–phonon interaction*). Being themselves bosons with an integer spin that is not necessarily zero, such pairs form a coherent superfluid condensate (see *Cooper pairs*). As for the charged Fermi particles (e.g., electrons in a *metal*) such a situation corresponds to the phenomenon of *superconductivity*. Superfluidity is possible for a Bose liquid (Bose gas) of charged bosons, e.g., bipolarons (see *Bipolaron mechanism of superconductivity*).

SUPERFLUIDITY CRITERION

See *Landau superfluidity criterion*.

SUPERFLUID PHASES OF ^{3}He (A-phase and B-phase)

Low-temperature phases of liquid ^{3}He discovered by D.D. Osheroff, R.C. Richardson, D.M. Lee in 1972 (Nobel Prize, 1996). *Phase transitions* of normal ^{3}He (N-phase) into its A-phase, and of the A-phase into the B-phase of ^{3}He were discovered in the solid ^{3}He melting curve ($P = 35$ atm) at temperatures of $T_c = 2.6$ mK and $T_{AB} = 2.07$ mK, respectively. When lowering the pressure, the temperature T_c separating natural ^{3}He from its superfluid state decreases to 0.9 mK at $P = 0$, while T_{AB} increases to 2.4 mK at $P = 20$ atm. In other words, there is a *multicritical point* present. The phase transition between the N-phase and either the A- or the B-phase is a *second-order phase transition*, while the AB-transition (between ^{3}He-A and ^{3}He-B) is a *first-order phase transition*. The T_c curve splits in an applied magnetic field, and the transition from the N-phase to the A-phase goes via another superfluid phase, the A_1-phase. Moreover, the B-phase is separated from the N-phase by a band of ^{3}He-A at the B- to N-phase boundary near the melting line (see Fig.).

The transition of ^{3}He into a superfluid state (see *Superfluidity*) is associated with the formation of a *Bose condensate* of Cooper pairs formed

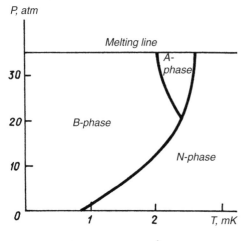

P, atm

Superfluid phases of ^{3}He.

by atoms of ^{3}He. L.P. Pitayevsky predicted the mechanism of pairing for ^{3}He as early as in 1959: *Cooper pairs* with a nonzero orbital angular momentum l form due to the van der Waals attraction. Theoretical and experimental studies of superfluidity of ^{3}He showed that *triplet pairing* takes place in ^{3}He (spin of pairs is $S = 1$ at $l = 1$). Superfluid states of the *Fermi liquid* with $l = 1$ were first studied by P.W. Anderson and P. Morel (1961), as well as by R. Balian and N.R. Werthamer (1963). Such states find their natural existence as the *A*-phase and the *B*-phase of superfluid ^{3}He, respectively. Atoms with their spins "up" or "down" couple in the *A*-phase while all the components of the triplet, $S_z = +1, 0, -1$, couple together in the *B*-phase. As a result, all the pairs have a projection $l = 1$ on a certain l axis ($l^2 = 1$) in the *A*-phase, while their spin projection on a certain other axis, V ($V^2 = 1$), is zero. The directions V and $-V$ are equivalent to each other for the latter. As for the *B*-phase, no selected quantization axes exist there for either the spin or the orbital angular momentum. The *order parameters* in ^{3}He-*A* and ^{3}He-*B* are, respectively:

$$A_{kj}^A = \Delta_A(T) V_k \Delta_j, \tag{1}$$

$$A_{kj}^B = \Delta_B(T) e^{i\varphi} R_{kj}(\boldsymbol{n}, v), \tag{2}$$

where $\boldsymbol{\Delta} = \boldsymbol{\Delta}' + i\boldsymbol{\Delta}''$, $\{\boldsymbol{l}, \boldsymbol{\Delta}', \boldsymbol{\Delta}''\}$ is a triplet of the mutually orthogonal vectors ($\boldsymbol{l} = \boldsymbol{\Delta}' \times \boldsymbol{\Delta}''$), R_{kj} is

a real orthogonal matrix that describes the rotations around a certain direction $\boldsymbol{n}$ ($\boldsymbol{n}^2 = 1$) through a certain angle v, and Δ_A and Δ_B are phenomenological parameters of the gap in the spectrum of *quasi-particles*. The order parameters (1) and (2) are constructed to take into account the degeneracy along the orientations of the spin axis V and the triplet of vectors $\{\boldsymbol{l}, \boldsymbol{\Delta}', \boldsymbol{\Delta}''\}$ in ^{3}He-*A*, and for degeneracy in phase φ and orientations of the spin space relative to the orbital space in ^{3}He-*B* (the degeneracy parameters are $\boldsymbol{n}$ and v).

The *spin–orbit interaction*, its role played by the *dipole–dipole interaction* of the magnetic moments of ^{3}He nuclei, as well as the presence of an external magnetic field, work to partially lift the degeneracy of the superfluid phase of ^{3}He. These effects combine to define the diverse physical and topological properties of ^{3}He. These involve the properties of a *superconductor*, a *magnetic substance*, and a *liquid crystal* simultaneously. From the theoretical point of view, superfluid ^{3}He is particularly interesting since it poses new fundamental problems in condensed matter physics.

SUPERFLUID VELOCITY

A gradient-invariant quantity defining the *current states* in a superconductor. The superconductor current density $\boldsymbol{j}_s$ is proportional to the superfluid velocity $\boldsymbol{V}_s$:

$$\boldsymbol{j}_s = n_s e \boldsymbol{V}_s,$$

$$\boldsymbol{V}_s = \frac{\hbar}{2m}\left(\nabla\varphi - \frac{2e^2}{\hbar c}\boldsymbol{A}\right),$$

where φ is the phase of the complex *order parameter* (macroscopic wave function of *Cooper pairs*), and $\boldsymbol{A}$ is the vector potential of the magnetic field (see *Superconductivity*).

SUPERHARD MATERIALS

Monocrystals, polycrystals, composite materials with a strength exceeding that of corundum (α-Al$_2$O$_3$) which rates 9 on the *Mohs hardness scale* (F. Mohs, 1811). The only superhard (SH) natural *mineral* is *diamond*. Superhard materials (SHMs) are synthesized by applying ultrahigh pressures (>5 GPa) at high temperatures, and then selecting the product through multistage enrichment, and various sorting techniques. SHMs feature high compression *strength* and *thermal conductivity*, low *friction constant*, and *specific heat*.

Table 1. Characteristics of monocrystalline synthetic diamond and cubic boron nitride

Characteristic	Diamond	Cubic boron nitride
Crystal structure	cubic	cubic
Lattice constant, nm	0.3567	0.3615
Minimum distance between atoms, nm	0.1544	0.1565
Density, theoretical (by X-rays), g/cm^3	3.515	3.48
Density, pycnometric, g/cm^3	3.23–3.60	3.44–3.49
Heat resistance, K	1123	1473
Number of atoms per unit cell	8	8
Number of atoms per cm^3	$1.76 \cdot 10^{23}$	$1.69 \cdot 10^{23}$
Coefficient of contraction from hexagonal to cubic structure	1.55	1.53

There are many SH nonmetallic and metal-like *carbides*, borides, and *nitrides*, the most common being *synthetic diamonds* and *cubic boron nitride* (cBN), see Table 1. A specific index of the physical and chemical properties of diamond-like materials is the directional *covalent bond* of atoms in their *crystal lattices*: the high strength of the carbon sp^3-hybrid configuration, and the maximum bonding density. The redistribution of electrons among the atoms of metals and nonmetals results in the formation of an ionic admixture in the *chemical bonds*. Compared to diamond, the valence electrons of SH carbides, borides, and nitrides of various transition metals exhibit lower stability, and the interatomic *ionic-covalent bonds* of these materials are also weaker. The yield limit for diamond, cubic *boron nitride*, and many other fragile hard materials has not yet been found.

The hardness of SH compounds is approximately proportional (with correlation coefficient of 0.8) to the ratio of the product of the interatomic bonding energy E_0 times the index of covalent bonding a_c divided by the square of the interatomic distance d (see Table 2). Crystalline SHMs are fragile, with microplasticity practically absent. They are obtained by *crystallization* at high static pressure (diamond, cBN), by hot isostatic *pressing*, by *sintering* in an electric discharge plasma or laser beam (polycrystals and *films* are obtained that way), in *shock waves* (crystals), by epitaxial growth of priming crystals from the gas phase at low pressures. By varying the reactant composition as well as the conditions of synthesis and separation, one may vary widely the physicochemical properties, size and shape of

Table 2. Hardness (GPa) and rigidity parameter $E_0 a_c / d^2$ for various compounds

Material	Hardness, GPa	Rigidity parameter $\dfrac{E_0 a_c}{d^2} \cdot 10^{16}$, eV/cm^2
BN	60.0	5.37
B$_4$C	45.0	4.39
TiB$_2$	33.7	2.42
SiC	33.0	2.92
Si$_3$N$_4$	32.3	2.80
BP	32.0	2.62
TiC	31.7	2.86
ZrC	29.5	2.72
YB$_2$	28.0	2.05
VC	24.8	2.52
ZrB$_2$	22.5	2.13
AlN	12.0	2.04

polycrystals, level of purity, and density of composites. Synthetic diamonds and other SHMs are mainly produced as powders of particles 0.03–0.1 μm in size. These are used to manufacture abrasive instruments, substrates for electronic circuitry, to produce polycrystalline superhard materials. Large synthetic diamonds of the ballas and black types, microcrystalline in nature, are also produced this way. Monocrystalline synthetic diamonds up to 10 mm in size are grown, and the ballas and black diamond are produced in the shape of plates and cylinders 100–200 mg in mass. Synthetic diamond (*dismite, carbonite*, CB) polycrystals are baked at high temperatures and pressures from finely powdered synthetic diamond, and the final product is cylinders 3–5 mm in di-

Table 3. Fracture toughness of superhard materials

Material	K_{1c}, MHz·m$^{1/2}$
Cubic boron nitride (monocrystal)	2.20
B$_4$C (sintered)	3.38
Elbor RM *	3.91
AlN + TiN (sintered)	4.20
Composite 12 *	4.30
Si$_3$N$_4$ (sintered)	4.33
Carbonite	4.64
Ballas diamond	5.50
Diamond (monocrystal)	5.70
Hexanite RM *	5.89
TiC + VC (sintered)	6.00
Dismite	6.38
Slavutich (sintered)	7.60
Cyborite	8.16
Tvesal (sintered)	8.50

* Produced by high-pressure synthesis (>5 GPa).

ameter and 4 mm high. High-pressure synthesis may produce diamond–hard alloy plates from fine powders, 8–13 mm in diameter, with an 0.9 mm thick diamond-saturated layer. cBN polycrystals (hexanite-R, belbor, elbor-R, cyborite, composite-10, teplonite) are made in the shape of plates 6–50 mm in diameter, 4–5 mm high or alternatively as hard alloy substrates with a cBN layer 0.8–1 mm thick.

Polycrystals and composites of SHMs are used to produce spindle teeth, lather mills, saws, drills and other cutting instruments. These are used to process hard tempered *steels* and *cast irons*, abrasive light alloys that contain silicon, viscous and corrosion active plastics, composites armored with high strength filaments. There are additional composite SHMs containing diamond (*slavutich*, *tvesals*, diamond containing tool material) based on hard-alloy and metal binding that even exceed diamond in their wear resistance.

High compression strength and wear resistance makes it possible to use superhard materials for machine parts, as well as instruments. There is a technical limitation specifying that fragile superhard materials may only be of limited use, depending on their *fracture toughness* (see Table 3). High values of optical parameters, thermal and electric conductivity, of superhard materials control

their effective use in electronics, electric technologies, instrument building. Superhard materials are widely used as *coatings* that are put on the surface of metal and nonmetal pieces by directional charging (rubbing-in), detonation and plasma *spraying*, galvanic precipitation, epitaxial growth. They all feature high wear strength (see *Wear*), corrosion and erosion strength (see *Corrosion* and *Surface erosion*) and a low friction constant. The class of superhard materials may be expanded using the principles of isovalent substitution to form triple and quadruple metal-ceramic compounds with the aid of *high pressure* and high temperature techniques.

SUPERHYPERFINE STRUCTURE

See *Hyperfine structure*.

SUPERLATTICE OF ALLOY, superstructure of alloy

A structure in an ordered *alloy* (either substitutinal or interstitial) where atoms of different types alternate in a regular order, forming a periodic lattice with a period exceeding that of the disordered alloy *crystal lattice*. Superlattice structures form below a certain temperature, called the *order–disorder phase transition* temperature, in cases when it is favorable in energy for the atoms of a given type to be surrounded by atoms of a different type. Often a superlattice structure forms as a result of a *second-order phase transition*. An example of a superlattice structure is found in the Cu–Zn alloy (so-called β-brass) in which the Cu and Zn atoms are ordinarily in a disordered state, distributed with equal probability among the body-centered cubic lattice sites. When, however, the alloy transforms to an ordered state, the atoms of one type occupy the sites at the apices, while those of the other type are found at the centers of these cubic cells, forming a superlattice. Similar superlattice structures are found in substitutional alloys of, e.g., the compositions Cu–Be, Cu–Pd, Ag–Mg, Fe–Al, Au–Zn, etc. An interstitial phase, isomorphous to those alloys, has its stoichiometric composition of the form Me$_2$X (Me is the metal, X is the nonmetal).

SUPERLATTICES

Solid-state structures in which, besides the periodic potential of their *crystal lattice*, there is another one-dimensional potential present, with a periodicity significantly larger than the lattice constant. Such semiconductor systems were first considered by L.V. Keldysh in 1962. *Composite superlattices* have alternate layers of crystalline structures of different types, and *doped superlattices* have periodic changes of impurity concentration. *Strained superlattices* with large discrepancies in their lattice constant (1–5%) are also known. Superlattice periods vary from a hundred to several thousand nanometers. One may obtain systems of such types using molecular beam epitaxy from the gas phase, and other state-of-the-art technologies.

Semiconducting superlattices are the ones most widely used and studied. They offer the unique possibility of constructing a customized *band structure* in a system. While the motion of current carriers transverse to the superlattice axis remains free, their motion along that axis is miniband in nature, and the operational characteristics of such minibands may be adjusted so as to modify the transport and optical properties of the system. Examples of this are systems containing alternating layers of GaSb and InAs, in which the top of the GaSb *valence band* is higher in energy than the bottom of the InAs *conduction band*. Another example is offered by GaAs with periodic doping that results in alternating *n*-type and *p*-type layers. Such superlattices provide the basis for constructing low-noise photodiodes, photomultipliers, photodetectors, light-emitting devices, including *semiconductor lasers*.

Metallic superlattices can be used to construct systems that have superconducting layers alternating with normal (systems with Nb/Cu, Pb/Cu) or metallic ones (V/Ni system). Here Nb, Pb and V are superconductors, Cu is a normal metal, and Ni is a band *ferromagnet*. Systems have also been constructed with alternating superconductor metals of different types (Nb/Zr, Ru/Ir). Such superlattices are of especial interest from the point of view of producing new superconducting systems. Because of changes in their crystalline structure in thin layers the properties of materials used for

them can change significantly. Therefore, it becomes possible to produce new structures not realizable in the materials themselves used to produce superlattices under ordinary conditions. An example of such a new system is Au/Ge with Au layers 1 nm thick and Ge layers 1.3 nm thick. It exhibits novel properties, not found in its initial components. The system becomes superconducting at $T_c = 1.5$ K, despite the fact that neither Au nor Ge display any *superconductivity* at atmospheric pressure.

Both critical currents and critical fields significantly change in *superconducting superlattices*. Due to the layered structure of the material, they become anisotropic, and their values increase. Such an effect is explained by their spatial inhomogeneity that induces *vortex pinning*. The anisotropic nature of the electron motions in superlattices that combine superconducting and normal semiconducting or metallic layers may be so strong that the interaction of superconducting layers may exhibit a *Josephson effect*. According to theoretical calculations, the spectrum of *quasiparticles* is a zero gap type in superlattices consisting of alternating superconducting and normal layers, provided their structure is sufficiently perfect.

Magnetic superlattices are associated with paramagnetic atoms which align their spins in an ordered manner (ferromagnets, antiferromagnets, ferrimagnets, standing spin waves) with a periodicity length which exceeds the crystallographic lattice constant.

SUPERLATTICE, SUPERCONDUCTING

See *Superconducting superlattices.*

SUPERLATTICE, SURFACE

See *Surface superlattice.*

SUPERLOCALIZATION OF STRAIN SHEAR BANDS

A particular type of *strain* of monocrystals that appears at high (pre-melting) temperatures (see *Melting temperature*). The strain under these conditions is mainly concentrated in high-temperature localized *slip bands* (HTLSB) where the shears reach 1000% and more. Outside the shears, the strain is a few parts per thousand. The HTLSB begins to manifest itself at $T > 0.6T_{\text{melting}}$, by its growth with increasing T. Their structure and properties differ considerably from those of the shear bands observed at the low-temperature localization of slip (see *Localization of plastic deformation*). The HTLSB shears are so pronounced that they form noticeable macroscopic steps at crystal faces, whereupon the degree of shear of the HTLSB increases, and the related step growth is accompanied by a noticeable broadening of the width of the comparatively narrow HTLSB (in crystals of LiF their width is 10 to 50 μm). As the strain ends and the sample cools, it is impossible to reveal the difference in the dislocation structure in regions within and beyond the HTLSB by using ordinary methods such as selective *etching*, *X-radiography*, *electron microscopy*, optical microscopy, and so on. The phenomenology of the representation of this superlocalization studied in various materials is the same: the profile of strain curves (see *Deformation diagram*) consist of jumps, *yield cusps*, and so on. However, the mechanism for the phenomenon requires more study *in situ* (during strain development), since *structure studies* of samples as the strain terminates are not very informative.

SUPERPARAMAGNETISM

A quasi-paramagnetic behavior observed in magnetically-weakly-coupled very small particles of magnetically ordered materials. In *ferromagnets* and *ferrimagnets* their spontaneous uniform *magnetization* M always has $2n$ ($n \geqslant 1$) possible orientations for a minimum of the free energy (two orientations for each of the n equivalent *easy magnetization axes*). A reorientation of M is related to overcoming a barrier height kV, where k is the *magnetic anisotropy* constant, and V is the particle volume. In the case of typical values k and $T \sim 100$ K, the barrier is of the order of the thermal energy $k_B T$ for volumes $\sim 10^{-27}$–10^{-23} m^3. A probability of thermally activated reorientation of M in particles of such a volume becomes of practical significance. For the values of V indicated above the particle is usually a *single-domain particle* with a magnetic moment $\mu = M \cdot V$ which reorients as a whole while conserving its magnitude. The magnetization of the ensemble of such particles is described by the *Langevin function* just as in the case of a classical paramagnet, but with the difference that in this case the role of the el-

ementary magnetic moment is not played by the moment of an individual ion, but rather by the much greater moment of the particle as a whole. As a result, the magnetic susceptibility of superparamagnetic particles can be very high.

For hyper-small antiferromagnetic particles, the disruption of the exact equivalency of their *magnetic sublattices*, and the lack of compensation of their magnetic moments, becomes probable. The particles acquire permanent magnetic moments, and superparamagnetism is characteristic for their ensemble (compare with *superantiferromagnetism*).

The concept of superparamagnetism is applicable in problems of *nondestructive testing techniques* of alloys with inclusions of a magnetic phase (*clusters*), and in studies of rock magnetism.

SUPERPLASTICITY

Capability of solids under certain conditions to undergo very large uniform distortions, to acquire high *strains* ($\delta > 80$–100%), and to withstand them without breaking up. This usually occurs at high temperatures. Cases are known when δ may reach 2000% and more. In contrast to *superconductivity* and *superfluidity*, superplasticity is not a state of matter resulting from a *phase transition*. In the 1930s, G. Pearson observed superplastic behavior in such *eutectics* as Pb–Sn and Bi–Sn. During the 1940s–1950s A.A. Bochvar and A.A. Presnyakov conducted the first systematic studies of superplasticity and coined the term. One may differentiate between the *superplasticity of a phase transition* and *structural superplasticity*. The former is observed during the cyclic thermal processing of a solid in the vicinity of its phase transition temperature. Then, the strain $\varepsilon \propto \sigma$, where σ is the applied external *stress*. Of particular importance is the structural superplasticity that appears when the conditions are absent for either *brittle failure* or *tough failure*. That failure results from quick formation and growth of a *neck* during dilation of the sample. Structural superplasticity manifests itself under the following conditions: (1) temperature $T > 0.4 T_{\text{melting}}$, where T_{melting} is the *melting temperature*; (2) the rate of deformation $\dot{\varepsilon} \approx \varepsilon \cdot 10^{-3}$ s^{-1}; (3) the crystalline grain size (see *Polycrystal*) is $d \approx 1$ μm. Moreover, the grains should not grow during the process of deformation. Certain eutectic alloys formed by binary or more complex systems in which the structural superplasticity is most clearly observed meet the latter condition easily: Al–33%Cu; Ag–28%Cu; Co–40%Al, etc. (mass percentage is indicated here). The principal input to *plastic deformation* in the case of superplasticity is provided by slip that develops along *grain boundaries*. Grains do not elongate, but maintain their equiaxial nature. No dislocation substructure appears in them, since lattice *dislocations*, which develop during the course of internal accommodation deformations in the grains, then pass through grain boundaries. These boundaries are in a specific excited state so that dislocations are absorbed by them, and no *texture* forms in the process. The number of grains in the sample cross-section diminishes during its dilation, which means that deformation is not of an affine nature. Modern theories of superplasticity yield certain expressions to describe the rate of deformation during structural superplasticity. This rate is more or less close to the following in its form:

$$\dot{\varepsilon} \sim \sigma^{1/m} d^{-A} \exp\left(-\frac{Q}{k_B T}\right).$$

Here m is the flow tension versus deformation rate sensitivity index. In the case of structural superplasticity, we have $0.5 < m < 1$ and $A = 2$–3; Q is the *activation energy*, its value falling between the activation energy for volume *diffusion* and that for diffusion along the grain boundaries. Superplasticity finds its technical application in low load pressure processing of pieces with an intricate shape.

SUPERSTRONG MAGNETIC FIELDS

Fields beyond $B \geqslant 50$–100 T in strength (the lower value is the maximum constant field attainable without using pulsed technology; the upper value is such that the shortest burst of such a field results in the destruction of metal solenoids). Superstrong magnetic fields are used in the physics of solids to study *galvanomagnetic effects*, *thermomagnetic phenomena*, *magnetooptical effects*, *magnetic phase transitions*. Absorption spectra and *cyclotron resonance* in solids, and the *Faraday effect* in the visible and the infrared, Zeeman energy level splittings, the *magnetoresistance* of

thin bismuth wires, etc., were all studied in magnetic fields up to 200 T.

Superstrong pulsed magnetic fields provide a source of quasi-hydrostatic pressures up to 10^{11} Pa and high energy densities. For example, the energy density of a magnetic field of 500–1000 T exceeds the binding energy of most solids, and the *magnetic pressure* reaches values found in the center of the Earth. Pulsed fields in the range of 50–80 T are used for pressure processing of metals, e.g., for magnetic pulse *welding* of metals.

Achieving superstrong magnetic fields is related to the problem of material *strength*. The magnetic pressure of a 50 T field ($P_H \approx B^2/(2\mu_0)$) reaches 10^9 Pa, which exceeds the strength characteristics of most metals. A high energy density is released in the surface layer of the solenoid material, and the resulting enormous magnetic pressures can lead to the destruction of that solenoid. Therefore, choosing adequate materials and judicious solenoid designs becomes one of the key problems in obtaining superstrong magnetic fields. Another problem lies with developing high-power critical current sources. Cooling the solid with cryogenic liquids (liquid nitrogen, neon, hydrogen) is a promising approach. Lowering its electric resistance suppresses heat release and also increases metallic strength. Combined magnetic systems that include cryogenic and *superconducting magnets* makes it possible to obtain stationary magnetic fields up to 50 T. Fields stronger than this are only obtained using pulse techniques, and the stronger the field, the shorter its duration. The *skin-effect*, however, hampers achieving superstrong magnetic fields for particularly short pulses: the current density increases and the heat transfer worsens. To reach fields up to 80 T use is often made of multitwisted single layer solenoids built from strong materials of high electrical conductivity, e.g., beryllium, chrome bronze, or *composite materials*. Superstrong magnetic fields also may be obtained by compressing the magnetic flux by high explosives. Such devices are called *explosion magnetic generators* and *magnetic cumulative generators*. The magnetic flux may also be compressed by electrodynamic forces, using the pressure of an external magnetic field, and fields up to 280–310 T may thereby be reached. Superstrong magnetic fields are measured by calibrated induction sensors (magnetic probes), and also by monitoring the Faraday or *Zeeman effect*.

SUPERSTRUCTURE, SEMICONDUCTOR

See *Semiconductor superstructure*.

SUPERSTRUCTURE VECTOR

A vector other than zero in the first *Brillouin zone* of an initial (more symmetric) *crystal structure*, such that its transformation properties coincide with the symmetry properties of the corresponding superstructure (*superlattice*). The superstructure vectors corresponding to the symmetry points of the first Brillouin zone characterize the macroscopically uniform ordered structures that may be stable with respect to the formation of antiphase domains. The superstructure vector is identified experimentally from the diffraction pattern, and the positions of the so-called *superstructure Bragg reflections*. An irreducible representation of the initial crystallographic *space group* that characterizes the highly symmetric phase corresponds to the superstructure vector. Using superstructure vectors makes it possible to reconstruct the spatial distribution of ordered elements (atoms, spins) at the sites (or interstitial locations) of the initial lattice.

SURFACE ACOUSTIC WAVES

Propagating elastic *strain* of a solid, which has an energy that is localized in the neighborhood of the surface. Surface acoustic waves are subject to the same wave equations as *bulk acoustic waves*, and are also governed by the boundary conditions of the surface. An example of these waves is provided by Rayleigh waves (J. Rayleigh, 1885), which occur at a plane boundary that separates a solid from vacuum, and whose displacement vector lies in the plane formed by the normal to the surface and the direction of propagation of the waves. The depth of penetration of a Rayleigh surface acoustic wave into the *substrate* is of the order of its wavelength, and the *phase velocity* is lower than that of its bulk counterpart ($v_{\text{Rayleigh}} \approx 0.9 v_{\perp}$), which ensures the stability of surface acoustic waves. There are no other surface acoustic waves at the boundary described above, because a wave that is polarized in the plane of the surface and satisfies the boundary condition, is a bulk acoustic wave.

Surface acoustic waves of the above polarization may exist at the surface, for different boundary conditions. Consider the surface loaded with a solid layer, in which the speed of a transverse surface acoustic wave is less than the speed of this mode in the substrate; this surface carries *Love waves* (A. Love). An electroacoustic *Gulyaev–Bleustein wave* (Yu.V. Gulyaev, 1969; J.L. Bleustein, 1968) travels along the surface of *piezoelectric materials*. Besides the components of transverse displacements, the latter wave has components of an electric field, which accompanies the deformations of the medium because of the piezoelectric effect (see *Piezoelectricity*). Besides purely transverse surface acoustic waves, which were discussed in the above examples, the surface may also carry surface acoustic waves of the Rayleigh kind. In anisotropic media, this division of the wave into two groups (purely transverse modes and Rayleigh ones) is possible only in high symmetry directions of propagation. Another specific feature of anisotropic media is that under certain conditions there occur unstable surface modes (evanescent surface acoustic waves), which decay as they propagate because of radiation of bulk acoustic waves ($v_{\text{surface}} > v_{\text{bulk}}$).

All above-mentioned modes of surface acoustic waves occur at the boundary between a solid and a liquid or gaseous medium with the difference that in this case the surface acoustic waves are subject to attenuation, which is caused by sound emission into the outside medium. Both radiative unstable modes of surface acoustic waves and nonradiative ones (*Stoneley waves*, R. Stoneley, 1924) may exist at the boundary between two solids. The Stoneley waves are, in fact, two coupled Rayleigh waves, which travel on each side of the boundary. Propagation of coupled surface acoustic waves at the boundary between two piezoelectric materials is possible even if there is no mechanical contact between the media; in this case, the coupled waves propagate because of the relation between electric components of the waves (*gap waves*).

The nature of the propagation of acoustic waves in nonuniform media is determined by the ratio between the wavelength and the sizes of inhomogeneities. In layered media, long-wavelength surface acoustic waves exhibit the same modes, as in homogeneous media, yet new modes arise with an increase of frequency. These new modes are related to the propagation of elastic deformations inside the layers. The simplest example is provided by *Lamb waves* (H. Lamb, 1917): the acoustic modes of a solid *plate* (the single layer), which exhibit a pronounced wave-guiding character (see *Acoustic waveguide*) at low frequencies; in the short-wavelength limit the Lamb waves become separated into a pair of weakly bound Rayleigh waves, which travel along both sides of the plate. The nonuniformity of the surface of the medium, which is due to, e.g., grooves and stripes, is responsible for the generation of a retarding layer on the surface in the long-wave limit. In the case of high-frequency oscillations, these structures may play the role of acoustic waveguides, acoustic baffles, and dispersion elements. Surface acoustic waves have found the widest application in *acousto-electronics*, where they are used for radio signal processing.

SURFACE-ACTIVE AGENTS, surfactants

Substances which exhibit a pronounced capability of getting adsorbed at *phase interfaces* and, by so doing, cause a considerable reduction of the *surface energy* (*surface tension*). In aqueous solutions, this ability is exhibited by various organic materials with a molecular structure that includes polar functional groups (OH^-, $COOH^-$, NH_2^-, SO_3^-, etc.) and long hydrocarbon chains (e.g., radicals of the formula RC_nH_{2n+1}, RC_6H_5, etc.). Organic substances, which form colloidal solutions in water (e.g., fatty and synthetic soaps), also exhibit high surface activity. When adsorbed on a surface, surface-active agents influence (depending on their structure) either hydrophilic or hydrophobic properties of the surface (see *Lyophilic and lyophobic behavior*). Thus, it is possible to gain control over the nature and the intensity of the effect of these agents upon the surface properties of *solids* by a suitable selection of surfactants. The clearly defined surface properties of the solutions of surface-active agents determine their uses: wetting agents (surfactants, which change the wettability of solid surfaces), agents for modification of stability of disperse systems: foams, emulsions, suspensions.

Surface-active agents have a wide range of application. Surfactants are used for flotation of ores,

in mining (added to the mud fluids at boring, they reduce the hardness of rocks), in metal-working industry (additives to lubricating oils), as modifiers of *crystallization*; are also used for modification of the conditions of deposition and properties of electroplated coatings.

SURFACE ACTIVITY

The ability of a material to reduce the free energy of an interface as a result of *adsorption* on it. In the case of surface-inactive materials the adsorption is either absent or negative (i.e. the *surface layer* contains less dissolved substance than the bulk of the solution). Surface activity depends on the molecular properties of the medium that provides the adsorbate, and on the chemical composition of the surface-active agents. Surface activity often increases with decreased solubility of the substance, since such a substance is more easily adsorbed from the solvent, in which it is less soluble. Surface activity is high, when the interaction between adsorbate and *solid surface* involves *chemisorption*, and the adsorbed substance is rigidly attached to the surface by strong chemical bonds.

SURFACE ALLOYING

See *Surface doping*.

SURFACE ATOMIC STRUCTURE

The order of arrangement of atoms in the outermost atomic planes of a *crystal*. An equilibrium surface is one having minimal values of *surface energy* or *surface tension*. The surface free energy is a function of surface orientation. The plot of surface energy versus orientation $\gamma(\theta)$ (*Wulf diagram*, see Fig. 1) at 0 K has an infinite number of minima in all the directions that are describable in terms of rational Miller indices (see *Crystallographic indices*). The less prominent minima become smeared with increased temperature, and a finite number of acute minima remains at a sufficiently high temperature. The tendency of the system to form a minimal energy surface may result in the decomposition of a planar surface into several faces of various orientations, whereby the increase of surface area is more than compensated for by the decrease in the specific free surface energy of the resulting faces.

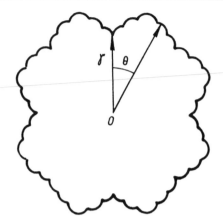

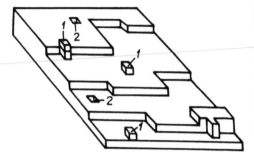

Fig. 2. Surface of a real crystal showing steps, adatoms (1), and vacancies (2).

Fig. 1. Wulf diagram; polar plot of surface energy versus orientation $\gamma(\theta)$ showing minima and maxima in varous directions.

Using the Wulf diagram, the surfaces may be divided into three types:

(1) *Singular surfaces*, to which the sharp minima on the $\gamma(\theta)$ plot correspond. These are low-index surfaces, which are bounded by atomic planes of indices equal to those of the surface.
(2) *Vicinal surfaces*, with orientations in direct proximity to the singular ones. Such surfaces often consist of the low-index facet terraces, connected by atomic steps with heights of one or several interplanar distances.
(3) *Nonsingular surfaces* comprise all the remainder; and sometimes they are called *atomically-rough surfaces* or *diffuse surfaces*. Usually they are combinations of planes with other indices, or combinations of the *facets*.

Atoms at the surface of the crystal have a smaller number of nearest neighbors than those within the volume, so the forces which affect them differ from the volume ones. This may lead to the reconstruction of the outermost atomic surfaces of the crystal. Such a *surface reconstruction* takes place mainly at the surfaces of covalent semiconductors, but they can also be encountered at planar surfaces of metals, e.g., at (100) surfaces of gold, platinum, iridium. As a result, *two-dimensional surface structures* form, including several upper layers, with periods which, as a rule, differ from the corresponding periods within the crystal volume. The *surface rearrangement* may occur either via the displacements of atoms from their normal positions in the lattice, or through the formation of orderly arranged *surface defects* (*vacancies*, adsorbed atoms). The surface reconstruction may be caused by impurity atoms dissolved in the crystal or adsorbed at the surface. Under the effect of an impurity not only can new two-dimensional surface structures appear, but in addition the type of surface may change considerably: a flat surface may be reconstructed into a step-like or faceted one, or vice versa.

When the system deviates from equilibrium during the course of growth, with increasing temperature, there is an increase of the surface defect concentration, and a number of fractures develop at the steps. A smooth close-packed face may transform to a rough one (see Fig. 2). The *surface roughness* of such a type is called *kinetic roughness*.

When the temperature is varied there may occur a bulk *order–disorder phase transition* or an order–order phase transition at the surface, when no bulk phase transition occurs within the volume of the crystal. These transitions may be either reversible or irreversible.

The main methods of surface structure investigation are *low-energy electron diffraction*, high-energy electron reflection-diffraction (see *Electron diffraction*), reflection and transmission diffraction *electron microscopy, scanning electron microscopy, X-ray diffraction*. The method of scanning tunnel microscopy has great prospects for the investigation of surface structure. See also *Solid surface*.

Structure of mechanically treated surface. Mechanical processing (cutting, turning, grinding, and polishing) of crystalline materials induces a significant distortion of the *crystal structure* of a treated surface. It is convenient to classify the distortions in relation to the *failure* mechanism of crystals: (1) *brittle failure*; (2) *quasi-brittle failure* in the presence of macroscopic plasticity $\delta = 0$; (3) quasi-brittle failure for $\delta > 0$; (4) *tough failure*.

In case (1) (diamond, SiC) the treated surface does not contain any traces of *plastic deformation*, and the removal of material during the processing takes place by means of the confluence of circular Hertz cracks. In case (2) (carbides, nitrides, oxides, semiconductor covalent crystals, and so on), the removal of material also occurs as a result of the confluence of cracks of various kinds, however, in this case a defect layer with a pronounced heterogeneous *strain* remains near the surface, with a small-grain polycrystalline or amorphous layer in close proximity to the surface (see *Amorphous state*). In the polycrystalline layer a *phase transition* can take place (e.g., in CdS). Below this is situated the host *crystal* with an increased *dislocation* density. The density of dislocations decreases as the distance from the surface increases. The untouched structure is at depths from a few μm to a few tens of μm. In case (3) (BCC metals at low temperatures including room temperature), the quasi-brittle failure facilitates the separation of shavings during processing in comparison with case (4) (metals with a close-packed lattice and BCC metals at high temperatures). In both cases, a disturbed subsurface layer is much thicker than in case (2), and can reach 1–2 μm. At the depth of a few μm below the surface, a highly fragmentized, crystallographically textured structure can form. As a result of mechanical polishing using the finest abrasive with a particle size of 1 μm, an amorphous layer can appear. The intensive absorption of embedding impurities during the treatment promotes the stabilization of the amorphous state. In steels, in the region of a treated surface, a series of phase transitions takes place in addition to plastic deformation.

SURFACE ATOM VIBRATIONS

See *Vibrations of surface atoms.*

SURFACE BAND STRUCTURE

The structure of *energy bands* of a semi-infinite *crystal*, which is bounded by a plane, under conditions of conservation of translational symmetry in directions parallel to the surface. Energy states are numbered with respect to components of the surface *quasi-momentum* parallel to the surface $k_\parallel$. The values of $k_\parallel$ are confined within the limits of a two-dimensional *Brillouin zone*, which is the projection of the Brillouin zone on the selected surface. The three-dimensional *band structure*, which is projected onto the two-dimensional Brillouin zone, retains the absolute gaps (energy regions, which have no eigenstates of the infinite crystal). Relative energy gaps, which exhibit discontinuities of the energy spectrum in a limited region of the three-dimensional Brillouin zone, may vanish or remain in the surface band structure, depending on the orientation of the projection plane. Besides volume band states, which are projected onto the surface, the surface band structure includes energy branches that are related to *surface electron states*, the wave functions of which are constructed using solutions that correspond to absolute and relative energy gaps.

Information on the details of the surface band structure is necessary for a quantitative description of the processes of *thermionic emission*, catalysis, *adsorption*, *adhesion*, etc. This information may be obtained through various methods of *electron spectroscopy* and *photoelectron spectroscopy*.

SURFACE BARRIER

Potential barrier, which arises in the neighborhood of a *solid surface* or near a solid–solid interface, solid–liquid interface, or solid–gas interface (see *Phase interface*). It prevents charge carriers from moving in directions which are perpendicular to the interface. An example is a Schottky barrier which is a junction between a layer of semiconductor material and a layer of metal characterized by hot current carriers. A surface barrier exerts a strong influence on current flow in semiconductors and *semiconductor devices*.

SURFACE BARRIER STRUCTURES

Multilayer structures, which consist of electrically conducting layers and thin dielectric (insulating) and semiconducting layers. The simplest

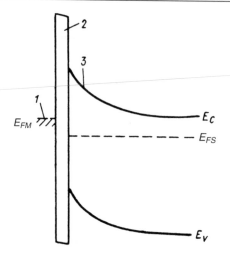

Fig. 1. Energy diagram of a surface barrier structure: 1, metal; 2, insulator; 3, semiconductor; E_{FM} and E_{FS} indicate the positions of the Fermi levels of the metal and the semiconductor, respectively, and E_C and E_V are the locations of the bottom of the conduction band and the top of the valence band, respectively.

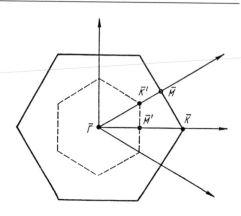

Fig. 2. The surface first Brillouin zones for two surface superstructures: Si (111)–(1 × 1) (solid line) and Si (111)–($\sqrt{3}$ × $\sqrt{3}$) (dashed line). High-symmetry points of these surface first Brillouin zones and some directions in the reciprocal three-dimensional space are also shown.

structure of this kind is the *metal–semiconductor junction*. Surface barrier structures exhibit rectifying properties. The specific feature of all surface barrier structures is that practically the entire applied voltage drop takes place in the semiconductor (see Fig. 1).

Besides that, the widths of contact conducting layers and dielectric layers of these structures are, as a rule, rather thin, so the light absorption can be neglected in these layers. Surface barrier structures find wide application in semiconductor electronics and photoelectronics.

SURFACE BRILLOUIN ZONE

This is the region of a two-dimensional reciprocal space (see *Reciprocal lattice*) with a center at the origin $\overline{\Gamma}$ (see Fig. 2) corresponding to the *Wigner–Seitz cell* of the direct lattice which involves a parallelogram system of equivalent points (a two-dimensional analog of the *Bravais lattice*) of the *surface superlattice* under consideration. The surface first Brillouin zone (see *Brillouin zone*) is a domain adequate for determining the *dispersion law* (periodic function of a two-dimensional wave vector) of a *surface quasi-particle*.

SURFACE CAPACITANCE

The *capacitance* of a condenser, which consists of a conducting electrode, and the near-surface *space charge region* of a semiconductor. High-frequency surface capacitance and low-frequency surface capacitance are distinguished. The surface capacitance is said to be low-frequency, when the modulation frequency ω of the applied low-amplitude voltage is considerably less than reversal periods of nonequilibrium charge carrier generation–recombination. Low-frequency surface capacitance consists of capacitances of *surface electron states*, which are connected in parallel, and the near-surface space charge region of a *semiconductor*.

High-frequency surface capacitance is realized when the magnitude of ω considerably exceeds the reversal periods of generation–recombination of nonequilibrium charge carriers. High-frequency surface capacitance is determined by the capacitance of the near-surface space charge region of a semiconductor.

Surface capacitances find wide application as adjustable capacitors (variable condensers, *varicaps*), in semiconductor electronics, and in *microelectronics*. The *surface capacitance method* is widely used in studies of generation–recombination characteristics of semiconductors, and for determination of the parameters of an insulator–semiconductor boundary.

SURFACE CHANNELING

Surface *channeling* is observed when high-velocity *ions* are scattered at glancing angles from a *monocrystal* surface. The motion of particles along potential valleys in the channeling mode is determined by the continuous plane potential (string potential), since the channeled particle undergoes correlated interactions with a large number of atoms along the plane or axis. When an ion moves at a glancing angle to the single crystal surface the scattering is due to the continuous potential valleys of atomic axes located along the trajectories of the ions in the neighborhood of the surface. The incident ions are scattered mostly by the potential valleys formed by atomic axes in the first (I) and second (II) layers at the monocrystal surface (see Figs. 1 and 2).

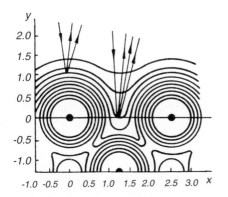

Fig. 1. Curves of equipotential surfaces (110) for the Ni $\langle 00\bar{1} \rangle$ direction.

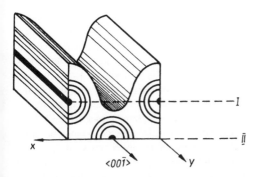

Fig. 2. Schematic of surface valley along the $\langle 00\bar{1} \rangle$ direction, showing equipotential surfaces in a continuous potential model.

An ion, which possesses transverse energy $E_1 = \theta^2 E$ (where E is total ion energy, θ is the angle of incidence on the crystal), is scattered by an equipotential surface that is formed by the continuous potentials of the first two layers of the monocrystal surface atoms (Figs. 1 and 2). By analogy with bulk channeling, the *critical angle of total external reflection* is introduced. At surface channeling the ion does not undergo collisions with small impact parameters, so little energy is lost by reflection from the surface. It has been proposed to use surface channeling for the purpose of changing the direction of propagation of a beam of high (or average) energy particles. The phenomenon discussed above is not unique to particular kinds of particles, and therefore can also take place at the reflection of light, electrons or *positrons*.

SURFACE, CLEAN

See *Atomically clean surface.*

SURFACE CONDUCTIVE CHANNEL

Electrically charged near-surface layer of increased electrical conductivity. It arises either because of the *injection* of mobile charge carriers from a material in contact with the semiconductor, or under action of the surface electric field perpendicular to the surface. Concentration and type of charge carriers in the surface conductive channel may vary according to the direction and value of the applied electric field. If the concentration of carriers in this channel is low, then processes of charge transport are influenced by inhomogeneity of the solid surface (see *Surface inhomogeneities*). As the density of free charge carriers increases, the thickness of surface conductive channel decreases, and *size effects* begin to influence the charge transport. Then when the channel thickness approaches the mean free path of electrons (holes), *surface scattering* begins to play a significant role. Besides bulk mechanisms of scattering, those typical of the surface alone take place: scattering by surface irregularities, by charged centers which are localized at *surface states* and in neighboring layers, by surface phonons (*surphons*). The effect of these mechanisms on charge transport increases with the decrease of the ratio between the conductor thickness and the mean free path of the charge carriers. On further decrease of the channel thickness, when it becomes comparable to the free charge carrier de Broglie wave length, then the charge carrier energy is quantized. Specific quantum effects manifest themselves in electron transport: change of effective mass of free charge carriers, initiation of anisotropy, oscillations of conductivity in a magnetic field, effects of carrier localization, *negative magnetoresistance*, onset of high mobilities. The ratio between characteristic lengths, which brings about quantization, is also enough to influence surface scattering; thus both size effects take place: processes of surface scattering proceed at quantization of the energy of free charge carriers. In such a situation, physical regularities of *transport phenomena* will differ from those observed for an unquantized electron gas. Quantization leads to increased efficiency of scattering by acoustic *phonons*. It changes the mobility $\mu(T)$ temperature dependence at scattering by Coulomb centers. In particular, the power n in the dependence $\mu \propto T^n$ changes from $3/2$ to 1. In the quantum limit, the mobility of carriers in a two-dimensional gas (see *Two-dimensional electron gas*) may tend with decreased temperature to a certain limiting minimum value, which does not depend on temperature of the sample or on the concentration of scattering centers.

SURFACE DECORATION

Method of detection of *point defects*, *dislocations*, steps of growth and other *defects* of the crystalline lattice through deposition of material at the surface of a crystal from gaseous or liquid phases, or by chemical methods. Material can be deposited in the form of particles at the positions of defects, and thus mark (decorate) them, and the decorated crystals are then examined by optical or *electron microscopy*. Surface decoration is used to study *crystallization* processes, the surface structure of *crystals*, *epitaxy*, and chemical reactions that take place at the *solid surface*. Unlike the methods of *etching* used for investigating defect structures and detecting spots where dislocations reach the surface of the crystal, the method of surface decoration can also provide information on the geometry of defects within the crystal.

Pictures of decorated thin voids, parallel to the (110) faces in ZnS crystals have been observed upon the application to the surface of Zn or Ag by the method of electrophoresis with successive annealing at 500 to 800 °C. Dislocations and twinning planes in ZnTe have been decorated by depositing Te via heating the crystals up to 1100 °C with the successive hardening. In CdS and CdSe crystals disc-like deposits formed by Cd, have been found, and also needle-like deposits of Cu and Ag sulfides. Vortices in superconductors and domains in magnets are observed by decorating with magnetic particles like iron powder.

SURFACE DEFECTS

Deviations from an ideal periodic arrangement of atoms on the surface of a material. Surface defects may be subdivided into point (zero dimensional), one-dimensional, and two-dimensional ones. *Point defects* include: *vacancies*, atoms shifted from their normal locations within the lattice, *impurity atoms* which substitute for atoms of the crystal, and *adatoms* which are inherent or impurity atoms located on the outer atomic plane. Surface defects may be arrayed in an ordered sequence, thus forming a surface structure. The periods of unit cells of such structures often differ from the corresponding periods in parallel crystallographic planes within the volume (see *Atomically clean surface*). A region of surface with the ideal periodicity of this surface structure may be considered as a *domain*. Domains shifted with respect to one another via a vector which is not a multiple of a surface structure vector are called *out-of-phase domains*. If the surface symmetry is lower than that of the unreconstructed crystal planes lying below, it is possible to form domains with surface structures which differ in orientation of their unit cells. For instance, on the surface with a square lattice, rectangular cells can be formed which are rotated by 90° with respect to one another, and on the surface of a hexagonal lattice there may form unit cells of three orientations in the form of parallelograms turned by ±120° relative to each other. *Domain boundaries* are associated with one-dimensional surface defects. Another type of one-dimensional surface defect is an *atomic step*. Examples of two-dimensional defects are *facets* or microfaces located obliquely to the surface plane. To a large extent surface defects influence the electronic properties of surfaces, as well as the processes of *adsorption*, catalysis, *diffusion*, the initial stages of growth, etc.

SURFACE DEGRADATION

Irreversible changes of the physical, chemical or other parameters of a surface towards their deterioration due to variations of structure, composition, geometrical relief, etc. The degradation of the surfaces of metallic, ceramic and other *construction materials* is usually associated with mechanical *wear*, *corrosion*, the appearance of microcracks upon exposure to intense irradiation, etc. For *semiconductor materials* surface degradation generally involves the degeneration of its electrophysical parameters such as the growth of *surface electron states* and *surface recombination* centers, the increase of *surface scattering* of current carriers which occurs due to the appearance of defects (see *Surface defects*), the *diffusion* of impurities, and uncontrollable surface oxidation. This degradation produces an irreversible deterioration of the properties of *semiconductor devices* such as the appearance of soft breakdown in ultra-high frequency diodes, of parasite channels in large-scale integrated circuits, the reduction in sensitivity of gas sensors, etc. Surface degradation can be costly since it reduces the reliability and stability of device operation; and to prevent it various types of protection can be used (*surface passivation*, conservation).

SURFACE DIFFUSION

Process of transport of surface atoms along a *solid surface*, through random displacements of atoms from one metastable position to another, which correspond to minima of the lattice potential. The *activation energy* of surface diffusion is sharply dependent on the atomic structure of the surface, and therefore has different values for different faces of the crystal. The activation energy of surface diffusion is usually several times smaller than that of bulk *diffusion*. Interaction of surface particles (which may belong to the solid under consideration or be foreign adsorbed particles) is responsible for the dependence of the surface diffusion coefficient on the surface concentration, and the correlation between the surface diffusion coefficient and changes in the surface layer structure. When passing from coatings, which are less than one molecular layer thick ($\theta < 1$, where θ is the relative surface concentration) to coatings of $\theta \approx 1$, the mechanism of surface diffusion changes. This change is related to the motion of adsorbed atoms of the upper layer along the *substrate* coated with a monolayer adsorbed film (so-called mechanism of the unrolling carpet). On further increase of θ, the generation and growth of small areas of adsorbate are possible, with a consequent *coalescence* and coagulation. The layers with closely packed particles exhibit group surface diffusion (coordinated transport of groups of atoms). Surface diffusion plays an important role in the processes of growth of crystals, catalysis, sintering, wetting, etc.

SURFACE DISLOCATIONS (introduced by G.I. Barenblatt, 1959)

Imaginary (fictitious) *dislocations*, which are spread over the free surface of a body, over the boundary between crystalline phases or over the surface, which separates mutually disoriented crystal blocks. The concept of "surface dislocations" was first used to describe the boundary conditions at the free surface and the surface of *pores* in crystals without *internal stresses*. It has also been used for the description of *grain boundaries* and coherent integration of crystals with different lattices, when the mismatch causes the generation of internal mechanical stresses. Unlike real (lattice) dislocations, surface dislocations are infinitesimal, i.e. the *Burgers vector* is very small. The introduction of surface dislocations is useful in solving problems of *elasticity theory* in connection with the development of numerical methods of energy minimization. The term "surface dislocations" is sometimes used in reference to dislocations that are located in the neighborhood of the free surface of a crystal whose properties are strongly influenced by *image forces*.

SURFACE DOPING, surface alloying

Modification of the surface or near-surface region of a solid, which results in changes of physical, chemical and mechanical properties of the treated regions. The term "*alloying*" is customarily used in reference to metallurgical processes, which consist in the modification of metallic compounds, whereas "*doping*" is the name given to the practice of modifying electrical and magnetic properties of a solid by incorporation of impurities in the host lattice; the lines of demarcation are not rigidly drawn between the two. We use here the term "surface doping" to include "surface alloying". There are various methods of surface doping of a solid: (1) adsorption of *impurity atoms* (*ions*) and molecules from gaseous and liquid media; (2) *adsorption* of atoms (molecules) from a molecular beam in vacuo; (3) *ion implantation*; (4) laser implantation; (5) electrochemical implantation; (6) *diffusion* of impurities, which reach the surface or the *phase interface* through the bulk of the material. Different methods of surface doping are responsible for modification of different properties of solids; on the whole, it is related to the condition of the impurities at the surface and in the near-surface region, namely: charge, coordination, energy state, degree of cooperation with other impurities and *defects*, etc. Surface doping plays an important role in semiconductor electronics because of microminiaturization of *semiconductor devices*, and application of thin layers and films in *integrated circuits*.

SURFACE DRIFT WAVES

Perturbations of an electromagnetic field which arise at the boundary between a current-carrying semiconductor (plasma) and a medium of either positive or negative relative high-frequency *dielectric constant*. They are borne along the surface at the speed of the drift of the charged particles, much slower than the speed of propagation of light in this medium. Therefore, the perturbation of the charge density takes place directly at the surface, and the amplitude of the electric field falls off exponentially on either side of the boundary. In the case of surface drift waves, the frequency of oscillation is related to the wave vector by a *dispersion law*, and the maximal component of the electric field is that directed along the motion of the particles. Since the motion of charges takes place under the action of external sources of electric energy, the rate of the drift may exceed the *phase velocity* of intrinsic excitations of the medium. In this case, the energy of surface drift waves is transformed into natural oscillations. Surface drift waves are widely used for the generation and amplification of electromagnetic and acoustic signals in solid state electronics.

SURFACE DYNAMICS

Branch of *crystal lattice dynamics* which considers the influence of the *solid surface* on the vibrations of the atoms of the bulk *crystal*. This influence is associated with the appearance of the free boundary and the variation of the *force constants* of the *vibrations of surface atoms*, with the appearance of *surface waves*. The force constants which determine the amplitudes of vibrations near the surface differ from those within the volume. They change with increased concentrations of *defects* near the surface. For a free surface with a distinctly expressed *crystal structure* the experiments of *low-energy electron diffraction* (LEED) and atomic beam scattering (e.g., helium atoms) are the main sources of information about the surface dynamics. Information about root-mean-square displacements of surface atoms, and the energy spectrum of surface vibrations, is extracted from these experiments. The root-mean-square displacements of surface atoms exhibit a noticeable anisotropy, and usually exceed the displacements in the bulk crystal.

SURFACE ELECTRICAL CONDUCTIVITY

1. The difference between the magnitudes of the total *electrical conductivity* of a sample recorded under different conditions: (1) when the surface electrostatic potential φ_s has a certain value, and (2) when it equals zero. The magnitude of surface conductivity is determined by the mobility of the charge carriers in the *space charge region*, and by the excess (or deficiency) of electron (ΔN) and hole (ΔP) charge carriers in the near-surface region:

$$\Delta N = \int_0^\infty \left[n(\varphi(z)) - n_0 \right] dz,$$

$$\Delta P = \int_0^\infty \left[p(\varphi(z)) - p_0 \right] dz,$$

where z is the coordinate, which is normal to the surface of the sample, $\varphi(z)$ is the potential in the space charge region, n_0 and p_0 are spatial concentrations, respectively, of electron and hole carriers at $\varphi_s = 0$. Under this definition, the coefficient of electrical surface conductivity depends on both the mobility of the charge carriers and their densities, as in the usual bulk conductivity. The dimensionality of specific surface conductivity is identical to that of total electrical conductivity. In order to draw a distinction between these quantities, the designation "siemens per square" is introduced (siemens (S) is the unit of conductivity, equal to Ω^{-1}, the reciprocal of resistance). The existence of ΔN and ΔP in the space charge region may be related to the presence of electric charge in *surface electron states*; the charge, which is induced in the surface charge region, is opposite in sign to the charge of the space charge region. The excess near-surface charge may also arise from an external electric field at the semiconductor surface (see *Field effect*). The influence of the field effect changes the space conductivity, which is used in engineering (*field-effect transistor*) and in investigations of surface properties of semiconductors (density and energy levels of surface electron states, mobilities of carriers in space charge region, etc.). Relative to bulk conductivity, surface conductivity increases when the surface charge region is enriched with majority carriers (*accumulation layer*) and minority carriers (*inversion layer*), and it decreases when the surface charge region is depleted of majority carriers (*depletion layer*).

2. Less often the term "surface conductivity" refers to the extrinsic conductivity along the solid surface, which is related to the existence of oxide layers (see *Oxidation*), and layers of adsorbed molecules (see *Adsorption*) on it. Another distinct kind of surface conductivity is *surface band conductivity*, which is due to motion of charge carriers in the energy band of surface electron states.

SURFACE ELECTRON STATES

States of electrons with wave functions localized in the neighborhood of an interface; surface electron states are specific for solid-state structures with phase boundaries that are sharp on the scale of interatomic distances. These states are responsible for the formation of the electric charge, which is concentrated within several planes of atoms; hence these surface electron states are of major importance for a whole class of *surface phenomena*.

The type of surface electron state is determined by the structure of the crystal matrix in the near-surface region. *Intrinsic surface electron states*

are formed if there exists translational symmetry along the interface. The most extensively studied intrinsic surface electron states are those at a free *atomically clean surface* of the solid in a vacuum (*Tamm levels, Shockley levels*). These states of the electrons are found at the surfaces of semiconductors, transition metals, noble metals and common (base) metals. Characteristic of surface states of this type is the presence of several allowed energy bands, both filled with electrons and empty. Intrinsic surface electron states include also resonance and sagging surface states in contact structures (see *Metal–semiconductor junction, Heterojunction*), where isoenergetic electron transitions from these states to allowed energy bands of media in contact are possible (through quantum tunneling effects). The nature of the energy spectrum of electrons in multilayer structures with thin crystalline layers (*superlattices*) also depends on quantum tunneling effects. When the layers become several interatomic distances thick (*quantum wells*), there is no more difference between surface and volume electron states. The electronic properties of such systems are determined by states of interfaces.

What all extrinsic surface electron states, which are associated with the presence of *surface defects*, have in common is a discrete energy spectrum due to the localization of the electron wave function in three dimensions. The degree of localization depends on the disposition of the defect with respect to the surface: the degree of localization is a maximum for a foreign atom, which is adsorbed on a free crystal surface, whereas a shallow *surface impurity state* exhibits a minimum degree of localization. Changes in the electronic structure of the surface at adsorption may be quite considerable (*chemisorption*). In this case, besides a change in the adatom energy spectrum, there occur both quantitative and qualitative transformations of the spectrum of intrinsic (Tamm) surface electron states, even for light coverages by the adsorbate. New ("induced") surface electron states may arise simultaneously with this process. The energy spectrum of these new states is specific to selected boundaries of crystal and adsorbate. This effect manifests itself in, e.g., a sharp difference between catalytic activities of different faces of the same crystal. Effects of rearrangement in the surface electron spectrum manifest themselves also at inner interfaces of contact structures. The following situation typically takes place: intrinsic surface electron states are displaced from the energy region close to the *Fermi level*, and electronic charge, which determines the height of potential barriers of the structure, becomes localized at impurity surface states. For high concentrations of impurity centers at the surface (above 10^{16} m^{-2}) the overlap of wave functions of electrons, which occupy impurity surface states, results in the generation of two-dimensional surface bands. *Magnetic surface levels* are a special case of surface electron states. An external magnetic field and an interface that is sharp and smooth on an optical scale are needed for their existence.

SURFACE ENERGY

The excess energy of a surface layer at a *phase interface*, compared to the specific energy inside the phase. The generation of this additional energy is related to the difference in intermolecular interaction in the bulk of the phase and at its boundary. An increase of surface area is accomplished through the transport of internal atoms to the surface layer. Work therefore must be performed to cause the surface to expand against the uncompensated forces of intermolecular interaction in the surface layer (cohesive forces). This work is equal to the free surface energy. It is experimentally determined from the change of the area of the boundary surface at the application of external forces $F = \int \sigma \, dS$, where σ is the specific surface energy, which is equal to *surface tension* in the case of liquid phases. Surface tension tends to reduce the surface energy to a minimum for a given volume. Therefore, a liquid of high surface energy assumes a spherical shape. The magnitude of surface energy depends on the temperature, as well as on the nature of the contacting phases and impurity inclusions.

In the case of solids, an important role is played by the anisotropy of surface energy (see *Anisotropy of crystals*). The magnitude of surface energy is strongly influenced by *dislocations* and other structural defects. The decrease of surface energy due to defects, may considerably reduce the melting temperature, ultimate strength and plasticity limit of solids.

The total surface energy is given by

$$E_{\mathrm{surf}} = \int \left[\sigma - T \left(\frac{\partial \sigma}{\partial T} \right) \right] dS,$$

where the second term on the right-hand side is the *bound energy*, i.e. the latent heat of formation of a unit surface area during the course of an irreversible thermodynamic process at temperature T, and $\partial \sigma / \partial T$ is the specific surface *entropy*. The free surface energy decreases significantly with increasing temperature under the action of *surface-active agents* (surfactants), whereas the total surface energy is nearly temperature-independent. In the neighborhood of the point of a *solid* (liquid)–gas phase transition (critical temperature) the distinction between neighboring phases decreases, and at the critical temperature there is no difference between liquid and gas. Among liquids, the highest values of surface energy are exhibited by melts of metals: Hg 0.484 N/m (at 300 K), Pt 1.82 N/m (at 2300 K); for comparison, H_2O has the surface energy 0.10 N/m (at 300 K). Surface energy has a profound effect on processes of *epitaxy*, effects related to adsorption, catalysis, mechanical strength, electrical properties, and it is used for the determination of specific features of phase boundaries.

SURFACE ENERGY OF SUPERCONDUCTORS

Energy which is related to the state of the boundary between the normal phase (N) and the superconducting phase (S). It is defined by the equation $\sigma_{ns} = \int_{-\infty}^{\infty} (f - f_n) \, dx$, where f is the free energy per unit volume, and f_n is the value of f at a large distance from the interface, e.g., deep inside the N-phase. The surface energy has a positive sign (*type I superconductor*), if the *Ginzburg–Landau parameter* $\kappa < 1/\sqrt{2}$; it has a negative sign (*type II superconductor*), if $\kappa > 1/\sqrt{2}$. The essential distinction between superconductors of the first and second kind manifests itself in the presence of the magnetic field. Type I superconductors undergo a *phase transition* to the normal state, when the field reaches its critical value B_c (*thermodynamic critical magnetic field*). The surface energy hinders the generation of the first nuclei of the N-phase, thereby providing the opportunity of the metastable conservation of the S-phase in fields which somewhat exceed B_c.

In the case of type II semiconductors, even prior to the attainment of the value of B_c by the field, the generation of "rods" of N-phase corresponding to *Abrikosov vortices* turns out to be thermodynamically advantageous above the lower critical field $B_{c1} < B_c$: the increase of the bulk energy being compensated for by the negative surface energy of the vortex (see *Superconductivity*).

SURFACE EROSION

Gradual breakdown of the surface layers of materials in a flux of liquid or gas, under mechanical action, or under the action of electrical discharges. Surface erosion is a complex of physical and physicochemical processes emerging as a result of the influence of the environment, *oxidation*, *strain hardening*, temperature and residual stress, brittle and fatigue *failure*, and so on. Mechanical surface erosion is a result of high-velocity collisions of (solid or liquid) particles of a condensed phase with material on the surface, accompanied by tearing out tiny particles of this substance. Surface erosion is enhanced by the large kinetic energy of the particles, the high hardness and angularity of solid particles, the *surface roughness* of the material, and the high temperature of its surface layer which reduces its *strength*. If the temperature of a metal during high-temperature surface erosion exceeds the melting point, then this erosion is a part of the process of mass carry-off, that is *ablation*. A variety of surface erosion is *cavitation surface erosion* that results from frequent hydraulic shocks arising from the cavitation (formation of gas bubbles) in the highly active liquid which is in contact with the surface, or close to it. A cavitation surface erosion can lead to comparatively uniform *wear*, *pitting*, formation of voids of diverse size, and even to full degradation of the surface (see also *Cavitation resistance*). Mechanical surface erosion also increases the surface roughness, which can induce the appearance of stress concentrators (see *Stress concentration*), and a related decrease of the *construction material strength*. This is a particular hazard for alloys and heterogeneous composite materials (see *Heterogeneous structure*) whose components possess different *erosion resistances*. *Electrical surface erosion* is the result of the action of electrical charges on the surface. Due to the appearance of localized origins of fusion and evaporation on a metal, as well as possible chemical and electrochemical processes, electrical erosion leads to tearing out the substance mass and "ulceration" of the surface. Nevertheless, there are cases when it can serve as a basis of some electrophysical methods of processing refractory metals and conductive compounds with a very pure treated surface (see *Pulse treatment of materials*).

In almost all cases the phenomena of erosion are harmful. In addition to changes of technological processes and design, the efficient means for counteracting surface erosion are diverse methods of *surface hardening*, applying protective *coatings*, etc. See also *Drop-impact erosion*.

SURFACE EXCITON

An *exciton* localized at the interface or at the surface of a semi-infinite *crystal* or *plate*. Molecular crystals may contain Frenkel surface excitons (mobile molecular excited states) and semiconductors may contain Wannier–Mott surface excitons (mobile weakly bound electron–hole pairs). The energy of a surface exciton in a crystal bounded by a flat surface depends on a discrete quantum number n, which characterizes the internal motion, and on the two-dimensional *quasimomentum* p, which is parallel to the surface (if translational symmetry exists along the surface). The difference in energy between the surface and volume excitons is determined by the properties of the near-surface layer of the crystal (change of properties of near-surface molecules in a molecular crystal, difference between electron–hole interactions in the bulk and near the surface). As shown by S.I. Pekar (1957), a surface exciton may exist in a molecular crystal, if the magnitude of surface distortion is larger than a certain value determined by the exciton band width. The probability of exciting molecules exponentially decays with distance from the crystal surface; the region of excitation localization is of the order of several lattice constants. Surface excitons manifest themselves in reflection and luminescence spectra of single crystals, and have been found in anthracene, naphthalene, tetracene. An exciton localized near a surface and related to an electromagnetic wave is called a *surface excitonic polariton* (see *Surface polaritons*). The depth of penetration of excitation into the crystal is of the order of the wavelength of light.

SURFACE FORCE

A force acting upon a *solid surface*. It is characterized by the force density f, the force which acts upon a unit surface area. Examples: the *pressure* P, which is exerted on the body by the surrounding liquid or gaseous medium: $f = -Pn$ (n is the outward normal to a surface); and a *thermoelastic force* which also acts upon a unit surface area (see *Thermoelasticity*).

SURFACE, FRESHLY PREPARED

See *Juvenile surface*.

SURFACE HARDENING

Set of procedures and techniques, which generate surface layers on work pieces of machines and mechanisms; these layers increase wear resistance (see *Wear*), fatigue resistance (see *Fatigue*), and *corrosion* resistance of the articles. Surface hardening is used in cases when the modification of properties of the whole article is either economically prohibitive or technically impossible; sometimes modification of the overall properties may be undesirable because it could impair the bulk properties, such as *plasticity*, *impact strength*, etc. Depending on the nature of the influence, layers of various thickness (from 0.1 μm to several millimeters) are exposed to *hardening*. This hardening involves modification of the chemical composition, grain refining, creation of appropriate *texture* and phase composition, generation of field of *residual stresses* (often compression ones), healing of *submicrocracks*, removal of microcracks, etc. Traditional methods of surface hardening include *nitriding*, *cementation*, surface plastic deformation, electric treatment, application of *coatings*. Today, these methods are supplemented by laser hardening, *plasma treatment*, irradiation with ion and electron beams, laser *alloying*, *ion implantation*, complicated techniques of thermal treatment, generation of regular surface textures, etc. Traditional methods of surface hardening were selected on an empirical basis and were aimed at the modification of a certain single property; they consist, e.g., in increasing hardness in order to elevate the wear resistance, or in the generation of a field of compression stresses for retardation of fatigue *cracks*. The quantitative improvement of operating parameters, which could be attained using traditional methods, ranged from 20–30% to 2–3 times. At the present time, understanding the physical mechanisms of the processes of *strain* and *failure* facilitates the development of precise complex techniques, which will not only increase the mean values of parameters by a factor of ten and more, but also markedly reduce their scatter. The latter circumstance is of particular importance for flexible industrial processes. Nowadays, the traditional finishing surface treatment (grinding and polishing) is not thought of as perfect finishing. Although it removes the most substantial stress concentrators (see *Stress concentration*) from the surface, the traditional finishing usually produces a layer with tensile stresses and unfavorable texture. Hardening of the near-surface layers results in changes in a number of effects, such as yield surface element and *yield cusp*, *Bauschinger effect*, Portevin–Le Chatelier effect, etc., which indicates that they are related to surface phenomena (see *Surface sound hardening*).

SURFACE IMPEDANCE

A quantity Z_s which characterizes the high-frequency electromagnetic properties of a good conductor (*metal*). It is defined by the relation $E_t = Z_s[n \times H_t]$, which connects the tangential components of the intensities of electric (E_t) and magnetic (H_t) fields at the boundary of a conductor (n is the unit vector of the normal to the boundary). This relation, *Leontovich boundary condition* (M.A. Leontovich, 1940), is used in cases where the wavelength, sizes of conductor, and radii of curvature of its surface are substantial in comparison with the thickness of the skin-depth (see *Skin-effect*). In this approximation, Z_s involves the dependence of the field outside the conductor and the parameters of the electrodynamic system to which it is connected on the properties of the material of the conductor. For an isotropic body $Z_s = 4\pi\omega\mu\beta\delta/c^2$, where ω is the field frequency (the field varies as $\exp(-i\omega t)$), μ is the magnetic permeability, δ is the penetration depth, β is a complex factor; e.g., in the case of the normal skin effect and low frequencies β is equal to $(1 - i)/2$, at high frequencies β is equal to $(-i)$, in the limit of the *anomalous skin-effect* (δ is then replaced by δ_a) β is equal to $2(1 - i\sqrt{3})/(3\sqrt{3})$.

In anisotropic crystals, isotropic crystals in a constant magnetic field, Z_S is a tensor in two dimensions, which has off-diagonal components. For a plane semiconducting layer the concept of surface impedance may be easily generalized to the case when the field penetration depth is comparable to the layer thickness; this allows Z_S to be used for the description of size dependences of high-frequency properties of metal *plates*, including *radio-frequency size effects*, excitations of slowly-decaying waves, etc. (see *Impedance*).

SURFACE IMPEDANCE OF A SUPERCONDUCTOR

Electrodynamical characteristic of a *superconductor*, defined by the relation $Z_S = R_S + iX_S = E_t/H_t$, where E_t, H_t are the tangential components of the electric and magnetic field vectors of an electromagnetic wave in the surface layer of the superconductor. The presence of a nonzero *impedance* is related to the transit-time effect of the superfluid condensate of *Cooper pairs*, which is equivalent to a certain induced resistance. The electric field, which is needed for the periodic acceleration of the condensate, acts also in normal excitations, which exist at a finite temperature. This causes dissipation of energy, which is proportional to the surface resistance R_S.

At high electromagnetic field frequencies $\hbar\omega \gg 2\Delta$ (Δ is the *energy gap*) the surface impedance of a superconductor is close in magnitude to the *surface impedance* of a metal in its normal state.

For $\hbar\omega < 2\Delta$ the quantity R_S decreases with decreasing temperature: having the value R_N in the normal state, tending to zero at $T \to 0$. This fact provides grounds for applying superconductors as materials for high-quality extra-high-frequency devices. In actual practice, however, there is always a certain finite residual resistance at $T \to 0$, which is due to defects at the superconductor surface (oxide layers, microcracks, non-superconducting inclusions, etc.).

The imaginary part of the surface impedance X_S, which characterizes the shift in resonance frequency of a superconducting cavity with respect to an ideally conducting one, remains finite also at $T \to 0$.

In the general case the surface impedance, which is calculated based on the microscopic theory of superconductors, depends on relations among ω, Δ, and T, and on the nature of the relationship between the current and the vector-potential in the superconductor.

SURFACE IMPURITY STATES

Local electronic levels of impurity centers at the crystal surface. The main factors that determine the energy spectrum of the impurity surface states are the lowering of the symmetry associated with the finite dimensions of a crystal, variation of the potential shape under the effect of electrical *image forces* of the impurity center potential, the interaction with surface atoms, and also features of the *surface band structure*. The characteristic localization length of the wave function of the impurity surface states outside the crystal is considerably smaller than the effective Bohr radius. From this there follows a reduction of the contribution to a wave function of components with zero orbital angular momentum. This effect is especially important for the ground state of small donor impurity centers that are located directly at the surface. Reduction of the symmetry in the finite crystal leads to a shift of the energy level of the impurity, and to the partial removal of degeneracy. The introduction of additional (surface) branches into the kinetic energy operator results in the appearance of new allowed energy levels in the spectrum of the impurity surface states, and in the transformation of the impurity states related to volume branches. These transform into quasi-stationary states because the localized electrons can then undergo constant-energy transitions into the intrinsic *surface electron state* band.

From the point of view of possible applications, the more important energy levels of impurity surface states lie in the principal *band gap* near the *Fermi level*. The presence or absence of *surface levels* in this energy range defines the electrical properties of junction structures. They are responsible for forming the potential barrier in a *metal–semiconductor junction*, define the carrier mobility in the subsurface range of the superconductor, prescribe the generation–recombination properties of the surface, etc.

SURFACE INHOMOGENEITIES

Spatial fluctuations of the distribution of defects of various types in the near-surface region or on the solid surface. Surface inhomogeneities produce a nonuniform distribution of properties on the surface, e.g., of electric charges, of *band bending* φ_s in semiconductors and insulators, etc.

The nonuniformity of electrophysical parameters arises during formation of a surface because of defects in the subsurface regions of semiconductors and on the insulator–semiconductor interface. These are generated by hardening, ion bombardment, heterogeneous *adsorption*, and mechanical impacts.

Surface inhomogeneities affect the properties of surface channels of M–I–S transistors, and this influence is the most important in the weak inversion region (close to the threshold voltage) when the concentration of shielded carriers in the channel is not high. The well-known phenomenon of an abrupt decrease of effective mobility in the channel with a decrease of φ_s is also due to surface inhomogeneities.

Since a typical recharge time of *surface states* τ strongly depends on φ_s, surface inhomogeneities lead to space fluctuations of τ. This implies an essential change in the frequency dependence of the capacitance and transverse electrical conductivity of structures with surface inhomogeneities, and this, in turn, can be used for diagnostics of the surface.

SURFACE IONIZATION

Thermal equilibrium *desorption* of atoms and molecules from *solid surfaces* in the form of ions. The most extensively studied case of thermal ionization involves the production of positively charged ions. The process readily occurs during the vacuum heating of a metal, which contains adsorbed (see *Adsorption*) atoms of an alkali metal at its surface. The ratio between the number of ions $n_\pm$ and the number of neutral atoms n_0, which have evaporated from a certain metal surface during the same period of time, is called the *degree of surface ionization* α. In the case of positively charged ions, α is of the form $\alpha_+ = n_+/n_0 = A_+ \exp[e(\varphi - V)/(k_B T)]$ (*Langmuir–Saha formula*). Here V is the ionization potential of evaporating electrons, $e\varphi$ is the electronic *work function* of a metal, T is the absolute temperature of

the surface, A_+ is the ratio of statistical weights of the states of positively charged ions and neutral atoms at the temperature T. If adsorbed atoms exhibit the *electron affinity* S, then they may be evaporated from the surface in the form of negatively charged ions. In this case we have the expression $\alpha_- = n_-/n_0 = A_- \exp[e(S - \varphi)/(k_B T)]$. It may be seen that surface ionization is a selective process, which depends on the relationship between V, S, and φ, and is rather effective for many atom–emitter combinations.

Nowadays, the method of surface ionization allows one to obtain currents of positive ions from elements with $V \leqslant 9$ eV (there are about 60 such elements), and currents of negative ions from elements (the number of such elements is about 20) with $S \geqslant 1$ eV. The method is effective for ionization of a number of molecules, including organic compounds, as well as various radicals, and clusters which are produced from molecules at the emitter surface. The values of $\alpha_\pm$, and of current densities in the ionization of particles of all compositions from free surfaces, are obtained by the application of ion-accelerating external electric fields. Surface ionization is used not only as an effective method of generation of ions, but also for investigations of the properties of surfaces, adsorbed layers and ionizing particles.

SURFACE, ISOENERGETIC

See *Isoenergetic surface*.

SURFACE, ISOFREQUENCY

See *Isofrequency surface*.

SURFACE LATTICES

Regular two-dimensional structures, which consist of adsorbed atoms at a crystal surface. *Commensurate lattices* and *incommensurate lattices* are observed in *films*, which are adsorbed at a pure surface or its reconstruction (see *Surface reconstruction*). The ratio between the lattice constant and those of the crystal surface is expressed in terms of rational numbers for commensurate lattices, and by irrational numbers for incommensurate lattices (at least for one of the incommensurate lattice constants). The main difference between commensurate and incommensurate lattices arises from the continuous depen-

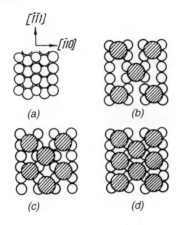

$[\bar{1}\bar{1}1]$

$[\bar{1}10]$

(a)

(b)

(c)

(d)

Surface lattices.

dence (ratio) between substrate lattice spacing and adsorbed atom parameters, which depends upon the extent to which the surface is coated with adsorbate. Fig. (a) shows a pure surface of the (112) face of tungsten (small circles), and the remaining Figs. (b, c, d) show different degrees of coverage by larger potassium atoms (cross hatched) of the top layer of tungsten atoms on the (112) face. The potassium arrangement is incommensurate with the tungsten crystal spacing along $\langle\bar{1}\bar{1}1\rangle$, which continuously changes with the degree of coverage. The spacing is commensurate along the $\langle\bar{1}10\rangle$ direction; and the crystal surface provides the coverage pattern for the adatoms. Incommensurate lattices are formed, as a rule, under conditions of a weak (compared with interactions between adatoms) charge pattern. In the neighborhood of a *phase transition* of a commensurate lattice to an incommensurate one, the latter has a complicated superstructure, which consists of regions of a commensurate lattice, separated by linear *stacking faults*: *domain walls* or static *solitons*. On leaving the phase transition point, solitons come closer together and an incommensurate lattice becomes practically uniform. At small wave vectors a commensurate lattice and an incommensurate one exhibit different excitation spectra: the spectrum of the commensurate lattice has a gap, whereas that of incommensurate lattice has a quasi-acoustic branch. As a result, a commensurate lattice exhibits ordinary long-range order (see *Long-range and short-range order*). At finite temperatures,

long-range order in an incommensurate lattice is destroyed by acoustic *phonons*. An incommensurate lattice exhibits so-called *quasi-remote order*, which is characterized by the decay of the correlation function of atomic displacement at long distances; this decay is not exponential as in the case of liquids, but rather is of an x^{-a} (power function) type.

SURFACE LAYER

Thin layer in the neighborhood of an interface between two *phases* (bodies, media); some properties of the material in the surface layer differ from bulk properties inside the material. This is because of the action of *surface forces*, which are specific to a particular phase boundary, such as dispersion and electrostatic forces, as well as structural forces, which are responsible for additional pressure in the submicrovolume of a liquid phase adjacent to an interface in thermodynamic equilibrium. Surface forces decay away from the phase boundaries in the bulk phase (characteristic spatial scale $l \sim 1$–10 nm).

The thickness, structure and state of the surface layer have an effect on operational and technological features of various structures, radio sets and other devices.

SURFACE LAYER, DISRUPTED

See *Disrupted surface layer*.

SURFACE LEVELS

Possible values of the energies of electrons in a crystal, which correspond to *surface electron states*. Lifetimes of electrons at surface levels are, as a rule, substantially less than those of volume (bulk) electrons. This is due to the extended number of possible quantum transitions between various types of surface electron states, as well as between surface and bulk states.

SURFACE MAGNETISM

A group of phenomena, which accompany the generation of ordered spins of surface and near-surface atoms of a *magnetic substance*. As a rule, surface *magnetization*, which exponentially decays away from the interface, is caused by a surface *second-order phase transition* at the temperature T_S, which is somewhat higher than the *Curie* point T_C of the material. As T approaches T_C, the depth of penetration of surface magnetization becomes infinite, accompanied by specific behavior of thermodynamic quantities. This transition, from a state with surface magnetization to one with the entire sample magnetized, is called an *extraordinary phase transition*. The existence of surface magnetism and its characteristics are determined by the structure of the surface and its prior treatment. The generation of surface magnetism is related to the sign of the parameter q in the equation for the surface component of the free energy of the material $F_s = (1/2)q\mu_0 M_s^2$, where M_s is the value of the magnetization at the surface. Surface magnetization arises for $q < 0$ provided $T_S - T_C \propto q^2$. The ordering of surface atom spins is a particular case of a second-order *surface phase transition*.

SURFACE MASS TRANSPORT

Transport of material along a *solid surface*, which is sustained either by a chemical potential gradient due to surface curvature ($\nabla \mu_s = \alpha \omega \nabla K$, where α is *surface tension*, ω is atomic volume, K is local curvature), or from the outside (by temperature gradients, electric field gradients, etc.). Mechanisms of mass transport are the following: surface or near-surface *diffusion*, diffusion in a gas, and evaporation–condensation. Surface mass transport influences the kinetics of formation of natural surface roughness, leveling of artificial irregularities, degradation of needle cathodes, decomposition of *microelectronics*, thin-film elements, etc. (see *Mass transport*).

SURFACE MOBILITY

Mobility of charge carriers, which are in a potential energy well near the surface, averaged over the width of the *space charge region*. Mobility is defined by the relation $\mu_s = \sigma_s/(e \Delta N)$, where σ_s is the *specific surface electrical conductivity*, ΔN is the near-surface excess of charge carriers, and e is the electron charge. The magnitude of the surface mobility is often lower than that of the bulk because of the additional scattering of charge carriers by the surface (see *Surface scattering*). At low temperatures and high values of the electric field at the surface, the magnitude of the surface mobility is strongly influenced by the quantization of the transverse motion of the carriers in the space charge region.

SURFACE OF A SOLID
See *Solid surface*.

SURFACE OF STRESS
See *Stress surface*.

SURFACE PASSIVATION

Prevention of chemical, electrical, or other interaction of a surface with the environment. A thin layer can be generated on the surface, which keeps the material of the *substrate* from chemical or other contact with the surrounding medium. Often this can be done through *oxidation* of the material, if the resulting film is continuous and prevents the possibility of contact with the surrounding medium. Foreign *coatings*, which exhibit high inertness to the surrounding medium (silicon nitride, aluminum dioxide, etc.), are applied if the intrinsic oxide of the material is permeable. The passivation of *atomically clean surfaces* of covalent compounds, which possess *dangling bonds*, is achieved through *adsorption* of molecules of those substances, which form chemical compounds with the substrate. Surface passivation of *semiconductor devices* is performed in a number of cases by applying a coat of organic lacquers, which are resistant to exposure to the atmosphere.

SURFACE PHASE

Crystallographic structure of an *atomically clean surface* of a *crystal*. Since the atoms at the surface are surrounded by a lesser number of closest neighbors than the atoms in the bulk, it can be thermodynamically advantageous for the surface atoms to assume a different type of arrangement, the symmetry of which is often different from the symmetry observed in the bulk of the crystal. As a result of the phenomenon of the so-called crystal *surface reconstruction* in ultrahigh vacuum, two-dimensional *surface superlattices* are generated, the unit cells of which exhibit constants, which are different from those of corresponding planes in the bulk of the crystal, and usually exceed "bulk unit cells" by several fold. Different faces of one and the same crystal are often characterized by different surface phases. In Si, e.g., the typical surface phase for the face (111) is a phase of the type 7×7, whereas the (100) face is typically characterized by a surface phase of the type 2×1, where the numbers indicate the ratio between the lengths of basis vectors of the surface unit cell and the corresponding lengths in the bulk. A change of temperature may cause *phase transitions* between various types of surface phases, as well as transitions from an ordered surface phase to a disordered one.

The main experimental methods for studying surface phases are *low-energy electron diffraction* and *scanning tunneling microscopy*.

SURFACE PHENOMENA in semiconductors

A group of electronic, optical and structural effects, which take place due to the presence of boundary surfaces (interfaces). Surface phenomena in semiconductors depend strongly on the state of the surface, the method of surface treatment, and the composition of adjacent gaseous media (see *Semiconductor surface*). Generation of surface phenomena in semiconductors is related to the existence of localized electron states at the surface, which are called *surface electron states*, and the change of the spectrum under external influences.

Field effect, surface recombination and *surface mobility* are among the fundamental surface phenomena in semiconductors. By virtue of the existence of surface electron states and a *contact potential difference*, a *space charge region* arises in the neighborhood of semiconductor surfaces, accompanied by the bending of the *valence band* and *conduction band* (see Fig.). The extent of the *band bending* is called *surface band bending*. The extent of this bending may be controlled by applying an electric field to the semiconductor. A field effect consists in a change of *electrical conductivity* of the semiconductor when an electric field is applied through the dielectric interlayer. This effect underlies the operating principle of field-effect unipolar transistors (see *Field-effect transistor, Schottky barrier*).

An important characteristic of surface phenomena in semiconductors is the *work function*, i.e. the minimum quantity of energy required to extract an electron from the surface of the semiconductor. It is equal to the height of the potential barrier that must be overcome by an electron in order to

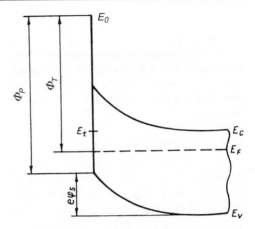

Energy diagram of a near-surface band of a semiconductor. E_0 is the energy of an electron in the vacuum, E_C, E_V and E_F are the energies at the bottom of the conduction band, at the top of the valence band, and the Fermi energy, respectively; $e\varphi_S$ is the surface band bending energy, Φ_T is the thermal work function of the semiconductor, Φ_P is the photoelectron work function of the semiconductor. E_t denotes the location of a discrete surface level.

escape into the vacuum. The *thermoelectric work function* and the *photoelectric work function* are singled out (see Fig.). The value of the work function depends on extent of surface band bending.

The annihilation of *electron–hole pairs* in the process of relaxation of a nonequilibrium state is called *surface recombination*. Effective recombination takes place at the *trapping* of electrons and holes by surface electron states. Surface recombination exerts a considerable influence on the operation of semiconducting *photoresistors* and photoconverters.

Surface mobility is lower than volume mobility because of additional mechanisms of surface scattering of electrons (see *Surface scattering*). The reduction of mobility becomes significant if the characteristic size of the region of electron motion (thickness of superconducting film or of near-surface space charge region) becomes comparable with electron mean free path.

The existence of surface electron states along with surface excitons, phonons and other *surface quasi-particles* leads to additional light absorption, which also takes place at free charge carriers, which are localized in narrow near-surface regions of space charge.

Structural surface phenomena include effects of rearrangement of crystal lattices at the surface of a semiconductor, ordering of multi-layer adsorbed films at the surface, and other related effects.

SURFACE PHOTOCAPACITANCE

The capacitance of a condenser, which is formed by a transparent or semitransparent conducting electrode, and the near-surface *space charge region* of a semiconductor under exposure to light. The important parameter, which characterizes surface photocapacitance, is the *light–dark capacitance ratio*, which is the ratio between the value of the capacitance during illumination and in the absence of illumination.

SURFACE PHOTOELECTROMOTIVE FORCE

Difference of surface electrostatic potentials of a semiconducting sample under illumination and in the dark. A surface photoelectromotive force consists of two components: barrier photoelectromotive force and surface charge photoelectromotive force. The initiation of a barrier photoelectromotive force is caused by the reduction of the potential barrier, which is due to an increase of the concentration of mobile charge carriers at illumination. The surface charge photoelectromotive force is generated by the *trapping* of surplus charge carriers (electrons or holes) by *surface electron states*. Initiation of the latter kind of photoelectromotive force does not require an initial *band bending* at the semiconductor surface.

The method of surface electromotive force is widely used for determining physical characteristics of semiconductor samples. Thus, measuring the spectral and photoresistance characteristics of this force permits one to determine important parameters, such as the diffusion length of *electron–hole pairs* and near-surface band bending (see *Surface phenomena in semiconductors*).

SURFACE PLASMON

Surface quasi-particle, which describes oscillations of surface charges and related electromagnetic fields at an interface. The amplitude of the field decreases sharply on either side of the surface where charges are localized. This definition corresponds to a surface plasmon: the *polariton*. The term "surface plasmon" is sometimes applied only to a plasma excitation obtained within the electrostatic limit (i.e. without regard for retardation). The surface plasmon frequency ω_{sp} for the metal–vacuum interface is given by the equation $\omega_{sp} = \omega_p/\sqrt{2}$, where ω_p is the volume *plasmon* frequency; surface plasmons are established by the electron beam and are observed in experiments on characteristic electron losses. A surface plasmon may be initiated by an external electromagnetic field under special conditions (*attenuated total internal reflection* method) or at a rough surface. Study of surface plasmon propagation provides information on surface properties.

SURFACE POLARITONS

Compound *surface quasi-particles* describing surface electromagnetic waves in *crystals* which interact with elementary excitations of various types. One distinguishes a *surface excitonic polariton*, *surface phononic polariton* (often simply referred to as a surface polariton), *surface plasmon-polariton*, depending on the particle with which the electromagnetic wave interacts. The existence conditions and *dispersion laws* of surface polaritons can be obtained by means of macroscopic electrodynamics. For a flat *phase interface* with dielectric constants ε_0 and ε_1 depending on the frequency ω, the dispersion law of surface polaritons follows from the relation

$$k_\parallel^2 = \frac{\varepsilon_0 \varepsilon_1}{\varepsilon_0 + \varepsilon_1} \left(\frac{\omega}{c}\right)^2,$$

where $k_\parallel$ is the *quasi-momentum* of the surface polariton parallel to the plane of interface, and c is the speed of light. Surface phononic polaritons can exist in the frequency range $\omega_{LO} < \omega < \omega_{TO}$ with negative $\varepsilon_\parallel(\omega)$ (here ω_{TO} and ω_{LO} are the frequencies of transverse and longitudinal phonons). For large $k_\parallel$, the frequency of surface polaritons is determined by the condition $\varepsilon_\parallel(\omega) = -\varepsilon_0$. In optical spectra, surface polaritons appear in the regions of phonon and exciton absorption bands. On a perfect insulator–vacuum interface the light does not directly excite surface polaritons, since the energy and momentum *conservation laws* do not hold simultaneously for surface polaritons and the external electromagnetic wave. For excitation and investigation of surface polaritons, special methods are used, e.g., *attenuated total internal reflection* or a *diffraction grating* deposited on the surface. The investigation of surface polaritons provides information on the properties of surface excitations and is an important investigation tool for solid surfaces and thin films.

SURFACE QUASI-PARTICLES

Elementary excitations specified by the presence of boundary surfaces in a system of many particles, e.g., surface quasi-particles in a semi-infinite crystal or surface quasi-particles on an insulator–semiconductor interface. Every type of *quasi-particle* in the solid bulk may have a surface analog under certain conditions. For example, there exist surface *phonons*, *excitons*, *magnons*, *plasmons*, etc. Surface quasi-particles are localized in the near-surface region, the wave function of surface quasi-particles decays when receding from the surface. From the localization condition it follows that the surface quasi-particles can exist in energy regions forbidden in the spectrum of bulk elementary excitations. Surface quasi-particles are characterized by the energy ε and quasi-momentum p connected by the dispersion law $\varepsilon(p)$, and by the spin. In a *crystal*, bounded by a plane with rational *Miller indices*, the quasi-momentum is two-dimensional. Surface quasi-particles may create and annihilate under external perturbations; interact with each other and with other types of quasi-particles. The existence conditions, energy spectrum, lifetimes, and mean free paths of the surface quasi-particles depend on the properties of the particular surface. The investigation of phenomena involving surface quasi-particles provides information about characteristics of the surface (its structure, level of perfection). Specific properties of surface quasi-particles are used in designing various devices, e.g., acoustoelectronic ones for surface waves, etc.

SURFACE RECOMBINATION

Annihilation of nonequilibrium *electron–hole pairs* at a semiconductor surface or at the boundary (see *Phase interface*) between the semiconductor and certain condensed media, or a gaseous phase. This takes place mostly through the intermediary of *surface electron states*. It consists in trapping an electron or a hole by a definite kind of surface electron state, which is accompanied by either radiative or nonradiative release of excess energy. It is often characterized by an effective *surface recombination rate S*. In the general case the process of surface recombination is described both in terms of recombination activity of surface electron states, and in terms of delivery of nonequilibrium charge carriers to the surface. At relatively low concentrations of surface electron states and *band bending* at the surface, the value of S is determined by the recombination activity of the surface; it depends on their concentration, and on cross-sections of trapping of electrons and holes by surface electron states, as well as on the surface band bending (see *Surface phenomena in semiconductors*). If the potential barrier is high or a considerable concentration of recombination surface electron states is observed, then the value of S is limited by the rate of *diffusion* of carriers through the layer of the *space charge region*, and therefore depends on the method of generation of nonequilibrium current carriers (see *Generation of current carriers*) in the neighborhood of the surface and in the bulk. The term "surface recombination" is also used in reference to neutralization of charged chemical radicals or ions in a liquid or gas, which are in contact with a *solid*.

SURFACE RECONSTRUCTION

A spontaneous change in the positioning of atoms in the surface layer or several adjacent layers with respect to their locations in an infinite *ideal crystal*. Surface reconstruction is a characteristic feature of semiconductors with *covalent bonding*. A surface hybridized state or band then forms at the free surface with *dangling bonds* (see *Tamm levels*), half-filled with electrons. This state is unfavorable in energy so the surface undergoes a transformation of the *Jahn–Teller effect* type which is called surface reconstruction. In essence, it consists in a specific displacement of surface atoms (2–3 atomic layers) that form a *superlattice* with a period which is a multiple of the bulk lattice period. During that transformation the initial hybridized state of broken bonds splits into two states, positioned outside the *band gap*, so that the surface energy is lowered as a result. Displacement of the surface atoms leads to deformation of hybridized orbitals directed into the crystal, and to raising the energies of the corresponding levels. As a result, there appear states filled with electrons in the band gap in the vicinity of the top of the *valence band*, associated with bonds directed inside the crystal. Similar unfilled states are split off from the bottom of the *conduction band*. The existence of these states has been demonstrated by data from *electron spectroscopy*.

To describe surface reconstruction one introduces two elementary translation vectors of the ideal (unreconstructed) surface, *a* and *b*. Experiments in *low-energy electron diffraction* demonstrate that *real surfaces* of C, Si, Ge may have translation periods *na* and *mb* where *n* and *m* are some numbers that are not necessarily integers. In that case it is assumed that a surface reconstruction of the $(n \times m)$ type exists.

Adsorption of atoms on the surface may significantly alter the nature of the surface reconstruction, since the adsorbed atoms enter into chemical bonds with hybridized orbitals of the broken bonds. The bonding state of the surface atom is then occupied by two electrons, which eliminates the very cause of surface reconstruction. That effect is demonstrated by experiments in which a monolayer of adsorbed atoms leads to the vanishing of the (2×1) superstructure on the reconstructed surface of C, Si, Ge.

In contrast to covalent solids, lowering of the surface energy in *metals* and *ionic crystals* is achieved through displacement of the surface plane without any change in their translation vectors. Such a phenomenon is called *surface relaxation*, and it is not equivalent to the Jahn–Teller effect. The displacements of atoms in this case will be small enough in that case, of the order of only several percent.

SURFACE RELAXATION

Variation of the crystal lattice constant in the direction normal to a solid surface for atoms in the surface layer and those distributed over several atomic layers below. This was initially observed by C. Davisson and L. Germer (1927) during *low-energy electron diffraction* experiments. Surface relaxation is associated both with the change of the electronic density at the surface and with the formation of a *dipole layer*, as well with the difference of the interatomic forces which affect the atoms located at the surface and within the bulk of a solid. Surface relaxation may be positive or negative (surface lattice constants are smaller than within the bulk). The relative value of the surface relaxation does not exceed 15%, it is usually within ±5%, and it decreases to zero over the thickness of several atomic layers. It may differ for different faces of the same *monocrystal*. Surface relaxation changes its magnitude and even its sign in the presence of adsorbed *coatings* on the surface. For example, one third of a monolayer of oxygen adsorbed at the (110) surface of nickel changes the surface relaxation from −4% to +1%, i.e. from −0.01 to +0.0025 nm. Surface relaxation is measured by the methods of low-energy electron diffraction and high-energy ion scattering, and the *orientation effects* in *Auger ion spectroscopy* may also be used.

SURFACE ROUGHNESS

Surface profile deviations from an atomically smooth contour (see *Atomically clean surface*). The natural and artificial roughness of a surface should be distinguished. *Natural roughness* appears on *crystals* which do not expose an equilibrium face, and results from anisotropy of the *surface energy*. As a consequence, the surface structural minimum and the surface energy minimum do occur together. With increasing temperature the growth of natural roughness is accelerated. *Artificial roughness* may be the result of factors distorting the surface (mechanical treatment, irradiation, etc.). When the temperature increases, the artificial roughness of a surface is smoothed by various mechanisms of *mass transport*. A *surface atomic roughness* may develop against a background of macroscopic roughness.

SURFACE SCATTERING

Change of the momentum and energy of a particle, which takes place at its collision with a *solid surface*. The term "surface scattering" is mostly used for describing the interaction of particles with intrinsic elementary excitations of the solid at its surface. Two reasons are responsible for this: first, *quasi-particles* of the crystal are more sensitive to the symmetry of its surface than the particles that are incident upon its surface from a vacuum; second, the scattering by *surface defects* determines the *kinetic phenomena* in *thin films* and near-surface layers, which are the working media in most devices of *solid-state quantum electronics.* Accordingly, two types of surface scattering study are distinguished.

I. Collision with an ideal surface. If the surface exhibits the natural translational symmetry of the mid-plane of the crystal, then the particle, which collides with it, conserves its component $P_\parallel$ of *quasi-momentum* parallel to the surface. For an arbitrary form of *dispersion law* $\varepsilon(P)$, the normal component $P_\perp$ may assume different values after the reflection of a particle from the surface, depending on the number of points at which the line $P_\parallel = \text{const}$ intersects the surface $\varepsilon(P) = \text{const}$ throughout the entire *reciprocal lattice.* If the surface exhibits translational symmetry, which is different from the symmetry of the mid-plane of the crystal, then its reciprocal lattice is characterized by its own set of two-dimensional vectors $k_\parallel$. On reflection from this surface, the change of the tangential component of the quasi-momentum to any of the $\hbar k_\parallel$ vectors is possible. An experimental study of the law of reflection of electrons provides information on $\varepsilon(P)$, and can lead to a direct determination of the set of vectors $k_\parallel$ that describe the *surface atomic structure.*

II. The importance for kinetics of the *relaxation* of energy and momentum at the expense of surface scattering. In the case of quasi-particles which obey an isotropic dispersion law, the scattering by an ideal surface does not result in relaxation of momentum and energy ($P_\parallel$ is conserved and $P_\perp$ changes its sign, which does not cause a change of the current along the surface). This kind of surface scattering is called *specular surface scattering.* Consideration of the scattering by surface defects upsets the specular nature of the scattering. In accordance with the type of potential of the defect, the following *surface scattering mechanisms* are distinguished:

- *Scattering by local surface irregularities*, the typical dimensions of which are smaller then the *mean free path.* If the surface is treated as an infinitely high *potential barrier* for the mobile charge carriers, then these irregularities are

scattering centers, which exhibit a potential that is infinite for a limited spatial region. The *kinetic coefficients* are expressed in terms of the scattering cross-section at a center, and the surface concentration of scattering centers. A surface of a random shape $z = U(\rho)$ is considered as a unit defect of the potential, which is given by the equation $V(r) = V_0 \delta(z) U(\rho)$ (where V_0 is the work function, z is the direction of the normal to the surface, ρ is a vector in the plane of the surface). All results are expressed in terms of the *correlation function* of irregularities $\langle U(\rho) U(\rho') \rangle$, where $\langle \ldots \rangle$ denotes averaging over the ensemble of random functions $U(\rho)$.

- *Scattering by charged centers*, which are randomly distributed over the surface or in the neighboring layer. The conductivity of this layer is insignificant (e.g., the insulating layer in *metal–insulator-semiconductor structures*), but the long-range Coulomb potential penetrates into the current-carrying region. The charged centers cause the generation of near-surface *band bending*. Therefore, the problem of describing the scattering by these centers falls into two parts: calculation of the potential of the center with regard for the nonuniform density of charge, and calculation of the amplitude of scattering by this potential.

What all mechanisms of surface scattering have in common is that the scattering of particles, which move at shallow angles to the surface, is specular. An important point is that this result holds true for an arbitrary dispersion law. The direct proof of specular reflection of glancing electrons has been provided by the discovery of *magnetic surface levels.*

Surface scattering results in a noticeable decrease of the effective mobility of carriers under the condition that the typical dimension of the region of their near-surface motion (the thickness of current-carrying region) is smaller than or of the same order of magnitude as the mean free path. Such conditions occur in thin films and wires, in narrow near-surface channels (see *Field effect, Inversion layer*), under conditions of the *skin-effect.* The contribution of different mechanisms of surface scattering to the effective mobility of charge carriers involves the dependences of the electrical conductivity and the Hall coefficient on the values of the layer thickness, temperature, and field.

The scattering by the short-range surface centers may result in a spin flip (see *Electron spin-flip scattering*). This effect of surface scattering is of importance for materials which exhibit a large *spin–orbit interaction*, and in studies of the surface *photovoltaic effect*. Spin flip in surface scattering was discovered during the course of experiments on electron paramagnetic resonance in thin films and metallic *small particles*.

The theoretical treatment of the surface scattering of *conduction electrons* is based on the establishment of boundary conditions for the *distribution function* in quasi-classical theories of *transport phenomena*, or on the method of the *quantum kinetic equation*.

The phenomenon of the cooling of nonequilibrium carriers in surface scattering has been observed experimentally. It is assumed that this cooling is related to the irradiation of *phonons* in surface scattering, i.e. to scattering by the surface imperfections, which are caused by lattice vibrations, or to the excitation of internal degrees of freedom of the molecules, which are adsorbed on the surface; it is also possible, that the cause of this effect is more direct: the cooling may be caused by electronic excitations to *surface states*.

SURFACE SCATTERING, ADIABATIC

See *Adiabatic surface scattering*.

SURFACE SMOOTHING, structure planarization

The process of leveling the underlying elements of relief of individual layers of laminated circuits by filling these elements with a certain material until a totally smooth surface is obtained. The necessity for carrying out such a smoothing operation becomes clear at the changeover to the production of submicron-scale *integrated circuits* with large and medium scales of integration. The leveling of concave surface elements may be performed by filling the surface regions with viscous fluids followed by hardening (after-bake), or by optimizing the number of processes of vacuum precipitation of *coatings*. Ion-plasma methods, in which the processes of precipitation and sputtering are combined, show promise. These methods permit controlling the relief of a growing *film* by

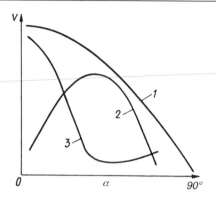

Dependence of the precipitation rate (curve 1), sputtering rate (curve 2), and total rate of film deposition (curve 3) on the angle α between the normal to the surface and the direction of motion of incident particles.

varying the total angular dependence of the precipitation rate (see Fig.), i.e. the dependence of the precipitation rate on the angle α between the normal to the surface and the direction of the motion of the particle flow. The angular dependence of the precipitation rate can be accurately approximated by application of the cosine law (curve 1 in Fig.). The angular dependence of the sputtering rate (curve 2 in Fig.) has a pronounced maximum at angles $50–70°$, and may change its character depending on conditions. The structure planarization is provided by attaining a suitable combination of the rates of precipitation and hardening, as well as by alternating the coating precipitation and sputtering processes.

SURFACE SOUND HARDENING

Hardening of the surfaces of articles with a deforming agent (strengthener) which vibrates at an acoustic or ultrasonic frequency. There are two kinds of surface sound hardening: (1) *impact hardening*, caulking (intermittent contact between the work piece and the block head; the tool vibrates along the normal to the surface); (2) *vibrational burnishing* (continuous permanent contact, the work piece and the strengthener are tightened together; the strengthener vibrates parallel to the surface, drawing a certain microscopic relief on it). Surface sound hardening falls into the category of methods, which use surface plastic deformation (see *Plastic deformation*), and is used as a finishing treatment in order to increase fatigue *strength*, *durability* and wear resistance of working parts of machines and mechanisms. Relative to methods which use low-frequency vibrations, the application of ultrasonic vibrations allows one to improve the qualitative characteristics of the process: to reduce the surface roughness, and obtain a considerable increase of residual compression stresses in the surface layer of articles. At impact hardening (see *Impact strength*), the sound waves penetrate into the metal, promoting favorable redistribution of *residual stresses*, particularly in welded structures; the waves also reduce the *stress concentration* in the neighborhood of *pores* and microcracks (see *Crack*), which results in an increase of *corrosion fatigue* resistance.

SURFACE SPECIFIC HEAT, surface heat capacity

The main contribution to the *specific heat* or heat capacity of a *solid* is surface vibrations. This contribution is proportional to the area of the *solid surface* and therefore may be experimentally determined only in the case of very *small particles* of a solid, when surface atoms account for an appreciable part of the total number of atoms of the solid. Since the energy spectrum of surface vibrations is, as a rule, shifted in frequency toward lower energies in comparison with spectrum of bulk vibrations, the surface specific heat is positive and observable only at fairly low temperatures. The computations of surface specific heat in the context of various models at low temperature lead to a quadratic form for the temperature dependence of this specific heat, in contrast to the cubic temperature dependence of a bulk solid. This quadratic temperature dependence of surface specific heat is, on the whole, qualitatively confirmed by experiments, yet quantitative experimental data exceed the theoretically predicted values by several fold. This discrepancy between theory and experiment is, presumably, related to the fact that the theory fails to account for a number of factors, such as *surface roughness*, *stacking faults*, contribution of the *conduction electrons* to the specific heat of small metallic particles, etc.

SURFACE STATES

States of *quasi-particles* of a solid, which are localized near its boundary with the vacuum, or other medium, and in the neighborhood of internal phase boundaries or grain contact interfaces. Surface states are characterized by particular values of energy and quasi-momentum (complex or purely imaginary), which preclude the possibility of the free propagation of quasi-particles in each direction from the surface, and thus are forbidden for corresponding volume quasi-particles. The most extensively studied surface states are *surface electron states*, surface *phonons, magnons, excitons, plasmons, polarons*, etc. (see *Surface quasiparticles*). Detecting and identifying surface states, determining their energetic, spatial, and dynamic characteristics provide important information on surface properties of both the electronic subsystem and the crystal matrix.

SURFACE STATES, MAGNETIC

See *Magnetic surface states* and *Magnetic surface levels*.

SURFACE STEPS

See *Atomic steps*.

SURFACE SUPERCONDUCTIVITY

The phenomenon of initiating *superconductivity* in the neighborhood of a *metal* surface, in a layer which is about one *coherence length* in thickness ($\xi \sim 10^{-5}$ cm); the initiation of superconductivity at the surface takes place under conditions of disruption of the superconductivity in the bulk of the sample by a strong magnetic field. Surface superconductivity is realized either in *type II superconductors* in the interval of fields which are parallel to the surface, and range from B_{c2} to B_{c3}, where B_{c2} is the *upper critical field* and $B_{c3} = 1.69 B_{c2}$ is the *third critical field*, or in *type I superconductors*, whose *Ginzburg–Landau parameter* κ ranges from 0.42 to 0.707, over the magnetic field interval between B_c and $1.69 B_1$, where B_c is the *thermodynamic critical magnetic field*, and B_1 is the *field of superconductor supercooling*. Experimentally the indicator of surface superconductivity is the vanishing of the resistance in strong fields, in which the *Meissner effect* does not yet take place.

The initial picture of surface superconductivity, as predicted by Saint-James and de Gennes

(1963), is related to nonsingular onset of an *order parameter* in the proximity of the surface in the form of a laminar current layer $j(z)$; the total current parallel to the surface is equal to zero: $\int_0^\infty j\,dz = 0$. At a later time, however, it was shown (I.O. Kulik, 1968) that this state corresponds only to the case of a field, which is exactly parallel to the surface, whereas under actual conditions the field always exhibits a finite deviation from parallelism (however small it may be), and the uniform state is unstable and gives way to a pattern of inclined vortices (see *Vortex lattices in superconductors*). The surface superconducting layer may carry a finite current only in the case when the inclined vortices are held fixed or pinned (*Vortex pinning*) at defects and inhomogeneities of the surface. This pattern of surface superconduction is confirmed by the sharp dependence of the current-carrying capacity of the surface layer (i.e. its critical field) on the nature of surface treatment.

SURFACE SUPERLATTICE

Type of ordered arrangement of surface atoms (ions, molecules), which is different from their arrangement in bulk planes of atoms parallel to the surface planes of the *crystal*. For the purposes of designation of a surface superlattice or superstructure, the following form of nomenclature is accepted: [name of crystal, *Miller indices* of crystal face, $(m \times n)$] (the numbers m and n indicate how many times the lattice constants of the superlattice are greater than the corresponding constants in the bulk crystal), e.g., Si (111)–(2×1). The ordered distribution of surface electron density (*charge density waves, spin density waves*, not accompanied by corresponding displacement of atomic nuclei) is designated in a similar manner. When the *unit cell* of a surface superlattice is a primitive one (occupied sites are located at its corners, and not inside it), then $(m \times n)$ can be prefixed with the letter p, e.g., Si (100)–$p(2 \times 2)$. If the surface unit cell under consideration is centered, then $(m \times n)$ can be prefixed with the letter c, e.g., LaB$_6$ (110)–$c(2 \times 2)$.

The most general nomenclature of surface superlattices uses the following matrix which relates the main surface superstructure vectors a, b

and the main vectors of an unrestructured surface a_0, b_0:

$$\begin{pmatrix} a \\ b \end{pmatrix} = \begin{pmatrix} m_{11} & m_{12} \\ m_{21} & m_{22} \end{pmatrix} \begin{pmatrix} a_0 \\ b_0 \end{pmatrix};$$

e.g., these matrix elements for the preceding examples are $m_{11} = m$, $m_{22} = n$, $m_{12} = m_{21} = 0$. The designation of adsorptive surface superlattice must include the specification of the adsorbate; if the unit cell is rotated by an angle φ with respect to the unit cell of the *substrate* surface then $(m \times n)$ in the designation of the superlattice is followed by $R\varphi$, e.g., Ge(111)–$(\sqrt{3} \times \sqrt{3})R30°$ Pb. According to another type of notation, the designation of a surface superlattice begins with a specification of the adsorbate, which is followed by the description of the substrate, e.g., $p(2 \times 2)S/Ni\,(100)$ (see *Atomic order* on a crystal surface).

SURFACES, VICINAL

See *Vicinal surfaces*.

SURFACE TENSION

The work necessary for the isothermal generation of a surface of unit area. An equivalent definition may be given as "the specific free energy of the surface layer of a liquid (*solid*), which is contiguous to its saturated vapor". An analogous quantity characterizes the *interfacial surface tension* at the boundary between two condensed phases. The *surface tension coefficient* γ is the force which acts in the surface layer of a condensed phase along a tangent to the surface of the phase along each unit of length of a line drawn in the surface. The coefficient γ has dimensionality J/m^2 or N/m. For various liquids, the order of γ ranges from 10^{-1} to 10^3 mJ/m^2. Examples of values of γ for several materials are: helium (at $-270°$C) 0.35, ethanol (20 °C) 22, water (20 °C) 73, mercury (20 °C) 485, osmium (2700 °C) 2500. Surface tension decreases with an increase of temperature, and it vanishes at the critical temperature. Static methods of investigation are based on the theory of capillarity (see *Capillary phenomena*). There exist methods in which the value of the surface tension is determined from the maximum pressure of bubbles and drops, which are separated from a capillary tube; certain techniques deduce γ from the height of capillary rise of a liquid, the weight of a separating drop, the separation of the curvature circle. Values of the surface tension of solids are often found using the *zero-creep method*, which is based on determining the magnitude of a load which, when applied (in combination with *surface tension forces*), does not change the surface area of the object under investigation.

The value of surface tension at the boundary between a conducting phase and an electrolyte has a pronounced dependence on the potential jump at this boundary (see *Electrocapillary phenomena*). Surface tension may be considerably decreased by the presence of surface-active impurities (see *Surface-active agents*). In thin layers ($\sim 10^{-7}$ m), surface tension exhibits a nonmonotonic dependence on the layer thickness.

SURFACE TREATMENT

Exposure of the *surface layer* of a solid to various kinds of physicochemical treatment, which results in modification of its structure, composition, free energy; upgrading the surface layer can improve the overall performance of a device. Mechanical, thermal and chemical kinds of surface treatment are recognized; there exist also combinations of these three types (chemical heat treatment, mechanochemical heat treatment, etc.). Mechanical surface treatment is based on modifying the content of structural defects in the surface layer; this may be achieved either by the removal of impaired layers and the creation of a certain surface topography (*abrading, honing,* etc.), or by plastic deformation: surface *cold working*, which is achieved, e.g., through shot blasting. The process of surface strain hardening brings about an increase of *yield limit, hardness, fatigue* resistance of the material and decrease of *plasticity, impact strength* and resistance to *strain* of the opposite sign. The depths of peening range from several hundreds of micrometers (shot peening) to tens of millimeters (impact caulking). Thermal surface treatment involves structural transformations, which take place in surface layers of the material under special heating conditions. The methods of thermal surface treatment include *surface hardening, annealing out of defects, amorphization*; these techniques are carried out through the agency of rapid radio-frequency, gas flame, electron-beam and ion-beam surface heating, as well as heating by intense fluxes of laser and ordinary optical radiation. Surface hardening attains high hardness of the surface layer with retention of softness and viscosity of the core of a steel article. Properties of the surface layer depend not only on the temperature, but also on the heating rate, duration of holding at the hardening temperature, the state of the structure before hardening. The operating characteristics are also strongly influenced by the degree of uniformity of hardening across the whole width of the layer, and by characteristics of the gradual transformation of a hardened layer into an unhardened core. Chemical surface treatment is based on modification of the chemical composition of the surface layer and its properties, which are limited by *adsorption* and *chemisorption* processes. The chemical treatment consists in the preparation of *coatings*, which are intended for protection and for consequent saturation (*diffusion* in solid state, *alloying* at melting) of the surface layer. Chemical surface treatment is performed using methods of chemical and electrolysis precipitation, detonation application of coatings, etc. The complex of various properties of surface layers allows one to employ combined techniques of surface treatment. Most widely used among them are: chemical heat surface treatment, which consists in diffusion saturation of the surface layer with adsorbed atoms (*borating, carburizing, nitriding, carbonitriding, chromonitriding,* etc.); the process involves generation of the *diffusion zone*, which consists of *solid solutions* and chemical compounds. Another popular method is chemical heat machining, whereby the properties of the generated surface layers are enhanced through the application of additional surface strain hardening (see *Plasma treatment, Ion-plasma treatment*).

SURFACE WAVES

Waves that travel along the boundary between two media (see *Phase interface*) or a medium and a vacuum, and decay rather rapidly with distance from the boundary. Most extensively studied are surface electromagnetic waves in *solids*, surface acoustic waves in solids, and magnetostatic waves in *ferromagnets*.

Surface electromagnetic waves in plasmas and magnetostatic waves (see *Spin waves*) in ferromagnets owe their existence to the fact that the high-frequency dielectric constant ε of a plasma and the high-frequency magnetic permeability μ of a ferromagnet may assume negative values over certain frequency ranges. As a rule, these waves are low-velocity ones, i.e. their phase and group velocities are smaller than the corresponding velocities of bulk waves. The amplitude of surface waves falls off as $\propto e^{-\gamma|x|}$ (decay starts at the phase boundary $x = 0$), and the depth of localization of the wave ($\propto 1/\gamma$) is frequency-dependent.

Surface acoustic waves in isotropic solids are characterized by a vector of particle displacement that lies in the *sagittal plane* (the plane containing the normal to the boundary and the wave vector). The velocity of surface acoustic waves is less than that of space waves (waves in the bulk of the phase). The depth of localization of surface acoustic waves is of the order of a wavelength. Surface acousto-electric waves, which decay slightly away from the interface, may travel in piezoelectric crystals; the existence of these waves is due to the piezoelectric effect (see *Piezoelectricity*).

SURFACTANT

See *Surface-active agents*.

SUSCEPTIBILITY

A parameter expressing the response of a system to an external applied field. For example, the electric polarization (usually called simply *polarization*, i.e. *dipole moment* per unit volume) P is related to an applied electric field E through the *dielectric susceptibility* $\kappa = P/(\varepsilon_0 E)$, and the *magnetization* (magnetic polarization) M is related to the applied magnetic field $H = B/\mu_0$ through the *magnetic susceptibility* $\chi = M/H$. These dimensionless susceptibilities are called *static susceptibilities*; they are ratios of a value induced by the field to the field value itself. At sufficiently low fields χ does not depend on the field (*linear response*). At high fields the response becomes non-linear and χ does depend on the field. In this case a *differential susceptibility* is introduced (e.g., $\chi_{dif} = dM/dH$). In *anisotropic media* (e.g., in crystals) the susceptibility is a tensor, the static susceptibility is a symmetric second rank tensor κ_{ij}, χ_{ij}, etc., so the magnetization and polarization depend both on the direction of the applied field, and on the directions of the symmetry axes of the crystal. If the field varies in time

and is non-uniform in space the susceptibility determines the non-local (retarding) connection between the response and the applied field, for times $t' < t$, e.g.,

$$P_\alpha(\mathbf{r}, t) = \int d\mathbf{r}' \int dt' \kappa_{\alpha\beta}(\mathbf{r}, \mathbf{r}'; t, t') E_\beta(\mathbf{r}', t').$$

In a spatially uniform medium with properties which do not depend on time, the susceptibility tensor depends only on the differences $(r - r')$ and $(t - t')$. Then the Fourier-component of the susceptibility $\chi_{\alpha\beta}(\mathbf{k}, \omega)$ which depends on the wave vector $\mathbf{k}$ (*space dispersion* of χ) and on the frequency ω (*time or frequency dispersion* of χ) is often used as the susceptibility itself; the connection between the Fourier-components $\mathbf{H}(\mathbf{k}, \omega)$ and $\mathbf{M}(\mathbf{k}, \omega)$ is local: $M_\alpha(\mathbf{k}, \omega) = \chi_{\alpha\beta}(\mathbf{k}, \omega) H_\beta(\mathbf{k}, \omega)$. Analogous expressions can be written for the dielectric case κ_{ij}. Unlike the static case, the *dynamic susceptibility* can have a complex value, and may contain symmetrical and antisymetrical parts. If the susceptibility has an antisymetrical part (e.g., $\kappa_{ik} \neq \kappa_{ki}$), the medium is called an optically active or *gyrotropic medium*. In an isotropic medium the susceptibility is a scalar. The imaginary part χ'' of the complex susceptibility $\chi = \chi' + i\chi''$ determines the energy dissipation in the medium; and the real part χ' is the susceptibility introduced above. Sometimes χ' is called the dispersion part and χ'' the absorption part. These real and imaginary parts are connected by the *Kramers–Kronig relations*. Onsager relations (see *Onsager theory*) also provide some restrictions on the components of the susceptibility tensors, in particular, in the most common general case (when a constant value of *magnetic flux density* B_0 is present within the medium) the following relation holds: $\kappa_{ik}(\mathbf{k}, \omega, \mathbf{B}_0) = \kappa_{ki}(-\mathbf{k}, \omega, -\mathbf{B}_0)$. In crystals with a space-periodic structure the *microscopic susceptibility* is the matrix $\kappa_{\alpha\beta}(\mathbf{k} + \mathbf{g}, \mathbf{k} + \mathbf{g}', \omega)$ in the space of the *reciprocal lattice* vectors $\mathbf{g}$ and $\mathbf{g}'$. This must be taken into account when considering the interaction with solids of electromagnetic waves (or charged particles) with characteristic wavelengths of the order or less than the period of the *crystal lattice*. However, in the long-wave approximation (see *Crystal optics*) it is possible to use a *macroscopic susceptibility* which has been averaged over a region considerably exceeding the volume of an elementary lattice cell, $\kappa_{\alpha\beta}(k, \omega) \equiv \kappa_{\alpha\beta}(k+0, k+0, \omega)$, which has the same meaning as the susceptibility of a uniform medium.

At a *phase transition* the susceptibility has singularities near the transition point. At a *first-order phase transition* the susceptibility exhibits a finite jump. At a *second-order phase transition* the type of singularity of the susceptibility depends on the character of the relationship between the external field G_i and the *order parameter* η describing this transition. If there is an *invariant* quantity $\eta \cdot G_i$ in the *thermodynamic potential* of the system, then the susceptibility is called the *critical susceptibility*, and it diverges at the transition point; $\chi \sim \tau^{-\nu}$, where $\nu > 0$ is a *critical index*, and τ is the distance to the transition point. If the relationship between η and G_i has another form, then the susceptibility is called a *non-critical susceptibility*. In the presence of the invariant $\eta^2 G_i$ the susceptibility is subjected to a finite jump in the transition point, and if the invariant is $\eta G_i G_j$ or $\eta^2 G_i G_j$ then the susceptibility is continuous but has a slope discontinuity (cusp).

SUZUKI ATMOSPHERE (H. Suzuki, 1952)

A raised concentration of one of the elements of an alloy at *stacking faults*. This is caused by a difference in the crystalline structure of the stacking faults and the bulk metal. The equilibrium *solubility* of the doping elements (see *Alloying*) is not the same for different lattices, hence the concentration of doping elements on stacking faults differs from the average in the alloy. Consequently, the interaction of doping elements with stacking faults is of a chemical nature. This was predicted by Suzuki to explain the *hardening* of alloys with a face-centered cubic lattice in which an elevated stacking fault density is present. The latter is the main obstacle opposing the development of a *plastic deformation*. The formation of a Suzuki atmosphere is accompanied by an increase of stacking fault energy, which stimulates the increase of their linear dimensions, thereby inducing a hardening effect. There are no direct data on the existence of Suzuki atmospheres, but experiments indicate that the formation of the atmosphere is associated with the observation of an increase of stacking fault length due to either doping or low-temperature *annealing*. Chemical interaction forces (~ 10 N/mm^2) are an order of magnitude smaller than the elastic forces inducing the formation of *Cottrell atmospheres*. However, in contrast to the latter, the former vary weakly with the temperature, and it is because of this that the chemical interaction forces conserve the effect of hardening up to higher temperatures. The Suzuki atmosphere determines to a considerable extent the mechanical behavior of alloys with face-centered lattices.

SWELLING by irradiation

An increase in the size of a material when it is irradiated by high doses of fast *neutrons*. It is characterized by a relative volume growth $S = (V - V_0)/V_0$, where V_0 is the initial and V is the final volume of a *solid* after irradiation. The value of S may reach 20%. Swelling is a consequence of the formation of vacancy *pores* or pores filled with chemically foreign atoms during the irradiation, with diameters that may reach several hundreds of nanometers, and concentrations up to 10^{16} cm^{-3}. The phenomenon of swelling is clearly expressed in *metals* into which atoms of inert gases are introduced (either resulting from nuclear reactions or *ion implantation*) since the latter feature a large positive dissolution energy (see *Solubility*). The nature of the swelling of an irradiated crystal (isotropic swelling or elongation in one direction) depends on the type of lattice. For example, cubic materials swell in an isotropic manner. Swelling is also commonly observed in polymers.

SWIHART WAVES in tunnel junctions
(J.C. Swihart, 1961)

Electromagnetic oscillations propagating through a *tunnel junction* between two superconductors. Such a junction may be considered a stripline, in which the wave velocity (the *Swihart velocity*) has the value $c_S = c[d/(\varepsilon \Lambda)]^{1/2}$, where c is the speed of light in vacuo, d is the thickness of an insulating gap between the superconductors, ε is the relative dielectric constant, $\Lambda = \lambda_1 + \lambda_2 + d$ is the overall *penetration depth of magnetic field* into the superconductor tunnel structure. A typical value of the Swihart velocity is $c_S \approx 10^9$ cm/s. Interacting with the wave density of the Josephson current $j = j_c \sin(\omega t - kx)$

(see *Josephson effects*) through the junction, the Swihart waves induce the appearance of certain stepwise features in the respective current–voltage characteristic (*Fiske steps*). Here the frequency ω and wave number k are determined by the voltage at the junction, and by the constant magnetic field applied to it.

SWIRL DEFECTS

A generic name for impurity, precipitation and segregation microdefects of growth with a swirl pattern that occur in single crystal ingots of *silicon*, independent of their specific origin. Following the commonly accepted classification one may identify A-, B-, C- and D-types of such defects. The presence of microdefects of a particular type in a material depends on the conditions of *monocrystal growth*. For example, the following microdefects were found to form in silicon produced by a crucible-free technique: evenly distributed large microdefects of A-type at a concentration of $N \sim 10^3$ cm^{-3} (crystal growth rates $V \leqslant 1$ mm/min); A- and B-type microdefects with striated distribution, $N_A \sim 10^6$–10^7 cm^{-3}, $N_B \sim 10^7$–10^{10} cm^{-3} ($V = 1$–4.5 mm/min); localized areas of types C- and D-type microdefects, $N \sim 10^{11}$–10^{12} cm^{-3} ($V > 5$ mm/min). As to crucible-grown single crystals of silicon, the rate of extraction controls the type of microdefects that they acquire and, due to a different axial temperature gradient, the density of these defects appears to be half as high. The largest A- and B-type microdefects exhibit a characteristic spiral distribution across the crystal cross-section, hence the name "swirl-defects". Studies have shown A-type defects to be dislocation ensembles (see *Dislocation*), while B-type defects consist of particles of a different *phase*, so that their formation is not accompanied by dislocations. C- and D-type microdefects are mainly interstitial in nature (see *Interstitial atoms*) controlled by the influence of the residual impurities *oxygen* and *carbon*. These elements apparently form the basis for such microdefects. It is only exceptionally pure crystals grown at high rates that feature *vacancy* microdefects of D-type.

SWITCH, HEAT
See *Heat switch*.

SWITCHING
See *Magnetic switching, Polarization switching*.

SWITCH, SUPERCONDUCTING
See *Superconducting switch* and *Superconducting group switch*.

SYMMETRIZED FUNCTIONS
Functions arranged to form the basis of an irreducible representation of a symmetry group. Usually it is convenient to symmetrize some initial function set beforehand using methods of *group theory*. With this intention in mind the application of group-theoretical rules can provide the needed linear combinations of initial functions. The use of symmetrized functions considerably simplifies calculations: providing the opportunity to identify possible non-zero matrix elements, to decompose secular equations, etc. Sometimes molecular orbital (see *Linear combinations of atomic orbitals*) symmetrized wave functions are called *group orbitals*. *Symmetric coordinates* used in the theory of vibrations, the Jahn–Teller effect, phase transitions, etc. are special cases of symmetrized functions.

SYMMETRY, AXIAL
See *Axial symmetry*.

SYMMETRY AXIS, axis of symmetry
An axis, rotation about which through a certain angle φ brings a physical body into coincidence with itself. If $\varphi = 2\pi/n$, where n is an integer, then the axis of symmetry is of nth order, and it is denoted by C_n. For example, a regular pyramid with a regular n-sided polygon as its base possesses an nth order symmetry axis that passes through the vertex, and is normal to the basal plane. A cube has three types of symmetry axes: three C_4 axes which pass through the centers of opposite faces, four C_3 axes which coincide with the diagonals, and six C_2 axes that pass through the midpoints of opposite edges.

SYMMETRY AXIS, ROTATIONAL
See *Rotation symmetry axis*.

SYMMETRY BREAKING
See *Spontaneous symmetry breaking*.

SYMMETRY CENTER

See *Center of symmetry.*

SYMMETRY CLASSES

The same as *Crystal classes.*

SYMMETRY GROUP

A group whose elements are symmetry operations (e.g., rotation, reflection, inverson) of a system (e.g., molecule, crystal lattice, Hamiltonian) (see *Group theory, Crystal symmetry*).

See also *Point groups of symmetry, Colored symmetry group, Defect symmetry group, Limiting symmetry groups, Local symmetry group, Magnetic symmetry group, Continuous symmetry transformation group.*

SYMMETRY, LOCAL

See *Local symmetry, Local symmetry group.*

SYMMETRY OF CRYSTALS

See *Crystal symmetry.*

SYMMETRY PLANE, reflection plane, mirror plane

A finite symmetry element, designated by m (mirror reflection), the symmetry operation involving the replacement of an object by its reflection in a plane (by its mirror image). Symmetry planes bisect all straight lines perpendicular to them that connect symmetrical (symmetrically identical) points of the mirror images.

SYNCHROTRON

A machine that accelerates charged particles in circular orbits, often at relativistic speeds.

SYNCHROTRON RADIATION, synchrotron emission

Electromagnetic radiation excited by deceleration of relativistic charged particles. The sources of synchrotron emission are mainly electrons (or other charged particles) in uniform circular motion established by electron accelerators and storage rings. Synchrotron emission was predicted by A. Schott (1912) and then observed in a *synchrotron* electron accelerator, hence its name "synchrotron radiation".

Applications of synchrotron radiation in solid state physics are growing because of its following properties which make it preferable to electromagnetic radiation from other sources:

(1) high intensity in a wide spectral range (most useful is the vacuum UV range where the synchrotron emission intensity is comparable with that of sources with discrete spectra, and the X-ray range where the synchrotron intensity exceeds that of the most powerful X-ray tubes by several orders of magnitude);

(2) a continuous spectrum;

(3) high and well established degree of polarization;

(4) high degree of directionality as a result of which the synchrotron emission intensity decreases inversely proportional to distance;

(5) the possibility of forming ultrashort pulses of duration down to 10^{-11} s;

(6) high vacuum in the source which is essential for examining *solid surfaces.*

With the help of synchrotron radiation spectral optical constants can be determined with high accuracy (reflection and absorption coefficients, *dielectric constants* with their real and imaginary parts), and also the spectrum of *luminescence* excitation and electron emission in the vacuum UV and X-ray ranges. Using a technique based on the angular dependence of *photoelectron emission* the detailed band and electronic structure of *surface electron states* (including impurity) of a number of *metals* and *semiconductors* were determined.

With the help of the *fine structure of X-ray absorption spectra*, the geometric structure of crystalline and noncrystalline solids, *polymers*, and biologically active macromolecules, as well as the details of *phase transitions* and photoinduced changes, have been studied. At this high level of X-ray intensity, including the dependence on the time (resolution up to 1 ns) and photon energy, *X-radiography, X-ray topography*, and *X-ray microscopy* are developing. The X-ray *lithography* of semiconductors using synchrotron radiation is of great practical value.

SYNERGISM (term introduced by H. Haken in the early 1970s)

The field of science that studies general regularities in processes of *self-organization* in various (physical, chemical, biological, etc.) nonequilibrium systems comprising, ordinarily, great numbers of objects (atoms, molecules, more complex subsystems), with their collective behavior giving rise to the formation of macroscopic spatial or temporal structures or systems. From another viewpoint, synergism (synergetics) concerns an interaction of elements when their combined effect exceeds the sum of their individual effects.

Synergism effects are observed in open dissipative systems: space–time *dissipative structures* develop as a result of instabilities due to external energy sources exceeding threshold values. The spontaneous origin of structures is considered in the synergism rationale as a nonequilibrium *order–disorder phase transition*. The analogy to equilibrium *critical phenomena* defines the terminology of synergism: *static space–time correlation, order parameter*, etc. The mathematical formulation of synergism uses nonlinear systems of equations for system order parameters. Many problems of solid state physics associated with synergetic effects are defined in terms of particle creation and annihilation processes (electrons, phonons, etc.), and are expressed in terms of probabilities of transitions from one state to the other. The relevant stochastic definition determines not only the changes of large-scale order parameters with time, but also *fluctuations* of these parameters.

Examples of synergetic effects in solid state physics are:

(1) The formation of a coherent electromagnetic wave in a *laser* (time correlation of acts of *induced radiation*). The effect depends critically on the pumping power, order parameter, electric field intensity.
(2) The formation of an electric field structure in a *semiconductor* due to the *Gunn effect*.
(3) The statistically effective clustering of defects produced by radiation (see *Radiation physics*).
(4) The formation of structures on a solid surface under exposure to laser radiation.

SYNGONY (fr. Gr. $\sigma \upsilon \nu$, together, and $\gamma o \nu \iota \alpha$, angle)

Crystallographic term synonymous with crystal family or crystal system, and commonly used in the Russian literature. There are four families in two dimensions (oblique, rectangular, square and hexagonal), and six families in three dimensions (triclinic, monoclinic, orthorhombic, tetrag-onal, hexagonal, and cubic). Trigonal, which is classified here under hexagonal, is often considered as a separate seventh syngony. See *Crystal system*.

SYNTHETIC MONOCRYSTALS

Artificially grown *crystals* of elements or chemical compounds of needed sizes or quality. Synthetic monocrystals of elements, various alloys, inorganic, organic and complex compounds are distinguished by special features pertaining to particular atomic, ionic or molecular structures. Programmed doping can provide crystals with predetermined types, distributions and numbers of defects (impurities (donors, acceptors), dislocations, etc.) to control a particular property (or set of properties). Monocrystals are grown from melts, solutions in melts (spontaneously and using a seed), aqueous and nonaqueous solutions at standard pressure p and temperature T, using hydrosolvothermal methods at elevated p and T (up to essentially supercritical values); from vapor-gas phase using sublimation, chemical transport, thermal or photoinduced (including resonance absorption) disproportionation of vapors of element-organic compounds, and many other methods (see *Monocrystal growth*).

Semiconducting synthetic crystals are basic materials for practically all devices of modern *microelectronics*, including *integrated optics*. These crystals are often used in the form of epitaxial films with monocrystalline structure. The films are obtained through various *epitaxy* methods (liquid state, gaseous phase, molecular-beam, etc., on various substrates) from the same material (*autoepitaxy*) or from different materials (*heteroepitaxy*).

Synthetic monocrystals of metals exhibit enhanced plasticity to ensure large drawing depth and intricate shape of items produced from them. *Heterodesmic crystals* (with two or more types of bonding) and *ionic crystals* include quartz (piezoelectric, optical, silicon glass). Available monocrystals are *semiconductor materials, active solid-state laser materials, electrooptic materials* and *nonlinear-optical crystals, pyroelectric materials*, acousto-optical materials (see *Acousto-optics*), optical materials (alkali halides), gems (emerald, amethyst, ruby), doubly refracting materials (see *Birefringence*), *superhard materials*, *magnetic materials, molecular crystals.* Growing synthetic monocrystals requires especially pure raw materials (often of standardized grades) and precise technological equipment that provides reliable operation during long continuous cycles (up to 1.5 years for optical quartz). Here, observance of strictly specified process parameters is mandatory, e.g., the pressure interval is from 10^{-8} Pa for molecular-beam epitaxy to many GPa to synthesize superhard crystals; temperatures are up to 3000 K, held constant with all attainable precision.

SYSTEM OF UNITS

See *Units of physical quantities.*

TAMM LEVELS (I.Ye. Tamm, 1932)

Energy levels related to *surface electron states* localized near the interface between media under the condition of conserved translational symmetry along this interface. The quantum number that denotes the Tamm levels is a component of the electron *quasi-momentum* $k_\parallel$ which is parallel to the interface. The necessary condition for the formation of Tamm levels is their simultaneous entering the energy bands of both adjacent media which are forbidden for electrons with the same $k_\parallel$. This means, physically, that there emerges a wave with certain values of the energy which propagates exclusively along the surface as a result of Bragg reflections. The energy spectrum of the Tamm levels is determined from the condition of an exponential decay of solutions of the Schrödinger equation for electrons with increasing distance from the surface. The Tamm levels are concentrated in the energy region near the absolute and relative gaps of the two-dimensional *surface band structure*. In the latter case, if a Tamm level occurs in the quasi-continuous spectral range then the wave function contains an admixture of non-decaying states which are called *Tamm resonances*; they can exist even in the range of the continuous spectrum, higher than the vacuum level. *Surface reconstruction* and *surface relaxation* can lead to the shifting, splitting, and disappearance of the Tamm levels. The observation of Tamm levels is possible in a high vacuum with the aid of *photoelectron spectroscopy* in the X-ray and UV ranges, the elastic reflection of electron, atom, and ion beams from the surface, and so on.

TANABE–SUGANO DIAGRAMS (Y. Tanabe, S. Sugano, 1954)

Diagrams which determine the positions of the energy levels of *ions* of the iron group with an unfilled $3d^n$ electronic shell ($2 \leqslant n \leqslant 8$) in a crystal field of cubic symmetry (see *Crystal field*). The energies of the terms $^{2s+1}\Gamma$ with multiplicity $2s + 1$ classified in respect to irreducible representations Γ of symmetry point group O_h are calculated in the framework of a single-configuration approach which takes into account the interelectron Coulomb interaction and the interactions of $3d$-electrons with the crystal field whose value is characterized by the $3d$-state splitting parameter $10Dq$. The energy of terms measured in units of the Racah parameter B is plotted against the ratio Dq/B, keeping constant the ratio of Racah parameters C/B of the interelectron interaction. The energy is reckoned from the principal state; the break point at the curves of Tanabe–Sugano diagrams responds to a change of the latter, i.e. the transition of the ion from the high-spin state to low-spin state as the crystal field strength increases. The Tanabe–Sugano diagrams are used for data processing involving optical measurements aimed at determining the nature of broad bands and narrow lines in the spectra of *transition metal* ions in crystals.

TANGENTIAL STRESS INTENSITY

A value equal to the square root of the second invariant of the *stress deviator* σ_{ik} (see *Elasticity theory invariants*):

$$T = \left[l_2 \mathrm{Dev}(\sigma) \right]^{1/2}$$

$$= \left(\frac{1}{6} \right)^{1/2} \left[(\sigma_{11} - \sigma_{22})^2 + (\sigma_{22} - \sigma_{33})^2 \right.$$

$$\left. + (\sigma_{11} - \sigma_{33})^2 + 6(\sigma_{12}^2 + \sigma_{23}^2 + \sigma_{13}^2) \right]^{1/2}.$$

It is proportional to the tangential stress t in the plane equally inclined to the principal axes of *stress tensor* (octahedral plane): $t = (2/3)^{1/2} T$.

TANTALUM, Ta

A chemical element of Group V of the periodic system with atomic number 73 and atomic mass 180.9479. Natural tantalum has one stable isotope ^{181}Ta and one radioactive isotope ^{180}Ta with half-life 10^{12} years; 15 radioactive isotopes have been obtained artificially. Outer shell electronic configuration is $4f^{14}5d^36s^2$. Successive ionization energies are 7.7, 16.2 and 30.5 eV. Atomic radius is 0.146 nm; radius of Ta^{5+} ion is 0.068 nm. Oxidation states are -1, $+1$, $+2$, $+3$, $+4$. Electronegativity is ≈ 1.68.

Compact tantalum is a *metal* of steel-gray color. It has a body-centered cubic lattice with parameter $a = 0.33025$ nm. Density is 16.65 g/cm^3 at 293 K; $T_{melting} = 3269$ K, $T_{boiling} = 5573$ K. Heat of melting is 28.89 kJ/mole, heat of evaporation is 7.54 kJ/mole; specific heat is 0.1391 kJ·kg^{-1}·K^{-1} (at 273 K), 0.14084 kJ·kg^{-1}·K^{-1} (at 373 K); coefficient of thermal conductivity is 54.428 W·m^{-1}·K^{-1} (at 293 K); temperature coefficient of linear expansion is $6.55 \cdot 10^{-6}$ K^{-1} (at 273 to 373 K). Electrical resistivity is 0.124 μΩ·m (at 291 K), 0.54 μΩ·m (at 1273 K), 0.87 μΩ·m (at 2273 K). Hardness is 0.882 GPa at purity 99.95%; hardness after electron-beam or zone melting is 0.686 to 0.882 GPa, at 1473 K hardness is 0.196 GPa. Ultimate strength of high purity tantalum is 0.186 to 0.225 GPa; yield limit is 0.184 GPa, relative elongation is 36 to 38%; relative constriction of the cross-section is around 90%. Depending on the impurity content the ultimate strength reaches 1.235 GPa. With increasing temperature ultimate strength is reduced to 0.049 GPa (at 1823 K) and to 0.035 GPa (at 2253 K). Adiabatic elastic moduli of tantalum are $c_{11} = 266.04$, $c_{12} = 160.98$, $c_{44} = 82.47$ (GPa) at 298 K. Elastic modulus is 186.2 GPa; shear modulus is 68.6 GPa. Transition from plastic state to brittle state does not occur until the temperature 23 K. Recrystallization temperature is 1323 to 1773 K depending on purity and on degree of deformation. Electronic work function is 4.1 eV. Tantalum is paramagnetic with magnetic susceptibility $0.849 \cdot 10^{-6}$ (at 298 K). Superconducting transition temperature T_c is 4.38 K. Cross-section for trapping thermal neutrons is 21.3 barn/atom. Due to its exceptional *corrosion resistance* tantalum is used for manufacturing chemical apparatus (heat exchangers, heaters, pipelines, reactors, etc.). Electrical capacitors of tantalum have very small dimensions and considerable capacitance values (see *Tantalum alloys*).

TANTALUM ALLOYS

Alloys based on *tantalum* (Ta). These alloys are usually prepared for the purpose of obtaining an increase in *strength*, in particular *high-temperature strength*, and *corrosion resistance*. *Solid-solution hardening* of Ta is achieved by *alloying* it with tungsten, molybdenum and rhenium. Additions of zirconium and hafnium together with carbon and other interstitial elements bring about precipitation hardening (see *Precipitation-hardened materials*). Also it undergoes combined *hardening* by several elements, as a result of which the *ultimate strength* of 740 MPa is attained (approximately 3 times higher than for unalloyed Ta), and its *plasticity* is reached at 196 °C. In order to increase its corrosion resistance Ta is doped with titanium, aluminum, chromium and beryllium. Tantalum alloys have a low thermal expansion coefficient, low elasticity of vapors, considerable acid resistance. These alloys are obtained by the method of *powder metallurgy* and melting in vacuo. Cold deformation substantially increases the alloy strength. Tantalum alloys in the form of ribbons, sheets, wire, etc., are used in electronic engineering, for manufacturing chemical equipment, and also in rocket engineering.

TEAR

See *Fissures*.

TECHNETIUM, Tc; obsolete name Masurium, Ma

A radioactive element of Group VII of periodic system with atomic number 43 and atomic mass 97.9072. There are no stable isotopes; 21 isotopes are known with mass numbers ranging from 90 to 110, and with half-life periods from 0.8 s to $4.2 \cdot 10^6$ years. No "primary" technetium is found in nature. Approximately $2.5 \cdot 10^{-13}$ g of the isotope ^{99m}Tc with half-life $2.12 \cdot 10^5$ years is contained in 1 g of uranium pitchblende ore. Outer shell electronic configuration is $4d^55s^2$. Successive ionization energies are 7.28, 15.26, 31.9 eV. Atomic radius is 0.136 nm, Tc^{7+} ion radius is 0.056 nm, of Tc^{4+} ion radius is 0.064 nm. Oxidation state is $+7$, $+3$, $+2$, -1, less often $+6$, $+5$, $+4$. Electronegativity is ≈ 1.54.

Technetium is a silvery-gray *metal* with a brown shade, slowly becoming dim in air. It has a hexagonal close-packed crystal lattice, space group $P6_3/mmc$ (D_{6h}^4), with parameters $a = 0.2740$ nm, $c = 0.4398$ nm at room temperature. In thin (less than 15.0 nm) films technetium has a face-centered cubic lattice, space group $Fm\bar{3}m$ (O_h^5), with parameter $a = 0.368$ nm. Density is 11.371 g/cm^3 at 298 K; $T_{melting} = 2445$ K, $T_{boiling} \approx 5010$ K. Heat of melting is ≈ 22 kJ/mole, heat of evaporation is 502 kJ/mole; specific heat is 0.243 kJ·kg^{-1}·K^{-1}; Debye temperature is ≈ 435 K; thermal conductivity coefficient is 50.2 W·m^{-1}·K^{-1}. Adiabatic elastic moduli of technetium monocrystal: $c_{11} \approx 437.36$, $c_{12} \approx 212.91$, $c_{13} \approx 159.2$, $c_{33} \approx 572.02$, $c_{44} \approx 115.83$ (in GPa) at 298 K and zero external pressure. Bulk modulus of polycrystalline technetium in temperature range 4.2 to 298 K varies within the limits 303 to 278 GPa, Young's modulus from 327 to 318 GPa, shear modulus from 124 to 122 GPa, Poisson ratio from 0.320 to 0.309, velocities of longitudinal and transverse elastic vibrations propagation respectively from 6410 to 6220 m/s and from 3300 to 3270 m/s. Coefficient of linear low-temperature electronic specific heat is 4.84 to 6.28 mJ·mole^{-1}·K^{-2}; specific resistivity of technetium polycrystal in temperature range 295 to 373 K varies from 140 to 690 nΩ·m. Work function of technetium polycrystal is 5.3 eV. Superconducting critical temperature is 7.8 K; critical magnetic field is 141 mT (0 K). Technetium is paramagnetic with molar magnetic susceptibility $+268\cdot10^{-6}$ CGS units (at 298 K). Technetium possesses high *corrosion resistance*, therefore it is prospective for *corrosion protection* of cooling ducts of nuclear reactors. It is used as a component of catalysts, and also for β-radiography, for strength determination of paper, tissues and other products from light materials; for gas ionization in gas discharge installations as a standard radiation source. ^{99m}Tc isotope with $t_{1/2} = 6.04$ h is effectively used in preparations for radio-isotope medical diagnostics as a source of γ-radiation with low, measurable dose; it is not toxic, but accumulates rapidly in the human organism, and after radioactive transformation it is completely eliminated by physiological processes.

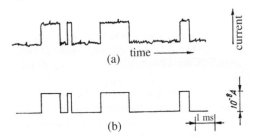

Representative sketch of telegraph noise current: (a) with superimposed random noise, and (b) after removing the latter.

TELEGRAPH NOISE, breakdown noise, explosion noise

Specific *fluctuations* of current observed in some solid-state devices, such as *diodes* and *transistors* at *semiconductor junctions, tunnel diodes*, and composition *resistors*. In a simple case telegraph noise is displayed as a bistable electronic signal of a step form and constant amplitude, but with randomly distributed time intervals between the steps, i.e. it resembles a random telegraph signal (see Fig.) If telegraph noise is amplified and reproduced through a loudspeaker, it reminds a person of the snaps of bursting corn grains, hence its name *popcorn noise*. Several known causes of telegraph noise are: random turning on and off of the surface current channel in a *p–n* junction, motion of *dislocations* through a *p–n* junction and, possibly, the presence of a *defect* of the metal sediment type in the region of the junction. Clarification of the nature of telegraph noise helps to improve the technology of *semiconductor device* manufacturing.

TELLURIUM (Lat. *tellus, telluris*, the earth), Te

A chemical element of Group VI of the periodic system with atomic number 52 and atomic mass 127.60. Natural tellurium is a mixture of eight stable isotopes with mass numbers 120, 122 to 126, 128, 130. Outer shell electronic configuration is $4d^{10}5s^25p^4$. Successive ionization energies are 9.01, 18.8, 31.0, 38.0, 66.0, 83.0 eV. Atomic radius is 0.144 nm; radius of Te^{2-} ion is 0.224 nm, of Te^{4+} ion 0.070 nm, of Te^{6+} ion 0.056 nm. Oxidation state is -2, $+4$, $+6$. Electronegativity is 2.05.

Massive tellurium is a glistening silvery-gray substance with metallic luster. It has a hexagonal crystal lattice with parameters $a = 0.4457$ nm, $c = 0.5929$ nm. Density is 6.22 g/cm^3 (at 293 K); $T_{\text{melting}} = 722.6$ K, $T_{\text{boiling}} = 1263$ K. The "amorphous" modification of tellurium (dark-brown powder), transforms irreversibly into the crystalline state upon heating. Heat of melting is 134.4 kJ/kg, heat of evaporation is 449.4 kJ/kg; specific heat is 0.2010 kJ·kg^{-1}·K^{-1} (at 298 K); coefficient of thermal conductivity is 5.86 W·m^{-1}·K^{-1} (at 293 K); temperature coefficient of linear expansion of polycrystalline tellurium is $\approx 16.5 \cdot 10^{-6}$ K^{-1} Adiabatic elastic moduli of tellurium of trigonal space group $P3_121$ (D_3^4) with unit cell parameters $a = 0.4495$ nm, $b = 0.374$ nm, $c = 0.5912$ nm, $d = 0.286$ nm and with angle $\alpha = 102.6°$ are $c_{11} = 32.8$; $c_{12} = 8.6$; $c_{44} = 31.4$; $c_{13} = 25.0$; $c_{33} = 72.2$; $c_{14} = 12.3$ (in GPa). Mohs hardness is 2.0 to 2.5; average microhardness is 0.568 GPa. Modulus of normal elasticity is 41.16 GPa. Monocrystals of tellurium oriented along (0001) undergo brittle failure at the stress 0.137 GPa.

Tellurium is a *semiconductor* with the band gap width 0.34 eV. Electrical conductivity of tellurium depends on its purity and on the degree of crystal perfection; in the purest samples it is ~ 2 (Ω·m)$^{-1}$ with the mobility of electrons 0.1700, and of holes 0.1200 m^2·V^{-1}·s^{-1}. During the course of melting tellurium transforms to the metallic state. Tellurium is diamagnetic with specific magnetic susceptibility $-0.3 \cdot 10^{-6}$ cm^3/g. Compounds of tellurium with metals are used in the television tubes, in dosimeters, in radiation counters, etc. In *metallurgy* tellurium serves as the alloying addition to lead for improving its mechanical properties.

TEMPERATURE

A physical (thermodynamic) quantity which characterizes the equilibrium state of a macroscopic *thermodynamic system*. The temperature can be regarded as a measure of the energy contained in a body: the higher the temperature, the greater its energy. It is the same and positive in all parts of an isolated system in an equilibrium state. If the system is not in equilibrium, then *heat transport* takes place, in accordance with the second law of thermodynamics, from regions at higher temperature to regions at lower temperature (the temperature decreases in the former and rises in the latter regions) until the system equilibrates at a final intermediate uniform temperature. In the general form, the temperature can be determined as the inverse value of the derivative of the material *entropy S* with respect to its energy: $1/T = dS/dE$ (at constant volume). This expression for T, called the *absolute temperature* or *thermodynamic temperature*, can take on values from zero (*absolute zero of temperature*) to infinity. It is possible to construct an absolute temperature scale if a certain empirical parameter $\tau(T)$ with an unambiguous relationship with T is known. Within the accuracy of an arbitrary factor that determines the scale, such a construction is carried out with the help of the equation

$$\frac{d\ln T}{d\tau} = -\frac{(\partial V/\partial \tau)_p}{(\partial Q/\partial p)_\tau},$$

where the experimentally observable quantities on the right-hand side are as follows: $(\partial V/\partial \tau)_p$ is the volume variation of the dependence of τ under constant pressure p; $(\partial Q/\partial p)_\tau$ is the amount of heat transferred to the body when varying p while keeping τ, and therefore T, constant.

Temperature is usually expressed in terms of the convenient and historically accepted scales of Celsius or Kelvin degrees, °C or K. In the international absolute thermodynamic scale (*Kelvin scale*) absolute zero is taken as zero, and the temperature of the *triple point* of water is accepted to be exactly 273.16 K. To convert into energy values, the temperature is multiplied by the *Boltzmann constant* $k_B = 1.3807 \cdot 10^{-16}$ erg/K $= 1.3807 \cdot 10^{-23}$ J/K. The commonly used *Celsius scale* (*centigrade scale*) where the temperature is designated as °C, sets the ice melting point at 0 °C, and 100 °C corresponds to the boiling point of water under normal pressure. In this scale, $T_C = T_K - 273.16$, with equal increments: $\Delta 1$ °C $= \Delta 1$ K. Old-fashioned scales are the *Réaumur* scale with $T_R = 0.8 T_C$, and the *Fahrenheit scale* $T_F = 32 + (9/5)T_C$.

In statistical physics the absolute temperature determines the filling of energy levels ε_i through the Boltzmann factor $\exp[-\varepsilon_i/(k_B T)]$ (see *Gibbs distribution, Boltzmann distribution*); the greater the energy, the less the filling. In classical physics (valid for solids at high temperatures) the absolute temperature is directly proportional to the mean kinetic energy of the atoms of the solid. In fact, the temperature is only defined for systems in a completely equilibrium state. Nevertheless, in many cases the concept of temperature can be used under incomplete equilibrium (*quasi-equilibrium states*). So, if the time for establishing equilibrium inside a subsystem is significantly less than that for equilibrating the temperature between the subsystems (even inside the same body) then one can speak about the individual temperature of each of these subsystems during finite durations of time (times much less than that needed for establishing equilibrium between the subsystems). It is in this sense that the concepts of *electron temperature, phonon temperature, magnon temperature, spin temperature*, and *impurity temperature* are used. For systems with discrete energy levels (e.g., for impurity centers in a crystal), there is a probability that for a sufficiently long time the population of levels with greater energy would exceed the population of lower energy levels. Such quasi-stationary states are described with the help of the concept of *negative temperature*. In the sense indicated above, a negative temperature is higher than a positive temperature, since the lower its absolute value, the greater the inverse population (the positive and negative temperature scales meet at $T = \pm\infty$). Systems with negative temperatures (e.g., spin systems) are the basis for quantum generators, oscillators and amplifiers (see *Laser, Maser, Quantum amplifier*).

TEMPERATURE CONDITIONS OF PLASTIC DEFORMATION

Plastic deformation and *pressure* treatment of crystalline materials including regions of hot, warm, and cold *strain*. The positions of these regions depend on the nature of interatomic bonds and the type of structure; this is determined using the characteristic temperature scale with parameter $\alpha = U/(k_B T_{\text{melting}})$ where U is the activation energy of the *dislocation* motion, and T_{melting} is the *melting temperature* (see Fig.).

Hot deformation of a material is carried out at a temperature above its *recrystallization* temperature T_R (region I), thus bringing about the

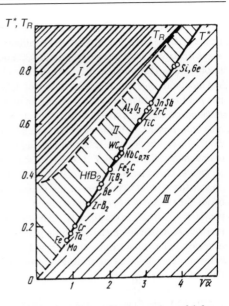

Temperature dependence of various regions of deformation of crystalline materials: I, II and III are regions of hot, warm and cold deformation, respectively; T^* is the characteristic deformation temperature; T_R is the recrystallization temperature.

formation of an equiaxial grain structure free of dislocations and *internal stresses*. Recrystallization can occur both directly during the process of shape variation (*dynamic recrystallization*) and during material heating between reductions (*static recrystallization*). The boundary between the regions of warm and cold deformation is the *characteristic temperature of deformation T^**. In this case $T^* \approx 0.22\sqrt{\alpha}$ as shown by the data plotted in the figure. The value T^* can be defined as the temperature of the sharp growth of the *yield limit* and the *hardness* for decreasing temperature. A *warm deformation* occurs in the temperature range $T^* < T < T_R$ (region II). In this region there is a nonequiaxial elongation in the stretching force direction, and grain structure is formed; inside the grains there is a cellular, dislocation permeated, and fragmented structure (see *Cellular structure, Fragmentation*) forms. The level of internal stresses rises as the degree of deformation grows up to a certain value; then this stabilizes or even reduces a little. A *cold deformation* takes place at temperatures $T < T^*$ (region III). As in the case of the warm deformation, a nonequiaxial grain structure is formed, however, inside the grains a cellular structure does not form, although a high density of randomly distributed dislocations is observed. The internal stresses grow rapidly and monotonically with increasing extent of deformation. In some materials deformation-stimulated *twinning of crystals* as well as the formation of *faults* takes place. Failure below temperature T^* exhibits a brittle or quasi-brittle nature; failure at temperatures higher than T^* is tough (see *Brittle failure, Quasi-brittle failure, Tough failure*). The greater the covalent character of the interatomic bonds (see *Covalent bond*), the higher the temperature T^*; in all cases $T_R > T^*$. The primary pressure treatment of metal ingots is performed in region I. The final treatment aimed to form an optimal combination of mechanical properties is implemented in region II.

TEMPERATURE CONDUCTIVITY FACTOR
See *Diffusivity*.

TEMPERATURE OF ELECTRONS
See *Electron temperature*.

TEMPERING

A type of *heat treatment* of alloys, carried out after *quenching*, and consisting of heating to a certain temperature with subsequent cooling. The term "tempering" is applicable mostly to the heat treatment of *steel*. Tempering of nonferrous metals is usually called *artificial aging* (see *Alloy aging*). Hardening by quenching causes steel to become not only hard, but also brittle (see *Alloy brittleness*), which is undesirable. Besides, high hardness hampers the finishing of parts. The tempering procedure is applied to decrease the brittleness and increase the *plasticity* of hardened steel. There is low (120–250 °C), medium (300–400 °C) and high temperature (450–650 °C) tempering. The tempering regime is selected to obtain the required ratio between the *strength* and the plasticity of steel.

TEMPORAL DIELECTRIC DISPERSION,
frequency dielectric dispersion

A dependence of the components of the *dielectric constant* tensor $\varepsilon_{\alpha\beta}$ on the frequency ω, caused by the retardation of the *linear response* of the

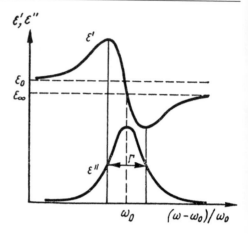

Temporal dielectric dispersion.

medium (in particular, of a *solid*) to a time dependent electromagnetic field. As a result of this delay the polarization of the medium $P(t)$, and consequently, the electric field induction $D(t)$ at a given moment of time t, due to the causality principle, is determined by the values of the external field $E(t)$ at all earlier moments of time. In an isotropic dielectric crystal (gas, liquid) in the long-wave limit $k \to 0$ the typical temporal dispersion has the form

$$\varepsilon(\omega) = 1 + \sum_{j \neq k} \frac{\omega_p^2 f_{jk}}{\omega_{jk}^2 - \omega(\omega + i\gamma_{jk})},$$

where ω_p is the electronic plasma frequency, ω_{jk} are the frequencies of the dipole-permitted band-to-band (level-to-level) transitions, and f_{jk} and γ_{jk} are the corresponding oscillator strengths and attenuation decrements, while the real $\varepsilon'(\omega)$ and imaginary $\varepsilon''(\omega)$ parts of the dielectric constant should satisfy the *Kramers–Kronig relations*. In Fig. the frequency dependences of ε' and ε'' are shown in the vicinity of one of resonances at $\omega = \omega_0$ with the decrement Γ.

TENSION, SURFACE
See *Surface tension*.

TENSORESISTIVE EFFECT
See *Piezoresistive effect*.

TENSOR OPERATOR, IRREDUCIBLE

See *Irreducible tensor operators*.

TERA... (fr. Gr. τερας, monster)

A prefix added to the name of a physical quantity to form a multiple value equal to 10^{12} of the original units. The symbolic designation is T; e.g., 1 TN (one teranewton) $= 10^{12}$ N.

TERBIUM, Tb

A chemical element of Group III of the periodic system with atomic number 65 and atomic mass 158.9254; it is one of the lanthanide elements. There are 19 isotopes known with mass numbers from 146 to 164. Natural terbium has one stable isotope ^{139}Tb. Outer shell electronic configuration is $4f^9 5d^0 6s^2$. Ionization energy is 6.74 eV. Atomic radius is ≈ 0.1776 nm; radius of Tb^{3+} ions is ≈ 0.0927 nm, of Tb^{4+} ion − 0.081 nm. Oxidation states are +3, +4. Electronegativity is ≈ 1.22.

Terbium is a silvery-white *metal* which exists in two polymorphics modifications (α-Tb, β-Tb). Below ≈ 1575 K α-Tb is stable with a hexagonal close-packed crystal lattice, space group $P6_3/mmc$ (D_{6h}^4) and lattice parameters $a = 0.3605$ nm, $c = 0.5697$ nm at room temperature Above 1575 K up to $T_{melting} \approx 1630$ K β-Tb with body-centered cubic lattice is stable, space group $Im\bar{3}m$ (O_h^9) and parameter $a = 0.402$ nm (extrapolated value) after hardening. Density is 8.234 g/cm^3 (at 298 K); $T_{boiling} \approx 3150$ K. Bonding energy of terbium is -4.1 eV/atom at 0 K. Heat of melting is ≈ 13 kJ/mole, heat of evaporation is 290 kJ/mole, heat of sublimation is 293 kJ/mole; specific heat is 0.179 kJ·kg^{-1}·K^{-1} at room temperature. Debye temperature is ≈ 173 K; linear thermal expansion coefficient of polycrystal is $\approx 8.5 \cdot 10^{-6}$ K^{-1}; thermal conductivity coefficient is 14 W·m^{-1}·K^{-1}. Adiabatic elastic moduli of α-Tb monocrystal: $c_{11} = 67.88$, $c_{12} = 24.32$, $c_{13} = 22.99$, $c_{33} = 72.25$, $c_{44} = 21.40$ (in GPa) at 300 K; isothermal bulk modulus is 39.9 GPa; Young's modulus is 57.5 GPa; shear modulus is 22.85 GPa; Poisson ratio is 0.261; ultimate tensile strength is 0.167 GPa. Brinell hardness is 0.686 GPa (terbium of 99.9% purity). Effective thermal neutron cross-section is 44 barn. Resistivity of terbium monocrystal is 1020 nΩ·m along the principal axis 6 and 1230 nΩ·m in the perpendicular direction (at 298 K); temperature coefficient of electric resistivity of polycrystalline terbium is 0.00091 K^{-1}. Work function from polycrystal of terbium is 3.09 eV. Two temperature regions of magnetic ordering are observed in terbium: in the temperature range T_C to T_N it is a helical antiferromagnet with helicoid axis along the principal axis 6, in the range 0 K to T_C it is a collinear ferromagnet; in both magnetic structures the magnetic moments lie in the basal planes of α-Tb. With the reduction of terbium purity to 99.8% T_N grows from 229 to 232 K. At the Curie point T_C ($218 < T_C < 222.5$ K) there occurs a first-order phase transition to the ferromagnetic state, when α-Tb crystal is subjected to small jumpwise distortions, reducing its symmetry to rhombic in the low-temperature region. Magnetic susceptibility of terbium is $115,000 \cdot 10^{-6}$ CGS units; nuclear magnetic moment of ^{159}Tb is 1.52 nuclear magnetons. Pure terbium is used for investigations of, and its alloys and compounds are prospective for the manufacture of *magnetic materials* and of *luminophors*.

TERM SPLITTING in crystal field

A partial or a total lifting of the *degeneracy* of the energy levels of *ions* with unfilled electron shells as a result of their presence in a *crystal lattice*. The qualitative nature of the energy level structure of paramagnetic ions is based on *group theory* considerations. The expansion of the irreducible representations of the rotation group corresponding to the ^{2S+1}L-terms or the (LSJ) multiplets of a free ion into the irreducible representations of a *point group* of *local symmetry* for the ion in the lattice defines the number and the type of states into which the given term (multiplet) splits in a *crystal field*. The splitting magnitudes of the multiplets increase with the increase of the ionicity of the *chemical bonds* in the lattice, and their order of magnitude reaches 10^4 cm^{-1} for transition ions with partly filled nd^N atomic shells, and 10^3 cm^{-1} for *lanthanide* and *actinide* ions with partly filled nf^N shells. To compute the term splittings, one can employ techniques of perturbation theory using various models of the crystal fields.

TESTING, NONDESTRUCTIVE

See *Nondestructive testing techniques*.

TETRACRITICAL POINT, quadruple critical point

A point at the *phase diagram* where four lines of *second-order phase transitions* coincide. Quadruple critical points occur on diagrams of "uniaxial mechanical stress versus temperature" of some crystals which undergo *structural phase transitions*, on compositional diagrams of some magnets, and so on. See also *Multicritical point*.

TETRAGONAL SYSTEM

Crystallographic system determined by the presence in crystals of rotational or inversion tetrad *symmetry axes*; referred to the tetragonal *crystal class*. The *unit cell* of crystals of the tetragonal system is a rectangular prism with parameters $a = b \neq c$ where the c/a can be either greater or less than unity. This system includes two *Bravais lattices*, seven *point groups*, and 68 *space groups*.

TEXTURE, crystallographic texture

The presence of a prevailing orientation of *crystallites* in *polycrystals* or of molecules in amorphous solids such as *polymers*. Texture appears in materials whose macrostructure elements demonstrate anisotropic physical properties under elastic stress, in electric and magnetic fields, with changes in temperature, and for various combinations of these actions. The texture is one of the most important factors determining the anisotropy of material properties. The formation of texture is accompanied by a reduction of the symmetry of material physical properties.

A texture is differentiated by its origin and type. There exists: (1) a *growth texture* formed during *crystallization*, various kinds of deposition, *condensation*, epitaxial growing (see *Epitaxy*), adsorption; (2) a *recrystallization texture* formed as a result of *annealing, phase transitions*, recrystallization in the field of actual forces, and so on; and (3) a *deformation texture* that results from plastic yield of material under treatment by *pressure*. In some cases, the texture is classified with respect to the type of prior treatment, e.g., rolling texture, wiring texture, deposition texture, and so on. The types of texture are the following: axial, spiral, limited, multicomponent, etc., which differ in the character of the orientations of crystallites (molecules) with respect to the principal directions of the material, such as the axis of a wire, longitudinal and transverse directions in rolled sheet, and so on. As a consequence of crystallographic texture (or by analogy with it), optical, magnetic, piezoelectric, etc. textures are considered. Textures characterize man-made as well as natural origin materials. Texture is studied by *X-radiography* (see *Texture diffractometer*), *electron diffraction analysis*, and *neutron diffractometry*; magnetic and optical methods are also employed. Texture is described either with the help of experimentally measured pole figures, or analytically using an *orientation distribution function*.

The orientation distribution function is designated in space by *Eulerian angles*. As a rule, this function cannot be measured directly, but it can be reconstructed from one or several experimentally measured pole figures. It can be represented in the form of an expansion in a series of generalized spherical functions. The expansion coefficient of the orientation distribution function is determined from the expansion coefficients of the measured pole figures. There exists a method of obtaining this distribution function with the help of its approximation by statistically justified *Gaussian distributions* obtained on the basis of the *central limit theorem* for the rotation group. Some other methods for calculating this distribution function have limited application. A knowledge of this function permits one to calculate properties (e.g., anisotropies) of materials, or to predict the behavior of textured material under various actions; to compute pole figures; to study analytically the physical processes leading to the formation of texture.

TEXTURE DIFFRACTOMETER

An apparatus for detecting an X-ray diffraction pattern of a *polycrystal* in order to determine its prevailing orientations in respect to macroscopic coordinates such as the drawing axis, the plane and direction of rolling, the direction of growth of film *coatings* or *minerals*. A texture diffractometer contains an *X-ray source, X-ray goniometer*, computer-controlled texture-attachment, detector equipped with instrument for measuring the intensity of diffracted radiation, computer-graphic device for data processing. A sample is fixed at the texture-attachment; its surface of interest is positioned in coincidence with the rotation axis of the goniometer; then the sample and the detector are set at reflection angles corresponding to a given crystallographic plane. In accordance with a given program, the sample placed at the texture-attachment rotates both in its own plane and about an axis in this plane (angle α), thus bringing about displacements of the X-ray beam spots or poles of reflection planes over the projection sphere. The measured intensity, which takes into account the defocusing and the beam absorption, is directly proportional to the spot or pole density. In the reflection experiment, the whole projection sphere is not available due to the strong absorption of radiation at large sample inclinations. The resulting missing information is obtained from transmission experiments using a rotation of a thinner sample at an angle α about the vertical axis lying in the sample plane. Microcomputer-assisted control of the texture diffractometer allows one to have complete information within a minimal duration of the experiment by choosing the optimal combinations of angular increments and exposure in regions with different pole density. The computer performs the accumulation of primary information, its correction, the plotting of *pole figures*, the quantitative analysis of the pole density distribution, and the calculation of the anisotropy of physical properties.

TEXTURE, MAGNETIC

See *Magnetic texture*.

TEXTURE OF LIQUID CRYSTALS

A configuration of the field of the order parameter; a visually observable pattern in a *liquid crystal*. The *texture* is determined by the character of the structure of the liquid crystal at the molecular level, by boundary conditions at the sample surface (see *Boundary effects*), the sample prehistory, and the type and the importance of external (mechanical, temperature, electromagnetic) actions.

Classified among principal textures in thin (5–10 μm) flat samples in the absence of external actions are: planar, homotropic, and schlieren textures in *nematic liquid crystals*; planar, "fingerprints", confocal textures in *cholesteric liquid*

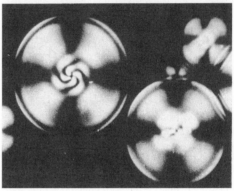

Texture in liquid crystals: polygonal texture in smectic A (above) and nematic with spherical drops (below).

crystals; planar, homotropic, and confocal textures in smectic liquid crystals. Ideal *planar textures* and *homotropic textures* do not contain any macroscopic defects; in the former case, the molecules are oriented parallel to the planar surfaces confining the sample, in the latter case they are normal to these surfaces. *Confocal textures* of liquid crystals are formed by a system of *confocal domains* in the form of conical bodies each of which contains a pair of defect lines, i.e. an ellipse and a hyperbola. Depending on the arrangement of these lines one can differentiate a *polygonal confocal texture* (ellipses in plane of boundary, see Fig.) and a *fan confocal texture* (ellipses normal to boundary). The presence of *disclinations* and *boojums* is specific in *schlieren textures*; in smectics the schlieren textures contain only disclinations. A *liquid-crystal texture "fingerprint"* in cholesterics is formed if a spiral twisting axis is situated along the sample surface. The textures which appear in external fields, are often of a dynamic type. These are represented by one- or two-dimensional modulation of the original (e.g., planar) texture, and form as a result of the motion of liquid-crystal material due to, e.g., a temperature gradient or charge transfer. One of the main dynamic textures is the *Kapustin–Williams domain texture* (see *Structural domain*) in the form of alternating dark and light strips related to cylindrical units of circulating matter. The observation of a texture in a microscope is a fundamental and reliable technique of identification of numerous liquid-crystal phases.

TEXTURE OF METALS

See *Structural texture of metals*.

THALLIUM, Tl

A chemical element of Group III of the periodic system with atomic number 81 and atomic mass 204.383. Natural thallium has 2 stable isotopes ^{203}Tl (29.52%) and ^{205}Tl (70.48%). Outer shell electronic configuration is $4f^{14}5d^{10}6s^26p^1$. Successive ionization energies are 6.106, 20.42, 29.8 eV. Atomic radius is 0.171 nm; radius of Tl$^+$ ion is 0.147 nm, Tl^{3+} is 0.095 nm. Oxidation state is $+1$, $+3$. Electronegativity of Tl$^+$ is 1.40 and of Tl^{3+} is ≈ 1.45.

Bulk thallium is a silvery-bluish *metal*, in air it is easily oxidized and quickly becomes dim. It exists in three allotropic modifications (α-Tl, β-Tl, γ-Tl). Under normal conditions α-Tl is stable; it has hexagonal close-packed lattice with parameters $a = 0.34564$ nm, $c = 0.55310$ nm (at 293 K), space group $P6_3/mmc$ (D_{6h}^4). Above 508.4 K up to $T_{\text{melting}} = 576.75$ K, β-Tl exists with body-centered cubic lattice, space group $Im\bar{3}m$ (O_h^9) with parameter $a = 0.3882$ nm at 535 K. Under high pressures (0.12 to 3.9 GPa) γ-Tl is formed with face-centered cubic lattice, space group $Fm\bar{3}m$ (O_h^5). Binding energy is 1.87 eV/atom (at 0 K). Density is 11.85 g/cm^3 (at 293 K); $T_{\text{boiling}} = 1730$ K. Heat of melting is 4.31 kJ/mole, heat of evaporation is 166 kJ/mole, heat of sublimation is 167.5 kJ/mole. Specific

heat is 0.136 kJ·kg^{-1}·K^{-1} (at 290 to 370 K). Debye temperature is 78.5 K, linear thermal expansion coefficient of thallium monocrystal is $36.3 \cdot 10^{-6}$ K^{-1} along the principal axis $\underline{6}$ and $26.1 \cdot 10^{-6}$ K^{-1} perpendicular to this axis at room temperature; thermal conductivity coefficient is 39.4 W·m^{-1}·K^{-1} at room temperature and normal pressure. Adiabatic elastic moduli of monocrystal: $c_{11} = 40.83$; $c_{12} = 35.40$; $c_{13} = 28.96$, $c_{33} = 52.91$, $c_{44} = 7.27$ (in GPa) at 295 K. Bulk modulus is 37.0 GPa, Young's modulus is 18.65 GPa, shear modulus is 6.58 GPa, Poisson ratio is 0.4160 (at 298 K), breaking strength is 0.0088 GPa. Brinell hardness is 3.0. Self-diffusion coefficient is $5.56 \cdot 10^{-16}$ m^2/s along the principal axis $\underline{6}$ and $7.81 \cdot 10^{-16}$ m^2/s perpendicular to this axis at 461 K. Ion-plasma frequency of thallium is 8.25 TGz, low-temperature linear term of molar electronic specific heat is 1.47 mJ·mole^{-1}·K^{-2}. Electrical resistivity of thallium is $0.164 \cdot 10^{-6}$ Ω·m (at 295 K), temperature coefficient of electrical resistivity is 0.00517 K^{-1}. Work function of polycrystal is 3.7 eV. Superconducting transition temperature T_c of thallium is 2.38 K, critical magnetic field is 17.8 mT, London magnetic field penetration depth is 92 nm (at 0 K). Molar magnetic susceptibility of thallium is $49 \cdot 10^{-6}$ CGS units at room temperature. Nuclear magnetic moment of isotope ^{205}Tl is $+1.612$ nuclear magnetons. Compounds of thallium are used in the production of materials for optical, luminescent and photoelectronic devices, and in high temperature superconductors. Thallium and its compounds are toxic.

THEORETICAL STRENGTH

Material *strength* calculated on the basis of analytical data on *interatomic interaction potentials* assuming an ideal defect-free solid structure and simultaneous tearing or *shear* over the whole surface of the interface. Estimates of the theoretical tensile stress of various materials are in the range $(0.05–0.2)E$, where E is *Young's modulus*. The theoretical shear strength is estimated to be in the range from $G/(2\pi)$ to $G/(10\pi)$, where G is the *shear modulus*. Taking into account the actual defect structure of solids, and also the influence of the thermal motion on the *failure* process, reduces the actual strength values significantly compared to the maximum values of theoretical strength calculated in defect-free solids at the temperature of absolute zero. The presence of discrepancies between the theoretical strength and the real strength stimulated the pursuit of studies of the influence of *defects* on strength properties of actual solids (including the role of *cracks* for the theoretical tensile strength, and the role of *dislocations* for the shear theoretical strength), as well as the development of technologies for obtaining materials whose strength approaches the theoretically calculated values. For some defect-free materials the actual strength is close to the theoretical value.

THERMAL ACTIVATION ANALYSIS

See *Activation energy*.

THERMAL CONDUCTIVITY, heat conductivity

A process of *heat transport* from more heated parts of a body to less heated parts which is not accompanied by the macroscopic movement of atoms of the medium. For a not very large temperature difference (relative variation of temperature T along a carrier *mean free path* l is small, i.e. $l(\mathrm{d}T/\mathrm{d}x) \ll T$), the *Fourier law* is valid:

$$q = -\hat{\kappa}\,\mathrm{grad}\,T, \tag{1}$$

where q is the heat flux vector, i.e. the amount of heat passed per unit time through a unit area oriented normal to vector q, κ is the *thermal conductivity coefficient* which in crystals is, in general, a symmetric second-rank tensor. The negative sign in Eq. (1) points out that the heat flux is directed against the temperature gradient, i.e. from hotter to cooler regions. The heat propagation in a body is described by the thermal conduction equation which in the absence of the heat sources takes the form:

$$\rho c_V \frac{\partial T}{\partial t} = \nabla \cdot q, \tag{2}$$

where ρ is the density of the material, and c_V is the *specific heat* at constant volume. In typical cases one can neglect the temperature dependence of parameters κ, ρ, and c_V, so Eqs. (1) and (2) combine to form the following linear second-order partial differential equation

$$\rho c_V \frac{\partial T}{\partial t} = -\kappa \nabla^2 T.$$

Even with this simplification, finding the function $T(t, r)$ becomes quite a complicated task when it

is necessary to take into account the dependence of the parameters of the problem on the coordinates. This nonlinear problem which takes into account the temperature dependencies of the parameters is soluble in closed form only in some special cases.

THERMAL CONDUCTIVITY COEFFICIENT, heat conductivity factor

A parameter which depends on the temperature and composition of the material medium which supports the *thermal conductivity* process. It is a coefficient of proportionality between the heat flux and the temperature gradient; having the dimensionality $\mathrm{W}\cdot\mathrm{m}^{-1}\cdot\mathrm{K}^{-1}$. In anisotropic bodies, the thermal conductivity coefficient is a second-rank tensor, which for the cubic crystals considered below reduces to a scalar value.

For gases, where the atoms possess only translational degrees of freedom, kinetic theory gives the following expression for the thermal conductivity coefficient κ:

$$\kappa = \frac{1}{3}c_V v\lambda, \tag{1}$$

where c_V is the *specific heat* of a unit volume at constant volume, v is the average velocity of the gas atoms (carriers of heat), and λ is the *mean free path* of atoms between collisions. In some cases, Eq. (1) can be used to estimate the thermal conductivity coefficient in other materials.

The heat transport carriers in solids are collective excitations: *crystal lattice vibrations* (phonons, κ_{ph}), *conduction electrons* (in *metals* and *semiconductors*, κ_{e}), and also oscillations of magnetization (*magnons*, κ_{m}) in magnetic materials. To a high degree of accuracy (disregarding interference effects) one can regard the total thermal conductivity coefficient κ as a sum of the three contributions κ_{ph}, κ_{e}, and κ_{m}:

$$\kappa = \kappa_{\mathrm{ph}} + \kappa_{\mathrm{e}} + \kappa_{\mathrm{m}}. \tag{2}$$

In metals the principal contribution arises from the conduction electrons (κ_{e}) while the other terms are 1 to 2 orders of magnitude lower in value. Therefore, the *thermal resistivity* W (inverse of thermal conductivity coefficient) can with sufficient accuracy be represented by the *Matthiessen rule* (1864):

$$W^{\mathrm{e}} = \frac{1}{\kappa_{\mathrm{e}}} = W_{\mathrm{d}}^{\mathrm{e}} + W_{\mathrm{ph}}^{\mathrm{e}}, \tag{3}$$

where the terms W_d^e and W_{ph}^e are related to the conduction electron *scattering* off lattice *defects* and phonons, respectively.

The value of W_d^e is related to the residual metallic electric resistivity ρ_{cond} (that is, the temperature-independent part of the total resistivity due to the scattering of electrons off lattice defects) by a simple expression called the *Wiedemann–Franz law*:

$$W_d^e = \frac{1}{LT}\rho_{cond},$$
$$L = \frac{\pi^2}{3}\left(\frac{k_B}{e}\right)^2, \tag{4}$$

where L is the *Lorenz number* for the case of degenerate electrons ($L = 2.445$ W$\cdot\Omega$/deg^2). In those crystals where the *Fermi surface* has a spherical shape,

$$\rho_{cond} = \frac{m}{ne^2}\frac{1}{\tau},$$

where m and n are the conduction electron *effective mass* and concentration, respectively, τ^{-1} is the part of their inverse lifetime which is determined by the scattering on defects. As can be seen from Eq. (4), the value of W_d^e grows $\sim 1/T$ as the temperature decreases, and at sufficiently low temperatures this becomes the dominant contribution to the electron thermal resistance (in all metals at this temperature the Wiedemann–Franz law is quite valid for the total thermal conductivity). The term W_{ph}^e in Eq. (3) at $T \ll \Theta$ (here Θ is the *Debye temperature*) grows with increasing temperature: $W_{ph}^e = BT^2$ (e.g., for copper the parameter is $B = 2.55 \cdot 10^{-7}$ m$\cdot$W$^{-1}\cdot$K^{-1}), and then at $T \gg \Theta$ it approaches its asymptotic constant value W_0^e, which, in fact, determines the thermal conduction of metals at high temperatures. There is an extrapolation formula for W_{ph}^e which is similar to the *Bloch–Grüneisen formula* for the electric resistance. For W_0^e the Wiedemann–Franz law is also valid:

$$W_0^e = \frac{1}{LT}\rho(T),$$
$$\rho(T) \sim T.$$

At intermediate temperatures $T \leqslant \Theta$, when W_{ph}^e becomes of the same order of magnitude as

W_d^e, the Wiedemann–Franz law no longer holds. In this temperature range the two terms in Eq. (3) describing the interference of conduction electron scattering off defects and phonons must sometimes be taken into account. The temperature dependence of total electron thermal conduction κ_e has the form of a curve with a maximum.

In *insulators* the phonons are the carriers of heat, and the thermal conductivity coefficient is determined only by the κ_{ph} term in Eq. (2). Solving the *kinetic equation* which determines the value of κ_{ph} is quite a complicated task since it is necessary to take into account various mechanisms of phonon–phonon scattering and *phonon scattering* off defects. Therefore, many solutions are known, but only in a qualitative form where the principal kinds of temperature dependence of κ_{ph} are characterized.

At $T \gg \Theta$ the phonon *specific heat* is constant, and the phonon mean free path has the dependence $\sim 1/T$. One can see from the Eq. (1) (where v is of the order of the sound velocity) that, as a result, $\kappa_e \sim 1/T$. At temperatures below the Debye temperature (but not very low), the main contribution to κ_{ph} comes from the processes of phonon scattering which are naturally divided into two kinds: the normal processes and the Umklapp processes. In *normal processes* the *quasi-momentum* is conserved, therefore, they themselves cannot provide a non-zero thermal resistance. In *Umklapp processes* the total quasi-momentum changes by the value $\hbar g$ where g is a vector of the *reciprocal lattice*. In this case the energies of two of the three phonons participating in the phonon scattering process would probably be comparable with the maximum photon energy. At $T \ll \Theta$, the *occupation numbers* of such phonons are exponentially small, so the corresponding value κ_{ph} becomes exponentially large, and growing as the temperature decreases: $\kappa_{ph} \sim \exp(\beta\Theta/T)$, where the parameter $\beta \approx 1/2$. The presence of *impurity atoms* leads to the situation wherein κ_{ph} grows again with decreasing temperature, but now according to the inverse power law: $\kappa_{ph} = A/T$, with A inversely proportional to the impurity concentration, and much greater than at $T \gg \Theta$. Other types of defects such as *dislocations*, twins (see *Twinning of crystals*), *grain boundaries*, and boundaries of crystals provide appreciable contributions at quite low temperatures; as a result, κ_{ph} goes to zero at $T \to 0$. If the scattering on dislocations is the determining mechanism of the momentum *relaxation* then $\kappa_{ph} \sim T^2$; for scattering on boundaries $\kappa_{ph} \sim T^3$ (this dependence occurs when the carrier mean free path for normal processes is smaller than the dimensions of the crystal or the grains).

In *semiconductors* it is necessary to take into account both phonon and electron contributions to the thermal conductivity coefficient. For the electron conductivity part one must keep in mind that the electron gas in semiconductors is non-degenerate, and the conduction can take place over *impurity bands*. There are various experimental methods for determining the thermal conductivity coefficient. The main experimental difficulties involve taking into account the heat radiation through the surface, and establishing a uniform temperature gradient along the sample.

THERMAL CONDUCTIVITY OF SUPERCONDUCTORS

The overall *thermal conductivity* of superconductors is composed of components from the electron subsystem (*thermal conductivity coefficient κ_e^s*), and the lattice thermal (phonon) conduction (its factor is κ_p^s). The principal relaxation mechanism for κ_e^s is the scattering on defects or impurities (κ_{ed}^s) and phonons (κ_{ep}^s), while for κ_p^s it is scattering off electrons (κ_{pe}^s), boundaries, and defects (κ_{pd}^s). It is approximately true that the components add as reciprocals: $(\kappa_e^s)^{-1} \approx (\kappa_{ep}^s)^{-1} + (\kappa_{ed}^s)^{-1}$, and $(\kappa_p^s)^{-1} \approx (\kappa_{pe}^s)^{-1} + (\kappa_{pd}^s)^{-1}$. The electron–electron and phonon–phonon interactions make an insignificant contribution at low temperatures to the thermal conductivity. As the superconducting state is mainly determined by the Cooper pairs of the electron subsystem, the trends of behavior of κ_e^s and κ_{pe}^s differ from those in normal metals, while κ_{pd}^s is the same as in a normal metal.

In pure (type I) and low impurity content type II superconductors the main contribution comes from κ_{ed}^s and κ_{ep}^s, respectively. The *Bardeen–Cooper–Schrieffer theory* predicts

$$\kappa_{ed}^s = \frac{2}{3} \frac{p_F^2 \tau_0}{\hbar^2 m_e} F(T),$$

where τ_0 is the normal metal electron *relaxation time*; $F(T) \to T$ at $T \to T_c$ and $F(T) \to e^{-\Delta/T}$ at $T \to 0$. The most significant *electron–phonon interaction* is in high-purity samples at $T \leqslant T_c$ where κ_{ep}^s also decreases with the decrease in temperature. The reduction of κ_e^s is caused by the decrease of the number of normal excitations at $T \to 0$.

In sufficiently "dirty" superconductors in the temperature range $T \geqslant (0.3–0.5)T_c$ the principal contribution is given by $\kappa_{pe}^s = BT^2 \Phi(T)$. As the temperature decreases, the function $\Phi(T)$ grows due to the increase of the phonon *mean free path* when the number of electron excitations decreases. Further decreases of temperature lead to phonon scattering off, mainly, defects and boundaries. As in the normal metal, $\kappa_{pd}^s \propto T^3$. Thus, the total κ_p^s passes through its maximum at $\kappa_{pd}^s \approx \kappa_{pe}^s$. Experimentally both these cases are observable in their pure forms as well as in various combinations.

In those cases when the weak coupling approximation does not hold, the dependences $\kappa_s(T)$ are slightly modified, so that in superconductors with strong electron–phonon coupling, at $T \sim T_c$, κ_{ed}^s drops more rapidly than predicted by the above formulae.

An interesting feature is also observed in the behavior of the thermal conductivity of the *intermediate state* when κ^s is noticeably reduced. At $T \ll T_c$ the reduction of κ^s correlates with the decrease of the phonon mean free path at the transition to the normal state. At higher temperatures, a specific mechanism of the electron reflection from the interface between normal and superconducting phases (*Andreev reflection*) is responsible for this effect. The electrons which come from the normal metal, at the interface, transform into positively charged holes and return back with the transferral of the extra charge to a condensate.

In superconductors, as in superfluid helium, the transport of heat can occur by means of convection. In this case, the current of the normal component $I_n \neq 0$, but the overall current $I = I_n + I_s = 0$. However, the ratio of convective to the ordinary thermal conduction is quite small, $\sim 10^{-4}–10^{-5}$.

In the cuprate superconductors (see *High-temperature superconductors*) there is observed a series of particular features in the temperature dependence of the thermal conductivity (a peak of κ^S at $T \leqslant T_c$, and a quadratic term at $T \ll T_c$) which have not yet been explained.

THERMAL DEFECTS

Crystal structure *defects* arising at the heating of samples followed by their subsequent cooling in a given system. The nature of defects in a material may differ, depending on the temperature, the duration of the heating, the cooling regime, the atmosphere, the material of crucibles or ampoules where heating takes place, the degree of perfection of the original material, and its impurity composition.

At a given temperature, there is established in a solid an equilibrium (for this temperature) concentration of *Frenkel pairs*, i.e. *vacancies* associated with *interstitial atoms*:

$$N_t = (N_{at} N') \left(\frac{v}{v'} \right)^{3r} \exp \left(-\frac{E}{2k_B T} \right),$$

where N_{at} is the concentration of regular atoms, N' is the concentration of interstices, v and v' are, respectively, the frequencies of atomic lattice vibrations in the absence or presence of defects, r is the coordination number, E is the energy of formation of a Frenkel pair, which in combination with the energy of vacancy migration produces the self-diffusion energy (from fractions of an electron volt to several eV). The indicated energies can be determined experimentally in studies of the equilibrium concentration of vacancies at several temperatures, as well as from the kinetics of *annealing* nonequilibrium vacancies. In pure *metals* and single-phase *alloys*, the stable thermal defects are those which emerge at fast cooling (quenching) from a higher temperature; also possible are the trapping of interstitial atoms by *impurity atoms*; the formation of *dislocations*, and so on. As a rule, the presence of defects favors *hardness*, *strength*, and *magnetic saturation*, and reduces *plasticity* and *creep*. There can occur a variation of the lattice constant, the electric resistance, and *diffuse scattering*.

In semiconductors the thermal defects affect the concentration and the lifetime of current carriers, optical, photoelectric and other properties.

The following processes may take place: (a) *precipitation* of impurities from a *solid solution* or, otherwise, a transition to the state of a solid solution; (b) *diffusion* of impurities from the crystal surface, the heating atmosphere, the material of crucibles and ampoules; (c) formation of clusters of vacancies or/and interstitial atoms with *impurity atoms*, inducing the loss or the acquisition of electric activity in the latter. Therefore, a final result of a *thermal treatment* depends on a large number of parameters and energy constants. In silicon and germanium with high oxygen contents, heating at 450–600 °C leads to the formation of oxygen-containing *donors* with two energy states in the *band gap*. At low oxygen content, a transition of *acceptors* (*copper, nickel, iron*) to an electrically active state prevails. In *n*-type germanium, donors of Group V lose their electrical activity through interaction with vacancies; this process can involve the reversal of the type of conductivity (*n–p* conversion). In binary compounds the most predominant defects are associated with the disturbance of the *stoichiometric composition*, and the appearance of vacancies in the sublattice of the more volatile component. Related to this is the phenomenon of *conductivity self-compensation*. The disturbance of stoichiometry and the appearance of nonequilibrium vacancies and interstitial atoms result from the thermal treatment of *alkali-halide crystals*.

THERMAL DESORPTION

The removal of adsorbed material from an absorbent surface at an increasing temperature. The thermal desorption rate is described by the *Arrhenius law*: $dn/dt = vn^x \exp[-E_D/(k_B T)]$ where n is the adsorbed particle concentration, E_D is the activation energy of *desorption*, v is the pre-exponential factor, x is an integer that reflects the kinetic order of desorption (e.g., $x = 2$ if the adsorbing atoms are desorbed as diatomic molecules). The composition of the desorption products is determined by the ratio of the desorption rates of different particles. Thus, the desorption of oxygen from noble *metals* takes place in the form of O_2 molecules; from *tungsten, molybdenum*, and *rhenium* it is in the form of atoms (in the case of small coatings); and in the form of oxides from *tantalum* and *niobium*. Thermal desorption of positive or negative ions is called, respectively, positive or negative *surface ionization*.

THERMAL DISCOLORATION

See *Photochromic effect*.

THERMAL DISPLACEMENTS OF ATOMS

Small displacements of atoms in a *crystal* related to their thermal motion about equilibrium positions. These displacements do not disturb the average periodicity of the *crystal lattice* (below the *melting temperature*). Disregarding the *anharmonic vibrations*, the distribution of probability of the atomic displacements in respect to the equilibrium position has a Gaussian profile. The mean square of displacement amplitudes $\langle u^2 \rangle$ serves as a measure of the magnitude of the thermal displacements of atoms. For a monatomic crystal this takes the form:

$$\langle u^2 \rangle = \frac{\hbar}{2MN} \int \frac{\nu(\omega)}{\omega} \coth\left(\frac{\hbar\omega}{2k_B T}\right) d\omega,$$

where M is the atomic mass, N is the number of atoms in crystal, T is the temperature (in energy units), and $\nu(\omega)$ is the phonon density of states. At high temperatures ($T \gg \Theta$, where Θ is the *Debye temperature*) the value of $\langle u^2 \rangle$ is proportional to the temperature: $\langle u^2 \rangle \sim \hbar^2 T/(M\Theta^2)$. At low temperatures ($T \ll \Theta$) the value $\langle u^2 \rangle$ approaches a constant (the square of the amplitude of *zero-point vibrations*): $\langle u^2 \rangle_0 \sim \hbar^2/(M\Theta)$. Typical values $\langle u^2 \rangle^{1/2}$ do not exceed 10–15% of the lattice parameter even at the melting temperature. However, in *quantum crystals* this parameter can be significantly larger. In crystals with symmetry (see *Crystal symmetry*) lower than cubic, there is anisotropy in the spacial distribution of thermal displacements of atoms, and this is particularly strong in *quasi-one-dimensional crystals* and *quasi-two-dimensional crystals*. Experimental values of $\langle u^2 \rangle$ can be found from, e.g., measurements of the *Debye–Waller factor* using the scattering of X-rays, neutrons, electrons, and γ-quanta by the crystal lattice.

THERMAL ELECTRON FIELD EMISSION,
thermally-assisted field emission

Emission of electrons by electrically conducting materials at high temperatures in a very strong external electric field (of strength $E \sim 10^9$ to 10^{10} V/m). The mechanism of thermal electron field emission includes the tunneling (see *Tunnel effect*) of electrons through the near-surface potential barrier composed of the main (below *Fermi level*) and thermally excited electron states, as well as the electrons escaping over the barrier. Upon decreasing the temperature, the thermal field electron emission transforms into a *field electron emission*. At decreasing E, the thermal field electron emission transforms into a *thermionic emission* produced by the field due to the Schottky effect (see *Schottky barrier*).

THERMAL-ENVIRONMENT RESISTANCE

The ability of a material to resist high-temperature *oxidation* (*gas corrosion*). The following parameters of thermal-environment resistance are used: specific changing of metal mass, thickness of oxidized film, depth of corrosion-induced damage. Thermal-environment resistance of a material is determined by the properties of the emergent oxide film: its continuity, density, and *plasticity*, *diffusion* rate of *metal* and oxygen ions within the film, correspondence between the metal and oxide lattices. The main alloying elements (see *Alloying*) that are used are *chromium*, *aluminum*, and *silicon*. The thermal-environment resistance sharply increases when strong oxide layers of Cr_2O_3, Al_2O_3, and double oxides with the spinel structure AB_2O_4, are formed at the surface of the material. The method of improving the thermal-environment resistance by applying a protective *coating* is wide spread. To protect heat tolerant nickel and cobalt alloys, operating at high temperatures and under high loads, against gas corrosion, use is made of aluminide coatings and more complex coatings based on it.

THERMAL-ENVIRONMENT RESISTANT
ALLOYS, scale-resistant alloys

Alloys based on nickel–chromium or iron–chromium, known for their *thermal-environment resistance* and characterized by their considerable resistance to gas *corrosion* at high temperature

(800–1100 °C) in the air and other gas media. The gas corrosion resistance depends on the chemical composition of the alloy, temperature, gas medium composition, period of operation, values of mechanical stresses and cyclic recurrence of loading. The combustion of raw oil or especially heavy fuel (black oil, etc.), which contains increased amounts of sulfur, *vanadium*, alkaline or alkaline-earth metal salts, etc., produces a gas medium that sharply deteriorates the *corrosion resistance* of alloys, and reduces the service life of products made of them. The corrosion displays itself to a lesser extent in purified fuel (e.g., kerosene, petroleum). Alloys with high contents of *chromium* or those subjected to special *alloying*, as well as the products with diffusion *coatings* created in the process of chromium calorizing or aluminum siliconizing, are known for higher gas corrosion resistance. Thermal environmental resistance of alloys is due to their high chromium content (18–25%) and additions of *aluminum*, while that of iron–chromium–nickel alloys also results from the increased content of *silicon* (up to 2–3%). In order to achieve the optimum combination of thermal environmental resistance, *high-temperature strength*, *plasticity* and weldability some other elements, in addition to chromium, are introduced into these alloys, namely, *molybdenum*, *tungsten*, *niobium*, *titanium*, *boron*, *cesium*, and others.

Thermal-environment resistant alloys can be classed as homogeneous and heterogeneous (see *Heterogeneous structure*). *Homogeneous alloys* are characterized by somewhat reduced mechanical properties and lower high-temperature strength, but possess good plasticity in the cold state, and high weldability. Their considerable stability against *oxidation* and gas corrosion at high temperature is due to the formation of an adhesive refractory oxide *film* (Cr_2O_3, Al_2O_3, $NiCr_2O_4$, etc.) which precludes the *diffusion* of oxygen into the metal, and of alloying elements from metal into the slag. Both high-temperature strength and thermal environmental resistance of alloys can be sharply degraded by additions of molybdenum and boron, and also the presence of sulfur in combination with *nickel* or *cobalt*, which causes the formation of fusible *eutectics* with melting temperature about 650 °C. *Heterogeneous alloys*, strengthened with titanium, aluminum and niobium, often along with the addition of refractory metals, exhibit high-temperature strength and thermal environmental resistance. They are, however, less weldable, tending to form hot *fractures* during the course of *welding* and *heat treatment*. These alloys must be welded under special conditions, using special modes of heat treatment. Thermal-environment resistant alloys are preferably manufactured by the methods of vacuum *metallurgy* and hot pressure treatment. They are used to manufacture combustion chambers, flue tubes, nozzle vanes of gas turbines, afterburners, elements of welded constructions, etc. (see also *High-temperature alloys*).

THERMAL-ENVIRONMENT RESISTANT MATERIALS

Materials known for their enhanced *thermal-environment resistance* in addition to *high-temperature strength*, *thermal-environment resistance* and specified thermal physical properties. According to their chemical and physical composition, and also some structural and technological features, the following classes of thermal-environment resistant materials are distinguished: ceramic materials (see *Ceramics*) based on refractory oxides and their compounds, or oxygen-free refractory compounds (*carbides*, borides, *nitrides*, silicides); metallic and ceramic-metallic materials; carbon-graphite materials; glass-ceramic and crystalline glass materials; *composite materials*. They also include refractory glass, heat-proof concrete, and cement. The operating parameters of these materials are constantly being improved because of their need in nuclear reactors, gas-turbine engines, etc. In addition, new areas of application have emerged: plasma chemistry, thermonuclear power engineering, etc.

THERMAL ETCHING GROOVE

Groove that forms at a high temperature along the line of intersection of a *polycrystal* surface with a *grain boundary*, or with a *twinning structure* boundary. The angle β of the sample surface relative to the tangent to the groove profile at its apex is determined by the formula $\beta = \arcsin(\alpha_B/2\alpha)$, where α and α_B are the *surface tension* of a crystal and grain boundary, respectively. For $\alpha_B/\alpha = 10^{-1}$, then $\beta \approx 5 \cdot 10^{-2}$, and thus the sides of a thermal etching groove are gently sloping. Concerning the presence of an equilibrium angle at the apex of a groove, a region of a surface of positive curvature is formed near it, where the chemical potential of atoms is higher than its value on a flat surface. This determines the fluxes of atoms from the groove apex and the kinetics of the groove propagation. The width h and depth d of a groove change with time according to a law that depends on the mechanism of *mass transport*. If the *surface diffusion* mechanism predominates, then $h_s = 4.6(Bt)^{1/4}$, $d_s = 0.973m(Bt)^{1/4}$, $m = \tan\beta$, $B = D_s\alpha\omega^2 n/(k_B T)$, where D_s is the surface self-diffusion coefficient, ω is the atomic volume, and n is the surface density of atoms. If volume *diffusion* dominates then $h_0 \sim d_0 \propto (D_0 t)^{1/3}$, where D_0 is the volume self-diffusion coefficient. The thermal etching groove fixes the position of boundaries and retards the collective *recrystallization* of near-surface grains. Data on the kinetics of the thermal etching groove propagation yield information on the self-diffusion coefficient and the value of α_B.

THERMAL EXPANSION

A change of body dimensions with a variation in *temperature* under constant external stress (*pressure P*). As a characteristic of the thermal expansion, there is a *temperature linear expansion coefficient* which is defined in an isotropic medium by the equation

$$\alpha = \frac{1}{L}\left(\frac{\partial L}{\partial T}\right)_{P=\text{const}},$$

where L is the body size. The thermal expansion in *crystals* is characterized by the second-rank symmetric tensor

$$\alpha_{ij} = \left(\frac{\partial \varepsilon_{ij}}{\partial T}\right)_{\sigma_{ij}=\text{const}}, \quad i, j = x, y, z,$$

where ε_{ij} is the *strain tensor* of the crystal, and σ_{ij} is the stress applied to the crystal. The *temperature volume expansion coefficient* is equal to

$$\beta = \frac{1}{V}\left(\frac{\partial V}{\partial T}\right)_P,$$

where V is the body volume. In crystals $\beta = \sum \alpha_{ii}$.

The thermodynamic relation which links tensor α_{ij} with the crystal Helmholtz free energy F has the form:

$$\alpha_{ij} = \frac{1}{V}\sum_{mn} S_{ijmn}\left[\frac{\partial}{\partial\varepsilon_{mn}}\frac{\partial}{\partial T}F\right],$$

where S_{ijmn} are the components of the crystal isothermal *compressibility* tensor (tensor of the fourth rank) which is the inverse of the tensor of elastic constants. In cubic crystals we have

$$\beta = -k\left[\frac{\partial}{\partial V}\frac{\partial}{\partial T}F\right],$$

where

$$k = -\frac{1}{V}\left(\frac{\partial V}{\partial P}\right)_T = 3(S_{1111} + 2S_{1122})$$

is the isothermal compressibility of a crystal. The thermal expansion of solids is caused by *anharmonic vibrations*. All the thermal expansion coefficients decrease to zero as $T \to 0$, which follows from the Nernst theorem. This theorem states that as absolute zero is approached the rate of change of entropy converges to zero, and the entropy itself approaches a minimum value. At high temperatures the thermal expansion coefficient reaches a constant value that is 10^{-4}–10^{-6} K^{-1}. As a rule, the linear expansion coefficient is positive. In most cases the dependence of the volume expansion coefficient β on the temperature is described quite well by the *Grüneisen rule*: $\beta = \gamma k C_V/V$ where C_V is the lattice *specific heat* at constant volume, and γ is the *Grüneisen constant*.

In analytical derivations of the Grüneisen rule it is usually assumed that all the frequencies ν of the crystal vibrations depend on the deformation through a single parameter. In this case, the Grüneisen constant is

$$\gamma = -\frac{V}{\nu}\frac{\partial\nu}{\partial V} = -\frac{\partial \ln\nu}{\partial \ln V}.$$

When the law of corresponding states holds, the Grüneisen constant is described by the formula $\gamma = [1/(k\Theta)]\partial\Theta/\partial P$, where Θ is the *Debye temperature*. For most simple solids the parameter γ is close to two.

At low temperatures the lattice specific heat C_V of crystals follows the Debye relation, being $\propto T^3$, while in glasses (and also in crystals in the presence of reorienting impurity centers) we have $C_V \propto T$. It follows from the Grüneisen rule that the same dependence on temperature also takes place for the coefficient β. In glasses the parameter γ is 1 to 2 orders of magnitude greater than in crystals. If it is possible to isolate a few subsystems in a crystal (e.g., the *crystal lattice* and the impurity centers) with their individual frequency-versus-deformation dependences, and the Grüneisen rule can be rewritten as a sum over these subsystems. In this case, the form of the temperature dependence of the total Grüneisen constant may be quite complicated, and even include a sign change. Of great practical interest are so-called *invar magnetic alloys* which have abnormally small or negative thermal expansion arising from corrections to the ordinary thermal expansion to take into account the magnetostriction reduction of the volume at heating. In the vicinity of a *second-order phase transition* point the expansion coefficient diverges as $(T - T_c)^{-\alpha}$ where T_c is the transition temperature, and the *critical index* α has the same value as the critical index for the specific heat under constant pressure ($\alpha \sim 0.1$). The experimental values of thermal expansion for crystals are determined by methods of *dilatometric analysis*.

THERMAL EXPANSION, ANOMALOUS
See *Anomalous thermal expansion*.

THERMAL FATIGUE
Fatigue resulting from cyclic alteration of thermal stresses which appear in a solid at a temperature change $\Delta T = T_2 - T_1$ if free expansion and free compression are restricted. The values of thermal stress and mechanical *strain* are, in this case, dependent on the constriction coefficient $R = -\Delta\varepsilon/(\alpha\Delta T)$ where $\Delta\varepsilon$ is the amplitude of preliminary mechanical strain, $\alpha\Delta T$ is the amplitude of thermal strain, α is the linear temperature expansion coefficient. The amplitude of apparent strain is determined by the equation $\Delta\varepsilon_{ap} = \Delta\varepsilon + \alpha\Delta T = (1 - R)\alpha\Delta T$. Depending on the coefficient R, the temperature change range ΔT, and the maximum temperature, there emerges in the solid either an elastic strain, an elastic-plastic strain, or an elastic-plastic *creep*. One may differentiate, respectively, the fatigue in the elastic range (*multicyclic fatigue*), in the elastic-plastic range

(*few-cyclic fatigue*), or in the range of the elastic-plastic creep (superposition of creep and fatigue). Tests are carried out under independent loading and heating (*thermomechanical fatigue*) and under load by restricting the heat deformation (thermal fatigue). The main characteristics of fatigue are its dependence on the number of cycles prior to crack formation N_f versus amplitudes of full strain ε_d, *plastic deformation* ε_{pd}, elastic strain ε_{ed}, irreversible strain ε_{ird}, strain under hard loading, and stress amplitude of cycle σ_d at soft loading (*Cyclic heat treatment*).

THERMAL INDICATORS
Substances which change their physical or physicochemical characteristics under the action of heat, and are utilized for detecting and monitoring changes in temperature (see *Thermography*). The two most widely used types of thermal indicators are luminophors and liquid crystals. The luminophor thermal indicators change their *luminescence* under the action of heat. Liquid crystal thermal indicators are cholesteric substances (see *Cholesteric liquid crystal*) oriented in a certain manner which reversibly change color under the action of heat.

THERMAL ION EMISSION
See *Thermoionic emission*.

THERMAL LOSS
See *Heat loss*.

THERMALLY ACTIVATED PROCESSES
See *Activation energy*.

THERMALLY STIMULATED ELECTRICAL CONDUCTIVITY
A nonequilibrium *electrical conductivity* arising at the heating of a sample containing *deep levels* previously occupied by charge carriers (sometimes called *thermally stimulated current*). As a rule, in studies of thermally stimulated conductivity the illumination of the sample at low temperature is used for nonequilibrium filling of the levels. The sample is heated at a low rate in the darkness. The charge carriers trapped at the levels release and exhibit peaks in the temperature dependence; from the peak positions one can determine the energies of the levels, their concentration

and trapping cross-sections of charge carriers. For determination of these values a *thermally stimulated luminescence* is also used. In some cases the studies of the thermally stimulated conductivity include measurements of the Hall effect (see *Galvanomagnetic effects*). In a sample with a specific resistivity less than 10^5 $\Omega \cdot$cm, the measurement of thermally stimulated conductivity is complicated by the high level of dark current. This difficulty is overcome by measuring the current, the *barrier capacitance* of *p–n* junctions, and the *Schottky barriers* shifted in the barrier direction (see *Capacitance spectroscopy*). When studying *insulators* the analysis of *thermally stimulated depolarization* is used.

THERMALLY STIMULATED LUMINESCENCE

The activation of the persistence of *phosphorescence* by heating. This is specific for luminophors. As a rule, this is seen as a light flash during the heating of the preliminary excited substance at a sufficient rate. The thermally stimulated luminescence is caused by that fact that the persistence in crystalline phosphors represents a recombination *luminescence*: the persistence is due to the accumulation of electrons and *holes* in traps during the excitation. After the excitation ends these gradually leave the traps under the action of heat and recombine. The heating activates this process and induces the sharp rise of the persistence, which then decays rapidly as the reserve of accumulated carriers depletes.

THERMAL METHODS OF STUDIES in solid state physics

They are used for studying the temperature (energy) parameters of a *thermodynamic system.*

Means and methods for measuring the *temperature* are developed in the branch of physics called *thermometry*. The techniques of temperature measurements are conventionally divided into contact methods (thermometry itself) and noncontact methods (radiation thermometry or *pyrometry*). The main components of apparatus designed for temperature measurements are the sensitive element where a thermometric property is implemented (alteration of electromotive force, electric resistance, pressure, volume, radiation intensity, etc.), and the measuring unit connected to the element by means of a communication line.

One of the principal tasks of thermal studies is the determination the distribution of temperature in space and time. Important factors are the thermal conductivity, enthalpy, and specific heat. By means of calorimetric analysis (see *Calorimetry*), the thermal characteristics of *phase transitions*, the annihilation and redistribution of *defects in crystals*, *sorption*, dissolution, etc., are investigated. The most informative devices for solid-state studies are differential scanning calorimeters, and apparatus for differential thermal analysis.

With the help of the experimental determination of the specific heat, enthalpy, thermal expansion coefficient, and isothermal compressibility for particular conditions, the parameters of the state of a solid can be calculated. Thermal methods of study are powerful instruments for investigating the integrated characteristics of processes in solids, thereby providing more complete information on the nature of the physical processes. Thermal methods of study are widely used in combination with other methods.

THERMAL MICROSCOPY

A method of studying *solids* with the help of heat waves, based on the interaction of thermal waves with structural elements and defects in solids. The thermal waves are excited in the solid by a modulated source without a direct contact: by a focused electron beam or *laser* radiation. The measured parameters are those of the thermal wave propagation process (attenuation, phase shift, reflection, interference, diffraction, etc.) as well as the physical parameters whose variations are caused by the interaction of thermal waves with the solid (acoustical waves in solids, sound waves in the medium in contact with the heating surface, temperature-sensitive acoustical properties of solids, IR radiation from a heated surface, and so on). To visualize the thermal inhomogeneities various detection techniques, including *fluorescence* of temperature-sensitive films, are used.

A *thermal microscope* contains a source of radiation, a scanning system, and a radiation receiver in the form of an IR ellipsoidal concave mirror, with the sample situated at its principal focus, and with a detector placed at another focus. The subject under investigation is irradiated by, e.g., an electron beam which is scanned over the surface to obtain a *raster* trace. The radiation scattered by the surface is collected, focused by a reflector, and directed through a filter upon a semiconductor radiation detector. The resolution of thermal microscopy depends on the heat transport constants of a given solid, the diameter of the incident beam, and the modulation frequency. Thermal microscopy is used for research purposes, and in the *defectoscopy* of solids.

THERMAL NEUTRON DIFFRACTION

Phenomenon of the scattering of neutrons with energies $E \sim 100$ eV in condensed media (crystals, liquids, biological structures), when the neutron *de Broglie wave* length ($\lambda \sim 10^{-8}$ cm) is comparable with the average separation between the atoms or molecules (see *Neutron diffractometry, Neutrons in solid state physics*).

THERMAL OXIDATION

Formation of an oxide layer at a solid *substrate* through exposure of the sample at elevated temperatures to an oxidizing atmosphere (see *Oxidation*). This is most widely used in silicon planar technology for the manufacture of masking, isolation, and passivation *coatings* (see *Surface smoothing*). As an oxidative medium either dry or wet oxygen can be used. The process of thermal oxidation takes place at the interface oxide–substrate where molecules of oxygen or water are brought to the site of the reaction (as a rule, by *diffusion* through the oxide layer). The rate of thermal oxidation grows with an increase of the temperature and the pressure of the oxidizer. In some cases catalyzing additives and illumination of the surface influence the oxidation rate. The thermal oxidation of *silicon* produces an interface with close to ideal electric physical parameters, including an insignificant quantity of *surface states*, and it exhibits a high free current carrier mobility.

THERMAL RADIATION, heat radiation

Electromagnetic radiation emitted by matter due to its internal energy at a certain *temperature*. The heat radiation exhibits a continuous spectrum whose maximum depends on the temperature of the emittor. With increasing temperature the total rate of emitted heat radiation energy P grows as the fourth power of the temperature: $P = \sigma T^4$, where σ is the *Stefan–Boltzmann constant*, and there is a shift of the maximum radiation value toward shorter wavelengths in accordance with *Wien's law*: $\lambda_{1\,\mathrm{max}} T_1 = \lambda_{2\,\mathrm{max}} T_2$. The thermal radiation is generated by the material under the conditions of an overall equilibrium for all nonradiative processes, e.g., for the energy interchange between electronic and vibrational motions in solids. The thermal radiation is in *thermodynamic equilibrium* with the material and with the ambient *equilibrium radiation*; the energy distribution spectrum is determined by the *Planck formula*. For one or more bodies in equilibrium with each other and with a radiation field inside an enclosure the Kirchhoff radiation law is valid, whereby these bodies have equal rates of emission and absorption. A good emitter is a good absorber.

THERMAL RESISTANCE, heat resistance

The ability of brittle materials to withstand *failure* during abrupt variations of temperature. The thermal resistance of materials (e.g., fireproofing, structural, and metal-ceramic materials, as well as those designed for fabricating structural elements to perform at elevated temperatures) depends on the kind of material, its structure, *thermal conductivity*, linear expansion coefficient, and other structural characteristics. The thermal resistance of a material depends significantly on its temperature gradient rate during a *cyclic heat treatment*, as well as on the kind of refrigerant. A criterion of thermal resistance is the number of cycles of temperature change (heat cycling) experienced prior to the appearance of a *crack* or overall decomposition of the sample during the heat cycling, or it is the magnitude of the temperature gradient which induces the appearance of crack. The index of thermal resistance is usually applied when comparing various kinds of materials according to their *fracture toughness* at sharp heat jumps.

THERMAL SPIKE in radiation physics of solids

A localized region in a *solid* with a raised (compared to a mean value) temperature of *atoms* (*ions*) which results from the onset of a *collision cascade* of atoms followed by the redistribution of the kinetic energy of atoms involved in mutual collisions, and collisions with surrounding atoms. In metals with a high electron *thermal conductivity*, the formation of heat (thermal) spikes is possible due to the fact that the characteristic time for the heat transfer between ions and electrons is $4m_e/m_i \sim 10^{-4}$ times less than that between ions (m_e and m_i are the masses of electrons and ions, respectively). A typical peak temperature of heat spikes in metals above that of the surroundings is about 10^3 K.

THERMAL TRANSPORT

See *Heat transport*.

THERMIONIC CATHODE

A filament in a vacuum tube or gas-filled counterpart. The main characteristics of a thermionic cathode are the value of the emission current at a given temperature, the *work function $e\varphi$*, the durability, the specific power of the filament, the power losses from heating the elements of the structure, and the radiation Q. The durability of a thermionic cathode depends on the reserve of active material that is expended under the action of the electron or *ion bombardment*, as a result of etching by residual gases, and so on. The efficiency of a thermionic cathode is defined as the ratio $e\varphi/Q$ of the work function to the filament power.

In respect to the type of filament, thermionic cathodes are divided into directly heated cathodes and equipotential cathodes. The most widespread among *directly heated thermionic cathodes* are those based on refractory metals (W, Ta). These have the minimal ratio $e\varphi/Q$ and operate at temperatures above 2000 K. A particular place among *thermionic equipotential cathodes* is occupied by effective thermionic cathodes such as the *oxide-coated cathode*. This is a semiconductor cathode, but it has a particular porous structure what stimulates extra electron transport through the pores in the cathode. The oxide-coated cathode is not a homogeneous system: there is a high-ohmic intermediate layer at the metal–semiconductor interface. The presence of inhomogeneities in oxide-coated cathodes leads to the transport of excess barium ions and empty oxygen sites (*vacancies*) during its working process. The presence of defects-vacancies in these cathodes, on the one hand, restricts the degree of its *doping* with barium, and on the other hand, provides an opportunity to carry out the doping, not by annealing in a barium at-

mosphere, but by using evaporation under high vacuum when it is mainly oxygen that the cathode evolves.

THERMIONIC EMISSION, thermoelectronic emission

The emission of electrons by heated *solids* (or liquids). For most *emitters* (i.e. bodies which emit electrons), the thermionic electron emission is observed at temperatures much higher than room temperature.

The simplest thermionic emission experiment can be performed with the help of a vacuum *diode*. At a given temperature, which is controlled by the current of the lamp filament, one can plot the dependence of the current passing through the diode on the anode voltage, that is the *current–voltage characteristic*. There are two well-pronounced regions of the current–voltage characteristics of diodes with metal cathodes. First, this is a region where the current passing through the diode is proportional to $V^{3/2}$ (*Langmuir law*, I. Langmuir, 1913); where V is the anode voltage. In this region the current is limited by the space charge. As the anode voltage grows the current becomes independent of it, reaching a saturation value when there is no longer any space charge in the cathode cavity. The current density in this case is determined by the cathode parameters and its temperature, in accordance with the *Richardson–Dushman equation* (O.W. Richardson, S. Dushman, 1923):

$$j = (1 - \bar{r})AT^2 \exp\left(-\frac{e\varphi}{k_B T}\right),$$

where $A = 4\pi e k_B^2 m/h^3 = 120$ A·cm^{-2}·K^{-2} is the *Richardson–Dushman constant*; r is the quantum-mechanical reflection coefficient for electrons passing from the emitter into the vacuum; $e\varphi$ is the *work function*. The work function is the principal parameter which determines the thermionic emission. This is weakly dependent on the temperature, but depends substantially on the condition of the emitter surface and the external electric field. For metal cathodes this dependence is due to the *Schottky external effect* whereby the the process of specular reflection reduces the work function by a factor dependent on the external electric field E near the cathode, and the *dielectric constants* ε and ε_0 of the emitter crystal lattice and vacuum, respectively.

The dependence of thermionic emission on the electric field of semiconductor cathodes is more complex. In addition to the Schottky external effect, in comparatively weak fields there is observed a rise of the emission current due to the enrichment of the subsurface region of the cathode by excess electrons. As a result, the cathodes of this type do not exhibit a clearly pronounced saturation region. With a further increase of the anode voltage, the electric field penetrates the semiconductor cathode and warms up the electron gas there. Hence the probability of electron emission from the semiconductor is no longer determined by the equilibrium Fermi–Dirac distribution function (see *Fermi–Dirac statistics*), but rather by a nonequilibrium one derived from the *kinetic equation*. Due to a strong dependence of the latter on the electric field strength, the thermionic emission grows rapidly with an increase of the anode voltage. Such a current rise may lead to a second region of space charge limited current which appears beyond the range of quasi-saturation.

THERMOACOUSTIC MICROSCOPY

A technique of representing as an image the near-surface thermomechanical characteristics of a solid, such as changes in crystallographic structure, phase composition, heat and strain parameters, mechanical damage, and so on. A focused laser or electron beam is scanned over the surface of the object under investigation, and modulated with a frequency of the order of hundreds kilohertz. The absorbed radiation induces local time-and-space periodic elastic stresses which, in turn, generate acoustic waves. These waves are detected by a piezoelectric *acoustic wave transducer* coupled to the sample. A signal at the transducer's input is used for forming a scanned and magnified image of the object with contrast furnished by the spatial variation of its heat and elastic properties. When studying semiconductors, the image contrast is also influenced by their electronic properties.

The amplitude of the video signal for a given density of absorbed energy is proportional to the value $B\alpha(\kappa\rho c\omega)^{-1/2}$ where B is the bulk *elastic modulus*, α is the *thermal expansion* coefficient, κ, ρ, and c are, respectively, the *thermal conductivity coefficient*, the specific *density*, the *specific*

Thermoacoustic microscopy.

heat of the solid material, and ω is the modulation frequency. The spatial resolution is determined by the diameter of the region where the elastic waves originate due to the periodic heating (i.e. the energy dissipation volume of the irradiating beam), and by the contribution of the thermal *diffusion* that spreads to a microvolume of thermoelastic expansion beyond the limits of the energy absorption region of size $d \cong [2\kappa/(\omega\rho c)]^{1/2}$ (≈ 1 to $10\ \mu m$).

Raster electron thermal acoustic microscopy provides unique information about the near-surface microstructure of a solid, which in many cases is unattainable by any other visualization method (see *Visualization of acoustic fields*). For example, in the electron microphotograph (see Fig., left) one can see the indenter-produced mechanical damage at the surface of a silicon plate. The same region imaged via thermal acoustic scanning (see Fig., right) in a raster electron microscope (see *Scanning electron microscope*) provides much more detailed information on subsurface microscopic inhomogeneities, including zones of elastic and *plastic deformation*, *shear* lines, and related *dislocations* and *stacking faults*.

THERMOCAPILLARY EFFECT

An effect caused by the motion of nonuniformly heated liquid due to the presence of a gradient of *surface tension* σ. Both $\nabla\sigma$ and the motion of the liquid stimulated by it are in the direction of decreasing temperature. The pressure distribution in the liquid layer of thickness z under the action of ∇T in direction x is given by the expression

$$\frac{dP}{dx} = \frac{3}{2z}\frac{d\sigma}{dT}\frac{dT}{dx}.$$

In the case of uniformly *spreading* liquid drops along a strip with ∇T from the cold end to the hot one, the boundary of the drop stops its motion. In the opposite direction, the spreading is accelerated in comparison with isothermal spreading. Individual drops of non-spreading liquids on a substrate with a ∇T are displaced as a whole under action of the thermocapillary effect toward the direction of the temperature decrease. During glass fusion the thermocapillary effect leads to failure of those crucible walls to which a heated glass moves. During a welding process the thermocapillary effect causes the formation of craters with the smelting of the basic metal. Surface convection of the liquid can, in addition to the thermocapillary effect, be caused by the presence of a material concentration gradient. Movements of liquid under the actions of temperature or concentration gradients are called *Marangoni effects*.

THERMOCOUPLE

A *temperature* transducer consisting of two electrical conducting elements of different types joined together. The working principle of a thermocouple involves the Seebeck effect (see *Thermoelectric phenomena*). Advantages of a thermocouple include the simplicity of construction, its small *specific heat*, and the rapidity with which it reaches equilibrium. Among its disadvantages are a strong dependence of thermoelectric power on the composition, homogeneity, and annealing conditions of its component materials. The thermocouple sensitivity reaches 80 μV/K. The most commonly used thermocouples are the following: Chromel/Au–0.02 at.% Fe and Chromel/Au–0.07 at.% Fe for the range 1–280 K (the substitution of a *superconductor* for the Chromel lead results in lowering the measurement limit to 0.05 K); copper–constantan (200–900 K); Chromel–Copel (220–1000 K); Chromel–Alumel (220–1500 K); platinum/rhodium–platinum (250–1900 K); tungsten–rhenium (300–2800 K); and tungsten–molybdenum (300–3100 K).

THERMODIELECTRIC EFFECT, Costa-Ribeiro effect

A charge separation which appears at the transition from one *state of matter* to another. As a result of the thermodielectric effect, different phases acquire electric charges of different signs. It is observable at *melting*, solidification, and *sublimation* of insulators (dielectrics), at the dissolving or *diffusion* into the solvent of one of the components. The thermodielectric effect was discovered by J. Costa-Ribeiro (1944) when studying *electrets*. In application to water solutions the effect was investigated by Workman and Reynolds and has been called the *Workman–Reynolds effect*. The thermodielectric current does not depend on the material thickness but depends only on the speed of the phase interface drift. In anisotropic crystals the current depends on the direction of growth of the solid phase, which is usually positively charged. The total amount of charge precipitated within the time of the boundary drift in a two-phase insulating system is proportional to the variation of the mass of one of phases: $Q = k\Delta m$ where $\Delta m = m_2 - m_1$, and k is the specific constant of the *insulator* material which is called the *thermodielectric coefficient*. The magnitude of the thermodielectric effect is determined by the voltage, the difference in densities of weakly bounded electrons, and the rate of charge penetration through the phase interface. The effect is utilized for the manufacture of electrets.

THERMODIFFUSION, Soret effect

A phenomenon of particle *diffusion* in the presence of a temperature gradient ∇T. In an homogeneous phase consisting of two components, a component separation takes place in the presence of the temperature gradient. To describe the thermodiffusion flux, a term proportional to the temperature gradient is added to the first Fick law:

$$j = -D\left(\nabla n + \frac{Q^*}{k_B T^2}n\nabla T\right),$$

where j is the diffusion flux density, D is the *diffusion coefficient*, n is the number of diffusing particles per unit volume, and Q^* is the heat of transport.

THERMODYNAMIC CRITICAL MAGNETIC FIELD

A magnitude of the external magnetic field strength $B = B_c$ at which a breakdown of the superconducting state occurs, either in the entire volume of a *type I superconductor* or in parts of it (see *Intermediate state*). The thermodynamic critical field is determined by the difference between the free energies of the normal *metal* and the superconductor (see *Thermodynamics of superconducting transition*), and it depends on the *energy gap* Δ in the spectrum of quasi-particles at a given temperature T: $B_c(T) \sim \Delta^2(T)$. In the vicinity of the critical temperature T_c of the superconducting transition $B_c(T) \to 0$ as $T \to T_c$ from below. The value of the critical field $B_c(T)$ is given by the relation $B_c(T) = \Phi_0/[2\sqrt{2}\pi\xi(T)\lambda(T)]$ where $\xi(T)$ is the *coherence length*, $\lambda(T)$ is the *penetration depth of the magnetic field* in the superconductor, and Φ_0 is the *flux quantum*.

THERMODYNAMIC EQUILIBRIUM

A state which a *thermodynamic system* spontaneously reaches, within a comparatively long time interval (compared to system *relaxation times*), characterized by an isothermal condition of either isolation from its environment, or contact with another sufficiently large system in a thermodynamic equilibrium state (thermal reservoir). At thermodynamic equilibrium all *irreversible processes* involving energy dissipation cease: i.e. *thermal conductivity*, *diffusion*, chemical reactions, and so on. Sufficient conditions of thermodynamic equilibrium (*conditions of stability*) can be obtained from the second law of thermodynamics. In a general case, a system is in thermodynamic equilibrium when the thermodynamic potential of the system, involving independent variables associated with external conditions, is a minimum, and the *entropy* is maximal.

THERMODYNAMIC PARAMETERS

The same as *State variables*.

THERMODYNAMIC PHASE

See *Phase*.

THERMODYNAMIC POTENTIALS

Values which characterize the properties of a *thermodynamic system* in its equilibrium state, and allow one to completely describe its behavior. The thermodynamic potentials are proportional to the system volume and do not depend on the way in which the system arrived at a given state (the integral around a closed loop is zero). For a closed system in the simplest case the thermodynamic potentials depend on two variables which might be either the *temperature* (T) and *pressure* (P), the temperature and volume (V), the *entropy* (S) and pressure, or the entropy and volume. For the variables S and V the thermodynamic potential is the *internal energy* (E), which is identical to the average energy of the system. The differential of the internal energy for arbitrary *irreversible processes*, or *reversible processes* which result in the system remaining in an equilibrium state, is given by

$$dE = T\,dS - P\,dV, \tag{1}$$

where the differential of the work $dW = P\,dV$. In variables S and P, the thermodynamic potential is the *enthalpy* H (sometimes called the *heat function* or *heat content*): $H = E + PV$, and its differential dH is

$$dH = T\,dS + V\,dP. \tag{2}$$

In terms of variables T and V, the *Helmholtz free energy* $F = E - TS$ serves as the thermodynamic potential,

$$dF = -S\,dT - P\,dV, \tag{3}$$

and for variables T and P the potential is *Gibbs free energy* $G = H - TS = F + PV$,

$$dG = -S\,dT + V\,dP. \tag{4}$$

If the system is self-developing then by the second law of thermodynamics its entropy cannot decrease, and eventually it reaches the state corresponding to an entropy maximum, i.e. a minimum of the thermodynamic potential which depends on the variables corresponding to the external conditions (e.g., at given T and P with other parameters being free, the system will go to the state with a minimum Gibbs free energy G).

In statistical physics, the free energy is expressed through the temperature and the energy of the system in a given state E_n with the help of the expression:

$$F = -k_\mathrm{B} T \ln \sum_n \exp\left(-\frac{E_n}{k_\mathrm{B} T}\right), \qquad (5)$$

which follows from the canonical ensemble distribution function (see *Gibbs distribution*).

For anisotropic solids (see *Anisotropy of crystals*) in which a *shear strain* exists, the value PV should be replaced (for a small strain) by $V_0 \sum_{ij} \sigma_{ij} u_{ji}$ where V_0 is the volume of the undeformed body (i.e. at $u_{ji} = 0$); $i, j = x, y, z$ are the Cartesian coordinates; σ_{ij} and u_{ji} are the *stress tensor* and *strain tensor* which, in general, might be functions of coordinates. In this case the differentials $P\,dV$ and $V\,dP$ in Eqs. (1)–(4) would be replaced by, respectively, $V_0 \sum_{ij} \sigma_{ij}\,du_{ji}$ and $V_0 \sum_{ij} u_{ji}\,d\sigma_{ij}$.

In addition to the indicated values, there may exist other parameters λ_k which determine states of the system (e.g., the electric, magnetic, gravitational fields, surface tension, etc.). In this case, in differential Eqs. (1)–(4) for the thermodynamic potentials, the terms $\sum_k d\lambda_k$ (here λ_k are certain functions of the state of the system) which are the same for all thermodynamic potentials, must be added. When the system under consideration can exchange matter with other systems (e.g., a thermostat) the dependence of the thermodynamic potential on the number of particles must be taken into account. Hence, in Eqs. (1)–(4) the terms $\sum_m \mu_m\,dN_m$ must be added (here dN_m is the change of the number of particles of the mth kind, μ_m is their *chemical potential*). It can be shown that the Gibbs energy is expressed through the factors μ_m and N_m in the form of the homogeneous function: $\Phi = \sum_m \mu_m N_m$. At the equilibrium of several phases (see *Phase equilibrium*), the chemical potentials of the same material in different phases are the same.

When considering systems with a variable number of particles it is convenient to introduce one more thermodynamic potential often referred to as the *grand potential*: $\Omega = E - TS - \Phi = F - \sum_m \mu_m N_m$. This is associated with the fact that using the *grand canonical ensemble* one can write for variable particle systems systems an expression which relates Ω with the temperature, the chemical potentials, and the energies of particles

in different states. Such an expression for Ω is identical in form to Eq. (5) which corresponds to systems with a constant number of particles. This allows one to carry out the analytical description of systems with variable numbers of particles.

THERMODYNAMIC POTENTIALS OF A BODY UNDER STRAIN

Functions which characterize the thermodynamic state of a body in the presence of elastic strain. The most important *thermodynamic potential* of a deformable body is the *internal energy* (E) which is determined by the expression $dE = T\,dS - \sigma_{ik}\,du_{ik}$. The latter is a mathematical expression of the first law of thermodynamics where T is the body temperature, S is the *entropy*, σ_{ik} is the elastic *stress tensor*, u_{ik} is the *strain tensor* (all the thermodynamic potentials are referred to a unit volume). Other potentials can be obtained from the internal energy with the help of Legendre transformations. For example, for *Gibbs free energy* G this transformation takes the form $G = E - TS - \sigma_{ik} u_{ik}$; hence, $dG = -S\,dT + u_{ik}\,d\sigma_{ik}$. The variables entering the expression for the exact differential of any thermodynamic potential are called its *natural variables*. For example, components u_{ik} are the natural variables of the internal energy, and σ_{ik} are those of the Gibbs potential. Formally, the calculation of various thermodynamic values is reduced to the computation of partial derivatives of the thermodynamic potential with respect to natural variables. In particular, $\sigma_{ik} = -(\partial E/\partial u_{ik})_S$, $u_{ik} = (\partial G/\partial\sigma_{ik})_T$. For exact calculations it is important to know the explicit expressions for thermodynamic potentials through the components of tensors u_{ik} and σ_{ik}. In most cases these are written in the form of a series which begins from the terms which are quadratic with respect to u_{ik} and σ_{ik}, i.e. items like $E_2 = \lambda_{iklm} u_{ik} u_{lm}$ and $G_2 = \mu_{iklm} \sigma_{ik} \sigma_{lm}$. The form of tensors λ, μ, and the tensor factors of the subsequent terms of the series are determined by requiring the invariance of the thermodynamic potential with respect to the symmetry group of an undeformed body. Therefore, those changes of the structure and physical state of a deformable body which cause an alteration of the *crystal symmetry* influence the form of the invariant series for the thermodynamic potentials. This underlies the role of the latter in revealing various features important in physical problems.

THERMODYNAMICS OF SUPERCONDUCTING TRANSITION

Relations which correspond the changes of *thermodynamic potentials*, and their derivatives with critical parameters, of a *superconductor* at the superconducting transition temperature T_c. For example, the difference in free energies between a normal *metal* and a *type I superconductor* is given by

$$F_N - F_S = V \frac{B_c^2}{2\mu_0},$$

where B_c is the *thermodynamic critical magnetic field*, V is the superconductor volume. The entropy difference is

$$S_S - S_N = V \frac{B_c}{\mu_0} \frac{\partial B_c}{\partial T}.$$

At $T < T_c$, $B_c > 0$ and $\partial B_c / \partial T < 0$, and consequently, $S_S < S_N$, i.e. the superconducting phase is more ordered. The latent heat of the transition $Q = T(S_S - S_N)$ is equal to zero at $T = T_c$, $B = 0$, and it is positive at $T < T_c$ when $B > 0$ (*first-order phase transition*). The jump of the *specific heat* at the transition is described well by the Rutgers formula (1936):

$$C_S - C_N = \frac{VT}{\mu_0} \left(\frac{\partial B_c}{\partial T} \right)^2_{T=T_c}.$$

By taking into account the volume change at the transition and the dependence of B_c on the pressure P allows one to obtain the volume *thermal expansion* coefficient $\alpha = (1/V)(\partial V/\partial T)$ and bulk modulus of elasticity $\kappa = -V(\partial P/\partial V)$:

$$\alpha_N - \alpha_S = \frac{1}{\mu_0} \frac{\partial B_c}{\partial T} \frac{\partial B_c}{\partial P},$$

$$\kappa_N - \kappa_S = \frac{\kappa^2}{\mu_0} \left(\frac{\partial B_c}{\partial P} \right)^2.$$

In *type II superconductors*, the superconducting phase transition is a *second-order phase transition* at the point $B = B_{c2}(T)$ for $T < T_c$ (see *Upper critical field*).

THERMODYNAMIC SYSTEM

A macroscopic body or a complex of macroscopic bodies which, in general, can exchange energy and matter with each other, or with surrounding bodies (environment). It consists of a great number of individual component particles (atoms, molecules). The state of a thermodynamic system can be completely described by *thermodynamic potentials*, i.e. by characteristic functions of volume V, *pressure* P, *temperature* T, *entropy* S, number of particles in the system N, and other macroscopic parameters x_i.

A thermodynamic system is in equilibrium (see *Thermodynamic equilibrium*) if the parameters of the system do not change with time, and there are no steady-state fluxes (of heat, matter, etc.). For thermodynamic systems in equilibrium one can introduce the concept of temperature as a *state parameter* which has the same value for all macroscopic regions or parts of the system. A thermodynamic system can be either homogeneous (single-phase) or heterogeneous (multiphase, see *Phase*). For a single-phase thermodynamic system, the number of independent thermodynamic degrees of freedom (*variance*) in the simplest case is equal to two (either T, P or T, V; S, V; S, P). Other parameters can be expressed through these independent ones with the help of *equations of state*. For multiphase systems with c components, the number of different phases φ that can be in equilibrium (see *Phase equilibrium*) as well as the number of thermodynamic degrees of freedom f are related by the *Gibbs' phase rule* $f = c - \varphi + 2$. Since all the phases are in equilibrium, the temperature, the pressure, and the chemical potentials of all components are the same. Thermodynamics takes into consideration: *isolated thermodynamic systems* which exchange neither energy nor matter with other systems; *closed systems* which can exchange energy but not matter with other systems; *open systems* which exchange both matter, and energy with other systems; as well as *adiabatic systems* in which there is no *heat transport* involving other systems.

THERMOELASTIC EFFECT

The appearance of elastic *stresses* and *strains* in an elastic material as a result of heating. A mathematical description of thermoelastic phenomena is achieved by introducing into the *elas-

ticity theory equations some terms related to the heating of a material, and into the *thermal conductivity* equations some terms related to the strain. Depending on the type of problem under consideration, thermoelastic effects are classified as static or dynamic, and as linear or nonlinear.

THERMOELASTICITY

1. A property of a material to acquire elastic *strains* due to stresses induced by a change of *temperature*. Heat thermoelastic stresses arise either from a nonuniform temperature distribution, or as a result of the dependence of the *thermal expansion* on the coordinates and (in noncubic crystals) in the presence of *texture*. However, not every temperature distribution over the bulk induces the appearance of thermoelastic stresses. In simply connected homogeneous continua free of external loads, for which the laws of elasticity and thermal expansion are linear, and in the absence of internal heat sources, an arbitrary stationary temperature distribution does not induce thermoelastic stresses. If such stresses result in the lack of elasticity then one speaks about the appearance of *thermoelastic strains*. It is necessary to take thermoelastic stresses into account when resolving some engineering and physical problems. These are responsible for *failure* caused by a heat impact, for thermal *fatigue*, they induce warping of a material, and they accelerate thermal cyclic *creep* (see *Cyclic heat treatment*).

2. A property of a series of crystals to undergo thermoelastic *martensitic transformations*. A particular feature of these reactions is the gradual growth of *martensite* crystals that takes place at cooling; the deeper the cooling, the more the growth. Heating produces a permanent decrease of the size of formed martensite crystals, leading to their eventual disappearance. Here a temperature factor is similar to a force factor in respect to its action. Thermoelastic martensitic transformations are the basis for the appearance of such properties as *shape memory effect* and *superelasticity*.

THERMOELASTIC MARTENSITE

A *martensite* formed in *alloys* where there is *thermoelastic phase equilibrium*.

THERMOELASTIC PHASE EQUILIBRIUM, Kurdyumov effect (G.V. Kurdyumov, L.V. Khandros, 1949)

A phenomenon of gradual growth (decay) of crystals of *martensite* at a change of temperature and/or *state of stress*. Thermoelastic phase equilibrium is explained in terms of the presence of equilibrium between the driving force of the transformation of chemical origin, and the counteraction of the elastic energy which is proportional to the martensite crystal size. The increase of the Helmholtz free energy difference (ΔF) (see *Thermodynamic potentials*) of the martensite phase and the host phase at a decrease in temperature leads to the growth of martensitic crystals. This growth ceases when the change of elastic energy ΔE becomes equal to ΔF. At heating, as a result of decreasing ΔF, equilibrium is achieved for smaller martensite crystals. A similar influence on the drift of phase interfaces arises from external stresses: martensitic crystals grow as the stress rises, and decrease and vanish as the stress diminishes. A necessary condition of thermoelastic phase equilibrium is the conservation of the interlinking or the coherence of the lattices of martensite and host at the *phase interface*, which takes place in the case of a small transformation motive force, small shear component of shape *strain*, insignificant bulk effects on the transformation, and high strength characteristics (see *Strength*) of the host and the martensite. The phenomenon of a thermoelastic phase equilibrium occurs in those alloys where the *martensitic transformations* satisfy the conditions indicated above (e.g., in Au–Cd, Cu–Al–Ni, Cu–Al–Mn, Ni–Ti, Cu–Zn–Al, and so on).

THERMOELECTRIC DOMAIN in metals

A nonuniform distribution of temperature and electric field which arises spontaneously along a thin homogeneous sample whose *current–voltage characteristic* has the shape of a letter N. The falling portion at the current–voltage characteristic occurs with Joule heating of the sample up to temperatures at which the electric resistance is determined, mainly by inelastic *electron scattering* (as a rule, by electron–phonon scattering) and hence it grows with an increase in the electric field. The relation between the sample temperature

$T = T(E)$ and the applied electric field strength E is determined by the condition of heat balance

$$f(T, j) = -j^2\sigma^{-1}(T) + d^{-1}q(T) = 0,$$

$$j = \sigma(T)E,$$

where j is the electric current density, $q(T)$ is the heat flux through a unit area of the sample of thickness d, $\sigma(T)$ is the *electrical conductivity* of the metal as a function of temperature, f is the heat yield per unit sample area. Hence, there is a falling portion at the current–voltage characteristic in the temperature range where $\partial(\sigma(T)q(T))/\partial T|_{T=T(E)} < 0$. If the sample length exceeds the critical length

$$L_{cr} = 2\pi\left[-\frac{\kappa\sigma d}{\partial(\sigma q)/\partial T}\right]^{1/2}\Bigg|_{T=T(E)},$$

where κ is the metal *thermal conductivity*, then for a given voltage applied to the sample, the uniform distributions of T and E are unstable, and there appear thermoelectric domains which move spontaneously through the sample with the speed $\sim 10^{-1}$–10^{-2} cm/s. In contrast to the domain instability in semiconductors (see *Gunn effect*), the thermoelectric domains in metals exist under conditions of local electrical neutrality.

In the limit $L \gg L_{cr}$, the appearance of thermoelectric domains leads to the stabilization of current in a sample. With the exponential (with respect to parameter L/L_{cr}) accuracy, the current j in the sample coincides with the critical current j_{cr} which is determined from the condition of "area equality":

$$\int_{T_1}^{T_2}\kappa(T)\left(d^{-1}q(T) - j_{cr}^2\sigma^{-1}(T)\right)dT = 0,$$

where T_1 and $T_2 > T_1$ are zeros of the function $f(T, j_{cr})$ which correspond to growing branches of the current–voltage characteristic. Here the thermoelectric domain has a trapezoidal shape: the sample is divided into two portions, i.e. a low-temperature part with $T = T_1$ (values T_1 are close to 4.2 K) and a "hot" part with $T = T_2 \sim \Theta \gg T_1$ (Θ is the *Debye temperature*) with narrow ($\sim L_{cr}$) intermediate regions between them. Wide range variations of potential difference V applied to the sample change only the length of

the "hot" portion of the thermoelectric domain leaving the magnitude of current constant in the sample. Spontaneous nucleation and decay of the thermoelectric domains in the sample occur at different values V and j which leads to *hysteresis* of the current–voltage characteristic. In that case when L is comparable to L_{cr}, the current stabilization does not take place but the current–voltage characteristic of the sample can consist of several independent branches and have self-intersecting points. As a rule, in an experiment $L_{cr} \sim 2$–3 cm, thermoelectric domains appear at $E \sim 1$ V/m, $j \sim 10^9$ A/m^2, their minimal temperature is 7–10 K, their maximal temperature is of the order of the Debye temperature, and the speed of their directed motion is $\sim 10^{-2}$ cm/s.

THERMOELECTRIC INSTABILITY OF PHOTOCURRENT

A phenomenon involving the impossibility of establishing a stationary photocurrent (see *Photoconductivity*) in an electric circuit with a permanent voltage source V, at a constant ambient temperature T, and an illumination intensity I in a certain range of these parameters. Within the range of instability periodic photocurrent oscillations are generated, and near the boundaries of the range of instability, quasi-periodic or random oscillations occur. The mechanism of thermoelectric instability of photocurrent is related to the presence in a semiconductor of two types of centers, i.e. centers with trapping levels M and recombination centers R. The trapping levels are free of electrons in the dark. Transitions 1 (see Fig. 1) of electrons to the conduction band under the action of light lead to filling the trapping levels (transitions 2, 3), as well as to the *recombination* of electrons and holes at levels R (transitions 4, 5). Owing to the yield of *Joule heat*, at values U, T, and I in the instability range, a small increase of any of them induces a sharp increase in the probability of transition 3 which is proportional to $\exp[-\varepsilon/(k_B T)]$ (here ε is the depth of the trapping level). These cause an increase of the electron concentration in the conduction band, the generation of extra Joule heat, and so on, i.e. to an avalanche depletion of trapping levels and a current spike. Subsequently the crystal cools, the trapping levels again become occupied by photoelectrons, and the cycle of *self-oscillations* repeats.

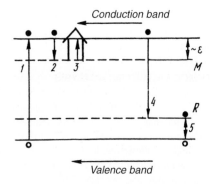

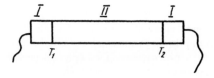

Fig. 2. Arrangement for measuring thermoelectric phenomena.

Fig. 1. Sketch of the transitions in the case of thermoelectric instability of photocurrent (solid points and open circles denote electrons and holes, respectively; arrows indicate electron transitions, wide arrow indicates exponentially growing thermal excitation in the region of instability).

Photoconductivity oscillations caused by the thermoelectric instability of photocurrent are observed in the crystals CdS, CdSe, CdIn$_2$S$_4$, GaAs, BaTiO$_3$, and many others. The oscillation amplitude varies under different conditions of illumination from 10^{-8} Å to decimal parts of an Å, and the period varies from 10^{-4} s to many hours. There are various auto-wave regimes of photocurrent, whereby a crystal is illuminated by light of intensity a little less than that needed to fall into the instability regime at a given T and U (so-called waiting regime). A subsequent local change of one of the controlling parameters leads to the development of an instability in the exciting portion of the crystal, and the propagation of this instability over the sample.

THERMOELECTRIC PHENOMENA

The appearance of an *electromotive force* and an *electric current* as a result of the nonuniform heating of a conductor, as well as the absorption and emission of excess heat (compared to Joule heat) when an electric current flows through a heterogeneous and anisotropic material. Thermoelectric phenomena include the Seebeck effect, the Peltier effect, and also the Thomson effect responsible for the operation of a *thermocouple* made of two different types of conducting materials I and II (see Fig. 2). If the temperatures at the connections (junctions of the thermocouple) T_1 and T_2 are the same no potential difference arises, i.e. $V_{k1} = V_{k2}$. If the temperatures differ, $T_1 \neq T_2$, then the *Seebeck effect thermomotive force* (Th.T. Seebeck, 1821) arises, inducing the *contact potential difference*, $V_{k1} \neq V_{k2}$, between the ends of thermocouple, and its associated electric field. The following causes can induce $U = V_{k1} - V_{k2}$: inequality caused by the temperature dependence $V_k(T)$, the pressure variation of current carriers along the thermocouple at nonuniform heating, and the electric field produced by the diffusion of current carriers associated with the involvement of current carriers in the *phonon* flux from the hot end to the cold end of the thermocouple. At small $T_2 - T_1$: $U = \alpha_{I,II}(T_2 - T_1)$ where $\alpha_{I,II}$ is the specific thermoelectric motive force (*thermoelectric coefficient, Seebeck coefficient*). If material II is a nondegenerate semiconductor than $\alpha_{I,II}$ is much greater than that of a pure metal thermocouple. This is explained by a stronger dependence of V_k and pressure on T in a nondegenerate gas than occurs in the case of *degeneracy*.

When passing a current I, *Peltier heat* Q_P (J. Peltier, 1834) is released at one of the junctions; and absorbed at the other. The value Q_P is given by $Q_P = P_{I,II} I t$, where $P_{I,II}$ is the *Peltier coefficient*, t is the time. The *Peltier effect* is explained by different average energies of current carriers in conductors I and II. As a result, when the carriers move from the conductor with greater energy to the conductor with less energy, they give their energy to the latter. If $T_2 \neq T_1$ then the current heats or cools conductor II, depending on the current direction. This effect is called the *Thomson effect* (W. Thomson, 1856) and is explained by the change of the average energy of the current carriers along conductor II due to its nonuniform heating and charge transport between the portions with different energies. The *Thomson heat* Q_T is determined as $Q_T = \tau(T_2 - T_1)t$, where τ is the *Thomson coefficient*. In some cases the Thomson and Peltier effect are called electrothermic effects. Like the thermoelectric motive force, they are greater in magnitude for a lack of degeneracy, compared to the case of degeneracy.

There appear *bulk gradient effects* related to the motion of current carriers in the internal heterogeneous field of a bulk nonuniform conductor. In anisotropic media, the Seebeck, Peltier, and Thomson *transverse thermoelectric effects* occur. The emission or absorption of heat in the bulk crystal at a change of current direction with respect to the crystallographic axes is called the *Bridgeman effect* (P.W. Bridgeman, 1929). The appearance of an electromotive force at $T_1 = T_2$ in semiconductors containing impurities or internal strains plus bipolar conductivity (see *Electrical conductivity*) and two oppositely directed temperature jumps with different temperature gradients ∇T is called the *Benedicks effect* (C. Benedicks, 1918). This electromotive force is equal to the ∇T dependent thermoelectric motive force difference at jumps of T.

Thermoelectric phenomena arise during hole or *electron heating* in a field. In particular, they are known as thermoelectric motive force, Peltier effect, and Benedicks effect for *hot electrons*. In the case of electron warming-up in a non-symmetric field in a homogeneous isotropic semiconductor (like lattice warming-up at the Benedicks effect), the *bigradient electromotive force of hot current carriers* arises.

THERMOELECTRONIC EMISSION

See *Thermionic emission*.

THERMOGRAPHY

A technique of visualization of thermal fields of diverse materials. The *thermal radiation* of a body is detected using an appropriate receiver and transformed into an electrical signal. In terms of the method of detection there are two principal types of thermographic devices: a contact apparatus and a non-contact one. For non-contact detection, special electronic devices called *IR imagers* are used. In contact thermography, special substances which change their color under the action of thermal fields (*thermal indicators*) serve for detection. The main areas of application of thermography are the *nondestructive testing techniques* of articles in technology, and medical diagnostics.

THERMOGRAVIMETRIC ANALYSIS

A physicochemical method which supplies heat to a material and determines the dependence of its change in mass on the temperature or the duration of the heating. A thermogravimetric analysis allows one to determine, from the magnitude and the rate of change of mass of a material, the degree and the rate of development of various physicochemical processes, such as *oxidation, desorption, evaporation*, and decomposition. In practice, a thermogravimetric analysis is performed in special automatic apparatus with detects simultaneously the mass change, the rate of this change, and the temperature at a given sample. *Thermogravimetry* is often combined with other methods, e.g., thermal analysis, and analysis of the composition of gas given off at heating.

THERMOIONIC EMISSION, thermal ion emission

The emission of positive or negative ions by heated bodies. Referred to as thermoionic emission are: the ionization of atoms of a *solid* during its *evaporation*; *ion emission* of impurity atoms (situated in lattice interstices, at grain boundaries, in dislocation channels, and which migrate to the surface by means of *diffusion*); the emission of ions of foreign substances located on the surface. Concerning the mechanism, thermoionic emission is a type of surface ionization (thermal equilibrium ionization) with the exception of those cases when the ionized matter is the product of dissociation of a compound with significant internal energy added to the surface. Such products are formed in reactions with an energy yield which becomes involved in the ionization process. The thermoionic emission current density is determined using the formulae for surface ionization while taking into account the conditions of the transport of particles to the surface, or their formation there as a result of reactions.

It is on the basis of thermoionic emission that efficient and economical ion sources with long working lifetimes are manufactured; their current density can reach the order of magnitude of hundreds of mA/cm^2.

THERMOMAGNETIC GENERATOR

A device for the direct transformation of heat energy into electrical energy based on the phenomenon of electromagnetic induction. The working principle of a thermomagnetic generator depends on the time variation of the *magnetization* M of the working material, i.e. a material with a strong dependence $M(T)$, by means of periodic heating and cooling (*thermal cycling*) in a permanent magnetic field. The idea of a thermomagnetic generator had been suggested by T.A. Edison (1887), but its practical implementation only became feasible in the mid-twentieth century when materials with the requisite properties became available.

The average value of the electromotive force induced in an electric bobbin with N coils per half-cycle is $\langle \varepsilon \rangle = kSN\Delta M(B)/T$, where S is the cross-section of working material, k is a coefficient of proportionality, $\Delta M(B)$ is the variation of M of the working material body in the magnetic field B in the range of thermal cycling δT, and T is the period of a half-cycle. Diverse *magnetic substances* are used as working materials. When applying the ferromagnets, the thermal cycling is carried out in the vicinity of the *Curie point* T_C; δT is of the order of tens of kelvins; B is close to B_S of engineering saturation (see *Saturation magnetization curve*). It is also possible to employ monocrystal ferromagnets in which the phenomenon of spin reorientation (SR) takes place, e.g., $NdCO_5$, with $T_{SR1} = 245$ K, $T_{SR2} = 285$ K. In this case $\delta T = T_{SR1} - T_{SR2}$, and B is of the order of the *demagnetization field*. Prospective magnets are those which at low temperatures undergo first-order *magnetic phase transitions* such as a antiferromagnet–ferromagnet transition. In this case, a value $\Delta M(B)$ is a maximum, and δT is close to the value of the *hysteresis* of the phase transition, $B \sim B_S$. The limiting value of the thermomagnetic generator efficiency is proportional to that of a Carnot cycle, which reaches 10% (e.g., for Ho, $T_{AF \leftrightarrow F} = 20$ K, $\delta T \sim 3$ K) at the particular output power density ~ 1 W/cm^3.

THERMOMAGNETIC INFORMATION RECORDING

The recording of individual points, images, or a hologram (see *Holography*) on *magnetic film* situated in a magnetic field which is less than the coercive force, and is antiparallel to the *magnetization*. The recording of information is performed by means of rapid heating of local portions of the film up to the temperature at which their *magnetic reversal* occurs abruptly. For the thermomagnetic recording of information one employs films with a rectangular *hysteresis* loop or with a microband *domain structure*. When performing this information recording a localized region of the film is heated up to the needed temperature with the help of a *laser*, an electron beam, or a light flash within a time interval less than 1 μs, and the magnetization is reversed. The reading is performed using the *Kerr effect*, the *Faraday effect*, or magnetic suspension (see *Magnetic powder patterns*). The density of the recorded information can reach 10^8 bit/mm^2, its erasure is performed by a magnetic field whose strength is higher than the strength of the field applied when recording. As a magnetic carrier it is used sputtered on substrate films of *amorphous magnetic materials* such as Tb–Fe, Gd–Fe, Tb–Fe–Co, Gd–Tb–Fe, Tb–Dy–Fe, or permalloy films (see *Soft magnetic materials*) with a microband domain structure. Employed in modern magnetooptical carriers are films based on *superlattices* of Co/Pt in which the number of layers is ~ 20, and the layer thickness is of the order of a nm. For magnetooptical carriers with multilayer Co/Pt films in which the principle of information recording is based on scanning optical microscopy of the nearest field (for overcoming the diffraction limit), the achieved density of recording information is about 700 Mbit/mm^2.

THERMOMAGNETIC MATERIALS

Ferromagnetic *alloys* with a strong dependence of saturation magnetization M_S on the temperature T in a given magnetic field. This property manifests itself in the vicinity of the *Curie point* T_C. In many of these alloys values of T_C fall on the range between 0 and 200 °C. As a rule, thermomagnetic materials are divided in two groups: *thermomagnetic (compensation) alloys*, such as Ni–Cu, Ni–Fe–Cr (*compensator*), Ni–Fe, and *multilayer thermomagnetic materials*. Thermomagnetic materials are applied, mainly, as magnetic shunts, or additional magnetic resistors. The introduction of such elements into *magnetic circuits* allows one to compensate for the changes of the magnetic flux in the circuits induced by the temperature changes of the electric resistance in the coils of a magnet, the air gap of a magnet, and so on. Thermomagnetic materials are used in relays whose response function depends on T. See also *Magnetic materials*.

THERMOMAGNETIC PHENOMENA

Phenomena which arise in a conductor with a temperature gradient as it is positioned in an external magnetic field. For the appearance of thermomagnetic phenomena, the magnetic field should have a component orthogonal to the temperature gradient. The fluxes of "hot" and "cold" *current carriers* moving from hot and cold ends of a sample towards each other are deflected by the magnetic field to opposite side walls of the sample. As a result one of the walls will be heated, and another wall cooled. The appearance of a thusly induced temperature gradient directed normally to the primary (applied from outside) gradient is the manifestation of the *Righi–Leduc effect* (A. Righi, S. Leduc, 1887). However, if the mobility depends on the energy of the current carriers then these oppositely directed fluxes would not be compensated. As a result, charge will accumulate at the side walls, thereby inducing an electric field orthogonal to the applied gradient; this field is called the *Nernst–Ettingshausen transverse field* (W. Nernst, A. Ettingshausen, 1886). In anisotropic conductors the field appears with an energy-independent mobility.

The Righi–Leduc effect and the *Nernst–Ettingshausen transverse effect* are called *transverse thermomagnetic effects*. These are odd in respect to the magnetic field direction, and their values are determined by the temperature change $\Delta T = -SB(T_1 - T_2)$ and the value of the transverse electric field $V_2 = -QB(T_1 - T_2)$ where B is the magnetic field; T is the absolute temperature (indices 1, 2 refer to heated and cooled regions of sample, respectively); the factor Q is called the *Nernst constant*; S is the *Righi–Leduc coefficient*.

In addition to transverse effects, there also exist *longitudinal thermomagnetic effects*. These are even in respect to the magnetic field direction. If the mobility of an isotropic conductor is dependent on the energy of the current carriers, or the conductor is anisotropic, then the fluxes of charged particles which flow in a direction normal to ∇T are turned by the magnetic field toward the direction along ∇T, and the magnitude of the *thermal electromotive force changes*. Thus, in a magnetic field an addition to the coefficient of thermal electromotive force α arises (which is even with respect to the magnetic field direction). This effect is called the *Nernst–Ettingshausen longitudinal effect*. Moreover, the electron *thermal conductivity* and, in turn, the heat flux along the original gradient both change; this is the so-called *Maggi–Righi–Leduc effect* (G. Maggi, A. Righi, S.A. Leduc, 1887). In small-size conductors the presence of the magnetic field can also change the primary temperature profile by inducing the dependence on coordinates of the applied temperature gradient in a homogeneous conductor. In this case thermoelectric currents, odd with respect to the magnetic field direction, emerge in the conductor. These currents are responsible for the appearance of odd (in respect to B) additions to the thermal electromotive force (*Nernst–Ettingshausen longitudinal even effect*), and an alteration of the primary heat flux (*Maggi–Righi–Leduc even effect*).

There are isothermal and adiabatic thermomagnetic phenomena. If the sample side walls are thermally isolated then there are no heat fluxes normal to the temperature gradient. In these cases the effects are classified as adiabatic ones. However, if there is a perfect heat transport to the environment, the transverse effects are missing, and the isothermal effects take place with values which differ from the those in adiabatic effects.

In the subsurface layers the temperatures of *phonons* and current carriers can fail to be coincident. In this case, the heat isolation of side walls does not result, in general, in the lack of heat fluxes in every subsystem (phonons and current carriers), and the character of thermomagnetic phenomena changes.

THERMOMAGNETIC TREATMENT

Heat treatment of metals and alloys in a *magnetic field*. A thermomagnetic treatment of *steel* increases its *impact strength* and mechanical strength properties. During the treatment the mechanical *strain* in the direction of the magnetic field improves in addition to the steel *strength*.

In practice, the influence of the thermomagnetic treatment on magnetic characteristics of *metals* is used to improve the properties of *thin films, ferrites*, alloys for *permanent magnets*. The action of an external field is observed if one of the phase components of the alloy is ferromagnetic. The magnetic field activates *phase transitions* which lead to the formation of the *phase* with higher *magnetization*, and expands the region of its presence. If the *martensitic transformation* in steel takes place in a magnetic field, then as a result of a noticeable increase in the *diffusion* of carbon atoms the *martensite* already formed during cooling undergoes a *tempering* and a partial decomposition with the carbon yield from a *solid solution*. Self-tempering of martensite in a magnetic field reduces the brittleness of fresh-hardened steel. These technologically important changes of the structural state of steel and its properties are achieved in magnetic fields with strengths up to 0.1 T.

THERMOMECHANICAL EFFECT

See *Mechanocaloric effect*.

THERMOMECHANICAL TREATMENT

A *heat treatment* of metals and alloys combined with a deforming process aimed to improve their properties. During a thermomechanical treatment the material structure reforms under the conditions of increased density and optimum distribution of crystal structure *defects* which nucleate at deforming and *phase transitions*. Deforming can be performed prior to the thermomechanical treatment, during the treatment, or after it, as well as both prior and subsequently. There are many technological schemes of thermomechanical treatment, including those involving multiple treatments. For a low-temperature thermomechanical treatment the material is deformed at a *temperature* a little below the *recrystallization* temperature, and for a high-temperature one the *plastic deformation* is carried out at a temperature above the recrystallization temperature. Conditions of plastic deforming and subsequent hardening are selected that produce stable structural states with an increased density of defects, and the presence of *polygonization*. This leads to the increase of the material *strength* in combination with sufficiently

high *plasticity, fatigue* resistance, and impact resistance.

THERMOMETRY (fr. Gr. $\theta\varepsilon\rho\mu\eta$, heat and $\mu\varepsilon\tau\rho\varepsilon\omega$, am measuring)

A branch of applied physics devoted to the measurement of *temperature*. Thermometry is closely related to metrology, it involves providing reliability and precision to temperature measurements: establishing temperature scales and standards, developing techniques for calibration and testing apparatus for variable temperature studies. The measurement of temperature entails a well-defined dependence on it of a particular property of a solid selected as a *thermometer*, and the selection of a temperature scale. The *absolute temperature scale* or Kelvin temperature scale selects absolute zero as 0 K, and chooses its unit increment of temperature as (1/273.16)th part of the thermodynamic temperature of the *triple point of water*. A primary thermometer is used for measuring the *thermodynamic temperature* (its *equation of state* can be written in closed form without unknown constants), and examples are gaseous, acoustical, noise, and total radiation types. With the aid of primary thermometers, extremely precise measurements can be made of standard fixed temperature points, and using these as benchmarks, the scales of practical thermometers can be graduated at these temperatures.

The international community has recommended the following thermometers as acceptable standards: a platinum resistance thermometer for the range from 13.81 K to 630.74 °C; a *thermocouple* with platinum–platinum/rhodium electrodes (10% rhodium) for the range from 630.74 °C to 1034.43 °C; and an optical pyrometer (see *Pyrometry*) for temperatures above 1064.43 °C. In the range 0.5–30 K a thermodynamic interpolation (gaseous or magnetic) procedure is accurate; at $T < 5$ K this is based on the vapor pressure of ^{4}He and ^{3}He. In the range 0.5 to 273 K the highly stable iron–rhodium (0.5 at.% Fe) resistance thermometer is applicable. The thermocouples Chromel/Au–0.02 at.% Fe and Chromel/Au–0.07 at.% Fe can be employed in the range 1–280 K. The methods of acoustic noise, magnetic thermometry, nuclear orientation, measurement of static nuclear susceptibility, and so on, are applicable at extremely low temperatures.

Semiconductor thermometry. Techniques for measuring temperature with the help of semiconductor devices, transducers whose main feature is a pronounced temperature dependence of one or more of its parameters. From the metrology viewpoint semiconductor temperature transducers, because of their specificity, are *secondary thermometers*, i.e. they require an initial calibration with subsequent verification by comparison with standard (primary) thermometers.

A high thermal sensitivity and thermoelectric power at the *p–n* junction which are specific for semiconductor materials (Ge, Si, GaAs, CdSb, and so on) stimulated implementation of two types of thermometers: *temperature sensitive resistors* and *thermal diodes* (transistors); the former being more common. A broad range of applications necessitates a series of partially contradicting requirements to which the *semiconductor resistance thermometers* must respond: accuracy and stability of readings, reliability and simplicity of operation, minimal dimensions, low inertia, independence of magnetic fields and some other physical influences, wide range, low cost, etc. Present day semiconductor technology is able to produce semiconductor resistance thermometers that satisfy the growing requirements of science and engineering.

An important aspect of semiconductor thermometry is *cryothermometry*. Semiconductor resistance thermometers are the most convenient for practical measurements of low (<100 K) and extremely low (<1 K) temperatures. Precise doping by, e.g., Ge performed by the nuclear transmutation method provides the possibility of fabricating semiconductor resistance thermometers for operation down to 0.01 K.

Magnetic thermometry. A method of determining the temperature in the vicinity of absolute zero based on the temperature dependence of the *magnetic susceptibility χ of paramagnets*. Those paramagnets whose susceptibility depends on the temperature through the simple Curie law $\chi = C/T$ are suitable for magnetic thermometry. One can determine the so-called *magnetic temperature T^** from a measurement of χ in a weak field, of the static nuclear susceptibility, and so on. These procedures are applicable at extremely low temperatures using the known value of the paramagnet

Curie constant C. In that temperature range where the Curie law is valid, T^* coincides with the thermodynamic temperature T. At lower temperatures deviations from the Curie law cause the magnetic temperature T^* to differ noticeably from the actual (thermodynamic) temperature T. In practice, the magnetic temperature is adjusted to the thermodynamic one using tables and curves composed on the basis of careful studies of the temperature dependence $\chi(T)$.

THERMOPLASTIC MATERIALS

Natural or artificial high molecular weight materials (resins) which soften at heating and harden at cooling, and which conserve these abilities after multiple heating and cooling cycles (see *Polymeric materials*). Thermoplastic materials are characterized by a *vitrification* temperature T_g and a *yield* temperature $T_f > T_g$. For $T < T_g$ solidification takes place. A softening, i.e. a transition to the viscous-fluid state occurs at $T > T_f$. Within the range $T_g < T < T_f$ high-elasticity state responses are specific for flexible polymer molecules of the material (see *Flexibility of polymeric chain*). The tendency for thermoplastic materials to be deformed irreversibly under the application of heat and pressure, and to conserve their form at $T < T_g$, is utilized for manufacturing diverse products, and for *information recording* by *thermoplastic media*. These thermoplastic media are thin layers whose particular feature is the capability of easy deformation up to a viscous-fluid state under the action of the field of the forces that form the latent image. The latent image is produced by the modulation of the electrostatic field forces (surface charge), by the optical field of the image, or by the output signal, e.g., electrons. Depending on this, thermoplastic media are divided into photothermoplastic and simply thermoplastic types. A developed image in the form of surface relief is fixed by rapid cooling of the thermoplastic medium below the vitrification temperature; the image is read with the help of special optical systems such as *schlieren-systems* (see *Optical defectoscopy*), which transform the phase contrast into an amplitude one, or with the help of illumination by a reference beam if a phase hologram is recorded (see *Holography*). Heating the thermoplastic medium to a higher temperature provides the *relaxation* of the forces forming the latent image, and a smoothing of the relief (deletion of record) under surface tension forces, thus permitting the repeated use of the thermoplastic medium for recording and reading images.

During the development process, there appear random surface strains ("frost-made pattern") at a uniformly charged layer of the thermoplastic medium, which constitute a noise source or a random raster; modulating this also allows one to record information. Thermoplastic media are considered the best reversible photosensitive media for recording thin phase holograms. These possess an extensive range of positive qualities such as the absence of grain structure, high sensitivity (5–100 $\mu J/cm^2$) comparable to the sensitivity of haloid-silver photoemulsions, a rapid and dry development process, capability for cycling, high resolution reaching 10^3 lines/mm in thin layers, and a high rate of recording and erasing of information.

Thermoplastic media were applied by G.Ch. Lichtenberg (1777) for observation of the action of electric charge on the surface of a softened layer of rosin (*Lichtenberg figures*). F. Fisher (1940) used thin layers of viscous-fluid oils deformed by static electric charge in a television system "Eidophor", which was then modified for using thermoplastic media.

THIN FILM GROWING

Production of thin layers (see *Films*) of crystalline or amorphous material on *substrates*. Some purposes of film growing are to create passivating, insulating or conducting *coatings*. This is done to form *heterojunctions*, to create multilayer structures for solid-state electronic and optical devices, to vary the mechanical properties of surface layers and perhaps increase their *hardness* and wear resistance, or to create a decorative coating (see *Surface decoration*). Films can be grown by various methods: deposition of material from vapor, liquid (melt or solution) and solid phases; deposition from molecular and ionic beams; by thermal or ionic scattering; with the help of chemical reactions at the substrate surface (*oxidation*, *anodization*, nitriding, formation of silicides, etc.), and by the electrolytic method. Depending on the method of growing and on their application, the thickness can be varied from one monolayer (Langmuir

monomolecular films) up to fractions of a millimeter (electrolytic coatings).

Historically, the formation of decorative and protective metal layers (nickel, gold, silver plating) from aqueous solution or electrolytes was the first thin film preparation method which found wide use. At present monocrystal films are widely grown on oriented (*epitaxy*) and non-oriented (*rheotaxy*) substrates. The convenience of *doping* films during the course of epitaxial growth facilitates the preparation of the epitaxial structures so widely used in semiconductor electronics.

THIN FILMS

Layers of material whose thickness is small in comparison with other dimensions. One could regard thin films as layers whose thickness is less than any characteristic length (screening length, *diffusion length*, *mean free path*, Maxwell *relaxation* length), or comparable to it. In this case *size effects* (scale effects) will manifest themselves in thin films, i.e. the dependence of their physical properties, as well as the *tunnel effect*, on the thickness. In this sense thin films are close in their properties to *thread-like crystals*. The size effects in the optics of thin films (see *Thin layer optics*), e.g., *light interference*, have been known since the time of Newton.

The size effect in the *electrical conductivity* of thin metal films was first observed by I. Stone (1898) and explained by J. Thomson (1901). Later the size effects in *thermal conductivity*, thermoelectromotive force (see *Thermoelectric phenomena*), and *galvanomagnetic effects* were discovered and investigated. Unlike the classical effects which are observable in samples whose thickness is comparable to the electron mean free path, *quantum size effects* are observable when the thickness is comparable to the electron *de Broglie wave* length λ. In this case, the role of quantizing the electron quasi-momentum becomes essential, and as a rule, this changes radically the thermodynamic and kinetic characteristics of the film. In many cases thin films are discontinuous (*insular films*). A limiting thin film case is *monomolecular films* (see *Langmuir–Blodgett films*) which had been studied in the experiments of A. Pockels (1891), and interpreted by T. Rayleigh (1899).

THIN-FILM SOLAR CELLS

Solid state devices used for the direct transformation of solar energy into electric current, *solar cells*. These consist of a few layers of materials with different properties. An active (working) region of a thin-film solar energy cell is a semiconductor film which incorporates an electric field (*semiconductor junction, heterojunction*, and so on). The direct transformation of solar energy into electrical energy takes place due to the separation of light-generated electron–hole pairs by the internal electric field, which results in the appearance of an *electromotive force* and an *electric current*. In thin-film solar energy cells semiconductor films with direct optical transitions are used (compounds $A^{III}B^V$ and $A^{II}B^{VI}$, amorphous hydrogenated *silicon*); they have a large *light absorption* coefficient. A typical thickness of the active region of such devices is from one to a few micrometers. The highest efficiency of thin-film solar energy cells (close to 30% in the order of magnitude under concentrated illumination) has now been achieved in devices based on the heterojunction $GaAs–Al_xGa_{1-x}As$.

THIN FILM SUPERCONDUCTIVITY

A group of *superconductivity* phenomena important both in theory and in practice. Indeed, almost all the nonstationary, nonequilibrium and nonlinear phenomena found in superconductivity, including those applicable to *superconductor electronics*, are found in thin film structures. Thin film superconductivity already starts in films $d \approx 1$ nm thick. However, the critical temperature T_c in extra thin layers is strongly perturbed by *fluctuations*. The longitudinal critical magnetic fields in films sharply increase for smaller d, and transverse ones result in the formation of quantized *Abrikosov vortices* in the film (for $d < \lambda$ it happens in films of every superconductor). These vortices interact with currents flowing through the film in a complex nonlinear manner. As for highly ohmic films, bound states of the vortex–antivortex type form in them, causing the topological *Kosterlitz–Thouless transition*. Effects of electron localization upon the film superconductivity also appear. In the presence of vortices, critical currents are controlled by the phenomenon of *vortex pinning*, while in their absence the same role is played by the kinetic depairing of the superconducting

Bose condensate (the latter occurs at current densities $\sim 10^7$ A/cm^2). Studying such films is also complicated by several factors: vortices can start to form in the Earth's magnetic field; the current distribution across the film cross-section is nonuniform; certain thermal effects take place, etc. Using thin films one may study in detail mechanisms for generating the *resistive state* while the overall superconducting state persists. Such mechanisms include the formation and melting of *vortex lattices*, the nonequilibrium dynamics of vortices, the formation of centers and lines of *phase slip* that can maintain their resistive state up to currents significantly exceeding the critical value, as well as effects involving the generation of electromagnetic oscillations, and nonlinear interactions with microwave fields. The critical temperature T_c and other parameters of superconducting films are sensitive to the conditions of *condensation*. At substrate temperatures close to that of liquid helium metastable amorphous condensates easily form, as well as new superconducting modifications of metals. Powder films are obtained via condensation in oxidizing media at low pressures. Co-evaporation of metals with insulators, semiconductors, or common metals yields granular films. These may be of different types, with granules linked together by weak *tunnel junctions* or bridge contacts. The physical properties of thin films depend on various phenomena: percolation (see *Percolation theory*), metal–insulator transitions, *Josephson effects*, Coulomb effects, and strong fluctuation effects. Thin film superconductivity opens the way to studying the effect of thermodynamic fluctuations of the *order parameter*, both far from the *phase transition* point and in the vicinity of T_c and B_c. The same applies to changes in the effective dimensionality of fluctuations under the effects of temperature and applied magnetic field.

THIN LAYER OPTICS

A branch of optics in which the optical properties of thin films of *solids* are studied. True thin layers are those with optical indices that depend on the thickness. From experimental observations, the limiting thickness where this dependence can yet be detected in *metals* does not exceed a few tens of nanometers, i.e. covers the thickness range where metal films are still somewhat transparent. The limiting thickness for *insulators* and *semiconductors* is 100–150 nm. The presence of interference is insufficient evidence for classifying films as truly "thin", when their *refractive index* and *absorption index* do not differ from those in the bulk material, and are independent of thickness.

Based on the macroscopic Maxwell equations, early phenomenological theories produced relations implying that the values of so-called *effective refractive indices* n_{eff}, calculated from *reflectance* and *transmittance* experimental data, grow rapidly as the effective layer thickness d_{eff} decreases, while the absorption index κ_{eff} in the region of very small thickness ($d < 100$ Å) has a maximum. Such a trend of n_{eff} leaves one perplexed, since as the thickness decreases, so does, typically, the density of the matter in the layer; therefore, one would expect a decrease of optical density and, accordingly, n_{eff}. Thus, phenomenological theories are unsuitable for the description of optical phenomena in thin layers. Other attempts have been made to provide a quantitative description of these phenomena. *Mie theory* (G. Mie, 1908) dealt with light scattering by colloidal particles; and *Maxwell–Garnett theory* (J. Maxwell, J.G. Garnett, 1904) described the granular structure of thin films (see *Granular films*) in the form of a three-dimensional colloid. Then, *David–Schopper theory* (E. David, 1939; H. Schopper, 1951) took into account the grain orientation in a *thin film*; and *Rosenberg theory* (G.V. Rosenberg) proceeded from the concept of a thin layer as a two-dimensional colloid. However, these approaches also failed to provide a satisfactory agreement between theory and experiment. Many believe that a proper microscopic theory of the optics of thin layers should take into account the various factors that determine these properties, namely: the role of the substrate surface, the way it fits on to the thin layer material, the shape and size of grains, their interaction, the free surface of grains, and so on. *Surface states* in an optical thin layer very likely play an important role in determining its optical properties.

THIRD CRITICAL FIELD, critical field of surface superconductivity

A magnetic field which destroys *superconductivity* in the vicinity of a plane surface of a superconductor. For the applied field parallel to the superconductor interface in a vacuum, the third critical field is given by $B_{c3} = 1.69B_{c2} = 1.69\sqrt{2}\kappa B_c$, where κ is the *Ginzburg–Landau parameter*, and B_c is the *thermodynamic critical magnetic field*. At the *upper critical field* B_{c2} the bulk superconductor goes normal, and in fields between B_{c2} and B_{c3} the superconductivity persists in the form of a surface superconducting layer of thickness equal to the *coherence length* which can carry a finite *critical current*. The region inside this layer is now in the normal state. Coating the superconducting surface with a normal *metal* reduces B_{c3} to a value very close to B_{c2} (see *Surface superconductivity*). For *type I superconductors*, the surface superconductivity is observed when $B_{c3} > B_c$, that is in the range $0.418 < \kappa < 0.707$. For values $\kappa < 0.418$, the field B_{c3} is less than B_c, and represents an overcooling field of the superconductor.

THOMSON EFFECT
See *Thermoelectric phenomena*.

THORIUM, Th

A radioactive element of Group III of periodic system with atomic number 90 and atomic mass 232.0381; it belongs to the *actinides*. Natural thorium consists almost entirely of the ^{232}Th isotope with half-life $1.39 \cdot 10^{10}$ years. Twenty four thorium isotopes are known with mass numbers 213 to 236. Outer shell electronic configuration is $6d^2 7s^2$. Oxidation state is $+4$, less often $+2$, $+3$. Atomic radius is 0.179, radius of Th^{4+} ion is 0.102 nm. Electronegativity is ≈ 1.27.

Bulk thorium is a light-gray *metal*. It exists in two polymorphous modifications (α-Th, β-Th). The low-temperature α-Th has a cubic close-packed lattice with period $a = 0.50851(7)$ nm (at 298 K). At 1633 K it transforms to β-Th with a body-centered cubic lattice and period $a = 0.411$ nm (at 1723 K). Density is 11.72 g/cm^3 (298 K); $T_{melting} = 2023$ K. Specific heat is 27.382 kJ·kg^{-1}·K^{-1} (300 K); coefficient of thermal conductivity is 37.6812 W·m^{-1}·K^{-1} (at 383 K); average linear thermal expansion coefficient is $11.55 \cdot 10^{-6}$ K^{-1} (at 303 to 473 K). Resistivity is 0.13 µΩ·m (at 293 K). Below 1.38 K thorium becomes a superconductor. Work function is 3.51 eV; magnetic susceptibility is $0.54 \cdot 10^{-6}$ emu/g (at 293 K). Total effective cross-section of adsorption and scattering of high-energy (3 to 10 MeV) neutrons is 7.2 barn, of thermal neutrons (0.025 eV) it is 20 barn. At room temperature normal elastic modulus is 68.81 GPa; shear modulus is 27.44 GPa; Poisson ratio is 0.27. The purest thorium iodide is of low strength, and is plastic at room temperature. In its strained and annealed state at 923 K its tensile strength is 0.119 GPa; relative elongation is 36%; relative constriction is 62%; Vickers hardness is 45 HV. Adiabatic elastic moduli of α-Th: $c_{11} = 75.3$, $c_{12} = 48.9$, $c_{44} = 47.9$ (in GPa) at 300 K. Less pure thorium, caked from electrolytic powder, strained and then annealed, has tensile strength 0.153 GPa, yield limit 0.078 GPa; relative elongation 35%; Vickers hardness 53.0 HV. Carbon dissolved in thorium increases its strength, and C contents of 0.05% makes it brittle below 273 K. Due to low solubility, additions of oxygen and nitrogen barely affect the mechanical properties. Thorium iodide becomes brittle. Thorium in powder form is pyrophoric in air and in oxygen. Thorium is used for *alloying* of Mg alloys. The dioxide ThO_2 is used as a fireproofing material.

THREAD-LIKE CRYSTALS, filamentary crystals, whiskers

Microscopic *monocrystals* with their length in one direction far exceeding their width in other directions (typical length 1–2 mm, diameter ~1–2 μm). Ordinarily filamentary crystals grow along certain crystallographic directions (e.g., along the normal to a close-packed face) and possess isometric cross-sections (hexagonal, square, etc.). Thin films and plates are also found (0.1–10 μm thick, 0.1–1 mm wide) (see also *Quasi-one-dimensional crystals*). Thread-like crystals grow from either gas or vapor via the so-called vapor–liquid–crystal mechanism (see *Crystallization from the gas (vapor) phase*). When these crystals grow from a solution or a solid phase, *screw dislocations* play an important role: either the apex or the base of the growing crystal has a growing step that reproduces itself as more matter reaches the thread-like crystal surface.

An important class of filamentary crystals discovered in 1991 is carbon nanotubes which are rolled up graphite sheets forming tubes, sometimes closed at the ends by fullerene (C_{60}) type half-spheres. The properties of nanotubes depend on tha axis about which the graphite sheet is wrapped, and two important configurations of carbon hexagons are called "armchair" and "zig-zag" types. Nanotubes can be multiwalled (concentric cylinders) with outer diameters between 2 and 25 nm, and single walled with a diameter between 1 and 2 nm. The electrical conductivity can range from metallic to semiconducting, and the mechanical strength is very high.

Filamentary crystals in general possess a number of unique properties: they have practically no *defects* and their *strength* can be close to the theoretical limit, hence it may exceed the strength of common crystals by a factor of 10^2–10^3. The latter feature allows studying the effects of strong elastic *strains* on various physical properties of crystals, such as electrical resistance. The broad thickness range and high chemical purity of thread-like crystals makes them useful for studying *size effects*. A number of instruments have been designed (miniature thermometers, strain-gauge transducers, Hall sensors, radiation instruments) in which filamentary crystals act as sensors. These crystals are used in electronics as highly efficient field-emission cathodes (see *Field-emission cathode*), and also to produce extra strong *composite materials*.

THRESHOLD ENERGY OF DEFECT FORMATION

Amount of energy transferred to an atom of a *crystal* to initiate starting the formation of a *Frenkel defect*. The experimental dependence of the cross-section for *defect* formation σ on the energy E of a foreign particle of type S (high-energy electron, proton, *neutron*, etc.) has a region of fast growth $\sigma(E)$, which approximately provides the threshold energy of defect formation E_d. The energy T transferred to the atom during an elastic collision by the foreign particle is considerably different from E, e.g., for nonrelativistic particles: $T \leqslant T_{max} = 4m\mu E/(m + \mu)^2$, where m, μ are the masses of the foreign particle and atom of the crystal. The values $T = E_d$ for different materials lie within the limits from several electron volts to ~100 eV: the corresponding threshold energy of the electrons is hundreds of keV. The cross-section of defect formation by the foreign particles

$$\sigma_S(E) = \int\limits_0^{T_{max}} dT \, W(T) \frac{d\sigma_S(E,T)}{dT},$$

where $d\sigma_S(E,T)$ is the differential cross-section of transfer to the crystal atom of the energy within the range from T to $T + dT$ from the particle S with energy E, and $W(T)$ is the probability of formation of a defect upon the transfer of energy T to the atom. The dependence $W(T)$ determined from the experimental curves $\sigma_S(E)$ is considerably different from the rough stepwise approximation, $W(T) = 1$ at $T > E_d$; $W(T) = 0$ at $T < E_d$.

THRESHOLD POTENTIAL SPECTROSCOPY

Method of investigating the composition and electronic structure of *solids*, based on studying the characteristic features of either secondary radiation from near-surface layer atom electronic excitation, or the reflected beam of primary ions which bombard the solid and undergo a change in energy. The excitation of atoms of the solid takes place at certain threshold values of the potential, the minimum value of which may be obtained from the equation $eV = E_{be} - e\varphi_c$, where E_{be} is the binding energy of an electron at the ground level with respect to the *Fermi level*, and φ_c is the work function of the cathode. The process of excitation involves the generation of a hole at the inner level, and the generation of two electrons (primary, and excited) above the Fermi level. *Relaxation* of this state may be accompanied by *recombination* of the hole with emission of an X-ray photon, or by the generation of an Auger electron with a probability close to 100% if the energy of the hole is less than 2 keV (see *Auger electron spectroscopy*). The energy band structure and composition of a solid may be determined through studying the energy distribution of Auger electrons, the primary electron elastic reflection, or soft X-ray radiation. A low-energy variety of threshold potential spectroscopy employs primary electrons of energy less than 10–15 eV; it is highly sensitive to physicochemical processes which involve a structural rearrangement of the density of valence electron and free electron states of the solid under consideration.

THROTTLING

A drop of the pressure in a gas, steam, or liquid without any thermal interchange with the environment, and without doing work. During throttling a temperature change takes place (see *Joule–Thomson effect*) which can be utilized for achieving *cryogenic temperatures*. A throttling process is also called a *Joule–Thompson process*.

THULIUM, Tm

Element of Group III of the periodic table with atomic number 69 and atomic weight 168.9342; belongs to yttrium subgroup of *lanthanides*. Natural thulium consists of one stable isotope ^{169}Tm. Outer shell electronic configuration $4f^{13}5d^06s^2$. Atomic radius 0.1743 nm; ionic radius of Tm^{3+} is 0.0869 nm. Oxidation state +3. Electronegativity $\approx$1.24.

Bulk thulium is a silvery-white *metal*. It has a hexagonal close-packed lattice at temperatures below 1793 K, space group $P6_3/mmc$ (D_{6h}^4); lattice constants are $a = 0.35375$ nm, $c = 0.55540$ nm at room temperature. The body-centered cubic lattice, space group $Im\bar{3}m$ (O_h^9), may arise above 1793 K (up to $T_{melting} = 1818$ K) with lattice constant $a = 0.392$ nm after hardening. Binding energy 2.6 eV/atom (at 0 K). Density 9.314 g/cm^3 at 298 K; $T_{boiling} \approx 2110$ K. Heat of melting 18.2 kJ/mole, heat of sublimation 247 kJ/mole, heat of evaporation 240 kJ/mole; specific heat 0.160 kJ$\cdot$kg$^{-1}\cdot$K^{-1} at room temperature; Debye temperature 167–200 K; coefficient of linear thermal expansion of thulium polycrystal is $\approx$12.5$\cdot$10^{-6} K^{-1}. Adiabatic crystal elastic moduli are: $c_{11} = 86.3$, $c_{12} = 31.0$, $c_{13} = 24.5$, $c_{33} = 84.0$ (GPa) at 298 K. Adiabatic bulk modulus is 44.5 GPa (300 K), isothermal bulk modulus is 39.7 GPa. Young's modulus is 74.8 GPa, rigidity is 30.4 GPa, Poisson ratio is 0.235. Brinell hardness is 0.539 GPa (for 99.0% pure thulium). Thulium is readily amenable to machining. Effective thermal neutron capture cross-section is 118 barn. The Sommerfeld coefficient of linear low-temperature electronic heat capacity is 14.41 mJ$\cdot$mole$^{-1}\cdot$K^{-2} (2.5–10.0 K). Resistivity of thulium polycrystal is $\approx$755 n$\Omega\cdot$m at room temperature. Temperature coefficient of resistance is 0.00195 K^{-1}. Work function of polycrystal is 3.12 eV. At the Néel temperature $T_N = 57$ K, thulium transforms from the paramagnetic to the antiferromagnetic state with a sinusoidal-shaped arrangement of spins directed parallel to the 6 principal axis. Possible additional magnetic structures are reported at $T_c \approx 35$ K, at $T_c = 21$ K, and over the temperature range 21–38 K. Magnetic susceptibility $+25600 \cdot 10^{-6}$ CGS units. Nuclear magnetic moment of ^{169}Tm isotope is 0.20 nuclear magnetons. Thulium is used as a getter in electrical vacuum devices and for scientific research purposes. The ^{170}Tm isotope is used as a portable source of X-radiation in defectoscopy and in medicine.

THYRISTOR (fr. Gr. ϑυρα, door)

A *semiconductor device* which contains four or more sequential, alternating sign conduction regions, e.g., a *p–n–p–n* transitor. In addition to anode and cathode, it is possible to introduce a third control electrode in one of the internal regions. Accordingly, there are a diode thyristor (*dynistor*) and a controlled thyristor. The thyristor *current–voltage characteristic* is determined by a combination of the actions of forward and reverse bias at *p–n* and *n–p* junctions, and characterized by a sharp transition between the states with low and high conductivity. Thyristors are employed as rectification and switching devices in electrical circuits, and in power control operations. The most common thyristors are silicon controlled rectifiers.

TICONAL (fr. titanium, cobalt, nickel and aluminum; known in Russia as YuNDKT)

Alloy based on the system Fe–Ni–Al–Co–Cu–Ti. A particular feature of ticonal is its high contents of *cobalt* (to 8%) and *titanium* (to 8%). Permanent magnets made of these alloys are obtained by *casting* and grinding. After a *thermomagnetic treatment* these become anisotropic *magnetic materials*. A large effect is achieved in the case of *monocrystals* or castings with an acicular *texture*. For the properties of ticonals, see Table 1 in *Hard magnetic materials*.

TIGHT BINDING METHOD

Approximation of tightly bound electrons; one of many methods for the theoretical calculation of the energy of a molecule or a *crystal*. According to the tight binding method, the wave function

of an electron $\psi(r)$ (in the *single-electron approximation*) is expressed in the form of a linear combination of atomic functions $\psi_a(r)$. It is supposed that when an electron is located near some atom with number n, then its potential $V(r)$ is mainly determined by the potential of this atom: $V_a(r - r_n)$, where r_n is the coordinate of the nucleus. The difference $V(r) - V_a(r - r_n)$ is a minor perturbation, providing a correction to the electron energy E with respect to the ionization potential of atom E_a. In the case of a crystal with a simple lattice, due to translational symmetry:

$$\psi(r) = A \sum_n \exp(ikr_n)\psi_a(r - r_n).$$

Then, to first order of perturbation theory, the energy is

$$E = E_a + \int \psi_a^*(r)\big[V(r) - V_a(r)\big]\psi_a(r)\,dr$$

$$+ \sum_{n \neq 0}' \int \psi_a^*(r)\big[V(r) - V_a(r)\big]\psi_a(r - r_n)\,dr$$

$$\times \exp(ikr_n).$$

In order to use the tight binding method it is necessary that the corrections to the energy $E - E_a$ for any k should remain much smaller than the difference between the ground state energy E and the first excited level. In other words, the widths of the permitted bands in the crystal should be much smaller than the widths of the forbidden gaps. In practice this condition is seldom fulfilled. It is valid for the *valence bands* of alkali-halide crystals, for crystals of inert gases, and for all inner shell bands; it may not be satisfied for the *conduction band*, and for higher energy bands. The improvement of the tight binding method in the many-electron approach leads to the Heitler–London–Heisenberg method (see *Heitler–London approximation*).

TIME INVERSION

Change in sign of the time variable t ($t \rightarrow -t$) in equations of motion. In mechanics the velocity v changes sign while the acceleration a does not, and in electromagnetism the magnetic field B changes sign while the electric field E does not.

The equations of motion (e.g., $F = ma$) are invariant with respect to this operation (with the simultaneous reversal of the magnetic field B direction). For $B = 0$, this extra symmetry of the Hamiltonian, $HK = KH$ (or $H = H^*$ if the spin interactions are disregarded), where K is the time inversion operator, means that both eigenfunctions ψ_i and $K\psi_i$ belong to the same energy level and, if these are orthogonal (as Bloch functions ψ_k and $K\psi_k \sim \psi_{-k}$ at any value of the quasi-wave vector k), an additional degeneracy is possible. That is, there appears a qualitative change of the band state energy dependence $E(k)$ on k; in particular, $E(k) = E(-k)$. Time inversion can also double the dimension of the matrix Hamiltonian of the *effective mass* method and, as with any new symmetry element, it can establish additional relationships between the matrix elements. There are some differences in the time inversion theory, depending on whether the spin interaction is disregarded or not. In particular, in the latter case, the bands in crystals with a center of symmetry are, by virtue of the *Kramers theorem*, even-fold degenerate at any k.

TIN (Lat. *stannum*), Sn

Chemical element of Group IV of the periodic system with atomic number 50 and atomic mass 118.710. Natural tin consists of ten isotopes with mass numbers 112, 114 to 120, 122 and 124. Isotope ^{124}Sn is weakly radioactive. Outer shell electron configuration is $4d^{10}5s^25p^2$. Successive ionization energies are 7.342, 14.628, 30.49, 39.4, 80.7 eV. Atomic radius is 0.140 nm; radius of Sn^{4+} ion is 0.071 nm; radius of Sn^{2+} ion is 0.093 nm. Oxidation state is $+2$, $+4$. Electronegativity is ≈ 1.85.

In the free form tin is a silvery-gray lustrous *metal*. It is polymorphous; below the temperature 286.4 K, α-Sn (*gray tin*) exists with a diamond type cubic structure $a = 0.64891$ nm (at 286 K). Above 286.4 K, β-Sn (*white tin*) is stable with a tetragonal crystal lattice, $a = 0.5832$ nm, $c = 0.3183$ nm (at 298 K). At the transition β-Sn $\rightarrow$ α-Sn, the specific volume increases considerably (to 25.6%), which results in tin collapsing to a gray powder. The process is sharply accelerated in the presence of α-Sn nucleation centers (see *Tin pest*). The density of β-Sn is 7.30 g/cm^3

(at 293 K), $T_{\text{melting}} = 505$ K, $T_{\text{boiling}} = 2543$ K. Heat of melting is 7.181 kJ/mole; heat of evaporation is 272.74 kJ/mole. Thermal linear expansion coefficient is $22.4 \cdot 10^{-6}$ K^{-1}; at 233 K, coefficient of heat conductivity is 65.32 W·m^{-1}·K^{-1}; specific heat is 0.2261 kJ·kg^{-1}·K^{-1}. Electrical resistivity is 115 nΩ·m (at 293 K). Mechanical properties of tin: tensile strength is 9.8 to 39.2 MPa, relative elongation 40%, relative constriction 75%, modulus of normal elasticity is 53.9 GPa. Brinell hardness is ≈ 40.5 MPa. Adiabatic elastic moduli of β-Sn: $c_{11} = 72.0$, $c_{12} = 58.5$, $c_{13} = 37.4$, $c_{33} = 88.0$, $c_{44} = 21.9$ GPa at 301 K.

Tin is a soft and plastic metal, easily rolled into thin sheets and foil. When infected with tin pest, tin is remelted. High-purity tin is a semiconductor engineering material.

TIN PEST

A polymorphic transformation (see *Structural phase transitions*) of white *tin* (β-Sn) to gray tin (α-Sn). On heating in air, α-Sn powder oxidizes and then compact β-Sn cannot be regenerated. The nucleated centers of α-Sn "infect" their environment, from which the transformation of tin proceeds further with the formation of "warts", the evolution of the latter to "tumors", and so on. Tin pest is a classical example of a polymorphic phase transformation in a solid. The temperature of the $\beta \leftrightarrow \alpha$ transition in pure Sn (99.9995%) is $+1.5\,^\circ$C; the latent heat of the transition at $0\,^\circ$C is 2.22 kJ/mole. As a result of the transformation of tin, the specific volume increases by 26%; therefore, the transformation always starts at the sample surface. The following *orientation relations* are valid from structural considerations: (001) planes of tetragonal β-Sn are parallel to (001) planes of cubic α-Sn, the [100] direction of β-Sn is parallel to the [110] direction of α-Sn. Tin pest requires prolonged incubation (sometimes months and years); and the α-Sn modification is obtained by *low-temperature deformation* with subsequent heating. A seed with particles of CdTe or InSb, which have a *diamond* type lattice (the same as α-Sn) and a *crystal lattice constant* close to α-Sn, can also be used. Tin pest formation accelerates under the action of ultrasound and neutron beam irradiation, and in the solution of Sn salts. Impurities of Al and Zn stimulate the nucleation of α-Sn centers, while Bi, Sb, and Pb hinder it. Preliminary annealing of Sn samples with small concentrations of Bi, Sb, and Pb additionally retards tin pest. Using the method of spreading Hg on Sn at $T < 0\,°C$, and the method of *acoustic emission* from Sn samples that were overheated after preliminary cooling below $0\,°C$, multiple nucleation of α-Sn centers was demonstrated. The incubation is needed for their transformation into growing centers. In super-high-purity Sn (99.9995%), the centers grow according to a dislocation mechanism; in less pure Sn, the mechanism of nucleation of two-dimensional centers is in effect.

TITANIUM, Ti

Element of Group IV of the periodic system with atomic number 22 and atomic mass 47.88. Natural titanium has 5 stable isotopes: ^{46}Ti (7.99%), ^{47}Ti (7.32%), ^{48}Ti (73.98%), ^{49}Ti (5.46%), ^{50}Ti (5.25%). Outer shell electronic configuration is $3d^2 4s^2$. Successive ionization energies are 6.83, 13.57, 28.14, 43.24 eV. Atomic radius is 0.1463 nm; radius of Ti^{4+} ion is 0.068 nm, of Ti^{3+} ion 0.076 nm, of Ti^{2+} ion 0.086 nm, of Ti^+ ion 0.095 nm. Oxidation state is $+4$, less often $+3$, $+2$. Electronegativity is ≈ 1.45.

Titanium is a *metal* of silvery-white color. It exists in two allotropic modifications (α-Ti, β-Ti). Below 1156 K α-Ti is stable with hexagonal close-packed crystal lattice, space group $P6_3/mmc$ (D_{6h}^4), with lattice parameters $a = 0.29507$ nm, $c = 0.46818$ nm at 294 K. Above 1156 K up to $T_{\text{melting}} = 1941$ K β-Ti is stable with body-centered cubic lattice, space group $Im\bar{3}m$ (O_h^9) and parameter $a = 0.33065$ nm (at 1173 K). β-Ti is not fixed by hardening, in the process of cooling there takes place the diffusion-free transformation β-Ti $\rightarrow \alpha'$-Ti. Density is 4.51 g/cm^3 (298 K); $T_{\text{boiling}} \approx 3530$ K. Bonding energy is -4.86 eV/atom at 0 K. The heat of $\alpha \rightarrow \beta$ transformation is 3.402 kJ/mole (at 101.33 kPa); heat of melting is ≈ 19.9 kJ/mole (at 101.33 kPa), heat of sublimation is ≈ 458 kJ/mole; heat of evaporation is 471 kJ/mole (at 101.33 kPa); the pressure of titanium saturated vapor is 0.0266 Pa at 1763 K; specific heat is 0.5225 kJ·kg^{-1}·K^{-1} at 298 K and 101.33 kPa. Debye temperature is ≈ 410 K; linear thermal expansion coefficient of α-Ti monocrystal: $\approx 5.7\cdot 10^{-6}$ K^{-1} along the

principal axis $\underline{6}$ and $10.1\cdot 10^{-6}$ K^{-1} perpendicularly to this axis at 300 K (in the temperature range 298 to 673 K), for β-Ti monocrystal: $12\cdot 10^{-6}$ K^{-1} in the temperature range 1173 to 1343 K; temperature conductivity coefficient is $6.2\cdot 10^{-6}$ m^2/s; thermal conductivity coefficient is 22.064 W·m^{-1}·K^{-1} (at 293 to 298 K). Adiabatic elastic moduli of α-Ti monocrystal: $c_{11} = 162.4$, $c_{12} = 92.0$, $c_{13} = 69.0$, $c_{33} = 180.7$, $c_{44} = 46.7$ (in GPa) at 298 K and zero external pressure; bulk modulus is ≈ 105.6 GPa, Young's modulus is 114.8 GPa, shear modulus is 43.5 GPa, Poisson ratio is ≈ 0.338 (at 298 K); ultimate tensile strength of high-purity titanium annealed in vacuum at 1073 K is ≈ 0.257 GPa, relative elongation is 55 to 70%, yield limit is ≈ 0.152 GPa, Brinell hardness is ≈ 0.6 GPa at room temperature. High-purity titanium is plastic, it is applicable for all the types of mechanical treatment (forging, rolling, stamping) in the hot and cold state. Titanium is not subjected to brittle failure down to cryogenic temperatures. At 1873 K the surface tension of solid titanium is 1.67 J/m^2, of liquid titanium 1.137 to 1.641 J/m^2. Self-diffusion coefficient of atoms in titanium polycrystal is $7.5\cdot 10^{-14}$ m^2/s at the melting temperature. Linear low-temperature electronic heat capacity coefficient is ≈ 3.44 mJ·mole^{-1}·K^{-2}. Density of states at Fermi level is 0.51 state·eV^{-1}·atom^{-1}·spin^{-1} at $T \approx 0$; parameter of electron–phonon interaction in titanium is ≈ 0.36. At 298 K resistivity of α-Ti monocrystal is 480.0 nΩ·m along the principal axis $\underline{6}$ and 453.5 nΩ·m in perpendicular to this axis; temperature coefficient of electrical resistivity of α-Ti polycrystal is 0.00546 K^{-1} at room temperature; resistivity of β-Ti is 1576 nΩ·m at 1173 K; Hall constant is $+0.95\cdot 10^{-11}$ m^3/C (at 283 K); thermoelectromotive force coefficient is $+30$ µV/C (at 1073 K). Coefficient of reflection of optical waves of 5.0 µm length is equal to 87.4%. Work function of titanium polycrystal is ≈ 3.9 eV. Superconducting transition temperature is 0.40 K; critical magnetic field is ≈ 0.78 mT at 0 K. Titanium is paramagnetic material; at room temperature molar magnetic susceptibility is $+168\cdot 10^{-6}$ along the principal axis $\underline{6}$ and $+143.5\cdot 10^{-6}$ CGS units perpendicular to this axis; nuclear magnetic moment of ^{47}Ti isotope is 0.787 nuclear magnetons. Titanium is widely used as light, strong and corrosion-resistant *construction material* in the aviation and space industries, in chemical machine building, in ferrous and non-ferrous *metallurgy*, in medical industry, in military and vacuum electronic engineering, etc.

TITANIUM ALLOYS

Alloys based on *titanium* containing alloying elements introduced for the improvement of technological and maintenance properties (see *Alloying*). The possible formation of stable and metastable solid solutions based on α (hexagonal close-packed) and β (body-centered cubic) polymorphous modifications and *intermetallic compounds* by the introduction of alloying elements is the physical basis of titanium alloy synthesis. The physicochemical and electrochemical properties of titanium alloys are determined by the phase compositions and structures that are created by *heat treatment* and *thermomechanical treatment*. According to their effect at the temperature of the polymorphous $\alpha \leftrightarrow \beta$ transformation, the alloying elements and impurities are subdivided into those stabilizing the *alpha phase* (Al, O, N, C), those stabilizing the *beta phase* (V, Mo, Fe, Cr, Mn), and neutral ones (Sn, Zr). The β-stabilizing elements, in their turn, are divided into eutectoid-forming and isomorphous ones, having respectively limited and unlimited *solubility* in β-titanium. Al, which is introduced into practically all the alloys in quantities of 2 to 8%, is the main alloying element of these. Al increases *strength*, *high-temperature strength* and the *elastic modulus*, while it simultaneously reduces the specific gravity of the alloys. At Al contents up to 5% *solid-solution hardening* plays the main role; at higher Al contents *dispersion hardening* due to separation of the ordered Ti_3Al phase is possible. Alloying of titanium alloys with elements stabilizing the β-phase in quantities exceeding the solubility limit in an α-solid solution leads to the appearance of the β-phase or of intermetallides in the structure (in alloys with eutectoid-forming elements). The compounds and temperature versus time modes of titanium alloys are selected in order to avoid the formation of intermetallides, which cause *embrittlement*. *Nitinol* (Ti–50 at.% Ni) which exhibits the *shape memory effect*, and *high-temperature alloys* based on the aluminides TiAl and Ti_3Al, which operate up to 800 to 1000 °C and are very brittle at room temperatures, are the best known among the precision titanium alloys. High *strength* and *corrosion resistance* favor the application of titanium alloys in aviation and space, chemical industry, power engineering, and ship-building industry.

TITANIZING

The diffusion saturation of a metal surface by *titanium*; a kind of *chemical heat treatment*. Titanizing results in higher *corrosion resistance* and *cavitation resistance*, *hardness* and wear resistance (see *Wear*). In industry it is applied to steels, cast iron, and nonferrous metals. When titanizing carbon-containing alloys, a layer of titanium carbide is formed at the surface (see *Carbides*). In the bulk of the *diffusion zone* the titanium carbide forms mainly along *grain boundaries*. Titanizing is performed by heating articles in the form of pastes, powders, salt solutions, and gas blends, as well as by thermal evaporation of titanium in vacuo.

TOMOGRAPHY (fr. Gr. $\tau o \mu o \varsigma$, cutting, and $\gamma \rho \alpha \varphi \omega$, am writing, drawing)

Techniques for studying the internal structure of materials based on a sequential, stepwise determination of its characteristics. Unlike transmission *endoscopy* and *defectoscopy* which give two-dimensional shadow images of three-dimensional objects, tomographic techniques provide substantially more contrast and a more realistic perspective. In routine *X-ray tomography* the radiation source and X-ray film move in parallel planes in different directions. In this case only one layer of the subject gives a clear image while the image of other layers is suppressed. In modern *computational microtomography* (e.g., CAT or *computerized axial tomography*) a narrow beam from the radiation source with a receiver, by means of parallel translations and rotations, irradiates the sample sequentially at different points of the layer, and at various angles, thus restoring with the help of the computer the bidimensional characteristics of this layer. The main principles of tomography were known from the last century. The development of computational tomography is related with its success in the area of medical diagnostics (see *Magnetic resonance imaging*). In solid state physics the application of tomography is mostly restricted to the study of model subjects. Tomography is a nondestructive method of testing and analysis.

TOMOGRAPHY, MICRO-
See *Computational microtomography.*

TOMOGRAPHY, MRI
See *Magnetic resonance imaging.*

TOMOGRAPHY OF MICROFIELDS

A method of restoration of the two-dimensional spatial distribution of electric and magnetic microfields over a solid surface which is based on measuring the deviations of an *electron probe* in the field under investigation, with a subsequent calculation of the field strength distribution. A focused electron probe goes through the regions of localization of the field under study (scattering field), deviates at every point of scanning under the action of the *Lorentz force*, and arrives at the attached two-coordinate detector. The coordinates and velocities of electrons emerging from the scattering field are determined experimentally. Then the problem of restoring the two-dimensional space distribution of electric or magnetic fields over the solid surface is reduced to solving a set of linear equations of a special form in respect to the scalar or vector potential by means of computer calculations. Using numerical differentiation of the potential function with the application of algorithms of regularization, one reconstructs the spatial configuration of the field, or the distribution of the absolute value of the field strength, or that of its components. When using this method the inaccuracy of measurements does not exceed 5%, with a spatial resolution of decimal parts of a micrometer. A method implemented on the basis of a raster microscopy system (see *Scanning electron microscopy*) coupled to a microcomputer (PC) allows one to study the two-dimensional scattering fields of electric and magnetic domains, fields over the surface of magnetic recording devices, and so on. The figure presents the calculated and restored contours of constant strength of the modulus of a magnetic scattering field over a recording head with a working clearance $2\delta = 3.2$ μm (noted in Fig.) for various indicated magnetomotive forces $(1, 2, \ldots, 6)$. See also *Computational microtomography.*

TOPOGRAPHY, X-RAY
See *X-ray topography.*

TOPOLOGICAL INDEX of a magnetic bubble domain

An integer s which designates the number of turns of the *magnetization* vector M in the middle of the magnetic domain wall of a *magnetic bubble domain* (MBD) when going counter-clockwise along the MBD perimeter. The topological index determines the properties of the MBD as a topological *magnetic soliton* with a topological charge (see *Topological inhomogeneity*) which, in turn, determines the degree of the reflection of the *magnetic film* plane onto the sphere $M^2 =$ const. For a *magnetic substance* with a strong *magnetic anisotropy* in the basal plane, the direction of M at different points of the boundary is the same, and the MBD with $s = 0$ is energetically favorable. For a Bloch *domain wall* with vertical *Bloch lines* $n = 2(s - 1)$, where n is the difference of numbers of Bloch lines of different signs. In the presence of *Bloch points* the value of the MBD topological index is not defined. The value s determines a set of static and, in particular, dynamic properties of the MBD (see *Rigid magnetic bubble domain, Magnetic domain wall dynamics*).

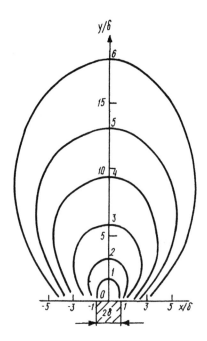

Constant field strength contours on recording head obtained by microfield tomography.

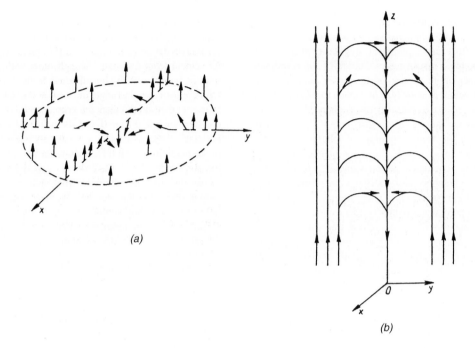

Nonsingular vortices: (a) point type in a plane, and (b) linear type in three dimensions.

TOPOLOGICAL INHOMOGENEITY, topological soliton, topological singularity

Spatially nonuniform configuration of *order parameter* fields that possesses a topological stability. Examples of topological inhomogeneities are *dislocations*, vortices (see *Abrikosov vortices*), *hedgehogs*, *disclinations*, *solitons*, and so on. There are two kinds of (point, linear, planar, and particle-like) topological inhomogeneities: those which involve either singular or nonsingular configurations of order parameter fields. A configuration is called singular if its order parameter field is either not defined or goes to zero at a certain number of points; this is related to a topological singularity. If the order parameter is defined and differs from zero then the corresponding configurations of the order parameters are called nonsingular; they describe a topological soliton. For example, the radial distribution of *magnetization M* in a *ferromagnet*, or the unit vector $n = M/|M|$, called a hedgehog, is a topological singularity: the vector $n = r/|r|$ is not defined at the origin of coordinates (at the point $r = 0$). This is an example of a *point singularity*. Examples of *linear singularities* are disclinations in nematics (see *Nematic liquid crystal*) and vortices in superfluid ^{4}He. Singular configurations of order parameter fields emerge for inhomogeneous conditions far from a topological singularity (at $|r| \to \infty$). Nonsingular topological inhomogeneities occur either when the homogeneous boundary conditions are at infinity, or in the presence of interactions breaking the degeneracy of the equilibrium state of the system. In the latter case, *magnetic domain walls* in magnets (see *Magnetic solitons*) or nematics (*planar topological singularities*) are examples. In the former case, examples could be a *nonsingular point vortex* in a planar magnet (i.e. with an order parameter in the form of a vector in a plane) or a nematic (Fig. (a)), as well as a *linear nonsingular vortex* or a *linear topological soliton* in the three-dimensional case (Fig. (b)). A point vortex in a planar model can be regarded as a section of a linear soliton in the three-dimensional case. An example of a *particle-like topological soliton* in the three-dimensional case is a ring-like, nonsingular vortex (a linear soliton bent into a ring).

To every topological singularity there corresponds a certain representation of the coordinate space in the space of the order parameter M. Depending on the type of a given topological inhomogeneity, an n-dimensional sphere S^n is selected (for an inhomogeneous distribution of the parameter n, far from a topological inhomogeneity, the order of n is 0, 1, and 2 for planar, linear, and point topological inhomogeneities, respectively; for a homogeneous distribution n is equal to 1, 2, and 3, respectively). Thus in the general case, to a topological singularity there correspond transformations $S^n \rightarrow M$. These transformations are characterized by the *homotopic groups*, with the help of which one can classify the singularities, resolve the problem of their topological stability, and indicate for them *topological invariants* or charges. It should be stressed that some *conservation laws* are valid for *topological charges*; this makes them similar to such quantities like electric charge, spin, and so on. Therefore, topology provides not only methods for studying physical systems, but also a language for describing the physical phenomena involving topological inhomogeneities or, in other words, the topological consequences of the theory expected to be observable in experiment.

TOPOLOGICAL TRANSITION

A transition of a material from one state to another which is accompanied by a change of the topological characteristics of the state. There are topological transitions of different types, such as the *Lifshits transition* due to alteration of the topology of the *Fermi surface*, and the (Berezinski)–*Kosterlitz–Thouless transition* involving the dissociation of connected states of *topological inhomogeneities*, i.e. vortices ("vortex molecules"). At this latter transition only the structure corresponding to *correlation functions*, etc., changes.

TOTAL EXTERNAL REFLECTION

The reflection of radiation (X-rays, electrons, neutrons, etc.) incident upon a given medium from air or a vacuum at grazing angles θ, which are smaller than a certain critical value θ_c. Similar to *total internal reflection*, practically all the energy of the incident beam is transferred to the reflected beam. The effect of total external reflection is described on the basis of *Snell's law*. If a plane wave is incident from a vacuum on a medium with *refractive index n*, then the refracted beam forms with the boundary the angle of $\cos\theta' = (\cos\theta)/n$. If $n < 1$, then there exists the critical angle $\theta_c = \cos^{-1} n$, such that $\theta' = 0$, and the refracted wave is cancelled. For $\theta < \theta_c$ there exists only one wave in the medium, which travels along the boundary; the amplitude of this wave decays exponentially inside the medium. Total external reflection of corpuscular radiation may be described with the help of the pseudopotential $U = E(1 - n^2)$ (see *Pseudopotential method*), which characterizes the interaction between radiation of energy E and the medium. If $n < 1$, then $U > 0$, and total external reflection may be treated as the reflection of particles of transverse energy $E_\perp < U$ ($\theta < \theta_c$) from a potential barrier, the value of the critical angle being $\theta_c = \sin^{-1}(U/E)^{1/2}$. In a *crystal* the field U exhibits spatial periodicity, and the reflection pattern becomes more complicated: total external reflection turns into *Bragg double reflection*.

Total external reflection is observed, e.g., for hard photons and thermal neutrons at small grazing angles ($\theta_c \sim \lambda$, where λ is the wavelength). This circumstance provides the basis for the operation of *neutron guides*. The refractive index of many materials has an imaginary value; if the incident radiation consists of extremely slow (ultracool) neutrons: $n^2 < 0$, then $U > E$, and total external reflection is observed for all values of θ. This effect is used for the transport and storage of slow neutrons. The operation of an *X-ray mirror* is based on total external reflection.

TOTAL INTERNAL REFLECTION

The reflection of electromagnetic radiation from the boundary between two transparent media when the radiation propagates from the medium of higher *refractive index* n_1 and the angle of incidence exceeds the critical angle i_{cr}, which is given by the condition $\sin i_{cr} = 1/n$, where n is the relative index of refraction $n = n_1/n_2$. In accordance with *Snell's law*, the value of i_{cr} corresponds to a refraction angle of $90°$. The electromagnetic energy at total internal reflection is completely returned to the denser medium at the angle of incidence, which is equal to the angle of reflection; the field in the other (optically less dense) medium exponentially decays with distance for the order of several wavelengths λ from the interface. The *reflectance* at total internal reflection is very close to 1. This property is utilized in reflecting prisms, light guides, and in investigations through the *attenuated total internal reflection* method.

TOTALLY SYMMETRIC VIBRATIONS of crystal lattice

Vibrations that involve no change of sign of normal coordinates during the course of all symmetry operations. During the process of total symmetric vibrations, all atoms that make up a *unit cell* exhibit rectilinear motion, which is symmetric with respect to the *center of symmetry* or the principal axis of the crystal. At any given instant of time all atoms are in the same phase of the motion. In a breathing mode all atoms move alternately in and out relative to the center of inversion, or a (principal) axis of inversion.

TOUGH CRACK, viscous crack, elastic crack

A *crack* which can develop into a *plastic deformation*; it can appear not only in the vicinity of a surface damage point, but also within the volume of a body. A tough crack develops by the formation of microcavities at its vertices, followed by their coalescence via the rupture of the bridges between them. As a result of the development of tough cracks, specific *pitted relief* can be formed. The energy of formation of the crack surfaces generally exceeds (by several orders) the *surface energy* of the body as a result of the energy spent in plastic deformation. The speed of tough crack growth is determined by macroprocesses in its plastic zone, and appears to be slower than the speed of *brittle crack* growth.

TOUGH FAILURE, elastic failure, ductility failure, viscous failure

A type of *failure*, accompanied by the development of intense *plastic deformation* at the site of breakdown. For single-axis tension it is displayed in the form of material rupture at the site of strong local contraction of the sample *neck*, the so-called

destruction with cup and cone. Tough failure is characteristic of all plastic metals, especially the soft ones (pure *iron*, *copper*, *aluminum*, etc.). The micromechanism of tough failure comprises nucleation, growth, and merging of pores with the formation of a *tough crack* with a blunt, rounded vertex in the center of the neck. The mechanism of viscous crack growth at the slow, subcritical stage lies in the formation of new pores in front of the *crack* front, the rupture of bridges between adjacent pores, and merging of the pores with the crack vertex. At the supercritical, fast stage two versions are possible: (1) the merging of pores gradually yields to the formation in front of the viscous crack of regions of local *spalling* and quasi-spalling, which give to the final fracture texture an appearance characteristic of *quasi-brittle failure*; (2) in the more plastic state of the metal the final failure occurs via the cut-off along the neck periphery. The transition from the stable to the destructive quasi-brittle stage in case (1) can be approximately described by the *Griffith–Orowan criterion* (see *Griffith theory*): $\sigma_f \approx (E\gamma_{eff}/c_{cr})^{1/2}$, where σ_f is the failure stress, E is *Young's modulus*, γ_{eff} is the effective *surface energy*, equal to the total energy expended by movement of the fracture per unit length, including the true surface energy of the metal γ and the work of plastic deformation at the crack vertex γ_{pl}, and c_{cr} is the viscous crack critical dimension. Ordinarily a criterion for σ_f is applied with the aid of the parameter γ_{eff} experimentally determined beforehand from the rupture of a sample with a known crack dimension c_{cr}. One can define the *fracture toughness* of a material according to the relation $Y_{1c} = K_{1c}^2/E$, where $Y_{1c} = 2\gamma_{eff}$ is the critical velocity of the energy release, K_{1c} is the critical intensity factor (*crack resistance factor*). At present there is no quantitative criterion for cut-off tough failure. Not every kind of failure involving a neck is related to the tough type. In many tensile tests of steels with small and medium contractions in the neck ($\psi \approx 10$ to 50%) the failure conditions are subject to the *microspalling criterion* of the deformed material. In spite of the existence of the neck, the presence of this type of destruction should be related to the quasi-brittle variety.

TOUGH FRACTURE, viscous fracture, ductile fracture

The *fracture* of a material during the course of *tough failure*. Macroscopically, tough fracture has the appearance of a fibrous structure; it is characteristic of materials with high resistance against the spreading of *cracks* and *spalling*. Microscopically, tough fracture is associated with the pitted nature of damage caused by the generation and coalescence of *pores*. Upon *tough failure*, the vertex of a slowly developing crack is rounded due to *plastic deformation*, and in its vicinity there appears a large number of cavities, during the formation of which second *phase* inclusions play an important role. The growth of cavities is caused by the "falling" of *dislocations* into them, i.e. by the transformation of lattice dislocations into cleaving ones. The crack moves by jumps from one cavity to the next accompanied by the destruction of bridges between them, thereby forming the *pitted relief* of the surface, where comparison of the surfaces separated by the crack shows that a ridge on one side corresponds to a ridge on the other, and a furrow lines up with a furrow.

TOUGHNESS, impact malleability

The ratio of the total work expended on the deformation and *failure* of a notched test bar to its cross-section at the point of the *notch* made under impact loading. Impact malleability is a complex characteristic, involving the whole set of strength properties of the material under consideration. Impact bending tests (see *Flexure*) are performed using pendulum pile drivers. Almost all of the work done by an impact is expended on *plastic deformation* of the test piece in the neighborhood of the notch, and on the formation and propagation of *cracks*.

Since the law of similarity is not obeyed during dynamic tests, the dimensions of impact specimens and test conditions must be rigidly controlled. The most commonly used test specimens are 5.5 cm long rods of square cross-section measuring 1×1 cm^2, with a U-shaped notch in the middle. One of the major challenges of impact tests consists in evaluating the susceptibility of *metals* to *brittle failure* in the context of their *cold brittleness*. This task is accomplished by constructing the temperature dependence of impact malleability, and determining the visco-brittle

transition temperature. The form of the dependence of impact malleability on the composition and structure of metals and alloys is similar to that of the characteristics of *fracture toughness* under static loading.

TPATT DIODE

The same as *Trapped plasma avalanche triggered transit diode*.

TRACERS, isotope tracers, labeled atoms

Radioactive atoms of various elements introduced into an object under study. The process of their movement through the medium is monitored by changes in the spatial distribution of the emitted radiation intensity (*autoradiography*), or with the aid of a *layer-by-layer analysis*. This underlies the isotope tracer method for studying *diffusion* and self-diffusion of atoms, as well as chemical reactions in *solids* and other (chemical, biological) materials. In some cases, the presence of an isotope can be detected using other methods unrelated to radioactivity. For instance, the lines in optical spectra of crystals containing *impurity atoms* can split upon introduction of various isotopes (e.g., carbon isotopes into silicon and semiconductors) in measurable quantities. Use can be made of stable isotopes as well (see also *Activation analysis*).

TRANSCRYSTALLINE FAILURE

A kind of *failure* of semicrystalline solids (see *Polycrystal*) at which a *crack* propagates over a body consisting of disoriented *crystallites* (grains), unlike the case of *intercrystalline failure* when the path of the main crack passes predominantly along the *grain boundaries*. At transcrystalline failure, the crack, as a rule, propagates along a crystallographic plane with small indices (in a body-centered cubic lattice this is the {001} plane). Transcrystalline failure is inherent in materials with a high interatomic coupling *strength* at grain boundaries due to the lack of atomic impurity *segregations* or precipitates of particles of small-disperse *phases* which produce local *microstresses*. Transcrystalline failure of *metals* and *alloys* may be either brittle (*spalling* failure), quasi-brittle, or tough (see *Brittle failure*, *Quasi-brittle failure*, and *Tough failure*). This failure is specific for metallic *construction materials*, since this is related to higher strength, *fracture toughness*, and *plasticity*, i.e. with the properties that are required to provide greater *reliability* to materials and structural elements.

TRANSDUCER, ACOUSTIVE WAVE
See *Acoustic wave transducer*.

TRANSFORM, EMERSLEBEN
See *Emersleben transform*.

TRANSIENT ABSORPTION

Rapidly relaxing optical absorption, which is induced in *solids* by pulses of radiation. The transient absorption is often initiated through pulses of electron accelerators and *lasers*. Time resolution at determinations of transient absorption is limited by the duration of the initiating pulse and is $\sim 10^{-14}$ s if lasers are used and $\sim 10^{-11}$ s if the absorption is initiated by electron accelerators. The ultimate resolving power in time at the determination of transient absorption spectra is set by the uncertainty relation, and is $\sim 10^{-15}$ s.

Transient absorption occurs in a number of dielectric and *semiconductor materials* (alkali halides, alkaline earth metals, silver halides, oxides, *gallium arsenide*, etc.). Transient absorption is related to electronic excitations (electrons, *holes*, *excitons*) and short-lived local electron and hole centers generated by the initiating pulse.

The most extensively studied kind of transient absorption is related to electronic transitions from the ground states of the short-lived *color centers* in *alkali-halide crystals*, e.g., H (*interstitial atom* of halide X^0) and V_k-centers (self-localized hole–molecular *ion* X_2^-) at temperatures high enough for these centers to be mobile. The investigation of transient absorption by these centers provides valuable information on the mechanism of their migration. The investigation by color centers is an efficient method for studying radiation processes. For example, the increase of transient absorption by Frenkel pairs of defects (see *Unstable Frenkel pair*) ($\sim 10^{-11}$ s in KCl and $\sim 10^{-10}$ s in CsI) has shown that the decomposition of an *exciton* into a pair of defects at low temperatures proceeds from the excited state of the *self-localized exciton*. The detection of short-lived components (with annihilation kinetics identical to those of short-lived Frenkel pair annihilation) in the transient absorption relaxation of self-localized excitons in KCl (10^{-7} s) was a proof of the occurrence of a *radiation-induced jolt*. The transient absorption due to electronic transition in a self-localized exciton, and due to transitions from relaxed excited states of color centers, permits the study the the energy band structure of excited states. Investigation of the transient absorption related to optical transitions of electrons and holes, in the *conduction band* and in the *valence band*, provides data on the energy band structure, lifetimes and thermalization times of carriers in bands. For instance, the absorption band, which peaks at 1.65 eV in CsI, is related to the transitions among nonoverlapping subbands of the split valence band, and relaxes in approximately 10^{-11} s (self-localization time of holes); the absorption band at 2 eV in gallium arsenide relaxes during the period of the order of 10^{-13}–10^{-14} s, which is due to the electron–electron interaction.

TRANSISTOR (abbr. of transfer and resistor)

A *semiconductor device* with three or more electrodes which is designed for the generation, amplification, and transformation of electromagnetic oscillations and electrical signals. Is was invented in 1948 and described analytically by Shockley, Bardeen and Brattain. A transistor is a solid state analogue of a triode vacuum tube. The main action of a transistor is the control of the flux of charge carriers between two electrodes (source or emitter and drain or collector) by means of the potential change at an intermediate (gate) electrode. In accordance with the characteristics of the charge transfer process, transistors are divided into unipolar (*field-effect transistors*) and bipolar transistors. *Unipolar transistors* are those in which the charge transfer is performed by a single type of charge carrier, i.e. by holes or electrons. In *bipolar transistors* the charge transfer takes place simultaneously via both holes and electrons.

A metal oxide semiconductor field effect transistor (MOSFET) has a metal electrode (e.g., Cu) just outside a semiconductor surface (usually Si) which is insulated from it by an oxide layer (often SiO_2). A gate voltage applied between the metal and the semiconductor causes charge carriers to collect at the interface of the oxide. The gate voltage can be used to control or modulate the flow of electrons from the source electrode to the drain electrode.

Depending on their function, transistors come in diverse types, and in terms of their application they are divided into a several groups which differ in their working parameters. The basic materials used to manufacture transistors are Si, Ge, GaAs, and InP incorporated in various *heterostructures*. A permanent improvement of the working characteristics of transistors has resulted from technological development, the inclusion of particular materials, and the development of new designs. For example, the production of the *high electron mobility transistor* became possible when the heterostructure GaAs–GaAlAs became available. A substantial increase in the response speed, and the upper limit of amplification and generation of SHF oscillations came about as a result of designing a *hot electron transistor*. At present transistors cover, in fact, all the SHF range.

See also *Bipolar transistor, Field-effect transistor, High electron mobility transistor, Hot electron transistor*.

TRANSITION METAL MAGNETISM

See *Magnetism of transition metals*.

TRANSITION METALS, transition elements

The group of elements with incomplete inner electron shells. In atoms of transition metals, d- or f-sublevels of the second or third (from the outside) electron shell are being filled with electrons; d- and f-transition metals (an alternative name is d- and f-transition elements) are distinguished. The transition series contain the atoms of the periodic table secondary subgroups. The first transition series, from atomic number 21 (Sc) to 30 (Zn), consists of $3d$-elements, and is in the fourth period of the periodic table. The transition metals of the fifth period are $4d$-elements from Y (atomic number 39) to Cd (48). In these two first transition series, $3d$- and $4d$-orbitals are being filled. The transition metals of the sixth period are $5d$-elements (filling the $5d$ orbitals) starting at La (57) and going from Hf (72) to Hg (80). There are also f-shell elements (f-orbitals are being filled: $4f$-orbitals in *lanthanides* and $5f$-orbitals in *actinides*). Except for Fr and Ra, all known elements

of the seventh period belong to transition metals (d- and f-block ones). The total number of transition metals exceeds 60. The outer or valence shells of the transition elements do not contain many electrons (usually from 1 to 3). Transition elements have certain properties in common, as well as many distinguishing features. For more details see entries for individual elements.

TRANSITION RADIATION (predicted by V.L. Ginzburg and I.M. Frank, 1945)

The radiation that occurs when a charged particle, which is moving at the constant velocity v, intersects the interface between two media with different values of the *dielectric constant* ε. Since the energy of the radiation is provided at the expense of the kinetic energy of the particle, the velocity of the latter is not constant in the strict sense. If, however, the energy of the particle is high compared with the losses by radiation, then the motion may be thought of as uniform and rectilinear.

A charged particle does not emit electromagnetic radiation during uniform rectilinear motion in vacuo. The field of the charged particle in a certain medium excites and polarizes the atoms of the medium, which then become sources of secondary electromagnetic waves that propagate in all directions. These secondary waves cancel each other if the medium is stationary and homogeneous and the velocity of the particle is smaller than the wave *phase velocity* in this medium. In the case of the violation of at least one of these conditions, the cancellation of secondary waves is not complete, and some radiation results. If the medium shows a certain nonuniformity with respect to electrodynamic characteristics (at the interface between two media), then the complete suppression of the secondary waves emitted by the atoms of the medium under the action of the traveling charged particle does not take place, and transition radiation occurs. The radiation is emitted into both hemispheres, because the electromagnetic field of the charged particle assumes different values in the media of different ε. The energy of transition radiation of a relativistic charge is concentrated mostly in the forward direction at small angles with respect to the incident particle velocity $\theta_\gamma \propto mc^2/E = 1/\gamma$; and mainly *photons* of the UV and

X-ray spectral regions are emitted; the spectrum of small-angle radiation extends to high frequencies: $\omega = \omega_0 \gamma$, where ω_0 is the plasma frequency of the electrons of the medium, $\omega_0^2 = Ne^2/(\varepsilon m)$, and γ is the Lorentz factor (ratio of the total energy to the rest energy). The energy losses by the transition radiation of an ultrarelativistic particle are proportional to its Lorentz factor:

$$W = \frac{1}{3} \frac{e^2 \omega_0 \gamma}{c}.$$

This circumstance allows using the transition radiation advantageously for detecting high-velocity charged particles: the transition radiation is usable for detecting particles in those energy regions where ionization and Cherenkov detectors (see *Nuclear radiation detectors*) are inefficient (i.e. in the ultrarelativistic energy region). The energy of the moving particle can be judged from measuring the *charged particle energy loss* by transition radiation, which increases linearly with respect to the energy of the particle. However, the intensity (the number of quanta) of radiation, which occurs at a single interface, is small (on the average, about $1/137$ of a quantum per charged particle). Therefore, a multi-interface transition radiation is used for the determination of the energy of extra-high-velocity particles. The transition radiation counter consists of about 1000 layers of material, which are interlaced with gas (or with a certain porous material, e.g., plastic foam). The plates are widely separated from one another: spacing between plates must exceed the so called *region of transition radiation generation* in vacuo, which is equal (in the order of magnitude) to the wavelength multiplied by γ^2, so that the intensities of radiations of individual plates of the pile will add.

A periodic medium, which consists of alternating plates of two different sets (the densities and, respectively, dielectric constants of the plates from different sets are sharply different), exhibits *resonance transition radiation*. Attempts have been made to use resonance transition radiation for obtaining quasi-chromatic coherent X-ray radiation. The main obstacle to be surmounted in this case is the low intensity of the resonance transition radiation. Resonance transition radiation with peaks in energy regions around 4 and 12 keV was generated by a *positron* beam of energy 4 GeV in a pile of 100 plates of mylar ($C_5H_4O_2$) separated by air layers.

TRANSITION RADIATION BY DISLOCATIONS

Acoustic radiation, which occurs when a *dislocation* intersects a surface with a discontinuity of *elastic moduli*, e.g., crosses the *grain boundary* of a *polycrystal*, reaches the crystal surface, etc. At the instant the dislocation crosses the boundary between media with different elastic characteristics, its elastic field undergoes a sharp rearrangement; as a result of this rearrangement part of the field "breaks away" from the dislocation and departs in the form of sound pulses. As the dislocation reaches the surface, a pulse of *surface waves* (*Rayleigh waves* or similar ones) arises, in addition to the pulses of bulk waves which are propagating inside the crystal. The intensity of transition radiation by dislocations is determined by the *Burgers vector* of the dislocation, and the value of its velocity at the moment of crossing the interface. The phenomenon of transition radiation associated with dislocations was theoretically predicted by V.D. Natsik (1968), and experimentally detected by V.S. Boiko et al. (1969) while studying the process of a cluster of twinning dislocations reaching a calcite monocrystal surface (see *Elastic twinning*). Transition radiation by dislocations is one of the important mechanisms of *acoustic emission* that occurs during the *plastic deformation* of crystals.

TRANSITION REGION between film and substrate

The region between a *substrate* and a *film*, which is generated during *thin film growing* of metals, *semiconductors* and *insulators*, and differs in properties from both substrate and film. The generation of the transition region at the growing film is related to the influence of the substrate, the effect of the growth processes, and changes in the original medium from which the film is grown. Of greatest interest is the transition region of epitaxial semiconductor films. Smearing of the film–interface boundary, which is due to the existence of the transition region, is detected by methods of *electron microscopy, electron diffraction analysis, X-radiography*, optical analysis, including *ellipsometry*, as well as by Auger spectroscopy, secondary ion mass spectrometry, spectrometry of ion back scattering. Electric conductivity and mobility of charge carriers are also indicative of the boundary smearing. The extent of the transition region in *semiconductor diodes* and *transistors* of the 1960–1970s reached several micrometers; the film itself being 10–20 μm thick, so the influence of the transition region could be neglected. With the advent of microelectronic devices (see *Microelectronics*), heterolaser thin-film structures called for reduction of the transition region width. *Gas phase epitaxy* provides doped layers, with a transition region less than 30 nm. The width of these layers obtained through *molecular beam epitaxy* in vacuo is less than 5 nm. The extent of this region in heteroepitaxial structures (see *Heteroepitaxy*) like silicon on sapphire reaches 1 μm.

The main reasons for the occurrence of the transition region are:

(1) the discrepancy between *crystal lattices* of the film and the substrate, including the change of lattice constant with concentration; the difference between the elastic moduli of the two lattices; the consequences of mechanical and chemical treatment of the substrate surface, contaminations; *diffusion* of impurities from the substrate; changes in the substrate under heat treatment;
(2) three-dimensional nucleation at the initial stage of the growth; change of the microrelief of the growth surface; dynamics and relaxation of defects; diffusion of impurities from the film;
(3) the period of nonstationarity in the growth mode; changes in temperature of the sources, the substrate and the impurity flux; autodoping through a gaseous or liquid phase; changes in the external force fields.

The decrease of the extent of the transition region was achieved through reducing the temperature of the growing process, creating the *vicinal surface* of growth, regulation and increase of purity of growing conditions. The possibility of exerting control over the transition region facilitated the production of *superlattices* and composition-modulated semiconductor layers (periodic structures).

TRANSLATION

Transport of an object in space for a certain distance *a* along a straight line called a translation axis. A translation is specified by a vector *a*. If as a result of translation a regular periodic lattice is brought into coincidence with itself then the action is a symmetry operation of the space group of the crystal (see *Crystal symmetry*). The translation operation is inherent in lattices or other objects which are periodic in one-, two-, or three-dimensional space. Examples are crystals and chain molecules of polymers.

TRANSLATION VIBRATIONS

See *Acoustic vibrations*.

TRANSMITTANCE

The ratio of the flux of energy (or particles) that passes through a medium without scattering to the total incident flux. Sometimes, provided that the scattering is particularly weak, transmittance is defined in a different way: it is then taken as a ratio of the total transmitted flux to the incident flux. In this latter case the transmittance is related to the *absorption coefficient A* and the *reflectance R* by $T + A + R = 1$.

TRANSMUTATION DOPING

The same as *Nuclear doping*.

TRANSPARENCY

A characteristic of a material defined by the ratio of the radiation flux that passes through the medium without changing its direction to the flux that enters the medium in the form of a parallel beam. The transparency is lower the more medium absorbs and scatters radiation.

TRANSPARENCY, ACOUSTIC

See *Acoustic self-induced transparency*.

TRANSPARENCY, SELF-INDUCED

See *Self-induced transparency, Acoustic self-induced transparency*.

TRANSPARENCY TO RADIATION

See *Radiation transparency*.

TRANSPORT, BALLISTIC

See *Ballistic transport*.

TRANSPORT OF HEAT

See *Heat transport*.

TRANSPORT OF MASS

See *Mass transport, Surface mass transport*.

TRANSPORT PHENOMENA

Irreversible processes in multiparticle systems related to the spatial transport of physical quantities under action of inhomogeneities or external fields. The evolution of the general nonequilibrium state has a double-stepwise character. Initially, very quickly, the localized equilibrium state is established within the time duration of spatially homogeneous *relaxation*. Its parameters evolve towards their equilibrium values in accordance with laws that follow from the transport equations, which are equations of macroscopic physics (see *Kinetic phenomena*).

Examples of transport phenomena are the following: *thermal conductivity* that corresponds to heat transport, viscous flow that is momentum transport (see *Viscosity*), *diffusion* that is material transport, *electrical conductivity* that is electric charge transport, *spin diffusion* that involves spin magnetic moment transport (see *Spin*).

A phenomenological (macroscopic) description of transport phenomena is reduced to a set of differential equations for a certain set of parameters. Examples of such equations are the Navier–Stokes hydrodynamic equations (L. Navier, 1821; G. Stokes, 1843, 1845), the heat equations, and the equations for diffusion in solids (Fick's laws). A linear response approximation is often made.

A microscopic description of transport phenomena is based on the use of *kinetic equations* (or equivalently, the mathematical apparatus of quantum kinetics) which govern the evolution of the system of particles or quasi-particles (molecules, electrons, phonons, magnons, etc.) responsible for the transport of the relevant quantities (mass, charge, energy, spin, etc.). In the case of condensed matter with *spontaneous symmetry breaking* (magneto-ordered crystals, *quantum liquids*), for an adequate description of transport phenomena it is necessary to utilize the equations

of motion for certain parameters related to the particular nature of the *order parameter* of the *phase* under consideration. In superfluid ^{4}He (see *Superfluidity*) this is the equation for the *superfluid velocity*; in spin systems it is the equation of motion for the local angle of spin rotation.

TRANSURANIUM ELEMENTS

Chemical elements beyond *uranium* ($Z = 92$) in the periodic table: *neptunium* Np (93), *plutonium* Pu (94), *americium* Am (95), *curium* Cm (96), *berkelium* Bk (97), *californium* Cf (98), *einsteinium* Es (99), *fermium* Fm (100), *mendelevium* Md (101), *nobelium* No (102), *lawrencium* Lr (103), *kurchatovium* Ku (104), etc., with atomic numbers up to 109. Traces of Np and Pu have been found in nature, formed from nuclear reactions of U with neutrons of the Earth's crust. All other transuranium elements are obtained artificially: by irradiation of U or of transuranium elements by neutrons, in nuclear reactions of Pb, U, or by bombarding transuranium elements with accelerated ions (He, C, O, Ne, Ar, etc.). All isotopes of transuranium elements are radioactive. Instability due to spontaneous fission and α-decay limits the possibilities of synthesis of new transuranium elements. The search for hypothetical comparatively long-living nuclides with $Z \sim 114$ protons, and the number $N \sim 184$ of neutrons, is underway. Transuranium elements with $Z = 93$ to 103 belong to the family of actinides and are f-elements: with increasing Z the electrons sequentially fill $5f$-orbitals (with some participation of $6d$-orbitals). The successive reduction of ionic radii (*actinide compression*) is observed. Transuranium elements of the actinide family are similar in their chemical properties, and have the oxidation state $+3$. Due to the weak binding of $5f$-electrons with the nucleus for Np, Pu and Am there is a variety of valence states (up to $+7$). Transuranium elements with $Z > 103$ belong, respectively, to periods IV, V, etc., of the periodic system. The element Ku is the chemical analogue of *hafnium*, and the element with $Z = 105$ is the analogue of *tantalum*.

The transuranium elements with $Z = 93$ to 99 are obtained in the metallic state, and their alloys and intermetallic compounds with many metals are known. Their crystal structure, phase transitions, electrical, magnetic and optical properties are known, and the solid state properties of *hydrides*, *carbides*, *nitrides*, oxides, silicides and other compounds of transuranium elements have been studied.

TRANSVERSE CONTRACTION

Contraction of a sample cross-sectional area during *mechanical testing of materials* for stretching. The relative transverse contraction of the sample is $\psi = \Delta S/S$, where ΔS is the reduction of the cross-area during *plastic deformation* by stretching, and S is the initial cross-section area. One may distinguish the uniform relative transverse contraction ψ_B and the relative transverse contraction localized in the *neck* ψ. In the first case ΔS is measured under a load corresponding to the *ultimate strength* of the material, i.e. up to the possible formation of the neck at the sample, and to its subsequent *failure*, in the second case ΔS is measured after the rupture of the sample (at the point of rupture). The quantities ψ_B and ψ characterize the *plasticity* of the material. The higher the values of ψ_B and ψ, the more plastic is the material. In the *theory of elasticity* the term transverse contraction means the reduction of the transverse dimension of the body during uniaxial elastic stretching (see *Poisson ratio*).

TRAPPED ELECTRONS

See *Magnetodynamic nonlinearity*.

TRAPPED PLASMA AVALANCHE TRIGGERED TRANSIT DIODE, TPATT diode

A *semiconductor diode* with a dynamic *negative resistance* resulting from periodic switching back and forth between the nonconducting and conducting states of the p^+-n-n^+ structure. The resistance in a TPATT diode lessens because its n-band is filled with the electron–hole plasma, and the dissipation of that plasma results in the restoration of the high resistance of the n-band. The electron–hole plasma forms when an electric field wave passes through the layer of depleted carriers, the strength of that wave being sufficient to produce *impact ionization* and an avalanche multiplication of carriers. The front of the impact ionization wave moves at a velocity exceeding the carrier drift velocity. To transform the source

energy of a TPATT diode circuit into VHF oscillations, an external circuit is necessary to ensure the phase locking, to a sufficient extent, of the higher-order harmonics of the current with respect to the fundamental. TPATT diodes are often used to generate VHF electromagnetic oscillations.

TRAPPING of current carriers

The passage of a free carrier to a localized state at a *defect* or impurity. At low temperatures when many carriers are trapped at attractive centers the cascade trapping mechanism dominates. At the limits of the temperature range where cascade trapping is effective, the one-phonon mechanism of trapping can manifest itself. At high temperatures *shallow levels* stop being effective and much of the trapping is to *deep levels* as a result of many-phonon transitions. The energy of the carrier bond at a deep center is determined by the short-range part of the potential, and in some cases the carrier is bound by a center having a charge of the same sign, despite the repulsive long-range Coulomb field. The trapping temperature is determined by the rate of energy loss, if the carriers approach the center faster than they are trapped by it; otherwise, the rate of trapping is limited by the rate of space *diffusion* of electrons to the center. The possibility of trapping at centers with levels in the *band gap* plays an important role both in radiation *recombination* and the *Auger effect* (see also *Traps for mobile particles*).

TRAPS for mobile particles

Defects in semiconductors and dielectrics that have *local electronic levels*, and are able to trap nonequilibrium electrons and holes (see *Trapping* of current carriers). Two types of traps are known together with their respective levels: *centers (levels) of attachment*, and *centers (levels) of recombination*. Depending on the concentration of holes, an electron entering a local level of a center from the *conduction band* may either become thermally excited so as to enter the conduction band, or recombine with a *hole* in the *valence band*. The first process dominates for attachment levels, and the second for recombination levels. Depending on the temperature, T, and the concentration of recombining particles, one and the same level may act as either an attachment or a recombination level. Quantitatively, the criterion for a local electronic level being of a certain type depends on the position of the *demarcation level* – the level of energy of a local electron center, where the probabilities of thermal excitation and *recombination* are equal to each other. The position of the demarcation level for electrons, measured downward from the bottom of the conduction band, E_{De}, is defined by the relation $E_{De} = -f_h - k_B T \ln[\gamma_e N_c/(\gamma_h N_v)]$, where F_h is the *Fermi quasi-level* for holes, N_c, N_v are the effective *densities of states* in the conduction and the valence bands, respectively, and γ_e and γ_h are probabilities of elementary acts of electron/hole trapping at the center, respectively. Levels positioned in the *band gap* of a semiconductor above the level E_{De} are attachment levels for electrons, and those positioned below E_{De} are recombination levels. Similarly the concept is introduced of hole levels of attachment – local electron levels positioned below the demarcation level for holes: $E_{Dh} = -F_e + k_B T \ln[\gamma_h N_v/(\gamma_e N_c)]$, where F_e is the Fermi quasi-level for electrons. *Defects in crystals* may also act as traps for moving *point defects*. The above separation of interaction processes into those of attachment and those of recombination is also valid for *vacancies* and *interstitial atoms* that are generated by *nuclear radiation* or other factors.

TRIBOELECTRICITY (fr. Gr. τριβω, rub or scrape, and ηλεκτρον, amber)

Generation of electric charges caused by the *friction* of solids being pressed together and moving against one another. Both of the solids being rubbed become electrified in the process, with the acquired charges equal in value but opposite in sign. Triboelectricity in *metals* and *semiconductors* results from electron transfer from the material with the lower *work function* to that with the higher one. In the case of contact between a metal and an *insulator* triboelectricity arises due to a metal–insulator electron transition. When two insulators rub against each other, both electronic and ionic *diffusion* may bring about the initiation of triboelectricity. Associated with the onset of triboelectricity is the fact that the solids being rubbed against each other become unequally heated, due to differences in their specific heats

and thermal conductivities, and this causes the transfer of carriers from the local inhomogeneities of the more heated surface. Triboelectricity may also be caused by mechanically stripping individual regions from the surface of pyroelectrics and piezoelectrics (see *Pyroelectricity* and *Piezoelectricity*).

TRIBOLUMINESCENCE

Luminescence, which occurs at the grinding, crushing or cleaving of *crystals*. A variety of reasons are responsible for triboluminescence; in a number of cases it is related to *photoluminescence* initiated by electric charges released during the cleavage of a crystal. In other cases, triboluminescence is caused by the motion of *dislocations* in a crystal under *strain*.

TRIBOMETRY (fr. Gr. τριβω, rub or scrape, and μετρεω, am measuring)

Methods used for measurements in tribology, that is measurements of an *external friction* force or external friction coefficient, external friction threshold, and the wear of friction surfaces. Two types of tribometry measurements are distinguished: laboratory tests which assess *friction* forces and wear resistance of materials under various conditions, and natural condition tests involving the evaluation of the amount of friction of a given working unit.

Samples used in laboratory tests involve single point or linear contacts (e.g., a sphere along a plane, two counter-rotating cylinders that rub along the generatrix); in applications there are also samples with small planar friction surfaces. These test specimens are used for determining specific friction forces and specific wear, i.e. quantities related to a unit surface of the actual contact.

TRICLINIC SYSTEM (fr. Gr. τρεις, three and κλινω, incline)

Crystallographic system characterized by a *unit cell* with the lowest degree of symmetry: $a \neq b \neq c$, $\alpha \neq \beta \neq \gamma \neq 90°$. The coordinate axes are directed along the unit cell axes, often selecting $c < a < b$. The triclinic system includes one primitive *Bravais lattice* and two *space groups*.

TRIGLYCINESULFATE,
$(NH_2CH_2COOH)_3 \cdot H_2SO_4$

Monoclinic water-soluble *ferroelectric*, with the *Curie point* $T_C = 322$ K. The T_C value increases to 333 K by deuteration. Triglycinesulfate gives rise to a series of isomorphous materials, which may be obtained by replacing the sulfate ion with selenate, fluoroberyllate, phosphate and fluorophosphate ions. Compounds which are isomorphous to triglycinesulfate form continuous sequences of solid solutions with one another. The substitution of a small part of glycine molecules by D or L optical isomers of α-alanine (which differs from glycine in the spatial arrangement of atoms) brings about steric effects, which hinder the repolarization process: triglycinesulfate undergoes impurity-induced conversion to a single-domain structure. The latter transformation is responsible for the asymmetry of the dielectric *hysteresis* loop, and facilitates the practical application of α-alanine-doped triglycinesulfate as targets of *pyroelectric radiation detectors*, which do not require the application of a variable magnetic field. Doped triglycinesulfate is comparatively one of the most sensitive *pyroelectric materials* for pyrovidicon targets, which provide television resolution of IR images in real time. Triglycinesulfate is a model ferroelectric, serving during the last 35 years as an object for the study of *ferroelectric domain structure* rearrangement kinetics, specific features of *phase transitions*, effect of impurities on properties of ferroelectric samples, etc. Of importance for engineering is the combined doping with alanine and chromium or copper, to optimize the pyroelectric characteristics.

TRIGONAL SYSTEM

Crystallographic system with a 3-fold axis (3 or $\bar{3}$) as a principal axis; it belongs to hexagonal syngony (see *Crystal system*). In the hexagonal coordinate system (see *Hexagonal system*), the 3 ($\bar{3}$)-axis is parallel to the z-axis, $a = b \neq c$, $\alpha = \beta = 90°$, $\gamma = 120°$; and in the rhombohedral coordinate system, $a = b = c$, $\alpha = \beta = \gamma \neq 90°$, with the 3 ($\bar{3}$)-axis directed along the principal body diagonal of the rhombohedral unit cell. The trigonal system involves one (primitive) *Bravais lattice*, 5 *point groups* and 18 *space groups*.

TRION, exciton ion

A *quasi-particle*, which is a bound complex formed by three charge carriers (see *Current carriers*) two of which carry a charge of like sign, e.g., a complex of an electron e and two *holes* h. The bonding energy of a trion depends on the ratio between the *effective masses* of the charge carriers forming the complex. Trions are formed in the *electron–hole plasma* in semiconductors, and during the decay of *X-ray excitons* in *ionic crystals*. Trions are observed in *luminescence* spectra of Ge and *cyclotron resonance* spectra of Si, as well as in spectra of luminescence excitation in NaCl.

TRIPLE CRITICAL POINT, tricritical point

The point on a *phase diagram*, at which either the lines of three *first-order phase transitions* meet, or two lines related to a *second-order phase transition* changes into the line of a first-order phase transition. An example of the former case is the triple point on a pressure versus temperature diagram at which the sublimation line, the melting (freezing) line and the vaporization (condensation) line intersect together, with all three phases (solid, liquid, gas) in equilibrium there. In the latter case the triple critical point is also called a *first-order phase transition critical point*, or a *Landau critical point*. *Susceptibility*, *specific heat* and a number of other physical quantities show anomalous behavior at the critical point; the specific heat anomaly is unsymmetrical: the specific heat diverges only when approaching the triple critical point from the ordered *phase*. Triple critical points are found in temperature–pressure, temperature–external field, and temperature–composition diagrams of many *ferroelectrics*, *ferromagnets* and *antiferromagnets*, crystals subject to *structural phase transitions*, *liquid crystals*, and *polymers*. See also *Multicritical point*.

TRIPLE POINT

A point on the *phase diagram*, which corresponds to the equilibrium coexistence of three phases of a material. It follows from the *Gibbs' phase rule* that no more than three phases may coexist in equilibrium in a single chemical substance (one-component system). These three phases (e.g., solid, liquid and gas, or as with sulfur, a liquid phase and two allotropic crystalline modifications)

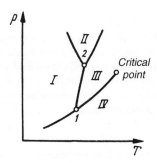

Pressure–temperature plot of a material exhibiting two triple points 1 and 2, showing the critical point and the phases: I, first crystalline modification; II, second crystalline modification; III, liquid state; and IV, gaseous state. The critical point (beyond which the liquid and gas phases are indistinguishable) is also indicated.

may coexist only at a particular temperature T_T and pressure p_T that specify the location of the triple point in the pressure–temperature diagram (see Fig.). For carbon dioxide, as an example, $T_T = 216.6$ K, $p_T = 5.16 \cdot 10^5$ Pa. The coordinates of the triple point for water, which is the principal reference point of the absolute thermodynamic temperature scale, are $T_T = 273.16$ K, $p_T = 4.68$ mm Hg (610 Pa).

TRIPLET EXCITONS

Delocalized *crystal* excitation states of unit spin ($S = 1$). Considerable study has been given to triplet excitons in *molecular crystals*. As a consequence of spin quantum number *selection rules* the probability of an optical transition from the ground singlet state to an excited triplet state is negligible (the transition is allowed by the *spin–orbit interaction*). Therefore, triplet excitons have relatively long lifetimes (10^{-4}–10 s depending on the crystal), and are observed for the most part in emission spectra (*phosphorescence*). The matrix elements of excitation transfer for the triplet states are determined by exchange integrals between electron states of different molecules; consequently, these matrix elements are small, and the bands are narrow. Due to the long lifetime of triplet excitons, it is possible to establish high concentrations of these states, and to observe transitions between spin sublevels under the action of microwave irradiation. Electron paramagnetic resonance spectra

of triplet excitons, as well as *optical detection of magnetic resonance* spectra, show a number of specific features caused by the passage of excitation across the lattice.

TRIPLET PAIRING

Generation of a bound state of two $S_i = 1/2$ Fermi particles with parallel spins, the total spin of the resulting state thus being $S = 1$. It is a triplet since the spin degeneracy $2S + 1 = 3$. By virtue of the Pauli exclusion principle, the wave function of the pair is antisymmetric with respect to exchange of the particles, and therefore triplet pairing takes place provided the value of the orbital angular momentum L of the pair is odd (in distinction from *singlet pairing* in an ordinary superconductor, when the total spin and orbital angular momenta of the *Cooper pairs* are, respectively, $S = 0$ and $L = 0$). Triplet pairing occurs, e.g., in *superfluid phases of* ^{3}He ($S = 1$, $L = 1$). There is evidence that the superconducting properties (see *Superconductivity*) of heavy-fermion systems (see *Heavy fermions*) ($CeCu_2Si_2$, UBe_{13}, UPt_3) may be due to triplet pairing, which proceeds presumably by a phononless attraction mechanism. In an isotropic system (superfluid ^{3}He), the role of the *order parameter* is played by the wave function of the pair, or by a Green's function, $\Psi_{\alpha\beta}(\boldsymbol{n})$, where $\boldsymbol{n}$ is a unit vector in *momentum space*, and α and β are spin indices. In the case of triplet pairing, $\Psi_{\alpha\beta}(\boldsymbol{n})$ is an odd function of $\boldsymbol{n}$, and is a symmetric spinor of rank two which may be expressed as

$$\Psi_{\alpha\beta}(\boldsymbol{n}) = \mathrm{i}\boldsymbol{d}(\boldsymbol{n})(\hat{\boldsymbol{\sigma}}\hat{\sigma}^y)_{\alpha\beta},$$

where $\hat{\boldsymbol{\sigma}} = \{\hat{\sigma}^x, \hat{\sigma}^y, \hat{\sigma}^z\}$ are Pauli matrices. The quantities $(\hat{\boldsymbol{\sigma}}\hat{\sigma}^y)_{\alpha\beta}$ in these equations make up the basic set for the decomposition of the $\Psi_{\alpha\beta}$ spinor in spin space, and the order parameter is the complex vector $\boldsymbol{d}(\boldsymbol{n})$. The order parameter $\boldsymbol{d}$ is a vector with respect to rotations in the spin space. When carrying out the construction of the superconductor order parameter, proper allowance must be made for the crystallographic symmetry. Different order parameters and hence different superconducting phases may occur in a given crystal class (e.g., cubic *crystals*). Some of these phases resemble superfluid ^{3}He phases in their properties, and besides that they have other uncommon features (e.g., a superconductor with triplet pairs may exhibit orbital and spin *ferromagnetism*).

TROOSTITE (after L.-J. Troost)

A structural constituent of *steel* and *cast iron*; a fine-grained mixture of *ferrite* and *cementite*. Troostite is formed at a decomposition of *austenite* over the lower region of its stability range (500–550 °C), and by medium-temperature (350–400 °C) *tempering* of hardened steel. Troostite after *quenching* (*primary troostite*) is an extremely fine-grained mixture of ferrite and cementite. The thickness of the ferrite laminae is less than 0.1 μm. The laminated structure of troostite reveals itself in electron microscope studies (see *Electron microscopy*); the dark regions of troostite against the background of light areas of *martensite* can be seen by an optical microscope. Tempered troostite (*secondary troostite*) has a ferritic matrix; fine crystals of cementite (mainly randomly shaped) are spread throughout the said matrix. The increased *hardness* of this type of troostite is due to the high dispersity of the cementite, and the distortions of the ferritic matrix lattice. Steel of the tempered troostite structure is characterized by a high ratio between the *elastic limit* and *ultimate strength*. This kind of steel is used to produce springs, etc.

TUBULAR DIFFUSION, pipe diffusion

The preferential *diffusion* of atoms along *dislocations*. This so-called pipe or tubular diffusion is a consequence of the increased mobility of atoms along a dislocation line compared to their mobility in the bulk *crystal*. This phenomenon may be formally represented as diffusion in a pipe of the material, the *diffusion coefficient* of which is much higher than that of the bulk material. Since the *activation energy* of diffusion assumes different values in the bulk and along dislocations (E_B and E_D, respectively), pipe diffusion manifests itself only in the region of relatively low temperatures, where the contribution of the bulk diffusion to the overall diffusion flow is insignificant. The central problem with this simplified interpretation is finding the radius r_0 of the above "pipe". In terms of a more rigorous treatment, the *edge dislocation* core is considered as consisting of a sequence of extra-plane atoms and a line of lattice vacancies in the neighborhood of this sequence. The major contribution to E_D is made by the generation of *interstitial atom*–vacancy pairs

which are capable of dissociation. The energy of migration of an interstitial atom along the sequence of *vacancies* is very low. Since diffusing atoms change places with the nearest lattice vacancies, the region of accelerated diffusion covers the cylinder-shaped region of radius r_0 on the order of interatomic distances.

TUNABLE LASERS

Laser sources that are tunable, that is, sources whose wavelength can be adjusted. Along with *nonlinear optics* devices (harmonic generators, generators of sum and difference frequencies, etc.), tunable lasers provide a solution to one of the principal problems of quantum electronics: to induce the emission of radiation at any wavelength of the optical region. The principle of operation of tunable lasers is based on a controlled variation of the threshold wavelength of radiation within the limits of the amplification band of a laser *active medium*. The principal characteristics of tunable lasers are: radiation width $\delta\nu_g$ and the frequency change range $\Delta\nu_g$ (many tunable lasers exhibit $\delta\nu_g = 10^3$–10^4 Hz over $\Delta\nu_g = 10^{13}$–10^{14} Hz).

There are three main types of tunable lasers: (1) laser sources with a *dispersion laser resonator*; (2) laser sources, which are based on shifting the amplification band over the spectral region; (3) *optical parametric oscillators*. Tunable lasers with a dispersion resonator are capable of frequency variations over most of the amplification band (lasers based on solutions of organic materials, *semiconductors*, *color centers*, etc.). A broadband (beyond 10^3 nm) shift of the amplification band is realized in certain types of *semiconductor lasers* on exposing the active medium to the action of a magnetic field, pressure, varying the temperature, etc. In parametric lasers based on *nonlinear media* a transformation of the pumping quantum into two emission quanta of lower energy occurs, provided that phase synchronization conditions are met (see *Second harmonic generation*). The smooth frequency change of such lasers is based on control over the wavelengths that obey the phase synchronization conditions, by rotation of the nonlinear element, or change of its *refractive index*. Highly promising are *sweep lasers*, i.e. lasers, which are tuned during the course of generation, particularly by simultaneously coordinating the variation of the resonator optical length (sweep lasers using dynamic modes). Most broad-band frequency variations are achieved by integrating tunable lasers and nonlinear optical frequency converters into a single system. Tunable lasers find wide application in laser spectroscopy, laser photochemistry, etc.

TUNGSTEN (Ger. *Wolfram*), W

A chemical element of Group VI of the periodic table, atomic number 74, atomic mass 183.85. Natural tungsten consists of a mixture of five stable isotopes ^{180}W (0.135%), ^{182}W (26.41%), ^{183}W (14.4%), ^{184}W (30.64%) and ^{186}W (28.41%). Configuration of outer electronic shells is $4f^{14}5d^46s^2$. First two ionization energies are 7.98, 17.7 eV; estimated 3rd, 4th, 5th, and 6th ionization energies are 24, 35, 48, 61 eV. Atomic radius is 0.137 nm; ionic radius of W^{4+} is 0.071, and of W^{6+} is 0.062 nm. Tungsten is chemically inert, oxidation states are $+2$ to $+6$, with $+6$ the most typical. Electronegativity is 1.40.

In the free form under normal conditions tungsten is a light-gray *metal*. It has a body-centered cubic lattice with parameter $a = 0.3165$ nm, space group $Im\bar{3}m$ (O_h^9). The density is 19.35 g/cm^3, $T_{\text{melting}} = 3693$ K, $T_{\text{boiling}} = 5953$ K, heat of melting is 32.25 kJ/mole; specific heat is 0.136 kJ·kg^{-1}·K^{-1} at 273 to 1273 K, Debye temperature is 405 K; linear thermal expansion coefficient is $4.6 \cdot 10^{-6}$ K^{-1} at room temperature, thermal conductivity coefficient is 168 W·m^{-1}·K^{-1}; adiabatic elastic moduli of tungsten monocrystal: $c_{11} = 523.27$, $c_{12} = 204.53$, $c_{44} = 160.72$ GPa at 300 K; Young's modulus is ≈402 GPa (at 298 K) for monocrystalline filament and ≈365 GPa for wire; rigidity (at 298 K) is ≈155 GPa for monocrystal and 137.0 GPa for polycrystal, Poisson ratio is ≈0.283 (at 298 K); tensile strength of a caked slab is 110 MPa, for forged rod it is 350 to 1500 MPa, for drawn wire it is 3800 MPa, for annealed and unannealed wire it is correspondingly 1100 and 1800 to 4150 MPa (depending on diameter); yield limit for the annealed wire (diameter 0.1 to 0.5 mm) is ≈770 and for unannealed wire (of the same diameter) it is 1490 MPa; Brinell hardness of a caked slab is ≈2250 and of a forged one it is ≈3750 MPa; self-diffusion coefficient of tungsten is $2.52 \cdot 10^{-15}$ m^2/s (at

2620 K); effective thermal neutron trapping cross-section is 19.2 barn/atom, ion-plasma frequency is 23.32 THz; linear low-temperature electronic specific heat coefficient is 0.95 mJ·mole^{-1}·K^{-2}; specific electrical resistivity is 53 nΩ·m (at 296 K); temperature coefficient of electrical resistance 0.00510 K^{-1}; reflection coefficient of the optical wavelength of 5 μm is 97%; work function of polycrystal is 4.54 eV; electron emission saturation current density is $1.5 \cdot 10^{-10}$ to 1690 mA/cm^2 (at 1103 to 2700 K); radiation power of filament supply is 18 to 245 W/cm^2 (at 1873 to 3303 K); superconducting transition temperature is ≈0.014 K, superconducting critical magnetic field is 0.11 mT (at 0 K). Tungsten is paramagnetic with molar magnetic susceptibility $+53.3 \cdot 10^{-6}$ cgs; nuclear magnetic moment of ^{183}W is 0.115 nuclear magnetons.

Tungsten is widely used for alloying steels and producing durable high-temperature alloys (see *Tungsten alloys*). Due to its refractoriness and low vapor pressure at high temperatures it serves as a material for light bulb filaments, and for parts in radioelectronics and X-ray engineering.

TUNGSTEN ALLOYS

Alloys based on *tungsten*; they are related to superhigh-temperature alloys. Tungsten is alloyed with *rhenium*, *molybdenum*, *nickel*, *tantalum*, *iron*, oxides, *carbides*, etc., to increase *high-temperature resistance* and *plasticity*, to improve processability and other properties. Tungsten alloys are formed by the methods of *powder metallurgy* and vacuum (arc or electron-beam) melting.

TUNNEL CONTACT

See *Tunnel junction*.

TUNNEL DIODE, Esaki diode (discovered by L. Esaki, 1957)

A *semiconductor diode* with a *current–voltage characteristic* curve having a negative slope over part of its operating range. This phenomenon of a semiconductor with a *negative resistance* arises from electron tunneling through the potential barrier that separates the *p*- and *n*-regions of the *semiconductor junction* (see *Interband tunneling*). The tunnel diode may be used as an amplifier, an oscillator, or in switching circuits, using frequencies up to 10^{12} Hz. A thin (~15 nm) *p–n* tunneling junction, which is sharp and transparent, is formed from doped *germanium, silicon,* and *gallium arsenide* prepared by the addition of large amounts ($>10^{19}$ cm^{-3}) of dopant atoms. The resulting *impurity band* is overlapped by the *conduction band* on the n-side and by the *valence band* on the p-side of the junction; with the *Fermi level* located within the impurity band. When forward bias is applied, the conduction band electrons tunnel into the empty energy states of the valence band (the device depends for its operation on majority carriers only), thereby being transferred to the other side of the p–n junction. The rapidity of response of the tunnel diode is determined by the time it takes for the potential barrier to adjust to the shape of the applied voltage (*Maxwell relaxation time*). The height of the potential barrier decreases with an increase of applied forward bias, which brings about the increase in the transparency to tunneling, and consequently to an increase of the current flow across the diode. On further increasing the forward bias, however, there are no more vacancies for electrons (and the current would vanish, were it not for the "surplus" current due to the tail of the *density of states* in the *band gap*); then the ordinary diffusion current is superimposed. The tunnel diode may operate close to helium temperatures, subject to slight changes of peak current due to the variation of the width of the gap. The tunnel diode exhibits *radiation resistance*. See also *Generation type semiconductor diodes.*

TUNNEL EFFECT

The quantum transition of a particle (or system of particles) between states that are separated by a *potential barrier* such that there is no direct path between these states that can be traversed by a particle (or system of particles) moving in accordance with the classical mechanical equations of motion. The tunnel effect is due to the fact that the wave function of a system is different from zero in regions which are forbidden from the classical motion standpoint, and the probability of the particle passing this region is given by the squared modulus of the wave function. In the limit of the quasi-classical approximation to quantum mechanics, the probability of *tunneling D* is exponentially small: $D \propto \exp[-(2/\hbar)|S|]$, where

the magnitude of the action integral $|S|$ of the tunneling particle is smaller, for lower heights U and narrower widths d of the barrier to be overcome. For a particle of energy E tunnelling through a one-dimensional potential barrier $U(x) > E$ that extends from $x = 0$ to $x = d$, the magnitude of the action integral is given by

$$|S| = \int_0^d \left[2m(U(x) - E)\right]^{1/2} dx.$$

TUNNEL EFFECTS IN SUPERCONDUCTORS

A number of phenomena related to the tunneling of electric charge across an *insulating* layer separating two *superconductors.* Since both *Cooper pairs* and *quasi-particles* act as charge carriers in superconductors, there are two types of charge tunneling. The tunneling transfer (see *Tunnel effect*) of Cooper pairs causes a nonzero electric current to flow even for a zero electric potential difference eV across the contact (see *Josephson effects*). The tunneling current of quasi-particles I arises only under the action of a voltage applied across the tunnel barrier. The form of the *current–voltage characteristic* is determined by the presence of gaps Δ_s, where $s = 1, 2$, in the spectra of elementary excitations and singularities in the densities of quasi-particle states $N_s(E)$ of the superconductor: $N_s(E) = \mathrm{Re}\{|E|/(E^2 - \Delta_s^2)^{1/2}\}$, where E is the excitation energy, counted from the *Fermi level.*

The calculation of the quasi-particle current may be carried out wih the aid of the *tunneling Hamiltonian.* This Hamiltonian is an operator describing quantum transitions of particles between two different phase space states. Given that p and q are quantum numbers of a particle in these phase spaces and a_p^+, a_p and b_q^+, b_q are respectively the creation and annihilation operators of particles (see *Second quantization*) in these spaces, the tunneling Hamiltonian is written as $H_T = \sum_{pq}\{T_{pq}a_p^+ b_q + T_{pq}^* b_q^+ a_p\}$, where T_{pq} is proportional to the amplitude of the tunneling electron current. The Hamiltonian is used to treat the problem of electron tunneling across the potential barrier between two *metals*, which are in either normal or superconducting states.

In accordance with general quantum-mechanical conventions, the tunnel current is defined as $I = e\langle dN/dt \rangle = -(ie/\hbar)\langle [N, H_T] \rangle$, where N is the electron number operator in one of the conductors, $N = \sum_p a_p^+ a_p$. In the limit of low potential barrier transparency the tunnel Hamiltonian approach provides an adequate description of the tunneling problem. To second order in T_{pq}, the tunnel current is given by

$$I = \frac{2\pi e}{\hbar} \langle |T_{pq}|^2 \rangle \int\limits_{-\infty}^{+\infty} dE\, N_1(E) N_2(E + eV)$$

$$\times [f(E) - f(E + eV)],$$

where $f(E) = \{1 + \exp[E/(k_B T)]\}^{-1}$. At $T = 0$, the charge transfer proceeds through the generation of a quasi-hole excitation ($E < 0$) in one superconductor and a quasi-particle excitation ($E > 0$) in the other, which is possible only if the inequality $eV > \Delta_1 + \Delta_2$ is satisfied. At $T \neq 0$, the energy threshold in the current–voltage characteristic becomes smeared due to tunnel transfer of thermally excited quasi-particles. At $eV = |\Delta_1 - \Delta_2|$, the singularities in the *densities of states* of both superconductors correspond to one and the same quasi-particle energy, and the current–voltage characteristic shows an inflection corresponding to the logarithmic singularity of the differential conductivity. If one of the conductors is in the normal state ($N_2(E) = N(0) =$ const), then the current–voltage characteristic no longer exhibits a logarithmic singularity, but it has a threshold at $eV = \Delta$. A measurement of the current–voltage characteristic of tunnel transfers is one of the methods for determining the *energy gap in superconductors*. Taken alone, the presence of singularities in the current–voltage characteristic is an important performance criterion for tunnel barriers.

TUNNEL EMISSION

The same as *Field electron emission*.

TUNNELING, INTERBAND

See *Interband tunneling*.

TUNNELING, MACROSCOPIC QUANTUM

See *Macroscopic quantum tunneling*.

TUNNELING MICROSCOPY

See *Scanning tunneling microscopy*.

TUNNELING PHENOMENA IN SOLIDS

Phenomena that are due to the quantum *tunnel effect*. Tunneling phenomena in solids are observed, as a rule, at low temperatures, and are classified into several groups depending on whether the tunneling system is an electron, an atom, or a macroscopic (many-particle) system. Electron tunneling is commonly investigated by employing structures especially designed for such studies. These structures involve various types of *semiconductor junction*: *point contacts*, metal–semiconductor and metal–semiconductor–metal contacts, etc. The conductivity of a tunnel contact depends on the properties of the bulk conductors (sides of the contact), and the insulating layer sandwiched between them. Therefore, determining the *current–voltage characteristic* of the contact makes it possible to study elementary excitations in both the conducting and the insulating materials by the application of *tunneling spectroscopy* (see Fig. 1). The generation of an *energy gap* Δ in a *quasi-particle* spectrum caused by the

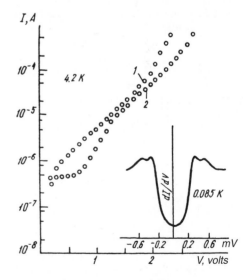

Fig. 1. Current–voltage characteristic at 4.2 K for the tunneling contacts Al–Al$_2$O$_3$–GeTe: (1) direct and (2) reverse bias. Inset: differential conductivity at the lower temperature 0.085 K.

transition of at least one of the sides to the superconducting state results in the suppression of the tunnel current at $eV < \Delta$, where V is the voltage across the contact. (The *Josephson effect* consists in the presence of a nondissipative current in the *tunnel junction* of two superconductors at $V = 0$.) The emission of *phonons* by nonequilibrium electrons opens inelastic channels for tunneling; the tunneling electron may lose its energy by emission of quasi-particles not only at the edge, but also in the tunneling interlayer. This makes it possible to study, e.g., fluctuations of molecular impurities in the barrier layer. The voltage shifts ($V \sim \hbar\omega/e$) commonly amount to tens of mV. The region of small shifts ($V \sim 0.2$ mV) is subject to commonly occurring anomalies, which consist in an abrupt decay or increase in the differential conductivity $G(V)$ (*zero anomaly*). This anomaly vanishes with increased temperature, and on exposure to strong enough magnetic fields. This abrupt increase in differential conductivity may be considered a consequence of tunneling accompanied by scattering off *magnetic moments* localized in the insulator (*Kondo scattering*, see *Kondo effect*). A similar phenomenon was also observed at tunneling across the *Schottky barrier* in *metal–semiconductor junctions*. Coulomb correlation in disordered conducting sides of a junction may be responsible for $G(V)$ passing through a minimum. The current–voltage characteristic of the contact depends on the type of potential barrier that separates the conductors. In multilayer *heterostructures*, one can produce barriers with complicated shapes. The states formed in these barriers are resonance states of the tunneling electrons. Clearance of the barrier, which occurs for certain shifts in V, results in a nonmonotonic V-dependence of I. Resonance states may also be generated in amorphous semiconductor interlayers, and in short channels of MIS-transistors (see *Metal–insulator–semiconductor structure*) through the agency of a *random potential*. In the latter case, such states give rise to a sequence of random spikes in the dependence of the channel conductivity G and the voltage applied to the gate. In addition to the above-listed features of the current–voltage characteristic, there exists a smooth nonlinear $I(V)$ dependence resulting from a decrease of the barrier with an increasing shift of V.

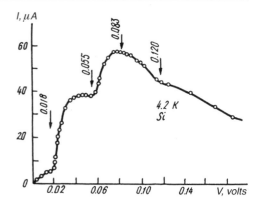

Fig. 2. Current–voltage characteristic for a Si tunneling diode at 4.2 K. The arrows indicate voltages corresponding to characteristic phonon frequencies.

The role of the tunnel interlayer may be played by a narrow *p–n* junction between heavily doped *p*- and *n*-regions of a semiconductor (Esaki *tunnel diode*). When the applied bias is small the current in the *Esaki diode* is due to *interband tunneling*. The application of a pronounced forward bias suppresses the current, for final states fall within the *band gap*. Because of its negative differential conductivity (see *Negative resistance*), the Esaki diode is of importance for applications in engineering (see Fig. 2).

A number of tunneling phenomena are related to electron tunneling in the bulk of a homogeneous material. The band-to-band tunneling in semiconductors may occur by the joint action of a strong electric field E and optical photons with energies less than the gap width E_g (*Franz–Keldysh effect*). Hence, the light absorption factor α is different from zero at $\hbar\omega < E_g$. The specific form of band-to-band tunneling may be observed in metals in the presence of a strong magnetic field. When following the quasi-classical orbit in *k*-space, an electron may pass into another orbit, some points of which lie close to those of the former orbit (*magnetic breakdown*).

The process of *atom tunneling* in a *crystal lattice* can be severely suppressed by the massiveness of atoms. Therefore, tunnel effects show themselves mainly for the lightest elements, whose atoms are in motion under the action of a sufficiently weak crystalline potential. Atom tunneling in, e.g., solid *helium* results in the delocalization of defects and the generation of their associated *impuriton* and *vacancion* bands. In amorphous materials (in glasses, see *Vitreous state of matter*), the tunnel transport of individual atomic groups turns out to be a factor, since the barriers between various system states are low. These processes appear to give rise to *two-level systems* in the energy spectra of glasses. The random scatter of two-level systems provides an explanation for *low-temperature anomalies* observed in kinetic and thermodynamic characteristics of glasses.

A number of phenomena in solids are due to the tunnel transport of many-particle systems. The phenomenon of *self-localization* is related to the tunneling of the electron–lattice system. Small *Josephson junctions* may exhibit tunneling between the superconducting condensate states that are characterized by different values of the *order parameter* phase (*macroscopic quantum tunneling*). Such tunneling brings about a finite and temperature-independent (in the low temperature limit) resistance of the Josephson contact.

TUNNELING SPECTROSCOPY

Investigations of energy spectra of solids carried out employing the *tunnel effect* at low temperatures. The resolving power of the method is determined by the thermal diffusion of the Fermi distribution (see *Fermi–Dirac statistics*) of electrons, and amounts to several units of $k_B T$. Elastic and inelastic tunnel effects are distinguished. Spectroscopy using *elastic tunneling* is based on electron tunneling across a potential barrier that takes place with the conservation of energy. The *current–voltage characteristic* of the tunnel contact in this case shows a structure that is representative of the relationship between the *density of electron states* and the electrode energy. By this means, it is possible to reconstruct the energy spectrum of *degenerate semiconductors*, *semimetals*, and *superconductors*; tunneling spectroscopy also permits the determination of certain features of the density of electronic states in normal *metals*. In the case of *tunnel junctions* between a superconductor and a normal metal, the differential conductivity, which is a function of the voltage at low enough temperatures, is directly proportional to the relation between the density of the electronic states

and the energy. In the case of superconductors subject to a strong enough *electron–phonon interaction* (at $T_c > 1.5$ K), these relationships are used for reconstructing the spectral function of the electron–phonon interaction (*Eliashberg function*). Spectroscopy employing *inelastic tunneling* makes possible the study of the spectral density of excitations of *quasi-particles* in the barrier layer (*insulator*). The tunneling, which involves the partial transfer of the electron energy to these excitations, opens up a new channel of electron penetration through the potential barrier provided that the excitation energy satisfies the inequality $\hbar\omega \leqslant eV$ (e is electron charge, where V is the potential difference across the contact). The dependences of the second derivative of the current with respect to voltage, which are called *tunneling spectra*, are directly proportional to the density of states of quasi-particle excitations in the barrier layer. In this manner it is possible to study the spectra of *phonons* in semiconductors and insulators, *vibrational spectra* of molecules that are adsorbed by the barrier layer, and other excitations. In order to identify specific features of a solid under spectroscopic investigation, the tunnel current is commonly subjected to magnetic fields, mechanical stresses, electromagnetic radiation, and other external actions.

TUNNELING SPECTROSCOPY OF SUPERCONDUCTORS

Investigation of spectra of *quasi-particles* by application of *tunnel effects in superconductors* that arise in layered structures of the metal–insulator–metal type, when one or both electrodes of such a sandwich are in the superconducting state. In order for a finite probability of quantum-mechanical electron transfer across the insulator to exist, its thickness must be very small ($\sim 10^{-9}$ m). Tunnel spectroscopy of superconductors has been made possible by the discovery of electron tunneling across oxide films (I. Giaever, 1960; Nobel Prize of 1973).

The experiment provides data on the *energy gap* and its dependence on temperature, magnetic field, pressure, impurities, and crystallographic orientation; also obtainable is information on phonon and electron spectra, and the mechanism of superconductivity. The investigation is generally concerned with thin-film systems prepared by high vacuum deposition. Besides naturally grown oxides, the insulator can be a thin semiconductor layer; *Schottky barriers* have also been used.

Tunnel spectroscopy has its origins in the relationship between the derivative of the tunneling current I with respect to the voltage across the barrier V (dI/dV), and the *density of states* $N(eV)$ of the superconductor:

$$\frac{dI_S/dV}{dI_N/dV} = N(eV) = \text{Re}\left\{\frac{eV}{\sqrt{(eV)^2 - \Delta^2(eV)}}\right\}.$$
(1)

When the voltage across the barrier corresponds to the magnitude of the energy gap given by $eV = \Delta_0 = \Delta(\Delta_0)$, the *current–voltage characteristic* has a readily measurable deep singularity. Recording the tunnel conductivity over a broader energy range provides the right-hand side of Eq. (1), while the real $\text{Re}\,\Delta(\omega)$ and the imaginary $\text{Im}\,\Delta(\omega)$ parts of the gap function $\Delta(\omega)$ are determined from the dispersion relation. The complex parameter $\Delta(\omega)$ contains much information on the electron–phonon interaction, which is responsible for the formation of the superconducting state. The equations of the microscopic theory of superconductors plus the experimentally measured quantities Δ_0 and $\Delta(\omega)$ permit the numerical reconstruction of the reduced spectral function (*Eliashberg expression*) $g(\omega) = \alpha^2(\omega)F(\omega)$ (where $\alpha^2(\omega)$ is the squared matrix element of the electron–phonon interaction, $F(\omega)$ is the phonon density of states) and the calculation of the critical temperature T_c, the critical magnetic fields B_{c1} and B_{c2}, and various electron–phonon interaction parameters. The details of the *phonon spectrum*, in particular *van Hove singularities*, have a direct impact on the experimental relationship of the second derivative of the tunnel current with respect to the voltage (d^2I/dV^2 versus V plot).

The superposition of the degenerate "quasi-electron" and "quasi-hole" superconductor states in structures consisting of thick ($d_S \sim 10^{-5}$ m) superconducting films results in oscillations of the tunnel conductivity of these structures. Measuring the period of these oscillations $\delta V = \pi \hbar v_F/d_S$ provides a simple means for determining the *Fermi velocity* v_F.

TUNNEL JUNCTION, tunnel contact

An electrical contact between two conductors established by virtue of quantum-mechanical electron tunneling through an intervening *potential barrier*. The transmittance of the barrier is of the order of $\exp(-A\Phi^{1/2}S)$, where $A \approx 1$ provided that the barrier height Φ and width S are expressed, respectively, in electron volts and nanometers; S does not ordinarily exceed several nanometers or tens of nanometers. The tunneling may occur through a thin vacuum gap (in a scanning tunneling microscope, see *Scanning tunneling microscopy*), through a thin insulating layer (in laminated metal–insulator–metal structures), through a thin semiconductor layer which is depleted of carriers (in Esaki *tunnel diodes*), or else through a metal–semiconductor diode with a *Schottky barrier*. The *current–voltage characteristic* curve of a tunnel junction is linear for $eV \ll \Phi$. For higher values of applied bias the nonlinearity of the current–voltage characteristic depends on the relationship between the effective height and width of the potential barrier, and the bias. For $eV \gg \Phi$, the current–voltage characteristic follows the *Fowler–Nordheim law* (R.H. Fowler, L. Nordheim, 1928) of *cold emission* in metal–insulator–metal contacts:

$$J = \frac{3.38 \cdot 10^8 E^2}{\Phi} \exp\left(-\frac{0.69\Phi^{3/2}}{E}\right),$$

where the tunnel current density J comes in units of A/cm^2, the intensity of the electric field in the barrier E is expressed in $V/\text{Å}$, and the barrier height Φ at zero bias is expressed in eV. Besides these features of the current–voltage characteristic, the mechanism of electron tunneling through the potential barrier is responsible for a strong nonlinearity of the current–voltage characteristic in the region of small biases, which is due to the energy gap in the semiconductor energy spectrum.

Of broad application in low-temperature physics and engineering are *superconducting tunnel junctions*. These are laminated structures consisting of three layers: superconductor–insulator–superconductor (S–I–S contact), and superconductor–insulator–normal metal (S–I–N contact). The conductivity of superconducting tunnel junctions is determined by electron tunneling (see *Tunnel effect*) through the insulator layer, which is no more than several nm thick. This junction exhibits a nonlinear current–voltage characteristic, since its conductivity is proportional to the energy dependence of the *density of states* of quasi-particle excitations in the superconductor. Applications of superconducting tunnel junctions are: for direct determinations of the *energy gap* Δ of semiconductors; for reconstructing the Eliashberg function (see *Eliashberg equations*) by utilizing the nonlinearities of current–voltage characteristic caused by the fact that Δ is energy dependent; and for nonlinear and switching elements in microelectronics. The S–I–S contacts exhibit *Josephson effects*. See also *Josephson junction*.

TURBOSTRATUM STACKING FAULTS (fr. Lat. *turbo*, vortex and *stratum*, layer)

Disruptions in the regular sequence of atomic layers in a crystal, which arise at the turning or displacing of these layers parallel to one another by an arbitrary amount. Turbostratum stacking faults are characteristic of layered structures exhibiting pronounced anisotropy (see *Anisotropy of mechanical properties*), like *graphite* and the graphite-like modification of *boron nitride*. Since the interlayer bonds in graphite-like compounds are weak, the concentration of turbostratum stacking faults γ in these structures may approach 1 ("*turbostratum*" structures). The quantity $(1 - \gamma)$ is referred to as the *degree of graphitization* of coal/graphite and related materials. The turbostratum structure is found in soot, coal (see *Active carbons*), low-temperature pyrocarbon and pyro-BN.

TVESAL

A superhard *composite material* composed of a tungsten–cobalt hard *alloy* and synthetic *diamonds*. The composition, structure, properties and application area are similar to *slavutich*. The manufacturing technology of tvesal excludes the dehardening of synthetic diamond grains during the thermal expansion of inclusions of metal-solvent.

TWINNING, ELASTIC

See *Elastic twinning*.

TWINNING, IRRATIONAL

See *Irrational twinning*.

TWINNING OF CRYSTALS

The coexistence in one solid of two identical *crystal structures* with different orientations, related by a certain *crystal symmetry* transformation: by the reflection in some crystallographic plane (*twinning plane*), by rotation around a certain crystallographic axis (*twinning axis*), by inversion with respect to a point (corresponding to *reflection twins* (see Fig.), *axial twins*, *inversion twins*). In the same crystal there can be several types of twinning at the same time, distinguished by their planes or axes (by *twinning laws*). The process of transformation of some uniform structure into one that is twinned is called the twinning of crystals. Often this twinning is accompanied by a change of the form of the crystal, e.g., the occurrence of reflection twins in calcite. Twinning with a variation of shape can be considered as a simple *shear*. Twinning can result in a sphere being transformed into an ellipsoid (*twinning ellipsoid*). The twinned atoms are displaced proportional to their distance from the twinning plane, and the direction of their displacement is called the *twinning direction*. If one of the twin structures occupies an insignificant part of the crystal, it is called a *twinning interlayer*, and the remaining part is the *source crystal*. A crystal pierced by parallel twinned interlayers is called a *polysynthetic twin*. Twins can appear during the course of crystal growth (*growth twins*), under the effect of external loads (*mechanical twins*), upon the heating of the deformed crystals (annealing or recrystallization twins), at the onset of phase transformations (especially *martensitic transformation* and *ferroelectric phase transition*

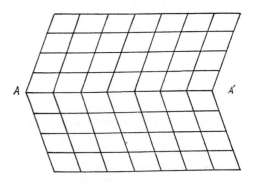

Sketch of the crystalline atomic planes of reflection twins; AA' indicates the twinning plane.

types), and under the effect of electric and magnetic fields. The twin components may be singled out by their physical properties which facilitates the experimental study of a *twinning structure*, and permits one to determine the relation of the *structural domains* (twins) with *domains* of other types. For instance, in some *ferroelastics* (Rochelle salt, gadolinium molybdate) the twins are at the same time *ferroelastic domains*. Mechanical twinning is one of the basic types of *plastic deformation* of crystals, and it becomes the prevailing type in cases of forced slipping (e.g., under dynamic loads, at low temperatures, in low-symmetry structures). In *metals* twinning is most common in hexagonal close packed lattices, it seldom appears in body-centered ones, and is even rarer in face-centered crystalline lattices. From the mechanical point of view twinning is polar, i.e. the shift may occur only to one side. The displacement to the opposite direction destroys the twin (*untwinning*). Mechanical twinning is a sequential process, and at the first stage an elastic twin appears (see *Elastic twinning*). The appearance of a macroscopic nucleus, which requires the participation of high local stresses, and the presence of concentrators of stresses (see *Stress concentration*), precede the appearance of a macroscopic twin. Having appeared, the macroscopic nucleus can elongate and grow quasi-statically in proportion to the increase in external load. At the second stage it intersects the crystal sample (under some loading) and becomes a residual interlayer (remaining in the crystal after the unloading). At the third stage this interlayer becomes thicker. Further development of the twin structure is due to the interactions of twins. Variation of the sign of the external load may bring about a reduction of the thickness of the twin interlayer. A twin which appears at high stress levels grows rapidly and as a result twinning can be accompanied by intensive *acoustic emission*. The theoretical description of the dynamics and kinetics of twinning is based on the *dislocation model of a twin*, where the twin replaces an aggregate of twinning *dislocations*. The theory provides a description of the basic stages of mechanical twinning, of dimensional changes, the formation of twins in external elastic fields, and other phenomena. As a rule, mechanical twinning is accompanied by the variation of the sample

shape, but twinning is also possible without any shape variation (e.g., the mechanical twinning of quartz). In this case the twin shift is absent and the twinning ellipsoid reduces to a sphere. The shape of the *unit cell* during this type of twinning is preserved, and the atomic structure of the twin components is distinguished only by the mutual displacement of the sublattices. The direction of the displacement of the twin boundary without a change of shape does not depend on the sign of the stresses. Some reflection and rotation twins, as well as all inversion twins, not only lack the twin shift, but their *elastic moduli* coincide (e.g., triglycinesulfate). In this case the external mechanical stresses cannot cause the transport of the twin boundary, but the boundary can displace, e.g., under the effect of an electric field.

TWINNING OF MARTENSITE

A method of *stress relaxation* which occurs at a *martensitic transformation* with a variation of the volume and shape of the transforming areas (or under the effect of external stresses). The changes during the twinning deformation which occur at the transition (see *Twinning of crystals*) are associated with a minimum of the elastic distortion energy, and the invariance of the habitus plane of the martensitic crystal. In this case the complex *martensite* substructure is formed, including a considerable number of *transformation twins*. Also the deformation produced twins must be differentiated from those formed after the completion of the martensite transition due to the effect of the external load on the crystal. In lamellar martensite crystals and in *iron alloys* the transformation twins are resolved in an *electron microscope* as a totality of thin parallel twin interlayers in a central region. The thickness of these interlayers ranges from several tenths of a nanometer to some tens of nanometers. The transformation twins in nonferrous alloys are larger and can be observed with the help of an optical microscope. In alloys on the basis of iron the martensite twins are preferentially formed according the twinning system $\{112\}\langle 111\rangle$. Besides, in carbon *steels* with a tetragonal martensite lattice twins are also observed corresponding to the system $\{011\}\langle 0\bar{1}1\rangle$. In some steels the $\{011\}\langle 0\bar{1}1\rangle$ twins have a relaxational nature, they form only after completion of the martensitic transformation during the course of heating the freshly-formed martensite, and thus they are not transformation twins. The twinning of carbon martensite according to the system $\{011\}\langle0\bar{1}1\rangle$ includes the motions of *carbon* atoms in octahedral *pores* of the crystal lattice along different axes of the unit cell. These movements are thermally activated so there can be an isothermal character to the twinning.

TWINNING PLANE SUPERCONDUCTIVITY

A phenomenon during which superconductivity in certain metals (*tin, niobium, indium*, etc.) appears to be localized in the vicinity of a twinning plane (TP), see *Twinning of crystals*. Twinning plane superconductivity (TPS) is characterized by a *critical temperature* somewhat higher than the critical temperature T_c of the corresponding bulk metal. However this increase in T_c is not very pronounced (0.1 K in In and 0.04 K in Sn) because of the *proximity effect* of the normal metal surrounding the TP. The superconducting *order parameter* then decreases as we move away from that plane, and the characteristic scale of that decrease is of the order of the *correlation length* ξ. It follows then that one cannot treat TPS as a purely two-dimensional effect. Twinning plane superconductivity develops while stronger bonding of electrons into Cooper pairs takes place in the vicinity of the plane (see *Cooper pairs*) due to the appearance of soft quasi-two-dimensional phonon modes close to the plane, and the corresponding changes in the electron spectrum. TP superconductivity in *type II superconductors* (e.g., niobium) located in a magnetic field may also be observed at temperatures below T_c. In fields somewhat above the bulk *upper critical field* $B_{c2}(T)$, TPS exists as far down as 0 K. The situation is different for *type I superconductors* (e.g., tin) where TPS appears in a quite narrow range around T_c. Moreover, the transition to TPS develops as a first order phase transition with clearly evident hysteresis. The weakening of the proximity effect, either near a close network of twins, or in small particles containing twins, may result in a noticeable increase of the critical temperature of the TPS. For example, an increase in the critical temperature from 3.7 K to 10 K was observed in tin due to twinning.

TWINNING STRUCTURE

The structure formed during the course of the *twinning of crystals*. It is determined by a series of factors: the value and speed of the external effect variation, its direction with respect to the crystallographic axes, the atomic structure of the twin boundary, the interactions of the twin interlayers with each other and with the *defects* of the lattice, the temperature, the features of the interaction processes of the twinning, *slip* and *failure* of the crystal, etc. At the initial stage of twinning there are non-interacting, incomplete wedge-like or lens-like twin inclusions (see *Twinning of crystals*). Depending on the *state of stress* parallel twin interlayers may appear, and also interlayers which grow along several systems of mutually intersecting planes. Crystals with a rhombohedral lattice (calcite, *bismuth, antimony*) may be twinned along three intersecting planes, hexagonal close packed metals along six of them, body-centered metals along 12 planes, etc. As a result, a framework of twin plates, dividing the crystal into fragments, may be formed. At the regions of contact of intersecting twins voids may be formed (e.g., *Rose channels* in calcite). Twinning structures which appear during the course of *martensitic transformations* and of ferroelectric phase transitions (see *Ferroelastics*) as a rule have the appearance of packets of thin wedges, which are then transformed to the packets of flat-parallel interlayers. The interface twin-matrix (*twinning boundary*), along which the crystal lattices of different orientations are conjoined, is the most important element of the twinning structures. The twinning boundary is a flat defect of the crystal lattice, and is always associated with additional *surface energy*. Data from physical measurements (*field*

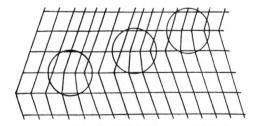

Sketch of an incoherent twin boundary; the circles enclose nuclei of twin dislocations.

ion microscopy) and computer simulations show that twinning boundaries are atomically sharp. If a boundary coincides with a plane of twinning, the location of atoms along it is regular (*coherent domain boundary*), and such a boundary does not cause macroscopic stresses within the crystal. A boundary, deviating from the twinning plane, consists of coherent sections, separated by steps of single-atom thickness, which are the nuclei of *twinning dislocations* (see Fig.). The broadening of the twinning interlayer occurs by the tangential movement of twinning dislocations along the boundary. The passage of each dislocation adds one interplanar distance to the thickening interlayer. *Noncoherent twinning boundaries* are the sources of high internal stresses. Near these boundaries the processes of generation and redistribution of defects are intensified, and *accommodation* regions appear. The twinning structure strongly influences the mechanical, electrical, superconducting, and other physical properties of crystals. Single crystals of the orthorhombic superconductor $YBa_2Cu_3O_{6.9}$ often exhibit twinning.

TWIST EFFECT in liquid crystals

An effect based on the optical properties of a *nematic liquid crystal* and a *Frederiks transition*. The liquid crystal has rod-like molecules with an electric dipole moment along the molecular axis. The end faces of a liquid crystal cell are treated so that the rod-shaped molecules are aligned parallel to the surface, but vertical at one face and horizontal at the opposite face. The molecules in-between the end faces undergo a gradual twist (*twist-structure*) of the liquid crystal *director n*, from a vertical to a horizontal alignment. An incident beam of linearly polarized light rotates through $\pi/2$ when the light traverses the liquid crystal located between two polaroids P_1 and P_2, as shown in Fig. (top). Therefore, the twist structure is opaque in parallel polaroids. When a magnetic or electric field of strength exceeding the threshold value of the Frederiks transition (a nematic has positive diamagnetic or dielectric anisotropy) is applied to the liquid crystal cell, the director *n* will align along the field as shown in Fig. (bottom). In this case the director does not rotate the light polarization plane, and the

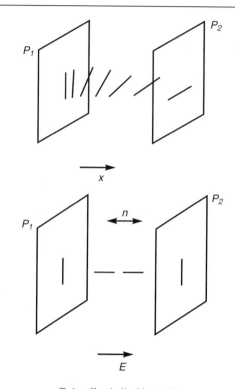

Twist effect in liquid crystals.

cell remains transparent in parallel polaroids, and opaque in crossed polaroids. The twist effect is used in the manufacture of *liquid-crystal displays*.

TWO-DIMENSIONAL CRYSTALLOGRAPHY

The 17 *crystal structures* (*space groups*) in two dimensions are grouped into four systems (syngonies): oblique ($a \neq b$, $\theta \neq 90°$), rectangular ($a \neq b$, $\theta = 90°$), square ($a = b$, $\theta = 90°$) and hexagonal ($a = b$, $\theta = 120°$). The rectangular system has both primitive and centered Bravais lattices, and the other three systems only have primitive lattices, so there are in all five Bravais lattices in two dimensions. The notation for a primitive space group symbol starts with the letter p, and that for a centered space group starts with c. The first three crystal systems mentioned above each have two point groups or (holohedries) associated with them, and the hexagonal system has four point groups, so in all there are 10 point groups or holohedries in two dimensions. Table 1 presents symmetry diagrams for the 15 primitive

Table 1. Diagrams of symmetry elements of the 17 two-dimensional space groups classified according to their crystal systems and their point groups. Both International and Schönflies symbols are given for the point groups. Primitive space groups start with the letter p, and centered space groups start with c.

Crystal system (syngony)	Space groups with lower symmetries	Space groups with higher symmetries
Oblique	1 (C_1) $p1$	2 (C_2) $p2$
Rectangular	m (C_{1h}) pm pg cm	$2mm$ (C_{2v}) $p2mm$ $p2mg$ $c2mm$ $p2gg$
Square	4 (C_4) $p4$	$4mm$ (C_{4h}) $p4mm$ $p4gm$
Hexagonal	3 (C_3) 6 (C_6) $p3$ $p6$	$3m$ (C_{3v}) $6mm$ (C_{6v}) $p3m1$ $p6mm$ $p31m$

and 2 centered space groups classified by their crystal systems and point groups.

TWO-DIMENSIONAL ELECTRON GAS

A system of electrons localized within a layer, whose thickness is less than their characteristic de Broglie wavelength. In such a layer the translational movement of the electrons is quantized. Two-dimensional (2D) motion along the layer is associated with each quantum level. In the narrow sense the twofold dimensionality of the system of electrons means that their motion with respect to only one (the lowest) level is displayed. Under specific conditions the two-dimensional electron gas appears in the *inversion layers* near the surface or interface of a semiconductor, within a thin semiconductor or metallic *film* (layer), near the surface of some liquid or solid *insulators*, etc. The possibility of controlling the parameters of a 2D electron gas over a wide range constitutes the basis of operation of a series of fast-response miniature *transistors* and other devices. Thermodynamic properties and kinetic phenomena which occur within the gas are in many respects different from those within the corresponding three-dimensional electron gas. For an ideal 2D gas with a parabolic dispersion law the *density of states* $g(\varepsilon)$ does not depend on energy within the limits of an allowed band, $g(\varepsilon) = m/(\pi \hbar^2)$. If a magnetic field $\boldsymbol{B}$ is applied perpendicular to the layer, then the energy spectrum of the ideal 2D electron gas is formed from two (taking into account spin) sets of discrete *Landau levels* (see *Quantizing magnetic field*), degenerate due to translation symmetry (see *Crystal symmetry*) and the density of states arising, respectively, from two sets of equidistant δ-like peaks. Due to such a form of $g(\varepsilon)$ the oscillations of the *magnetic susceptibility* and the conductivity resulting from the variation of the magnetic field in the low temperature region appear to be sharper than those in three-dimensional systems. It is also clear that the non-ideality of the two-dimensional electron gas essentially influences its properties, especially within a magnetic field. In most cases it is possible to increase the concentration n of the electron gas by one or two orders of magnitude by changing the difference between the potentials of the 2D-layer and the metallic electrode where the neutralizing charge is located (characteristic values of n for electrons over a thick film of *helium* are 10^7 to 10^9 cm^{-2} and in the vicinity of interfaces in the *heterostructures* on the basis of $Ga_x Al_{1-x} As$ it is of the order of 10^{10} to 10^{12} cm^{-2}). Therefore, the 2D electron gas is a convenient subject for the investigation of the effects of the electron–electron interaction. It is essential to note that although the electronic gas is two-dimensional, the electromagnetic field created by the electrons and exerting its influence over them is three-dimensional. One of the consequences of this fact is the comparable "weakness" of the screening in this 2D gas; the screened potential of the point charge at large distances r in the 2D-layer decreases as r^{-3} (the effective *screening radius* is the *Bohr radius* within the medium). Also the dispersion law of the plasma waves is much different from the three-dimensional case: at low wave numbers q the frequency is given by

$$\omega(q) = \left(\frac{ne^2}{m\varepsilon}\right)^{1/2} q^{1/2}$$

(ε is the mean *dielectric constant* of the medium). The electron–electron interaction in the 2D-system in the absence of Fermi degeneracy at very low temperatures leads to spatial ordering of the electrons or to the formation of a *Wigner crystal*. The absence or low concentration of the built-in charged centers, which are inevitably present in three-dimensional systems, is favorable for this. *Wigner crystallization* appears when the ratio of the characteristic energy of the Coulomb interaction and the heat energy $e^2(\pi n)^{1/2}/(\varepsilon k_B T)$ is close to 140, when a triangular lattice is formed

(at large distances with a finite temperature the density correlator falls by a power law). In a degenerate 2D system the interaction of the electrons may result in the formation of systems like *charge density waves* or *spin density waves*.

By varying the electric field pressing the two-dimensional electron gas to the surface of a dielectric or a semiconductor, in the region where electrons are located, it is possible to change their interaction with the vibrations of the medium and with defects localized in the vicinity of the surface. For a sufficiently strong interaction with the vibrations there appear selflocalized states of the electrons, similar to polaron states. In addition the simultaneous *self-localization* of many electrons is possible (it is observed, in particular, for a 2D electron gas over the surface of liquid helium, see *Levitating electrons*). Due to the reduced dimensionality of the 2D-system at low temperatures there are quantum corrections to the conductivity σ which are caused by the weak disorder (*weak localization effects*), and they produce a logarithmic reduction of σ with the lowering of the temperature. In addition to σ, the electron–electron scattering in the presence of scattering at *defects* (see *Current carrier scattering*) has a similar temperature dependence. The specifics of 2D systems are also displayed for strong disorder, where the temperature and frequency dependencies of the *hopping conductivity* with a variable jump length differ from those in three-dimensional systems. The interactions of electrons with one another, with defects and with vibrations of a medium in the presence of a strong transverse magnetic field, serve to remove the degeneracies of the Landau levels and make the energy spectrum of the electrons continuous rather than discrete, and this is among the most interesting 2D system results. The localization of the states is related to the tails of energy bands formed within the field of defects, and also to the electron–electron correlations which bring about such important phenomena as the *integral and fractional quantum Hall effect* (see *Quantum Hall effect*). In sufficiently perfect structures the reconstruction of the energy spectrum and of the states within the magnetic field is in turn associated with the electron-electron interaction, and therefore it influences the character of the scattering at vibrations and defects. It results in particular concentration dependences of *cyclotron resonance* spectra, and influences the magnetic conductivity in the limit of low concentrations, $n \ll |eB|/(2\pi\hbar)$, when the two-dimensional electron gas is non-degenerate.

TWO-DIMENSIONAL LATTICE MODELS

Models of statistical physics where the space variable is confined to discrete values in a plane. The first accurate solution was found by L. Onsager (1944) for the *Ising model*. There are a small number of two-dimensional lattice models which have been solved exactly, at least in the thermodynamic limit. They are: the two-dimensional Ising model, ferroelectric model, eight-vertex and three-spin models. An exact solution permits one to check many postulates of a general theory, and to determine the application limits of approximate methods. All known two-dimensional lattice models involve a physical interaction of restricted radius, provide a qualitative representation, and sometimes quantitative parameters, of real systems. They are good for describing systems with strong horizontal and weak vertical interactions: lamellar *magnetic substances* (e.g., K_2NiF_4, Rb_2MnF_4), liquid *helium* films, *superconducting films*, *smectic liquid crystal* films, monolayers of adsorbed atoms, and *charge density waves*. There is a correspondence between two-dimensional lattice models and two-dimensional models of quantum field theory.

For the majority of two-dimensional lattice models the solutions have been obtained only in the absence of an external field, which does not allow us to give a complete description of critical behavior and to calculate the scaling function (see *Scaling invariance hypothesis*). An exception is the ferroelectric model in the presence of an electric field. The nature of the interaction and its symmetry are the determining factors for the selection of a model to represent a real system. A brief description of several two-dimensional lattice models and the conditions of their realization is given below.

Models with pairwise interactions:

1. *Gauss model (free field)*.
2. *Discrete Gauss model*. It is used for describing a system of atoms adsorbed at the surface of metals with a high ratio of the two substrate periods.

3. *Cabrera model* (N. Cabrera). It describes the fluctuations of a crystalline surface. The variable is the height of a protruding column on a surface with the corresponding site number.
4. *XY-model (planar magnetic)*. It is used for the description of superfluid ^{4}He and superconductors.
5. *Potts model*. Describes lattices with *vacancies*. Corresponds to planar magnetic medium with an anisotropy axis of magnitude equal to the number of spin components.
6. Models with Z_q symmetry allow discrete flat rotations through the angles θ_j. The *Ashkin–Teller model* described by two Ising spins is the best known example.

Vertex models. The statistical weight of each model is given by the configuration of spins at the corresponding faces which belong to each particular vertex. If the statistical weight is decomposed into a product of statistical weights with a *pairwise interaction* then the net result for the original system is a set of non-interacting subsystems with pairwise interactions. The following models are related to the vertex ones:

1. *Eight-vertex model* (8V-model). It is exemplified by atomic hydrogen adsorbed on tungsten.
2. *Baxter model* (R.J. Baxter, 1971). Corresponds to a specific parametrization of the eight-vertex model.
3. *Six-vertex model*. A model which describes ice.
4. *Model of rigid hexagons* (triangular *lattice gas*).

For each of the two-dimensional lattice models a description of critical behavior has been obtained. For some models the difference between the system phases is determined by the behavior of a correlator at large distances. The *critical indices* are determined and their connection with the interaction parameters is established. There is a correspondence between some of the models described above, and other models involving the presence of a dual transformation which converts one model into another.

TWO-LEVEL WEAKLY COUPLED SYSTEM

System of two nearby energy wells in which the occupation oscillates between the wells. The term usually applies to a quantum-mechanical system with two discrete energy level populations which can appreciably vary due to thermal fluctuations or applied external forces, whereas transitions involving other levels are not appreciably effected. Such phenomena as *optical nutation* (oscillations of the population difference of two atomic levels), *photon echo*, coherent radiation of optical generators, etc., are described in terms of two-level systems which are very weakly coupled with one another and undergo transitions under the influence of thermal *fluctuations* or of external fields (electromagnetic or elastic). Two-level systems with closely spaced (separated by $\sim 10^{-4}$ eV) levels each corresponding to a different configuration and able to transform to one another through tunneling transitions (see *Tunneling phenomena*) occupy a special place in the physics of amorphous solids. An atom (or electron) located in a potential well, close to symmetrical in shape, with two minima (see Fig.) separated by a low ($\sim 10^{-4}$ eV) energy potential barrier, is the simplest two-level system. Two closely spaced energy levels (designated by solid lines) correspond to the atom located in the vicinity of the right (E_1) or left (E_2) minimum, respectively. The configuration can change via thermally activated jumps over the energy barrier, as well as via tunnelling transitions through the barrier. The latter determine the atomic frequency of interchange at the low temperatures when thermally activated transitions are suppressed. The two-level system may be associated not only with a single atom, but also with a group of atoms. Many *low-temperature anomalies* of amorphous solids can be explained by the presence of two-level systems. The microscopic structure of such two-level systems in amorphous solids remains unclear.

TWO-PHOTON LIGHT ABSORPTION

Light absorption with the simultaneous participation of two *photons* which have a combined total energy equal to the resonance frequency of the system, in particular, of the *crystal*. For *insulators* and *semiconductors* the total energy may be equal to or greater than the *band gap*. It is observed in the interaction with matter of very intensive light sources such as lasers, for which the polarization is presented in the form of a series

$$P = \alpha E + \beta E^2 + \cdots.$$

The coefficients α and β characterize respectively the linear and non-linear (quadratic, etc.) polarizability of the material. The first term describes the common phenomena of linear optics and the additional terms ones describe those of *nonlinear optics*. The two-photon light absorption is determined by the second term so the probability of the process, i.e. the intensity of the light flux is proportional to E^2. For anisotropic crystals this absorption is also anisotropic.

TWO-WAVE APPROXIMATION

An approximation in the theory of the *dynamic radiation scattering* by crystals. It is used when there are two sites of the *reciprocal lattice* 0 (000) and H (hkl) in the vicinity of the *Ewald sphere* (see *X-ray structure analysis*). Due to this the basic dynamic scattering equations divide into two systems of equations (for two states of polarization) each of which has two amplitudes of the displacement (electric induction), one for the transmitted plane wave (D_0) with wave vector $\mathbf{k}_0$, and one for a diffracted plane wave (D_h) and with wave vector $\mathbf{k}_h = \mathbf{k}_0 + \mathbf{h}$, where $\mathbf{h}$ is the reciprocal lattice vector corresponding to the reflecting planes (hkl). The nontrivial solution of these basic equations (*dispersion equations*) involves two allowed wave vectors $\mathbf{k}_{0j}$ ($j = 1, 2$) whose origins describe a two-sheeted dispersion surface in the reciprocal lattice. The cross-section cut through this surface by the scattering surface is a hyperbola with two

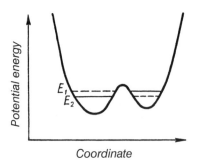

Potential energy dependence of a double potential well on the coordinate. Energy level E_1 corresponds to localization (with an exponential probability) in the right-hand well, and level E_2 is for localization in the ground state of the left-hand well.

branches, and the minimal distance between them is equal to the reciprocal of the *extinction length*. The positions of the origins of the two wave vectors k_{0j} on the dispersion plane are determined by the boundary conditions, and these establish the relation between the amplitudes D_{0j} and D_{hj}. As a result, two Bloch wave fields ($j = 1, 2$) are formed within the crystal from the plane waves:

$$D_j(r) = \exp\big[2\pi i \, (vt - k_{0j} \cdot r)\big]$$
$$\times \big[D_{0j} + D_{hj} \exp(-2\pi i \, h \cdot r)\big],$$

which have amplitudes that are periodic functions of the coordinate r (v is frequency, t is time). The maxima of these amplitudes are located in planes parallel to the planes $h \cdot r = n + \text{const}$ (n is an integer), i.e. the wave field is a standing wave with maxima and minima that lie in planes parallel to the reflecting planes (hkl). It follows from the solution of the basic equations and the dispersion equation that the phases of D_0 and D_h coincide in one wave field, and in the second wave field they differ in phase by π, i.e. the minima of one field coincide with the maxima of the second one, and vice versa. This circumstance plays an important role in the phenomenon of the *anomalous passage of X-rays*. Taking into account the boundary conditions for the amplitudes of the incoming and diffracted waves in accordance with the geometrical conditions of *diffraction of waves* provides formulae for explaining the interference phenomena of both *Laue diffraction* and *Bragg diffraction*.

TYPE I SUPERCONDUCTORS

Superconductors featuring the property of *absolute diamagnetism* (see *Meissner effect*) in applied magnetic fields B whose strength does not exceed a certain critical, temperature dependent value B_c (see *Thermodynamic critical magnetic field*). Type I superconductors comprise all pure *metals*, except for *niobium*. Superconductivity in type I superconductors vanishes in higher applied magnetic fields $B > B_c(T)$.

For certain shapes of type I superconductors a partial destruction of superconductivity may take place within the bulk sample even at $B < B_c(T)$ (see *Intermediate state*). In a nonzero magnetic field $B \neq 0$ the transition of a type I superconductor from its normal to its superconducting state

(the N–S transition) that follows the lowering of its temperature to $T = T_c(B)$ (see *Critical temperature of superconductors*) is a *first-order phase transition*, accompanied by release of a latent heat

$$Q = \frac{T B_c}{\mu_0} \left| \frac{dB_c}{dT} \right|.$$

During a reverse S–N transition, latent heat is absorbed, and, provided the sample is thermally insulated, its temperature decreases. In the absence of a magnetic field ($B = 0$) the N–S (or S–N) transition at point T_c is a *second-order phase transition* that is accompanied by a jump in the *specific heat* of the sample without any latent heat. In the case of type I superconductors in a magnetic field, the interface between the normal and the superconducting phases (the N–S interface) has a certain positive *surface energy* that prevents the magnetic flux from penetrating into the bulk volume of the superconductor. This results from the suppression of the superconducting *order parameter* in the neighborhood of the N–S interface due to the *proximity effect*. It develops within a distance of the order of the *coherence length*, ξ, a distance which exceeds the *penetration depth of magnetic field* λ in a type I superconductor. For $\xi \gg \lambda$ the surface energy per unit surface area is $\gamma \approx \xi B_c^2 / (2\mu_0)$. The *Ginzburg–Landau parameter* in a type I superconductors is $\kappa = \lambda / \xi < 1/\sqrt{2}$.

TYPE II SUPERCONDUCTORS

Materials retaining their superconducting properties over a wide range of magnetic fields, while *magnetic flux* partially penetrates the bulk of the sample (the partial *Meissner effect*). Among the type II superconductors are disordered *alloys* and solid solutions of *metals*; *intermetallic compounds*, doped *degenerate semiconductors* and semimetals; *amorphous superconductors*, conducting organic and metal-oxide compounds (see *Organic conductors and superconductors*, *High-temperature superconductors*) and *niobium*. As for the *absolute diamagnetism*, it is observed only in type II superconductors in weak enough magnetic fields, with their strength remaining below the *lower critical field*, $B_{c1} = \Phi_0 \ln(\lambda/\xi)/(4\pi\lambda^2)$, where Φ_0 is the *flux quantum*, and λ and ξ are the *penetration depth of magnetic field* and the *coherence length*, respectively. Since $\lambda > \xi$ for type II superconductors, the B_{c1} field is smaller than the *thermodynamic critical magnetic field* $B_c = \Phi_0/(2\pi\sqrt{2}\xi\lambda)$. Within the range $B_{c1} < B_c < B_{c2}$, where $B_{c2} = \Phi_0/(2\pi\xi^2)$ is the *upper critical field*, there exists a *mixed state* called the *Shubnikov phase* of type II superconductors in which the magnetic field penetrates the sample bulk in the form of *Abrikosov vortices* (vortex filaments). Fig. 1 presents the dependence of B_{c1} and B_{c2} on temperature, and Figs. 2 and 3 the dependence of the *magnetic flux density* B_{in} and *magnetization* M inside the superconductor on the applied field B_{app}, as observed in type II superconductors at $T < T_c$. Along the full length of the $B_{c2}(T)$ curve the transformation from the normal state to the superconducting state and back is a *second-order phase transition*. The *surface energy* of the interface between the normal and the superconducting phases is negative for type II superconductors. Provided $\lambda \gg \xi$, this energy is equal to $\gamma = -\lambda B_c^2$ per unit surface area of the interface (compare that result to the one found for *type I superconductors*). Therefore, for field strengths $B_{app} > B_{c1}$ it is energetically advantageous for the magnetic field to penetrate the bulk sample in the form of individual vortices, each carrying a single flux quantum Φ_0 and covering a maximum surface area of the normal to superconductor interface. With increasing B_{app} the number of such vortices in a type II superconductor increases, their mutual interaction (repulsion) increases (see *Vortex lattices*), and at $B_{app} = B_{c2}$ the normal material vortex cores (their radius equal to ξ) finally almost touch. Superconductivity in the bulk of type II superconductor then vanishes. However, up to $B_{app} = 1.69B_{c2} = B_{c3}$ (*third critical field*), the presence of the superconducting phase still exists at its surface.

As for an ideally pure type II superconductor (e.g., single crystal niobium or a stoichiometric A15 compound), the transport current $\boldsymbol{j}$ flowing through it at an angle to $\boldsymbol{B}_{app}$ in a mixed state ($B_{c1} < B_c < B_{c2}$) results in the motion of vortices under the action of the *Lorentz force* that is transverse to both $\boldsymbol{B}_{app}$ and $\boldsymbol{j}$, so that a finite ohmic resistance develops (see *Resistive state*). Thus, a superconducting nondissipative current is only

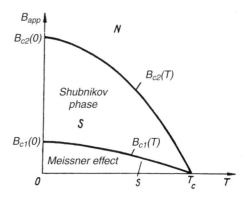

Fig. 1.

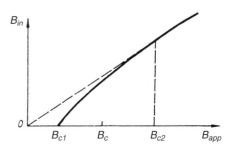

Fig. 2.

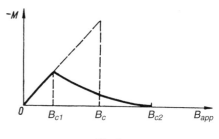

Fig. 3.

possible in non-ideal (e.g., "doped") type II superconductors when the vortices are pinned at various defects of the *crystal lattice* (see *Vortex pinning*). One of the important characteristics of type II superconductors is their *Ginzburg–Landau parameter* $\kappa = \lambda/\xi$, which defines the ratio $B_{c2}/B_{c1} = 2\kappa^2/(\ln\kappa)$. There are many practical applications for type II superconductors with $\kappa \gg 1$.

ULTIMATE RESISTANCE
See *Ultimate strength*.

ULTIMATE STRENGTH, ultimate resistance

The limit stress corresponding to the maximum load, i.e. to the maximum carrying ability of a sample during a short-term *mechanical testing of materials*. The value of ultimate strength depends on the type of *strain* being tested. Ultimate strength is defined for stretching (σ_B), compression (σ_{BCOM}), and *flexure* (σ_{BF}). The basic measuring unit is the pascal (Pa), which equals $1\,\text{N}/\text{m}^2$. In plastic materials (see *Plasticity*) undergoing tensile tests, the *deformation diagram* which plots load P versus strain ε has a maximum, which corresponds to σ_B. The maximum arises from the simultaneous processes of *strain hardening* and *transverse contraction* of the sample. At the continuation of a deformation after reaching the ultimate strength, a *neck* is formed on the sample and *failure* occurs. In materials of low plasticity the ultimate strength corresponds to the *breaking stress*, in this case by this the failure may be observed at low stresses and for insignificant values of the *plastic deformation*.

ULTRASONIC DELAY LINES

Devices that provide a time delay after the conversion of electrical signals into acoustic vibrations. The time delay is realized due to the low (10^4–10^6 cm/s) velocity of propagation of elastic vibrations in the sound conductor. The main components of solid-state ultrasonic delay lines are *acoustic wave transducers*, and a conductor of sound. These delay lines may operate either "by transmission" or "by reflection". In the former case, separate input and output acoustic wave transducers are used, whereas in the latter case the generation and detection of the acoustic vibrations in the sound conductor are performed by

one and the same transducer. Depending on the type of vibrations employed, three main groups of solid-state ultrasonic delay lines may be distinguished: lines operating by *bulk acoustic waves*, lines employing *surface acoustic waves*, and ultrasonic delay lines of the waveguide type. These delay lines may also be conventionally classified by their purpose; in this case, the following types of lines are distinguished: calibration lines, which feature a definite time interval between input and output pulses or between several output radio frequency pulses; variable length lines; multidrop lines used for generating coded signals and as matched filters for processing these signals; dispersion lines which have a delay time that is a function of the signal frequency are used for generating and compressing signals with pulse-frequency self-modulation.

The key parameters of solid-state ultrasonic delay lines are delay time, operating frequency, bandwidth, operating losses, spurious signal level, and temperature coefficient of delay. Dispersion ultrasonic delay lines are characterized also by their *pulse compression factor* (product of pulse duration and bandwidth), and by the height of side lobes of the compressed signal. The types of materials used for producing sound conductors includes isotropic materials (fused quartz, special varieties of glass, magnesium alloys, etc.), as well as perfect *monocrystals* of quartz, lithium *niobate*, ruby, sapphire, yttrium *iron garnet*, *yttrium–aluminum garnet*, etc.

ULTRASONIC TREATMENT

Subjection of a material to ultrasonic vibrations. Ultrasonic treatment of *solids* incorporates mechanical dimensioning (drilling, profiling, grinding), application of ultrasound for plastic working of metals (*pressing*, rolling, *drawing*), surface hardening, ultrasonic welding and ultrasonic soldering, crystallization of melts in the field of ultrasonic vibrations, ultrasonic cleaning and dispersing (see Dispersion) solids in liquids. In the process of shaping the material through the agency of a mechanical ultrasonic treatment, the particles of the abrasive material, which are suspended between the shaping tool and the machined object, are raised to a high speed through the agency of ultrasonic vibrations. Collisions of these particles against the material promote the controlled erosion of the latter. Exposure of the site of strain to ultrasonic vibrations during the course of plastic working of metals permits a considerable reduction of contact friction of the tool against the wrought material, and brings down the level of internal stresses in the processed material. Ultrasonic treatment of annealed material brings about hardening, and when the material had been strengthened prior to ultrasonic treatment, ultrasonic vibrations bring about the redistribution of internal stresses. This treatment raises the rate of diffusion processes in solids (see Diffusion). In the course of welding and soldering, ultrasonic vibrations are used for decomposing and removing scale from the surfaces of the parts to be joined, to increase the temperature and impact pressure at the junction. The action of ultrasound during the process of crystallization of a melt favors degassing of the latter, and facilitates obtaining a homogeneous fine-grained crystal structure. Ultrasonic cleaning of surfaces consists in immersing the material to be purified into a liquid, which undergoes cavitation caused by an ultrasonic wave. Liquids which are subjected to cavitation under the action of ultrasound, are also used for dispersing solids.

ULTRASOUND

Elastic vibrations and waves that are generally coherent and fall within the frequency range of $(1.5–2)\cdot10^4$ to $(10^{12}–10^{13})$ Hz: the lower and upper limits of this interval are respectively the high-frequency threshold of audibility for a human ear, and the frequency corresponding to the sound wavelength that equals the interatomic distance of a solid. The physical nature of ultrasound does not differ from that of sound, and therefore the boundary separating the frequency ranges of sound and

ultrasound vibrations is conventional. The major sources of ultrasound in solids are magnetostrictive and piezoelectric transducers, which convert electromagnetic energy to mechanical (acoustic) energy using the effects of magnetostriction and piezoelectricity, respectively (see Acoustic wave transducer).

Methods based on the application of ultrasound are used in various areas of solid state physics, in particular, for determining elastic moduli and dissipative characteristics of crystals (see, e.g., Ultrasound attenuation), for studying interactions of phonons with conduction electrons, magnons and other quasi-particles (see Acoustic methods of studying solids). Ultrasound is also used in defectoscopy and acoustic microscopy of solids. The employment of ultrasound in engineering gave rise to new trends in electronics and optics, namely, acousto-electronics and acousto-optics.

ULTRASOUND ABSORPTION BY DISLOCATIONS

Scattering of energy of ultrasonic waves within crystals at dislocations moving under their influence. Dislocations transfer the energy obtained through interactions with ultrasonic waves to the phonon (see Phonons) electron, magnon (see Magnon) and other subsystems of elementary excitations within crystals. The efficiency of energy transfer to one or another subsystem is determined by the properties of crystals (by the nature of the interatomic bonds, by the phonon, electronic and other spectra, by the presence of impurity atoms and other defects, etc.), as well by external conditions (temperature, external electrical and magnetic fields, etc.). Originally the absorption of ultrasound by dislocations was observed by W.T. Read (1940) in the course of the investigation of the orientation dependence of the absorption of ultrasound in zinc monocrystals. He had found that this absorption reached a maximum value when the ultrasonic vibration produces the maximum shear stresses in the slip planes of the most mobile dislocations. Theoretical descriptions of ultrasonic absorption by dislocations had been provided on the basis of the string model of dislocations (see Granato–Lücke model, Dislocation string) and the model of bends (A. Seeger, 1955). In the theory of Granato and Lücke the dislocation segment is considered as a section of a string,

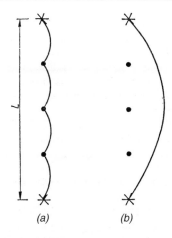

(a)   (b)

Fig. 1. Motion of a dislocation fixed by weak (•) and strong (∗) pinning centers under the influence of an ultrasonic wave: (a) linear mode of motion while still pinned to both strong and weak pinning centers; and (b) motion after break-off from weak pinning centers.

fixed by one of the two types of stationary pinning centers sketched in Fig. 1: by strong pinning centers at the ends of a section (nodes of the *dislocation grid* play this role), and by weak pinning centers in the gap between the strong ones (*point defects*, impurities, vacancies, etc.). The vibratory motion of a dislocation in a field of alternating stress $\tau = \tau_0 \sin \omega t$, actually in the slip plane, is described by an equation similar to that of a vibrating string

$$M \frac{\partial^2 u}{\partial t^2} + B \frac{\partial u}{\partial t} - C \frac{\partial^2 u}{\partial x^2} = b\tau,$$

where M is the effective mass per unit dislocation length, B is a damping coefficient, C is the linear tension of the dislocation, b is the modulus of the *Burgers vector* of the actual slip system, $u(x, t)$ is the displacement of a dislocation element, x is the direction along the dislocation, τ_0 is the shear stress amplitude, and ω is the angular frequency of the ultrasound. The damping properties of the medium are determined by the spectra of elementary excitations within the crystal, and by the nature of their interaction with dislocations. If τ_0 is smaller than some critical value τ_c (*stress of dislocation break-off*), the dislocation segments perform the vibratory motions

shown in Fig. 1(a). In this case the coefficient of absorption of ultrasound by dislocations does not depend on τ_0, and is called *dynamic absorption of ultrasound by dislocations* (linear). The frequency dependence of this dynamic absorption has a broad maximum at $\omega_m = \pi^2 C/(Bl^2)$, where for a qualitative description usually a uniform distribution of weak pinning centers along the dislocation is assumed. At this frequency ω_m the dynamic absorption coefficient α is

$$\alpha = \frac{k_1 \Lambda l^2 \omega_m}{v},$$

where Λ is the density of dislocations, v is the *sound velocity*, the k_j here and below are dimensionless constants, and l is the length of a dislocation segment between the weak pinning center.

When $\omega \ll \omega_m$ we have

$$\alpha = \frac{k_2 \Lambda l^4 B \omega^2}{Cv}.$$

For $\omega \gg \omega_m$ the dynamic absorption coefficient reaches its maximum value with no dependence on l:

$$\alpha = \frac{k_3 \Lambda C}{Bv}.$$

If $\tau_0 > \tau_c$, then during the first (third) quarter of the vibration period the weak dislocations break off from the weak pinning centers, causing a jump-like increase of the dislocation strain $\varepsilon_d(\tau)$ (Figs. 1(b), 2). During the second (fourth) quarter the forces of linear tension cause the dislocation segments to return to their initial positions. On the curve of stress τ versus strain ε_d a closed *hysteresis* loop is obtained (Fig. 2), corresponding to additional energy losses during each semi-period that are proportional to the area of the loop. This is called *hysteresis absorption of ultrasound by dislocations*, and it arises from the non-linear absorption of ultrasound by dislocations. For small values, $\tau_0 \ll \Gamma$ (Γ is the characteristic level of stress), and with an exponential distribution of segments by length, the hysteresis absorption may be expressed in the following form:

$$\alpha = k_4 \frac{\Lambda L^3 \Gamma \omega}{v Cl \tau_0} \exp\left(-\frac{\Gamma}{\tau_0}\right),$$

where L is the distance between the strong pinning centers. In its pure form the hysteresis mechanism

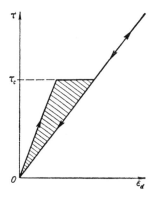

Fig. 2. Relationship between dislocation deformation and stress during half of the period of ultrasonic vibrations.

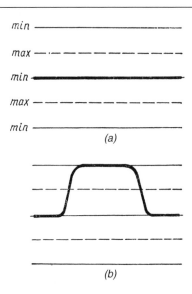

Fig. 3. Dislocation within the Peierls relief: (a) position of dislocation at mechanical equilibrium in a valley of the potential contour at $T = 0$ K; (b) pair of bends on the dislocation for $T > 0$ K.

can be displayed only at $T = 0$ K in the absence of thermal (and quantum) *fluctuations*, which assist the dislocation to break off from the weak pinning centers. At $T > 0$ K the break-off process becomes partly relaxational, and can be detected by the presence of relaxational peaks in the temperature dependences of the absorption (*Hasiguti peak*). These peaks are particular cases of relaxational peaks of ultrasound absorption, determined by the condition

$$\omega t_r \simeq 1, \qquad t_r = t_0 \exp\left(\frac{E}{k_B T}\right).$$

Here t_r is the *relaxation time* of the process and E is its *activation energy*. The string model does not take into account the discreteness of the crystal lattice structure. Because of the presence of regularly interlacing rows of atoms the linear energy of a dislocation is a periodic function of its displacement within the crystal. At $T = 0$ the dislocation seeks the position of minimum energy in the valley of a potential contour (so-called *Peierls relief*, Fig. 3(a)). As the temperature increases thermal fluctuations are able to displace some of the dislocations from one valley of the potential contour to another valley. In this way two bends of opposite sign (pair of bends) can be formed on the dislocation (Fig. 3(b)). The process of forming pairs of bends at dislocations is a relaxational one, and the relaxation time t_r depends exponentially on the temperature. With respect to this relaxational process the peaks on the temperature dependence

plots had been initially detected by P.G. Bordoni (1949).

ULTRASOUND, ANOMALOUS PENETRATION

See *Anomalous penetration of ultrasound.*

ULTRASOUND ATTENUATION

in superconductors

Decrease of the intensity of acoustic oscillations in a superconductor associated with the transfer of their energy to electronic excitations and to the lattice. It occurs according to the law $I = I_0 \exp(-\alpha x)$, where I, α are the intensity and coefficient of absorption of ultrasound, respectively. The transition to the superconducting state in a metal involves a change in the coefficient of ultrasound absorption. In the *Bardeen–Cooper–Schrieffer theory* (BCS theory) the temperature dependence of the relative coefficient of electronic absorption of longitudinal ultrasound has the form

$$\frac{\alpha_S}{\alpha_N} = \frac{2}{\exp[\Delta/(k_B T)] + 1}, \qquad (1)$$

where Δ is the superconductor *energy gap*, and α_S/α_N is the ratio of the absorption coefficient of ultrasound in the superconductor to that in the normal metal. For transverse ultrasound, a jump-like decrease of α_S/α_N is observed at the superconducting transition, which is related to the screening of the electromagnetic interaction of transverse ultrasound with normal excitations in a superconductor by the superconductor condensate. The investigation of the temperature and orientation dependences of α_S/α_N provides information on the value, temperature dependence, and anisotropy of the energy gap in the superconductor.

However, there exist several wide classes of superconducting compounds (*heavy fermions*, A-15 compounds (Nb_3Ge, V_3Si, etc.), Chevrel phases, etc.) where the temperature dependence of α_S fails to follow Eq. (1). This is related to the particular features of their electronic structures (e.g., the presence of narrow peaks in the *density of states* near the *Fermi level*).

ULTRAVIOLET PHOTOEMISSION SPECTROSCOPY (UPS)

Method of obtaining energy distributions of photoelectrons (photoelectron spectra) that are emitted by atoms, molecules, and solids on exposure of the latter to monochromatic electromagnetic radiation. The radiation generally employed is the resonance emission produced by a gas discharge; with energies in the range $h\nu = 10$–40 eV. The most commonly used sources are He I ($h\nu = 21.2$ eV) and He II ($h\nu = 40.8$ eV). The radiation of this energy is capable of exciting only valence shell electrons, yet the natural *line (level) width* amounts to several meV, in contrast to other radiation sources. With the aid of a suitably designed spectrometer it is possible to determine the *band structure* of solids by the measurement of their photoemission spectra.

Excitation of electrons in a crystal may arise from particular points (or regions) of the *Brillouin zone*. The energy spectrum of excited electrons is obtained by considering plausible band-to-band *direct transitions* between all pairs of occupied and vacant bands. This spectrum depends on the energy of the exciting radiation, and particular features of the energy band structure of initial and final electron states in solids. Under certain conditions, a band-to-band *indirect transition* may also

take place. These transitions may be conditioned by *surface plasmons*, by *surface electron states* in *metals* and *semiconductors*, and by *phonons*. Since the thickness of the near-surface layer analyzable by photoemission spectroscopy amounts to only several monatomic layers, the method is surface-sensitive. UV photoemission spectroscopy shows great promise for studying the processes of *adsorption* and catalysis. The changes in energy band structure of valence shell electrons that are caused by interactions between adsorbed atoms and the substrate manifest themselves in the UV spectrum.

In the case of monocrystalline substrates, the method of UV angle resolved photoemission spectroscopy (ARPES, dependence on photoelectron emission angle, or on angle of incidence of the electromagnetic radiation) permits analyzing not only the electronic structure of the adsorbed layers, but also the atomic one as well. The most efficient for angular resolution in the UV is *synchrotron radiation* which is characterized by high stability, linear polarization, and a broad range of energy.

UMKLAPP PROCESSES, U-processes (fr. Germ. *Umklapp*, tossing over)

Scattering events of *quasi-particles*, whereby the total *quasi-momentum* (expressed in $\hbar$ units) of all quasi-particles after scattering differs from its total value before scattering by a *reciprocal lattice* vector b. The reason for the nonconservation of the quasi-momentum has to do with the *periodic potential* of the crystal, which provides the background in which the scattering takes place. Scattering events which conserve the total quasi-momentum are called normal processes (*N-processes*).

The division of scattering events into Umklapp and normal processes is determined by the selection of the *momentum space* unit cell. To explain the difference between U- and N-processes let us consider the absorption of a *phonon* with quasi-momentum q by an electron with quasi-momentum k_1. Both q and k_1 are inside the first *Brillouin zone* (FBZ). After scattering, in the final state one has the electron with quasi-momentum k_2. If the sum $k_1 + q$ is outside the FBZ, then the momentum conservation law has

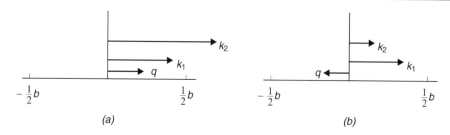

Umklapp and normal processes. (a) U-process, (b) N-process.

the form $k_2 = q + k_1 + b$, and the U-process takes place. If the sum $q + k_1$ is inside the FBZ, then the momentum conservation law has the form $k_2 = q + k_1$, and the N-process takes place. The U- and N-processes are shown schematically in Figs. (a) and (b), respectively.

The occurrence of Umklapp processes is indicative of the fact that the scattering of quasiparticles involves not only quasi-momentum exchange between particles (e.g., within an electron–phonon system), but also the transfer of quasi-momentum to the crystal as a whole, i.e. to those degrees of freedom, which are responsible for the "rigid crystal" motion. Hence Umklapp processes result in dissipation of the momentum of the system of quasi-particles, and they are responsible for the thermal and electrical resistance in pure crystals (Peierls, 1931). The U-processes are also important for investigations of X-ray, Möss-bauer, neutron, etc., *scattering* by the crystal (Laue points). See also *Conservation laws in solid state physics*.

UNIAXIAL CRYSTALS

Crystals with a single *optical axis*, along which a light beam does not experience double refraction. The optical axis in this case coincides with the principal *symmetry axis* (see *Crystal symmetry*). In a uniaxial crystal, one of the *refractive indices*, n_o (ordinary one), is independent of the direction, while the other, n_e (extraordinary one), is directionally-dependent. In *optically positive crystals*, $n_o - n_e > 0$, in *optically negative crystals* $n_o - n_e < 0$.

UNIAXIAL MAGNETICS

See *Magnetic anisotropy*.

UNIDIRECTIONAL ANISOTROPY

See *Magnetic anisotropy*.

UNIFORM COMPRESSION

Strain of a body under the effect of uniformly distributed compressing (extending) stresses (p) along the body surface. Under elastic uniform compression the *stress tensor* assumes the form $\sigma_{kl} = -p\delta_{kl}$, where δ_{kl} is the Kronecker delta symbol. *Hooke's law* is written in the form $u_{ij} = -ps_{ijkl}\delta_{kl} = -ps_{ijkk}$ ($i, j, k, l = 1, 2, 3$), where u_{ij} are the components of *strain tensor* and s_{ijkl} are the components of *tensor of elastic pliability*. Under elastic uniform compression, a reduction of the strain tensor u_{ij} occurs: $u_{ii} = -p[s_{11} + s_{22} + s_{33} + 2(s_{12} + s_{23} + s_{31})]$, where s_{jk} are the elastic pliabilities in matrix *Voigt notation*. The uniform elastic compression (expansion) of elastic isotropic solids and *crystals* with a lattice of cubic symmetry brings about only the alteration of the material volume without a change of shape. In the case of crystals with lower symmetry, a change in shape also takes place. The change of volume under unit compression stress given by

$$-\frac{1}{p}\frac{\Delta v}{v} = s_{11} + s_{22} + s_{33} + 2(s_{12} + s_{23} + s_{31})$$

is an elastic characteristic of a solid called the *bulk compressibility* (see *Bulk modulus*). The conditions of uniform compression or close to uniform compression can prevail during the *high pressure* treatment of materials.

UNIFORM STRAIN

Strain that is the same for all points of a deformed body. Planes and straight lines remain planes and straight lines after such a strain; parallel planes and parallel straight lines remain parallel. Upon the application of uniform strain a sphere becomes an ellipsoid. When stretching a *rod* of constant cross-section, uniform strain is observed in the elastic strain region and at the beginning of plastic flow. The condition for the uniformity of the *plastic deformation* to be disturbed, and its localization to occur (see *Localization of plastic deformation*), is the equality of the stress and the *strain hardening*. Microscopic inhomogeneity appears as early as at the first stages of the plastic deformation as a result of the strain localization in *slip planes* of dislocations.

UNIPOLARITY

The phenomenon of the non-invariance of properties of an object with respect to a change of ionicity of a certain unipolar axis. For example, if a *crystal* of formula MX consists of alternating parallel plane layers of ions M^+ and X^-, and the spacings between the layers alternate between $a + \delta$ and $a - \delta$, where $\delta \ll a$, then the crystal lattice exhibits a nonzero electric *dipole moment* since the distances from M to X and from X to M along the direction perpendicular to the atomic planes differ from each other. Unipolar properties may be exhibited by crystals which lack a center of inversion. The phenomenon of unipolarity is responsible for the piezoelectric effect (see *Piezoelectricity*), the electret effect (see *Electrets*), the *photovoltaic effect*, etc.

UNIT CELL

A parallelepiped in a crystal *lattice* constructed by three noncoplanar *translations*. The set of three linear (a, b, c) and three angular (α, β, γ) values determining a unit cell are called the *unit cell parameters* (cf. *Crystal lattice constants*). As a rule, a unit cell is chosen so that it can represent the *crystal symmetry* in the most convenient fashion; ordinarily it has a maximal number of right angles and possesses a minimal volume. Space is filled by the replication of unit cells.

UNIT CELL, QUASI-EXPANDED

See *Quasi-expanded unit cell*.

UNITS OF PHYSICAL QUANTITIES

Values of physical quantities adopted as benchmarks to compare and measure quantities of identical nature. A few physical units may be chosen arbitrarily and independently of each other, but relationships exist between various quantities, prescribed either by their definition or through physical laws. There are also practical matters to consider such as the feasibility of the exact replication of standards, and the accuracy of comparisons during the measurement process. Therefore, the independence of physical units is assigned to as few quantities as possible (*basic units*), expressing the others in terms of them using physical relationships (*derived units*). An individual dimension is attributed to each basic unit, while dimensionality formulae explicitly describing relationships between basic and derived units provide the dimensions of the latter. The sizes (scale) of the basic units have been selected so that commonly encountered values are neither too small nor too large.

The set of basic and the derived physical units pertaining to an area of physical phenomena and measurement forms a *system of units*. The separation of into basic and derived units is to some extent arbitrary. Some theoreticians adopt a system of units in which they set $\hbar = c = 1$. Besides, some fields of science and technology use specialized systems of units with relative independence, and this can make it difficult to compare results of scientific research. That is why the acceptance of an *International System of Units* (SI) which embraces all the areas of physical phenomena and includes many of the formerly used units, has been a major success. This system was adopted by the XI General Conference on Units and Weights (ICUW) in 1960, and further updated by later ICUWs.

New units are developed following the rules of formation of *coherent derived units*, some having their own names. These may, in their turn, be used to form additional new units. The SI electric and magnetic units are chosen in accordance with the rationalized electromagnetic field equations. The *MKS system* (meter-kilometer-second) is an integral part of the SI system. To determine a specific physical quantity its numerical value (and respective physical constants) should be reduced to SI. In that sense, the *CGS system*, in particular, the

Table 1. Relationships between SI and CGS systems of units

Variable	SI unit		Conversion factor	CGS unit	
	Name	Notation		Name	Notation
Basic units					
Length	meter	m	10^{-2}	centimeter	cm
Mass, m	kilogram	kg	10^{-3}	gram	g
Time, t	second	s	1	second	s
Electric current, I	ampere	A	$10/c^a \approx 3.33564 \cdot 10^{-10}$	not a basic unit in CGS	
Thermodynamic temperature, T	kelvin	K	1	kelvin	K
Amount of matter	mole	mole, mol	1	mole	mole, mol
Luminous power	candela	cd	1	candela	cd
Additional units					
Plane angle	radian	rad	1	radian	rad
Solid angle, Ω	steradian	sr	1	steradian	ster
Derived units					
Area, A	square meter	m^2	10^{-4}	square centimeter	cm^2
Volume, V	cubic meter	m^3	10^{-6}	cubic centimeter	cm^3
Velocity, v	meter per second	m/s	10^{-2}	centimeter per second	cm/s
Acceleration, a	meter per second squared	m/s^2	10^{-2}	centimeter per second squared	cm/s^2
Angular velocity, ω	radian per second	rad/s	1	radian per second	rad/s
Wavelength, λ	meter	m	10^{-2}	centimeter	cm
Wave number	inverse meter	m^{-1}	10^2	inverse centimeter	cm^{-1}
Period, T	second	s	1	second	s
Frequency, v or f	hertz	Hz	1	hertz	Hz
Frequency of rotation, ω	inverse second	s^{-1}	1	inverse second	s^{-1}
Density, ρ	kilogram per cubic meter	kg/m^3	10^3	gram per cubic centimeter	g/cm^3
Momentum, p	kilogram-meter per second	$kg \cdot m \cdot s^{-1}$	10^{-5}	gram centimeter per second	$g \cdot cm \cdot s^{-1}$
Moment of momentum	kilogram-meter squared per second	$kg \cdot m^2 \cdot s^{-1}$	10^{-7}	gram centimeter squared per second	$g \cdot cm^2 \cdot s^{-1}$
Moment of inertia, I	kilogram-meter squared	$kg \cdot m^2$	10^{-7}	gram centimeter squared	$g \cdot cm^2$
Force, weight, F	newton	N	10^{-5}	dyne	dyn
Force moment	newton meter	$N \cdot m$	10^{-7}	dyne centimeter	$dyn \cdot cm$
Force momentum	newton second	$N \cdot s$	10^{-5}	dyne second	$dyn \cdot s$
Stiffness, surface tension	newton per meter	N/m	10^{-3}	dyne per centimeter	dyn/cm
Stress, pressure, elastic modulus	pascal	Pa	10^{-1}	dyne per square centimeter	dyn/cm^2
Dynamic viscosity	pascal second	$Pa \cdot s$	10^{-1}	poise	P
Kinematic viscosity	square meter per second	m^2/s	10^{-4}	stokes	St

[a]Hereinafter the value of c in the "Conversion factor" column is equal to the speed of light in vacuo in CGS units, i.e. $c = 2.99792 \cdot 10^{10} \approx 3 \cdot 10^{10}$ cm/s.

Table 1. (Continued)

Variable	SI unit		Conversion factor	CGS unit	
	Name	Notation		Name	Notation
Energy, work, W, amount of heat, Q	joule	J	10^{-7}	erg	erg
Power, energy flux, P	watt	W	10^{-7}	erg per second	erg/s
Sound intensity	watt per square meter	W/m^2	10^{-3}	erg per second per square centimeter	$erg \cdot s^{-1} \cdot cm^{-2}$
Acoustic resistance	pascal second per cubic meter	$Pa \cdot s \cdot m^{-3}$	10^5	dyne second per centimeter to the power of five	$dyn \cdot s \cdot cm^{-5}$
Temperature coefficient (factor)	inverse kelvin	K^{-1}	1	inverse kelvin	K^{-1}
Heat flux	watt	W	10^{-7}	erg per second	erg/s
Thermal conductivity	watt per meter per kelvin	$W \cdot m^{-1} \cdot K^{-1}$	10^{-5}	erg per second per centimeter per kelvin	$erg \cdot s^{-1} \cdot cm^{-1} \cdot K^{-1}$
Heat transfer factor	watt per square meter per kelvin	$W \cdot m^{-2} \cdot K^{-1}$	10^{-3}	erg per second per square centimeter per kelvin	$erg \cdot s^{-1} \cdot cm^{-2} \cdot K^{-1}$
Entropy, S	joule per kelvin	J/K	10^{-7}	erg per kelvin	erg/K
Specific heat, C	joule per kilogram per kelvin	$J \cdot kg^{-1} \cdot K^{-1}$	10^{-4}	erg per gram per kelvin	$erg \cdot g^{-1} \cdot K^{-1}$
Heat content	joule per kilogram	J/kg	10^{-4}	erg per gram	erg/g
Diffusion coefficient (diffusivity)	square meter per second	m^2/s	10^{-4}	square centimeter per second	cm^2/s
Molar internal energy	joule per mole	J/mole	10^{-7}	erg per mole	erg/mole
Molar entropy, chemical potential	joule per mole per kelvin	$J \cdot mol^{-1} \cdot K^{-1}$	10^{-7}	erg per mole per kelvin	$erg \cdot mol^{-1} \cdot K^{-1}$
Molar concentration	mole per cubic meter	$mole/m^3$	10^6	mole per cubic centimeter	$mole/cm^3$
Electric charge, Q	coulomb	C	$10/c \approx 3.33564 \cdot 10^{-10}$	statcoulomb	–
Electric space charge density, ρ	coulomb per cubic meter	C/m^3	$10^7/c \approx 3.33564 \cdot 10^{-4}$	statcoulomb per cubic centimeter	–
Electric surface charge density, σ	coulomb per square meter	C/m^2	$10^5/c \approx 3.33564 \cdot 10^{-6}$	CGS unit	–
Electric potential, electromotive force, V	volt	V	$c/10^8 \approx 2.99792 \cdot 10^2$ ≈ 300	statvolt	–
Electric field strength, E	volt per meter	V/m	$c/10^6 \approx 2.99792 \cdot 10^4$	statvolt per centimeter	–

Table 1. (Continued)

Variable	SI unit		Conversion factor	CGS unit	
	Name	Notation		Name	Notation
Electric displacement (electric induction), D	coulomb per meter square	C/m^2	$10^5/(4\pi c)$ $\approx 2.65442 \cdot 10^{-7}$	statvolt per centimeter	–
Capacitance, C	farad	F	$10^9/c^2$ $\approx 1.11265 \cdot 10^{-12}$	centimeter	cm
Electric current density, J	ampere per square meter	A/m^2	$10^5/c$ $\approx 3.33564 \cdot 10^{-6}$	statamp per square centimeter	–
Electric resistance, R	ohm	Ω	$10^{-9}c^2$ $\approx 8.98755 \cdot 10^{11}$	second per centimeter	–
Electric resistivity, ρ	ohm meter	$\Omega \cdot m$	$10^{-11}c^2$ $\approx 8.98755 \cdot 10^9$	CGS unit	–
Electric conductivity, σ	siemens	S $(\Omega^{-1}m^{-1})$	$10^9/c^2$ $\approx 1.11265 \cdot 10^{-12}$	CGS unit	–
Current carrier mobility, μ	square meter per volt per second	$m^2 \cdot V^{-1} \cdot s^{-1}$	$10^4/c$ $\approx 3.33564 \cdot 10^{-7}$	CGS unit	–
Electrochemical equivalent	kilogram per coulomb	kg/C	$10^{-4}c$ $\approx 2.99792 \cdot 10^6$	CGS unit	–
Magnetic flux	weber	Wb	10^{-8}	maxwell	Mx
Magnetic flux density, B	tesla	T	10^{-4}	gauss	G or Gs
Magnetic vector potential, A	tesla meter	$T \cdot m$	10^{-6}	CGS unit	–
Magnetomotive force	ampere	A	$10/(4\pi)$ ≈ 0.795775	gilbert	Gi or Gb
Magnetic field strength, H	ampere per meter	A/m	$10^3/(4\pi)$ ≈ 79.5775	oersted	Oe
Magnetization, M	ampere per meter	A/m	$10^3/(4\pi)$ ≈ 79.5775	CGS unit	–
Inductance, L	henry	H	10^{-9}	CGS unit	(cm)
Reluctance	ampere per weber	A/Wb	$10^9/(4\pi)$ $\approx 79.5775 \cdot 10^6$	CGS unit	(Gb/Mx)
Intensity of radiation (radiant intensity)	watt per steradian	W/sr	10^{-7}	erg per second per steradian	$erg \cdot s^{-1} \cdot sr^{-1}$
Irradiance, radiant emittance	watt per square meter	W/m^2	10^{-3}	erg per second per square centimeter	$erg \cdot s^{-1} \cdot cm^{-2}$
Luminant intensity (radiance)	watt per steradian per square meter	$W \cdot sr^{-1} \cdot m^{-2}$	10^{-3}	erg per second per steradian per square centimeter	$erg \cdot s^{-1} \cdot sr^{-1} \cdot cm^{-2}$
Luminous flux	lumen	lm	1	lumen	lm
Luminous energy	lumen second	$lm \cdot s$	1	lumen second	$lm \cdot s$
Luminous emittance	lumen per square meter	lm/m^2	10^4	radphot (lumen per square centimeter)	rph

Table 1. (Continued)

Variable	SI unit		Conversion factor	CGS unit	
	Name	Notation		Name	Notation
Illuminance	lux	lx	10^4	phot	ph
Luminance	candela per square meter	cd/m^2	10^4	stilb	sb
Absorbed radiation dose	gray	Gy	10^{-4}	erg per gram	erg/g
Radiation dose power	gray per second	Gy/s	10^{-4}	erg per second per gram	$erg \cdot s^{-1} \cdot g^{-1}$
Ionizing particle flux density	inverse second by inverse square meter	$s^{-1} \cdot m^{-2}$	10^4	inverse second by inverse square centimeter	$s^{-1} \cdot cm^{-2}$
Nuclide activity in a radioactive source	becquerel	Bq	1	inverse second	s^{-1}
Equivalent radiation dose	sievert	Sv	10^{-4}	erg per gram	erg/g
Exposure dose (for X- and gamma rays)	coulomb per kilogram	C/kg	$10^4/c$ $\approx 3.33564 \cdot 10^{-7}$	statcoulomb per gram	–
Base-10 logarithm of a dimensionless ratio of a physical variable to a reference value	bell (usually expressed as a decibel, dB)	B	–	–	–

so-called symmetric CGS system (*Gauss system* for electric and magnetic variables) is a system in its own right with the basic physical units centimeter (cm), gram (g) and second (s). In this system the *dielectric constant* and *magnetic permeability* are both dimensionless and equal to unity in vacuum: $\varepsilon_0 = \mu_0 = 1$. The electric units have no official names, and the magnetic units for B and H are gauss and oersted, respectively. Having specified the thermodynamic temperature, the amount of matter and the light intensity (illumination unit) as additional basic units to complement the above, we obtain a CGS system applicable to all physical phenomena. Due to the coherence between the SI and the CGS systems, formulae and expressions written in either one can be transformed into the other. Both systems are absolute, since length, mass and time are among their basic physical units. For measuring mechanical quantities in engineering sometimes use is made of an *MKS system* with the basic units meter (m), kilogram of force (kgf) and second (s).

The choice of the values of basic units remains to some extent arbitrary, and this prescribes the values of universal constants, with physical relationships between them. If these constants are set equal to unity then one arrives at a *dimensionless system of units* that leaves no freedom in choosing the values for any particular unit. Since there are dimensionless relationships between various universal constants, only some of them may be equated to unity, and that choice depends on the particular area of study. Such systems are known as *natural systems of units*. The equations of physics acquire a simple form in them, so such systems are often used for the construction of physical theories, but they are generally not practical for use in experimental measurements.

In addition to the physical units belonging to standard systems, there exist a number of nonsystem units. Examples are the decimal multiple and fractional units (deci-, centi-, deca-, hecto-, etc.) employed with numbering systems having other

Table 2. Relations between SI units and some non-system units

Variable	Non-system unit		Conversion factor
	Name	Notation	to system unit
Length	angstrom	Å	10^{-10} m
	micron	μm	10^{-6} m
Mass, m	atomic mass unit	u	$\approx 1.66057 \cdot 10^{-27}$ kg
Plane angle	degree	$\ldots^{\circ}$	$\pi/180$
			$\approx 1.745329\ldots\cdot10^{-2}$ rad
	minute	$\ldots'$	$\pi/10800$
			$\approx 2.908882\ldots\cdot10^{-4}$ rad
	second	$\ldots''$	$\pi/648000$
			$\approx 5.848137\ldots\cdot10^{-6}$ rad
Volume, V	liter	l	10^{-3} m^3
Rotation frequency, ω	revolutions per second	rev/s	2π rad/s
	revolutions per minute	rev/min	$2\pi/60 \approx 0.104719$ rad/s
Force, F	kilogram force	kgf	9.80665 N
Pressure, P	kilogram force per square centimeter	kgf/cm^2	98066.5 N
	millimeter of mercury	mm Hg	133.322 Pa
	torr	torr	133.322 Pa
	atmosphere	atm	$1.01325 \cdot 10^5$ Pa
	bar	bar	10^5 Pa
Stress	kilogram force per square millimeter	kgf/mm^2	$9.80665 \cdot 10^6$ Pa
Energy	electronvolt	eV	$\approx 1.60219 \cdot 10^{-19}$ J
Temperature	degree Celsius	°C	T [K] $- 273.15$
Quantity of heat	calorie	cal	4.1868 J
Electrical resistivity	ohm-square millimeter per meter	$\Omega \cdot$mm^2/m	$10^{-6}\Omega\cdot$m
Magnetomotive force	ampere turn	At	1
Luminous intensity	(candle)		1.005
Absorbed radiation dose	rad	rad, rd	10^{-2} Gy
Equivalent radiation dose	roentgen-equivalent-man	rem	10^{-2} Gy
Exposure radiation dose of X- and gamma rays	roentgen	R	$2.58 \cdot 10^{-4}$ Gy
Nuclide activity in a radioactive source	curie	Ci	$3.700 \cdot 10^{10}$ s^{-1}
Natural logarithm of a dimensionless ratio of a physical variable to a reference value	neper	Np	0.8686 B = 8.686 dB

than a decimal base (minute, hour, etc.), and those that do not corresponding to units from any standard system (atmosphere, calorie, mm Hg, etc.). The latter are still convenient in specialized areas of scientific knowledge, or are retained as a tribute to historical traditions. It is quite convenient to transform the numerical values of physical quantities and various constants from one system to another, or reduce them to SI using Tables 1 and 2. Because of the coherence between the two systems the form of physical relationships is also accordingly transformed.

UNSHARED ELECTRON PAIRS

Correlated electron pairs in nonbonding *orbitals* (usually of the same atom). For normal atomic coordination unshared pairs do not participate in *chemical bonds*. Of most interest are unshared pairs that form the top of the *valence band* in vitreous compounds (see *Vitreous state of matter*), which are associated with p-electron state elements of Group VI of the periodic table. These provide easy recoordination of atoms in the structure, and determine many properties of oxide and chalcogenide glasses (see *Chalcogenide materials*).

UNSTABLE FRENKEL PAIR

A *Frenkel defect* whose components, a vacancy V and an interstitial atom I produced at the displacement of the atom from its regular lattice site, are fairly close together and have a high probability of annihilation. Many Frenkel defects which are radiation-induced in *metals, insulators* or *semiconductors* are unstable. The development of *pulse spectrometry* provided methods for determining the characteristics of short-lived (lifetime less than 1 s) Frenkel unstable pairs in various crystals (see *Transient absorption*). The *defects* are induced in the materials through an initiating radiation pulse of nanosecond or picosecond duration. Most of these defects subsequently annihilate; those relatively few which have not undergone annihilation either exist in the form in which they were created during the initial event, or they become converted into defects which are stable under the conditions of the experiment. In *alkali-halide crystals* the primary products of the decay of electronic excitations are neutral (F and H) or charged (α and I) pairs of defects (see *Color centers*). The ratio between the quantities of charged and neutral pairs (induced by short radiation pulses) is determined by the properties of the material. In chlorides, e.g., most primary pairs are neutral, whereas in iodides most primary pairs are charged. The majority of primary defects are annealed after formation. The number of recombined centers and the rate of *annealing* are determined by the temperature of the sample during the course of experiment. For example, in KCl the quantity of F-centers reduces by a factor of 3 in 100 ns at 100 K, but hardly changes at all at

300 K. The duration of the annealing of F-centers at 300 K is mostly in the range of milliseconds. In KI at 300 K the annealing of F-centers takes place partially in the time range of microseconds, whereas the duration of annealing of α-centers is completed in the time range of milliseconds. The efficiency of the process of decay of electronic excitations into pairs of structural defects is high. Thus, the generation of an individual $F-H$ pair in alkali-halide crystals requires not more than 50 eV of the absorbed radiation energy (15–20 eV are required to produce an *exciton*), whereas the overall energy accumulation for a single pair reaches 10^3–10^6 eV. This means that for each created defect only 10^{-1}–10^{-4} survive. A similar situation is observed also for other materials, in particular for crystals of the *fluorides* of alkaline-earth metals. The primary defects induced in silicon have short lifetimes (less than 1 μs), which is temperature-dependent. Only 10^{-4}–10^{-6} defects remain after the lapse of 1 s. The major fraction of the defects, which are induced through the agency of a radiation pulse, turns out to be short-lived. The high value of *radiation resistance* of these materials, which is determined by stationary experiments of the above types, is evidence for the low stability of induced primary defects, but not for a low efficiency of their creation. The efficiency of the annealing of Frenkel unstable pairs is determined by the functional distribution of the components of pairs in terms of the distances between them; this distribution function is determined by the conditions of defect generation and the annealing mechanism. Depending on the temperature of the sample, the correlated pairs under irradiation become annealed either through tunneling processes, or for the most part, through thermally activated motion of the mobile component of the pair. As this takes place, some portion of the pairs may become spatially separated and then transform into stable defects.

UNZIPPING

The process of sequential detachment of *dislocations* from *point defects* in the crystal lattice (e.g., *impurity atoms*) which pin them.

Unzipping is observable when dislocation motion resulting from external stress is simultaneously retarded by many local defects. The mechanical or fluctuational dislodging of a defect from a dislocation, and the advancement of the latter at this location, helps to alleviate the conditions of strain from neighboring defects. As a result an avalanche-like process of detachment, either thermally activated or activation-free, referred to as an *unzipping wave*, can propagate along the dislocation line in both directions from the initial breach. Computer simulations show that the unzipping wave travels through the dislocation line in each direction for about 6–10 point defect spacings, and then decays. The onset of unzipping is facilitated by an increase of defect density, or by a decrease of the temperature.

UPPER CRITICAL FIELD, second critical field

The value of external magnetic field beyond which the superconductive state is destroyed in the bulk of a *type II superconductor*. The upper critical field $B_{c2}(T)$ decreases with increasing temperature (T) and tends to zero as T approaches the critical temperature (T_c). In the range of applied fields $B_{c1}(T) < B_{app} < B_{c2}(T)$, where B_{c1} is the *lower critical field*, there is a partial penetration of the magnetic flux into the bulk of a type II superconductor (see *Mixed state, Shubnikov phase*) in the form of *Abrikosov vortices*, each of which carries the *flux quantum* Φ_0. According to the *Ginzburg–Landau theory of superconductivity* at $B_{app} = B_{c2}(T)$, there is a *second-order phase transition* from the superconducting to the normal state, and $B_{c2}(T)$ has the value

$$B_{c2}(T) = \frac{\Phi_0}{2\pi \xi^2(T)} \equiv \kappa \sqrt{2} \cdot B_c(T),$$

where $\xi(T)$ is the *coherence length*, $\kappa = \lambda/\xi$ is the *Ginzburg–Landau parameter*, λ is the *penetration depth of magnetic field*, and $B_c(T)$ is the *thermodynamic critical magnetic field*. In many type II superconductors (e.g., cuprates) $\kappa \gg 1$ and $B_{c2}(T) \gg B_c(T)$.

UPS

See *Ultraviolet photoemission spectroscopy*.

URANIUM, U

A radioactive element of Group III of the periodic system; with atomic number 92 and atomic weight 238.0289, belongs to the *actinide* group. Natural uranium consists of three isotopes: ^{238}U

(99.275%), ^{235}U (0.720%) and ^{234}U (0.005%). Outer shell electronic configuration is $5f^3 6d^1 7s^2$. Atomic radius 0.153 nm, ionic radii 0.097 nm (U^{4+}), 0.080 nm (U^{6+}). Uranium compounds occur in oxidation states from $+3$ to $+6$, the most typical states being $+4$ and $+6$. Electronegativity value ~ 1.27.

Uranium is a silvery-white lustrous *metal*. It occurs in three allotropic modifications (α-U, β-U, γ-U). α-U has an orthorhombic crystal structure with lattice constants: $a = 0.28542$ nm, $b = 0.58692$ nm, $c = 0.49563$ nm (at 298 K). At 941 K, α-U converts to β-U, which is tetragonal with lattice constants $a = 1.0759$ nm, $c = 0.5656$ nm (at 993 K). Body-centered cubic γ-U (lattice constant $a = 3524$ nm at 1078 K) is stable above $T = 1078$ K. The density of α-U is 19.05 g/cm^3 at room temperature, $T_{melting} = 1405$ K, $T_{boiling} = 4093$ K at atmospheric pressure. The heats of α-U $\rightleftarrows$ β-U and β-U $\rightleftarrows$ γ-U transformations are respectively 2.93 kJ/mole and 4.81 kJ/mole. Heat of melting 19.89 kJ/mole, heat of evaporation ~ 469 kJ/mole. Specific heat at constant pressure is 26.8 J/mole (298 K). Mean coefficients of thermal expansion of α-U along a, b and c axes are respectively $32.9 \cdot 10^{-6}$ K^{-1}, $-6.3 \cdot 10^{-6}$ K^{-1}, $27.6 \cdot 10^{-6}$ K^{-1} (at 293–773 K). Coefficient of thermal conductivity is ≈ 25.1208 W·m^{-1}·K^{-1} at room temperature (increases with increased temperature). Electric resistance of α-U depends on the crystallographic direction; averaged resistivity of high-purity polycrystalline sample is ≈ 0.30 μΩ·m at room temperature, and rises to 0.54 μΩ·m at 873 K. Molar magnetic susceptibility is $414 \cdot 10^{-6}$ at standard conditions. Superconducting critical temperature is 0.68 K for α-U, 1.8 K for γ-U. The Debye temperature is 207 K. Electronic work function of the uranium polycrystal is 3.74 eV. Thermal neutron capture cross-section is 7.59 barn. Young's modulus of α-U is anisotropic, with polycrystalline value 204.96 GPa; shear modulus 83.36 GPa; Poisson ratio 0.23. Adiabatic elastic moduli of α-U: $c_{11} = 214.74$, $c_{12} = 46.94$, $c_{13} = 21.77$, $c_{22} = 198.57$, $c_{23} = 107.91$, $c_{33} = 267.11$, $c_{44} = 124.44$, $c_{55} = 73.42$, $c_{66} = 74.33$ (all values are GPa) at 298 K. The value of Vickers hardness (HV) of α-U is 200 at room temperature, but decreases to ≈ 12 HV at 873 K. The transformation

α-U $\rightarrow$ β-U increases hardness from $\approx$10 HV to $\approx$30 HV. The rupture strength of annealed α-U (0.02% C) is $\approx$0.412 GPa at 293 K, it increases with temperature to 0.480 GPa at 373 K, and then almost linearly decreases to 0.108 GPa at 873 K. The yield limit is 0.255 GPa at 373 K, 0.088 GPa at 873 K, the relative elongation is 8% at 293 K, 26% at 873 K, the relative contraction is 11% at 293 K, 65% at 873 K. The increase of carbon content to 0.2% increases the strength and yield limits of uranium. All mechanical characteristics of uranium are sharply dependent on the presence of impurities and pretreatment. The creep of uranium is particularly dependent on cyclic temperature changes, which is related to extra thermal stresses which arise due to large differences between the coefficients of thermal expansion along different crystallographic directions of α-U. A characteristic property of uranium is elongation of polycrystalline α-U rods, which exhibit texture along the [010] axis for repeated heatings and coolings. The fission of uranium atoms gives rise to isotopes of neon and krypton, which results in swelling of the metal, a property which is highly undesirable for nuclear fuel applications. Metallic uranium, uranium-based alloys and uranium dioxide are used as nuclear fuel, and for the preparation of *plutonium.*

VACANCION

A *quasi-particle* emerging from a *vacancy* in a *crystal lattice* as a result of quantum tunneling (see *Tunneling phenomena*); a particular case of a *defecton*.

VACANCY (fr. Lat. *vacans*, empty)

Point defect in a crystal: an unoccupied atomic site of a crystal lattice.

Under increased temperature (T) *crystals* contain equilibrium numbers of vacancies that appear due to thermal *fluctuations* by local evaporation of atoms to or from the surface or interstitial positions. The vacancy concentration C_V is determined by the energy of their formation E_V in accordance with the expression

$$C_V = \exp\left(\frac{\Delta S}{k_B}\right) \exp\left(-\frac{E_V}{k_B T}\right),$$

where ΔS is the difference in *entropy* of the crystal in the presence of and the absence of the vacancy. Near the melting temperature C_V is of the order of 10^{-3}. Excessive nonequilibrium vacancies for a given temperature can be obtained by *quenching*, deformation, irradiation by high-energy particles, or by ultrasound. The presence of vacancies in a crystal can considerably affect its physical properties. Vacancies are centers for the *current carrier scattering*. In non-metallic crystals the electrically active vacancies influence the concentration of mobile charge carriers. The introduction of vacancies leads to the appearance of *light absorption* bands and *luminescence*, and depending on the presence of unpaired electrons, it leads to the appearance of EPR lines, etc. Vacancies can facilitate the transfer of atoms between lattice points. *Diffusion* controlled processes of *alloy ordering, alloy aging, creep, homogenization, sintering*, etc., are associated with vacancy motion.

Vacancies can aggregate with each other or with *impurity atoms*, resulting in the formation of *bivacancies* as well as larger aggregates of vacancies (see *Point defect clustering*). This plays a role in the mechanisms of *swelling* of solids. Complexes of vacancies with impurities have reduced mobility compared to pure vacancies. With increasing temperature, nonequilibrium vacancies migrate to sites which can be internal or external crystal surfaces or *dislocations*.

VACANCY–DISLOCATION INTERACTION

The forces of attraction or repulsion acting on a *vacancy* arising from elastic and (in *ionic crystals*) electrostatic fields created by a *dislocation*. Vacancies cause elastic distortions of the lattice characterized by a *dilatation* ΔV of the order of an atomic volume, and in ionic crystals a vacancy can carry a charge, so some forces that affect vacancies come from dislocation fields. In a cylindrical coordinate system r, ϑ associated with an *edge dislocation* (ϑ is measured from the *slip plane*)

$$P = \frac{1+\nu}{1-\nu}\frac{\sin\vartheta}{r}\frac{Gb}{3\pi},$$

so the energy $W = P\Delta V$ of the elastic vacancy–dislocation interaction at point (r, ϑ) is

$$W = \frac{Gb}{3\pi}\frac{1+\nu}{1-\nu}\Delta V \frac{\sin\vartheta}{r},$$

where G is the *shear modulus*, ν is *Poisson ratio*, and b is the *Burgers vector*. The vacancy is attracted to the compressed region near the dislocation ($\vartheta \approx \pi/2$), and it is repulsed from the stretched region ($\vartheta \approx 3\pi/2$). At $r = b$ the value of W is usually a fraction of an electronvolt, so that at the distance $r > r_0 = (3–5)b$ it is comparable to the thermal energy of lattice vibrations at room temperature, and the atmosphere of vacancies is washed out. From $r < r_0$ the equilibrium concentration of vacancies $c(r, \vartheta) = c_0 \exp[W(r, \vartheta)/(k_B T)]$, where c_0 is the concentration of vacancies far from the dislocation. The electrical interaction is weaker (with the exception of charged vacancies in ionic crystals) near charged *dislocation steps*.

Under an excess concentration of vacancies in the bulk of the crystal (e.g., after irradiation by fast particles), a drift diffusion flow of vacancies to dislocations is established. In the dislocation core the vacancies are connected to the edge of the extra plane, which causes *creeping-over of the dislocation*. For a sufficiently high vacancy oversaturation the formation of vacancy *pores* at a dislocation is possible. Since in the isotropic *theory of elasticity*, a helical screw dislocation possesses no dilatation field, a single vacancy does not interact with it elastically. The elastic dipole moment of a *divacancy* interacts with the tangential stresses of a *screw dislocation*. At split dislocations, in addition to the elastic interaction of vacancies with *partial dislocations*, there is also their chemical interaction with the surface of a *stacking fault*, with energies of fractions of an electron volt.

VACUUM ELECTRONICS, emission electronics

Branch of physics covering the studies and practical applications of physical phenomena related to the emission by a solid or liquid of electrons, ions, or neutral particles through a vacuum or gaseous medium (see *Particle emission*).

VACUUM EVAPORATION

A method of applying a *coating* to the surface of a product in vacuo. It consists of *evaporating* the coating material at high temperatures and under pressure below 1 Pa (usually 10^{-2}–10^{-3} Pa), with subsequent *condensation* on a cold surface. Heating is performed by electric current running along the vaporizer heating coil, its turns including small pieces or thin short wires (up to 8 mm in length) of the material to be evaporated. Depending on the vaporizer power, the wire diameter varies from 0.2 mm to 0.8 mm. The thickness of the applied layer is determined by the amount of evaporating material, and by the distance from the coated surface to the vaporizer. To obtain especially pure *films*, the evaporation is carried out in vacuo at the pressure 10^{-7}–10^{-8} Pa. Advantages of this method are its applicability to any material of a product or coating, its operation at low voltage (10–20 V), and the possibility of emitting the material on either side from the vaporizer.

VACUUM STATE OF A SYSTEM

Ground state of a macroscopic quantum system, i.e. the state with the minimum possible energy.

The vacuum state of a system can be degenerate (see *Degeneracy*). Systems with a continuous parameter describing a degenerate state are of special interest, such as *superconductors*, superfluid *helium* (see *Superfluidity*), *ferromagnets* and *antiferromagnets* in the exchange approximation (parameters are polar and azimuthal angles). The spectrum of *quasi-particles* in continuous order parameter systems has as many nonactivated (gapless) branches as there are continuous order parameters (see *Goldstone theorem*), if the interactions between the quasi-particles act at short range.

VALENCE (f. Lat. *valens*, strong)

Ability of an atom to form *chemical bonds* with one or more neighboring atoms. The valence of an atom depends mainly on the structure of its outer electronic shell because the inner shell electrons do not take part in forming chemical bonds. Ionic bonds involving the transfer of electrons are formed between cations that are monovalent (e.g., Na^+, K^+), divalent (e.g., Mg^{2+}, Ca^{2+}), trivalent (e.g., Al^{3+}) or higher in valency with anions that are monovalent (F^-, Cl^-), divalent (e.g., O^{2-}) or higher in valency. The charge of an ion is called its oxidation state. Covalent bonds involve the sharing of electrons between atoms. Carbon, which has a valence of four, can form covalent bonds that are single (e.g., CH_4), double (e.g., CO_2) or triple (e.g., acetylene C_2H_2). Some chemical bonds are partly ionic and partly covalent in character. There are also cases of coordination compounds which do not fit the above simple models, and an example is the anion $[CCo_8(CO)_{18}]^{2-}$ in which the carbon atom is bonded to eight atoms of cobalt.

VALENCE BAND

Energy band of a *crystal* filled with valence electrons of the constituent atoms. In *insulators* and *semiconductors* the valence band in the ground state is completely filled with electrons, which are thus not capable of taking part in electron transport phenomena. However, thermal, electromagnetic, and other excitations, or the presence of lattice defects (acceptors) can create holes in the valence band, which provides for hole conduction. The overlapping of hole bands, observed in some semiconductors, can play an important role. In *metals* the valence band is full, and the high electrical conductivity arises from delocalized electrons in the *conduction band* which lies above the *band gap*.

VALENCE, INTERMEDIATE

See *Intermediate valence*.

VALENCE, VARIABLE

See *Variable-valence pair*.

VANADIUM, V

A chemical element of Group V of the periodic system of elements; with atomic number 23, atomic mass 50.9415. Natural vanadium consists of stable isotopes ^{51}V (99.75%) and ^{50}V (0.25%). Electronic configuration of external shells $3s^2 3p^6 3d^3 4s^2$. Successive ionization energies are, respectively, 6.740, 14.66, 29.32, 46.71, 65.2 eV. Atomic radius is 0.134 nm, radius of V^{2+} ion is 0.072 nm; of V^{3+} ion 0.067 nm; of V^{4+} ion 0.061 nm; of V^{5+} ion 0.04 nm. Oxidation state is +2 to +5. Electronegativity is ≈ 1.7. In a free form vanadium is a silvery-gray *metal*. It has a body-centered cubic lattice with parameter $a = 0.30282$ nm; density 6.11 g/cm^3 (at 20 °C); $T_{melting} = 2190$ K, $T_{boiling} = 3580$ K; coefficient of thermal conductivity is 30.99 W·m^{-1}·K^{-1} at 373 K and 36.84 W·m^{-1}·K^{-1} at 773 K; specific heat is 462.5 J·kg^{-1}·K^{-1} at 250 K; heat of melting is 345.4 kJ/kg; heat of evaporation is 9002 kJ/kg. Cross-section for thermal neutron capture is 4.5 barn/atom. Mechanical properties depend considerably on the presence of impurities, and depending on contents of additions it is possible to get vanadium from brittle and hard to plastic and soft. Brinell hardness is 628 MPa. Ultimate strength is 118 MPa. Relative elongation depending on purity 17 to 45%, cross-section reduction 25 to 75–85%. Bend angle without destruction of pure vanadium is 180°. Elasticity modulus is 137 GPa, rigidity is 46.4 GPa, compressibility modulus is 542.7 GPa. Adiabatic elastic moduli of vanadium: $c_{11} = 230.98$; $c_{12} = 120.17$; $c_{44} = 43.76$ (GPa) at 300 K. Resistivity of vanadium of 99.9% purity is 0.226 µΩ·cm (299 K). Work function is 3.79 eV. Vanadium is a superconductor below the transition temperature of 5.3 K. Paramagnetic properties are characteristic for vanadium. *Vanadium alloys* are widely used. The A15 compound with gallium V$_3$Ga has the superconducting transition temperature of 14.5 K. Vanadium is used in nuclear power and electronic engineering.

VANADIUM ALLOYS

Alloys based on the element *vanadium*. They have low density, high *corrosion resistance* in liquid *alkali metals*, good technological effectiveness (processability, weldability), high low-temperature *plasticity*, small cross-section for the capture of thermal and fast (0.1–1.0 MeV) neutrons. According to specific strength in the range 500–1000 °C, vanadium alloys exceed *titanium alloys, iron alloys, nickel alloys, cobalt* alloys and *niobium* alloys. Vanadium alloys with *gallium* and *silicon* exhibit *superconductivity*. They are prospective materials for the first wall of thermonuclear reactors. Vanadium alloys with niobium and titanium are the basis for complex triple and multicomponent alloys. Vanadium alloys are obtained by induction or arc melting in vacuo, or in a protective atmosphere. After primary hot *pressing* or *forging* the majority of vanadium alloys can be subjected to rolling without heating.

VAN DER WAALS FORCES (J.D. van der Waals, 1873)

Forces of interaction between *atoms, ions, molecules*, aside from the basic ionic, covalent, and other bonding interactions. They arise from the mutual influence of interacting particles causing distortions of their electronic shells even when a direct overlap of electron wavefunctions is negligible. By carrying out a multipole expansion of the Coulomb interaction $\widehat{V}(r, r')$ for the electrons and nuclei (e.g., of two atoms without intrinsic dipole, quadrupole or higher moments) terms proportional to $1/R^3$, $1/R^4$, $1/R^6$, are obtained, where R is the distance between nuclei. Correction to the energy in the second order of perturbation theory has the form

$$\Delta E = \sum_i \frac{|V_{0i}|^2}{E_0 - E_i},$$

where E_0, E_i are the energies of the ground and excited states of two noninteracting atoms, and V_{0i} are matrix elements of the interaction $\widehat{V}$ between the states $0, i$. Thus the interaction energy of the atom is reduced to a sum of terms C/R^6, C'/R^8, C''/R^{12}. In some approximation for two identical atoms $C = (3/2)\alpha^2 I$, where $\alpha = \alpha_1 = \alpha_2$ is the polarizability of the atoms, and I is their ionization potential.

VAN HOVE SINGULARITIES (L. Van Hove, 1953)

Singularities in the *density of states* $g(E)$ of solids, associated with infinities of the derivative $dg(E)/dE$ at certain values of energy $E = E_n$. For solids

$$g(E) = \sum_j \int_{\omega_j(p)=E} \frac{dS}{|v_j(p)|},$$

where $\omega_j(p)$ is the energy of a *quasi-particle* of jth type with *quasi-momentum* p; and $v_j(p) = \partial \omega_j(p)/\partial p$ is the group velocity (assume $\hbar = 1$). The integration is performed over an *isoenergetic surface* $\omega_j(p) = E$ in momentum space, with dS being its differential area element inside the *Brillouin zone*. The values $E_n = \omega_j(p_n)$ correspond to such p_n at which $v(p_n)$ is equal to zero. The existence of Van Hove singularities is connected with the fact that $\omega(p)$ in solids is a periodic function of p. These singularities occur at points of change in topology of an isoenergy surface $\omega_i(p) = E$ (transition from a closed to an open surface with a change of E, etc.). At E values close to E_n the density of states $g(E)$ is described by the expression

$$g(E) - g(E_n) \sim (m_1 m_2 m_3)^{1/2} |E - E_n|^{1/2},$$

where m_1, m_2, m_3 are the diagonal components of the *effective mass* tensor of the quasi-particle.

The significance of Van Hove singularities is determined by the role of the density of states $g(E)$ in the thermodynamic and kinetic properties of solids. These singularities influence the scattering *cross-sections* of *neutrons* and *gamma rays* in solids, the *light absorption* and ultrasound absorption, and the formation of bound states of quasiparticles. Van Hove singularities of the electronic *density of states* determine several phenomena in solids, such as anomalies of the *specific heat* and *compressibility* at the transition of order $2\frac{1}{2}$ (*Lifshits transition*) in a *metal* in the normal state, nonlinear dependence of the temperature of the superconducting transition (see *Superconductivity*), and of the dependence of the critical magnetic fields of a superconductor (see *Meissner effect*) on the pressure and the concentration of impurities. Van Hove singularities can be enhanced if one or more of the effective masses becomes infinite simultaneously with $v(p_n) = 0$.

VAN VLECK MECHANISM, Kronig–Van Vleck mechanism (J.H. Van Vleck, R. Kronig, 1939)

One of the effective mechanisms for *spin–lattice interactions* of *magnetic ions* in condensed matter.

The *Van Vleck mechanism* arises from modulation of the crystalline electric fields (see *Crystal field*) at the magnetic ion sites due to lattice vibrations. The fluctuating electric field influences the spins due to the *spin–orbit interaction*, and induces transitions between the spin states of the magnetic ions. In the limit of low concentration the spin–phonon interaction constants, which serve as measures of the efficiency of the Van Vleck mechanism, do not depend on the concentration. For ions of the iron group typical values of these spin–phonon constants are in the range from 0.1 to 10 cm^{-1}. The Van Vleck mechanism is used for the quantitative description of *spin–lattice relaxation* of paramagnetic ions in nonmetallic crystals, of *acoustic paramagnetic resonance*, and of pressure effects in the spectra of *electron paramagnetic resonance*.

VAN VLECK PARAMAGNETISM (J.H. Van Vleck, 1932)

Polarizational paramagnetism, magnetism of atoms (ions) arising from the magnetic polarization of their excited states by an applied magnetic field. Van Vleck paramagnetism is commonly observed in ions which have no *magnetic moments* in their ground states, and materials of this type are called Van Vleck paramagnets. Unlike the usual orientational spin *paramagnetism*, the Van Vleck type is practically independent of the temperature. A theoretical expression for the *magnetic susceptibility* $\Delta\chi$ associated with this paramagnetism has the form

$$\Delta\chi \approx \sum{}' \frac{|(0|\widehat{M}_z|n)|^2}{(E_n - E_0)},$$

where the summation is performed over the excited states n, E_n is the nth energy level, $\widehat{M}_z$ is the operator of the magnetic moment projection, and $n = 0$ is the ground state term. Materials containing atoms of *rare-earth elements* with $(0|\widehat{M}_z|0) = 0$ are particular examples of Van Vleck paramagnets.

VAPORIZATION

The *phase transition* from a *condensed state of matter* (either liquid or solid) to the gaseous state. Vaporization takes place as a *first-order phase transition*. Vaporization is called *evaporation*, if it occurs at the free surface of a liquid, whereas vaporization, which takes place at a *solid surface*, is called *sublimation*; vaporization in the bulk liquid is called *boiling*.

VARACTOR, also called varactor diode
(B.V. Wool, 1954)

A *semiconductor device* with a capacitance which depends on the applied voltage, that can be used in *parametric semiconductor amplifiers* of extra-high frequency oscillations. These parametric amplifiers exhibit a low noise level at the condition $f_0 \ll f$, where f_0 is the operating frequency, and f is the upper bound of the frequency region, within which the varactor exhibits *negative resistance*. Modern varactors produced from *gallium arsenide* exhibit $f \sim 10^3$ GHz, which provides parametric semiconductor amplifiers of *noise temperature* $T_N = 30$–50 K without cooling, and $T_N = 10$–30 K when cooling the *diodes* with liquefied nitrogen over extra- and ultrahigh frequency wave bands.

VARIABLE-VALENCE PAIR

A pair of charged *defects* in chalcogenide glasses (see *Chalcogenide materials*) which are produced by a change of the coordination number of the chalcogen atom due to the use of *unshared electron pairs* for the generation of new bonds.

VARIANCE (fr. Lat. *vario*, am changing)

Number of degrees of freedom of a *thermodynamic system*, i.e. the number of independent system parameters (pressure, temperature, concentration of components, etc.) which can be changed (varied) within some limits without changing the phase equilibrium (see *Gibbs' phase rule*).

Another definition of variance used in mathematical statistics is the expression

$$\sum_i \frac{(y_i - y_m)^2}{n}$$

taken in the limit $n \to \infty$, where y_m is the mean value of y_i.

VARIBAND SEMICONDUCTORS

Crystals whose energy *band structure* parameters (*band gap*, inter-valley energy distances in the *conduction band*, of spin–orbit splitting in the *valence band*), *effective masses* of carriers and the *dielectric constant* vary smoothly along one or more directions within the crystal. Gradients of applied pressure, smoothly varying internal mechanical stresses, temperature gradients, and positional dependent *solid solution* composition can be the causes of the change. Variband semiconductor are sometimes called *variband structures* or smooth *heterojunctions*.

The structures GaAs–Al$_x$Ga$_{1-x}$As with gradually changing x and therefore with a gradient of the *electron affinity* and of the band gap E_g are typical examples of variband semiconductors. Due to the presence of a built-in quasi-electric field in these regions the drift distance of nonequilibrium carriers can exceed the diffusion length L. This difference is especially pronounced for high radiation *recombination* efficiency when the contribution of *photon drift* is added to the transport of carriers. Variband structures have been utilized in *optoelectronics*, and a variband layer has been inserted between the emitter and active area in double *heterostructures* of *light emitting diodes*. Photoelectric transformers based on variband structures have enhanced values of the collection coefficient and of efficiency due to reduction of surface (see *Surface recombination*) and volume recombination losses. Photodiode structures with an illuminated variband layer of width $d < L$ where the value of E_g changes from E_g^{max} to E_g^{min} function as wide-band receivers sensitive in the spectral range from E_g^{max} to E_g^{min}. At $d \gg L$ the selective photoreceivers are obtained. *Avalanche photodiodes* with a smoothly varying multiplication coefficient, injection *photodiodes* with a controllable area of spectral photosensibility, *magnetodiodes* with enhanced voltage sensitivity are also examples of variband structure applications.

Since the positions of different maxima in the conduction band with respect to the top of the valence band can change with the composition, there can be a change of the absolute energy minima in the conduction band at some point within the crystal. So, in $Al_x Ga_{1-x} As$ at $x < 0.4$ the minimum energy corresponds to a Γ-valley, whereas at $x \geqslant 0.4$ the X-minimum is absolute (so-called $\Gamma - X$ *junction*). In other words, the direct-gap type band structure is replaced by an indirect-gap variety, and this introduces a coordinate dependence of the absorption coefficient, of edge *luminescence* intensity, and other effects. In the area close to a $\Gamma - X$ transition the redistribution of carriers between Γ- and X-valleys occurs. The other important effect of the spatial dependence of a conduction band extremum is the variation with the coordinate of the depths of impurity levels associated with different valleys, relative to the band edge.

In variband semiconductors there exist built in *quasi-electric and quasi-magnetic fields*. They are realized under the following conditions: the position of a dispersion curve minimum depends on the coordinate, for example, it shifts from the center of the *Brillouin zone* to its boundaries upon changing the composition of the solid solution, and *isoenergetic surfaces* are ellipsoids of rotation. The built-in quasi-electrical fields cause the diffusion-drift character of the motion of the unbalanced carrier in variband semiconductors, which brings about the change of efficiency of the

drift length, and also of the surface recombination rate.

This establishes the advantages of these materials in band engineering and optoelectronic devices. Variband semiconductors are also favorable for highly-sensitive *strain gauges* and photoreceivers with internal amplification (see *Optical radiation detectors*).

VARICAP (fr. variable and capacity)

Semiconductor device utilizing non-linear properties of the *barrier capacitance* of a diode. Upon application to the *p–n* transition layer of a direct bias voltage to an *semiconductor junction* ($n-p$ junction with the positive potential on the side of the *p*-region) the width of the bulk charge layer is reduced and its capacity increases. Upon application of a reverse bias the capacity of the junction decreases. Varicaps are widely used as variable capacitors (in the range 10^{-2} to 10^2 pF) and as nonlinear capacitors (see *Varactor*).

VARICOND (fr. variable and condenser)

Ferroelectric capacitor (see *Ferroelectric ceramics*) with a pronounced non-linear dependence of the capacity on the intensity of the applied electric field resulting from the non-linearity of the *dielectric constant* (solid solutions of Ba (Ti, Zr, Sn) and (Ba, Sr)TiO_3 are commonly used) in the vicinity of the *Curie point* in both the ferroelectric as well as the *paraelectric phases*. Variconds are manufactured in bulk and film form, they have a high specific capacitance ($10-10^3$ pF/mm^3), small dimensions ($1-10$ mm^3) and control voltages, and a wide range of operational frequencies. Nominal values of the capacitance of variconds lie in the range from several pF to several μF and change by a factor of $1.5-20$ (depending on frequency and temperature) for a control voltage of several tens of volts. Variconds are characterized by high mechanical strength, stability to vibrations, moisture resistance radiation stability, and almost unlimited operational life. But time and temperature instability of their capacity and unacceptable *dielectric losses* restrict their practical applications. Variconds are used in automatics, electronics, UHF engineering (including hybrid integral circuits) – for contactless remote control, for parametric power amplification, also for current and voltage stabilization, frequency phase, and signal transformation, etc.

There is a need for developing new ferroelectric ceramics and films which combine high non-linearity with low dielectric losses over a wide frequency range, together with stabile properties.

VARISTOR (fr. variable and resistor)

A two electrode semiconductor *resistor*, whose electrical resistance is a non-linear function that drops with an increase of the applied voltage. A varistor is a *semiconductor* in the form of a powder (usually *silicon carbide*) pressed into a resin or a liquid glass. The contact resistance between grains is determined by phenomena at the random network of *semiconductor junctions* being formed.

VAVILOV LAWS (S.I. Vavilov, 1927)

The laws which establish the relationships between the luminescence energy yield (see *luminescence yield*), and the mutual spectral positions of the exciting monochromatic light and the emitted *luminescence* light. The first law states that the absolute value of the *photoluminescence* energy yield can not exceed one unit. The second law asserts that for excitation of the Stokes' type (see *Stokes' rule*) the photoluminescence energy yield is proportional to the wavelength (*quantum yield* is constant), it saturates at the approach to the region where the exciting and irradiated spectra overlap (see *Anti-Stokes luminescence*), and then sharply decreases beyond this point. The lower the temperature, the sharper the decrease in yield.

VEGARD RULE (L. Vegard, 1928)

Relationship between the composition of a *solid solution* and the dimensions of its crystallographic unit cell. In the first approximation the dimensions of the *unit cell* (or the interatomic distances) vary according the linear law $a = xa_1 + (1 - x)a_2$, where x is the mole fraction of the solvent, and a_1 and a_2, respectively, are the dimensions of the unit cell of the solvent and the dissolved matter. The measured parameters of the unit cell often differ from those calculated according the Vegard rule, and ordinarily they are smaller than the calculated ones. This has been attributed to the chemical affinity of the solid solution components, their elasticity parameters, and their thermodynamic properties. In spite of this, the Vegard rule helps to establish the concentration limits over which solid solutions exist, and the change in solubility with reduction in the temperature.

VELOCITY LIMIT OF DOMAIN WALL

See *Domain wall velocity limit*.

VENER SPOTS

Appearance on a screen of spots formed by particles emitted from the *sputtered* surface of a monocrystal. They result from the increased yield of *atoms* (ions) which under *ion bombardment* escape from the crystal in directions with low Miller indices. The appearance of Vener spots is associated with the propagation of *focusons*. In addition, due to the energy losses, the crystal surface is reached mostly by low-energy atoms (ions) that can either recoil in the direction along a channel (e.g., $\langle 100 \rangle$ in a cubic crystal), or knock out an atom from the surface layer of the crystal by a head-on collision (e.g., in the $\langle 111 \rangle$ direction), which also can induce the formation of Vener spots.

VERDET CONSTANT (M. Verdet), specific magnetic rotation

A coefficient V characterizing the magnitude of the angle of rotation θ of the plane of polarization of linearly polarized light propagating a distance d through a medium parallel to the direction of an applied magnetic field B. The Faraday effect equation is $\theta = VBd$. For non-magnetic transparent *crystals* V has a value from 10^{-2} to 10^3 rad$\cdot$T$^{-1}\cdot$m^{-1}, and depends on the temperature, on the *current carrier* concentration and on the radiation wavelength. For low magnetic fields $B < 10$ T as a rule it does not depend on B.

VERTEX DISLOCATIONS

A version of a *partial dislocation* in a face-centered cubic structure. Vertex dislocations are formed as a result of *dislocation* splitting, accompanied by the transition of one of the partial dislocations to the adjacent *slip plane*, or as a result of reactions between the split dislocations moving in the intersecting slipping planes. Since the *Burgers vector* of a vertex dislocation does not lie in a slipping plane, it is immobile and plays an important role as a barrier for the motion of other dislocations.

VERTICAL TRANSITIONS

See *Direct transitions*.

VESICULAR MATERIALS

See *Nonsilver photography*.

VIBRATIONAL SPECTRA of solid solutions

These spectra characterize the dynamics of *crystals* that have a configurational disordering in one or more sublattices. The vibrational spectra appear in the form of narrow bands with about the same width as those in *ideal crystals* (although frequency shifts with concentration in *solid solutions* are not small). This permits one to describe the vibrational spectra of solid solutions in terms of a small number of effective oscillators, the latter transforming in limiting cases into the long-wave optical vibrations of the corresponding perfect crystals.

The nature of the optical lattice spectrum of a solid solution depends upon the degree of difference in masses and in *force constants*. If these quantities are not appreciably different then the response of a solid solution to an external electromagnetic field corresponds to that of a crystal with the parameters averaged over its composition (*virtual crystal*). The optical spectrum has the same number of active bands as the spectrum of each of its constituent components. When the relative concentrations change the bands shift from the position of one component to that of the other. Exemplifying this behavior are the *phonon spectra* of AB_xC_{1-x} solid solutions that can be fairly well described in terms of a single effective oscillator (*single-mode spectral behavior*). If the parameters differ substantially then the spectrum consists of several bands, each of them corresponding to one of the constituent components, and having an intensity defined by the component content in the solid solution (*multi-mode spectral behavior*). When the content of a corresponding component decreases sufficiently then its band turns into the band of *local vibrations* of an impurity atom.

The following methods are employed to compute the vibrational spectra of solid solutions: virtual crystal model, *cluster approximation*, *coherent potential method*, averaged *t*-matrix method, etc.

VIBRATIONS

Repetitive changes with time, t, of some quantity, for instance, a coordinate and velocity (e.g., vibrations of atoms in solids, including those giving rise to *sound*, mechanical *vibrations of rods and plates*, oscillations of electric and magnetic field intensities, electromagnetic oscillations, etc.). An important class is periodic oscillations, for which the changes of magnitude of a variable u recur over a particular time interval T which is termed the *oscillation period*: $u(t + T) = u(t)$. The reciprocal quantity $v = 1/T$ is called the *frequency* and $\omega = 2\pi v$ is called the *angular frequency*. The *oscillation amplitude* is its maximum magnitude, and its *phase* is the value of the argument of a periodically changing function counted with respect to an arbitrarily chosen origin of coordinates. Of particular importance are *harmonic oscillations* described by a sinusoidal function:

$$u(t) = A \sin(\omega t + \varphi_0), \qquad (1)$$

where A is the amplitude and φ_0 its phase at the initial time $t = 0$. Any variations of the quantity u, including random ones, within an arbitrary finite time interval t, can be represented by a superposition of harmonic oscillations with different ω, φ_0 and A values. Two oscillations having closely similar frequencies and phases are termed *coherent oscillations*. Actual oscillations may not conform to Eq. (1), but sometimes can be approximated by this equation, with φ_0 slowly changing with time (changes during the time T should be small compared to π). For longer times $\tau \gg T$, when the change of φ_0 becomes of the order of π, the coherence of the oscillation becomes disrupted. The particular time interval τ_0, when φ_0 approaches or reaches π, is termed the *coherence time*.

The rate of vibration damping in the presence of *friction* is determined by the logarithmic *decrement* that equals the natural logarithm of the ratio of two successive maxima of the oscillating quantity. If the system vibrations occur with what is termed *negative friction* (which is possible in the presence of external factors, e.g., when the system parameters are oscillating), the decrement is negative, with its absolute value called an *increment*. Oscillations can be excited in systems under the action of external forces (termed *forced vibrations*), or due to system parameters which change periodically (*parametric vibrations*), and vibrations may also occur in autonomous systems (with no outside agents acting). In conservative systems periodic vibrations are only possible if one neglects small energy transfers. In non-conservative systems, called *dissipative systems*, in which dissipative forces of friction are acting, the amplitude of the vibrations is damped, and the only possible stationary state is the state of rest. *Self-oscillations* are possible in non-conservative nonlinear systems. The nature of possible vibrations of a system depends upon the individual system parameters (masses, coordinates, and forces of interaction in a system composed of atomic particles; density, *elastic modulus*, shape and size of an elastic medium; *inductance*, capacitance, and resistance of an electric circuit, etc.). A certain set of *free (intrinsic) vibrations* of a system exists called *eigenfrequencies* ("*natural*" *frequencies*); the system performs vibrations at such frequencies in the absence of external forces, provided that it receives an initial impulse. For instance, free vibrations of a string with fixed ends are such that an integer or half-integer number of wavelengths fits into the length of the string, and all frequencies of free vibrations are multiples of the minimum one termed the *fundamental frequency*. Free oscillations in linear systems are called *normal modes* (see *Normal vibrations*). If the frequency of an external force coincides with a system eigenfrequency then a pronounced amplification of forced vibrations occurs (*resonance*). The free vibrations of bound systems are called *bound vibrations* (e.g., interacting systems possessing a single degree of freedom each).

VIBRATIONS, INTRAMOLECULAR

See *Intramolecular vibrations*.

VIBRATIONS, LOCAL AND QUASI-LOCAL

See *Local vibrations, Quasi-local vibrations*.

VIBRATIONS, NORMAL

See *Normal vibrations*.

VIBRATIONS OF CRYSTAL LATTICE

See *Crystal lattice vibrations, Acoustic vibrations, Anharmonic vibrations, Antisymmetric vibrations, Bending vibrations, Optical vibrations, Silent lattice vibrations, Totally symmetric vibrations, Zero-point vibrations*.

VIBRATIONS OF RODS AND PLATES

The main types of free *vibrations* that can exist in *rods* are: *stretching vibrations* (along the rod axis), *bending vibrations* (perpendicular to the axis), and *torsional vibrations* (around the axis). In all these cases, an integral number of half-wavelengths fits into the length, l, of a rod with either fixed or free ends: $l = n\lambda/2$, where $n = 1, 2, \ldots$. The *natural frequencies of vibration*, ω_n, form a discrete spectrum. For stretching vibrations of small amplitude, $\omega_n = (\pi n/l)(E/\rho)^{1/2}$, where E is *Young's modulus*, and ρ is the *density*. For larger n, dispersion corrections should be taken into account; thus, for a round rod of radius $R \ll l$, a factor $1 - (I\pi\nu R/\lambda)^2$ (ν is *Poisson ratio*) should be added to the expression for ω_n. For torsional vibrations, $\omega_n = (\pi n/l)[D/(\rho I)]^{1/2}$ (D is the rod's *torsional rigidity*), where I is the *moment of inertia* of the cross-section with respect to the *center of mass*; in particular, $\omega_n = (\pi n/l)(\mu/\rho)^{1/2}$ for a round rod (μ is the *shear modulus*). In the case of bending vibrations, the frequencies ω_n do not make a harmonic series because of strong *dispersion* (the velocity of bending waves is proportional to $\omega^{1/2}$), but rather depend non-linearly on n: $\omega_n = (A_n/l^2)[EI_0/(\rho S)]^{1/2}$ ($A_1 = 133$, $A_2 = 339, \ldots$; S is the cross-section area, and I_0 is the moment of inertia of the cross-section with respect to the neutral axis lying in the cross-sectional plane, see *Flexural waves*).

In thin *plates* (films), both longitudinal extension–contraction vibrations and transverse bending vibrations can occur, with displacements parallel and perpendicular to the plane of the plate, respectively. The vibrational spectrum and the distribution of displacements depend explicitly on the shape of the plate, and on the boundary conditions. For instance, for a square plate having hinge-supported edges, the eigenfrequencies of longitudinal vibrations oriented along one of the plate sides will be: $\omega_n = (\pi n/l)[E/\rho(1 - \nu^2)]^{1/2}$, and the frequencies of transverse vibrations will be: $\omega_{mn} = (m^2 + n^2)(\pi^2 h/S)\{E/[12\rho(1 - \nu^2)]\}^{1/2}$, $m, n = 0, 1, 2, \ldots$. Corresponding to each eigenfrequency there is a particular configuration of *nodal lines* along which the displacements vanish.

VIBRATIONS OF SURFACE ATOMS

Vibrations which differ from those of atoms within the bulk *crystal* by their amplitudes (root-mean-square displacements), *Debye temperature*, and frequency spectrum. For instance, the ratio of the root-mean-square displacements of atoms in the first two surface layers to the corresponding values in the crystal bulk, as determined by *low-energy electron diffraction*, amounts to the following values for several crystals: Nd, 2.65; Pb, 2.43; Bi, 2.42; Ag, 2.16; Pt, 2.12; Pd, 1.95; Cr, 1.8; Ni, 1.77; Ir, 1.63; V, 1.52; Rh, 1.35. Similar values exist for the ratio of the Debye temperature in the bulk of a crystal to its value at the surface. The phenomenon of increased amplitudes of surface atom vibrations was discovered by V.E. Lashkarev and G.A. Kuzmin (1934), and by S.G. Kalashnikov and O.I. Zamsha (1939). Enhanced amplitudes of surface atom vibrations were also found for simple chemical compounds. The initially conjectured "softening" of the frequencies of surface vibrations was not always confirmed by subsequent investigations. For instance, the force constant for the Ni (001) surface plane equals or exceeds by 20% that in the bulk, according to a number of reports. Changes of the energy of surface atom vibrations have an important bearing on some technological processes such as *sintering* and crystal *melting*.

VIBRONIC ANHARMONICITY

Terms of higher order than quadratic in the potential for nuclear displacements in a vibrating solid lattice. The usual *adiabatic potential* contains only quadratic terms that result in harmonic linear displacements of the atoms through the *vibronic interaction*. Vibronic anharmonicity contributes to thermal expansion, coupling between different phonons, thermal resistivity, and other effects.

VIBRONIC EFFECTS IN OPTICS

Effects in the spectroscopy of the optical region which occur as a result of *vibronic interactions*.

Due to the complex spectrum of *vibronic states* the observed optical spectra of electronic transitions which arise from electronically degenerate or pseudo-degenerate terms contribute to the *fine structure* and the lineshape of broad bands (envelope of the elementary transitions). For example, for an $A \rightarrow E$ transition which produces two peaks, the distance between them increases with increasing *Jahn–Teller effect* stabilization energy, and the depth of the dip decreases with increasing contribution of fully-symmetrical oscillations. For other types of transition the line shape is more complex; but it never coincides with the single-peak curve of a transition between nondegenerate terms. The fine structure repeats the spectrum of the vibronic state. It is possible to observe the splitting of *phononless lines* which are due to *inversion splitting*, and also the appearance of local *resonances* and pseudo-resonances for the impurity states in crystals as a result of a polymode Jahn–Teller effect. With the help of the moment method for investigating optical spectra influenced by external effects it is possible to determine the vibronic constants and other parameters of the system. Qualitatively new optical properties occur since electronic degeneracy and *pseudodegeneracy* can cause the polyatomic system electron distribution to be lower in symmetry than that of the nuclear frame. Highly symmetrical systems can acquire an abnormal anisotropy of polarizability and of hyper-polarizabilities which manifest themselves through abnormal depolarization of the *Rayleigh scattering of light* and *hyper-Rayleigh scattering* of light, through *birefringence* in external fields (*anomalous Kerr effect* and *anomalous Cotton–Mouton effect*), and through purely rotational absorption and *Raman scattering of light* by molecules.

VIBRONIC EFFECTS IN RADIO-FREQUENCY SPECTROSCOPY

Effects of the *vibronic interaction* found in radio-frequency spectroscopy which are especially important in *electron paramagnetic resonance* (EPR).

In the case of weak vibronic interactions (small vibronic constants) when the *inversion splitting* δ is high, an EPR at sufficiently low temperature corresponds to the ground *vibronic state* (dynamic spectrum) for which the orbital part of the g-factor, g_2 is multiplied by the vibronic reduction factor. In the case of the $E-E$ problem of the *Jahn–Teller effect* (ground vibronic *doublet* ^2E) in cubic systems $g_{\pm} = g_s + g_2 \pm q g_2 f$, where g_s is the

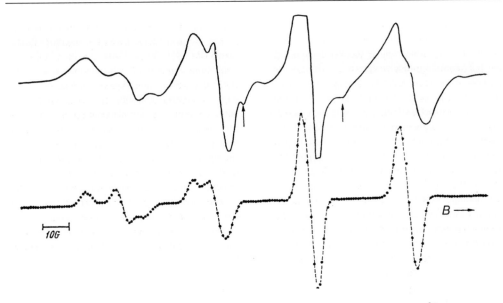

Vibronic effects in radio-frequency spectroscopy, showing the comparison of an experimental CaO:Cu^{2+} EPR spectrum (above) with a calculated spectrum (below).

spin part, $f = [1 - 3(l^2m^2 + l^2n^2 + m^2n^2)]^{1/2}$ (l, m, n are the direction cosines of the applied magnetic field $\boldsymbol{B}$) and the vibronic reduction factor q varies from 1 (neglecting vibronic interactions) to 1/2 (limit of strong vibronic interactions). In the limiting case $g_2 \beta B \gg \delta$ (β is Bohr magneton) the magnetic field mixes the ground doublet with a vibronic singlet and shifts the system into one of its minima so that EPR corresponds to the Jahn–Teller static deformation $g_\| = g_s + 2g_2$; $g_\perp = g_s + g_2/2$. Additional splittings of the vibronic doublet under the influence of random crystalline lattice distortions $\overline{\Delta}$ together with the effects of spin relaxation make the EPR spectrum more complex. As a result, at low temperature for $g_2 \beta B/\delta < 0.1$ and $\overline{\Delta}/\delta < 0.1$ a dynamic spectrum is observed. At $g_2 \beta B/\delta \geqslant 5$ or $\overline{\Delta}/\delta \geqslant 5$ the spectrum becomes static. In the intermediate range ($0.1 < g_2 \beta B/\delta \leqslant 5$ or $0.1 < \overline{\Delta}/\delta \leqslant 5$) one of the dynamic spectral peaks is broadened and displaced, the angular dependence acquires a special character, and the hyperfine structure for an arbitrary direction of $\boldsymbol{B}$ is very complicated. With increasing temperature the increase in δ is accompanied by a transition from a static to a dynamic

spectrum, and the occupation of the excited orbital singlet plus relaxation gradually reduce the whole spectrum to a single isotropic line. The figure compares the experimental EPR spectrum (above) of the CaO:Cu^{2+} system for $\boldsymbol{B} \parallel [111]$ (arrows indicate forbidden quadrupole lines) measured at $T = 1.3$ K and $v = 8.7$ GHz with a calculated spectrum (below) which takes into account vibronic effects with $\overline{\Delta}/\delta = 0.67$ and $q = 0.5$. Transitions between the lowest vibronic states lead to the absorption of microwaves.

VIBRONIC INSTABILITY

Absence of an *adiabatic potential* minimum of a polyatomic system with a highly-symmetrical configuration Q_0 of atomic nuclei in the direction of symmetrized displacements Q, arising from appreciable mixing of one electronic state with another under the influence of *vibronic interactions* (see *Jahn–Teller effect*, *Renner effect*). For a non-degenerate ground state a maximum of the adiabatic potential along the Q direction can appear only as a result of vibronic mixing with excited electronic states. This can be a source of dynamic instability of polyatomic systems, and predicts the existence of stable excited states. Vibron instability in the direction of symmetrized nuclear displacements Q along which the highly symmetric polyatomic system is deformed with the formation of *dipole moments* is commonly called a *dipole instability*. Usually this instability brings mutually compensated dipole moments to several equivalent minima of the adiabatic potential. The observed properties of such a system depend on how its parameters are related to the dipole moments averaged over all the minima. In gases and liquids the dipolar instability can be displayed in the anomalies of the temperature dependence of the polarizability, in the purely rotational radiation absorption: in *Rayleigh scattering of light* and *hyper-Rayleigh scattering* of light, in the *Kerr effect* and the *Cotton–Mouton effect*; in intermolecular interactions. The ordering of dipole-unstable centers can bring about in a crystal *spontaneous polarization* and *ferroelectricity* (see *Vibronic theory of ferroelectricity*).

VIBRONIC INTERACTIONS

A part of the electron–nuclear interaction $V(r, Q)$ in the Hamiltonian of a polyatomic system, corresponding to the displacements of nuclei Q from their initial positions $Q = Q_0$, most often from a point of high symmetry (r are the coordinates of the electrons, Q are the symmetrized nuclear coordinates). If V is expanded in a series with respect to small displacements $Q - Q_0$:

$$V(r, Q) = V(r, Q_0) + \sum_{\alpha} \left(\frac{\partial V}{\partial Q_\alpha} \right)_0 (Q_\alpha - A_{\alpha 0})$$

$$+ \frac{1}{2} \sum_{\alpha, \beta} \left(\frac{\partial^2 V}{\partial Q_\alpha \partial Q_\beta} \right)_0$$

$$\times (Q_\alpha - A_{\alpha 0})(Q_\beta - Q_{\beta 0}) + \cdots,$$

then the expansion terms, except for the zero-order one, represent linear, quadratic, etc., vibronic interactions. When the electronic energy (obtained from zero-order term $V(r, Q)$) is degenerate then the influence of the these interactions is especially important, and it introduces some new effects. Similar effects appear in the absence of degeneracy when the mixing of electronic states of different energies is sufficiently pronounced. The *vibronic constant*, introduced as a vibronic

interaction parameter, designates the electronic matrix element (diagonal or off-diagonal) coefficients $(\partial V / \partial Q_\alpha)_0$, $(\partial^2 V / \partial Q_\alpha^2)_0$, etc., in the corresponding terms of the vibronic interaction (similarly diagonal or non-diagonal, linear, quadratic, etc.). A non-diagonal vibronic constant is a measure of the mixing of electronic states of the corresponding interactions. The diagonal linear vibronic constant $(\partial V_{ii} / \partial Q_\alpha)_{Q_\alpha = Q_{\alpha 0}}$ is equal to the force with which the electrons at the ith electronic state deform the nuclear configuration under consideration $Q_{\alpha 0}$ to the direction of the symmetrized displacement Q_α. From the diagonal matrix elements it is possible, with the help of *group theory*, to select the force constant, as well as the vibronic constant corresponding to the symmetry. The quantum theory of polyatomic systems in which (unlike the simple *adiabatic approximation*) both diagonal and off-diagonal matrix elements are taken into account, is called vibronic interaction theory. In the case of electronic degeneracy, or *pseudodegeneracy*, the vibronic interaction theory displays a range of new effects and regularities in the physics and chemistry of molecules and crystals (see *Jahn–Teller effect*, *Vibronic theory of ferroelectricity*, *Vibronic effects in optics* and *Vibrionic effects in radio-frequency spectroscopy*, *inversion splitting*, etc.).

VIBRONIC REDUCTION, Ham–Bersuker effect
(I.B. Bersuker, F.S. Ham)

Effect of reduction (suppression) by the *vibronic interaction* of low-symmetry electronic excitations, which remove the degeneracy of a Jahn–Teller ground state (see *Jahn–Teller effect*), compared to level splitting in the absence of vibronic coupling. This is described by the so-called *vibronic reduction factors* which are coefficients of matrix elements of electron operators acting on degenerate electronic states. These reduction factors show to what extent external perturbations of the electronic ground state become less effective when vibrations are taken into account. These reduction factors are determined by the parameters of the Jahn–Teller effect, are constant for a given system, are independent of the nature of a disturbance, and depend only on its symmetry. Vibronic reduction factors of first- and second-order perturbation theory are distinguished.

VIBRONIC STATES

Discrete *electronic-vibrational states* of a polyatomic system defined by taking into account the mixing of electronic states of the same degenerate term, or of other terms, with the help of the *vibronic interaction*.

In order to determine the vibronic states, instead of confining ones attention to one oscillation level of the *adiabatic approximation*, it is necessary to solve a system of coupled equations (depending on the number of mixed electronic states). It is typical for this system of equations that basic vibronic states are obtained which have the same symmetry and multiplicity as the source electronic state, i.e. for the electronic E-term the vibronic state should be a doublet, for the T-state a vibronic triplet, etc. Wave functions of the vibronic states are called *vibronic functions*. If f electronic states contribute and the wavefunctions φ_i are extensively mixed then the vibronic function $\Psi(r, Q)$ has the form of a linear combination

$$\Psi(r, Q) = \sum_{i=1}^{f} \varphi_i(r, Q) \chi_i(Q),$$

where the coefficients $\chi_i(Q)$ are functions of nuclear coordinates determined from the solution of the system of f coupled equations. In the absence of mixing $f = 1$ and $\Psi(r, Q) = \varphi(r, Q)\chi(Q)$, a result conforming to the adiabatic approximation, with $\chi(Q)$ here playing the role of the vibrational wave function. The spectroscopic properties of a polyatomic system arising from vibronic states are called *vibronic spectra* (see *Vibronic effects in optics* and *Vibronic effects in radio-frequency spectroscopy*).

VIBRONIC THEORY OF FERROELECTRICITY

Microscopic theory of the origin of spontaneous polarization (see *Ferroelectricity*) and ferroelectric (and other) *structural phase transitions*, based at the assumption of the dipolar instability (*vibronic instability*) of the corresponding crystal centers and their interaction (see *Jahn–Teller effect*). In various versions the spontaneous polarization appears as a result of the vibronic instability of the lattice with respect to the corresponding longitudinal optical oscillations. The wave vector

dependence of the frequency ω_q of these oscillations is

$$\omega_q^2 = \tilde{\omega}_q^2 [1 + \Pi_q(\omega_q)],$$

where $\tilde{\omega}_q$ is the frequency without taking account of the *vibronic interaction* (i.e. of the *electron–phonon interaction*), and Π_q is the polarization operator in the Dyson equation for the phonon *Green's function* $D = D_0 + D_0 \Pi D$. When calculated by taking into account the mixing of the states of the *valence band* and *conduction band* by the vibronic interaction, the values of $\Pi_q(\omega_q)$ decrease with changing temperature (*soft mode*). For some q at low temperatures $1 + \Pi_q < 0$, so ω_q^2 can be negative, and the lattice loses its stability with respect to the corresponding nuclear displacements so the structural phase transition occurs. It has been shown that $\tilde{\omega}_q^2 > 0$, so that zero and negative values of ω_q^2 can occur due to the vibronic interaction. The latter is the origin of structural phase transitions in crystals. These same vibronic interactions in higher orders (see *Vibronic anharmonicity*) lead to the positivity of ω_q^2 at higher temperatures, and also provide stability to the the paraelectric phase. In the language of the *adiabatic potential*, e.g., for BaTiO$_3$ taking into account the pseudo-Jahn–Teller effect in the sublattice of Ti^{4+} ions, it is possible to explain the origin of the the observed *ferroelectric phase transitions*, to evaluate the temperatures of the transitions, and also the magnitude of the polarization. The theory predicts that only the low-temperature rombohedral phase should be completely ordered, the orthorhombic and tetragonal phases should be disordered in one or two directions, respectively, and the paraelectric phase should be completely disordered. These conclusions are confirmed by precise experiments of electron paramagnetic resonance at impurities, and by *diffuse scattering of X-rays*. The theory provides descriptions of some additional features of *ferroelectrics*.

VICALLOY

High coercivity alloy based on the Fe–Co–V system. The high coercive state is achieved after *quenching* and *tempering* vicalloy-1 (10% of V), and also after hardening, cold deformation and tempering vicalloy-2 (12–14% of V). Before tempering the alloys are deformable, afterwards they possess high *hardness*. Magnetic properties can be improved by *thermomechanical treatment* (these properties are given in Table 1 of *Hard magnetic materials*).

VICINAL SURFACES (fr. Lat. *vicinalis*, adjacent, near)

Surfaces that are close in their orientation to faces with low *crystallographic indices*. Vicinal surfaces retain their gently sloping sections above the main face formed during the growth of crystals (see *Crystallization*) and they are called *vicinals*. Different forms of vicinals correspond to various faces which conform to *crystal symmetry* directions. There are two points of view concerning the nature of vicinal surfaces: (1) they are planes with high indices forming small angles (up to several degrees) with low-index faces; (2) they are faces with low indices which have deviated from their ideal position in the crystalline lattice. Especially prepared atomically pure vicinal surfaces of crystals usually consist of terraces of low-index faces connected by *atomic steps* which are one or more interplanar distances high. In some cases vicinal surfaces consist of *facets*. At atomically pure vicinal surfaces (110) of *germanium* some reversible reconstructions of these surfaces relative to step ones with terraces can be observed.

VILLARI EFFECT (E. Villari, 1865)

Magnetoelastic effect or variation of *magnetization M* in magnetically ordered materials under the effect of external elastic stresses. Thermodynamically it is the inverse of *magnetostriction*, and it is caused by the relative variations of the *spin–orbit interactions, magnetic dipole interactions* and *exchange interactions* at deformations of the crystal. In some magnetic materials the magnetoelastic effect changes sign at a particular value of the magnetic field B (*Villari point*). At the Villari point the magnetostriction λ, to which magnetoelastic effect is connected with the following expression:

$$\left(\frac{\partial M}{\partial \sigma}\right)_{B,T} = -\left(\frac{\partial \lambda}{\partial B}\right)_{\sigma,T},$$

also changes its sign, where σ designates unilateral elastic stress acting in the direction of the measured magnetostriction.

VIRTUAL CRYSTAL

A *crystal*, the values of whose parameters (lengths of interatomic bonds, ionic character of bonds, atomic potentials, masses, etc.) coincide with the average values of the corresponding substitutional *solid solution*. Replacement of the real crystal by the virtual one corresponds to the *virtual crystal approximation* (random statistical fluctuations are not taken into account). This approximation permits one to find the regions of the spectrum where excitations in the crystal might occur, but it is not suitable for the description of *elementary excitation spectra of disordered solids* near boundaries or edges (see *Anderson model*).

VIRTUAL FERROELECTRICS

Materials which remain *paraelectric* down to the temperature of 0 K (in the absence of external effects) but have a *dielectric constant* which increases with the decrease in temperature.

As is the case with true *ferroelectrics*, these materials are characterized by an abnormally high dielectric constant, which depends on the temperature, and is associated with the presence of phonon *soft modes*, occurring due to the partial compensation of forces of long-range attraction of ions in different unit cells (polarization forces), and by forces of short-range repulsion of ions which belong to the same unit cell. Compensation of the polarization and repulsive forces in virtual ferroelectrics, unlike real ones, is never complete, so the soft mode frequencies do not tend to zero at any temperature, and no *ferroelectric phase transition* takes place. The properties of virtual ferroelectrics are similar to those of true ferroelectrics in the *paraelectric phase*. For example, the dielectric constant at high temperatures satisfies the *Curie–Weiss law*, but with increasing temperature the dependence $\varepsilon(T)$ becomes greater, and it tends to a finite limit at $T \to 0$. The presence of quantum effects or *zero-point vibrations*, which bring lattice atom displacements to a finite value (unlike classical theory) and, more specifically, to the finite value consistent with the repulsive forces even at the lowest temperatures (see *Barrett formula*), provide a physical mechanism to explain the stabilization of the paraelectric phase with decreasing temperature. Virtual ferroelectrics can be transformed to real ferroelectrics by the application of external pressure which changes the interaction constants of the atoms, as is the case with $KTaO_3$ and $SrTiO_3$ upon application of axial pressure. On the other hand, *uniform compression* can produce an increase of the soft mode frequency, and in some cases it transforms a true ferroelectric into a virtual one, as happens, for example, in crystals of potassium dihydrogenphosphate KH_2PO_4. It is also possible to induce the ferroelectric phase transition in virtual ferroelectrics by the introduction of impurities (see *Ferroelectricity induced by impurities*).

VIRTUAL PHONONS

A term used for the quantum-mechanical description (according to perturbation theory) of the interaction of quanta of *crystal lattice vibrations* (*phonons*) with one another or with some other elementary excitations (e.g., with *conduction electrons*). In this interaction process the energy *conservation law* may not hold. Virtual phonons, with postulated momentary existence, can contribute to the renormalization of the frequency spectrum due to the *anharmonic vibrations* of a crystal, and also to the formation of collective quasi-particle oscillations in the crystal (see *Collective excitations*). It is often convenient to represent *quasi-particles* in the crystal as a complex of an initial (bare) elementary excitation in an ideal lattice and a cloud of virtual phonons. The concept of virtual phonons is also convenient for describing the indirect interaction of elementary excitations caused by their coupling to the lattice vibrations. In particular, the exchange of virtual phonons can bring about the attraction between two conduction electrons which leads to the formation of a Cooper pair (see *Superconductivity*).

VIRTUAL TRANSITION

A concept relating to (possibly hypothetical) intermediate quantum states in perturbation calculations of the probability of a system to tundergo a transition due to a perturbation V from an initial quantum state i to a final state f.

First-order perturbation theory involves the direct matrix element V_{if}, whereas second- and higher-order perturbation approximations involve pairs of matrix elements V_{mf} and V_{im}, where m is

any possible ("intermediate") state of the system, sometimes called a *virtual state*. The intermediate transitions V_{mf} and V_{im} that enter the calculation are referred to as virtual transitions. Unlike an actual transition, the energy *conservation law* does not hold for an individual virtual transition, but it is valid for the totality of all virtual transitions leading to the actual transition $i \to f$. In solid-state physics the quantum states of the unperturbed system are often described by the numbers of participating particles and *quasi-particles* of different types (*photons, phonons*, free electrons and *holes, plasmons*, etc.), while transitions between different states are described in terms of the creation and annihilation of particles. The same description is valid for virtual transitions. Non-conservation of energy under virtual transitions implies a short lifetime of the *virtual particles* in conformity with the uncertainty relation, $\Delta E \cdot \Delta t \approx \hbar$. In *Feynman diagrams* the lines having no unattached ends correspond to virtual particles. For example, the probability of an electron jump under the influence of the *electron–phonon interaction* between two closely located *impurity atoms* with energies E_1 and $E_2 = E_1 + \Delta E$ in the first order of perturbation depends on the creation of an actual phonon with energy $E_{ph} = \Delta E$. In the second order of the perturbation the jump can be represented as a process involving two *virtual phonons* (one virtual phonon with energy E_a is created, and the other with energy E_b is annihilated) subject to the energy condition $E_a - E_b = \Delta E$.

VISCOSITY

A tendency of a fluid to resist the motion of one part of it with respect to other parts, or with respect to a boundary surface. *Newton's basic law of viscous flow* is $\tau = \eta \, dv/dz$, where τ is the tangential stress, causing the *shear* of layers with respect to one another, η is the *(dynamic) viscosity* (measured in Pa·s), and dv/dz is a gradient of flow velocity in the direction perpendicular to the plane of laminar flow. The reciprocal $1/\eta$ is called fluidity. The *kinematic viscosity* is determined as η/ρ, where ρ is the material density. Kinematic viscosity is measured in m^2/s. In liquids and solids viscosity depends on the interaction force and on the mobility of the molecules and atoms. In gases the distances between molecules are considerably greater than the radius of action of the intermolecular forces, and the viscosity arises from the molecular exchange between the layers with different speeds of motion. In solids viscosity can be considered as the resistance to *plastic deformation* expressed in terms of work per unit cross-section or per unit volume during the deformation process. In solids undergoing impact tests the *toughness* as well as the *fracture thoughness* or *crack resistance* are often estimated. For determining the viscosity of liquids and gases special devices called *viscometers* are often used. In solids at high temperatures the viscosity arises from diffusion mechanisms of flow, and at lower temperatures by different mechanisms of plastic flow, e.g., by dislocational *plasticity* in crystalline bodies, by growth, and also by movement of the free volume in amorphous metallic alloys (see *Plastic deformation of amorphous metallic alloys*). With the onset of flow in a solid the value of dv/dz in Newton's law is equal to the velocity of the shearing deformation $\partial\varepsilon/\partial t \equiv \dot\varepsilon$ and $\tau = \eta\dot\varepsilon$. This law is approximately satisfied for diffusion flow of solids, in this case $\eta \propto d^\alpha k_B T/(\omega D)$, where d is the dimension of a grain, $\alpha = 1$ to 3, ω is the atomic volume, D is the coefficient of volume self-diffusion or of boundary diffusion (see *Diffusion*), depending on the mechanism which determines the flow. In the presence of dislocational mechanisms of flow the relation between τ and $\dot\varepsilon$ is non-linear, i.e. Newton's law is not satisfied.

VISCOSITY, DIELECTRIC

See *Dielectric viscosity*.

VISUALIZATION OF ACOUSTIC FIELDS

Totality of methods for acquiring visual pictures of the spatial distribution of values characterizing sonic fields.

Two methods have been widely used for acoustic field visualization in solids, namely mechanical or electronic scanning of the surface under investigation by a sound receiver or buffer; and recording the light flux phase modulation which occurs when light passes through a region where the index of refraction changes under the influence of the sound wave. For example, in the thermal *Tepler method* (A. Tepler, 1864) only deviated rays are

recorded, and illumination at the screen is generally proportional to the deviation angles, so the regions with variable refraction index are displayed. In the *phase contrast method* when the phase-modulated light beam passes through a quarter-wave plate the amplitude modulation which occurs provides a visual image.

Surface acoustic waves can be visualized in various ways: by scanning the light beam over the investigated surface using a beam diameter which is less than the acoustic wavelength; by varying the angle of light reflection from the surface crimped under the effect of sound; by transforming the light intensity modulation with the help of a shadowing wedge; by recording the distribution of the electrical fields which accompany the surface waves in *piezoelectric materials*; by the use of optical holographic interference, when several optical holograms of an oscillating body are recorded at the same photoplate (see *Holography in solid media*). Interference bands appear at the reconstructed image corresponding to the distribution of the oscillation amplitudes at the surface.

VITREOUS STATE OF MATTER, glass

The solid state (see *States of matter*) of an amorphous material attained by cooling the melt (see *Amorphous state*). The transformation of a supercooled liquid into a glass occurs within a narrow temperature interval ΔT_g as shown in Fig. 1,

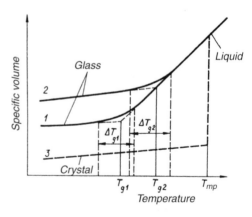

Fig. 1. Dependence of the specific volume of glass on the temperature for a slower cooling rate (curve 1), and a faster cooling rate (curve 2). A typical crystallization plot (curve 3) is also shown.

where curve 2 is for more rapid cooling than curve 1. Thus the width ΔT_g and location of the vitrifying point $T_g < T_{mp}$ both depend upon the rate of cooling. The process of transformation of a melt into a glass is referred to as *vitrification*, and the temperature T_g, at which this transformation occurs, is given the name *vitrification temperature*. A typical crystallization behaviour (curve 3) is shown for comparison. Materials which form melts capable of transforming into glasses are spoken of as *vitrescent (glass-forming) materials*. The vitrification process is treated as a *phase transition*, with the name *liquid–glass transition* (see *Vitrification*). An amorphous solid, which may turn into a supercooled liquid on cooling, is obtainable not only from a melt, but also by the *amorphization* of crystalline solids, or by precipitation of molecular, atomic, or plasma flows on substrates. Therefore, the materials conventionally referred to as glasses are amorphous solids exhibiting a liquid–glass transition, regardless of the method of preparation. The structure and properties of a material in the vitreous state depend on the method of preparation (see Fig. 1).

The vitreous state of matter is found in elements (C, S, Se, As, etc.), oxides (SiO_2, B_2O_3, GeO_2, V_2O_5, etc.), aqueous solutions of H_3PO_4, $HClO_4$, H_2SO_4, etc., chalcogenides (see *Chalcogenide materials*) of various elements (As, P, Ge, etc.), certain halides and carbonates, various metal *alloys*. There exists also a broad class of organic and *polymeric glasses*, which are composed mostly of *high molecular weight compounds* (see *High molecular weight glasses*). In terms of electronic properties, glasses are classed into three groups: insulators (the majority of oxide, organic and polymeric glasses), semiconductors (see *Amorphous semiconductors*), and metals (see *Metallic glasses*).

A supercooled liquid is metastable and exists within the stability region of a crystalline solid. Therefore, vitrification may occur only if the temperature of the melt (below the *crystallization* temperature) decreases so rapidly that the melt does not have enough time to crystallize during cooling to the temperature below T_g. Each type of vitrescent compound features a characteristic critical rate of cooling such that the vitrification is possible only above this temperature. For the majority of metallic alloys, the critical rate of cooling ranges from 10^5 to 10^6 K/s, and silicate glasses are characterized by critical cooling rates of 10^{-2} K/s.

Certain high molecular weight compounds, with molecular structures incompatible with translational symmetry (like the case of *noncrystalline clusters*), do not exist in crystalline states. The spatial distributions of atoms in the supercooled liquid and in the glass differ only slightly from one another, as shown by *structure studies*. The principal distinction between these spatial distributions is that the supercooled liquid exhibits an ergodic structure, and glass a nonergodic one. The structure is spoken of as ergodic if the physical quantities (dependent on coordinates and velocities of atoms) assume equal values when averaged over time and over phase space. In particular, the time average of the correlation of positions of a certain population of n atoms is equal to the ensemble average of configurations of all populations involving n atoms at a fixed instant of time. The *nonergodicity* (frozenness) of a glass structure is indicative of the rarity of thermally activated atomic migrations. Therefore, substantial reconfigurations do not occur over macroscopically long times.

The atomic structure of glasses, which is not easily observable by direct methods, still remains unknown in many respects. Several models of the structure of vitreous state of matter have been devised. These models include: (a) a *random continuous network* with *covalent bonds*; (b) *random close-packing model*; (c) *model of interpenetrating molecular globules*, used for describing macromolecular glasses; (d) the *polycluster model* (see *Polycluster amorphous solids, Lebedev crystallite hypothesis*). The glass structure is not in equilibrium, and becomes rearranged within limits specified by the diffusional mobility of the atoms. Subjecting glasses of complex structure (including high molecular weight polymer glasses) to isochronal *annealing* brings about, as a rule, several stages of structural *relaxation*, each stage related to a reconfiguration of certain types of structural members. Each stage of structural relaxation is characterized by a temperature T_k and *relaxation time* τ_k (index $k = 1, 2, \ldots$ enumerates relaxation stages). The most severe structural

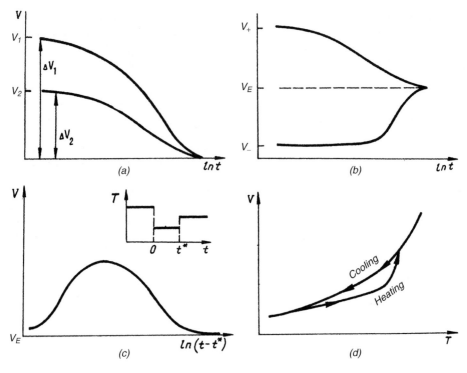

Fig. 2. Characteristics of the relaxation volume V of glass arising from the lack of equilibrium: (a) nonlinearity of relaxation rate, (b) asymmetry of relaxation, (c) memory effect, and (d) hysteresis for cyclic heating–cooling.

changes in glasses take place in the neighborhood of T_g. Atomic rearrangements responsible for these changes have been given the name *alpha motions*, and the related relaxation is spoken of as *alpha relaxation*. The irreversible rearrangements at $T < T_g$ have been called *beta relaxations*. Microscopic models of α- and β-rearrangements are at present inadequate. Experimental data indicate that the temperature dependences of relaxation times follow the form of an *Arrhenius law*. The phenomenological description of the relaxation of macroscopic quantities under annealing in terms of several "elementary" relaxation processes is provided by a set of equations of the form

$$\Delta V = \sum_k \Delta V_k,$$

$$\frac{\mathrm{d}\Delta V_k}{\mathrm{d}t} = -\frac{1}{\tau_k}\Delta V_k,$$

$$\tau_k = \tau_k(T\Delta V),$$

where ΔV is the relaxing volume of a body, ΔV_k is the partial change of volume due to relaxation rearrangements of the k type. Equations similar to those above are also specified for other macroscopic quantities. The coefficients in these equations are phenomenological, and determined by correlation with experimental data. The microscopic theory of relaxation phenomena in glasses is still far from completion. Stretched exponential relaxation behaviour has also been observed.

Processes of structural relaxation in glasses from an initial nonequilibrium volume V_i to a final quasi-equilibrium volume V_E at a temperature T display several general features which are independent of the type of glass and the details of its structure, and Fig. 2 presents some of these regularities. Fig. 2(a) shows how the deviation ΔV_i of a nonequilibrium volume V_i from V_E relaxes back to the value V_E in a non-exponential manner (i.e. it is nonlinear on a semilogarithmic plot). Fig. 2(b) illustrates the asymmetry of the relaxation, i.e. the difference in behavior of the kinetic curves approaching quasi-equilibrium from nonequilibrium volumes that are smaller V_- and larger V_+ than the value V_E which is designated by the dashed line. Fig. 2(c) illustrates the memory effect, which is representative of the sensitivity of the relaxation process to the prehistory of the sample, and Fig. 2(d) presents a typical hysteresis plot of the variation of $V(T)$ during a temperature cycle.

Glasses share a number of common distinctive properties, which include: lower density compared to their crystalline analogues: this difference in density amounts to 1–3% for metallic glasses and reaches 15% in the case of nonmetallic ones; enhanced *microhardness* and *strength*; low-temperature anomalies of *specific heat*, *thermal conductivity*, and *sound absorption*, which are due to the existence of *two-level systems* (see *Low-temperature anomalies*); high value of *internal friction*. Due to the lack of translational structural symmetry, glasses show electron and phonon properties that differ significantly from those of crystalline solids (see *Anderson localization*, *Amorphous semiconductors*, *Metallic glasses*, *Elementary excitation spectra of disordered solids*). Glasses find applications in many areas of engineering and manufacturing.

VITRIFICATION, liquid–glass transition

The transformation of a supercooled liquid into a glass, a process accompanied by an abrupt increase of the *viscosity* and density, a decrease of the temperature coefficient of expansion, and an abrupt change of the *specific heat*, all of which occurs within a narrow temperature range. The temperature T_g, at which the vitrification takes place, is referred to as the *vitrification temperature*. The glass forming process resembles a *second-order phase transition*, but does not formally constitute this type of transitions, as evidenced, in particular, by the violation of the *Prigogine–Defay criterion* (I. Prigogine, R. Defay), whereby for second-order transitions the abrupt changes of heat capacity Δc_p, the *thermal expansion* coefficient $\Delta \alpha$, and *compressibility* coefficient Δk are related by an equation of the form

$$\frac{\Delta c_p \Delta k}{(\Delta \alpha)^2 T V} = 1$$

(T is *temperature*, V is sample volume). Besides that, the liquid–glass transition is not totally reversible due to the complicated relaxation processes in the temperature region of the transition (see *Vitreous state of matter*), and the nature of the transition depends strongly on the rate at which the temperature changes. The vitrification process has been described in the context of the *free volume model*, the polycluster model (see *Polycluster amorphous solids*), and by the application of various phenomenological approaches.

VOIGT AVERAGING of elastic constants

Approximate method of calculating *elastic moduli* of single-phase isotropic *polycrystals* in terms of elastic moduli of *monocrystals*. The calculation is performed by expressing the *elastic modulus tensor* of the monocrystal in a random coordinate system, and averaging the result over all the possible orientation of its crystalline axes.

For the *bulk modulus K* the Voigt averaging leads to $K_V = (1/9)[(c_{11} + c_{22} + c_{33}) + 2(c_{12} + c_{23} + c_{31})]$, and for the *shear modulus* μ we obtain $\mu_V = (1/15)[(c_{11} + c_{22} + c_{33}) - (c_{12} + c_{23} + c_{31}) + 3(c_{44} + c_{55} + c_{66})]$, where c_{ik} are elastic moduli of the monocrystal in the *Voigt notation*. These relations are valid for all the *crystal symmetry* classes. Actual values of the K and μ moduli for polycrystals are, as a rule, to some extent lower than those calculated by Voigt averaging, that is $K < K_V$ and $\mu < \mu_V$, so K_V and μ_V may be considered as limiting (maximum) values of the K and μ moduli.

VOIGT EFFECT (W. Voigt, 1899)

Birefringence in an optically isotropic material located in a magnetic field $\boldsymbol{B}$ for propagation of radiation in a direction, perpendicular to the direction of $\boldsymbol{B}$. It was investigated in collaboration with E. Wiecher for the D-lines of sodium vapor. The Voigt effect was also observed by K. Majorana (1902) in colloidal solutions of iron, and in 1907 E. Cotton and H. Mouton established the magnetic birefringence for organic materials (*Cotton–Mouton effect*). The phase difference δ of the components of radiation, polarized along and perpendicular to the direction $\boldsymbol{B}$, is proportional to $\boldsymbol{B}^2$, and increases linearly with the sample thickness. Measured values of δ are relatively low. In *semiconductors* the Voigt effect of free current carriers is small compared to the *Faraday effect*, $\delta/\theta \sim 10^{-2}$, where θ is the Faraday rotation angle. In optically isotropic semiconductors with anisotropic *isoenergetic surfaces* the Voigt effect of free carriers is anisotropic and, as is the case of *cyclotron resonance*, may be used for determining the anisotropy of the *effective mass* of charge carriers. Due to the smallness of δ, the Voigt effect is only studied in certain favorable semiconductors (Ge, InAs, InSb).

VOIGT NOTATION (W. Voigt, 1910)

Notation of the components of *elasticity theory* tensors, whereby to each pair of tensor indices (ij), the designated Cartesian coordinates ij correlate with the corresponding Voigt index k according to the following rule: $(11) \to 1$, $(22) \to 2$, $(33) \to 3$, $(23) = (32) \to 4$, $(31) = (13) \to 5$, $(12) = (21) \to 6$. The *strain tensor* u_{ij}, *stress tensor* σ_{ij} and *elastic modulus tensor* c_{ijkl} of three-dimensional x, y, z Cartesian space appear, respectively, in the six-dimensional Voigt notation space as the corresponding vectors u_α, σ_α and matrix $c_{\alpha\beta}$. Due to its compactness the Voigt notation is widely used in elasticity theory, although it does have some drawbacks. In particular, in order to replace the tensor designation according to Cartesian indices (e.g., in *Hooke's law*, or in the expression for the *strain* energy) by one according to Voigt indices, it is necessary to introduce additional multipliers (e.g., 2). Besides this, the Voigt formulation indices are not invariant under rotations of the coordinate system.

VOLUME FORCE

See *Bulk force*.

VOLUMETRY (fr. Lat. *volumen*, volume, and Gr. $\mu\varepsilon\tau\rho\varepsilon\omega$, am measuring)

Methods for measuring the specific volume and its variations, accompanying different chemical processes.

The following methods are applicable for *solids*: pycnometric, hydrostatic, flotational and dilatometric. The *pycnometric method* is based on the determination of the volume of a liquid, displaced from a special vessel (pycnometer) as the solid sample is submerged into it (accuracy does not exceed 0.1%). The *method of hydrostatic weighing* involves the subsequent weighing of the sample in two media: in air and in the operating liquid. The accuracy of the determination of the specific volume of the sample depends essentially on the accuracy of measuring the media *densities*, which depend on *pressure* and temperature. For careful thermostatic control of the media and sample the accuracy reaches 0.01%. The *method of differential weighing* is more accurate: the small changes of specific weight are measured by subsequent weighing of two samples (reference and investigated ones) in two media simultaneously. For a controled temperature difference between the samples the measurement attains the accuracy 0.001%. The *flotational method* is based on the flotation of the body, submerged in a liquid, when through a change in the contents or temperature their specific volumes are equalized. The flotation method, which is simple to carry out, requires a highly accurate determination of the temperature dependence of the working liquid density for each composition, and it is not applicable to solids with specific volumes less than 0.25 cm^3/g. The accuracy of this method is 0.01 to 0.001%. For solids interacting with working liquids, the isotropic specific volume changes which occur during the course of different processes can be measured with the help of a *dilatometer*. For solids with anisotropic properties the dilatometric measurements should be performed along three orthogonal axes.

VORTEX FILAMENTS

See *Abrikosov vortices*.

VORTEX, JOSEPHSON

See *Josephson vortex*.

VORTEX LATTICES in superconductors

Lattices formed by *Abrikosov vortices* in a superconductor placed in an external magnetic field.

Vortex lattices exist in bulk *type II superconductors*, superconducting films in perpendicular (two-dimensional vortex lattice) and parallel (one-dimensional vortex lattice) magnetic fields, in the surface layer of a superconductor, in an oblique magnetic field (*Kulik's vortices*), and in superconducting tunnel contacts (*Josephson vortices*). Typical vortex lattices are two-dimensional arrays of

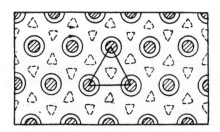

Fig. 1. Abrikosov vortex lattice.

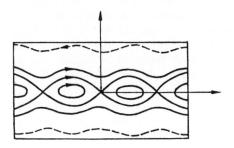

Fig. 2. Vortex state within a thin film.

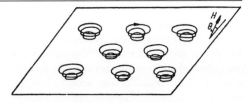

Fig. 3. Oblique vortices in a superconductor.

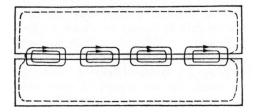

Fig. 4. Josephson vortices.

points; and tracing around one of them changes the *order parameter* phase by 2π, while the flux through each lattice unit cell is equal to the *flux quantum* $\Phi_0 = h/2e$.

The most important case of a vortex lattice is the *mixed state* in type II superconductors which exists in the interval of applied fields between the *lower critical field* (B_{c1}) and *upper critical field* (B_{c2}). In this case the vortices form the regular triangle lattice shown in Fig. 1 with a spacing which depends on the magnetic field through the *quantization of flux* condition $B = n\Phi_0$, where n is the density of vortices (reciprocal area of vortex lattice unit cell). Very thin superconducting films in a perpendicular field are always in the mixed state ($B_{c1} \sim 0$) with a triangular Abrikosov lattice. In the case of a film in a parallel field the periodic distribution of B is in the form of a one-dimensional chain of vortices as shown in Fig. 2, which appears at a thickness $d < 1.6\xi$, where ξ is the *coherence length* of the superconductor. In a field exceeding B_{c2}, where the bulk material goes normal, superconductivity persists near the surface in the form of a surface layer (see *Surface superconductivity*), pierced by the oblique vortices shown in Fig. 3 which are arranged at the vertices of of nonequilateral triangles oriented in the direction of the projection of the field to the surface. In the case of a contact between two superconducting metals, separated by a thin insulating layer (*tunnel junction*) and placed in a magnetic field parallel to the surface, in the junction plane there appears a one-dimensional chain of Josephson vortices, as shown in Fig. 4.

In the absence of an externally applied current the vortex lattice is stationary, i.e. immobile in space. When transport current of density J is caused to flow perpendicular to the field B (i.e. vortex direction) the *Lorentz force* $J \times B$ on the vortices starts the lattice moving in a direction perpendicular to both J and B (if the sample is nonuniform, then for small currents the lattice can remain immobile due to the *vortex pinning*). The moving vortex state is called the *resistive state* because it corresponds to the appearance of electric fields and a finite resistance in the superconductor. In the case of a tunnel junction the energy dissipation through the motion of vortices is very low, and various dynamic phenomena can take place (see *Josephson effect*).

VORTEX PINNING

The phenomenon of pinning of *Abrikosov vortices* at inhomogeneities of chemical composition or imperfections (defects) of crystal structure in *type II superconductors*. The *bulk pinning force*

F_p determines the *critical current* density in magnetic fields and balances the *Lorentz force*, which acts on the *vortex lattice* as the transport current passes through the superconductor: $F_p = j_c \times B$, where j_c is the critical current density, B is the magnetic flux density. The bulk pinning force is the sum of elementary interactions of isolated vortices with defects. With allowance made for collective interaction of the vortex lattice with the ensemble of random pinning centers, the equation for the bulk pinning force takes the form $F_p = [n_p \langle f_p^2 \rangle / V_c]^{1/2}$, where n_p is the number of pinning centers, f_p is the value of the *unit pinning force*, and the angle brackets denote averaging over the correlation volume V_c, within which the vortex lattice retains short-range order. In the context of linear *elasticity theory*, the dimensions of the correlation volume are expressed via the *elastic moduli* of the vortex lattice.

In the event of a strong interaction with a pinning center, the short-range order of the vortex lattice within the correlation volume is violated, and the collective pinning becomes a single-particle type process. Three varieties of interaction between vortices and defects, which lead to the pinning of vortices, are distinguished: elastic interaction, interaction with the vortex core, and magnetic interaction.

The elastic interaction is due to the change of volume and elastic moduli of a material in the process of a normal-to-superconducting transition. As a result of this change, the normal vortex core becomes more dense and stiff compared to the superconducting regions of the body. Therefore, elastic force fields arise around the vortex core, which are capable of effectively interacting with the stress fields of various crystal lattice defects.

The interaction with the vortex core is determined by the local changes of energy of the superconducting condensate in the neighborhood of the defect, and is characteristic of the pinning of vortices at *grain boundaries*, micropores, *pile-ups of dislocations* (dislocation walls), and second-phase inclusions.

The magnetic interaction is most pronounced in those superconductors, where the dimensions of pinning centers considerably exceed the London *penetration depth of magnetic field*, and is related to the abrupt change of the magnetic flux density or magnetic moment at the transit of a vortex across the interface between two regions (e.g., the pinning of vortices at a superconductor surface).

Flux flow can involve vortices moving past weakly pinning centers, or past other vortices that are already pinned. This flow can take place at higher temperatures where the thermal energy overcomes pinning forces.

VORTEX, QUANTUM

See *Quantum vortices, Dynamics of quantum vortices*.

VORTEX STATE

See *Mixed state*.

VORTICES, INCLINED

See *Inclined vortices*.

WAKE POTENTIAL

A potential which is produced by a charged particle moving in a medium, and which appears behind in the region already traversed by the particle (in its "wake") due to the medium polarization. The corresponding electric charge is referred to as the *wake charge*. The emergence of the wake potential affects the passage of a beam of charged particles through a medium, since each particle is acted upon by the wake potential generated by preceding particles. That influences, e.g., the characteristics of the dissociation due to their ionization when a molecular beam passes through a material. Oscillations of the wake charge density result in an additional ionization added to that directly produced by the charged particle. This secondary ionization proceeds along the whole track axis for a certain period of time after the particle has passed, but it is retarded with respect to the primary ionization. Because of the wake potential the track of a charged particle appears to be a linear source of secondary electrons. The difference between the *dielectric constant* in the particle track and that of the surrounding space or medium results in the *Raman scattering of light* from the wake charge oscillations.

WALKER MODES (L.R. Walker, 1953)

Natural *magnetostatic oscillations* of a saturated isotropic ferromagnetic sample that is spheroid-like in shape. Walker modes are named after L.R. Walker who was the first to calculate them. These modes are characterized by three indices that are descriptive of the radial and azimuthal dependencies of the magnetostatic potential amplitude, and the shrinkage of *magnetization* along the spheroid rotation axis. The spectrum of Walker modes for a slightly anisotropic *magnetic substance* lies within the frequency range from γB_0 to $\gamma (B_0 + \mu_0 M_0)$, where μ_0 is the *magnetic permeability of vacuum*, γ the is *gyromagnetic ratio*, B_0 is the value of the constant external magnetic field, M_0 is the saturation magnetization. The uniform precession of magnetization is one of the types of Walker modes. Walker modes are experimentally observed as resonance lines of electromagnetic energy absorption of a sample located in the nonuniform magnetic field of a microwave waveguide or *resonator* (cavity).

WALKER SOLUTION (L.R. Walker, 1953)

The exact one-dimensional solution of the *Landau–Lifshits equation* involving *Gilbert's relaxation term*. The Walker solution describes a planar *magnetic domain wall*, which is set in motion at a speed v under the action of an external magnetic field B, the direction of the latter being parallel to the *easy magnetization axis*. The solution holds true for a *ferromagnet* with a *magnetic anisotropy* of the form $W_a = -\beta M_z^2/2 + \beta' M_x^2/2$. The Walker solution is defined by the equations

$$\frac{M_z}{M_0} = \pm\tanh\left[\frac{x - vt}{\Delta(\varphi_0)}\right],$$

$\varphi_0 = \varphi_0(v) = \text{const}$, where $\tan \varphi = M_y/M_x$, M_0 is the saturation magnetization, $\Delta(\varphi_0) = \Delta/(1 + \varepsilon \cos^2 \varphi_0)^{1/2}$, $\varepsilon = (4\pi + \beta')/\beta$, Δ is the thickness of the wall at rest. The relationship between φ_0, v and B is given by the equations

$$v = \gamma M_0 \Delta \varepsilon \sin \varphi_0 \cos \varphi_0 \left(1 + \varepsilon \cos^2 \varphi_0\right)^{-1/2}$$

or

$$v = \frac{\gamma B \Delta}{\alpha}\left\{1 + \frac{\varepsilon}{2}\left(1 \pm \left[1 - \left(\frac{B}{B_c}\right)^2\right]^{1/2}\right)\right\}^{-1/2},$$

where $B_c = \alpha\beta\varepsilon M_0/2$ is the *Walker critical field*. In the case of $\pi/4 < \varphi_0 < \pi/2$ and a plus sign, the above equations describe the motion of a stable

Bloch wall, whereas the condition $0 < \varphi_0 < \pi/4$ for the former equation and a minus sign in the latter equation correspond to an unstable *Néel wall*. At $B = B_c$, φ_0 assumes the value $\varphi_0 = \pi/4$ and $v = v_c = \varepsilon\beta\Delta M_0/2$ is the *Walker critical velocity*. For a nondissipative medium, the Walker solution holds true at $B = 0$ and describes the simplest *magnetic soliton*, with the velocity v the free parameter of the solution; the value of velocity in this case is bounded by the *Walker limiting velocity* $v_W = \gamma M_0(\alpha\beta)^{1/2}((1+\varepsilon)^{1/2} - 1)$ (it should be noted that $v_W > v_c$). The domain wall in real magnetic materials is multidimensional because of *domain wall twisting*, the presence of Bloch lines, etc., and therefore, the Walker solution often provides an inadequate description of experimental data (in particular, the velocity does not reach the critical value v_c, see *Domain wall peak velocity*). The agreement between the Walker solution and experiment is better, the stronger the anisotropy in the basal plane.

WALLER MECHANISM (I. Waller, 1932)

One of the channels of magnetization relaxation in *paramagnets* (see *Magnetic relaxation*).

The Waller mechanism is determined by the coupling of *spins* with *phonons*, which occurs due to changes of the magnetic dipole interactions between paramagnetic ions and lattice vibrations. The *selection rules* for the Waller mechanism allow relaxation transitions between Zeeman sublevels (see *Zeeman effect*) of a single paramagnetic ion or several ions, with some processes involving the simultaneous flipping of two spin moments. In the case of a dilute paramagnetic spin, the rate of relaxation due to the Waller mechanism is proportional to the paramagnetic ion concentration n, and for sufficiently small n it gives way to the more effective concentration-independent *relaxation* processes. In concentrated paramagnetic salts, the Waller mechanism can appear together with other mechanisms, caused by *spin–spin interactions* of a nonmagnetic nature (typically, by *exchange interactions*). In most cases other relaxation processes such as the Van Vleck mechanism are more dominant than the Waller mechanism, so the latter is rarely invoked.

WANNIER FUNCTIONS (G. Wannier, 1937)

Wave functions $\varphi(r)$ in the coordinate space defined in terms of an expansion of Bloch functions Ψ_{kn} (see *Bloch theorem*) ψ_{kn}:

$$\varphi_n(r - R) = \frac{1}{\sqrt{N}} \sum_k e^{-ikR} \psi_{kn}(r),$$

where the summation is over all the allowed *quasi-wave vectors* k. The inverse transformation permits Bloch functions to be expressed in terms of Wannier functions:

$$\psi_{kn} = \frac{1}{\sqrt{N}} \sum_k e^{ikR} \varphi_n(r - R),$$

where R designates sites of the direct lattice (*Bravais lattice*). Wannier functions are orthonormal: $\int \varphi_n^*(r)\varphi_{n'}(r)\,d\tau = \delta_{nn'}$, and they are convenient for the expression of electronic wave functions in the presence of various perturbations of a *periodic potential*.

WANNIER–MOTT EXCITON (G. Wannier, N. Mott, 1937)

Weakly bound state of electron and hole in *semiconductors* and *insulators* occurring when the interparticle distance of an *exciton* is far greater than the lattice constant. The energy of a bound Wannier–Mott exciton, in contrast to that of the strongly bound *Frenkel exciton*, is determined by the structure of the *conduction band* and *valence band*, as well as by the dielectric properties of the medium. The electron and hole are bound by the Coulomb force of attraction. If they are so widely separated that the background atomic lattice structure variations can be neglected, then the interaction potential can be written in the form

$$V(r) = -\frac{e^2}{\varepsilon r}, \tag{1}$$

where r is the distance between the electron and hole, e is the charge, ε is the macroscopic high-frequency (short wavelength) *dielectric constant*. In this case the exciton problem is similar to the problem of a hydrogen-like atom (Wannier, 1937); the factor ε^{-1} was introduced by Mott (1938). The potential (1) produces an infinite series of hydrogen-like exciton bound states (Rydberg series) with the energies:

$$E_n^{ex}(k) = -\frac{e^4\mu}{2\hbar^2\varepsilon^2 n^2} + \frac{\hbar^2 k^2}{2M_{ex}}, \qquad (2)$$

where $n = 1, 2, \ldots$ is the principal quantum number, μ and M_{ex} are respectively the reduced (optical) mass and the mass of the translational exciton motion, and $\hbar k$ is the momentum of the center of mass. The bound exciton state thus has a continuous energy spectrum associated with the translational motion of the exciton as a free particle. This motion transfers energy but does not transfer charge. The energy level scheme of a Wannier–Mott exciton (2) is shown in the figure.

The wave function of a Wannier–Mott exciton with $n = 1$ without including the normalization factor has the form

$$\psi(k, R, r) \approx \exp\left(ikR - \frac{r}{a_{ex}}\right), \qquad (3)$$

where R is the coordinate of the center of mass, and $a_{ex} = \varepsilon\hbar^2/(\mu e^2)$ is the exciton *Bohr radius*. For excited states $\exp(-r/a_{ex})$ in Eq. (3) should be replaced by the corresponding Coulomb wave function $F_{nlm}(r/a_{ex})$ where l, m are orbital and magnetic quantum numbers. The bound energy of a Wannier–Mott exciton with $n = 1$ is counted down from the continuum energy band ($n = \infty$) where the states of the exciton are dissociated (see the shaded area in the figure); this value is often called the *excitonic Rydberg energy*:

$$E_{ex} = \frac{\mu e^4}{2\hbar^2\varepsilon^2} = \frac{e^2}{2\varepsilon a_{ex}} = \frac{\hbar^2}{2\mu a_{ex}^2}. \qquad (4)$$

The energy E_{ex} in semiconductors can be very small, e.g., $E_{ex} \approx 4$ meV in GaAs for which $\mu \approx 0.06m_0$ (m_0 is free electron mass), $\varepsilon \approx 12$, $a_{ex} \approx 10$ nm. A Wannier–Mott exciton can appear at any critical point of the band structure, and the criterion for its existence is $\nabla_k E_e = \nabla_k E_h$, i.e. the equality of the group velocities of the electron and hole. Excitons can be created during optical transitions, with particular energies and features that follow from the *conservation law* of energy and momentum. Discrete hydrogen-like spectra of excitons are observed at the direct photon absorption (see Fig.) in semiconductors with a direct energy gap. This absorption spectrum was initially found

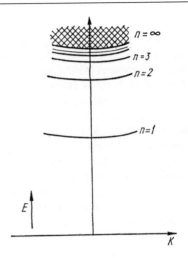

Hydrogen-like energies in a semiconductor with a direct energy gap E_g. Wannier–Mott excitons are excited upon absorption of photons with energies $h\nu = E_g - E_n^{ex}(0)$.

and investigated in Cu_2O crystals. If the extrema of the conduction and valence bands in momentum space are separated as, for example, in crystals of Ge and Si, Wannier–Mott excitons are excited by light in an indirect manner, namely with the participation of phonons. These excitons play an important role in explaining optical, magnetic, photoelectric, collective and other properties of semiconductors and insulators.

WATSON SPHERE (R.E. Watson, 1958)

A sphere surrounding a fragment of a *crystal lattice* used for calculations in the *cluster approximation*, or surrounding a molecule under consideration using the method of scattered waves. Within the Watson sphere the potential is selected in the form of a *muffin-tin potential*, and outside the sphere the potential vanishes. When calculating charged *clusters*, a compensating charge is placed on the Watson sphere.

WAVE DIFFRACTION

See *Diffraction of waves.*

WAVE FRONT

The surface generated by all of the points where a wave has the same phase at a particular moment of time. Wave propagation occurs in the direction normal to the wave front, and can be re-garded as wave front motion through the medium. Spherical, plane and other waves are differentiated according to the shapes of their wave fronts. This is a basic notion in *crystal optics* and *holography* (see *Wave front reversal*).

WAVE FRONT REVERSAL

Generation of a conjugate wave beam. Two wave beams E_1 and E_2 are conjugate ($E_2 \sim E_1^*$) or possess mutually reversed wave fronts, if the beams propagate toward each other and their fronts coincide. Wave front reversal of coherent

beams takes place in *nonlinear media*: in *induced light scattering*, and in multiple beam parametric interaction when synchronism conditions are sat-isfied (see *Optical parametric oscillator*). For the case of stimulated scattering (e.g., *Brillouin scatter-ing*) the wave E_1 induces oppositely scattered ra-diation E_2 in a nonlinear medium 1 (Fig. (a)). Inci-dent wave E_1 passes through nonlinear medium 3 and has its wave front distorted. Their interference field records shifted dynamic (real time) holo-grams 2; owing to diffraction from these holo-grams, the scattered radiation is amplified. Here, the component with wave front reversal $E_2 \sim E_1^*$

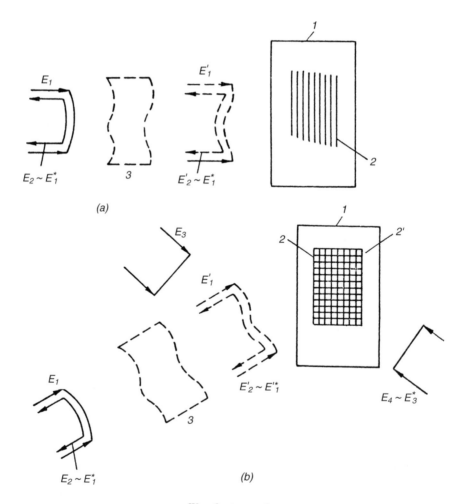

Wave front reversal.

is amplified with the largest gain by virtue of its large overlap with E_1.

In the case of multiple beam interactions (Fig. (b)), two conjugate pump waves $E_4 \sim E_3^*$ are used, mostly with planar wave fronts, as shown. Signal waves E_1 with E_3 and/or E_4 in nonlinear medium 1 record dynamic hologram 2 (and/or $2'$), while E_4 (and/or E_3) transforms by diffracting on these holograms into a wave with wave front reversal $E_2 \sim E_1^*$. Nonlinear solids, e.g., photorefractive crystals (see *Holography*), *fiber glass*, and *semiconductors* bring about an efficient wave front reversal of continuous and pulsed laser beams, especially, in a *holographic laser*. These devices are called *wave front reversal mirrors*, since they not only "reflect" the incident wave, but also compensate the optical inhomogeneities in its path during the flyback (Fig. (b), 3). This unique property of wave front reversal finds applications in laser devices, homing systems, for image transmission through the atmosphere, in multimodal optical fibers, and so on.

WAVEGUIDE, ACOUSTIC

See *Acoustic waveguide*.

WAVE-LIKE MODULATION in liquid crystals

Formation of a configuration of periodically bent or curved liquid-crystalline layers (in *smectic liquid crystals* and *cholesteric liquid crystals*) due to external effects, such as mechanical *uniform strain*, electric or magnetic fields; wave-like modulation appears as a result of instability of the layer system towards disturbance of macroscopic uniformity. Thus, upon stretching a smectic sample perpendicular to the layers, when a deformation γ_0 achieves the critical value γ_c there arises a disturbance in the form of a periodical bending of the layers with wave vector $q \sim (\lambda d)^{-1/2}$, where d is the thickness of the sample, $\lambda = (K/B)^{1/2} \ll d$ is the characteristic length, and K and B are elasticity coefficients (see *Elasticity of liquid crystals*). These layers try to fill the free space with the help of the wave-like modulation in order to preserve the value of their thickness and reduce the elastic stretching energy.

WAVE PACKET REPRESENTATIONS OF WAVE FUNCTIONS

Complete sets of wave functions of *quasiparticles* in crystals, which are derived from the original basis (e.g., the wave functions in the momentum representation, $|\kappa\rangle$) through the so-called *wave packet transformation*

$$|x, k\rangle = \sum_{\kappa \in V_k} A(\kappa \cdot x)|\kappa\rangle,$$

where V_k is a certain volume in the *Brillouin zone* in the neighborhood of the wave vector k, which includes n states κ, $A(\kappa \cdot x)$ is the weighting function of the wave packet transformation, x is the radius-vector of the wave packet center. The completeness of the set of wave functions $|x, k\rangle$ means that k assumes N/n values (N is the number of *unit cells*) and x assumes n values at the sites of the lattice, which is the reciprocal of the k lattice (see *Reciprocal lattice*). The wave packet itself can be a pulse-like or localized travelling waveform consisting of very rapid, short-wavelength, oscillations enclosed by an envelope shape function, e.g., a Gaussian:

$$\exp\left[-\frac{(k' - k_0')^2}{2(\Delta k')^2}\right],$$

several or many wavelengths wide. In the case of free particles, the wave packet representation of wave functions leads to wave packets of the de Broglie and Gaussian form, and in the case of $n = NA \propto \mathrm{e}^{\mathrm{i} \kappa \cdot x}$ this representation of wave functions is identical to the *site representation of wave functions*. Since $|x, k\rangle$ are states of finite uncertainty in coordinates and momenta, the wave packet representation may serve as a convenient basis for the description of inhomogeneous systems in the context of the *quantum kinetic equation*. Various versions of the wave packet representation method of wave functions are used in theoretical studies of localized particles such as *excitons*, electrons, phonons.

WAVES, SURFACE

See *Surface waves, Surface acoustic waves, Surface drift waves*.

WAVE VECTOR GROUP

The totality of elements of a *space group*, that leave the wave vector invariant. The wave vector group is an important subgroup of the spatial symmetry group, because through it the dynamics of carriers in separate valleys of the energy *band structure* of a crystal are connected. With the help of the wave vector group the Bloch functions are formed (see *Bloch theorem*) for the *effective mass* Hamiltonian, the *selection rules* for optical transitions are established, and reactions to external disturbances are taken into consideration. Theoretically it is often possible to neglect translations, and then instead of the wave vector group the term *factor-group* can be used.

WAVE VECTOR STAR

A set of gk vectors, where g is an element of the *point group* of a *space group*, and k is a wave vector. A wave vector star establishes the symmetry of an energy band (see *Band theory*) since $E(k) = E(gk)$. Star vectors in semiconductors may indicate possible equivalent regions of increased *density of states* (valleys; see *Many-valley semiconductors*), but their specific position depends on the particular crystal (compare n-Ge and n-Si). See also *Wave vector group*.

WEAK FERROMAGNETISM

See *Antiferromagnetism*.

WEAK LOCALIZATION

Group of phenomena involving *quantum interference of conduction electrons* in *solids* with a finite electrical conductivity σ at $T \to 0$. We are dealing with a wide class of systems (*amorphous metals and metallic alloys*, disordered *metals*, *degenerate semiconductors*, various two-dimensional conducting structures) at temperatures sufficiently low so that σ is determined by electron scattering off static random potentials localized, e.g., at randomly distributed impurities. With sufficiently strong disorder, quantum interference results in localization of electronic states (see *Anderson localization*). Weak localization occurs in the presence of weak disorder when $p_F l \gg \hbar$ where p_F is the electron Fermi momentum (see *Fermi velocity*), and l is the *mean free path*. The interference results in a quantum correction for σ which is small relative to the parameter $\hbar / p_F l$, but with an unusual dependence on the temperature and external fields. The physical basis of weak localization

is the interference of probability amplitudes of an electron which returns to a given point after passing along the same path in different directions (see *Quantum interference phenomena*).

If magnetic flux Φ penetrates a hollow metal cylinder then the difference in wave phases passed through for a transit n times along the same path in different directions is equal to $2\pi n \Phi / \Phi_0$ where $\Phi_0 = \pi \hbar / e$ is the magnetic flux quantum. Therefore $\sigma(\Phi)$ is an oscillating function of the flux with the period Φ_0. The *anomalous magnetoresistance* has the same nature. The magnetic field B effects the interference of those paths through which the flux is Φ_0. Since the magnitude of the flux differs along different paths, instead of oscillations there is a noticeable monotonic dependence $\sigma(B)$ at such small B values where classical effects are negligible. The magnitude and the sign of weak localization effects are determined by inelastic *relaxation* processes, as well as by the *spin–spin relaxation* and spin–orbit relaxation (see *Spin–orbit interaction*) of electron spins, and this enables one to study these processes quantitatively.

WEAK SUPERCONDUCTIVITY

A number of phenomena and effects in spatially nonuniform superconducting systems containing weak superconducting contacts (weak links or *tunnel junctions, superconducting microbridges*, etc.). The basis of weak superconductivity is *Josephson effects* which appear not only in tunnel contacts, but also in any superconducting system where the *order parameter* is attenuated over short distances $L \leqslant \xi(T)$ (ξ is the *coherence length*). Weak superconducting contacts or weak links (both individual and grouped) are widely used in *superconductor electronics*. The simplest systems are independent contacts and *superconducting quantum interference devices* (SQUIDs) containing one or two weak links. In weak superconductivity a basic feature is the space and time phase coherence of the wave functions of superconducting electrons at macroscopic distances and times, specifying the *quantization of flux* and quantum interference. This provides the opportunity to study the quantum behavior of macroscopic systems in the presence of dissipation, in particular of

macroscopic quantum tunneling, and also (at sufficiently low temperatures) the quantum electrodynamics of superconducting circuits. *Josephson
media* are of special interest. Thermodynamic and
kinetic properties of such media have been intensively studied from the viewpoint of *topological
transitions*, of localization of electrons in superconductors, Coulomb effects, *percolation theory*.
Under certain conditions these systems undergo
random quantum *fluctuations*, macroscopic quantum tunneling and interference, in particular Bloch
oscillations in spatially periodic lattices. These oscillations display a new effect, the appearance of
voltage steps at current values $I_{m,n} = 2\nu e(m/n)$,
where m, n are integers, on the *current–voltage
characteristic* of the contact in the presence of
direct and alternating currents with frequency ν.
Along with Josephson effects and the *quantum
Hall effect*, the new macroscopic interference effect creates the "metrology triad" related to each
other through universal constants of current, voltage and frequency.

Also referred to as weak superconductivity is
an extensive class of effects in spatially homogeneous superconducting systems (films, filaments),
such as *resistive state* mechanisms, nonequilibrium and nonlinear effects arising in regions with
small direct and alternating currents and voltages,
and representing great scientific and practical interest.

WEAR

A change of dimensions (size, shape, mass)
of a solid and of the condition of a *solid surface* because of *strain* and *failure* of the surface
layer under the action of *external friction*. The wear
of a solid can result from interaction with another solid during their relative displacement under compression load, or from interaction with
a moving liquid or gas (e.g., cavitation wear).
The main kinds of wear in interactions between
solids are *abrasive wear* and *contact-fatigue wear*.
Types of wear processes are repeated deformation at elastic or plastic contact, microcutting, and
deep tearing. At elastic contact, a failure results
from slow *crack* growth during *fatigue*. At plastic
contact, the stress concentrators (see *Stress concentration*) form cracks due to repeated nonuniform *plastic deformation* in the subsurface layer,

accompanied by nonuniform heat emission. Crack
merging causes the separation of wear particles.
Microcutting arises when a solid particle, viz., an
abrasive or peened (see *Strain hardening*) wear
particle, is embedded deeply enough into the surface. Deep tearing out occurs when rubbing surfaces fasten due to molecular interactions. As a
rule, several processes are concurrent during wear;
their relative role depends on the friction conditions (load, speed, lubrication, type of frictional
contact) and the properties of the friction pair
components. Of greatest importance among them
are *hardness* (*strength*), *plasticity*, failure mechanism, structure, phase composition, *thermal conductivity*, the chemical interaction between one another and with the environment (see *Corrosion*).
Worn surfaces develop specific nonequilibrium
structures. The wear process is often described in
the context of nonequilibrium thermodynamics.

WEAVON (fr. weave)

Stable motion of a charged particle in the
regime of planar *channeling*. In channeling along
planes an atomic plane is "perceived" by the channeling particle as a continuous charged plane. The
averaged potential of the plane atoms depends
only on a single coordinate, the distance from
the plane. The structure of the energy levels of
the transverse motion of electrons is shown in
the figure. In the potential wells there are several
quasi-discrete levels with negative energy of transverse motion ("transverse" energy). The broadening of the lower levels due to the periodicity of
the plane potential (*band broadening*) is exponentially small while, e.g., the uppermost negative
level may spread into a wide band. The separate
bands with positive transverse energy are well-
defined only at small transverse energies, at large
energies they form a continuum spectrum. With
increasing total particle energy the number of discrete levels grows proportional to $\gamma^{1/2}$, where γ is
the *Lorentz factor* (ratio of total energy to rest energy). For non-relativistic electrons there is only
one level of quasi-bound transverse motion, for
electrons with energy $E \sim 50$ MeV the number
of such levels $N \sim 10$, for $E \sim 5$ GeV, $N \sim 100$,
etc. A weavon can be regarded as a state of a one-

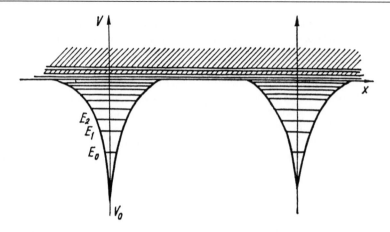

Potential wells and energy levels associated with a weavon.

dimensional atom. A negatively charged particle sort of "weaves" itself into a crystalline plane.

WEDGE-SHAPED CRACK

A sharp-pointed *crack* with an opening converging to an interatomic distance. The concept of a micro-crack as a starting point of a subsequent material *failure* was first introduced in the *Griffith theory*. Still the crack had the shape of an ellipse in that theory, since at that time one could only compute the elastic energy for such a shape. The concept of a wedge-shaped crack having an opening meeting at an interatomic distance was first introduced by P.A. Rebinder in the 1930s. Nowadays the notion of wedge-shaped nucleating microcracks is universally accepted. A refined determination of the shape of a wedge-like crack using the formulation of *elasticity theory* was made by K.H.G. Ashbee, G.I. Barenblatt, and some other researchers. The crack profile was computed, and its edges were shown to have a curved pattern near the opening. During a perfectly brittle failure (see *Brittle failure* of solids) the crack point (tip) remains sharp at the atomic level. In other cases the point gets blunted, which brings about a growth of breaking stress. See also *Failure*.

WEIGERT EFFECT

See *Polarization holography*.

WEISS INDICES

See *Crystallographic indices*.

WEISS THEORY (F. Weiss, 1907)

A theory explaining the magnetic properties of *ferromagnets*.

It is based on two hypotheses. According to the first one, in the temperature range from 0 K to the *Curie point* θ_C in ferromagnets there exists spontaneous *magnetization* which depends on the temperature T, attains maximum values at $T \to 0$ K, decreases with increasing T, and vanishes at $T \geqslant \theta_C$. The temperature θ_C is called the *ferromagnetic Curie point*, and above it the Curie–Weiss law $\chi \sim 1/(T - \theta_C)$ of paramagnetism is obeyed by the susceptibility. Its value is determined mainly by the *exchange interaction*, and in different ferromagnets it varies from fractions of a kelvin to several hundred kelvin. The existence of spontaneous magnetization in ferromagnets is explained by the *molecular field* theory proposed by Weiss. The absence of macroscopic magnetization of ferromagnets in the natural state (without an external magnetic field) is explained by the presence of *domains* in the sample (second hypothesis). According to this hypothesis, the effect of an applied magnetic field on a ferromagnet lies in the redistribution of the fractions of positively and negatively oriented domains. This can occur by the motion of *domain walls* between individual domains, and also by the flipping or rotation of the magnetization within individual domains. These two basic processes determine the character of the nonlinear magnetization curve of ferromagnets, and the existence of a *hysteresis* loop. Both hypotheses have been completely substantiated.

WELDING

A technical process for producing single-piece (welded) joints of homogeneous metallic and non-metallic (plastics, ceramics, glass) materials, as well as combinations of heterogeneous materials (steel–aluminum, titanium–ceramics, etc.). A joint produced by welding is characterized by continuous structural bonding and monolithic formation. Atomic (molecular) bonding between the elementary particles of the connected pieces forms these joints. To weld pieces together, they should be brought into direct contact, i.e. to within distances of the order of an atomic radius (10^{-8} cm). The surface atoms of the two solids may interact directly, with chemical bonding accompanied by *diffusion* taking place.

The existing techniques of welding may be divided into two basic groups: fusion welding (welding in a liquid phase), and pressure welding (welding in a solid phase). During *fusion welding*, the material of the joined pieces forms a single structure. That happens without application of any external forces, by way of melting, mutual mixing, and further *crystallization* of the metal in the seam. During *solid phase welding*, pieces are joined together without melting by the application of considerable *pressure* (compression). Welding that combines both processes is also possible. Considering the basic features of the process of forming a welded joint, such as the form of energy used as well as the technical and technological features of welding, one may divide the available forms of welding into three classes. They are thermal (electric arc, solar radiation, electroslag, electron-beam, laser, thermite, etc.), thermomechanical (contact, diffusion, pressure gas, furnace pressure, etc.), and mechanical (cold, explosion, ultrasonic, etc.). More than a hundred different techniques of welding are currently available, and new technological developments and upgrades of technologies to produce welded joints continuously appear with the passage of time.

Welding makes it possible to join metal pieces of widely varying thickness: from several micrometers to 2 m and more in a single operation. The ambient media used to weld separate pieces or whole units also varies widely: beside the common terrestrial environment, welding may be carried out in outer space under the conditions of zero gravity and deep vacuum (see below), as well as underwater, at depths of 60 m and more. Outer space offers the best environment for electronic beam, solar radiation, diffusion and cold welding. Metals underwater are mainly welded by an electric arc using a melting electrode.

From the operational point of view existing welding techniques may be divided into manual, mechanical (semiautomatic) and automatic. Currently, welding robots are used more and more widely. Welding gave birth to the techniques of *cladding* when layers of substances with differing physical and mechanical properties are built up onto the surfaces of various pieces and units at high temperatures and pressures. This is the so-called welding-on technology that is similar to the techniques for covering pieces with protective and hardening *coatings*.

To heat pieces during welding, one needs to utilize specific energy sources, such as slag baths, electronic beams, plasma arcs, etc. Welded seams feature certain specific micrometallurgical properties. These particularities were combined together to launch a new branch of high-quality *metallurgy – electric metallurgy*.

Currently the processes of electric arc fusion welding and contact welding are the leading techniques in producing welded joints. The scope and application range for the new special welding techniques (e.g., by electronic beam in vacuo, by laser beam, by diffusion, friction, etc.) are also on the rise.

Welding or permanently joining together metals in a vacuum such as in an outer space environment may proceed in a solid phase (*diffusion welding, pressure welding*, or *explosion welding*), in a liquid phase (fusion welding, soldering), and by evaporation and *condensation* of material on to a solid substrate (coating). All these processes result in strong bonding between the atoms and molecules of the pieces joined, so that a physical boundary between the attached pieces vanishes or acquires the properties of an intercrystalline boundary in polycrystalline bodies. The specifics of welding in outer space immediately follow from the environmental conditions one finds there: lack of gravity, plus a deep vacuum that is accompanied by an extremely high evaporation rate. It was as early as the 1960s when S.P. Korolyov and B.E. Paton proposed the idea of industrial scale welding in outer space as a means to assemble and repair large space station constructions. The first experiments were conducted by the cosmonauts V.N. Kubasov and G.S. Shonin on board the "Soyuz-6" spacecraft (October 1969). They successfully tested electronic beam (see *Electron beam technologies*) and electric arc fusion welding (see *Laser technologies*) on titanium and stainless steel.

WELDING DIFFUSION

See *Diffusion welding*.

WENTZEL–KRAMERS–BRILLOUIN APPROXIMATION (WKB)

See *Quasi-classical approximation*.

WERTHEIM EFFECT (C. Wertheim)

The development of a torsional *strain* in a conductor carrying an electric current and located in a longitudinal magnetic field. It can be considered as a *Joule effect* (*longitudinal magnetic striction*) involving the interaction of the magnetic field of the current with the applied magnetic field. The Wertheim effect attains particularly high values in alloys of *rare-earth elements*. It is the basic principle behind highly sensitive emitters and receivers of ultrasound. It is also known as the *Wiedemann effect* and was discovered in ferromagnets by G. Wiedemann (1858).

WETTING

A set of processes which determines the properties of a liquid–solid, or a liquid–liquid interface (surface contact). Wetting is caused by intermolecular interaction forces on the *phase interface*, and is the effect of *adhesion* at a surface. One can distinguish *contact wetting* and *immersion wetting*. The contact type means wetting at a three-phase boundary, e.g., solid, liquid and gas (vacuum or some other liquid). In this case the wetting is characterized by an *edge angle* value θ_0. When $0 \leqslant \theta_0 \leqslant \pi/2$ the liquid is called a *wetting liquid* (*lyophilic*), and when $\pi/2 \leqslant \theta_0 \leqslant \pi$ it is *non-wetting* (*lyophobic*) (see *Lyophilic and lyophobic*

behavior). In the case $\theta_0 = 0$ the wetting is called *absolute (full) wetting*, and if $\theta_0 = \pi$ it is *absolute non-wetting*. Wetting is called immersional when the solid (or other liquid) is fully dipped into the liquid. If the solid is smoothly lifted out from the wetting liquid the *solid surface* remains covered with a liquid layer, but if the liquid is non-wetting then the solid is not so covered.

The *contact angle* changes due to *diffusion* processes, the formation of chemical bonds at the contact, and so on. *Young's equation* (T. Young, 1805) is satisfied during the equilibration of a liquid drop on a smooth, homogeneous, non-deforming plane: $\cos \theta_0 = (\sigma_{sg} - \sigma_{sl})/\sigma_{lg}$, where σ_{sg}, σ_{sl}, σ_{lg} are respectively the surface tension (interphase) coefficients (see *Surface tension*) on the boundary solid–gas, solid–liquid, and liquid–gas.

WETTING ANGLE

The same as *Edge angle*.

WHISKERS

See *Thread-like crystals*.

WIDMANSTÄTTEN STRUCTURE
(A. Widmanstätten, 1808)

A structure of *alloys* characterized by the crystallographically oriented arrangement of secondary phase *crystals* within the main matrix.

A Widmanstätten structure is detected on the surface of an *etched* sample by the appearance of figures of various geometric shapes: squares, triangles, rhombi, etc. It was originally observed by W. Tompson in the course of etching of Pallas iron (1804), by A. Widmanstätten in iron–nickel meteorites (1808) and by N.T. Belyaev in carbon *steels*. This structure forms under accelerated cooling of cast or strongly overheated steel with coarse austenite grains (see *Austenite*). In this case the primary products of austenite decomposition are preferentially located along crystallographic planes, conditioned by the crystallogeometrical correspondence between the lattices of decomposition products and that of the Widmanstätten structure matrix; and as a rule it depends on the *carbon* content and the cooling rate. In pre-eutectic steels (less than 0.8% C) a Widmanstätten structure is formed from the free ferrite extracted along and within *grain boundaries* in the form of plates and needles, and in over-eutectic steels as a result of the deposition of secondary *cementite* plates. Steels with a course-grained Widmanstätten structure are distinguished by low *impact strength* and by increased brittleness (see *Alloy brittleness*) which are to a large extent eliminated by phase *recrystallization*. A Widmanstätten structure is also formed upon the decomposition (see *Alloy decomposition*) of an oversaturated *solid solution*, and at a diffusion-less *martensitic transformation* when the *martensite* crystals are oriented along definite crystallographic planes of the matrix phase. The analysis of the conditions of Widmanstätten structure occurrence clarifies the *phase transition* mechanisms involved in the *heat treatment* of alloys.

WIEDEMANN EFFECT

See *Wertheim effect*.

WIEDEMANN–FRANZ LAW (G. Wiedemann, R. Franz, 1853)

Correlation connecting the electronic *thermal conductivity* κ and the *electrical conductivity* σ in solids.

The law states that for *metals* $\kappa/\sigma = LT$ where L is a constant and T is the absolute temperature. L. Lorenz found (1882) that L is a universal constant L_O called the *Lorenz number*, which according to the electronic theory of metals has the value $L_O = (\pi^2/3)(k_B/e)^2 = 2.45 \cdot 10^{-8}$ W$\cdot\Omega\cdot$K^{-2}. The Wiedemann–Franz law is valid under the condition of elastic *electron scattering* (no change in magnitude of electron energy upon scattering). In non-cubic crystals the Wiedemann–Franz law is satisfied individually for every pair of the corresponding tensor components of electrical conductivity σ_{ik} and of electronic thermal conductivity κ_{ik}. For non-elastic scattering the Wiedemann–Franz law does not hold.

This law has been experimentally confirmed for the majority of metals at high (room temperature and above) temperatures, when the scattering of electrons by *crystal lattice vibrations* is sufficiently elastic, and at the low temperatures when elastic scattering off impurities prevails. There are some exceptions (Be, Mn) without any adequate explanation.

In *semiconductors* the Wiedemann–Franz law is subject to some modification since in most cases

the carriers are non-degenerate so all free carriers participate in the transport processes (in metals only those located at the *Fermi surface* contribute). The presence of more than one type of charge carrier is possible. In the case of bipolar conduction performed by *conduction electrons* and *holes*, most of the heat transfer arises from the generation of *electron–hole pairs* at the hot end of the sample and their recombination at the cold end. As a result $\kappa/\sigma \neq L_O T$, and there is an effective Lorenz number L which depends on the mechanism of carrier scattering. Besides, for bipolar conductivity L depends also on the *band gap* width E_g, on the temperature, and on the partial conductivities of electrons and holes σ_n and σ_p ($\sigma_n + \sigma_p = \sigma$). For non-degenerate semiconductors under elastic scattering L is

$$L = \left(\frac{k_B}{e}\right)^2 \left\{\left(r + \frac{5}{2}\right)\right.$$
$$\left. + \left[2\left(r + \frac{5}{2}\right) + \frac{E_g}{k_B T}\right]^2 \frac{\sigma_n \sigma_p}{\sigma^2}\right\},$$

where r is the power in the power law dependence of the scattering time on the carrier energy, e.g., for scattering by acoustic *phonons* $r = -1/2$, and for scattering by ionized impurities $r = 3/2$ (see *Brooks–Herring formula*). The formula assumes the same scattering mechanism for electrons and holes, if this is not the case then it becomes even more complex. For the case of inelastic scattering of carriers (in particular, scattering off optical phonons at low temperatures), and also for an arbitrary degeneracy of carriers, the quantity L has a complex dependence on the temperature. Nevertheless, in many cases these factors do not change the order of magnitude of L so it is generally feasible to use the Wiedemann–Franz law for estimating the electronic thermal conductivity from measured values of the electrical conductivity.

WIGNER CRYSTAL (E.P. Wigner, 1934)

Crystalline state of a system of electrons within a uniformly distributed, neutralizing positively charged medium. Suppose that there are n electrons in a unit volume, neutralized by a uniformly distributed positive charge (*jellium model*). If each electron is localized within the volume $1/n$ the energy lost per electron is proportional to $n^{2/3}$; while the gain in correlation energy is proportional to $n^{1/3}$. At low electron density the latter is dominant and makes localization of the electrons energetically favorable. A *highly-doped semiconductor* as well as a compensated semiconductor, in which the Bohr radius is large compared with the distance between centers, can serve as good approximation to such a situation. An *electronic crystal* is considered as a collection of equal-sized spherical cells with one electron localized in each, and negligible interactions between cells. Experimentally the Wigner crystal was originally observed by Grimes and Adams for electrons on the surface of liquid helium (see *Levitating electrons*).

WIGNER–SEITZ CELL

A primitive *unit cell* which has the symmetry of the point group of the crystal *Bravais lattice*.

A Wigner–Seitz cell is a polyhedron within which all points are closer to its center than they are to any other *crystal lattice* point. To construct the Wigner–Seitz cell around a particular lattice point called its center, straight line segments are drawn from the center to all closest neighbor

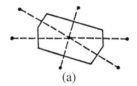

(a)

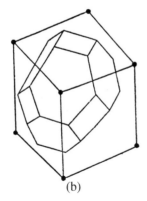

(b)

Wigner–Seitz cell: (a) for a two-dimensional oblique Bravais lattice, and (b) for a three-dimensional body-centered Bravais lattice (truncated octahedron).

lattice points. Then one constructs perpendicular planes through the midpoint of each line segment, and these planes intersect to enclose a volume V_O called the Wigner–Seitz cell (see Fig.). It is a polyhedron of minimum volume $V_O = V_{TOT}/N_{TOT}$ where the overall crystal volume V_{TOT} contains N_{TOT} lattice points. The remaining Wigner–Seitz cells of the crystal can be obtained from this constructed one by translations via a lattice vector. For a cubic Bravais lattice with a high *point group of symmetry* (FCC or BCC) the Wigner–Seitz cell is often replaced by a sphere of the same volume (*Wigner–Seitz sphere*). The subdivision of a crystal into Wigner–Seitz cells allows one to take into account the lattice symmetry in the best possible manner. These cells are used in calculations of electronic structure and bond energies of metals (see *Wigner–Seitz method*), in the theory of electronic crystallization (see *Wigner crystal*), and also for the study of other problems of solid-state physics. For systematization of electronic states and for a *band structure* description one makes use of the Wigner–Seitz cell in *reciprocal space* (*k-space*) where it is called the first *Brillouin zone*. In theoretical crystallography a Wigner–Seitz cell is sometimes called a *Voronoi–Dirichlet domain* (see *Two-dimensional crystallography*).

WIGNER–SEITZ METHOD (E. Wigner, F. Seitz, 1933)

A method for calculating the energy spectrum, wave functions and bonding energy of electrons in *alkali metals*. It was originally used to calculate the bottom of the *conduction band* of metallic *sodium*. In the Wigner–Seitz method the metal is considered as a collection of *Wigner–Seitz cells*, the centers of which coincide with the sites of the *crystal lattice*. To obtain the wavefunction the Shrödinger equation (with proper boundary conditions) is solved within a *unit cell*, and then extended to the whole crystal using the *Bloch theorem*: $\Psi_k(r + R) = e^{ikR}\Psi_k(r)$. This single-electron Schrödinger equation is based on the following assumptions. (1) There is a single *conduction electron* within each cell. (2) Wigner–Seitz cells are overall electrically neutral, so the Coulomb interaction between electrons from neighboring cells can be neglected. (3) The potential $U(r)$ within the cell is assumed to be spherically symmetric; it is small outside the ionic shell and coincides with the ion potential within the ionic core. As a result, for the Bloch function of the electron $\Psi_k(r) = u_k(r)e^{-ikr}$ an equation is obtained which at $k = 0$ coincides with the free atom Shrödinger equation

$$\left[\frac{(p + \hbar k)^2}{2M} + U(r)\right]u_k(r) = E_k u_k(r).$$

The solution of this equation will be the true wave function of the *electron in a crystal* if the requirement is met that the normal derivative of $\Psi_k(r)$ be zero at the cell boundary: $\hat{n} \cdot \nabla \Psi_k(r)|_{\text{boundary}} = 0$. This condition plays a principal role in the Wigner–Seitz method. It provides the band character of the spectrum and reduces the energy relative to that of a free atom. This equation with the prescribed boundary condition is solved (at $k = 0$) using standard methods by expanding the wavefunction $u(r)$ in terms of spherical harmonics $u(r) = \sum_{lm} Y_{lm}(\theta, \varphi)\chi_l(r)$. When explicit calculations are carried out the Wigner–Seitz cell is often approximated by a sphere of equal volume. In this case, the boundary condition has the simple form $\partial u_0/\partial r|_{r=r_S} = 0$, and only the term with $l = 0$ remains in the expansion of $u(r)$. The Wigner–Seitz method does not take into account electron correlations, exchange, or Coulomb effects in metals. Nevertheless, the method provides results which agree well with experiment for the calculation of the *cohesion* energy and other electronic properties of alkali metals which have almost spherical *Fermi surfaces* (Na, K). The method was also used by Wigner for electronic crystallization theory (see *Wigner crystal*).

WIND OF PHONONS

See *Phonon wind*.

WITTKE–GARRIDO GROUPS (O. Wittke, I. Garrido, 1959)

Groups which map the dependence of a property (color) of a system on a group element. They describe the symmetry of crystallographic polyhedra with colored facets. This group is determined by its subgroup, which preserves the fixed color (property). Wittke–Garrido groups along with the usual ones also contain the set of colored symmetry elements (see *Colored symmetry*).

There are 279 *colored symmetry groups* which is equal to the number of all the subgroups of the 32 point symmetry groups. Thirty two of them are uncolored, and of the remaining 247 colored groups there are 35 which are determined by the color-preserving non-equivalent subgroups, distinguished only by the rotational symmetry elements through an angle of 30° or 45° around the common axis. Of $211 + 1$ (classical group C_1), 139 groups are determined as invariant, and 73 are noninvariant subgroups of the symmetry groups (proper Wittke–Garrido groups). Of the 139 invariant groups, 58 of them are two-colored (*Shubnikov groups*), 18 are polycolored (*Belov groups*), etc. The Wittke–Garrido groups can be used in the investigation of the distribution of *defects* in crystals, in the study of *phase transitions*, etc.

WKB APPROXIMATION

See *Quasi-classical approximation*.

WORK FUNCTION

The energy expended to remove an electron from either a *solid* or liquid material into a vacuum with zero kinetic energy. When an electron enters a solid from the vacuum, an energy equal to the work function is released and converted into kinetic energy of the remaining particles. The work function from a *metal* equals the difference between the electron energy in vacuo and its *Fermi energy*. In semiconductors and insulators, it is usually measured from the bottom of the *conduction band* (the *internal work function*). If it is measured from the Fermi energy it is called the *external work function*. Part of the work function value results from the presence of an *electric double layer* at the outer boundary of the solid that forms because of the different environments in which the surface atoms find themselves on the side of the solid and on the side of the vacuum or other contacting medium. The double layer forms in metals due to the shift of the center of gravity of the electron clouds of atoms in the surface plane of the solid with respect to the positions of their nuclei. The depth of the electric double layer in semi-conductors and insulators significantly exceeds the atom-to-atom distance. The *dipole moment* of a surface double layer plays a major role in determining the value of the work function.

The work function is a characteristic of a *solid surface*. Faces of the same *crystal* formed by different crystallographic planes or covered by different atoms feature differing work functions. For example, the work function from the tungsten (110) face is 5.3 eV, while that from the (111) face is 4.4 eV.

The work function may change significantly due to *adsorption* of various atoms and molecules at the surface. Atoms of metals with low ionization energy (e.g., Cs) acquire a dipole moment pointed toward the vacuum during the process of adsorption. In certain cases this may lower the work function by $\sim$1 eV.

The absolute value of the work function is measured by the amount of heat that must be brought to the body to keep its temperature constant when a thermionic current is being emitted. Other evaluations use the temperature dependence and the total value of the thermionic current. When dealing with metals and degenerate semiconductors the same evaluation may be done using the concept of the red limit (low-energy limit) of *photoelectron emission*.

WORK HARDENING, cold hardening

Hardening of *metals* and *alloys* (see *Hardening*) with the help of *plastic deformation*, accompanied by changes in physical and mechanical properties and, sometimes, in phase composition of the materials. An indicator of work hardening is the value of the latent energy that characterizes an increase of the internal energy of a metal because of structural changes in the material of the matrix, and other *phases*. An increase of internal energy in pure metals and *solid solutions* takes place due to the increase of the *defect* density, which also can be accompanied by a proportionate increase of the specific volume. Under aggravated *strain* this proportionality may fail because of the formation of *submicrocracks*, *pores*, *fissures*, etc., i.e. because of overpeening. *Overpeening* is the main cause of *embrittlement* of a deformed metal, and reduction

of its *construction material strength*. Work hardening proper and *phase work hardening* are distinguished: phase work hardening is caused by external loads (e.g., by rolling, *drawing*) or by *phase transitions*. Heating a metal that has been exposed to work hardening causes *dehardening (restoration of crystals, polygonization, recrystallization)*. Work hardening often leads to an increase of *strength* and a decrease of *plasticity* of a metal. Work hardening finds application in metal working, e.g., in operations of surface finishing, hardening through cold rolling, drawing, etc.

WORK OF DEFORMATION

Work performed by external forces during deformation of a solid. In general the work of deformation $d\omega$ of a unit volume of solid is

$$d\omega = \sigma_{ij}\, d\varepsilon_{ij},$$

where σ_{ij} are components of the *stress tensor*, and $d\varepsilon_{ij}$ are the components of the *strain tensor*. During single axis extension under load p, $d\omega = p\, d\varepsilon$, where $d\varepsilon$ is the relative elongation of the specimen. In the case of pure *shear* one has $d\omega = \tau\, d\gamma$ where τ and γ are the shear stress and strain, respectively (see *Shear strain*). For a complex *state of stress* the work of deformation may be expressed via the invariants of stress and strain (see *Invariants of elasticity theory*):

$$d\omega = \sigma_e\, d\varepsilon_e + 3\sigma\, d\varepsilon,$$

where the first term is the work of changing the specimen shape which can be expressed via the invariants of shear stress σ_e and strain ε_e. The latter two have the following appearance in Cartesian coordinates:

$$\sigma_e = \frac{1}{\sqrt{2}}\big[(\sigma_{xx} - \sigma_{yy})^2 + (\sigma_{yy} - \sigma_{zz})^2 \\ + (\sigma_{zz} - \sigma_{xx})^2 \\ + 6(\sigma_{xy}^2 + \sigma_{yz}^2 + \sigma_{zx}^2)\big]^{1/2},$$

$$\varepsilon_e = \frac{\sqrt{2}}{3}\big[(\varepsilon_{xx} - \varepsilon_{yy})^2 + (\varepsilon_{yy} - \varepsilon_{zz})^2 \\ + (\varepsilon_{zz} - \varepsilon_{xx})^2 \\ + 6(\varepsilon_{xy}^2 + \varepsilon_{yz}^2 + \varepsilon_{zx}^2)\big]^{1/2}.$$

The second term corresponds to the work of changing the specimen volume, and is expressed via the average (hydrostatic) pressure

$$\sigma = \frac{1}{3}(\sigma_{xx} + \sigma_{yy} + \sigma_{zz})$$

and the relative change of the specimen volume

$$\varepsilon = (\varepsilon_{xx} + \varepsilon_{yy} + \varepsilon_{zz}).$$

If the volume does not change during deformation ($d\varepsilon = 0$) and the *deformation diagram* is known ($\sigma_e = f(\varepsilon_e)$), the specific work for *uniform strain* may be calculated by integrating the deformation curve:

$$\omega_n = \int_{\varepsilon_e}^{\varepsilon_1} \sigma_e \, d\varepsilon_e.$$

The above expressions hold for elastic *strain* (see *Elasticity*) and for *plastic deformation*. The work of elastic deformation is unambiguously defined by the points of the displacements of the deformed specimen, while the work of plastic deformation also depends on the path of the deformation.

In the course of doing elastic deformation work, the energy from external forces transforms into potential energy of the deformed specimen.

The energy per unit volume of an isotropic elastically deformed specimen (see *Isotropy of elasticity*) is

$$\omega = \frac{E}{2(1+v)}\left(\varepsilon_{ij}^2 + \frac{v}{1-2v}\varepsilon_{kk}^2\right),$$

where E and v are *Young's modulus* and the *Poisson ratio*. For an anisotropic crystal specimen (see *Anisotropy of elasticity*) we have for the energy

$$\omega = \frac{1}{2}c_{ijkl}\varepsilon_{ij}\varepsilon_{kl},$$

where c_{ijkl} are the *elastic moduli* of the crystal. When the external load is lifted, this energy transforms into elastic vibrational energy, and subsequently, as a result of damping, it dissipates as heat. During plastic deformation a far greater part of the energy (90–95%) transforms into heat during the process of deformation itself, and the remainder is absorbed by defects (*dislocations, disclinations, point defects*, residual elastic stresses at *grain boundaries*, etc.), formed during deformation. That latter energy is called accumulated (stored) energy from the deformation. A very small amount of energy is released as *acoustic emission* (sound) during the process of deformation.

X-RADIOGRAPHY

Totality of methods for investigating solids, and also amorphous substances and liquids, associated with the application of X-rays. X-radiography can be subdivided into *X-ray structure analysis*, *X-ray spectrum analysis* (including *Auger spectroscopy*, *X-ray electronic spectroscopy*, etc.), X-ray defectoscopy (see *Defectoscopy*). See also *High-temperature X-radiography*, *Low-temperature X-radiography*, *X-radiography in a magnetic field*.

X-RADIOGRAPHY AT HIGH TEMPERATURES

See *High-temperature X-radiography*.

X-RADIOGRAPHY AT LOW TEMPERATURES

See *Low-temperature X-radiography*.

X-RADIOGRAPHY IN A MAGNETIC FIELD

X-ray diffraction methods for studying the problems of solid state physics and physics of magnetic phenomena associated with the effect of an applied *magnetic field*. X-radiography is applied in magnetic fields up to several tesla, generated with the help of *permanent magnets*, electromagnets, and *superconducting magnets*. One of the main tasks of X-radiography is the investigation of thermodynamic, kinetic and morphologic aspects of first-order and second-order *magnetic phase transitions*, and plotting the results on field versus temperature plots ($B–T$ diagrams) of magnetic substances. Also investigated is the forced *magnetostriction* of ferro-, antiferro- and ferrimagnets, and also the evolution of processes of *magnetic domain structures* in magnetic fields. Studies have been made concerning the deformation of magnetic structures of magnetically ordered solids under the effect of a magnetic field. With the help of precision methods of X-radiography in a magnetic field the influence of the field on the structural instability of superconductors is studied. The investigation of *liquid crystal* structure is a specific branch of X-radiography in a magnetic field; the role of the magnetic field in these investigations involves the orientation of the molecules in a specific direction.

X-RAY ACOUSTIC RESONANCE

Resonance suppression of the *anomalous passage of X-rays* by a transverse ultrasonic wave with wavelength λ which is equal to the *extinction length* l_E. The direction of ultrasound propagation divides the angle between the direct and reflected X-ray beams in half. In a perfect crystal under the conditions of *Bragg diffraction* two X-ray Bloch states with different values of the absorption coefficient are formed. The state with an absorption coefficient close to zero reaches the output surface of a sufficiently thick crystal. Scattering into the second state from the ultrasonic displacements occurs, as a result of which the absorption coefficient of the anomalously transmitted wave increases. The condition of resonance $\lambda = l_E$ follows from the *quasi-wave vector* conservation law for such scattering. Typical resonance values of $\lambda = 10$ to 100 μm. At resonance the sensitivity to acoustic deformations increases by 2 to 3 orders of magnitude (deformations of $\sim 10^{-9}$ reduce the intensity of an X-ray beam by several times). See also *Diffraction of radiations in ultrasonic field*.

X-RAY ANOMALOUS PASSAGE

See *Anomalous passage of X-rays*.

X-RAY CAMERA

A device used in *X-ray structure analysis* for the photographic recording of the diffraction pattern of X-rays scattered from a sample. Common structural elements of an X-ray camera are the following: the device forming the primary beam, a sample holder with an alignment mechanism, and a photographic cassette. An X-ray camera is used to obtain *X-ray diffraction patterns* from monocrystalline, polycrystalline, amorphous, and vitreous materials. In an X-ray camera both the characteristic radiation and that of the continuous spectrum are used. Several types of X-ray cameras may be distinguished: those for investigations at ordinary temperatures and pressures, high- and low-temperature cameras, those for investigations under *high pressure*, etc.

X-RAY DEFECTOSCOPY

See *Defectoscopy*.

X-RAY DIFFRACTION

The phenomenon of the coherent scattering of X-rays (photons) within *crystals* (see *Bragg diffraction*, *Laue diffraction*), liquids and other condensed media under the conditions where the photon wavelength is comparable with the average distance between the atoms (molecules) of the medium. X-ray diffraction is the basic method for investigating the atomic structure of materials (see *X-ray structure analysis*, *X-radiography*, *X-ray spectrum analysis*, *X-ray topography*).

X-RAY DIFFRACTION ANALYSIS

See *X-ray structure analysis*.

X-RAY DIFFRACTION PATTERN, X-rayogram

A photographic image of a diffraction pattern of scattered X-rays fixed at a light-sensitive film or plate. In an X-rayogram, the diffraction picture appears at a noticeable solid angle with the simultaneous recording of a large number of reflections. X-rayograms allow analyzing diffraction patterns with a complex scattering geometry. According to the degree of blackening of the X-rayogram lines, one can judge the intensity of corresponding diffracted X-rays. An X-rayogram is used for *X-ray structure analysis*, and for the *X-ray spectrum-analysis* of materials. See also *Asterism*.

X-RAY DIFFRACTOMETER

A device for measuring the intensity distribution of X-rays scattered in different directions with respect to the primary incident X-ray beam. It is used for the *X-ray structure analysis* of crystalline, amorphous, and liquid materials. High accuracy in determining the angular coordinates and the intensities of the X-rays scattered in various particular directions is characteristic of such devices (compared to *X-ray cameras*). The basic elements of an X-ray diffractometer are as follows: source of X-rays (e.g., *X-ray tube*), *X-ray goniometer* where the sample under investigation is located; recording system which consists of the X-ray detector (scintillation, proportional and semiconductor counters), and the electronic units which process the information provided by the detector; as well as sensors of the angular position of the sample and detector. The measurement of the angular radiation distribution diffracted at the sample is carried out by varying the relative positions of the X-ray source, sample and detector (see Fig. in *X-ray diffractometry*), and thereby establishing the diffracted beam direction and intensity with the help of the recording system. See also *Texture diffractometer*, *X-ray diffractometry*.

X-RAY DIFFRACTOMETRY

Basic method of recording the diffraction pattern in *X-ray structure analysis* based on the use of an X-ray detector. A Geiger–Müller counter, scintillation counter, proportional counter, etc., are used as a detector. Recording of the diffraction pattern is carried out through sequential motions of the detector. Position-sensitive detectors are used to fix simultaneously the intensities and positions of several reflections.

An *X-ray diffractometer* is a device for carrying out X-ray diffractometry. To obtain reflections from atomic planes with different crystallographic indices hkl with the aid of monochromatic radiation, the detector and the sample are simultaneously rotated in order to fulfill the *Bragg law*. In most cases use is made of the *Bragg–Brentano geometry of measurements* (W.L. Bragg, J.C.M. Brentano) (see Fig.), whereby the detector rotation angle 2θ is twice the sample rotation angle θ.

Sometimes Bragg reflections with different interplane distances d_{hkl} are obtained with polychromatic X-rays using X-ray recording with the

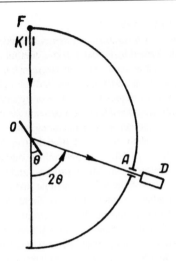

Sketch of the Bragg–Brentano geometry of measurements using an X-ray diffractometer. The symbols designate: A, restricting slit at the detector; D, detector; F, focus of the X-ray tube; K, collimator restricting the primary beam; O, sample; θ, sample rotation angle.

help of a multichannel amplitude analyzer (*energy-dispersion X-ray diffractometry*). At fixed detector positions, the intensity distribution is presented as a function of wavelength λ with maxima used to calculate the corresponding d_{hkl} from the Bragg law.

According to their purpose, X-ray diffractometers are subdivided into the following types: those of general purpose (study of both poly- and monocrystalline samples); monocrystal ones (determining new structures and refinement of known structures, investigating structural variations in *monocrystals* during external influences); polycrystalline devices; texture diffractometers (investigation of sample structure); those intended for the investigation of *thin films*. In the latter case the amplification of the reflection intensity, the survey geometry with a fixed sample having an operating surface (the plane of film) subtending a small angle (1 to 10°) to the primary beam, and collimation of the diffracted beam are performed by two systems of restricting slits (so-called *Soller slits*), horizontal and vertical.

The main tasks of X-ray diffractometry. Using *phase analysis*, particular crystal phases are identified (qualitative phase analysis) and their concentration in the sample is determined (quantitative phase analysis). Identification is based on the comparison of interplane distances d_{hkl} and intensities of reflections I_{hkl} in the *X-ray diffraction pattern* of the sample with the d_{hkl} and I_{hkl} in X-rayograms of reference materials which are tabulated in X-ray diffractometry card files. Quantitative phase analysis is based on the proportional dependence of the integrated intensity of the reflections on the concentration of specified phases. Phase analysis is used for investigating *phase transitions* (polymorphous transformations, transitions from the crystalline to the amorphous state); phase *size effects* (phase transformations caused by the smallness of the particle dimensions); for the analysis of source materials before carrying out physicochemical studies.

Unit cell parameters are determined from the d_{hkl} values of the diffraction patterns. The relative measurement error is $\delta d/d = -\cot\theta\,\delta\theta$, where $\delta\theta$ is the error of Bragg angle evaluation. In solid state physics, X-ray diffractometry is used to identify the terms of isomorphous series; to calculate concentrations of isomorphous inclusions in *solid solutions* (commonly the *Vegard rule* is used); to investigate phase transitions (accompanied by an abrupt change of volume); to determine heat expansion tensors and elastic tensors, etc.

The determinations of *coherent scattering regions* and of structural microdistortions are carried out according to the diffraction profiles of the reflections. The coherent scattering region size Λ (in the direction perpendicular to the crystallographic plane of the reflection) is calculated with a knowledge of the half-width of the reflection $2\Delta\theta_\Lambda$, when the broadening arises from the smallness of dimension Λ, according to the formula $\Lambda = K\lambda/(2\Delta\theta_\Lambda\cos\theta)$, where K is a coefficient close to 1 which depends on the shape of the coherent scattering region. Broadening of the reflection caused by the scattering along the interplane distances $\Delta d/d$ caused by the microdistortions of structure, by nonuniformity of the sample, by the internal macrostresses, is $2\Delta\theta_{\Delta d/d} = f\tan\theta$, where f depends on concentration and on the type of defects, on the structure of the crystal, and on the crystallographic reflection plane. The width of the diffraction profile at half-height is assumed to lie under the half-width of the reflection. The dimensions of the coherent scattering region, functions of its distribution according to the sizes and microdistortions of structure, are calculated with the help of harmonic analysis of the reflection diffraction profile. Information about static and dynamic displacements of atoms with respect to their ideal geometric lattice sites within the structure is obtained by the weakening of the integrated intensities of reflections due to the line broadening coefficient (*Debye–Waller factor*), and also by taking into account the *diffuse scattering of X-rays*.

The investigation of the sample *texture* proceeds by obtaining the distribution (at the sphere) of the densities of normals to crystallographic planes (so-called *pole figure*). It is used to determine the preferable orientation of the *crystallites* for the plastic treatment of metals, of *crystallization* in orienting fields, and during the course of *spraying* thin films.

X-RAY DIFFRACTOMETRY WITH ENERGY DISPERSION

See *Energy-dispersion X-ray diffractometry.*

X-RAY EMISSION SPECTROSCOPY

The field of X-ray spectroscopy associated with investigations of X-ray characteristic radiation which appears at the transitions of electrons from outer filled levels of atoms (or molecules) to vacancies in their deep-lying levels. Due to the *selection rules* the investigation of the intensity distribution in X-ray emission bands (transitions from the valence band to inner shell levels of atoms within the crystal) allows the study of partial *densities of states*. The method possesses high sensitivity with respect to the environmemt of the given radiating atom, and allows establishing the local symmetry of atomic sites, the constituent elements, and the types of chemical bonds. From the chemical shift of lines that arise from transitions involving filled levels, it is possible to determine the form of charge transfer between the atoms of the components of the alloy or chemical compound.

X-RAY EXCITONS

The core *excitons* or high-energy excited states of solids formed by a *hole* in some internal (core) atomic electron shell, with an electron bound to the resultant ion (atom in this excited state). X-ray excitons decay mainly by radiation or by an Auger process (see *Auger effect*), and optical excitons as well as *trions* can be found among their decay products. Due to their short lifetime (on the order of 10^{-13} to 10^{-10} s) determined mainly by the probability of an Auger process, and due to weak overlap of the wave functions of the inner shells of adjacent atoms in solids, these excitons are practically immobile. X-ray excitons are identified by a narrow band in the *X-ray absorption spectra* in solids, located at energies somewhat below the ionization energies of the corresponding inner electron shells, and also by specific decay products (*photons*, Auger electrons of particular energy, trions).

X-RAY FINE STRUCTURE

See *Fine structure of X-ray absorption spectra.*

X-RAY GONIOMETER

A device to determine or establish simultaneously the direction of a diffracted X-ray beam at a sample under investigation, and also the angular orientation of the sample with respect to the direction of the primary incident beam; this device is a basic component of *X-ray diffractometers.* An X-ray goniometer equipped with a film cassette is an *X-ray camera* intended for obtaining *X-ray diffraction patterns*, where it is possible to establish the sites of a *reciprocal lattice* of the crystal under study. There are X-ray goniometers that allow fixing the sites of the reciprocal lattice lying in one plane (X-ray goniometers of the Weissenberg design, cameras for reciprocal lattice photography) and also special-purpose X-ray goniometers (e.g., a texture type).

X-RAY INTERFEROMETRY

Application area of solid state physics where X-ray interference patterns are obtained with the aid of *interferometers* and other phase sensitive devices, and investigations are carried out with them. Coherent beams for *X-ray interferometers* are obtained by the reflection of X-rays from atomic

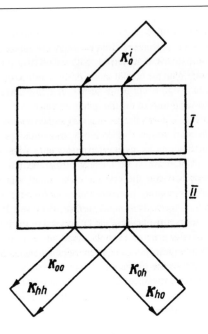

Fig. 1

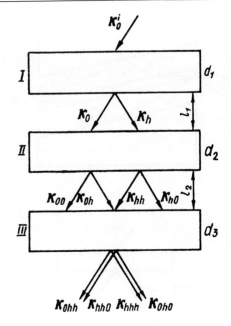

Fig. 2

planes of *crystals*. Different types of X-ray interferometers are known: there are double- and
triple-monocrystal interferometers where, respectively, two or three blocks are cut out of the
same crystal without their separation, based on the
same mounting, as shown in Figs. 1 and 2; there
are also triple-block, double-crystal interferometers, the two latter crystalline blocks of which are
monocrystalline and have the same basis, and the
first one is separated, as indicated in Fig. 3, etc.

X-ray interferometers are also distinguished
according to the type of reflection: Laue reflection (L) or Bragg reflection (B). Fig. 1 shows
an LL-interferometer, Figs. 2 and 3 present interferometers with LLL-reflections; Fig. 4 shows a
BBB-interferometer, and Fig. 5 illustrates an interferometer with mixed LBL-reflections. Many interferometers of novel types have been created,
including multiblock ones (with four or more
blocks) and also multiwave ones. The principle
of operation of X-ray interferometers will be explained using the examples of the LLL- (see
Fig. 2) and LL-interferometers (see Fig. 1).

Fig. 2 shows a triple monocrystal X-ray interferometer with the reflecting planes of all the

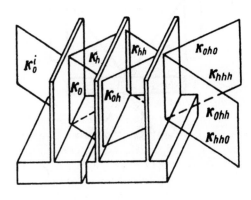

Fig. 3

crystals perpendicular to the surfaces of the input
and output X-radiation. The primary wave with the
wave vector K_0^i is incident at crystal I at the Bragg
reflection angle. The waves with wave vectors K_0
and K_h emerge from crystal I and fall at crystal II
in the directions of incidence and reflection. From
crystal II the waves with wave vectors K_{00}, K_{0h},
K_{hh} and K_{h0} emerge and fall in the same directions but at the surface of crystal III. The waves
with wave vectors K_{0h} and K_{hh} superimpose and

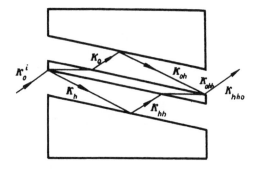

Fig. 4

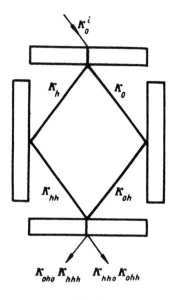

Fig. 5

ter is not fulfilled the conditions of parallelism of the wave vectors are violated, and there appear periodic distributions of intensity called *Moiré patterns*. With the help of those, it is possible to detect even insignificant bends of the reflecting surfaces and variations of the interplane distance.

Fig. 1 shows the ray diagram in the two-crystal LL-interferometer. Two-block monolith system can function as an interferometer only when the incident beam is sufficiently broad and the distance between the crystals is so small that the waves emerging from crystal I superimpose with one another at the input surface of crystal II. If the reflecting planes of the last crystal of the interferometer are rotated to the small angle α around the axis perpendicular to the input surface, with respect to the planes of the other crystals, or the interplane distances of this crystal are changed to Δd, then in the beams coming out of the interferometer Moiré pictures appear with respective periods

$$\Delta_\varepsilon = \frac{d}{\varepsilon} \quad \text{and} \quad \Delta_d = \frac{d^2}{\Delta d} \cos\theta,$$

where θ is Bragg angle (see *Bragg law*). The minimum measured value of the angle ε and relative variations of interplane distance $\Delta d/d$ are restricted by the maximum achievable value of the Moiré picture period, which in its turn is restricted by maximum attainable cross-section of the primary coherent beam ($\Delta_\varepsilon = 1$ cm corresponds to $\varepsilon \approx 10^{-8}$ rad, and $\Delta_d \approx 1$ cm to $\Delta d/d \approx 10^{-8}$). Devices have been created and X-ray interferometry methods have been developed for the performance of metrology investigations with ultrahigh accuracy: *X-ray length meters* (*ångström-meters*), *refracrometers* (see *Refractometry*). Combinations of X-ray and optical interferometers for absolute measurement of the lattice constants, and of the other fundamental constants to the accuracy of hundredth parts of ångström units, are especially effective.

X-RAY LASER, raser

A device for the generation and amplification of coherent X-rays; a *raser* (Roentgen ray Amplification by Stimulated Emission of Radiation). An *X-ray laser based on free electrons* involves the stimulated radiation of relativistic free electrons

interfere, and therefore, further on there are waves which are a result of this interference which will be considered like these waves. Due to the superimposition of these waves the waves with vectors K_{hh0}, K_{0h0}, K_{0hh} and K_{hhh} emerge from crystal III. In ideal interferometers where orientations of the reflecting planes and interplane distances between the crystals are strictly the same ($l_1 = l_2$) the conditions $K_{hh0} \parallel K_{0hh}$ and $K_{0h0} \parallel K_{hhh}$ are satisfied, and therefore, the intensities of interference beams emerging from the ideal interferometer have a uniform distribution. If at least one of the requirements of ideality of the interferome-

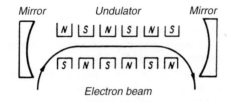

Mirror   Undulator   Mirror

Electron beam

A scheme of the X-ray free electron laser showing the undulator, mirrors and electron beam.

undergoing, together with their forward motion, an additional vibrational movement brought about by the pumping action of space-periodic magnetic (magnetostatic pumping) or electric (electric pumping) fields, or of an intense electromagnetic wave incident on an electron beam (electromagnetic pumping). Pumping of the first type occurs in undulators and "ubitrons", pumping of the second type may take place during the *channeling* of electrons in *crystals*. Electromagnetic pumping is used for the transformation of high-power long-wave SHF radiation to short-wave radiation (IR, optical, UV) through its scattering by a relativistic electron beam. One may make use of "wigglers" of different types (multipole or spiral "snakes") (see Fig.).

Multipole "snakes" may have a flat or spiral form. Thus, the trajectories in undulators may be divided into two classes: (1) flat ones, the simplest example of these being sinusoidal trajectories, and (2) trajectories close to a helical form. Both types serve to provide high intensities in a narrow spectral region (see Fig.).

A typical feature of X-ray lasers using free electrons, which distinguishes them from the devices of classical electronics, is that the use of relativistic effects provides a possibility for the generation of short-wave radiation in macroscopic systems, and also a possibility of a smooth adjustment of the frequency over a broad range with the help of the variation of macroscopic parameters: such as the electron beam energy, or the period of the magnetic field of pumping.

An X-ray laser with nuclear pumping consists of a cylindrical system of thin metallic fibers surrounding a nuclear explosion device. Thermal X-radiation which results from the nuclear fission stimulates X-ray emission from the atoms of fibers

during several seconds. The formation of the directed X-ray laser radiation is brought about by the choice of a cylindrical geometry for the radiating rods. The divergence of the beam is determined by the ratio of the fiber dimensions d/L (L is rod length, d its diameter), and by diffraction effects in accordance with the expression

$$\theta^2 \sim \left\{ \frac{d^2}{L^2} + \frac{\lambda^2}{d^2} \right\},$$

where λ is the laser radiation wavelength. The divergence can be quite modest, e.g., at a distance of 4000 km an X-ray laser illuminates a spot with a 200 m diameter. It is possible for such a device to generate a pulse of soft X-rays at energies of about 1 keV.

The idea of creating a *gamma ray laser* (*gaser*) requires the presence of comparatively long-lived metastable quantum levels of the excited nuclei. The conserved energy from the pumping action is stored at these levels. The energy "drop" is brought about by the excitation of transitions from metastable levels to closely spaced upper levels with the subsequent transition of the nucleus back to its ground state. Additional excitation can be carried out by electromagnetic radiation (of radio frequency or X-ray range). Gaser is a term taken from the American literature, it is an acronym for the expression "Gamma-quantum Amplification by Stimulated Emission of Radiation", which is similar to the rationale for the terms *maser, laser* and *raser*.

There is a speculation that a gamma laser could possess tremendous radiation power; e.g., using the metastable excited state of the ^{57}Fe nucleus it could generate hard electromagnetic radiation with $\lambda \sim 0.1$ nm. It shall be noted that a developed gamma laser would be an amplifier of spontaneous radiation, but the problem of creating resonator mirrors for a true laser has yet to be solved for both gamma radiation and X-rays.

X-RAY LUMINESCENCE

Luminescence that arises from the action of X-rays. X-ray luminescence is usually associated with *recombination* of the charge carriers created within the material upon the absorption of X-rays. *Impurity atoms* and ions, as well self-localized carriers (usually *holes*), can also be *luminescent centers.* X-ray luminescence is used for the detection of X-rays, and for the visualization of images created by X-rays. The most often used *X-ray luminophors* are: $CaWO_4$, $ZnS \cdot CdS$-Ag, $Ba_3(PO_4)_2$-Eu, CsI-Na, NaI-Tl.

X-RAY MICROSCOPY

Totality of methods of investigation of the microstructure of materials with the help of X-rays. A device for applying the methods of X-ray microscopy is called an *X-ray microscope*. The basic elements of the microscope are as follows: source of X-rays (*X-ray tube*, generator of synchrotron radiation); system of X-ray optical elements (*X-ray mirrors*, diaphragms, zone plates, etc.); image recording system (photographic film, fluorescent screen, image transformer, device for image processing and computer display). The image in X-ray microscopy is obtained according to the diagram (see Fig.): S is the source of X-radiation, O is the object, D is the image detector, CZP is condenser zone plate, MZP is microzone plate.

Depending on the method of forming the image, the following methods of X-ray microscopy can be distinguished:

- *Contact X-ray microscopy*, Fig. (a), is based on obtaining a shadow image the natural size of the object in contact with the photographic plate. The image reflects the difference in the adsorption of the X-radiation at various positions of the object. Resolution is limited by the degree of granularity of the photographic material, and can reach 0.1 μm. It is used mainly for investigations of biological specimens.

- *Projection X-ray microscopy*, Fig. (b), is based on image formation by a divergent beam of X-radiation passing through the object and emerging from the microfocus (diameter 0.1 to 1 μm) of the X-ray tube in combination with the point diaphragm. The mechanism of image formation is the same, but due to the divergence of the X-ray beam it is possible to achieve a significant enlargement of the image.

- *Diffraction X-ray microscopy*, Fig. (c), provides natural size images of sufficiently perfect crystalline samples via X-radiation diffracted by a system of reflecting planes. Diffracted radiation usually carries information, not necessarily on the object itself (e.g., on a structural defect) but on its field of long-range deformations, which displace the diffracted beam from accurate Bragg positions, and this serves as a source of contrast (see *X-ray topography*). Diffraction X-ray microscopy is used in solid state physics and in materials science to observe separate structural defects, to monitor the structural perfection of the crystals, to investigate deformed surface layers, etc.

- In *reflection X-ray microscopy*, Fig. (d), image A' of the point of object A is formed by an X-ray beam reflected from one or several X-ray mirrors or concave reflectors operating at *total*

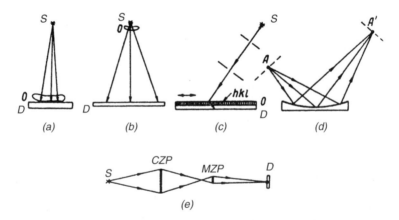

Types of X-ray microscopy: (a) contact, (b) projection, (c) diffraction, (d) reflection, and (e) image transmission.

external reflection at grazing incidence, or from curved perfect monocrystals. Magnification of the reflection X-ray microscope may reach 10^5. Its resolution strongly depends on the quality of the mirror polishing, and in the soft X-ray region ($\lambda = 0.4$ to 40 nm) it may reach the diffraction limit; present reflection X-ray microscopes have a resolution of the order of 1 μm.

- *X-ray microscopy with image transmission* is based on the use of zone plates which serve for focusing and enlargement of the image, Fig. (e). Magnification reaches 10^3, exposure time when utilizing powerful sources is several seconds, resolution is about 0.1 μm.
- In *scanning X-ray microscopy* the object is scanned by the X-ray beam sharply focused with the help of the condenser zone plate. The spatial resolution, determined by the size of the focal point, reaches about 50 nm (in principle it may be taken down to 10 nm).

X-ray microscopy is used to supplement the techniques of optical and *electron microscopy*. Methods of X-ray microscopy allow the observation of individual defects, the investigation of surface processes, obtaining holographic images of microobjects, etc.

X-RAY MIRROR

A device for reflecting and focusing X-rays. Existing X-ray mirrors are based mainly on *total external reflection*. In the range of frequencies ω for which the absorption of X-rays in the material is not high the refractive index has the form $n = 1 - \omega_p^2/(2\omega^2)$, where ω_p is the plasma frequency of electrons in the solid. Total external reflection is observed at grazing incidence of the beams at the mirror surface at angles $\theta < \theta_c = \omega_p/\omega$. For photons with energy from 1 to 10 keV, the angle $\theta_c \approx 10^{-2}$ to 10^{-3} rad. Therefore, the X-ray mirror, e.g., in X-ray telescopes, is made in the shape of a ring section of an oblong conic or parabolic surface with its axis directed to the source, and it focuses its radiation to the detector.

Another type of X-ray mirror is a sufficiently *perfect crystal*. When radiation at the wavelength λ is incident at the flat surface of the crystal under diffraction conditions, then in the narrow range of angles near the Bragg angle (see *Bragg law*) there is interference amplification of the waves scattered by these planes, and almost the entire reflected part of the incident beam emerges in the direction of reflection. The focusing of radiation by such an X-ray mirror is brought about by the uniform bend of the crystal.

In addition to the above cases, a separated atomic plane (two-dimensional crystal) may possess the properties of an X-ray mirror for a series of frequencies close to the resonance frequencies of its atoms. In this case the incident and scattered waves are mutually extinguished, and all the energy is passed on to the reflected beam.

X-RAYOGRAM

See *X-ray diffraction pattern*.

X-RAY SCATTERING

Change in the characteristics of X-ray flux (intensity, direction of propagation, frequency, polarization) during its interaction with matter. X-ray scattering is largely independent of the *state of matter* of various physical bodies, being controlled mainly by the interaction between the X-ray photons and electrons. Incoherent X-ray scattering by a weakly bound electron is determined by the *Compton effect*. The *scattering amplitude* $f(s)$ for an individual atom is given by $f(s) = \int \rho(r) \exp(-isr)\, dr$, where $s = 2\pi/\lambda$ is the diffraction vector. Its value is calculated via a Fourier-transform of the electron cloud density, $\rho(r)$. When the photon frequency ν is close to the natural frequencies, ν_{0i} corresponding to core level (inner electron) transitions, the effects of *anomalous dispersion* must be taken into account, e.g., the dependence of the real and the imaginary parts of $f(s)$ on ν.

To describe X-ray scattering by *solids*, which are mostly crystalline in structure, various other factors that influence the intensity are introduced beside $f(s)$, such as structural, thermal, absorption, polarization, etc., types. The first of these (see *Structure amplitude*) accounts for the phase shifts generated by various periodicities in the solid. The thermal factor describes the effect of *phonons* that weaken the diffraction maxima, their angular position, θ, being prescribed by the *Bragg law*: $2d_{hkl} \sin\theta = \lambda$, where d_{hkl} is the interplanar distance.

Two principal approaches are used to account for the intensity of X-ray scattering by crystalline atomic lattices. The first or kinetic approach based on the first *Born approximation* (neglect of scattered wave amplitude) is applicable to small and strongly perturbed crystals (see *Elastic scattering of radiation*). Such an approach is used to interpret X-ray scattering by polycrystalline targets containing structural perturbations of the first and second kind (after Krivoglaz). To describe the effect of the latter upon the intensity of the $I_R(\theta)$ maxima a static *Debye–Waller factor* is introduced, of the form $\exp(-W)$, which varies within the interval $0 \ll e^{-W} < 1$ for perturbations of the first kind, while the corresponding factor tends to zero for perturbations of the second kind. Defects weaken and displace the scattering maxima, and also result in the *diffuse scattering of X-rays*, its anomalies yielding information on the nature of the perturbations and on their spatial distribution.

The second or dynamic approach (C.G. Darwin, 1913; P.P. Ewald, 1917; M. Laue, 1925) assumes an interaction between the incident and the reflected rays to form wave fields in an *ideal crystal* that "adjust" to the medium periodicity d_{hkl} itself. In the simplest case the wave field represents a Bloch wave, $D = D_1 \exp[ik_0 r] + D_2 \exp[-i(k_0 + b_h)r]$, where D_1 and D_2 are the respective amplitudes of the incident (wave number k_0) and the reflected waves, and $b_h = 2\pi/d_{hkl}$. Two types of wave fields may form, one of them having its crest at atomic sites, and the other between the lattice sites. The second field is responsible for the *Borrman effect* (see *Anomalous passage of X-rays*) during which the absorption factor for X-rays sharply drops. The dynamical theory of X-ray scattering also explains the effect of pendulum beats of the two wave fields (see *Pendular oscillations of intensity*) that serves as the basis of the most accurate state-of-the-art techniques for determining $f(s)$. The *dynamic scattering of X-rays* is extremely sensitive to weak perturbations of the medium periodicity, and is used both to obtain images of different structures in reflected beams (*diffraction topography*), and to extract the integral characteristics of the structural perfection of *monocrystals*. The theory of X-ray scattering by perturbed crystals serves as the basis for perfecting detailed experimental techniques for identifying weak structural perturbations of various origins in solid bodies. It has been found that

Krivoglaz' ideas on the effects of defects of the first and second kind upon X-ray scattering retain their overall validity during dynamical scattering as well.

X-RAY SCATTERING, SMALL-ANGLE

See *Small-angle scattering of X-rays*.

X-RAY SOURCES

Converters of various kinds of radiation into the electromagnetic type in the spectral range between γ- and ultraviolet sources within the wavelength range from 10^{-3} to 10^2 nm. The principal X-radiation source with wavelength $\lambda \sim 0.1$ nm for *X-ray structure analysis* and other solid state physics methods is an *X-ray tube*. This is a high-voltage electron tube (diode or triode) where the radiation is generated by the braking of accelerated electrons (*Bremsstrahlung*) in the target material (anode of the tube). In hot cathode tubes (*electron X-ray tubes*) free electrons appear due to *thermionic emission*; in cold cathode tubes (*ionic X-ray tubes*) they appear due to *field electron emission*. Depending on the anode material and the accelerating voltage, the X-ray tubes can serve as sources of both a continuous ("white") and a characteristic (discrete) X-ray spectrum. Used as powerful continuous X-radiation sources are electron circular accelerators (synchrotrons, etc.) and storage rings of electron accelerators. In them, an electron with energy E moving in a circular orbit of radius R radiates a quasi-continuous electromagnetic spectrum with the intensity maximum at the wavelength $\lambda_{\max} \sim R/E^3$ (at $E = 1$ GeV, $R = 0.3$ m, $\lambda_{\max} \sim 0.07$ nm). The *synchrotron radiation* thus obtained is polarized in the plane of the orbit of the electron motion, and directed along a tangent to the orbit. Radioisotope X-radiation sources use the phenomenon of K-capture radioactivity, with the characteristic lines of the element appearing as a result of the decay of a particular radioactive nuclide (e.g., ^{55}Fe).

X-ray *defectoscopy* uses, in addition to X-ray tubes, linear accelerators of electrons and betatrons as sources of short-wave X-radiation.

X-RAY SPECTROGRAPH

A device for recording emission and adsorption *X-ray spectra* by a light-sensitive film or plate. The main elements of a spectrograph are: source of X-rays (*X-ray tube*); dispersing element that decomposes the radiation of the X-ray tube into a spectrum of spread out frequencies; cassette for the light-sensitive film at which the spectrum is recorded. A monocrystalline plate or diffraction grating may serve as a dispersing element. All the components of the spectrograph are positioned to satisfy the requirements established by the conditions of X-ray diffraction (see *Bragg law*).

X-RAY SPECTROMETER

A device that is similar to an *X-ray spectrograph* but uses ionization recording of the *X-ray spectrum*. A schematic diagram (see Fig.) of an X-ray spectrometer provides the relative positions of the basic elements (source of X-rays, dispersing element, and detector) which satisfy the requirements of the *Bragg law*. The following types of counters are commonly used as detectors: Geiger

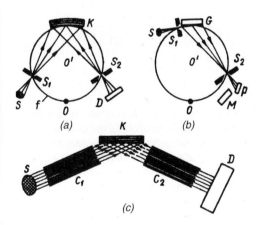

Ray diagrams of three X-ray spectrometers: (a) focusing spectrometer with crystal-analyzer K; (b) focusing spectrometer with diffraction grating G; (c) spectrometer with a flat crystal K and a Soller collimator (C_1 and C_2). The symbols designate: D, detector; f, focal circumference; M, secondary electron multiplier; O, center of circumference along which the crystal is curved, or the center of the concave surface of the lattice; O', center of circle f; P, photocathode; S, radiation source; S_1 and S_2, slits.

counter, flow-through proportional counter, scintillation counter, semiconductor counter or photocathodes with a secondary electron multiplier of the open type or with a channel multiplier. For the solution of special problems a synchrotron can serve as a source of X-rays. For *X-ray structure analysis diffraction-free spectrometers* are used where X-rays of the sample are recorded directly, and analytical lines of the spectrum are separated by a single- or multichanneled amplitude pulse analyzer (see Fig.). See *Crystal diffraction spectrometers*.

X-RAY SPECTRUM

Dependence of the intensity of X-radiation on its wavelength (i.e. on the photon energy). The *X-ray emission spectrum* appears during the interaction of accelerated charged particles or X-ray photons with inner electron shells of atoms. When a charged particle decellerates in the target material there appear X-rays with a *continuous (Bremsstrahlung or braking) spectrum* and with a sharp short-wave boundary $\lambda_0 = hc/(eV)$, where e is the charge of the bombarding particle, and V is the potential difference across the particle. The intensity of the continuous spectrum is proportional to the energy of the particles and to the atomic number of the target atoms, and it is inversely proportional to mass of the bombarding particles.

X-rays with a *discrete characteristic spectrum* appear superimposed on the continuous Bremsstrahlung background, and arise from transitions of atoms from an initial higher energy state E_i to a final lower energy state E_f. At each such individual transition an X-ray quantum with energy $h\nu = E_i - E_f$ is emitted. State E_f (corresponding to a missing electron) may be created by a charged particle (*primary spectra*) or by a photon (*secondary* or *fluorescent spectra*) of sufficient energy. If both states are characterized by the absence of an electron at one of the inner electron shells of the atom, the transition leads to the appearance of the so-called *internal X-ray emission lines*. The positions of these internal X-ray spectrum lines depend on the atomic number of the target. The intensity I_{if} of the energy emitted by these lines is determined by a number of factors: by the probability of ionization P_i of ith level of the atom, by the populations n_i and vacancies n_f

of levels i and f, respectively, by the probability P_{if} of inducing the transition $E_i \rightarrow E_f$, and by the energy $h\nu$ of the photon that is emitted as a result of this transition: $I_{if} \propto P_i n_i n_f P_{if} h\nu$. In the case of a massive target the intensity also depends on the properties of the anode material, on the angle of incidence the bombarding beam at the target surface, and on the output angle of the X-rays.

The charge of the emitting atom and the geometry of its surroundings influence the energy state and the structure of the spectral lines. If state E_f is characterized by the absence of an electron in the full part of the *valence band*, then the transitions $E_i \rightarrow E_f$ give the so-called *terminal lines* or *X-ray emission bands*. In their dependence on the atomic shell in which the electron vacancy corresponding to E_f is present the sequence of X-ray emission lines are designated by the series K, L, M, etc., corresponding to the principal quantum numbers $n = 1, 2, 3$, etc., of the ground level from which the photoelectron is knocked-out by the photon, thereby creating the above-mentioned electron vacancy. The core electron levels of atoms have symmetries that depend on their orbital substates ($1s$ for K; $2s$ and $2p$ for L; $3s$, $3p$, $3d$ for M, etc.). The atomic *selection rules* in the emission bands of the different series influence the partial *densities of states* of various symmetry types. In addition, parameters of the emission bands are influenced to a considerable extent by the character of the chemical bonds in solids. The characteristic X-ray radiation spectrum which appears at the bombardment of a solid by high-energy heavy ions contains information about the distribution of the radiating atoms according to the multiplicity of their internal ionization.

The diagram of the dependence of the spectral intensity of the continuous (near the quantum boundary) or of the characteristic (near the excitation potential) emission spectrum on the energy of electrons incident on the material is called the *isochromat curve of the X-ray spectrum*. When the electron beam is incident on the solid, the long-range structure of isochromat curves in their initial region (up to 30 to 50 eV) is determined by the presence of empty electronic states of the solid; the long-range structure of the isochromat curves of the characteristic spectrum (up to 600 to 1000 eV) is determined by the configurations of atoms of the solid.

The *X-ray absorption spectrum* provides the dependence of X-ray intensity of the continuous spectrum, after passing through the adsorber, on its wavelength (i.e. on the photon energy). While passing through the absorber, its atoms absorb X-ray photons of specific energies. During this absorption process a photon, in the course of its collision with an atom, excites it to the state E_f. To accomplish this the photon energy must fulfill the condition $h\nu > E_f - E_i$. A photoelectron produced by the excitation of an atom is taken up by the empty valence states of the solid (in compliance with selection rules), or by its crystal lattice.

The X-ray absorption spectrum may also be divided into three regions: main absorption edge, initial region of absorption, and long-range (extended) *fine structure of X-ray absorption spectrum*. The energy position of the main absorption edge (point of bending) maps the energy distance between the maximum density of the core level ($n = 1, 2, 3, \ldots$) and the closest vacant states, to which a transition is permitted by the selection rules. The charge of the atom whose spectrum is studied, the geometry of its nearby surroundings, and the nature of the chemical bond in the solid under investigation influence the position of the main absorption edge. The structure of the initial absorption region, which often begins at the main edge and is spread beyond it up to a distance of 30 eV, displays the structure of vacant partial densities of states of the solid (according to the selection rules). By comparing at the common scale of the X-ray absorption spectrum of different series of lines it is possible to obtain the structure of the unfilled part of the valence band distorted by the probability of the transition to, and by the width of, the internal level (and also by the resolution of the device). The charge of the atom, whose spectrum is studied, the geometry of its nearby surroundings, the distribution of the density of vacant states of the solid, and the nature of its chemical bonds, all influence the structure and position of the extrema of the initial region of adsorption. The long-range fine structure of the X-ray absorption spectrum, which extends toward short wavelengths up to the distance of 600 to 1000 eV, is determined by the spatial distribution of the matrix of photoelectron dissipation by the atoms of the solid. From the analysis of this spectrum it is possible to obtain some information on the atomic structure of the solid, and in the case of a multicomponent non-crystalline adsorber, it also depends on the geometry of the nearest neighbor surroundings of atoms of each component separately, and on the quantity and types of atoms in these nearby surroundings.

X-RAY SPECTRUM ANALYSIS

Totality of the physical methods of finding the chemical composition of a material by its adsorption and emission *X-ray spectra*. *Emission X-ray spectrum analysis* is carried out by primary and secondary (fluorescent) spectra excited by electrons, ions or X-ray photons. *Qualitative X-ray spectrum analysis* is based on the detection of characteristic lines of the elements in the emission spectrum of the material under investigation. The energies of these lines are related to the atomic numbers of the elements by *Moseley's law* (H. Moseley, 1913–1914). The sensitivity depends on the ionization cross-section of the atomic shell, on the radiation fluorescence yield, on the ratio of the intensities of the analytical line and the background under it, on the chemical composition of the material, on statistical fluctuations of the intensity, and also on instrumental factors, on the method of excitation, of recording and dispersion of radiation to the spectrum. *Quantitave emission X-ray spectrum analysis* is based on the relation between the intensity of the analytical line of the element A being determined I_i^A and its concentration within the sample C^A. Depending on the method of defining the relative intensity of the line χ_i^A, external or internal standards can be used. In the former case, χ_i^A is the ratio of intensities I_i^A of the analyzed and standard samples, in the second one it is ratio of the intensities I_i^A and I_j^B of the analytical line j of element B artificially introduced into the sample with a precisely known concentration. Element B is selected according to the closeness of its comparison line to the analytical line of element A. In *X-ray fluorescent analysis* the intensity of X-rays scattered by the analyzed sample is also used as an internal standard. The dependence of χ_i^A on C^A is initially established either theoretically, or experimentally for samples of the known composition. There are two types of theoretical connection equations. The first type are based on a physical model of excitation and adsorption of X-rays in the material. Equations of connection with fundamental parameters are the most general type, which allow using only pure elements as the standard. In these equations, χ_i^A is related to C^A by a function depending on the chemical composition of the material, and on the parameters of its atoms (fluorescence yields, energy of lines, probabilities of emission, coefficients of adsorption and scattering, etc.). The connection equations of the second type, called phenomenological or regression equations, are expansions of I_i^A in power series with coefficients determined for standard samples with compositions satisfying a so-called "diagram of orthogonal correlations". The connection equations are solved on a computer by iteration.

Methods of X-ray spectral analysis with photon and electron probe spectral excitation are the most effective and well developed. X-ray fluorescent analysis is noted for its speed, accuracy (achieving the accuracy of standard chemical analysis), possibility of analyzing materials (liquids and solid layers with a thickness more than 10 nm) without their destruction or decomposition. The sensitivity is 10^{-1} to $10^{-3}\%$ in the atomic number range $9 \leqslant Z \leqslant 20$, and 10^{-3} to $10^{-4}\%$ in the range $20 \leqslant Z \leqslant 92$. The thickness of the investigated layer of the solid in its dependence on chemical composition and on analytical line energy is 1 to 10^3 μm.

Electron-probe X-ray spectrum analysis allows the determination of elements from boron to uranium with a sensitivity of 10^{-1} to $10^{-3}\%$ for a microvolume of material, with a size depending on its chemical composition and the energy of the exciting ions, of $2 \cdot 10^{-1}$ to 10^3 μm, and also permits one to obtain the "image" of a surface layer of the material in monochromatic X-rays, reflected and absorbed electrons.

The method finds its application in investigations of the chemical composition of inclusions, *films*, protective *coatings*, elements of *microelectronics*, in the investigation of *diffusion* processes, in plotting *phase diagrams*, etc. In investigating the composition of thin layers of a material with thickness of 10^{-3} to 10^{-4} μm, X-ray spectral analysis with ionic excitation of the spectrum has the best sensitivity (10^{-4} to $n \cdot 10^{-8}\%$) in the range of atomic numbers $5 \leqslant Z \leqslant 92$.

Absorption X-ray spectrum analysis has low sensitivity and therefore rather restricted possibilities for evaluating the chemical composition of solids. In this method, the content of an element is determined by the jump of adsorption of the continuous or characteristic spectrum in the vicinity of the adsorption edge, or by the reduction of intensity of X-rays upon the adsorption. The relation of the analytical signal with C^A is established with the aid of standard samples. *X-ray spectrograph, X-ray spectrometer* and other devices are used in X-ray spectral analysis. See also *Microprobe X-ray spectrum analysis.*

X-RAY SPECTRUM, MICROPROBE

See *Microprobe X-ray spectrum analysis.*

X-RAY STRUCTURE ANALYSIS, X-ray diffraction analysis

Main method for investigating the structure of solids based on *X-ray diffraction* (M. Laue, 1912). X-rays of a wavelength on the order of 0.1 nm, i.e. of the order of an interatomic distance, are used in X-ray structure analysis. An *X-ray tube*, or electron gun where X-rays are generated as a result of quantum transitions in the target atoms, or from an anode bombarded by accelerated electrons, is the source of monochromatic characteristic X-rays. These X-rays superimposed on a continuous spectrum (*braking radiation* or *Bremsstrahlung*) are obtained with the help of X-ray tubes, and also of various accelerating systems.

Synchrotron radiation, whose source is relativistic charged particles circulating in storage rings, distinguished by high intensity and directivity, is used for X-ray structure analysis. From the continuous spectrum of synchronous radiation, specified wavelengths are selected by single-crystal monochromators.

Interactions of X-rays with matter consists of adsorption, e.g., of fluorescent excitation radiation, formation of Auger electrons, as well as elastic (coherent) and inelastic (incoherent or Compton) scattering. In X-ray structure analysis ordinarily the elastic scattering of X-rays by electrons is used. The amplitude of the wave scattered by all the electrons of an individual atom is called the *atomic form factor,* and the amplitude of a wave scattered by all the atoms in a unit cell is called

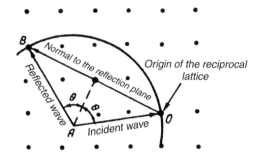

Ewald construction for X-ray structure analysis.

the *geometric structure factor.* The latter involves a sum over terms for the individual atoms of the unit cell, with the term for each atom proportional to its respective atomic form factor (see *Structure amplitude*). Geometric structure factors depend on the type of crystal structure.

Diffraction of X-rays at the crystal lattice is described by the Laue (see *Laue conditions*) and Bragg (see *Bragg law*) equations. The figure shows the *Ewald construction* (P. Ewald, 1916) in reciprocal space (*k-space*) with a circle (more generally, a sphere called the *Ewald sphere*) centered at point A drawn through point O with radius $k = 2\pi/\lambda$, where λ is the X-ray wavelength. When the circle also passes through another point B of the reciprocal lattice then there is a diffracted (reflected) wave in the $A \rightarrow B$ direction at the angle 2θ relative to the incident wave direction AO, and the distance O to B is a reciprocal lattice vector also of length $2\pi/\lambda$. From the Ewald construction there follows the essence of all three basic methods of X-ray structure analysis: in the *Laue method* (stationary monocrystal, continuous spectrum) a site of the reciprocal lattice is measured out to the Ewald sphere so that the value $2\pi/\lambda$ varies over a wide range; in the *rotating crystal method* (a monocrystal rotates with respect to axis $[u, v, w]$, monochromatic spectrum) the site is taken out to the sphere at particular orientations of the reciprocal lattice with respect to the incident wave; in the *Debye–Scherrer–Hull method* or *powder method* (polycrystal, characteristic spectrum) due to the presence of many different orientations some of the points always lie on the Ewald sphere. The establishment of the crystal structure from its diffraction pattern is the fundamental task of X-ray structure analysis. Methods for solving this problem reduce to plotting the three-dimensional distribution of electron density in the unit cell (*Fourier synthesis*), to the application of a "trial-and-error method" (minimization of differences between measured diffraction line intensities and those calculated from a structure model), etc. All methods of deciphering structures are individualized since the experiment gives only the modulus of the *structure amplitudes*, with the phase remaining indeterminate (*Mössbauerography* can, in principle, determine the phase). With the help of X-ray structure analysis it is possible to solve a large number of application problems in solid state physics, such as those involving imperfections of the crystal, atomic vibrations, phase transformations, textures, etc.

X-RAY TOPOGRAPHY

Method for recording images of *defects in crystals* via their scattered X-ray intensity distribution in space. Radiation from the source is incident on the *crystal*, it undergoes Bragg scattering, and the intensity distribution is recorded on a photographic film (or photographic plate). Defects cause a local variation of the lattice parameter, distortion and deflection of the crystallographic planes which are deviated from their proper Bragg reflection positions, leading to large local variations in the X-ray intensity distribution. As the angular area of the Bragg scattering is small, small deformations of the crystal may cause large variations of Bragg reflections. This allows observation of the conformation of individual *dislocations* (at densities up to 10^6 cm^{-2}), nonuniformities in the distribution of interstitial atoms, scratches, mosaic and domain structures of crystals, and other defects. Using different reflections, it is possible to obtain different projections, which permits a reliable determination of the geometrical characteristics of defects. Compared to electronic transmission microscopy the methods of X-ray topography have less resolution, but they have an advantage of applicability to the investigation of large crystals. X-ray topography is used for investigating dislocation dynamics, for nondestructive monitoring of the structural perfection of monocrystals, for investigation of the mechanism of plastic deformation, of aging, and of other such processes. See also *Microprobe X-ray defect structure analysis, Microradiography, Dislocation contrast*.

X-RAY TUBE

A device for generating X-radiation. The simplest X-ray tube consists of a sealed glass or ceramic container with gas pressure of 10^{-3} to 10^{-4} Pa, with cathode and anode units fixed within it at some distance from each other. The generation of X-radiation is achieved by bombardment of the anode with the electron beam emitted by the cathode, and accelerated by the electric field that is applied between the two electrodes. The structural features of the X-ray tube depend on their application: for X-ray defectoscopy apparatus, for *X-ray structure analysis* instruments, for *X-ray spectrum analysis* apparatus.

Y y

YIELD

A property of solids deformed plastically or viscously under the action of stress. This is the condition for the initiation of flow in a solid. Yield is determined by a closed surface $f(\sigma_{ij}) = 0$ in the six-dimensional space of the components of the *stress tensor* σ_{ij}, each point of which represents a certain *state of stress*. There are elastic *strains* inside this surface (see *Elasticity*); and at the surface itself *plastic deformations* take place. An analytical contour of the onset of yield is plotted on the basis of one of three hypotheses: the hypothesis of the initial tangent stress; the hypothesis of the most reduced stress; and the hypothesis of the "energetic" condition at the onset of *plasticity*. The yield elements in crystalline materials are represented by crystallographic *slip planes* with a certain *slip* direction, referred to as *slip systems*. If one considers the entire complex of yield onset for body-centered cubic *monocrystals* with slip systems $\{110\}\langle 111\rangle$ or $\{112\}\langle 111\rangle$ for certain orientations using the superposition of contours, and one supposes that the original slip system ceases its action when *hardening*, and another system then starts to act, such a consideration predicts the presence of the "easy" deformation orientations $\{001\}\langle 110\rangle$ in the slip system $\{110\}\langle 110\rangle$. This orientation is unique; for then the contour of yield onset is the only possible one since the traces of the yield surface intersect only once. There is no "easy" deformation orientation under action of slip systems of the kind $\{112\}\langle 111\rangle$. These considerations are supported by experiment.

In *amorphous metals and metallic alloys*, the slip elements act either at the scale of cluster units or in the macroscopic bulk of the sample, involving so-called excessive free volume in the yield process (see *Plastic deformation of covalent crystals*).

YIELD CUSP

Abrupt increase of flow stress at the initial stage of *plastic deformation* during the *mechanical testing of materials*, followed by an abrupt decrease of the flow stress. A yield cusp appears on a *deformation diagram* as a maximum (*upper yield limit*) at the initial stage of deformation. This change of flow stress may be related to the pinning of *dislocations* arising from interactions between dislocations and *impurity atoms*. In this case, the upper yield limit corresponds to the stress needed to detach dislocations from impurity distributions. Later on, with an increase of the density of mobile dislocations, the flow stress decreases to its *lower yield limit* value. More generally, the phenomenon of yield cusp is explainable in terms of the dynamics of the motion and the increase of dislocations. The appearance of a yield cusp is a consequence of the rapid increase of the rate of plastic deformation (by virtue of the intensive *pile-up of dislocations* at a low initial dislocation density) up to values which exceed the rate of the testing device. The generation of a yield cusp is facilitated by a low initial dislocation density, their high aggregation rate, and the strong dependence between this rate and the applied stress. The initial section of a deformation diagram with yield cusp may be calculated theoretically within the context of available *dislocation models of plastic deformation*.

YIELD LIMIT, yield point

Stress at which irreversible *plastic deformation* starts to develop, caused by the breakup and movement of crystalline structure *defects*, *dislocations* and *disclinations*, in crystalline materials, and of the free volume in amorphous solids (see *Plastic deformation of amorphous metallic alloys*). In materials which exhibit a cusp in their *deformation diagram* (the *yield cusp*) both an *upper yield limit* and a *lower yield limit* are observed. The upper yield limit (maximum on the initial stage of a deformation diagram) corresponds to the stress at which the velocity of plastic deformation equals the velocity of the testing machine, and then for higher stresses exceeds it (e.g., due to a sharp increase of the mobile dislocation density). The lower (or physical) yield limit (minimum on the deformation diagram) corresponds to the stress at which the velocity of the sample plastic deformation and the velocity of the testing machine coincide, and the deformation may develop for some time at a constant stress, corresponding to a plateau of yield on the deformation diagram. The notion of a *conditional yield limit* σ_ε, which corresponds to the stress at which plastic deformation achieves some fixed value ε expressed in %, is widely used in several fields of engineering. For instance, $\sigma_{0.2}$ corresponds to the stress at which plastic deformation is 0.2%.

YIELD LIMIT ANOMALY

See *Low-temperature yield limit anomalies.*

YIELD POINT

See *Yield limit.*

YOUNG'S MODULUS, modulus of elongation
(T. Young, 1807)

In the linear *elasticity theory* of isotropic solids, the ratio E of stress P (force per unit area) to *strain* $\Delta L/L$ along the axis of loading of a rod or other regular solid of length L is called Young's modulus: $E = PL/\Delta L$. Young's modulus is a function of the *elasticity moduli* (elastic stiffness constants) c_{ij} of the material, and for a cubic crystal we have

$$E = \frac{(c_{11} - c_{12})(c_{11} + 2c_{12})}{c_{11} + c_{12}}.$$

This can be expressed in terms of the *bulk modulus* $K = (c_{11} + 2c_{12})/3$ and the *shear modulus* $\mu = (c_{11} - c_{12})/2$ as follows:

$$E = \frac{9K\mu}{3K + \mu}.$$

E is measured in pascals (newtons per square meter), and some representative values are: Al, $E = 70$ GPa, W, 360; Fe, 210; Cu, 110; cast iron, 120; brass, 91; steel, 200; and quartz, 54. These values

are only approximate since they depend on the purity and the pretreatment of the sample.

YOUNG'S MODULUS MAGNETIC EFFECT
(ΔE-effect)

Variation of *Young's modulus* E under the influence of an applied magnetic field in ferromagnetic materials (see *Ferromagnet*). In the absence of external mechanical stress and of a magnetic field, the direction of the inherent *magnetization* depends on the energy of the *magnetic anisotropy* and corresponds to the directions of easy magnetization $\langle 100 \rangle$. The application of an external stress σ by itself which can be static (measuring E in stretching experiments) or dynamic (as in method of *internal friction*) influences the position of the magnetization vector. The dimensions of a solid in this case, besides the variation of a purely elastic origin (Δl_{el}), will change due to the *magnetostriction* expansion Δl_{M} caused by the variation of the orientation of the magnetization vector. Upon application of a strong magnetic field (state of saturation) the probability of changing the orientation of *domains* by a weak stress is excluded. When ΔE is determined as a difference between the modulus E in the saturated state $E_{\mathrm{s}} \propto \sigma/\Delta l_{\mathrm{el}}$ and the modulus in the demagnetized state $E_0 \propto \sigma/(\Delta l_{\mathrm{el}} + \Delta l_{\mathrm{M}})$, the variation ΔE is positive and may reach about 20%. If σ preferably causes the motion of 90-degree *magnetic domain walls* (the case of high anisotropy energy and low internal stresses, e.g., a crystal of the *nickel* type):

$$\frac{\Delta E}{E_0} = \frac{3}{5} \frac{\chi_0 \lambda_{100}^2 E_{\mathrm{s}}}{J_{\mathrm{s}}^2},$$

where χ_0 is the initial *magnetic susceptibility* and λ_{100} is the magnetostriction constant for saturation in the direction $\langle 100 \rangle$. To determine the *shear modulus* from tests on *rod twisting*, the corresponding variation of the shear modulus is observed (ΔG-effect). Dynamic tests at low frequencies detect an E variation associated with *relaxation* due to microvortex currents, and this may give some contribution to ΔE.

YTTERBIUM, Yb

Chemical element of Group III of the periodic system with atomic number 70 and atomic mass 173.04; it belongs to *lanthanides*. Natural ytterbium consists of 7 stable isotopes with mass numbers 168, 170 to 174 and 176, among which ^{174}Yb (31.84%) is the most abundant and ^{168}Yb (0.135%) is the least common. Electronic configuration of outer shells is $4f^{14}5d^{0}6s^{2}$. Successive ionization energies are 6.254, 12.17, 25.5 eV. Atomic radius is 0.193 nm. Oxidation state is +3, more rarely +2. Electronegativity is 1.2.

In the free form, ytterbium is a soft plastic silvery-white *metal*. The face-centered cubic crystal lattice of α-Yb with parameter $a = 0.5483$ nm at 1070 K transforms into β-Yb with a body-centered cubic lattice with parameter $a = 0.444$ nm. Density of α-Yb is 6.96 g/cm^3; $T_{\text{melting}} = 1094$ K; $T_{\text{boiling}} = 1484$ K; heat of melting 7.66 kJ/mole; heat of sublimation 144.1 kJ/mole; heat of evaporation is 130.3 kJ/mole; specific heat $c_p = 26.8$ J·mole^{-1}·K^{-1}; Debye temperature is 118 K. Electrical resistivity is 0.27 $\mu\Omega$·m (at 298 K); temperature coefficient of electrical resistance is $1.3 \cdot 10^{-3}$ K^{-1}. Ytterbium is a *paramagnet* with magnetic susceptibility $\chi = 0.41 \cdot 10^{-9}$. Hall constant is $+377 \cdot 10^{-10}$ m^3/C. Temperature coefficient of linear expansion is $29.9 \cdot 10^{-6}$ K^{-1} (298 to 973 K). Brinell hardness is 196 MPa, modulus of normal elasticity is 17.8 GPa; shear modulus is 6.97 GPa.

Metallic ytterbium is used as a getter in electronic vacuum devices. It is used as an additive to special *alloys* and fireproof materials. Additions of ytterbium serve as activators in *luminophors* and crystal phosphors.

YTTRIUM, Y

Chemical element of Group III of the periodic system with atomic number 39 and atomic mass 88.906; it belongs to the second transition series elements. Natural yttrium has one stable isotope ^{89}Y. Outer shell electronic configuration is $4d^{1}5s^{2}$. Successive ionization energies are 6.217, 12.24, 20.52 eV. Atomic radius is 0.181; radius of Y^{+3} ion is 0.092 nm. Oxidation state is +3. Electronegativity is 1.20.

In a free form, yttrium is a silvery-white *metal*. It has two allotropic modifications: α-Y

and β-Y. α-Y with hexagonal close-packed lattice is stable, its parameters: $a = 0.36451$ nm, $c = 0.57305$ nm at 298 K, space group $P6_3/mmc$ (D_{6h}^4). At a temperature above 1760 K it transforms into β-Y with body-centered cubic lattice with parameter $a = 0.411$ nm, space group $Im\overline{3}m$ (O_h^9). Density of α-Y is 4.469 g/cm^3 (293 K); $T_{\text{melting}} = 1800$ K; $T_{\text{boiling}} = 3595$ K. Binding energy is 4.387 eV/atom at 0 K. Heat of melting is 10.1 kJ/mole; heat of evaporation is 335 kJ/mole; specific heat is 0.299 kJ·kg^{-1}·K^{-1}; Debye temperature is 256 K. Linear thermal expansion coefficient of monocrystal is $19.2 \cdot 10^{-6}$ K^{-1} along the principal crystallographic axis $\underline{6}$ and $4.6 \cdot 10^{-6}$ K^{-1} perpendicular to it at 300 K; coefficient of heat conductivity is 14.7 W·m^{-1}·K^{-1} at room temperature. Adiabatic coefficients of elastic rigidity of α-Y monocrystal: $c_{11} = 79.51$, $c_{12} = 30.58$, $c_{13} = 29.07$, $c_{33} = 78.70$, $c_{44} = 25.13$ GPa at 300 K; bulk modulus is 41.3 GPa at 275 K, Young's modulus is 64.1 GPa (at 275 K), shear modulus 25.8 GPa (at 275 K), Poisson ratio is 0.241 (at 275 K); ultimate strength is 0.309 GPa, yield limit is 0.172 GPa; relative elongation is 35%. Brinell hardness is 400 MPa, Vickers hardness is 38 in HV units. Thermal neutron capture cross-section is 1.31 barn. Linear low-temperature electronic heat capacity (Sommerfeld coefficient) is 10.23 mJ·mole^{-1}·K^{-2}; electric resistivity of α-Y is 350 nΩ·m along the principal axis $\underline{6}$ and 720 nΩ·m^1 perpendicular to it (at 298 K). Hall constant is $-7.7 \cdot 10^{-11}$ m^3/C at room temperature; polycrystal work function is 3.07 eV; superconducting transition temperature at the pressure 16 GPa is 2.7 K; molar magnetic susceptibility of yttrium is $+184 \cdot 10^{-6}$ CGS units along the principal axis $\underline{6}$, and $+196 \cdot 10^{-6}$ CGS units perpendicular to it at room temperature; nuclear magnetic moment of ^{89}Y isotope is 0.137 nuclear magnetons.

Yttrium is used as a structural material in the nuclear industry. Additions of yttrium to *aluminum alloys* increase their *strength*, addition to *vanadium* improves its *plasticity*. Yttrium is one of the components of *luminophors*. Yttrium *ferrites* are used in radio electronics. Many compounds of yttrium are laser materials. The layered cuprate $YBa_2Cu_3O_7$ was the first compound to go superconducting (1987) above the temperature 77 K of liquid nitrogen ($T_c = 95$ K, see *High-temperature superconductivity*).

YTTRIUM–ALUMINUM GARNET, $Y_3Al_5O_{12}$

It is structurally a garnet, with cubic *space group* $Ia\bar{3}d$ (O_h^{10}); the *unit cell* contains eight formula units, i.e. 160 ions. The Y^{3+} ions are coordinated with eight oxygen atoms and occupy Wyckoff position c, the Al^{3+} ions are located at positions a and d with octahedral and tetrahedral coordination, respectively; oxygen ions occupy general position h with no site symmetry. Yttrium–aluminum garnet crystals have the following physical parameters: $T_{\text{melting}} = 2200$ K; thermal conductivity 0.11 to 0.14 $W \cdot cm^{-1} \cdot K^{-1}$, heat capacity 0.145 $cal \cdot g^{-1} \cdot K^{-1}$; density is 4.55 g/cm^3, Mohs hardness is 8.5; speed of sound is $8.56 \cdot 10^5$ cm/s along $[001]_l$, dielectric constant $\varepsilon_0 = 11.7$ ($\varepsilon_\infty = 3.5$); optical transparency is 0.2 to 6 μm. Yttrium–aluminum garnet is used as a laser material. For this purpose small amounts of the ions Nd^{3+}, Cr^{3+}, Tb^{3+}, Ho^{3+}, Yb^{3+}, Er^{3+} are introduced into the garnet crystals during their growth by the Czochralsky method, the most common being Nd^{3+} which yields pink-violet crystals with the coherent radiation wavelength 1.06 μm. See also *Active solid-state laser materials*.

ZEEMAN EFFECT (P. Zeeman, 1896)

Splitting of energy levels in an external magnetic field B that removes orbital and spin degeneracies by the quantized projection of the system *magnetic moment* μ associated with the total angular momentum $J = L + S$ along the magnetic field direction, where L and S are the total orbital and spin angular momenta, respectively, in the units of $\hbar$. In the magnetic field B the system acquires the *Zeeman energy* $E = -\mu_J B$, where $\mu_J = -g_J \mu_B M_J$ is the projection of μ along the B direction; g_J is the dimensionless *g-factor* of the system; μ_B is the Bohr *magneton*; and M_J is the projection of J along the B direction. For a system of electrons in an atom or ion with levels described in the Russell–Saunders or LS coupling approximation, the magnetic moment is the vector sum of the system total spin magnetic moment $\mu_S = -g_S \mu_B S$ and its orbital magnetic moment $\mu_L = -g_L \mu_B L$, where the g-factors of the free electron spin and of its orbital moment are $g_S = 2.00232$, $g_L = 1$. Since $g_S \neq g_L$, the direction of the resultant angular momentum of the electrons $J = L + S$ does not coincide with that of the total magnetic moment $\mu = \mu_L + \mu_S$. In this case,

$$g_J = \{ J(J+1)(g_L + g_S)$$
$$+ [L(L+1) - S(S+1)](g_L - g_S) \}$$
$$\times [2J(J+1)]^{-1}.$$

Assuming $g_S = 2$, $g_L = 1$, this formula reduces to the well-known expression for the *Landé g-factor*. In solids, the anisotropic interaction of an impurity atom (ion) with its environment may partially cancel the degeneracy along the J direction. The orbital motion L can quantize along the crystalline electric fields, which partially or totally uncouple L from S, a phenomenon called *quench-*

ing. The Zeeman effect of the remaining degenerate levels also becomes anisotropic, and is described by an effective spin $\overline{S}$, such that the degeneracy equals $2\overline{S} + 1$, and by the Hamiltonian $H = \sum g_{ij} \mu_B B_i \overline{S}_j$, with the symmetric g-tensor ($g_{ij} = g_{ji}$) introduced. In the case of cubic symmetry, the g-factor is isotropic.

In some cases, for degenerate levels, the value $\mu_J \to 0$ in the limit $B \to 0$ (as occurs, e.g., with $B \perp J$, when the J direction is fixed by an axial *crystal field* in uniaxial crystals). As a result, the Zeeman effect can be quadratic ($\propto B^2$ for ions with even numbers of electrons) or cubic ($\propto B^3$ for ions with odd numbers of electrons) in B. This effect resembles the *nuclear Zeeman pseudoeffect*, where the spin splitting of neutrons moving in a

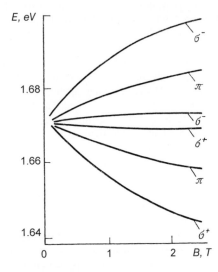

Magnetic field dependence of the giant spin splitting Γ_6–Γ_8 of exciton reflection spectra in Cd$_{0.95}$Mn$_{0.05}$Te, observed for σ^+, σ^- and π light polarizations at the temperature 1.97 K.

crystal is amplified owing to nuclear interactions with spin-polarized crystal nuclei in the magnetic field.

Internal fields can also appear in solids, and when they are added to the applied field B which has created them, they influence the Zeeman effect. This influence is particularly pronounced when excitons are present in *semimagnetic semiconductors*. In this case the strong *exchange interaction* of charge carriers with the ensemble of *magnetic ions* can be described in the approximation of an average *molecular field* that is proportional to the sample magnetization. The latter is created by the field B, while the Zeeman effect of energy bands is displayed as a giant spin splitting with the effective g-factor $g_{\text{eff}} \gg 2$ (see figure). The Zeeman effect can be observed both by the splitting of *light absorption* and *luminescence* spectral lines, and by various *magnetic resonance* methods.

ZEEMAN EFFECT, NUCLEAR

See *Nuclear Zeeman effect*.

ZENER MODEL of ferromagnetism of transition metals

A model initially suggested by C. Zener (1951) as an explanation of the connection between the *ferromagnetism* and electrical conductivity of *transition metal* oxides with *intermediate valence*. It was assumed within the framework of the model that the substitution, e.g., of Ca^{2+} for La^{3+} in $LaMnO_3$, results in the appearance of the Mn^{4+} ion instead of Mn^{3+} which can capture an electron from one of the adjacent Mn^{3+} ions. The repeated hopping motion of this electron interchange determines the finite conductivity of the sample. It also gives rise to the ferromagnetic ordering of the spins of the electrons belonging to atoms at crystal lattice sites. The reason is that, according to *Hund's rule*, the spin of an atom in its ground state should have a maximum value. To meet this condition during the passage of an electron from atom to atom, the spins of these atoms and of the hopping electron should be aligned. The electron moving from ion to ion is referred to as a *Zener electron*.

D.M. Edwards (1970) proposed a modified Zener model for explaining the ferromagnetism of transition metals with a lattice structure of identical atoms, some of them containing x d-electrons,

and the remainder having $x + 1$, where $1 \leqslant x \leqslant 4$. For a more than half full *conduction band* we have $5 \leqslant x \leqslant 8$, and the same reasoning can be applied to holes (see *Band theory*). This approach is applicable to the case of more than one d-electron per atom. The modified Zener model is a generalization of the *Hubbard model*, and the s–d exchange *Shubin–Vonsovsky model* (see *Ferromagnetism of metals and alloys*). L.C. Bartel (1973) calculated the spectrum of magnetic excitations in the *random phase approximation* within the Zener model, and showed that, in contrast to the one-band Hubbard model, the spectrum contains an additional *spin wave* optical branch.

The Zener model and related theories became very popular with the discovery of colossal magnetoresistance in the 1990s.

ZENER RELAXATION (C. Zener, 1947)

A relaxation peak (see *Relaxation*) on plots of the temperature dependence of *internal friction* in *substitutional alloys*, originally observed in α-brass *monocrystals*. The theory of the Zener effect is based on a model of the reorientation of pairs of nearest neighbor *substitutional atoms*. Elastic distortion fields in face-centered and body-centered cubic crystals have symmetries other than spherical, which provide an energy stimulus for orienting the atom pairs in the field of external stresses. The mechanism of the Zener peak in substitutional alloys is similar to that of the Snoek peak in *interstitial alloys*. The effect is anisotropic and depends on the substitutional element concentration c ($\propto c^2$), and on the temperature T (i.e. $\propto (T - T_c)^{-1}$), where T_c is the ordering temperature. It is used to study parameters of *diffusion*, substitution, *vacancies*, and aging (see *Alloy aging*), as well as the establishment of long-range order in alloys (see *Long-range and short-range order*).

ZEOLITES (fr. Gr. $\zeta\varepsilon\omega$, to boil, and $\lambda\iota\vartheta o\varsigma$, stone)

Natural and synthetic porous *crystals* (hydrated *aluminosilicates*) with developed internal surfaces, which reach surface areas of 500 to 700 m^2/g. The structure of zeolites is permeated by a three-dimensional network of channels, often of complex configuration. The diameters of channels in various individual zeolites are fixed, ranging from 0.3 to 1 nm, allowing *adsorption* of different molecular substances, which in natural zeolites is usually water. The chemical composition of zeolites is given by the general formula $M_n Al_n Si_m O_{2(n+m)} \cdot l H_2 O$, where M is an alkali or alkaline-earth element. In synthetic zeolites the replacement of Al by Ga or B, and also of Si by Ge or P is possible. *Silicalites* are known; they are pure or almost pure silicon zeolites (with $n = 0$), with some resemblence to crystalline, highly porous SiO_2. Silicates are hydrophobic (see *Lyophilic and lyophobic behavior*); they are used as catalysts for the production of synthetic fuels. More generally, zeolites are used as *molecular sieves*. Zeolites are colorless crystals with sizes of several millimeters (natural ones) or of several micrometers (synthetic ones). Density is 1.8 to 2.3 g/cm^3, thermal expansion is similar to quartz. Mohs hardness of natural zeolites is 4 to 6, *refractive index* is 1.47 to 1.52. In industrial applications zeolites are granulated with a bonding agent. The bulk density of granules is 0.5 to 0.8 g/cm^3, in the dehydrated state the *specific heat* is 0.83 to 1.0 J·g^{-1}·K^{-1}, *thermal conductivity* is 2.1 W·m^{-1}·K^{-1}. Upon mild heating (150 to 350°C) zeolites give off water without disturbing their structure. Heating at 400 to 1000°C disintegrates the structure.

The large surface of zeolites allows one to investigate the macroscopic physical parameters of adsorbed *films*, e.g., the specific heat. In particular, adsorbed ^{4}He, ^{3}He, H_2, CH_4, H_2O and other substances have been investigated in zeolites of various structures. *Phase transitions* were observed, which were interpreted as *melting* (freezing) of the adsorbate. As a rule, the transition takes place at lower temperatures compared to the case of these free substances. Structural investigations by diffraction methods show that at low temperatures the "solid" adsorbed phase is incommensurate with the substrate, i.e. there is no *epitaxy*. At temperatures above $T_{melting}$ of the adsorbate the molecular mobility takes place by diffusional jumps between minima of the interaction potential "adsorbed molecule–zeolite surface". Intermolecular interactions at high densities of adsorbate filling lead to a noticeable smearing of the diffraction minima, which is clear from the diminished intensity of diffraction lines associated with the adsorbed phase. In contrast to this the diffractograms of some hydrated zeolites appear almost the same as those of dehydrated ones. These experimental results help to clarify the nature of the phase transitions in zeolites that are associated with incommensurability, i.e. the transition from dynamic disordering of the adsorbate above $T_{melting}$ to static disordering below $T_{melting}$. Zeolites are used as catalysts, as components of detergents, as *ionites* (ion exchangers), and as fillers.

ZERO-FIELD ENERGY LEVEL SPLITTING

Energy level splittings of paramagnetic ions (nuclei) in the absence of an external magnetic field. The magnitude of the zero field splitting depends on the *local symmetry group*, interaction with the crystalline electric field, the *spin–orbit interaction*, *hyperfine interactions*, and *exchange interactions* with other ions. As for atomic nuclei, their zero field splitting results from interactions of their electric quadrupole moments with *electric field gradients* (see *Nuclear quadrupole interaction*). For example, the fivefold orbitally degenerate d-electron level splits into a doublet and a triplet in a cubic *crystal field* and into five singlets in a rhombic field. To estimate the sequence of terms in d^n, f^n-configurations *Tanabe–Sugano diagrams* are used. The splitting of the initial levels is measured using *zero-field magnetic resonance, paramagnetic resonance, nuclear quadrupole resonance*, nuclear gamma ray resonance (see *Mössbauer effect*), and optical techniques. It is described by a generalized *spin Hamiltonian*, and by the parameters of the crystal field.

ZERO-FIELD MAGNETIC RESONANCE

A microwave or radio frequency (RF) power absorption by a system of *paramagnetic centers* (ions, nuclei) in the absence of an external constant applied magnetic field B. When $B = 0$, the energy level splitting is caused by the interaction of the total angular momentum J of the center with the crystal field (*fine structure*), by *hyperfine interactions* (A), *exchange interactions* and quadrupole interactions (Q). The resonance is detected by sweeping the microwave or RF frequency. To obtain an undistorted recording of the *line shape*, a modulation is applied in the form of turning the magnetic field on/off with an amplitude 1–20 mT at low switching frequencies

(≤1–10 kHz). Zero-field magnetic resonance includes observations of the electron paramagnetic resonance (EPR) of intrinsic and impurity centers with electronic spin $S \geqslant 1$ at $B = 0$, *nuclear quadrupole resonance*, and *optical detection of magnetic resonance*. The advantages of zero-field are the improved accuracy of determination of the zero field splitting constants D and E (elimination of B-dependent line broadening and crystal orientation dependence), an opportunity to work with powders rather than crystals; the simplicity of the spectra, ease of spectral processing; and coincidence of lines of magnetically nonequivalent centers. The disadvantages of this method are the lack of information on the symmetry and the orientation of the center axes, and on the individual components of the crystalline field and the A-, D-, and Q-tensors.

ZERO-PHONON SPECTRAL LINE

See *Phononless spectral line*.

ZERO-POINT VIBRATIONS of a crystal lattice

Vibrations of the atoms of a *crystal* which are characteristic of its state at $T = 0$ K. Zero-point vibrations are a purely quantum effect related to the fact that the energy of the ground state of a harmonic oscillator differs from zero. A characteristic of zero-point vibrations is the mean square of the atomic displacement amplitude $\langle u^2 \rangle$, which in the *harmonic approximation* for a monatomic crystal is given by

$$\langle u^2 \rangle = \frac{\hbar}{2MN} \sum_{k,\alpha} \frac{1}{\omega_\alpha(k)},$$

where M is the atomic mass, N is the number of atoms in the crystal, $\omega_\alpha(k)$ is the *phonon* frequency for the αth branch, and the wave vector is k. The value of $\langle u^2 \rangle^{1/2}$ in an ordinary (classical) crystal is significantly smaller than its lattice constant, while in quantum crystals (such as H or He) the value of $\langle u^2 \rangle^{1/2}$ may reach 30% of the lattice constant. The *de Boer parameter* is a measure of the quantum nature of a crystal.

ZERO SPIN OSCILLATIONS

Quantum contribution to the ground state energy of a spin system undergoing low amplitude oscillations about its classical equilibrium positions. Zero spin oscillations take place when the system symmetry is lowered (e.g., due to the *magnetic dipole interaction*) so no component of the total spin is conserved. Like *zero-point vibrations* of an oscillator, zero spin oscillations result from the uncertainty relation. Unlike the oscillator, the zero spin oscillation energy is negative, i.e. the quantum effects lower the ground state energy. The corresponding contribution to the *magnetization* is also negative, i.e. the latter remains lower than its nominal value for any external magnetic field amplitude due to the zero spin oscillations.

ZIMAN–FABER THEORY (J. Ziman, 1961; T.E. Faber, 1965)

Theory of electronic properties of *liquid metals* whereby their electronic structure is described by the *nearly-free electron approximation* with a spherically symmetric *Fermi surface*. The electron scattering by the pseudopotentials of atoms (see *Pseudopotential method*) is considered as weak, and the *Born approximation* is applicable. The resistivity ρ is described by the expression

$$\rho = \frac{3\pi}{\hbar e^2 v_F^2 \Omega} \int_0^{2k_F} \frac{|v(q)|^2 S(q) q^3 \, dq}{4k_F^4},$$

where $v(q)$ is the Fourier transform of the atom pseudopotential, Ω is the system volume, and $S(q)$ is the *structure factor*

$$S(q) = \frac{1}{N} \int \left[1 + e^{iqR} \right]^2 P(R) \, d^3 x,$$

where $P(R)$ is the pair distribution function. In *binary alloys*, the quantity $|v(q)|^2 S(q)$ in the integrand of the above equation is replaced by its average value:

$$\overline{|v(q)|^2 S(q)} = c_1 |v_1(q)|^2 \{c_2 + c_1 S_{11}(q)\}$$
$$+ c_2 |v_2(q)|^2 \{c_1 + c_2 S_{22}(q)\}$$
$$+ c_1 c_2 \left(v_1^*(q) v_2(q) + v_1(q) v_2^*(q) \right)$$
$$\times \{S_{12}(q) - 1\},$$

where c_i and $v_i(q)$ are the concentration and potential of the ith component, and $S_{ij}(q)$ is the partial atomic factor for the i–j pair. R. Evans has improved the Ziman theory by replacing $v(q)$ with the t-matrix:

$$v_j(q) \rightarrow t_j(q)$$

$$= -\frac{2\pi \hbar^3}{m(2mE_F)^{1/2}\Omega}$$

$$\times \sum_l (2l+1)\sin \eta_l^j (E_F)$$

$$\times \exp[i\eta_l^j (E_F)]P_l(\cos\theta),$$

where $\eta_l^j (E_F)$ is the phase shift for scattering with the angular momentum l, calculated at the *Fermi energy* E_F. With this replacement the expressions for ρ and $S(q)$ become applicable for the description of scattering processes with strong interactions in liquid transition metals and alloys. According to Ziman, the temperature dependence of the resistivity is determined by the structure factor $S(q)$. For instance, the temperature factor of resistance depends on the position of $2k_F$ relative to q_p, and the position of the main peak of the structure factor $S(q)$. With increasing temperature the structure factor peak broadens, and all $S(q)$ values near q_p grow smaller. If the alloy is characterized by the value $2k_F \approx q_p$, then its temperature coefficient of resistance is expected to be negative. On the other hand, if the value $2k_F$ is far from q_p then all $S(q)$ values increase, and the temperature coefficient of resistance should be positive. Thus the Ziman–Faber theory makes it possible to explain the electrical resistance behavior of liquid and *amorphous metals and metallic alloys*, namely the temperature dependence of the resistance, the origin of its negative temperature coefficient, and the continuity of the variation in electrical resistance at the transition from the liquid to the *amorphous state*.

ZINC, Zn

A chemical element of Group II of the periodic system with atomic number 30 and atomic mass 65.39. There are 5 stable isotopes: ^{64}Zn (48.6%), ^{66}Zn (27.9%), ^{67}Zn (4.1%), ^{68}Zn (18.8%) and ^{70}Zn (0.6%). Outer shell electronic configuration is $3d^{10}4s^2$, ionization energies are 9.391, 17.96, 39.70 eV. Atomic radius is 0.134 nm; radius of Zn^{2+} ion is 0.074 nm. Oxidation state is +2. Electronegativity is $\sim$1.60.

Zinc is a bluish-white *metal*. It has a hexagonal close-packed lattice, space group $P6_3/mmc$ (D_{6h}^4), $a = 0.26649$ nm, $c = 0.49469$ nm (at 298 K). Density is 7.131 g/cm^3 (at 298 K); density of liquid zinc is 6.66 g/cm^3. $T_{melting} = 692.73$ K, $T_{boiling} = 1180$ K. Binding energy is -1.35 eV/atom at 0 K. Heat of melting is 7.29 kJ/mole, heat of sublimation is 131.43 kJ/mole, heat of evaporation is 115 kJ/mole, specific heat is 461 J·kg^{-1}·K^{-1} (at 693 K, constant pressure), Debye temperature is $\sim$327 K. Linear thermal expansion coefficient of monocrystal is 63.5·10^{-6} K^{-1} along the principal axis $\underline{6}$ and 13.2·10^{-6} K^{-1} perpendicularly to this axis at 300 K, and at 60 K, respectively, we have 55·10^{-6} K^{-1} and 2·10^{-6} K^{-1}. Coefficient of thermal conductivity is 111 W·m^{-1}·K^{-1} (at 293 K). Adiabatic elastic moduli of zinc monocrystal: $c_{11} = 163.68$, $c_{12} = 36.4$, $c_{13} = 53.0$, $c_{33} = 63.47$, $c_{44} = 38.79$ (in GPa) at 295 K; bulk modulus is 68.8 to 72.5 GPa, Young's modulus is $\sim$114 GPa; shear modulus is $\sim$46 GPa; Poisson ratio is $\sim$0.23 (at 298 K); tensile strength is $\sim$0.22 GPa, relative elongation is $\sim$45%. Brinell hardness is $\sim$0.44 GPa; yield limit of common zinc is 0.00039 GPa, and of thread-like crystal ("whisker") it is 0.38 GPa. Zinc of high purity (99.999%) is plastic and can easily be drawn to a thin wire. Zinc of 99.9% purity in the cold state is brittle, and at 370 to 420 K it is plastic and may be easily rolled to sheets and foil with a thickness hundredths of a millimeter. With increasing temperature within the range 470 to 520 K, zinc becomes brittle again and transforms to a powder. The recrystallization temperature is 293 K. Self-diffusion coefficient is 2.50·10^{-14} m^2/s along the principal axis $\underline{6}$ and 1.29·10^{-14} m^2/s perpendicularly to this axis at 554 K. Ion-plasma frequency is 13.34 THz, linear low-temperature electronic specific heat (Sommerfeld coefficient) is $\sim$0.62 mJ·mole^{-1}·K^{-2}. Electrical resistivity of zinc monocrystal is 55.8 nΩ·m along the principal axis $\underline{6}$ and 53.8 nΩ·m perpendicular to this axis, temperature coefficient of electrical resistance is 0.00414 K^{-1} along the principal axis $\underline{6}$ and 0.00406 K^{-1} perpendicular to this axis (at room temperature and normal pressure), Hall constant is +2.7·10^{-11} m^3/C (+14.4·10^{-11} m^3/C for external magnetic field applied along principal axis $\underline{6}$ at room temperature); coefficient of the absolute thermal electromotive force is $+0.4$ μV/C along the principal axis $\underline{6}$ and $+2.1$ μV/C perpendicular to this axis. Coefficient of reflection of the optical waves of wavelength 5.0 μm is 97.9%. Work function of zinc polycrystal is 4.24 eV. Superconducting transition temperature is 0.850 K; critical magnetic field is 5.4 mT (at 0 K). Zinc is diamagnetic with molar magnetic susceptibility $-11.4 \cdot 10^{-6}$ CGS units at 300 K (value along principal axis $\underline{6}$ is 1.4 times higher than in perpendicular direction); nuclear magnetic moment of ^{67}Zn isotope is 0.874 nuclear magnetons. Thermal neutron capture cross-section is 1.1 barn.

Zinc is used for *corrosion protection* (see *Zinc plating*), for the production of many alloys, e.g., with copper (*brass*, see *Copper alloys*). Zinc compounds are used for the preparation of paints, of electrodes in chemical current sources, etc.

ZINC PLATING

Application of layer of *zinc* at the surface of metallic (mainly steel and cast iron) products to protect them from *corrosion*. The thickness of the layer is often 20 to 40 μm. Hot, electrolytic, and diffusion methods of zinc plating are used, and also zinc plating by metallization. *Hot zinc plating* is the most widespread, when the products are submerged into the tank with molten zinc at ~ 723 K. *Electrolytic zinc plating* is carried out in acid or alkali electrolytes. *Diffusion zinc plating* is performed by the vapor-phase method at 573 to 1123 K; the *coating* formed in this case is an iron–zinc alloy. Metallization is accomplished using special guns provided with devices for melting the metal and spraying it along the surface of the product being treated. See also *Chemical heat treatment*.

ZINTL PHASES (after German scientist E. Zintl, 1931)

Compounds with the AB formula, where component A is always an *alkali metal* (Li or Na). When the *atomic radii* of the components have the ratio $R_A/R_B < 1$, the BCC CsCl-type *crystal structure* appears, and when $R_A/R_B > 1$, then, as a rule, the structure is the FCC NaCl-type. The homogeneity regions of Zintl phases in double crystal systems are extremely narrow.

ZIRCONIUM, Zr

Chemical element of Group IV of the periodic system with atomic number 40 and atomic mass 91.224. There are 5 stable isotopes: ^{90}Zr (51.45%), ^{91}Zr (11.32%), ^{92}Zr (17.19%), ^{94}Zr (17.28%) and ^{96}Zr (2.76%). Outer shell electronic configuration is $4d^2 5s^2$. Ionization energies are 6.835, 12.92, 24.8, 33.97, 82.3 eV. Atomic radius is 0.158 nm; radius of Zr^{4+} ion is 0.079 nm. Oxidation state is $+4$, less often $+2$, $+3$. Electronegativity is ~ 1.36.

Zirconium is a silvery-white *metal*. It exists in two allotropic modifications (α-Zr, β-Zr). Below 1135 K α-Zr is stable with hexagonal closely-packed crystal lattice, space group $P6_3/mmc$ (D_{6h}^4), $a = 0.32317$ nm, $c = 0.51476$ nm (at 302 K). Above 1135 K up to $T_{melting} = 2125$ K β-Zr is stable with body-centered cubic lattice, space group $Im\bar{3}m$ (O_h^9), $a = 0.3620$ nm (at 1140 K). At pressures above 6.08 GPa the hexagonal ω-phase of zirconium is formed with $a = 0.5036$ nm, $c = 0.3109$ nm (in metastable state after pressure removal). Density of α-Zr is 6.49 g/cm^3 (at 293 K), of β-Zr 6.36 g/cm^3 (at 1136 K), $T_{boiling} = 4650$ K. Binding energy is -6.316 eV/atom at 0 K. Heat of $\alpha \leftrightarrow \beta$ transformation is 2.98 kJ/mole; heat of melting is 19.3 kJ/mole; heat of sublimation is 523 kJ/mole; heat of evaporation is 536 kJ/mole; specific heat is 0.290 kJ·kg^{-1}·K^{-1} (298 to 373 K). Debye temperature is 291 K; linear thermal expansion coefficient of zirconium monocrystal is $7.36 \cdot 10^{-6}$ along the principal axis $\underline{6}$ and $4.99 \cdot 10^{-6}$ K^{-1} perpendicular to this axis (at 300 K); thermal conductivity coefficient is 20.93 W·m^{-1}·K^{-1} (293 to 323 K); adiabatic elastic moduli of zirconium monocrystal: $c_{11} = 143.68$, $c_{12} = 73.04$, $c_{13} = 65.88$, $c_{33} = 165.17$, $c_{44} = 32.14$ (in GPa) at 298 K and zero external pressure; bulk modulus is ~ 110 GPa; Young's modulus is 95.6 GPa; shear modulus is 35.2 GPa, Poisson ratio is ~ 0.355 (at 298 K); tensile strength of annealed zirconium is ~ 0.255 GPa (at relative elongation 20 to 25%), yield limit is ~ 0.087 GPa. Vickers hardness is ~ 0.97 GPa at room temperature. High-purity zirconium is plastic, and is readily subjected to hot and cold mechanical treatment (rolling, stamping, forging); recrystallization temperature is 435 K. Surface tension of liquid zirconium at $T_{melting} = 1.48$ J/m^2. Self-diffusion coefficient of atoms in zirconium monocrystal at $T_{\text{melting}} = 3.4 \cdot 10^{-14}$ m^2/s. Effective thermal neutron trapping cross-section is 0.185 barn. Bulk modulus of interacting electron gas in zirconium is 23.1 GPa (at 4 K). Linear low-temperature electronic specific heat is $\sim$2.785 mJ·mole^{-1}·K^{-2}. Electrical resistivity is 415 nΩ·m at room temperature for α-Zr polycrystal and 1247 nΩ·m at 1427 K for β-Zr; temperature coefficient of the electrical resistance is 0.0044 K^{-1} (at 273 to 373 K); Hall constant is $+2.27 \cdot 10^{-11}$ m^3/C (at 300 K); coefficient of thermal electromotive force is 2 μV/K at room temperature, electronic work function from β-Zr polycrystal is 4.0 eV. Superconducting transition temperature is 0.61 K; critical magnetic field is 4.7 mT (at 0 K). Zirconium is paramagnetic with molar magnetic susceptibility $+129 \cdot 10^{-6}$ in CGS units at room temperature (value along principal axis $\underline{6}$ is higher than in perpendicular direction); nuclear magnetic moment of ^{91}Zr isotope is 1.298 nuclear magnetons.

Zirconium is a preferred *construction material* for the active zone of thermal neutron reactors, due to its combination of favorable physical mechanical parameters, and its high *corrosion resistance* in many aggressive media.

ZONE MELTING

A technique of crystal growth (see *Crystallization*), or for the *zone refinement* of a material. The source material is placed into a boat (horizontal option), into an ampoule, or is pressed into the shape of a cylindrical rod (horizontal or vertical option). When using zone melting to grow crystals, a seed is placed at one end of the boat or ampoule, or put in contact with the rod. With the help of a concentrated source of thermal flux or heat (induction, laser, xenon lamp, etc.), a melted zone is created within the source material. The zone is then transported at a specified speed along the boat, ampoule or rod, which results in *recrystallization* (in case of zone refinement) or *monocrystal* growth from the seed.

The prevailing option with rod melting is *non-crucible zone melting*. The melted zone with vertical positioning is held in place by *surface tension* forces, and the zone dimensions are limited by the conditions of its stability. For the horizontal rod position, an electromagnetic field is used to maintain the zone. Unlike crucible methods (see *Monocrystal growth*), the non-crucible method provides a higher purity of the final product crystal due to the absence of an interaction between the melt and the material forming the crucible.

ZONE REFINEMENT

A technique for purifying materials containing impurities with the aid of *zone melting*. The impurity collects in the molten zone and, through gradual displacement of the region of melting, is transferred from one end of the ingot to the other, thereby providing purification of the material. The dependence of the impurity distribution, $C(x)$, on the distance x along the recrystallized part of the ingot can be expressed as

$$C(x) = C_0 \left[1 - (1-k) \exp\left(-\frac{kx}{b} \right) \right],$$

where C_0 is the initial concentration of the impurity, b is the zone length, and k is the dimensionless *interphase impurity distribution coefficient*. For $k > 1$ the impurity is forced to the far end of slab, and for $k < 1$ to the beginning of it. In order to increase the effectiveness of the refinement one can either process several zones simultaneously, or one zone repeatedly. The degree of purification, C_∞, which is achievable in theory with a large enough number of repetitions, does not depend on the number of passages, but rather depends on the value of the interphase impurity distribution coefficient. For example, for $C_0 = 1$ and $b/L = 0.1$ (L is the ingot length), C_∞ varies from 10^{-3} for $k = 0.5$ to 10^{-25} for $k = 0.01$. Zone refinement is used to purify materials for final use, as well as to provide intermediate compounds.

ZUBAREV METHOD (named after D.N. Zubarev)

A way to describe nonequilibrium statistical ensembles that is based on the introduction of the *nonequilibrium statistical operator*.

Subject Index

All terms in this Subject Index are arranged alphabetically. There are many cross-references between them. If the term corresponds to the title of a separate article in the main text, then its page number is emphasized by **boldface**.

band structure, 93, 100, 101, **102**, 144, 214, 240, 290, 390, 401, 437, 518, 535, 560, 587, 588, 637, 675, 715, 716, 778, 850, 1007, 1040, 1056, 1112, 1128, 1130, 1232, 1305, 1321, 1327, 1435, 1451, 1474, 1480

band structure computation, *see* linear methods of band structure computation

band structure computation methods, 95, **102**, 102, 290, 683, 716, 865, 941

band structure, surface, *see* surface band structure

band theory, 101, **102**, 237, 272, 290, 357, 368, 395, 538, 581, 638, 712, 872, 925, 1060, 1083, 1090, 1093, 1096, 1189, 1232, 1474, 1506

band theory relativistic effects, *see* relativistic effects in band theory of metals

band-to-band recombination, 1122

band-to-band tunneling, *see* interband tunneling

band-type domain, 755

band, valence, *see* valence band

bands of allowed energies, 87

bands of microwave frequencies, *see* microwave frequency bands

bands of slip, *see* slip bands

bands with secondary shear, 1208

Bardeen barrier, **104**

Bardeen–Cooper–Schrieffer theory, 60, 61, **104**, 115, 123, 149, 184, 212, 230, 409, 425, 478, 498, 507, 522, 523, 555, 576, 662, 907, 1174, 1256, 1258, 1308, 1309, 1311, 1369, 1434

Bardeen–Herring sources, 867

Bardeen law, 268

barium, **105**, 574

barium–sodium niobate, 881

barium titanate, 19, **105**, 335, 450, 455, 881, 1045, 1121

Barkhausen effect, **106**, 336, 759, 785, 791, 1033

Barkhausen jump, 106

Barkhausen noise, **106**, 884

Barkhausen pulses, 1033

Barnett effect, **106**, 358, 800

barodiffusional effect, **106**, 491, 681

Barrett formula, **107**, 1460

barrier, Bardeen, *see* Bardeen barrier

barrier capacitance, **107**, 150, 833, 1164, 1375, 1452

barrier injection transit time diode, **107**, 311, 516, 875, 1179

barrier, local, *see* local barriers for dislocations

barrier, Lomer–Cottrell, *see* Lomer–Cottrell barrier

barrier, potential, *see* potential barrier

barrier, Schottky, *see* Schottky barrier

barrier structures, surface, *see* surface barrier structures

barrier, surface, *see* surface barrier

BARRITT diode, *see* barrier injection transit time diode

basal (coordinate) faces, 108

basal plane, **108**, 238

basalt, **108**

basic units, 1437, 1438

Bauschinger effect, **108**, 1336

Baxter model, 1427

BCS criterion for superconductivity, 231

BCS relationship, 105

BCS theory, *see* Bardeen–Cooper–Schrieffer theory

B-defects, 724

beams of high-energy electrons, 279

Beer–Lambert law, 3, **108**, 914

Beilby layer, **109**

Belousov–Zhabotinsky reaction, 333

Belov groups, 1481

Belov theorem, 179

Bénard cells, 333

bending moment, 475

bending of particles, *see* particle bending

bending vibrations, **109**, 1455

Benedicks effect, 1386

Bennett condition, 758

Berezinski–Kosterlitz–Thouless transition, *see* Kosterlitz–Thouless transition

Berezinski phase, 685

berkelium, 26, **109**, 1409

Berkovich hardness, 608

Bernal cavities, 1113

Bernal cells, 1113

Bernal's fluid model, *see* random close-packing model

Bernal's liquid model, 825

berthollides, 845

beryl, 511

beryllium, **109**, 110, 511, 595, 1036

beryllium alloys, **110**

beryllium bronze, 215

beta phase, **111**, 134, 591, 1233, 1399

beta relaxations, 1464

beta spectroscopy, 397

Bethe ansatz, 61

Bethe lattice, **111**, 533, 592

Bethe–Peierls approximation, **111**

Bethe splitting, **112**

Betti reciprocity theorem, **112**

biaxial crystals, **112**, 168, 204, 253, 356, 431, 1022, 1256

biaxial magnetic, 752

bicritical point, **112**, 866

bidentate ligands, 711

biexciton, **112**, 129, 234, 410, 423, 424, 566, 916, 944, 1040, 1301

bifurcations, 347

bigradient electromotive force of hot current carriers, 1386

bimetals, **113**

binary alloy system, 113

contact corrosion, **208**, 725
contact-fatigue wear, 1475
contact interaction, 597
contact, nonohmic, *see* nonohmic contact
contact, ohmic, *see* ohmic contact
contact, point, *see* point contacts
contact potential difference, 104, 150, 173, **208**, 370,
    616, 1181, 1251, 1341, 1385
contact problem of elasticity theory, **208**, 209
contact stresses, **208**
contact wetting, 1478
contact X-ray microscopy, 1492
continuity conditions, 1282
continuous acoustic emission, 7
continuous alloy decomposition, 42
continuous (Bremsstrahlung or braking) spectrum, 1495
continuous potential, 164
continuous symmetry transformation group, 1353
continuous symmetry transformation groups, **209**, 509,
    525, 637, 1272
continuum approximation, 209
continuum integral, *see* path integral
continuum mechanics, 8, 143, **209**, 361, 418, 689, 794,
    1283, 1289
continuum theory of defects, **211**, 958, 1010, 1246
contraction, transverse, *see* transverse contraction
conversion, 517
conversion electrons, 397
converter, 906
converter, optical frequency, *see* optical frequency
    converter
convolution of line shape, *see* line shape convolution
Conwell–Weisskopf formula, 265
cooling, adiabatic demagnetization, *see* adiabatic
    demagnetization cooling
cooling, adiabatic paramagnetic, *see* adiabatic
    paraelectric cooling
cooling agent, **211**, 234, 235, 736
cooling, desorption, *see* desorption cooling
cooling duct, **211**
Cooper effect, 104, **212**, 576, 783, 1313
Cooper pair, kinetic decoupling, *see* kinetic decoupling
    of Cooper pairs
Cooper pairs, 4, 21, 37, 60, 61, 67, 104, 115, 124, 129,
    166, 184, 190, **212**, 226, 230, 268, 392, 409, 424,
    446, 475, 477, 507, 522, 526, 530, 536, 576, 662,
    671, 679, 747, 783, 806, 819, 830, 886, 934, 949,
    961, 969, 976, 982, 1015, 1071, 1080, 1094,
    1125, 1174, 1207, 1258, 1301, 1303, 1307, 1311,
    1313, 1317, 1318, 1337, 1413, 1416, 1423
cooperative luminescence, 82, **212**, 739, 740
coordinates, Eulerian, *see* Eulerian coordinates
coordinates, Lagrange, *see* Lagrange coordinates
coordinates, rotating, *see* rotating coordinate system

coordination compounds, **213**, 559, 711
coordination number, 91, 172, 213
coordination polyhedron, 1038
coordination sphere, 43, 91, 111, 181, 189, **213**, 237,
    258, 619, 728, 911, 935, 1037, 1038, 1206, 1252,
    1262
copolymer, 937, 1042
copper, 80, 134, 140, 211, **214**, 220, 293, 504, 564, 627,
    656, 736, 892, 991, 1059, 1370, 1403
copper alloys, 45, **214**, 220, 518, 1510
corals, 514
Corbino disk, **215**
Corbino disk geometry, 501
cordierite, 514
correlation, **215**
correlation energy, 202, **215**, 289, 440, 551, 668, 858,
    892
correlation energy of a center, **216**
correlation function, 59, **216**, 217, 347, 476, 494, 515,
    533, 681, 687, 717, 728, 855, 892, 965, 1050,
    1159, 1345, 1402
correlation function of random field, **217**, 1115
correlation in alloys, 43, 1138
correlation length, 68, **217**, 227, 228, 304, 350, 441,
    523, 629, 659, 679, 772, 962, 963, 972, 1159,
    1168, 1307, 1423
correlation parameters, 935
correlation radius, 217, 728, 1159, *see* correlation
    length
correlation time, 121, **217**, 494
correlator, 216
corrosion, 50, 52, 83, 208, 214, **218**, 220, 426, 469, 525,
    528, 626, 647, 705, 725, 750, 804, 813, 814, 857,
    860, 883, 943, 1233, 1275, 1302, 1320, 1331,
    1336, 1371, 1475, 1510
corrosion, contact, *see* contact corrosion
corrosion cracking, 8, **218**, 219, 220
corrosion fatigue, **219**, 220, 408, 438, 1347
corrosion, intercrystalline, *see* intercrystalline corrosion
corrosion, local, *see* local corrosion
corrosion-mechanical strength, 220
corrosion of metals, **219**, 1010
corrosion potential, 218
corrosion protection, 73, 218, 219, **220**, 1359, 1510
corrosion resistance, 42, 46, 47, 80, 95, 173, 176, 177,
    195, 219, **220**, 333, 379, 407, 646, 647, 650, 653,
    654, 705, 750, 805, 814, 880, 883, 1021, 1053,
    1129, 1215, 1230, 1275, 1279, 1358, 1359, 1372,
    1399, 1449, 1511
corrosion under stress, **220**, 816
corrosive, 508
corundum, 511
Costa-Ribeiro effect, *see* thermodielectric effect

Mandelshtam–Brillouin scattering, *see* Brillouin
   scattering
manganese, 157, 564, 750, **806**
manganese spar, 513
manganin, 215
manometers, 572
many-configuration approximation, 510, 551, **807**, 1074
many-soliton solutions, 1244
many-valley semiconductors, 101, 262, 264, 310, 385,
   588, 589, 635, 636, 639, **808**, 847, 883, 982,
   1007, 1156, 1190, 1202, 1220, 1238, 1474
many-wave approximation, 610, **808**
many-well potential, **809**
Marangoni effects, 1379
margin of plasticity, 137, 283, **809**
margin of safety, **809**
martensite, 95, 154, 415, 548, 655, 701, 782, **809**, 810,
   811, 940, 975, 1019, 1060, 1166, 1206, 1316,
   1383, 1389, 1414, 1422, 1479
martensite, thermoelastic, *see* thermoelastic martensite
martensite twinning, *see* twinning of martensite
martensitic point, 811
martensitic polytypes, 8, **810**, 1044
martensitic transformation, 7, 95, 135, 186, 306, 328,
   367, 415, 599, 654, 809, 810, **811**, 940, 1019,
   1044, 1059, 1119, 1166, 1206, 1248, 1316, 1383,
   1389, 1421–1423, 1479
maser, 392, 706, 778, **812**, 901, 1072, 1140, 1141,
   1361, 1491
mask, 1137
mass, *see* effective mass
mass-analyzers, 813
mass concentration, 198
mass crystallization, 247
mass force, 141
mass operator of quantum mechanics, 186, 351, 536,
   812, 1033
mass spectrometry, 180, 469, 470, 497, 648, 699, **812**,
   1003, 1166
mass spectrometry via secondary ions, *see* secondary
   ion mass spectrometry
mass spectrum, 812
mass transfer, anomalous, *see* anomalous mass transfer
mass transport, 158, 226, 518, 551, 595, 609, 708, 746,
   747, **813**, 1026, 1051, 1064, 1219, 1340, 1344,
   1373
mass transport, surface, *see* surface mass transport
master equation, 1079
masurium, *see* technetium
Matano–Boltzmann method, 628
material coordinates, 689
material mechanical testing, *see* mechanical testing of
   materials
material reliability, *see* reliability of materials

material tensor, 269
materials, acousto-electronics, *see* acousto-electronics
   materials
materials, amorphous magnetic, *see* amorphous
   magnetic materials
materials, anisotropic magnetic, *see* anisotropic
   magnetic materials
materials, antiemission, *see* antiemission materials
materials, antifriction, *see* antifriction materials
materials, bimetallic, *see* bimetals
materials, casting, *see* casting materials
materials, chalcogenide, *see* chalcogenide materials
materials, composite, *see* composite materials
materials, construction, *see* construction materials
materials, electrooptic, *see* electrooptic materials
materials, high-temperature, *see* high-temperature
   materials
materials, hyperelastic, *see* hyperelastic material
materials, hypoelastic, *see* hypoelastic material
materials, insulating, *see* insulating materials
materials, magnetic, *see* magnetic materials
materials, magnetostrictive, *see* magnetostrictive
   materials
materials, nonmagnetic, *see* nonmagnetic materials
materials of membranes, *see* membrane materials
materials of powder metallurgy, *see* strength and
   plasticity of powder metallurgy materials
materials, photographic, *see* photographic materials
materials, piezoelectric, *see* piezoelectric materials
materials, polymeric, *see* polymeric materials
materials, precipitation-hardened, *see*
   precipitation-hardened materials
materials, pyroelectric, *see* pyroelectric materials
materials, radiation-absorbing, *see* radiation-absorbing
   materials
materials, refractory, *see* refractory materials
materials science, 140, 437, **814**, 827, 832, 860, 1085
materials, semiconductor, *see* semiconductor materials
materials, soft magnetic, *see* soft magnetic materials
materials, solid-state laser active, *see* active solid-state
   laser materials
materials, sound-absorbing, *see* sound-absorbing
   materials
materials, superhard, *see* superhard materials
materials, thermal-environment resistant, *see*
   thermal-environment resistant materials
materials, thermomagnetic, *see* thermomagnetic
   materials
materials, thermoplastic, *see* thermoplastic materials
Mathieu equation, 966
matrix, 186
matrix diffusion ratios, 480
matrix isolation, **814**
Matthias rule, **815**

superlocalization of strain shear bands, **1321**
supermalloy, 571
superparamagnetism, 457, 1218, 1303, **1322**
superplasticity, 35, 529, 726, 1016, 1018, **1322**
superplasticity of a phase transition, 1322
superposition approximations, 681
supersaturated solid solution, 1236
superstep, 326
superstrong magnetic fields, 428, **1323**
superstrong pulsed magnetic fields, 1323
superstructure, 1202
superstructure Bragg reflections, 1324
superstructure of alloy, *see* superlattice of alloy
superstructure, semiconductor, *see* semiconductor
    superstructure
superstructure vector, **1324**
supplementary light waves, *see* additional light waves
supporting diode, 310
surface acoustic waves, 10, 17, 19–22, 25, 224, 498,
    881, 888, 1112, 1226, 1243, 1302, **1324**, 1350,
    1431, 1462
surface-active agents, 37, 279, 281, 329, 376, 593, 626,
    742, 817, 1289, **1325**, 1334, 1349
surface activity, **1325**
surface alloying, *see* surface doping
surface atom vibrations, *see* vibrations of surface atoms
surface atomic roughness, 1344
surface atomic structure, 1136, **1325**, 1345
surface, atomically clean, *see* atomically clean surface
surface band bending, 1341
surface band conductivity, 1333
surface band structure, **1327**, 1337, 1357
surface barrier, **1327**
surface-barrier silicon detectors, 906
surface barrier structures, **1327**
surface Brillouin zone, **1328**
surface capacitance, 1203, **1328**
surface capacitance method, 1328
surface channeling, 165, 955, **1329**
surface, characteristic, *see* characteristic surface
surface, clean, *see* atomically clean surface
surface coalescence, 182
surface conductive channel, 165, **1329**
surface decoration, 961, **1330**, 1391
surface defects, 427, 640, 1326, **1330**, 1331, 1333, 1345
surface degradation, **1331**
surface diffusion, 248, 306, 468, 469, 1243, **1331**, 1373
surface dislocations, 325, **1331**
surface doping, 699, **1331**
surface drift waves, **1332**
surface dynamics, **1332**
surface electrical conductivity, 1203, **1332**
surface electromagnetic waves, 1349

surface electron states, 104, 604, 823, 850, 1162, 1203,
    1210, 1251, 1327, 1328, 1331, **1333**, 1337, 1339,
    1341–1343, 1347, 1353, 1357, 1435
surface energy, 2, 49, 92, 139, 207, 229, 251, 328, 329,
    366, 376, 406, 433, 522, 537, 626, 630, 704, 746,
    835, 841, 857, 909, 941, 971, 988, 1051, 1055,
    1170, 1175, 1242, 1252, 1273, 1287, 1289, 1293,
    1325, **1334**, 1344, 1403, 1423, 1429, 1430
surface energy of superconductors, **1334**
surface erosion, 118, 346, 375, 841, 1320, **1335**
surface exciton, 423, **1335**
surface excitonic polariton, 1335, 1342
surface, Fermi, *see* Fermi surface
surface force, 112, 361, 471, 1010, 1155, **1336**, 1339
surface, freshly prepared, *see* juvenile surface
surface hardening, 189, 546, 699, 1335, **1336**, 1349,
    1432
surface hydration, *see* hydration of surfaces
surface impedance, 3, 72, 273, 605, 677, 783, 784, 797,
    1109, 1140, 1271, **1336**, 1337
surface impedance of a superconductor, 1306, **1337**
surface impurity states, 1333, **1337**
surface inhomogeneities, 1329, **1338**
surface ionization, 468, 642, 955, **1338**, 1371
surface, isoenergetic, *see* isoenergetic surface
surface, isofrequency, *see* isofrequency surface
surface, juvenile, *see* juvenile surface
surface Langmuir waves, 1239
surface lattices, 88, 977, **1338**
surface layer, 109, 417, 1325, **1339**, 1349
surface layer, disrupted, *see* disrupted surface layer
surface levels, 707, 992, 1210, 1243, 1337, **1339**
surface levels, magnetic, *see* magnetic surface levels
surface magnetism, **1339**
surface mass transport, 813, **1340**
surface mobility, 847, **1340**, 1341
surface modification, ion-stimulated, *see* ion-stimulated
    surface modification
surface of a solid, *see* solid surface
surface of normals, 253
surface of semiconductor, *see* semiconductor surface
surface of stress, *see* stress surface
surface optical spectroscopy, *see* optical spectroscopy
    of surfaces
surface passivation, 220, 653, 943, 1120, 1331, **1340**
surface phase, 88, **1340**
surface phase transition, 1340
surface phenomena, 733, 850, 1203, 1243, 1333, **1341**
surface phenomena in semiconductors, 466, 988, 1342,
    1343
surface phononic polariton, 1342
surface photocapacitance, **1342**
surface photoelectric effect, 995
surface photoelectromotive force, 673, **1342**

ISBN 0-12-561465-9

9 780125 614658

PERIODIC SYSTEM

PERIOD	I	II	III	IV	V
1	**H** 1 1.00794±7 $1s^1$ 1 Hydrogen				
2	**Li** 3 6.941±2 $2s^1$ 1 2 Lithium	**Be** 4 9.01218±1 $2s^2$ 2 2 Beryllium	**B** 5 10.811±5 $2s^2 2p^1$ 3 2 Boron	**C** 6 12.011±1 $2s^2 2p^2$ 4 2 Carbon	**N** 14.0067±1 Nitrogen
3	**Na** 11 22.98977±1 $3s^1$ 1 8 2 Sodium	**Mg** 12 24.305±1 $3s^2$ 2 8 2 Magnesium	**Al** 13 26.98154±1 $3s^2 3p^1$ 3 8 2 Aluminum	**Si** 14 28.0855±3 $3s^2 3p^2$ 4 8 2 Silicon	**P** 30.97376±1 Phosphorus
4	**K** 19 39.0983±1 $4s^1$ 1 8 8 2 Potassium	**Ca** 20 40.078±4 $4s^2$ 2 8 8 2 Calcium	21 **Sc** 2 9 8 2 $3d^1 4s^2$ 44.95591±1 Scandium	22 **Ti** 2 10 8 2 $3d^2 4s^2$ 47.88±3 Titanium	23 2 11 8 2 $3d^3 4s^2$
	29 **Cu** 1 18 8 2 $3d^{10} 4s^1$ 63.546±3 Copper	30 **Zn** 2 18 8 2 $3d^{10} 4s^2$ 65.39±2 Zinc	**Ga** 31 69.723±4 $4s^2 4p^1$ 3 18 8 2 Gallium	**Ge** 32 72.59±3 $4s^2 4p^2$ 4 18 8 2 Germanium	**As** 74.9216±1 Arsenic
5	**Rb** 37 85.4678±3 $5s^1$ 1 8 18 8 2 Rubidium	**Sr** 38 87.62±1 $5s^2$ 2 8 18 8 2 Strontium	39 **Y** 2 9 18 8 2 $4d^1 5s^2$ 88.9059±1 Yttrium	40 **Zr** 2 10 18 8 2 $4d^2 5s^2$ 91.224±2 Zirconium	41 1 12 18 8 2 $4d^4 5s^1$
	47 **Ag** 1 18 18 8 2 $4d^{10} 5s^1$ 107.8682±3 Silver	48 **Cd** 2 18 18 8 2 $4d^{10} 5s^2$ 112.41±1 Cadmium	**In** 49 114.82±1 $5s^2 5p^1$ 3 18 18 8 2 Indium	**Sn** 50 118.710±7 $5s^2 5p^2$ 4 18 18 8 2 Tin	**Sb** 121.75±3 Antimony
6	**Cs** 55 132.9054±1 $6s^1$ 1 8 18 18 8 2 Cesium	**Ba** 56 137.33±1 $6s^2$ 2 8 18 18 8 2 Barium	57 **La*** 2 9 18 18 8 2 $5d^1 6s^2$ 138.9055±3 Lanthanum	72 **Hf** 2 10 32 18 8 2 $5d^2 6s^2$ 178.49±3 Hafnium	73 2 11 32 18 8 2 $5d^3 6s^2$
	79 **Au** 1 18 32 18 8 2 $5d^{10} 6s^1$ 196.9665±1 Gold	80 **Hg** 2 18 32 18 8 2 $5d^{10} 6s^2$ 200.59±3 Mercury	**Tl** 81 204.383±1 $6s^2 6p^1$ 3 18 32 18 8 2 Thallium	**Pb** 82 207.2±1 $6s^2 6p^2$ 4 18 32 18 8 2 Lead	**Bi** 208.9804±1 Bismuth
7	**Fr** 87 223.0197 $7s^1$ 1 8 18 32 18 8 2 Francium	**Ra** 88 226.0254 $7s^2$ 2 8 18 32 18 8 2 Radium	89 **Ac**** 2 9 18 32 18 8 2 $6d^1 7s^2$ 227.0278 Actinium	104 **Ku** 2 10 32 32 18 8 2 $6d^2 7s^2$ (261) Kurchatovium	105 2 11 32 32 18 8 2 $6d^3 7s^2$

*

Ce 58 2 9 19 18 8 2 140.12±1 $4f^1 5d^1 6s^2$ Cerium	**Pr** 59 2 8 21 18 8 2 140.9077±1 $4f^3 6s^2$ Praseodymium	**Nd** 60 2 8 22 18 8 2 144.24±3 $4f^4 6s^2$ Neodymium	**Pm** 61 2 8 23 18 8 2 144.9128 $4f^5 6s^2$ Promethium	**Sm** 62 2 8 24 18 8 2 150.36±3 $4f^6 6s^2$ Samarium	**Eu** 63 2 8 25 18 8 2 151.96±1 $4f^7 6s^2$ Europium	**G** 157.25±3 Gadoli

**

Th 90 2 10 18 32 18 8 2 232.0381±1 $6d^2 7s^2$ Thorium	**Pa** 91 2 9 20 32 18 8 2 231.0359 $5f^2 6d^1 7s^2$ Protactinium	**U** 92 2 9 21 32 18 8 2 238.0289±1 $5f^3 6d^1 7s^2$ Uranium	**Np** 93 2 9 22 32 18 8 2 237.0482 $5f^4 6d^1 7s^2$ Neptunium	**Pu** 94 2 8 24 32 18 8 2 244.0642 $5f^6 7s^2$ Plutonium	**Am** 95 2 8 25 32 18 8 2 243.0614 $5f^7 7s^2$ Americium	**C** 247.0703 Curium